ANNUAL REVIEW OF BIOCHEMISTRY

EDITORIAL COMMITTEE (1983)

ANNUAL REVIEW OF BIOCHEMISTRY

VOLUME 52, 1983

ESMOND E. SNELL, *Editor*
University of Texas at Austin

PAUL D. BOYER, *Associate Editor*
University of California, Los Angeles

ALTON MEISTER, *Associate Editor*
Cornell University Medical College

CHARLES C. RICHARDSON, *Associate Editor*
Harvard Medical School

ANNUAL REVIEWS INC.   4139 EL CAMINO WAY   PALO ALTO, CALIFORNIA 94306 USA

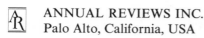
International Standard Serial Number 0066-4154
International Standard Book Number: 0-8243-0852-2
Library of Congress Catalog Card Number: 32-25093

Annual Review and publication titles are registered trademarks of Annual Reviews Inc.

Annual Reviews Inc. and the Editors of its publications assume no responsibility for the statements expressed by the contributors to this *Review*.

PREFACE

In its 52-year history as one of the most influential publications in biomedical science, the *Annual Review of Biochemistry* has had only three editors, J. Murray Luck, Paul D. Boyer (now serving as associate editor), and Esmond E. Snell. Fortunately, all three men are still vital members of the editorial committee that guides the series.

However, the present volume is the final one to be published during the editorship of Dr. Snell [although because of the *Review*'s two-year planning cycle, Volume 54 (1985) will be the last planned under his chairmanship of the editorial committee]. During Dr. Snell's 15-year term as editor (he also served as acting editor in 1963–1964), the volume—and Annual Reviews Inc. itself—joyously celebrated its 50th anniversary in 1981. Also during this period, several new *Annual Review* series were sired to fill the needs of those subdisciplines of biochemistry that had grown into clearly defined fields of their own—namely the *Annual Reviews* of *Biophysics and Bioengineering, Neuroscience,* and *Nutrition.* The first volume of the most recent addition, the *Annual Review of Immunology,* was published earlier this year. None of these offspring has undermined the preeminent position of the *Annual Review of Biochemistry*; highly qualified authors continue generously to contribute to the *Review.* As Associate Editor Boyer commented in the preface to Volume 50, those who contributed to the first volume back in 1932 could have had little idea how enormous the field of biochemical research would become in the ensuing half century. This exponential growth will surely continue, and further subdisciplines of biochemistry will gain their own *Annual Review* volumes.

Happily for Annual Reviews Inc. as a whole, however, this year does not signal a farewell to Dr. Snell, who will continue to serve on the Board of Directors (which he chaired while serving as President of Annual Reviews Inc. in 1973–1975). In this way, he will be able to keep a close watch over the progress of the series he has served so devotedly and with such editorial acumen. We thank him heartily for his lasting contribution to the *Annual Review of Biochemistry.* Dr. Snell can leave the editorial chair with the assurance that this prestigious series has earned increased international respect under his guidance.

Regular readers will notice that the present volume lacks the customary author index. Typesetting problems encountered in mid-production re-

quired that we engage the services of Aberdeen University Press of Glasgow, Scotland, who are to be thanked for completing the book at extremely short notice. Unfortunately, the computer program needed to generate the author index was not transferable to our new typesetter. An attempt to develop the author index computer program or to index by hand would have greatly delayed the publication of this volume. We thus postponed the author index in order to maintain our July publication date. It will appear next year in Volume 53 (1984).

ALISTER BRASS
EDITOR-IN-CHIEF
ANNUAL REVIEWS INC.

Annual Review of Biochemistry
Volume 52, 1983

CONTENTS

LONG AGO AND FAR AWAY, *Luis F. Leloir*   1

STRUCTURE AND CATALYSIS OF ENZYMES, *William N. Lipscomb*   17

ARCHITECTURE OF PROKARYOTIC RIBOSOMES, *H. G. Wittmann*   35

AFFINITY LABELING OF PURINE NUCLEOTIDE SITES IN PROTEINS,
    Roberta F. Colman   67

DNA METHYLATION AND GENE ACTIVITY, *Walter Doerfler*   93

COMPARATIVE BIOCHEMISTRY OF PHOTOSYNTHETIC LIGHT-
    HARVESTING SYSTEMS, *A. N. Glazer*   125

ADENYLATE CYCLASE–COUPLED BETA-ANDRENERGIC RECEPTORS:
    STRUCTURE AND MECHANISMS OF ACTIVATION AND
    DESENSITIZATION, *Robert J. Lefkowitz, Jeffrey M. Stadel, and
    Marc G. Caron*   159

BIOCHEMISTRY OF SULFUR-CONTAINING AMINO ACIDS, *Arthur J. L.
    Cooper*   187

LIPOPROTEIN METABOLISM IN THE MACROPHAGE: IMPLICATIONS FOR
    CHOLESTEROL DEPOSITION IN ATHEROSCLEROSIS, *Michael S.
    Brown and Joseph L. Goldstein*   223

DYNAMICS OF PROTEINS: ELEMENTS AND FUNCTION, *M. Karplus and
    J. A. McCammon*   263

CELLULAR ONCOGENES AND RETROVIRUSES, *J. Michael Bishop*   301

LEUKOTRIENES, *Sven Hammarström*   355

MECHANISM OF FREE ENERGY COUPLING IN ACTIVE TRANSPORT,
    Charles Tanford   379

VITAMIN D: RECENT ADVANCES, *Hector F. DeLuca and Heinrich K.
    Schnoes*   411

THE PATHWAY OF EUKARYOTIC MRNA FORMATION, *Joseph R.
    Nevins*   441

THE GENE STRUCTURE AND REPLICATION OF INFLUENZA VIRUS,
    Robert A. Lamb and Purnell W. Choppin   467

RIBULOSE-1,5-BISPHOSPHATE CARBOXYLASE-OXYGENASE, *Henry M.
    Miziorko and George H. Lorimer*   507

FATTY ACID SYNTHESIS AND ITS REGULATION, *Salih J. Wakil, James K. Stoops, and Vasudev C. Joshi*   537

PROKARYOTIC DNA REPLICATION SYSTEMS, *Nancy G. Nossal*   581

GLUCONEOGENESIS AND RELATED ASPECTS OF GLYCOLYSIS, *H. G. Hers and L. Hue*   617

HUMAN PLASMA PROTEINASE INHIBITORS, *J. Travis and G. S. Salvesen*   655

GLUTATHIONE, *Alton Meister and Mary E. Anderson*   711

CELL SURFACE INTERACTIONS WITH EXTRACELLULAR MATERIALS, *Kenneth M. Yamada*   761

PROTON ATPASES: STRUCTURE AND MECHANISM, *L. Mario Amzel and Peter L. Pedersen*   801

PENICILLIN-BINDING PROTEINS AND THE MECHANISM OF ACTION OF β-LACTAM ANTIBIOTICS, *David J. Waxman and Jack L. Strominger*   825

A MOLECULAR DESCRIPTION OF NERVE TERMINAL FUNCTION, *Louis F. Reichardt and Regis B. Kelly*   871

INDEXES
Subject Index   927
Cumulative Index of Contributing Authors, Volumes 48–52   940
Cumulative Index of Chapter Titles, Volumes 48–52   943

SOME RELATED ARTICLES IN OTHER *ANNUAL REVIEWS*

From the *Annual Review of Biophysics and Bioengineering*, Volume 12 (1983)

Interactions of Water With Nonpolar Solutes, A. Hvidt

Coupling of Proton Flux to the Hydrolysis and Synthesis of ATP, J. H. Wang

Protozoan and Related Photoreceptors: Molecular Aspects, P.-S. Song

Intracellular Measurements of Ion Activities, R. Y. Tsien

Neutron Protein Crystallography: Advances in Methods and Applications, A. A. Kossiakoff

Theoretical Studies of Protein Folding, N. Gō

Mechanisms of Assembly and Disassembly of Microtubules, J. J. Correia and R. C. Williams, Jr.

Structural Studies of Protein-Nucleic Acid Interactions, D. H. Ohlendorf and B. W. Matthews

Thermodynamics of Protein-Ligand Interactions: Calorimetric Approaches, H.-J. Hinz

Sodium Channel Gating: Models, Mimics, and Modifiers, R. J. French and R. Horn

Protein and Nucleic Acid Sequence Database Systems, B. C. Orcutt, D. G. George, and M. O. Dayhoff

Acetylcholine Receptor-Controlled Ion Translocation: Chemical Kinetic Investigations of the Mechanism, G. P. Hess, D. J. Cash, and H. Aoshima

Dynamics of tRNA, R. Rigler and W. Wintermeyer

From the *Annual Review of Genetics*, Volume 16 (1982)

Attenuation in Amino Acid Biosynthetic Operons, C. Yanofsky and R. Kolter

Genetic Control of Nitrogen Assimilation in Bacteria, B. Magasanik

DNA Uptake in Haemophilus Transformation, S. H. Goodgal

Genetic Defects in Human Purine and Pyrimidine Metabolism, J. E. Seegmiller and G. R. Boss

Strand Transfer in Homologous Genetic Recombination, C. M. Radding

Molecular Genetics of Yeast Mating Type, K. A. Nasmyth

From the *Annual Review of Immunology*, Volume 1 (1983)

Getting Started 50 Years Ago—Experiences, Perspectives, and Problems of the First 21 Years, E. A. Kabat

Cellular Mechanisms of Immunologic Tolerance, G. J. V. Nossal

Structural Basis of Antibody Function, D. R. Davies and H. Metzger

Regulation of B-Cell Growth and Differentiation by Soluble Factors, M. Howard and W. E. Paul

From the *Annual Review of Medicine*, Volume 34 (1983)

Insulin Receptors and Insulin Resistance, J. S. Flier

Structural Variants of Human Growth Hormones: Biochemical, Genetic, and Clinical Aspects, R. K. Chawla, J. S. Parks, and D. Rudman

From the *Annual Review of Microbiology*, Volume 37 (1983)

Yeast DNA Plasmids, N. Gunge

Structure, Assembly, and Function of Cell Walls of Gram-Positive Bacteria, G. D. Shockman and J. F. Barrett

Role of Proton Motive Force in Sensory Transduction in Bacteria, B. L. Taylor

From the *Annual Review of Neuroscience*, Volume 6 (1983)

The Classification of Dopamine Receptors: Relationship to Radioligand Binding, I. Creese, D. R. Sibley, M. W. Hamblin, and S. E. Leff

From the *Annual Review of Pharmacology and Toxicology*, Volume 23 (1983)

The Endorphins: A Growing Family of Pharmacologically Pertinent Peptides, F. E. Bloom

Structure-Activity Relationships of Dopamine Agonists, J. G. Cannon

Mechanisms of Selective Action of Pyrethroid Insecticides, J. E. Casida, D. W. Gammon, A. H. Glickman, and L. J. Lawrence

Suicidal Destruction of Cytochrome P-450 During Oxidative Drug Metabolism, P. R. Ortiz de Montellano and M. A. Correia

Superoxide Radical: An Endogenous Toxicant, I. Fridovich

Biosynthesis of the Enkephalins and Enkephalin-Containing Polypeptides, R. V. Lewis and A. S. Stern

The Activity of Sulfonamides and Anions Against the Carbonic Anhydrases of Animals, Plants, and Bacteria, T. H. Maren and G. Sanyal

From the *Annual Review of Physical Chemistry*, Volume 34 (1983)

Electronic States and Luminescence of Nucleic Acid Systems, P. R. Callis

From the *Annual Review of Physiology*, Volume 45 (1983)

The Red Cell Calcium Pump, H. J. Schatzmann

Calcium Channels in Excitable Cell Membranes, R. W. Tsien

Calcium Transport Proteins, Calcium Absorption, and Vitamin D, R. H. Wasserman and C. S. Fullmer

New Neurotrophic Factors, Y. A. Barde, D. Edgar, and H. Thoenen

Antifreeze Peptides and Glycopeptides in Cold-Water Fishes, A. L. DeVries

CNS Peptides and Glucoregulation, L. Frohman

Lipid Digestion and Absorption, M. Carey, D. Small, and C. Bliss

Heterogeneity of Apolipoprotein B and the Metabolism of Lipoproteins in Plasma, J. Kane

Bile Acid Synthesis, G. Salen, S. Shefer

ERRATA

ANNUAL REVIEW OF BIOCHEMISTRY, Volume 50 (1981)

In *In Vivo Chemical Modification of Proteins* (*Post-Translational Modification*), by Finn Wold:

On page 811, Ref. 62 should read, "Lederer, F...." (not Lederer, P.). On page 1117, in the Author Index, that reference (cited on page 792) should be listed with the other references to Lederer, F.

ANNUAL REVIEW OF BIOCHEMISTRY, Volume 51 (1982)

In *Enzymes of the Renin-Angiotensin System and Their Inhibitors*, by M. A. Ondetti and D. W. Cushman:

On page 286, paragraph two, the statement in which the definition of the International Unit of renin is equated with that of the Goldblatt Unit is incorrect. The Goldblatt Unit is an "animal unit" depending on the (variable) response in a dog. The International Unit of human renin is defined by the International Reference Preparation of Renin, human, for Bioassay, established by the World Health Organization in 1974.[1] Ampoules of this International Reference Preparation, each containing 0.1 IU of human renin, are available on request from the National Institute for Biological Standards and Control, Hampstead, London NW3 6RB, England.

[1] Bangham, D. R., Robertson, I., Robertson, J. I., Robinson, C. J., Tree, M. 1975. *Clin. Sci. Mol. Med.* 48: (Suppl. 2) 135–59

Luis Thelvis

Ann. Rev. Biochem. 1983. 52: 1–15

FAR AWAY AND LONG AGO

Luis F. Leloir

Instituto de Investigaciones Bioquimicas, "Fundacion Campomar" and Facultad de Ciencias Exactas y Naturales, Buenos Aires, Argentina

CONTENTS

B. A. HOUSSAY'S INSTITUTE OF PHYSIOLOGY 2
INITIATION IN BIOCHEMISTRY 3
FATTY ACIDS 4
AN ADVENTURE IN HYPERTENSION 5
A STAY IN USA 7
FUNDACION CAMPOMAR 8
DOLICHOL DERIVATIVES 13
WHY RESEARCH 14

Biochemistry and I were born and grew at about the same time. Before the turn of the century some organic chemists and physiologists had lain the bases of Biochemistry. In 1906 two journals dealing with it appeared, the *Biochemische Zeitschrift* and the *Biochemical Journal.* The *Journal of Biological Chemistry* had started publication only one year before. In 1906 Arthur Harden and W. J. Young were able to separate "yeast juice into a residue and filtrate, each of which was itself incapable of setting up the alcoholic fermentation of glucose, whereas, when they were reunited the mixture produced almost as active fermentation as the original juice." This finding occurred only nine years after Edward Buchner had prepared a cell free yeast juice capable of fermentation. This line of work led eventually to the discovery of the multitude of enzymes, coenzymes, and intermediates of cell metabolism. In 1906 Tswett published the first description of chromatography.

Another important event (from my point of view) occurred in 1906. This was my birth in Paris, 81 Avenue Victor Hugo, just a few blocks away from the Arc de Triomphe.

The growth of Biochemistry was rapid; in a few decades most of the vitamins, hormones, enzymes, and coenzymes were discovered but at the

1

time of writing this essay it is showing signs of dismemberment. Molecular Biology, Cell Biology, Chemical Genetics, etc have risen from it and surely there will be others. I reached the age of 76 thanks to some clever arterial repair work carried out by Michael Debakey in Houston.

I have borrowed the title of this essay from a delightful book by W. H. Hudson (1) that describes the wild life in the country near Buenos Aires. Hudson describes the same scenery and the same animals, flamingoes, armadillos, caranchos, vizcachas, etc that I saw in my infancy. It seems that both of us were interested in animal life and understanding nature, but while I became convinced that scientific knowledge and technology would be good for mankind, Hudson (2) had some doubts and he expressed them as follows: "Ah yes, we are all seeking after happiness in the wrong way. It was with us once, and ours, but we despised it, for it was only the old common happiness which Nature gives to all her children, and we went away from it in search of another grander kind of happiness, which some dreamer—Bacon or another—assured us we would find. We had only to conquer Nature, but how weary and sad we are getting!! The old joy in life and gaiety of heart have vanished."

B. A. HOUSSAY'S INSTITUTE OF PHYSIOLOGY

When I was two years old my Argentine parents brought me to Buenos Aires, where, after going through the studies and examinations necessary for graduating as an M.D. at the University of Buenos Aires in 1932, I worked at the Hospital of the University (Hospital de Clinicas) for about two years. I was never satisfied with what we did for the patients. Looking back on those times, I realize how profoundly medicine has changed since then. Medical treatment was in those days only slightly better than that exemplified by the French story, in which the doctor ordered, "Today we shall bleed all those on the left side of the ward and give a purgative to all those on the right side."

When I practiced medicine, except for surgery, digitalis, and a few other active remedies, we could do little for our patients. Antibiotics, psychoactive drugs, and all the new therapeutic agents were unknown. It was therefore not strange in 1932 that a young doctor such as I should try to join efforts with those who were trying to advance medical knowledge. The most active research laboratory in town was that directed by Dr. Bernardo A. Houssay, professor of physiology. In his work on the role of the pituitary gland on carbohydrate metabolism, he obtained some very novel findings, for which he, along with Carl and Gerty Cori, was awarded the Nobel Prize in Physiology and Medicine in 1947. Dr. Houssay suggested that I could do my thesis work under his direction and proposed several topics. My choice was the role of the adrenals in carbohydrate metabolism. My first task was to learn how to measure blood sugar with the method of Hagedorn & Jensen. It was my first experience in a research laboratory. My ignorance in chemistry was unfathomable; for that reason I decided to follow some courses in the Faculty of Sciences.

Houssay helped me a lot. Not only did he do the brain work but he also carried out most of the adrenalectomies on the dogs. Houssay made daily rounds in the Institute and often left messages on minute pieces of paper. It was apparently through him that I learned to be economical. Even now, I usually write manuscripts on half sheets of paper already used on one side. Young people now are spendthrifts and would scandalize Houssay as they do me. The thesis work was awarded the Annual Prize of the Faculty for the best thesis, but it was undoubtedly Houssay's merit, not mine. Our close association lasted until Dr. Houssay's death in 1970. During all those years we saw each other daily and I could appreciate his cyclopean work in favor of Argentine science.

My enthusiasm for research increased gradually and, without noticing the change, I began to put in more hours at the laboratory and less at the hospital. I could do so because I did not have to earn my living with medical work. My great-grandparents came to Argentina, some from France, others from Spain, and bought land when it was cheap but still unsafe from the incursions of the Indians. Later these lands produced the cereal and grains and the cattle that brought riches to the country and to the pioneers who worked on them. These circumstances allowed me to devote myself to research when it was very difficult or impossible to find a full-time position for research.

It was a great privilege to be associated with Houssay. He worked very hard all his life trying to modernize the teaching of medicine, as well as directing his numerous students. His interest in research was very wide. Endocrinology was his main concern but he also ventured into many other aspects of physiology and biochemistry. He made intensive efforts to promote science. He was for many years president of the Argentine Association for the Progress of Science and later president of the Argentine Research Council. At times his efforts were very successful, but at other times the government was against him because of his outspoken manner and his liberal views.

INITIATION IN BIOCHEMISTRY (3)

After I finished my thesis work, Dr. Houssay advised me to work some time abroad. In consultation, with Dr. V. Deulofeu, professor of biochemistry, and Dr. R. de Meio, I decided that a good place would be the Biochemical Laboratory of Cambridge University, directed by Sir Frederick Gowland Hopkins, who had received the Nobel Prize in 1929, together with Eijkman, for "his discovery of growth stimulating vitamins." Cambridge was then at the peak of its glory with Rutherford, Dirac, and other scientific giants in the Department of Physics. Biochemistry was also excellent with Hopkins, the father of English Biochemistry, at the helm of the biochemical laboratory, and David Keilin, the discoverer of the cytochromes, in the Department of Parasitology. Arriving in Cambridge thirsty for knowledge, I began immediately to work under the direction of Malcolm Dixon on the effect of cyanide and pyrophosphate on succinic acid dehydrogenase. After that I worked with Norman L. Edson on ketogenesis using liver slices. Edson had worked with Hans Krebs, whom he admired greatly. When Edson returned to his home country, New Zealand, I worked with David Green on the purification and properties of β-hydroxybutyrate dehydrogenase. The atmosphere in the biochemical laboratory was very stimulating because of the many talented people there, such as Marjorie Stephenson, one of the pioneers in bacterial biochemistry, Norman Pirie, who had crystallized tobacco mosaic virus, Robin Hill, who was well known for his work on photosynthesis (the Hill effect), Joseph Needham, who started chemical embryology and ended up as an orientalist, Dorothy Moyle Needham, an expert in muscle chemistry, and many others.

FATTY ACIDS (3, 4)

After my year at Cambridge, I returned to the Institute of Physiology in Buenos Aires, where I became associated with J. M. Muñoz. I never enjoyed working by myself so I was glad I could work with him. He had a rather original personality, and degrees in Medicine and Chemistry. Not content with these titles, he also earned a degree in Dentistry. Actually he did so not merely to increase his knowledge, but more importantly in order to become the professor of physiology in the school of dentistry. That type of ludicrous requirement was not too rare at our university. As a way to avoid competition from other graduates, it was of course detrimental to the university and the students.

Dr. Muñoz had been estimating ethanol, and had a reliable method with a beautiful, small distillation apparatus. Therefore we decided to work on ethanol metabolism, with interesting results that were published in the *Biochemical Journal*.

After our alcoholic adventure we took advantage of the same distillation apparatus for estimating volatile fatty acids. Muñoz found that the recovery was quantitative if the boiling point of the samples was increased by the addition of sodium sulfate. The volatile fatty acids (butyric acid was generally used) were then measured by oxidation with dichromate.

After working with tissue slices, we turned our attention to tissue homogenates, a rather ambitious project since fatty acid oxidation was believed to occur only in intact cells. Our approach was different from previous ones because we planned to measure fatty acid disappearance while other workers had relied on measurements of oxygen uptake with or without fatty acids. We obtained some homogenates that consumed butyric acid, but in general our experiments failed. Learning from our failures, we managed to find conditions under which our liver homogenates were active. It was a question mostly of working fast and in the cold. Because we wanted to fractionate the extracts, we had to cool our centrifuge. We did this with an automobile inner tube filled with freezing mixture and wrapped around the head of an old pulley-driven centrifuge. In this way we were able to separate a cell free particulate fraction and a supernatant fluid. The particulate fraction was found to require several factors for oxidizing fatty acids. These factors were cytochrome C, a C_4 dicarboxylic acid (we usually used fumaric acid) and adenosin phosphate (ATP was active but we used adenylic acid because it was easier to obtain the pure substance). We suspected that mitochondria were involved in fatty acid oxidation but could not prove it. In those days mitochondria were known to cytologists but practically unknown to biochemists. Claude had just started centrifugal fractionation and separated tissue homogenates into fractions containing mainly nuclei, mitochondria, or smaller constituents (microsomes).

AN ADVENTURE IN HYPERTENSION (5, 6, 7)

While we were working with Dr. Muñoz on fatty acid oxidation, J. C. Fasciolo was experimenting on the mechanism of renal hypertension. Under Houssay's guidance, he followed the discovery of Harry Golblatt, who had found that if the renal artery of dogs was mechanically constricted the animals developed a permanent hypertension. This had been a milestone because it provided an experimental method of producing high blood pressure. Fasciolo's contribution consisted in grafting one of those kidneys in a normal dog and observing the changes in blood pressure. The result was an increase in blood pressure, which proved that the effect was due to some substance that the constricted kidney poured into the blood. At that stage, I became involved. Dr. J. C. Fasciolo, E. Braun Menendez, J. M. Muñoz, and I formed a team that successfully clarified the problem. Dr. A. Taquini also collaborated in many experiments.

An early observation was that from the blood of the constricted kidney a substance that produced a transient rise in blood pressure could be extracted with acetone. The effect did not last as long as that produced by grafting a constricted kidney. In 1898 Tigersted & Bergman had discovered a pressor substance they called renin, which could be extracted from kidneys. We found that the effect of renin was clearly different from that produced by acetone extracts of the venous blood of constricted kidneys.

We then carried out a variety of experiments to obtain the in vitro production of the pressor substance. Incubations of kidney extracts or slices with blood plasma under aerobic or anaerobic atmosphere yielded only inactive extracts. One day Dr. Braun Menendez suggested that I should incubate renin with blood plasma. I told him I had incubated kidney extracts that contained renin with blood with negative results. I said I had done this experiment many times but that if it made him happy I might test renin. The incubation gave rise to the formation of a pressor substance soluble in aqueous acetone; we soon found that the properties were the same as those of the pressor compound found in the constricted kidney blood. We worked feverishly and obtained in a short time much information on the new substance, which we called hypertensin. We suggested that renin acted as an enzyme and released hypertensin from a precursor found in blood plasma. We also detected an enzyme in tissues and blood that destroyed hypertensin; this explained my previous failure to detect hypertensin formation. If the incubation time was too long or the destroying enzyme too active, all the hypertensin formed would disappear. The action of renin was found to have an interesting specificity, thus dog renin acted on dog serum, human renin on human serum, and so on, but they did not cross-react.

While we were working busily in Buenos Aires, others were conducting similar experiments elsewhere. In the laboratories of Eli Lilly in Indianapolis, Irwin Page and co-workers observed that purified renin showed pressor activity when injected into the blood but was inactive as a vasoconstrictor when tested on a dog's tail perfused with Ringer solution. They suggested that renin required a blood protein as activator. In fact they found that addition of blood plasma restored vasoconstrictor activity. Page et al proposed the name angiotensin for what we used to call hypertensin. Although our paper was already published (1939) when we learned of the work of the other group we were considerably depressed because we could not claim a discovery but only a co-discovery. Looking back now after so many years I view those feelings as quite infantile. However, such incidents happen quite often in research, and they affect workers even more experienced than we were at that time.

For several years, both groups tried to impose the names they had proposed. We used the terms hypertensin, hypertensinogen and hypertensinase, while the Indianapolis group referred to angiotensin and renin activator. Finally Dr. Braun Menendez and Page agreed on a Solomonic solution and proposed the names angiotensin and angiotensinogen. These are the names generally used today.

The purification and determination of the structure of angiotensin seemed an insurmountable problem in those days. We knew it was a polypeptide because it became inactive on treatment with proteolytic enzymes but the methods were too primitive to obtain much more information. Determining the sequence of a polypeptide was conceived only by stretching a wild dream.

It is now known that renin splits a leucyl-leucyl bond in angiotensinogen to yield the decapeptide angiotensin I, which is inactive, but when acted upon by a converting enzyme, it loses the dipeptide histidyl-leucine to yield angiotensin II, which is very active. The latter peptide may lose an aspartic acid residue to yield angiotensin III, which is less active than II. Much was accomplished after we abandoned the problem. The great advances came possibly because of the progress of the methods for amino acid analysis and peptide synthesis.

My incursion in hypertension research lasted only about one year but it was one of the most productive years in my career. Two important factors in the success were the congenial atmosphere and the personal quality of my teammates. They had very different personalities but worked together successfully. E. Braun Menendez was full of energy, enthusiasm and entrepreneurial ability; J. Muñoz had a very original personality and many unique ideas; good-humored Juan Carlos Fasciolo always told jokes or funny stories, but he also performed his work seriously, intelligently and efficiently. All were very clever and diligent. We had quite a lot of fun with our work. After successful experiments, I often said "You see, nothing can resist systematic research." But after failed experiments, I looked tired and depressed, so Dr. Fasciolo would make fun of me, saying "You see, nobody can resist systematic research." However, we worked hard; the rate of research was limited only by the availability of dogs for measuring pressor substances. We also used toads for measuring vasoconstrictor substances but with little success. They always gave confusing results.

A STAY IN U.S.A. (3)

Our work in the Institute of Physiology was interrupted in 1943 by unexpected and disagreeable events. Dr. Houssay never mixed up in politics but had signed an apparently innocent letter that appeared in the newspapers and also carried the signature of many of the most important persons of the country. The letter asked for "constitutional normality, effective democracy, and American solidarity." The government reacted completely out of proportion, and decreed the dismissal of all the signers who had positions in the state institutions. Many of the best university professors lost their posts. Dr. Houssay was one of them. Most of the members of the Institute of Physiology resigned in protest and disbanded. Days of confusion and preoccupation followed. It was finally decided to continue working, not in the university but in a private institution that had to be organized from scratch. Since I was always in a hurry to do experiments and not lose time, I thought it would be a good moment to work abroad for some time. This decision coincided with an important and fortunate event in my life, the start of a happy marriage. My wife and I decided to travel to the United States. Commercial planes were still two-motored and flew only by day; thus after several stops in various parts of the American continent, we finally arrived in New York. Since I had no previous arrangements, I had to look for a place in which to work. One place that was highly regarded then was Cori's laboratory in St. Louis. The Coris had just published a thorough and careful study on the crystallization and preparation of phosphorylase.

After looking around a few days in New York we traveled to St. Louis, where Carl Cori accepted me kindly in the laboratory. There he arranged that I should collaborate with Ed Hunter on the formation of citric acid. Thus for six months I had the privilege of working in a place full of traditions, where I met daily with Carl and Gerty Cori, Sidney Colowick, Arda Green, and other outstanding scientists. To broaden my outlook, I spent time in New York, where I arranged to work again with David E. Green. He had two rooms in the College of Physicians and Surgeons of Columbia University, and a small group of collaborators including Sarah Ratner, Eugene Knox, and Paul Stumpf. For some time we worked on the purification of aminotransferases and we were able to separate the alanine from the aspartate-aminotransferase. Green was always full of ideas and sometimes came in with a list of all the processes that had to be clarified. The list was impressive but I did not know how to attack any of those problems. When I told him about the fatty acid oxidation system that needed so many co-factors and that might be localized in mitochondria, he looked at me with a rather incredulous smile. He did not suspect then that these organelles would occupy a great part of his future.

One of the important things I learned from Green was that if one could find a place to work, one should be able to form a research group by asking around for the required salaries, equipment, and chemicals. Indeed, that was just what I did upon returning to Argentina, where I formed a small research group that grew slowly and gave rise to the Fundacion Campomar.

FUNDACION CAMPOMAR (3, 8, 9, 10, 11)

After my stay in the United States, I returned to the Institute of Physiology. Dr. Houssay had been reinstated and was trying to put the Institute together again. For some time I worked by myself and endeavored to start a small research team. The first person I attracted was Dr. Ranwel Caputto of the University of Cordoba, who had just returned from the Biochemical Laboratory in Cambridge. He had worked with Malcolm Dixon, and had succeeded in crystallizing glyceraldehyde dehydrogenase. Later he found that this enzyme is more active on glyceraldehyde phosphate, which is probably the natural substrate. The second to join the group was microbiologist Raul Trucco. The idea in selecting him was to continue the work on fatty acid oxidation but with bacterial enzymes.

I learned in 1946 that Dr. Houssay had been approached by Jaime Campomar, one of the owners of an important textile industry. Campomar wanted to finance a research institution specializing in biochemistry. I suspect that there were few candidates for director of the new institution; thus, even though Dr. Houssay may not have been very convinced that I could be successful in the enterprise, he suggested me.

As things were being organized, we continued to work in a cellar in the faculty of medicine, but only until Dr. Houssay was again removed from his position as professor of physiology and director of the Institute of Physiology, on the pretext that he was overage. This new abuse produced a great commotion in the faculty and most of us decided to leave. If the installations and equipment were poor at the faculty, they were disastrous in the laboratory into which we moved. This was the Instituto de Biologia y Medicina Experimental, a private institution that had been created when Dr. Houssay was first dismissed from the university. Shortly afterwards we rented a small, four-room house beside the above mentioned institute and adapted it for our laboratory. Others joined the group then, including Carlos Cardini, who had been for several years professor at the University of Tucuman, N. Mittelman, who had good training in protein fractionation, and A. C. Paladini, the youngest of all, who joined us with a fellowship. Mr. Campomar's annual contribution of 100,000 Argentine pesos was equivalent to about US $25,000. It was a very generous gift. With it we installed the laboratory, acquired some equipment, and paid some salaries.

After several false starts we happened to step into a field that turned out to be rather fertile. Dr. Caputto had told us that he had done some experiments that indicated that mammary gland homogenates produced lactose when incubated with glycogen. This seemed an interesting finding because the biosynthesis of saccharides was a rather unexplored field. At that time the only disaccharide that had been synthesized in vitro was sucrose. The enzyme used was of bacterial origin and the reactants were glucose 1-phosphate and fructose. We carried out many experiments with the mammary gland extracts but they gave generally ambiguous results, mainly because we used only very primitive methods for lactose detection, such as osazone formation. Discouraged, we thought of another approach—to find an enzyme that would catalyze the reversible synthesis of lactose. For this purpose we obtained a culture of *Saccharomyces fragilis,* a yeast that utilizes lactose. With this yeast it was found that oxygen uptake was faster when lactose was the substrate than with a mixture of glucose and galactose. At first we thought we would find some direct pathway for lactose utilization but after a time we concluded that it was a question of selective permeability, but we never investigated the point exhaustively. We did find a very active lactase and therefore directed our attention to galactose utilization. We obtained evidence indicating that galactose was phosphorylated and therefore the corresponding enzyme was studied. The reaction product was found to be galactose 1-phosphate and the study of its utilization led us to the detection of glucose 1, 6-diphosphate and uridine diphosphate glucose. We prepared glucose 1-phosphate and galactose 1-phosphate by chemical synthesis and observed that they were utilized when we incubated them with enzymes from galactose-adapted yeast. The disappearance was increased by the addition of a heated yeast extract. At first we thought that there was only one factor in the heated extract that activated the reactions with the two esters but after more work and many confusing experiments we realized that two thermostable factors were involved. One was necessary for the conversion of galactose 1-phosphate into glucose 1-phosphate, which requires the inversion of the hydroxyl group at carbon four. The other cofactor was required for the formation of glucose 6-phosphate from glucose 1-phosphate, that is, a change of position of the phosphate group.

Quantitative methods of estimating each of the two co-factors with the appropriate enzymes could be developed, and the next step was to purify the active substances. This was not as easy as it is now with all the current methods of fractionation. The procedures then available consisted of the separation of the barium, calcium, or mercury salts.

The co-factor of the phospho-glucomutase reaction had properties similar to those of fructose 1, 6-diphosphate; in fact, our partially purified preparations were enriched in it, but it was possible to destroy the fructose ester, leaving the active substance intact. Finally the co-factor was obtained pure, which turned out to be glucose 1, 6-diphosphate.

Studies on the other co-factor were also advancing. Purification by fractionation of the mercury salts and charcoal adsorption had yielded concentrates that adsorbed light at 260 nm. The spectrum was similar to that of adenosine but had some differences. For quite some time we were unable to discover the identity of the 260 nm absorbing substance until Dr. Caputto came in one morning with the last issue of the *Journal of Biological Chemistry* and showed us a spectrum identical to ours. This spectrum was that of uridine. Things became easier because we could measure the substance with a spectrophotometer instead of the enzymatic tests.

In addition to uridine the co-factor was found to contain glucose and two phosphates. Identification of the glucose was carried out by paper chromatography, a technique that we had just started to use. We employed a very inadequate solvent, but the results were correct. Only later were good solvents for sugar separations described. The presence of uridine in a cofactor was rather novel because in other compounds (ATP, NAD, FAD) the nucleoside present was adenosine. The occurrence of a sugar derivative combined with a nucleoside was also novel. The mechanism of action of uridine diphosphate glucose (UDPG) began to become clear when we found that the enzyme preparations active in converting galactose 1-phosphate into glucose 1-phosphate could also convert reversibly the glucose moiety of UDPG into galactose. Later Kalckar and coworkers found that NAD was necessary in this reaction; it seemed likely that the inversion of hydroxyl four occurred by an oxidoreduction reaction.

Dr. Paladini detected a new compound similar to UDPG by applying paper chromatography to yeast extracts. When very alkaline solvents were used, the sugar moiety of UDPG became separated from the nucleotide as could be ascertained by the noncoincidence of ultraviolet absorption and the color given by reagents for sugar detection. The products of decomposition were identified as uridine 5′-phosphate and cyclic glucose1, 2-monophosphate. Besides this compound, crude preparations of UDPG gave another substance that was not decomposed in alkaline solvents. It seemed the same as UDPG but with a different sugar moiety. This substance turned out to be uridine diphosphate acetylglucosamine. Since it has no hydroxyl group at position 2 of the sugar, it cannot form a cyclic phosphate. At about the same time James Park discovered other uridine nucleotides containing acetylmuramic acid plus some amino acids that are now known to be intermediates in the formation of peptidoglycan, a component of the cell walls of bacteria.

Paper chromatography of crude UDPG preparations allowed us to detect still another sugar nucleotide. The guanosine moiety could be identified easily because of its typical ultraviolet spectrum with a peak at 260 nm and a shoulder at 280 nm. It was studied mostly by Dr. Enrico Cabib, a bright young man who had joined our team as a fellow replacing Paladini, who had gone to work with L. Craig at the Rockefeller Foundation in New York. He had a good sense of humor and was a hard worker. Except for minor episodes that occur in all human groups, the atmosphere in the laboratory was very pleasant and all our time was dedicated to research. We had no lectures to give, no committees, and no forces pulling us away from the laboratory.

In 1957, Mr. Campomar's death left our institute without resources. Before disbanding we played our last card and asked for a grant from the National Institute of Health of the United States. We had little hopes of success, but to our surprise the grant was approved, and we continued our work.

To detect any unknown reaction involving UDPG, we used the same method of estimation that we had used for developing the isolation procedure, which consisted in measuring the activating action on the galactose

1-phosphate-glucose 1-phosphate transformation. We used this method in experiments with yeast extracts with various additions and found that glucose-6-phosphate produced an increase in UDPG disappearance. With Dr. Cabib we traced the effect to the synthesis of trehalose-6-phosphate, an ester that W. T. Morgan had isolated several years earlier as a product of the action of yeast enzymes.

A similar procedure with wheat germ enzymes allowed us, with C. Cardini and J. Chiriboga, to detect the biosynthesis of sucrose. In the course of studies on the later process, we detected another enzyme that makes sucrose phosphate.

Hexosamine metabolism was the object of considerable research. We became interested in this field because of the discovery of UDP-acetylglucosamine. We wanted to find a role for this compound. One obvious role was the formation of acetylgalactosamine made likely by H. Pontis, who isolated UDP-acetylgalactosamine from liver. Another function of UDP-acetylglucosamine that seemed obvious was the formation of chitin. In fact, years later L. Glaser obtained a transfer of acetylglucosamine to chitin. With C. Cardini we studied the formation of glucosamine phosphate from glutamine and hexosephosphate.

In 1957 we entered another field. It was found that liver extracts could catalyze the formation of glycogen from UDPG. In those days it was firmly believed that glycogen was formed from glucose 1-phosphate. However some penetrating minds had expressed doubts about this. For instance Earl Sutherland had reasoned that it was difficult to understand why epinephrine, which activated glycogen phosphorylase in vitro, should always produce glycogenolysis in vivo. Herman Niemeyer went further and suggested that glycogen was formed not from glucose-1-phosphate but from UDP-glucose. Our finding was therefore not totally unexpected but nevertheless helped considerably in the clarification of the mechanism of glycogen synthesis and degradation. Soon after, we detected the activating effect of glucose-6-phosphate, a finding which gave new impetus to the work on regulation of glycogen metabolism. The studies on glycogen synthetase were carried out with C. Cardini, H. Carminatti, S. Goldemberg and J. M. Olavarria. In 1958 the government offered us a large house which had been a girls' school, where we are still working. Funds were scarce, so we moved into the new house after minimal adaptation. A short time thereafter the National Research Council was created and we became associated with the faculty of sciences of the University of Buenos Aires. As new workers joined the laboratory, it lost the romantic flavor it used to have.

Research on glycogen biosynthesis was extended to include starch. It seemed obvious that the synthetic mechanism should be the same or very similar. After trying many plant enzyme preparations with negative results, we detected activity in transferring glucose from UDPG in some starch grains. The activity seemed low and we reasoned it might be because we used the wrong substrate. G. Khorana had then developed rather simple chemical methods for the synthesis of sugar nucleotides. We attracted a new collaborator, trained in organic chemistry, Dr. E. Recondo. The idea, which at times seemed quite silly, was to synthesize nucleotides of glucose and different bases, and to test them with the starch enzyme. The most available starting materials were AMP and glucose 1-phosphate, so Dr. Recondo synthesized ADPG first. The enzymatic tests showed that this sugar nucleotide was active as donor and better than UDPG. We were not sure that the result was not an artifact until Dr. J. Espada detected in plant material an enzyme that led to the synthesis of ADPG. Later other workers found that it was the substrate for glycogen synthesis in bacteria.

Another aspect of glycogen synthesis which interested us and Dr. J. Mordoh, A. J. Parodi and C. Krisman was the high molecular weight glycogen that can be extracted from tissues by avoiding treatment with acid or alkali. These preparations have molecular weights ranging from 10 to 1,000 million daltons and give a characteristic aspect on examination with the electron microscope. Similar preparations could be obtained in vitro with glycogen synthetase and UDPG but not with glycogen phosphorylase working in the synthetic direction. The make up of high molecular weight glycogen still remains rather mysterious.

DOLICHOL DERIVATIVES (12, 13, 14)

In 1965 a rather important development in the field of polysaccharide biosynthesis was the simultaneous discovery of lipid intermediates by two groups of workers: P. Robbins et al in Boston, who were investigating the biosynthesis of lipopolysaccharides by *Salmonella,* and J. Strominger in Madison, who was working on the formation of the cell wall peptidoglycan in *Staphylococcus.* Both groups detected the formation of liposoluble intermediates identified as polyprenyl phosphate-sugars. Our colleague Marcelo Dankert, who worked in Dr. Robbins' group, returned to our laboratory and told us about these compounds. He was very enthusiastic about the new field. We happened to have some crude preparations of glycogen synthetase, and radioactive UDPG was always easy to find around the laboratory. With N. Behrens we mixed these two substances, incubated them, added *n*-butanol, as Dankert taught us, separated the organic phase, and counted its radioactivity. The incubated sample had only a few counts more than the control. The experiment was repeated with the same results, but after optimizing the conditions, we obtained a considerable incorporation of glucose into a substance soluble in some organic solvents.

The next advance was the detection of an activation of the reaction by some liver extracts. In contrast to our previous work, in which water soluble substances were involved, the activator was soluble in certain organic solvents. We purified considerably the activating substance of liver extracts, but we could not obtain the amounts necessary for analysis. Judging from properties of the substance it seemed that a polyprenol phosphate was a good bet. Therefore dolichol, a 20-isoprene unit, saturated polyprenol present in liver, was isolated and phosphorylated chemically. The reaction product was active as an acceptor of glucose and had the same properties as the activating substance of liver. It was concluded therefore that the reaction product was dolichol-phosphate-glucose.

Here we had a new compound of unknown function that might be an artifact. Finding a role for it was not easy. Quite a lot of time was spent in trying to find out if it had a role in collagen synthesis, since this is the only glycoprotein that contains glucose. The results were negative. By chance it was found that dolichol-phosphate-glucose disappeared when incubated with liver extracts and that a new substance was formed. The role of this new substance remained rather mysterious until it was observed that it could be dissolved in chloroform-methanol-water $1:1:0.3$. Mixtures of the same solvents in other proportions did not solubilize it. This was a small but important finding. Once the right solvent was found, the new substance could be fractionated on anion exchange columns and thus obtained pure. It was decomposed by mild acid treatment yielding a lipidic part and another hydrophylic part. The water soluble part had the properties of an oligosaccharide, which is now known to contain 3 glucoses, 9 mannoses, and 2 acetylglucosamine molecules; the lipophylic part appeared to be dolichol-diphosphate. The sugar moiety was found to be transferred to protein. This part of the work was done in collaboration with A. Parodi, R. Staneloni, and H. Carminatti.

Since then the field has attracted many brilliant investigators and has progressed rapidly.

WHY RESEARCH

I do not know how it happened that I followed a scientific career. It was not a family tradition, because my parents and brothers were involved mainly in rural activities. My father had graduated as a lawyer but did not practice. In our home there were always a lot of books on the most varied subjects and I had the opportunity to acquire information on natural phenomena. I suppose the most important factor in determining my future was that I received a set of genes that gave me the required negative and positive abilities.

Among the negative abilities I might mention that my musical ear was very poor so that I could not become a composer or a musician. In most sports I was mediocre so that was another activity that did not attract me too much. My lack of oratorical ability closed the door to politics and law. I was a bad practising physician because I was never sure of the diagnosis or of the treatment.

These negative conditions were accompanied presumably by others not so negative: great curiosity in understanding natural phenomena, normal or slightly subnormal capacity for work, average intelligence, and excellent capacity for teamwork. Probably the most important thing was my opportunity to spend my days in the laboratory and to do many experiments. Most failed but a few succeeded either due to pure good luck or to having made the right mistake.

Nearly 50 years have elapsed since I started to research. These have been years of fairly hard work but with many agreeable moments. Research has many aspects that make it an attractive venture. One is the intellectual pleasure of discovering previously unknown facts. There are also human aspects worth mentioning. Some of the most pleasant periods in my career were those in which I worked with enthusiastic, clever people who had a good sense of humor. The discussion of research problems with such people is always a most stimulating experience.

The less agreeable part of research, the routine work that accompanies most experiments, is more than compensated for by the interesting aspects, which include meeting and sometimes winning the friendship of people of superior intellect from different parts of the world. The balance is clearly positive.

Literature Cited

1. Hudson, W. H. 1939. *Far Away and Long Ago*. London/Toronto: Dent. 337 pp.
2. Hudson, W. H. 1904. *The Purple Land*, p. 261. London/Toronto: Dent
3. Leloir, L. F. 1974. *5th Lynen Lecture in Biology and Chemistry of Eucaryotic Cell Surfaces, Miami Winter Symp.*, 7:1–19
4. Leloir, L. F. 1948. *Enzymologia* 12:263–76
5. Leloir, L. F. 1946. *Spec. Publ. NY Acad. Sci.* 111:60–76
6. Braun Menendez, E., Fasciolo, J. C., Leloir, L. F., Muñoz, J. M., Taquini, A. C. 1946. In *Renal Hypertension*, Illinois: Thomas. 451 pp.
7. Fasciolo, J. C. 1977. *Hypertension*, ed. J. Genest, E. Koiw, O. Kachels. New York: McGraw-Hill. 134 pp.
8. Leloir, L. F. 1972. In *Biochemistry of the Glycosidic Linkage*, ed. R. Piras, H. Pontis, pp. 1–18. New York: Academic
9. Leloir, L. F. 1971. *Science* 172:1299–1303
10. Leloir, L. F., Paladini, A. C. 1983. In *Comprehensive Biochemistry (History of Biochemistry)*. Amsterdam: Elsevier Biomedical. Submitted for publication
11. Leloir, L. F. 1964. The Fourth Hopkins Memorial Lecture. *Biochem. J.* 91:1–9
12. Behrens, N. H. 1974. See Ref. 3, pp. 159–80
13. Parodi, A. J., Leloir, L. F. 1979. *Biochim. Biophys. Acta* 559:1–37
14. Staneloni, R. J., Leloir, L. F. 1982. *CRC Crit. Rev. Biochem.* 12:289–326

Ann. Rev. Biochem. 1983. 52: 17–34

STRUCTURE AND CATALYSIS OF ENZYMES

William N. Lipscomb

Gibbs Chemical Laboratories, Harvard University, 12 Oxford Street, Cambridge, Massachusetts 02138

CONTENTS

PERSPECTIVES AND SUMMARY 17
REACTIVITIES IN SOLUTION AND IN VARIOUS CRYSTAL
    STRUCTURES 18
ZINC ENZYMES 20
  Carbonic Anhydrase EC 4.2.1.1 20
  Carboxypeptidase A (CPA) EC 3.4.17.1 22
  Thermolysin EC 3.4.24.4 24
  Liver Alcohol Dehydrogenase (EC 1.1.1.1) 25
  Comments on Zn^{+2} Enzymes 26
ALLOSTERIC ENZYME STRUCTURES 27
  Introduction 27
  Aspartate Transcarbamylase (EC 2.1.3.2) 27
  Phosphofructokinase (EC 2.7.1.11) 29
  Phosphorylases a and b (EC 2.4.1.1) 30

PERSPECTIVES AND SUMMARY

About 60 enzyme structures are known from X-ray diffraction and chemical sequence studies to atomic resolution. One class of enzymes, the serine proteinases, was treated in two of the most recent reviews (1–3). In this review, I concentrate on four enzymes containing zinc as an essential cofactor and then briefly on allosteric enzymes. This introduction and the first section provides some general background for the orientation of the reader.

A large recent increase occurred in power and scope of these studies. Improved procedures for obtaining new crystalline enzymes included amphiphilic molecules in place of detergents for solubilizing and crystallizing membrane bound proteins (4, 5). Improved methods include the use of synchrotron radiation (6), area counters (7, 8, 8a), low temperature tech-

17

niques (9), advanced use of the fast Fourier transform (10) including restraints in refinement (11), and computer graphics (12). On the theoretical side are improved methods for deducing structure from sequence (13), studies of protein dynamics (14), and the use of model reactions of small molecules and chemical theory for mechanistic deductions (15). The near future looks bright indeed, if adequate support can be found.

Now that three-dimensional structures are increasingly available since 20 years ago, the accomplishments are about an average of those who thought (or feared) that everything might be explained and those who believed that these results would be totally irrelevant. It is sensible to think of the structures of enzymes and their complexes as one of the best starting points for studies of folding, dynamics and mechanism of catalysis. It is still my hope that the advances briefly referred to above, together with the library of results at the Brookhaven Data Bank, will encourage many biochemists to add X-ray diffraction as one of the standard instrumental methods. At least, the known structures could be used for studies of derivatives, complexes, and mutants.

Most of the structural investigations of enzymes were followed by results on structures of complexes with substrates, products, or inhibitors. With very few exceptions, e.g. the binding of glucose-1-phosphate to phosphorylase b, these further structures were consistent with atomic shifts of about 1Å or less to reach a transition state, intermediate, or product. The choice of reactive groups was severely limited by these structural results, and the stereochemistry is apparent. However, substantial ambiguities inevitably remain in the mechanisms. Essentially all other methods of biochemistry are then required to aid in the resolution of these ambiguities. And, as we shall exemplify, aspects of all known enzymatic reactions remain unresolved.

After a brief section on reactivities of enzyme molecules in solution and in crystals, the active sites of some well-studied Zn-enzymes are discussed, including carboxypeptidase A, carbonic anhydrase, thermolysin and liver alcohol dehydrogenase. In the last section I turn briefly to the allosteric enzymes aspartate transcarbamylase, phosphofructokinase, and phosphorylases a and b.

REACTIVITIES IN SOLUTION AND IN VARIOUS CRYSTAL STRUCTURES

Protein crystals differ from crystals of most compounds of chemistry. They are extremely fragile, and contain on the average (16) about 43% by volume of water (range, 27 to 65%). Enzymes, as opposed to small flexible peptides, have well defined structures that, however, are not rigid. Nevertheless, most enzymes give X-ray diffraction data to atomic resolution, although certain regions may show more flexibility, which appears as disorder, than others. Usually, when a substrate binds to the crystalline enzyme, the complex shows activity comparable to that in solution unless the active site is blocked by intermolecular contacts. When a large conformational change of the enzyme occurs on binding this change affects intermolecular contacts, and the crystal is usually destroyed. I now turn to a few examples.

In tetragonal lysozyme crystals (17), intermolecular contacts interfere with the last two subsites E and F, where the six sugar rings of the substrate are designated as A-F. Nevertheless, the binding of inhibitors at the other four subsites gave a good starting point for a three-dimensional view of the mechanism thought to proceed through a stabilized carbonium ion intermediate. The molecular conformation in the triclinic form (18) of lysozyme is the same as that in the tetragonal form, except for regions involved in intermolecular contacts. Even for these regions, the expected time scale of changes of conformation in solution is short compared to times of substrate turnover.

In another enzyme, subtilisin Carlsberg, the k_{cat} and K_m values are only moderately affected by crystallization (19).

On the other hand, in glycogen phosphorylase a and phosphorylase b the maximum rates are decreased by about a factor of 10 to 100 upon crystallization (20), while K_m values are not much changed. The presently known molecular structures of both phosphorylases a (21) and b (22, 23) in the crystals are in the less active T molecular form. Although substrates produce some local conformational changes, the crystals are destroyed before the active R form is attained.

Carboxypeptidase A was the most studied example of a reduced activity in one crystalline phase. It is known (24) that the crystalline phase with 1/300 of the activity of that in solution has cell parameters a = 50.9, b = 57.9, c = 45.0A, and $\beta = 94°40'$. These crystals are probably the ones studied extensively by Vallee's group (25–33). In particular, the arsanilazo-Tyr 248 derivative is yellow in the crystals and red in solution at pH 8.2. However, the crystals used in the presently available X-ray diffraction study have different cell parameters, a = 51.41, b = 59.89, c = 47.19Å and $\beta = 97°35'$, and show the same behavior in the crystal as in solution when the arsanilazo-Tyr 248 derivative is examined (34): red at pH 8.2 and yellow at pH 7.4. It is thus illogical that the mechanistic proposals from studies of these latter crystals, which show 1/3 of the activity in solution (reduced by loss of substrate activation in the crystal), were criticized so extensively on the basis of chemical studies of a different crystalline phase, which has only 1/300 of the solution activity. In addition, the conclusion that reduced activity in a crystalline state is caused by a change in the conformation of the enzyme has a more likely explanation: that the active site is partially blocked by intermolecular contacts.

As more structures are available, so are answers to the solution-solid state discussion (35). Three environments are known for chymotrypsin molecules in crystals, two have already been noted for lysozyme, and three are available for the complex of carboxypeptidase A with inhibitors. For example, the two molecules in independent environments in the complex of carboxypeptidase A (36) with the inhibitor from the potato show rms (root mean square) deviations of 0.26 Å for main chain atoms and 0.36 Å for all atoms. These values are comparable with the experimental errors at 2.5 Å resolution. In all of these comparisons intermolecular interactions change the structures by amounts that can be expected to revert to the solution conformations in times that are short compared with turnovers. Most changes involve single-bond rotations, and formation of intermolecular salt links or hydrophobic contacts. Thus, if the activity of an enzyme is greatly reduced in one crystalline phase, the same enzyme may be active in a different crystalline phase. It would be unwise to ignore the three-dimensional structural information of an enzyme, whether or not the crystals show activity toward substrates.

ZINC ENZYMES

Carbonic Anhydrase EC 4.2.1.1 (37)

The reversible hydration of CO_2 is catalyzed in isoenzyme II(C) at a maximal turnover of 10^6 s^{-1} at 25°, and is favored above pH 7 while the proton assisted dehydration of HCO_3^- is favored below pH 7. In the forward direction, proton loss is rate limiting at low and moderate buffer concentrations; thus buffer is required for rapid catalysis in order to facilitate proton transfer (38–41). Although the crystal structure (42) also allows Glu 106 as a potential catalytic group, most investigators including the structural workers (43) strongly favored Zn bound H_2O or OH^- as either an inner sphere or outer sphere catalytic unit. The low pK_a is associated with the neutrality of imidazole ligands to the protein, and with the deep partially hydrophobic pocket in which the Zn^{+2} is bound. The most probable mechanism is inner sphere, based on $L_3Zn^{+2}OH^-$ where a lone pair of OH^- attacks the carbon of CO_2, in the hydration direction. Here, L is His, and in the resting enzyme the Zn^{+2} is four coordinated.

Interest in increased coordination to 5 when a neutral inhibitor is bound arose when Khalifah (44) showed that in the B enzyme imidazole inhibits the catalytic hydration of CO_2, and when a nitrogen of bound imidazole was then found at a distance of 2.7 Å from the otherwise four-coordinated Zn^{+2} in carbonic anhydrase at pH 8.7 (42, 43). If the principle of electroneutrality [relatively constant charge density (45)] is added to the four-five coordination change (46), a sequence of stages can be proposed for the hydration of CO_2 at pH 7.5, where the active site is mostly $L_3Zn^{+2}OH^-$.

1. CO_2 binds near Zn^{+2},
2. the new CO bond is formed to give HCO_3^- with its OH on Zn^{+2};
3. proton transfer gives Zn^{+2} bound HCO_3^- with its OH not on Zn;
4. H_2O adds to Zn^{+2} in this complex;
5. departure of HCO_3^- may be aided by ionization of this H_2O, as
6. proton transfer occurs to a group on the enzyme, then to buffer and then to solvent.

The binding of CO_2 at pH 5.5 does not influence the 2341 cm^{-1} asymmetric stretching mode sufficiently to require binding of one oxygen to the Zn^{+2} ion (47). While this result is consistent with binding of CO_2 near, but not at, Zn^{+2}, this site may be nonproductive (44). Moreover, one cannot exclude the momentary bonding of CO_2 to Zn^{+2} in a transient catalytic complex. However, the high pH form, $L_3Zn^{+2}OH^-$, would be less likely to bind oxygen of CO_2 than would the low pH $L_3Zn^{+2}OH_2$ form of the enzyme.

Step 2 is assumed here to occur while Zn^{+2} remains four-coordinated, so that the pKa remains low, maximizing the percentage of $L_3Zn^{+2}OH^-$ form of the enzyme that reacts with CO_2. While it is true that nucleophilicity of this Zn^{+2} bound OH^- would be increased by a change to five coordination (46), say by the addition of water, the concentration of the active form of the enzyme would than be reduced. This is a delicate balance that needs further study.

The proton transfer in Step 3 could be facilitated by Zn-bound H_2O in a five-coordinated complex assigned to Step 4. However, molecular orbital calculations (J. V. Ortiz, W. N. Lipscomb, unpublished study) suggest a low barrier for the direct, nonfacilitated proton transfer within bound HCO_3^-. The negative charge [APS distribution (48)] is about equally distributed among the three oxygens of HCO_3^-. Here, the charge density on Zn^{+2} is not raised as much as it is when OH^- adds. Thus the addition of H_2O in Step 4 may be easier when HCO_3^- is bound than it would be when OH^- is bound (see above).

The departure of HCO_3^- is also aided by this tendency toward equal charge distribution among oxygens in HCO_3^-, and its loss is further facilitated if the fifth ligand (H_2O) loses a proton (J. V. Ortiz, W. N. Lipscomb, unpublished study).

As noted above, this 6th step, proton transfer, is facilitated by buffer (38–41). Lindskog, et al (37) propose at least two steps in proton transfer, first from the catalytic group (EH^+) to the proton transfer group (HE^+), which may be His 64 in isoenzyme II, and then transfer to buffer.

The recent discovery of uncompetitive inhibition of CO_2 hydration by chloride ion (37, 46) was interpreted in two ways. Lindskog, et al (37) regard the inhibition by anions as requiring the protonated form of the enzyme EH^+. The anion would have little effect at high pH and low CO_2 concentrations where E is dominant, but would bind to EH^+ as it accumulates in the steady state at low pH and high CO_2 and buffer concentrations. The resulting effect on k_{cat} would appear as uncompetitive inhibition. On the other hand Pocker & Deitz (46) suggest that anions bind to a fifth coordination site on Zn^{+2} as the charge density on Zn^{+2} is relieved by attack of oxygen of $L_3Zn^{+2}OH^-$ on the carbon of CO_2, which has neither oxygen bound to Zn^{+2}. Clearly, the recent discovery of anion inhibition at high pH has opened a new approach to the mechanism.

Evidence that intramolecular H^+ exchange and the buffer dependent H^+ transfer are independent of the CO_2-HCO_3^- exchange pathway is that the rates of ^{13}C NMR exchange are the same in H_2O and in D_2O (49, 50). This CO_2-HCO_3^- exchange is competitively inhibited by Cl^-, indicating that a ternary complex among enzyme, substrate, and inhibitor is not formed at pH 6.8. This result is consistent with the conclusion (37) that the uncompetitive inhibition at high pH is due to accumulation of EH^+ in the steady state.

Independent rates were obtained for this CO_2-HCO_3^- exchange and the exchange between CO_2 and water by examination of ^{18}O and ^{13}C exchanges at equilibrium (51). Because ^{12}C and ^{13}C exchange occurs on those molecules having ^{18}O, the ^{18}O from one substrate molecule must remain in the active site so that this oxygen can enter another substrate molecule. The conclusion is that both intramolecular and buffer-mediated H^+ exchange are required for rapid removal of oxygen as well as protons from the active site in the dehydration direction.

While the simplified sequence 1–6 given above is favored, ambiguities remain on a fundamental level. Even an outer sphere reaction is still possible, as are dissociative mechanisms rather than changes between four and five coordination. Participation of Glu 106 in the reaction is not excluded. However, the studies of recent years have focused mostly on the inner sphere zinc-hydroxyl reaction for hydration of CO_2 and the zinc ion catalyzed dehydration of HCO_3^-. Finally, the role of divalent ions, such as SO_4^{-2}, in the effective pKa's of the reaction is more important (37, 45, 52) than had previously been thought.

Carboxypeptidase A (CPA) EC 3.4.17.1

The active site for cleavage of the C-terminal amino acid of a peptide substrate has been defined from X-ray diffraction studies of the enzyme (53), and of its complexes with Gly-Tyr (54), (–)-2-benzyl-3-p-methoxybenzoylpropionic acid (55) (an ester analog), and the 39 amino acid inhibitor from the potato (36, 56). The single water molecule on Zn is displaced by the carbonyl oxygen of the ester analog, and by both the carbonyl oxygen and the NH_2 terminus of Gly-Tyr. In the potato inhibitor the C-terminal Gly 39 was cleaved, leaving a product with a newly formed carboxylate with one oxygen on Zn.

These studies indicate that the two atoms of the scissile peptide or ester bond are near Glu 270, which could attack the carbonyl carbon to make an anhydride (57) or could promote the attack of water on this carbonyl carbon, and Tyr 248, which could donate a proton to the leaving NH as it becomes NH_2. These models suggest that in the initial stage of the reaction, the Zn polarizes the carbonyl group, making the carbonyl carbon more susceptible to attack, although the Zn could function in deacylation if the anhydride intermediate is formed (58, 59). Alternatively, the participation of a Zn-hydroxyl group in initial attack of the carbonyl carbon of the substrate is not out of the question (53, 59).

However, an anhydride intermediate, which, in view of the structure should be at Glu 270, was isolated at –40° for the ester substrate O-(trans-p-chlorocinnamoyl)-L-β-phenyllactate (60, 61). Acylation depends at –40°C on pKa's of 7.65 for the Zn^{+2} enzyme and 6.33 for the Co^{+2} enzyme. This former pKa was assigned to the L_3ZnOH_2 ionization, where L refers to the ligands from the protein (His 69, Glu 72, and His 196). It would be of considerable interest to discover whether a covalent intermediate occurs for other ester substrates, whether peptide substrates use this anhydride pathway at a greater cost of free energy than that for esters, and which rate steps dominate for esters and peptides at various temperatures.

Use of ^{17}O broadening of Co^{+2} resonance in the anhydride at about pH 7.5 indicates (62) that Co^{+2} binds to both H_2O (or OH^-) and to the carbonyl oxygen derived from substrate in formation (from enzyme and substrate) of the mixed anhydride with Glu 270. It is not yet certain that the anhydride formed from Zn^{+2} enzyme and substrate also has two nonprotein ligands to Zn^{+2}, nor that there are two nonprotein ligands in the active complex as it undergoes deacylation. Theoretical models of this active site intermediate indicate that a carbonyl oxygen becomes unbound as the $ZnOH_2$ ionization occurs, (R. Staudigl, W. N. Lipscomb, unpublished study), and an X-ray diffraction study of the complex of apo CPA with Gly-Tyr indicates that the carbonyl group of this dipeptide binds to Arg 127, displaced by only about 1Å from the now-vacant Zn^{+2} site (63). Moreover, in the holoenzyme, a lowering of coordination number would lower the pKa of H_2O on Zn^{+2}. Hence, deacylation may proceed by this shift of the carbonyl oxygen from Zn^{+2} to Arg 127, thus promoting ionization of the Zn-bound water.

A kinetically indistinguishable kinetic scheme is the "reverse protonation" mechanism of Mock (64), in which Tyr 248 in its anionic form catalyzes the attack of H_2O on the carbonyl carbon of the substrate. The concentration of active enzyme is down by about a factor of 10^3 over that of the mechanisms described above. This proposal was used to interpret the retention near pH 6.5 of the peptidase activity of the nitro-Tyr 248 enzyme (64). This activity was previously thought to be very low (65). Another less-specified proposal is characterized by the binding of Tyr 248 anion to Zn^{+2}. These studies are confined to the arsanilazo-Tyr 248 derivative (26–33), which at low substrate concentrations has only about 1/3 of the activity of the unmodified enzyme (66). However, at pH 7.5 (the pKa of the arsanilazo-Tyr 248 interaction with Zn^{+2}) the unmodified enzyme has no measurable binding of Tyr 248 to the Zn^{+2} ion (67). Even when crystals are grown at pH 8.5, no bonding of Tyr 248 can be detected (G. Shoham, W. N. Lipscomb, study in progress). It is therefore unlikely that this Tyr-Zn^{+2} interaction is a property of the unmodified enzyme; thus, it is not related to the catalytic activity. It remains therefore probable that the nucleophile is Glu 270, or that Glu 270 promotes the attack of water, in the formation of the first tetrahedral intermediate of the reaction of carboxypeptidase A with substrates.

Thermolysin (EC 3.4.24.4)

The modes of binding of extended substrates were illuminated by X-ray diffraction studies of inhibitor-enzyme complexes of thermolysin (68–71). Of these, the hydroxamates (72), which have inhibition constants as low as 10^{-6} M, bind in the anionic form RCO N⁻OH (or its resonance hybrid). Here, the carbonyl oxygen and hydroxyl oxygen are the fourth and fifth ligands to Zn^{+2}, in addition to the three protein ligands, His 142, His 146, and Glu 166. Moreover, the N-hydroxyl group of the bound hydroxamate forms a hydrogen bond to Glu 143, and is close to a probable position suggested earlier for H_2O attack promoted by Glu 143. Thus substrates are believed to be hydrolyzed by general acid-base catalysis, where Glu 143 promotes the attack of water and His⁺231 is the proton donor (71). The H_2O that attacks the carbonyl carbon of the substrate is postulated to bind to both Glu 143 and Zn^{+2}. [A similarly bound H_2O was suggested for the carboxypeptidase A mechanism on theoretical grounds (73).]

The Zn^{+2} sites are very similar in thermolysin and carboxypeptidase A, while the protein structures are very different. The main difference from a mechanistic point of view is that Glu 143 is far enough from the carbonyl carbon of substrate to accommodate an intervening water molecule in thermolysin, while Glu 270 is in Van der Waals' contact with the carbonyl carbon of Gly-Tyr in carboxypeptidase A. In thermolysin the general base pathway, assisted by Zn^{+2}, which binds both the water molecule and the substrate's carbonyl oxygen, is perhaps preferred. Although sufficient distortion might occur to make an anhydride intermediate, this pathway seems less favorable, and the mechanism seems to be less ambiguous in thermolysin than in carboxypeptidase A.

Liver Alcohol Dehydrogenase EC 1.1.1.1

In horse liver alcohol dehydrogenase (LADH), the indication from X-ray diffraction studies (74, 75) of inner sphere bonding to Zn^{+2} by oxygen of substrate was strengthened by a study of the ternary complex (76). In contrast, the electron spin resonance evidence (77) for outer sphere binding of substrate was questioned (78) because of weakness of the signal. More recently, this ESR evidence (77) was attributed to a sample in which Co^{+2} substitution occurred at the structural Zn^{+2} site instead of the catalytic site (79). The occurrence of an inner sphere complex is also in agreement with kinetic evidence in the reduction of trans-4-(N,N'-dimethylamino) cinnamaldehyde (80). The same substrate leads to the order: (a) activation of the aldehyde for hydride attack by binding of oxygen to Zn^{+2}, (b) hydride transfer; (c) proton transfer to the resulting Zn-bound alcoholate ion (81); and (d) product release (82). Whether the hydride transfer from NADH occurs as such, or is broken into steps is still an open question (83, 84).

The Zn^{+2} of the active site is coordinated to Cys 46, His 67, and Cys 174 of the enzyme, and to a water molecule. The pKa of either this water (85) or His 51 (86) changes from 9.6 to 7.6 when NAD^+ binds. The proposal that the Zn^{+2} becomes five coordinated as the substrate ROH binds (87), is supported by the five-coordinated complex observed when 1-10 phenanthroline binds as an inhibitor (85). When both OH^- and ROH are presumed to be bound to Zn^{+2} the OH^- is stabilized by a hydrogen bond to Ser 48, which is hydrogen bonded to His 51. Hydride transfer to NAD^+ from the CH_2 group of ROH, which loses its proton to OH^-, then completes the oxidation to aldehyde. While this five-coordinated intermediate is reasonable, it is not yet proved as a unique mechanism. However, the enzyme in which Co^{+2} was substituted for Zn^{+2} at the active site (88), shows an increase, when NAD^+ binds, from four to five ligands: three from the protein and probably two water molecules, as deduced from zero field splitting parameters in the electron paramagnetic resonance (89). The ternary complex (Co^{+2} enzyme, NAD^+ and pyrazole) also shows five-coordinated metal in these EPR experiments. The kinetic data, while interpretable on the basis of four-coordinated Zn^{+2} in the native enzyme (90), were reevaluated in terms of five-coordinated Zn^{+2} (86). Whether the coordination is maintained at five, or changes between five and four, and whether the pKa is due to His 51 or water on Zn^{+2} remain as open questions, which may soon be answered. At least the X-ray diffraction results of the ternary complex of LADH with NADH and dimethylsulfoxide established that there are conformational changes, and that the Zn^{+2} is four coordinated to the three protein ligands and DMSO.

Comments on Zn^{+2} Enzymes

Because of differences in ligands to Zn^{+2} and the presence of potential nucleophiles other than ZnOH or proton donors other than $ZnOH_2$, few generalizations are apparent for the four Zn enzymes briefly discussed here. In the three hydrolytic enzymes the pK_a of H_2O on Zn varies from about 7 for carbonic anhydrase, which also has three neutral His ligands to Zn^{+2} in a deep pocket, to much higher values in carboxypeptidase A and thermolysin. Also, a high value is expected in liver alcohol dehydrogenase in the absence of NAD^+, because of the two Cys^- ligands.

In all of these examples most probably H_2O, not OH^-, is displaced from Zn^{+2} by the substrate. This initial binding may proceed through increase of coordination number followed by loss of H_2O. Similarly, loss of a negatively charged group from Zn^{+2} probably proceeds by addition of H_2O as a fifth ligand followed by loss of a proton from this H_2O depending on its pKa and the pH. These changes from four to five to four coordination of Zn^+ appear to be common among these enzymes.

Most probably, Zn^{+2} delivers OH^- to the carbon of CO_2 in carbonic anhydrase and to the mixed-anhydride intermediate of an ester substrate of carboxypeptidase A, and assists in general base catalysis by thermolysin. The Zn^{+2} ion also serves a polarization function: activating nucleophilic attack in the initial stage of carboxypeptidase reaction, and activating the substrates of liver alcohol dehydrogenase. Although Zn^{+2} is not unique in activities of these enzymes, only adjacent and nearby doubly ionized metals of the periodic table such as Co^{+2} and Ni^{+2} show comparable activity towards natural substrates; and when they do, it is yet to be established that their coordination is the same as that for Zn^{+2}. Functional metals may be restricted in carbonic anhydrase to those that yield a low pKa for the metal bound water, and in carboxypeptidase A to metals that have a high pKa for displaceable water in the resting enzyme, and a low pKa for a water molecule in a hydrophobic environment as the substrate is bound and the reaction proceeds. The choice of mechanism then depends on how the local environment affects the Zn^{+2}-water ionization, and also very much on the presence and spacial distribution of other catalytic groups, which are proton donors, nucleophiles, or general base catalysts.

ALLOSTERIC ENZYME STRUCTURES

Introduction

Because detailed structures of both active (R) and inactive (T) forms of no allosteric enzyme are yet available, this section is a brief progress report. Within a year or so, detailed structural information will likely be available on aspartate transcarbamylase (91, 92) of *Escherichia coli,* phosphofructokinase (93, 94) of *Bacillus stearothermophyllus,* and glycogen phosphorylases a (21) and b (22, 23) of rabbit muscle. At present, at least one detailed structure is known for each of these allosteric enzymes, and some indication was found for the very large conformational changes of each of them.

Aspartate Transcarbamylase (EC 2.1.3.2)

This enzyme catalyzes the reaction of carbamyl phosphate and aspartate to yield carbamyl aspartate and phosphate. Both substrates produce homotropic (sigmoidal) kinetics. Moreover, allosteric inhibition by cytidine triphosphate (CTP) and activation by ATP occur (95–97). The catalytic chain c (34,000 daltons) and regulatory chain r (17,000 daltons) assemble to c_3 and r_2 and then to the holoenzyme c_6r_6 (98–100) of symmetry D_3. The active site is shared between adjacent c chains within each catalytic trimer.

In the unliganded, or CTP-liganded, enzyme the two c_3 subunits are in contact (101). When the potent inhibitor *N*-phosphonacetyl-L-aspartate (PALA) (102) is bound at saturating concentration, the c_3 units move apart as indicated by changes in unit cell parameters (92) at pH 5.8 and in low angle X-ray scattering (103) at pH 8.3. The X-ray diffraction study shows that the two c_3 units move apart by 11–12 Å, and reorient relatively by 8–9°, while each regulatory dimer reorients by 14–15° about its two-fold axis. These large changes preserve the crrc units, where the two c's are in separate c_3's and the two r's are within a regulatory dimer. These crrc units, apparent from the structure (100), were suggested as important functional units (104, 105) in the cooperative behavior of this enzyme. A recent calculation (R. B. Altman, J. E. Ladner, W. N. Lipscomb, study in progress) of the low angle X-ray scattering curves from the T and R structures at pH 5.8 indicates excellent agreement with the observed curves (103) at pH 8.3. Hence, the quaternary structural changes that occur when the T to R transition is induced by *N*-phosphonacetyl-L-aspartate (PALA) are very similar at these two pH values.

Details of the associated tertiary structural changes, however, are still being obtained from the refinement of the PALA-ligated enzyme. Therefore, detailed molecular interpretations are not yet available for the increase of activity by about a factor of five upon activation from the T form to the R form, and for the regulation by CTP and ATP. The molecular pathways for these interactions occur over a distance of at least 60 Å from regulatory to catalytic sites, and of 22 Å between catalytic sites within a catalytic trimer.

Negative cooperativity between r chains in the r_2 subunit (105–109) is associated with disruption of coupling between N-terminal regions as the first molecule of allosteric effector binds to r_2 in the enzyme (110). The structure of the enzyme in the absence of substrate or allosteric effectors shows full D_3 symmetry, and the r's are symmetry equivalent in the r_2 units of the unliganded enzyme. When CTP binds to one r unit, the N-terminal strand of this unit moves toward CTP; when ATP binds, this strand moves away from the ATP binding site. These motions perturb the N-terminal region of the adjacent r chain within r_2.

Nucleoside triphosphates also bind to the isolated catalytic trimer in a negatively cooperative manner: the first molecule lowers the affinity of the second and third effector molecules by 10–100 fold (109). The most obvious constraint on the c . . . c interactions is the salt link, Arg 54-Glu 86, which, when broken, may stabilize the previously symmetry-related salt links (110).

These c . . . c interactions within c_3 units, plus the crrc allosteric units, plus the c_3 . . . c_3 interactions in the T form of the enzyme make the cooperativity somewhat more complex than a simple two-state model, which is a first approximation. The effects of these unusual quaternary constraints were studied in hybrids in which mutant or chemically modified c_3 subunits have replaced the wild-type c_3 (111).

A further complication, or "insurmountable opportunity,"[1] is that various chemical or genetic modifications can greatly reduce homotropic interactions, although the inhibition by CTP and activation by ATP are retained. These results imply separate pathways or structural mechanisms for homotropic and heterotropic effects (112–117). At least one of these modifications, nitration of Tyr 232(?), has been located in the provisional sequence (W. H. Konigsberg, personal communication) by S. M. Landfear of Harvard, who isolated the modified peptide.

Finally, the T to R transition lowers Km for Asp from 90 mM to 7 mM (114), and hence may be associated with apparent conformation changes in the active site of the enzyme (J. E. Ladner, J. P. Kitchell, K. W. Volz, H. M. Ke, W. N. Lipscomb, study in progress).

---

[1] With apologies to Walt Kelly and "Pogo."

Phosphofructokinase (EC 2.7.1.11)

The enzyme from *B. stearothermophilus,* under study by X-ray diffraction (93, 94), is a tetramer of molecular weight 4 × 33,900 daltons, containing 4 × 316 amino acids (118). It is somewhat different from the yeast and mammalian enzymes. Catalysis of the key control step of glycolysis, the phosphorylation of fructose-6-phosphate by ATP to form fructose-1, 6-biphosphate and ADP, is under allosteric control: the kinetics are sigmoidal depending on the substrate F-6-P, and this cooperative behavior in the bacterial enzyme is inhibited by phosphoenolpyruvate (PEP) and stimulated by ADP. The two-state model applies; the R form, induced by F-6-P and favored by ADP, binds F-6-P more strongly (lower K_m) than does the T state (higher K_m), while V_{max} and k_{cat} are about the same for these two states of the enzyme (119).

The structure of the more active (R) form, crystallized in the presence of F-6-P, was solved to atomic resolution (93, 94). Transition to the T form by removal of the activating ligands destroys these crystals. However, crystals of the T form have been obtained in the presence of the allosteric inhibitor PEP or 2-phosphoglycollate (a weaker inhibitor, which is more stable toward hydrolysis than is PEP). The T structure was solved to about 6 Å resolution, which is sufficient to show that the relative shifts of the subunits move a helix toward the active site region as the R to T transformation occurs, so that binding of the substrate is reduced, in qualitative accord with the K_m values (P. R. Evans, private communication).

The active site consists of two adjacent regions that bind F-6-P and ADP. ATP probably binds in this ADP site, where an ATP analog, 5'-adenylimidodiphosphate, binds. The regulatory site binds either ADP or PEP, provided that inorganic phosphate is not present. The active site lies in a cleft between two domains of a subunit in such a way that the phosphate of F-6-P is shared between subunits. One domain binds primarily ATP, while the other binds F-6-P, except for two arginines of an adjacent subunit that bind to the phosphate of F-6-P. The effector site lies in a different cleft between subunits. It is the diphosphate and Mg^{+2} of ADP that are buried in this cleft, rather than the adenine. One might expect that this sharing of an active or effector site between domains or subunit will be discovered in other, but not all, multi-subunit enzymes. This feature allows the substrate or effector to lock the quaternary, and some aspects of tertiary, structure into an active or inactive arrangement of subunits or domains.

Relative movement of domains is well established in the kinases, and recently was found in citrate synthase (120) where the more open forms are probably associated with reactant assembly and product release, while the relatively closed forms promote catalysis. Movements of subunits and tertiary structure were described in detail for hemoglobin by Perutz and co-workers (121, 121a). However, at present there are few unifying generalizations that can be drawn from these conformational changes.

Phosphorylases a and b (EC 2.4.1.1)

The rabbit skeletal muscle enzyme in the b (nonphosphorylated) form is inactive except in the presence of the allosteric activators, AMP, IMP, or analogs of AMP, and is inhibited by ATP, ADP, glucose, or glucose-6-phosphate. In the active form, Ser 14 has been phosphorylated by phosphorylase kinase. The activity of phosphorylase a is increased by AMP and decreased by glucose. Integration of these aspects of glycogen metabolism into a network of regulatory pathways has recently been reviewed (122).

Three dimensional structures are now available for the T form of phosphorylase a (21), inhibited by glucose, and the T form of phosphorylase b (23, 23) in the presence of IMP. The sequence of the 841 amino acids was carried out by Titani et al (123). Pyridoxal 5'-phosphate as a cofactor is bound as a Schiff's base to Lys 679, and the enzyme remains active when this base is reduced by sodium borohydride (124).

In the T form of the enzyme, the catalytic site is 32 Å from the allosteric AMP site and 30 Å from the glycogen (attachment) site, while these last two are 39 Å from one another. The nucleoside site is 10–12 Å from the catalytic site in the general direction of the glycogen site. The enzyme is a dimer, having separations of 45 Å between the allosteric and catalytic sites on separate subunits. While Ser 14, phosphorylated in the a form of the enzyme, is near the interface between subunits, the mechanism by which it affects the T $\leftrightarrows$ R equilibrium is not understood (21).

From the studies of binding of G-1-P and of the cyclic G-1,2-P it has been suggested that G-1-P binds in a nonproductive mode (23). While the sugar rings of these two inhibitors bind in about the same region, the phosphate group of G-1-P is about 2.4 Å away from that of G-1,2-P. The earlier suggestion (22) that some reorientation of G-1-P occurs for productive binding is supported by this other position for phosphate found in G-1,2-P. This new position for phosphate is near the 5'-phosphate of pyridoxal phospate, and thus supports the view that this 5'-phosphate acts as a proton shuttle to promote attack of inorganic phosphate on α-1,4-glucoside polymer to split off α-D-glucose-1-P. This alternative orientation of G-1-P, obtained from model building, allows attack of inorganic phosphate at C1 with retention of configuration after cleavage of the C1-O1 bond. The 5'-phosphate of pyridoxal phosphate can function as a dianion (124) to promote and stabilize a carbonium-oxonium ion intermediate, and His 376 can serve as a general base for the phosphorylation reaction (23). Whether these views of the mechanism are correct will be further tested when crystals of the R form of the enzyme become available.

At the allosteric nucleoside site, 10–12 Å from the active site, several inhibitors including adenine, caffeine, adenosine, inosine, ATP, or FMN were studied structurally and thermodynamically in glucose-inhibited phosphorylase a (125). These inhibitors form an intercalative complex in which the ring system of the inhibitor is stacked between the side chains Phe 285 and Tyr 612. This binding mode stabilizes the 282–286 region so that α-D-glucose binds more readily. The thermodynamic aspects are unusual: ΔC_p of association of the inhibitor FMN increases from a normally large negative value (126) at 15°C to an abnormally small negative value at 35°C.

Because the previous review of allosteric behavior, as judged from conformational changes in the crystals of the T form, is so recent and complete, the reader is referred to that review (21). Large conformational changes will be seen in the R form, when the structures are solved, in view of the destruction of the crystals of the T form by high concentrations of glucose-1-phosphate (127). Finally, a new review has appeared (128).

ACKNOWLEDGMENTS

I wish to thank S. Lindskog for a communication on carbonic anhydrase, and the National Institutes of Health (GMO6920) for support.

Literature Cited

1. Blow, D. M., Steitz, T. A. 1970. Ann. Rev. Biochem. 39:63–100
2. Kraut, J. 1977. Ann. Rev. Biochem. 46:331–58
3. Fletterick, R. J., Madsen, N. B. 1980. Ann. Rev. Biochem. 49:31–61
4. Michel, H., Oesterhelt, D. 1980. Proc. Natl. Acad. Sci. USA 77:1283–85
5. Garavito, R. M., Rosenbusch, J. P. 1980. J. Cell. Biol. 86:377–87
6. Rosenbaum, G., Holmes, K. C., Witz, J. 1970. Nature 230:434–37
7. Arndt, U. V., Champness, J. N., Phizackerley, R. P., Wonacott, A. J. 1973. J. Appl. Crystallogr. 6:457–63
8. Cork, C., Fehr, D., Hamlin, R., Vernon, W., Xuong, N. H., Perez-Mendez, V. 1974. J. Appl. Crystallogr. 6:457–63
8a. Cork, C., Fehr, D., Hamlin, R., Vernon, W., Xuong, N. H., Perez-Mendez, V. 1974. Acta Crystallogr. Sect. A 31:702–03
9. Bartunik, H. D., Schubert, P. 1982. J. Appl. Cryst 15:227–31
10. Ten Eyck, L. F. 1977. Acta Crystallogr. Sect. A 33:486–92
11. Konnert, J. H., Hendrickson, W. A. 1980. Acta Crystallogr. Sect. A 36:344–50
12. Britton, E. G., Lipscomb, J. S., Pique, M. E. 1978. Comput. Graphics 12(3):222–27
13. Ptitsyn, O. B., Finkelstein, A. V. 1980. Q. Rev. Biophys. 13:339–86
14. Karplus, M., McCammon, J. A. 1981. CRC Crit. Rev. Biochem. 9:293–349
15. Jencks, W. P. 1975. Adv. Enzymol. Rel. Areas. Mol. Biol. 43:219–410
16. Matthews, B. W. 1968. J. Mol. Biol. 33:491–501
17. Blake, C. C. F., Mair, G. A., North, A. T. C., Phillips, D. C., Sharma, V. R. 1967. Proc. R. Soc London Ser. B 167:365–77
18. Moult, J., Yonath, A., Traub, W., Smilansky, A., Podjarny, A., Rabinovich, D., Saya, A. 1976. J. Mol. Biol. 100:179–95
19. Tuchsen, E., Ottesen, M. 1977. Carlsberg Res. Commun. 42:407–20
20. Kavinski, P. J., Madsen, N. B. 1976. J. Biol. Chem. 251:6852–59

21. Fletterick, R. J., Madsen, N. B. 1980. *Ann. Rev. Biochem.* 49:31–61
22. Johnson, L. N., Jenkins, J. A., Wilson, K. S., Stura, E. A., Zanotti, G. 1980. *J. Mol. Biol.* 140:565–80
23. Jenkins, J. A., Johnson, L. N., Stuart, D. I., Stura, E. A., Wilson, K. S., Zanotti, G. 1981. *Philos. Trans. R. Soc. London Ser. B* 293:23–41
24. Quiocho, F. A., Richards, F. M. 1966. *Biochemistry* 5:4062–76
25. Vallee, B. L., Riordan, J. F., Johansen, J. T., Livingston, D. M. 1971. *Cold Spring Harbor Symp. Quant. Biol.* 36: 517–31
26. Johansen, J. T., Vallee, B. L. 1971. *Proc. Natl. Acad. Sci. USA* 68:2532–35
27. Johansen, J. T., Livingston, D. M., Vallee, B. L. 1972. *Biochemistry* 11: 2584–88
28. Johansen, J. T., Vallee, B. L. 1973. *Proc. Natl. Acad. Sci. USA* 70:2006–10
29. Riordan, J. F., Muszynska, G. 1974. *Biochem. Biophys. Res. Commun.* 57: 447–51
30. Harrison, L. W., Auld, D. S., Vallee, B. L. 1975. *Proc. Natl. Acad. Sci. USA* 72:3930–33
31. Harrison, L. W., Auld, D. S., Vallee, B. L. 1975. *Proc. Natl. Acad. Sci. USA* 72:4356–60
32. Spilburg, C. A., Bethune, J. L., Vallee, B. L., Scheraga, H. A. 1980. *Biochemistry* 19:759–66
33. Schuele, R. K., Van Wart, H. E., Vallee, B. L., Scheraga, H. A. 1980. *Biochemistry* 19:759–66
34. Quiocho, F. A., McMurray, C. H., Lipscomb, W. N. 1972. *Proc. Natl. Acad. Sci. USA* 69:2850–54
35. Rupley, J. A. 1969. The comparison of protein structure in the crystal and in solution. In *Biological Macromolecules Series*, ed. S. N. Timasheff, G. D. Rasman, II:291–352. New York: Dekker.
36. Rees, D. C., Lipscomb, W. N. 1982. *J. Mol. Biol.* 160:475–98
37. Lindskog, S., Ibrahim, S. A., Jonsson, B.-H., Simonsson, I. 1982. Carbonic anhydrase: structure, kinetics, and mechanism. In *Coordination Chemistry of Metalloenzymes in Hydrolytic and Oxidative Processes*, ed. I. Bertini, R. A. Drago, C. Luchinat. NATO Advanced Study Inst. Dordrecht, Netherlands: Reidel. In press
38. Khalifah, R. G. 1973. *Proc. Natl. Acad. Sci. USA* 70:1986–89
39. Prince, R. H., Woolley, P. R. 1973. *Bioorg. Chem.* 2:337–44
40. Lindskog, S., Coleman, J. E. 1973. *Proc. Natl. Acad. Sci. USA* 70:2502–8
41. Jonsson, B.-H., Steiner, H., Lindskog, S. 1976. *FEBS Lett.* 64:310–14
42. Kannan, K. K., Notstrand, B., Fridborg, K., Lovgren, S., Ohlsson, A., Petef, M. 1975. *Proc. Natl. Acad. Sci. USA* 72:51–55
43. Kannan, K. K., Petef, M., Fridborg, K., Cid-Dresdner, H., Lovgren, S. 1977. *FEBS Lett.* 73:115–19
44. Khalifah, R. G. 1971. *J. Biol. Chem.* 246:2561–73
45. Koenig, W. H., Brown, R. D. III, Jacob, G. S. 1980. The pH-independence of carbonic anhydrase activity: apparent pK_a due to inhibition by HSO_4^- In *Biophysics and Physiology of Carbon Dioxide*, ed. C. Bauer, H. Bartels, pp. 238–43. New York: Springer
46. Pocker, Y., Deitz, T. L. 1981. *J. Am. Chem. Soc.* 103:3949–51
47. Riepe, M. E., Wang, J. H. 1968. *J. Biol. Chem.* 243:2279–87
48. Armstrong, D. R., Perkins, P. G., Stewart, J. J. P. 1973. *J. Chem. Soc. Dalton Trans.* pp. 838–40
49. Koenig, S. H., Brown, R. D., London, R. E., Needham, T. E., Matwiyoff, N. A. 1974. *Pure Appl. Chem.* 40:103–13
50. Simonsson, I., Jonsson, B.-H., Lindskog, S. 1979. *Eur. J. Biochem.* 93:409–17
51. Silverman, D. N., Tu, C. K., Lindskog, S., Wynns, G. C. 1979. *J. Am. Chem. Soc.* 101:6734–40
52. Bertini, I., Canti, G., Luchinat, C., Scozzafava, A. 1977. *Biochem. Biophys. Res. Commun.* 78:158–60
53. Lipscomb, W. N., Hartsuck, J. A., Reeke, G. N. Jr., Quiocho, F. A., Bethge, P. H., et al. 1968. *Brookhaven Symp. Biol.* 21:24–90
54. Rees, D. C., Lewis, M., Honzatko, R. B., Lipscomb, W. N., Hardman, K. D. 1981. *Proc. Natl. Acad. Sci. USA* 78:3408–12
55. Rees, D. C., Honzatko, R. B., Lipscomb, W. N. 1980. *Proc. Natl. Acad. Sci. USA* 77:3288–91
56. Rees, D. C., Lipscomb, W. N. 1980. *Proc. Natl. Acad. Sci. USA* 77:4633–37
57. Reeke, G. N., Hartsuck, J. A., Ludwig, M. L., Quiocho, F. A., Steitz, T. A., Lipscomb, W. N. 1967. *Proc. Natl. Acad. Sci. USA* 58:2220–26
58. Lipscomb, W. N. 1974. *Tetrahedron* 30:1725–32
59. Lipscomb, W. N. 1980. *Proc. Natl. Acad. Sci. USA* 77:3875–78
60. Makinen, M. W., Yamamura, K., Kaiser, E. T. 1976. *Proc. Natl. Acad. Sci. USA* 73:3882–86

61. Makinen, M. W., Kuo, L. C., Dymowski, J. J., Jaffer, S. 1979. *J. Biol. Chem.* 254:356–66
62. Kuo, L. C., Makinen, M. W. 1982. *J. Biol. Chem.* 257:24–27
63. Rees, D. C., Lipscomb, W. N. 1983. *Proc. Natl. Acad. Sci. USA* In press
64. Mock, W. L., Chen, J.-T. 1980. *Arch. Biochem. Biophys.* 203:542–52
65. Riordan, J. F., Sokolovsky, M., Vallee, B. L. 1967. *Biochemistry* 6:3609–17
66. Urdea, M. S., Legg, J. I. 1979. *J. Biol. Chem.* 254:1868–74
67. Rees, D. C., Lipscomb, W. N. 1981. *Proc. Natl. Acad. Sci. USA* 78:5455–59
68. Weaver, L. H., Kester, W. R., Matthews, B. W. 1977. *J. Mol. Biol.* 114:119–32
69. Kester, W. R., Matthews, B. W. 1977. *Biochemistry* 16:2506–16
70. Bolognesi, M. C., Matthews, B. W. 1979. *J. Biol. Chem.* 254:634–39
71. Holmes, M. A., Matthews, B. W. 1981. *Biochemistry* 20:6912–20
72. Nishimo, H., Powers, J. C. 1978. *Biochemistry* 17:2846–50
73. Nakagawa, S., Umeyama, H., Kitaura, K., Morokuma, K. 1981. *Chem. Pharm. Bull.* 29:1–6
74. Brändén, C.-I., Jornvall, H., Eklund, H., Furugren, B. 1975. Alcohol dehydrogenases. *Enzymes* 11:103–90
75. Brändén, C.-I., Eklund, H. 1978. Coenzyme-induced conformational changes and substrate binding in liver alcohol dehydrogenase. In *Molecular Interactions of Proteins, Ciba Found. Symp.*, pp. 63–80
76. Eklund, H., Samama, J.-P., Wallén, L., Brändén, C.-I., Ålseson, Å., Jones, T. A. 1981. *J. Mol. Biol.* 146:561–87
77. Sloan, D. L., Young, J. M., Mildvan, A. S. 1975. *Biochemistry* 14:1998–2008
78. Burton, D. R., Forsen, S., Karlstrom, G., Dweck, R. A. 1979. Proton relaxation enhancement (PRE) in biochemistry: a critical survey. In *Progress in nmr Spectroscopy*, ed. J. W. Emsley, J. Feeney, L. H. Sutcliffe, 13:1–45. Oxford: Pergamon
79. Anderson, I., Maret, W., Zeppezauer, M., Brown, R. D. III, Koenig, S. H. 1981. *Biochemistry* 20:3424–32
80. Dunn, M. F., Biellman, J.-F., Branlant, G. 1975. *Biochemistry* 14:3176–82
81. Brändén, C.-I., Eklund, H. 1980. Structure and mechanisms of liver alcohol dehydrogenase, lactate dehydrogenase and glyceraldehyde-3-phosphate dehydrogenase. In *Dehydrogenases Requiring Nicotinamide Coenzymes, Experi-entia Suppl.*, ed. J. Jeffery, 36:40–84 Basel, Birkhäuser
82. Morris, R. G., Saliman, G., Dunn, M. F. 1980. *Biochemistry* 19:725–31
83. van Eikeren, P., Grier, D. L. 1977. *J. Am. Chem. Soc.* 99:8057–60
84. Klinman, J. P. 1981. *CRC Crit. Rev. Biochem.* 10:39–78
85. Bowie, T., Brändén, C.-I. 1977. *Eur. J. Biochem.* 77:173–79
86. Cook, P. F., Cleland, W. W. 1981. *Biochemistry* 20:1805–16
87. Dworschack, R. T., Plapp, B. V. 1977. *Biochemistry* 16:111–16
88. Maret, W., Andersson, I., Dietrich, H., Schneider-Bernlohr, H., Einarsson, R., Zeppezauer, M. 1979. *Eur. J. Biochem.* 98:501–12
89. Makinen, M. W., Yim, M. B. 1981. *Proc. Natl. Acad. Sci. USA* 78:6221–25
90. Kvassman, J., Petersson, G. 1980. *Eur. J. Biochem.* 103:565–75
91. Ladner, J. E., Kitchell, J. P., Honzatko, R. B., Ke, H. M., Volz, K. W., et al. 1982. *Proc. Natl. Acad. Sci. USA* 79:3125–28
92. Monaco, H. L., Crawford, J. L., Lipscomb, W. N. 1978. *Proc. Natl. Acad. Sci. USA* 75:5276–80
93. Evans, P. R., Farrantz, G. H., Hudson, P. J. 1981. *Philos. Trans. R. Soc. London Ser. B* 293:53–62
94. Evans, P. R., Hudson, P. J. 1979. *Nature* 279:500–4
95. Yates, R. A., Pardee, A. B. 1956. *J. Biol. Chem.* 221:757–70
96. Gerhart, J. C., Pardee, A. B. 1962. *J. Biol. Chem.* 237:891–96
97. Kantrowitz, E. R., Pastra-Landis, S. C., Lipscomb, W. N. 1980. *Trends Bioch. Sci.* 5:124–28, 150–53
98. Weber, K. 1968. *Nature* 218:1116–19
99. Wiley, D. C., Lipscomb, W. N. 1968. *nature* 218:1119–21
100. Wiley, D. C., Evans, D. R., Warren, S. G., McMurray, C. H., Edwards, B. F. P., et al. 1972. *Cold Spring Harbor Symp. Quant. Biol.* 36:285–90
101. Honzatko, R. B., Crawford, J. L., Monaco, H. L., Ladner, J. E., Edwards, B. F. P., et al. 1982. *J. Mol. Biol.* 160:219–64
102. Collins, K. D., Stark, G. R. 1971. *J. Biol. Chem.* 246:6599–6605
103. Moody, M. F., Vachette, P., Foote, A. M. 1979. *J. Mol. Biol.* 133:517–32
104. Markus, G., McClintock, D. K., Bussel, J. B. 1971. *J. Biol. Chem* 246:762–71
105. Chan, W. W.–C. 1975. *J. Biol. Chem.* 250:668–74
106. Buckman, T. 1970. *Biochemistry* 9:3255–65

107. Gray, C. W., Chamberlain, M. J., Gray, D. M. 1973. *J. Biol. Chem.* 248:6071–79
108. Wu, C.-W., Hammes, G. G. 1973. *Biochemistry* 12:1400–8
109. Suter, P., Rosenbusch, J. P. 1977. *J. Biol. Chem.* 252:8136–41
110. Honzatko, R. B., Lipscomb, W. N. 1982. *J. Mol. Biol.* 160:265–86
111. Gibbons, I., Flatgaard, J. E., Schachman, H. K. 1975. *Proc. Natl. Acad. Sci. USA* 72:4298–4302
112. Weitzmann, P. D., Wilson, I. B. 1966. *J. Biol. Chem.* 241:5481–88
113. Hammes, G. G., Wu, C.-W. 1971. *Science* 172:1205–11
114. Kerbiriou, D., Hervé, G. 1972. *J. Mol. Biol.* 64:379–92
115. Kerbiriou, D., Hervé, G. 1973. *J. Mol. Biol.* 78:687–702
116. Kantrowitz, E. R., Lipscomb, W. N. 1977. *J. Biol. Chem.* 252:2873–80
117. Landfear, S. M., Evans, D. R., Lipscomb, W. N. 1978. *Proc. Natl. Acad. Sci. USA* 75:2654–58
118. Hudson, P. J., Hengartner, H., Kolb, E., Harris, J. I. 1979. The primary structure of phosphofructokinase from B. Stearothermophilus. In *Proc. 12th FEBS Symp. 2,* ed. W. Pfeil, E. Hofmann, H. Aurich, pp. 341–48. Oxford: Pergamon
119. Blagney, D., Buc, H., Monod, J. 1968. *J. Mol. Biol.* 31:13–35
120. Remington, S., Wiegland, G., Huber, R. 1982. *J. Mol. Biol.* 158:111–52
121. Baldwin, J., Chothia, C. 1979. *J. Mol. Biol.* 129:175–220
121a. Perutz, M. F. 1980. *Proc. R. Soc. London Ser. B* 208:135–62
122. Cohen, P., Embi, N., Foulkes, G., Hardie, G., Nimmo, G., et al. 1979. The role of protein phosphorylation in the coordinated control of intermediary metabolism. In *From Gene to Protein: Information Transfer in Normal and Abnormal Cells,* ed. T. R. Russell, K. Brew, H. Faber, J. Schultz, pp. 463–81. New York: Academic
123. Titani, K., Koide, A., Hermann, J., Ericsson, L. H., Kumar, S., et al. 1977. *Proc. Natl. Acad. Sci. USA* 74:4762–66
124. Helmreich, E. J. M., Klein, H. W. 1980. *Angew. Chem. Int. Ed. Engl.* 19:441–55
125. Sprang, S., Fletterich, R., Stern, M., Yang, D., Madsen, N., Sturtevant, J. M. 1982. *Biochemistry.* 21:2036–48
126. Sturtevant, J. M. 1977. *Proc. Natl. Acad. Sci. USA* 74:2236–40
127. Madsen, N. B., Kasvinsky, P. J., Fletterick, R. J. 1978. *J. Biol. Chem.* 253:9097–9101
128. Fletterick, R. J., Sprang, S. R. 1982. *Acc. Chem. Res.* 15:361–69

Ann. Rev. Biochem. 1983. 52:35–65

ARCHITECTURE OF PROKARYOTIC RIBOSOMES

H. G. Wittmann[1]

Max-Planck-Institut für Molekulare Genetik, D-1000 Berlin-Dahlem, Germany

CONTENTS

PERSPECTIVES AND SUMMARY 35
SHAPE AND SIZE OF RIBOSOMAL PARTICLES 37
    Small Angle Scattering 37
    Electron Microscopy.. 38
    Diffraction 40
SPATIAL ARRANGEMENT OF THE PROTEINS IN SITU 43
    Immune Electron Microscopy 43
    Neutron Scattering.. 46
    Fluorescence.. 48
    Protein-Protein Cross-links 49
    Protein Complexes... 51
    Accessibility 51
    Assembly Maps 52
SPATIAL ARRANGEMENT OF THE RNAs IN SITU 55
    Electron Microscopy.. 56
    RNA-RNA Cross-links.. 56
    RNA-Protein Cross-links 57
    Protein Binding Sites on the RNA 58
    Fragmentation of Subunits 59
    Spatial Packing of the RNAs 59
CONCLUDING REMARKS 60

PERSPECTIVES AND SUMMARY

The biosynthesis of proteins is a complex process that takes place on the ribosome and proceeds in numerous steps. Extensive studies during the last two decades have led to a formal description of this process and its many steps. However, our knowledge of the structure-function correlation is still rudimentary. For instance little is known as to which ribosomal compo-

[1]I thank Drs. R. Brimacombe, B. Hardesty, W. E. Hill, K. H. Nierhaus, G. Stöffler, M. Stöffler-Meilicke, and A. Yonath for helpful comments on the manuscript.

35

0066-4154/83/0701-0035$02.00

nents are involved in the initiation, elongation, and termination of protein biosynthesis, or what happens at the molecular level on the ribosome. The main reason for this deficiency has been a lack of information on the ribosomal structure at a sufficiently detailed molecular level to enable these and other interesting questions to be answered.

Without insight into the structure of the ribosome and its components at high resolution our knowledge of the process of protein biosynthesis will remain at the descriptive level, and many important questions (e.g. on the molecular mechanism of the translocation process or the peptidyltransferase reaction, and on the interaction of the ribosome with factors, mRNA, tRNA, and antibiotics) will remain unanswered or at a hypothetical stage.

Since it was realized in the 1960s that ribosomes are complex particles consisting of many individual proteins and several RNA molecules, great effort has been made to isolate and characterize the ribosomal components by chemical, physical, and immunological methods. These investigations concentrated mainly on the *Escherichia coli* ribosome and led to the determination of the amino acid sequences of all 53 proteins as well as the primary and secondary structures of the three RNA molecules from this ribosome. Furthermore, the secondary and tertiary structures and the shapes of the ribosomal proteins were extensively investigated [see (1) for a recent review on the components of bacterial ribosomes].

A battery of techniques was used in the last ten years in the attempt to reveal the architecture of the ribosomal particles, i.e. their shape and the spatial arrangement of their RNA and protein components in situ. The most widely used techniques include electron microscopy combined with immunological methods, for the elucidation of the shape of the ribosomal subunits and the location of proteins on their surface (2, 3); neutron scattering for measuring the distances between the mass centers of proteins in situ (4); protein-protein cross-linking with bifunctional reagents for revealing which proteins are neighbors within the ribosomal particles, or which proteins are located at the interface between the two subunits when the latter are associated in the 70S ribosome (5); RNA-protein cross-linking to elucidate which RNA region is close to a given protein in situ (6); determination of the RNA regions to which individual proteins bind, thus also providing information on RNA-protein neighborhoods (7); measurement of singlet-singlet energy transfer between two different fluorescent dyes attached to specific ribosomal sites, providing data on the distance between the sites (8); construction of assembly maps, showing the interaction of protein during the assembly process and giving indirect topographical information (9, 10).

Among the techniques just mentioned, immune electron microscopy, neutron scattering, and cross-linking have contributed most to our present knowledge on the architecture of the *E. coli* ribosome. There is good agreement among the results from these methods when the most recent data are compared. This review concentrates on the results obtained by the various approaches since the last article on this topic in the *Annual Review of Biochemistry* (11), and it describes our current knowledge of the internal and external architecture of prokaryotic ribosomes, especially of the *E. coli* ribosome and its subunits.

SHAPE AND SIZE OF RIBOSOMAL PARTICLES

Small-Angle Scattering

Early small-angle X-ray studies and results of hydrodynamic measurements [summarized in (12)] showed the ribosomal subunits to be best approximated as ellipsoids. The data from X-ray scattering of 30S particles were best approximated by those calculated for an oblate ellipsoid having dimensions of 5.5 X 22 X 22 nm. The data from the 50S subunit were fit to those calculated for a 11.5 X 23 X 23 nm ellipsoid. There is considerable discrepancy between these X-ray results and those emanating from electron microscopy for the 30S subunit, whereas there is reasonable agreement for the 50S particle. Subsequent small-angle neutron scattering studies using the contrast-variation method with a spherical-harmonic analysis also showed that the 50S subunit results are consistent with a trilobal hemispherical model similar to that seen in electron micrographs (13).

Using models derived from electron micrographs it is possible to calculate the scattering curves that would be expected from such particles. A comparison of the experimental curves with the calculated curves provides information on models that are not consistent with scattering results, but it cannot predict a unique model. Efforts of this nature (14, 15) showed that shapes of the ribosomal subunits other than ellipsoidal structures can provide data that approximate the experimental data quite well. In this way the measured neutron scattering curve was found to be similar to that calculated for a 30S particle as observed by electron microscopy (15).

Neutron scattering studies gave some insight into the distribution of the protein and RNA moieties within the ribosome and its subunits. In the 30S particle the radius of gyration for the RNA is about 6 nm and for the protein moiety about 8 nm. The distribution of RNA and protein is approximately concentric but not homogeneous, the RNA being mainly located towards the center and the proteins mainly towards the periphery of the particle. The centers of mass for the RNA and protein moieties roughly coincide (15–18). It was also reported that the scattering data for the 16S RNA within the 30S particle (15) are consistent with a V-like structure as seen on electron micrographs of isolated 16S RNA molecules (19) and of small subunits from which most of the proteins have been removed (20, 21).

After some initially controversial results (13, 22, 23), there is now general agreement that in the 50S particle there is no large separation (less than 3 nm) between the mass centers of the RNA and protein, and that the distribution of the two moieties within the subunit is nonuniform, but less than in the 30S particle. Based on the large difference between the radii of gyration for the RNA and proteins (5–6 nm and 9–10 nm, respectively) it was concluded that the RNA is predominantly in the inner zone, whereas the proteins are distributed concentrically mainly in the outer regions of the 50S subunit (13, 15, 25, 26).

In the 70S particle the radii of gyration are 7 nm for the RNA and 10 nm for the protein moiety, respectively (18). From a comparison of the scattering data from the 70S ribosome and its individual free subunits, it can be concluded that no drastic conformational change of the subunits occurs when they associate to form the 70S particle.

Electron Microscopy

In principle, electron microscopy is the most direct method for studying the overall shape and the gross structure of ribosomes and their subunits. Consequently, it has been widely applied for this purpose with a variety of methods for the preparation of the samples (e.g. positive and negative staining, single and double layer technique, air and freeze drying, shadow casting) and with different electron microscopical techniques (e.g. conventional bright field transmission EM, dark field EM, scanning transmission EM, three-dimensional EM).

Ribosomes are relatively small and highly hydrated particles that can undergo distortions during dehydration and are easily damaged by electron beam irradiation. Because of this and the subjective interpretation of the images by the various electron microscopists, it is not too surprising that divergent models for the *E. coli* ribosome and its subunits were proposed (2, 3, 27–33). This is illustrated for the small and large subunits in Figure 1 and for the 70S ribosome in Figure 2.

In spite of the differences between the various models there is general agreement that the small subunit is a prolate asymmetric particle with approximate dimensions of 23 X 11 nm. It consists of a "head" comprising about one third, and a "base" or "body" comprising about two thirds of the particle, and it has a partition ("collar" or "neck") between the two parts as well as one or two protuberances ("lobe" or "platform"). As seen in Figure 1 the differences between the various models concern the degree of asymmetry of the subunit, the number, size, and shape of the protuberances and the depth of the partition between the "head" and the "body."

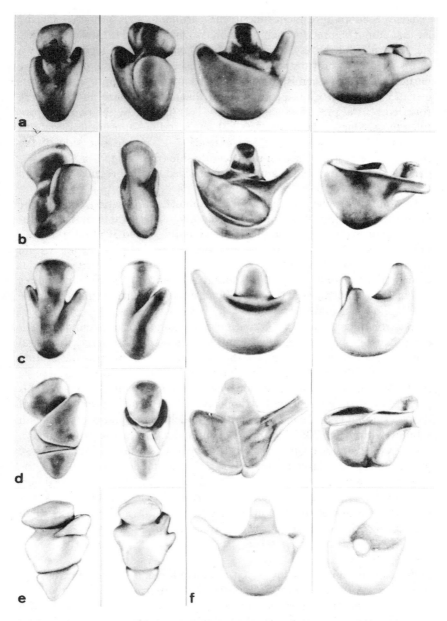

Figure 1   Models of *E. coli* ribosomal subunits. (*a*) Stöffler et al (33); (*b*) Lake (3); (*c*) Boublik (29); (*d*) Vasiliev et al (27, 28); (*e*) Korn et al (30); (*f*) Spiess (31). The views were selected in such a way that the most characteristic features of the models can be seen.

There is more agreement among the various models proposed for the large subunit (Figure 1), which is seen in two main views: the quasi-symmetric "crown" and the asymmetric "kidney" or "crescent" forms. The two forms represent the front and side view, respectively, of the subunit. The particle consists of a hemispherically shaped main body (approximate diameter: 23 nm) with three protuberances that differ in size and shape. A rounded central protuberance (the "nose") is flanked by the two side protuberances. One of them (the "stalk") has a rodlike appearance, contains proteins L7/L12 (see below) and extends 8–12 nm to one side of the particle. The protuberance on the opposite side (the "ridge" or "L1-shoulder") is somewhat rounded and is characterized by the presence of protein L1.

In addition to the contours of the subunits, information can also be obtained concerning their internal structure by three-dimensional reconstruction of electron micrographs of individually stained 30S and 50S subunits. In this way an indication for the location of the RNA chains within the subunits was obtained (34). From these studies it can further be concluded that the stalk of the 50S subunit is oriented as shown in Figures 1a,b,d. The handedness of the 30S subunit has only been determined for the model shown in Figure 1b (35).

Electron micrographs of 70S ribosomes show more or less rounded particles (diameter of approximately 23 nm) lacking highly distinctive structural features. As depicted in Figure 2 the various models differ in the mutual orientation of the small and large subunits within the 70S ribosome. In three models the small subunit lies obliquely and diagonally with the head of this particle located between the L1-shoulder and the central protuberance of the 50S subunit (Fig. 2a,b,d). In another model the orientation of the small subunit within the 70S ribosome is lengthwise across the crown region between the two side protuberances, with the head of the small subunit close to the stalk of the large subunit (Fig. 2c). The models also differ in the interface between the two subunits. This discrepancy arises from the differences in the shapes of the subunits as proposed by the various models, but in all the 70S models the protuberances (lobe, platform, etc.) of the small subunit face the notched side of the large subunit.

Diffraction

Diffraction techniques, such as X-ray crystallography, three-dimensional image reconstruction of optical diffraction patterns of electron micrographs, and neutron diffraction are very powerful methods to establish a detailed model of the ribosome. The techniques for three-dimensional structure determination of macromolecules have recently enabled rapid advances, especially in the effectiveness of collecting, processing, and

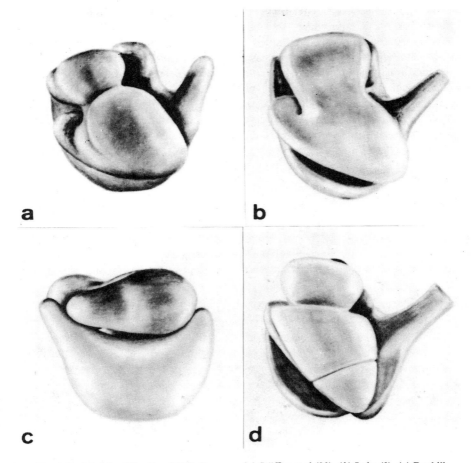

Figure 2   Models of the *E. coli* 70S ribosome. (*a*) Stöffler et al (33); (*b*) Lake (3); (*c*) Boublik (29); (*d*) Vasiliev (unpublished).

analyzing crystallographic data. However, since ribosomes are not only of enormous size but also asymmetric and somewhat flexible in structure, the determination of their three-dimensional structure at the molecular level is expected to be extremely difficult and will require the development of highly advanced techniques.

The use of diffraction methods depends on the availability either of three-dimensional crystals or, if these cannot be obtained, of two-dimensional ordered arrays or helical structures. Thus, much effort has been devoted to the crystallization of ribosomes and their subunits.

TWO-DIMENSIONAL ORDERED ARRAYS   Two-dimensional periodic organization of eukaryotic ribosomes occurs under special conditions in vivo. Ordered sheets were obtained from various sources, such as *Entamoeba histolytica* (36), chick embryos (37, 38), lizards (39, 40), and brain of senile humans (41). Also, helical and two-dimensional arrays of the *E. coli* ribosomal subunits were recently obtained (42, 43). These were analyzed using three-dimensional image reconstruction techniques.

The parameters of a P21 lattice of small crystalline sheets of the large ribosomal subunits of *E. coli* are $a = 33 \pm 3$ nm, $b = 33 \pm 3$ nm, and $\gamma = 123°$. Three-dimensional reconstruction studies performed to a resolution of 4.5 nm on a single electron micrograph revealed the outer contour of the particle.

THREE-DIMENSIONAL CRYSTALS   The in vitro growth of three-dimensional crystals of ribosomes and ribosomal subunits has proved to be extremely difficult. Only recently were three-dimensional crystals of varying degrees of order obtained for two intact ribosomal particles from bacterial sources (Figure 3), namely from the *E. coli* 70S ribosome (44), and the large

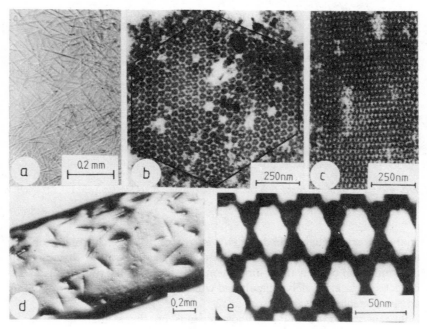

Figure 3   (*a*) Crystals of *E. coli* 70S ribosomes; (*b* and *c*) electron micrographs of sections through crystals shown in *a* in two orthogonal directions (44); (*d* and *e*) crystals and computed filtered image of a section through a crystal of *B. stearothermophilus* 50S large ribosomal subunits (47, 48); *d* and *e* are related to two different crystal forms.

subunit of *Bacillus stearothermophilus* (45–48). Among those, at least one crystal form of *B. stearothermophilus* 50S subunits seems to be suitable for X-ray crystallographic studies since the diffraction patterns of cross-linked single crystals extend to about 0.9 nm resolution with spacings observed at 15.4 nm and 26.1 nm.

Three-dimensional image reconstruction studies were performed on positively stained thin sections of several crystal forms of *B. stearothermophilus* 50S subunits. Unit-cell constants as well as three-dimensional representation of the stain distribution for the ribosomal particles were obtained. It was found that the portion of the subunit interacting with uranyl acetate (presumably the rRNA) is distributed in a nonuniform manner within a bilobal structure, whereas the ribosomal components interacting with phospho-tungstic acid (presumably the proteins) are located mainly on the surface and are involved in interactions between the crystalline particles (K. Leonard, B. Tesche, H. G. Wittmann, A. Yonath; to be published).

SPATIAL ARRANGEMENT OF THE PROTEINS IN SITU

Immune Electron Microscopy

The availability of specific antibodies against *E. coli* ribosomal proteins, combined with the total lack of immunological cross-reaction among the individual proteins (49, 50), has allowed immune electron microscopic investigations in which the attachment site of an antibody specifically directed against a given individual protein is localized on the surface of the ribosomal particle by electron microscopy.

Since the first application of immune electron microscopy to ribosomes (51), many studies on the localization of ribosomal proteins on the small and large subunits have been carried out [see (2) and (3) for reviews]. Some of the earlier results were recently corrected, the errors arising mainly from impurities present in several antisera. The current maps with the location of the proteins on the small and large subunits are illustrated in Figures 4 and 5, respectively. There is relatively good agreement between the results of the two principal groups conducting immune electron microscopic studies on *E. coli* ribosomes (2, 3, 52–57).

Although there is general agreement that the stalk contains the proteins L7/L12, a discrepancy still exists as to whether the four copies of these proteins are exclusively located in the stalk (57, 58) or also in the second lateral protuberance (29). However, most of the biochemical, physical, and immunological data are consistent with the following hypothesis. The four elongated L7/L12 molecules are present in the stalk, and lie parallel to each other. Their C-termini point away from the 50S particle, whereas their *N*-termini are located at the base of the stalk and bind to protein

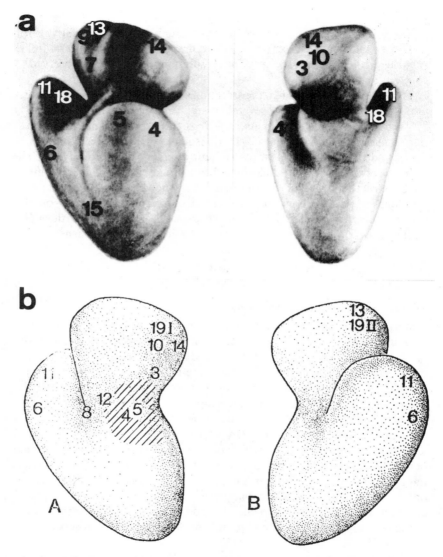

Figure 4 Mapping of proteins on the *E. coli* 30S subunit by immune electron microscopy. (*a*) Stöffler et al (56); (*b*) Lake et al (52, 54).

L10, which, in turn, binds to the 23S RNA. The results supporting this hypothesis are summarized in (59).

Electron microscopic studies were also performed to localize functional domains on the ribosome, e.g. the binding sites for mRNA, tRNA, antibiot-

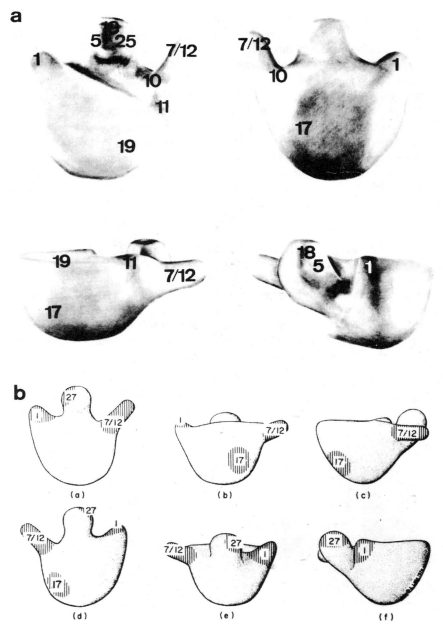

Figure 5 Mapping of proteins on the *E. coli* 50S subunit by immune electron microscopy. (*a*) Stöffler et al (57); (*b*) Lake et al (53).

ics, initiation and elongation factors as well as the ribosomal region where the peptidyltransferase is located. Information concerning these domains derives either from the direct electron microscopic visualization of ribosomal attachment sites of antibodies raised against mRNA, tRNA, factors, etc. or from a more indirect approach aiming at the localization of binding sites for antibiotics on the ribosome. An antibiotic or its affinity analog is bound to ribosomes, and its binding site is then mapped by electron microscopy with antibodies against this antibiotic. In this way the ribosomal binding sites of puromycin (60–62) and chloramphenicol (61) were mapped. Since both antibiotics act on the peptidyltransferase, information about the location of the peptidyltransferase center on the large ribosomal subunit can be obtained (61). It is located in the valley between the central protuberance and the L1-shoulder. Figure 6 shows a summary of the results on the mapping of functional domains on the small and large *E. coli* ribosomal subunits.

The following domains have been localized: the mRNA binding region (55, 63); the 3' end of tRNA and the decoding region (64); binding sites for the initiation factor IF-3 (55, 63) and the elongation factor EF-G (65); binding sites for chloramphenicol (61) and puromycin (60–62), indicating the location of the peptidyltransferase center; the thiostrepton binding site (55, 61); the exit domain for the nascent polypeptide chain (66). The localization of these functional domains yields illustrative information as to where the various steps in protein biosynthesis take place on the ribosome.

Neutron Scattering

In contrast to immune electron microscopy, which yields information on the location of ribosomal components at the surface of the particle, neutron

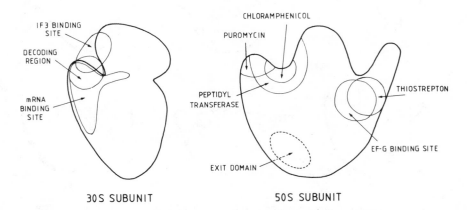

Figure 6 Functional domains on *E. coli* ribosomal subunits as determined by immune electron microscopy. For references see text.

scattering is a method by which information also on the internal architecture of the ribosome can be obtained. This method was applied to the determination of the distances between proteins and of their shapes in situ. Individual deuterated ribosomal components are isolated from *E. coli* cells grown in D_2O; 30S subunits are then reconstituted, in which only the two proteins under study are deuterated, whereas all other components are protonated (4). By extensive neutron scattering studies with small ribosomal subunits containing different pairs of deuterated proteins in an otherwise protonated particle, the distances between the mass centers of gravity have been measured for 14 proteins (4, 67). When the spatial arrangement of these proteins as measured by triangulation (Figure 7) is compared with the corresponding location of the proteins as determined by immune electron microscopy, a very good agreement between the two different sets of data is obtained. This mutually corroborates the results from the two methods and, together with other data, allows a rather reliable map for the locations of many proteins within the 30S subunit to be constructed.

In neutron scattering studies with the *E. coli* large subunits a different approach is used: 50S particles are reconstituted in which all the ribosomal components are deuterated, with the exception of the two protonated proteins under study. When the ribosomal RNAs are isolated from bacteria grown in 76% D_2O and the ribosomal proteins from cells grown in 84% D_2O, then the 50S subunits reconstituted from these components have the same neutron densities as 100% D_2O. The matching of the deuterated ribosomal components in undiluted D_2O gives a highly improved signal-to-noise ratio for the interference pattern of the neutrons scattered by the two

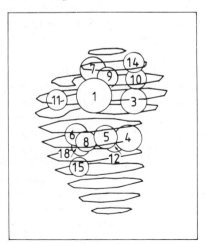

Figure 7   Map of proteins within the *E. coli* 30S subunit as determined by neutron scattering studies (4, 67).

protonated proteins in the deuterated matrix. Thus the distances between the mass centers of gravity for three protein pairs have so far been measured (68, 69).

This method allows not only the measurement of the distance between the mass centers of gravity of a pair of proteins, but also the direct determination of the shape of individual proteins in situ when only one (instead of two) protonated protein(s) is incorporated into the deuterated 50S particle. In this way the shapes of five proteins were determined (69). When their shapes within the ribosomal particle and as individual proteins in solution were compared no significant differences were found. This demonstrates that no drastic conformational changes of the proteins occur during their assembly into the ribosomal subunit.

Fluorescence

Individual ribosomal (protein and/or RNA) components are labeled with different fluorescent dyes, and subunits are then reconstituted in which one component is labeled with one dye (donor) and another component with a different dye (acceptor). By means of singlet-singlet energy transfer measurements it is possible to determine the distance between the fluorescent labels. The early studies with this technique were concerned with distance measurements between labeled proteins within the 30S particle (8). In recent years, besides continued studies with 30S proteins (70–72), fluorescent measurements were also carried out with the 50S proteins L6 (73), L10 (74, 75), and L7/L12 (74–77) as well as with 5S, 16S, and 23S RNAs labeled at their 3' ends (78–80). Furthermore, fluorescence was used to locate the positions of tRNAs (80–85), the initiation factor IF-3 (86, 87), and the antibiotics streptomycin (88) and erythromycin (89) on the ribosome.

When the results from the fluorescent studies are compared with those from neutron scattering, immune electron microscopy, and cross-linking, there is a good or at least moderate fit in many cases, whereas in others the results cannot be reconciled at all. A possible reason for the discrepancy is the fact that the degree of energy transfer between the two fluorescent dyes depends not only on the distance between the donor and acceptor molecules, but also on other factors, such as the relative orientation of the dyes. However, in spite of these and other limitations, e.g. the relatively narrow range in which energy transfer is possible, the fluorescence method can give very useful topographical data. This is especially true when a single dye is bound to a specific residue in a protein, e.g. its only SH-group. In this way measurements between very specific points within the ribosomal particles are possible.

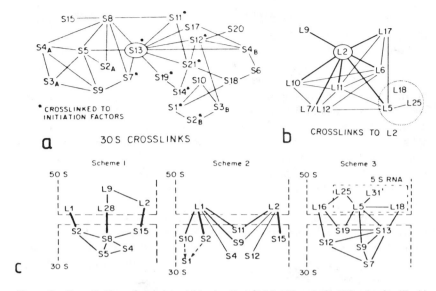

Figure 8   Cross-linking of proteins within the *E. coli* (*a*) 30S and (*b*) 50S subunits (5); (*c*) protein neighborhoods at the subunit interface (96). Scheme 1 shows the cross-links found in highest amount, scheme 2 those between 50S proteins L1 and L2 and several 30S proteins, and scheme 3 those between 5S RNA binding proteins and 30S proteins.

Protein-Protein Cross-links

The application of bifunctional reagents for the identification of neighboring proteins within the ribosome was the earliest method used to probe ribosomal topography. A wide variety of bifunctional reagents has been used, many of which are imidoesters. The most frequently applied reagent in recent years was 2-iminothiolane (5), which can also be used for RNA-protein cross-linking (90). The identification of the cross-linked proteins can be made by diagonal gel electrophoresis (91) if the bifunctional reagent is cleavable, whereas with noncleavable reagents the cross-linked proteins have mainly been identified by immunological techniques (e.g. 92).

Extensive cross-linking studies were carried out with both subunits as well as with the 70S ribosome by several groups (93–95), the main contribution being by Traut et al (5, 96–98). These results are summarized in Figure 8, which represents two-dimensional maps of the cross-linking data. There is a rather good agreement between the cross-linking data on the one hand and the results from neutron scattering and immune electron microscopy on the other. In general, proteins that can be cross-linked are also found to be close to each other by the other two methods. In those cases where two proteins are neighbors but have not been found cross-linked, it is likely that there are no suitable amino acid residues, e.g. lysine or cysteine, in positions accessible to the cross-linking reagent.

A number of cross-links, e.g. between S5–S8, S7–S9, S6–S18, and S13–S19, were found by several groups with different reagents and in high yield, whereas other protein pairs were only identified with one reagent and in low yield. In the latter cases, in particular when the cross-links are at variance with the results from other methods, the question arises as to how valid the cross-linking data are. It could be that such cross-links derive from a small subpopulation of ribosomes that are in a functionally inactive conformation. This possibility could be ruled out in a few cases (99, 100) by incorporating the cross-linked protein pair (instead of the two single proteins) into the subunit, and demonstrating that the function of the particles was not impaired.

In addition to data on the 30S and 50S subunits, cross-linking studies also provide topographical results on the interface between the two subunits when they are associated within the 70S ribosome (96, 97). A summary of the current data is given in Figure 8, which shows that a relatively large number of proteins are present at the interface. The cross-links L1–S2, L2–S15, and L28–S8 were found in the highest yield. Furthermore, proteins L1 and L2 are cross-linked to five and four 30S proteins, respectively. Proteins L5 and L18, which are associated with the 5S RNA (101), have been found cross-linked to three 30S proteins, namely S9, S13, and S19. From this it can be concluded that the 5S RNA complex is located at the interface. This conclusion is in good agreement with immune electron microscopical results, which place the three 30S proteins in the head of the small subunit and the 5S RNA binding proteins in the central protuberance of the large subunit. As shown in Figure 2 the head and the central protuberance are in close proximity within the 70S ribosome.

Since the primary sequences of all *E. coli* ribosomal proteins are known (1, 102, 103) it is possible to determine the exact positions of the amino acid residues within the protein chains that are cross-linked by a bifunctional reagent. In this way, position 93 in protein S8 could be cross-linked to position 166 in S5 (104), i.e. these two amino acid residues within the 30S subunit cannot be further apart than 8 Å, which is the span of the bifunctional reagent used. The validity of this finding was demonstrated by control experiments with a mutant of S5 lacking residue 166. As expected, no cross-linking occurs in this case (104).

The results from the identifications of the cross-links at the amino acid level become especially important when they can be combined with data on the tertiary structure of the cross-linked proteins as elucidated by X-ray analysis. The first example to be obtained is the combination of a cross-link between proteins L7/L12 (100) with crystallographic data on these proteins (105). More such cases can be expected when the tertiary structures become available for those ribosomal proteins whose X-ray structure analysis is in progress (105–110).

Protein Complexes

During the isolation of individual proteins, a number of specific complexes between two or more proteins were isolated. The most intensively studied complexes are those containing proteins L7/L12, which can be isolated as a dimer (111), as a tetramer (112), or as a pentameric complex between four L7/L12 molecules and the protein L10 (113–115). As shown by immune electron microscopic studies the four copies of L7/L12 form the "stalk" of the 50S subunit (116), and they bind to protein L10 with their N-termini [see (59) for further discussion].

In addition to the $(L7/L12)_4$–L10 complex, a number of other protein-protein complexes have been isolated (117, 118). They elute from the column at positions different from those of their individual components. It is interesting that some of them, e.g. the S13–S19 complex, bind to their cognate rRNA, whereas the individual proteins do not (114).

Another way to study protein-protein interaction is by mixing individual proteins. Some of these combinations, but not others, lead to the formation of protein complexes with association constants of 10^4–10^6. Several complexes were identified in this way: S3–S4, S3–S5, S3–S5–S10, S4–S5, S4–S20, S5–S8, and S5–S10 (119, 120). Most of them are consistent with the results from neutron scattering, protein crosslinking, and immune electron microscopy, which supports the notion that they are genuine complexes reflecting topographical neighborhoods and specific interactions between the corresponding proteins within the 30S subunit.

Accessibility

A variety of probes, such as antibodies (121), trypsin (122–125), kethoxal (126), fluorescent reagents (127), N-ethyl-maleimide (128–130), and iodination by carrier-bound lactoperoxidase (131, 132), has been used to identify proteins at the surface of the ribosomal subunits, and most proteins were found to be accessible to one or more of these probes. When the iodination of proteins in the isolated subunits and in the 70S ribosome was compared, seven proteins were found to be labeled to a greater and four to a lesser extent in the subunits as opposed to the ribosome, which indicates a conformational change during the association of the subunits (131). These results are consistent with those from other studies using enzymatic iodination (132).

Iodination experiments were also carried out to reveal neighborhoods between proteins within the 30S subunit (133). Partially reconstituted particles containing the 16S RNA and a different number of proteins (up to 11) were chemically iodinated, and the protection of individual proteins from iodination by the presence of other proteins was analyzed. In this way 30 protein pairs were identified in which one protein protects another from iodination, and the proteins within a given pair were interpreted as neighbors in situ. These results in general agree with those from other topographical approaches, such as cross-linking, neutron scattering, immune electron microscopy, and fluorescence studies. From additional experiments with iodination of proteins in the subunits it was concluded that the *E. coli* 30S population isolated by standard procedures is conformationally heterogeneous (134).

In addition to studies on the accessibility of proteins to various probes, several enzymes and reagents have also been used for identification of accessible RNA regions. The most widely used reagent for these studies is kethoxal, which reacts with guanines in single-stranded RNA regions. About two dozens guanine residues, mainly from the middle to the 3' end of the 16S RNA are modified by kethoxal in the intact 30S particle (135). When the two ribosomal subunits are associated, six guanines in the middle region (positions 673–817) and six near the 3' end (pos. 1165–1516) are protected from modification by the reagent, whereas three guanines (pos. 1052–1067) become more reactive (136), indicating a conformational change affecting this region. Similar results were obtained using RNase T1 as a probe in studies on 30S subunits in the free and associated state (137).

Assembly Maps

SMALL SUBUNIT   Incubation of 16S RNA with the proteins extracted from the small subunit of the *E. coli* ribosome under reconstitution conditions leads to the formation of a 30S particle that is physically and functionally indistinguishable from the native subunit (138). This process, which consists of at least three steps, has been extensively studied from a structural and functional point of view. The earlier results are summarized in (9), and more recent studies on the 30S assembly process have been described (139–148). In brief, approximately two-thirds of the 30S proteins bind at low temperature to the 16S RNA and form an intermediate particle sedimenting at 21S. Upon heating to 37°C the 21S particle undergoes a conformational change leading to a more compact form that sediments at 26S, without any increase in the number of proteins contained in the particle. As a result of the conformational change new binding sites are created, and the remaining one-third of the proteins can bind even at low temperature, thus leading to intact and functionally active 30S particles.

The analysis of the sequence of protein incorporation into the intermediate particles, and of the interdependence between the individual proteins during the assembly process allowed the construction of an assembly map (9, 149). Figure 9*a* illustrates the interactions of the 30S proteins with the

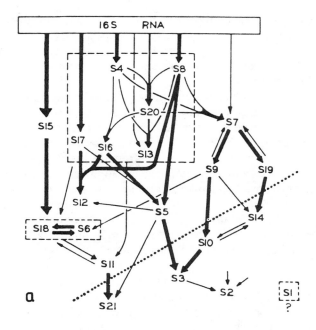

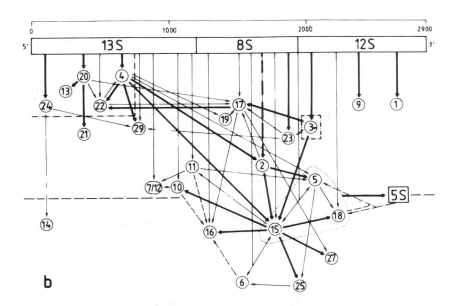

Figure 9   (*a*) Assembly map of the 30S (9), and (*b*) of the 50S subunit (160).

16S RNA and with each other. Under the conditions used for the studies on the assembly map, seven proteins bind independently and directly to the 16S RNA. Only after these primary binding proteins become attached to the RNA is it possible for other proteins (the secondary binding proteins) to bind. Some proteins, e.g. S3, S6, S10, S14, S18, and S21, even need the presence of the secondary binding proteins prior to their own binding.

The interdependence shown by the assembly map may reflect parameters other than direct protein-protein contact, and therefore two proteins connected by an arrow in the assembly map need not necessarily be neighbors in situ. However, comparison of the data from the assembly map on the one hand with results from other topological approaches, such as immune electron microscopy, cross-linking, and neutron scattering, on the other hand, yields a general agreement. Proteins interacting according to the assembly map have usually been found by the other methods to be close within the 30S subunit. Therefore, the assembly map can give (although somewhat indirect) topological information.

LARGE SUBUNIT   Although the in vitro assembly of the large subunit from the *B. stearothermophilus* ribosome can be accomplished under reconstitution conditions similar to those for the *E. coli* small subunit (150), considerable effort was necessary to find corresponding conditions under which the total reconstitution of the large ribosomal subunit from *E. coli* was possible (151). In contrast to the single step incubation for 30S subunits, the in vitro assembly of *E. coli* 50S subunit requires two incubations, the first at 44°C and 4 mM Mg^{++} and the second at 50°C and 20 mM Mg^{++} (151–152). The assembly process occurs in at least four steps, leading to three intermediate particles sedimenting with 33S, 41S, and 48S. Incubation of the 23S RNA, the 5S RNA, and the proteins derived from the 50S subunit initially results in a 33S particle consisting of the two RNAs and about two-thirds of the proteins. Upon heating at 44°C, the sedimentation coefficient increases from 33S to 41S, suggesting a drastic conformational tightening of the particle. To this 41S particle additional proteins are then bound, and the sedimentation coefficient increases to 48S. Upon heating at 50°C, the 48S particle is converted into the functionally active 50S subunit (153–155).

In spite of the large number (about 20) of proteins that bind to the 23S RNA in the first step, only two "initiator proteins" (L24 and L3) are able to initiate the assembly process, and only 5 out of the 20 proteins within the 33S particle are necessary and sufficient to achieve formation of the 41S particle at 44°C. For 2 of the 5 proteins, namely L20 and L24, it was shown that they are absolutely essential for this step, but they can be removed after formation of the 41S particle without impairing the later assembly steps or the full activity of the resulting 50S subunit. Therefore, it was concluded that these proteins are mere assembly proteins without a function in the mature 50S particle (156, 157). Interestingly, the proteins that are essential for the early assembly bind close to the 5' end of the 23S RNA, whereas those proteins assembled at a later stage bind to the 3' half of the RNA strand. This finding suggests an assembly gradient, which indicates that the assembly process starts while the ribosomal RNA chain is still being synthesized, and that the assembly progress in vivo depends on the availability of the growing RNA strand (158).

The 50S assembly map (Figure 9b), as derived from the study of the protein interdependence during the assembly process (159–161), shows a similar hierarchy for the incorporation of the proteins into the intermediate particles as described above for the 30S assembly. Firstly, the primary binding proteins bind to the 23S RNA, and their presence on the particle is necessary for the binding of the secondary binding proteins, etc. As shown in Figure 9b there are at least two assembly domains around proteins L20 and L15, respectively. Most proteins in the L20 domain are important for the early assembly process, whereas the majority of those in the L15 domain are late assembly proteins and are involved in ribosomal functions, such as peptidyltransferase activity or binding of elongation factors. There is general agreement between the 50S assembly map on the one hand and topological results, e.g. from studies on protein cross-linking and on the successive removal of proteins from the 50S particle by increasing salt concentrations, on the other hand. Furthermore, there is some correlation between the assembly map and the organization of genes for 50S proteins within certain operons (160, 161). A possible explanation for this surprising result is the formation of preformed complexes consisting of proteins coded by genes within the same transcriptional unit. The incorporation of preformed protein complexes instead of individual proteins into the intermediate particle would accelerate the assembly process and increase its cooperativity.

SPATIAL ARRANGEMENT OF THE RNAs IN SITU

Rapid progress has been made in determining the primary and secondary structure of ribosomal RNAs from various species (reviewed in 1, 6, 162–164). The models for the secondary structure of the E. coli 16S (165–169) and 23S (170–172) RNAs as proposed by several groups using different techniques are in almost complete agreement (6). However, the next step, the elucidation of the spatial arrangement of the RNA strands within their ribosomal subunits, has so far caused considerable difficulties and is still in its early stage.

Electron Microscopy

Attempts have been made to determine the shape of isolated 5S (173, 174), 16S (174–177), and 23S (174, 176–178) RNA molecules by direct visualization in the electron microscope. Both the 16S and the 23S RNAs were seen in a specific and compact structure in some studies (175, 178). According to these results the isolated 16S RNA has a V-like structure with one arm somewhat thicker and longer than the other (175). The isolated 23S RNA has a size and shape that can be accommodated within the "crown" or "kidney" views of the intact 50S subunit (178). Based on these observations it was concluded that in both subunits the RNAs form a structural skeleton to which the proteins bind without causing major conformational changes in the RNA framework, and a model for the small subunit has been proposed (179). However, the conclusion that the shape and size of the ribosomal RNAs are very similar in situ and as isolated moieties was challenged by recent electron microscopical results (177); it is also in disagreement with other physical studies (143, 180, 181).

Immune electron microscopy was used to determine the location of some characteristic regions of the ribosomal RNA strands within their cognate subunits. By means of antibodies against the modified nucleoside N^6, N^6-dimethyladenosine, which occurs at positions 24 and 25 from the 3' end of the 16S RNA, this region could be localized near the partition between the head and the body of the 30S subunit (55, 182). Similarly, the 7-methylguanosine at position 526 of the 16S RNA was also mapped near the junction of the upper one-third and the lower two-thirds of the 30S particle and maximally distant from the platform (183). Furthermore, the 3' ends of 16S RNA (184–186), of 23S RNA (28, 79), and of 5S RNA (79, 187) as well as the 5' end of the 16S RNA (188) were localized on the small and large ribosomal sub-units of *E. coli*, respectively (Figure 10). The results from the various groups are in good agreement with each other, and the localized regions can be used as fixed points in the attempts to elucidate the spatial arrangement of the ribosomal RNAs within their subunits (see below).

RNA-RNA Cross-links

Among the various approaches so far used, cross-linking between different regions of the RNA strand in situ is the most direct and promising one. The goal is the identification of enough cross-links between RNA regions "outside" the secondary structure to enable the RNA strand to be folded into three dimensions in the ribosomal model. To this end, after treatment of intact 30S or 50S subunits with suitable cross-linking reagents, the RNA is extracted and partially digested with nucleases. The fragments are separated by two-dimensional electrophoresis and analyzed for cross-links. In

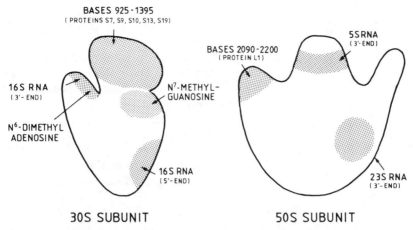

BASES 925-1395
(PROTEINS S7, S9, S10, S13, S19)

5S RNA
(3'- END)

BASES 2090-2200
(PROTEIN L1)

16S RNA
(3'- END)

N⁷-METHYL-
GUANOSINE

N⁶-DIMETHYL
ADENOSINE

16S RNA
(5'- END)

23S RNA
(3'- END)

30S SUBUNIT   50S SUBUNIT

Figure 10   Location of RNA regions in the small and large *E. coli* subunits. For references see text.

this way a number of intra-RNA cross-links were identified (170, 189) but most of those found so far connect RNA regions that are within elements of the secondary structure. However, recently the first cross-links were localized from which some conclusions about the three-dimensional arrangement of the 23S RNA strand in situ can be drawn (190).

Intra-RNA cross-linking studies on the relatively small 5S RNA molecule have yielded two cross-links within the secondary structure (191, 192) and one between residues G-41 and G-72, which reflects a tertiary structural interaction (193). This latter result has been used to construct an improved three-dimensional model of the 5S RNA molecule (193).

Intra-RNA cross-links within the 16S RNA caused by intercalating psoralen derivatives were investigated by electron microscopy (194–197), in order to identify the approximate positions along the 16S RNA molecule of the cross-linked regions. It will be useful to improve the resolution of the identification by isolation of RNA fragments containing the cross-links, and by identifying the cross-linked nucleotides by sequence analysis.

RNA-Protein Cross-links

The identification of RNA-protein cross-links is a very useful method for obtaining information concerning the spatial arrangement of the RNA molecules within their subunits. If on the one hand the locations of the proteins on the subunit are known, e.g. by immune electron microscopy and neutron scattering, and if on the other hand the RNA regions neighboring individual proteins are identified by RNA-protein cross-linking studies, then it should be possible to trace the RNA strand within the particle. In this way one should be able to reveal for instance which regions of the 16S RNA are in the various parts (head, neck, body, etc) of the *E. coli* 30S subunit.

In order to obtain this type of information it is not sufficient to identify those proteins that can be cross-linked to the intact RNA strand, as was done in the early studies [reviewed in (7)], but it is necessary to determine, at the nucleotide level, the precise RNA region to which a given protein has been cross-linked. This is in principle possible since the primary structures of the ribosomal RNAs are known. However, because of technical difficulties in the performance of these analyses, only relatively few identifications have so far been accomplished. The identification of the position of the amino acid residue cross-linked to the RNA strand has proved to be even more difficult [see (6) for a summary and discussion of recent results].

A variety of cross-linking reagents has been used, as well as a number of methods for the isolation and analysis of the cross-linked RNA and protein regions (6). Ultraviolet irradiation (198–200), periodate oxidation (201, 202), and many bifunctional reagents, such as diepoxybutane (203), bis-(2-chloroethyl)-amine (204), iminothiolane combined with UV-irradiation (90), and several aromatic azido derivatives (205–208) were applied for cross-linking studies. In this way many of the ribosomal proteins were found to be cross-linked to the 16S or 23S RNA (7). This shows that many, if not all, of the proteins are in close neighborhood to the RNA within the subunits. It should be noted that results from cross-linking studies give purely topographical information, and it cannot be concluded from these studies that the cross-linked protein and RNA regions are associated by RNA-protein interactions. This latter type of information can be derived from investigations in which individual ribosomal proteins are bound to their cognate RNAs and the binding sites are identified.

Protein Binding Sites on the RNA

Many of the individual ribosomal proteins bind independently and specifically to the isolated 5S, 16S, and 23S RNA. The precise number of these primary binding proteins depends on the methods by which the proteins and the RNAs are isolated. By mild nuclease treatment of the complex between the RNA and the individual protein bound to it, the binding site on the RNA protected by this protein can be isolated and identified by sequence analysis. Since these studies were reviewed elsewhere (1, 6, 7) only a brief summary of the results is given here.

The length of the binding site varies by a factor of ten depending on the individual protein bound to the RNA. It ranges from 50–60 nucleotides for protein S8 or L11 to 400–500 nucleotides for proteins S4 or L24. When the binding sites are compared with the map of the secondary structure of the 16S or 23S RNA, it can clearly be seen that they correspond closely to structural domains, i.e. stems, loops, etc. With some proteins, e.g. S4 and L24, the RNA fragments protected by the bound proteins consist of non-contiguous sequences held together by long-range interactions. It is interesting that these interactions are so strong that they can survive the removal of the protein, and identical fragments can be isolated by mild nuclease digestion of naked RNA. This indicates that the secondary (and probably also the tertiary) structure of the RNA is the stabilizing force. A given protein can associate with one or more of the structural domains, and especially in the latter case it helps in stabilization of the ribosomal architecture.

Fragmentation of Subunits

A further useful method to obtain information on the neighborhood between proteins and RNA regions within the ribosome is to treat intact subunits (or reconstitution intermediates lacking some proteins) with nuclease under mild conditions and to isolate the resulting ribonucleoprotein fragments of various sizes by sucrose gradient centrifugation or by gel electrophoresis. The proteins within the fragments are identified by two-dimensional electrophoresis and the RNA regions by sequence analysis. The results of this approach, which has been applied more successfully to the analysis of the 30S than of the 50S subunit, have been summarized (10, 209). The limitation of the method, namely the ease of dissociation of the ribonucleoprotein complexes and the concomitant loss of the specificity of the interactions, was recently overcome to a large extent by a variation of the procedure (210). By the improved method a few large RNP fragments from the small subunit were isolated and analyzed (211, 212). The results agree with those from binding studies (7) showing the specificity of the approach and also give evidence for the existence of long-range interactions between various regions of the RNA strand that are far apart in the primary sequence.

Spatial Packing of the RNAs Within Their Subunits

If the data from all the approaches described above are combined, then it becomes possible to determine the spatial packing of the RNA strands in situ to a certain extent. A summary of the present knowledge is given in Figure 10. In addition to the locations shown in this figure, the approximate positions of several additional RNA regions can be inferred, e.g. of those regions (nucleotides 25–550) that interact with protein S4 or of those (nucleotides 550–800) that are close to proteins S6, S8, S15, and S18. Both

RNA regions are located in the body of the 30S subunit, and the latter region is closer to the platform than the former one. However, additional data are necessary for a precise localization of these and other RNA regions in situ.

CONCLUDING REMARKS

In a recent review (1) the known primary structures of all the *E. coli* ribosomal proteins and RNAs were summarized, and the development of reliable models for the secondary structure of the three RNA molecules was described. The present article represents a sequel to the previous review and deals with the problem of how the individual proteins and RNA molecules are arranged in situ, and it shows how the results from a variety of physical, chemical, and immunological studies have already given some insight into the architecture of the *E. coli* ribosome. This is especially true for the small ribosomal subunit for which recent results, in particular those from immune electron microscopy and neutron scattering, yield a gratifying agreement on the spatial arrangement of many proteins within the particle. In addition, the locations of some RNA regions, e.g. the 5' and 3' ends, and of several functional domains on the ribosomal subunits have been directly determined. Other RNA regions can be mapped indirectly in those cases where it is known which proteins lie close to them in situ, and if the positions of these proteins within the particle have been located. The three-dimensional packing of the RNA chains in situ is being revealed gradually by intra-RNA cross-linking and other methods.

Future research will lead to a refinement of the current model for the architecture of the *E. coli* ribosome. This can be accomplished by further application of the various approaches used so far. Of particular importance in this context are immune electron microscopy, neutron scattering, cross-linking, and fluorescence studies. It can be expected that these investigations will eventually lead to a reliable model of the ribosome in which the spatial arrangements of the proteins and of the RNA chains within their cognate subunits are known. Furthermore, the three-dimensional structure at a high resolution will be determined by X-ray diffraction analysis of many crystallized ribosomal proteins.

In conclusion, enormous and rapid progress has been made in elucidating the structure of the *E. coli* ribosome, but there still remains much to be done in order to reveal its architecture at a level which is detailed enough to understand, in molecular terms, the function of the ribosome and the involvement of its many components in the complex process of protein biosynthesis.

Literature Cited

1. Wittmann, H. G. 1982. *Ann. Rev. Biochem.* 51:155–82
2. Stöffler, G., Bald, R., Kastner, B., Lührmann, R., Stöffler-Meilicke, M., et al. 1979. In *Ribosomes: Structure, Function and Genetics,* ed. G. Chambliss, G. R. Craven, J. Davies, K. Davis, L. Kahan, M. Nomura, pp. 171–205. Baltimore: Univ. Park Press
3. Lake, J. A. 1979. See Ref. 2, pp. 207–36
4. Moore, P. M. 1979. See Ref. 2, pp. 111–33
5. Traut, R. R., Lambert, J. M., Boileau, G., Kenny, J. W. 1979. See Ref. 2, pp. 89–110
6. Brimacombe, R., Maly, P., Zwieb, C. 1983. *Prog. Nucleic Acid Res. Mol. Biol.* 28: In press
7. Zimmermann, R. A. 1979. See Ref. 2, pp. 135–69
8. Cantor, C. R., Huang, K. H., Fairclough, R. 1974. In *Ribosomes,* ed. M. Nomura, A. Tissières, P. Lengyel, pp. 587–99. Long Island, NY: Cold Spring Harbor Lab.
9. Nomura, M., Held, W. A. 1974. See Ref. 8, pp. 193–223
10. Nierhaus, K. H. 1982. *Curr. Top. Microbiol. Immunol.* 97:81–155
11. Brimacombe, R., Stöffler, G., Wittmann, H. G. 1978. *Ann. Rev. Biochem.* 47:217–49
12. Van Holde, K. E., Hill, W. E. 1974. See Ref. 8, pp. 53–91
13. Stuhrmann, H. B., Haas, J., Ibel, K., De Wolf, B., Koch, M. H. J., et al. 1976. *Proc. Natl. Acad. Sci. USA* 73:2379–83
14. Hill, W. E., Fassenden, R. J. 1974. *J. Mol. Biol.* 90:719–26
15. Serdyuk, I. N., Grenader, A. K., Zaccai, G. 1979. *J. Mol. Biol.* 135:691–707
16. Moore, P. B., Engelman, D. M., Schoenborn, B. P. 1975. *J. Mol. Biol.* 91:101–20
17. Beaudry, P., Peterson, H. V., Grunberg-Manago, M., Jacrot, B. 1976. *Biochem. Biophys. Res. Commun.* 72: 391–97
18. Stuhrmann, H. B., Koch, M. H. J., Parfait, R., Haas, J., Ibel, K., Crichton, R. R. 1978. *J. Mol. Biol.* 119:203–12
19. Vasiliev, V. D., Selivanova, O. M., Koteliansky, V. E. 1978. *FEBS Lett.* 95:273–76
20. Vasiliev, V. D., Koteliansky, V. E., Shatsky, I. N., Rezapkin, G. V. 1977. *FEBS Lett.* 84:43–47
21. Vasiliev, V. D., Koteliansky, V. E., Rezapkin, G. V. 1977. *FEBS Lett.* 79:170–74
22. Moore, P. B., Engelman, D. M., Schoenborn, B. P. 1974. *Proc. Natl. Acad. Sci. USA* 71:172–76
23. Serdyuk, I. N., Grenader, A. K. 1975. *FEBS Lett.* 59:133–36
24. Deleted in proof
25. Crichton, R. R., Engelman, D. M., Haas, J., Koch, M. H. J., Moore, P. B., et al. 1977. *Proc. Natl. Acad. Sci. USA* 74:5547–50
26. Serdyuk, I. N., Grenader, A. K., Koteliansky, V. E. 1977. *Eur. J. Biochem.* 79:504–8
27. Vasiliev, V. D. 1974. *Acta Biol. Med. Ger.* 33:779–93
28. Shatsky, I. N., Evstafieva, A. G., Bystrova, T. F., Bogdanov, A. A., Vasiliev, V. D. 1980. *FEBS Lett.* 122:251–55
29. Boublik, M. 1982. In *Anatomy of Escherichia coli,* ed. N. Nanninga. London: Academic. In press
30. Korn, A. P., Spitnik-Elson, P., Elson, D. 1982. *J. Biol. Chem.* 257:7155–60
31. Spiess, E. 1978. *FEBS Lett.* 91:289–92
32. Kastner, B., Stöffler-Meilicke, M., Stöffler, G. 1981. *Proc. Natl. Acad. Sci. USA* 1981. 78:6652–56
33. Kastner, B., Stöffler-Meilicke, M., Stöffler, G. 1982. *Electron Microscopy, Int. Congr. Electr. Microscopy, Hamburg,* 3:105–6
34. Hoppe, W. 1982. See Ref. 33, 1:97–106
35. Leonard, K. R., Lake, J. A. 1979. *J. Mol. Biol.* 129:155–63
36. Kress, Y., Wittner, M., Rosenbaum, R. M. 1971. *J. Cell. Biol.* 49:773–84
37. Byers, B. 1967. *J. Mol. Biol.* 26:155–67
38. Barbieri, M. 1982. *J. Theor. Biol.* 91: 545–601
39. Taddei, C. 1972. *Exp. Cell. Res.* 70: 285–92
40. Kühlbrandt, W., Unwin, P. N. T. 1982. *J. Mol. Biol.* 156:431–48
41. O'Brien, L., Shelley, K., Towfighi, J., McPherson, A. 1980. *Proc. Natl. Acad. Sci. USA* 77:2260–64
42. Clark, M. W., Hammons, M., Langer, J. A., Lake, J. A. 1979. *J. Mol. Biol.* 135:507–12
43. Clark, M. W., Leonard, K., Lake, J. A. 1982. *Science* 216:999–1001
44. Wittmann, H. G., Müssig, J., Piefke, J., Gewitz, H. S., Rheinberger, H. J., Yonath, A. 1982. *FEBS Lett.* 146: 217–20
45. Yonath, A., Müssig, J., Tesche, B., Lorenz, S., Erdmann, V. A., Wittmann, H. G. 1980. *Biochem. Int.* 1:428–35
46. Yonath, A., Khavitch, G., Tesche, B., Müssig, J., Lorenz, S., Erdmann, V. A.,

Wittmann, H. G. 1982. *Biochem. Int.* 5:629–35
47. Yonath, A., Müssig, J., Wittmann, H. G. 1982. *J. Cell Biochem.* 19:145–55
48. Leonard, K. R., Arad, T., Tesche, B., Erdmann, V. A., Wittmann, H. G., Yonath, A. E. 1982. See Ref. 33, pp. 9–15
49. Stöffler, G., Wittmann, H. G. 1971. *Proc. Natl. Acad. Sci. USA* 68:2283–87
50. Stöffler, G., Wittmann, H. G. 1971. *J. Mol. Biol.* 62:407–9
51. Wabl, M. R. 1973. *Elektronenmikroskopische Lokalisierung von Proteinen auf der Oberfläche ribosomaler Untereinheiten von Escherichia coli mittels spezifischer Antikörper.* PhD thesis. Freie Univ., Berlin
52. Kahan, L., Winkelmann, D. A., Lake, J. A. 1981. *J. Mol. Biol.* 145:193–214
53. Lake, J. A., Strycharz, W. A. 1981. *J. Mol. Biol.* 153:979–92
54. Winkelmann, D. A., Kahan, L., Lake, J. A. 1982. *Proc. Natl. Acad. Sci. USA* 79:3111–15
55. Stöffler, G., Stöffler-Meilicke, M. 1981. In *International Cell Biology*, ed. H. G. Schweiger, pp. 93–102. Heidelberg: Springer
56. Stöffler-Meilicke, M., Stöffler, G. 1982. See Ref. 33, pp. 99–100
57. Noah, M., Stöffler-Meilicke, M., Stöffler, G. 1982. See Ref. 33, pp. 101–2
58. Strycharz, W. A., Nomura, M., Lake, J. A. 1978. *J. Mol. Biol.* 126:123–40
59. Liljas, A. 1982. *Prog. Biophys. Mol. Biol.* In press
60. McKuskie-Olsen, H., Grant, P. G., Glitz, D. G., Cooperman, B. S. 1980. *Proc. Natl. Acad. Sci. USA* 77:890–94
61. Stöffler, G., Bald, R., Lührmann, R., Tischendorf, G., Stöffler-Meilicke, M. 1980. In *Electron Microscopy*, ed. P. Bredero, W. de Priester, pp. 566–67. Leiden: Eur. Congr. Electr. Microscopy
62. Lührmann, R., Bald, R., Stöffler-Meilicke, M., Stöffler, G. 1981. *Proc. Natl. Acad. Sci. USA* 78:7276–80
63. Lührmann, R., Stöffler-Meilicke, M., Dieckhoff, J., Tischendorf, G., Stöffler, G. 1980. See Ref. 61, pp. 568–69
64. Keren-Zur, M., Boublik, M., Ofengand, J. 1979. *Proc. Natl. Acad. Sci. USA* 76:1054–58
65. Girshovich, A. S., Kurtskhalla, T. V., Ovchinnikov, Y. A., Vasiliev, V. D. 1981. *FEBS Lett.* 130:54–59
66. Bernabeau, C., Lake, J. A. 1982. *Proc. Natl. Acad. Sci. USA* 79:3111–15
67. Moore, P. B., Engelman, D. M., Langer, J. A., Ramakrishnan, V. R., Schindler, D. G., et al. 1983. In *Basic*

Life Sciences Series, ed. B. P. Schoenborn. New York: Plenum. In press
68. Nierhaus, K. H., Lietzke, R., May, R. P., Nowotny, V., Schulze, H., et al. 1983. *Proc. Natl. Acad. Sci. USA.* In press
69. Nierhaus, K. H., Lietzke, R., Nowotny, V., Schulze, H., Wurmbach, P., et al. 1982. *Europhysics J., Phys. Status Solidi B.* In press
70. Epe, B., Woolley, P., Steinhäuser, K., Littlechild, J. 1982. *Eur. J. Biochem.* 129:211–19
71. Odom, O. W., Robbins, D., Kramer, G., Hardesty, B., Subramanian, A. R. 1983. *Arch. Biochem. Biophys.* In press
72. Chu Gloria, Y., Cantor, C. G. 1979. *Nucleic Acids Res.* 6:2363–79
73. Steinhäuser, K. G., Woolley, P., Epe, B., Dijk, J. 1982. *Eur. J. Biochem.* 127:587–95
74. Zantema, A., Maassen, J. A., Möller, W. 1982. *Biochemistry* 21:3069–76
75. Zantema, A., Maassen, J. A., Kriek, J., Möller, W. 1982. *Biochemistry* 21:3077–82
76. Lee, C. C., Cantor, C. R., Wittmann-Liebold, B. 1981. *J. Biol. Chem.* 256:41–48
77. Lee, C. C., Wells, B. D., Fairclough, R. H., Cantor, R. C. 1981. *J. Biol. Chem.* 256:49–53
78. Odom, O. W., Robbins, D. J., Lynch, J., Dottavio-Martin, D., Kramer, G., Hardesty, B. 1980. *Biochemistry* 19:5947–53
79. Stöffler-Meilicke, M., Stöffler, G., Odom, O. W., Zinn, A., Kramer, G., Hardesty, B. 1981. *Proc. Natl. Acad. Sci. USA* 78:5538–42
80. Robbins, D., Odom, O. W., Lynch, J., Kramer, G., Hardesty, B., Liou, R., Ofengand, J. 1981. *Biochemistry* 20:5301–9
81. Fairclough, R. H., Cantor, C. R., Wintermeyer, W., Zachau, H. G. 1979. *J. Mol. Biol.* 132:557–73
82. Fairclough, R. H., Cantor, C. R. 1979. *J. Mol. Biol.* 132:575–86
83. Fairclough, R. H., Cantor, C. R. 1979. *J. Mol. Biol.* 132:587–601
84. Wintermeyer, W., Gualerzi, C. 1981. *Biophys. Struct. Mechan.* 7:287–89
85. Johnson, A. E., Adkins, H. J., Matthews, E. A., Cantor, C. R. 1982. *J. Mol. Biol.* 139:113–39
86. Box, R., Woolley, P., Pon, C. 1981. *Eur. J. Biol.* 116:93–99
87. Weiel, J., Hershey, J. W. B. 1981. *Biochemistry* 20:5859–65
88. Langlois, R., Lee, C. C., Cantor, C. R.,

Vince, R., Pestka, S. 1976. *J. Mol. Biol.* 106:297–313
89. Hall, J., Davis, J. P., Cantor, R. R. 1977. *Arch. Biochem. Biophys.* 179:121–30
90. Wower, I., Wower, J., Meincke, M., Brimacombe, R. 1981. *Nucleic Acids Res.* 9:4285–4302
91. Sommer, A., Traut, R. R. 1975. *Proc. Natl. Acad. Sci. USA* 71:3946–50
92. Lutter, L. C., Bode, U., Kurland, C. G., Stöffler, G. 1974. *Mol. Gen. Genet.* 129:167–76
93. Kurland, C. G. 1977. *Ann. Rev. Biochem.* 46:173–200
94. Expert-Bezancon, A., Barritault, D., Milet, M., Guerin, M. F., Hayes, D. H. 1977. *J. Mol. Biol.* 112:603–29
95. Peretz, H., Towbin, H., Elson, D. 1976. *Eur. J. Biochem.* 63:83–92
96. Lambert, J. M., Traut, R. R. 1981. *J. Mol. Biol.* 149:451–76
97. Cover, J. A., Lambert, J. M., Normann, C. M., Traut, R. R. 1981. *Biochemistry* 20:2843–52
98. Boileau, G., Sommer, A., Traut, R. R. 1981. *J. Biol. Chem.* 256:8222–27
99. Lutter, L. C., Kurland, C. G. 1973. *Nature New Biol.* 243:15–17
100. Maassen, J. A., Schop, E. N., Möller, W. 1981. *Biochemistry* 20:1020–25
101. Erdmann, V. A. 1976. *Prog. Nucleic Mol. Biol.* 18:45–90
102. Wittmann, H. G., Littlechild, J. A., Wittmann-Liebold, B. 1979. See Ref. 2, pp. 51–88
103. Wittmann-Liebold, B. 1982. In *Methods of Protein Sequence Analysis,* ed. M. Elzinga, pp. 27–63. Clifton, NJ: Humana Press Inc.
104. Allen, G., Capasso, R., Gualerzi, C. 1979. *J. Biol. Chem.* 254:9800–6
105. Leijonmarck, M., Eriksson, S., Liljas, A. 1980. *Nature* 286:824–26
106. Appelt, K., Dijk, J., Epp, O. 1979. *FEBS Lett.* 103:66–70
107. Liljas, A., Newcomer, M. E. 1981. *J. Mol. Biol.* 153:393–98
108. Appelt, K., Dijk, J., Reinhardt, R., Sanhuesa, S., White, S. W., et al. 1981. *J. Biol. Chem.* 256:11787–90
109. Appelt, K., White, S. W., Wilson, K. 1983. *J. Biol. Chem.* In press
110. Dijk, J., White, S. W., Wilson, K., Appelt, K. 1983. *J. Biol. Chem.* In press
111. Möller, W., Groene, A., Terhorst, C., Amons, R. 1972. *Eur. J. Biochem.* 25:5–12
112. Dijk, J., Georgalis, Y., Labischinski, H., Wills, P. R. 1983. *Biochemistry.* In press

113. Pettersson, I., Hardy, S. J. S., Liljas, A. 1976. *FEBS Lett.* 64:135–38
114. Dijk, J., Littlechild, J., Garrett, R. A. 1977. *FEBS Lett.* 77:295–300
115. Gudkov, A. T., Tumananova, L. G., Venyminov, S. Y., Khechinashvilli, N. N. 1978. *FEBS Lett.* 93:215–18
116. Strycharz, W. A., Nomura, M., Lake, J. A. 1978. *J. Mol. Biol.* 126:123–40
117. Dijk, J., Littlechild, J. 1978. *Methods Enzymol.* 59:481–502
118. Wystup, G., Teraoka, H., Schulze, H., Hampl, H., Nierhaus, K. H. 1979. *Eur. J. Biochem.* 100:101–3
119. Aune, K. C. 1977. *Arch. Biochem. Biophys.* 180:172–77
120. Tindall, S. H., Aune, K. C. 1981. *Biochemistry* 20:4861–66
121. Stöffler, G., Wittmann, H. G. 1977. In *Molecular Mechanism of Protein Biosynthesis,* ed. H. Weissbach, S. Pestka, pp. 117–202. New York: Academic
122. Craven, G. R., Gupta, V. 1970. *Proc. Natl. Acad. Sci. USA* 67:1329–36
123. Chang, F. N., Flaks, J. G. 1971. *J. Mol. Biol.* 61:387–400
124. Spitnik-Elson, P., Breiman, A. 1971. *Biochim. Biophys. Acta* 254:457–67
125. Crichton, R. R., Wittmann, H. G. 1971. *Mol. Gen. Genet.* 114:95–105
126. Benkov, K., Delihas, N. 1974. *Biochem. Biophys. Res. Commun.* 60:901–8
127. Spitnik-Elson, P., Schechter, N., Abramovitz, R., Elson, D. 1976. *Biochemistry* 15:5246–53
128. Ginzburg, I., Zamir, A. 1976. *J. Mol. Biol.* 100:387–98
129. Modolell, J., Vazquez, D. 1976. In *Ribosomes and RNA Metabolism,* ed. J. Zelinka, J. Balan, 2:259–68. Bratislava: Publ. House Slov. Acad. Sci.
130. Ghosh, N., Moore, P. B. 1979. *Eur. J. Biochem.* 93:147–56
131. Michalski, C. J., Sells, B. H. 1975. *Eur. J. Biochem.* 52:385–89
132. Litman, D. J., Beekman, A., Cantor, C. R. 1976. *Arch. Biochem. Biophys.* 174:523–31
133. Changchien, L. M., Craven, G. R. 1977. *J. Mol. Biol.* 133:103–22
134. Lam, M. K. T., Changchien, L. M., Craven, G. R. 1979. *J. Mol. Biol.* 128:561–75
135. Noller, H. F. 1974. *Biochemistry* 13:4694–703
136. Chapman, N. M., Noller, H. F. 1977. *J. Mol. Biol.* 109:131–49
137. Santer, M., Shane, S. 1977. *J. Bacteriol.* 130:900–10
138. Traub, P., Nomura, M. 1968. *Proc. Natl. Acad. Sci. USA* 59:777–84

139. Dunn, J. M., Wong, K. P. 1979. *J. Biol. Chem.* 254:7705–11
140. Dunn, J. M., Wong, K. P. 1979. *J. Biol. Chem.* 254:7712–16
141. Dunn, J. M., Wong, K. P. 1979. *Biochemistry* 18:4380–85
142. Tam, M. F., Hill, W. E. 1981. *Biochemistry* 20:6480–84
143. Tam, M. F., Dodd, J. A., Hill, W. E. 1981. *J. Biol. Chem.* 256:6430–34
144. Tam, M. F., Hill, W. E. 1981. *Biochem. Int.* 3:655–62
145. Bogdanov, A. A., Zimmermann, R. A., Wang, C. C., Ford, N. C. 1978. *Science* 202:999–1001
146. Barritault, D., Guerin, M. F., Hayes, D. H. 1979. *Eur. J. Biochem.* 98:567–71
147. Changchien, L. M., Craven, G. R. 1978. *J. Mol. Biol.* 125:43–56
148. Zagorska, L., Skopinska, A., Klita, S., Szafranski, P. 1980. *Biochem. Biophys. Res. Commun.* 95:1152–59
149. Mizushima, S., Nomura, M. 1970. *Nature* 226:1214–18
150. Nomura, M., Erdmann, V. A. 1970. *Nature* 228:744–48
151. Nierhaus, K. H., Dohme, F. 1974. *Proc. Natl. Acad. Sci. USA* 71:4713–17
152. Dohme, F., Nierhaus, K. H. 1976. *J. Mol. Biol.* 107:585–99
153. Dohme, F., Nierhaus, K. H. 1976. *Proc. Natl. Acad. Sci. USA* 73:2221–25
154. Sieber, G., Nierhaus, K. H. 1978. *Biochemistry* 17:3505–11
155. Sieber, G., Tesche, B., Nierhaus, K. H. 1980. *Eur. J. Biochem.* 106:515–23
156. Spillmann, S., Nierhaus, K. H. 1978. *J. Biol. Chem.* 253:7047–50
157. Nowotny, V., Nierhaus, K. H. 1980. *J. Mol. Biol.* 137:391–99
158. Spillmann, S., Dohme, F., Nierhaus, K. H. 1977. *J. Mol. Biol.* 115:513–23
159. Roth, H. E., Nierhaus, K. H. 1980. *Eur. J. Biochem.* 103:95–98
160. Röhl, R., Roth, H. E., Nierhaus, K. H. 1982. *Hoppe-Seyler's Z. Physiol. Chem.* 363:143–57
161. Röhl, R., Nierhaus, K. H. 1982. *Proc. Natl. Acad. Sci. USA* 79:729–33
162. Noller, H. F. 1979. See Ref. 2, pp. 3–22
163. Noller, H. F., Woese, C. R. 1981. *Science* 212:403–11
164. Brimacombe, R. 1982. *Biochem. Soc. Symp.* 47:49–60
165. Woese, C. R., Magrum, L. J., Gupta, R., Siegel, R. B., Stahl, D. A., et al. 1980. *Nucleic Acids Res.* 8:2275–93
166. Glotz, C., Brimacombe, R. 1980. *Nucleic Acids Res.* 8:2377–95
167. Zwieb, C., Glotz, C., Brimacombe, R. 1981. *Nucleic Acids Res.* 9:3621–40
168. Stiegler, P., Carbon, P., Zuker, M., Ebel, J. P., Ehresmann, C. 1980. *CR Acad. Sci. Ser. D* 291:937–40
169. Stiegler, P., Carbon, P., Zuker, M., Ebel, J. P., Ehresmann, C. 1981. *Nucleic Acids Res.* 9:2153–72
170. Glotz, C., Zwieb, C., Brimacombe, R., Edwards, K., Kössel, H. 1981. *Nucleic Acids Res.* 9:3287–306
171. Branlant, C., Krol, A., Machatt, M. A., Pouyet, J., Ebel, J. P., et al. 1981. *Nucleic Acids. Res.* 9:4303–24
172. Noller, H. F., Kop, J., Wheaton, V., Brosius, J., Gutell, R. R., et al. 1981. *Nucleic Acids Res.* 9:6167–89
173. Tesche, B., Schmiady, H., Lorenz, S., Erdmann, V. A. 1980. See ref. 61, pp. 534–35
174. Sieber, G., Tesche, B., Nierhaus, K. H. 1980. *Eur. J. Biochem.* 106:515–23
175. Vasiliev, V. D., Salivanova, O. M., Koteliansky, V. E. 1978. *FEBS Lett.* 95:273–76
176. Edlind, T. D., Bassel, A. R. 1980. *J. Bacteriol.* 141:365–73
177. Boublik, M., Robakis, N., Hellmann, W., Wall, J. S. 1982. *Eur. J. Cell Biol.* 27:177–84
178. Vasiliev, V. D., Zalite, O. M. 1980. *FEBS Lett.* 121:101–4
179. Spirin, A. S., Serdyuk, I. N., Shpungin, J. L., Vasiliev, V. D. 1979. *Proc. Natl. Acad. Sci. USA* 76:4867–71
180. Robakis, N., Boublik, M. 1981. *Biochem. Biophys. Res. Commun.* 103:1401–8
181. Tam, M. F., Dodd, J. A., Hill, W. E. 1981. *FEBS Lett.* 130:217–20
182. Politz, S. M., Glitz, D. G. 1977. *Proc. Natl. Acad. Sci. USA* 74:1468–72
183. Trempe, M. R., Ohgi, K., Glitz, D. G. 1982. *J. Biol. Chem.* 257:9822–29
184. McKuskie-Olson, H., Glitz, D. G. 1979. *Proc. Natl. Acad. Sci. USA* 76:3769–73
185. Shatsky, I. N., Mochalova, L. V., Kojouharova, M. S., Bogdanov, A. A., Vasiliev, V. D. 1979. *J. Mol. Biol.* 133:501–5
186. Lührmann, R., Stöffler-Meilicke, M. Stöffler, G. 1981. *Mol. Gen. Genet.* 182:369–76
187. Shatsky, I. N., Estafieva, A. G., Bystrova, T. F., Bogdanov, A. A., Vasiliev, V. D. 1980. *FEBS Lett.* 121:97–100
188. Mochalova, L. V., Shatsky, I. N., Bogdanov, A. A., Vasiliev, V. D. 1982. *J. Mol. Biol.* 159:637–50
189. Zwieb, C., Brimacombe, R. 1980. *Nucleic Acids Res.* 8:2397–2411
190. Stiege, W., Zwieb, C., Brimacombe, R. 1982. *Nucleic Acids Res.* 10:7211–29

191. Wagner, R., Garrett, R. A. 1978. *Nucleic Acids Res.* 5:4065–76
192. Rabin, D., Crothers, D. M. 1979. *Nucleic Acids Res.* 7:689–703
193. Hancock, J., Wagner, R. A. 1982. *Nucleic Acids Res.* 10:1257–69
194. Wollenzien, P., Hearst, J. E., Thammana, P., Cantor, C. R. 1979. *J. Mol. Biol.* 135:255–69
195. Thammana, P., Cantor, C. R., Wollenzien, P. L., Hearst, J. E. 1979. *J. Mol. Biol.* 135:271–83
196. Wollenzien, P. L., Cantor, C. R. 1982. *J. Mol. Biol.* 159:151–66
197. Wollenzien, P. L., Cantor, C. R. 1982. *Proc. Natl. Acad. Sci. USA* 79:3940–44
198. Möller, K., Brimacombe, R. 1975. *Mol. Gen. Genet.* 141:343–55
199. Ehresmann, B., Reinbolt, J., Backendorf, C., Tritsch, D., Ebel, J. P. 1976. *FEBS Lett.* 67:316–19
200. Baca, O. G., Bodley, J. W. 1976. *Biochem. Biophys. Res. Commun.* 70:1091–96
201. Kenner, R. A. 1973. *Biochem. Biophys. Res. Commun.* 51:932–38
202. Czernilofsky, A. P., Kurland, C. G., Stöffler, G. 1975. *FEBS Lett.* 58:281–84
203. Bäumert, H. G., Sköld, S. E., Kurland,

C. G. 1978. *Eur. J. Biochem.* 89:353–59
204. Ulmer, E., Meinke, M., Ross, A., Fink, G., Brimacombe, R. 1978. *Mol. Gen. Genet.* 160:183–93
205. Fink, G., Fasold, H., Rommel, W., Brimacombe, R. 1980. *Anal. Biochem.* 108:394–401
206. Rinke, J., Meinke, M., Brimacombe, R., Fink, G., Rommel, W., Fasold, H. 1980. *J. Mol. Biol.* 137:301–14
207. Millon, R., Olomucki, M., LeGall, J. Y., Golinska, B., Ebel, J. P., Ehresmann, B. 1980. *Eur. J. Biochem.* 110:485–92
208. Millon, R., Ebel, J. P., LeGoffic, F., Ehresmann, B. 1981. *Biochem. Biophys. Res. Commun.* 101:784–91
209. Brimacombe, R., Nierhaus, K. H., Garrett, R. A., Wittmann, H. G. 1976. *Prog. Nucleic Acid Res. Mol. Biol.* 18:1–44
210. Spitnik-Elson, P., Elson, D. 1979. *Methods Enzymol.* 59:461–81
211. Spitnik-Elson, P., Elson, D., Avital, S., Abramowitz, R. 1982. *Nucleic Acids Res.* 10:1995–2006
212. Spitnik-Elson, P., Elson, D., Avital, S., Abramowitz, R. 1982. *Nucleic Acids Res.* 10:4483–92

Ann. Rev. Biochem. 1983. 52:67–91
Copyright © 1982 by Annual Reviews Inc. All rights reserved

AFFINITY LABELING OF PURINE NUCLEOTIDE SITES IN PROTEINS

Roberta F. Colman

Department of Chemistry, University of Delaware, Newark, Delaware 19711

CONTENTS

PERSPECTIVES AND SUMMARY 67
PERIODATE-OXIDIZED NUCLEOTIDES 69
ALKYL HALIDE DERIVATIVES OF PURINE NUCLEOTIDES 72
PHOTOREACTIVE PURINE NUCLEOTIDE ANALOGS 74
FLUOROSULFONYLBENZOYL ANALOGS OF NUCLEOTIDES 76
   General Considerations 76
   Glutamate Dehydrogenase 80
   Pyruvate Kinase 84
   ADP Receptor Protein of Platelets 86
CONCLUDING REMARKS 87

PERSPECTIVES AND SUMMARY

Purine nucleotides have a multifunctional role in cellular metabolism; for example, they are directly involved in most kinase reactions, they are the precursors of RNA and DNA, they function as regulators of many allosteric enzymes, they are involved in the aggregation of platelets, and they participate in dehydrogenase reactions as part of the NAD or NADP molecules. Analogs of nucleotides, many of which bind reversibly to proteins, have been used in a wide variety of experiments to probe the function of nucleotides in enzymatic reactions, as well as the structure and environment of the nucleotide binding sites. For example, the compounds adenylyl methylenediphosphonate (1) and adenylyl imidodiphosphate (2, 3), in which the β,γ-bridge oxygen of ATP is replaced by $-CH_2-$ or $-NH-$, respectively, are stable to enzymatic hydrolysis and solvolysis. These have provided tools to assess the relative importance of nucleotide binding and

67

hydrolysis in promoting various catalytic reactions and cellular processes. The diastereomeric thiophosphate analogs of nucleotides have been essential elements in investigations of the stereochemistry of enzymatic reactions (4, 5). Fluorescent derivatives, such as 1,N^6-ethenoadenosine triphosphate (6, 7) have been used to measure the binding of these nucleotides by enzymes (8), to report on the types of amino acid residues constituting the site and the spatial arrangement of the nucleotide within that site (9, 10), and to measure the distance between sites on an enzyme by energy transfer (11). Spin-labeled analogs, such as N^6-(2,2,6,6-tetramethylpiperidin-4-yl-1-oxyl) adenosine 5'-monophosphate (12) and adenosine 5'-diphosphate-4(2,2,6,6-tetramethylpiperidine-1-oxyl) (13) have been used to probe the local conformation of their binding sites and the distances between sites. Extended analogs that retain the terminal pyrimidine and imidazole rings, such as linear-benzoadenine nucleotides, made possible the evaluation of the size of the space available for the purine in several enzymes (14). The synthesis, chemical characteristics, and applications of these reversibly binding nucleotide analogs have been thoroughly reviewed (see 14–16). This chapter focuses instead on a special class of purine nucleotide analogs: those that contain reactive functional groups and therefore function as covalent affinity labels of the specific nucleotide binding sites of proteins.

Chemical modification contributed much to identification of amino acid residues in the active sites of enzymes; however, it is difficult to limit the covalent modification of a protein. Affinity labeling may be particularly suitable for specific modification of the protein binding sites of the purine nucleotides, because such amino acid residues are not expected to be unusually reactive unless they are direct participants in the catalytic reaction. Baker et al (17) pointed out the strategies used in affinity labeling, and Schoellman & Shaw demonstrated admirably the strategies for chymotrypsin (18) and trypsin (19). Baker (20) and Shaw (21) also wrote reviews of affinity labeling. More recently, an entire book has devoted itself to the subject (22). In designating a given compound as a specific site-directed reagent, three criteria should optimally be fulfilled:

1. There should be kinetic evidence for the initial formation of a reversible enzyme-inhibitor complex prior to irreversible inactivation. The existence of such a complex is indicated by a "rate saturation effect," in which the rate of inactivation is proportional to the reagent concentration until the enzyme site is saturated with reagent; further increases in the reagent concentration do not enhance the inactivation rate.
2. The extent of modification of amino acid residues produced by the nucleotide analog should be more limited than that produced by a structurally unrelated reagent with the same functional group.

3. The presence of the natural purine nucleotide or a reversible inhibitor that binds to the same site should decrease the rate of inactivation by the reagent.

In selecting a purine nucleotide for use as an affinity label, it is desirable for the compound to be structurally close to the natural nucleotide; hence, it should have the purine and ribose moieties and the appropriate negatively charged group, since these are features that are frequently important in directing the binding of the nucleotide. The compound should be soluble in water and reasonably stable over the range of pH and buffer conditions likely to be used for reactions with enzyme. The active functional group should be capable of reacting with many different types of amino acids, because in most cases the contributing residues to the binding site are not known in advance. The functional group should not be too bulky, so that the mode of binding of the nucleotide analog closely approximates that of the natural compound. Finally, the compound should produce a stoichiometric, stable, isolable product of reaction with the enzyme, so that the modified amino acid residue can be identified. Several types of purine nucleotide analogs are discussed, and the extent to which they conform to these ideal characteristics is evaluated.

PERIODATE-OXIDIZED NUCLEOTIDES

It has long been known that oxidation of ribonucleotides by periodate results in cleavage of the 2',3'-cis-diol to yield the corresponding dialdehyde derivative, but only recently has the structure of the periodate-oxidized ATP been characterized carefully (23). Aldehydes are capable of reacting with primary amines, such as the ϵ-amino group of lysine or the α-amino group of the protein N-terminal and this knowledge has provided the major impetus behind the increased use since 1976 of the dialdehyde nucleotide analogs as affinity labels of nucleotide sites in enzymes. Easterbrook-Smith et al (24) established for pyruvate carboxylase that, in the absence of a reducing agent, the 2',3'-dialdehyde derivative of ATP (oATP) behaved as a reversible linear competitive inhibitor with respect to MgATP. When NaBH$_4$ was added, irreversible inactivation was observed upon incubation of pyruvate carboxylase with oATP; these results were consistent with reduction of a Schiff base. The covalent reaction was limited; extrapolation to 100% inactivation indicated the incorporation of about one mole of radioactive oATP, and the coenzyme MgATP protected the enzyme against inactivation. This suggested that oATP reacts in the region of the metal-nucleotide binding site of that enzyme. Although a few reports indicate that periodate-oxidized nucleotides can react relatively nonspecifically with lysine residues in proteins (25, 26), for a growing number of enzymes the case for affinity labeling by this class of compounds is reasonably well established.

The 2',3'-dialdehyde derivative of ATP acts as an affinity label of the ATP site of the latent ATPase from *Mycobacterium phlei* (27). The rate constant for inactivation exhibits a nonlinear dependence on the concentration of oATP, indicating the reversible formation of an enzyme-reagent complex prior to the irreversible inactivation. ATP provides specific protection against inactivation, which is proportional to the incorporation of about one mole [^{3}H]-oATP per mole enzyme. The catalytic site of the rat liver succinyl CoA synthetase also is modified by a periodate-oxidized nucleotide: the 2',3'-dialdehyde analog of GDP (28). In this case, rate saturation kinetics was observed when the rate constant for inactivation was determined as a function of oGDP concentration, and complete inactivation was correlated with the incorporation of approximately one mole of reagent per mole of enzyme of 80,000 daltons. The 80,000-dalton enzyme species is a heterodimer of 46,500 and 33,500 subunits and, because radioactive oGDP was located in both types of subunits of the modified enzyme, it was proposed that the nucleotide binding site of succinyl CoA synthetase includes components from both subunits. Recently, the allosteric ADP activator site of NAD-dependent isocitrate dehydrogenase was specifically modified by periodate-oxidized ADP (M. M. King, R. F. Colman, 1983, *Biochemistry*, in press). The concentration dependence of inactivation by oADP in the presence of varying $MnSO_4$ concentrations revealed a consistent K_I of 23 μM for the enzyme and unchelated oADP, the same form of the natural ADP that binds to isocitrate dehydrogenase. Up to one mole of radioactive oADP was incorporated per average enzyme subunit and the affinity label was distributed among the three types of dissimilar subunits of NAD-dependent isocitrate dehydrogenase as shown by isoelectric focusing in polyacrylamide gels. Thus in several cases periodate-oxidized nucleotides were used effectively to evaluate the location of a nucleotide site in a multisubunit enzyme.

Upon treatment of NADP with periodate, only the ribose bound to the nicotinamide is converted to the corresponding 2',3'-dialdehyde. This periodate-oxidized NADP (oNADP) functions as a specific modifier of the coenzyme site of 6-phosphogluconate dehydrogenase from *Candida utilis* (29–31). The presence of the substrate (6-phosphogluconate), while not protecting against inactivation, alters from 2 to 1 the number of subunits of this dimeric enzyme that must be labeled to produce an inactive enzyme. The substrate may induce a conformational change leading to negative cooperativity between the subunits, an interaction reflected in the reaction of the oNADP affinity label (31). More recently, periodate-oxidized NADP was reduced enzymatically to the corresponding dialdehyde derivative of

NADPH (M. Mas, R. F. Colman, submitted for publication), and this nucleotide analog proved to be an affinity label for the pig heart NADP-dependent isocitrate dehydrogenase.

The product of reaction of the various enzymes with periodate-oxidized nucleotides is assumed usually to be a Schiff base resulting from nucleophilic attack of one of the dialdehydes by the ϵ-amino group of lysine; however, only a few authors have presented direct evidence for this product. Easterbrook-Smith et al (24) showed that after $NaBH_4$ reduction of oATP-modified pyruvate carboxylase, a total enzymatic digest yielded a product that co-migrated upon paper chromatography with a standard prepared by reaction of lysine with oATP. Dallocchio et al (30) synthesized from α-N-acetyllysine and glyceraldehyde the two reaction products expected from reaction (to form initially a Schiff base) of a lysyl residue with either the 2'-aldehyde or the 3'-aldehyde of a periodate-oxidized nucleotide, followed by reduction with $NaBH_4$ and acid hydrolysis. These acid-stable products had the same mobility on thin layer chromatography as those derived from oNADP-modified 6-phosphogluconate dehydrogenase (30). The same synthetic, acid-stable lysine products were shown more recently to be separable and quantifiable on an amino acid analyzer (M. M. King, R. F. Colman, 1983, *Biochemistry*, in press).

Although substantial evidence for a few enzymes indicates that reaction of periodate-oxidized nucleotides involves formation of a Schiff base with lysine, in an increasing number of cases the product is clearly not a Schiff base. Periodate-oxidized ATP inactivates phosphofructokinase (32); however, despite evidence for reaction with lysine residues, the measured lysine content of modified enzyme following acid hydrolysis showed no decrease, even after reduction with $NaBH_4$. These results contradict the expectations for a reducible Schiff base. Instead, the product may be a dihydroxymorpholino derivative, which would not be affected by $NaBH_4$. A similar dihydroxymorpholine product was proposed in the affinity labeling of phosphorylase kinase by oATP (33) to account for the long-term instability of the nucleotide-enzyme linkage either with or without reduction by $NaBH_4$. Primary amines promote the β-elimination of the phosphate fragment from nucleoside 5'-phosphates (34, 35). In the inactivation of mitochondrial ATPase by periodate-oxidized ATP, incorporation of $[2,8\text{-}^3H]$-oADP was much greater than was incorporation of $[\alpha\text{-}^{32}P]$-oATP (36). Elimination of the triphosphate group of oATP bound to mitochrondial ATPase was suggested initially to yield a stable conjugated Schiff base with lysine that was not reducible by $NaBH_4$ (37), but an alternate reaction scheme involving a dihydroxymorpholino derivative of the β-elimination product was proposed more recently (36). In the case of affinity labeling of the ADP activator site of the NAD-dependent isocitrate dehydrogenase, incorporation of $[8\text{-}^{14}C]$-oADP occurred concomitant with inactivation, but only about 3% as much incorporation of $[^{32}P]$-oADP was observed, suggesting that the enzyme-bound product involved loss of the pyrophosphoryl group (M. M. King, R. F. Colman, 1983, *Biochemistry*, in press). For isocitrate dehydrogenase, no significant specific incorporation of tritium into the oADP-modified enzyme after treatment with $[^{3}H]$-NaBH$_4$ was found; these and other results were consistent with the formation of a 4',5'-didehydro-2',3'-dihydroxymorpholino derivative of oADP and the ε-amino group of lysine in this enzyme.

How does the class of periodate-oxidized nucleotides rate as a group of potential affinity labels of nucleotide sites of proteins? Structurally they are quite close to the natural nucleotides, having the purine ring of the parent compound and the appropriate number of phosphoryl groups (at least before reaction with the enzyme); only the original ribose ring is perturbed by the periodate oxidation reaction, and many enzymes can tolerate structural changes in the ribose moiety of the nucleotide. The solubility of the periodate-oxidized analogs does not differ appreciably from that of their parent compounds. The application of these compounds is limited primarily by the restricted types of residues with which they react, and by the characteristics of the reaction products. The 2',3'-dialdehydes react in proteins only with lysine residues and potentially with α-amino groups of the N-terminal residue. Thus the periodate-oxidized nucleotides are useful as affinity labels only for those enzymes in which lysine residues happen to be located in the region of the ribose binding site. Furthermore, in the case of enzyme products other than a Schiff base reducible by NaBH$_4$, the instability of the product to acid hydrolysis (32, 33) and possibly to prolonged proteolytic digestion and peptide isolation procedures may limit the value of those compounds as tools for identifying the particular labeled amino acid residue within the primary sequence of an enzyme.

ALKYL HALIDE DERIVATIVES OF PURINE NUCLEOTIDES

Alkyl halides are among the most frequently used compounds for chemically modifying proteins. As exemplified by iodoacetamide, they have the potential to react with the nucleophilic side chains of many types of amino acids including cysteine, histidine, lysine, methionine, glutamic, and aspartic acids; in most of these cases reasonably stable, isolable products are expected (38–40). A purine nucleotide derivative employing an alkyl halide as the electrophilic moiety is a good choice for an affinity label. As long as the purine moiety determines the specificity, the probability is high that one of the susceptible amino acids is in the vicinity of the binding site. The adenosine 5'-, 2'-, and 3'-(2-bromoethyl)-phosphates were synthesized and characterized (41, 42), and kinetic evidence suggested that the 5'-derivative might function as an affinity label of the ADP activator site of the NAD-dependent isocitrate dehydrogenase (41). These compounds have the favorable attributes of a negative charge at neutral pH, which might be important in directing the binding of nucleotides; high solubility in water; and stability over a wide range of pH and buffer conditions likely to be used for chemical modification experiments. The reaction rates of the bromoethyl-AMPs with cysteine, lysine, histidine, and tyrosine were measured and the product of the reaction with cysteine was isolated and characterized (42).

A variety of other purine nucleotide derivatives employing an alkyl halide as the electrophilic moiety were introduced recently (43, 44), including adenosine-5'-chloromethane pyrophosphonate, adenosine-5'-(β-chloroethylphosphate), adenosine-5'-chloromethane phosphonate, adenosine-5'-(β-bromoethanepyrophosphonate), and adenosine-5'-(β-bromoethanephosphonate). These promising compounds were tested as affinity labels of histone kinase (45, 46), phosphorylase b (47, 48), tryptophanyl-tRNA synthetase (43), leucyl-tRNA synthetase (49), and cAMP-dependent protein kinase (44). Phosphorylase b is inactivated by adenosine-5'-chloromethylphosphonate, with a nonlinear dependence of the rate constant on reagent concentration (47); however, the maximum rate is very slow, exhibiting a $t_{1/2}$ about 4 days. Reaction may have occurred at the AMP activating site; however, despite the incorporation of only 1.3–1.5 moles of the analog per enzyme subunit and the isolation of two modified peptides (48), it is not clear that the reaction is truly specific or that these are the only two peptides that are modified. The kinetic evidence for affinity labeling of tryptophanyl-tRNA synthetase by adenosine-5'-(β-chloroethylphosphate) is better (43), and the minimum half-life at saturating reagent concentration is 12.6 minutes. However, the full potential of these compounds has not yet been evaluated thoroughly. Berghäuser & Geller reported on the preparation of adenosine-5'-(2,3-dibromohydrogen succinate) and provided preliminary evidence for the specific modification of myokinase and glyceraldehyde-3-phosphate dehydrogenase (50), but detailed studies of these reactions have not been conducted.

A reactive AMP derivative was synthesized with the functional group linked to the 6-NH$_2$ group of the purine ring N^6-p-bromoacetamino benzyl adenosine-5'-phosphate (51). Covalent incorporation of one mole of this reagent per mole phosphorylase b leads to activation, suggesting that the AMP activating site of the enzyme was labeled, but the modified amino acid has not been identified. Other evidence showed the affinity labeling of phosphoglycerate kinase (52) and glyceraldehyde-3-phosphate dehydrogenase (53) by this compound; thus, it may prove to be broadly applicable as an affinity label of nucleotide sites.

A major limitation in the usefulness of the alkyl halide derivatives as affinity labels is their inherently low reactivity. Dahl & McKinley-McKee

(54) measured the reaction rates of a variety of alkyl halides with thiolate anions. Bednar & Colman (42) reported pH-independent rate constants for reaction at 25° of several amino acids with the bromoethyl-AMP analogs: 4.0×10^{-4} M^{-1} sec^{-1} for reaction with cysteine, 5.2×10^{-6} M^{-1} sec^{-1} for reaction with N-acetyl lysine, and 3.0×10^{-7} M^{-1} sec^{-1} for reaction with N-acetyl histidine. The bromoethyl-AMP analogs have relatively low reactivity because they are not activated by a carboxyl group or a ketone group α to the bromine. Comparison of the reactivity of several chemical modification reagents with thiol groups illustrates this point. The pH-independent reaction rate of bromoethyl-AMP, bromoacetate (55), and 3-bromo-2-ketoglutarate (56) are 3×10^{-4} M^{-1} sec^{-1}, 2.7 M^{-1} sec^{-1}, and 2.2×10^4 M^{-1} sec^{-1}, respectively. Each reagent differs by approximately four orders of magnitude in intrinsic reactivity. In order for a reagent that contains a highly reactive electrophilic group, such as 3-bromo-2-ketoglutarate to function as a specific affinity label, it would not only have to exhibit enhanced affinity for the active site but also exhibit decreased access toward reactive groups on the surface of the enzyme. The low intrinsic reactivity of the bromoethyl AMPs, on the other hand, leads to a very low probability of nonspecific modification; reaction at a reasonable rate would require the presence of an entropic activation that could result from the specific binding of the analogs in the ligand-binding site. Noncovalent ligand-binding (57) brought about rate enhancements of 10^{10}–10^{13}. The existence of a significant rate enhancement would fulfill one criterion of an affinity label, since it would provide evidence for a specific interaction between the electrophilic group of the nucleotide analog and a nucleophilic group on the enzyme. The inactivation of tryptophanyl-tRNA synthetase by adenosine 5'-(2-chloroethyl) phosphate has a k_{max}/K_{inact} of 4.0 M^{-1} sec^{-1} at pH 7.2 (43). The identified group is not yet determined. If the most intrinsically reactive amino acid (cysteine) were modified, the rate enhancement for this modification reaction would be greater than five orders of magnitude, as based on a comparison with the reaction rates of bromoethyl-AMP. If a tyrosine were modified, the rate enhancement would be more than 10 orders of magnitude. Studies of the reactions of the alkyl halide derivatives of purine nucleotides may yield a better understanding of the mechanism and importance of entropic rate enhancement in enzyme-catalyzed reactions.

PHOTOREACTIVE PURINE NUCLEOTIDE ANALOGS

The approach of photoaffinity labeling has been applied increasingly to the exploration of nucleotide sites in proteins. The photoaffinity analogs have two important advantages. First, they remain chemically unreactive until they are deliberately activated, thus allowing the investigator to study the reversible interaction of the parent compound with the protein. Second, the usual reactive groups that are photogenerated (carbenes and nitrenes) react relatively indiscriminately with a wide variety of chemical groups and thus are less restricted in the amino acids with which they can react than are the affinity labels with electrophilic functional groups. However, the application of photoaffinity labeling is not without compensatory difficulties. Bayley & Knowles (58) and Chowdhry & Westheimer (59) have thoroughly reviewed the general principles, some of the experimental problems, and approaches to confronting them.

Brunswick & Cooperman used diazomalonyl derivatives of cAMP to label the allosteric site of rabbit muscle phosphofructokinase (60, 61); typically the photoreactive compounds labeled 16–35% of the cAMP binding sites on the enzyme, but the yields could be improved by continuous replacement of the photoaffinity reagent. The 8-azido nucleotide analogs developed by Haley et al are probably the most commonly used among the photoaffinity nucleotide analogs. The 8-azido cAMP was first used to label membrane proteins of human red cells and a partially purified protein kinase from bovine brain (62–65); typically about 14% of the measurable cAMP sites were covalently labeled. In a few exceptional cases, the photoreaction of 8-azido nucleotide analogs led to almost stoichiometric incorporation: the reaction of 8-azido-cAMP with the cAMP-dependent protein kinase holoenzyme (66) and of 8-azido-AMP with the allosteric inhibitory site of fructose-1,6-bisphosphatase (67). More recently 8-azido-GTP was used as a probe of tubulin-GTP interactions (68, 69), but corrections had to be made for nonspecific incorporation, and more typical, low incorporation was observed. The compound 8-azido-1,N^6-ethenoadenosine 3'-, 5'-cyclic monophosphate was used to insert a fluorescent probe into protein kinases (70). The azido analogs of nucleotides are formally close in structure to the parent compounds. However, introduction of a substituent in the 8-position may well cause the compound to prefer the syn conformation, whereas the parent nucleotide exists predominantly in the anticonformation (16).

The compound arylazido-β-alanine ATP, in which the ATP molecule is modified by the addition of an N-(4-azido-2-nitrophenyl)-amino acyl group to the 3'-hydroxyl group of the ribose, was examined as a photoaffinity label for myosin ATPase (71–73). The photoreactive substituent is extremely bulky and labeling may occur at sites adjacent to (rather than within) the binding site of the natural nucleotides. Analogous derivatives of NAD (74, 75) and NADP (76), in which the arylazido-β-alanine is linked to the 3'-hydroxyl of the nicotinamide ribose, were prepared for potential use as photoaffinity labels of the coenzyme sites of dehydrogenases.

Several mono- and dinucleotides containing a 4-azido phenyl group coupled to the 5'-phosphate were synthesized (77). These compounds are structurally well-designed for binding to the appropriate enzyme nucleotide sites, but as yet experience is too limited to evaluate their usefulness. A similar photoreactive analog of a nucleoside triphosphate, ATP-γ-p-azidoanilidate, was prepared and covalently bound to phenylalanyl and tryptophanyl-tRNA synthetases upon illumination (78). An even smaller photolabile substituent was used in the compounds guanosine 5'-O-(2-azidodiphosphate) and guanosine 5'-O-(3-azidotriphosphate) and these were tested for their interaction with the elongation factor G of polypeptide synthesis (79). A newly described type of nucleotide photoaffinity label is 3'-O-(4-benzoyl)-benzoyl ATP, in which the benzophenone group can be excited to a diradical triplet state intermediate (80). This compound seems to be reasonably effective in photolabeling mitochondrial F_1-ATPase.

The use of highly reactive photoaffinity labels, despite occasional success, has certain inherent problems, because upon irradiation the label tends to react with the solvent as well as with any amino acid adjacent to the compound. The reagent may therefore react incompletely with several amino acid residues, making it difficult to ascertain which residues are actually involved in binding the purine nucleotide. The structures of the reaction products of the enzyme are not known in many cases, and some of the linkages may well be chemically labile. In most cases, the amount of radioactivity incorporated at a specific site is relatively low so that identification of the reacted amino acid is difficult. The frequently encountered problems in the application of photoaffinity labeling are well described in Standring & Knowles' study of the reaction of lactate dehydrogenase with the carbene derived from the 3-diazirino analog of NAD (81). The most effective applications of the nucleotide photoaffinity labels may be in the qualitative identification of proteins containing particular nucleotide binding sites, and the localization of such sites in proteins with dissimilar subunits, rather than in the stoichiometric labeling of purified proteins with the goal of characterizing the modified proteins.

FLUOROSULFONYLBENZOYL ANALOGS OF NUCLEOTIDES

General Considerations

The fluorosulfonylbenzoyl analogs of nucleotides proved to be compounds appropriate for the stoichiometric, specific modification of nucleotide sites of a wide variety of proteins. The prototype of these compounds is 5'-p-fluorosulfonylbenzoyl adenosine (5'-FSBA) prepared by reaction of p-fluorosulfonylbenzoyl chloride with adenosine (82, 83). This compound might be considered an analog of ADP, ATP, NAD, or NADH. In addition to the adenine and the ribose moieties, it has a carbonyl group adjacent to the 5'-position that is structurally similar to the first phosphoryl group of the naturally occurring purine nucleotides. If the molecule is arranged in an extended conformation, the sulfonyl fluoride moiety may be located in a position analogous to the terminal phosphate of ATP or to the 5'-position of the ribose adjacent to the nicotinamide group of NAD. The sulfonyl fluoride is a reactive functional group that can act as an electrophilic agent in covalent reactions with several classes of amino acids, including tyrosine, lysine, histidine, serine (84), and cysteine (85–87). Because of the broad range of reactions in which the sulfonyl fluoride participates, there is a reasonable probability of reaction within any particular nucleotide site. The 5'-p-fluorosulfonyl adenosine thus far reacts specifically with the NADH regulatory site of glutamate dehydrogenase (82, 88), the coenzyme binding site of malate dehydrogenase (89), 3α, 20β-hydroxysteroid dehydrogenase (90), and 17β-estradiol dehydrogenase (91), as well as the nucleotide binding sites of several kinases including rabbit muscle (85, 92) and yeast pyruvate kinase (93), phosphofructokinase (94–96), cAMP-dependent protein kinase (97–99), cGMP-dependent protein kinase (100, 101), casein kinase II (102), and epidermal growth-factor-stimulated protein kinase (103). In addition, it reacts specifically and covalently with such diverse enzymes as the mitochondrial F_1-ATPase (104–106), chloroplast ATPase (107), platelet actin and myosin (108), RNA polymerase (109), glutamine synthetase (110), carbamyl phosphate synthetase (111), luciferase (112), 5-oxo-prolinase (113) and acetyl coenzyme A carboxylase (113a). Furthermore, it reacts with an ADP receptor protein of platelet membranes (114, 115). Thus, the first to be synthesized of this class of fluorosulfonylbenzoyl analogs of nucleotides already has wide applicability.

Related compounds have also been prepared and characterized. The 3'-p-fluorosulfonylbenzoyl adenosine (83, 116) is not as close an analog of ATP, ADP, and NAD as is 5'-FSBA. The 3'-hydroxyl group of ribose is generally free in purine nucleotides that bind to the substrate and regulatory sites of enzymes; thus, for those enzymes that can tolerate the added bulk in this region, 3'-FSBA may function in accordance with Baker's definition of an exo-alkylating agent: one that participates in covalent bond formation with an enzymic nucleophilic group located in a position immediately adjacent to if not actually within the purine nucleotide binding site (20). It might be expected that 3'-FSBA would prove complementary to 5'-FSBA as an affinity label for adenine nucleotide sites in enzymes; this seems to be the case for glutamate dehydrogenase (83, 116). The compound 5'-p-fluorosulfonylbenzoyl guanosine (5'-FSBG), in which guanine replaces the adenine in 5'-FSBA, was synthesized (117) and shown to react specifically with GTP binding sites in glutamate dehydrogenase (118), phosphoenol-pyruvate carboxykinase (119) and pyruvate kinase (86). Recently the preparation of 5'-p-fluorosulfonylbenzoyl inosine was reported (113) and the compound was shown to react with 5-oxo-prolinase.

Fluorescent analogs of natural biochemical compounds as well as fluorescent labeling agents are valuable in probing the environment of binding sites in proteins and in elucidating distances between defined site markers on proteins by energy transfer (120–122). Fluorescent nucleotide alkylating agents thus provide effective tools for introducing into a defined nucleotide binding site a covalent fluorescent probe that can then be used to examine the properties of that site and its interactions with other sites of the protein. Two fluorescent nucleotide analogs with fluorosulfonylbenzoyl substituents are thus far reported. The 5'-p-fluorosulfonylbenzoyl-1,N^6-ethenoadenosine (5'-FSBϵA) has an excitation maximum at 308 nm and a fluorescence emission peak at 412 nm (123); this compound reacts at the active site of pyruvate kinase (123, 124) and at a GTP regulatory site of glutamate dehydrogenase (125). The 5'-p-fluorosulfonylbenzoyl-2-aza-1,N^6-etheno-adenosine, like its parent compound, exhibits an excitation maximum at 356 nm and an emission peak at 490 nm; it reacts at the cAMP site of phosphofructokinase (126). Because of the differences in their spectral properties, these two fluorescent nucleotide alkylating agents are expected to have differential applicability in probing the nucleotide sites of proteins.

Of the amino acids that can react with the fluorosulfonylbenzoyl nucleotide analogs, both tyrosine and lysine reacted with 5'-FSBA or p-fluorosulfonylbenzoic acid to give stable derivatives (88, 89, 93, 104). Upon treatment with 6N HCl, the ester linkage between the benzoyl and nucleoside moieties is hydrolyzed to yield the acid-stable products N^ϵ-(4-carboxybenzenesulfonyl)lysine   (CBS-lys)   and O-(4-carboxybenzene-sulfonyl)tyrosine (CBS-tyr). Both are relatively hydrophobic as compared to most amino acids, but they have a net negative charge at neutral pH. Because of this unusual combination of properties, these two amino acid derivatives may be separated from the normal complement of amino acids present in protein hydrolysates by thin layer electrophoresis at pH 6.4, followed by thin layer chromatography using butanol-acetic acid-water as solvent. Under these conditions, only glutamate, aspartate, CBS-lys, and CBS-tyr are negatively charged, and the latter two distinguish themselves readily from aspartate and glutamate, as well as from each other, by their chromatographic behavior (88, 89). CBS-lys and CBS-tyr were separated and quantified as distinct peaks in the region of tyrosine and phenylalanine on an amino acid analyzer (93, 104); their extinction coefficients at their ultraviolet absorption maxima and their color constants for reaction with ninhydrin were determined (88). Lysine was shown to be modified by 5'-FSBA in several proteins, (88, 89, 93, 96, 97, 101, 112) and tyrosine to be the site of attack in others (85, 88, 93, 104). Evidence for reaction of 5'-FSBA, 5'-FSBG, and 5'-FSBϵA with cysteinyl residues in proteins was based on the loss of free sulfhydryl groups and the reactivation of the modified enzymes by treatment with dithiothreitol (85, 86, 113, 123, 124, 127), although a stable thiolsulfonate derivative of cysteine has not yet been isolated. In the cases of reaction of rabbit muscle pyruvate kinase with 5'-FSBA and 5'-FSBG (85, 86) and of myosin subfragment 1 with 5'-FSBA (127), it was proposed that a cysteine reacts initially with the fluorosulfonyl-benzoyl nucleoside to yield a thiolsulfonate; subsequently a second neighboring cysteine displaces the sulfinic acid moiety by attacking at the cysteinyl sulfur of the thiolsulfonate and forming a disulfide. The result is that the final enzyme product does not contain the elements of the reagent but has one disulfide per mole of reacted enzyme. Parsons et al (128) described the reaction of thiolsulfonate with thiols as a synthetic route to disulfides. A model reaction of 6 μmoles free cysteine with 3 μmoles 5'-FSBA established the feasibility of this reaction sequence (127). The reactivation by dithiothreitol of pyruvate kinase and myosin that was inactivated by fluorosulfonylbenzoyl nucleosides may have important implications for earlier studies on the modification of enzymes by 5'-FSBA. In several cases, dithiothreitol or mercaptoethanol quenched the reaction between 5'-FSBA and the enzyme. More recent results suggest the need to ascertain the effect of added thiol on the activity, extent of reagent incorporation and free sulfhydryls regenerated for these modified enzymes. Serine is another possible site of sulfonylation by these reagents. However, despite explicit attempts to evaluate serine as the site of attack in at least two enzymes (88, 123), no evidence has yet demonstrated reaction of serine in proteins with the fluorosulfonylbenzoyl nucleosides.

Fluorosulfonylbenzoyl derivatives, when dissolved in aqueous buffer solutions, hydrolyze with release of fluoride ion. The rates of hydrolysis depend on the pH and the buffer used (83). For example, the half-life for decomposition of 5'-FSBA has been measured as 32 min at pH 8.6 in 20-mM potassium barbital buffer (93); $t_{1/2} = 43$ min for decomposition of 5'-FSBϵA at pH 8.0 in 20-mM potassium barbital buffer (125); and $t_{1/2} = 90$ min and 223 min for hydrolysis of 5'-FSBG at pH 7.93 and pH 7.65, respectively (86). The stability of these compounds increases dramatically below pH 7.6. In analyzing the kinetics of modification of an enzyme by these reagents, corrections can be made for the effects of reagent decomposition under the conditions used (86, 93, 125). Because of the analogs' limited stability and water solubility, most studies with the fluorosulfonylbenzoyl nucleotide analogs involved the initial preparation of solutions of the compounds in organic solvents; the addition of small volumes of the solvents into a reaction mixture containing 1–15% ethanol, dimethylformamide or dimethylsulfoxide initiated the reactions. The need for these organic solvents may be a disadvantage in the use of these compounds. It is important to ascertain whether the presence of even low levels of the solvents causes any significant perturbations of the structure or activity of the enzyme (e.g. 89).

Although the ester linkage between the benzoyl and nucleoside moieties of the fluorosulfonylbenzoyl nucleosides is reasonably stable under the conditions usually employed for protein modification reaction, it decreases below pH 6 and above pH 9. This characteristic is important in the design of experiments aimed at quantitating the incorporation of reagent and at isolating peptides derived from modified enzyme. Most often the stoichiometry of incorporation was ascertained using radioactive reagent in which the label was located in the purine moiety (83, 117, 125); this type of radioactive label is satisfactory if the pH is maintained in the neutral range. An alternative synthesis starting with [carboxy-^{14}C]-p-aminobenzoic acid, which leads to 5'-FSBA with ^{14}C in the benzoyl moiety was described (129), and the benzoyl-labeled compound was used for the isolation of radioactive peptides, even under acidic or basic conditions (104), as well as for the quantification of CBS-lys and CBS-tyr in acid hydrolysates of modified protein (93). The purification of peptides from proteins labeled with these nucleotide analogs sometimes exploited the unusual characteristics conferred on the peptides because of the introduction of the sulfonylbenzoyl nucleoside group. For example, a modified peptide from cAMP-dependent protein kinase was purified by chromatography on DEAE-Sephadex before and after mild alkaline hydrolysis that specifically cleaved the ester bond of the reagent, generating an extra negative charge on the carboxybenzoyl-sulfonyl-labeled peptide and thereby increasing its affinity for the ion exchange resin (97). A peptide from pyruvate kinase modified with 5'-FSBA was isolated by column chromatography on polyacrylamide covalently substituted with dihydroxyborylphenyl groups (130, 131). The procedure involved complex formation between the cis-diol moiety of the nucleosidyl peptide and the dihydroxyboryl group of the resin; thus, the nucleosidyl peptide was bound reversibly to the column, while most other peptides were eluted in the void volume. The range of information obtained by the use of the fluorosulfonylbenzoyl nucleosides as affinity labels of nucleotide sites is best illustrated by describing in more detail the reactions of a few specific proteins with these compounds.

Glutamate Dehydrogenase

Bovine liver glutamate dehydrogenase has several purine nucleotide sites, including for each enzyme subunit a site for the activator ADP, two sites for the inhibitor GTP, and two sites for NADH (118, 132–135). It was proposed that high concentrations of NADH bind to a site distinct from the active site (132, 136) In the controversy over whether the inhibitory NADH site is identical with or distinguishable from the site occupied by the activator ADP (136–140), affinity labeling might be a useful approach. Incubation of glutamate dehydrogenase (0.6 mg/ml) with 0.3–0.6 mM 5'-p-fluorosulfonylbenzoyl adenosine at pH 8 leads to the time-dependent incorporation of up to 1.1 moles reagent per peptide chain (82). The resultant enzyme loses its normal ability to be inhibited by NADH, but retains full activity as measured in the absence of allosteric ligands, is activated normally by ADP, and is still inhibited 93% by GTP. The rate constant for loss of NADH inhibition upon incubation with 5'-FSBA is decreased considerably by the addition of relatively high concentrations of NADH, such as would be needed to bind to the inhibitory site or by NADH plus GTP. However, neither the substrate α-ketoglutarate, nor the coenzymes NAD and NADPH, nor the regulators ADP and GTP alone (82, 88) provide protection against modification. This indicates that the ADP and NADH sites cannot be identical and that 5'-FSBA attacks exclusively the second inhibitory NADH site of glutamate dehydrogenase. Upon reaction at a protein concentration of 0.6 mg/ml, the NADH inhibition is abolished when an average of only 0.5 moles 5'-SBA/peptide chain are incorporated, although as much as one mole reagent/peptide chain can react; whereas, upon reaction at a protein concentration of 2.0 mg/ml, only 0.53 mole radioactive reagent are incorporated, and the enzyme also becomes unresponsive to NADH inhibition (82, 88). Both CBS-tyr and CBS-lys were detected as the only unusual amino acids in this modified glutamate dehydrogenase; together, they accounted for the total incorporation of radioactivity into the enzyme. As a function of time of incubation of the enzyme with 5'-FSBA the product ratio of CBS-tyr to (CBS-tyr plus CBS-lys) remains essentially constant at 0.47, with 0.25 mole CBS-tyr and 0.28 mole CBS-lys detected upon complete loss of NADH inhibition. Both tyrosine and lysine are thus present in the NADH inhibitory site and covalent modification of either residue on 3 of the 6 peptides of the catalytically active hexameric enzyme is sufficient to eliminate NADH inhibition (88).

In contrast, the guanosine analog, 5'-FSBG reacts at an inhibitory site of glutamate dehydrogenase (117, 118). Incubation of the enzyme with 2-mM 5'-FSBG produces no change in the maximum velocity when measured without modifiers, implying that this compound also does not react at the active site. When assayed in the presence of 1-μM GTP, native glutamate dehydrogenase exhibits a maximum velocity that is only 6% of that measured in the absence of modifiers. 5'-FSBG produces a progressive decrease in the ability of the enzyme to be inhibited by GTP, as indicated by an increase in the velocity measured in the presence of a constant concentration of GTP. The rate constant for reaction of glutamate dehydrogenase with 5'-FSBG, as measured by the diminished sensitivity to GTP inhibition, is specifically decreased by low concentrations of GTP in the presence of reduced coenzyme, but not by substrates, NADH alone or ADP either with or without NADH. The modified enzyme binds only 1 mole of GTP per peptide chain in the presence or absence of NADH (as compared to the 2 moles of GTP bound by native enzyme in the presence of NADH), implying that reaction with 5'-FSBG eliminates one of the allosteric sites for GTP. In contrast, the modification reaction does not alter the ability of the enzyme to be inhibited by high concentrations of NADH, suggesting that the NADH regulatory site is not the site of attack by 5'-FSBG. The extent of covalent incorporation of radioactive 5'-SBG is directly proportional to the percentage decrease in GTP inhibition; a maximum alteration in the sensitivity to GTP is observed when about 2 moles 5'-SBG are incorporated per enzyme subunit. Enzyme treated with radioactive 5'-FSBG under the same conditions but with GTP and reduced coenzyme present exhibits no change in GTP inhibition; this protected enzyme contains only 1.0 mole of radioactive reagent per peptide chain. Thus, the extra mole of 5'-SBG incorporated in the absence of GTP and reduced coenzyme is probably responsible for the loss of inhibition by added GTP. Glutamate dehydrogenase that reacted with two moles 5'-SBG per peptide chain contains equal amounts of CBS-tyr and CBS-lys; in contrast, the protected enzyme prepared in the presence of GTP and reduced coenzyme yields only CBS-lys (141). Modification of a tyrosine residue by 5'-FSBG may be responsible for the decreased sensitivity of the enzyme to GTP inhibition.

With the goal of preparing a modified glutamate dehydrogenase with a fluorescent probe covalently linked at a defined nucleotide site, Jacobson & Colman examined the reaction of the enzyme with the fluorescent analog, 5'-FSBεA (125). The enzyme is reversibly inhibited by the fluorescent nucleotide 1,N^6-etheno-ATP and reacts irreversibly with the corresponding affinity label 5'-p-fluorosulfonylbenzoyl-1,N^6-ethenoadenosine. As in the case of 5'-FSBG, this reaction does not inactivate the enzyme as measured in the absence of modifiers. However, a time-dependent increase is observed in the catalytic activity when measured in the presence of the allosteric inhibitor GTP, which allows the determination of a rate constant for the reaction with 5'-FSBεA. A nonlinear dependence of the reaction rate on reagent concentration suggests a reversible binding (K_D = 1.4 mM) prior to irreversible modification. GTP or 1,N^6-etheno-ATP causes a specific decrease in the rate constant, and a combination of GTP in the presence of reduced coenzyme provides complete protection. As compared to native glutamate dehydrogenase, modified enzyme exhibits a decreased affinity for and diminished maximum inhibition by saturating concentrations of GTP, a decreased maximum extent of activation with no change in affinity for ADP and a normal ability to be inhibited by high NADH concentrations. Only one mole of GTP is reversibly bound per peptide chain of the modified enzyme as contrasted with the two found by native enzyme. These results imply that, as in the case of 5'-FSBG, reaction with 5'-p-fluorosulfonyl benzoyl-1,N^6-ethenoadenosine eliminates one of the natural GTP sites. It appears that reaction occurs at a GTP site and that alteration of the N-1 and 6-NH$_2$ positions of the adenine ring leads to recognition by glutamate dehydrogenase as an inhibitory nucleotide. However, reaction of the enzyme with 5'-FSBεA is more specific than in the case of 5'-FSBG: only 1.28 moles of the fluorescent analog are incorporated per peptide chain at 100% change in sensitivity to GTP (125). Modification of tyrosine increases as a function of time of incubation of enzyme with 5'-FSBεA, the amount of CBS-tyr being proportional to the change in GTP inhibition (M. A. Jacobson, R. F. Colman, manuscript in preparation). The results suggest that the fluorescently labeled tyrosine residue is in a GTP binding site of glutamate dehydrogenase.

   The fluorescent properties of 5'-FSBεA and of 5'-SBεA-modified glutamate dehydrogenase were investigated. The compound ethenoadenosine is highly fluorescent with a quantum yield of 0.54. In contrast, a quantum yield of only 0.01 was measured for 5'-FSBεA in aqueous solution. A conformation in which the purine base and the benzoyl moiety are stacked may account for the apparent internal quenching of the fluorescence. To provide models for environments in which the 5'-FSBεA might exist bound to the enzyme, investigators measured the quantum yields of ethenoadenosine and 5'-FSBεA in solvents of different polarity. The quantum yields of 5'-FSBεA increased 4-fold, while that of ethenoadenosine remained the same, as the dielectric constant was decreased from 78 to 10. The decrease in polarity of the solvent can interfere with the ring interactions in 5'-FSBεA, thus producing a partially extended conformation in the molecule. The quantum yield of 5'-SBεA covalently bound to glutamate dehydrogenase was measured in a buffer that stabilized the native conformation of the enzyme. A value of 0.011 was found, suggesting that the fluorescent probe exists in the same conformation on the enzyme as it does in aqueous solution (M. A. Jacobson, R. F. Colman, manuscript in preparation). The compound 2'(3')-O-trinitrophenyl-ADP functions as an activator of glutamate dehydrogenase, competing with the natural activator ADP. The overlap between the absorption spectrum of 2'(3')-O-trinitrophenyl-ADP and the emission spectrum of 5'-SBεA covalently linked to a GTP site was used to estimate the distance between the activator and inhibitor sites of glutamate dehydrogenase (M. A. Jacobson, R. F. Colman, manuscript in preparation).

   These investigations of the reactions of glutamate dehydrogenase with the three fluorosulfonylbenzoyl nucleosides indicate the specificity of the enzyme for the purine moiety of the reagent and illustrate how several of these nucleotide analogs are used to probe different nucleotide sites in a single enzyme. They also demonstrate the importance of the ability to isolate stable enzyme derivatives that can be characterized completely with respect to their kinetic, regulatory and ligand-binding properties.

Pyruvate Kinase

The reactions of 5'-FSBA, 5'-FSBG, and 5'-FSBϵA with pyruvate kinase provide examples of the manner in which these nucleotide analogs aid the exploration of the active site of an enzyme. Rabbit muscle pyruvate kinase exhibits a wide tolerance for nucleotides, using such nucleotides as ADP, GDP, IDP, or UDP as coenzymes in the catalytic reaction (142). However, it is still unestablished whether these several nucleotides occupy the same or distinct sites. One approach might be the comparison of the reactions of the nucleotide binding site with several purine nucleotide analogs that have the same reactive functional group but differ in the purine moiety. Rabbit muscle pyruvate kinase is inactivated readily by the adenosine analog, 5'-FSBA (85); similarly, a rapid and specific inactivation is observed when adenosine is replaced by guanosine (86) or by ethenoadenosine (123, 124). Since p-fluorosulfonylbenzoic acid (which contains the same reactive functional group) does not inactivate the enzyme at a measurable rate (92), the marked inactivation by 5'-FSBA, 5'-FSBG, and 5'-FSBϵA must be attributable to interactions of the purine moieties of the structures at particular sites on the enzyme.

The inactivation of rabbit muscle pyruvate kinase proceeds with biphasic kinetics upon incubation with 2-mM 5'-FSBA at pH 7.4 (85). Addition of 20-mM dithiothreitol to the incubation mixture produces a partial reactivation. The inactivation of pyruvate kinase by 5'-FSBA includes three events: a fast dithiothreitol-reactivatable reaction yielding a partially active enzyme with 67% residual activity and two slow first-order reactions yielding fully inactive enzyme. One of the slow reactions is sensitive to dithiothreitol reactivation and the other is not. The two dithiothreitol-sensitive reactions are attributable to the formation of disulfides of cysteine and do not lead to the covalent incorporation of reagent. The fast dithiothreitol-sensitive reaction leading to partial inactivation results from modification of cysteine residues at the noncatalytic nucleotide site, which crystallographic studies demonstrated (143, 144). The dithiothreitol-insensitive component of the inactivation is proportional to the incorporation of about 1 mole of radioactive reagent per enzyme subunit and is attributed to the modification of a critical tyrosine residue. ADP plus Mg^{2+} provides almost complete protection against the dithiothreitol-insensitive inactivation, suggesting that the tyrosine is located within the catalytic metal-nucleotide binding site. A tryptic peptide containing the CBS-tyrosine was purified by chromatography on a column of dihydroxyborylphenyl-polyacrylamide, followed by high pressure liquid chromatography (131). Determination of k_{obs} for the DTT-insensitive reaction as a function of pH indicates a fivefold increase in the rate constant as the pH is raised from 7.0 to 8.0 (145). A plot of $1/k_{obs}$ versus (H^+) yields a straight line and allows the calculation of a pKa value of 7.94 for a single ionizable amino acid residue that reacts with 5'-FSBA (145). If this pK reflects the ionization of the critical tyrosine, it is abnormally low compared with typical values for tyrosine residues in proteins (146). A positively charged group in the vicinity of the critical tyrosine would tend to decrease its pK, and indeed evidence has been presented for the location of positively charged groups such as lysine and arginine residues in the active site of pyruvate kinase (147–149).

The inactivation of pyruvate kinase by the guanosine analog 5'-FSBG also proceeds with biphasic kinetics, the first phase leading to partially active enzyme of 45% residual activity (86). At pH 8, the inactivation is substantially reversed by dithiothreitol. For both phases of the reaction, Mg^{2+} plus either ADP or GDP provides the best protection, which suggests that the reaction occurs in the region of the active metal-nucleotide binding site. Inactivation correlates well with the loss of two free sulfhydryl groups per enzyme subunit and the restoration of activity correlates with the regeneration of two free sulfhydryls after treatment of modified enzyme with dithiothreitol. The inactivation of pyruvate kinase at pH 8 proceeds by formation of the thiolsulfonate followed by rapid displacement of the sulfinic acid moiety by a second cysteine to yield a disulfide. A negative cooperativity in the interaction of subunits with 5'-FSBG might best account for the biphasic inactivation kinetics in this case (86). It is likely that 5'-FSBA and 5'-FSBG do not react with the same cysteinyl residue in pyruvate kinase (86, 145). In contrast, examination of the reaction of pyruvate kinase with 5'-FSBG over the pH range from 7.0–8.0 reveals a dithiothreitol-insensitive inactivation that is remarkably similar to that caused by 5'-FSBA: the rates are essentially identical and yield a pKa value of 7.94 for a single essential ionizable residue (145). It would not be surprising to identify the same modified tyrosyl residue in pyruvate kinase with 5'-FSBG used as the inactivator.

The fluorescent compound, 5'-FSBϵA produces biphasic inactivation of pyruvate kinase: one phase can be related to disulfide formation and the other (dithiothreitol-insensitive inactivation) to covalent incorporation of about one mole reagent per mole enzyme subunit (123, 124). The pH dependence of the rate constant for dithiothreitol-insensitive inactivation by 5'-FSBϵA is easily distinguishable from those of the 5'-FSBA and 5'-FSBG reactions. The pK of the group reactive toward 5'-FSBεA is 8.90. Furthermore, careful analysis of the 5'-SBεA-modified enzyme eliminated the possibility of modification of tyrosine or lysine (123, 124). Although these purine nucleotide analogs may react in the general region of the metal-nucleotide site of pyruvate kinase, they may be oriented somewhat differently and do not all attack the same residues.

ADP Receptor Protein of Platelets

The effect of 5'-p-fluorosulfonylbenzoyl adenosine on human platelets shows how a purine nucleotide affinity label is used to identify and probe the function of a nucleotide-binding protein in such a complex system as an intact cell. Adenosine diphosphate induces human platelets to change shape from discs to spiculated spheres followed by platelet aggregation and granule secretion (150). The detailed mechanism by which ADP participates in these platelet functions is unknown, but may be initiated by the binding of ADP to a protein on the exterior surface of the platelet. Nachman & Ferris (151), who investigated the binding of radioactive ADP to isolated human platelet membranes, and Lipps et al (152), who studied the binding of ADP to intact platelets as well as to platelet membranes, presented evidence for the existence of such sites. Fibrinogen binding by platelets is essential for ADP-induced aggregation (153–163). If 5'-FSBA is to serve as an affinity label for the platelet ADP receptor, it must be demonstrated that the nucleotide analog interferes with the normal effects of ADP on platelets. Bennett et al (114) showed that 5'-FSBA is a potent inhibitor of the ADP-induced platelet shape change. A K_I value of 11 μM was estimated for the interaction of 5'-FSBA with the cell. Treatment of either gel-filtered or washed platelets (115) with 5'-FSBA prevents ADP-induced aggregation. The extent of binding of ^{125}I-fibrinogen and aggregation decreased to a degree related to the incorporation of 5'-SBA into platelets, indicating that 5'-FSBA could inhibit the ADP-induced exposure of fibrinogen receptors, which is necessary for aggregation (115).

Incubation of isolated platelet membranes with radioactive 5'-FSBA leads to a time-dependent covalent incorporation. The inclusion of ADP in the incubation mixture inhibits the incorporation of reagent more than 90%, whereas GDP, adenosine and epinephrine are much less effective, suggesting that incorporation occurs at specific ADP sites on the membranes (114). Radioactivity was found to be incorporated into four protein components of the isolated platelet membrane, as demonstrated by polyacrylamide gel electrophoresis in the presence of sodium dodecylsulfate; these components have molecular weights of 43,000, 100,000, 135,000, and 200,000 (114). The incorporation of radioactivity into all four peaks decreases when the incubations are conducted in the presence of unlabeled

ADP, suggesting that labeling of all these proteins involves specific reaction at nucleotide sites. Two of the labeled components, those with molecular weights of 200,000 and 43,000, correspond in size to myosin and actin; these proteins react with 5'-FSBA (108). The function and identity of the 135,000-dalton species are still unknown. Experiments such as these, in which 5'-FSBA is incubated with isolated platelet membrane preparations, do not distinguish readily between labeling of proteins originating on the inside or outside of the cell.

In contrast, when intact platelets are incubated with radioactive 5'-FSBA and the membranes are isolated subsequent to the modification reaction, a strikingly different labeling pattern emerges: a single component of 100,000 daltons is modified under these conditions (108, 114). Since 5'-FSBA does not penetrate platelets without prior hydrolysis (114), these results indicate that the protein of molecular weight 100,000 is the only 5'-FSBA reactive protein located on the exterior surface of the intact platelet. The intact cells appear not to have myosin, actin, or the 135,000-dalton protein available on the cell surface; rather, these become available during procedures that perturb the integrity of the platelet membrane or during the isolation of membranes which can form "inside-out" as well as "right-side-out" vesicles. The 100,000-dalton protein seems to be the best candidate for an ADP receptor protein of platelets: it incorporates radioactive 5'-FSBA at concentrations that inhibit ADP-induced platelet shape change, fibrinogen binding, and aggregation; the protein is protected by ADP against incorporation of 5'-FSBA; the protein is located on the exterior surface of the platelet membrane and is labeled in intact platelets. Researchers are devoting efforts to the purification of this protein.

CONCLUDING REMARKS

The technique of affinity labeling provides an extremely powerful approach to the identification of specific purine nucleotide sites within purified proteins, as well as to the recognition of nucleotide-binding proteins in complex systems. By specific occupation of a purine nucleotide site, an affinity label can be used to assess the role of that site in the catalytic function, in the regulation or in the subunit interactions of a homogeneous protein. The use of affinity labeling in a more complex system, such as an intact cell, can be used to evaluate the involvement of a purine nucleotide binding protein in a multistep process. The ideal characteristics for a purine nucleotide affinity label set forth at the beginning of this review are not met entirely by any of the compounds described here. Although not all of the potential purine nucleotide affinity labels extant were included, this conclusion also extends to those compounds not specifically considered herein. Potential affinity labels that are reasonably stable under the likely conditions for reaction with enzymes may not be sufficiently reactive to modify certain proteins. Compounds that are negatively charged at neutral pH and structurally closest to the natural nucleotide structures may not have a functional group capable of reaction with the full range of amino acids located in the vicinity of such binding sites. Finally, compounds that meet the criteria of general reactivity and structural similarity to the natural nucleotides may not yield products of reaction with enzymes that are sufficiently stable to allow satisfactory characterization of a stoichiometrically modified enzyme. Despite the limitations in the existing compounds, their application in the affinity labeling of many particular proteins has revealed a surprising amount of new information on the nucleotide-binding sites of these proteins. The future development of additional analogs with improved characteristics and with reactive functional groups at different positions of the purine or ribose moieties will contribute further to our understanding of the structure and role of purine nucleotide sites in proteins.

ACKNOWLEDGMENTS

Work from the author's laboratory reported in this review was supported by USPHS Grant GM 21200 and by NSF Grant PCM-8201969.

Literature Cited

1. Moos, C., Alpert, N. R., Myers, T. C. 1960. *Arch. Biochem. Biophys.* 88: 183–92
2. Yount, R. G., Babcock, D., Ojala, D., Ballantyne, W. 1971. *Biochemistry* 10: 2484–89
3. Yount, R. G., Ojala, D., Babcock, D. 1971. *Biochemistry* 10:2490–96
4. Eckstein, F., Goody, R. S. 1976. *Biochemistry* 15:1685–91
5. Burgers, P. M. J., Eckstein, F. 1979. *J. Biol. Chem.* 254:6889–93
6. Secrist, J. A., Barrio, J. R., Leonard, N. J., Weber, G. 1972. *Biochemistry* 11: 3499–506
7. Roberts, J. E., Aizono, Y., Sonenberg, M., Swislocki, N. I. 1975. *Bioorganic Chemistry* 4:181–87
8. Barrio, J. R., Secrist, J. A., Chien, Y., Taylor, P. J., Robinson, J. L., Leonard, N. J. 1973. *FEBS Lett.* 29:215–18
9. Vandenbunder, B., Morange, M., Buc, H. 1976. *Proc. Natl. Acad. Sci. USA* 73:2696–700
10. Miyata, H., Asai, H. 1981. *J. Biochem.* 90:133–39
11. Cerione, R. A., Hammes, G. G. 1982. *Biochemistry* 21:745–52
12. Busby, S. J. W., Hemminga, M. A., Radda, G. K., Trommer, W. E., Wenzel, H. 1976. *Eur. J. Biochem.* 63:33–38
13. Zantema, A., DeSmet, M.-J., Robillard, G. T. 1979. *Eur. J. Biochem.* 96:465–76
14. Leonard, N. J. 1982. *Acc. Chem. Res.* 15:128–35
15. Yount, R. G. 1975. *Adv. Enzymol.* 43: 1–56
16. Scheit, K. H. 1980. *Nucleotide Analogs: Synthesis and Biological Function.* New York: Wiley.
17. Baker, B. R., Lee, W. W., Tong, E., Ross, L. O. 1961. *J. Am. Chem. Soc.* 83:3713–17
18. Schoellman, G., Shaw, E. 1963. *Biochemistry* 2:252–55
19. Shaw, E., Mares-Guia, M., Cohen, W. 1965. *Biochemistry* 4:2219–24
20. Baker, B. R. 1976. *Design of Active Site-Directed Irreversible Enzyme Inhibitors.* New York: Wiley.
21. Shaw, E. 1970. *Enzymes* 1:91–146
22. Jakoby, W. B., Wilchek, M., eds. 1977. *Affinity Labeling, Methods in Enzymology,* Vol. 46. New York: Academic.
23. Lowe, P. N., Beechey, R. B. 1982. *Bioorg. Chem.* 11:55–71

24. Easterbrook-Smith, S. B., Wallace, J. C., Keech, D. B. 1976. *Eur. J. Biochem.* 62:125–30
25. Mehler, A. H., Kim, J. P., Olsen, A. A. 1981. *Arch. Biochem. Biophys.* 212: 475–82
26. Hinrichs, M. V., Eyzaguirre, J. 1982. *Biochim. Biophys. Acta* 704:177–85
27. Kumar, G., Kalra, V. K., Brodie, A. F. 1979. *J. Biol. Chem.* 254:1964–71
28. Ball, D. J., Nishimura, J. S. 1980. *J. Biol. Chem.* 255:10805–12
29. Rippa, M., Signorini, M., Signori, R., Dallocchio, F. 1975. *FEBS Lett.* 51: 281–83
30. Dallocchio, F., Negrini, R., Signorini, M., Rippa, M. 1976. *Biochim. Biophys Acta* 429:629–34
31. Rippa, M., Bellini, T., Signorini, M., Dallocchio, F. 1979. *Arch. Biochem. Biophys.* 196:619–23
32. Gregory, M. R., Kaiser, E. T. 1979. *Arch. Biochem. Biophys.* 196:199–208
33. King, M. M., Carlson, G. M. 1981. *Biochemistry* 20:4382–87
34. Khym, J. X., Cohn, W. E. 1961. *J. Biol. Chem.* 236:PC9–PC10
35. Schwartz, D. E., Gilham, P. T. 1972. *J. Am. Chem. Soc.* 94:8921–22
36. Lowe, P. N., Beechey, R. B. 1982. *Biochemistry* 21:4073–82
37. Lowe, P. N., Baum, H., Beechey, R. B. 1979. *Biochem. Soc. Trans.* 7:1133–36
38. Gundlach, H. G., Stein, W. H., Moore, S. 1959. *J. Biol. Chem.* 234:1754–60
39. Gundlach, H. G., Moore, S., Stein, W. H. 1959. *J. Biol. Chem.* 234:1761–64
40. Gomi, T., Fujioka, M. 1982. *Biochemistry* 21:4171–76
41. Roy, S., Colman, R. F. 1980. *J. Biol. Chem.* 255:7417–520
42. Bednar, R. A., Colman, R. F. 1982. *J. Protein Chem.* 1:203–24
43. Kovaleva, G. K., Ivanov, L. L., Madoyan, I. A., Favorova, O. O., Severin, E. S., et al. 1978. *Biochemistry-USSR* 43:419–26
44. Grivennikov, I. A., Bulargina, T. V., Khropov, Y. V., Gulyaev, N. N., Severin, E. S. 1979. *Biochemistry-USSR* 44:605–13
45. Gulyaev, N. N., Tunitskaya, V. L., Baranova, L. A., Nesterova, M. V., Murtuzaev, L. M., Severin, E. S. 1976. *Biochemistry-USSR* 41:1014–20
46. Severin, E. S., Nesterova, M. V., Gulyaev, N. N., Shylapnikov, S. V. 1976. *Adv. Enzyme Regul.* 14:407–44
47. Skolysheva, L. K., Vul'fson, P. L., Gulyaev, N. N., Severin, E. S. 1978. *Biochemistry-USSR* 43:1914–23

48. Mikhailova, L. I., Vul'fson, P. L., Skolysheva, L. K., Agalarova, M. B., Severin, E. S. 1978. *Biochemistry-USSR* 43:2016–21
49. Krauspe, R., Kovaleva, G. K., Gulyaev, N. N., Baranova, L. A., Agalarova, M. B., et al. 1978. *Biochemistry-USSR* 43:656–61
50. Berghäuser, J., Geller, A. 1974. *FEBS Lett.* 38:254–56
51. Eguchi, C., Suzuki, K., Imahori, K. 1977. *J. Biochem.* 81:1401–11
52. Suzuki, K., Eguchi, C., Imahori, K. 1977. *J. Biochem.* 81:1393–99
53. Suzuki, K., Eguchi, C., Imahori, K. 1977. *J. Biochem.* 81:1147–54
54. Dahl, K. H., McKinley-McKee, J. S. 1981. *Bioorg. Chem.* 10:329–41
55. Dickens, F. 1933. *Biochem. J.* 27: 1142–51
56. Bednar, R. A., Hartman, F. C., Colman, R. F. 1982. *Biochemistry* 21: 3681–89
57. Jencks, W. P., 1980. In *Mol. Biology, Biochemistry and Biophysics*, eds. F. Chapeville, A.-L. Haemi, 32:1–25. New York: Springer.
58. Bayley, H., Knowles, J. R. 1977. See Ref. 22, pp. 69–114
59. Chowdhry, V., Westheimer, F. 1979. *Ann. Rev. Biochem.* 48:293–325
60. Brunswick, D. J., Cooperman, B. S. 1971. *Proc. Natl. Acad. Sci. USA* 68: 1801–4
61. Brunswick, D. J., Cooperman, B. S. 1973. *Biochemistry* 12:4074–79, 4079–84
62. Haley, B. E. 1975. *Biochemistry* 14:3852–57
63. Haley, B. E. 1977. See Ref. 22, pp. 339–46
64. Pomerantz, A. H., Rudolph, S. A., Haley, B. E., Greengard, P. 1975. *Biochemistry* 14:3858–62
65. Czarnecki, J., Geahlen, R., Haley, B. E. 1979. *Methods Enzymol.* 56:642–53
66. Taylor, S. S., Kerlarage, A. R. 1982. *Fed. Proc.* 41:660
67. Marcus, F., Haley, B. E. 1979. *J. Biol. Chem.* 254:259–61
68. Geahlen, R. L., Haley, B. E. 1977 *Proc. Natl. Acad. Sci. USA* 74:4375–77
69. Geahlen, R. L., Haley, B. E. 1979. *J. Biol. Chem.* 254:1982–87
70. Dreyfuss, G., Schwartz, L., Blout, E. R., Barrio, J. R., Liu, F. T., Leonard, N. J. 1978. *Proc. Natl. Acad. Sci. USA* 75:1199–1203
71. Jeng, S. J., Guillory, R. J. 1975. *J. Supramol. Struct.* 3:448–68
72. Guillory, R. J., Jeng, S. J. 1977. See Ref. 22, pp. 259–88

73. Szilagyi, L., Balint, M., Sreter, F. A., Gergely, J. 1979. *Biochem. Biophys. Res. Commun.* 87:936–45
74. Chen, S., Guillory, R. J. 1977. *J. Biol. Chem.* 252:8990–9001
75. Chen, S., Guillory, R. J. 1979. *J. Biol. Chem.* 254:7220–27
76. Chen, S., Guillory, R. J. 1980. *J. Biol. Chem.* 255:2445–53
77. DeRiemer, L. A., Meares, C. F. 1981. *Biochemistry* 20:1606–12
78. Grachev, M. A., Knorre, D. G., Lavrik, O. I. 1981. *Biology Reviews, Soviet Scientific Reviews, Sect. D,* ed. V. P. Skulachev, 2:107–43
79. Chladek, S., Quiggle, K., Chinali, G., Kohut, J. III, Ofengand, J. 1977. *Biochemistry* 16:4312–19
80. Williams, N., Coleman, P. S. 1982. *J. Biol. Chem.* 257:2834–41
81. Standring, D. N., Knowles, J. R. 1980. *Biochemistry* 19:2811–16
82. Pal, P. K., Wechter, W. J., Colman, R. F. 1975. *J. Biol. Chem.* 250:8140–47
83. Colman, R. F., Pal, P. K., Wyatt, J. L. 1977. See Ref. 22, pp. 240–49
84. Paulos, R. L., Price, P. A. 1974. *J. Biol. Chem.* 249:1453–57
85. Annamalai, A. E., Colman, R. F. 1981. *J. Biol. Chem.* 256:276–83
86. Tomich, J. M., Marti, C., Colman, R. F. 1981. *Biochemistry* 20:6711–20
87. Togashi, C. T., Reisler, E. 1982. *J. Biol. Chem.* 257:10112–18
88. Saradambal, K. V., Bednar, R. A., Colman, R. F. 1981. *J. Biol. Chem.* 256:11866–72
89. Roy, S., Colman, R. F., 1979. *Biochemistry* 18:4683–90
90. Sweet, F., Samant, B. R. 1981. *Biochemistry* 20:5170–73
91. Tobias, B., Strickler, R. C. 1981. *Biochemistry* 20:5546–49
92. Wyatt, J. L., Colman, R. F. 1977. *Biochemistry* 16:1333–42
93. Likos, J. J., Hess, B., Colman, R. F. 1980. *J. Biol. Chem.* 19:9388–98
94. Mansour, T. E., Colman, R. F. 1978. *Biochem. Biophys. Res. Commun.* 81:1370–76
95. Pettigrew, D. W., Frieden, C. 1978. *J. Biol. Chem.* 253:3623–27
96. Weng, L., Heinrickson, R. L., Mansour, T. E. 1980. *J. Biol. Chem.* 255:1492–96
97. Zoller, M. J., Taylor, S. S. 1979. *J. Biol. Chem.* 254:8363–68
98. Hixson, C. S., Krebs, E. G. 1979. *J. Biol. Chem.* 254:7509–14
99. Zoller, M. J., Nelson, N. C., Taylor, S. S. 1981. *J. Biol. Chem.* 256:10837–42
100. Hixson, C. S., Krebs, E. G. 1981. *J. Biol. Chem.* 256:1122–27
101. Hashimoto, E., Takio, K., Krebs, E. G. 1982. *J. Biol. Chem.* 257:727–33
102. Hathaway, G. M., Zoller, M. J., Traugh, J. A. 1981. *J. Biol. Chem.* 256:11442–46
103. Buhrow, S. A., Cohen, S., Staros, J. V. 1982. *J. Biol. Chem.* 257:4019–22
104. Esch, F. S., Allison, W. S. 1978. *J. Biol. Chem.* 253:6100–6
105. DiPietro, A., Godinot, C., Martin, J.-C., Gautheron, D. C. 1979. *Biochemistry* 18:1738–45
106. DiPietro, A., Godinot, C., Gautheron, D. C. 1981. *Biochemistry* 20:6312–18
107. DeBenedetti, E., Jagendorf, A. 1979. *Biochem. Biophys. Res. Commun.* 96:440–46
108. Bennett, J. S., Vilaire, G., Colman, R. F., Colman, R. W. 1981. *J. Biol. Chem.* 256:1185–90
109. Kumar, S. A., Mooney, C., Krakow, J. S. 1977. *Fed. Proc.* 36:882
110. Foster, W. B., Griffith, M. J., Kingdon, H. S. 1981. *J. Biol. Chem.* 256:882–86
111. Boettcher, B. R., Meister, A. 1980. *J. Biol. Chem.* 255:7129–33
112. Lee, Y., Esch, F. S., DeLuca, M. A. 1981. *Biochemistry* 20:1253–56
113. Williamson, J. M., Meister, A. 1982. *J. Biol. Chem.* 257:9161–72
113a. Chen, S.-L., Kim, K.-H. 1982. *J. Biol. Chem.* 257:9953–57
114. Bennett, J. S., Colman, R. F., Colman, R. W. 1978. *J. Biol. Chem.* 253:7346–54
115. Figures, W. R., Niewiarowski, S., Morinelli, T. A., Colman, R. F., Colman, R. W. 1981. *J. Biol. Chem.* 256:7789–95
116. Pal, P. K., Wechter, W. J., Colman, R. F. 1975. *Biochemistry* 14:707–15
117. Pal, P. K., Reischer, R. J., Wechter, W. J., Colman, R. F. 1978. *J. Biol. Chem.* 253:6644–46
118. Pal, P. K., Colman, R. F. 1979. *Biochemistry* 18:838–45
119. Jadus, M., Hanson, R. W., Colman, R. F. 1981. *Biochem. Biophys. Res. Commun.* 101:884–92
120. Brand, L., Witholt, B. 1967. *Methods Enzymol.* 11:776–856
121. Horton, H. R., Koshland, D. E. 1967. *Methods Enzymol.* 11:856–70
122. Stryer, L. 1978. *Ann. Rev. Biochem.* 47:819–46
123. Likos, J. J., Colman, R. F. 1981. *Biochemistry* 20:491–99
124. Tomich, J. M., Colman, R. F. 1982. *Fed. Proc.* 41:1176
125. Jacobson, M. A., Colman, R. F. 1982. *Biochemistry* 21:2177–86
126. Craig, D. W., Hammes, G. G. 1980. *Biochemistry* 19:330–34

127. Togashi, C. T., Reisler, E. 1982. *J. Biol. Chem.* 257:10112–18
128. Parsons, T. F., Buckman, J. D., Pearson, D. E., Lamar, F. 1965. *J. Org. Chem.* 30:1923–26
129. Esch, F. S., Allison, W. S. 1978. *Anal. Biochem.* 84:642–45
130. Annamalai, A. E., Pal, P. K., Colman, R. F. 1979. *Anal. Biochem.* 99:85–91
131. Annamalai, A. E., Colman, R. F. 1982. *Int. Congr. Biochem., 12th,* Perth, Australia, p. 315 (abstr.)
132. Krause, J., Buhner, M., Sund, H. 1974. *Eur. J. Biochem.* 41:593–602
133. Sund, H., Markau, K., Koberstein, R. 1975. *Subunits in Biological Systems C.* 7:225–87. New York: Dekker.
134. Goldin, D. R., Frieden, C. 1972. *Curr. Top. Cell. Regul.* 4:77–117
135. Fisher, H. F. 1973. *Adv. Enzymol. Relat. Areas Mol. Biol.* 39:369–417
136. Frieden, C. 1963. *J. Biol. Chem.* 238:3286–99
137. Pantaloni, D., Dessen, P. 1969. *Eur. J. Biochem.* 11:510–19
138. Cross, D. G., Fisher, H. F. 1970. *J. Biol. Chem.* 245:2612–21
139. Koberstein, R., Krause, J., Sund, H. 1974. *Eur. J. Biochem.* 41:593–602
140. Pantaloni, D., Lecuyer, B. 1973. *Eur. J. Biochem.* 40:381–401
141. Pal, P. K., Colman, R. F. 1979. *Int. Congr. Biochem., 11th,* Toronto, Canada, p. 301 (abstr.)
142. Plowman, K. M., Krall, A. R. 1965. *Biochemistry* 4:2809–14
143. Stammers, D. K., Muirhead, H. 1975. *J. Mol. Biol.* 95:213–25
144. Stuart, D. I., Levine, M., Muirhead, H., Stammers, D. K. 1979. *J. Mol. Biol.* 134:109–42
145. Annamalai, A. E., Tomich, J. M., Mas, M. T., Colman, R. F. 1982. *Archives Biochem. Biophys.* 219:47–57
146. Cohn, E. J., Edsall, J. T. 1943. *Proteins, Amino Acids and Peptides.* p. 445. New York: Reinhold.
147. Hollenberg, P. F., Flashner, M., Coon, M. J. 1971. *J. Biol. Chem.* 246:946–63
148. Johnson, G. S., Deal, W. C. Jr. 1970. *J. Biol. Chem.* 246:238–48
149. Cardemil, E., Eyzaguire, J. 1979. *Arch. Biochem. Biophys.* 192:533–38
150. Marcus, A. J. 1982. In *Hemostasis and Thrombosis: Basic Principles and Clinical Practice,* eds. R. W. Colman, J. Hirsh, V. J. Marder, E. W. Salzman, pp. 380–89. Philadelphia: Lippincott
151. Nachman, R. L., Ferris, B. 1974. *J. Biol. Chem.* 249:704–10
152. Lips, J. P. M., Sixma, J. J., Schiphorst, M. E. 1980. *Biochem. Biophys. Acta* 628:451–67
153. Cross, M. J. 1964. *Thromb. Diath. Haemorrh.* 12:524–27
154. McLean, J. R., Maxwell, R. E., Hartler, D. 1964. *Nature* 202:605–6
155. Tangen, O., Berman, H. J., Marfey, P. 1971. *Thromb, Diath. Haemorrh.* 25:268–87
156. Niewiarowski, S., Regoeczi, E., Mustard, J. F. 1972. *Ann. NY Acad. Sci.* 201:72–83
157. Mustard, J. F., Perry, D. W., Ardlie, N. G., Packham, M. A. 1972. *Br. J. Haematol.* 22:103–4
158. Marguerie, G. A., Plow, E. F., Edgington, T. S. 1979. *J. Biol. Chem.* 254:5357–63
159. Marguerie, G. A., Edgington, T. S., Plow, E. F. 1980. *J. Biol. Chem.* 255:154–61
160. Bennett, J. S., Vilaire, G. 1979. *J. Clin. Invest.* 64:1393–401
161. Hawiger, J., Parkinson, S., Timmons, S. 1980. *Nature* 283:195–97
162. Peerschke, E. I., Zucker, M. B., Grant, R. A., Egan, J. J., Johnson, M. 1980. *Blood* 55:841–47
163. Kornecki, E., Niewiarowski, S., Morinelli, T. A., Kloczewiak, M. 1981. *J. Biol. Chem.* 256:5696–701

Ann. Rev. Biochem. 1983: 52:93–124

DNA METHYLATION AND GENE ACTIVITY

Walter Doerfler

Institute of Genetics, University of Cologne, D-5000 Cologne, Germany

CONTENTS

PERSPECTIVES AND SUMMARY 93
BIOCHEMISTRY OF DNA METHYLATION 95
  Occurrence of Modified Bases 95
  Maintenance and de novo Methylations 96
  Methods to Determine Methylated Bases 99
  Distribution of 5-Methylcytosine 100
  DNA Methylation and Structural Changes of DNA 102
THE BIOLOGICAL FUNCTION OF DNA METHYLATION 103
  DNA Methylation—a Regulatory Signal in Eukaryotic Gene Expression 103
  Correlations 105
  The Drosophila Problem 109
  Treatment of Cells with 5-Azacytidine Can Lead to the Activation of Previously
     Dormant Genes 110
  DNA Methylation at Specific Sites is Causally Related to Gene Inactivation 112
  Additional Observations 115
  Overall Conclusions 116
  DNA Methylation and Genetic Recombination 117
  A Model of Herpes Simplex Virus Latency 117
IN SEARCH OF MORE COMPLEX GENETIC CODING SIGNALS 118
OUTLOOK 118

PERSPECTIVES AND SUMMARY

DNA methylation and its functional significance, particularly in eukaryotic systems, have become very active areas of research. Although the biochemistry of DNA methylation has been studied in prokaryotic systems in great detail (1–7), the biological role of DNA modifications, particularly in eukaryotes, remains to be examined in great depth. A possible function in gene regulation was suggested in 1964 (5). In prokaryotes, it has long been apparent that methylation of specific DNA sequences renders the DNA refractory to the action of most restriction endonucleases that recognize

0066-4154/83/0701-0093$02.00

these specific sites (1–4). Whereas most restriction endonucleases are unable to cleave at methylated sequences, some require the presence of methylated bases. The enzyme *Dpn* I (from *Diplococcus pneumoniae*), for example, cuts at the sequence GATC only when the internal A-residue is methylated (8). Thus, it is important to recognize that methylated bases can assume positive or negative signal values relative to the functions they control or affect. The interference of methylated DNA sequences with the activity of restriction endonucleases constitutes the most clearly documented example yet of an effect of DNA modification on DNA-protein interactions. Other striking biochemical consequences of cytosine methylation are increased binding of lac repressor to a mutant (5-mC) operator (9) and reduced transcription of the arabinose operon upon methylation of a promoter sequence (10). Adenine modification also regulates the expression of a DNA-modifying function of bacteriophage Mu (11, 63). Regulation of gene expression in eukaryotes is a fascinating topic in molecular biology, with interesting implications for differentiation and for many other areas of biology, for example virology, immunology, genetic disease, and oncology. In this review, major emphasis is placed on examining the evidence that implicates DNA methylation in transcriptional regulation, especially in eukaryotes. Other possible functions of DNA methylation are also discussed.

The function that methylated bases in DNA exert seems to depend decisively on the base involved and on its highly specific position in regulatory sequences of genes or in the vicinity thereof. One of the major challenges in molecular biology is the elucidation of more complex coding principles, which play important roles as regulatory signals in DNA. It is probably too simplistic to search for these signals exclusively in specific DNA sequences; they may also be found in complex structures of DNA. In recent years, it has become increasingly apparent that the double-stranded DNA molecule can assume a number of structures that differ strikingly from the classical B form of DNA (for review see 12–14). It is uncertain to what extent these structures occur naturally and whether and how they can be implicated in special functions of DNA. Recent evidence from studies on a synthetic poly(dG·dC)·poly(dG·dC) polynucleotide suggests that the transition from the right-handed B form to the left-handed Z form (15–18) can be facilitated and stabilized by 5-mC (19–21) or by 7-methylguanine (22) in the alternating polynucleotide poly(dG·dC)·poly(dG·dC). The formation of nucleosome core particles stabilizes the B form. In contrast, the Z form of DNA seems to prevent nucleosome formation (23). Z-DNA has also been shown for molecules of sequence poly (dT·dG)·poly(dC·dA) (24). A sequence of 50 alternating dT and dG residues within one of the introns of a human cardiac muscle actin gene has been reported (25). A probe specific for poly (dT·dG) sequences revealed that potential Z-DNA-forming sequences are highly repeated in the human genome. If naturally occurring DNAs indeed contain the Z form of DNA, as antibody studies suggest (26–29), the possibility that the function of methylated bases in DNA could also be exerted via specific and localized structural alterations of DNA will have to be critically examined.

In the past few years, evidence has been accumulating rapidly for the implication of DNA methylation in a number of biologic processes (for reviews see 30–38). DNA methylation affects DNA-protein interactions, protects DNA against restriction endonucleases, regulates gene expression in eukaryotes, enhances mutation and recombination, can affect the structure of synthetic polynucleotides and probably of DNA as well, and may influence DNA replication, virus latency, and differentiation. Although the stringency of the evidence for some of these putative functions of DNA methylation may vary in quality, it is beyond doubt that DNA methylation impinges upon DNA-protein interactions in positive or negative ways, and thus may affect many different biochemical reactions. DNA methylation can be considered a modulator of these interactions and often acts as a long-term signal, since demethylations seem complicated to achieve by the cell. Since DNA-protein interactions are at the core of many different gene functions, it is not surprising that DNA methylation has been plausibly, sometimes convincingly, implicated in a number of different mechanisms. Nevertheless, destructive analyses of seemingly familiar concepts will be attempted in order to gain an improved understanding of this complex signal in DNA.

The burgeoning and rapidly expanding field of DNA methylation has prompted a series of reviews within the last three years focusing on different aspects of this topic (30–38). I have therefore decided—at the risk of unintended omissions—not to repeat all the details covered in these previous reviews, but rather to concentrate on recent developments and functional aspects of DNA methylation with particular emphasis on eukaryotic systems. The reader should not expect encyclopedic coverage of the entire literature, but a treatise on selected topics. Admittedly, I have not suppressed my bias toward the advantages that viral systems offer in the in-depth study of certain phenomena and their underlying mechanisms in eukaryotic systems.

BIOCHEMISTRY OF DNA METHYLATION

Occurrence of Modified Bases

Apart from extensive structural changes of DNA, modifications of bases have been envisaged as potential signals, although the precise role of the modifications has not yet been unraveled. In 1925, the base 5-methylcytosine (5-mC) was first crystallized as 5-methylcytosine picrate in the chemistry department of Yale University (39). According to that study, "This salt occurs in exceedingly minute, lath-like crystals. They possess a most brilliant luster and are a golden yellow." Methylated bases in DNA were described already in the late 1940s and early 1950s (40–42). The DNA of prokaryotes contains the modified bases N^6-methyladenine (6-mA) and 5-methylcytosine (5-mC). Other modified bases in DNA have apparently not been found in appreciable quantities although they may occur in minute amounts. Mycoplasmal DNA contains 0.2% to 2% 6-mA and 5.8% 5-mC (42a). The DNAs of higher eukaryotes contain 5-mC as a modified base (41, 43–47) and the DNAs of certain green algae (48–50), protozoa (51–52), and insects (53) also contain low levels of 6-mA (0.03 mol%). The percentage of 5-mC in eukaryotic DNA varies over a wide range. It is about 0.03 mol% in some insects such as the mosquito *Aedes albopictus,* 2 to 8 mol% in mammals, and can be as high as 50 mol% in higher plants (43, 44, 46, 47, 53–57). The DNAs of lower eukaryotes like *Drosophila* (58, 59, 89, 200, 200a) or nematodes appear to contain very few, if any, methylated bases although analyses with restriction endonucleases may reveal minute amounts of 5-mC in decisive sequences. In spite of intensive investigations, it cannot be ruled out that the DNAs of these species contain methylated bases in concentrations below the levels of detectability by current methods. Perhaps, these few modified bases are located in functionally important sequences. It should also be mentioned that in the DNA of the T–even bacteriophages, a highly modified base, glycosylated hydroxymethylcytosine, occurs in genetically determined patterns (60–62). In bacteriophage Mu the mom gene controls an unusual DNA modification [N^6-(1-acetamido)-adenine]. Expression of the mom function is positively regulated by DNA methylation (63, 63a).

Maintenance and de novo Methylations

Patterns of DNA methylation are inheritable (36, 64–69) and are maintained from cell division to cell division by the action of DNA methyltransferases (Figure 1). All biochemical evidence indicates that DNA is methylated in an early postreplicative step (70). It is only the newly synthesized strand that becomes methylated (67). Thus DNA replication is a prerequisite for changing patterns of DNA methylation. Demethylating activities have been found in nuclear extracts in mammalian cells (70a).

It also appears possible that specific inhibition of DNA methyltransferases is required to generate new unmethylated sites in DNA. Maintenance DNA methyltransferases recognize hemimethylated DNA as their specific substrate (69), i.e. DNA that is methylated in only one strand, and methylate bases in the newly synthesized DNA strand in an antiparallel, mirror-

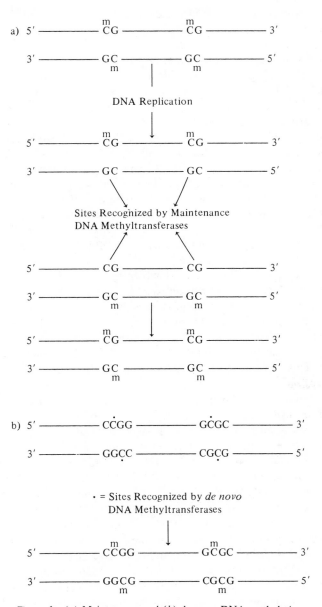

Figure 1   (*a*) Maintenance and (*b*) de novo DNA methylation.

like sense. Methylation patterns are inherited in a semiconservative fashion (64–66). In this way, specific methylation patterns of DNA can be perpetuated in the cell population.

There are also de novo DNA methyltransferases that impose a sequence specific pattern of DNA methylation on DNA not previously or only partly methylated. A relatively large number of prokaryotic DNA methyltransferases has been characterized and purified, e.g. from *Haemophilus parainfluenzae* (*Hpa*II; 71, 72), from *H. haemolyticus* (*Hha*I; 71, 73), from *Thermus thermophilus* HB8, *Thermus aquaticus* YTI, and *Caryophanon latum* L (74), from *Escherichia coli* SK (N_1 and G_{II}; 75), or from *Bacillus subtilis* or phage-infected *B. subtilis* (76). All these enzymes are de novo DNA methyltransferases that modify specific sequences in unmethylated DNA. Maintenance as well as de novo DNA methyltransferases have also been isolated from eukaryotic sources (77–85, for review see 31, 33). The DNA methyltransferase in Friend erythroleukemia cells is bound to linker DNA in condensed regions of chromatin. It is conceivable that chromatin proteins play a role in regulating the access of DNA methyltransferases to specific DNA sites (85, 185). The preferred substrate for the in vitro reaction of a mouse ascites DNA methyltransferase is hemimethylated DNA (69). This enzyme methylates exclusively cytosin residues at the dinucleotide 5'-CpG-3'. Interestingly, the restriction endonuclease from *E. coli* K is a multifunctional enzyme that can also efficiently methylate hemimethylated DNA requiring S-adenosylmethionine, adenosine triphosphate, and Mg^{2+} for the reaction. This methylation function can apparently modify different sites in DNA (86).

Viral systems often play useful roles in research on the molecular biology of eukaryotes. Results from work on a viral system can also provide convincing evidence for the existence of de novo methylation in mammalian cells. The DNA of human adenoviruses is not methylated, or methylated to a very limited extent, while the DNA of human cells, on which these viruses are propagated, contains 3.5–4.4% of 5-mC (87–90). Furthermore, human cellular DNA incorporated into a symmetric recombinant of adenovirus type 12 containing viral and cellular DNA covalently linked is not detectably methylated either (91). In contrast, adenoviral DNA integrated into the genome of transformed hamster cells or hamster tumor cells (for review see 37, 92) is methylated in highly specific patterns (93–97, 189). Thus, viral DNA must be methylated de novo at some time after infection and probably during or after insertion into the host genome. These findings also point to a chromosomal location and/or site of action of DNA methyltransferases. In Friend erythroleukemia cells, the bulk of DNA methyltransferase activity is bound to linker DNA in condensed regions of chromatin (85). DNA methyltransferase activities capable of de novo methylating adenovirus DNA have been extracted from nuclei of human KB cells. Differences in this activity between uninfected and Ad2-infected cells have not been observed (90, 98). It is therefore likely that the fact that adenovirion DNA is not significantly methylated in KB cells (87–90) can best be explained by assuming that DNA methyltransferases are normally sequestered into the chromatin "compartment" of the nucleus. Thus, they have limited access to viral DNA replicating in the nucleoplasm. More complicated explanations are also possible.

Methods to Determine Methylated Bases

For the precise quantification of 5-mC and other modified bases, a number of very sensitive techniques have been developed. Chromatographic procedures (99), two-dimensional chromatographic, and electrophoretic separations (87), gas chromatography (100), mass spectrometry (101, 102), and high-pressure liquid chromatography (102a, 102b, 103) have been successfully applied to the detection of 5-mC after total acid hydrolysis of DNA. Enzymatic hydrolysis of DNA is an additional method to degrade DNA (104). The sensitivity of the HPLC method may still be enhanced when one uses radioactively labeled DNA.

When 5-mC is covalently bound to bovine serum albumin, it can be used to raise antibodies against 5-mC in rabbits. These antibodies have proved useful, though possibly not entirely specific tools in the detection of small amounts of 5-mC (105–108). However, restriction enzymes (for review see 109) have assumed a pivotal role in the determination of methylation patterns, an option that has not been afforded by purely chemical or physical techniques. For this development, an incidental discovery by Waalwijk & Flavell (110) was of decisive importance. These authors established that the restriction endonuclease pair *Hpa*II (from *H. parainfluenzae*) and *Msp*I (from *Moraxella* species) are isoschizomers that cleave at CCGG sites. Whenever the internal C-residue of the sequence is modified to 5'-C^{m}CGG-3' by methylation, the endonuclease *Hpa*II cannot cut whereas *Msp*I does cut at such sites. Conversely, methylation of the external C, ^{m}CCGG, renders *Msp*I refractory, while the endonuclease *Hpa*II remains fully active (111). The availability of this isoschizomeric pair of restriction enzymes with a complementary refractoriness for methylation facilitates the elucidation of methylation patterns in specific genes. In brief, the DNA to be investigated is cleaved either with the restriction endonuclease *Hpa*II or *Msp*I, and the fragments are separated by electrophoresis in an appropriate gel system and transferred to nitrocellulose filters by the Southern technique (112). Using cloned cellular or virus genes as specific ^{32}P-labelled hybridization probes (113), the distribution of specific DNA fragments can be determined. Depending on the cleavage patterns generated by the *Hpa*II or *Msp*I restriction endonuclease, conclusions can be drawn about the state of methylation at all CCGG sites in a given gene. The only precondition for this type of analysis is the availability of a molecular clone of the gene to be investigated. The method can be improved by using two-dimensional DNA electrophoresis (59, 113a).

A high percentage of all 5-mC residues in eukaryotic DNA occurs in CpG sequences (44, 114–117); only a limited amount of all CpG groups occurs in the 5'-CCGG-3' sequence. Thus, differential cleavage by the *Hpa*II-*Msp*I or SmaI-XmaI restriction endonuclease pair will reveal only part of all 5-mC groups present in a given gene. For special sequence constellations, the restriction endonuclease pair *Sau*3A GATC and *Taq*I (TCGA) can be used on a similar basis, namely for the sequence GATCGA, which constitutes an overlap of both sites. A methylated TaqI site (GAT ^{m}CGA) is still sensitive to cleavage; a methylated Sau3A site (GAT^{m}CGA) is, however, rendered resistant (118). *Taq*I is also inhibited by 6-mA. In prokaryotes, 6-mA can be detected by cleavage with the isoschizomeric restriction endonuclease pair *Dpn*I (*D. pneumoniae*) and *Dpn*II, which cleave at GATC sites. *Dpn*II is inhibited by 6-mA; *Dpn*I only cleaves methylated DNA. There are several additional restriction endonucleases that are sensitive to the methylation of a ^{m}CpG site (for review see 37).

The presence of the base 5-mC in DNA can also be detected by the Maxam-Gilbert nucleotide-sequencing method (119, 120), since 5-mC does not appear in the pyrimidine cleavage patterns of sequencing gels. This is because 5-mC is less reactive with hydrazine than are cytosine or thymine. The presence of 5-mC can then be ascertained by sequencing of the complementary strand. The restriction enzyme analysis of methylation patterns yields necessarily a somewhat limited answer, because there are more possible methylated CpG dinucleotide combinations than there are currently isoschizomeric restriction endonuclease pairs. Therefore, it would be highly desirable to have available a sequencing technique that would allow one to determine the total pattern of methylation in situ. Such methods are not yet available.

Distribution of 5-Methylcytosine

As mentioned above, 5-mC appears to be the only modified base in DNA of eukaryotes and occurs predominantly in the sequence CpG (114–117), but it has also been found in the sequences CpA, CpT, and CpC (121). As discussed below, actively transcribed regions of the chromosome are undermethylated or unmethylated, inactive regions are strongly methylated (for review see 30–38). In animal cell DNA, an average of 70% of all CpG sequences are methylated. Active nuclear regions of DNA that exhibit increased sensitivity toward DNase I (122, 123) show 30–40% of the CpG sequences to be methylated (116).

Since the sequence CpG is the main site of methylation in eukaryotes, it may be of decisive importance that this dinucleotide is statistically under-represented in most higher eukaryotes (124) and many of their viruses (125). Perhaps the CpG dinucleotide occurs predominantly in functionally crucial positions in DNA sequences exerting a regulatory role. For some genes or groups of genes it has been recognized that frequency clusters of the dinucleotide CpG can be found close to the promoter/leader and/or 5′ regions of these genes (90, 126, 228). In adenovirus type 12 DNA, there is a very striking cluster of CpG dinucleotides and of interdigitating CCGG and GCGC sites at or close to the 5′ end of the E1a region (37, 90, 97), which is thought to control the expression of all other early regions of this viral genome (127, 128). Thus a decisive element in gene regulation may lie in the strategic positioning of regulatory nucleotide sequences that can be methylated at crucial sites.

If methylated bases served a regulatory or modulating role in various gene functions, one would expect that, in addition to the mechanism of DNA methylation, demethylating activities would also exist. Apart from inducible dealkylating mechanisms of a highly specialized nature in proka-ryotes and in mammalian cells (129, 130), such demethylases are just being recognized in eukaryotic cells (70a). Demethylations, at least in part, are a consequence of DNA replication in the absence of concomitant or subse-quent methylation of the hemimethylated DNA formed in the course of DNA replication. In this case, mechanisms regulating the activity of main-tenance DNA methyltransferases might assume a pivotal role in DNA methylation and demethylation. I consider it an important research project to elucidate these control mechanisms and also to search for active DNA demethylation. The cytidine analog 5-azacytidine (131, 131a), which cannot be methylated, inhibits DNA methylation (132–135). There is convincing evidence that 5-azacytidine can be incorporated into replicating DNA and, in this state, can inhibit the activity of DNA methyltransferases perhaps by irreversibly binding these enzymes (132–135). I discuss below the way in which 5-azacytidine can be used to activate previously inactive genes by locally and specifically inhibiting DNA methyltransferases.

It should be mentioned parenthetically that incubation of DNA with S-adenosyl-L-methionine can lead to base modifications with formation of 7-methylguanine and 3-methyladenine (136). Thus, S-adenosyl-L-methio-nine may prove mutagenic and carcinogenic.

A series of experiments revealed that the patterns of DNA methylation can be transmitted from cell generation to cell generation, i.e. that patterns of methylation are stably inherited (68, 137, 138). In these experiments, cloned viral or eukaryotic genes were methylated using prokaryotic DNA methyltransferases and subsequently transfected or microinjected into eukaryotic cells. Some of the methylated sites remained stably methylated as late as 30 generations after transfection, although a few sites had apparently lost the methyl groups. Stability of methylation patterns can be explained by maintenance methyltransferases recognizing specifically hemimethylated DNA. However, there may be more complex recognition mechanisms operating that account for the preservation of highly specific patterns of methylation.

In vitro methylation of previously unmethylated DNA, e.g. viral DNAs, provides a clear example of de novo methylation by prokaryotic or even eukaryotic DNA methyltransferases. Such in vitro methylated DNA preparations have served useful roles in studies on the role of DNA methylation in gene regulation (139–142, 228).

DNA Methylation and Structural Changes of DNA

It is an interesting and significant question whether DNA methylation at highly specific sites affects essential biological functions directly or via structural changes of DNA that are elicited or stabilized by DNA methylation. DNA can assume many different conformations depending on experimental conditions (12–14). Theoretically, several different structures of DNA have to be considered (for review see 12–14); three families of DNA helices have been examined in detail: A, B, and Z. In solutions containing high concentrations of salt, $poly(dG \cdot dC) \cdot poly(dG \cdot dC)$ can form a left-handed Z structure. For $poly(dG \cdot m^5dC) \cdot poly(dG \cdot m^5dC)$ the transition from the B to the Z form is observed at close to physiological salt concentrations (19–22). Di- and polyvalent ions are very effective in causing this transition, as well as a variety of in part exotic ionic conditions (29). The sequence $dG \cdot m^5dC$ can have a very striking effect on the B-Z transition. The B-Z transition is facilitated by alternating $dC \cdot dG$, or more generally by alternating pyrimidine-purine tracks, whereas the sequence dCdCdGdG may favor transition from the B to the A form of DNA (143). It is presumed that the latter sequence might sterically inhibit the B-Z conversion. Since certain DNA sequences can cause specific structural alterations, it will be important to determine under which conditions these transitions can occur in the living cell and which modifications might stabilize certain transitions. DNA methylation at highly specific sites could have this effect. On the other hand, a methylated base with the methyl group protruding into the major groove of B DNA can efficiently provide a signal in its own right.

THE BIOLOGICAL FUNCTION OF DNA METHYLATION

Not merely for the sake of analogy, but for more profound reasons, the encoding of information in DNA can be compared with that in language (144). Methylation of certain bases in highly specific positions in DNA resembles alterations or modifications in language. Minor modifications can drastically alter the meaning of language and equally of a signal in DNA (e.g. English: step → stop; German: *Achtung* = high esteem → *Ächtung* = proscription). If one accepts the possibility that methylated bases somehow modulate DNA-protein interactions, as has been clearly demonstrated for the activity of restriction endonucleases (1–4), it is plausible that DNA methylations in highly specific sequences and at unique sites can affect a host of biological functions. In that sense, the relatively simple signal of a methylated base could have positive or negative values. I shall summarize evidence for DNA methylation and its role as a regulatory signal in eukaryotic gene expression, in inactivating the X chromosome, in stimulating genetic recombination of bacteriophage λ, and in virus latency.

For some of the listed mechanisms, the accumulated evidence begins to be rather convincing, for others it is still tentative. In this discussion, I do not separate strictly prokaryotic and eukaryotic systems, but rather, I present examples from both systems.

DNA Methylation—a Regulatory Signal in Eukaryotic Gene Expression

The regulation of gene expression in eukaryotes is most likely to involve highly specific DNA-protein interactions. Proteins binding to specific sequences in eukaryotes are just being recognized (e.g. 145); their regulatory function is still uncertain. DNA methylation at specific sequences may modulate these interactions; usually DNA methylation of C residues in specific sequences is associated with the inactive state of a gene. Chromatin structure as well probably plays an important part in the regulation of gene expression (for review see 146), and nucleosome structure may in itself be responsive to DNA methylation (23).

Hypotheses relating DNA methylation and gene activity evolved from observations on the differential states of DNA methylation in developing organisms and in different tissues (64–67, 146–148). In a given gene that was actively expressed in one organ, the CCGG sites were found to be under- or unmethylated, and the DNA in the same gene in other organs not expressing this gene was completely methylated at the CCGG sites. These sites were preferentially investigated, because of the availability of the isoschizomeric restriction endonuclease pair *Hpa*II and *Msp*I (110) that permits the determination of the state of methylation at all CCGG sites. In many genes the state of methylation at these sites appears to have special significance. In this way, inverse correlations between the extent of DNA methylation at CCGG sites in specific genes and the degree to which these genes are expressed have originally been established in several different systems (93–97, 150–154). In some cases, these inverse correlations have been extended to other sequences containing CG dinucleotides (96, 97). An inverse relationship between the extents of DNA methylation and gene expression could obviously indicate cause or consequence of gene inactivation.

Investigations directed toward the core of the problem used in vitro methylated cloned genes that were microinjected into the nucleus of oocytes of *Xenopus laevis* (139, 140, 228) or into mammalian cells in culture (141, 142). Methylated genes were not expressed in these experiments. The use of sequence specific de novo DNA methyltransferases from prokaryotic organisms allowed the determination of highly specific sequences presumably involved in gene regulation (139–142). Other sequences, when methylated, did not exhibit any effect on gene expression (155, 156). These data accumulating from work in different systems and in different laboratories provided strong direct evidence for the notion of methylated sequences at highly specific sites playing an important part in the regulation of gene expression. And yet, there are still examples (157, 157a) that do not seem to conform to this straight-forward scheme. Of course, it can always be argued that, for some genes, certain unknown sequences are susceptible to DNA methylation and that these specific sites have not been modified in experiments failing to demonstrate the usual inverse correlations.

Further convincing evidence comes from work with the cytidine analog 5-azacytidine. This compound can be incorporated into replicating DNA (132–135), and because of its chemical structure (N in the 5-position, instead of C) cannot be methylated. However, the main inhibitory role of 5-azacytidine toward DNA methylation appears to emanate from its ability to trap DNA methyltransferases and thus block their action (132–135). In a number of different systems it has been possible to activate previously dormant genes and trigger their expression (see below). The original report on the gene activating property of 5-azacytidine described the activation of a complex set of cellular functions leading to in vitro differentiation of mouse fibroblasts to twitching muscle cells (158). It is not to be expected that all dormant cellular genes can be turned on by treatment of cells with 5-azacytidine. It is argued below that DNA methylation at highly specific sites, which could be different for different genes, plays a role in the long-term inactivation of genes. Moreover, absence of DNA methylation appears a necessary but not sufficient precondition for gene activation (96, 150).

A crucial mechanism like gene activity is probably subject to multifaceted regulatory mechanisms, DNA methylation constituting only one important parameter. Thus, depending on the stringency of inactivation for a given gene, 5-azacytidine treatment may or may not lead to the activation of a certain gene or set of genes. As with the use of any drug or inhibitory analog certain reservations remain with respect to the interpretation of results adduced using this cytidine analog. It may affect cellular metabolism in several different ways, and there is no evidence ascertaining that the inhibition of DNA methyltransferases is in fact the only or the major influence on gene activity that this analog can exert.

For a complete understanding of the role of DNA methylation in gene regulation, methods will be required to determine the state of methylation at possibly all the CG sites in a gene and its adjacent sequences. Improved in vitro transcription systems will also be required to evaluate the effect of methylation in different CG sites. Presently available in vitro transcription systems do not seem to respond to DNA methylation.

In ascribing 5-mC residues a major role in gene regulation, one will also have to deal with those systems in which the DNA allegedly or actually lacks this modified base. At present, there is no positive evidence for the occurrence of significant amounts of 5-mC in *Drosophila,* in nematodes or some of the other lower eukaryotes, although minute amounts of 5-mC may be present in these DNAs. Several possibilities are discussed to reconcile this apparent discrepancy with the role that DNA methylation seems to play in higher eukaryotic systems and perhaps also for some prokaryotic genes (10, 11 63, 63a).

I now examine in detail the evidence available to substantiate the notion ascribing to DNA methylation a decisive function in the regulation of gene expression.

Correlations

Inverse correlations between the degree of DNA methylation of certain genes and the extent to which these genes are expressed have now been established in many viral (93–97, 152–154, 159–167, 174) and nonviral eukaryotic systems (116, 147–150, 168–173, 175). Tissue-specific patterns of the distribution of 5-mC continue to be elucidated in many different systems (176–179). In general, the experimental approach has been based on the use of the isoschizomeric restriction endonuclease pair *Msp*I (M. species I) and *Hpa*II (H. *parainfluenzae* II) (Table 1). A survey of other methylation-sensitive restriction endonucleases was presented in a previous review (37). Although for many of these enzymes isoschizomers are not available, they can still be used in the analysis of methylation patterns of specific genes as long as the gene under investigation is available as a molecular clone. The clone is then grown up in a methylation-deficient host to assure its unmethylated state, and this DNA, upon cleavage with the 5-mC-sensitive enzyme, is then coelectrophoresed as an internal marker with the cleaved cellular DNA to be investigated. The blotted DNA is next hybridized with the ^{32}P-labeled cloned DNA as a probe. By comparing the restriction patterns obtained with cloned and cellular DNAs one will be able to deduce the extent of methylation at a number of different sites in a particular gene. In this way, it has been possible to demonstrate that, in an organism, distinct genes not expressed in certain organs are hypermethylated, whereas the same genes are hypomethylated or unmethylated in those organs in which they are specifically expressed.

Gene expression involves the concerted activity of a number of functions; the DNA-dependent RNA polymerases are of prime importance in this event. It was therefore significant to ask whether not only genes known to be transcribed by RNA polymerase II were subject to control in one way or another by DNA methylation, but also those requiring the activity of RNA polymerase I. Inverse correlations between DNA methylation and gene expression have been described for these genes as well, in particular for the ribosomal DNA of *Xenopus* and mouse (180–185). It appears that the location of methylated CpG sequences is also decisive in this system (183–185). Unmethylated rDNA genes are hypersensitive to DNase I in mouse liver nuclei (184). It has been speculated that during *Xenopus* development a nuclear protein interferes specifically with DNA methylation at specific sites (185). In the elucidation of the function that DNA methylation has in gene regulation in higher eukaryotes, viral systems have played a very significant role (37, 93–97, 139–141, 152–156). The virion DNA that is encapsidated in the virus particle is not methylated to a detectable extent in many of the DNA viruses (87–90, 186–188). This generalization is supported by data using acid or enzymatic hydrolysis of DNA followed by two-dimensional electrophoretic and chromatographic analyses, high-pressure liquid chromatography (87–90), or by using restriction enzyme analysis (93–95, 189) of viral DNAs. There is at least one very interesting

Table 1 Activity of Restriction Endonucleases *Hpa*II and *Msp*I at methylated or unmethylated CCGG sites[a]

Restriction endonuclease	CCGG	$\overset{m}{CCGG}$	$\overset{m}{CCGG}$
*Hpa*II	+	+	−
*Msp*I	+	−	+

[a] This scheme is derived from data by Waalwijk & Flavell (110) and Sneider (111). Hemimethylated DNA, i.e. DNA carrying 5-mC in only one strand, is nicked by *Msp*I in the unmethylated strand (175).

exception to this general rule: the DNA of frog virus 3 (FV3), a herpes-like virus, is heavily methylated, containing over 20% of 5-mC instead of C residues, whereas the DNA of the host fathead minnow cells contains only 6–8% 5-mC (190). The viral DNA replicates in the nucleus, and at this stage remains unmethylated. Upon transport into the cytoplasm, where mature virions are assembled, the DNA becomes heavily methylated (A. Granoff, personal communication). There is preliminary evidence that infection with FV3 alters host cell RNA polymerase II to render it capable of transcribing methylated viral DNA (D. B. Willis, personal communication).

On the other hand, viral DNA integrated into the host genome by covalent phosphodiester bonds in transformed or virus-induced tumor cells exhibits highly specific methylation patterns (37, 93–97, 155–156). Free intracellular viral DNA in productively or abortively infected cells is not methylated (95). This observation documents in a most direct way that mammalian cells can de novo methylate foreign and perhaps their own DNA. The existence of de novo DNA methyltransferases has been demonstrated in a number of systems (77–85, 90, 98). Mammalian (rodent) cells transformed by human adenoviruses have been analyzed in detail and have proved useful tools in studies on the regulation of viral gene expression. The patterns of viral DNA integration have been determined in some 70 adenovirus type-2- or adenovirus type-12-transformed rodent cell lines or in adenovirus type-12-induced tumors or tumor lines (for review see 92). In most of these cell lines, the viral genes expressed late in a productive infection are permanently turned off and highly or completely methylated at the CCGG or GCGC sites. In contrast, some or all of the early viral genes are expressed in transformed cell lines and these regions of the inserted viral genomes are undermethylated (37, 93–97, 139, 140, 191–193, 193a). This example has allowed the establishment of a convincing inverse correlation between DNA methylation and gene expression. Recently, it has become necessary to map precisely the CCGG sequences in the early regions of the inserted viral genome in a number of cell lines (97). The data summarized in Figure 2 and results from other systems (185; H. Cedar, personal communication) indicate that DNA methylation at the 5' end and/or close to the promoter region of a gene correlates with the absence of gene expression. Perhaps, only methylated bases in highly specific sequences at certain sites can assume the regulatory functions.

The complexity of the interdependence between DNA methylation and gene expression can be documented by investigations on the extent of DNA methylation of inserted viral genomes in adenovirus type-12-induced tumors and in cell lines originating from these tumors (96, 194, 195). The integrated Ad12 DNA in these tumors is not methylated, or is methylated

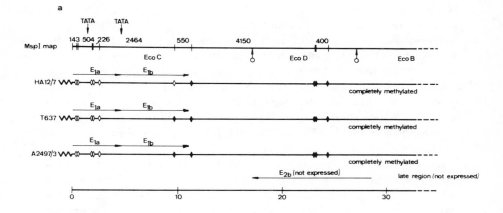

Figure 2   Functional maps of the left (*a*) and the right (*b*) halves of the Ad12 DNA molecules integrated in cell lines HA 12/7, T637, and A2497-3. The horizontal lines represent the Ad12 genomes (*straight lines*) integrated into the genomes of cell lines (*jagged lines*) as indicated. The *Msp*I maps of the left (*a*) and right (*b*) ends of adenovirus type 12 DNA are presented in the top line. Vertical bars on and figures above the horizontal lines indicate the locations of the *Msp*I sites and the size of some of the *Msp*I fragments, respectively. The vertical arrows designate *Eco*RI sites. TATA marks the locations of presumptive Goldberg-Hogness signals in the E1a region. The unmethylated (*open diamonds*) and methylated (*closed diamonds*) CCGG sites in the integrated adenovirus type 12 genomes in the lines HA12/7, T637, and A2497-3 are also indicated. The horizontal arrows indicate map positions and direction of transcription of the individual early regions in each of the cell lines. Presence of an arrow indicates expression of corresponding regions of the adenovirus type 12 genome. *Msp*I sites to the right of the *Eco*RI-D/B and to the left of the *Eco*RI-E/F junctions have not been mapped. The early regions of adenovirus type 12 DNA and a fractional length scale have also been indicated. This figure was taken from Kruczek & Doerfler (97).

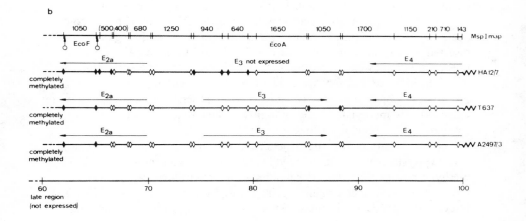

only to a very limited extent, at the CCGG and GCGC sites (96). Nevertheless, the viral DNA in these tumors is either not expressed at all, or only at a very low level (96, 194). Thus, there must be additional, more complex parameters other than DNA methylation that ascertain the inactivation of integrated viral genes. These parameters are still unknown.

It would also appear that DNA methylation might be used as a regulatory signal whenever a gene or a group of genes has to be turned off permanently, or for a long time. When cell lines in culture were established from the adenovirus type-12-induced tumors and passaged several times, an increase in the extent of viral DNA methylation was observed (96). It appears that this increase followed a certain pattern, in that certain sites in the inserted viral DNA were methylated earlier than others (96, 195). This shift in the patterns of methylation did not seem to be accompanied by altered expression of the viral DNA. It must still be determined whether and to what extent the altered culture conditions, i.e. the transition from animal to cell culture, can affect levels of DNA methylation. Furthermore, it has been demonstrated that the levels of viral DNA methylation in those morphological revertants of cell line T637 (196–198), which still contain one or half a copy of adenovirus type 12 DNA, have increased. Concomitantly, the extent to which the integrated viral DNA is expressed in the revertants is reduced relative to the parent T637 line (192, 198).

In considering correlations between DNA methylation patterns and gene activity, it is necessary to emphasize that hypomethylation and gene activity also correlate with increased DNase sensitivity and altered chromatin structure at sites of increased gene expression (199). It will be very interesting to elucidate in what way precisely these different parameters, which can be determined by current technology, are related. The dependence of chromatin structure on certain patterns of DNA methylation has only rarely been investigated (199a).

Obviously, the evidence reviewed in this section represents only correlations between gene activity and DNA methylation. Based on the information gleaned from the experiments discussed, one can argue that DNA methylation may be the cause or the consequence of the absence of gene expression. More decisive experimental evidence must be adduced to distinguish between these alternatives.

The Drosophila Problem

Results correlating DNA methylation and gene inactivation have come mainly from studies on higher eukaryotes. It is currently thought that the DNA of *Drosophila* cells does not contain 5-mC at levels detectable by chemical methods (59, 89, 200, 200a). It may be an open question whether the DNA in polytene chromosomes in *Drosophila* also lacks the modified base. At this stage, the following possibilities require further study:

(*a*) There may be an exceedingly low amount of 5-mC in highly strategic positions of *Drosophila* DNA. These rare modified bases might then still suffice to exert a regulatory function. These rare modified bases could be detectable by restriction analysis and blotting using highly specific probes.

(*b*) Gene regulation in *Drosophila* may be totally different from that in higher eukaryotes and not require modified bases at all.

(*c*) If the role of 5-mC in gene regulation were to be exerted via structural alterations of DNA, e.g. by facilitating the transition from the B to the A or to the Z conformation or by fixing DNA structure in either configuration, perhaps the presence of certain sequences alternating frequently in purine and pyrimidine bases might render the occurrence of 5-mC in large amounts unnecessary.

(*d*) Other, not-yet-identified modified bases may play an important role in *Drosophila* DNA.

Considerably more analytical work will be required to clarify these tantalizing questions.

Treatment of Cells with 5-Azacytidine Can Lead to the Activation of Previously Dormant Genes

The 5-mC analog 5-azacytidine (131, 131a) can be incorporated into DNA but cannot be methylated (132). This analog specifically inhibits methylation of cytosine residues in DNA and RNA (201, 202). Increasing levels of 5-azacytidine incorporated inhibit the action of DNA methyltransferases (132, 133). It therefore appears likely that the decrease in DNA methylation observed in 5-azacytidine-treated cells is due to the inhibitory effect of 5-azacytidine on DNA methyltransferases. Treatment of Friend erythroleukemia cells with 5-azacytidine or 5-aza-2′-deoxycytidine leads to a rapid decrease of the activity of DNA methyltransferases and to hypomethylation of DNA (132). Inhibition of DNA synthesis in these experiments abolishes the loss of DNA methyltransferase activity. About 0.3% substitution of 5-azacytidine for cytidine in DNA reduces the DNA methyltransferase level by >95% (133). The 5-azacytidine effect is totally reversible. In this system, 5-azacytidine also functions as weak inducer of erythroid differentiation (133). There is evidence that 5-azacytidine and related cytosine analogs do not exert significant mutagenic action in eukaryotic cells (203). Thus, there is no support for a mutational basis for 5-azacytidine-induced biological effects. The analog has, however, been shown to cause neoplastic transformation of mouse cells (204, 205) and to be weakly mutagenic in microorganisms (205) and viruses (206).

When mouse fibroblasts are treated with 5-azacytidine at a narrow dose range, the formation of functional muscle cells, chondrocytes, and adipocytes can be induced in culture (158, 207–211). It has also been demonstrated that genes encoded on an inactive X chromosome in mouse-human somatic cell hybrids generated by cell fusion can be reactivated by treating these hybrids with 5-azacytidine (212, 213). The human hypoxanthine-guanine-phosphoribosyltransferase gene from structurally normal inactive human X chromosomes has been induced in 0.1–8% of the surviving cells (214). Maximal induction effects are observed when cells are treated in late S phase. The inactive X chromosome replicates late in the hybrid cells. In muscle cells induced by 5-azacytidine or other analogs with modifications in the 5 position of the pyrimidine ring (e.g. 5-aza-2'-deoxycytidine, pseudoisocytidine, or 5-fluoro-2'-deoxycytidine), the DNA has been found to be undermethylated 48 hours after treatment (209). These data imply that the inhibition of maintenance methylation can lead to highly significant demethylations in certain genes and to the activation of these genes.

Thymidine kinase minus (tk⁻) Chinese hamster cells, which were resistant to bromodeoxyuridine (215), were exposed to 5-azacytidine. This treatment led to a massive conversion of cells to the tk⁺ state (216). This interesting observation may necessitate reinvestigation of many cellular mutants for their actual mutant genotype, or else for the possibility that certain genes or groups of genes had been permanently turned off by methylation at highly specific sites.

Similar observations about the induction of genetic activity by 5-azacytidine and concomitant changes in the patterns of methylation have been made for the metallothionine I gene in a mouse thymoma cell line (217). Normally, this gene is expressed in the presence of either cadmium or glucocorticoids.

Again, work using viral systems has produced very convincing evidence for the gene-activating role engendered in cells treated with 5-azacytidine. Chicken cells harbor the transcriptionally active ev-3 and the inactive ev-1 endogenous retroviral loci. DNA sequences in the ev-3 region are undermethylated, preferentially sensitive to DNase I, and contain nuclease hypersensitive sites (199). Treatment of cells with 5-azacytidine causes hypomethylation and transcriptional activation of the ev-1 locus (162, 199). This induction was stable for at least 10 generations. Similarly, 5-azacytidine-activation studies have been reported with mouse cells transformed with *Herpes simplex* virus containing the viral thymidine kinase in an inactive state. Upon two days of exposure to 5-azacytidine (10 μM), the thymidine kinase gene was maximally expressed (218). The inactive thymidine kinase gene was methylated, the reactivated gene unmethylated. Similar observations have been presented for Moloney murine leukemia virus in murine embryonal carcinoma cells. The inactive proviral genomes have been activated by treatment with 5-azacytidine and have been subsequently found to lose 5-mC (219).

As described previously, the late viral genes in adenovirus type-12-transformed hamster cells are usually not expressed and these DNA sequences are completely methylated at CCGG sites (93, 94, 97, 191, 192). Treatment of cell line T637, one of the adenovirus type-12-transformed hamster cell lines with 5-azacytidine leads to specific demethylations in the late region of the viral genome. These demethylation patterns are different in more than 30 separate clones isolated and analyzed individually (90, 220). There is, however, no evidence that late viral genes are reactivated by 5-azacytidine in these clones. This finding lends support to the notion that absence of methylation is a necessary but not sufficient precondition for gene activity. In adenovirus type-12-transformed hamster cells, other, hitherto unknown, factors are missing for late viral gene expression. It will be interesting to substantiate whether 5-azacytidine treatment of adenovirus type-12-transformed cells can indeed lead to demethylations predominantly at certain sites of the viral genome (220).

Quite a number of examples relating 5-azacytidine treatment of cells and gene activation attest to the general mechanism: Inhibition of maintenance methylation in growing cell cultures by a series of analogs of cytidine causes induction of certain genes. Since absence of methylation is only one of the factors prerequisite for gene activity, not all genes can possibly respond to this treatment. In the following section, I present further, more direct evidence to corroborate the notion that DNA methylation at specific sites and gene inactivation are causally related.

DNA Methylation at Specific Sites is Causally Related to Gene Inactivation

Microinjection experiments using either *X. laevis* oocytes or mammalian cells as recipients and unmethylated or in vitro methylated, cloned DNA of viral or eukaryotic origin reveal that genes methylated in specific sequences are transcriptionally inactivated upon transfer to recipient cells (139–142, 155, 221). These data provide the first conclusive evidence for a causal role of DNA methylation in gene regulation. In vitro transcription systems have so far not yielded differences in the rate of transcription of methylated and unmethylated DNA. The factors recognizing the structural alteration due to the methyl group may not be present in currently used in vitro systems. In many of the microinjection experiments, the DNA was methylated at the CCGG sites using the prokaryotic *Hpa*II DNA methyltransferase (71, 72). This enzyme was chosen because of its ability to modify the sequence that had been implied to play a role in gene regulation in many experiments correlating DNA methylation and gene inactivation (see above). Alternatively, newly developed plasmid cloning vectors that can replicate in both *E. coli* and *H. parainfluenzae*, might be useful in specifically methylating cloned genes at the CCGG sites (222).

In addition, it has been shown that in vitro methylation with the *Hpa*II or *Hha*I DNA methyltransferases inactivates the transforming activity of the cloned Moloney sarcoma viral DNA (221).

There is now very good evidence demonstrating that oocytes from *X. laevis* can faithfully express a number of eukaryotic genes (223, 224) upon microinjection into the nuclei. The oocyte system thus appears capable of recognizing as well as mammalian cells the regulatory signal constituted by methylated sequences.

The E2a region of adenovirus type 2 DNA encodes the virus-specific DNA-binding protein required for viral DNA replication. In three lines of adenovirus type-2-transformed hamster cells HE1, HE2, and HE3, multiple copies of the major part of the viral genome persist in an integrated state (225). Cell lines HE2 and HE3 do not express the DNA-binding protein whereas line HE1 does so (193). All three cell lines also contain the intact E2a region of adenovirus type 2 DNA and a functional (late) promoter (155, 226). Thus, there is no a priori reason for this gene not to be expressed. In cell line HE1, all CCGG (*Hpa*II/*Msp*I) sites in the E2a region remain unmethylated. Conversely, in lines HE2 and HE3 lacking expression of the E2a region, all *Hpa*II sites are methylated (95). The cloned E2a region of adenovirus type 2 DNA, the *Hind*III A fragment in pBR322, was methylated in vitro by using *Hpa*II DNA methyltransferase or was left unmethylated. In vitro methylation did not break or nick supercoiled circular DNA. Methylated or unmethylated DNA was then microinjected into the nuclei of *X. laevis* oocytes, and the subsequent synthesis of virus-specific RNA was monitored. In vitro methylated DNA remained in the methylated state for 24 h on microinjection into nuclei of *Xenopus* oocytes; unmethylated DNA remained unmethylated. When the injected DNA had been methylated by using HpaII DNA methyltransferase, Ad2-specific RNA was not synthesized as late as 24 h after microinjection. Unmethylated DNA was readily expressed into virus-specific RNA (139, 140). As an internal control, unmethylated histone genes (h22 DNA) from sea urchin were microinjected together with methylated E2a DNA from adenovirus type 2. Adenovirus type-2-specific RNA was not found; h22 DNA-specific RNA was readily detected. This finding ruled out nonspecific inhibitory effects in the methylated DNA preparation. It was also shown that transcription of the unmethylated *Hind*III A fragment of adenovirus type 2 DNA in *Xenopus* oocytes was initiated on the late promoter of the E2a region (140). The same promoter was used in productively infected KB cells. Methylation by BsuRI methylase (GGCC) did not inactivate the *Hind*III A fragment (155). In the cell lines HE1, HE2, and HE3, the GGCC sites in the E2a region are not methylated at all, hence it was not expected that these sites had any regulatory significance (155). These results provide direct evidence for the notion that methylated sequences at highly specific sites are involved in the regulation of gene expression. The actual nature of the regulatory signal is not yet understood. A detailed description of all the results on DNA methylation using the adenovirus system has been published elsewhere (227).

Similar experiments were performed in other systems as well. Plasmid DNA methylated at the CCGG sites and injected into oocytes of *X. laevis* preserved its state of methylation through replication. Unmethylated sites remained in this state in progeny molecules (138). Hemimethylated DNA became fully methylated. In this system inheritance of methylation did not require integration of foreign DNA into the host chromosome.

Furthermore, cloned genes for the T antigen of SV40 DNA or for the thymidine kinase of *Herpes simplex* virus were methylated with the *Eco*RI DNA methyltransferase, which modifies the internal adenosine in the sequence GAATTC (141). DNA was injected into nuclei of thymidine-kinase-deficient (tk⁻) hamster cells. The methylation of the SV40 T antigen gene at up to 24 different sites did not affect its expression in mammalian cells. Conversely, methylation of a single site upstream from the cap site inhibited expression of the tk gene (141). These findings are consistent with the idea that DNA methylations at highly specific sites have signal values for different genes. It is considered an unusual finding that adenosine methylation has such striking effects.

When the only *Hpa*II site in intact SV40 DNA, which is located close to the late leader, is methylated, late viral genes are not expressed in *X. laevis* oocytes, whereas the early viral genes are expressed normally (228). Similarly, in vitro methylation of the AGPRT gene by the *Hpa*II methylase leads to transcriptional inactivation in mouse-L cells (142). The transforming activity of Moloney sarcoma virus DNA was reduced by in vitro methylation with HpaII DNA methyltransferase. The transforming efficiency was partly restored when transformed cells were treated with 5-azacytidine (221).

These results, derived from very different systems and laboratories, build a rather strong case, perhaps even proof, for the direct involvement of DNA methylation at highly specific sites in the control of gene expression. Experiments are now in progress in several laboratories to methylate selectively the 5'- or the 3'-end of a gene or its promoter region exclusively and to examine differential effects of selective site methylations on gene activity. There is evidence from several systems that methylation at the 5' end and/or at the promoter region (97) may constitute the decisive signal (see above and Figure 2).

It is conceivable that a methylated base at each individual site has specific modulator functions. Much more refined analyses are needed to unravel these possibly very complicated functional implications.

Additional Observations

In the chicken genome and in the genomes of other avian species, clusters of middle repetitive DNA sequences are present. The bulk of these clustered repetitive sequences is heavily methylated regardless of the chicken tissue. The same patterns of methylation have been noted in DNA from thirty individual animals. The distribution of methylation clusters appears to be regional; long methylated DNA sequences are interrupted by hypomethylated stretches (177, 229). Very high levels of DNA methylation have also been found in repetitive DNA sequences and in particular in inverted repetitive DNA sequences in mouse P815 mastocytoma cells. The inverted repeats contain about 50% more methylated bases than do normal repeat sequences (230).

In hepatocellular carcinomas, hypomethylation correlated with malignant transformation of cells (231). This finding is in agreement with that made in adenovirus type-12-induced hamster tumors (96). However, I consider it premature to attempt to establish correlations between DNA methylation and malignant transformation.

Transcription of the J chain gene is initiated in the differentiation of mouse B cells. The expression of this gene correlates with a loss of methyl groups in this gene, whereas nonexpressed J chain genes in embryos or lymphomas are heavily methylated. A similar correlation exists for the C_{2B} sequences; the $C\mu$ sequences (μ chains of the constant region of immunoglobulins) are undermethylated in all stages of differentiation (169).

In the green alga *Chlamydomonas,* DNA methylation can play a dual function. Firstly, it protects DNA from the restriction system in *Chlamydomonas* (83, 232, 233). Secondly, DNA methylation controls the transcriptional activity of certain algal genes. Site-specific DNA methyltransferases in *Chlamydomonas* have been isolated and characterized (83, 234). In *Chlamydomonas*, chloroplast genes are maternally inherited. They are transmitted from the female parent to all progeny, while the corresponding genes from the male parent are lost. It has been postulated that this type of inheritance is regulated by selective methylation. Recently, it was demonstrated that only the chloroplast DNA of female gametes is methylated at CCGG sites. The chloroplast DNAs of male gametes and of vegetative cells of both mating types remain unmethylated. Thus, methylation appears to be sex determined and provides the molecular basis for maternal inheritance. Methylation was detected by use of 5-mC-specific antibodies (233a). Sager et al (235) isolated mutants of *Chlamydomonas* (mat-1) in which methylation also occurs in the chloroplast DNA of male gametes. Thus, methylation patterns in this system are genetically controlled. In the recently described mutant *me-1*, extensive methylation of the cytosine residues in chloroplast DNA has been found in vegetative cells of both mating types. The CpG dinucleotide is extensively methylated, CpC is methylated occasionally. Interestingly, extensive methylation has been reported at the GpC dinucleotide (236). These authors suggest that chloroplast DNA methylation may be insufficient to account for inheritance patterns in *Chlamydomonas* chloroplast DNA. However, if the methylation-restriction system in *Chlamydomonas* is site specific as in bacteria, then the specific sites (as yet unidentified) may not be methylated in the vegetative cells.

Early in embryonic development, one of the two X chromosomes is inactivated in the somatic cells of the normal female in mammals (237). In 1975, Holliday & Pugh (64) and Riggs (65) proposed that DNA methylation may play a role in this event. Mohandas et al (212, 213) provided direct evidence in support of this hypothesis. They prepared a mouse-human somatic cell hybrid that was deficient in hypoxanthine-guanine phosphoribosyl transferase (HGPRTase) and contained an inactive human X chromosome. These hybrid cells were treated with the cytidine analog 5-azacytidine and were subsequently assayed for reactivation of the X chromosome and the expression of HGPRTase located on the human X chromosome. HGPRTase-positive clones were about 10^3 times more frequent than in untreated cells and the HGPRTase synthesized was shown to be human. In some clones other X-chromosome-linked functions were also expressed, namely glucose-6-phosphate dehydrogenase or phosphoglycerate kinase. Since 5-azacytidine treatment leads to hypomethylation of DNA (132), it is likely that DNA methylation is somehow responsible for the inactivation of the X chromosome. Recently, these data have been refined by extracting DNA from mouse-human or hamster-human hybrids containing inactive human X chromosomes that had been either induced by 5-azacytidine or left uninduced, and by using these DNA preparations in transformation of HGPRT⁻ cells. DNA from cells containing the activated human X-chromosome transformed cells to the HGPRT⁺ phenotype; DNA from noninduced cells did not (238). Results from another laboratory (239, 240) indicate that the extent of changes in methylation patterns in DNA sequences located on the X chromosome upon activation of this chromosome may depend on the DNA probe used, i.e. on the genes investigated.

Overall Conclusions

There is now a substantial amount of information leading to the conclusion that DNA methylation at sites specific possibly for each gene is somehow causally related to the regulation of gene expression in higher eukaryotes. This regulatory function is most likely to be exerted via specific DNA-protein interactions; structural changes of DNA may also be involved. At the molecular level, the regulatory signal 5-mC is not understood. Many intriguing problems remain to be solved:

1. Which 5-mC residues are the functionally significant ones?
2. What are the rules in positioning 5-mC residues relative to other regulatory signals?
3. What structural changes of DNA can 5-mC elicit and/or stabilize, and are structural changes of DNA decisive as regulatory signals?
4. How does 5-mC in DNA affect DNA-protein interactions?
5. What functions other than transcription can be affected by 5-mC?
6. How can we determine all 5-mC residues in or in the vicinity of a gene and establish their functions? A method is needed urgently.

Lastly, there are genes that do not appear to subscribe to regulation by DNA methylation, e.g. genes in most free viral genomes (for review see 227). Studies on the methylation pattern of the $\alpha2$ collagen gene in DNA from five cell types of the chicken reveal that expression of the $\alpha2$ collagen gene seems to be independent of the levels of methylation at CCGG sites (157). Thus, DNA methylation is an element in long-term gene inactivation; it may not be invariably used as a regulatory signal, and it may occasionally assume other functions as well.

DNA Methylation and Genetic Recombination

Lambda bacteriophage DNA can contain varying amounts of 5-mC depending on the strain of *E. coli* it is grown on (1). Thus, in prokaryotes, viral DNA is methylated by the DNA methyltransferases of the host. This observation contrasts with conditions in eukaryotic systems; most DNA viruses replicating in the nucleus lack methylated bases. For bacteriophage lambda, correlations have been described between decreased frequencies of 5-mC and increased genetic recombination and unusual sensitivity of viral DNA to endonuclease S1 (241). Hemimethylated CC^A_TGG sequences are possibly required for enhanced recombination, and necessary but not sufficient for S1 sensitivity. Further work will be necessary to define clearly the role that methylated sequences can play in recombination.

A Model of Herpes Simplex Virus Latency

A lymphoblastoid T-cell line can be persistently infected with *H. simplex* virus type 1 (HSV-1) (242). The nonproductive state of these cells can be maintained by concanavalin A, but reversed with phytohemagglutinin. Similar to the situation in Epstein-Barr virus-transformed cells (152), the viral DNA in nonproducing cells is extensively methylated. In producer cells, however, the viral DNA is not methylated (243).

IN SEARCH OF MORE COMPLEX GENETIC CODING SIGNALS

The endeavor to understand the functional meaning of 5-mC in DNA is only part of a more extensive effort to recognize and decipher more complex genetic codes. Recently, I raised the question whether recognition models gleaned from a completely different field, i.e. linguistics, could serve as a guide (144). Striking similarities exist between the genetic code and human languages and their linear or more complex representations. These analogies may point to an inherent, very basic interrelationship connecting code and language. The nature of this connection will have to be sought perhaps in the realm of brain function. It is conceivable and perhaps even likely that, in addition to the familiar linear genetic code, more complex codes are operative that might be superimposed on the DNA sequences that code for polypeptides and/or may also be found in "noncoding" or in repetitive DNA sequences, provided these sequences can store sufficient information. This latter idea is reflected in the similarities between some languages and the primary code. Other languages exhibiting more complex structural elements are postulated to have their correlations in more complex coding elements. Some well-documented examples make this idea less hypothetical than it may appear at first sight (144).

Thus, there is a pool of resources of structural and evolutionary elements available in linguistics that may prove useful in the structural analyses of more complicated coding principles in genetics. Naturally, these analyses cannot be sensibly performed without the availability of solid genetic evidence. Nevertheless, an available set of principles may be useful as a guide in the analysis of genetic data, in particular of a host of sequence data that can easily be subjected to computer analyses. With some effort, structural and perhaps even evolutionary principles known from linguistic research might be testable for their applicability in genetic research. This comparison may appear farfetched, and it will require considerable effort before one can decide whether this model will prove useful in genetic analyses.

OUTLOOK

The evidence summarized in this review demonstrates that methylated bases in DNA, in particular 5-mC in eukaryotic DNA, have a signal function with plurivalent potential. From work in many different laboratories employing very different systems, convincing evidence has emerged that implicates DNA methylation in the regulation of gene expression at least in the higher eukaryotes. Many different experimental approaches have helped to establish a causal relation between DNA methylation and gene inactivation. It is still advisable to view this correlation with a sense of caution in that only certain groups of genes might be responsive to this regulatory signal. Moreover, it is not certain whether gene regulation in lower eukaryotes is also subject to this regulatory mechanism. In prokaryotes, only a few investigations have so far focused on this problem. Major problems remain for future research:

1. The determination of patterns of methylation at all 5'-CpG-3' sites of a gene and its regulatory sequences.
2. The elucidation of the functional significance, if any, of each methylated base in DNA.
3. Studies on the regulation of DNA methylation. If DNA methylation does indeed play a decisive role in the regulation of gene activity, DNA methyltransferases (maintenance or de novo) and factors affecting their activity must be ascribed a very important role.
4. Can DNA be actively demethylated by demethylating enzymes or proteins? Recent evidence (70a) on demethylating activities from mammalian cells will encourage this search.

At this point, one may view DNA methylation as a pluripotent signal that may assume different meanings depending on sites and sequences involved. It is interesting to recall that methylations of other macromolecules, RNA, protein carbohydrates, etc, also seem to have striking functional consequences [for a recent summary, see (38)]. A particularly interesting case is that of transmethylation reactions in specific proteins (244–246) affecting, for example, the chemotactic behavior of cells (244, 245). Thus, minor modifications—if an added methyl group can be considered a minor alteration—of macromolecules can have far-reaching functional consequences. Future research on chemical or structural modifications of DNA will undoubtedly lead to a better understanding of the coding capacity of DNA and may help to unravel more complex genetic coding principles. So far, we have seen perhaps only the tip of an iceberg.[1]

ACKNOWLEDGMENTS

I am indebted to many colleagues for providing preprints of unpublished work and for the permission to cite work in progress. I should like to thank Birgit Kierspel for typing this manuscript. Work in my laboratory has been supported by the Deutsche Forschungsgemeinschaft through SFB 74-C1 and by the Ministry of Science and Research of the State of North Rhine-Westfalia.

[1] Two significant recent developments should be mentioned. It was reported that 5-azacytidine selectively increased γ-globin synthesis in a patient with β^+ thalassemia and hypomethylation in bone-marrow DNA at the γ-globin and ε-globin genes (247). Also, ultimate chemical carcinogens of diverse chemical nature were shown to inhibit in vitro methylation of hemimethylated DNA (248).

Literature Cited

1. Arber, W. 1974. *Prog. Nucleic Acid Res. Mol. Biol.* 14:1–37
2. Arber, W., Linn, S. 1969. *Ann. Rev. Biochem.* 38:467–500
3. Meselson, M., Yuan, R., Heywood, J. 1972. *Ann. Rev. Biochem.* 41:447–66
4. Boyer, H. W. 1971. *Ann. Rev. Microbiol.* 25:153–76
5. Srinivasan, P. R., Borek, E. 1964. *Science* 145:548–53
6. Gold, M., Hurwitz, J. 1964. *J. Biol. Chem.* 239:3858–65
7. Fujimoto, D., Srinivasan, P. R., Borek, E. 1965. *Biochemistry* 4:2849–55
8. Lacks, S., Greenberg, B. 1975. *J. Biol. Chem.* 250:4060–66
9. Fisher, E. F., Caruthers, M. H. 1979. *Nucleic Acids Res.* 7:401–16
10. Horwitz, A., Wilcox, G. 1982. Personal Communication
11. Kahmann, R. 1982. *Cold Spring Harbor Symp. Quant. Biol.* 47:639–46
12. *Cold Spring Harbor Symp. Quant. Biol.* 1982. 47: In press
13. Dickerson, R. E., Drew, H. R., Conner, B. N., Wing, R. M., Fratini, A. V., Kopka, M. L. 1982. *Science* 216:475–85
14. Zimmerman, S. B. 1982. *Ann. Rev. Biochem.* 51:395–427
15. Pohl, F. M., Jovin, T. M. 1972. *J. Mol. Biol.* 67:375–96
16. Wang, A. H.-J., Quigley, G. J., Kolpak, F. J., Crawford, J. L., van Boom, J. H., et al. 1979. *Nature* 282:680–86
17. Drew, H., Takano, T., Tanaka, S., Itakura, K., Dickerson, R. E. 1980. *Nature* 286:567–73
18. Arnott, S., Chandrasekaran, R., Birdsall, D. L., Leslie, A. G. W., Ratliff, R. L. 1980. *Nature* 283:743–45
19. Behe, M., Felsenfeld, G. 1981. *Proc. Natl. Acad. Sci. USA* 78:1619–23
20. Behe, M., Zimmerman, S., Felsenfeld, G. 1981. *Nature* 293:233–35
21. Klysik, J., Stirdivant, S. M., Singleton, C. K., Zacharias, W., Wells, R. D. 1983. *J. Mol. Biol.* In press
22. Möller, A., Nordheim, A., Nichols, S. R., Rich, A. 1981. *Proc. Natl. Acad. Sci. USA* 78:4777–81
23. Nickol, J., Behe, M., Felsenfeld, G. 1982. *Proc. Natl. Acad. Sci. USA* 79:1771–75
24. Crawford, J. L., Kolpak, F. J., Wang, A. H.-J., Quigley, G. J., van Boom, J. H., et al. 1980. *Proc. Natl. Acad. Sci. USA* 77:4016–20
25. Hamada, H., Kakunaga, T. 1982. *Nature* 298:396–98
26. Lafer, E. M., Möller, A., Nordheim, A., Stollar, B. D., Rich, A. 1981. *Proc. Natl. Acad. Sci. USA* 78:3546–50
27. Nordheim, A., Pardue, M. L., Lafer, E. M., Möller, A., Stoller, B. D., Rich, A. 1981. *Nature* 294:417–22
28. Nordheim, A., Peck, L. J., Lafer, E. M., Stollar, B. D., Wang, J. C., Rich, A. 1982 *Cold Spring Harbor Symp. Quant. Biol.* 47:93–100
29. Jovin, T. M., van de Sande, J. H., Zarling, D. A., Arndt-Jovin, D. J., Eckstein, F., et al. 1982. *Cold Spring Harbor Symp. Quant. Biol.* 47:143–54
30. Razin, A., Riggs, A. D. 1980. *Science* 210:604–10
31. Drahovsky, D., Boehm, T. L. J. 1980. *Int. J. Biochem.* 12:523–28
32. Burdon, R. H., Adams, R. L. P. 1980. *Trends Biochem. Sci.* 5:294–97
33. Hattman, S. 1981. *Enzymes* 14:517–47
34. Razin, A., Friedman, J. 1981. *Prog. Nucleic Acid Res. Mol. Biol.* 25:33–52
35. Ehrlich, M., Wang, R. Y.-H. 1981. *Science* 212:1350–57
36. Wigler, M. H. 1981. *Cell* 24:285–86
37. Doerfler, W. 1981. *J. Gen. Virol.* 57:1–20
38. Usdin, E., Borchardt, R. T., Creveling, C. R., eds. 1982. *Biochemistry of 5-Adenosylmethionine and Related Compounds.* London: Macmillan. 760 pp.
39. Johnson, T. B., Coghill, R. D. 1925. *J. Am. Chem. Soc.* 47:2838–44
40. Hotchkiss, R. D. 1948. *J. Biol. Chem.* 175:315–32
41. Wyatt, G. R. 1951. *Biochem. J.* 48:581–84
42. Sinsheimer, R. L. 1955. *J. Biol. Chem.* 215:579–83
42a. Razin, A., Razin, S. 1980. *Nucleic Acids Res.* 8:1383–90
43. Shapiro, H. S., Chargaff, E. 1960. *Biochim. Biophys. Acta* 39:68–82
44. Doskočil, J., Šorm, F. 1962. *Biochim. Biophys. Acta* 55:953–59
45. Doskočil, J., Šormova, Z. 1965. *Biochim. Biophys. Acta* 95:513–15
46. Vanyushin, B. F., Belozersky, A. N., Kokurina, N. A., Kadirova, D. X. 1968. *Nature* 218:1066–67
47. Vanyushin, B. F., Tkacheva, S. G., Belozersky, A. N. 1970. *Nature* 225:948–49
48. Pakhomova, M. V., Zaitseva, G. N., Belozersky, A. N. 1968. *Dokl. Akad. Nauk SSSR* 182:712–14
49. Hattman, S., Kenny, C., Berger, L., Pratt, K. 1978. *J. Bacteriol.* 135:1156–67
50. Burton, W. G., Grabowy, C. T., Sager, R. 1979. *Proc. Natl. Acad. Sci. USA* 76:1390–94

51. Gorovsky, M. A., Hattman, S., Pleger, G. L. 1973. *J. Cell Biol.* 56:697–701
52. Cummings, D. J., Tait, A., Goddard, J. M. 1974. *Biochim. Biophys. Acta* 374:1–11
53. Adams, R. L. P., McKay, E. L., Craig, L. M., Burdon, R. H. 1979. *Biochim. Biophys. Acta* 563:72–81
54. Dunn, D. B., Smith, J. D. 1958. *Biochem. J.* 68:627–36
55. Drozhdenyuk, A. P., Sulimova, G. E., Vanyushin, B. F. 1977. *Biochimia* 42:1439–44
56. Thomas, A. J., Sherratt, H. S. A. 1956. *Biochem. J.* 62:1–4
57. Deumling, B. 1981. *Proc. Natl. Acad. Sci. USA* 78:338–42
58. Kurnick, N. B., Herskowitz, I. H. 1952. *J. Cell Comp. Physiol.* 39:281–99
59. Smith, S. S., Thomas, C. C. Jr. 1981. *Gene* 13:395–408
60. Cohen, S. S. 1968. *Virus-Induced Enzymes.* New York: Columbia Univ. Press
61. Mathews, C. K., Brown, F., Cohen, S. S. 1964. *J. Biol. Chem.* 239:2957–63
62. Josse, J., Kornberg, A. 1962. *J. Biol. Chem.* 237:1968–76
63. Hattman, S., Goradia, M., Monaghan, C., Bukhari, A. I. 1982. *Cold Spring Harbor Symp. Quant. Biol.* 47:647–53
63a. Hattman, S. 1982. *Proc. Natl. Acad. Sci. USA* 79:5518–21
64. Holliday, R., Pugh, J. E. 1975. *Science* 187:226–32
65. Riggs, A. D. 1975. *Cytogenet. Cell Genet.* 14:9–25
66. Sager, R., Kitchin, R. 1975. *Science* 189:426–33
67. Bird, A. P. 1978. *J. Mol. Biol.* 118:49–60
68. Wigler, M., Levy, D., Perucho, M. 1981. *Cell* 24:33–40
69. Gruenbaum, Y., Cedar, H., Razin, A. 1982. *Nature* 295:620–22
70. Burdon, R. H., Adams, R. L. P. 1969. *Biochim. Biophys. Acta* 174:322–29
70a. Gjerset, R. A., Martin, D. W., Jr. 1982. *J. Biol. Chem.* 257:8581–83
71. Mann, M. B., Smith, H. O. 1977. *Nucleic Acids Res.* 4:4211–21
72. Quint, A., Cedar, H. 1981. *Nucleic Acids Res.* 9:633–46
73. Roberts, R. J., Meyers, P. A., Morrison, A., Murray, K. 1976. *J. Mol. Biol.* 103:199–208
74. McClelland, M. 1981. *Nucleic Acids Res.* 9:6795–804
75. Nikolskaya, I. I., Lopatina, N. G., Anikeicheva, N. V., Debov, S. S. 1979. *Nucleic Acids Res.* 7:517–28
76. Jentsch, S., Günthert, U., Trautner, T. A. 1981. *Nucleic Acids Res.* 9:2753–59
77. Simon, D., Grunert, F., von Acken, U., Döring, H.-P., Kröger, H. 1978. *Nucleic Acids Res.* 5:2153–67
78. Turnbull, J. F., Adams, R. L. P. 1976. *Nucleic Acids Res.* 3:677–95
79. Browne, M. J., Turnbull, J. F., McKay, E. L., Adams, R. L. P., Burdon, R. H. 1977. *Nucleic Acids Res.* 4:1039–45
80. Adams, R. L. P., McKay, E. L., Craig, L. M., Burdon, R. H. 1979. *Biochim. Biophys. Acta* 561:345–57
81. Sneider, T. W., Teague, W. M., Rogachevsky, L. M. 1975. *Nucleic Acids Res.* 2:1685–1700
82. Roy, P. H., Weissbach, A. 1975. *Nucleic Acids Res.* 2:1669–84
83. Sano, H., Sager, R. 1980. *Eur. J. Biochem.* 105:471–80
84. Bromberg, S., Pratt, K., Hattman, S. 1982. *J. Bacteriol.* 150:993–96
85. Creusot, F., Christman, J. K. 1981. *Nucleic Acids Res.* 9:5359–81
86. Burckhardt, J., Weisemann, J., Yuan, R. 1981. *J. Biol. Chem.* 256:4024–32
87. Günthert, U., Schweiger, M., Stupp, M., Doerfler, W. 1976. *Proc. Natl. Acad. Sci. USA* 73:3923–27
88. von Acken, U., Simon, D., Grunert, F., Döring, H.-P., Kröger, H. 1979. *Virology* 99:152–57
89. Eick, D., Fritz, H.-J., Doerfler, W. 1982. Submitted for publication
90. Doerfler, W., Kruczek, I., Eick, D., Vardimon, L., Kron, B. 1982. *Cold Spring Harbor Symp. Quant. Biol.* 47:593–503
91. Deuring, R., Klotz, G., Doerfler, W. 1981. *Proc. Natl. Acad. Sci. USA* 78:3142–46
92. Doerfler, W. 1982. *Curr. Top. Microbiol. Immunol.* 101:127–94
93. Sutter, D., Doerfler, W. 1979. *Cold Spring Harbor Symp. Quant. Biol.* 44:565–68
94. Sutter, D., Doerfler, W. 1980. *Proc. Natl. Acad. Sci. USA* 77:253–56
95. Vardimon, L., Neumann, R., Kuhlmann, I., Sutter, D., Doerfler, W. 1980. *Nucleic Acids Res.* 8:2461–73
96. Kuhlmann, I., Doerfler, W. 1982. *Virology* 118:169–80
97. Kruczek, I., Doerfler, W. 1982. *EMBO J.* 1:409–14
98. Kron, B., Eick, D., Doerfler, W. 1983. In preparation
99. Randerath, K., Randerath, E. 1968. *Methods Enzymol.* 12:323–47
100. Razin, A., Sedat, J. 1977. *Anal. Biochem.* 77:370–77
101. Razin, A., Cedar, H. 1977. *Proc. Natl. Acad. Sci. USA* 74:2725–28
102. Singer, J., Roberts-Ems, J., Luthardt,

F. W., Riggs, A. D. 1979. *Nucleic Acids Res.* 7:2369–85
102a. Fritz, H.-J., Belagaje, R., Brown, E. L., Fritz, R. H., Jones, R. A., et al. 1978. *Biochemistry* 17:1257–67
102b. Kuo, K. C., McCune, R. A., Gehrke, C. W. 1980. *Nucleic Acids Res.* 8: 4763–76
103. Fritz, H.-J., Eick, D., Werr, W. 1982. In *Chemical and Enzymatic Synthesis of Gene Fragments—A Laboratory Manual*, ed. H. G. Gassen, A. Lang, pp. 199–223. Weinheim: Chemie
104. Ford, J. P., Coca-Prados, M., Hsu, M.-T. 1980. *J. Biol. Chem.* 255:7544–47
105. Erlanger, B. F., Beiser, S. M. 1964. *Proc. Natl. Acad. Sci. USA* 52:68–74
106. Eastman, E. M., Goodman, R. M., Erlanger, B. F., Miller, O. J. 1980. *Chromosoma* 79:225–39
107. Lubit, B. W., Pham, T. D., Miller, O. J., Erlanger, B. F. 1976. *Cell* 9:503–9
108. Sano, H., Royer, H. D., Sager, R. 1980. *Proc. Natl. Acad. Sci. USA* 77:3581–85
109. Roberts, R. J. 1982. *Nucleic Acids Res.* 10:r117–r44
110. Waalwijk, C., Flavell, R. A. 1978. *Nucleic Acids Res.* 5:3231–36
111. Sneider, T. W. 1980. *Nucleic Acids Res.* 8:3829–40
112. Southern, E. M. 1975. *J. Mol. Biol.* 98:503–17
113. Wahl, G. M., Stern, M., Stark, G. R. 1979. *Proc. Natl. Acad. Sci. USA* 76:3683–87
113a. Yee, T., Inouye, M. 1982. *J. Mol. Biol.* 154:181–96
114. Grippo, P., Iaccarino, M., Parisi, E., Scarano, E. 1968. *J. Mol. Biol.* 36:195–208
115. Gautier, F., Bünemann, H., Grotjahn, L. 1977. *Eur. J. Biochem.* 80:175–83
116. Naveh-Many, T., Cedar, H. 1981. *Proc. Natl. Acad. Sci. USA* 78:4246–50
117. Manes, C., Menzel, P. 1981. *Nature* 293:589–90
118. Streeck, R. E. 1980. *Gene* 12:267–75
119. Maxam, A. M., Gilbert, W. 1977. *Proc. Natl. Acad. Sci. USA* 74:560–64
120. Ohmori, H., Tomizawa, J. I., Maxam, A. M. 1978. *Nucleic Acids Res.* 5: 1479–85
121. Gruenbaum, Y., Naveh-Many, T., Cedar, H., Razin, A. 1981. *Nature* 292: 860–62
122. Weintraub, H., Groudine, M. 1976. *Science* 193:848–56
123. Garel, A., Zolan, M., Axel, R. 1977. *Proc. Natl. Acad. Sci. USA* 74:4867–71
124. Subak-Sharpe, H., Bürk, R. R., Crawford, L. V., Morrison, J. M., Hay, J., Keir, H. M. 1966. *Cold Spring Harbor Symp. Quant. Biol.* 31:737–48
125. Morrison, J. M., Keir, H. M.,Subak-Sharpe, H., Crawford, L. V. 1967. *J. Gen. Virol.* 1:101–8
126. Felsenfeld, G., Nickol, J., McGhee, J., Behe, M. 1982. *Cold Spring Harbor Symp. Quant. Biol.* 47:577–84
127. Roberts, R. J., Bing-Dong, J., Bullock, P. A., Gelinas, R. E., Gingeras, T. R., et al. 1982. *Cold Spring Harbor Symp. Quant. Biol.* 47:1025–37
128. Berk, A. J., Lee, F., Harrison, T., Williams, J., Sharp, P. A. 1979. *Cell* 17: 935–44
129. Karran, P., Lindahl, T., Griffin, B. 1979. *Nature* 280:76–77
130. Olsson, M., Lindahl, T. 1980. *J. Biol. Chem.* 255:10569–71
131. Piskala, A., Šorm, F. 1964, cited in Pačes, V., Doskočil, J., Šorm, F. 1968. *Biochim. Biophys. Acta* 161:352–60
131a. Raška, K., Jurovčik, M., Fučik, V., Tykva, R., Šormova, Z., Šorm, F. 1966. *Coll. Czech. Chem. Commun.* 31:2809–15
132. Jones, P. A., Taylor, S. M. 1981. *Nucleic Acids Res.* 9:2933–47
133. Creusot, F., Acs, G., Christman, J. K. 1982. *J. Biol. Chem.* 257:2041–48
134. Christman, J. K., Weich, N., Schoenbrun, B., Schneiderman, N., Acs, G. 1980. *J. Cell Biol.* 86:366–70
135. Jones, P. A., Taylor, S. M., Wilson, V. L. 1983. In *Recent Results in Cancer Res.* Heidelberg: Springer. 84:202–11
136. Rydberg, B., Lindahl, T. 1982. *EMBO J.* 1:211–16
137. Stein, R., Gruenbaum, Y., Pollack, Y., Razin, A., Cedar, H. 1982. *Proc. Natl. Acad. Sci. USA* 79:61–65
138. Harland, R. M. 1982. *Proc. Natl. Acad. Sci. USA* 79:2323–27
139. Vardimon, L., Kuhlmann, I., Cedar, H., Doerfler, W. 1981. *Eur. J. Cell Biol.* 25:13–15
140. Vardimon, L., Kressmann, A., Cedar, H., Maechler, M., Doerfler, W. 1982. *Proc. Natl. Acad. Sci. USA* 79:1073–77
141. Waechter, D. E., Baserga, R. 1982. *Proc. Natl. Acad. Sci. USA* 79:1106–10
142. Stein, R., Razin, A., Cedar, H. 1982. *Proc. Natl. Acad. Sci. USA* 79:3418–22
143. Conner, B. N., Takano, T., Tanaka, S., Itakura, K., Dickerson, R. E. 1982. *Nature* 295:294–99
144. Doerfler, W. 1982. *Medical Hypotheses* 9:563–79
145. Jack, R. S., Megan, T., Gehring, W. J. 1982. *Cold Spring Harbor Symp. Quant. Biol.* 47:483–91
146. McGhee, J. D., Felsenfeld, G. 1980. *Ann. Rev. Biochem.* 49:1115–56

147. Christman, J. K., Price, P., Pedrinan, L., Acs, G. 1977. *Eur. J. Biochem.* 81:53–61
148. McGhee, J. D., Ginder, G. D. 1979. *Nature* 280:419–20
149. Shen, C.-K. J., Maniatis, T. 1980. *Proc. Natl. Acad. Sci. USA* 77:6634–38
150. Van der Ploeg, L. H. T., Flavell, R. A. 1980. *Cell* 19:947–58
150a. Smith, S. S., Yu, J. C., Chen, C. W. 1982. *Nucleic Acids Res.* 10:4305–20
151. Mandel, J. L., Chambon, P. 1979. *Nucleic Acids Res.* 7:2081–103
151a. Weintraub, H., Larsen, A., Groudine, M. 1981. *Cell* 24:333–44
152. Desrosiers, R. C., Mulder, C., Fleckenstein, B. 1979. *Proc. Natl. Acad. Sci. USA* 76:3839–43
153. Cohen, J. C. 1980. *Cell* 19:653–62
154. Guntaka, R. V., Rao, P. Y., Mitsialis, S. A., Katz, R. 1980. *J. Virol.* 34:569–72
155. Vardimon, L., Günthert, U., Doerfler, W. 1983. *Mol. Cell. Biol.* 2:1574–80
156. Doerfler, W., Vardimon, L., Kruczek, I., Eick, D., Kron, B., Kuhlmann, I. 1983. In *Gene Transfer and Cancer*, ed. N. L. Sternberg, M. L. Pearson. New York: Raven
157. McKeon, C., Ohkubo, H., Pastan, I., de Crombrugghe, B. 1982. *Cell* 29:203–10
157a. Gerber-Huber, S., May, F. E. B., Westley, B. R., Felbar, B. K., Hosbach, H. A., et al. 1983. Submitted for publication
158. Constantinides, P. G., Jones, P. A., Gevers, W. 1977. *Nature* 267:364–66
159. Kintner, C., Sugden, B. 1981. *J. Virol.* 38:305–16
160. Harbers, K., Schnieke, A., Stuhlmann, H., Jähner, D., Jaenisch, R. 1981. *Proc. Natl. Acad. Sci. USA* 78:7609–13
161. Breznik, T., Cohen, C. 1982. *Nature* 295:255–57
162. Conklin, K. F., Coffin, J. M., Robinson, H. L., Groudine, M., Eisenman, R. 1982. *Mol. Cell. Biol.* 2:638–52
163. Hynes, N. E., Rahmsdorf, U., Kennedy, N., Fabiani, L., Michalides, R., et al. 1981. *Gene* 15:307–17
164. Stuhlmann, H., Jähner, D., Jaenisch, R. 1981. *Cell* 26:221–32
165. Desrosiers, R. C. 1982. *J. Virol.* 43:427–35
166. Katz, R. A., Mitsialis, S. A., Guntaka, R. V. 1983. *J. Gen. Virol.* In press
167. Jähner, D., Stuhlmann, H., Stewart, C. L., Harbers, K., Löhler, J., et al. 1982. *Nature* 298:623–28
168. Nakhasi, H. L., Lynch, K. R., Dolan, K. P., Unterman, R. D., Feigelson, P. 1981. *Proc. Natl. Acad. Sci. USA* 78:834–37
169. Yagi, M., Koshland, M. E. 1981. *Proc.*

Natl. Acad. Sci. USA 78:4907–11
170. Rogers, J., Wall, R. 1981. *Proc. Natl. Acad. Sci. USA* 78:7497–501
171. Hjelle, B. L., Phillips, J. A., III, Seeburg, P. H. 1982. *Nucleic Acids Res.* 10:3459–74
172. Marcaud, L., Reynaud, C.-A., Therwath, A., Scherrer, K. 1981. *Nucleic Acids Res.* 9:1841–51
173. Waalwijk, C., Flavell, R. A. 1978. *Nucleic Acids Res.* 5:4631–41
174. Subramanian, K. N. 1982. *Nucleic Acids. Res.* 10:3475–86
175. Gruenbaum, Y., Cedar, H., Razin, A. 1981. *Nucleic Acids Res.* 9:2509–15
176. Sturm, K. S., Taylor, J. H. 1981. *Nucleic Acids Res.* 9:4537–46
177. Sobieski, D. A., Eden, F. C. 1981. *Nucleic Acids Res.* 9:6001–15
178. Pages, M., Roizes, G. 1982. *Nucleic Acids Res.* 10:565–76
179. Ehrlich, M., Gama-Sosa, M. A., Huang, L.-H., Midgett, R. M., Kuo, K. C., McCune, R. A., Gehrke, C. 1982. *Nucleic Acids Res.* 10:2709–21
180. Bird, A. P., Southern, E. M. 1978. *J. Mol. Biol.* 118:27–47
181. Bird, A. P., Taggart, M. H. 1980. *Nucleic Acids Res.* 8:1485–97
182. Bird, A. P., Taggart, M. H., Smith, B. A. 1979. *Cell* 17:889–901
183. Bird, A., Taggart, M., Macleod, D. 1981. *Cell* 26:381–90
184. Bird, A. P., Taggart, M. H., Gehring, C. A. 1981. *J. Mol. Biol.* 152:1–17
185. La Volpe, A., Taggart, M., Macleod, D., Bird, A. 1982. *Cold Spring Harbor Symp. Quant. Biol.* 47:585–92
186. Kaye, A. M., Winocour, E. 1967. *J. Mol. Biol.* 24:475–78
187. Low, M., Hay, J., Keir, H. M. 1969. *J. Mol. Biol.* 46:205–7
188. Soeda, E., Arrand, J. R., Smolar, N., Griffin, B. E. 1979. *Cell* 17:357–70
189. Sutter, D., Westphal, M., Doerfler, W. 1978. *Cell* 14:569–85
190. Willis, D. B., Granoff, A. 1980. *Virology* 107:250–57
191. Ortin, J., Scheidtmann, K. H., Greenberg, R., Westphal, M., Doerfler, W. 1976. *J. Virol.* 20:355–72
192. Schirm, S., Doerfler, W. 1981. *J. Virol.* 39:694–702
193. Esche, H. 1982. *J. Virol.* 41:1076–82
193a. Esche, H., Siegmann, B. 1982. *J. Gen. Virol.* 60:99–113
194. Kuhlmann, I., Achten, S., Rudolph, R., Doerfler, W. 1982. *EMBO J.* 1:79–86
195. Kuhlmann, I., Doerfler, W. 1983. In preparation
196. Groneberg, J., Doerfler, W. 1979. *Int. J. Cancer* 24:67–74

197. Groneberg, J., Sutter, D., Soboll, H., Doerfler, W. 1978. *J. Gen. Virol.* 40:635–45
198. Eick, D., Stabel, S., Doerfler, W. 1980. *J. Virol.* 36:41–9
199. Groudine, M., Eisenman, R., Weintraub, H. 1981. *Nature* 292:311–17
199a. McGhee, J. D., Wood, W. I., Dolan, M., Engel, J. D., Felsenfeld, G. 1981. *Cell* 27:45–55
200. Smith, S. S., Reilly, J. G., Thomas, C. A., Jr. 1981. *ICN-UCLA Symp. Mol. Cell. Biol.*, ed. D. Brown, C. F. Fox, 23:635–45
200a. Urieli-Shoval, S., Gruenbaum, Y., Sedat, J., Razin, A. 1982. *FEBS Lett.* 146:148–52
201. Friedman, S. 1979. *Biochem. Biophys. Res. Commun.* 89:1328–33
202. Lu, L.-J. W., Randerath, K. 1980. *Cancer Res.* 40:2701–5
203. Landolph, J. R., Jones, P. A. 1982. *Cancer Res.* 42:817–23
204. Benedict, W. F., Banerjee, A., Gardner, A., Jones, P. A. 1977. *Cancer Res.* 37:2202–8
205. Marquardt, H. 1977. *Cancer* 40:1930–34
206. Halle, S. 1968. *J. Virol.* 2:1228–29
207. Constantinides, P. G., Taylor, S. M., Jones, P. A. 1978. *Dev. Biol.* 66:57–71
208. Taylor, S. M., Jones, P. A. 1979. *Cell* 17:771–79
209. Jones, P. A., Taylor, S. M. 1980. *Cell* 20:85–93
210. Taylor, S. M., Jones, P. A. 1982. *J. Cell Physiol.* 111:187–94
211. Sager, R., Kovac, P. 1982. *Proc. Natl. Acad. Sci. USA* 79:480–84
212. Mohandas, T., Sparkes, R. S., Shapiro, L. J. 1981. *Science* 211:393–96
213. Mohandas, T., Sparkes, R. S., Hellkuhl, B., Grzeschik, K. H., Shapiro, L. J. 1980. *Proc. Natl. Acad. Sci. USA* 77:6759–63
214. Jones, P. A., Taylor, S. M., Mohandas, T., Shapiro, L. J. 1982. *Proc. Natl. Acad. Sci. USA* 79:1215–19
215. Brown, M. M. M., Clive, D. 1977. *Mutat. Res.* 53:116
216. Harris, M. 1982. *Cell* 29:483–92
217. Compere, S. J., Palmiter, R. D. 1981. *Cell* 25:233–40
218. Clough, D. W., Kunkel, L. M., Davidson, R. L. 1982. *Science* 216:70–73
219. Stewart, C. L., Stuhlmann, H., Jähner, D., Jaenisch, R. 1982. *Proc. Natl. Acad. Sci. USA* 79:4098–102
220. Eick, D., Doerfler, W. 1982. Unpublished data
221. McGeady, M. L., Ascione, R., van de Woude, G. F. 1982. *Cold Spring Harbor Tumor Virus Meeting* (Abstr.)
222. Danner, D. B., Pifer, M. L. 1982. *Gene* 18:101–5
223. Gurdon, J. B., Lane, C. D., Woodland, H. R., Marbaix, G. 1971. *Nature* 233:177–82
224. McKnight, S. L., Gavis, E. R., Kingsbury, R., Axel, R. 1981. *Cell* 25:385–98
225. Vardimon, L., Doerfler, W. 1981. *J. Mol. Biol.* 147:227–46
226. Vardimon, L., Renz, D., Doerfler, W. 1983. In *Recent Results in Cancer Res.* Heidelberg: Springer. 84:90–102
227. Doerfler, W. 1983. *Cur. Top. Microbiol. Immunol.* In press
228. Fradin, A., Manley, J. L., Prives, C. L. 1982. *Proc. Natl. Acad. Sci. USA* 79:5142–46
229. Eden, F. C., Musti, A. M., Sobieski, D. A. 1981. *J. Mol. Biol.* 148:129–51
230. Drahovsky, D., Boehm, T. L. J., Kreis, W. 1979. *Biochim. Biophys. Acta* 563: 28–35
231. Lepeyre, J. N., Becker, F. F. 1979. *Biochem. Biophys. Res. Commun.* 87:698–705
232. Burton, W. G., Grabowy, C. T., Sager, R. 1979. *Proc. Natl. Acad. Sci. USA* 76:1390–94
233. Royer, H. D., Sager, R. 1979. *Proc. Natl. Acad. Sci. USA* 76:5794–98
233a. Sano, H., Royer, H.-D., Sager, R. 1980. *Proc. Natl. Acad. Sci. USA* 77:3581–85
234. Sano, H., Grabowy, C., Sager, R. 1981. *Proc. Natl. Acad. Sci. USA* 78:3118–22
235. Sager, R., Grabowy, C., Sano, H. 1981. *Cell* 24:41–47
236. Bolen, P. L., Grant, D. M., Swinton, D., Boynton, J. E., Gillham, N. W. 1982. *Cell* 28:335–43
237. Lyon, M. F. 1961. *Nature* 190:372–73
238. Venolia, L., Gartler, S. M., Wassman, E. R., Yen, P., Mohandas, T., Shapiro, L. J. 1982. *Proc. Natl. Acad. Sci. USA* 79:2352–54
239. Wolf, S. F., Migeon, B. R. 1982. *Nature* 295:667–71
240. Migeon, B. R., Wolf, S. F., Mareni, C., Axelman, J. 1982. *Cell* 29:595–600
241. Korba, B. E., Hays, J. B. 1982. *Cell* 28:531–41
242. Hammer, S. M., Richter, B. S., Hirsch, M. S. 1981. *J. Immunol.* 127:144–48
243. Youssoufian, H., Hammer, S. M., Hirsch, M. S., Mulder, C. 1982. *Proc. Natl. Acad. Sci. USA* 79:2207–10
244. Adler, J. 1975. *Ann. Rev. Biochem.* 44:341–56
245. Koshland, D. E., Jr. 1979. *Physiol. Rev.* 59:811–62
246. Paik, W. K., Kim, S. 1980. *Protein Methylation.* New York: Wiley. 282 pp.
247. Ley, T. J., DeSimone, J., Anagnou, N. P., Keller, G. H., Humphries, P. K., Turner, P. H., et al. 1982. *N. Engl. J. Med.* 307:1469—75
248. Wilson, V. L., Jones, P. A. 1983. *Cell* 32:239–46

Ann. Rev. Biochem. 1983. 52:125–57
Copyright © 1983 by Annual Reviews Inc. All rights reserved.

COMPARATIVE BIOCHEMISTRY OF PHOTOSYNTHETIC LIGHT-HARVESTING SYSTEMS

A. N. Glazer

Department of Microbiology and Immunology, University of California, Berkeley, California 94720

CONTENTS

PERSPECTIVES AND SUMMARY 125
ANOXYGENIC PHOTOSYNTHESIS 127
   Green Bacteria 127
   Purple Bacteria 132
OXYGENIC PHOTOSYNTHESIS 135
   Green Algae and Higher Plants 136
   Cyanobacteria (Blue-Green Algae) 145
   Prochlorophyta 147
   Algal Light-Harvesting Photosynthetic Accessory Pigments 147
CONCLUDING REMARKS 152

PERSPECTIVES AND SUMMARY

In photosynthetic organisms, light provides the energy for the production of ATP and the generation of reducing power. The first step in this process is the absorption of photons by an array of light-harvesting protein-pigment complexes ("antenna complexes"). The energy migrates from one pigment molecule in the antenna to the next in $\sim 10^{-13}$ s and arrives at a photoactive center, the reaction center, after about 10^{-11} s. The reaction center is contained within an integral membrane protein of the photosynthetic lamellae. Excitation of a special pigment species (D, donor) within the reaction center leads to a charge separation with the formation of D^+A^-, where A^- is called the primary acceptor. Electron transfer from A^- to secondary electron acceptor species occurs at a faster rate than charge recombination to $D \cdot A$, which would lead to the dissipation of the excitation energy. Hence, energy

125

trapping in the reaction center is essentially irreversible. Appropriate positioning of the electron acceptors results in electron transport in one direction across the membrane and proton transport in the opposite direction with consequent generation of an electric field across the membrane. Collapse of this electrochemical gradient by proton transport through the membrane-bound coupling factor ATPase leads to the formation of ATP. Depending on the particular reaction center, the overall electron flow may be cyclic, essentially when D^+ is reduced by recapture of its own electron after the electron's passage through a chain of electron carriers, or noncyclic, when the electron is captured by an acceptor such as ferredoxin or NADPH, and D^+ is reduced by an electron from an external donor, such as water in photosystem II of green plants.

Bacteriochlorophyll (BChl) species function as reaction center pigments in green and purple bacteria; chlorophyll (Chl) species play this role in cyanobacteria (blue-green algae), the prochlorophytes, and in all photosynthetic eukaryotes. Much more information is now available on the spectroscopic properties of reaction center species than on their chemistry. Hopefully, this situation will soon change with the recent crystallization of the photosynthetic reaction center from *Rhodopseudomonas viridis* in a form diffracting to beyond 2.5 Å (1).

Green and purple bacteria perform anoxygenic photosynthesis, so named because this photochemical process does not lead to the production of O_2. In these bacteria only one type of reaction center is present. The D^+/D redox potential for the reaction center pigment of green bacteria, P_{840}, lies at $+0.25$ V (2), that for P_{870} of purple bacteria at 0.44 V (3). ATP is produced as a direct consequence of photosynthesis in these organisms. The modes of NADPH generation remain to be established definitively (4).

Organisms performing oxygenic photosynthesis invariably possess two photosystems, photosystem I (PSI) and photosystem II (PSII). The reaction center of PSII, P_{680}, has a redox potential of $> +0.81$ V. P_{680}^+ is a powerful oxidant and abstracts electrons from water with the evolution of oxygen. Excitation of P_{700}, the reaction center of PSI with a redox potential of $\sim +0.5$ V (5), mediates a cyclic electron flow resulting in ATP synthesis. It is generally believed that when PSII and PSI are excited simultaneously, electrons abstracted by PSII from water are transported to PSI via plastoquinone to replace PSI electrons used for the reduction of ferredoxin and thence NADP. The noncyclic electron flow between PSII and PSI is accompanied by ATP synthesis. This series of events is depicted by the classical Z scheme of photosynthetic electron transport (6). Evidence for an alternative scheme has been presented by Arnon and colleagues (7, 8), in which the reduction of ferredoxin and NADP, as well as ATP generation, are performed by PSII. The sole function of PSI is envisaged to be production of ATP by cyclic photophosphorylation. In this formulation PSI is seen to be closely analogous to the single anoxygenic type of photosystem possessed by the green and purple bacteria.

This review focuses on the structure of the antennae in photosynthetic organisms. These antennae vary greatly in size. At one extreme are those of the green bacteria with as many as 2000 molecules of light-harvesting BChl per reaction center; at the other are those of the purple bacteria with as few as 60. In the middle lie the PSII and PSI of green plants with 200–300 pigment molecules per reaction center. Whereas the reaction center complexes are invariably contained within membranes, antenna complexes are not always intrinsic membrane components. The antennae of green bacteria, cyanobacteria, and cryptomonads form structures attached to the outside of the thylakoid membrane; the location of the water-soluble peridinin-Chl a-protein complexes of dinoflagellates remains to be determined. The available information on the components of these antennae and their modulation by environmental conditions is considered here.

Several recent findings modify drastically our view of the structure of the photosynthetic apparatus. These include: (a) the relative ratios of PSI and PSII vary over a factor of ~ 8 in different photosynthetic membranes; (b) PSI is largely if not completely excluded from the grana partition regions of higher plant chloroplasts, whereas PSII is greatly enriched in these regions; (c) Chl b, traditionally assigned to PSII, is a significant component of the antenna of PSI as well, (d) the role of carotenoids as light-harvesting pigments is receiving greater appreciation; and (e) chemically distinct chlorophylls additional to the classical Chl a and Chl b are present in light-harvesting antenna and reaction center complexes.

ANOXYGENIC PHOTOSYNTHESIS

Green Bacteria

The green bacteria are classified in two families: the Chlorobiaceae (which includes green- and brown-colored species) and the Chloroflexaceae. The Chlorobiaceae are strict photolithotrophs and obligate anaerobes, whereas the Chloroflexaceae are photoorganotrophs and facultative anaerobes. The two families have in common an unusual organization of the photosynthetic apparatus: the majority of the antenna pigments are located in chlorosomes (9) [formerly chlorobium vesicles (10)] which are ovoid bag-like structures appressed to the cytoplasmic membrane, whereas the reaction centers are located in the cytoplasmic membrane (11). Green-colored Chlorobiaceae possess light-harvesting BChl c or d and the carotenoids chlorobactene and OH-chlorobactene, whereas the brown-colored species contain light-harvesting BChl e and the carotenoids isorenieratene and β-isorenieratene. The

Chloroflexaceae contain light-harvesting BChl *c* and γ and β-carotene (12). All species contain BChl *a* as reaction center pigment with an associated small BChl *a* antenna (11). The organization of the light-harvesting machinery has been examined in members of each family: *Chlorobium limicola* (13) and *Prosthecochloris aestuarii* (Chlorobiaceae) and *Chloroflexus aurantiacus* (Chloroflexaceae) (13–15). Fowler et al (2) showed that the photosynthetic units of the Chlorobiaceae contain 1000–1500 BChl *c, d,* or *e,* and about 100 BChl *a* molecules per reaction center. This enormous antenna enables these green bacteria to grow well in weak light. Clayton (16) observed, "A culture growing in a 30-liter jug, illuminated only by a 25-W tungsten lamp, soon resembles liquid spinach." In *C. aurantiacus* the size of the photosynthetic unit is of the same order of magnitude (<300) as that found in purple bacteria (9).

Detailed electron microscopic studies by Staehelin et al (13) have revealed structural features of the chlorosome of *C. limicola* illustrated in Figure 1. These oblong structures vary in size from 40 × 70 nm to 100 × 260 nm and exhibit no particular mutual orientation. The chlorosome core is surrounded by a ～3 nm thick featureless single-layered envelope. Rod elements, 10 nm in diameter, 10–30 in number, are closely packed within the core and extend to the full length of the chlorosome. A crystalline baseplate (5–6 nm thick) connects the chlorosome to the cytoplasmic membrane. The ridges of the baseplate have a repeating distance of 6 nm and a granular substructure with a periodicity of ～3.3 nm. A distinctive feature of the chlorosome attachment site is the presence of 20–30 very large

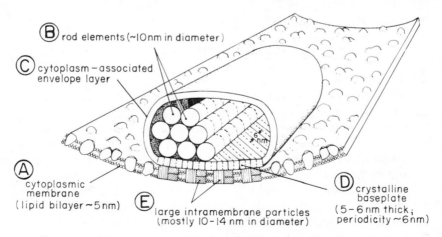

Ⓑ rod elements (~10nm in diameter)

Ⓒ cytoplasm – associated envelope layer

Ⓐ cytoplasmic membrane (lipid bilayer ~5nm)

Ⓔ large intramembrane particles (mostly 10-14 nm in diameter)

Ⓓ crystalline baseplate (5-6 nm thick; periodicity ~6nm)

Figure 1 Model of a *Chlorobium limicola* chlorosome and its associated cytoplasmic membrane, based primarily on results of freeze-fracture electron microscopy. Redrawn with minor modifications from Figure 15 of Ref. (13).

($\sim$12.5 nm diameter) intramembrane particles (13). The chlorosome of *C. aurantiacus* (Chloroflexaceae) is similar to that of *Chlorobium,* but smaller—35 nm wide, 100 nm long, $\sim$12 nm high (17).

Feick et al (18) recently devised a procedure for the parallel isolation of cytoplasmic membrane and purified chlorosome fractions from *C. aurantiacus* with no cross-contamination. In accord with earlier results on other green bacteria (see 11), reaction center BChl *a* and antenna BChl *a* (806 nm, 865 nm) species were found in the cytoplasmic membrane. Since 10–14% of the 865-nm peak was photobleachable, this peak is attributable both to an antenna BChl *a* species and to reaction center BChl *a*. No BChl *c* was found in the cytoplasmic membrane. The chlorosome fraction showed absorption maxima at 450 nm (contributed by carotenoids and BChl *a* Soret band), 742 nm (BChl *c*), and 792 nm (BChl *a*), the last two at the intensity ratio of 25:1 (18). Spectroscopic studies indicated that the BChl *c* was highly organized, with transition dipoles more or less parallel to the long axis of the chlorosome, and transferred energy efficiently to the 792-nm BChl *a* (19). The energy transfer sequence in *C. aurantiacus* is: 742 nm BChl *c* → 792-nm BChl *a* → 805-nm BChl *a* → 865-nm BChl *a* → 865-nm BChl *a* reaction center (19).

Purified chlorosomes have a surprisingly simple polypeptide composition; only three components, of 15, 12, and 6 kilodaltons (kD), present in comparable amounts were detected (18). An 8-kD polypeptide was a prominent component of the cytoplasmic membrane in phototrophically but not in chemotrophically grown cells. This polypeptide is absent from purified chlorosomes and consequently is a candidate for a specific antenna BChl *a*-associated protein in the cytoplasmic membrane (18).

The foregoing results are consistent with the following tentative assignment of specific components of the photosynthetic apparatus to the morphological entities shown in Figure 1 (13). The rod structures within the chlorosome core are assumed to be BChl *c*-protein complexes whereas the crystalline baseplate is believed to be made up of a BChl *a*-protein complex. Betti et al (19) suggest that in *C. aurantiacus,* the chlorosome-associated 792-nm BChl *a* may form the baseplate, but it is not evident that the BChl *a* component(s) absorbing at 805 and 865 nm could not fulfill that role. The large particles (12–14 nm) in the region of the cytoplasmic membrane underlying the baseplate are proposed to be complexes containing a reaction center associated with noncrystalline light-harvesting BChl *a* antenna (11). Since there are numerous such particles in contact with each baseplate, Staehelin et al (13) proposed two functions for the baseplate: transfer of excitation energy to reaction centers and distribution of this energy among the reaction centers.

When cells of the chlorobiacean green bacteria, *Chlorobium* and *Pros-*

thecochloris, are broken by sonication at alkaline pH ($\sim$pH 10), a water-soluble BChl *a*-protein with an absorption maximum at 809 nm is released (20). This protein can be readily crystallized and its structure has been determined at 2.8 Å resolution by X-ray diffraction (21, 22). Since this is the only light-harvesting protein whose structure is known at high resolution, its properties will be discussed in some detail. The protein from *P. aestuarii* is a trimer of $\sim$135 kD made up of three identical subunits tightly packed around a three-fold symmetry axis. The trimer cannot be dissociated into monomers without using denaturing conditions. Each subunit is made up of about 354 amino acid residues and seven BChl *a* molecules, adding up to a total molecular weight of 45,000. The polypeptide chain of each subunit forms an extensive 15-strand β-sheet that wraps around the BChl *a* molecules shielding them effectively from the external aqueous environment (Figure 2).

The position and orientation of each BChl *a* molecule are governed mainly by specific interactions with the protein rather than those with other BChl molecules. BChl-BChl interactions are limited to hydrophobic interactions between phytyl tails. The phytyl tails constitute an inner hydrophobic core of the molecule. Inspection of the structure indicates that the folding of the subunit must be a cooperative process involving both the protein and the bacteriochlorophyll. The Mg^{2+} atoms of five of the seven bacteriochlorophylls are ligated to histidyl residues, one to a water molecule, and one to a main chain carbonyl oxygen. The average center-to-center distance between porphine rings is 12 Å for nearest neighbors. The seven BChl *a* chromophores are so close together that they act as a single unit in absorbing and emitting photons. The closest distance between bacteriochlorophylls in adjacent subunits is 24 Å. Even at this distance, significant strong interactions between certain bacteriochlorophylls in adjacent subunits are expected to produce pertubations in their spectra (22). Olson (11) proposed that the crystalline chlorosome baseplate in *Chlorobium,* described by Staehelin et al (13; see above, and Figure 1), is a two-dimensional crystal in the trigonal space group $P3_1$ of the BChl *a*-protein.

Matthews et al (22) believe that the *Prosthecochloris* BChl *a*-protein provides a generally applicable model for the organization of chlorophyll in photosynthetic membranes, and this plausible view has gained much acceptance. However, it is likely that the *Prosthecochloris* BChl *a*-protein typifies but one kind of a chlorophyll-protein complex. For example, Feick et al (18) reported that isolated *Chloroflexus* chlorosomes contain 600 nmol BChl *c*/mg of protein. If one were to assume that all of the protein in the chlorosome were involved in interaction with BChl *c,* this ratio would require the binding of 24 BChl *c* molecules per 40,000 g of protein as compared to the seven BChl *a* contained within each 38,600 dalton subunit

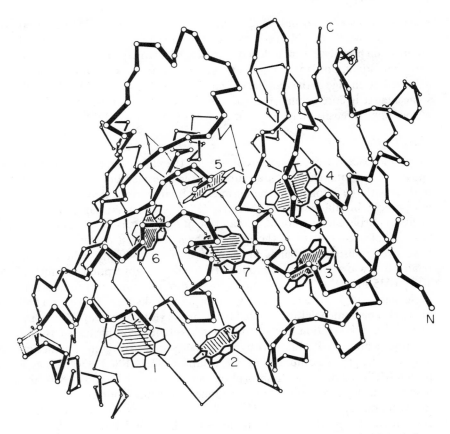

Figure 2   Polypeptide backbone of the *Prosthecochloris aestuarii* bacteriochlorophyll *a*-protein subunit viewed from the 3-fold axis, which is horizontal, toward the exterior of the protein. For clarity the phytyl chains and other bacteriochlorophyll ring substituents of the seven bacteriochlorophylls enfolded within the subunit have been omitted. Reproduced with permission from B. W. Matthews et al (22).

of the trimeric BChl *a*-protein. A similar BChl *c*/protein ratio was obtained in an earlier study in a different green bacterium by Cruden & Stanier (23). The manner in which the BChl *c* is organized is clearly different from that determined for BChl *a*.

The BChl *c* content of green bacterial cells varies inversely with the light intensity under which the cells are grown (9). For *C. aurantiacus,* the BChl *c* content is ～40 times higher in cells grown at 240 lux than in those grown at 54,000 lux (9). However, the ratios of BChl *c*/BChl *a* (absorbing at 790 nm) of isolated chlorosomes, of BChl *a*/reaction center in the cytoplasmic membrane, and of reaction center/mg of cytoplasmic membrane protein remained constant and independent of light intensity (18). The measured length and width of chlorosomes and the area of the two-dimensional chlorosome baseplate were independent of light intensity. These data establish that adaptation to changes in light intensity in *C. aurantiacus* is achieved by change in the number of chlorosomes (18, see also 24), although some change in core volume, evidenced by increase in chlorosome thickness at low light intensity, has not been rigorously excluded (15).

Purple Bacteria

In purple bacteria grown anaerobically in the light, the photosynthetic apparatus is contained within an intracytoplasmic membrane system continuous with the cytoplasmic membrane. Such infoldings exhibit distinctive morphology in different purple bacteria (25–27). The light harvesting pigments in purple bacteria are BChl *a* and BChl *b* and a variety of carotenoids (28). Whole cell spectra of those purple bacteria which produce BChl *a* have strong absorption bands in the region of 800–870 nm, whereas the relatively few organisms that produce BChl *b* show absorption maxima at 1015–1035 nm (27). The carotenoids vary from one organism to another; these pigments give rise to distinctive absorption peaks between 420 and 560 nm (27). The nomenclature adopted for light-harvesting protein-BChl complexes and for reaction centers designates antenna complexes as B and reaction centers as P, followed by a number(s) approximating the actual long wavelength maximum(a) of the complex. Thus B800–850 defines a protein-BChl complex with absorption maxima at ~800 and ~850 nm. Some purple bacteria contain only one type of antenna complex in addition to a reaction center, such as B875 in *Rhodospirillum rubrum,* while others contain two different antenna complexes e.g. B875 and B800-850 in *Rhodopseudomonas sphaeroides* (see Table 1; 29). Purple bacteria adapt to variation in incident light intensity by varying (*a*) the amount of intracytoplasmic membrane, (*b*) the number of photosynthetic units per given area of membrane, and (*c*) the size of the photosynthetic unit (total amount of BChl per photochemical reaction center). Response (*c*) is not seen in purple bacteria with only one major antenna complex.

The structure of the reaction center complex has been examined most extensively in *R. sphaeroides* (5, 30–33, 38). The reaction center complex is made up of three polypeptide subunits: L, M, and H. Chemical modification and immunological studies have shown that all three polypeptides span the membrane (5, 39). The H subunit can be removed by treatment of the complex with lauryldimethylamine oxide in $LiClO_4$; the residual LM complex retains photochemical activity and all of the pigments (5). The H subunit plays a role in the binding of a secondary quinone that serves as an electron acceptor from the primary quinone bound tightly to the reaction center complex (see Table 1) (40).

Table 1 Properties of the reaction center and light-harvesting components in *R. sphaeroides*

Reaction center complex (5, 30–33)	
λ_{max}	870, 800, 760 nm
Molecular weight	~95,000
Polypeptide composition	L, M, H in a molar ratio of 1:1:1
Molecular weight SDS-PAGE[a]	L 21,000, M 24,000, H 28,000
amino acid analysis	L 28,000, M 32,000, H 34,000
Other constituents	4 BChl *a*, 2 BPheo, 1 carotenoid[b], 1 ubiquinone (tightly bound), 1 Fe^{2+}
B875 Light-harvesting complex	
λ_{max}	875 nm
Polypeptide composition and molecular weight[c]	LH1 12,000; LH3 8,000
Other constituents	2 BChl *a*, 2 carotenoid
B800–950 Light-harvesting complex	
λ_{max}	850, 800 nm
Polypeptide composition and molecular weight[c]	LH2 10,000; LH3 8,000
Other constituents	2 BChl *a*, 2 carotenoid

[a] Abbreviations used are SDS-PAGE, polyacrylamide gel electrophoresis in presence of SDS; BChl *a*, bacteriochlorophyll *a*, BPheo, bacteriophaeophytin; LH, Light-harvesting polypeptide.

[b] In several strains of *R. sphaeroides*, the major reaction center carotenoid is spheroidene (34–36). It is interesting to note that in *R. sphaeroides* Ga, the reaction center carotenoid is chloroxanthin, while the major carotenoid in this organism is neurosporene (36).

[c] Determined by SDS-PAGE (37).

During formation of photosynthetic membranes, reaction center complexes with their associated B875 antennae are formed first and the B800–850 complexes added subsequently (41–43). In *R. sphaeroides* Ga, for example, B875 is always present at a fixed molar ratio of ~30 moles B875 BChl per mole of reaction center, whereas the amount of B800–850 varies inversely with the light intensity to which the cells are exposed, reaching ~300 moles B800–850 BChl per mole of reaction center (44). From these data, and from the compositions of B800–850 and B875 (Table 1), the maximum size of the photosynthetic unit, which contains ~300 BChl per reaction center protein, in this organism is ~2.5 × 10^6.

All of the BChl complexes contain carotenoids as well. The carotenoids protect the photosynthetic apparatus from photo-oxidative damage by quenching triplet BChl and singlet oxygen, $^1O_2$ (36). Carotenoidless mutants are killed by exposure to light and oxygen (45). The carotenoids also contribute to the harvesting of light energy between 400 and 550 nm. Fluorescence measurements on B875 and B800–850 indicate that the energy absorbed by the carotenoids is transferred to BChl *a* with an efficiency of 55–70% (46). For comparison, the quantum efficiency of photochemistry (electrons transferred per quantum absorbed) is >98% for isolated *R.*

sphaeroides reaction centers and ~90% for light absorbed by antenna BChl *a* in photosynthetic membrane fragments (16, 47).

How is the light-harvesting apparatus organized in purple bacteria? Upon mechanical disruption of *R. sphaeroides* cells, most of the photosynthetic apparatus is recovered in vesicles called chromatophores (48) that are derived from the intracytoplasmic membranes. Dissociation of chromatophores in cold lithium dodecylsulfate and subsequent polyacrylamide gel electrophoresis leads to the separation of a photoactive LM complex, B800–850 and B875 antenna complexes, and a family of higher complexes with increasing B800–850/B875 ratios (37). Spectroscopic measurements indicate a pathway of energy transfer to be B800–850 → B875 → reaction center, and suggest a "lake" arrangement for these complexes; B875 complexes are thought to surround and interconnect several reaction centers, and the B875 complexes are in turn surrounded by arrays of B800–850 complexes (49, 50).

In *R. rubrum,* which contains a single antenna complex, there are only ~30 BChl per photochemical reaction center as compared to ~300 in *R. sphaeroides,* and the organization appears to be simpler. It should be noted, however, that picosecond difference spectroscopy indicates that the antenna system in *R. rubrum* contains ~5% of a BChl with an absorption maximum at ~905 nm. The B905 minor fraction accepts energy from B875 (51). A 7×10^5 dalton complex, corresponding to a "photoreaction unit," has been purified from chromatophores solubilized by a mixture of cholate and deoxycholate (52). The complex showed a reaction center activity equivalent to that of chromatophores and contained 33 BChl per photochemical reaction center. It was made up of ~10 different polypeptide species with molecular weights ranging from 38,000 to 10,000, but contained only a very small amount of phospholipid (52). X-Ray diffraction studies indicate that the arrangement of such photoreaction units in chromatophore membranes is random (53).

The BChl *a* complexes of *Rhodopseudomonas capsulata* have also been examined carefully. The properties of the reaction center complex from this organism are similar to those of *R. sphaeroides* reaction center (54). The B875 antenna complex, present in a molar ratio of 25:1 to the reaction center, contains one BChl *a* and one carotenoid per polypeptide of 12,000 (55, 56). Only one type of polypeptide appears to be present in this complex. The B800–850 antenna complex contains three different polypeptides in a molar ratio of 1:1:1 (57). The apparent molecular weights of these polypeptides, as determined by sodium dodecylsulfate (SDS)-polyacrylamide gel electrophoresis, are 14, 10, and 8 kD, respectively (58). However, calculations based on amino acid analyses indicate lower molecular weights of 12, 9.3, and 5.1 kD respectively (57). These data suggest that the smallest polypeptide exists as a dimer in SDS solutions. No pigment is associated with the largest polypeptide, and its role is not known. Two BChl a appear to be bound to the 10-kD polypeptide and one BChl a and one carotenoid to the smallest polypeptide (55, 59). The organization of this complex in the membrane has yet to be established, but the smallest aggregate that retains the native spectroscopic characteristics is either a trimer or tetramer (57).

To date the sequence of only one light-harvesting BChl protein has been determined—that of the B870 protein purified from a carotenoidless mutant of *R. rubrum* G-9 (60). This polypeptide, which may account for up to 50% of chromatophore membrane protein (61), is soluble in chloroform/methanol (1:1), and this unusual property aids in its purification (60). The sequence of this 52-residue (M_r 6106) polypeptide is:

5   10   15

N^f-Met-Trp-Arg-Ile-Trp-Gln-Leu-Phe-Asp-Pro-Arg-Gln-Ala-Leu-Val-
20   25   30

Gly-Leu-Ala-Thr-Phe-Leu-Phe-Val-Leu-Ala-Leu-Leu-Ile-His-Phe-
35   40   45

Ile-Leu-Leu-Ser-Thr-Glu-Arg-Phe-Asn-Trp-Leu-Glu-Gly-Ala-Ser-
50

Thr-Lys-Pro-Val-Gln-Thr-Ser-COOH

where N^f-Met is N-formylmethionine (62).

Interestingly, SDS-polyacrylamide gel electrophoresis gave an apparent molecular weight of 14 kD for this polypeptide, and sedimentation equilibrium analysis in 60% formic acid gave a value of 12 kD (60). It appears that this polypeptide exists as a dimer under these conditions. The wide discrepancy between the actual and apparent molecular weights of this membrane polypeptide indicates that accurate calculations of the physical size of supramolecular light-harvesting assemblies will have to await determination of the amino acid sequences of the constituent polypeptides.

OXYGENIC PHOTOSYNTHESIS

All photosynthetic eukaryotes perform oxygen-evolving photosynthesis and contain PSII and PSI. These properties are shared by two groups of prokaryotes, cyanobacteria (blue-green algae) and the prochlorophytes. The structure of the reaction center complexes appears to be conserved among all of these organisms. In contrast, there is a wide variety of antenna complexes with distinctive compositions and spectroscopic properties.

Therefore, the discussion of the photosynthetic apparatus of green algae and higher plants includes consideration of reaction center complexes, whereas discussion of that of cyanobacteria and various algae emphasizes antenna complexes.

Green Algae and Higher Plants

The number of PSI and PSII reaction centers and the size of the chlorophyll antennae surrounding these centers have been estimated for a variety of plants and algae from spectroscopic measurements on intact cells, chloroplasts, and membrane preparations. These measurements lead to two important observations. The relative numbers of PSI and PSII reaction centers vary widely in different plants (63, 64) and within the same plant grown under different conditions. For example, the ratio of reaction centers of PSI to those of PSII is ~0.7 in the spinach chloroplast and ~0.3 in developing pea leaves (63). In cyanobacteria, the ratio depends on the light conditions under which the cells are grown; ratios of the reaction centers of PSI to those of PSII approaching 4 have been determined (65, 66). The second observation is that the antenna size per reaction center varies significantly with culture conditions, or with the stage of chloroplast development (67). Over a hundred chlorophyll molecules are contained within the antenna for each PSI and PSII (68).

In green algae and green plants both the reaction centers and the antenna pigments are integral components of the thylakoid membranes. Thornber et al (69) and Ogawa et al (70) were the first to demonstrate that when such membranes were dissolved in anionic detergents and subjected to polyacrylamide gel electrophoresis, much of the chlorophyll coelectrophoresed with distinct polypeptides although some free chlorophyll was released. The conditions for both membrane solubilization and electrophoresis have been significantly modified since that time. In general, "improvement" has been defined as a decrease in the amount of chlorophyll migrating as free pigment (71–74).

Electrophoretic analyses have been performed on subchloroplast particles greatly enriched in PSII (75) or PSI (76, 77) activity, or on membranes from mutants deficient in one of the reaction center components (78–86), or lacking an antenna complex (83, 87–90). Such studies are all consistent with the view that chlorophyll in green plant photosynthetic membranes is localized within three macromolecular complexes: a PSI complex containing P700, a PSII complex containing P680, and a "light-harvesting chlorophyll a/b protein," also called LHC or LHCP (72, 74, 91–93). The degree to which the native complex structure is retained, and hence the composition of the components isolated, depends critically on the method of purification.

The results obtained with the SDS-polyacrylamide gel electrophoresis systems of Anderson (73, 74) and of Machold et al (83) are described briefly here as examples of widely used analyses. Table 2 shows the results obtained with the Anderson system. Six chlorophyll-binding complexes are seen in these gels: CP1, a PSI complex, and CP1a, a larger complex containing CP1, CP2, believed to be the RCII-containing core of PSII; LHCP[3], the light-harvesting chlorophyll a/b protein; LHCP[2] and LHCP[1], believed to be oligomers of LHCP[3].

The system of Machold et al (83) (see Table 3) resolves ten chlorophyll-protein bands and reveals a greater degree of complexity in the chlorophyll a/b-containing components. At least two different Chl a/b-polypeptides were resolved. These electrophoretic analyses reveal the minimum level of complexity of the protein-chlorophyll assemblies of thylakoid membranes.

PHOTOSYSTEM I   Dissociation of thylakoid membranes from a variety of sources with SDS under mild conditions results in the solubilization of a P700-Chl a protein (equivalent to CP1 or Chl_a-P1 in Tables 2 and 3) containing some 40 Chl a and 1–2 β-carotenes per P700 (77, 93, 97). This

Table 2 Chlorophyll-protein complexes of thylakoids from spinach (*Spinacia oleracea* L.) Comparison with corresponding complexes from barley thylakoids[a]

| | Chlorophyll distribution | | | | |
| | Spinach | | Barley | | |
Component[b]	Percent	$\dfrac{Chl\ a}{Chl\ b}$	Wild type (%)	Chlorina–f2[c] (%)	Comments on complexes from spinach
CP1a (Chl_a–P1*)	14	>20	6	—	P700-chlorophyll a-protein. Apoprotein has mol wt of 68,000
CP1 (Chl_aP1)	11	Chl a	17	28	As for CP1a
LHCP[1] ($Chl_{a/b}$–P2**)	12	1.09	9	—	Component of LHC
LHCP[2] ($Chl_{a/b}$–P2*)	10	1.14	7	—	Component of LHC
CPa	10	3.66	8	25	P680 reaction center complex (?). Apoprotein has mol wt of 42,500
LHCP[3] ($Chl_{a/b}$–P2)	31	1.28	37	—	Chl a/b-protein of LHC. Apoproteins have mol wts of 25,000 and 23,000
Free chlorophyll	11	Chl a	16	47	

[a] Data from refs. 95, 96.
[b] As resolved by the sodium dodecyl sulfate-polyacrylamide gel electrophoresis system of Anderson (73, 74). Designations in brackets refer to the corresponding barley thylakoid complexes described by Machold et al (83) (see Table 3).
[c] Mutant lacking chlorophyll b (87).

protein has been described as the "heart of photosystem I" (97). SDS-polyacrylamide gel electrophoresis of the fully denatured complex shows an apoprotein of ~ 70 kD (77); occasionally the apoprotein band is resolved into two polypeptides of very similar apparent molecular weights (98, 99). From cross-linking experiments, Nelson (cited in 5) concluded that the P700-Chl a protein contained one P700 per two ~ 70-kD polypeptides.

When thylakoid membranes are dissociated under mild conditions with nonionic detergents, much larger P700-containing complexes are isolated with up to 130 Chl a/P700 (76). One such complex, obtained from Swiss chard, containing ~ 80 Chl a/P700 has been well characterized by Nelson

Table 3 Chlorophyll proteins of thylakoids from barley (*Hordeum vulgare* L.)[a]

Chlorophyll proteins[b]	Complexes[c]	Apparent mol wt $\times 10^{-3}$	Comments
	Chl_a-P1**		
	Chl_a-P1**		
Chl_a-P1		110	P700[d]–Chl a-protein. Apoprotein has mol wt of 69,000
	$Chl_{a/b}-P2***$	107	Minor component in LHC[e] pattern
	$Chl_{a/b}-P2**$	71	Prominent component in LHC pattern
	$Chl_{a/b}-P2*$	50	Minor component in LHC pattern
Chl_a-P2		49	Apoprotein has mol wt of 45,000
Chl_a-P3		41	Apoprotein has mol wt of 40,000 P680 reaction center complex (?)
$Chl_{a/b}-P1$		32	Not found in mutant *chlorina*–f2 that lacks Chl b.
$Chl_{a/b}-P2$		29	Major component in LHC pattern. Major apoproteins have mol wts of 24,000 and 23,000. Not found in mutant *chlorina*–f2.

[a] As resolved by the sodium dodecylsulfate-polyacrylamide gel electrophoresis system of Machold et al (83, 86).

[b] The term "chlorophyll protein" is used to describe chlorophyll-containing bands of unique composition. Where several chlorophyll-containing bands can be shown to consist of identical polypeptides, the band with the lowest apparent molecular weight, assumed to be the monomer, is called a chlorophyll protein. The subscripts a or a/b denote the presence of Chl a or Chl a and b, respectively. For complete discussion of nomenclature, see (83).

[c] $Chl_{a/b}-P2*$, $-P2**$, and $-P2***$ are considered to be multimers of $Chl_{a/b}-P2$ on the basis of absorption spectra and polypeptide composition of these complexes and of purified LHC. Chl_a-P1** and $-P1*$ are higher order complexes of Chl_a-P1.

[d] P700 and P680 designate the reaction centers of PSI and PSII, respectively.

[e] LHC represents the light-harvesting Chl a/b-protein complex from barley purified by the method of Burke et al (94). This complex contains $\sim 50\%$ of the chlorophyll of wild-type thylakoids and most of the Chl b.

and co-workers (100). This complex was made up of six subunits, I–VI, with apparent molecular weights of 70, 25, 20, 18, 16, and 8 kD (100, 101). A molar ratio of 2:1:1:1:1 was proposed for subunits I, II, III, IV, and V. Subunit I was identical with the ~70-kD apoprotein discussed above; subunits IV, V, and VI were tentatively identified as iron-sulfur proteins and believed to correspond to the PSI electron acceptors X, and centers A and B, identified spectroscopically (for review see 77). A simpler PSI complex, isolated from barley thylakoids, contained 90–110 Chl a/P700. This complex was made up of the 100-kD CP1 (Table 2) and two iron-sulfur proteins of 18.3 and 15.2 kD (85). Mullet et al (76) developed a procedure for the preparation of particles enriched in PSI activity by very mild extraction of membranes with Triton X-100. Such particles from barley thylakoids contain 110 Chl a and b/P700 (PSI-110) and have a ratio of Chl a/Chl b equal to 18. When these particles are treated further with Triton, particles containing only Chl a bound to CP1, at a ratio of 65 Chl a/P700, are obtained and at least three polypeptides are lost from the complex. Analysis of a mutant of barley missing chlorophyll b showed the loss of six polypeptides from the thylakoid membranes; three of these could be assigned to the light-harvesting complex (LHC), and three (22-, 21-, and 20-kD) to the PSI-110 particle (102). A similar conclusion was reached in a study of chlorophyll-protein complexes in *Chlamydomonas reinhardtii* (103). It appears therefore that a chlorophyll a/b-protein antenna, containing Chl a and Chl b at a ratio of 6.5:1, is to be associated with CP1 in the intact membrane.

Genetic studies (78, 80, 82, 84) confirm that PSI is indeed a multicomponent assembly. Girard et al (80) investigated 25 *Chlamydomonas* mutants lacking or deficient in PSI reaction centers. These mutants belonged to 13 complementation groups scattered throughout the nuclear genome. The thylakoid membranes of all of these mutants were lacking in the apoprotein of CP1 and in at least six low molecular weight polypeptides. When the procedure of Mullet et al (see above; 76) was applied to *Chlamydomonas,* a PSI-110 preparation was obtained that contained ~25% of the total chlorophyll (80). All of the polypeptides missing from the PSI-deficient mutants are major components of this PSI preparation. Since all of the mutations lead to the absence of the seven polypeptides from the thylakoid membrane, the results indicate that the P700-Chl a protein is a part of a multisubunit complex whose assembly requires the presence of all of these constituents.

PHOTOSYSTEM II   Preparations enriched in PSII activity, and free of PSI, have been obtained by density gradient fractionation of thylakoid membranes treated with such detergents as Triton X-100 (e.g. 104, 105) or digitonin (106–108). These preparations do not perform the water-splitting reaction, but can be assayed for PSII activity by using artificial electron donors and acceptors, e.g. 1,5-diphenylcarbazide as donor and dichlorophenol indophenol as acceptor.

Analyses of PSII mutants in barley (82) indicated that CHl_a-P2 and Chl $_a$-P3 (Table 3) are derived from PSII, and that Chl_a-P3 most probably contains reaction center II (RCII) (83). Similar conclusions were reached in studies of PSII mutants in *Chlamydomonas reinhardtii,* where two chlorophyll-protein complexes, CPIII and CPIV, were found to be associated with PSII (79). The apparent molecular weights of the apopolypeptides of these two complexes were 50 and 47 kD, respectively (79). Protein, chlorophyll $a,$ and β-carotene were present in a molar ratio of $1:4:0.4$ in CPIII, and $1:5:1.3$ in CPIV. From the amount of chlorophyll associated with CPIII and CPIV, it was calculated that roughly one copy of each complex was present for a photosynthetic unit of 400–500 chlorophylls (79). Analysis of *C. reinhardtii* mutants in PSII in a manner similar to that described above for PSI, demonstrated that the reaction center of PSII was contained within a complex made up of five polypeptides of 50, 47, 32, 14, and 3 kD (80, 81). Moreover, in the same study it was shown that other polypeptides of 19, 24, and 34 kD were associated with the water-splitting reaction (81). A photoactive PSII complex purified by Satoh (108, 109) from spinach contained only 43-, 27-, and 6.5-kD polypeptides.

The function of each of the polypeptides in the complexes described above have yet to be determined. The functional and spectroscopic properties of PSII reaction center preparations were thoroughly reviewed recently (5, 110–113).

LIGHT-HARVESTING CHLOROPHYLL a/b-PROTEIN COMPLEX (LHC) The LHC encompasses $\sim 50\%$ of the chlorophyll of mature chloroplasts. An enrichment in Chl b is a distinctive hallmark of this complex; in spinach LHC, Chl a/b is 1.1 compared with 2.8 in the parent thylakoid membranes (114). A current procedure for the preparation of highly purified LHC involves Triton X-100 solubilization of thylakoid membranes, sucrose gradient fractionation, and Mg^{2+} induced aggregation of the complex (94). LHC purified in this manner contains several polypeptides in the 20–30-kD range: barley LHC, 27, 25, and 24 kD (102); pea LHC, 28.5, 26, 24.5, and 24 kD (114); spinach LHC, 26 and 23.5 kD (114); and *Chlamydomonas,* 39.5, 26.5, and 25 kD (103). These polypeptides are present in varying proportions to each other. Studies of their biosynthesis indicate that they differ in primary structure (115). LHC exists in the membrane as a multisubunit complex, perhaps as large as an octamer (116, 117). It is not evident whether the different polypeptides are present within each multimer or whether there are several LHC complexes with distinctive compositions.

Spinach LHC purified by the procedure outlined above (94) contained ~7 molecules of Chl *a* and *b*, ~5 molecules of diacyl lipid, and ~1 carotenoid per 26 kD of polypeptide (114). In general, lipid appears to be tightly bound to LHC components (118–121). LHCP[1], LHCP[2], and LHCP[3] from *Nicotiana tabacum,* prepared by SDS-polyacrylamide gel electrophoresis (see Table 2), all contained lipid (119). In LHCP[3], polar lipids represented ~25% of the chlorophyll content. Interestingly the diacyl lipid composition of the oligomer of LHC, LHCP[1], included a high concentration of phosphatidyl glycerol containing 3-trans-hexadecenoic acid, a selective enrichment relative to the content of this lipid in the thylakoid membranes (119). Treatment of LHCP[1] with phospholipase A_2 converted it to the monomer $LHCP_3$ (120). These observations suggest that interactions with specific lipids may be important to the assembly of LHC within thylakoid membranes.

ORGANIZATION OF THE CHLOROPHYLL-PROTEIN COMPLEXES IN THYLAKOIDS   Mechanical fractionation of thylakoids leads to a separation of a "heavy fraction" derived from grana stacks and a "light fraction" derived from stroma lamellae (122). The heavy fraction is somewhat depleted in PSI activity, enriched in PSII activity, and markedly enriched in LHC, as deduced from the Chl *a/b* ratio. The light fraction is highly enriched in PSI activity, and displays some PSII activity. When chloroplasts are disrupted by suspension in a low-salt medium and the thylakoids sheared by passage through a press, two classes of vesicles can be fractionated in a two-phase partition system: inside-out vesicles obtained from appressed membranes in the partition regions of grana stacks (see Figure 3, *A* and *B*) and right-side-out vesicles obtained from stroma lamellae and end membranes (Figure 3, *A* and *B;* 123–125). The intrinsic properties of thylakoid membranes appear to be such that after disruption, fragments seal to form right-side-out vesicles unless they are in an appressed state (124, 125). Electrophoretic analysis of the chlorophyll-protein complexes present in the two classes of vesicles showed that PSII and LHC were highly enriched in the partition regions, but were also present in the stroma-exposed membrane regions, whereas PSI appeared to be largely, if not entirely localized in the stroma-exposed membrane regions (126–128). This distribution is illustrated in Figure 3 *C.* These observations are consistent with spectroscopic studies, which indicate that LHC transfers energy primarily to PSII (129).

Insight into the organization of PSII and LHC within the partition regions has come from electron microscopy studies (116, 130). These indi-

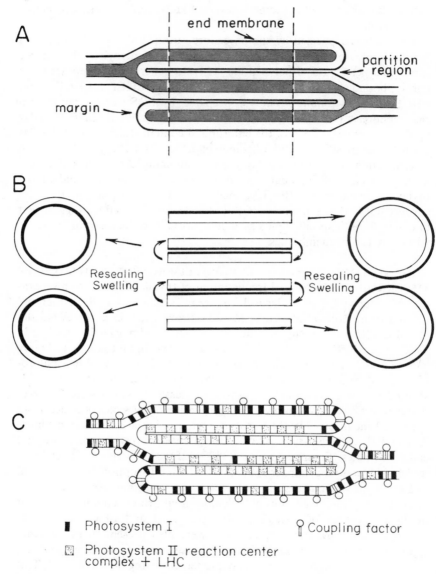

Figure 3 (*A*) Schematic representation of appressed thylakoid membranes (grana partitions) and exposed thylakoid membrane regions (stroma thylakoids, and grana end membranes and margins). Space external to the thylakoids is shaded. (*B*) Origin of "inside-out" and "right-side-out" vesicles obtained upon shearing of thylakoid membranes along planes such as those indicated by the dotted lines in *A*. Stroma thylakoids give rise to right side-out vesicles. Based on scheme 1 of Ref. (125) and Figure 11 of Ref. (124). (*C*) Localization of PSI and of PSII-LHC complexes in grana-containing chloroplasts. Based on Figure 2 of Ref. (128).

cate that LHC complexes, 60–80 Å in diameter (116, 117), reside partially within the protoplasmic stacked surface of the thylakoid membranes in the partition region as illustrated in Figure 4. The LHC complexes are associated with PSII complexes that are exposed at the endoplasmic surface, and which may traverse the bilayer (Figure 4; 96). Spectroscopic studies indicate that several PSII reaction centers share a common antenna, and that this interaction is dependent on membrane appression (131). Numerous studies also indicate that LHC contributes to the formation of grana stacks (92, 132–134), although grana are formed in the absence of intact LHC (135). These observations are compatible with the organization suggested by electron microscopy (Figure 4) which indicates that important interactions between PSII-LHC complexes can take place efficiently between complexes in different membranes when these are in the appressed state. It appears that LHC does not mediate energy transfer between PSII centers within the plane of a single thylakoid membrane, since sharing of antennae between PSII units is not seen in de-stacked thylakoids (83).

Present evidence shows that the interaction between PSI and PSII in higher plants and green algae is primarily at the level of electron transport and that direct sharing of excitation energy between the two photosystems is minor relative to the energy absorbed directly by each photosystem (128).

MULTIPLICITY OF CHEMICALLY DIFFERENT CHLOROPHYLL CHROMOPHORES   Until recently, it was accepted that there were only two chlorophyll species in green plants: 2-vinyl, 4-ethyl Chl *a,* and 2-vinyl, 4-ethyl Chl *b,* which differed from Chl *a* by the presence of a formyl group instead of a methyl group at position 3 of the macrocycle. Chl *a* was believed to function both as antenna pigment and in the conversion of

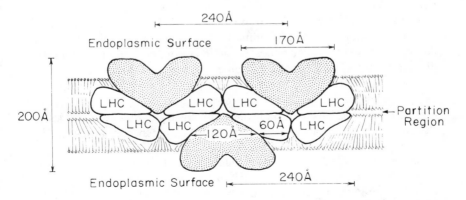

Figure 4   Model of the thylakoid membrane from the grana of barley chloroplasts based on freeze-fracture electron microscopy. The stippled areas represent PSII particles. Redrawn with minor modifications from Figure 34 of Ref. (116).

excitation energy to electron flow as a component of the reaction center (136). It had been noted that the absorption spectrum of photosynthetic lamellae in the Chl absorption region resulted from multiple overlapping absorption bands (137). This heterogeneity in absorption behavior was attributed to the unique environments in which the Chl chromophores were located within the lipoproteins of the thylakoid membrane (22) and also to the electromagnetic interactions between the Chl chromophores themselves (138, 139).

The above views require drastic revision to accommodate the demonstration that multiple chemically different Chl a and Chl b species are universally present in green plants (140, 141, see also 142). There are at least four Chl a species: two of these, Chl a (E432F664) and Chl a (E436F670) are referred to as "short wavelength" or SW Chl a species; the other two, Chl a (E443F670) and Chl a (E446F674) as the "long wavelength" or LW Chl a species, where E refers to the Soret excitation maximum and F to the fluorescence emission maximum in ether at 77 K (143). The classical Fischer-Stern 2-vinyl, 4-ethyl Chl a, presumably Chl a (E446F674), accounts for some 60–80% of the total Chl a, Chl a (E432F664) for 6–8%, and the balance is divided between Chl a (E436F670) and Chl a (E443F670) (141). Spectroscopic evidence suggests that the SW Chl a species may differ from the LW Chl a species in the cyclopentanone ring portion of the structure; their spectra are similar to those of 10-OH Chl a lactone (140). However, rigorous proof of structure for the less abundant Chl a components must still be presented.

The SW Chl a species are enriched in PSI and PSII complexes, whereas the LW Chl a species are enriched in LHC (144). Chl b of green plant thylakoids is also made up of four different species: SW Chl b (E465) and Chl b (E470), and LW Chl b (E475) and Chl b (E485) (143).

Studies of PSI preparations from spinach and *Scenedesmus* suggest that the Chl chromophore corresponding to P700 differs from the Chl a species described above (145). The putative reaction center chromophore, Chl RCI, differed from Chl a in fluorescence quantum yield and redox behavior, and gave a molecular weight 35 units higher than that of Chl a, as determined by [252]Cf-plasma desorption mass spectroscopy (145).

Much work remains to be done on the chemistry of these Chl species, but the evident implications are summarized well by Rebeiz et al (143): "It is possible to visualize the side-chain modification of the chlorophyll chromophores as one way of tagging a particular chlorophyll molecule for a particular site and a particular orientation within its macromolecular environment in the thylakoid membrane. On the other hand, modifications of the cyclopentanone ring may result in differences in the basic photochemistry of the chlorophyll chromophore which may be dictated by some specific functional role."

Cyanobacteria (Blue-Green Algae)

Cyanobacteria and red algae do not contain Chl *b*. As much as 50% of the light-harvesting capacity of cyanobacterial and red algal cells resides in a family of colored proteins called phycobiliproteins. These proteins are assembled in vivo into macromolecular aggregates, phycobilisomes, which are attached to the outer surface of the photosynthetic lamellae. Phycobilisomes range in size from $\sim 7 \times 10^6$ daltons in cyanobacteria to perhaps twice that size in certain red algal chloroplasts, and vary considerably in their gross morphology. They can be isolated intact as judged by spectroscopic criteria and electron microscopy (146, 147). The properties of phycobilisomes (148–150) and phycobiliproteins (150–153) have been reviewed in detail recently. These reviews should be consulted for extensive literature surveys and only the most recent advances are summarized here.

Phycobilisomes consist entirely of proteins: phycobiliproteins and proteins that function in the assembly of the particle. Quantitatively, the major biliprotein are the red phycoerythrin (λ_{max} 565 nm) and the blue phycocyanin ($\lambda_{max} \sim 620$ nm) and allophycocyanin (λ_{max} 650 nm). The monomer of each of these proteins is made up of two dissimilar polypeptide chains, α and β, of 17–22 kD. In addition there are quantitatively minor biliproteins, allophycocyanin B and a high molecular weight biliprotein (also called allophycocyanin I), that absorb at 670 nm. The prosthetic groups of the phycobiliproteins are open-chain tetrapyrroles, phycoerythrobilin in phycoerythrin, and phycocyanobilin in the other biliproteins, covalently attached to the polypeptide chain by thioether linkage.

The following energy transfer pathway has been established in phycobilisomes (149, 150): Phycobilisomes donate energy primarily to PSII (154, 155):

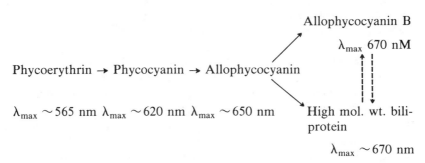

The hemidiscoidal phycobilisomes present in many cyanobacteria (156–158) and certain red algae (159) have been particularly favorable objects for ultrastructural analysis. These particles have two morphologically distinct domains: a core made up either of three (156–159) or two (157) (see Figure 5) cylindrical objects, from which six rods made up of stacked discs (120 × 60 Å) extend in a hemidiscoidal array. In phycobilisomes containing both phycoerythrin and phycocyanin, the peripheral discs of the rod substructures contain phycoerythrin, the discs proximal to the core contain phycocyanin, whereas the core itself contains allophycocyanin (156, 158–163). From studies of mutants (158, 163–165), and of chromatic adaptation (156, 158, 166), it is evident that the structure is assembled in a stepwise manner outward from the thylakoid membrane. The phycobilisome appears to be anchored to the thylakoid membrane through a segment of the high molecular weight biliprotein. This biliprotein ranges from 75–120 kD in phycobilisomes of different organisms (149, 150). In *Synechococcus* 6301, this biliprotein has a molecular weight of 75,000 and one copy is present in each of the two core cylinders (Figure 5c).

The *Synechococcus* 6301 phycobilisome, illustrated in Figure 5, is the only phycobilisome for which nearly complete structural information is available (164, 165, 167–174). This is a simple phycobilisome; it contains only phycocyanin and allophycocyanin as major biliproteins. Each of the core cylinders is made up of four "trimeric" allophycocyanin complexes. A remarkable feature of this structure is that allophycocyanin participates as a common building block within four subelements of differing composition, as shown in Figure 5d. Complexes 3 and 4 are released upon dissociation of phycobilisomes under appropriate conditions, whereas complexes 1 and 2 are recovered in an ∼560-kD core-rod junction complex and are only separated after mild tryptic degradation of this complex (170–172). A different terminal energy acceptor is present in complex 2 and in complex 3. In complex 3, it is the allophycocyanin B α subunit (indicated by the asterisk in Figure 5d, complex 3), as is evident from a comparison of the fluorescence emission properties of complexes 3 and 4. In complex 2, the terminal energy acceptor is a phycocyanobilin-bearing tryptic fragment of the 75-kD polypeptide remaining with this complex (25–27).

The assembly and properties of the rods also offer a number of points of interest. Each $(\alpha\beta)_6$ phycocyanin disc in the rod substructure is associated with one copy of a specific uncolored "linker polypeptide" (165, 168, 169). Each linker polypeptide fulfills three functions: it mediates the assembly of the biliprotein into the appropriate aggregate, concomitantly modifies the spectroscopic properties of the phycocyanin, and, by means of a projecting segment, determines the location of the particular aggregate within the rod structure (see Figure 5A) (176, 177). The spectroscopic properties of the rod components are thus such that they favor flow of energy toward the core (Figure 5A), despite the single biliprotein makeup of the rod substructures. Each $(\alpha\beta)_6$ phycocyanin disc contains 18 bilin chromophores and each allophycocyanin trimer 6 bilins (151).

The modification of this particle in cyanobacteria that face changes in light quality or light intensity illustrates the economical nature of structural adaptation made possible by the modular nature of the phycobilisome. Two groups of chromatic adapters among cyanobacteria have been described (177). Group II chromatic adapters do not make phycoerythrin in red light, and do not change their rate of phycocyanin synthesis. Group III chromatic adapters cease to make phycoerythrin in red light but express a second set of phycocyanin genes in addition to those expressed in green light (178). In both groups of organisms the rate of allophycocyanin synthesis is unaffected by light quality (177). The rod substructures of phycobilisomes of group II organisms grown in red light are shorter than those from cells grown in green light, whereas those of phycobilisomes from group III chromatic adapters show no change in length—the increase in the number of phycocyanin discs compensates for the loss of phycoerythrin discs (156, 158, 166). In all of these phycobilisomes, the structure of the core domains is unaltered (156, 158, 166). *Synechococcus* 6301, which makes no phycoerythrin, adjusts to conditions of low light intensity or those that promote phycocyanin synthesis by making phycobilisomes with longer rods (165). Thus the adaptive changes in the complex structure of the phycobilisome are restricted to the components at the periphery of the rods.

Prochlorophyta

The prochlorophyta are the only group of prokaryotes other than the cyanobacteria that perform oxygen-evolving photosynthesis (4, 179). These organisms have been described only recently, presumably because of their rather narrow ecological distribution and close resemblance to cyanobacteria (179). *Prochloron* spp. are found in association with certain didemnid ascidians (179) and have not yet been induced to reproduce successfully in laboratory culture. The cells have a typical prokaryotic ultrastructure and have carotenoids characteristic of cyanobacteria. However, they possess both Chl *a* and Chl *b* and show some stacking of the thylakoids (179). Components analogous to the P700-Chl *a*-protein and the Chl *a/b*-protein of green plants were demonstrated by SDS-polyacrylamide gel electrophoresis of *Prochloron* sp. photosynthetic lamellae (180). The prochlorophytes show affinities with cyanobacteria (180, 181) and with green algae (180) and more data are needed before it can be resolved whether these organisms are the long-sought evolutionary precursors of green plant chloroplasts.

Algal Light-Harvesting Photosynthetic Accessory Pigments

Table 4 lists the major light-harvesting pigments other than Chl *a* present in various classes of algae. Chl c_1 and c_2 are particularly widely distributed. Chl c_1 and c_2 differ from Chl *a* and *b* in that they are porphyrin rather

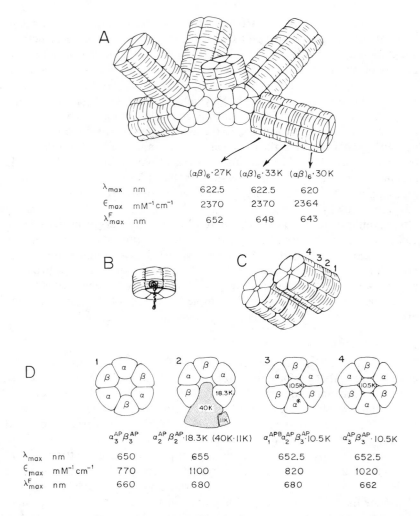

		$(\alpha\beta)_6 \cdot 27K$	$(\alpha\beta)_6 \cdot 33K$	$(\alpha\beta)_6 \cdot 30K$
λ_{max}	nm	622.5	622.5	620
ϵ_{max}	$mM^{-1}cm^{-1}$	2370	2370	2364
λ^F_{max}	nm	652	648	643

		$\alpha_3^{AP}\beta_3^{AP}$	$\alpha_2^{AP}\beta_2^{AP}\cdot18.3K\ (40K\cdot11K)$	$\alpha_1^{APB}\alpha_2^{AP}\beta_3^{AP}10.5K$	$\alpha_3^{AP}\beta_3^{AP}\cdot10.5K$
λ_{max}	nm	650	655	652.5	652.5
ϵ_{max}	$mM^{-1}cm^{-1}$	770	1100	820	1020
λ^F_{max}	nm	660	680	680	662

Figure 5 Structure of the phycobilisome of an unicellular cyanobacterium *Synechococcus* 6301 (*Anacystis nidulans*). Stacked 120 × 60 Å hexameric phycocyanin complexes form the rod substructures (*A*), which are attached in a hemidiscoidal array to two core cylinders (157, 164). From the end proximal to the core, the phycocyanin discs contain "linker polypeptides" of 27K, 33K, and 30K, respectively (164, 165, 168, 169). A portion of each linker polypeptide mediates the assembly of the hexameric disc while the remainder projects in a polar manner from the disc, as shown in *B,* and functions in the attachment of the disc to the adjacent component in the structure (169, 175, 176). In *C,* the core cylinders are shown in an isometric view. Each of the core cylinders is a stack of four 110 × 30 Å allophycocyanin-containing complexes (164, 170, 172, 173). Complexes 3 and 4 have been purified from dissociated phycobilisomes; complexes 1 and 2 are isolated together within an 18 S complex (171–173). Separation of 1 from 2 involves mild tryptic degradation resulting in the breakdown of a 75K

than chlorin derivatives with ring IV unsaturated, containing an acrylic rather than a propionic acid side-chain at C_7 and not esterified to phytol or other alcohols. Chl c_1 has an ethyl and c_2 a vinyl group at C_4 in ring II (188, 189). In some diatoms and dinoflagellates Chl c is present in amounts about equal to Chl a. In other diatoms and dinoflagellates, as well as in chrysomonads, cryptomonads, and brown algae, its amount ranges from 20–50% of that of Chl a (186).

It is now generally recognized that carotenoids serve both a light-harvesting and a photoprotective function in all photosynthetic organisms. However, carotenoids do not make a quantitatively major contribution to light harvesting in green plants. In certain algal species, such a contribution of carotenoids, in the region of 460–580 nm, is very important (185, 190). In some dinoflagellates, in diatoms, and brown algae, fucoxanthin (191), the dominant carotenoid, functions with high efficiency in light harvesting for photosynthesis (e.g. 192). Peridinin (193) plays this role in many dinoflagellates (194), and it was recently demonstrated that siphonaxanthin (195) makes a major contribution to light absorption in certain green algae that live in deep coastal waters where their photosynthesis is sustained mostly by green light (183, 184).

DINOFLAGELLATES   The photosynthetic action spectrum of dinoflagellate chloroplasts is dominated by the contributions of Chl a (438, 672 nm), Chl c_2 (460, 630 nm) and peridinin (470–560 nm) (196). Whereas the Chl a and Chl c_2 are located in integral membrane complexes that require detergent for solubilization (197, 198), peridinin is recovered within a water-soluble peridinin-Chl a-protein complex (199). The best studied of such peridinin-containing proteins are those from *Glenodinium* sp. and *Amphidinium carterae*. The native protein from *Glenodinium* sp. has a molecular weight of 35,500 and contains peridinin and Chl a in a molar ratio of ~4:1. On treatment with 1% SDS, the protein dissociates to subunits of 15 kD (200). The protein from *Amphidinium carterae* is 39.2 kD and consists of a 31.8-kD apoprotein associated with 9 peridinin and 2 Chl a molecules (201). Spectroscopic studies on these proteins (202, 203), recently reviewed by Song (204), show 100% efficient energy transfer from peridinin to Chl a. A model has been suggested for the spatial arrangement of the pigments within the peridinin-Chl a-protein and speculation has been

←————————————

phycocyanobilin-containing polypeptide, present at one copy per core cylinder (171, 172). Two fragments of that polypeptide, 40K and 11K, are recovered in complex 2. The spectroscopic properties are tabulated for each phycobilisome subcomplex; λ_{max} and λ^F_{max} represent the wavelengths of maximum absorption and fluorescence emission, respectively; ϵ_{max} is the extinction coefficient. Abbreviations are $(\alpha\beta)_6.27K$ is the complex of a hexamer of phycocyanin with a 27-kD linker polypeptide; α^{AP}, β^{AP} are the α and β subunits of allophycocyanin; α^{APB} is the α subunit of allophycocyanin B.

Table 4   Major light-harvesting pigments other than Chl a present in various algal classes

Algae class	Pigments
Chlorophyceae (green algae)	Chl b, siphonaxanthin[a]
Dinophyceae (dinoflagellates)[b]	Peridinin and Chl c_2[c] or fucoxanthin, Chl c_1 and Chl c_2
Cryptophyceae (cryptomonads)	Phycobiliproteins and Chl c_2[d]
Chrysophyceae (golden algae)	Fucoxanthin, Chl c_1 and Chl c_2
Bacillariophyceae (diatoms)	Fucoxanthin, Chl c_1 and Chl c_2
Phaeophyceae (brown algae)	Fucoxanthin, Chl c_1 and Chl c_2
Rhodophyceae (red algae)	Phycobiliproteins

[a] Siphonaxanthin is characteristic of many siphonous green algae (182) and of certain chlorophycean seaweeds living deep in coastal waters (183, 184), but is not a significant light-harvesting pigment in the majority of green algae. In the marine green alga *Caulerpa cactoides*, siphonaxanthin was found to be associated with the light-harvesting Chl a/b-protein (185).
[b] *Gymnodinium cyaneum* Hu was found to contain a phycocyanin-like protein (187).
[c] *Exuviella cassubica* contains both Chl c_1 and c_2 (186).
[d] *Chroomonas mesostigmatica* contains both Chl c_1 and c_2 (186).

offered on the mechanism of the carotenoid to Chl a energy transfer process (203–205). The manner in which these proteins interact with the integral light-harvesting components of the thylakoid membranes remains to be established.

Significant increases in the cell content of the peridinin-Chl a protein were seen in cells of *Glenodinium* sp. grown at low light intensities. Increases were also seen in the level of an unidentified Chl a component(s) of the thylakoid membrane (196, 206). Consequently, an unambiguous analysis of the molecular mechanism of intensity adaptation in this dinoflagellate cannot yet be made.

CRYPTOMONADS   Little is known about the molecular organization of the photosynthetic apparatus in cryptomonads. Comparison of the spectra of isolated pigments with photosynthetic action spectra shows that Chl a and c_2 and phycobiliproteins function as light-harvesting species in these organisms (207, 208). The cryptomonad phycobiliproteins differ in several important respects from those of cyanobacteria and red algae. In general, only a single type of phycobiliprotein is present. These proteins fall into two classes: phycoerythrins (λ_{max} 560–570nm) that carry only phycoerythrobilin chromophores, and phycocyanins (λ_{max} 610–650 nm) that carry phycocyanobilin chromophores as well as other bilins whose structures are not established (151). The spectra of proteins within each class differ depending on the source organism (151, 209). In vitro, the cryptomonad biliproteins exist either as a monomer, $\alpha\beta$, or as dimers (210–215). The structure $\alpha\alpha'\beta_2$ has been proposed for dimeric cryptomonad biliproteins, but the possibility that two proteins, $\alpha_2\beta_2$ and $\alpha'_2\beta_2$, are present must be considered (213, 216). The amino-terminal sequence of the β subunit of a cryptomonad phycocyanin was found to be homologous to that of the β subunit of a rhodophytan phycoerythrin (217). The α subunit sequence does not appear to be closely related to the sequences of the corresponding subunits of cyanobacterial or rhodophytan biliproteins (217). Morever, the α subunit of the cryptomonad phycobiliproteins is small, 9–10 kD, as compared to 17–20 kD, for the α subunits of the other biliproteins (151). The cryptomonad biliproteins are located within the thylakoids, but their in vivo organization is not known (218, 219).

DIATOMS   In addition to Chl a and β-carotene, diatoms contain Chl c_1 and c_2 and fucoxanthin as major pigments. From a study of the action spectra for photosynthesis, Mann & Myers (192) assigned fucoxanthin and Chl c to PSII. Holdsworth & Arshad (220) isolated a large complex that displayed PSII activity from the diatom *Phaeodactylum tricornutum*. This complex contained Chl a: Chl c: fucoxanthin in a molar ratio of 2:1:1. Gugliemelli et al (221) isolated such a complex from *P. tricornutum* by a different procedure and obtained a preparation containing Chl a: Chl c: fucoxanthin in a molar ratio of 1:1:4, which retained energy transfer from carotenoids to Chl a.

BROWN ALGAE   Dissociation of the thylakoids of the brown seaweed *Acrocarpia paniculata* with Triton X-100 and subsequent fractionation led to the isolation of a P700-Chl a-protein complex with a molar ratio of P700:38 Chl a: 4 β-carotene molecules, similar to the P700-containing complex isolated with Triton X-100 from higher plants and green algae. Two light-harvesting complexes were also isolated: an orange-brown complex containing Chl a: Chl c_2: fucoxanthin in a ratio of 2:1:2, and a green complex containing Chl a: Chl c_1: Chl c_2: violaxanthin in a ratio of 8:1:1:1 (222, 223). A Chl a-protein (λ^F_{max} 694 nm), with properties suggestive of a PSII reaction center complex was also isolated (224). Treatment of the Chl a-Chl c_2-fucoxanthin protein complex with cholate led to the separation of a fucoxanthin-protein, a Chl a-protein (λ_{max} 672 nm), and a Chl c_2-protein. The M_r of these pigment-proteins ranged from 13,000–16,000 (224). A proposed model for the arrangement of these complexes assigns the Chl a-Chl c_2-fucoxanthin protein to the antenna of PSII and the Chl a-Chl c_1-Chl c_2-violaxanthin protein to the antenna of PSI (224). The results obtained with other brown algae are in general agreement with those obtained with *A. paniculata* (225, 226).

RED ALGAE The phycobilisomes of red algae were reviewed earlier with those of cyanobacteria.

CONCLUDING REMARKS

This review attempts to provide a broad coverage of the available information on the structure and assembly of the conponents of all types of photosynthetic light-harvesting systems with the exception of the unusual bacteriorhodopsin-containing purple membrane of halobacteria (227). For most of these systems, structural studies have lagged behind studies of function and mechanism. This is largely a consequence of the finding that the reaction centers and most types of antennae are integral membrane complexes. Rapid progress can be anticipated as a consequence of increasing sophistication in the handling of membrane proteins and of the advances in molecular genetics.

ACKNOWLEDGMENTS

This review was written during the tenure of a John Simon Guggenheim Memorial Foundation fellowship for which the author wishes to express his thanks. He is also grateful to Dr. Mel Y. Okamura for a preprint of Reference 5.

Literature Cited

1. Michel, H. 1982. *J. Mol. Biol.* 158: 567–72
2. Fowler, C. F., Nugent, A., Fuller, R. C. 1971. *Proc. Natl. Acad. Sci. USA* 68:2278–82
3. Prince, R., Dutton, P. L. 1978. In *The Photosynthetic Bacteria,* ed. R. K. Clayton, W. R. Sistrom, pp. 439–53. New York: Plenum
4. Stanier, R. Y., Pfennig, N., Truper, H. G. 1981. In *The Prokaryotes,* ed. M. P. Starr, H. Stolp, H. G. Truper, A. Balows, H. G. Schlegel, 1:197–211. Berlin: Springer
5. Okamura, M. Y., Feher, G., Nelson, N. 1982. In *Photosynthesis: Energy Conversion by Plants and Bacteria,* ed. Govindjee, 1:195–272. New York: Academic
6. Trebst, A. 1974. *Ann. Rev. Plant Physiol.* 25:423–58
7. Arnon, D. I., Tsujimoto, H. Y., Tang, G. M. -S. 1980. *Proc. Natl. Acad. Sci. USA* 78:2942–46
8. Arnon, D. I., Tsujimoto, H. Y., Tang, G. M. -S. 1981. In *Photosynthesis II. Electron Transport and Photophosphorylation,* ed. G. Akoyunoglou, pp. 7–18. Philadelphia: Balaban Int. Sci. Serv.
9. Pierson, B. K., Castenholtz, R. W. 1978. See Ref. 3, pp. 179–97
10. Cohen-Bazire, G. 1963. In *Bacterial Photosynthesis,* ed. H. Gest, A. San Pietro, L. P. Vernon, pp. 89–110. Yellow Springs, Ohio: Antioch
11. Olson, J. M. 1980. *Biochim. Biophys. Acta* 594:33–51
12. Pfennig, N. 1978. See Ref. 3, pp. 3–18
13. Staehelin, L. A., Golecki, R. J., Drews, G. 1980. *Biochim. Biophys. Acta* 589: 30–45
14. Schmidt, K. 1980. *Arch. Microbiol.* 124: 21–31
15. Sprague, S., Staehelin, L. A., DiBartolomeis, M. J., Fuller, R. C. 1981. *J. Bacteriol.* 147:1032–39
16. Clayton, R. K. 1980. *Photosynthesis: Physical Mechanisms and Chemical Patterns,* p. 48. Cambridge: Cambridge Univ. Press. 281 pp.
17. Staehelin, L. A., Golecki, J. R., Fuller, R. C., Drews, G. 1978. *Arch. Microbiol.* 119:269–77
18. Feick, R. G., Fitzpatrick, M., Fuller, R. C. 1982. *J. Bacteriol.* 150:905–15
19. Betti, J. A., Blankenship, R. E., Natarajan, L. V., Dickinson, L. C., Fuller, R.

C. 1982. *Biochim. Biophys. Acta* 680: 194–201
20. Olson, J. M. 1978. See Ref. 3, pp. 161–78
21. Fenna, R. E., Matthews, B. W., Olson, J. M., Shaw, E. K. 1974. *J. Mol. Biol.* 84:231–40
22. Matthews, B. W., Fenna, R. E., Bolognesi, M. C., Schmid, M. F., Olson, J. M. 1979. *J. Mol. Biol.* 131:259–85
23. Cruden, D. L., Stanier, R. Y. 1970. *Arch. Mikrobiol.* 72:115–34
24. Schmidt, K., Maarzahl, M., Mayer, F. 1980. *Arch. Microbiol.* 127:87–97
25. Oelze, J., Drews, G. 1972. *Biochim. Biophys. Acta* 265:209–39
26. Drews, G., Oelze, J. 1981. *Adv. Microb. Physiol.* 22:1–92
27. Trüper, H. G., Pfennig, N. 1981. See Ref. 4, pp. 299–312
28. Trüper, H. G., Pfennig, N. 1978. See Ref. 3, pp. 19–27
29. Cogdell, R. J., Thornber, J. P. 1980. *FEBS Lett.* 122:1–8
30. Reed, D. W., Clayton, R. K. 1968. *Biochem. Biophys. Res. Commun.* 30: 471–75
31. Clayton, R. K., Wang, R. T. 1971. *Methods Enzymol.* 23:696–704
32. Feher, G. 1971. *Photochem. Photobiol.* 14:373–87
33. Feher, G., Okamura, M. Y. 1978. See Ref. 3, pp. 349–86
34. Beugeling, T., Slooten, L., Barelds-van de Beck, P. G. M. M. 1972. *Biochim. Biophys. Acta* 328:328–33
35. Jolchine, G., Reiss-Husson, F. 1974. *FEBS Lett.* 40:5–8
36. Cogdell, R. J., Parson, W. W., Kerr, M. A. 1976. *Biochim. Biophys. Acta* 430: 83–93
37. Broglie, R. M., Hunter, C. N., Delepelaire, P., Niederman, R. A., Chua, N.-H., Clayton, R. K. 1980. *Proc. Natl. Acad. Sci. USA* 77:87–91
38. Olson, J. M., Thornber, J. P. 1979. In *Membrane Proteins in Energy Transduction*, ed. R. A. Capaldi, pp. 279–340. New York: Marcel Dekker
39. Bachmann, R. C., Gillies, K., Takemoto, J. Y. 1981. *Biochemistry* 20:4590–96
40. Debus, R. J., Okamura, M. Y., Feher, G. 1981. *Biophys. J.* 33:19a
41. Sistrom, W. R. 1978. See Ref. 3, pp. 841–48
42. Takemoto, J., Huang Kao, M. Y. C. 1977. *J. Bacteriol.* 129:1102–9
43. Niederman, R. A., Mallon, D. E., Langen, J. J. 1976. *Biochim. Biophys. Acta* 440:429–47

44. Drews, G. 1978. *Curr. Topics Bioenerg.* 8:161–207
45. Sistrom, W. R., Griffiths, M., Stanier, R. Y. 1956. *J. Cell. Comp. Physiol.* 48: 473–515
46. Hunter, C. N., Niederman, R. A., Clayton, R. K. 1981. In *Photosynthesis III. Structure and Molecular Organization of the Photosynthetic Apparatus*, ed. G. Akoyunoglou, pp. 539–45. Philadelphia: Balaban Int. Sci. Serv.
47. Wraight, C. A., Clayton, R. K. 1974. *Biochim. Biophys. Acta* 333:246–60
48. Schachman, H. K., Pardee, A. B., Stanier, R. Y. 1952. *Arch Biochem. Biophys.* 38:245–60
49. Campillo, A. J., Hyer, R. C., Monger, T. G., Parson, W. W., Shapiro, S. L. 1977. *Proc. Natl. Acad. Sci. USA* 74: 1997–2001
50. Monger, T. G., Parson, W. W. 1977. *Biochim. Biophys. Acta* 460:393–407
51. Razjivin, A. P., Danielius, R. V., Gadonas, R. A., Borisov, A. Yu., Piskarskas, A. S. 1982. *FEBS Lett.* 143:40–44
52. Nishi, N., Kataoka, M., Soe, G., Kakuno, T., Ueki, T. et al. 1979. *J. Biochem.* 86:1211–24
53. Kataoka, M., Ueki, T. 1981. *J. Biochem.* 89:71–78
54. Nieth, K. F., Drews, G., Feick, R. 1975. *Arch. Microbiol.* 105:43–45
55. Feick, R., Drews, G. 1978. *Biochim. Biophys. Acta* 501:499–512
56. Schumacher, A., Drews, G. 1978. *Biochim. Biophys. Acta* 501:183–94
57. Shiozawa, J., Cuendet, P., Zuber, H., Drews, G. 1981. See Ref. 46, pp. 427–34
58. Shiozawa, J. A., Cuendet, P. A., Drews, G., Zuber, H. 1980. *Eur. J. Biochem.* 111:455–60
59. Feick, R., Drews, G. 1979. *Z. Naturforsch. Teil C* 34:196–99
60. Cuendet, P. A., Zuber, H. 1977. *FEBS Lett.* 79:96–100
61. Tonn, S. J., Gogel, G. E., Loach, P. A. 1977. *Biochemistry* 16:877–85
62. Brunisholz, R. A., Cuendet, P. A., Theiler, R., Zuber, H. 1981. *FEBS Lett.* 129:150–54
63. Melis, A., Brown, J. S. 1980. *Proc. Natl. Acad. Sci. USA* 77:4712–16
64. Falkowski, P. G., Owens, T. G., Ley, A. C., Mauzerall, D. C. 1981. *Plant Physiol.* 68:969–73
65. Kawamura, N., Mimuro, M., Fujita, Y. 1979. *Plant Cell Physiol.* 20:697–705
66. Myers, J., Graham, J-R., Wang, R. T. 1980. *Plant Physiol.* 66:1144–49
67. Akoyunoglou, G., Argyroudi-Akoyunoglou, J. H. 1978. In *Photosynthetic Oxygen Evolution*, ed. H.

Metzner, pp. 453–88. London: Academic
68. Raven, J. A., Beardall, J. 1982. *Plant Cell Environ.* 5:117–24
69. Thornber, J. P., Gregory, R. P. F., Smith, C. A., Leggett-Bailey, J. 1967. *Biochemistry* 6:391–96
70. Ogawa, T., Obata, F., Shibata, K. 1966. *Biochim. Biophys. Acta* 112:223–34
71. Markwell, J. P., Miles, G. D., Boggs, R. T., Thornber, J. P. 1979. *FEBS Lett.* 99:11–14
72. Thornber, J. P., Markwell, J. P., Reinman, S. 1979. *Photochem. Photobiol.* 29:1205–16
73. Anderson, J. M., Waldron, J. C., Thorne, S. W. 1978. *FEBS Lett.* 92:227–33
74. Anderson, J. M. 1980. *Biochim. Biophys. Acta* 591:113–26
75. Henry, L. E. A., Møller, B. L. 1981. *Carlsberg Res. Commun.* 46:227–42
76. Mullet, J. E., Burke, J. E., Arntzen, C. 1980. *Plant Physiol.* 65:814–22
77. Malkin, R. 1982. *Ann. Rev. Plant Physiol.* 33:455–79
78. Chua, N. -H., Bennoun, P. 1975. *Proc. Natl. Acad. Sci. USA* 72:2175–79
79. Delepelaire, P., Chua, N. -H. 1979. *Proc. Natl. Acad. Sci. USA* 76:111–15
80. Girard, J., Chua, N. -H., Bennoun, P., Schmidt, G., Delosme, M. 1980. *Curr. Genet.* 2:215–21
81. Bennoun, P., Diner, B. A., Wollman, F.-A., Schmidt, G., Chua, N. -H. 1981. See Ref. 46, pp. 839–49
82. von Wettstein, D., Møller, B. L., Høyer-Hansen, G., Simpson, D. 1982. In *On the Origins of Chloroplasts*, ed. J. A. Schiff, pp. 243–55. New York: Elsevier
83. Machold, O., Simpson, D. J., Møller, B. L. 1979. *Carlsberg Res. Commun.* 44:235–54
84. Hiller, R. G., Møller, B. L., Høyer-Hansen, G. 1980. *Carlsberg Res. Commun.* 45:315–28
85. Møller, B. L., Høyer-Hansen, G., Hiller, R. G. 1981. See Ref. 46, pp. 245–56
86. Machold, O., Simpson, D. J., Møller, B. L., Meister, A. 1981. See Ref. 46, pp. 357–63
87. Thornber, J. P., Highkin, H. R. 1974. *Eur. J. Biochem.* 41:109–16
88. Henriques, F., Park, R. B. 1975. *Plant Physiol.* 55:763–67
89. Machold, O., Meister, A., Sagromsky, H., Høyer-Hansen, G., von Wettstein, D. 1977. *Photosynthetica* 11:200–206
90. Burke, J. J., Steinback, K. E., Arntzen, C. J. 1979. *Plant Physiol.* 63:237–43

91. Boardman, N. K., Anderson, J. M., Goodchild, D. J. 1978. *Curr. Topics Bioenerg.* 8:35–109
92. Arntzen, C. J. 1978. *Curr. Topics Bioenerg.* 8:111–60
93. Hiller, R. G., Goodchild, D. J. 1981. In *The Biochemistry of Plants*, ed. M. D. Hatch, N. K. Boardman. 8:1–49. New York: Academic
94. Burke, J. J., Ditto, C. L., Arntzen, C. J. 1978. *Arch Biochem. Biophys.* 187:252–63
95. Boardman, N. K., Anderson, J. M. 1978. In *Chloroplast Development*, eds. G. Akoyunoglou, J. M. Argyroudi-Akoyunoglou, pp. 1–14. Amsterdam: Elsevier
96. Andersson, B., Anderson, J. M., Ryrie, I. J. 1982. *Eur. J. Biochem.* 123:465–72
97. Thornber, J. P. 1975. *Ann. Rev. Plant Physiol.* 26:127–58
98. Bar-Nun, S., Schantz, R., Ohad, I. 1977. *Biochim. Biophys. Acta* 459:451–67
99. Chua, N. -H., Matlin, K., Bennoun, P. 1975. *J. Cell Biol.* 67:361–77
100. Bengis, C., Nelson, N. 1975. *J. Biol. Chem.* 250:2783–88
101. Bengis, C., Nelson, N. 1977. *J. Biol. Chem.* 252:4564–69
102. Bellemare, G., Bartlett, S. G., Chua, N.-H. 1982. *J. Biol. Chem.* 257:7762–67
103. Wollman, F. -A., Bennoun, P. 1982. *Biochim. Biophys. Acta* 680:352–67
104. Vernon, L. P., Ke, B., Satoh, K. 1966. *Brookhaven Symp. Biol.* 19:102–14
105. Vernon, L. P., Klein, S. M. 1975. *Ann. NY Acad. Sci.* 244:281–96
106. Wessels, J. S. C., Borchert, M. T. 1978. *Biochim. Biophys. Acta* 503:78–93
107. Satoh, K., Butler, W. L. 1979. *Plant Physiol.* 61:373–79
108. Satoh, K. 1979. *Biochim. Biophys. Acta* 546:84–92
109. Satoh, K. 1980. *Proc. 5th Int. Cong. Photosynthesis*, Kallithea-Kassandra, Halkidiki, Greece, 7–13 Sept.
110. Amesz, J., Duysens, L. N. M. 1977. In *Primary Processes of Photosynthesis*, ed. J. Barber, pp. 151–85. Amsterdam: Elsevier
111. Velthuys, B. R. 1980. *Ann. Rev. Plant Physiol.* 31:545–67
112. Ke, B. 1981. *Isr. J. Chem.* 21:283–90
113. Amesz, J. 1981. *Progress in Botany* 43:49–63
114. Ryrie, I. J., Anderson, J. M., Goodchild, D. J. 1980. *Eur. J. Biochem.* 107:345–54
115. Schmidt, G. W., Bartlett, S. G., Grossman, A. R., Cashmore, A. R., Chua, N.-H. 1981. *J. Cell Biol.* 91:468–78

116. Simpson, D. J. 1979. *Carlsberg Res. Commun.* 44:305–36
117. Li, J., Hollingshead, C. 1982. *Biophys. J.* 37:363–70
118. Rawyler, A., Henry, L. E. A., Siegenthaler, P. -A. 1980. *Carlsberg Res. Commun.* 45:443–51
119. Trémolières, A., Dubacq, J. -P., Ambard-Bretteville, F., Rémy, R. 1981. *FEBS Lett.* 130:27–31
120. Rémy, R., Trémolières, A., Duval, J. C., Ambard-Bretteville, F., Dubacq, J. P. 1982. *FEBS Lett.* 137:271–75
121. Selstam, E. 1981. See Ref. 46, pp. 631–34
122. Sane, P. V., Goodchild, D. J., Park, R. B. 1970. *Biochim. Biophys. Acta* 216: 162–78
123. Åkerlund, H. -E., Andersson, B., Albertsson, P. -Å. 1976. *Biochim. Biophys. Acta* 449:525–35
124. Andersson, B., Simpson, D. J., Høyer-Hansen, G. 1978. *Carlsberg Res. Commun.* 43:77–89
125. Andersson, B., Sundby, C., Albertsson, P.-Å. 1980. *Biochim. Biophys. Acta* 599:391–402
126. Andersson, B., Anderson, J. M. 1980. *Biochim. Biophys. Acta* 593:427–40
127. Anderson, J. M., Andersson, B. 1981. See Ref. 46, pp. 23–31
128. Anderson, J. M. 1981. *FEBS Lett.* 124:1–10
129. Joliot, P. 1965. *Biochim. Biophys. Acta* 102:116–34
130. Miller, K. R. 1981. In *Electron Microscopy in Biology*, ed. J. D. Griffith, pp. 1–30. New York: J. Wiley
131. Joliot, P., Bennoun, P., Joliot, A. 1973. *Biochim. Biophys. Acta* 305:317–28
132. Gerola, P. D. 1981. *Physiol. Veg.* 19: 565–80
133. Anderson, J. M. 1982. *Photobiochem. Photobiophys.* 3:225–41
134. Mullet, J. E., Baldwin, T. O., Arntzen, C. J. 1981. See Ref. 46, pp. 577–82
135. Goodchild, D. J., Highkin, H. R., Boardman, N. K. 1966. *Exp. Cell Res.* 43:684–88
136. Katz, J. J., Norris, J. R., Shipman, L. L., Thurnauer, M. C., Wasielewski, M. R. 1978. *Ann. Rev. Biophys. Bioeng.* 7:393–434
137. Brown, J. S. 1972. *Ann. Rev. Plant Physiol.* 23:73–86
138. Strouse, C. E. 1976. In *Chemistry*, ed. D. Lippard, 21:159–77. New York: Wiley
139. Katz, J. J., Norris, J. R., Shipman, L. L. 1976. *Brookhaven Symp. Biol.* 28:16–55
140. Rebeiz, C. A., Belanger, F. C., McCarthy, S. A., Freyssinet, G., Dugan, J. X., et al. 1981. In *Photosynthesis V. Chloroplast Development*, ed. G. Akoyunoglou, pp. 197–212. Philadelphia: Balaban Int. Sci. Serv.
141. Rebeiz, C. A. 1982. *Chemtech* 12:52–63
142. Bazzaz, M. B., Brereton, R. G. 1982. *FEBS Lett.* 138:104–8
143. Rebeiz, C. A., Belanger, F. C., Freyssinet, G., Saab, D. G. 1980. *Biochim. Biophys. Acta* 490:234–47
144. Freyssinet. G., Rebeiz, C. A., Fenton, J. M., Khanna, R., Govindjee. 1980. *Photochem. Photobiophys.* 1:203–12
145. Dörnemann, D., Senger, H. 1982. *Photochem. Photobiol.* 35:821–26
146. Gantt, E., Lipschultz, C. A., Grabowski, J., Zimmerman, B. K. 1979. *Plant Physiol.* 63:615–20
147. Gantt, E., Lipschultz, C. A. 1972. *J. Cell Biol.* 54:313–24
148. Gantt, E. 1980. *Int. Rev. Cytol.* 66: 45–80
149. Gantt, E. 1981. *Ann. Rev. Plant Physiol.* 32:327–47
150. Glazer, A. N. 1982. *Ann. Rev. Microbiol.* 36:173–98
151. Glazer, A. N. 1981. See Ref. 93, pp. 51–96
152. Glazer, A. N. 1980. In *The Evolution of Protein Structure and Function*, ed. D. S. Sigman, M. A. B. Brazier, pp. 221–44. New York: Academic
153. Scheer, H. 1981. *Angew. Chem. Int. Ed. Engl.* 20:241–61
154. Wang, T. W., Stevens, C. L. R., Myers, J. 1977. *Photochem. Photobiol.* 25: 103–8
155. Mimuro, M., Fujita, Y. 1978. *Biochim. Biophys. Acta* 504:406–12
156. Bryant, D. A., Guglielmi, G., Tandeau de Marsac, N., Castets, A. -M., Cohen-Bazire, G. 1979. *Arch. Microbiol.* 123: 113–27
157. Glazer, A. N., Williams, R. C., Yamanaka, G., Schachman, H. K. 1979. *Proc. Natl. Acad. Sci. USA* 76: 6162–66
158. Williams, R. C., Gingrich, J. C., Glazer, A. N. 1980. *J. Cell Biol.* 85:558–66
159. Mörschel, E., Koller, K.-P., Wehrmeyer, W., Schneider, H. 1977. *Cytobiologie* 16:118–29
160. Gantt, E., Lipschultz, C. A., Zilinskas, B. A. 1976. *Brookhaven Symp. Biol.* 28:347–57
161. Koller, K.-P., Wehrmeyer, W., Mörschel, E. 1978. *Eur. J. Biochem.* 91:57–63
162. Mörschel, E., Wehrmeyer, W., Koller, K.-P. 1980. *Eur. J. Cell Biol.* 21:319–27
163. Gingrich, J. C., Blaha, L. K., Glazer, A. N. 1982. *J. Cell Biol.* 92:261–68

164. Yamanaka, G., Glazer, A. N., Williams, R. C. 1980. *J. Biol. Chem.* 255: 11004–10
165. Yamanaka, G., Glazer, A. N. 1981. *Arch. Microbiol.* 130:23–30
166. Rosinski, J., Hainfeld, J. F., Rigbi, M., Siegelman, H. W. 1981. *Ann. Bot.* 47: 1–12
167. Yamanaka, G., Glazer, A. N., Williams, R. C. 1978. *J. Biol. Chem.* 253: 8303–10
168. Yamanaka, G., Glazer, A. N. 1980. *Arch. Microbiol.* 124:39–47
169. Lundell, D. J., Williams, R. C., Glazer, A. N. 1981. *J. Biol. Chem.* 256:3580–92
170. Lundell, D. J., Yamanaka, G., Glazer, A. N. 1981. *J. Cell Biol.* 91:315–19
171. Yamanaka, G., Lundell, D. J., Glazer, A. N. 1982. *J. Biol. Chem.* 257:4077–86
172. Lundell, D. J., Glazer, A. N. 1983. *J. Biol. Chem.* 258:894–901
173. Lundell, D. J., Glazer, A. N. 1983. *J. Biol. Chem.* 258:902–8
174. Lundell, D. J., Glazer, A. N. 1982. *J. Biol. Chem.* 256:12600–6
175. Yu, M. H., Glazer, A. N., Williams, R. C. 1981. *J. Biol. Chem.* 256:3130–36
176. Yu, M. H., Glazer, A. N. 1982. *J. Biol. Chem.* 257:3429–33
177. Tandeau de Marsac, N. 1977. *J. Bacteriol.* 130:82–91
178. Bryant, D. A., Cohen-Bazire, G. 1981. *Eur. J. Biochem.* 119:415–24
179. Lewin, R. A. 1981. See Ref. 4, pp. 257–66
180. Withers, N. W., Alberte, R. S., Lewin, R. A., Thornber, J. P., Britton, G., Goodwin, T. W. 1978. *Proc. Natl. Acad. Sci. USA* 75:2301–5
181. Seewaldt, E., Stackebrandt, E. 1982. *Nature* 295:618–20
182. Kleinig, H. 1979. *J. Phycol.* 5:281–84
183. Yokohama, Y., Kageyama, A. 1977. *Bot. Mar.* 20:433–36
184. Kageyama, A., Yokohama, Y., Shimura, S., Ikawa, T. 1977. *Plant Cell Physiol.* 18:477–80
185. Anderson, J. M., Barrett, J., Thorne, S. W. 1981. See Ref. 46, pp. 301–15
186. Jeffrey, S. W. 1976. *J. Phycol.* 12:349–54
187. Honjun, H., Minjuan, Y., Xiankong, Z. 1980. *Kexue Tongbao* 25:882–84
188. Dougherty, R. C., Strain, H. H., Svec, W. A., Uphaus, R. A., Katz, J. J. 1966. *J. Am. Chem. Soc.* 88:5037–38
189. Strain, H. H., Cope, B. T., McDonald, G. N., Svec, W. A., Katz, J. J. 1971. *Phytochemistry* 10:1109–14
190. Halldal, P. 1970. In *Photobiology of Microorganisms,* ed. P. Halldal, pp. 17–55. London: Wiley
191. Jensen, A. 1964. *Acta Chem. Scand.* 18:2005–7
192. Mann, J. E., Myers, J. 1968. *Plant Physiol.* 43:1991–95
193. Strain, H. H., Svec, W. A., Aitzetmuller, K., Grandolfo, M. C., Katz, J. J., et al. 1971. *J. Am. Chem. Soc.* 93: 1823–25
194. Siegelman, H., Kycia, J. H., Haxo, F. T. 1976. *Brookhaven Symp. Biol.* 28: 162–69
195. Strain, H. H., Svec, W. A., Aitzetmuller, K., Cope, B. T., Harkness, A. L., Katz, J. J. 1971. *Org. Mass. Spectrom.* 5:565–72
196. Prézelin, B. B., Ley, A. C., Haxo, F. T. 1976. *Planta* 130:251–56
197. Prézelin, B. B., Alberte, R. S. 1978. *Proc. Natl. Acad. Sci. USA* 75:1801–4
198. Prézelin, B. B., Boczar, B. A. 1981. See Ref. 46, pp. 417–26
199. Siegelman, H. W., Kycia, J. H., Haxo, F. T. 1976. *Brookhaven Symp. Biol.* 28:162–69
200. Prézelin, B. B., Haxo, F. T. 1976. *Planta* 128:133–41
201. Haxo, F. T., Kycia, J. H., Somers, G. F., Bennett, A., Siegelman, H. W. 1976. *Plant Physiol.* 57:297–303
202. Song, P.-S., Koka, P., Prézelin, B. B., Haxo, F. T. 1976. *Biochemistry* 15:4422–27
203. Koka, P., Song, P.-S. 1977. *Biochim. Biophys. Acta* 495:220–31
204. Song, P.-S. 1978. *Trends Biochem. Sci.* 3:25–27
205. Mauzerall, D. 1981. *Israel J. Chem.* 21:321–24
206. Prézelin, B. B. 1976. *Planta* 130:225–33
207. Haxo, F. T., Fork, D. C. 1959. *Nature* 184:1051–52
208. Haxo, F. T. 1960. In *Comparative Biochemistry of Photoreactive Pigments,* ed. M. B. Allen, pp. 339–60. New York: Academic
209. Gantt, E. 1979. In *Biochemistry and Physiology of Protozoa,* ed. H. S. Hutner, M. Lewandowsky, pp. 121–37. New York: Academic. 2nd ed.
210. Brooks, C., Gantt, E. 1973. *Arch. Microbiol.* 88:193–204
211. MacColl, R., Habig, W., Berns, D. S. 1973. *J. Biol. Chem.* 248:7080–86
212. Glazer, A. N., Cohen-Bazire, G. 1975. *Arch. Microbiol.* 204:29–32
213. Mörschel, E., Wehrmeyer, W. 1975. *Arch. Microbiol.* 105:153–58
214. Mörschel, E., Wehrmeyer, W. 1977. *Arch. Microbiol.* 113:83–89
215. MacColl, R., Berns, D. S., Gibbons, O. 1976. *Arch. Biochem. Biophys.* 177: 265–75

216. Jung, J., Song, P. -S., Paxton, R. J.,
     Edelstein, M. S., Swanson, R., Hazen,
     E. E. Jr. 1980. *Biochemistry* 19:24–32
217. Glazer, A. N., Apell, G. S. 1977. *FEMS
     Lett.* 1:113–16
218. Gantt, E., Edwards, M. R., Provasoli,
     L. 1971. *J. Cell Biol.* 14:433–44
219. Gantt, E. 1980. In *Dinoflagellates:
     Form and Function,* ed. E. Cox, pp.
     381–405. Amsterdam: Elsevier
220. Holdsworth, E. S., Arshad, J. H. 1977.
     Arch. Biochem. Biophys. 183:361–73
221. Gugliemelli, L. A., Dutton, H. J., Jur-
     sinic, P. A., Siegelman, H. W. 1981.

Photochem. Photobiol. 33:903–7
222. Anderson, J. M., Barrett, J. 1979. *Ciba
     Found. Symp.* 61:81–104
223. Barrett, J., Anderson, J. M. 1980. *Bio-
     chim. Biophys. Acta* 590:309–23
224. Barrett, J., Thorne, S. W. 1981. See Ref.
     46, pp. 347–56
225. Barrett, J., Anderson, J. M. 1977. *Plant
     Sci. Lett.* 9:275–83
226. Kirk, J. T. O. 1977. *Plant Sci. Lett.*
     9:373–80
227. Stoeckenius, W., Bogomolni, R. A.
     1982. *Ann. Rev. Biochem.* 51:587–616

Ann. Rev. Biochem. 1983. 52:159–86

ADENYLATE CYCLASE-COUPLED BETA-ADRENERGIC RECEPTORS:
Structure and Mechanisms of Activation and Desensitization

Robert J. Lefkowitz, Jeffrey M. Stadel, and Marc G. Caron

Howard Hughes Medical Institute Research Laboratories, Duke University Medical Center, Departments of Medicine (Cardiology) and Biochemistry, Durham, North Carolina 27710

CONTENTS

PERSPECTIVES AND SUMMARY 159
ADRENERGIC RECEPTORS 160
LIGAND-BINDING STUDIES 162
STRUCTURE OF BETA$_1$- AND BETA$_2$-ADRENERGIC RECEPTORS 163
   Affinity and Photoaffinity Labels for the Beta-Adrenergic Receptor 164
   Purification of the Receptors 171
AGONIST-SPECIFIC BINDING PROPERTIES AND THE MECHANISMS OF
    ACTIVATION OF ADENYLATE CYCLASE 174
RECONSTITUTION STUDIES OF BETA-ADRENERGIC RECEPTORS 180
DESENSITIZATION 181

PERSPECTIVES AND SUMMARY

Although hormone and drug receptors have been studied by pharmacologists and physiologists through most of this century, only within the past decade have the properties and regulation of these structures become the focus of intensive interest by biochemists. This change has been brought about largely by the development of ligand-binding techniques that have permitted the direct assay of the binding function of the receptor macromolecules. This has facilitated the study of their molecular properties and

159

has paved the way for their purification and biochemical characterization. Of the membrane-bound hormone and drug receptors, few have attracted more attention than the adenylate cyclase-coupled beta-adrenergic receptors. They have become the premier model system and the focus of interest for diverse groups of scientists interested in the nature and regulation of receptors and the mechanisms by which they accomplish transmission of biological signals across the plasma membrane. There are several reasons for this: (a) their ubiquity and the diversity of the physiological responses they mediate; (b) the close coupling of the receptors to a well-defined biochemical effector unit, the enzyme adenylate cyclase, which has facilitated the use of the beta-adrenergic receptor adenylate cyclase system as a model for dissecting the biochemical mechanisms of hormone receptor-effector coupling; and (c) the important therapeutic consequences of stimulation or blockade of these receptors by pharmacological means, which have implications for the therapy of a wide variety of human illnesses.

An enormous amount of information about these receptors has become available since their initially successful identification by ligand-binding studies in 1974. The receptor (R) binding subunits have been identified and purified from several tissues. The individual effector units such as the guanine nucleotide regulatory protein (N) and the enzyme adenylate cyclase (C), to which the receptors are coupled in the plasma membrane, have been identified and the N unit has been purified. The sequence of molecular events involved in the interaction of these components that leads ultimately to activation of the adenylate cyclase has been partially elucidated, as have a number of factors which regulate these interactions, e.g. agonist-induced desensitization. At each stage of these investigations new insights have been preceded by the development of new technologies such as ligand-binding methods, quantitative approaches to analysis of such data, affinity chromatography and photoaffinity labelling to name but a few.

Future research will focus on the characterization of purified components, their interactions in defined reconstituted systems, and the mechanisms by which they are regulated. Although these tasks will be formidable (individual components must be purified $\sim$ 100,000 fold from membranes), application of genetic engineering and monoclonal antibody technologies offer new approaches and strategies. Previous experience suggests that insights and principles discerned from the beta-adrenergic receptor system may be of fairly general significance for understanding the biochemical basis for hormone- and drug-induced transmembrane signaling mechanisms.

ADRENERGIC RECEPTORS

Catecholamines, such as the endogenous hormone adrenaline, the neurotransmitter noradrenaline and a wide variety of synthetic congeners exert their physiological and pharmacological effects by interacting with specific receptor structures localized to the plasma membranes of responsive cells. Pharmacological approaches have delineated several distinct classes of such receptors that are defined in terms of the potency series of various agonists for stimulating or antagonists for inhibiting responses mediated by the various receptor types (1). The two major classes of receptors are termed alpha- and beta-adrenergic receptors. Each of these two classes of adrenergic receptors has been further subdivided into two subclasses termed $beta_1$ and $beta_2$ (2) and $alpha_1$ and $alpha_2$ (3, 4), respectively. Some pharmacological properties and physiological functions of these receptor subtypes are summarized in Table 1.

$Beta_1$- and $beta_2$-adrenergic receptors stimulate the membrane-bound enzyme complex adenylate cyclase, which catalyzes the formation of the second messenger cyclic AMP from substrate ATP (5). $Alpha_2$-adrenergic receptors often appear to inhibit the enzyme (6). $Alpha_1$-adrenergic recep-

Table 1 Some properties of adrenergic receptor subtypes

| Receptor subtype | Drug specificity | | Location and effect | Mechanism of action |
	Agonists	Antagonists		
$Alpha_1$	Order of potency: Epi $\geqslant$ NE $>$ ISO[a] Specific agonists: methoxamine phenylephrine	Prazosin	Generally postsynaptic (vasoconstriction)	Alter cellular Ca^{2+} fluxes
$Alpha_2$	Order of potency: Epi $\geqslant$ NE $>$ ISO Specific agonists: clonidine tramazoline	Yohimbine Rauwolscine	Presynaptic (feedback inhibition of norepinephrine release from sympathetic nerves) Postsynaptic (vasoconstriction in certain vascular beds) Also human & rabbit platelets, hamster & human adipocytes	Often inhibit adenylate cyclase
$Beta_1$[b]	ISO $>$ NE $\cong$ Epi	Atenolol Betaxolol Metoprolol Practolol	Mammalian heart (contractility & rate), adipose tissue (lipolysis), avian erythrocytes	Stimulate adenylate cyclase
$Beta_2$[b]	ISO $>$ Epi $\gg$ NE	ICI 118551 Butoxamine	Vascular, uterine and bronchial smooth muscle, (relaxation) amphibian erythrocytes	Stimulate adenylate cyclase

[a] Epi = epinephrine, NE = norepinephrine, ISO = isoproterenol.
[b] See Table 2 for a more complete listing of the tissue distribution of beta-adrenergic receptor subtypes.

tors in general do not appear to mediate changes in adenylate cyclase activity. At a biochemical level less is known about alpha-adrenergic than beta-adrenergic receptors. Information about alpha receptors has been summarized elsewhere (7, 8).

LIGAND-BINDING STUDIES

Study of the biochemical properties of any membrane-bound hormone or drug receptor requires the ability to measure the receptor in membranes and solubilized or purified fractions. The simplest approach is to assay the binding function of the receptor macromolecule by ligand-binding techniques. Successful direct radioligand-binding methods for studying the beta-adrenergic receptors were developed only in 1974. The initial ligands were (–)[^{3}H]dihydroalprenolol (9), ($\pm$)[^{125}I]iodohydroxybenzylpindolol (10), and ($\pm$)[^{3}H]propranolol (11). Since these studies, a number of antagonist and agonist radioligands have been developed permitting direct assay of the receptors under a wide variety of circumstances (12). For each of these radioligands, the binding specificity, stereoselectivity, and saturability are all as expected for binding to the physiologically relevant beta-adrenergic receptors. The uses of such ligand-binding approaches include: characterization of the pharmacological binding properties of receptors in intact cells, isolated membranes, and soluble preparations; tagging the receptors through purification procedures (see below); studies of regulation of the number of the receptors in tissues under a variety of physiological and pathophysiological circumstances (reviewed in 13–15); studies comparing the pharmacology of the mechanisms of receptor-effector coupling (see below); and studies of beta$_1$- and beta$_2$-adrenergic receptor subtypes. In general, antagonist radioligands have been more generally useful than have agonist radioligands. All of the antagonist radioligands available are non-subtype selective, i.e. they bind with equal affinity to the beta$_1$- and beta$_2$-adrenergic receptors. However, by constructing competition curves of the binding of one of the nonsubtype selective radioligand antagonists by a subtype selective drug and analyzing the data in one of a variety of ways, it has been possible to quantitate the numbers of beta$_1$ and beta$_2$ adrenergic receptors in various tissues (16–20). Of the several methods of analysis that have been used for this purpose, the nonlinear least-squares curve fitting by computer-assisted methods appears to be the most satisfactory and valid (19, 20). Such experiments reveal that beta$_1$- and beta$_2$-adrenergic receptors often coexist within the same tissue (18). The distribution of the two subtypes of beta-adrenergic receptors in some representative tissues is summarized in Table 2.

Table 2 Tissue distribution of beta-adrenergic receptor subtypes determined by radioligand binding techniques[a]

Tissue	Species	Beta$_1$	Beta$_2$
Heart (left	Rat	83–100	0–17
ventricle)	Dog	100	0
	Frog	15	85
	Rabbit	95	5
	Guinea pig	100	0
	Human	85	15
Lung	Rat	20	80
	Rabbit	80	20
	Bovine	25	75
Placenta	Human	75	25
Uterus	Rat—estrogen dominated	20	80
	Rat—progesterone dominated	0	100
Brain	Rat—cerebral cortex	65	35
	Rat—cerebellum	0	100
	Rat—striatum	65	35
	Rat—limbic forebrain	55	45
Spleen	Rat	35	65
Erythrocyte	Rat	0	100
	Amphibian	0	100
	Avian	100	0

[a]Compiled from references (18, 120) and data from the authors' laboratories.

STRUCTURE OF BETA$_1$- AND BETA$_2$-ADRENERGIC RECEPTORS

Elucidation of the structure and molecular properties of hormone and drug receptors is an important goal in understanding the mechanisms of hormone and drug action at a biochemical level. Two approaches have led to current understanding of the structure of the beta-adrenergic receptors, purification of the receptor-binding macromolecules, and photoaffinity labeling of these receptor structures. The two techniques are complementary and together provide the potential for gaining information about the molecular properties of the receptor macromolecules.

Much has been learned about beta-adrenergic receptor structure and function by studying simple model systems. Erythrocytes of amphibian, avian, and some mammalian species have generally been used for this purpose. The plasma membranes of such nucleated erythrocytes contain beta-adrenergic receptors coupled to adenylate cyclase. There is evidence that receptor control of the enzyme, in turn, modulates ion fluxes across the cell membrane via cyclic AMP dependent mechanisms (21). Since these cells can be obtained as homogeneous populations, and plasma membrane fractions can be easily prepared, numerous investigators have used them as convenient sources of the receptors. The beta-adrenergic receptors of amphibian erythrocytes and rat erythrocytes and reticulocytes appear to be of the beta$_2$ subtype, whereas those of avian species are beta$_1$ in nature (Table 2). Pharmacological differences do exist between the beta$_1$ receptor subtype found in avian erythrocytes and that of mammalian species (22) and between the beta$_2$ receptor subtype of amphibian erythrocytes and mammalian tissues (23). The structural basis for these differences in pharmacological specificity, which relate primarily to the binding of various selective synthetic antagonists, remains to be elucidated. Nonetheless, these cells have provided excellent simple model systems for the initial characterization of these receptors.

Affinity and Photoaffinity Labels for the Beta-Adrenergic Receptor

A number of different affinity and photoaffinity probes for the beta-adrenergic receptor have been described in recent years. These are tabulated in Table 3. NHNB-NBE (24) and BrAlpM (25) are affinity ligands based on modification of the structure of the standard beta antagonists propranolol and alprenolol by inclusion of a bromacetyl moiety. Acebutalol-azide (26), para-azidobenzylpindolol (27), cyanopindolol-azide (28), and the isomers of para-azidobenzylcarazolol (29, 30) are all photoaffinity reagents in which the beta-adrenergic antagonist structure has been modified to include the highly reactive azide moiety, which upon irradiation forms a highly reactive nitrene that can insert covalently into a variety of chemical bonds. Para-aminobenzylcarazolol, a precursor of the photoactive p-azidobenzylcarazolol, is a beta-adrenergic antagonist of high affinity containing a free amine moiety that has been crosslinked into the receptor sites with a photoactive bifunctional crosslinker (31).

Several criteria are necessary for a compound to be of general usefulness as an affinity probe. The compound must: (a) possess high affinity for the site to be labeled; (b) possess a high degree of specificity; (c) be labeled to high specific radioactivity; and (d) possess a reactive group that can be readily covalently incorporated into the binding site of interest. Several of the compounds listed in Table 3 fulfill all these criteria. However, some of the compounds listed have been of only limited usefulness in the characterization of beta-adrenergic receptors. For example, acebutolol-azide has not yet been described in radioactive form and its very modest affinity ($K_D \approx$ 3–4 $\times$ 10^{-7}M) suggests that it would not be a particularly effective probe for studying the beta-adrenergic receptor in any case (26). BrAlpM, a bromoacetylated menthane derivative of the antagonist alprenolol, has been labeled to a specific activity of $\sim$40 Ci/mmol with tritium and possesses reasonable affinity ($K_D \approx 500$ pM) but because of low reactivity has been found to incorporate very marginally into partially purified receptor preparations (25). Similarly, NHNB-NBE, the first irreversible probe described, is of relatively low affinity ($K_D \approx 2 \times 10^{-7}$M) and has been radiolabeled to only low specific activity ($\sim$5.5 Ci/mmol) with tritium. This compound was reported to incorporate into peptides of $M_r \sim 37,000$ and 41,000 in both intact turkey erythrocytes (beta$_1$-adrenergic receptors) and intact L6 muscle cells (beta$_2$-adrenergic receptors) (24). However, the putative receptor bands on radioautographs of SDS-PAGE were only partially protected by high concentrations of the antagonist hydroxybenzylpindolol, thus making the results difficult to interpret with confidence.

In contrast, [^{125}I]para-azidobenzylpindolol, [^{125}I]cyanopindolol-azide, the tritiated and iodinated forms of para-azidobenzylcarazolol and radioiodinated para-aminobenzylcarazolol, appear to represent valuable and generally applicable tools for photoaffinity labeling of the beta-adrenergic receptors. The structures and some of the essential properties of these compounds are shown in Table 3. All of the above compounds possess high affinity for binding to the beta-adrenergic receptor ($K_D \approx 5$–900 pM); they show a high degree of specificity for the receptor; with the exception of [^{3}H]para-azidobenzylcarazolol (40 Ci/mmol) (30) all are labeled with [^{125}I] at theoretical specific activity ($\sim$2000 Ci/mmol) and all compounds possess functional groups that can be covalently incorporated into protein by photolysis or crosslinking.

One of the most important criteria for validating the affinity or photoaffinity labeling approach for the identification of any biological macromolecule is the demonstration that incorporation of the ligand into the macromolecule is specifically blocked by a variety of ligands that specifically occupy the receptor in question. In the case of the beta-adrenergic receptor, using para-azidobenzylcarazolol, para-aminobenzylcarazolol, para-azidobenzylpindolol, and cyanopindolol-azide, it was demonstrated that both beta-adrenergic agonists and antagonists are capable of blocking incorporation of the ligands into specific peptides. An example of this is shown in Figures 1 and 2. Here, erythrocyte membranes from the frog (beta$_2$-adrenergic receptors) (Figure 1) and from the turkey (beta$_1$-adrenergic receptors) (Figure 2) were labeled with [^{125}I]para-azidobenzylcarazolol and photolyzed. Covalent incorporation can be observed into a polypeptide of M_r 58,000 (Figure 1, left panel) for the beta$_2$-receptor of frog erythrocytes and into two polypeptides of M_r 40,000 and 45,000 (Figure 2, left panel) for the beta$_1$-receptor system of the turkey erythrocytes.

As shown in the center panels, incorporation into the various polypep-

Table 3 Characteristics of several beta-adrenergic affinity ligands

Compound		Structure	Functional group and mode of incorporation	Radioactive label	Specific radioactivity (Ci/mmol)	Apparent dissociation constant (pM)[c]	Reference
Name	Trivial name						
N-[2-hydroxy-3-(1-naphthyloxy)-propyl]-N'-bromacetyl-ethylenidia-nine	NHNB-NBE		Bromacetyl affinity	^{3}H	~5	200,000	24
Bromoacetylated menthane derivatives of alprenolol	BrAlpM[a]		Bromacetyl affinity	^{3}H	~40	500	25
p-Azidobenzylpindolol	pABP		Arylazide photoaffinity	^{125}I	~2000	300–900	27
Acebutalol-azide			Arylazide	—		390,000	26

Compound	Abbreviation	Structure	Method	Isotope	K_d		Ref.
(4-Azidobenzimidyl)-3,3-dimethyl-6-hydroxy-7-(2-cyano-3-iodionated-4-yloxy)-1,4-diazaheptane	Cyano-pindolol-azide-1	$R_1 = $ ... $R_2 = $ ... azide-1 X = NH; azide-2 X = O	Arylazide photoaffinity	^{125}I	~2000	40-45	28
1-(4-azidobenzoyl-3,3-dimethyl-6-hydroxy-7-(2-cyano-3-iodionated-4-yloxy)-1,4-diazaheptane	Cyano-pindolol-azide-2						
15-[4'-azidobenzyl] carazolol	pABC		Arylazide photoaffinity	^{3}H	~40	100	30
15-[4'-azidobenzyl] carazolo	pABC[b]	[125I] pAIBC: X = H; Y = I / [125I] pABIC: X = I; Y = H	Arylazide photoaffinity	^{125}I	~2000	5-20	29
15[4'-aminobenzyl] carazolol	pAMBC[b]	[125I] pAMIBC: X = H; Y = I / [125I] pAMBIC: X = I; Y = H	Amine crosslinking	^{125}I	~2000	5	31

[a] For BrAlpM the presence of two R_2 substituents indicates that two different compounds were synthesized and characterized. [125I] pAIBC: [125I] 15-[3'-iodo-4'-azidobenzyl] carazolol; [125I] pABIC: [125I] 15-[4' azidobenzyl]-1-iodocarazolol; [125I] pAMIBC: [125I] 15-[3'-iodo-4'-aminobenzyl] carazolol.

[b] For the two compounds pABC and pAMBC, two different radioiodinated isotopomers for each compound have been prepared. [125I] pAMBIC: [125I] 15-[4'aminobenzyl]-1-iodocarazolol.

[c] In the case of the bromoacetyl affinity reagents a true dissociation constant cannot be calculated since binding of the reagent tends to become irreversible with time and true equilibrium conditions do not exist.

PHOTOAFFINITY LABELING AND PURIFICATION OF THE FROG
ERYTHROCYTE β_2 ADRENERGIC RECEPTOR

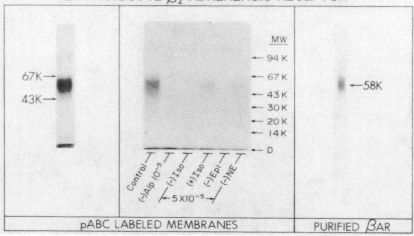

Figure 1   Identification of the beta$_2$-adrenergic receptor-binding subunit of frog erythrocytes by photoaffinity labeling and purification. Left and center panels: Aliquots of frog erythrocyte membranes (equivalent of 300 μg protein/SDS-PAGE lane) were diluted with 25 mM Tris-HCl, 2 mM MgCl$_2$, pH 7.4 at 25°C to a receptor concentration of 20–30 pM ($\sim$ 15 ml) and incubated under dim light for 90 min with [^{125}I]p-azidobenzylcarazolol ([^{125}I]pABC) (20–30 pM) alone (left panel and control lane of center panel) or with (–)alprenolol, 10^{-5} M (Alp); or 5 $\times$ 10^{-5} M of the following drugs: (–)isoproterenol [(–)Iso], (+)isoproterenol (+)Iso], (–)epinephrine [(–)Epi], or (–)norepinephrine [(–)NE]. After incubation membranes were washed twice with the above buffer containing 0.5% fatty-acid-free bovine serum albumin and a last time with buffer alone by centrifugation at 40,000 $\times$ g for 15 min. The pellets were resuspended in 15 ml of 25 mm Tris-HCl, 2 mM MgCl$_2$, pH 7.4 at 25°C and photolyzed for 90 sec, 12 cm from a Hanovia 450-W medium pressure mercury arc lamp filtered through 5 mm of Pyrex glass. After photolysis, the suspension was again sedimented at 40,000 $\times$ g for 15 min and the pellet resuspended in 10% SDS, 10% glycerol, 5% beta-mercaptoethanol, 50 mM Tris-HCl, pH 6.8 and incubated at room temperature for 30–60 min and run on SDS-PAGE as described previously (29). The gel was subsequently dried and an autoradiogram developed by exposing on Kodak XAR-5 film with one intensifying screen. Data in center panel are taken from (29). Right panel: For comparison a purified beta$_2$-adrenergic receptor preparation from frog erythrocyte membranes obtained by one affinity chromatography step on Sepharose-alprenolol and two sequential high performance liquid chromatography steps on steric exclusion columns as described previously (31) was iodinated by chloramine T and Na[^{125}I] and electrophoresed on a 12% SDS polyacrylamide slab gel and an autoradiogram developed as described above. Other conditions were as described in (31). Preparations containing only this polypeptide, as evidenced by radioiodination, bound adrenergic ligands with affinity and specificity identical to the receptor in solubilized preparations.

tides can be blocked with appropriate beta$_1$- or beta$_2$-adrenergic specificity and stereoselectivity. Thus, the antagonist (–)alprenolol (10 μM) completely blocked incorporation into all receptor polypeptides as did 50 μM (–)isoproterenol (an agonist). However, the (+)isomer of isoproterenol was

PHOTOAFFINITY LABELING AND PURIFICATION OF THE TURKEY
ERYTHROCYTE β_1 ADRENERGIC RECEPTOR

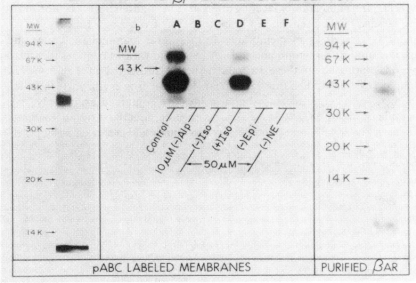

Figure 2   Identification of the beta$_1$-adrenergic receptor binding subunit of turkey ery-throcytes by photoaffinity labeling and purification. Left and center panels: Aliquots of turkey erythrocyte membranes (equivalent of ~ 300 μg protein/SDS-PAGE lane) were diluted in 75 mM Tris-HCl, 25 mM MgCl$_2$, pH 7.4 at 25°C to a receptor concentration of 20–30 pM ($\sim$ 15 ml total volume) and incubated under dim light with [^{125}I]p-azidobenzylcarazolol (20–30 pM) alone or in the presence of 10 μM (–)alprenolol (left panel). In the center panel incuba-tions were either alone or in the presence of the stated concentrations of the various adrenergic drugs indicated. Abbrevations are the same as in Figure 1. All procedures were exactly as described in the legend to Figure 1 and as previously described (29). Data for left and center panels were taken from (29). Right panel: A purified beta$_1$ adrenergic receptor preparation from turkey erythrocyte membranes obtained by an affinity chromatography pass on Se-pharose-alprenolol and two sequential high performance liquid chromatography steps on steric exclusion columns as described previously (35) was iodinated, electrophoresed on a 10% SDS-polyacrylamide slab gel and an autoradiogram developed. Data (right panel) are taken from (35).

much less effective in preventing covalent labeling of the polypeptides in agreement with the stereoselective interaction of catecholamines with their receptors. Moreover, whereas (–)epinephrine and (–)norepinephrine were equipotent in blocking incorporation at the beta$_1$-receptor system of the turkey erythrocyte membranes (Figure 2), (–)epinephrine was slightly more potent at the beta$_2$-adrenergic receptor system (Figure 1), a situation ex-pected from the known pharmacology of ligand-binding and physiological response for these two subtypes of receptors (Table 1). For comparison, on the right-hand panels of Figures 1 and 2, autoradiograms of SDS-PAGE of purified receptor preparations from both frog and turkey erythrocytes are shown, which are obtained by the purification procedures described below.

Using the photoaffinity labeling approach, beta-adrenergic receptor-binding peptides have been labeled in erythrocytes from frogs, turkeys, ducks, and rats as well as in membranes from several mammalian tissues and species, and in S49 lymphoma cells. A list of the various peptides identified by this approach in these various tissues is presented in Table 4. The simplest system studied thus far appears to be the frog erythrocyte in which the beta$_2$-adrenergic receptor-binding sites are contained exclusively on a peptide of about 58,000 M_r (32). This peptide, identified by photoaffinity labeling (29, 30, 33) and photoaffinity crosslinking (31) as well as by purifi-

Table 4 Beta-adrenergic receptor polypeptides identified by affinity or photoaffinity labeling and purification techniques

| Species | Tissue | Receptor subtype | Peptides identified | |
			Affinity labeling	Purification
Frog	Erythrocytes	Beta$_2$	58,000 (29, 31) 63,000 (33)	58,000 (31–32)
	Heart	Beta$_2$	62,000 (34)	
Turkey	Erythrocytes	Beta$_1$	45,000[a]; 40,000 (29, 35) 43,500 (33) 50,000[a]; 40,000 (28) 41,000; 37,000 (24)	45,000[a]; 40,000 (35)
Duck	Erythrocytes	Beta$_1$	48,500[a]; 45,000 (27, 33)	
Pigeon	Erythrocytes	Beta$_1$	52,000; 45,000[a] (33)	
Rat	Erythrocytes	Beta$_2$	65,000; 53,000; 39,000[a] (29) 62,500 (33)	
	Reticulocytes	Beta$_2$	65,000; 53,000; 39,000[a] (29)	
	Lung	Beta$_1$ (20%) Beta$_2$ (80%)	62,000; 47,000; 36,000 (29, 37)	
	Heart	Beta$_1$	62,000; 55,000[a] (34)	
Rabbit	Lung	Beta$_1$ (80%) Beta$_2$ (20%)	65,000; 45,000; 38,000 (29)	
	Skeletal muscle	Beta$_2$ (80%) Beta$_1$ (20%)	54,000; 48,000; 44,000[a] (36)	
	Heart	Beta$_1$	62,000; 55,000[a] (34)	
Dog	Lung	Beta$_1$ (20%) Beta$_2$ (80%)	52,000; 39,000[a] (36)	
	Heart	Beta$_1$	62,000; 55,000[a] (34)	
Guinea Pig	Lung	Beta$_1$ (20%) Beta$_2$ (80%)	62,000; 52,000[a]; 40,000[a] (36)	
Cultured cells (mouse)	S49 lymphoma	Beta$_2$	65,000; 55,000 (38)	
Human	Heart	Beta$_1$	62,000; 55,000[a] (34)	

[a] Denotes minor components.

cation methods (31, 32) (see below), generally migrates as a broad band centered around M_r 58,000 on SDS-PAGE. The reasons for this apparent micro-heterogeneity are not clear, but they may well relate to heterogeneity of carbohydrate moieties present on the glycoprotein receptor. In contrast, in avian erythrocytes [turkey (29, 35), pigeon (33), duck (27)] containing beta$_1$-adrenergic receptors, two peptides are labeled having molecular weights of 38,000–40,000 and 45,000–50,000, respectively. The smaller molecular weight peptide appears to be present in a 3–4:1 ratio as compared with the larger migrating peptide (28, 29). The same two peptides have been purified to apparent homogeneity from turkey erythrocytes (35).

In contrast, for all mammalian beta$_1$- and beta$_2$-adrenergic receptors examined thus far (Table 4) (29, 33–38), either two or three peptides have been found in varying ratios. The peptides are of 62,000–65,000, 47,000–53,000, and 35,000–40,000 M_r (Table 4). However, inclusion of certain protease inhibitors, especially those affecting metalloproteases such as EDTA or EGTA leads to enhanced labeling of the largest, i.e. $\sim$ 62,000 M_r peptide. Of particular interest are the very similar patterns observed in the rat and rabbit lung systems, which contain respectively 80% beta$_2$- and 80% beta$_1$-adrenergic receptors. Protection experiments with subtype selective agents indicate that the labeled peptides in the rat lung membranes are of the beta$_2$ subtype whereas those from the rabbit lung membranes appear to be all of the beta$_1$ subtype (36, 37). In rat lung membranes, the beta$_2$ selective antagonist ICI 118551 blocks covalent labeling more potently than the beta$_1$ selective antagonist betaxolol and the agonist (–)epinephrine is more potent than (–)norepinephrine. In rabbit lung, the reverse situation is observed for the antagonists, and (–)epinephrine and (–)norepinephrine are equipotent, typical of a beta$_1$ pharmacology (36). These results suggest that even though in some tissues beta-adrenergic receptor polypeptides may be very similar in their migration pattern on SDS-PAGE, they nonetheless display different pharmacological characteristics. More rigorous assessment of the degree of identity or nonidentity between these peptides will have to await further comparison by peptide mapping and amino acid sequencing of the purified or photoaffinity labeled peptides. The photoaffinity probes recently developed should facilitate these types of studies. The photoaffinity probes have already proved to be useful tools for studying purified receptor peptides (see below). Possible explanations for the apparent size heterogeneity of beta-adrenergic receptor ligand-binding peptides are discussed below.

Purification of the Receptors

Successful purification of beta-adrenergic receptors from several sources has been accomplished, and in each case the key appears to be the use of affinity chromatography. The most useful tool is an affinity support involving Se-pharose-immobilized alprenolol (39, 40). Alprenolol is a potent ($K_D \approx 1$–2 nM) beta-adrenergic antagonist. Plasma membranes have generally been solubilized with digitonin, since this has been the only detergent that solu-bilizes the receptors in a form permitting their subsequent interaction with specific ligands (40, 41). Numerous attempts to solubilize the receptors with a variety of other detergents have led to complete loss of receptor-binding activity. Affinity chromatography of digitonin-solubilized beta-adrenergic receptors on Sepharose-alprenolol gels leads to a 100–1000-fold purification depending on the system in question and whether elution is performed in a step or gradient fashion using competing beta-adrenergic ligand (32, 35, 40). It has been documented that both adsorption to and elution from such Sepharose-alprenolol gels are highly biospecific phenomena that reflect the inherent specificity and stereoselectivity of beta-adrenergic receptor interac-tions. This procedure is thus a good example of true affinity chromatogra-phy (40). By coupling repeated affinity chromatography cycles with ion exchange (32) or affinity chromatography and several high performance liquid chromatography steps (31, 35), the receptors from several sources have been purified to apparent homogeneity. In the case of the frog and turkey erythrocyte the same one or two peptides that can be labeled by the photoaffinity techniques in membranes are the ones that are purified by the above-described techniques (Figure 1, right panel; Figure 2, right panel) (29–31, 35).

The only way, short of complete reconstitution, of validating that a purified protein is in fact the beta-adrenergic receptor is to document that the ligand-binding properties of the pure protein are identical to those of the receptors observed in intact cells or membranes. Each of the purified beta-adrenergic receptor peptides has been shown to bind beta-adrenergic ligands with the appropriate beta$_2$- or beta$_1$-adrenergic specificity (32, 35). In each of these cases it has been documented that the pure peptides bind beta-adrenergic ligands with specific activities in the range of 12,000–18,000 pmol/mg, which generally represents an approximate 80–100,000-fold pu-rification from crude membrane fractions (31, 32, 35).

Several other reports of purification of turkey erythrocyte beta-adrener-gic receptors have appeared using affinity chromatography (42) or im-munoaffinity chromatography (43). However, these reports contained no documentation that any of the purified peptides bound ligands with a beta-adrenergic specificity, making the results difficult to interpret. As ob-served in Table 4, in those cases where beta-adrenergic receptors have been both purified and photoaffinity labeled in membranes or in purified prepara-tions, the results are in excellent accord. This correlation validates both approaches.

Another approach to gain information about the structural properties of beta-adrenergic receptors is use of the radiation inactivation technique. This technique estimates the functional target size of a protein under non-denaturing conditions. Radiation-inactivation studies of the beta$_2$-adrenergic receptor of frog erythrocyte membranes in situ and of highly purified preparations indicate that both appear to have a target size of about 54,000 daltons, in excellent agreement with the results obtained by purification and photoaffinity labeling as described above (44). These results, together with consideration of the specific activity of purified receptor preparations, suggest that a single binding site exists per subunit of M_r 54–58,000.

With membranes and purified receptor preparations of turkey erythrocytes, an estimate of 50,000–55,000 for the functional target size of the receptor has been obtained by radiation inactivation studies (44). This accords relatively well with the size of the polypeptides identified by photoaffinity and purification in turkey erythrocytes (cf Figure 2). However, it had been initially reported that the beta-adrenergic receptor of the turkey erythrocyte membranes had an apparent size of 90,000 (45). The reasons for these differences are not readily apparent. The receptor does not exist as a dimer of nonidentical subunits of M_r 40,000 and 45,000, because these two forms of the receptor can be separated by isoelectric focusing or high performance liquid chromatography (35) and the peptides shown to bind beta-adrenergic drugs with identical beta$_1$ specificity. Thus, these two peptides appear to represent two different forms of the beta$_1$ receptor.

Several major questions concerning beta-adrenergic structure are the following:

1. Do beta$_1$- and beta$_2$-adrenergic receptors have basically similar or different structures? The data available thus far suggest that in mammalian systems peptides of comparable size contain both beta$_1$- and beta$_2$-adrenergic receptor-binding sites. Table 4 shows that in every case the largest identified peptide is of $M_r \approx$ 60,000.

2. What is the relationship of the multiple receptor-binding peptides observed by both photoaffinity labeling and purification techniques? Because each of the several identified peptides appears to contain the appropriate ligand-binding sites, these may represent isoreceptors or various forms of processed receptors, e.g. receptors differing in their extent of glycosylation; or they may be caused by proteolysis during the preparation of samples for analysis. Since the proportion of receptor migrating on SDS-PAGE with larger apparent M_r is increased by alterations in sample preparation techniques or inclusion of batteries of certain protease inhibitors, some form of proteolysis is likely responsible for the heterogeneity observed. More rigorous comparison of structures will have to await amino acid sequence data for each of these peptides. However, to date no information is available

concerning the amino acid composition or sequence of the active binding site of the receptors. Beta-adrenergic receptors do, however, appear to be glycoproteins by virtue of their strong interaction with lectin-Sepharose resins (46).

In summary, current data suggest that in mammalian systems as well as in the frog, both beta$_1$- and beta$_2$-adrenergic receptor-binding sites reside on peptides of M_r 58–62,000. These are readily degraded by endogenous proteases to ligand-binding fragments of M_r 50–55,000, 38–40,000, and 30,000. The avian erythrocyte may be an exception, because as yet no evidence for peptides larger than $\sim$ 40–50,000 has been obtained.

AGONIST-SPECIFIC BINDING PROPERTIES AND THE MECHANISMS OF ACTIVATION OF ADENYLATE CYCLASE

Over the past several years ligand-binding studies as well as a variety of biochemical approaches have been brought to bear on the problem of understanding how the beta-adrenergic receptor interacts with the other components of the hormone responsive adenylate cyclase system resulting in activation of the enzyme. In addition to the receptors (R) these other components include the catalytic moiety of the enzyme (C) and the nucleotide regulatory protein (N or G/F). Attempts at purifying the catalytic unit have progressed very slowly, apparently owing to the instability of this entity when it is separated from the nucleotide regulatory protein. In contrast, the nucleotide regulatory protein has been purified from two sources, rabbit liver and the turkey erythrocyte, and a good deal is now known about its structure (47, 48). This regulatory component is composed of two major subunits of molecular weights $M_r = 35,000$ and 45,000. The properties of this protein have been reviewed in some detail elsewhere (49–51). As discussed below this regulatory protein is necessary to functionally couple agonist occupancy of the beta-adrenergic receptors with activation of the catalytic moiety. The actions of guanine nucleotides, which are absolutely required for hormone stimulation of the enzyme, appear to be mediated through this protein.

A major goal of investigators of adenylate cyclase has been to understand the sequence of molecular events required for transmembrane signaling. This process is initiated by agonist binding to the receptors and thus results in the stimulation of the catalytic moiety. The study of receptor-binding properties through the use of specific radioligands has contributed in a major way to the evolution of thinking about this problem. Guanine nucleotides not only modulate the catalytic activity of adenylate cyclase but also exert specific regulatory effects on agonist binding to beta-adrenergic receptors (52, 53). Thus, when examined in ligand-binding assays, guanine nucleotides decreased the overall affinity of agonist binding while having virtually no effect on antagonist binding. Moreover, the extent to which the affinity of binding of an agonist was decreased by guanine nucleotides was found to vary in direct proportion to its intrinsic activity, i.e. its maximum ability to stimulate adenylate cyclase (53). Subsequently, detailed computer modeling of ligand-binding data suggested an explanation for these phenomena. Distinctly different results were obtained in experiments in which various agonists and antagonists were used to inhibit competitively the binding of radiolabeled antagonists such as (–)[^{3}H]dihydroalprenolol to receptors in plasma membrane fractions. Whereas antagonist competition curves are uniphasic (slope factors or "pseudo-Hill coefficients" $\cong$ 1), agonist competition curves are shallow (slope factor $<$ 1) reflecting complex agonist-receptor interactions that can be resolved into two distinct affinity states (R_H and R_L) having respectively high (K_H) and low (K_L) affinity for agonists (54–56). Guanine nucleotides appear to convert all the high into low affinity state receptors. Moreover, the extent to which an agonist can stimulate the activity of the adenylate cyclase correlates with the percentage of the total receptor population stabilized in the "high affinity state" by the agonist (% R_H), or the ratio of the agonist's affinities for the two states of the receptor, K_L/K_H (54). Thus, both % R_H and K_L/K_H are directly related to the intrinsic activity of the agonist. As noted above, antagonists generally do not distinguish these different affinity states of the receptor.

Computer modeling methods have also been used to demonstrate that a "ternary complex" model can fit a wide variety of ligand-binding data obtained in beta-adrenergic receptor systems (57).

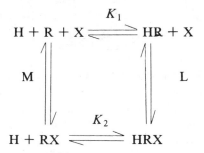

In such a scheme, H stands for hormone or agonist, R for the beta-adrenergic receptor and X for some putative additional membrane component. When binding data are analyzed with a curve fitter based on this model, the major correlate of the intrinsic activity of various agonists is the parameter *L*, which is the association constant for the reaction HR + X $\rightleftharpoons$ HRX.

Thus, agonists stabilize the ternary complex HRX. This modeling approach is independent of the nature of the component X, but since guanine nucleotides uniquely destabilize the agonist-induced ternary complex, it was suggested that X is a guanine nucleotide binding protein (56).

The availability of an agonist radioligand that binds to the beta-adrenergic receptors [³H]hydroxybenzylisoproterenol (57, 58) has led to further insights into the fundamental differences between agonist and antagonist binding to the beta receptors, and has allowed direct biochemical confirmation of the ternary complex model. Direct binding assays show that [³H]HBI labels predominantly the high affinity form of the receptor (R_H) described above (56, 58). Important results have been obtained by comparing the molecular properties of beta-adrenergic receptors solubilized from membranes after prelabeling with the agonist [³H]hydroxybenzylisoproterenol, or the antagonists [³H]dihydroalprenolol or [¹²⁵I]hydroxybenzylpindolol (59, 60, 61). Following solubilization with digitonin, the agonist-prelabeled receptor (the form of the receptor that binds agonists with high affinity) elutes from an ACA34 gel filtration column with a larger apparent molecular weight than the antagonist-prelabeled receptors. Moreover, guanine nucleotides can revert the larger molecular weight form of the receptor stabilized by agonist to the smaller form (61), consistent with the notion of guanine nucleotide-mediated interconversion of receptor states derived from computer modeling of agonist competition binding data.

The larger molecular weight form of the receptor, stabilized by agonists, is attributed to association of the guanine nucleotide regulatory protein with the agonist-receptor complex. This is demonstrated by covalently tagging the 45,000 M_r subunit of the regulatory protein in the presence of cholera toxin (62). Cholera toxin activates adenylate cyclase by catalyzing the incorporation of ADP ribose from NAD^+ into the nucleotide regulatory protein (47, 48, 63, 64, 65). If [³²P]NAD^+ is utilized as a cosubstrate for this toxin, a radioactive marker can be covalently incorporated into the regulatory protein prior to performing gel filtration experiments such as those described above. The [³²P]-labeled regulatory protein coelutes specifically with the larger form of the receptor, the formation of which is promoted uniquely by agonists. These data indicate that the ternary complex form of the beta-adrenergic receptor is likely a complex of hormone, receptor, and the nucleotide regulatory protein. This complex requires magnesium for its stability (66, 67), and cannot be formed in the absence of divalent cations. Studies using mutants of the S49 lymphoma cells that lack a functional N protein (cyc⁻ mutant) provide further evidence for the composition of this complex. Agonists do not promote a high affinity state of the receptor in membranes prepared from these cells (68). However, reconstitution of guanine nucleotide dependent agonist activation of adenylate cyclase in these membranes by insertion of the N protein into the membranes also restores the ability of agonists to promote formation of a high affinity complex with the beta-adrenergic receptor (69).

The unique properties of agonist interaction with beta-adrenergic receptors have also been probed by exploiting the observation that agonist but not antagonist binding produces a conformational change in the receptor-adenylate cyclase complex that appears to expose a critical sulfhydryl group that can be alkylated by N-ethylmaleimide (NEM) (70). Alkylation of this sulfhydryl inactivates the receptors in the sense that they can no longer bind radiolabeled ligands. Recent studies indicate that the effect of NEM is to "lock" the agonist very tightly into the receptor in the HRN complex, thus preventing subsequent assay of the receptors by radioligands (71, 72). The greater the intrinsic activity of the agonist the greater the rate at which it promotes NEM "inactivation" of the receptor (73, 74). Some uncertainty exists as to whether the critical sulfhydryl group is located on the receptor, on the N protein or both. In any case, the N protein is required in order to form the complex sensitive to the actions of NEM; agonists do not promote the NEM inactivation of receptors in the S49 cyc⁻ mutant that lacks the N protein (75). Moreover, GTP, which reverts the high affinity HRN complex to the low affinity form, also prevents the ability of agonists to induce the NEM inactivation of the receptor (75). Interestingly, only 50–60% of the receptors can be maximally "inactivated" by agonist plus NEM. The reasons for this are unclear at present.

Formation of the HRN complex represents a crucial intermediate step in the pathway to hormonal activation of the cyclase via the beta-adrenergic receptor as depicted in Figure 3. This model indicates that agonist binding to the receptor (Step 1) promotes the formation of the ternary complex HRN (Step 2). The formation of this "high affinity" complex facilitates the exchange of GTP for GDP on the regulatory protein (Steps 3 and 4) (76, 77). The binding of the guanine nucleotide triphosphate activates the nucleotide regulatory protein and releases N by destabilizing the HRN ternary complex. It is the free N protein (or at least one or more of its subunits) now charged with stimulatory GTP which appears to activate the catalytic moiety of the enzyme (Step 5), thus forming an $N_{GTP}C$ complex representing the active form of the enzyme (78–80). Moreover, this complex also appears to contain a GTPase activity (Step 6) which, by hydrolyzing GTP to GDP, terminates each cycle of adenylate cyclase activation (Step 7) (81, 82). Thus, nonhydrolyzable analogs of GTP such as Gpp(NH)p or GTP-γS lead to persistent activation of the enzyme, because they are only slowly hydrolyzed.

Similarly, the action of cholera toxin to irreversibly activate adenylate cyclase is attributable to its ability to catalyze a covalent modification (ADP ribosylation) of the nucleotide regulatory protein. This cholera toxin catalyzed reaction inhibits the GTPase, and thus renders the action of GTP to stimulate the enzyme essentially irreversible (83). The site of ADP-ribysylation is on the 45,000 M_r peptide but not the 35,000 M_r subunit of the regulatory protein (47). Consistent with the scheme described above are the observations that the intrinsic activities of a series of agonists for stimulation of adenylate cyclase, GTPase and GDP release, are identical (84, 85).

The fundamental concept in this mechanism is the action of the guanine nucleotide regulatory protein as a shuttle that conveys information from the agonist-occupied receptor to the catalytic moiety of the enzyme. Several lines of investigation provide support for this idea. In one approach the ternary complex (HRN) was formed initially in frog erythrocyte membranes by preincubation with agonists (86). This complex was solubilized with lubrol PX, a nonionic detergent and adsorbed to wheat germ lectin-Sepharose by virtue of interaction with the receptor, which is a glycoprotein. Because guanine nucleotide destabilizes the ternary complex, the N protein could then be eluted specifically from the lectin Sepharose with the nonhydrolyzable GTP analog, GTPγS. The GTPγS-N complex eluted from the resin could then be reconstituted with a suitable acceptor, in this case catalytic units of solubilized turkey erythrocyte adenylate cyclase. When the frog erythrocyte membranes were first preincubated with an antagonist or with an agonist, followed by GTP that disrupts the ternary complex, and the experiment then carried through, very little reconstitutive activity was found in the GTPγS eluate of the lectin resin. This is because the N protein does not ordinarily interact with the lectin-Sepharose except by virtue of its specific interaction with the agonist-occupied receptor,

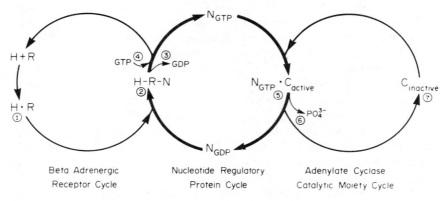

Figure 3 Schematic model for activation of adenylate cyclase by beta-adrenergic agonists. H = hormone or agonist drug, R = beta-adrenergic receptor, N = nucleotide regulatory protein, C = catalytic moiety of adenylate cyclase.

which can then be adsorbed. This experiment provides support not only for the concept of the ternary complex HRN as an intermediate in the adenylate cyclase reaction scheme, but also for the identity of the nucleotide regulatory protein that associates with the receptor and activates the catalytic moiety.

In a somewhat parallel line of experimentation, detergent solubilized fractions of R and N were prepared that were free of functional C (87). The mixing of these soluble fractions was followed by detergent removal and phospholipid treatment, which resulted in the reconstitution of fruitful R-N communication. Hormone-dependent activation of N was then assayed by fusion of the reconstituted system to S49 cyc⁻ mutant cells and subsequent assay of adenylate cyclase activity. The kinetics of R and N interactions were studied by varying the proportion of the soluble components in the reconstitution system. The agonist promoted activation of N showed a first-order dependence with respect to R but a zero-order dependence on the concentration of N. These studies indicate that R acts catalytically to activate N and represent an advance over previous kinetic experiments, which had been interpreted in the framework of agonist-receptor complexes interacting directly with the adenylate cyclase (88). Thus, both sets of reconstitution experiments support the notion that the guanine nucleotide regulatory protein acts as a shuttle interacting on the one hand with the receptor and on the other with the catalytic moiety of the enzyme.

The scheme shown in Figure 3 forms the basis for current research into receptor-cyclase communication. Researchers are focusing particularly on the two key intermediates that link the three components in the model, i.e. HRN and $N_{GTP}C$. It is necessary to define in molecular terms the rate-limiting step for cyclase activation. It has been suggested that this crucial step is the activation of the N protein that is catalyzed by the agonist-occupied receptor through the formation of the high affinity HRN complex. The activation requires the binding of a guanine nucleotide triphosphate to N. Preliminary evidence suggests that the binding of the nucleotide not only destabilizes the HRN complex but may also cause dissociation of the 45,000 M_r from the 35,000 M_r subunit of N. The GTP-bound 45,000 M_r subunit then associates with C to form the active holoenzyme adenylate cyclase. In some systems, however, GDP release from the regulatory protein appears to be the controlling step.

Another question is the nature of the GTPase that serves as the "turn off" mechanism for adenylate cyclase. Is this activity intrinsic to the N protein itself or is the association of N with C required to trigger this enzymatic activity? Other issues yet to be resolved are the role of allosteric effectors such as metal ions in receptor cyclase coupling, the manner in which the lipid environment of the membrane influences transmembrane signaling, and the purification of the catalytic moiety of adenylate cyclase. More direct approaches with well-characterized isolated components will be necessary to precisely define the molecular events involved in receptor-cyclase coupling.

RECONSTITUTION STUDIES
OF BETA-ADRENERGIC RECEPTORS

Hormone and drug receptors perform two essential functions. First, they bind biologically active molecules. Second, they transduce this binding into a biochemical and physiological response. As receptors are purified, ligand-binding studies can assess continually the functionality of the ligand-binding site. However, the ability of the receptor to activate biological responses requires some other form of assay, i.e. a reconstitution assay. In the case of the beta-adrenergic receptor, the obvious assay is one in which a purified receptor-binding component is reconstituted within an artificial or natural lipid milieu with the other components of the system (N and C) and confers upon these components sensitivity to stimulation by catecholamines with a typical beta-adrenergic specificity. The requirement for a lipid membrane in which the various components of the beta receptor-coupled adenylate cyclase system can interact has been stressed repeatedly. In fact, the solubilization of such systems with any detergent leads invariably to a complete loss of hormone sensitivity of the enzyme. Thus, a structured membrane environment is required for successful reconstitution and functioning of the beta-adrenergic receptor adenylate cyclase system. Reconstitution studies must answer the question of whether the various isolated beta-adrenergic receptor peptides thus far identified also contain all the functional units necessary for interaction with the nucleotide regulatory protein and activation of adenylate cyclase. It is possible that other protein components are required. However, if one of these isolated beta-adrenergic receptor binding sites could be shown to reconstitute the catecholamine responsiveness of a model system this would indicate that both the ligand-binding and activating functions are present on the same macromolecule. This goal has yet to be achieved.

To date, efforts at reconstitution have utilized a variety of unpurified systems. Schramm and colleagues (89, 90) have shown that beta-adrenergic receptors in intact cells or isolated membranes can be coupled to the adenylate cyclase system of acceptor cells by performing fusions using Sendai virus or polyethyleneglycol. By chemically inactivating the N and C components of the donor membranes or cells, it could be shown that their beta-adrenergic receptors became functionally coupled to the cyclase system of the acceptor cells or membranes. These results confirmed biochemical studies that indicated that the beta-adrenergic receptor component was in fact distinct from the more distal components of the adenylate cyclase system (91, 92). More recently, several groups have reconstituted solubilized beta-adrenergic receptors from turkey or rat erythrocytes with unpurified nucleotide regulatory proteins and catalytic moieties to form functioning catecholamine responsive systems (87, 93, 94). All such cases, however, used crude receptor preparations. Thus, although these studies demonstrate the feasibility of reconstituting solubilized receptors and other components of the system within lipid vesicles, they have shed light neither on the structural features of the receptor required for these interactions nor on the nature of the interactions themselves. However these preliminary studies point the way to what will hopefully be successful reconstitution of highly purified receptor macromolecules.

DESENSITIZATION

One of the most striking insights about receptors in general and beta-adrenergic receptors in particular to emerge from recent studies is the very dynamic nature of their regulation by a wide variety of influences. These have been reviewed in detail elsewhere (14, 95, 96). One highly studied aspect of the regulation of beta-adrenergic receptors has particularly far-reaching implications. Phenomena referred to as "desensitization," "tolerance," "tachyphylaxis," or "refractoriness" (97), undoubtedly mediated by diverse mechanisms, all exhibit upon exposure of cells to beta-adrenergic catecholamines a progressive, often rapid, loss in subsequent responsiveness of the adenylate cyclase system to further catecholamine stimulation. In some cases desensitization is highly specific. Thus, subsequent responsiveness of the enzyme to catecholamines may diminish after catecholamine exposure, whereas responsiveness of the enzyme to other effectors such as prostaglandins or the fluoride ion remains unaffected. Such desensitization phenomena are often referred to as "homologous" (98). In contrast, exposure of a system to one hormone, such as a prostaglandin, may render the cell refractory to further stimulation by all classes of hormonal activators and even to nonspecific activators of the enzyme such as fluoride or Gpp(NH)p. This type of desensitization has been referred to as "heterologous" (98). Although the molecular mechanisms of these regulatory phenomena remain to be elucidated fully, recent findings indicate that alterations in the beta-adrenergic receptors underlie homologous forms of desensitization (99–101). Thus, several types of reconstitution experiments indicate normally functioning N and C components in homologous forms of desensitization. Alterations in more distal components such as the nucleotide regulatory proteins probably also contribute to heterologous forms of desensitization (101).

Two very distinct pathways for agonist-promoted beta-adrenergic receptor alterations leading to desensitization have been described. The first pathway has been extensively studied in frog erythrocytes (54, 97, 102, 103, 104) as well as in various types of cultured cells, in particular astrocytoma cells (105). This pathway involves an agonist-promoted internalization of the receptor through an endocytotic process. This process occurs in several identifiable steps (105, 106). Occupancy of the receptors by agonists (not antagonists) in some systems leads to a rapid "uncoupling" of the receptors within the plasma membranes (54, 105, 107). This is apparent functionally as a decrease in the ability of the receptors to stimulate adenylate cyclase. It can also be seen as a change in the agonist-binding properties of the receptor. The uncoupled receptors are less able to form the ternary HRN complex (56). As a result, agonist binding to these receptors is of uniformly low affinity and is unresponsive to guanine nucleotides (54, 105, 107). This can be measured as a rightward shift and steepening of an agonist competition curve, or as a marked decrease in [^{3}H] agonist, i.e. [^{3}H]hydroxybenzylisoproterenol binding (107). Both alterations reflect the decreased ability of the agonist to stabilize the high affinity form of the receptor. At this early stage of desensitization the number of receptors in the plasma membrane is not yet altered (105). After their formation, these uncoupled receptors can be found in what appear to be internalized vesicles in which they are sequestered away from the cell surface and can no longer be assayed in plasma membrane fractions (102, 108). These receptors, now lost from the cell surface (109, 110), are sometimes referred to as "down regulated" (111). Earlier reports of the presence of a small percentage of these "internalized" receptors in the soluble fraction of the cell (103, 104), probably failed to note the presence of light vesicles in the supernatant fractions used for the binding assays. These internalized vesicles contain a large proportion of all the down regulated receptors after desensitization but appear to be devoid of other plasma membrane markers and even the N and C components of the adenylate cyclase (102). When studied in the frog erythrocyte system, the receptors within the vesicles appear not to be degraded and have normal mobility on SDS gels as assessed by photoaffinity labeling techniques (102).

When agonists are removed from the environment of the cells, the receptors likely recycle to the cell surface and are reinserted into the plasma membrane where they resume normal function. Further evidence for the nondegradation of the internalized receptors is the normal reappearance of receptors at the cell surface upon removal of the agonist, even when protein synthesis is blocked (105, 112, 112a) or when tunicamycin is present, an antibiotic that interferes with the glycosylation of newly synthesized proteins (113). This pathway of beta-adrenergic receptor internalization is somewhat analogous to the receptor recycling pathways recently described for lipoproteins and polypeptide hormone receptors (114). However, this pathway may not necessarily involve coated pits, at least in frog erythrocytes. The nature of the agonist-promoted alteration in the receptors that triggers initially their uncoupling and subsequently their internalization in sequestered vesicles away from other components of the adenylate cyclase system remains to be defined.

A very different form of agonist-promoted receptor alteration leading to desensitization has been described in turkey erythrocytes (115, 116). These cells become refractory to catecholamine stimulation upon incubation with agonists, but there is no down regulation of receptor number or internalization of the receptors. There is also a small but significant decrease in F$^-$-stimulated enzyme activity (115, 116). However, these cells show receptor "uncoupling" when incubated with catecholamines. This is apparent as a decreased ability of agonists to stimulate the cyclase through the receptor, as well as by a right-shifted agonist competition curve in binding experiments, which indicates decreased formation of the high affinity HRN complex (116). The ability of the agonist-occupied receptors to promote release of GDP from the N proteins is also impaired (117). Interestingly, the entire picture in the turkey erythrocyte, i.e. the desensitization as well as the receptor alterations, can be reproduced by incubating the cells with cyclic nucleotide analogs such as 8-bromocyclic AMP or dibutyryl cyclic AMP (116). Cyclic AMP appears to mediate desensitization in certain other cells as well (118).

When beta-adrenergic receptors from plasma membranes of control turkey erythrocytes and desensitized cells are photoaffinity-labeled with [125I] para-azidobenzylcarazolol, striking differences are observed in the mobility of the photolabeled receptors on SDS polyacrylamide gels. Both of the receptor peptides, M_r 40,000 and 50,000, move with somewhat larger apparent molecular weights of about 42,000 and 53,000, respectively, after desensitization. In addition, the smaller peptide may now be visualized as a doublet. Moreover, if desensitization is blocked with a beta-adrenergic antagonist such as propranolol, the mobility change of the receptor peptides is also blocked. Thus, the agonist-induced desensitization in these cells is associated with an agonist-promoted alteration in the structure of the receptor that is manifested by an altered mobility on SDS polyacrylamide gel electrophoresis.

The following experiments tested the hypothesis that phosphorylation of the beta-adrenergic receptor is responsible for the observed desensitization of the receptor and its altered mobility on SDS gels. Intact turkey erythrocytes were preincubated with [32P$_i$] to label the intracellular pool of ATP. These cells were subsequently desensitized by incubation with isoproterenol. The receptor was then purified to homogeneity using affinity chromatography and HPLC methods, and the purified receptor subjected to SDS polyacrylamide gel electrophoresis. Desensitization was associated with a marked stimulation of phosphorylation of both receptor peptides as assessed by radioautography. Moreover, the phosphorylated receptor showed the same characteristic changes on SDS-PAGE as originally discerned by PABC photoaffinity labeling. The beta antagonist propranolol blocked these agonist-promoted changes. These findings establish that, at least in the model turkey erythrocyte system, desensitization is associated with phosphorylation of the beta-adrenergic receptors (119). This is the first documented example of a structural alteration in a component of the adenylate cyclase system contributing to a desensitization process.

Considered together, the frog and turkey erythrocyte model systems are examples of two very different mechanisms by which beta-adrenergic receptors can be "uncoupled" from adenylate cyclase during desensitization. In one case, the receptor is sequestered away from the other components of the adenylate cyclase system in special vesicles. In the other, the receptors remain in the plasma membrane but are structurally altered and thus uncoupled from the effector adenylate cyclase. These findings underscore the divergent mechanisms that have evolved to regulate receptor function.

Literature Cited

1. Ahlquist, R. P. 1948. *Am. J. Physiol.* 153:586–600
2. Lands, A. M., Arnold, A., McAuliff, J. P., Luduena, F. P., Braun, T. G. 1964. *Nature* 214:597–98
3. Starke, K. 1977. *Revs. Physiol. Biochem. Pharmacol.* 77:1–124
4. Langer, S. Z. 1974. *Biochem. Pharmacol.* 23:1793–800
5. Sutherland, E. W., Robison, G. A., Butcher, R. W. 1968. *Circulation* 37: 279–89
6. Michel, T., Hoffman, B., Lefkowitz, R. J. 1982. *Biochemical Actions of Hormones*, ed. G. Litwack, pp. 43–68. New York: Academic. 362 pp.
7. Hoffman, B. B., Lefkowitz, R. J. 1980. *N. Engl. J. Med.* 302:1390–96
8. Hoffman, B. B., Lefkowitz, R. J. 1982. *J. Cardiovasc. Pharmacol.* 4:514–18
9. Lefkowitz, R. J., Mukherjee, C., Coverstone, M., Caron, M. G. 1974. *Biochem. Biophys. Res. Commun.* 60:703–09
10. Aurbach, G. D., Fedak, S. A., Woodard, C. J., Palmer, J. S., Hauser, D., Troxler, F. 1974. *Science* 186:1223–25
11. Atlas, D., Steer, M. L., Levitzki, A. 1974. *Proc. Natl. Acad. Sci. USA* 71: 4246–48
12. Burgisser, E., Lefkowitz, R. J. 1982. *Methods in Neurobiology,* ed. P. J. Marangos. New York: Academic. In press
13. Hoffman, B. B., Lefkowitz, R. J. 1980. *Ann. Rev. Pharmacol. Toxicol.* 20:581–608
14. Davies, A. O., Lefkowitz, R. J. 1981. *Receptor Regulation,* Ser. B, ed. R. J. Lefkowitz, 13:83–121. London: Churchill-Livingstone. 253 pp.
15. Motulsky, H. J., Insel, P. A. 1982. *N. Engl. J. Med.* 307:18–29
16. Rugg, E. L., Barnett, D. B., Nahorski, S. R. 1978. *Mol. Pharmacol.* 14:996–1005
17. Minneman, K. P., Hegstrand, L. R., Molinoff, P. B. 1979. *Mol. Pharmacol.* 16:34–46
18. Minneman, K. P., Pittman, R. N., Molinoff, P. B. 1981. *Ann. Rev. Neurosci.* 4:419–61
19. Hancock, A., De Lean, A., Lefkowitz, R. J. 1979. *Mol. Pharmacol.* 16:1–9
20. De Lean, A., Hancock, A., Lefkowitz, R. J. 1982. *Mol. Pharmacol.* 21:5–16
21. Gardner, J. D., Klaeveman, J. P., Bilezikian, J. P., Aurbach, G. D. 1973. *J. Biol. Chem.* 248:5590–96
22. Minneman, K. P., Weiland, G. A., Molinoff, P. B. 1980. *Mol. Pharmacol.* 17:1–7
23. Dickinson, K. E., Nahorski, S. R. 1981. *Eur. J. Pharmacol.* 94:43–52
24. Atlas, D., Levitzki, A. 1978. *Nature* 272:370–71
25. Pitha, J., Zjawiony, J., Nasrin, N., Lefkowitz, R. J., Caron, M. G. 1980. *Life Sci.* 27:1791–98

26. Wrenn, S. M. Jr., Homcy, C. J. 1980. *Proc. Natl. Acad. Sci. USA* 77:4449–53
27. Rashidbaigi, A., Ruoho, A. E. 1981. *Proc. Natl. Acad. Sci. USA* 78:1609–13
28. Burgermeister, W., Hekman, M., Helmreich, E. J. 1982. *J. Biol. Chem.* 257:5306–11
29. Lavin, T. N., Nambi, P., Heald, S. L., Jeffs, P. W., Lefkowitz, R. J., Caron, M. G. 1982. *J. Biol. Chem.* 257:12332–40
30. Lavin, T. N., Heald, S. L., Jeffs, P. W., Shorr, R. G. L., Lefkowitz, R. J., Caron, M. G. 1981. *J. Biol. Chem.* 256:1944–950
31. Shorr, R. G. L., Heald, S. L., Jeffs, P. W., Lavin, T. N., Strohsacker, M. W., et al. 1982. *Proc. Natl. Acad. Sci. USA* 79:2778–82
32. Shorr, R. G. L., Lefkowitz, R. J., Caron, M. 1981. *J. Biol. Chem.* 256:5820–26
33. Rashidbaigi, A., Ruoho, A. E. 1982. *Biochem. Biophys. Res. Commun.* 106:139–48
34. Stiles, G., Strasser, R., Lavin, T., Jones, L., Caron, M. G., Lefkowitz, R. J. 1983. Submitted for publication
35. Shorr, R. G. L., Strohsacker, M. W., Lavin, T. N., Lefkowitz, R. J., Caron, M. G. 1982. *J. Biol. Chem.* 257:12341–50
36. Nambi, P., Lavin, T., Caron, M. G., Lefkowitz, R. J. 1983. In preparation
37. Benovic, J., Stiles, G., Lefkowitz, R. J., Caron, M. G. *Biochem. Biophys. Res. Commun.* 110:504–11
38. Rashidbaigi, A., Ruoho, A. E., Green, D., Clark, R. B. 1982. *Fed. Proc.* 41:1327 (Abstr.).
39. Vauquelin, G., Geynet, P., Hanoune, J., Strosberg, A. D. 1977. *Proc. Natl. Acad. Sci. USA* 74:3710–14
40. Caron, M. G., Srinivasan, Y., Pitha, J., Kiolek, K., Lefkowitz, R. J. 1979. *J. Biol. Chem.* 254:2923–27
41. Caron, M. G., Lefkowitz, R. J. 1976. *J. Biol. Chem.* 251:2374–84
42. Durieu-Trautmann, O., Delavier-Klutchko, C., Andre, C., Vauquelin, G., Strosberg, A. D. 1980. *J. Supramol. Struct.* 13:411–19
43. Fraser, C. N., Venter, J. C. 1980. *Proc. Natl. Acad. Sci. USA* 77:7034–38
44. Shorr, R. G. L., Strohsacker, M., Kempner, E., Lefkowitz, R. J., Caron, M. G. 1982. Submitted for publication
45. Nielsen, T. D., Lad, T. N., Preston, S., Kempner, E., Schlegel, W., Rodbell, M. 1981. *Proc. Natl. Acad. Sci. USA* 78:722–26
46. Stadel, J. M., Shorr, R. G. L., Limbird, L. E., Lefkowitz, R. J. 1981. *J. Biol. Chem.* 256:8718–23
47. Northup, J. K., Sternweis, P. C., Smigel, M. D., Schleifer, L. S., Ross, E. M., Gilman, A. G. 1980. *Proc. Natl. Acad. Sci. USA* 77:6516–20
48. Hanski, E., Sternweis, P. C., Northup, J. K., Dromerick, A. W., Gilman, A. G. 1981. *J. Biol. Chem.* 256:2911–19
49. Ross, E. M., Gilman, A. G. 1980. *Ann. Rev. Biochem.* 48:533–64
50. Stadel, J. M., De Lean, A., Lefkowitz, R. J. 1982. *Adv. Enzymol.* 53:1–43
51. Spiegel, A. M., Downs, R. W. Jr. 1981. *Endocrinol. Rev.* 2:275–305
52. Maguire, M. E., Van Arsdale, T. M., Gilman, A. G. 1976. *Mol. Pharmacol.* 12:335–38
53. Lefkowitz, R. J., Mullikin, D., Caron, M. G. 1976. *J. Biol. Chem.* 251:4686–91
54. Kent, R. S., De Lean, A., Lefkowitz, R. J. 1980. *Mol. Pharmacol.* 17:14–23
55. Stadel, J. M., De Lean, A., Lefkowitz, R. J. 1980. *J. Biol. Chem.* 255:1436–42
56. De Lean, A., Stadel, J. M., Lefkowitz, R. J. 1980. *J. Biol. Chem.* 255:7108–16
57. Lefkowitz, R. J., Williams, L. T. 1977. *Proc. Natl. Acad. Sci. USA* 74:515–19
58. Williams, L. T., Lefkowitz, R. J. 1977. *J. Biol. Chem.* 252:7202–9
59. Limbird, L. E., Lefkowitz, R. J. 1978. *Proc. Natl. Acad. Sci. USA* 75:228–32
60. Limbird, L. E., Hickey, A. R., Lefkowitz, R. J. 1979. *J. Biol. Chem.* 254:2677–83
61. Limbird, L. E., Gill, D. M., Stadel, J. M., Hickey, A. R., Lefkowitz, R. J. 1980. *J. Biol. Chem.* 255:1854–61
62. Limbird, L. E., Gill, D. M., Lefkowitz, R. J. 1980. *Proc. Natl. Acad. Sci. USA* 77:775–79
63. Cassel, D., Pfeuffer, T. 1978. *Proc. Natl. Acad. Sci. USA* 75:2669–73
64. Gill, D. M., Meren, R. 1978. *Proc. Natl. Acad. Sci. USA* 75:3050–54
65. Johnson, G. L., Kaslow, H. R., Bourne, H. R. 1978. *J. Biol. Chem.* 253:7120–27
66. Williams, L. T., Mullikin, D., Lefkowitz, R. J. 1978. *J. Biol. Chem.* 253:2984–89
67. Bird, S. V., Maguire, M. E. 1978. *J. Biol. Chem.* 253:8826–32
68. Maguire, M. E., Ross, E. M., Gilman, A. G. 1977. *Adv. Cyclic Nucleotide Res.* 8:1–83
69. Sternweis, P. C., Gilman, A. G. 1979. *J. Biol. Chem.* 254:3333–40
70. Bottari, S., Vauquelin, G., Durieu-Trautman, O., Delavier-Klutchko, C., Strosberg, A. D. 1979. *Biochem. Biophys. Res. Commun.* 86:1311–19
71. Heidenreich, K. A., Weiland, G. A.,

Molinoff, P. B. 1982. *J. Biol. Chem.* 257:804–10
72. Korner, M., Gilon, C., Schramm, M. 1982. *J. Biol. Chem.* 257:3389–96
73. Vauquelin, G., Bottari, S., Strosberg, A. D. 1980. *Mol. Pharmacol.* 17:163–71
74. Vauquelin, G., Maguire, M. E. 1980. *Mol. Pharmacol.* 18:362–69
75. Vauquelin, G., Bottari, S., Andre, C., Jacobsson, D., Strosberg, A. D. 1980. *Proc. Natl. Acad. Sci. USA* 77:3801–5
76. Cassel, D., Selinger, Z. 1978. *Proc. Natl. Acad. Sci. USA* 75:4155–59
77. Lad, P. M., Nielsen, T. B., Preston, M. S., Rodbell, M. 1980. *J. Biol. Chem.* 255:988–95
78. Pfeuffer, T. 1977. *J. Biol. Chem.* 252:7224–30
79. Ross, E. M., Howlett, A. C., Ferguson, K. M., Gilman, A. G. 1978. *J. Biol. Chem.* 253:6401–9
80. Pfeuffer, T. 1979. *FEBS Lett.* 101: 85–89
81. Cassel, D., Selinger, Z. 1975. *Biochim. Biophys. Acta* 452:538–45
82. Cassel, D., Eckstein, F., Larve, M., Selinger, Z. 1979. *J. Biol. Chem.* 254:9835–41
83. Cassel, D., Selinger, Z. 1977. *Proc. Natl. Acad. Sci. USA* 74:3307–11
84. Pike, L. J., Lefkowitz, R. J. 1980. *J. Biol. Chem.* 255:6860–66
85. Pike, L. J., Lefkowitz, R. J. 1981. *J. Biol. Chem.* 256:2207–12
86. Stadel, J. M., Lefkowitz, R. J. 1982. *Curr. Top. Membr. Trans.* In press
87. Citri, Y., Schramm, M. 1980. *Nature* 287:297–301
88. Tolkovsky, A. M., Levitzki, A. 1978. *Biochemistry* 17:3795–810
89. Orly, J., Schramm, M. 1976. *Proc. Natl. Acad. Sci. USA* 73:4410–14
90. Schramm, M., Orly, J., Eimerl, S., Korner, M. 1977. *Nature* 268:310–15
91. Limbird, L. E., Lefkowitz, R. J. 1977. *J. Biol. Chem.* 252:799–802
92. Haga, T., Haga, K., Gilman, A. G. 1977. *J. Biol. Chem.* 252:5776–82
93. Peterson, S. E., Ross, E. M. 1982. *Fed. Proc.* 41:1411 (Abstr.)
94. Fleming, J. W., Ross, E. M. 1980. *J. Cyclic Nucleotide Res.* 6:407–19
95. Lefkowitz, R. J. 1982. *Am. J. Physiol.* 243:E43–E47
96. Lefkowitz, R. J. 1979. *Ann. Intern. Med.* 91:450–58
97. Lefkowitz, R. J., Wessels, M., Stadel, J. M. 1980. *Curr. Top. Cell. Regul.* 17:205–30
98. Su, Y. F., Cubeddu, L., Perkins, J. P. 1976. *J. Cyclic Nucleotide Res.* 2: 257–70
99. Pike, L. J., Lefkowitz, R. J. 1980. *Biochim. Biophys. Acta* 632:354–65
100. Iyengar, R., Bhat, M. K., Riser, M. E., Birnbaumer, L. 1981. *J. Biol. Chem.* 256:4810–15
101. Kassis, S., Fishman, P. H. 1982. *J. Biol. Chem.* 257:5312–18
102. Stadel, J. M., Strulovici, B., Nambi, P., Caron, M. G., Lefkowitz, R. J. 1983. *J. Biol. Chem.* In press
103. Chuang, D. M., Costa, E. 1979. *Proc. Natl. Acad. Sci. USA* 76:3025–28
104. Chuang, D. M., Kinnier, W. J., Farber, L., Costa, E. 1980. *Mol. Pharmacol.* 18:348–55
105. Su, Y. F., Harden, T. K., Perkins, J. P. 1980. *J. Biol. Chem.* 255:7410–19
106. Perkins, J. P., Waldo, G. L., Harden, T. K. 1982. *Fed. Proc.* 41:1327 (Abstr.)
107. Wessels, M. R., Mullikin, D., Lefkowitz, R. J. 1979. *Mol. Pharmacol.* 16:10–20
108. Harden, T. K., Cotton, C. U., Waldo, G. L., Lutton, J. K., Perkins, J. P. 1980. *Science* 210:441–43
109. Mukherjee, C., Caron, M. G., Lefkowitz, R. J. 1975. *Proc. Natl. Acad. Sci. USA* 72:1945–49
110. Mickey, J. V., Tate, R., Lefkowitz, R. J. 1975. *J. Biol. Chem.* 250:5727–29
111. Lefkowitz, R. J., Mukherjee, C., Caron, M. G., Limbird, L. E., Alexander, R. W., et al 1976. *Recent Prog. Horm. Res.* 32:597–632
112. Mukherjee, C., Caron, M. G., Lefkowitz, R. J. 1976. *Endocrinology* 99: 343–53
112a. Doss, R. C., Perkins, J. P., Harden, T. K. 1981. *J. Biol. Chem.* 256:12281–286
113. Doss, R. C., Harden, T. K., Perkins, J. P. 1982. *Fed. Proc.* 41:1534 (Abstr.)
114. Brown, M. S., Goldstein, J. L. 1979. *Proc. Natl. Acad. Sci. USA* 76:3330–37
115. Hoffman, B. B., Mullikin-Kilpatrick, D., Lefkowitz, R. J. 1979. *J. Cyclic Nucleotide Res.* 5:355–66
116. Stadel, J., Mullikin-Kilpatrick, D., Dukes, D., Lefkowitz, R. J. 1981. *J. Cyclic Nucleotides Res.* 7:37–47
117. Briggs, M., Stadel, J., Lefkowitz, R. J. 1982. Submitted for publication
118. Moylan, R. D., Barovsky, K., Brooker, G. 1982. *J. Biol. Chem.* 257:4947–50
119. Stadel, J., Nambi, P., Shorr, R., Caron, M., Lefkowitz, R. J. 1983. *Proc. Natl. Acad. Sci. USA.* In press
120. Nahorski, S. R. 1981. *Towards Understanding Receptors,* ed. J. W. Lamble, pp. 71–77. Amsterdam: Elsevier/North Holland Biomedical

Ann. Rev. Biochem. 1983. 52:187–222
Copyright © 1983 by Annual Reviews Inc. All rights reserved

BIOCHEMISTRY OF SULFUR-CONTAINING AMINO ACIDS

Arthur J. L. Cooper

Departments of Neurology and Biochemistry, Cornell University Medical College, 1300 York Avenue, New York, New York 10021

CONTENTS

PERSPECTIVES AND SUMMARY 187
THE SULFUR CYCLE... 188
ASSIMILATION OF SULFUR INTO AMINO ACIDS IN
        MICROORGANISMS 188
METHIONINE FORMATION IN MICROORGANISMS.................................. 190
TRANSSULFURATION PATHWAY OF METHIONINE DEGRADATION 191
TRANSAMINATIVE PATHWAY OF METHIONINE CATABOLISM 193
S-ADENOSYLMETHIONINE METABOLISM.. 195
    General Considerations 195
    Salvage of Methionine from 5'-Methylthioadenosine and α-Keto-γ-
        Methiolbutyrate 196

CYSTEINE METABOLISM.. 198
    General Considerations 198
    Enzymes that Transfer Sulfane Sulfur... 204
    Comments on the Metabolism of 3-Mercaptopyruvate 207

α-KETO ACID ANALOGS OF METHIONINE, CYST(E)INE AND
        HOMOCYST(E)INE... 209
CYSTEINE- AND HOMOCYSTEINE-CARBONYL ADDUCTS.......................... 211
ROLE OF CYSTEINE IN PIGMENT FORMATION 212
INBORN ERRORS OF SULFUR-AMINO ACID METABOLISM 213
L-METHIONINE AND L-CYST(E)INE REQUIREMENT
        OF CANCER CELLS 215
INDUSTRIAL APPLICATIONS.. 215

PERSPECTIVES AND SUMMARY

The literature on sulfur amino acid metabolism is too vast for a short chapter to cover in great depth. I attempt here a brief overview with references to many specialized review articles. This review emphasizes aspects of sulfur amino acid metabolism elucidated in the last ten years, in

187

0066-4154/83/0701-0187$02.00

particular aspects not generally covered in biochemistry texts, e.g. transaminative pathways of methionine metabolism. A selected list of reviews is given in references 1–15. References to reviews on glutathione are covered in the chapter by A. Meister in this volume (1a).

THE SULFUR CYCLE

Most biochemists are familiar with the outlines of the carbon, nitrogen, and sulfur cycles. Although the kinetics are not fully understood, there is no doubt that sulfur is recycled through the biosphere in considerable amounts (16–18). As with other cycles, a variety of organisms plays a role in both assimilation and dissimilation reactions (Figure 1). Mammals dissimilate sulfur by breakdown of sulfur amino acids, but cannot assimilate, i.e. meet their requirements with, inorganic sulfur, and must instead rely principally on ingested methionine and cysteine.

ASSIMILATION OF SULFUR INTO AMINO ACIDS IN MICROORGANISMS

Reductive dissimilation of sulfate occurs in certain obligatory anaerobic bacteria, such as *Desulfovibrio* and *Desulfotomaculum*. This reduction pathway produces hydrogen sulfide that is largely lost to the environment. The sulfate dissimilatory bacteria oxidize organic compounds or

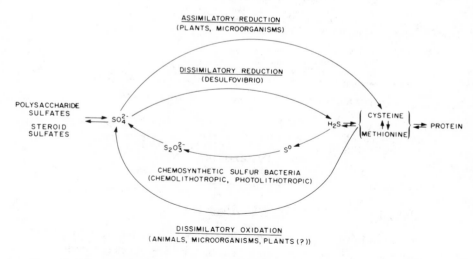

Figure 1   Schematic representation of flow of sulfur through the biosphere, modified from (16, 18).

molecular hydrogen by using metabolites of sulfate as a terminal electron acceptor in a manner similar to the way aerobes utilize oxygen as an acceptor (6, 16). Reduction of sulfate can also result in assimilation rather than loss of the reduced form of sulfur (Figure 1). Plants and many microorganisms, but not higher animals, have evolved mechanisms for sulfate assimilation. Reductive assimilation of sulfate, i.e. incorporation of sulfate sulfur into thiol groups of amino acids and other bioorganic compounds, occurs via two major enzymatic routes: the APS (adenosine 5'-phosphosulfate) pathway and the PAPS (adenosine 3'-phosphate-5'-phosphosulfate) pathway. In the APS assimilation pathway (Equation 1), which requires ferredoxin, glutathione (GSH) reacts with APS to yield the thiosulfate, $GSSO_3^-$ (a Bunte salt), which is converted in turn to GSS^- via a thiosulfate reductive reaction; GSS^- reacts with O-acetylserine to yield cysteine. In some organisms, other -SH compounds can partially replace GSH as a thiol carrier (18):

$$\text{SO}_4^{2-} \xrightarrow[\text{PP}_i]{\text{ATP}} \text{APS} \xrightarrow[\text{5'-AMP}]{\text{GS}^-} [\text{GSSO}_3^-] \xrightarrow[\text{Fer}_\text{red}\ \text{Fer}_\text{ox}]{} \text{GSS}^- \xrightarrow[O\text{-acetylserine}\ \text{acetate}]{\text{GS}^-} \text{L-cysteine.} \quad 1.$$

In the PAPS pathway, O-acetylserine reacts directly with HS^- (Equation 2):

$$\text{SO}_4^{2-} \xrightarrow[\text{PP}_i]{\text{ATP}} \text{APS} \xrightarrow[]{\text{ATP}\ \text{ADP}} \text{PAPS} \xrightarrow[\text{NADPH,H}^+\ \text{NADP}^+]{\text{PAP}} \text{HSO}_3^- \xrightarrow[]{} \text{HS}^- \xrightarrow[O\text{-acetylserine}\ \text{acetate}]{} \text{L-CYS.} \quad 2.$$

In yeast, conversion of PAPS to sulfite requires thioredoxin, but the exact role of thioredoxin is not yet fully elucidated (18). It was originally thought that thioredoxin was also a cofactor in the PAPS pathway of *Escherichia coli*, but in *E. coli* mutants lacking this protein, glutaredoxin can substitute (19). The APS pathway has been found in *Chlorella, Euglena,* spinach chloroplasts, and *Lemna* (duckweed) (e.g. 19–23). The PAPS system has been studied most extensively in yeast (24), *E. coli* (25), and *Salmonella typhimurium* (26).

Alternative routes for assimilation of sulfide are also known (Equations 3–5):

O-Acetylhomoserine + H_2S → L-homocysteine + acetate.   3.

O-Succinylhomoserine + H_2S → L-homocysteine + succinate.   4.

O-Phosphohomoserine + H_2S → L-homocysteine + P_i.   5.

O-Acetylhomoserine is the preferred substrate in spinach (27) and fungi (28). O-Succinylhomoserine is the preferred substrate in *E. coli* (28). *Chlorella* and *Lemnis* preferentially utilize O-phosphohomoserine, although *Chlorella* uses the O-malonyl-, O-oxalyl- and O-succinylhomoserine derivatives as well (29).

In 1957, Schlossmann & Lynen (30) showed that yeast contains an enzyme that catalyzes the reversible conversion of L-serine to L-cysteine (Equation 6):

$$\text{L-serine} + H_2S \leftrightarrows \text{L-cysteine} + H_2O. \qquad\qquad 6.$$

Serine sulfhydrase (cysteine synthase) was found subsequently to be widely distributed in nature, including mammalian tissues, and to have a very wide specificity (cf 31). In fact, there are similarities between cysteine synthase and cystathione β-synthase (32, 33); because (*a*) vertebrates do not possess a sulfate-reducing system, (*b*) in vivo incorporation of sulfide (S^{2-}) into cysteine is of minor importance, and (*c*) equilibrium lies far toward serine formation (34), the function of the enzyme in vertebrates is probably to synthesize cystathionine. Thus, cysteine synthase may be regarded as a variant cystathionine β-synthase; both enzymes are "C_3-specific β-replacing lyases" (33).

METHIONINE FORMATION IN MICROORGANISMS

In microorganisms, the major route to methionine is via cysteine and homocysteine (Equation 7). In vertebrates the reverse is true, i.e. methionine serves as a precursor of homocysteine and cysteine sulfur. A pivotal difference in the two groups is the fate of cystathionine. In microorganisms, β-cystathionase splits a C–S bond resulting in formation of homocysteine (Equation 7). In mammals, γ-cystathionase splits the other C–S bond resulting in formation of cysteine.

$$
\begin{array}{l}
\qquad\qquad\qquad\qquad\qquad\qquad \text{cysteine} \\
\qquad\qquad\qquad\qquad\qquad\qquad \downarrow \\
\text{Aspartate} \rightarrow \text{homoserine} \rightarrow O\text{-succinylhomoserine} \rightarrow \text{cystathionine} \\
\qquad\qquad\qquad\qquad (\text{or } O\text{-acetylhomoserine}) \qquad\quad \downarrow \\
\qquad\qquad\qquad\qquad\qquad\qquad\qquad\qquad\qquad \text{homocysteine} \\
\qquad\qquad\qquad\qquad\qquad\qquad\qquad\qquad\qquad \downarrow \\
\qquad\qquad\qquad\qquad\qquad\qquad\qquad\qquad\qquad \text{methionine.} \qquad 7.
\end{array}
$$

Flavin (35) and Woods et al (36) have presented detailed reviews of methionine biosynthesis in microorganisms. A few points are discussed here. In *E. coli, S. typhimurium,* and other enteric bacteria, the immediate precursor of cystathionine is O-succinylhomoserine (37). In *Neurospora crassa* (38) and *Aspergillus nidulans* (38) the immediate precursor is O-acetylhomoserine. Methylation of homocysteine in microorganisms can occur via two distinct pathways. Equation 8 shows the vitamin B_{12}-dependent homocysteine transmethylation reaction. This reaction requires a reducing system and can also utilize S-adenosylmethionine in place of methyltetrahydrofolate as methyl donor; 2-mercaptoethanol can act as a methyl acceptor, although less readily than homocysteine (39). Equation 9 shows the vitamin B_{12}-independent reaction. This reaction requires P_i and is stimulated by Mg^{2+}.

$$5\text{–}CH_3\text{–}H_4PteGLU + \text{L–homocysteine} \xrightarrow{B_{12}} \text{L–methionine} + H_4PteGLU.$$
$$8.$$

$$5\text{–}CH_3\text{–}H_4PteGLU_{3\text{–}7} + \text{L–homocysteine} \xrightarrow{P_i, \; Mg^{2+}} \text{L–methionine} +$$
$$H_4PteGLU_{3\text{–}7}. \qquad\qquad\qquad 9.$$

Many microorganisms and plants can synthesize homocysteine both by direct incorporation of H_2S (Equations 3–5) or by the cystathionine pathway (Equation 7). There has been some debate as to which is the most important route to methionine in these organisms (cf 35, 40). In plants (40) and in *A. nidulans* (41) the transsulfuration pathway probably predominates. On the other hand, recent evidence suggests that the major route for homocysteine synthesis in *Brevibacterium flavum* is via reaction of H_2S with O-acetylhomoserine (Equation 3; 42). Detailed reviews of methionine metabolism in plants and microorganisms have recently been published (43, 44).

TRANSSULFURATION PATHWAY OF METHIONINE DEGRADATION

Before 1940, cyst(e)ine was regarded as an essential dietary amino acid in mammals; later it was concluded that methionine rather than cysteine is an essential amino acid, and that cysteine sulfur could be derived from methionine (45). However, in the presence of adequate methyl donors, the methionine requirement may be met solely from ingested homocysteine (46). Moreover, it has recently been found in several species that cyst(e)ine is not indispensable after all. For example, it can supply 50% of the total dietary sulfur amino acid requirement in growing beagle dogs (47). Sturman et al were unable to detect cystathionase activity in the livers of human fetuses (48) and concluded that cyst(e)ine is indeed an essential amino acid for the human fetus (48, 49).

The transsulfuration pathway (Figure 2) is now well characterized (e.g. 2, 8, 12, 31, 46) so that only a few points are discussed here. Figure 2 shows that methionine and homocysteine are readily interconvertible. This has led some authors to use the term "methionine cycle." Mudd (46) concluded that, via this cycle, "methionine is controlled to just the right level for transmethylation reactions and polyamine synthesis." Evidently, much of the homocysteine skeleton is efficiently re-utilized. Mudd & Pool (50) calculated that the methyl group is conserved for an average of 1.9 turnovers in men, 1.5 in women. Krebs et al (51) investigated methionine metabolism and concluded that levels are controlled to a fine degree at the level of homocysteine: when methionine is needed, homocysteine is re-methylated by methyltetrahydrofolate; when methionine is in excess, catabolism of homocysteine via the γ-cystathionase reaction is accelerated.

Figure 2 shows two salvage pathways for methionine, methionine synthase and betaine-homocysteine methyltransferase. Some texts also state that dimethylthetin and dimethyl-β-propiothetin (cf 52) are methyl donors,

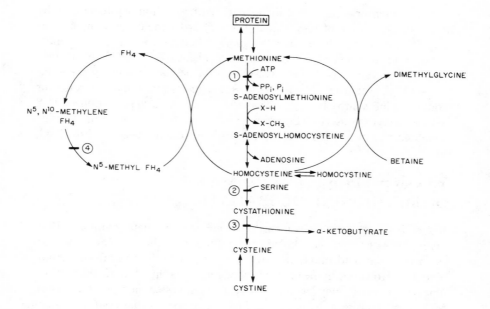

Figure 2 Transsulfuration pathway to cysteine. The solid bars and numbers indicate enzymes known to be absent or modified in various metabolic defects of man.

but the significance of this reaction is not clear. A third pathway (salvage from 5'-methylthioadenosine) is discussed later. Some authors include a desulfhydrase reaction (Equation 10) as a route for metabolism of homocysteine. However, the enzyme responsible, homocysteine desulfhydrase, is identical to γ-cystathionase (53), and the reaction is probably of little or no significance in mammalian tissues.

$$\text{L-Homocysteine} + H_2O \rightarrow \alpha\text{-ketobutyrate} + NH_3 + H_2S. \qquad 10.$$

TRANSAMINATIVE PATHWAY OF METHIONINE CATABOLISM

Large amounts of methionine are toxic, whether derived from the diet (e.g. 54–56) or from metabolic blocks in diseases such as liver dysfunction (57, 58) and some forms of hypermethionemia (e.g. 59). In patients with liver disease the transsulfuration pathway is diminished (60) leading to elevated mercaptans in blood (61) and in the breath (58, 61, 62). Benevenga (63) concluded that the metabolic basis for methionine toxicity cannot be attributed to catabolism via the transsulfuration pathway but is related to metabolism of the methyl moiety. Subsequently, Case & Benevenga (64, 65) showed that methionine can be extensively catabolized by a pathway independent of S-adenosylmethionine formation and that formaldehyde and formate were two intermediates in the oxidation of the methyl carbon by this pathway. From detailed experiments with labeled methionine, Benevenga and colleagues suggested a "transaminative" pathway of methionine catabolism in which methanethiol is a breakdown product (65–71; Figure 3).[1] Canellakis & Tarver (74) first showed that α-keto-γ-methiolbutyrate is a better precursor of methanethiol than is methionine.

A decrease in the transsulfuration pathway relative to the transamination pathway may explain the increase in methyl mercaptan and dimethyl disulfide in patients with liver disease. Moreover, the similarity in tissue damage brought about by both excess methionine and by 3-methylthiopropionate suggests that methionine toxicity is related to methanethiol and H_2S, both of which are extremely poisonous (69, 70). Steele & Benevenga (69) pointed out several lines of evidence indicating that the transaminative pathway is operative at physiological levels of methionine intake. Dimethyldisulfide is an attractant pheromone in hamster vaginal secretions (75). Low levels of methanethiol and dimethylsulfide can be detected in the

---

[1]Steele and Benevenga have also shown that ethionine is metabolized, at least in part, via a transaminative pathway in rat liver (72). This finding may explain some of the toxicity associated with ethionine; a major metabolite of this pathway is the markedly toxic 3-ethylthiopropionate (73).

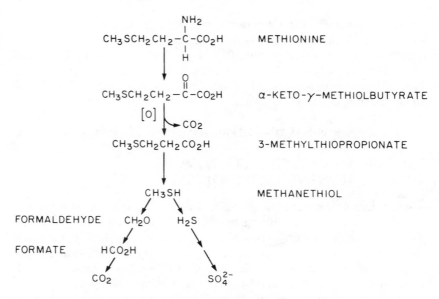

Figure 3   Transaminative pathway of methionine metabolism, modified from (65, 67).

breath of normal individuals (62). Toohey (76) discovered a thioalkane, tentatively identified as methanethiol, that is required for cell division in certain cell lines.

Rat tissue preparations can catalyze transamination between methionine and α-ketoglutarate (77), and methionine is a substrate of purified preparations of rat liver glutamine transaminase (78), rat kidney glutamine transaminase (79), and rat liver asparagine transaminase (80). Ikeda et al (81) demonstrated that a leucine transaminase isolated from rat liver mitochondria can transaminate methionine. However, the transaminase(s) responsible for the formation of α-keto-γ-methiolbutyrate in vivo remain unknown. The glutamine transaminases are probably not responsible. The much higher levels of glutamine in rat tissues, coupled with the virtually irreversible nature of the reaction with glutamine should ensure that the reaction is directed toward glutamine utilization (82, 83). Furthermore, pyruvate and especially α-ketoglutarate are poor substrates of the enzymes. The glutamine transaminases may act to spare the carbon skeleton of α-keto-γ-methiolbutyrate, arising from nonspecific transamination of methionine, at the expense of "nonessential" glutamine; without such a mechanism an excessive loss of essential methionine, via degradation of α-keto-γ-methiolbutyrate, might occur (78, 79, 82, 83). Recent evidence suggests that α-keto-γ-methiolbutyrate is oxidatively decarboxylated by branch-chain α-keto acid dehydrogenase (84). The presence of trace amounts of β-methylmercaptopropionaldehyde (methional) in foodstuffs (85) suggests that nonoxidative decarboxylation of α-keto-γ-methiolbuty-rate may also occur. Evidently, salvage relative to degradation depends on availability of glutamine and competition from endogenous α-keto acids. Interestingly, glutamine is rapidly metabolized by rat hepatocytes in the presence of α-keto-γ-methiolbutyrate but not in the presence of pyruvate (86). Moreover, very recent evidence suggests that the glutamine transaminase reaction acts to salvage α-keto-γ-methiolbutyrate formed from 5'-methylthioadenosine (see below).

Using a gas chromatographic–mass spectrometric determination of the quinoxalinol derivative, Kaji et al (87) showed that α-keto-γ-methiolbuty-rate occurs in trace amounts in normal urine. The urinary output of the α-keto acid increased after oral loading with D- or L-methionine. Kaji et al (62, 87) also showed that exhalation of dimethylsulfide is much greater following oral loading of D-methionine than with L-methionine. The authors suggested that the mechanism of Benevenga is more important for the metabolism of D-methionine than for L-methionine (87). [Presumably, α-keto-γ-methiolbutyrate is formed from D-methionine by the action of D-amino acid oxidase; D-methionine is an excellent substrate of this enzyme (88)]. Hydrogen sulfide, one of the proposed intermediates of the trans-aminative pathway, may also arise enzymatically from cyst(e)ine or from 3-mercaptopyruvate (see below). Although not detected in mammalian tissues, H_2S is present in trace amounts in flatus (89). The relative importance of the transaminative pathway vs the transsulfuration pathway of L-methionine breakdown must await further study. Careful methyl balance studies by Mudd and colleagues (e.g. 46) suggest that the transsulfuration pathway greatly predominates. However, impairment of the transsulfuration pathway, as in liver disease (57, 58) and/or in portacaval shunting (90), probably leads to a relative increase in the transaminative pathway.

S-ADENOSYLMETHIONINE METABOLISM

General Considerations

Following (a) the realization of the biological importance of 5'-methylthio-adenosine (MTA) in 1952 (91) and (b) the discovery of S-adenosylmethio-nine (SAM) in 1953 (92), Tabor et al showed that a number of microorgan-isms utilize SAM to convert putrescine to spermine and to spermidine (93). At the same time, Shapiro & Mather (94) showed that SAM could be rapidly degraded to MTA (and homoserine lactone) and then to 5-methyl-thioribose (MTR) by *Aerobacter aerogenes*. Later, Pegg & Williams-Ashman (95) showed that the rat ventral prostate has an enzyme system that catalyzes the SAM-mediated conversion of putrescine to spermidine.

Figure 4 gives a scheme of some of the known metabolic routes of SAM in prokaryotes (cf 96). Cleavage reactions involving all three of the S–C bonds are known. By far the most numerous are those reactions in which cleavage results in methylation (Figure 4; 12). Although methylation of relatively small molecules has long been known, the importance of SAM in the methylation of phospholipids (97), proteins (98), polysaccharides (99), and nucleic acids (100) is becoming increasingly apparent. In a few cases, S–C cleavage results in transfer of the adenosyl portion to enzyme protein (e.g. 101). Finally, breakage of a S–C bond with transfer of the 3-amino-3-carboxypropyl group to tRNA also occurs (102). Hydrolysis of SAM to MTA and homoserine lactone (Figure 4) is a special case of cleavage of this third S-C bond.

Salvage of Methionine from 5'-Methylthioadenosine and α-Keto-γ-Methiolbutyrate

Prokaryotes generally convert MTA to MTR and adenosine. However, mammalian cells convert MTA to 5'-methylthioribose 1-phosphate (MTRP) and adenine in a reaction catalyzed by 5'-methylthioadenosine phosphorylase (103, 104). The enzyme is present in a number of rat organs (104, 105) and in human prostate (106) and placenta (107). A single example of the enzyme occurring in a prokaryote has been recorded, i.e. in the extreme thermophile *Caldariella acidophila* grown optimally at 87°C (108). Until very recently very little was known about the metabolism of MTR and

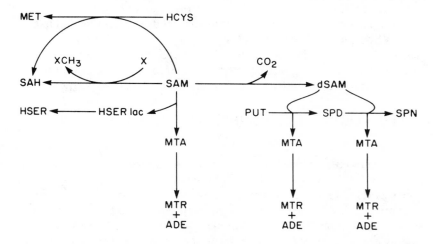

Figure 4   Some of the known metabolic routes of S-adenosylmethionine in prokaryotes. SAM, S-adenosylmethionine; SAH, S-adenosylhomocysteine; HCYS, homocysteine; dSAM, decarboxylated S-adenosylmethionine; SPD, spermidine; SPN, spermine; PUT, putrescine; HSER, homoserine; HSER lac, homoserine lactone; MTA, 5'-methylthioadenosine; MTR, 5-methylthioribose; ADE, adenine. X = methyl acceptor. Modified from (96).

MTRP, although in 1964 Schlenk & Ehninger (109) showed that the carbon and sulfur of the thiomethyl group of MTA is efficiently incorporated back into SAM by *Candida utilis*. Then, in 1981, Backlund & Smith (110) showed that MTA is converted to methionine by rat liver extracts. Carbons from the ribose portion, carbon and hydrogen of the methyl group, and the sulfur of MTA are all incorporated into methionine (110, 111). The authors proposed that the "pathway appears to be a significant salvage pathway for methionine synthesis in mammals, and may be necessary for removal of 5'-methylthioadenosine produced as by-product of polyamine biosynthesis." Apparently, the pathway is as follows:

$$\text{A.A. K.A.}$$
$$\text{MTA} \rightarrow \text{MTRP} \rightarrow \rightarrow \rightarrow \alpha\text{KMB} \rightarrow \text{MET.} \qquad \qquad 11.$$

MTRP is converted to α-keto-γ-methiolbutyrate (αKMB) by an unknown mechanism (111, 111a). α-Keto-γ-methiolbutyrate is then converted to methionine by transamination with a suitable amino acid donor (A. A., Equation 11). Glutamine and asparagine are the preferred donors. α-Keto-γ-methiolbutyrate is an excellent substrate of rat kidney and liver glutamine transaminases (78, 79) and a moderately good substrate of rat liver asparagine transaminase (80). The findings of Backlund et al (110, 111) support the earlier suggestion that one role of the glutamine transaminase is to salvage α-keto-γ-methiolbutyrate (78, 79, 82, 83).

Ethylene promotes ripening of fruit and is derived from methionine via 1-aminocyclopropane-1-carboxylate (ACC) (112). Much of the pathway in the apple has been recently elucidated (113; Equations 12, 13). Methionine is efficiently salvaged from 5'-methylthioribose (MTR) (113; Equation 12). Presumably, the mechanism is similar to that of the methionine salvage from MTRP in rat liver (110, 111) but α-keto-γ-methiolbutyrate has not yet been identified as a precursor in the apple salvage pathway:

$$
\begin{array}{c}
\quad \quad \quad \overset{\displaystyle H \quad H}{\underset{\displaystyle C}{\diagdown\diagup}} \\
\text{Met} \rightarrow \text{SAM} \quad \quad \overset{\diagup \diagdown}{H - C - C - NH_2.} \\
\uparrow \quad \quad \downarrow \quad \quad \quad H \quad CO_2H \\
\text{MTR} \leftarrow \text{MTA} \quad \quad \quad (\text{ACC})
\end{array}
\qquad 12.
$$

$$
\overset{O_2}{\text{ACC} \rightarrow CH_2{=}CH_2 + CO_2 + HCO_2H + NH_3.} \qquad 13.
$$

CYSTEINE METABOLISM

General Considerations

Cysteine participates in an extremely complex series of metabolic reactions. Figure 5 presents a general scheme. Cysteine is incorporated into most proteins and glutathione. It is also a precursor of CoA and can be readily oxidized to cystine; however, the intracellular cysteine : cystine ratio is very high (114).

Cystine is converted to thiocystine in reaction mixtures containing γ-cystathionase (115; Figure 5; see also below), but this pathway is of minor metabolic importance. Cavallini et al (116) showed that, following administration of [^{35}S]cystine to rats, kidney extracts and urine contained ^{35}S-labeled taurine, hypotaurine, thiotaurine, thiazolidine carboxylate, and cysteine sulfinate. However, interpretation of the results is complicated because the DL-form of [^{35}S]cystine was used and some of the labeled products may have arisen nonenzymatically during the isolation procedure. The compounds mercaptoacetate-cysteine mixed disulfide [S-(carboxymethylthio)cysteine] and 3-mercaptolactate-cysteine mixed disulfide [S-(2-hydroxy-2-carboxyethylthio)cysteine] occur naturally (117). They may

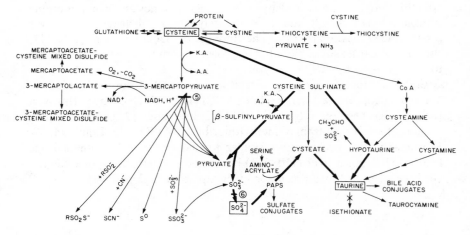

Figure 5   Cysteine metabolism in mammals. Solid bars indicate enzymes known to be absent or modified in certain defects of man. K.A. and A.A. represent α-keto acid substrate and amino acid product, respectively, of the cysteine and cysteine sulfinate transaminases. β-Sulfinylpyruvate is in square brackets because its existence in solution is not yet demonstrated. The relative flow of sulfur through the various pathways depends on factors such as species, sex, age, organ, and nutritional status; in general, the major flow of sulfur is through the pathways depicted with bold arrows. Sulfate and taurine are encased in order to emphasize that they are the major catabolites in urine and tissue, respectively. For more details of the interaction between thiosulfate and sulfite, see Figure 7.

arise directly from cystine (117), but are more likely formed from 3-mercaptopyruvate (Figure 5). The available evidence suggests that dietary cystine is reduced rapidly to cysteine (13) and that the major metabolic fate of cystine is conversion to cysteine (118); for a discussion of possible mechanisms for this conversion see (118).

Despite the metabolic importance of cysteine, its concentration in normal tissues is low, generally in the range of 10–100 μM (119, 120).[2] This low level may be a protection against the high reactivity of cysteine. On the other hand, the concentration of glutathione in animal tissues is much higher (0.5–10 mM) and it has long been supposed that one of the many functions of glutathione is to store cysteine (121, 122). Interestingly, glutathione levels in plasma are very low, $\sim$ 25 μM (123), but the concentration in red blood cells is much higher. Fahey et al (124) have developed a sensitive fluorescence technique for the estimation of biological thiols. In accord with earlier reports, the concentration of cysteine and glutathione in human red blood cells was found to be 8–10 μM and 2.3–2.8 mM, respectively (124).[3]

Two major pathways of cysteine breakdown in mammals are known (Figure 5), i.e. the direct oxidation (cysteine sulfinate) pathway and the transamination (3-mercaptopyruvate) pathway. However, some of the minor pathways are discussed first. Many authors still include a pathway in which cysteine is converted to pyruvate by liver "cysteine desulfhydrase" (Equation 14):

$$\text{Cysteine} + \text{H}_2\text{O} \rightarrow \text{pyruvate} + \text{NH}_3 + \text{H}_2\text{S}. \qquad 14.$$

While bacterial cysteine desulfhydrases are well characterized, no such enzyme has ever been isolated from mammalian tissues. Cavallini and co-workers (127, 128) attributed the apparent cysteine desulfhydrase activity of rat liver to γ-cystathionase. Thus, cystine present in reaction mixtures containing cysteine is converted to thiocysteine, pyruvate, and ammonia (Equation 15). If the reaction is carried out in the presence of cystine only, cysteine is produced (Equation 16):

$$\text{Cystine} + \text{H}_2\text{O} \rightarrow \text{pyruvate} + \text{NH}_3 + \text{thiocysteine}. \qquad 15.$$

$$\text{Thiocysteine} + \text{cystine} \rightarrow \text{thiocystine} + \text{cysteine}. \qquad 16.$$

[2]A compilation of concentrations of methionine, SAM, adenosylhomocysteine, cysteine, cystine, and acid soluble thiols in rat tissues has recently been published (120).

[3]The only other thiol detectable was ergothioneine (140 μM). This compound, a betaine, was first isolated from an ergot infection of rye grain in 1909 (125) and first shown to be present in human blood in 1927 (126); its origin in human blood may be due to ingested ergothioneine.

If, however, the reaction mixture contains appreciable cysteine, cystine is produced. The following reactions were thought to occur (Equations 17,18):

$$\text{Thiocysteine} \rightleftharpoons \text{cysteine} + S°. \qquad\qquad 17.$$

$$2\ \text{Cysteine} + S° \rightleftharpoons \text{cystine} + H_2S. \qquad\qquad 18.$$

Later Jollès-Bergeret & Chatagner (129) showed that γ-cystathionase catalyzes conversion of cysteine to pyruvate and ammonia even in the presence of excess 2-mercaptoethanol, i.e. under conditions in which cystine formation is negligible. These experiments suggest that rat liver γ-cystathionase catalyzes direct desulfhydration of cysteine. However, more work is needed to establish this point. More recently, Yamanishi & Tuboi (130) investigated the γ-cystathionase reaction on cyst(e)ine and showed that $S°$ is not free in solution but is enzyme-bound as a labile sulfane (130). The following reactions were proposed:

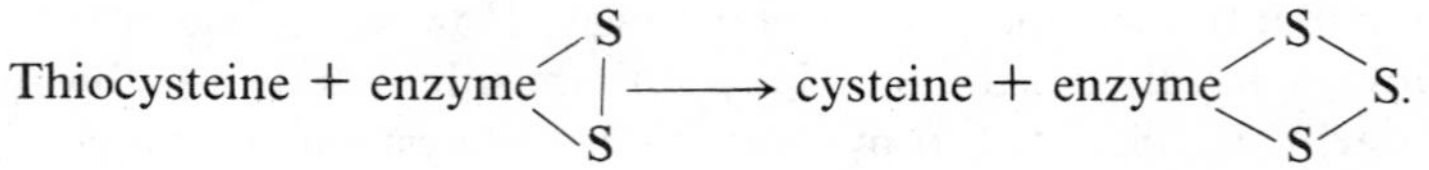

$$\text{Thiocysteine} + \text{enzyme}\begin{smallmatrix} S \\ | \\ S \end{smallmatrix} \longrightarrow \text{cysteine} + \text{enzyme}\begin{smallmatrix} S \\ \diagdown \\ S \end{smallmatrix} S. \qquad 19.$$

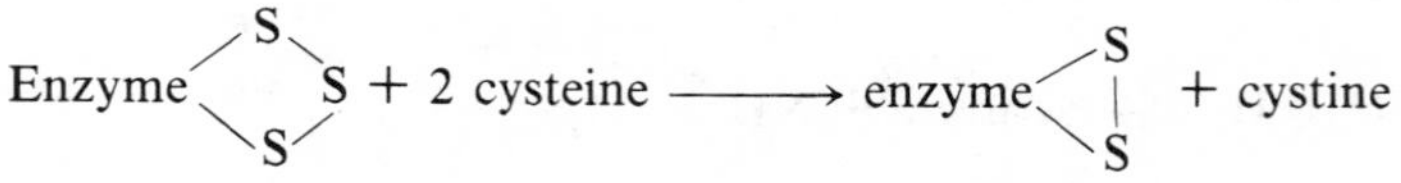

$$\text{Enzyme}\begin{smallmatrix} S \\ \diagdown \\ S \end{smallmatrix} S + 2\ \text{cysteine} \longrightarrow \text{enzyme}\begin{smallmatrix} S \\ | \\ S \end{smallmatrix} + \text{cystine} \qquad 20.$$
$$+ S^{2-} + 2H^+.$$

Other –SH compounds, such as dithiothreitol, also promote removal of sulfur from the sulfane-enzyme (130). Note that both cystine and cysteine are substrates in the overall proposed reaction sequence, but cystine is regenerated. Thus, the earlier observations that mammalian organs possess a "cysteine desulfhydrase" activity is explained by the sum of reactions 15, 19, and 20 catalyzed by γ-cystathionase.

Cysteamine is present in low concentrations in rat heart (131). However, there is no evidence that cysteamine is derived by direct decarboxylation of cysteine (132). Huxtable proposed that cysteamine is derived via a complex series of reactions from CoA catabolism (133) but the pathway awaits experimental validation (132). [For recent discussion of the biochemistry of other S-containing co-factors, i.e. biotin, lipoic acid, and 8α-(S-L-cysteinyl)-riboflavin, see appropriate chapters in (15)].

Recently, Scandurra et al (134) described a pathway to cysteamine from lanthionine. The authors demonstrated that lanthionine is enzymatically degraded to aminoethylcysteine and then to cysteamine by beef kidney cortex slices (134). Lanthionine is present in acid hydrolysates of many alkali-treated proteins [see references quoted by Dowling & Maclaren (135)]; however, lanthionine is thought to arise in these preparations via nonenzymatic decomposition and cross-linking reactions with cysteine (135). Since mammalian tissues normally contain little if any free lanthionine, the pathway of lanthionine to cysteamine is probably of little metabolic importance in mammals. [Insects contain lanthionine as part of the free amino acid pool (136), but the metabolism of lanthionine in these animals has not been elucidated.]

Federici et al (137) compared cysteine oxygenase and cysteamine oxygenase activities in a number of mammalian organs and showed that optimal cysteamine oxygenase activity was at least comparable to and in many cases greater than cysteine oxygenase activity. The authors suggested that the cysteamine pathway to taurine may be more important than previously suspected (137). The product of the cysteamine oxygenase reaction (hypotaurine) is oxidized readily and nonenzymatically to taurine, but it has been difficult to demonstrate whether this reaction is also enzyme-mediated. However, hypotaurine oxidase has recently been found in various rat tissues (138) and in ox retina (139).

Many texts indicate that taurine is converted to isethionate $[(SO_3^-)CH_2$ $CH_2OH]$, but careful isotope studies have been unable to detect such a pathway in mammalian organs (140). Possibly, isethionate in mammalian organs is at least in part of microbial origin (140). Interestingly, the concentrations of isethionate and taurine in the squid giant axon are remarkably high; 150 mM and $\geq$ 75 mM, respectively (141). Yet even in the squid axon, no evidence for conversion of taurine to isethionate was obtained, and the exact pathway for isethionate formation must await further study (141).

Cysteate and cystamine, in addition to hypotaurine, are potential precursors of taurine (Figure 5). Cysteate may arise by oxidation of cysteine sulfinate (probably a minor pathway) or via the PAPS transferase reaction (Figure 5). Cavallini and colleagues showed that cystamine, the disulfide of cysteamine, is a substrate of diamine oxygenase (142); the product is a cyclic imine (cystaldimine; 1,2-dehydrodithiomorpholine), which is then enzymatically cleaved by pig kidney extracts giving rise to a variety of products such as thiocysteamine, hypotaurine, thiotaurine, and taurine (143–145; Figure 6).

Equation 21 shows the sequence leading to thiocysteamine and cysteamine. Thiocysteamine and cysteamine are presumably the precursors of thiotaurine and taurine, respectively. However, thiocysteamine is unstable

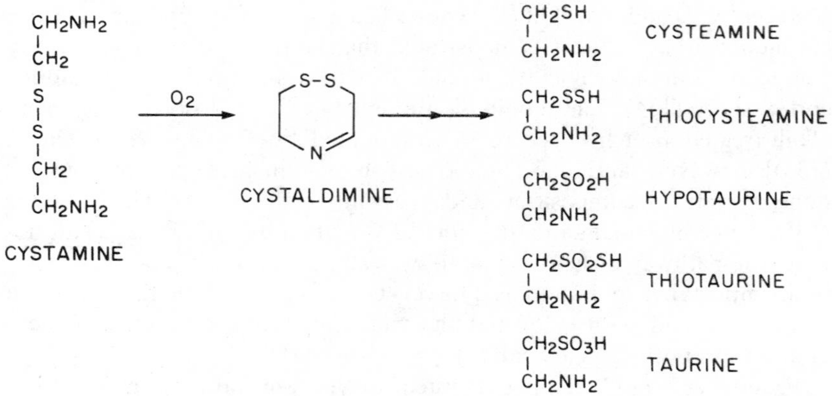

$$\begin{array}{l} CH_2NH_2 \\ | \\ CH_2 \\ | \\ S \\ | \\ S \\ | \\ CH_2 \\ | \\ CH_2NH_2 \end{array}$$

CYSTAMINE

$$\xrightarrow{O_2}$$

(cyclic structure) CYSTALDIMINE

$$\longrightarrow$$

$$\begin{array}{l} CH_2SH \\ | \\ CH_2NH_2 \end{array}$$ CYSTEAMINE

$$\begin{array}{l} CH_2SSH \\ | \\ CH_2NH_2 \end{array}$$ THIOCYSTEAMINE

$$\begin{array}{l} CH_2SO_2H \\ | \\ CH_2NH_2 \end{array}$$ HYPOTAURINE

$$\begin{array}{l} CH_2SO_2SH \\ | \\ CH_2NH_2 \end{array}$$ THIOTAURINE

$$\begin{array}{l} CH_2SO_3H \\ | \\ CH_2NH_2 \end{array}$$ TAURINE

Figure 6   Enzymatic conversions of cystamine (see 142–145).

and readily loses sulfur. Thus cystamine is readily converted to cysteamine, ammonia, a "C-2" fragment (probably glycolaldehyde or glyoxal) and labile sulfur (145). The labile sulfur is either incorporated into proteins or participates in transsulfuration reactions.

$$\text{Cystamine} \xrightarrow{\;\;\overset{NH_4^+}{\diagup}\;\;} \text{cystaldimine} \xrightarrow{\;\;\overset{\text{"}C_2\text{"}}{\diagup}\;\;} \text{thiocysteamine} \xrightarrow{\;\;\overset{S}{\diagup}\;\;} \text{cysteamine.} \qquad 21.$$

Several organisms are capable of producing taurine from sulfite or sulfate via sulfation reactions leading to cysteate; for a review of this and other pathways to taurine see (146).[4] Martin and colleagues showed that $^{35}SO_4^{2-}$ in the chick could be incorporated directly into taurine without label appearing in cysteine (150). Subsequently they showed that the pathway was also present in rat liver (151). The sulfation is due to the combined action of serine dehydratase and PAPS transferase (151; Equation 22):

$$SO_4^{2-} \xrightarrow{\;ATP\;} APS \xrightarrow{\;ATP\;} PAPS \overset{\text{aminoacrylate} \leftarrow \text{serine}}{\underset{\displaystyle + \, PAP}{\xrightarrow{\quad\downarrow\quad}}} \text{cysteate} \xrightarrow{\;O_2\;} \text{Taurine.} \quad 22.$$

[4]This review also describes the occurrence and biochemistry of some unusual taurine analogs, i.e. thiotaurine, taurocyamine, hypotaurocyamine, phosphotaurocyamine, and hypophosphotaurocyamine. Taurocyamine (*N*-guanidinotaurine) was first discovered in polycheate worms (147) but was subsequently found to be a constituent of rat urine (148) and of rat tissues ($\cong 20\,\mu M$)(149). In the chick embryo yolk sac, cysteate may arise from cysteine and sulfite via a reaction catalyzed by cysteine lyase (149a).

Most texts state that cysteine is oxidized to cysteine sulfinate, which is then converted to taurine, either by (a) decarboxylation to hypotaurine followed by oxidation, or (b) oxidation to cysteate followed by decarboxylation [see appropriate chapters in (11, 14, 15, 131, 133) and Figure 5]. The enzyme activities vary widely in different tissues and from species to species, but in general, pathway (a) is thought to greatly predominate. Interestingly, taurine is now considered an essential amino acid for the cat (152). Some authors also regard taurine as a dietary essential amino acid for the infant but not for the human adult (153, 154).

Almost 30 years ago Kearney & Singer showed that extracts of *Proteus vulgaris* contain (a) a cysteine sulfinate-α-ketoglutarate transaminase activity (155), and (b) a cysteine sulfinate dehydrogenase activity (155a). Since this time, textbooks have shown a pathway in which β-sulfinylpyruvate is formed via transamination of cysteine sulfinate, which in turn is desulfinated to pyruvate. However, Kearney & Singer showed that desulfination was extremely rapid and nonenzymatic (155). Later, it was shown that transamination results in stoichiometric formation of sulfite (156). β-Sulfinylpyruvate has not yet been prepared. Meister suggested [discussion in (151)] that desulfination may occur when the substrate is bound at the transaminase active site and that appreciable β-sulfinylpyruvate might not be formed free in solution. Previous findings support this hypothesis: John & Fasella (158) showed that glutamate-aspartate transaminase catalyzes elimination of sulfate from L-serine-O-sulfate; competing transamination was much slower. At the same time, slow inactivation of enzyme occurred because of alkylation with aminoacrylate (the product of α,β-elimination) (158). Cavallini et al (159) showed that cysteine sulfinate also inactivates glutamate-aspartate transaminase and that the rate of inactivation increases in the presence of α-ketoglutarate. Presumably, in the absence of α-ketoglutarate, the enzyme catalyzes an α,β-elimination to aminacrylate and sulfite, but slow transamination to the pyridoxamine P-form of the enzyme results in an enzyme resistant to inactivation; the presence of α-ketoglutarate ensures turnover to the susceptible pyridoxal P-form of the enzyme (159). Thiosulfate protects against inactivation by both serine-O-sulfate and cysteine sulfinate by trapping enzyme-generated aminoacrylate with the resultant formation of alanine thiosulfonate (Equation 23; 159). Cavallini et al speculated that a role of thiosulfate may be to trap the potentially toxic aminoacrylate, generated in several enzyme-mediated reactions, as nontoxic alanine thiosulfonate (159). The naturally occurring alanine thiosulfonate may also arise via the 3-mercaptopyruvate transsulfurase reaction (see below). The PAPS reaction (Equation 22) also removes aminoacrylate.

$$XCH_2CH(NH_2)CO_2H \rightarrow CH_2{=}C(NH_2)CO_2H + XH \qquad 23.$$

$$\downarrow HSSO_3H$$

$$HO_3SSCH_2CH(NH_2)CO_2H.$$

There has been some controversy as to whether cysteine sulfinate trans-aminase and glutamate-aspartate transaminase are identical. Recasens & Mandel (157) purified two cysteine sulfinate transaminases from rat brain to homogeneity; both are very active with aspartate but the authors were still unable to prove whether or not their purified enzymes were the same as soluble and mitochondrial forms of glutamate-aspartate transaminase. A similar controversy ensued over whether or not brain glutamate decar-boxylase and cysteine sulfinate decarboxylase are identical enzymes. How-ever, Spears & Martin (160) recently purified three enzymes with cysteine sulfinate decarboxylase activity from pig brain; decarboxylase I and II are specific for cysteine sulfinate whereas decarboxylase III is identical to glutamate decarboxylase. Wu (161) also recently purified distinct L-cys-teate/L-cysteine sulfinate and glutamate decarboxylases from bovine brain. It is also of interest that cysteine sulfinate can be converted to pyruvate and sulfite (presumably via β-sulfinylpyruvate or the corresponding α-imino acid) in a reaction catalyzed by a dehydrogenase in *P. vulgaris* (155a).

Taurine is abundant in animal tissues; this abundance may be due in part to its relatively slow turnover (e.g. 162). Certain bacteria can catalyze transamination between taurine and α-ketoglutarate (163), but since taurine is relatively inert in mammalian tissues, this reaction is probably of little importance in mammals. Interestingly, the taurine transaminase purified from *C. acidophila* is much more reactive with hypotaurine than with taurine (163). Recently, a mammalian hypotaurine-pyruvate (α-ketogluta-rate) transaminase was described (164). Transamination results in stoichi-ometric formation of acetaldehyde and sulfite (or sulfate). Evidently sulfinoacetaldehyde is extremely unstable. The possibility that desulfination occurs at the transaminase active site was also discussed (163).

Enzymes that Transfer Sulfane Sulfur

Transfer reactions of bivalent, or sulfane, sulfur are catalyzed by at least three separate enzymes: rhodanese (thiosulfate thiotransferase), thiosulfate reductase, and 3-mercaptopyruvate sulfurtransferase (for review see 5, 6, 165–169). These enzymes participate in the metabolism of cysteine (and possibly cystine) and other sulfur-containing compounds of low molecular weight (5, 6, 165, 167), detoxify sulfide (170), and detoxify cyanide (165, 171). In addition, rhodanese may be responsible for restoring labile sulfur in succinate dehydrogenase (172, 173); rhodanese and 3-mercaptopyruvate sulfurtransferase may be involved in the formation of the iron-sulfur chromophores of ferredoxin (174, 175).

RHODANESE   In 1933, Lang (171) described an enzyme, which he called rhodanese,[5] that converted cyanide to the less toxic thiocyanate. Subsequently, the enzyme was purified and crystallized. The enzyme is widespread, composed of two small (M_r 19,000) subunits, and possesses wide specificity (165, 176). Equation 24 summarizes the rhodanese reaction where $x = 0$, 1, or 2; $A = O$, H, or R; $Y^- = $ thiophilic anion; $(ASO_xS)^-$ = sulfane donor.

$$(ASO_xS)^- \quad \rightarrow \quad \text{rhodanese} \quad \rightarrow \quad YS^- \qquad\qquad 24.$$
$$(ASO_x)^- \quad \rightarrow \quad \text{rhodanese–S} \quad \rightarrow \quad Y^-.$$

The reaction is of the double displacement type with the enzyme acting as a sulfur carrier (165). Szczepkowski & Wood (170) described an interesting example of a thiane transfer, which may also be of biological importance. Thus, the rhodanese reaction can be coupled to the γ-cystathionase reaction on cystine. The resultant thiocystine (Equation 16) (and possibly thiocysteine; equation 15) acts as a sulfane donor to a suitable acceptor (170; Equation 25). Thiocystine reacts more rapidly with rhodanese than does thiosulfate (170):

$$\text{Thiocystine} + Y^- \rightarrow \text{cystine} + YS^-. \qquad\qquad 25.$$

Szczepkowski & Wood suggested that the combined action of γ-cysta-thionase and rhodanese makes possible the efficient utilization of cystine sulfur in tissues without the appearance of sulfide ion (170).

THIOSULFATE REDUCTASE   Koj and colleagues (166, 177–179) discovered this enzyme, which utilizes glutathione (Equation 26), and determined its subcellular distribution in tissues:

$$SSO_3^{2-} + 2GSH \rightarrow GSSG + H_2S + SO_3^{2-}. \qquad\qquad 26.$$

It is known that inner-labeled thiosulfate $(S \cdot {}^{35}SO_3)^{2-}$ is converted to ${}^{35}SO_4^{2-}$ more rapidly than outer-labeled thiosulfate $({}^{35}S \cdot SO_3)^{2-}$ [(178) and

---

[5]Lang named the enzyme rhodanese from the German word for thiocyanide, *Rhodanid*. von Euler suggested the suffix "-ese" for enzymes named for the product of the reaction. Rhodanese is the only example where this nomenclature is used. All other enzymes named for the product are referred to as synthases or synthetases.

references quoted therein]. Moreover, thiosulfate labeled in both sulfur atoms is formed when $(^{35}S \cdot SO_3)^{2-}$ is incubated with rat liver mitochondria (178). In order to explain these observations, Koj et al proposed a thiosulfate cycle (Figure 7) in which thiosulfate reductase (Equation 1; Figure 7) and rhodanese (Equation 2; Figure 7) are key enzymes (179). The cycle provides a means whereby sulfur derived from cyst(e)ine, cysteine sulfinate or 3-mercaptopyruvate is readily incorporated into sulfate.

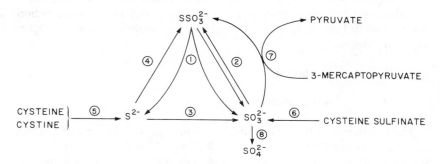

Figure 7 Thiosulfate cycle in animal tissues: (1) Thiol-dependent reduction of thiosulfate (rhodanese or thiosulfate reductase). (2) Sulfite exchange with inner atom of thiosulfate (rhodanese). (3) Sulfide oxidation to sulfite (nonenzymatic, sulfide oxidase). (4) Sulfide oxidation to thiosulfate (nonenzymatic, sulfide oxidase). (5) Sulfide production from both cysteine and cystine in presence of γ-cystathionase. (6) Sulfite production from cysteine sulfinate (enzymatic transamination-desulfination). (7) Thiosulfate formation by sulfur transfer from 3-mercaptopyruvate to sulfite (3-mercaptopyruvate sulfurtransferase). (8) Sulfite oxidation to sulfate (sulfite oxidase). Adapted from (178); the author added enzymatic step 7.

3-MERCAPTOPYRUVATE SULFURTRANSFERASE   Meister et al (180) discovered this enzyme in 1954. Attempting to demonstrate the enzymatic transamination of 3-mercaptopyruvate with a suitable amino donor, they consistently noted the appearance of alanine, rather than cysteine, on paper chromotograms. Subsequently the authors showed that rat liver possesses an enzyme that desulfurates 3-mercaptopyruvate. Elemental sulfur was shown to be the initial product; this finding was confirmed with the purified enzyme (181, 182). The enzyme is eventually inhibited by such accumulation (180, 182) but in the presence of a suitable acceptor, transsulfuration occurs (180–186). Thus, sulfur is transferred from 3-mercaptopyruvate to sulfite (183), cysteine sulfinate (183) and cyanide (184) to form thiosulfate, alanine thiosulfonate (= S-sulfocysteine) and thiocyanate, respectively. The mechanism of the transfer to cyanide has been studied in detail (182). In the presence of 2-mercaptoethanol, H_2S is generated (180) but the significance of this reaction is unclear. As noted above, sulfur can also be transferred to ferredoxin. Finally, Lipsett et al (187) demonstrated an absolute requirement for 3-mercaptopyruvate in the in vitro thiolation of *Escherichia coli* tRNA. 3-Mercaptopyruvate sulfurtransferase is widespread in animal tissues (167, 180, 188).

Compared to the cysteine sulfinate pathway, relatively little work has been carried out on the 3-mercaptopyruvate pathway. Direct evidence suggests that cysteine participates in transamination reactions (77–80, 180, 189–192). Glutamine and asparagine transaminases (78–80, 192) are capable of transaminating cysteine with a suitable donor. More recently mitochondrial cysteine-glutamate transaminase has been shown to be identical to glutamate-aspartate transaminase (189). A cysteine-glutamate transaminase not identical to glutamate-aspartate transaminase has also been described (191). The reverse reaction, conversion of 3-mercaptopyruvate to L-cysteine by transamination, has also been noted (78–80, 189–192). Cooper and Meister suggested that glutamine transaminase may act to salvage 3-mercaptopyruvate in addition to α-keto-γ-mercaptobutyrate (79, 80, 82, 83).

Although the 3-mercaptopyruvate pathway of cysteine metabolism has received relatively little attention, strong evidence indicates it is important in vivo. Thus, [^{35}S]alanine thiosulfonate was detected in rat kidney and urine following administration of [^{35}S]cystine (116). After administration of ^{35}SO$_4^{2-}$ to mutants of *A. nidulans,* Nakamura & Sato (193) noted that [^{35}S]alanine thiosulfonate accumulated. The authors concluded that alanine thiosulfonate is an obligatory intermediate in sulfate assimilation in these organisms. Mudd et al (194) noted that a child who eventually died with severe neurological disorders had a deficiency of sulfite oxidase (blockage 6, Figure 5) that led to the excretion of large amounts of sulfite, thiosulfate and alanine thiosulfonate. In another case, Crawhall et al described an inborn error in which large amounts of 3-mercaptolactate-cysteine-mixed disulfide were excreted in the urine (195–197). The disease was later recognized as being due to a deficiency of 3-mercaptopyruvate sulfurtransferase (blockage 5, Figure 5; 198, 199). Although in several cases the disease led to mental retardation, two sisters excreted large amounts of 3-mercaptolactate-cysteine-mixed disulfide, but were of normal intelligence (200).

Comments on the Metabolism of 3-Mercaptopyruvate

Thirty years ago Meister demonstrated 3-mercaptopyruvate to be an excellent substrate of lactate dehydrogenase (201), a finding that Kun (202) subsequently verified. 3-Mercaptolactate, in the form of its mixed disulfide with cysteine, is a normal constituent of human urine (117) and is increased in the urine of patients with mercaptolactate-cysteine disulfiduria (195–197; see above). Kobayashi (117) showed that pig kidney and liver extracts can convert cystine to 3-mercaptolactate- and mercaptoacetate-cysteine mixed disulfides. Presumably the reason for mixed disulfide formation is that thiols formed in minor amounts and oxidized in the extracellular space have a greater probability of forming a mixed disulfide with cysteine than of forming a symmetrical disulfide (167). It has been difficult to detect 3-mercaptopyruvate in biological samples because of its lability, but recently Hannestad et al (199) described a gas chromatographic method that relies on derivatization of carboxyl, thiol, and keto groups and showed that 3-mercaptopyruvate, (presumably as a mixed disulfide) is a normal trace constituent in human urine. A patient with 3-mercaptolactate-cysteine disulfiduria had a slightly elevated urine concentration of 3-mercaptopyruvate (199). Sörbo and colleagues have also developed methods for the determination of other cysteine- and 3-mercaptopyruvate-derived metabolites in urine. Interestingly, human subjects excrete $\cong$ 20–30 mmol of sulfur per day, of which at least 80% is inorganic sulfate (167, 203). Values for mercaptolactate (204), mercaptoacetate (204), N-acetylcysteine (204), thiosulfate (nonsmokers) (205), and thiocyanate (206) were 37, 9, 31, 32, and 44 μmol/day or < 1% of total urinary sulfur.[6] Kågedal and Källenberg devised an extremely sensitive method for the detection of mercaptoacetate and N-acetylcysteine and report average values in normal human urine of 6.4 and 30.7 μmol/liter, respectively (210). The origin of N-acetylcysteine is not clear but may be related to mercapturic acid metabolism (211); presumably mercaptoacetate and its mixed disulfide arise via oxidative decarboxylation of 3-mercaptopyruvate (or its mixed disulfide with cysteine). Ubuka & Yao (212) showed that, in the absence of catalase, cystine is converted in part to mercaptoacetate-cysteine mixed disulfide by L-amino acid oxidase. It is doubtful that this reaction is responsible for the in vivo formation of mercaptoacetate; most likely the enzyme responsible is an α-keto acid dehydrogenase . Recently, mercaptoacetate was shown to inhibit 3-hydroxybutyrate dehydrogenase (213); such an inhibition may account in part for the deleterious effects associated with sulfituria or 3-mercaptolactate-cysteine disulfiduria.

Since cyanide is extremely toxic and unlikely to arise via metabolic reactions, one might wonder about the origin of urinary thiocyanate. Cya-

---

[6]A portion of inorganic sulfate is activated to 3'-phosphoadenosine-5'-phosphosulfate (PAPS); sulfotransferases transfer "activated sulfate" to a large number of high and low molecular weight substrates (207; Figure 5). Thus, dermatan, keratan, heparan, and chondroitan sulfates are formed in this manner as are lipid- and polysaccharide-sulfate complexes (207). Sulfate conjugates of phenols, aliphatic alcohols, steroids, amino sugars, and choline also occur and many are found in trace amounts in the urine (208). Hypothiocyanate (OSCN⁻) has been found in saliva (209). Apparently, hypothiocyanate and possibly higher peroxythiocyanates result from the action of lactoperoxidase on thiocyanate; the higher peroxythiocyanates are thought to be antimicrobial (209).

nide is a low-level, widespread pollutant and present, for example, in inhaled cigarette smoke and in fire atmospheres, particularly of plastics. High levels of thiocyanate and cyanide have been noted in the blood of cigarette smokers and of smoke-inhalation victims (214, 215).

The relative importance of the 3-mercaptopyruvate and cysteine sulfinate pathways of cysteine metabolism remains in doubt. Stipanuk (216) concluded from careful studies of the fate of label derived from dietary L-[^{35}S]cysteine that, in the rat, the major pathway for cysteine metabolism is the cysteine sulfinate pathway. On the other hand, Krijgsheld et al (217) showed recently that oral administration of both D- and L-cysteine to rats resulted in 55% and 33% recovery of sulfur as sulfate in the urine within 24 hours. The sulfate derived from D-cysteine could not have come from cysteine sulfinate; D-cysteine is not a substrate of cysteine dioxygenase (218). Presumably, D-amino acid oxidase converts D-cysteine to 3-mercaptopyruvate, which can readily yield thiosulfate via transsulfuration; thiosulfate can then react through the thiosulfate cycle to yield sulfate (see Figures 5 and 7). The experiments of Krijgsheld et al (217) also suggest that transamination of 3-mercaptopyruvate in vivo is not very extensive. Thus no increase in taurine levels was noted following administration of D-cysteine; in contrast, L-cysteine administration markedly raised taurine levels (217). Possibly, as with methionine, the α-keto acid pathway is more important for the metabolism of the D-isomer of cysteine. Evidently the capacity to metabolize 3-mercaptopyruvate in the rat is very great, so that the limiting step in the 3-mercaptopyruvate pathway would appear to be the transamination of cysteine. Nevertheless, even though the 3-mercaptopyruvate pathway under normal conditions, i.e. L-cysteine breakdown, appears to be quantitatively a minor route, it is of major importance for the transfer of thiolane sulfur, both as a normal metabolic process and as a defense against toxins such as cyanide.

α-KETO ACID ANALOGS OF METHIONINE, CYST(E)INE AND HOMOCYST(E)INE

Waelsch & Borek (219) first prepared α-keto-γ-methiolbutyrate in 1939 by oxidizing D-methionine with kidney slices. α-Keto-γ-methiolbutyrate may be more readily obtained by incubating L-methionine with L-amino acid oxidase and catalase (220). The α-keto acid has also been chemically synthesized via hydrolysis of the appropriate azlactone (221). The sodium salt is relatively stable and the α-keto acid exhibits typical α-keto acid behavior (222). 3-Mercaptopyruvate (the α-keto acid analog of cysteine) cannot be made by oxidation of L-cysteine with L-amino acid oxidase, although cysteine is a substrate (192). Oxidation gave rise to a number of products, one of which was 3-mercaptopyruvate-cysteine disulfide (192). However, if the reaction was carried out in the presence of lactate dehydrogenase and NADH, 3-mercaptolactate was formed in good yield (192). 3-Mercaptopyruvate has been prepared as the ammonium salt by reacting β-chloro- or bromopyruvate with ammonium sulfide (180, 223). 3-Mercaptopyruvate is more reactive than most α-keto acids and readily undergoes aldol condensation particularly at alkaline pH; it is more stable under acidic conditions (192). Kumler & Kun noted that 3-mercaptopyruvate exhibits an anomalous carboxyl pK_a value and infrared spectrum (5, 202, 224). Despite negligible absorbance of aqueous solutions in the region 285–330 nm, it was concluded that 3-mercaptopyruvate exists largely in an enolic form (5, 224). However, 3-mercaptopyruvate was shown recently to exist in solution in equilibrium with a cyclic dithiane (Figure 8; 192). Most likely, this dithiane formation accounts for the anomalous kinetics noted for the reaction of 3-mercaptopyruvate with rat liver glutamine transaminase (78) and with lactate dehydrogenase (225).

The diketo acid analog of cystine, 3-mercaptopyruvate disulfide, has been chemically prepared by oxidation of 3-mercaptopyruvate with iodine (180, 223). The 2,4-dinitrophenylhydrazone of 3-mercaptopyruvate can also be oxidized to the corresponding disulfide (180, 223). However, treatment of 3-mercaptopyruvate disulfide with 2,4-dinitrophenylhydrazine results in a product with only one hydrazone linkage (180). Cysteine is a substrate of L-amino acid oxidase. The initial product is 3-mercaptopyruvate-cysteine mixed disulfide (212, 226–228). There is some evidence that this molecule is also a substrate of L-amino acid oxidase (212, 227). However, Ricci et al (228) presented evidence that 3-mercaptopyruvate-cysteine mixed disulfide cyclizes to a seven-membered dicarboxyl dihydro-dithiazine ring, analogous to the "cystaldimine" structure formed from cystamine (Figure 6).

The α-keto acid analog of homocysteine has not been isolated, although both homocysteine and homocystine are substrates of L-amino acid oxidase (226, 229). Cooper & Meister showed that α-keto-γ-mercaptobutyrate is unstable in solution; however, the corresponding 2,4-

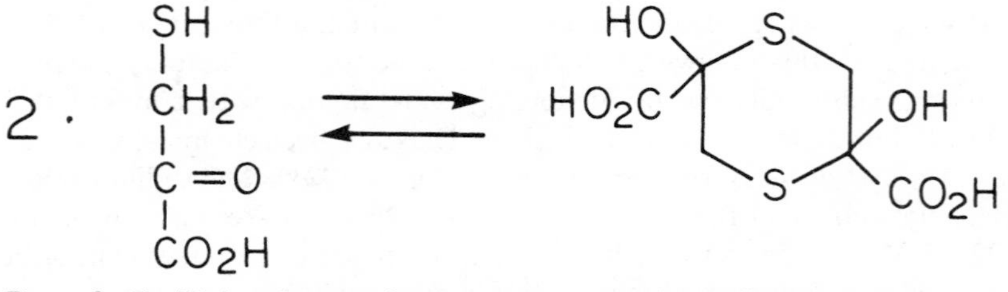

Figure 8 Equilibrium between 3-mercaptopyruvate and 2,5-dihydroxy-1,4-dithiane-2,5-dicarboxylate.

dinitrophenylhydrazone and its disulfide can be prepared (229). Oxidation of DL-homocysteine with L-amino acid oxidase gives rise to at least seven products, five of which have been tentatively identified, respectively, as α-keto-γ-mercaptobutyrate, the mono and diketo analogs of homocystine, and the mono and diketo analogs of homolanthionine (229). Apparently, some of the imine derived from homocysteine (the initial oxidation product) hydrolyzes to the α-keto acid, but another portion undergoes spontaneous β,γ-elimination of H_2S followed by γ-addition of RS^- giving rise to "keto" homolanthionine derivatives. Similar nonenzymatic β,γ-elimination-γ-addition reactions were previously noted with the imine derived from methionine sulfoximine (230). These nonenzymatic γ-exchange reactions are similar to the enzymatic γ-exchanges catalyzed by cystathionine γ-synthase and methionase (231), except that in these examples the imine nitrogen is part of an enzyme-bound pyridoxal P-ketimine. Homolanthionine [$S(CH_2CH_2CHNH_2CO_2H)_2$] is a rare amino acid found to accumulate in methionine-requiring mutants of $E.\ coli$ (232) and in the urine of patients with homocystinuria (233). Presumably homolanthionine arises at least in part via an enzymatic (or nonenzymatic) γ-replacement reaction. α-Hydroxy-γ-mercaptobutyrate-homocysteine mixed disulfide is also present in the urine of individuals with homocystinuria (233, 234). Presumably, homocysteine is transaminated in vivo to the corresponding α-keto acid, which is then reduced to α-hydroxy-γ-mercaptobutyrate. In support of this hypothesis is the recent finding that homocysteine can be transaminated in vitro; the corresponding α-keto acid is also a substrate of lactate dehydrogenase (229).

CYSTEINE- AND HOMOCYSTEINE-CARBONYL ADDUCTS

It has been known for more than 45 years that cysteine reacts with ketones, such as pyruvate, to yield crystalline hemithioketals (235, 236). In the reaction with reactive aldehydes, such as glyoxylate (237) and pyridoxal P (e.g. 238), water is lost and a cyclic thiazolidine is formed. L-Cysteine-α-keto acid hemithioketals can be chromatographed on paper, and can be detected with ninhydrin; they are substrates of L-amino acid oxidase (192). Homocysteine is also known to react with pyridoxal-P, presumably to yield a thiazine. Recently, Dewhurst & Griffiths (239) reported that they had crystallized this thiazine and confirmed its structure, but gave no details. Homocysteine-glyoxylate hemithioacetal and homocysteine-pyruvate hemithioketal have recently been prepared (229). Homocysteine-α-keto acid hemithioketals are relatively stable and can also be detected with ninhydrin following paper chromatography (229).

Both cysteine (e.g. 240) and homocysteine (e.g. 241, 242) inhibit pyridoxal-P enzymes, presumably by forming cyclic adducts at the active site. Homocysteine appears to inhibit enzymes such as GABA transaminase (241) and glutamate decarboxylase (242), both by competition with substrate and by forming an inhibitor/cofactor complex. High concentrations of cyst(e)ine (54, 243–245) and homocyst(e)ine (54, 246–249) are toxic to the central nervous system. Part of this toxicity may stem from adduct formation with essential carbonyls (e.g. with pyridoxal-P enzymes).

Collagen disease often occurs in patients with homocystinura. Jackson (250) reported that homocysteine reacts with formaldehyde and suggested that in homocystinurics, high levels of homocysteine will react with the aldehyde groups of procollagen; such a blockage will impede cross-linking, thus destabilizing the collagen. In support of this theory, homocysteine was shown to react directly with collagen in vitro, destroying the aldimine bridges formed between lysine and allysine (251).

Accelerated arteriosclerosis is a feature of several disorders of sulfur amino acid metabolism (e.g. 252). Several authors have attempted (with varying success) to produce an animal model of this disease by injection of DL-homocysteine lactone (252–256) or L-homocystine (257, 258).[7] The mechanism of arteriosclerosis in this animal model remains unknown but may also be related to S-adduct formation with essential carbonyls.

Many linear addition complexes of glyoxylate and –SH compounds [thioacetals; i.e. $R–S–C(OH)HCO_2H$] are substrates of L-α-hydroxy acid oxidase (260). On the other hand, when cyclization occurs with loss of water, the resultant thiazolidine is a substrate of D-amino acid oxidase (261, 262). Hamilton et al (261) suggested that the D-amino acid oxidase-catalyzed conversion of cysteamine + glyoxylate to Δ^2-thiazoline-2-carboxylate, via a thiazolidine intermediate, may be of physiological importance (261). Δ^2-Thiazoline-2-carboxylate has recently been shown to inhibit dopamine β-hydroxylase (263).

ROLE OF CYSTEINE IN PIGMENT FORMATION

Glutathione and cysteine are important for the synthesis of several types of pigments (for review see 264, 265). Tyrosine, in the presence of cysteine and tyrosinase (266) or peroxidase (267), gives rise to various cysteinyldopas. [If the oxidation of tyrosine is carried out in the presence of glutathione, glutathionedopas are generated that are enzymatically cleaved to cysteinyl-

---

[7]Homocysteine thiolactone is readily hydrolyzed to homocysteine in alkali or by pig liver carboxyesterase (259). At high concentrations the lactone readily forms a diketopiperazine (259). Therefore, some caution is needed when using the thiolactone in animal studies (259).

dopas (265)]. Generally, the major product is 5-(S)-cysteinyldopa with smaller amounts of 2-(S)-cysteinyldopa, 6-(S)-cysteinyldopa, 2,5-(S,S)-dicysteinyldopa, 3,5,6-(S,S,S)-tricysteinyldopa, and dihydrobenzothiazines (264–267). Interestingly, 2,5-(S,S)-dicysteinyldopa was first isolated from eyes of the alligator gar, *Lepisosteus spatula* (268). It has been assumed that 5-(S)-cysteinyldopa is produced by the nonenzymatic 1,4-nucleophilic addition of cysteine to dopaquinone (267) (equation 27). However, Nkpa & Chedekel (269) could not obtain evidence for such a mechanism from model reactions and suggested a free radical mechanism involving nucleophilic attack on the semiquinone species (269):

$$\text{Tyr} \xrightarrow{\text{O}_2,\text{ enz}} \text{dopa} \xrightarrow{\text{O}_2,\text{ enz}} (\text{dopaquinone}) \xrightarrow{+ \text{ cys; 1,4-addition}} 5\text{-(S)-cysteinyldopa.} \qquad 27.$$

Prota et al (270) showed that the 5-(S)-, 2-(S)-, and 2,5-(S,S)-cysteinyldopas are excreted in the urine of patients with melanomas. However, the 5-(S)-isomer was shown recently to be a normal constituent in urine (271).

Generally, three types of pigments are regarded as arising from the oxidation of tyrosine: (*a*) eumalins (black-brown), (*b*) phaeomalins (yellow-reddish brown), and (*c*) trichochromes (yellow-red). The eumalins contain little sulfur and are not derived from cysteinyldopas; the phaeomalins and trichochromes are derived from the oxidation of cysteinyldopas (Equation 28). The structures of several trichochromes, which occur in small quantities in red hair, have been elucidated (265). For a discussion of the occurrence of the three pigments in the various human racial types, see (264).

$$\text{Cysteinyldopa} \xrightarrow{\text{O}_2,\text{ tyrosinase}} \text{benzothiazinylalanine} \xrightarrow{} \begin{array}{c} \text{phaeomalins} \\ + \\ \text{trichochromes.} \end{array} \qquad 28.$$
$$\text{intermediates}$$

INBORN ERRORS OF SULFUR AMINO ACID METABOLISM

Sulfituria and 3-mercaptopyruvate-cysteine disulfiduria were cited above. Several other diseases in which sulfur amino acid metabolism is affected either directly or indirectly are known (Table 1; for review see 7, 8, 46, 169, 251, 272–279). There are at least four causes of homocystinuria. Homocystinuria I is caused by a defect in cystathionine β-synthase; the other types are caused by defects of the remethylation cycle. Two types of cystathionine β-synthase deficiency are known, a type that responds to B_6 treatment and another that does not. Similarly, B_6-responsive and B_6-unresponsive forms of γ-cystathionase deficiency have been described. Note that both methylmalonyl CoA mutase and N^5-methyltetrahydrofolate-homocysteine transmethylase require methylcobalamin as a cofactor. Therefore, methylmalonic aciduria is also a feature of homocystinuria types III and IV. Patients with 3-mercaptopyruvate cysteine disulfiduria (169, 195–199), sulfituria (194), homocystinuria (233, 234), and cystathioninuria (280) excrete a large number of unusual sulfur-containing amino acids. In addition to those already mentioned, may be added 5-amino-4-imidazolecarboxamide-5'-S-homocysteinylriboside (AICHR), a previously unknown amino acid found in the urine of homocystinuric patients (233). By 1972, Ohmori et al (234) identified 25 new sulfur amino acids in the urine from various species including eleven from the urine of homocystinuric patients.

Cystinosis is caused by an impairment of cystine transport (e.g. 279; and references quoted therein). Very recent evidence suggests that lysosome function is compromised in this disease (281, 282). Thus, cystinotic fibroblasts can be loaded with cystine by incubation in a medium containing glutathione-cysteine mixed disulfide; on removal of the disulfide from the medium, cysteine is very slow to clear from the lysosomes, compared to

Table 1 Inborn errors affecting sulfur amino acid metabolism in man

Name	Defect[a]	Number in Figure 2 or 5	Selected references
Hypermethioninemia	Liver methionine adenosyltransferase ↓[b]	1	276, 279
Homocystinuria (I)	Cystathionine β-synthase ↓	2	8, 46, 273, 274, 276–279
Cystathioninuria	γ-Cystathionase ↓	3	8, 46, 273, 275, 276, 278, 279
Cystinosis	Impairment of cystine transport		273, 274, 279, 281, 282
Homocystinuria (II)	N^5,N^{10}-Methylenetetrahydrofolate reductase ↓	4	8, 46, 274, 277, 279
3-Mercaptopyruvate-cysteine disulfiduria	3-Mercaptopyruvate sulfurtransferase ↓	5	8, 169, 276, 279
Sulfituria (sulfocysteinuria)	Sulfite oxidase ↓	6	273, 275, 279
Homocystinuria (III)	Low N^5-methyltetrahydrofolate-homocysteine transmethylase activity because of inability to synthesize cobalamin cofactor		46, 277, 279
Homocystinuria (IV)	Defective intestinal B_{12} absorption from gut leading to low N^5-methyltetrahydrofolate-homocysteine transmethylase activity		46, 277, 279
Methionine maladsorption syndrome	Inability to absorb methionine from gut		276

[a] Down-pointing arrow denotes low or absent enzyme activity.
[b] May occur also in cystathioninuria, tyrosinemia, and fructose intolerance.

control fibroblasts or to fibroblasts obtained from heterozygous individuals (281). Similarly, cystinotic leukocytes can be loaded with cystine by exposure to cystine dimethyl ester; again, clearance of cystine from the lysosomes of cystinotic leukocytes was shown to be much slower than in control leukocytes (282).

L-METHIONINE AND L-CYST(E)INE REQUIREMENT OF CANCER CELLS

A survey of six studies showed that in all cases normal human and rodent cell lines in culture can grow adequately in the presence of homocysteine instead of methionine (283). In contrast, malignant rodent cells and some malignant human cell lines have an absolute requirement for methionine (283, 284). Kreis et al (284) showed that in two human embryonic fibroblast lines (one line that had an absolute requirement for methionine and another that did not) growth was completely retarded by the addition of L-methionase to the medium. Growth was partially restored to the homocysteine-utilizing line, but not to the line with a requirement for methionine, by addition of D-homocystine or L-homocysteine thiolactone (284). The mechanism of rescue is unknown, but the data suggest that "methionine starvation" in vivo may have some therapeutic value (283).

Uren & Lazarus (285) have reviewed the cyst(e)ine requirements of certain malignant cell lines (285). Some cells are susceptible to enzymatic cyst(e)ine depletion in vitro. However, enzyme therapy in vivo has yet to be demonstrated because of rapid clearance of enzyme from the blood (285). Still, the possibility remains that some cysteine-utilizing enzyme(s) or in vivo inhibitors of γ-cystathionase may be of therapeutic value. Recent work suggests that propargylglycine is active against γ-cystathionase in rats in vivo (286), but whether this compound or related compounds is of clinical use remains to be seen.

INDUSTRIAL APPLICATIONS

Soybean protein is a major feed for livestock. It is heralded as a low-cost meat substitute (287) and as an ingredient in milk substitutes in infant formulas (288). However, although the quality of soybean protein is high compared to other vegetable proteins, it is limiting in L-methionine (289). Direct addition of L-methionine is expensive whereas addition of DL-methionine is less so. However, addition of L-methionine or of DL-methionine to foodstuffs results in unappetizing flavors caused by bacterial degradation and release of volatile sulfides (290, 291). Furthermore, despite earlier reports to the contrary, D-methionine may not be well utilized in man (289).[8] Not surprisingly, researchers are putting a great deal of effort into developing a cheap, palatable methionine analog that can be converted readily to methionine in vivo. One possibility is N-acetylmethionine, which has been shown to effectively replace L-methionine in the diet of rats (296), adult humans (289), and one-year-old fasting infants (297). However, the compound that has received most attention is hydroxymethionine analog, or HMA, (DL-α-hydroxy-γ-methiolbutyrate, calcium salt; Monsanto Chemical Co.). HMA is now manufactured in large quantities as a supplement to livestock feed.

In 1932 Block & Jackson reported that the zinc salt of HMA can promote growth (298). Subsequently, the DL-isomer was shown to promote growth in a number of species (47, 299), and recently the growth response to the D- and L-isomers in chicks has been investigated (299). It was found that efficacy in isosulfurous amino acid formulations was in the order L-HMA $<$ DL-HMA $<$ D-HMA $<$ D-MET $\leq$ L-MET (299). Apparently, the D- and L-HMA isomers are converted to α-keto-γ-methiolbutyrate in vivo, by D- and L-α-hydroxy acid dehydrogenase, respectively (299). Interestingly, it has long been known that α-keto-γ-methiolbutyrate can replace methionine in the diets of rats (221). Langer showed that in the rat, HMA is converted to α-keto-γ-methiolbutyrate, which is then transaminated to L-methionine (300). Glutamine, and to a lesser extent asparagine, were required for the transamination reaction (300). As discussed above, α-keto-γ-methiolbutyrate is a good substrate of rat liver and kidney glutamine transaminases and of rat liver asparagine transaminase (78, 80).

S-Carboxymethylcysteine (Mucodyne, Berk Pharmaceuticals) is an orally effective mucolytic agent used clinically for the treatment of respiratory disorders (301). Following oral administration of ^{35}S-carboxymethylcysteine to rats, label is concentrated in mucus-producing organs including the prostate in the male and the cervix in the female (301).

ACKNOWLEDGMENTS

Some of the work referred to from the author's laboratory was supported by US Public Health Service grant AM 16739. The author is a recipient of a US Public Health Service Career Development Award NS 00343. I wish to thank Ms. Susan M. Hall for preparing the manuscript and Drs. Owen Griffith and Alton Meister for their helpful suggestions.

[8]Earlier results suggested that D-methionine could effectively replace L-methionine in the diet in mice (292), rats (293), and man (294). More recent studies indicate that puppies can also utilize D-methionine (47). However, Zezulka & Calloway (289) were unable to demonstrate efficient utilization of D-methionine in man, even in the presence of choline (to prevent loss of methyl groups) and sulfate (to spare methionine; cf 295).

Literature Cited

1a. Meister, A., Anderson, M. E. 1983. *Ann. Rev. Biochem.* 52:711–60
1. Black, S. 1963. *Ann. Rev. Biochem.* 32:399–418
2. Meister, A. 1965. *Biochemistry of the Amino Acids.* pp. 74–82, 757–818 New York: Academic
3. Truffa-Bachi, P., Cohen, G. N. 1968. *Ann. Rev. Biochem.* 37:79–108
4. Trudinger, P. A. 1969. *Adv. Microb. Physiol.* 3:111–58
5. Kun, E. 1969. In *Metabolic Pathways.* ed. D. M. Greenberg, 3:375–401. New York: Academic, 3rd ed.
6. Roy, A. B., Trudinger, P. A. 1970. *The Biochemistry of Inorganic Compounds of Sulphur.* Cambridge: Cambridge University Press.
7. Muth, O. H.; Oldfield, J. E., eds. 1970. *Sulfur in Nutrition.* Westport, Conn: Avi
8. Scriver, D. R., Rosenberg, L. E. 1973. *Amino Acid Metabolism and its Disorders,* pp. 207–33. Philadelphia: Saunders
9. Schiff, J. A., Hodson, R. C. 1973. *Ann. Rev. Plant Physiol.* 24:381–414
10. Bender, D. A. 1975. *Amino Acid Metabolism,* pp. 112–42. New York: Wiley
11. Greenberg, D. M., ed. 1975. *Metabolic pathways.* Vol. 7, *Metabolism of Sulfur Compounds.* New York: Academic
12. Salvatore, F., Borek, E., Zappia, V., Williams-Ashman, H. G., Schlenk, F., eds. 1977. *The Biochemistry of Adenosylmethionine.* New York: Columbia Univ. Press
13. Jocelyn, P. C. 1972. *Biochemistry of the SH Group.* New York: Academic
14. *Ciba Found. Symp. 72, Sulphur in Biology* 1980. Amsterdam: Excerpta Medica
15. Cavallini, D., Gaull, G. E., Zappia, V., eds. 1980. *Natural Sulfur Compounds: Novel Biochemical and Structural Aspects.* New York: Plenum
16. Siegel, L. M. 1975. See Ref. 11, pp. 217–86
17. Kelly, D. P. 1980. See Ref. 14, pp. 3–18
18. Schiff, J. A. 1980. See Ref. 14, pp. 49–69
19. Tsang, M. L.-S. 1981. *J. Bacteriol.* 146:1059–66
20. Abrams, W. R., Schiff, J. A. 1973. *Arch. Microbiol.* 94:1–10
21. Tsang, M. L.-S., Schiff, J. A. 1976. *Plant Cell Physiol.* 17:1209–20
22. Tsang, M. L.-S., Schiff, J. A. 1978. *Plant Sci. Lett.* 11:177–83
23. Brunhold, C., Schmidt, A. 1978. *Plant Physiol.* 61:342–47
24. Wilson, L. G., Bierer, D. 1976. *Biochem. J.* 158:255–70
25. Tsang, M. L.-S., Schiff, J. A. 1976. *J. Bacteriol.* 125:923–33
26. Kredich, N. M., Hulanicka, M. D., Hallquist, S. G. 1980. See Ref. 14, pp. 87–99
27. Wiebers, J. L., Garner, H. R. 1967. *J. Biol. Chem.* 242:5644–49
28. Giovanelli, J., Mudd, S. H. 1967. *Biochem. Biophys. Res. Commun.* 27:150–6
29. Datko, A. H., Mudd, S. H., Giovanelli, J. 1977 *J. Biol. Chem.* 252:3436–45
30. Schlossmann, K., Lynen, F. 1957. *Biochem. Z.* 328:591–94
31. Greenberg, D. M. 1975. See Ref. 11, pp. 505–28
32. Braunstein, A. E., Goryachenkova, E. V., Lac, N. D. 1969. *Biochim. Biophys. Acta* 171:366–8
33. Braunstein, A. E., Goryachenkova, E. V., Tolosa, E. A., Willhardt, I. H., Yefremova, L. L. 1971. *Biochim. Biophys. Acta* 242:247–60
34. Huovinen, J. A., Gustafsson, B. E. 1967. *Biochem. Biophys. Res. Comm.* 136:441–47
35. Flavin, M. 1975. See Ref. 11, pp. 457–503
36. Woods, D. D., Foster, M. A., Guest, J. R. 1965. In *Transmethylation and Methionine Biosynthesis,* ed. S. K. Shapiro, F. Schlenk, pp. 138–56. Chicago: Chicago
37. Rowbury, R. J., Woods, D. D. 1964. *J. Gen. Microbiol.* 36:341–58
38. Kerr, D. S., Flavin, M. 1970. *J. Biol. Chem.* 245:1842–55
39. Rosenthal, S., Smith, L. C., Buchanan, J. M. 1965. *J. Biol. Chem.* 240:836–43
40. Giovanelli, J., Mudd, S. H., Datko, A. M. 1980. See Ref. 15, pp. 81–92
41. Paszewski, A., Grabski, J. 1975. *J. Bacteriol.* 124:893–904
42. Ozaki, H., Shiio, J. 1982. *J. Biochem.* 91:1163–71
43. Bright, S. W. J., Lea, P. J., Miflin, B. J. 1980. See Ref. 14, pp. 101–17
44. Giovanelli, J., Mudd, S. H., Datko, A. H. 1980. In *The Biochemistry of Plants,* ed. B. J. Miflin, Vol. 5. New York: Academic
45. du Vigneaud, V. 1952. *A Trail of Research in Sulfur Chemistry.* Ithaca: Cornell Univ. Press
46. Mudd, S. H. 1980. See Ref. 14, pp. 239–58
47. Burns, R. A., Milner, J. A. 1981. *J. Nutr.* 111:2117–24

48. Sturman, J. A., Gaull, G., Raiha, N. C. R. 1970. *Science* 169:74–76
49. Sturman, J. A. 1980. See Ref. 15, pp. 107–119
50. Mudd, S. H., Pool, J. H. 1975. *Metab. Clin. Exp.* 24:721–35
51. Krebs, H. A., Hems, R., Tyler, B. 1976. *Biochem. J.* 158:341–53
52. Ferger, M. F., du Vigneaud, V. 1950. *J. Biol. Chem.* 185:53–57
53. Riosin, M.-P., Chatagner, F. 1969. *Bull. Soc. Chim. Biol.* 51:481–93
54. Harper, A. E., Benevenga, N. J., Wohlhueter, R. M. 1970. *Physiol. Rev.* 50:428–58
55. Hardwick, D. F., Applegarth, D. A., Cockroft, D. M., Ross, P. M., Calder, R. J. 1970. *Metab. Clin. Exp.* 19:381–91
56. Benevenga, N. J., Yeh, M.-H., Lalich, J. J. 1976. *J. Nutr.* 106:1714–20
57. Rosen, H. M., Yoshimura, N., Hodgman, J. M., Fischer, J. E. 1977. *Gastroenterology* 72:483–87
58. Chen, S., Zieve, L., Mahadevan, V. 1970. *J. Lab. Clin. Med.* 75:628–35
59. Finkelstein, J. D. 1975. See Ref. 11, pp. 547–97
60. Horowitz, J. H., Rypins, E. B., Henderson, J. M., Heymsfield, S. B., Moffit, S. D., et al. 1981. *Gastroenterology* 81:668–75
61. Zieve, L., Doizaki, W. M., Zieve, F. J. 1974. *J. Lab. Clin. Med.* 83:16–28
62. Kaji, H., Hisamura, M., Saito, N., Murao, M. 1978. *Clin. Chim. Acta* 85:279–84
63. Benevenga, N. J. 1974. *J. Agric. Food Chem.* 22:2–9
64. Case, G. L., Benevenga, N. J. 1976. *J. Nutr.* 106:1721–36
65. Case, G. L., Benevenga, N. J. 1977. *J. Nutr.* 107:1665–76
66. Mitchell, A. D., Benevenga, N. J. 1978. *J. Nutr.* 108:67–78
67. Steele, R. D., Benevenga, N. J. 1978. *J. Biol. Chem.* 253:7844–50
68. Everett, G. B., Mitchell, A. D., Benevenga, N. J. 1979. *J. Nutr.* 109:597–605
69. Steele, R. D., Benevenga, N. J. 1979. *J. Biol. Chem.* 254:8885–90
70. Steele, R. D., Barber, T. A., Lalich, J. J., Benevenga, N. J. 1979. *J. Nutr.* 109:1739–51
71. Dixon, J. L., Benevenga, N. J. 1980. *Biochem. Biophys. Res. Commun.* 97:939–46
72. Steele, R. D., Benevenga, N. J. 1979. *Cancer Res.* 39:3935–41
73. Steele, R. D. 1982. *J. Nutr.* 112:118–25
74. Canellakis, E. S., Tarver, H. 1953. *Arch. Biochem. Biophys.* 42:387–98, 446–55
75. Singer, A. G., Agosta, W. C., O'Connell, R. J., Pfaffmann, C., Bowen, D. V., Field, F. H. 1976. *Science* 191:948–50
76. Toohey, J. I. 1977. *Biochem. Biophys. Res. Commun.* 78:1273–80
77. Cammarata, P. S., Cohen, P. P. 1950. *J. Biol. Chem.* 187:439–52
78. Cooper, A. J. L., Meister, A. 1972. *Biochemistry* 11:661–71
79. Cooper, A. J. L., Meister, A. 1974. *J. Biol. Chem.* 249:2554–61
80. Cooper, A. J. L. 1977. *J. Biol. Chem.* 252:2032–38
81. Ikeda, T., Konishi, Y., Ichihara, A. 1976. *Biochim. Biophys. Acta* 445:622–31
82. Cooper, A. J. L., Meister, A. 1977. *CRC Crit. Rev. Biochem.* 4:281–303
83. Cooper, A. J. L., Meister, A. 1981. *Comp. Biochem. Physiol. B.* 69:137–45
84. Livesey, G. 1981. In *Metabolism and Clinical Implications of Branched Chain Amino and Ketoacids,* ed M. Walser, J. R. Williamson, pp. 143–48. New York:Elsevier/North Holland
85. Hoshika, Y. 1982. *J. Chromatogr.* 237:439–45
86. Lund, P. 1980. *FEBS Lett.* 117:K86–92
87. Kaji, H., Saito, N., Murao, M., Ishimoto, M., Kondo, H., et al 1980. *J. Chromatogr.* 221:145–48
88. Meister, A., Wellner, D. 1963. *Enzymes* 7:609–48
89. Kirk, E. 1949. *Gastroenterology* 12:782–94
90. Benjamin, L. E., Steele, R. D. 1981. *Am. J. Physiol.* 241:G503–8
91. Smith, R. L., Schlenk, F. 1952. *Arch. Biochem. Biophys.* 38:159–65, 167–75
92. Cantoni, G. L. 1953. *J. Biol. Chem.* 204:403–16
93. Tabor, H., Rosenthal, S. M., Tabor, C. W. 1958. *J. Biol. Chem.* 233:907–14
94. Shapiro, S. K., Mather, A. N. 1958. *J. Biol. Chem.* 233:631–33
95. Pegg, A. E., Williams-Ashman, H. G. 1969. *J. Biol. Chem.* 244:682–93
96. Shapiro, S. K., Ferro, A. J. 1977. See Ref. 12, pp. 58–76
97. Mozzi, R., Andreoli, V., Porcellati, G. 1980. See Ref. 15, pp. 41–54
98. Oliva, A., Galletti, P., Zappia, V., Paik, W. K., Kim, S. 1980. See Ref. 15, pp. 55–66
99. Ballou, C. E. 1977. See Ref. 12, pp. 435–50
100. Salvatore, F., Traboni, C., Colonna, A., Ciliberto, G., Paolella, G., Cimino, F. 1980. See Ref. 15, pp. 25–40
101. Knappe, J., Schmitt, T. 1976. *Biochem. Biophys. Res. Commun.* 71:1110–17
102. Nishimura, S. 1977. See Ref. 12, pp. 510–20

103. Garbers, D. L. 1978. *Biochim. Biophys. Acta* 523:82–93
104. Zappia, V., Cartenì-Farina, M., Cacciapuoti, G., Oliva, A., Gambacorta, A. 1980. See Ref. 15, pp. 133–48
105. Ferro, A. J., Wrobel, N. C., Nicolette, J. A. 1979. *Biochim. Biophys. Acta* 570:65–73
106. Zappia, V., Oliva, A., Cacciapuoti, G., Galletti, P., Mignucci, G., Cartenì-Farina, M. 1978. *Biochem. J.* 175:1043–50
107. Cacciapuoti, G., Oliva, A., Zappia, V. 1978. *Int. J. Biochem.* 9:35–41
108. Cartenì-Farina, M., Oliva, A., Romeo, G., Napolitano, G., DeRosa, M., et al. 1979. *Eur. J. Biochem.* 101:317–24
109. Schlenk, F., Ehninger, D. J. 1964. *Arch. Biochem. Biophys.* 106:95–100
110. Backlund, P. S. Jr., Smith, R. A. 1981. *J. Biol. Chem.* 256:1533–35
111. Backlund, P. S. Jr., Chang, C. P., Smith, R. A. 1982. *J. Biol. Chem.* 257:4196–202
111a. Trackman, P. C., Abeles, R. H. 1981. *Biochem. Biophys. Res. Commun.* 103:1238–44
112. Amrhein, N., Schneebeck, D., Skorupka, H., Tophof, S. 1981. *Naturwissenschaften* 68:619–20
113. Yung, K. H., Yang, S. F., Schlenk, F. 1982. *Biochem. Biophys. Res. Commun.* 104:771–77
114. Crawhall, J. C., Segal, S. 1967. *Biochem. J.* 105:891–96
115. Abdolrasulnia, R., Wood, J. L. 1980. See Ref. 15, pp. 483–91
116. Cavallini, D., De Marco, C., Mondovì, B., Tentori, L. 1960. *J. Chromatogr.* 3:20–24
117. Kobayashi, K. 1970. *Physiol. Chem. Phys.* 2:455–66
118. Singer, T. P. 1975. See Ref. 11, pp. 535–46
119. Gaitonde, M. K. 1967. *Biochem. J.* 104:627–33
120. Finkelstein, J. P., Kyle, W. E., Harris, B. J., Martin, J. J. 1982. *J. Nutr.* 112:1011–18
121. Meister, A., Tate, S. S. 1976. *Ann. Rev. Biochem.* 45:559–604
122. Griffith, O. W., Meister, A. 1979. *Proc. Natl. Acad. Sci. USA* 76:5606–10
123. Anderson, M. E., Meister, A. 1980. *J. Biol. Chem.* 255:9530–33
124. Fahey, R. C., Newton, G. L., Dorian, R., Kosower, E. M. 1981. *Anal. Biochem.* 111:357–65
125. Tanret, C. 1909. *C. R. Acad. Sci.* 149:222–4
126. Eagles, B. A., Johnson, T. B. 1927. *J. Am. Chem. Soc.* 49:575–80
127. Cavallini, D., Mondovì, B., De Marco, C., Scioscia-Santoro, A. 1962. *Enzymologia* 24:253–66
128. Mondovì, B., Scioscia-Santoro, A., Cavallini, D. 1963. *Arch. Biochem. Biophys.* 101:363–64
129. Jollès-Bergeret, B., Chatagner, F. 1964. *Arch. Biochem. Biophys.* 105:640–41
130. Yamanishi, T., Tuboi, S. 1981. *J. Biochem.* 89:1913–21
131. Huxtable, R., Bressler, R. 1976. In *Taurine,* ed. R. J. Huxtable, A. Barbeau, pp. 45–57. New York: Raven
132. Huxtable, R. J. 1980. See Ref. 15, pp. 277–93
133. Huxtable, R. J. 1978. In *Taurine and Neurological Disorders,* ed. A. Barbeau, R. J. Huxtable, pp. 5–17. New York: Raven
134. Scandurra, R., Consalvi, V., De Marco, C., Politi, L., Cavallini, D. 1980. See Ref. 15, pp. 345–52
135. Dowling, L. M., Maclaren, J. A. 1965. *Biochim. Biophys. Acta* 100:293–4
136. Rao, D. R., Ennor, A. H., Thorpe, B. 1966. *Biochem. Biophys. Res. Commun.* 22:163–68
137. Federici, G., Ricci, G., Santoro, L., Antonucci, A., Cavallini, D. 1980. See Ref. 15, pp. 187–93
138. Pierre, Y., Loriette, C., Chatagner, F. 1980. See Ref. 15, pp. 195–200
139. Macaione, S., Di Giorgio, R. M., De Luca, G. 1980. See Ref. 15, pp. 265–76
140. Fellman, J. H., Roth, E. S., Fujita, T. S. 1978. See Ref. 133, pp. 19–24
141. Hoskin, F. C. G., Noonan, P. K. 1980. See Ref. 15, pp. 253–63
142. Cavallini, D., De Marco, C., Mondovì, B. 1956. *Experientia* 12:377–79
143. Cavallini, D., De Marco, C., Mondovì, B. 1957. *Biochim. Biophys. Acta* 24:353–58
144. Cavallini, D., De Marco, C., Mondovì, B. 1961. *Enzymologia* 23:101–10
145. De Marco, C., Bombardieri, G., Riva, F., Durré, S., Cavallini, D. 1965. *Biochim. Biophys. Acta* 100:89–97
146. Jacobsen, J. G., Smith, L. H. Jr. 1968. *Physiol. Rev.* 48:424–511
147. Van Thoai, N., Robin, Y. 1954. *Biochim. Biophys. Acta* 13:533–36
148. Van Thoai, N., Roche, J., Olumucki, A. 1954. *Biochim. Biophys. Acta* 14:448
149. Mori, A., Hiramatsu, M., Takahashi, K., Kohsaka, M. 1975. *Comp. Biochem. Physiol. B* 51:143–44
149a. Tolosa, É. A., Maslova, R. N., Goryachenkova, E. V. 1975. *Biokhimiya* 40:248–55
150. Sass, N. L., Martin, W. G. 1972. *Proc. Soc. Exp. Biol. Med.* 139:755–61
151. Martin, W. G., Sass, N. L., Hill, L., Tarka, S., Truex, R. 1972. *Proc. Soc. Exp. Biol. Med.* 141:632–38

152. Hayes, K. C., Carey, R. E., Schmidt, S. Y. 1975. *Science* 188:949–51
153. Gaul, G. E., Rassin, D. K., Räihä, N. C. R., Heinonen, K. 1977. *J. Pediatr.* 90:348–55
154. Rigo, J., Santerre, J. 1977. *Biol. Neonat.* 32:73–6
155. Kearney, E. B., Singer, T. P. 1953. *Biochim. Biophys. Acta* 11:270–89
155a. Singer, T. P., Kearney, E. B. 1953. *Biochim. Biophys. Acta* 11:290–99
156. Leinweber, F.-J., Monty, K. J. 1962. *Anal. Biochem.* 4:252–56
157. Recasens, M., Mandel, P. 1980. See Ref. 14, pp. 259–70
158. John, R. A., Fasella, P. 1969. *Biochemistry* 8:4477–82
159. Cavallini, D., Federici, G., Bossa, F., Granata, F. 1973. *Eur. J. Biochem.* 39:301–4
160. Spears, R. M., Martin, D. L. 1982. *J. Neurochem.* 38:985–91
161. Wu, J.-W. 1982. *Proc. Natl. Acad. Sci. USA* 79:4270–74
162. Sturman, J. A., Hepner, G. W., Hoffman, A. F., Thomas, P. J. 1976. See Ref. 131, pp. 21–33
163. Tanaka, H., Toyama, S., Tsukahara, H., Soda, K. 1974. *FEBS Lett.* 45:111–13
164. Fellman, J. H., Roth, E. S. 1982. In *Taurine in Nutrition and Neurology*, ed. R. J. Huxtable, H. Pasantes-Morales, pp. 99–113. New York: Plenum
165. Westley, J. 1973. *Adv. Enzymol.* 39:327–68
166. Koj, A. 1980. See Ref. 15, pp. 493–503
167. Sorbö, B. 1975. See Ref. 11, pp. 433–56
168. Sorbö, B., Hannestad, P., Lundquist, J., Mårtensson, J., Öhman, S. 1980. See Ref. 15, pp. 463–70
169. Mårtensson, J. 1981. *Studies on Human Sulfur Metabolism with Emphasis on Catabolic Conditions and Mercaptolactate Cysteine Disulfiduria.* Medical Diss. No. 19. Linköping Univ., Linköping, Sweden
170. Szczepkowski, T. W., Wood, J. L. 1967. *Biochim. Biophys. Acta* 139:469–78
171. Lang, K. 1933. *Biochem. Z.* 259:243–56
172. Pagani, S., Canella, C., Cerletti, P., Pecci, L. 1975. *FEBS Lett.* 51:112–15
173. Bonomi, F., Pagani, S., Cerletti, P., Cannella, C. 1977. *Eur. J. Biochem.* 72:17–24
174. Finazzi-Agrò, A., Canella, C., Graziani, M. T., Cavallini, D. 1971. *FEBS Lett.* 16:172–74
175. Taniguchi, T., Kimura, T. 1974. *Biochim. Biophys. Acta* 364:284–95
176. Sorbö, B. H. 1953. *Acta Chem. Scand.* 7:1129–36, 1137–45
177. Koj, A. 1968. *Acta Biochim. Pol.* 15:161–69
178. Koj, A., Frendo, J., Janik, Z. 1967. *Biochem. J.* 103:791–95
179. Koj, A., Frendo, J., Wojczak, L. 1975. *FEBS Lett.* 57:42–46
180. Meister, A., Fraser, P. E., Tice, S. V. 1954. *J. Biol. Chem.* 206:561–75
181. Hylin, J. W., Wood, J. L. 1959. *J. Biol. Chem.* 234:2141–44
182. Jarabak, R., Westley, J. 1980. *Biochemistry* 19:900–4
183. Sörbo, B. 1957. *Biochim. Biophys. Acta* 24:324–29
184. Kun, E., Fanshier, D. W. 1959. *Biochim. Biophys. Acta* 32:338–48
185. Ubuka, T., Yuasa, S., Ishimoto, Y., Shimomura, M. 1977. *Physiol. Chem. Phys.* 9:241–46
186. Ishimoto, Y. 1979. *Physiol. Chem. Phys.* 11:189–91
187. Lipsett, M. N., Norton, J. S., Peterkovsky, A. 1967. *Biochemistry* 6:855–60
188. Van Den Hammer, C. J. A., Morell, A. G., Scheinberg, I. M. 1967. *J. Biol. Chem.* 242:2514–16
189. Ubuka, T., Umemura, S., Ishimoto, Y., Shimomura, M. 1977. *Physiol. Chem. Phys.* 9:91–96
190. Ubuka, T., Umemura, S., Yuasa, S., Kinuta, M., Watanabe, K. 1978. *Physiol. Chem. Phys.* 10:483–500
191. Ip, M. P. C., Thibert, R. J., Schmidt, D. E. Jr. 1977. *Can. J. Biochem.* 55:958–64
192. Cooper, A. J. L., Haber, M. T., Meister, A. 1982. *J. Biol. Chem.* 257:816–26
193. Nakamura, T., Sato, R. 1963. *Biochem. J.* 86:328–35
194. Mudd, S. H., Irreverre, F., Laster, L. 1967. *Science* 156:1599–1602
195. Crawhall, J. C., Parker, R., Sneddon, W., Young, E. P., Ampola, M. G., et al. 1968. *Science* 160:419–20
196. Crawhall, J. C., Bir, K., Purkiss, P., Stanbury, J. B. 1971. *Biochem. Med.* 5:109–15
197. Crawhall, J. C. 1974. In *Heritable Disorders of Amino Acid Metabolism: Patterns of Clinical Expression and Genetic Variation*, ed. W. L. Nyhan, pp. 467–76. New York: Wiley
198. Shih, V. E., Carney, M. M., Fitzgerald, L., Monedjikova, V. 1977. *Pediatr. Res.* 11:464 (Abstr.)
199. Hannestad, U., Mårtensson, J., Sjödahl, R., Sörbo, B. 1981. *Biochem. Med.* 26:106–14
200. Niederweiser, A., Giliberti, P., Baerlocher, K. 1973. *Clin. Chim. Acta* 43:405–16

201. Meister, A. 1952. *J. Biol. Chem.* 197:309–17
202. Kun, E. 1957. *Biochim. Biophys. Acta* 25:135–37
203. Lundquist, P., Mårtensson, J., Sörbo, B., Öhman, S. 1980. *Clin. Chem.* 26:1178–81
204. Hannestad, U., Sörbo, B. 1979. *Clin. Chim. Acta* 95:189–200
205. Sörbo, B., Öhman, S. 1978. *Scand. J. Clin. Lab. Invest.* 38:521–27
206. Lundquist, P., Mårtensson, J., Sörbo, B., Öhman, S. 1979. *Clin. Chem.* 25:678–81
207. De Meio, R. H. 1975. See Ref. 11, pp. 287–358
208. Boström, H. 1965. *Scand. J. Clin. Lab. Invest.* 17 (Suppl. 86):33–52
209. Pruitt, K. M., Tenovuo, J., Andrews, R. W., McKane, T. 1982. *Biochemistry* 21:562–67
210. Kågedal, B., Källberg, M. 1982. *J. Chromatogr.* 229:409–15
211. Green, R. M., Elce, J. S. 1975. *Biochem. J.* 147:283–89
212. Ubuka, T., Yao, K. 1973. *Biochem. Biophys. Res. Commun.* 55:1305–10
213. Bauché, F., Sabourault, D., Giudicelli, Y., Nordmann, J., Nordmann, R. 1982. *Biochem. J.* 206:53–59
214. Butts, W. C., Kuehneman, M., Widdowson, G. M. 1974. *Clin. Chem.* 20:1344–48
215. Symington, I. S., Anderson, R. A., Oliver, J. S., Thomson, I., Harland, W. A., Kerr, T. W. 1978. *Lancet* 2:91–92
216. Stipanuk, M. H. 1979. *J. Nutr.* 109:2126–39
217. Krijgsheld, K. R., Glazenburg, E. J., Scholtens, E., Mulder, G. J. 1981. *Biochim. Biophys. Acta* 677:7–12
218. Ewetz, L., Sörbo, B. 1966. *Biochim. Biophys. Acta* 128:296–305
219. Waelsch, H., Borek, E. 1939. *J. Am. Chem. Soc.* 61:2252
220. Meister, A. 1957. *Methods Enzymol.* 3:404–14
221. Cahill, W. M., Rudolph, G. G. 1942. *J. Biol. Chem.* 145:201–5
222. Cooper, A. J. L., Redfield, A. G. 1974. *J. Biol. Chem.* 250:527–32
223. Parrod, J. 1942. *C. R. Acad. Sci. Ser. A* 215:146–48
224. Kumler, D., Kun, E. 1958. *Biochim. Biophys. Acta* 27:464–68
225. Pensa, B., Costa, M., Colosimo, A., Cavallini, D. 1982. *Mol. Cell Biochem.* 44:107–12
226. Chen, S. S., Walgate, J. H., Duerre, J. A. 1971. *Arch. Biochem. Biophys.* 146:54–63
227. Ubuka, T., Ishimoto, Y., Kasahara, K. 1975. *Anal. Biochem.* 67:66–73
228. Ricci, G., Federici, G., Achilli, M., Matarese, R. M., Cavallini, D. 1981. *Physiol. Chem. Phys.* 13:341–46
229. Cooper, A. J. L. and Meister, A. 1982. Manuscript in preparation
230. Cooper, A. J. L., Stephani, R. S., Meister, A. 1976. *J. Biol. Chem.* 251:6674–82
231. Ito, S., Nakamura, T., Eguchi, Y. 1975. *J. Biochem.* 78:1105–1107
232. Huang, H. T. 1963. *Biochemistry* 2:296–98
233. Perry, T. L. 1971. See Ref. 274, pp. 224–31
234. Ohmori, S., Kodama, H., Ikegami, T., Mizuhara, S., Oura, T., et al. 1972. *Physiol. Chem. Phys.* 4:286–94
235. Schubert, M. P. 1936. *J. Biol. Chem.* 114:341–50
236. Schubert, M. P. 1937. *J. Biol. Chem.* 121:539–48
237. Fourneau, J. P., Efimovsky, O., Gaignault, J. C., Jacquier, R., LeRidant, C. 1971. *C. R. Acad. Sci. Ser. C* 272:1515–17
238. Schonbeck, N. D., Skalski, M., Schafer, J. A. 1975. *J. Biol. Chem.* 250:5343–51
239. Dewhurst, I. C., Griffiths, R. 1981. *Biochem. Soc. Trans.* 9:426
240. Schirch, L., Mason, M. 1963. *J. Biol. Chem.* 237:2578–81
241. Tunnicliff, G., Ngo, T. T. 1977. *Can. J. Biochem.* 55:1013–18
242. Taberner, P. V., Paerce, M. J., Watkins, J. C. 1977. *Biochem. Pharmacol.* 26:345–49
243. Birnbaum, S. M., Winitz, M., Greenstein, J. P. 1957. *Arch. Biochem. Biophys.* 72:428–36
244. Olney, J. W., Ho, O. L., Rhee, V. 1971. *Exp. Brain Res.* 14:61–76
245. Karlsen, R. L., Grofova, I., Malthe-Sørenssen, D., Fonnum, F. 1981. *Brain Res.* 208:167–80
246. Folbergrová, J. 1974. *Brain Res.* 81:443–54
247. Folbergrová, J. 1975. *J. Neurochem.* 24:15–20
248. Blennow, G., Folbergrová, J., Nilsson, B., Siesjö, B. K. 1979. *Brain Res.* 179:129–46
249. Hurd, R. W., Hammond, E. J., Wilder, B. J. 1981. *Brain Res.* 209:250–54
250. Jackson, S. H. 1973. *Clin. Chim. Acta* 45:215–17
251. Bailey, A. J. 1975. In *Inborn Errors in Skin, Hair, and Connective Tissue,* ed. J. B. Holton, J. T. Ireland, pp. 105–18. Baltimore: Univ. Park

252. McCully, K. S., Ragsdale, B. D. 1970. *Am. J. Pathol.* 61:1–11
253. McCully, K. S., Wilson, R. B. 1975. *Atherosclerosis* 22:215–27
254. Donahue, S., Sturman, J. A., Gaull, G. 1974. *Am. J. Pathol.* 77:167–174
255. Makheja, A. N., Bombard, A. T., Randazzo, R. L., Bailey, J. M. 1978. *Atherosclerosis* 29:105–12
256. Reddy, G. S. R., Wilcken, D. E. L. 1982. *Metab. Clin. Exp.* 31:778–83
257. Harker, L. A., Slichter, S. J., Scott, C. R., Ross, R. 1974. *N. Eng. J. Med.* 291:537–43
258. Harker, L. A., Ross, R., Slichter, S. J., Scott, C. R. 1976. *J. Clin. Invest.* 58:731–41
259. Dudman, N. P. B., Wilcken, D. E. L. 1982. *Biochem. Med.* 27:244–53
260. Brush, E. J., Hamilton, G. A. 1981. *Biochem. Biophys. Res. Commun.* 103:1194–1200
261. Hamilton, G. A., Buckthal, D. J., Mortensen, R. M., Zerby, K. W. 1979. *Proc. Natl. Acad. Sci. USA* 76:2625–19
262. Fitzpatrick, N. F., Massey, V. 1982. *J. Biol. Chem.* 257:1166–71
263. Naber, N., Venkatesan, P. P., Hamilton, G. A. 1982. *Biochem. Biophys. Res. Commun.* 107:374–80
264. Prota, G., Thomson, R. H. 1976. *Endeavour* 35:32–38
265. Prota, G. 1980. See Ref. 15, pp. 391–97
266. Ito, S., Prota, G. 1977. *Experientia* 33:1118–19
267. Ito, S., Fujita, K. 1981. *Biochim. Biophys. Acta* 672:151–57
268. Ito, S., Nicol, J. A. C. 1975. *Tetrahedron Lett.* pp. 3287–90
269. Nkpa, N. N., Chedekel, M. R. 1981. *J. Org. Chem.* 46:213–15
270. Prota, G., Rorsman, H., Rosengren, A.-M., Rosengren, E. 1977. *Experientia* 33:720–21
271. Morishima, T., Hanawa, S. 1981. *Acta Derm. Venereol.* 61:149–50
272. Hsia, D. Y.-Y. 1966. *Inborn Errors of Metabolism.* Chicago: Year Book
273. Nyhan, W. L., ed. 1967. *Amino Acid Metabolism and Genetic Variation.* New York: McGraw-Hill
274. Carson, N. A. J., Raine, D. N., eds. 1971. *Inherited Disorders of Sulphur Metabolism.* Baltimore: Williams & Wilkins. 312pp.
275. Stanbury, J. B., Wyngaarden, J. B., Fredrickson, D. S. 1972. *Metabolic Basis of Inherited Disease.* New York: McGraw-Hill

276. Nyhan, W. L., ed. 1974. See Ref. 197, pp. 395–467
277. Vinken, P. J., Bruyn, G. W., eds. 1977. *Handbook of Clinical Neurology,* Vols. 28, 29. *Metabolic and Deficiency Diseases of the Central Nervous System.* Parts II and III. New York: North-Holland/American Elsevier
278. Sperling, O., DeVries, A., Tiqva, P., eds. 1978. *Inborn Errors of Metabolism in Man,* Part 1. Basel: Karger
279. Wellner, D., Meister, A. 1981. *Ann. Rev. Biochem.* 50:911–68
280. Kodama, H., Yao, K., Kobayashi, K., Hirayama, K., Fujii, Y., et al. 1969. *Physiol. Chem. Phys.* 1:72–76
281. Jonas, A. J., Greene, A. A., Smith, M. L., Schneider, J. A. 1982. *Proc. Natl. Acad. Sci.* 79:4442–45
282. Steinherz, R., Tietze, F., Gahl, W. A., Triche, T. J., Chiang, H., et al. 1982. *Proc. Natl. Acad. Sci. USA* 79:4446–50
283. Kreis, W. 1979. *Cancer Treat. Rep.* 63:1069–1072
284. Kreis, W., Baker, A., Ryan, V., Bertasso, A. 1980. *Cancer Res.* 40:634–41
285. Uren, J. R., Lazarus, H. 1979. *Cancer Treatm. Rep.* 63:1073–79
286. Kodama, H., Sasaki, K., Agata, T. 1982. *Biochem. Int.* 4:195–200
287. Hamdy, M. M. 1974. *J. Am. Oil. Chem. Soc.* 51:85A–90A
288. Fomon, S. J. 1959. *Pediatrics* 24:577–84
289. Zezulka, A. Y., Calloway, D. H. 1976. *J. Nutr.* 106:1286–91
290. Damico, R. 1975. *J. Agr. Food Chem.* 23:30-33.
291. Bookwalter, G. N., Warner, K., Anderson, R. A., Mustakas, G. C., Griffin, E. L. Jr. 1975. *J. Food Sci.* 40:266–70
292. Bauer, C. D., Berg, C. P. 1943. *J. Nutr.* 26:51–63
293. Wretlind, K. A. J., Rose, W. C. 1950. *J. Biol. Chem.* 187:697–705
294. Rose, W. C., Coon, M. J., Lochart, H. B., Lambert, G. F. 1955. *J. Biol. Chem.* 215:101–10
295. Smith, J. T. 1973. *J. Nutr.* 103:1008–11
296. Boggs, R. W., Rotruck, J. T., Damico, R. A. 1975. *J. Nutr.* 105:326–30
297. Steginck, L. D., Filer, L. J. Jr. 1982. *J. Nutr.* 112:597–603
298. Block, R. J., Jackson, R. W. 1932. *J. Biol. Chem.* 97:cvi–cvii
299. Baker, D. H., Boebel, K. P. 1980. *J. Nutr.* 110:959–64
300. Langer, B. W. Jr. 1965. *Biochem. J.* 95:683–87
301. Bodmer, J. L., Waring, R. H. 1981. *Biochem. Soc. Trans.* 9:549–50.

NOTE ADDED IN PROOF: The two rat brain cysteine sulfinate transaminases have now been shown to be identical to soluble and mitochondrial aspartate transaminase. Recasens, M., Benezra, R., Basset, P., Mandel, P. 1980. *Biochemistry* 19:4583–89

Ann. Review Biochem. 1983. 52:223–61

LIPOPROTEIN METABOLISM IN THE MACROPHAGE: Implications for Cholesterol Deposition in Atherosclerosis[1]

Michael S. Brown and Joseph L. Goldstein

Departments of Molecular Genetics and Internal Medicine, University of Texas Health Science Center at Dallas, Dallas, Texas 75235

CONTENTS

PERSPECTIVES AND SUMMARY 224
UPTAKE OF LIPOPROTEIN-BOUND CHOLESTEROL BY
    MACROPHAGES 226
    Receptor for Acetyl-LDL 227
    Receptor for LDL/Dextran Sulfate Complexes 235
    Receptor for β-VLDL 237
    Receptors for Cholesteryl Ester/Protein Complexes from
        Atherosclerotic Plaques 241
PROCESSING AND STORAGE OF LIPOPROTEIN-BOUND CHOLESTEROL
    BY MACROPHAGES 243
    Endocytosis and Lysosomal Hydrolysis 243
    Cytoplasmic Re-esterification and Hydrolysis of Lipoprotein-Derived Cholesteryl
        Esters 244
    The Cholesteryl Ester Cycle 247
SECRETION OF CHOLESTEROL AND APOPROTEIN E BY
    MACROPHAGES 249
    Cholesterol Excretion Dependent on Cholesterol Acceptors 249
    Synthesis and Secretion of Apoprotein E in Response to Cholesterol Loading 252
IMPLICATIONS FOR FOAM CELL FORMATION
    IN ATHEROSCLEROSIS 255
    The Foam Cell in Familial Hypercholesterolemia 257

[1]Abbreviations used: ACAT, acyl-CoA:cholesterol acyltransferase; apo, apoprotein; FH, familial hypercholesterolemia; HDL, high density lipoprotein; HDL_c, a cholesterol-induced form of HDL containing apoprotein E in addition to apoprotein A-I; IDL, intermediate density lipoproteins; LDL, low density lipoprotein; LCAT, lecithin:cholesterol acyltransferase; β-VLDL, β-migrating very low density lipoproteins; WHHL rabbit, Watanabe Heritable Hyperlipidemic rabbit.

0066-4154/83/0701-0223$02.00

PERSPECTIVES AND SUMMARY

Atherosclerotic plaques are filled with scavenger cells that have ingested large amounts of cholesterol and have become so stuffed with cholesteryl ester that they are converted into foam cells (1, 2). Most of these foam cells arise either from resident macrophages of the artery wall or from blood monocytes that enter the wall at sites of endothelial damage. Macrophages ingest and degrade cholesterol-carrying plasma lipoproteins that have leaked through damaged endothelium and penetrated into the tissue of the wall. When macrophages take up more lipoprotein cholesterol than they can excrete, the cholesterol is stored in the cytoplasm in the form of cholesteryl ester droplets. These droplets give the cytoplasm a foamy appearance in the electron microscope, thus accounting for the term foam cell.

The atherosclerotic plaque is a complicated structure. In addition to cholesterol-filled macrophages, the structure contains large numbers of proliferating smooth muscle cells and a large amount of extracellular material that includes sulfated glycosaminoglycans, collagen, fibrin, and cholesterol (3). Some of the smooth muscle cells contain cholesteryl ester droplets that resemble those of macrophage foam cells. In order to unravel such a complicated structure, in recent years scientists have begun to study the specialized properties of each of the cell types that comprise the lesion. For example, endothelial cells and smooth muscle cells were propagated in vitro, and their analyses identified several distinctive properties that might contribute to the initiation of atherosclerosis (reviewed in 3).

The macrophage, too, has come under study. Extensive investigations over the past five years disclosed that macrophages, isolated from the peritoneal cavity of mice and from the blood of man, possess mechanisms that allow them to take up and digest cholesterol-containing lipoproteins, to store the sterol, and to excrete it in large amounts when conditions permit (4–8). These mechanisms differ from those in other cell types, such as cultured fibroblasts and smooth muscle cells. Awareness of these special mechanisms for lipoprotein uptake made possible the conversion of macrophages into foam cells in vitro (4, 8). These studies shed new light on the possible mechanism for foam cell formation in vivo.

The uptake of lipoprotein-bound cholesterol in macrophages occurs through the process of receptor-mediated endocytosis (4–7). The initial event is the binding of the lipoprotein to a cell surface receptor. Although macrophages express few receptors for normal plasma lipoproteins, they exhibit abundant receptors for lipoproteins that have been altered by chemical derivitization (4) or by complexing with other molecules (5, 7). In addition, macrophages have receptors for at least one type of abnormal lipoprotein that accumulates spontaneously in plasma in hyperlipidemic states (6).

Most of the cholesterol in plasma lipoproteins is in the form of cholesteryl esters. Macrophages process these esters in a series of sequential reactions that take place in two cellular compartments (8, 9). Immediately after they enter the macrophage via receptor-mediated endocytosis, lipoprotein-bound cholesteryl esters are delivered to lysosomes (first cellular compartment) where they are hydrolyzed by an acid lipase. The liberated cholesterol crosses the lysosomal membrane and enters the cytoplasm (second cellular compartment) where it is re-esterified by a microsomal enzyme and stored in the cytoplasm as cholesteryl ester droplets.

The two-compartment pathway allows quantitative assay of the cellular uptake of cholesterol-rich lipoproteins without the need for radiolabeled lipoproteins. When incubated in the usual medium containing normal serum, macrophages do not take up lipoproteins at a high rate, and hence they do not synthesize cholesteryl esters (4, 8). Thus, when [^{14}C]oleate is added to the culture medium, the cells do not incorporate it into cholesteryl [^{14}C]oleate. However, when the cells are presented with a lipoprotein that they can ingest, cholesterol is liberated and then re-esterified, and this leads to a 100- to 200-fold increase in the rate of incorporation of [^{14}C]oleate into cholesteryl [^{14}C]oleate (4, 8). All of the cholesterol-rich lipoproteins that enter macrophages were found to enhance cholesteryl ester synthesis in this fashion and hence stimulation of cholesteryl [^{14}C]oleate synthesis is used as a functional assay to measure lipoprotein uptake (4–8).

The cholesteryl esters stored in the cytoplasm of macrophage foam cells undergo a continual cycle of hydrolysis and re-esterification (9). Hydrolysis is mediated by a nonlysosomal esterase distinct from the lysosomal acid lipase. Re-esterification is mediated by a membrane-bound enzyme that transfers a fatty acid from fatty acyl coenzyme A to cholesterol. When the extracellular fluid contains a substance, such as high density lipoprotein (HDL), that is capable of binding cholesterol, the free cholesterol is not re-esterified or stored, but is excreted from the cell. When no cholesterol acceptor is available, the free cholesterol is re-esterified for storage, and the cycle of hydrolysis and re-esterification continues (9).

If macrophages metabolize lipoprotein cholesterol in the body as they do in tissue culture, then the cholesterol that they excrete may have two metabolic fates: (a) some of it may be transported directly to the liver where it is excreted from the body (the so called "reverse cholesterol transport") (10); and (b) some of it may be transferred to other lipoproteins, such as low density lipoprotein (LDL), that deliver it both to liver and to extrahepatic tissues for use in the synthesis of new plasma membranes and steroid hormones (11, 12). When macrophages excrete cholesterol, they simultaneously synthesize and secrete large amounts of apoprotein E (13, 14), a component of plasma lipoproteins that binds avidly to lipoprotein receptors. Secreted apo E and secreted cholesterol may associate with the

HDL present in the medium to produce a lipoprotein called HDL_c. When injected intravenously into animals, HDL_c is taken up rapidly by lipoprotein receptors on the surface of hepatocytes (11, 12). Thus, apo E may be synthesized by cholesterol-loaded macrophages in order to target the secreted cholesterol to the liver, thereby facilitating "reverse cholesterol transport" (14).

In this article, we review studies carried out over the last five years that have led to these new insights into the mechanisms for cholesterol uptake, storage, and excretion by macrophages. While the data were obtained almost exclusively from in vitro systems, they have important implications for macrophage function in the body and suggest how macrophages might go awry during the formation of foam cells in the atherosclerotic plaque.

UPTAKE OF LIPOPROTEIN-BOUND CHOLESTEROL BY MACROPHAGES

Macrophages can take up large amounts of cholesterol by two mechanisms: (a) by phagocytosis of whole cells or fragments of membranes containing cholesterol; or (b) by receptor-mediated endocytosis of plasma lipoproteins either in solution or complexed in insoluble form with other tissue constituents. The factors governing phagocytosis were discussed elsewhere (15). In this section we review the various systems for receptor-mediated endocytosis of cholesterol-containing lipoproteins.

The initial studies on receptor-mediated endocytosis of lipoproteins by macrophages, reported in 1979 by Goldstein et al (4), were carried out to resolve a paradox that emerged from studies of the LDL receptor. LDL receptors are present on a variety of nonmacrophage cells grown in tissue culture or taken directly from the body. The LDL receptors mediate the uptake and degradation of LDL by body cells and hence are an important determinant of the plasma LDL-cholesterol level (11). Subjects with homozygous familial hypercholesterolemia have a genetically determined total or near total deficiency of LDL receptors. Plasma LDL cannot penetrate into their cells with normal efficiency, and as a result the plasma LDL level rises. Despite their deficiency of LDL receptors, subjects with homozygous familial hypercholesterolemia nevertheless accumulate LDL-derived cholesteryl esters in macrophage foam cells at several sites in the body, notably in the arterial wall, causing atheromas, and in tendons, causing xanthomas (16). This clinical observation suggested that macrophages have some alternative mechanism for taking up LDL-cholesterol distinct from the LDL receptor. However, in vitro tissue macrophages take up native LDL at extremely slow rates and do not accumulate excessive cholesteryl esters, even when exposed to high concentrations of LDL for prolonged periods of time (4). These paradoxical findings led to a search for altered forms of LDL that could be internalized by macrophages at rapid rates.

Receptor for Acetyl-LDL

The first plasma lipoprotein demonstrated to enter macrophages by receptor-mediated endocytosis was human LDL that had been reacted with acetic anhydride in vitro to form acetyl-LDL (4). These studies were conducted with monolayers of resident mouse peritoneal macrophages isolated by the classic techniques developed by Cohn and co-workers (reviewed in 17). Unlike most other cell types, normal tissue macrophages from the mouse and other species express few if any receptors for native LDL (4–6). When incubated with ^{125}I-labeled LDL in vitro, mouse peritoneal macrophages internalize only minimal amounts of the lipoprotein and do not increase cellular cholesterol content (4, 8).[2] In contrast, LDL that has been modified by chemical acetylation is taken up with extremely high efficiency by macrophages, resulting in massive cholesterol accumulation within the cells (4, 8).

BIOCHEMICAL PROPERTIES OF THE ACETYL-LDL RECEPTOR   Studies with ^{125}I-labeled acetyl-LDL showed that the rapid uptake by mouse macrophages is mediated by an initial binding of the lipoprotein to a limited number of high affinity binding sites (20,000–40,000 sites/cell) that recognize acetyl-LDL but not native LDL (4, 18). Binding leads to rapid internalization of acetyl-LDL by endocytosis and delivery to lysosomes. Within 60 min, virtually all of the cell-bound ^{125}I-acetyl-LDL is hydrolyzed and the label is excreted from the cell in the form of ^{125}I-monoiodotyrosine (4). The receptor for acetyl-LDL is just beginning to be characterized biochemically. It is not yet clear whether it is a single molecular entity or is comprised of several different molecular species, each of which is capable of binding acetyl-LDL and mediating its rapid internalization by the cell. All of the surface binding sites for ^{125}I-acetyl-LDL are destroyed when the cells are treated briefly with low concentrations of trypsin or pronase (4), suggesting that all of the receptors are composed of protein. Half-maximal binding of ^{125}I-acetyl-LDL is achieved at an acetyl-LDL concentration of 5 μg pro-

[2]Although small amounts of ^{125}I-LDL are taken up and degraded by mouse peritoneal macrophages, this uptake does not appear to be mediated by the classic LDL receptor in that it is competitively inhibited nonspecifically by lipoproteins, such as acetyl-LDL [see Figure 2 A in (5)] and typical HDL (24), which do not bind to the LDL receptor. The nature of this nonspecific uptake process for ^{125}I-LDL by tissue macrophages is not clear; it may be related to the ability of lipoproteins to bind nonspecifically to a site on cell membranes that recognizes multiple lipoproteins, i.e. LDL, HDL, methyl-LDL, and acetyl-LDL (93, 105).

tein/ml at 4°C and 25 μg protein/ml at 37°C (4). Binding is not inhibited by EDTA (4), indicating that divalent cations are not essential.

Using the mouse macrophage cell line P388D$_1$ as a source of receptor, Via, et al (19) reported the partial characterization of a solubilized membrane protein that, after a 300- to 400-fold purification, shows the same affinity and binding specificity as does the acetyl-LDL receptor of intact cells. The detergent-receptor complex has a $M_r = 283,000$, an isoelectric point of 5.9, and a sedimentation coefficient of 6.55.

DISTRIBUTION OF THE ACETYL-LDL RECEPTOR ON DIFFERENT CELL TYPES   The acetyl-LDL receptor has been found on macrophages from every source and species so far tested. These include resident peritoneal macrophages from mice (4), rats (4), and dogs (20); Kupffer cells from guinea pigs (4) and rats (21); monocyte-derived macrophages from humans (4, 18, 22, 23); and established lines of mouse macrophage tumors such as IC21 cells (24), J774 cells (25), and P388D$_1$ cells (19). Activated and inflammatory macrophages produced by intraperitoneal injection of mice with a variety of agents (including thioglycollate, fetal calf serum, phytohemagglutinin, BCG, *Corynebacterium parvum,* and pyran copolymer) express roughly the same amount of acetyl-LDL receptor activity as do unstimulated resident macrophages (26). This is in contrast to other receptors, such as those for mannose-conjugated proteins, which vary markedly in number after several of these treatments (26). Conditioned medium from human lymphocyte cultures stimulated by concanavalin A reduces the ability of macrophages to degrade malondialdehyde-treated LDL, a lipoprotein that enters the cell via the acetyl-LDL receptor (27; see below). This suggests that lymphocytes produce a substance that suppresses the function of the acetyl-LDL receptor.

Pitas et al (28) provided a particularly striking demonstration of the cell specificity of the acetyl-LDL receptor. They made mixed cultures of human fibroblasts and mouse peritoneal macrophages and incubated them with lipoproteins that had been rendered fluorescent through incorporation of the lipophilic fluorescent dye 3,3'-dioctadecylindocarbocyanine. When the 3,3'-dioctadecylindocarbocyanine was incorporated into acetoacetylated LDL, which binds to the acetyl-LDL receptor but not to the LDL receptor, the macrophages became intensely fluorescent but the interspersed fibroblasts did not (28).

In contrast to the LDL receptor of nonmacrophage cells whose number is suppressed when cellular cholesterol accumulates to high levels (29), acetyl-LDL receptors remain constant in number even when macrophages have accumulated massive amounts of cholesterol (4). As a result of their failure to suppress the production of acetyl-LDL receptors, macrophages incubated continuously with acetyl-LDL take up so much cholesterol that they are converted into foam cells in vitro (4, 8; see below).

In contrast to its apparently universal expression in macrophages, the acetyl-LDL receptor is generally absent from nonmacrophage cells, including cultured human fibroblasts, cultured human and bovine smooth muscle cells, freshly isolated human lymphocytes, human lymphoblasts, mouse Y-1 adrenal cells, and Chinese hamster ovary cells (4, 18). The one exception is cultured bovine endothelial cells, which express a small number of acetyl-LDL receptors and degrade ^{125}I-acetyl-LDL at 6% of the rate of resident mouse peritoneal macrophages (30). Endothelial cells are known to share other properties with macrophages, such as the presence of lipoprotein lipase (31, 32) and the ability to present antigens to T lymphocytes in an immunogenic form (30).

In contrast to tissue macrophages, which express acetyl-LDL receptors but virtually no LDL receptors, monocytes freshly isolated from the blood of normal subjects express receptors for both native LDL and acetyl-LDL (4, 18, 22, 23, 33, 34). After 5 days of culture in vitro, the activity of the acetyl-LDL receptor increases by as much as 20-fold and markedly exceeds (by more than 10-fold) the activity of the LDL receptor (33, 34). Cultured malignant macrophages such as J774 cells (25) and IC21 cells (24) express low levels of LDL receptors and high levels of acetyl-LDL receptors. Monocytes cultured from the blood of subjects with the homozygous form of familial hypercholesterolemia display normal acetyl-LDL receptor activity despite their genetic deficiency of receptors for native LDL (18, 34).

Figure 1 demonstrates the all-or-none difference in the ability of cultured human fibroblasts and mouse peritoneal macrophages to take up and degrade ^{125}I-acetyl-LDL and ^{125}I-LDL. This difference between acetyl-LDL receptors and LDL receptors is one of the most striking biologic differences between macrophage and nonmacrophage cells and implies an important role for the acetyl-LDL receptor in macrophage function in vivo.

LIGAND SPECIFICITY OF THE ACETYL-LDL RECEPTOR   Acetylation of LDL removes positive charges from the ϵ-amino groups of lysine and thereby converts a weakly anionic lipoprotein into a strongly anionic one (35). The acetyl-LDL loses its ability to bind to the classic LDL receptor of nonmacrophage cells, but it remains precipitable by antibodies to native LDL (35). The enhanced net negative charge of acetyl-LDL is responsible for its binding to the macrophage acetyl-LDL receptor (4). Other chemical modifications that abolish positive lysine residues and increase LDL's net negative charge also convert the lipoprotein into a ligand for the acetyl-LDL receptor. Such ligands include acetoacetylated LDL (20), maleylated LDL (4), succinylated LDL (4), and malondialdehyde-treated LDL (18,

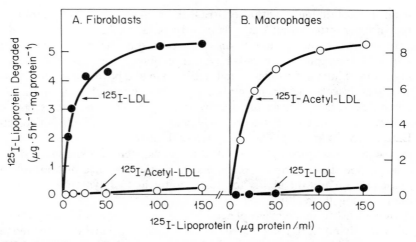

Figure 1   Differences in the uptake and degradation of ^{125}I-LDL (●) and ^{125}I-acetyl-LDL (O) by human fibroblasts (*A*) and mouse peritoneal macrophages (*B*). Monolayers of growing fibroblasts were incubated in lipoprotein-deficient serum for 48 hr prior to the experiments to induce maximal LDL receptors (29). Monolayers of freshly isolated macrophages were studied without prior incubation in lipoprotein-deficient serum (4). For degradation assays, each monolayer received medium containing 10% lipoprotein-deficient serum and the indicated concentration of either human ^{125}I-LDL (●) or ^{125}I-acetyl-LDL (O). After incubation for 5 hr at 37°C, the amount of ^{125}I-labeled acid-soluble material released into the medium was measured (4).

22). Reductive methylation of LDL, which modifies the lysine residues but does not remove their positive charge, fails to convert LDL into a species that will bind to the acetyl-LDL receptor (18).

Haberland et al (36) studied the stoichiometry of lysine modification of LDL. Using malondialdehyde as a lysine-modifying agent, they showed that binding to the acetyl-LDL receptor in human monocyte-derived macrophages occurs only above a threshold of 30 moles of malondialdehyde incorporated per mole of LDL protein. Additional incorporation of malondialdehyde (up to 60 moles/mole of LDL) produced no further increase in lipoprotein uptake. In contrast to this sharp threshold effect for ligand binding to the acetyl-LDL receptor, the ability of native LDL to bind to the classic LDL receptor was reduced gradually and in proportion to the number of malondialdehyde residues incorporated. High affinity binding of LDL to the LDL receptor disappeared entirely when 20 moles of malondialdehyde were incorporated per mole of LDL. When the incorporation of malondialdehyde was between 20 and 30 moles per mole of LDL, the particle would bind neither to the LDL receptor nor to the acetyl-LDL receptor (36).

The identity of the negatively charged residues on acetyl-LDL that mediate binding to the acetyl-LDL receptor is not known. LDL is a complex particle that contains negatively charged lipids as well as amino acids and carbohydrates. To simplify analysis of ligand-receptor interactions, experiments were performed with less complex polyanionic ligands that bind to the acetyl-LDL receptor and thus compete for the binding, uptake, and degradation of ^{125}I-acetyl-LDL. In general, acetylation of other proteins (such as albumin, gamma globulin, α-1-antitrypsin, transferrin, ferritin, ovalbumin, histones, ovomucoid, α-1-acid glycoprotein, and HDL) does not convert them into ligands for the acetyl-LDL receptor (37). However, maleylated albumin binds with high affinity (4). In contrast to acetylation, reaction with the dicarboxylic acid maleate not only removes positive charges on lysine residues but also adds additional negatively charged residues in the form of carboxyl groups. These experiments suggest that most native proteins, such as albumin, do not contain a sufficient number or arrangement of negatively charged residues to bind to the acetyl-LDL receptor, even when all of the positive charges on the available lysine residues have been obliterated. However, the addition of new negative charges in the form of maleate converts the molecule into a binding moiety. The unique aspect of LDL is that it contains sufficient negatively charged residues so that elimination of the positive lysine residues induces binding to the acetyl-LDL receptor without a requirement for additional negative charges. HDL behaves like albumin in that it requires maleylation in order to be recognized by the acetyl-LDL receptor (37).

The acetyl-LDL receptor also recognizes compounds in which the negative charges reside on noncarboxyl moieties (4, 18), such as sulfate (e.g. polyvinyl sulfate, dextran sulfate, and fucoidin) or phosphate (e.g. polyinosinic acid and polyxanthinylic acid). All binding polyanions have a high molecular weight. Low molecular weight polyanions (e.g. ATP and GTP) do not bind, as judged by their inability to compete for the uptake of ^{125}I-acetyl-LDL (37).

Table 1 lists a large number of compounds that were tested for binding to the acetyl-LDL receptor. Testing was performed by measuring the ability of each molecule to compete with ^{125}I-acetyl-LDL for uptake and degradation by the mouse peritoneal macrophage receptor. Multiple negative charges are necessary but not sufficient for receptor binding. Certain contrasts are striking. For example, certain polypurines (such as polyinosinic acid, polyguanylic acid, and polyxanthinylic acid) compete effectively for the binding of ^{125}I-acetyl-LDL, while another polypurine (polyadenylic acid) does not compete. Adenylic acid differs from the first three purines in that it has an amino group in place of a keto group at carbon 6. However, polyguanylic acid, which has an amino group at carbon 2, is recognized by the acetyl-LDL receptor. Since the simple presence of an amino group is not sufficient to prevent binding, the configuration of the polymer is probably the important feature. Two polypyrimidines (polycytidylic acid and polyuridylic acid) do not compete for the binding of ^{125}I-acetyl-LDL. Other polyanions that do not compete for receptor binding include sulfated polysaccharides of relatively low molecular weight (such as heparin and chondroitin sulfate), polycolominic acid, polyphosphate chains containing up to 65 phosphates, and polyglutamic acid (4, 18).

From the results of the inhibitor studies shown in Table 1, we conclude that binding to the acetyl-LDL receptor requires not only a large number of negatively charged residues but also an arrangement that produces a high density of charge within specific regions of the molecule (18).

FUNCTION OF THE ACETYL-LDL RECEPTOR IN VIVO   Evidence that the acetyl-LDL receptor is expressed in macrophages in vivo comes from experiments in which ^{125}I-labeled acetyl-LDL (4), ^{125}I-labeled acetoacetylated LDL (20, 38) and ^{125}I-labeled succinylated LDL (39) were administered intravenously to mice, dogs, and rats, respectively. The modified lipoproteins were cleared from the plasma within minutes by macrophages (Kupffer cells) of the liver (38). The hepatic uptake of intravenously administered ^{125}I-acetyl-LDL in mice was blocked when the animals were injected

Table 1 ^{125}I-Acetyl-LDL binding site in mouse peritoneal macrophages

Effective competitors		Ineffective competitors
Negatively charged compounds		Negatively charged compounds
Polyvinyl sulfate	1 µg/ml[a]	Polyadenylic acid, polycytidylic acid, poly-
Polyinosinic acid	1.5	polyuridylic acid, heparin, chondroitin
Polyguanylic acid	1.5	sulfates A and C, phosvitin, colominic
Poly G:I (1:1)	2	acid (polysialic acid), polyphosphates
Polyxanthinylic acid	3	($n = 65$), poly(D-glutamic acid)
Dextran sulfate	3	Positively charged compounds
Fucoidin	10	Lysozyme, spermine
Bovine sulfatides	10	Glycoproteins
Carragheenan	20	Mannan (yeast), thyroglobulin, fetuin,
Maleylated LDL	20	orosomucoid, asialoorosomucoid
Poly I: poly C	50	Others
Maleylated albumin	100	Acetylated proteins, including albumin,
Maleylated HDL	150	γ-globulin, α-1-antitrypsin, transferrin,
		ovalbumin, histones, ovomucoid, α-1-acid
		glycoprotein, and HDL
		Methylated LDL

[a] Values refer to the concentrations required for 50% inhibition of ^{125}I-acetyl-LDL binding.

simultaneously with fucoidin (4), confirming that the uptake was mediated by a saturable receptor with a specificity similar to the binding site demonstrated in vitro.

The normal ligand for the acetyl-LDL receptor in vivo, if any, is unknown. It has not yet been demonstrated that the receptor participates in the clearance of endogenous plasma lipoproteins, as opposed to injected modified lipoproteins. It is unlikely that acetyl-LDL would be formed extracellularly in the body since its biologic formation requires acetyl-CoA or some other donor of active acetate whose occurrence in extracellular fluid has not been demonstrated. Other functionally equivalent types of lysine modification reactions might take place extracellularly, however. For example, malondialdehyde is known to be secreted by platelets and macrophages as a by-product of the oxidation of arachidonic acid (40, 41). As originally shown by Fogelman, et al. (22), malondialdehyde can react with the lysines of LDL and convert the lipoprotein to a form that is recognized with high affinity by macrophages. Subsequent studies showed that the malondialdehyde-modified LDL was taken up by the acetyl-LDL receptor (18, 33). Chemical formation of malondialdehyde-LDL in vitro requires concentrations of malondialdehyde above 1 mM (18, 22, 33). This is several orders of magnitude higher than the malondialdehyde concentrations that are likely to occur in the body, even in platelet thrombi (42). Nevertheless, some local factor(s) operating within damaged tissues or platelet thrombi might increase the susceptibility of LDL to reaction with malondialdehyde.

Malondialdehyde is not the only substance derived from arachidonic acid that might modify the lysines of LDL. Table 2 lists several theoretical reactions that could occur between the ϵ-amino group of lysines in LDL and one or more metabolites generated normally in vivo during the conversion of arachidonic acid to thromboxanes and leukotrienes. Each of these reactions would convert LDL to a more negative form that could be recognized by the acetyl-LDL receptor.

In addition to the arachidonic acid-derived aldehydes listed in Table 2, there are several other naturally occurring aldehydes that might react with the lysine residues of LDL in extracellular fluids. One of these is glucose. Glucosylated-LDL was shown to be present in the plasma of patients with diabetes mellitus (43–45). When LDL is reacted with glucose in vitro, the lysine residues are modified to such an extent that the LDL loses its ability to bind to the native LDL receptor of fibroblasts (44, 46, 47). However, this modified LDL is not taken up by macrophages via the acetyl-LDL receptor, perhaps because an insufficient number of lysine residues is accessible to modification by glucose (36, 44).

In addition to chemical derivitization, the lysines of LDL could be altered enzymatically through the action of lysyl oxidase, an extracellular enzyme

Table 2 Arachidonic acid metabolites that could theoretically react with lysine residues of LDL so as to eliminate their positive charge

Pathway of Arachidonic acid metabolism	Reactive Metabolite	Chemical reaction with Lysine	Example
Cyclooxygenase	Malondialdehyde	Imination (Schiff base)	
Cyclooxygenase	Prostaglandin A	Michael addition	
Lipoxygenase	5-HPETE	Oxidation by active oxygen species	
Lipoxygenase	Leukotriene A$_4$	Alkylation	

This table was prepared by J. R. Falck. 5-HPETE, 5-hydroxyperoxyeicosatetraenoic acid.

that oxidizes the ϵ-amino group of certain lysine residues to allysine, an aldehyde intermediate that is essential to the cross-linking of collagen fibrils (48). Whether such oxidation of LDL's lysines would lead to its uptake by the acetyl-LDL receptor is unknown. Lysyl oxidase occurs in highest concentrations in the extracellular fluid of tendons and aorta (48). Interestingly, these are precisely the sites at which the most pronounced accumulation of LDL-cholesterol occurs in macrophages in vivo (16). A role for lysyl oxidase is rendered less likely by the belief that the enzyme is specific for collagen and elastin (48); it has not been shown to react with LDL. Oxidation of protein-bound lysine residues can also occur chemically in the presence of hydrogen peroxide, a peroxidase enzyme, and a hydrogen acceptor (49).

A recent exciting development was the demonstration by Henriksen et al (25) that cultured endothelial cells can convert human LDL into a form that is recognized by the acetyl-LDL receptor of macrophages. This modification follows incubation of human LDL with any one of several types of cultured cells: rabbit aortic endothelial cells, human umbilical vein endothelial cells, or aortic smooth muscle cells from guinea pig or swine (25, 50). It is not produced by the incubation of LDL with conditioned medium from these cultures or by incubation with intact cultured fibroblasts or red blood cells. The modified LDL has an increased density, owing to a decrease in its content of free and esterified cholesterol (50). It also has an enhanced electrophoretic mobility, presumably due to an increased negative charge (25). The chemical nature of this modification is unknown; malondialdehyde was not detectable (25).

Although the acetyl-LDL receptor was discovered on the basis of its ability to bind modified LDL, its function in vivo may involve pathways other than lipoprotein metabolism. Maleylated proteins, which enter macrophages through the acetyl-LDL receptor (4), were shown to stimulate markedly the secretion of three neutral proteases: neutral caseinase, plasminogen activator, and cytolytic proteinase (51). These findings suggest that the acetyl-LDL receptor may play a regulatory role in the inflammatory response. In this regard, it is striking that many of the ligands for the acetyl-LDL receptor are potent stimulators of interferon secretion, e.g. polyinosinic acid, polyvinyl sulfate, dextran sulfate, fucoidin, and polyinosinic acid:polycytidylic acid (52). Whether the activity of these agents is dependent upon binding to the acetyl-LDL receptor has not been explored.

Receptor for LDL/Dextran Sulfate Complexes

LDL has long been known to form soluble and insoluble complexes with sulfated polysaccharides, both naturally occurring and synthetic (53). In 1979 Basu et al (5) reported that complexes of LDL and dextran sulfate are taken up with high affinity by mouse peritoneal macrophages, apparently as a result of binding to saturable and high affinity surface receptor sites. As with acetyl-LDL, the protein component of the ^{125}I-LDL/dextran sulfate complexes is rapidly degraded in lysosomes and the cholesterol is retained by the cell for processing and storage (5; see below).

The LDL/dextran sulfate binding site has so far been found only on macrophages, including not only mouse peritoneal macrophages but also cultured mouse IC21 cells (24) and human monocyte-derived macrophages (27, 33, 34). A similar uptake does not occur in other cell types such as human fibroblasts (5). In fact, dextran sulfate inhibits the receptor-mediated endocytosis of LDL in fibroblasts because LDL that is complexed to dextran sulfate can no longer bind to the native LDL receptor (29).

The stoichiometry of the LDL/dextran sulfate complex entering macrophages is not known. When the two ligands are present in the culture medium in a 10:1 mass ratio, the maximal rate of ^{125}I-LDL degradation occurs at an LDL concentration of 250 μg/ml and a dextran sulfate concentration of 25 μg/ml (5). Recently, the converse experiment was performed, and the results demonstrated that LDL greatly increases the uptake of [^{3}H]dextran sulfate by macrophages (54), confirming that the two substances enter the cell as a complex.

Whereas dextran sulfate of $M_r = 500,000$ is effective in promoting the uptake of LDL, dextran sulfate of lower M_r (i.e. 40,000) is ineffective (5). A variety of naturally occurring sulfated glycosaminoglycans (including chondroitin-6-sulfate, chondroitin-4-sulfate, dermatan sulfate, heparan sulfate, keratan sulfate, and heparin) do not stimulate uptake and degradation of ^{125}I-LDL (5), even though many of these compounds are known to form tight complexes with LDL (53). Although purified heparin is not effective, a heparin-containing proteoglycan isolated from rat skin, $M_r > 900,000$ (55), stimulates the uptake and degradation of human ^{125}I-LDL in mouse macrophages by two-fold (56). These data suggest that stimulation of LDL uptake in macrophages requires the formation of large complexes.

The nature of the cell surface binding site for LDL/dextran sulfate is not known. Monocyte-derived macrophages isolated from the blood of subjects with homozygous familial hypercholesterolemia take up and degrade ^{125}I-LDL/dextran sulfate complexes with high affinity and at a normal rate under conditions in which no high affinity uptake and degradation of native ^{125}I-LDL is detected, thus providing genetic evidence that the binding site is distinct from the LDL receptor (34). Competition studies show that the binding site for LDL/dextran sulfate is also distinct from the acetyl-LDL binding site (5). Thus, uptake and degradation of ^{125}I-LDL/dextran sulfate is not inhibited by incubation with polyinosinic acid up to concentrations of 500 μg/ml, although uptake and degradation of ^{125}I-acetyl-LDL is completely inhibited at < 5 μg/ml (5). Similar differential inhibitory results are obtained with fucoidin, an inhibitor of acetyl-LDL binding (5). Polycations, such as spermine, spermidine, and putrescine, prevent the formation of the ^{125}I-LDL/dextran sulfate complex in vitro and thereby prevent dextran sulfate from stimulating ^{125}I-LDL degradation by intact macrophages (5). The potency of these agents declines in the following order: spermine $>$ spermidine $>$ putrescine.

Since the arterial wall and other interstitial spaces contain large amounts of sulfated proteoglycans (57), it is tempting to speculate that LDL forms complexes with these proteoglycans in vivo and that these complexes are then taken up by macrophages in a manner analogous to the uptake of LDL/dextran sulfate complexes in vitro. Whether such an event does occur in vivo remains speculative (see below).

Receptor for β-VLDL

In 1980, Goldstein et al (6) reported that mouse peritoneal macrophages express a surface binding site that mediates the uptake and lysosomal degradation of β-migrating very low density lipoproteins (β-VLDL). To date, β-VLDL are the only naturally occurring plasma lipoproteins known to bind to macrophage receptors (6, 58). Although these lipoproteins are not normally present in detectable amounts in the plasma of humans or animals, Mahley et al (58, 59) showed them to accumulate in the plasma of a variety of species, including dogs, rats, rabbits, and monkeys, when the animals are fed a high cholesterol diet. β-VLDL are also present in plasma of humans with the genetic disease familial dysbetalipoproteinemia, also called type 3 hyperlipoproteinemia (60).

Upon ultracentrifugation of plasma, β-VLDL are found in the same density range as VLDL (d < 1.006 g/ml). In contrast to normal VLDL, which have a triglyceride-rich core, the β-VLDL have a core composed largely of cholesteryl ester. Whereas normal VLDL contain apoproteins B, E, and C, β-VLDL particles contain predominantly apo B and apo E, with markedly reduced amounts of apo C. In contrast to normal VLDL, which have pre-β mobility on agarose gel electrophoresis, the β-VLDL particles show β mobility (59, 61).

β-VLDL particles are thought to represent exaggerated forms of remnant particles normally created during the catabolism of the triglyceride-rich lipoproteins, chylomicrons and VLDL. After the triglycerides of chylomicrons and VLDL are hydrolyzed by lipoprotein lipase, the particles are converted into cholesteryl ester-rich remnant lipoproteins that are rapidly removed by the liver (11, 61). In cholesterol-fed animals and in patients with familial dysbetalipoproteinemia, the normal hepatic clearance mechanism is either overloaded or functions inefficiently. As a result, the remnant particles remain in plasma where they grow in size and become even further enriched in cholesteryl esters to form β-VLDL (60, 61).

Mouse peritoneal macrophages have surface receptors that specifically bind β-VLDL but not VLDL (6, 58). Similar receptors are present on macrophages derived from human monocytes (58, 62). The β-VLDL receptor differs from the acetyl-LDL receptor since it does not bind acetyl-LDL, fucoidin, or polyinosinic acid, all of which bind to the acetyl-LDL receptor (6).

BIOCHEMICAL PROPERTIES OF THE β-VLDL RECEPTOR   In many respects the β-VLDL receptor resembles the LDL receptor of human fibroblasts. For example, both receptors are under feedback regulation. When fibroblasts are induced to accumulate large amounts of cholesterol, they suppress production of LDL receptors (29). Similarly, when mouse peritoneal macrophages (6) or human monocyte-derived macrophages (62) are loaded with cholesterol, the number of β-VLDL receptors is reduced. However, there are sufficient differences to indicate that the macrophage β-VLDL receptor and the fibroblast LDL receptor are probably different molecules. Thus, although the LDL and β-VLDL receptors both bind β-VLDL (58), binding of ^{125}I-labeled β-VLDL to the LDL receptor of human fibroblasts is competitively inhibited by excess unlabeled LDL, while binding of ^{125}I-labeled β-VLDL to the β-VLDL receptor in mouse peritoneal macrophages is not inhibited significantly by LDL (6, 58). Monocytes from patients with familial hypercholesterolemia cannot produce LDL receptors because of a defect in the gene for the LDL receptor (63). Yet, these mutant monocytes produce normal amounts of β-VLDL receptors (58, 62). Additional evidence for two different receptors comes from comparative studies of the effect of exogenously added apo E on the uptake of VLDL by the β-VLDL receptor of macrophages and by the LDL receptor of fibroblasts (see below).

LIGAND SPECIFICITY OF THE β-VLDL RECEPTOR   Based on Scatchard analysis, the Kd for ^{125}I-β-VLDL binding to mouse peritoneal macrophages at 4°C is about 1 μg protein/ml (64). The component of β-VLDL recognized by the macrophage receptor has been difficult to elucidate. When the proteins of β-VLDL are modified by reductive methylation, binding to the β-VLDL receptor is abolished, as judged by the loss of the ability to stimulate cholesteryl ester formation in macrophages (64). This experiment strongly suggests that the protein component is involved in binding. As mentioned above, the proteins of β-VLDL consist almost entirely of apoproteins B and E. Yet, lipoproteins containing only apo B (LDL) or only apo E (apo E-HDL$_c$) fail to inhibit the uptake and degradation of ^{125}I-β-

VLDL by the macrophage β-VLDL receptor (6, 58, 65), suggesting that neither apo B or apo E alone mediates binding. Moreover, normal ^{125}I-VLDL, which contain both apo B and apo E, fail to bind to the macrophage β-VLDL receptor (6). Thus, it has not been possible to determine which, if any, of the proteins on β-VLDL mediate binding. Binding of ^{125}I-β-VLDL is not competitively inhibited by ligands for other known macrophage receptors, such as yeast mannans (6).

Fainaru et al (66) recently subfractionated dog β-VLDL particles by agarose gel chromatography into two discrete populations. One population, designated Fraction I, was larger in diameter (90–300 nm) than Fraction II (20–70 nm). On paper electrophoresis, Fraction I remained at the origin, whereas Fraction II had β-mobility. Both Fractions I and II contained predominantly apoproteins B and E. However, Fraction I contained an equal mixture of high and low M_r forms of apoprotein B, whereas Fraction II contained only the high M_r form. On the basis of these findings, Fainaru et al suggested that the Fraction I component of canine β-VLDL was of intestinal origin, i.e. it was derived from intestinal chylomicrons. In contrast, Fraction II particles were postulated to be of hepatic origin (66). This hypothesis was supported by the finding that fasting for 48 hr led to a near disappearance of Fraction I particles from plasma, but did not lower the amount of Fraction II in dogs previously fed cholesterol. Both Fractions I and II bound to the β-VLDL receptor on mouse peritoneal macrophages, as evidenced by their ability to deliver cholesterol to cells and thereby stimulate the incorporation of [^{14}C]oleate into cholesteryl [^{14}C]oleate (66). Fraction I was 3- to 15-fold more active than Fraction II in stimulating cholesteryl ester synthesis when added at comparable cholesterol levels, suggesting that the β-VLDL receptor may interact preferentially with intestinally derived remnant particles.

Plasma from two patients with familial dysbetalipoproteinemia were also found to contain two fractions of β-VLDL (66). These fractions corresponded in size and composition to Fractions I and II of dog plasma. In both the dog and human samples, the Fraction I lipoprotein had a much lower protein-to-cholesterol ratio than did Fraction II. As with the canine samples, the human Fraction I material was several-fold more potent than Fraction II in stimulating cholesteryl ester synthesis in mouse peritoneal macrophages (66).

Gianturco et al (67) showed that chylomicrons and VLDL isolated from hypertriglyceridemic patients bind to the β-VLDL receptor of mouse peritoneal macrophages. Chylomicrons and VLDL from normal humans failed to bind or be taken up by the macrophages. Uptake of ^{125}I-VLDL from the hypertriglyceridemic patients was not inhibited competitively by acetyl-LDL, but was inhibited competitively by rabbit β-VLDL to a much greater degree than by LDL, suggesting that the hypertriglyceridemic VLDL particles were binding to the β-VLDL receptor. Uptake of hypertriglyceridemic VLDL by the cells delivered triglyceride and produced a massive increase in the triglyceride content of the macrophages. The apoproteins of hypertriglyceridemic VLDL that mediate binding to the β-VLDL receptor are unknown. No difference in protein composition was found between normal VLDL, which does not bind to this binding site, and hypertriglyceridemic VLDL, which binds with high affinity (67).

The addition of exogenous apo E to hypertriglyceridemic VLDL markedly limits its ability to be taken up by the β-VLDL receptor of macrophages (68). On the other hand, the addition of apo E to normal VLDL, which does not bind to either the β-VLDL receptor of macrophages or the LDL receptor of fibroblasts, results in a particle that is efficiently internalized by the LDL fibroblast receptor (69). Thus, incorporation of exogenous apo E inhibited binding to the β-VLDL receptor and stimulated binding to the LDL receptor. The mechanism for these effects is unknown.

Although hypertriglyceridemic VLDL and β-VLDL appear to bind to the same site on the macrophage surface and although the uptake mechanisms appear to be similar, the final result of the uptake process is different in the case of the two different lipoproteins. Since β-VLDL contain predominantly cholesteryl ester, uptake of this particle causes cholesteryl ester accumulation in the cells (6; see below). In contrast, uptake of hypertriglyceridemic VLDL leads to degradation of the internalized triglyceride within lysosomes, release of fatty acids, and re-esterification of the fatty acids within the cytoplasm to form triglycerides (67). Thus, although in both cases the cells develop Oil Red-O-positive inclusions, the lipids comprising these inclusions differ.

FUNCTION OF THE β-VLDL RECEPTOR IN VIVO   As mentioned above, the β-VLDL receptor appears designed to bind remnants of chylomicron or VLDL metabolism that have circulated in plasma for an abnormally long time and are excessively rich in cholesteryl esters or triglycerides. The macrophage receptor thus may function as a backup mechanism to clear remnant lipoproteins when they are not removed promptly by their normal receptors on hepatocytes or other cells. Evidence that the macrophage receptor functions in vivo is so far indirect. One line of evidence stems from the observation by Mahley (59) that cholesterol-fed dogs develop massive cholesteryl ester deposition in macrophages throughout the body. This deposition becomes marked only when the plasma cholesterol level exceeds 750 mg/dl. Similarly, β-VLDL becomes a predominant lipoprotein only when the plasma cholesterol level exceeds 750 mg/dl. The simultaneous appearance of circulating β-VLDL and cholesterol-loaded macrophages suggests that β-VLDL may be depositing its cholesterol in macrophages, presumably as a result of uptake via the β-VLDL receptor.

The second line of evidence for function of the β-VLDL receptor in vivo comes from studies in which ^{125}I-β-VLDL are injected into animals. These β-VLDL particles are removed from the circulation with extreme rapidity (70–72). Canine Fraction I particles are removed much more rapidly than Fraction II particles (66). The plasma clearance of β-VLDL is markedly reduced when the β-VLDL are reductively methylated, a reaction known to prevent the binding of β-VLDL to the macrophage receptor (64). Interpretation of these in vivo clearance experiments is clouded, however, because in addition to binding to the macrophage β-VLDL receptor, these particles also bind to LDL receptors which are present on hepatocytes and in extrahepatic tissues (58, 72). Binding of β-VLDL to the LDL receptor as well as to the β-VLDL receptor is inhibited by reductive methylation (64, 73). Thus, the rapid disappearance of β-VLDL from the plasma could be a consequence of its binding to LDL receptors on hepatocytes, rather than to β-VLDL receptors on macrophages. Indeed, Kovanen et al (72) showed that feeding of cholesterol to rabbits results in a marked diminution in the number of β-VLDL binding sites on liver membranes and a corresponding decrease in the rate of uptake of β-VLDL by the liver in vivo. The in vitro binding sites measured in these studies represented LDL receptors rather than β-VLDL receptors of macrophages, because the binding of ^{125}I-β-VLDL was susceptible to inhibition by LDL (72). Thus, at present, it is not possible to conclude definitively that β-VLDL receptors on macrophages participate in the clearance of β-VLDL from plasma.

Receptors for Cholesteryl Ester/Protein Complexes from Atherosclerotic Plaques

In addition to the receptors for modified or abnormal plasma lipoproteins, mouse peritoneal macrophages express binding sites that mediate the uptake of cholesteryl ester/protein complexes isolated from atherosclerotic plaques of human aortas (7). This uptake leads to a marked stimulation of cholesteryl [^{14}C]oleate synthesis and a marked cellular accumulation of cholesteryl esters (7; see below).

The biochemical elements of the aortic cholesteryl ester/protein complexes that promote macrophage cholesteryl ester accumlation have been defined in only a preliminary way. The bulk of the active complexes were excluded on a Biogel A-50m column and floated in the ultracentrifuge in the density range of 1.006–1.063 g/ml, indicating that they consisted of large lipid/protein aggregates (7). The complexes contained 19% free cholesterol, 35% esterified cholesterol, 11% phospholipid, 4% triglyceride, and 31% protein (7), a composition roughly similar to that of LDL. It is likely that the active complexes also contained sulfated glycosaminoglycans as found by Berenson and co-workers in studies of similar material (57, 74). The protein component of the complexes was important for uptake by macrophages since pronase treatment abolished their ability to stimulate cholesterol esterification (7).

Some of the aortic cholesteryl ester/protein complexes were retained on an anti-apoprotein B affinity column, which suggested that they contained immunoreactive apo B (7). However, the material that did not adhere to the affinity column was just as active as the retained material in stimulating cellular cholesteryl [^{14}C]oleate synthesis (7), showing that apo B was not required, at least in an immunoreactive form.

The aortic cholesteryl ester/protein complexes appeared to enter macrophages by binding to high affinity saturable sites on the cell surface. Uptake was prevented in an apparent competitive fashion by certain polyanions, such as polyinosinic acid, dextran sulfate, and fucoidin, but not by polycytidylic acid (7). This pattern of competition is similar to that displayed by the receptor for acetyl-LDL (4, 18). In direct competition experiments, however, the aortic extract competed to only a small degree for the uptake and degradation of ^{125}I-acetyl-LDL (7). It was concluded that the binding site for the aortic complexes was similar but not identical to that for acetyl-LDL (7).

Macrophages showed specificity at two levels in interacting with the aortic complexes. First, there was ligand specificity in that other types of cholesteryl ester/protein complexes, such as those isolated from liver and adrenal glands, were not taken up by macrophages, as monitored by their inability to stimulate cellular cholesteryl [^{14}C]oleate synthesis when added to the medium at comparable cholesterol concentrations (7). Second, there was receptor specificity in that other cell types such as cultured human fibroblasts failed to take up the cholesteryl ester-rich particles of the aortic extract (7).

In early studies, Werb & Cohn (75, 76) showed that mouse peritoneal macrophages take up particulate complexes formed by sonicating mixtures of albumin and cholesteryl ester. Goldstein et al (7) formed such complexes in the presence of ^{125}I-labeled albumin and showed that the complexes bound with high affinity to a surface binding site at 4°C. Surface binding was inhibited completely by polyinosinic acid and fucoidin, partially by acetyl-LDL, and not all by polycytidylic acid (7, 77). However, the albumin/cholesteryl ester complexes did not inhibit the binding of ^{125}I-acetyl-LDL (7). Thus, the albumin/cholesteryl ester complexes, like the aortic complexes, bind to a high affinity site that resembles but is not identical to the acetyl-LDL receptor. Interestingly, ^{125}I-albumin, in the absence of cholesteryl ester, was not taken up or degraded with high affinity by the macrophages (7).

The simplest interpretation of the above experiments is that macrophages express a family of receptors that recognize cholesteryl ester/protein complexes. One of these is the receptor for acetyl-LDL. Other receptors bind cholesteryl ester/protein complexes from aorta and cholesteryl ester/albumin complexes, but not acetyl-LDL. These receptors share the striking property that binding and uptake are inhibited by polyinosinic acid (but not by polycytidylic acid) and by sulfated polysaccharides such as fucoidin and dextran sulfate. However, cross-competition studies indicate that the receptors for acetyl-LDL differ from those for cholesteryl ester/protein complexes.

PROCESSING AND STORAGE OF LIPOPROTEIN-BOUND CHOLESTEROL BY MACROPHAGES

All of the lipoproteins that bind to receptors on the surface of macrophages appear to be processed in a similar fashion (4–7). All are rapidly internalized at 37°C, apparently by endocytosis, and then delivered to lysosomes where the protein and cholesteryl ester components are hydrolyzed. The liberated cholesterol is released from the lysosome and subsequently re-esterified in the cytoplasm or excreted from the cell, depending on whether or not a cholesterol acceptor is present in the culture medium. The most detailed studies were conducted with acetyl-LDL as a model (4, 8, 9, 18, 78), and these are described below.

Endocytosis and Lysosomal Hydrolysis

When ^{125}I-labeled acetyl-LDL is allowed to bind to the macrophage surface at 4°C and the cells are subsequently warmed to 37°C, the surface-bound lipoprotein is internalized and degraded with extreme efficiency (4). Within 30 min at 37°C, 50% of the initial cell-bound radioactivity is degraded and the major degradation product, ^{125}I-monoiodotyrosine, is excreted from the cell (4).

In the presence of low concentrations of the lysosomal inhibitor chloroquine (20–75 μM), the endocytic uptake of the ^{125}I-acetyl-LDL continues, but degradation is blocked (4, 8). As a result, large amounts of undigested ^{125}I-acetyl-LDL accumulate within the cell, demonstrating that degradation normally occurs in lysosomes. When binding of the ^{125}I-acetyl-LDL to the surface receptor is inhibited by compounds such as fucoidin and polyinosinic acid, cellular uptake and degradation are reduced by more than 95%, indicating that uptake and degradation require binding of the ^{125}I-acetyl-LDL to its receptor (4, 18).

In an extensive series of studies, Brown et al (8, 9) traced the fate of the cholesteryl ester component of acetyl-LDL within macrophages. These studies were carried out with the use of reconstituted lipoproteins, prepared according to the technique of Krieger et al (79). By this technique acetyl-LDL particles are lyophilized in the presence of starch and extracted with heptane to remove cholesteryl ester and free cholesterol. The core of the lipoprotein is then reconstituted by addition of [³H]cholesteryl linoleate in heptane. The resultant particle, designated r-[³H-cholesteryl linoleate]acetyl-LDL, is taken up by macrophages in the same way as acetyl-LDL, and the [³H]cholesteryl ester component is hydrolyzed (8). In the presence of chloroquine, hydrolysis of the [³H]cholesteryl linoleate is inhibited, and unhydrolyzed [³H]cholesteryl linoleate accumulates within the cell (8). Double-label studies using ¹²⁵I-acetyl-LDL and r-[³H-cholesteryl linoleate]acetyl-LDL indicate that the protein and cholesteryl ester components are taken up and hydrolyzed in amounts that are proportional to their occurrence in the lipoprtein particle, i.e. the particle appears to be taken up and delivered to lysosomes intact (8).

Cytoplasmic Re-esterification and Hydrolysis of Lipoprotein-Derived Cholesteryl Esters

The studies with r-[³H-cholesteryl linoleate]acetyl-LDL show that the [³H]cholesterol released after lysosomal hydrolysis does not remain as free cholesterol, but rather is rapidly re-esterified (8). The re-esterification reaction was demonstrated by two approaches. First, macrophages were incubated with r-[³H-cholesteryl linoleate]acetyl-LDL in the presence of unlabeled oleate. At intervals the cells were extracted and the [³H]cholesteryl esters were fractionated on silver nitrate-coated thin layer chromatogram sheets, which separate cholesteryl oleate and cholesteryl linoleate. At early time points (less than 2 hr), most of the cellular [³H]cholesteryl esters were composed of [³H]cholesteryl linoleate. Some free [³H]cholesterol also was formed. By 2 hr, the levels of intact [³H]cholesteryl linoleate and free [³H]cholesterol had reached a steady state plateau. Subsequently, there was a marked accumulation of [³H]cholesteryl oleate, indicating hydrolysis of the entering [³H]cholesteryl linoleate with subsequent re-esterification of the free [³H]cholesterol to form [³H]cholesteryl oleate. In the presence of chloroquine, the hydrolysis of [³H]cholesteryl lineoleate was inhibited. As a result, [³H]cholesteryl linoleate continued to accumulate in the cells without reaching a plateau, and there was no generation of free [³H]cholesterol or of [³H]cholesteryl oleate (8).

The second approach used to demonstrate the hydrolysis and re-esterification mechanism consisted of incubating macrophages with r-[³H-cholesteryl linolate]acetyl-LDL in the presence of [¹⁴C]oleate (8). During the initial 2 hr period, all of the rise in cholesteryl esters within the cell could be attributed to [³H]cholesteryl linoleate. At this point the cholesteryl ester fraction contained little [¹⁴C]oleate. However, after 2 hr the rise in [³H]cholesterol in the ester fraction was paralleled by an equimolar rise in the [¹⁴C]oleate content of the cholesteryl fraction. Thus, after 2 hr all of the increase in cholesteryl ester represented newly synthesized cholesteryl oleate (8).

The above two experiments documented that the cholesteryl esters of acetyl-LDL (predominantly cholesteryl linoleate) are hydrolyzed in lysosomes and that the resultant free cholesterol is rapidly re-esterified (primarily with oleate). Figure 2 shows the marked stimulation in the synthesis (Panel *A*) and accumulation (Panel *B*) of cholesteryl esters that occurs in macrophages incubated with acetyl-LDL. Fucoidin, which inhibits binding to the acetyl-LDL receptor, prevents the uptake of acetyl-LDL and thereby prevents the lipoprotein from delivering cholesterol to the cells (Figure 2*B*).

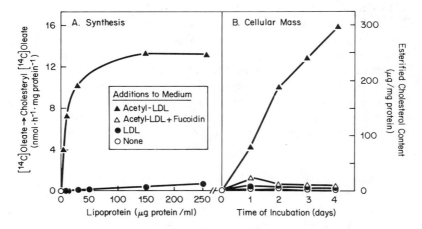

Figure 2 Accumulation of cholesteryl esters in mouse peritoneal macrophages incubated with acetyl-LDL. *Panel A:* Stimulation of cholesteryl [¹⁴C]oleate formation. Monolayers of macrophages were incubated for 2 days in lipoprotein-deficient serum and then with the indicated concentration of either native LDL (●) or acetyl-LDL (▲) 5 hr at 37°C. Each monolayer was then pulse-labeled for 2 hr with 0.1 mM [¹⁴C]oleate bound to albumin and the cellular content of cholesteryl [¹⁴C]oleate was measured by thin layer chromatography (8). *Panel B:* Time course of cholesteryl ester accumulation. Monolayers of macrophages received medium containing lipoprotein-deficient serum and one of the following additions: O, none; ●, 25 μg protein/ml of native LDL; ▲, 25 μg protein/ml of acetyl-LDL; or Δ, 25 μg protein/ml of acetyl-LDL plus 50 μg/ml of fucoidin. At the indicated time, the monolayers were harvested and their content of esterified cholesterol was measured by gas-liquid chromatography (8).

The mechanism of the cholesterol re-esterification reaction was studied in cell-free homogenates prepared from macrophages incubated in the absence or presence of acetyl-LDL (8). The cell-free homogenates were incubated with [^{14}C]oleate in the presence of ATP, magnesium, and coenzyme A. Homogenates of the cells that had been incubated with acetyl-LDL incorporated [^{14}C]oleate into cholesteryl [^{14}C]oleate at a rate that was 20-fold higher than that of homogenates of cells not incubated with acetyl-LDL. The cholesterol esterifying enzyme was associated with membranes, since it was recovered in the pellet after centrifugation at 100,000 X g. The reaction was totally dependent on ATP and coenzyme A, indicating that it was mediated by an acyl-CoA : cholesterol acyltransferase (ACAT) enzyme (8).

The cholesterol derived from lysosomal hydrolysis appears to be the component of acetyl-LDL that stimulates the ACAT enzyme, but the mechanism of this stimulation is not known. A similar stimulation of ACAT activity occurs when human fibroblasts take up and hydrolyze native LDL and thereby liberate free cholesterol (29). The activity of the ACAT enzyme can be enhanced in a similar fashion if intact cells are simply incubated with cholesterol or oxygenated derivatives of cholesterol that are dissolved in ethanol (29). However, the stimulation of the ACAT reaction is not simply a result of the provision of excess cholesterol substrate, since addition of cholesterol to the membranes in vitro does not reproduce the stimulation observed when acetyl-LDL or cholesterol is incubated with intact cells in vivo (8). In fibroblasts the stimulation of ACAT activity is not blocked by cycloheximide, indicating that LDL-derived cholesterol does not induce synthesis of new enzyme molecules, but rather activates pre-existing ACAT (29). Experiments with cycloheximide are not yet reported in macrophages.

Stimulation of ACAT in macrophages requires lysosomal hydrolysis of the acetyl-LDL-derived cholesteryl esters. If the acetyl-LDL is allowed to accumulate intact in lysosomes in the presence of chloroquine, no stimulation of ACAT activity occurs (6–8). However, if the cells are then washed free of extracellular acetyl-LDL and of chloroquine, the acetyl-LDL that has accumulated in lysosomes is hydrolyzed and the liberated cholesterol stimulates the ACAT reaction (6, 7). Similar effects of chloroquine on cholesteryl ester metabolism were obtained in macrophages incubated with β-VLDL (6) and human aortic extracts prepared from atherosclerotic plaques (7).

The cholesteryl esters that are synthesized by the ACAT enzyme accumulate in the cytoplasm of the cell as cholesteryl ester droplets. By electron microscopy, the cytoplasmic cholesteryl esters are not surrounded by a typical bilaminar membrane, but rather appear as discrete lipid droplets (8).

Figure 3 shows a polarized light micrograph of macrophages that were incubated with acetyl-LDL. At low power, the cytoplasm of the cells is packed with birefringent crystals of cholesteryl ester. The morphologic appearance of these cells, which were produced in culture dishes in vitro, is remarkably similar to that of macrophage foam cells that occur in vivo in atherosclerotic plaques.

The Cholesteryl Ester Cycle

Figure 4 shows a two-compartment model proposed by Brown, Ho & Goldstein (9) to account for the metabolism of cholesteryl esters in mouse peritoneal macrophages. Lipoprotein-bound cholesteryl esters that enter the macrophage via receptor-mediated endocytosis are delivered to lysosomes (first cellular compartment), where the cholesteryl esters are hydrolyzed. The liberated cholesterol leaves the lysosomes and enters the cytoplasm (second cellular compartment), where it has two fates. Some of the cholesterol is immediately excreted. The remainder of the excess cholesterol is re-esterified by the ACAT enzyme and accumulates in the cytoplasm as cholesteryl ester droplets. Although these droplets appear inert in the light and electron microscope, their cholesteryl esters are in a dynamic

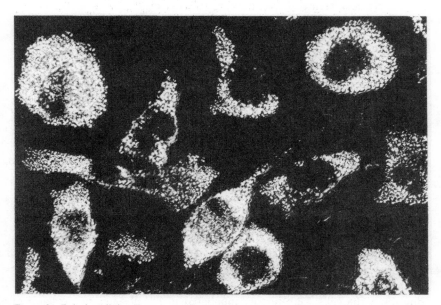

Figure 3 Polarized light microscopy of macrophages incubated with acetyl-LDL. Monolayers of mouse peritoneal macrophages were incubated for 3 days with 25 μg protein/ml of acetyl-LDL (8). The cytoplasm is packed with birefringent crystals of cholesteryl ester. *Magnification*: X 1000. Richard G. W. Anderson kindly provided this photograph.

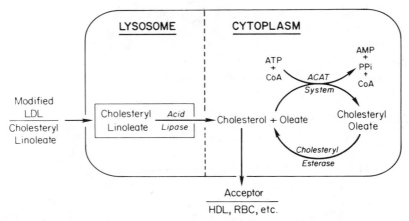

Figure 4   Two-compartment model for cholesteryl ester metabolism in macrophages, illustrating the cytoplasmic cholesteryl ester cycle. The salient features of the model are discussed in the text. RBC, red blood cells. Reproduced with permission from the *Journal of Biological Chemistry* (9).

state, continually undergoing a cycle of hydrolysis and re-esterification, which has been termed the cholesteryl ester cycle.

The cholesteryl ester cycle was demonstrated in several types of double-label experiments (9). Macrophages were incubated with acetyl-LDL in the presence of [³H]oleate, so that the cells accumulated cytoplasmic cholesteryl ester droplets containing cholesteryl [³H]oleate. The cells were then placed in medium free of acetyl-LDL and containing [¹⁴C]oleate in place of [³H]oleate. Under these conditions the cellular content of cholesteryl [³H]oleate declined steadily, falling by about 50% in 24 hr. This loss in cholesteryl [³H]oleate was balanced almost exactly by an increase in the content of cholesteryl [¹⁴C]oleate. Thus, although the total cholesteryl oleate content of the cells did not change significantly, there was a continual hydrolysis of the cholesteryl [³H]oleate and re-esterification of the cholesterol with [¹⁴C]oleate (9).

Hydrolysis of the cytoplasmic cholesteryl [³H]oleate was shown to be mediated by an enzyme distinct from the lysosomal acid lipase (9). Macrophages were induced to form radiolabeled cytoplasmic cholesteryl esters by incubation with *r*-[³H-cholesteryl linoleate]acetyl-LDL. After hydrolysis and re-esterification, the [³H]cholesterol accumulated in the cytoplasm as [³H]cholesteryl oleate. The acetyl-LDL was then withdrawn from the medium and the cells were further incubated in the absence or presence of chloroquine and in the presence of HDL. In the absence of chloroquine, the cytoplasmic [³H]cholesteryl esters were hydrolyzed, and the [³H]cholesterol was excreted from the cell. In the presence of chloroquine, hydrolysis and excretion were unaffected.

If the culture medium does not contain HDL or another acceptor for cholesterol, the free cholesterol released from the hydrolysis of cytoplasmic cholesteryl esters is re-esterified, completing the cholesteryl ester cycle (Figure 4). The re-esterification reaction appears to be catalyzed by the same microsomal ACAT enzyme that esterifies the lipoprotein-derived cholesteryl esters. This conclusion is based on experiments with progesterone, a known inhibitor of the microsomal ACAT reaction (80). These experiments employed a double-labeled protocol (9). Macrophages were incubated with acetyl-LDL in the presence of [^{3}H]oleate, so that they formed cytoplasmic cholesteryl esters containing cholesteryl [^{3}H]oleate. They were then switched to medium containing [^{14}C]oleate. In the absence of progesterone, the cholesteryl [^{3}H]oleate esters were hydrolyzed, and cholesteryl [^{14}C]oleate esters were synthesized, with no change in the total amount of cholesteryl ester. In the presence of progesterone, hydrolysis of the cholesteryl [^{3}H]oleate proceeded as before. However, progesterone prevented the re-esterification with [^{14}C]oleate. As a result, cholesteryl [^{14}C]oleate did not accumulate, and the total amount of cellular cholesteryl esters fell (9).

SECRETION OF CHOLESTEROL AND APOPROTEIN E BY MACROPHAGES

Cholesterol Excretion Dependent on Cholesterol Acceptors

Macrophages that have stored cytoplasmic cholesteryl esters after incubation with acetyl-LDL cannot secrete this cholesterol when incubated in serum-free medium. In the absence of serum, the cholesteryl esters are continually hydrolyzed and the cholesterol is re-esterified, as mentioned above. This cholesteryl ester cycle is interrupted in a striking fashion when the culture medium contains serum or other substances that are capable of binding cholesterol (9). Werb & Cohn (75, 76) originally showed that an acceptor molecule is required for the excretion of cholesterol by macrophages. They studied cells that were cholesterol-loaded by incubation with albumin/cholesteryl ester complexes. While a variety of cholesterol acceptors can perform this removal function, the most detailed studies were carried out with human HDL$_3$ with a density of 1.125–1.215 g/ml (9, 78).

Direct demonstration of HDL's action in promoting cholesterol excretion from macrophages has come from mass measurements of the free and esterified cholesterol content of cells that had been loaded with cholesterol by incubation with acetyl-LDL. Ho et al (78) showed that HDL produced a marked decline in the cellular content of esterified cholesterol, the level falling from 250 to 50 µg sterol/mg cellular protein within 24 hr, while the free cholesterol content dropped only slightly, from 30 to 20 µg sterol/mg cellular protein. The cholesteryl ester content of the cells decreased progressively with increasing HDL concentrations until it reached a minimum at an HDL concentration of 100–200 μg protein/ml. Although the bulk of the HDL-mediated decline was attributable to a decline in esterified cholesterol, all of the cholesterol that was excreted into the culture medium was in the free (i.e. nonesterified) form (78). This conclusion was based on experiments in which macrophages were incubated with r-[³H-cholesteryl linoleate]acetyl-LDL so that the cytoplasmic cholesteryl esters were radiolabeled. The subsequent addition of HDL led to a marked decline in the cellular content of [³H]cholesteryl esters, yet all of this sterol was recovered in the medium as nonesterified [³H]cholesterol (9, 78).

The above findings suggest that macrophages are unable to excrete cholesteryl esters and therefore excrete only free cholesterol. As discussed above, the cell's cholesteryl esters are continually undergoing hydrolysis and re-esterification in a cholesteryl ester cycle. HDL could increase the generation of free cholesterol by one of two means: (a) it could increase the rate of hydrolysis of cholesteryl esters; or (b) it could prevent the re-esterification reaction (Figure 4). These possibilities were distinguished in a series of double-label experiments (9). Macrophages were incubated with acetyl-LDL in the presence of [³H]oleate so that they accumulated cytoplasmic cholesteryl [³H]oleate. After loading, the [³H]oleate was replaced with an equal concentration of [¹⁴C]oleate. As described above, when this second incubation was performed in the absence of HDL, a decline in cholesteryl [³H]oleate was balanced by an increase in cholesteryl [¹⁴C]oleate, indicating that hydrolysis and re-esterification were occurring at equal rates. When HDL was present in the medium, the cellular content of cholesteryl [³H]oleate declined at the same rate, but [¹⁴C]oleate was no longer incorporated into cholesteryl [¹⁴C]oleate. This experiment demonstrated that HDL causes a net hydrolysis of cholesteryl esters by preventing the re-esterification reaction and not by stimulating the hydrolysis reaction. Consistent with this interpretation was an experiment in which macrophages were incubated with acetyl-LDL and then further incubated in medium with or without HDL. The cells were harvested for measurement of in vitro ACAT activity. The cells incubated with HDL had a 60% reduction in the activity of the cholesterol-esterifying enzyme (9).

The mechanism by which HDL blocks cholesterol re-esterification is not known precisely, but it must be related to the known ability of HDL to remove cholesterol from cell membranes (81–83). It seems likely that HDL removes free cholesterol from the macrophage plasma membrane and this depletion allows the membrane to bind the cholesterol that is generated within the cytoplasm from the hydrolysis of cholesteryl esters. The shift of this hydrolyzed cholesterol from the cytoplasm to the cell membrane would prevent its re-esterification. After binding to the plasma membrane, the cholesterol would be removed by HDL and this would facilitate a continuous net hydrolysis and excretion of cholesterol (Figure 4). When no cholesterol acceptor is available, the plasma membrane becomes saturated with cholesterol. The cytoplasmic cholesterol cannot enter the plasma membrane and therefore it is re-esterified by the ACAT enzyme.

In the absence of a cholesterol acceptor, about 50% of the stored cholesteryl esters are hydrolyzed and re-esterified each day in a type of futile cycle (9). Inasmuch as the ACAT enzyme uses a fatty acyl-CoA derivative that requires ATP for its synthesis, each turn of the cholesteryl ester cycle has the net effect of breaking down one molecule of ATP to AMP and pyrophosphate (Figure 4).

HDL is an effective acceptor for membrane cholesterol because its outer shell contains a low ratio of cholesterol to phospholipid as compared with most cell membranes (10). The cholesterol : phospholipid ratio is kept low in vivo by the combined action of the plasma enzyme lecithin : cholesterol acyl-CoA transferase (LCAT), which esterifies the surface cholesterol of HDL and initiates its transfer into the core of HDL (10), and of a plasma cholesteryl ester transfer protein, which promotes the transfer of cholesteryl esters from HDL to other plasma lipoproteins (85). Under the conditions of the macrophage experiments in vitro, HDL can accept cholesterol from cell membranes without a requirement for LCAT (9, 78). Presumably if the incubations were continued long enough in the absence of LCAT, the surface of HDL would become saturated with free cholesterol and such HDL would cease to be an effective cholesterol acceptor. In vitro, HDL has not been shown to become saturated with cholesterol, perhaps because the amount of HDL present in the medium is large relative to the amount of cholesterol that is excreted. Large amounts of HDL are used because the experiments were designed to maximize the rate of cholesterol removal.

In addition to HDL, other substances can remove cholesterol from cholesterol-loaded macrophages. For example, the lipoprotein-deficient fraction of serum (density > 1.215 g/ml) is quite effective (78). The two major proteins of this fraction, albumin and gamma globulin, have no cholesterol-removing activity when tested as isolated proteins (78). It is possible that the cholesterol-removing activity of lipoprotein-deficient serum resides in trace amounts of the apoproteins of plasma lipoproteins, such as apo A-I, which are known to be present in that fraction (84). High concentrations of casein or thyroglobulin (i.e. above 1 mg/ml) can also remove cholesterol from cholesterol-overloaded macrophages (78). The activity of these two proteins persists after lipid extraction, suggesting that the protein component itself, and not attached phospholipid, is responsible. Presumably these proteins have a small ability to bind cholesterol that is sufficient to promote cholesterol excretion when the proteins are present at high concentrations. Human erythrocytes have an exceptionally potent ability to remove cholesterol from cholesterol-overloaded macrophages (78). Five μl of packed washed erythrocytes per ml of culture medium were as effective as maximal concentrations of whole serum or HDL.

At least two lipophilic agents do not appear to remove cholesterol from cholesterol-overloaded macrophages. One is human LDL and the other is a liposome preparation composed of egg phosphatidylcholine with or without sphingomyelin (78). LDL has a much higher ratio of cholesterol to phospholipid than does HDL, and this presumably accounts for its inability to remove large amounts of cholesterol from cells. The inability of phospholipid liposomes to facilitate removal is more difficult to understand. Such liposomes bind cholesterol, and in other cell systems they are effective in cholesterol removal (86). The failure of the liposomes in the macrophage experiments may be due to the sensitivity of these cells to the toxic effects of phosphatidylcholine. Concentrations of phosphatidylcholine above 200 μg/ml were not tolerated by the macrophages and hence could not be tested (78).

Macrophages that have been loaded with cholesteryl ester by incubation with β-VLDL have the same requirements for cholesterol excretion as do the cells loaded with acetyl-LDL. Thus, HDL reduces the synthesis of cholesteryl esters and lessens their accumulation in cells incubated with β-VLDL (65).

Synthesis and Secretion of Apoprotein E in Response to Cholesterol Loading

Overloading macrophages with cholesterol stimulates the cells to synthesize and secrete large amounts of apo E, a normal constituent of plasma lipoproteins. This surprising observation was made initially by Basu et al (13, 14), who incubated mouse peritoneal macrophages with acetyl-LDL and then pulse-labeled the cells with [^{35}S]methionine. The cholesterol-loaded cells secreted a protein that floated in the ultracentrifuge at a density $<$ 1.215 g/ml, indicating that it was complexed with lipid. The molecular weight of this protein (35,000) and its isoelectric point corresponded to that of authentic mouse plasma apo E. Moreover, the ^{35}S-labeled apoprotein was precipitated by an antibody directed against apo E (13, 14). In the absence of cholesterol loading, macrophages synthesized and secreted small amounts of apo E. Secretion was stimulated 24-fold when the cells were subjected to prior incubation with acetyl-LDL or other sources of cholesterol (14). Replacement of the cholesteryl esters of acetyl-LDL with triglycerides abolished the ability of these lipoproteins to stimulate apo E secretion (14). Under conditions of maximal cholesterol loading, the synthesized apo E represented more than 10% of the total protein secreted by macrophages and more than 2% of the total protein synthesized (13, 14).

The identity of macrophage apo E was confirmed by genetic studies performed with human monocytes (14). These studies took advantage of the known genetic polymorphism of apo E in humans. The apo E locus in humans comprises three common alleles, apo ϵ^2, ϵ^3, and ϵ^4, whose products (E-2, E-3, and E-4) are distinguishable by isoelectric focussing (87, 88). The apo E secreted by human monocytes contained more sialic acid residues than the corresponding plasma apo E and therefore exhibited a more acidic isoelectric point on two-dimensional gels (14). However, a correspondence between monocyte and plasma apo E became apparent when the proteins from both sources were treated with neuraminidase to remove the charge heterogeneity created by sialic acid. After such neuraminidase treatment, it could be shown that blood monocytes from subjects homozygous for the ϵ^3 allele produced a protein that corresponded in size and isoelectric point to the E-3 protein isolated from the plasma of the same individual (14). Similarly, blood monocytes from an individual heterozygous for the ϵ^2 and ϵ^3 alleles synthesized and secreted proteins corresponding to both E-2 and E-3 (14).

Electron microscopic studies with negative staining techniques revealed that the macrophage apo E was secreted in the form of disc-like structures that measured 180 Å in diameter and 30 Å in width (14). These discs resembled the phospholipid/protein discs known to be secreted by perfused rat liver and designated as "nascent HDL" (89). Nascent HDL contains apo A-I as well as apo E (89), but the macrophage discs contain only apo E (14). Nascent HDL consists of a lamellar bilayer of phospholipid. The proteins are thought to circumscribe the edges of the discs, thereby stabilizing the structure (90). In plasma, these lamellar discs are converted into spherical pseudomicellar particles by the insertion of cholesteryl ester between the lamellae (91). The esters are synthesized by the plasma enzyme LCAT, which transfers a fatty acid from the 2-position of lecithin to the 3-hydroxyl position of cholesterol. The free cholesterol substrate for the LCAT reaction can come from the cholesterol on the surface of the nascent HDL particle or it can come from free cholesterol extracted from plasma membranes or from other lipoproteins (10).

The apo E/phospholipid discs secreted by macrophages do not carry large amounts of cholesterol out of the cell. As described above, cholesterol-loaded macrophages secrete large amounts of cholesterol only when an exogenous cholesterol acceptor, such as HDL, is added to the culture medium. Yet the cells synthesize and secrete nearly as much apo E in the absence of HDL as in the presence of HDL (13). In the absence of HDL, the apo E/phospholipid discs are secreted, but there is little secretion of cholesterol (13, 78), indicating that the discs do not carry significant amounts of cholesterol out of the cell. The converse is also true. Cholesterol excretion occurs even when secretion of apo E is blocked by incubation of the cells with monensin (91a), an ionophore that prevents the movement of secreted proteins from the Golgi apparatus to the plasma membrane (92). Thus, the fundamental mechanisms for secretion of apo E and cholesterol are different.

The above results suggest that secretion of apo E is neither necessary nor sufficient to mediate the excretion of large amounts of cholesterol by macrophages. The apo E must perform some other function. To integrate this information, Basu et al (14) proposed the model shown in Figure 5. In this model, cholesterol leaves the macrophages by binding to the surface of HDL. Apo E is secreted by the cells through an independent mechanism in the form of phospholipid discs. In the presence of plasma LCAT, cholesterol adsorbed to HDL is esterified. During this process, apo E may transfer from the disc to the HDL particle; alternatively, the cholesteryl esters and apo A-I of HDL may transfer to the core and surface of the apo E/phospholipid discs, respectively. By either mechanism, a large spherical particle would be formed that contains a core of cholesteryl esters and a coat containing apoproteins A-I and E (Figure 5).

Mahley (59) described a lipoprotein particle fitting the above description. This lipoprotein, termed HDL$_c$, is present in small amounts in plasma of normal man and most animals. Its concentration rises markedly after cholesterol feeding, especially in dogs, rabbits, and swine (59). HDL$_c$ contains a core of cholesteryl ester and a coat of apo A-I and apo E. When administered intravenously to normal animals, HDL$_c$ is rapidly taken up by hepatocytes through receptor-mediated endocytosis that is triggered by receptors that bind apo E (12, 59). This uptake suggests that the function of

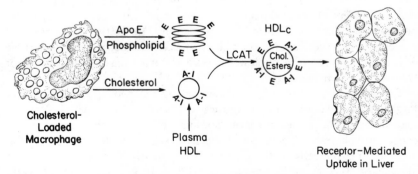

Figure 5 Working model for the role of apo E secretion in mediating "reverse cholesterol transport," i.e. the transport of cholesterol from cholesterol-loaded macrophages to the liver. The salient features of the model are discussed in the text. Adapted with permission from the *Journal of Biological Chemistry* (14).

HDL$_c$ is to transport cholesterol to the liver. However, the origin of plasma HDL$_c$ is unknown.

Since cholesterol-loaded macrophages secrete both apo E/phospholipid discs and cholesterol in the presence of HDL, it is attractive to speculate that these secreted products are assembled extracellularly into HDL$_c$, as depicted in Figure 5. The HDL$_c$ would then carry the excreted cholesterol to the liver, where receptor-mediated uptake would occur by virtue of the apo E component. Inasmuch as the actual formation of HDL$_c$ has not been demonstrated in macrophage cultures, the model in Figure 5 must be considered hypothetical. Nevertheless, the model has heuristic value in that it unites many of the ideas and observations of Glomset & Norum (10) on the role of LCAT in plasma cholesteryl ester formation, the findings of Mahley, Innerarity, and co-workers (12, 93) on the role of HDL$_c$ in delivering cholesterol to the liver, and the work of Brown, Goldstein and co-workers (4, 11, 14) on cholesterol metabolism in macrophages. Additional work will be necessary to determine whether the assembly of HDL$_c$ is accomplished in the manner suggested in Figure 5.

IMPLICATIONS FOR FOAM CELL FORMATION IN ATHEROSCLEROSIS

The in vitro studies of isolated macrophages disclosed many new properties of these scavenger cells relevant to cholesterol metabolism, and raised new questions about the role of macrophages in atherosclerosis. The studies demonstrated that macrophages are specifically adapted for internalizing, storing, and secreting large amounts of lipoprotein-cholesterol. Several types of lipoprotein receptors were identified on the surface of macrophages, and a two-compartment pathway for the storage of cytoplasmic cholesteryl ester droplets was defined. The extraordinary ability of the cholesterol-loaded macrophage to excrete large amounts of cholesterol and apo E was also disclosed.

The evidence that these mechanisms function in vivo is so far indirect. When plasma lipoprotein levels rise, macrophages of the artery wall accumulate cytoplasmic cholesteryl ester droplets (1, 2) that are morphologically identical to those of foam cells created in vitro (8). Such cholesteryl ester droplets are rich in cholesteryl oleate (8, 94, 95), indicating that they represent re-esterified cholesterol and suggesting that the two-compartment model shown in Figure 4 operates in vivo. Moreover, the foam cells of atherosclerotic plaques in experimental animals (95–97) and humans (94, 98) were shown to incorporate [^{14}C]oleate into cholesteryl [^{14}C]oleate at a high rate. Again, this finding is explained by the two-compartment model delineated in the in vitro macrophage studies.

There is also growing evidence to indicate that macrophages in vivo express the same types of lipoprotein receptors that were demonstrated in vitro. Foam cells from atherosclerotic plaques of cholesterol-fed rabbits were recently shown to contain active acetyl-LDL receptors and β-VLDL receptors (99). Direct demonstration that these and other receptors function in the uptake of plasma lipoprotein-cholesterol in the arterial wall will require treatment of animals with specific inhibitors that prevent this uptake and thereby prevent foam cell formation in vivo.

The plasma concentration of apo E rises in cholesterol-fed animals, and apo E-containing lipoproteins, such as HDL_c, appear in the plasma (59). Even though cholesterol-loaded macrophages can synthesize apo E in vitro (13, 14), one cannot yet conclude that these cells are an important source for plasma HDL_c in vivo. In this regard, it would be instructive to measure the synthesis of apo E in explants of aortas from atherosclerotic animals and humans.

The finding that HDL facilitates cholesterol excretion by macrophages (9, 78) may be relevant to the epidemiologic observation that high levels of plasma HDL are correlated with a reduced frequency of atherosclerotic complications in man (100). The excretion of cholesterol by macrophages in the artery wall in vivo may be limited by the availability of HDL. The concentration of HDL in the arterial wall is unknown, as is the route by which it enters and leaves. Whether an increase in the plasma level of HDL would lead to a higher arterial level of HDL and whether this would speed the removal of cholesterol from macrophages is a question that seems worthy of study.

The scavenging of lipoprotein-cholesterol by macrophages appears to be a protective mechanism that functions to rid the interstitial space of excessive lipoproteins. By this formulation, foam cell formation in atherosclerosis would result when this protective mechanism becomes overwhelmed, either because the amount of plasma lipoprotein-cholesterol that enters the arterial wall is too great for the macrophages to process, or because the ability of the macrophages to excrete cholesterol becomes limited. Entry of lipoproteins into the arterial wall could be controlled by: (*a*) lowering the lipoprotein level in plasma; or (*b*) improving the integrity of the endothelium. Excretion of cholesterol from macrophages might be enhanced by: (*a*) increasing the concentration of HDL in plasma (and presumably its concentration in the arterial wall); or (*b*) improving the ability of HDL to act as a cholesterol acceptor by increasing the efficiency of the LCAT and cholesteryl ester transfer protein reactions that lower the cholesterol content of HDL.

The question arises as to whether macrophages make any contribution to the pathogenesis of atherosclerosis other than by scavenging cholesterol.

Macrophages produce factors that stimulate the growth of smooth muscle cells, which form a major part of the bulk of the atherosclerotic plaque (3, 101). Macrophages also synthesize and secrete lipoprotein lipase (31, 32), which might liberate toxic fatty acids and triglycerides from plasma lipoproteins locally within the artery wall. Macrophages also secrete a host of other biologically active molecules, such as prostaglandins and proteases (102). The influence of macrophage cholesterol accumulation on all of these secretory events needs to be explored. Maleylated albumin, which enters macrophages through the acetyl-LDL receptor (4), stimulates the secretion of several proteases (51). If lipoproteins act similarly, then they might trigger a vicious cycle in vivo in which lipoprotein entry activates macrophages to secrete factors that lead to additional damage to the artery wall, which in turn leads to additional lipoprotein entry, etc.

The Foam Cell in Familial Hypercholesterolemia

The studies of lipoprotein metabolism in macrophages were initiated to explain the paradoxical finding that familial hypercholesterolemia (FH) patients whose cells lack receptors for LDL can nevertheless accumulate lipoprotein-derived cholesteryl esters in macrophages (4). The studies revealed a variety of receptors that might mediate the macrophage uptake of lipoprotein cholesterol in these patients. As mentioned above, only one of these receptors operates on a naturally occurring lipoprotein and that is the receptor for β-VLDL (6, 58). Recent studies of FH in man and in rabbits have begun to provide evidence that particles resembling β-VLDL are present in the circulation of affected individuals and that these particles may be an important source of macrophage cholesteryl esters.

The rabbit studies have dealt with a strain known as Watanabe Heritable Hyperlipidemic (WHHL) rabbits (103). These rabbits have a mutation in the gene for the LDL receptor that is analogous to the mutation in human FH. When present in the homozygous form, this mutation leads to a near complete deficiency of LDL receptors in tissues such as liver, adrenal, and cultured fibroblasts (104–106). As a result of this deficiency, LDL is removed slowly from the circulation and accumulates to massive levels in plasma (107). In these respects the homozygous WHHL rabbits resemble humans with homozygous FH (16).

In addition to binding LDL, the LDL receptor is known to bind particles such as β-VLDL and intermediate density lipoproteins (IDL) that contain apo E as well as apo B (11). IDL particles are remnant lipoproteins formed during the metabolism of VLDL in man and animals on normal diets. In normal rabbits IDL are rapidly removed from the circulation in the liver, apparently by binding to LDL receptors (108). When ^{125}I-labeled VLDL is injected into the circulation of homozygous WHHL rabbits, the VLDL is converted to IDL by lipoprotein lipase, but the IDL is not removed normally from the circulation (108). Hence, these mutant rabbits accumulate cholesterol-rich particles in the VLDL and IDL density classes as well as in LDL (109). Similar findings were made in studies of the turnover of ^{125}I-VLDL in humans with homozygous FH (110). The VLDL and IDL that accumulate in WHHL plasma are similar, though not identical, to the β-VLDL particles that accumulate in cholesterol-fed animals (109), especially to the Fraction II subfraction of these particles (66).

The cholesterol-rich VLDL and IDL particles from WHHL rabbits bind to a receptor on macrophages that appears to be the same as the β-VLDL receptor. As a result of this binding, VLDL and IDL particles from WHHL rabbits stimulate cholesteryl oleate synthesis and storage (Table 3).

The above data raise the possibility that cholesterol-rich VLDL and IDL particles, in addition to modified LDL, may constitute major sources of cholesterol in the atherosclerotic foam cells of FH homozygotes and perhaps of heterozygotes as well (although elevated IDL levels have not been well documented in heterozygotes). Zilversmit (111) proposed that IDL particles and other forms of remnant lipoproteins are the primary cause of atherosclerosis in cholesterol-fed animals; these lipoproteins may contribute to the atherosclerosis in FH as well.

Table 3 Stimulation of cholesteryl ester formation in mouse peritoneal macrophages by lipoproteins from normal and WHHL rabbits[a]

| Source of lipoproteins | Lipoprotein fraction added to medium | Concentration in medium | | [14C]oleate → cholesteryl [14C] oleate (nmol/mg protein) |
		Protein (μg/ml)	Cholesterol (μg/ml)	
—	None	0	0	0.26
Normal rabbit (2% cholesterol diet)	β-VLDL (d < 1.006)	5	50	16.5
		30	300	61.3
Normal rabbit (chow diet)	VLDL (d < 1.006)	60	50	0.24
		360	300	0.30
	IDL (d 1.006–1.019)	40	50	2.4
		240	300	8.9
	LDL (d 1.019–1.063)	40	50	0.30
		240	300	0.90
WHHL rabbit (chow diet)	VLDL (d < 1.006)	14	50	15.0
		84	300	25.4
	IDL (d 1.006–1.019)	20	50	3.7
		120	300	12.6
	LDL (d 1.019–1.063)	40	50	0.15
		240	300	0.68

[a] Each monolayer of mouse peritoneal macrophages received 0.6 ml medium containing 0.2 mM [14C]-oleate bound to albumin and the indicated concentration of the indicated lipoprotein fraction. After incubation for 7.5 hr at 37°C, the cellular content of cholesteryl [14C]oleate was determined by thin layer chromatography (8). The addition of either polyinosinic acid (30 μg/ml) or fucoidin (100 μg/ml) did not inhibit the formation of cholesteryl [14C]oleate in these experiments.

Among FH patients (both heterozygotes and homozygotes), there is considerable variation in the rate of progression of atherosclerosis, despite uniformly elevated LDL levels. The suggestion was made that those FH heterozygotes who have low HDL levels are more susceptible to atherosclerosis than those who have higher HDL levels (112, 113). VLDL, the precursor of IDL, is known to vary inversely with HDL levels. FH heterozygotes with low HDL levels may also have high IDL levels and the high IDL level may be the aggravating factor in atherosclerosis rather than the low HDL level.

The studies of macrophage lipoprotein metabolism have raised many questions concerning the role of lipoproteins in atherosclerosis. Further studies should throw new light on the biochemical mechanisms responsible for foam cell formation.

ACKNOWLEDGMENTS

The authors' experimental work described in this review was supported by a research grant from the National Institutes of Health (HL-20948).

Literature Cited

1. Fowler, S., Shio, H., Haley, N. J. 1979. *Lab. Invest.* 41:372–78
2. Schaffner, T., Taylor, K., Bartucci, E. J., Fischer-Dzoga, K., Beeson, J. H., et al. 1980. *Am. J. Path.* 100:57–80
3. Ross, R., Glomset, J. A. 1976. *N. Engl. J. Med.* 295:369–76, 420–25
4. Goldstein, J. L., Ho, Y. K., Basu, S. K., Brown, M. S. 1979. *Proc. Natl. Acad. Sci USA* 76:333–37
5. Basu, S. K., Brown, M. S., Ho, Y. K., Goldstein, J. L. 1979. *J. Biol. Chem.* 254:7141–46
6. Goldstein, J. L., Ho, Y. K., Brown, M. S., Innerarity, T. L., Mahley, R. W. 1980. *J. Biol. Chem.* 255:1839–48
7. Goldstein, J. L., Hoff, H. F., Ho, Y. K., Basu, S. K., Brown, M. S. 1981. *Arteriosclerosis* 1:210–26
8. Brown, M. S., Goldstein, J. L., Krieger, M., Ho, Y. K., Anderson, R. G. W. 1979. *J. Cell Biol.* 82:597–613
9. Brown, M. S., Ho, Y. K., Goldstein, J. L. 1980. *J. Biol. Chem.* 255:9344–52
10. Glomset, J. A., Norum, K. R. 1973. *Adv. Lipid Res.* 11:1–65
11. Brown, M. S., Kovanen, P. T., Goldstein, J. L. 1981. *Science* 212:628–35
12. Mahley, R. W. 1981. *Diabetes* 30(Suppl. 2):60–65
13. Basu, S. K., Brown, M. S., Ho, Y. K., Havel, R. J., Goldstein, J. L. 1981. *Proc. Natl. Acad. Sci. USA* 78:7545–49
14. Basu, S. K., Ho, Y. K., Brown, M. S., Bilheimer, D. W., Anderson, R. G. W., et al. 1982. *J. Biol. Chem.* 257:9788–95
15. Silverstein, S. C., Steinman, R. M., Cohn, Z. A. 1977. *Ann. Rev. Biochem.* 46:669–722
16. Goldstein, J. L., Brown, M. S. 1982. *Med. Clin. North Am.* 66:335–62
17. Edelson, P. J., Cohn, Z. A. 1976. In *In Vitro Methods in Cell-Mediated and Tumor Immunity,* ed. B. R. Bloom, J. R. David, pp. 333–40. New York: Academic
18. Brown, M. S., Basu, S. K., Falck, J. R., Ho, Y. K., Goldstein, J. L. 1980. *J. Supramol. Struct.* 13:67–81
19. Via, D. P., Dresel, H. A., Gotto, A. M. Jr. 1982. *Circulation* 66:II–37
20. Mahley, R. W., Innerarity, T. L., Weisgraber, K. H., Oh, S. Y. 1979. *J. Clin. Invest.* 64:743–50
21. Van Berkel, T. J. C., Nagelkerke, J. F., Kruijt, J. K. 1981. *FEBS Lett.* 132:61–66
22. Fogelman, A. M., Schechter, I., Seager, J., Hokom, M., Child, J. S., et al. 1980. *Proc. Natl. Acad. Sci. USA* 77:2214–18
23. Traber, M. G., Kayden, H. J. 1980. *Proc. Natl. Acad. Sci. USA* 77:5466–70
24. Traber, M. G., Defendi, V., Kayden, H. J. 1981. *J. Exp. Med.* 154:1852–67
25. Henriksen, T., Mahoney, E. M., Stein-

berg, D. 1981. *Proc. Natl. Acad. Sci. USA* 78:6499–6503

26. Imber, M. J., Pizzo, S. V., Johnson, W. J., Adams, D. O. 1982. *J. Biol. Chem.* 257:5129–35

27. Fogelman, A. M., Seager, J., Haberland, M. E., Hokom, M., Tanaka, R., et al. 1982. *Proc. Natl. Acad. Sci. USA* 79:922–26

28. Pitas, R. E., Innerarity, T. L., Weinstein, J. N., Mahley, R. W. 1981. *Arteriosclerosis* 1:177–85

29. Goldstein, J. L., Brown, M. S. 1977. *Ann. Rev. Biochem.* 46:897–930

30. Stein, O., Stein, Y. 1980. *Biochim. Biophys. Acta* 620:631–35

31. Mahoney, E. M., Khoo, J. C., Steinberg, D. 1982. *Proc. Natl. Acad. Sci. USA* 79:1639–42

32. Chait, A., Iverius, P.-H., Brunzell, J. D. 1982. *J. Clin. Invest.* 69:490–93

33. Fogelman, A. M., Haberland, M. E., Seager, J., Hokom, M., Edwards, P. A. 1981. *J. Lipid Res.* 22:1131–41

34. Knight, B. L., Soutar, A. K. 1982. *Eur. J. Biochem.* 125:407–13

35. Basu, S. K., Goldstein, J. L., Anderson, R. G. W., Brown, M. S. 1976. *Proc. Natl. Acad. Sci. USA* 73:3178–82

36. Haberland, M. E., Fogelman, A. M., Edwards, P. A. 1982. *Proc. Natl. Acad. Sci. USA* 79:1712–16

37. Basu, S. K., Ho, Y. K., Brown, M. S., Goldstein, J. L. Unpublished observations

38. Mahley, R. W., Weisgraber, K. H., Innerarity, T. L., Windmueller, H. G. 1979. *Proc. Natl. Acad. Sci. USA* 76:1746–50

39. Chao, Y. S., Chen, G. C., Windler, E., Kane, J. P., Havel, R. J. 1979. *Fed. Proc.* 38:896 (Abstr.)

40. Samuelsson, B., Goldyne, M., Granström, E., Hamberg, M., Hammarström, S., Malmsten, C. 1978. *Ann. Rev. Biochem.* 47:997–1029

41. Stossel, T. P., Mason, R. J., Smith, A. L. 1974. *J. Clin. Invest.* 54:638–45

42. Smith, J. B., Ingerman, C. M., Silver, M. J. 1976. *J. Lab. Clin. Med.* 88:167–72

43. Schleicher, E., Deufel, T., Wieland, O. H. 1981. *FEBS Lett.* 129:1–4

44. Witzum, J. L., Mahoney, E. M., Branks, M. J., Fisher, M., Elam, R., et al. 1982. *Diabetes* 31:283–91

45. Kim, H.-J., Kurup, I. V. 1982. *Metabolism* 31:348–53

46. Gonen, B., Baenziger, J., Schonfeld, G., Jacobson, D., Farrar, P. 1981. *Diabetes* 30:875–78

47. Sasaki, J., Cottam, G. L. 1982. *Biochem. Biophys. Res. Comm.* 104:977–83

48. Siegel, R. C. 1979. *Int. Rev. Connect. Tissue Res.* 8:73–118

49. Stahmann, M. A., Spencer, A. K. 1977. *Biopolymers* 16:1299–306

50. Henriksen, T., Mahoney, E. M., Steinberg, D. 1983. *Ann. NY Acad. Sci.* 401:102–6

51. Johnson, W. J., Pizzo, S. V., Imber, M. J., Adams, D. O. 1982. *Science* 218:574–76

52. DeClercq, E., Eckstein, F., Merigan, T. C. 1970. *Ann. NY Acad. Sci.* 173:444–61

53. Iverius, P.-H. 1972. *J. Biol. Chem.* 247:2607–13

54. Kielian, M. C., Cohn, Z. A. 1982. *J. Cell Biol.* 93:875–82

55. Robinson, H. C., Horner, A. A., Hook, M., Ogren, S., Lindahl, U. 1978. *J. Biol. Chem.* 253:6687–93

56. Basu, S. K., Horner, A. A., Brown, M. S., Goldstein, J. L. Unpublished observations

57. Srinivasan, S. R., Dolan, P., Radhakrishnamurthy, B., Parganokar, P. S., Berenson, G. S. 1975. *Biochim. Biophys. Acta* 388:58–70

58. Mahley, R. W., Innerarity, T. L., Brown, M. S., Ho, Y. K., Goldstein, J. L. 1980. *J. Lipid Res.* 21:970–80

59. Mahley, R. W. 1979. *Atherosclerosis Rev.* 5:1–34

60. Brown, M. S., Goldstein, J. L., Fredrickson, D. S. 1983. In *The Metabolic Basis of Inherited Disease,* ed. J. B. Stanbury, J. B. Wyngaarden, D. S. Fredrickson, J. L. Goldstein, M. S. Brown, pp. 655–671. New York: McGraw-Hill

61. Mahley, R. W. 1982. *Med. Clin. North Am.* 66:375–402

62. Van Lenten, B. J., Fogelman, A. M., Hokom, M. M., Benson, L., Haberland, M. E., et al. 1983. *J. Biol. Chem.* In press

63. Bilheimer, D. W., Ho, Y. K., Brown, M. S., Anderson, R. G. W., Goldstein, J. L. 1978. *J. Clin. Invest.* 61:678–96

64. Innerarity, T. L., Mahley, R. W. 1980. In *Drugs Affecting Lipid Metabolism,* ed. R. Fumagalli, D. Kritchevsky, R. Paoletti, pp. 53–60. Amsterdam: Elsevier/North-Holland Biochemical

65. Innerarity, T. L., Pitas, R. E., Mahley, R, W. 1982. *Arteriosclerosis* 2:114–24

66. Fainaru, M., Mahley, R. W., Hamilton, R. L., Innerarity, T. L. 1982. *J. Lipid Res.* 23:702–14

67. Gianturco, S. H., Bradley, W. A., Gotto, A. M. Jr., Morrisett, J. D.,

Peavy, D. L. 1982. *J. Clin. Invest.* 70:168–78
68. Gianturco, S. H., Gotto, A. M. Jr, Hwang, S. L. C., Karlin, J. B., Lin, A. H. Y. et al. 1983. *J. Biol. Chem.* In press
69. Gianturco, S. H., Brown, F. B., Gotto, A. M. Jr., Bradley, W. A. 1982. *J. Lipid Res.* 23:984–93
70. Ross, A. C., Zilversmit, D. B. 1977. *J. Lipid Res.* 18:169–81
71. Kushwaha, R. W., Hazzard, W. R. 1978. *Biochim. Biophys. Acta* 528: 176–89
72. Kovanen, P. T., Brown, M. S., Basu, S. K., Bilheimer, D. W., Goldstein, J. L. 1981. *Proc. Natl. Acad. Sci. USA* 78:1396–400
73. Weisgraber, K. H., Innerarity, T. L., Mahley, R. W. 1978. *J. Biol. Chem.* 253:9053–62
74. Woodard, J. F., Srinivasan, S. R., Zimny, M. L., Radhakrishnamurthy, B., Berenson, G. S. 1976. *Lab. Invest.* 34:516–21
75. Werb, Z., Cohn, Z. A. 1972. *J. Exp. Med.* 135:21–44
76. Werb, Z., Cohn, Z. A. 1971. *J. Exp. Med.* 134:1545–69
77. Ho, Y. K., Goldstein, J. L., Brown, M. S. Unpublished observations
78. Ho, Y. K., Brown, M. S., Goldstein, J. L. 1980. *J. Lipid. Res.* 21:391–98
79. Krieger, M., Brown, M. S., Faust, J. R., Goldstein, J. L. 1978. *J. Biol. Chem.* 253:4093–4101
80. Goldstein, J. L., Faust, J. R., Dygos, J. H., Chorvat, R. J., Brown, M. S. 1978. *Proc. Natl. Acad. Sci. USA* 75:1877–81
81. Bailey, J. M. 1973. In *Atherogenesis: Initiating Factors,* Ciba Found. Symp. 12 (NS):63–92
82. Stein, Y., Glangeaud, M. C., Fainaru, M., Stein, O. 1975. *Biochim. Biophys. Acta* 380:106–18
83. Bates, S. R., Rothblat, G. H. 1974. *Biochim. Biophys. Acta* 360:38–55
84. Oram, J. F., Albers, J. J., Cheung, M. C., Bierman, E. L. 1981. *J. Biol. Chem.* 256:8348–56
85. Chajek, T., Fielding, C. J. 1978. *Proc. Natl. Acad. Sci. USA* 75:3445–49
86. Phillips, M. C., McLean, L. R., Stoudt, G. W., Rothblat, G. H. 1980. *Atherosclerosis* 36:409–22
87. Utermann, G., Langenbeck, U., Beisiegel, U., Weber, W. 1980. *Am. J. Hum. Genet.* 32:339–47
88. Zannis, V. I., Breslow, J. L. 1981. *Biochemistry* 20:1033–41
89. Hamilton, R. L. 1978. In *Disturbances in Lipid and Lipoprotein Metabolism,* ed. J. Dietschy, A. M. Gotto Jr., J. A.

Ontko, pp. 155–71. Baltimore: American Physiology Society
90. Small, D. M. 1977. *N. Engl. J. Med.* 297:873–77, 924–29
91. Havel, R. J., Goldstein, J. L., Brown, M. S. 1980. In *Metabolic Control and Disease* ed. P. K. Bondy, L. E. Rosenberg, pp. 393–494. Philadelphia: Saunders. 8th ed.
91a. Basu, S. K., Goldstein, J. L., Brown, M. S. 1983. *Science* In press
92. Tartakoff, A., Vassalli, P. 1978. *J. Cell Biol.* 79:694–707
93. Hui, D. Y., Innerarity, T. L., Mahley, R. W. 1981. *J. Biol. Chem.* 256:5646–55
94. Smith, E. B. 1974. *Adv. Lipid Res.* 12:1–49
95. St. Clair, R. W. 1976. *Atherosclerosis Rev.* 1:61–117
96. Day, A. J., Phil, D., Wahlqvist, M. L. 1968. *Circ. Res.* 23:779–88
97. Hashimoto, S., Dayton, S., Alfin-Slater, R. B., Bui, P. T., Baker, N., et al. 1974. *Circ. Res.* 34:176–83
98. Wahlqvist, M. L., Day, A. L., Tume, R. K. 1969. *Circ. Res.* 24:123–30
99. Pitas, R. E., Innerarity, T. L., Mahley, R. W. 1983. *Arteriosclerosis* 3:2–12
100. Miller, G. J. 1980. *Ann. Rev. Med.* 31:97–108
101. Ross, R. 1981. *Arteriosclerosis* 1:293–311
102. Nathan, C. F., Murray, H. W., Cohn, Z. A. 1980. *N. Engl. J. Med.* 303:622–26
103. Watanabe, Y. 1980. *Atherosclerosis* 36:261–68
104. Tanzawa, K., Shimada, Y., Kuroda, M., Tsujita, Y., Arai, M., et al. 1980. *FEBS Lett.* 118:81–84
105. Kita, T., Brown, M. S., Watanabe, Y., Goldstein, J. L. 1981. *Proc. Natl. Acad. Sci. USA* 78:2268–72
106. Attie, A. D., Pittman, R. C., Watanabe, Y., Steinberg, D. 1981. *J. Biol. Chem.* 256:9789–92
107. Bilheimer, D. W., Watanabe, Y., Kita, T. 1982. *Proc. Natl. Acad. Sci. USA* 79:3305–09
108. Kita, T., Brown, M. S., Bilheimer, D. W., Goldstein, J. L. 1982. *Proc. Natl. Acad. Sci. USA* 79:5693–97
109. Havel, R. J., Kita, T., Kotite, L., Kane, J. P., Hamilton, R. L., et al. 1982. *Arteriosclerosis* 2:467–74
110. Soutar, A. K., Myant, N. B., Thompson, G. R. 1982. *Atherosclerosis* 43:217–31
111. Zilversmit, D. B. 1979. *Circulation* 60:473–85
112. Streja, D., Steiner, G., Kwiterovich, P. O. 1978. *Ann. Intern. Med.* 89:871–80
113. Hirobe, K., Matsuzawa, Y., Ishikawa, K., Tarui, S., Yamamoto, A., et al. 1982. *Atherosclerosis* 44:201–10

Ann. Rev. Biochem. 1983. 53:263–300

DYNAMICS OF PROTEINS: ELEMENTS AND FUNCTION

M. Karplus and J. A. McCammon

Department of Chemistry, Harvard University, Cambridge, Massachusetts 02138

Department of Chemistry, Fleming Building, University of Houston, Houston, Texas 77004

CONTENTS

INTRODUCTION 263
OVERVIEW 265
DYNAMICS METHODOLOGY 268
   Molecular Dynamics 270
   Stochastic Dynamics 270
   Harmonic Dynamics 272
   Activated Dynamics 272
   Simplified Model Dynamics 273
ATOMIC FLUCTUATIONS 273
   Mean-Square Fluctuations and Temperature Factors 273
   Time-Dependence: Local and Collective Effects 277
   Biological Function 278
SIDECHAIN MOTIONS 279
   Tyrosines in PTI 279
   Ligand-Protein Interaction in Myoglobin 283
   Exterior Sidechain and Loop Motions 287
RIGID BODY MOTIONS 288
   Hinge Bending 288
   Quaternary Structural Change 290
α-HELIX MOTION: HARMONIC AND SIMPLIFIED
     MODEL DYNAMICS 291
PERSPECTIVE 292

INTRODUCTION

The classic view of proteins has been static in character, primarily because of the dominant role of the information provided by high-resolution X-ray crystallography for these very complex systems. The intrinsic beauty and

263

0066-4154/83/0701-0263$02.00

remarkable detail of the drawings of protein structures led to an image in which each protein atom is fixed in place; an article on lysozyme by Phillips (1), the books by Dickerson & Geis (2), and by Perutz & Fermi (3), and the review by Richardson (4) give striking examples. Stating clearly the static viewpoint, Tanford (5) suggested that as a result of packing considerations "the structure of native proteins must be quite rigid." Phillips (6) wrote recently ". . . the period 1965–75 may be described as the decade of the rigid macromolecule. Brass models of double helical DNA and a variety of protein molecules dominated the scene and much of the thinking."

Most attempts to explain enzyme function have been based on the examination of the average structure obtained from crystallography; e.g. the high specificity of enzymes for their substrates has been likened to the complementarity of two pieces of a jigsaw puzzle. Cases in which conformational changes were known from X-ray data to be induced by ligand or substrate binding (e.g. the allosteric transition in hemoglobin) were generally treated as abrupt transitions between otherwise static structures.

The static view of protein structure is being replaced by a dynamic picture. The atoms of which the protein is composed are recognized to be in a state of constant motion at ordinary temperatures. From the X-ray structure of a protein, the average atomic positions are obtained, but the atoms exhibit fluidlike motions of sizable amplitudes around these average positions. Crystallographers have acceded to this viewpoint and have come so far as to sometimes emphasize the parts of a protein molecule they do not see in a crystal structure as evidence of motion or disorder (7).

The new understanding of protein dynamics subsumes the static picture in that use of the average positions still allows discussion of many aspects of protein function in the language of structural chemistry. However, the recognition of the importance of fluctuations opens the way for more sophisticated and accurate interpretations of protein function. The dynamic picture incorporates a variety of phenomena known to be involved in the biological activity of proteins, but whose detailed description was not possible under the static view. Transient packing defects due to atomic motions play an essential role in the penetration of oxygen to the heme-binding site in myoglobin and hemoglobin (8, 9). Functional interactions of flexible ligands with their binding sites often require conformational adjustments in both the ligand and the binding protein; the ligands involved include drugs, hormones, and enzyme substrates (10, 11). The structural changes in the binding proteins regulate the activity of many of these molecules through induced fit and allosteric effects (12–14). The chemical transformations of substrates by enzymes typically involve significant atomic displacements in the enzyme-substrate complexes. The mechanisms and rates of such transformations are sensitive to the dynamic properties of these systems; for example, the differences in the vibrational modes of the initial and transition states affect the free energies of activation and catalytic rates (13–15). Electron transfer processes may depend strongly on vibronic coupling and fluctuations that alter the distance between the donor and acceptor (16–19). The relative motion of distinct structural domains is important in the activities of myosin (20–22), other enzymes (23–25), and antibody molecules (26–28), as well as in the assembly of supramolecular structures such as viruses (29).

Any attempt to understand the function of proteins requires an investigation of the dynamics of the structural fluctuations and their relation to activity and conformational change. The review deals primarily with theoretical approaches to protein dynamics. This rapidly developing field of study is founded on efforts to supplement our understanding of protein structure with concepts and techniques from modern chemical theory, including reaction dynamics and quantum and statistical mechanics. From a knowledge of the potential energy surface, the forces on the component atoms can be calculated and used to determine phase space trajectories for a protein molecule at a given temperature. Such molecular dynamics simulations, which have been successfully applied to gases and liquids containing a large number of atoms, provide information concerning the thermodynamic properties and the time-dependence of processes in the system of interest (30). More generally, statistical mechanical techniques have succeeded very well in characterizing molecular motion and chemical reaction in condensed phases (31–33). The application of these methods to proteins is natural in that proteins contain many atoms, are densely packed, and function typically in liquid environments (34).

In this review we present first a brief overview of the wide range of motions that occur in proteins. We then outline the methods that can be used to study the various motions, and review the results obtained so far. We emphasize the role of the motions in the biological activity and compare with experiments where data exist. We conclude with an outlook for the future of this new and exciting field.

A number of reviews of the theory of protein dynamics has already appeared (35–39a). Specific aspects of protein dynamics, including the rapidly growing body of experimental data, have been reviewed (36, 40–50). The proceedings of a Ciba Foundation meeting (March 2–4, 1982) devoted to *Protein Motion and Its Relation to Function* are to be published (51). Two detailed reviews surveyed protein folding recently from the structural (52) and dynamic viewpoints (53).

OVERVIEW

Globular proteins have a wide variety of internal motions. They can be classified for convenience in terms of their amplitude, energy, and time scale, or by their structural type. Table 1 lists the ranges involved for these quantities; Careri, Fasella & Gratton (54) give a complementary summary. One expects an increase in one quantity (e.g. the amplitude of the fluctuation) to correspond to an increase in the others (e.g. a larger energy and a longer time scale). This is often true, but not always. Some motions are slow because they are complex, involving the correlated displacements of many atoms. An example might be partial-to-total unfolding transitions, in which the correlation of amplitude, energy, and time scale is expected to hold. However, in much more localized events, often involving small displacements of a few atoms, the motion is slow because of a high activation barrier; an example is the aromatic ring flips in certain proteins (55–60). In this case the macroscopic rate constant can be very slow ($k \sim 1$ sec^{-1} at 300°K), not because an individual event is slow (a ring flip occurs in $\sim 10^{-12}$ sec), but because the probability is very small ($\sim 10^{-12}$) that a ring has sufficient energy to get over an activation barrier on the order of 16 Kcal.

At any given time, a typical protein exhibits a wide variety of motions; they range from irregular elastic deformations of the whole protein driven by collisions with solvent molecules to chaotic librations of interior groups driven by random collisions with neighboring atoms in the protein. Considering only typical motions at physiological temperatures, the smallest effective dynamical units in proteins are those that behave nearly as rigid bodies because of their internal covalent bonding. Examples include the phenyl group in the side chain of phenylalanine, the isopropyl group in the side chains of valine or leucine, and the amide groups of the protein backbone. Except for the methyl rotations in the isopropyl group, these units display only relatively small internal motions owing to the high energy cost associated with deformations of bond lengths, bond angles, or dihedral angles about multiple bonds. The important motions in proteins involve relative displacements of such groups associated with torsional motions about the

Table 1 Classification of internal motions of globular proteins

Scales of motions (300°K)	
Amplitude	0.01 to 100 Å
Energy	0.1 to 100 Kcal
Time	10^{-15} to 10^3 sec

Types of motions	
Local	Atom fluctuations, side chain oscillations, loop and "arm" displacements
Rigid body	Helices, domains, subunits
Large-scale	Opening fluctuation, folding and unfolding
Collective	Elastic-body modes, coupled atom fluctuations, soliton and other non-linear motional contributions

rotationally permissive single bonds that link the groups together. High frequency vibrations occur within the local group, but these are not of primary importance in the relative displacements.

Most groups in a protein are tightly encaged by atoms of the protein or of the surrounding solvent. At very short times ($\lesssim 10^{-13}$ s), such a group may display a rattling motion in its cage, but such motions are of relatively small amplitude ($\lesssim 0.2$ Å). More substantial displacements of the group occur over longer time intervals; these displacements involve concomitant displacements of the cage atoms. Broadly speaking, such "collective" motions may have a local or rigid-body character. The former involves changes of the cage structure and relative displacements of neighboring groups, while the latter involves relative displacements of different regions of the protein but only small changes on a local scale.

The presence of such motional freedom implies that a native protein at room temperature samples a range of conformations. Most are in the general neighborhood of the average structure, but at any given moment an individual protein molecule is likely to differ significantly from the average structure. This in no way implies that the X-ray structure, which corresponds to the average in the crystal, is not important. Rather, it suggests that fluctuations about that average are likely to play a role in protein function. In a protein, as in any polymeric system in which rigidity is not supplied by covalent cross-links, significant fluctuations cannot be avoided; they must, therefore, have been taken into account in the evolutionary development.

Although the existence of the fluctuations is now well established, our understanding of their biological role in specific areas is incomplete. Both conformational and energy fluctuations with local to global character are expected to be important. In a protein, as in other nonrigid condensed systems, structural changes arise from correlated fluctuations. Perturbations, such as ligand binding, that produce tertiary or quaternary alterations do so by introducing forces that bias the fluctuations in such a way that the protein makes a transition from one structure to another. Alternatively, the fluctuations can be regarded as searching out the path or paths along which the transition takes place.

In considering the internal motions of proteins, one must separate the dynamic from the thermodynamic aspects; in the latter, the presence of flexibility is important (e.g. entropy of binding), while in the former the directionality and time scale play a role. Another way of categorizing the two is that in the second, equilibrium behavior is the sole concern, while in the first, the dynamics is the essential element. In certain cases, some aspects of the dynamics may be unimportant because they proceed on a time scale that is much faster than the phenomenon of interest. An example might be the fast local relaxation of atoms involved in a much slower hinge bending motion; here only the time scale of the latter would be expected to be involved in determining a rate process, though the nature of the former would be of considerable interest. In other situations, the detailed aspects of the atomic fluctuations may be the essential factor.

DYNAMICS METHODOLOGY

To study theoretically the dynamics of a macromolecular system, one must have a knowledge of the potential energy surface, the energy of the system as a function of the atomic coordinates. The potential energy can be used directly to determine the relative stabilities of the different possible structures of the system (30). The forces acting on the atoms of the systems are obtained from the first derivatives of the potential with respect to the atom positions. These forces can be used to calculate dynamical properties of the system, e.g. by solving Newton's equations of motion to determine how the atomic positions change with time (30, 31, 61). From the second derivatives of the potential surface, the force constants for small displacements can be evaluated and used to find the normal modes (62); this serves as the basis for an alternative approach to the dynamics in the harmonic limit (62, 63).

Although quantum mechanical calculations can provide potential surfaces for small molecules, empirical energy functions of the molecular mechanics type (64–67) are the only possible source of such information for proteins and their solvent surroundings. Since most of the motions that occur at ordinary temperatures leave the bond lengths and bond angles of the polypeptide chains near their equilibrium values, which appear not to vary significantly throughout the protein (e.g. the standard dimensions of the peptide group first proposed by Pauling et al in 1951; 68), the energy-function representation of the bonding can be hoped to have an accuracy on the order of that achieved in the vibrational analysis of small molecules. Where globular proteins differ from small molecules is that the contacts among nonbonded atoms play an essential role in the potential energy of the folded or native structure. From the success of the pioneering conformational studies of Ramachandran et al in 1963 (69), which used hardsphere nonbonded radii, it is likely that relatively simple functions (Lennard-Jones nonbonded potentials supplemented by a special hydrogen-bonding term and electrostatic interactions) can adequately describe the interactions involved.

The energy function used for proteins are generally composed of terms representing bonds, bond angles, torsional angles, van der Waals interactions, electrostatic interactions, and hydrogen bonds. The resulting expression has the form (64–67, 70):

$$E(\mathbf{R}) = \frac{1}{2} \sum_{\text{bonds}} K_b(b - b_0)^2 + \frac{1}{2} \sum_{\substack{\text{bond} \\ \text{angles}}} K_\theta (\theta - \theta_0)^2 \qquad\qquad 1.$$

$$+ \frac{1}{2} \sum_{\text{torsional}} K_\phi [1 + \cos (n\phi - \delta)]$$

$$+ \sum_{\substack{nb \text{ pairs} \\ r < 8 \text{ Å}}} \frac{A}{r^{12}} - \frac{C}{r^6} + \frac{q_1 q_2}{Dr} + \sum_{\substack{\text{H} \\ \text{bonds}}} \frac{A'}{r^{12}} - \frac{C'}{r^{10}} \; .$$

The energy is a function of the Cartesian coordinate set, $\mathbf{R}$, specifying the positions of all the atoms involved, but the calculation is carried out by first evaluating the internal coordinates for bonds (b), bond angles (θ), dihedral angles (ϕ), and interparticle distances (r) for any given geometry, $\mathbf{R}$, and using them to evaluate the contributions to Equation 1, which depends on the bonding energy parameters K_b, K_θ, K_ϕ, Lennard-Jones parameters A and C, atomic charges q_i, dielectric constant D, hydrogen-bond parameters A' and C', and geometrical reference values b_0, θ_0, n, and δ. For most protein atoms an extended atom representation is used; i.e., one extended atom replaces a nonhydrogen atom and any hydrogens bonded to it. However, although the earliest studies employed the extended atom representation for all hydrogens, present calculations treat hydrogen-bonding hydrogens explicitly and generally use a more accurate function to represent hydrogen bonding interactions (e.g. angular terms are included) than that given in Equation 1 (70).

Given a potential-energy function, one may take any of a variety of approaches to study protein dynamics. The most exact and detailed information is provided by molecular-dynamics simulations, in which one uses a computer to solve the Newtonian equations of motion for the atoms of the protein and any surrounding solvent (70–73). With currently available computers, it is possible to simulate the dynamics of small proteins for up to a few hundred ps. Such periods are long enough to characterize completely the librations of small groups in the protein and to determine the dominant contributions to the atomic fluctuations. To study slower and more complex processes in proteins, it is generally necessary to use methods other than straightforward molecular dynamics simulation. A variety of dynamical approaches, such as stochastic dynamics (74–78), harmonic dynamics (63, 79–81), and activated dynamics (59, 82–86), can be introduced to study particular problems.

Molecular Dynamics

To begin a dynamical simulation, one must have an initial set of atomic coordinates and velocities. These are obtained from the X-ray coordinates of the protein by a preliminary calculation that serves to equilibrate the system (72, 73). The X-ray structure is first refined using an energy-minimization algorithm to relieve local stresses caused by nonbonded atomic overlaps, bond length distortions, etc. The protein atoms are then assigned velocities at random from a Maxwellian distribution corresponding to a low temperature, and a dynamical simulation is performed for a period of a few ps. The equilibration is continued by alternating new velocity assignments, chosen from Maxwellian distributions corresponding to successively increased temperatures, with similar intervals of dynamical relaxation. The temperature, T, for this microcanonical ensemble is measured in terms of the mean kinetic energy for the system composed of N atoms as:

$$\frac{1}{2} \sum_{i=1}^{N} m_i <v_i^2> = \frac{3}{2} N k_B T. \qquad 2.$$

In this equation, m_i and $<v_i^2>$ are the mass and average velocity squared of the ith atom, and k_B is the Boltzmann constant. Any residual overall translational and rotational motion can be removed to simplify analysis of the subsequent conformational fluctuations. The equilibration period is considered finished when no systematic changes in the temperature are evident over a time of about 10 ps (slow fluctuations could be confused with continued relaxation over shorter intervals). It is necessary also to check that the atomic momenta obey a Maxwellian distribution and that different regions of the protein have the same average temperature. The actual dynamical simulation results (coordinates and velocities for all the atoms as a function of time) for determining the equilibrium properties of the protein are then obtained by continuing to integrate the equations of motion for the desired length of time.

Several different algorithms for integrating the equations of motion in Cartesian coordinates are used in protein molecular dynamics calculations. Most common are the Gear predictor-corrector algorithm, familiar from small molecule trajectory calculations (72) and the Verlet algorithm, widely used in statistical mechanical simulations (87).

Stochastic Dynamics

In certain cases it is advantageous to simplify the dynamical treatment by separating the system under study into two parts. One part is that whose dynamics are to be examined and the other serves as a heat bath for the first;

this could be a protein in a solvent or one portion of a protein with the surrounding protein serving as the heat bath. In such an analysis (e.g. of a tyrosine sidechain in a protein) the displacement of the part whose dynamics is to be studied relative to its neighbors is presumed to be analogous to molecular diffusion in a liquid or solid. The allowed range of motion can be characterized by an effective potential-energy function termed the "potential of mean force" (30, 72); this potential corresponds to the free energy of displacement of the elements being studied in the average field due to surrounding bath atoms. The motion of the group under study is determined largely by the time variation of its nonbonded interactions with the neighboring atoms. These interactions produce randomly varying forces that act to speed or slow the motion of the group in a given direction. In favorable cases, these dynamical effects can be represented by a set of Langevin equations of motion (30, 72, 74). For a particle in one dimension, we can write:

$$m\frac{d^2x}{dt^2} = F(x) - f\frac{dx}{dt} + R(t) \qquad\qquad 3.$$

Here, m and x are the mass and position of the particle, respectively, and t is the time; thus, the term on the left is simply the acceleration of the particle. The term $F(x)$ represents the systematic force on the particle derived from the potential of mean force. The terms $-f dx/dt$ and $R(t)$ represent the effects of the varying forces caused by the bath acting on the particle; the first term is the average frictional force caused by the motion of the particle relative to its surroundings (f is the friction coefficient), and $R(t)$ represents the remaining randomly fluctuating force. The Langevin equation and its generalized forms are phenomenological in character but they are consistent with more detailed models for the atomic dynamics.

The Langevin equation also provides a useful focal point in the discussion of large-scale motions (88, 89). For displacements of whole sections of polypeptide chain away from protein surface (local denaturation), the terms corresponding to the one on the left of Equation 3 are typically negligible in comparison to the others (78). The motion then has no inertial character and the chain displacements have the particularly erratic character of Brownian motion. For elastic deformations of the overall protein shape, such as those involved in interdomain or hinge-bending motions, the potential of mean force may have a simple Hooke's law or springlike character (88). Finally, the larger-scale structural changes involved in protein folding (e.g. the coming together of two helices connected by a coil region to form part of the native structure) are also likely to have Brownian character (90, 91).

Harmonic Dynamics

Harmonic dynamics provides an alternative approach to the dynamics of a protein or one of its constituent elements (e.g. an α-helix). Early attempts to examine dynamical properties of proteins or their fragments used the harmonic approximation. They were motivated by vibrational spectroscopic studies (92), in which the calculation of normal mode frequencies from empirical potential functions has long been a standard step in the assignment of infrared spectra (62). One assumes that the vibrational displacements of the atoms from their equilibrium positions are small enough that the potential energy can be approximated as a sum of terms that are quadratic in the displacements. The coefficients of these quadratic terms form a matrix of force constants which, together with the atomic masses, can be used to set up a matrix equation for the vibrational modes of the molecule (62). For a molecule composed of N atoms, $3N$-6 eigenvalues provide the internal vibrational frequencies of the molecule; the associated eigenvectors give the directions and relative amplitudes of the atomic displacements in each normal mode.

Although the harmonic model may be incomplete because of the contribution of anharmonic terms to the potential energy (Equation 1), it is nevertheless of considerable importance because it serves as a first approximation for which the theory is highly developed. Further, the harmonic model is essential for some quantum mechanical treatments of vibrational contributions to the heat capacity and free energy (81, 93) and for certain approaches to unimolecular reactions (94).

Activated Dynamics

Enzyme catalyzed reactions generally involve some processes in which the rate is limited by an energy barrier. In many cases the phenomenological time scale of such activated events is a microsecond or longer. Such processes that are intrinsically fast but occur rarely (i.e. with an average frequency much less than 10^{11} sec^{-1}) are not observed often enough for adequate characterization in an ordinary molecular dynamics simulation. To study such processes, alternative dynamical methods can be employed.

It is often possible to identify the particular character of the structural change involved (e.g. the reaction path) and then to approximate the associated energy changes. In the adiabatic-mapping approach, one calculates the minimized energy of the protein consistent with a given structural change (57, 95, 96). Minimization allows the remainder of the protein to relax in response to the assumed structural change, so that the resulting energy provides a rough approximation to the potential of mean force. Accurate potentials of mean force can be calculated by means of specialized molecular-dynamics calculations (59), but the computational requirements are greater. To analyze the time-dependence of the process, the potential of mean force is incorporated into a model for the dynamics such as the familiar transition state theory (57, 83). A more detailed understanding of the process can be obtained by analyzing trajectories chosen to sample the barrier region (59, 85, 86). The trajectory analysis displays the space and time correlations of the atomic motions involved and provides experimentally accessible quantities such as rate constants and activation energies.

Simplified Model Dynamics

To simulate processes that are intrinsically complicated (i.e. that involve the sampling of many configurations), it is sometimes possible to use simplified models for the structure and energetics of the protein. In one model of this kind, each residue in the protein is represented by a single interaction center and these centers are linked by virtual bonds (97, 98). The energy function for this model is obtained by averaging interresidue interactions over all the local atomic configurations within each residue (78, 98–100). Thus, the model incorporates the assumption of separated time scales for local and overall chain motions. The reduced number of degrees of freedom allows rapid calculation of the energy and forces, so that significantly longer dynamical simulations are possible than with a more detailed model. Such an approach may be particularly useful for studying local unfolding or folding of proteins and their secondary structural elements (78).

ATOMIC FLUCTUATIONS

Figure 1 gives a qualitative picture of the fluctuations observed in the molecular dynamics simulation of the basic pancreatic trypsin inhibitor (PTI), a small protein with 58 amino acids and 454 heavy atoms; only the α-carbon atoms plus the three disulfide bonds are shown. The left-hand drawing represents the X-ray structure and the right-hand drawing an instantaneous picture of the equilibrated structure after 3 ps (71). The two structures are very similar, but there are small differences throughout. The largest displacements appear in the C-terminal end, which interacts with a neighboring molecule in the crystal, and in the loop in the lower left, which has rather weak interactions with the rest of the molecule. Corresponding behavior and deviations from the X-ray structure would be observed in "snap shots" taken at any other time during the simulation.

Mean-Square Fluctuations and Temperature Factors

A more quantitative measure of the motions is obtained from the mean-square fluctuations of the atoms from their average positions. These can be

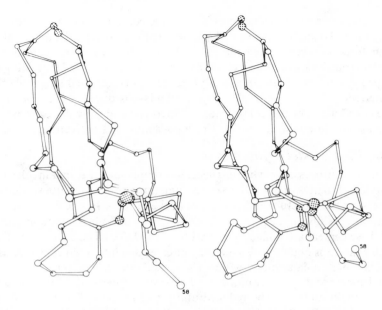

Figure 1   Drawing of α-carbon skeleton plus S–S bonds of PTI; left-hand drawing is the X-ray structure and right-hand drawing is a typical "snapshot" during the simulation.

related to the atomic temperature or Debye-Waller factors, B, determined in an X-ray diffraction study of a protein crystal (101–105). The mean-square positional fluctuation, $<\Delta r^2>_{\text{dyn}}$, with the assumption of isotropic and harmonic motion can be written:

$$<\Delta r^2>_{\text{dyn}} = \frac{3B}{8\pi^2} - <\Delta r^2>_{\text{dis}}. \qquad 4.$$

$<\Delta r^2>_{\text{dis}}$ is the contribution to B from lattice disorder and other effects that are difficult to evaluate experimentally. For a number of proteins at ambient temperatures (101–105), the measured value of $(3B/8\pi^2)$ averaged over all of the nonsurface atoms of the protein is in the range 0.48–0.58 Å^2. Comparison of this result with the mean value of $<\Delta r^2>_{\text{dyn}}$ from protein simulations (0.28–0.36 Å^2; 101, 106, 107) suggests that the nonmotional contribution to the B-factor $<\Delta r^2>_{\text{dis}}$, is in the range 0.20–0.25 Å^2. The only experimental estimate of $<\Delta r^2>_{\text{dis}}$ is from Mössbauer data for the heme iron in myoglobin (102); for that one atom a somewhat smaller value (0.14 Å^2) was obtained. Thus, in the cases examined, approximately half of the experimental B-factor is associated with thermal fluctuations in the atomic positions and half with other sources. However, some protein crystals, particularly those with a high percentage of water, appear to have a larger disorder contribution (109).

There is generally an increase in the magnitude of the experimental and theoretical fluctuations with distance from the center of the molecule. The magnitudes of the rms fluctuations range from ~0.4 Å for backbone atoms to ~1.5 Å for the ends of long sidechains. The hydrogen-bonded secondary structural elements (α-helices, β-sheets) tend to have smaller fluctuations than the random coil parts of the protein (106, 108). The magnitude of the fluctuations vary widely throughout the protein interior, suggesting that the system is inhomogeneous and that some regions are considerably more flexible than others.

To examine the importance of bond length and bond angle fluctuations, simulations were performed on PTI in which the bond lengths or both the bond lengths and the bond angles were fixed at their average values (73). It was found that use of fixed bond lengths (normal fluctuations ± 0.03 Å) does not significantly alter the dynamical properties on a time scale longer than 0.05 ps, but that constraint of the bond angles (normal fluctuations, $\pm 5°$) reduces the mean amplitude of the atomic motions by a factor of two. This result demonstrates that in a closely packed system, such as a protein in its native configuration, the excluded volume effects of repulsive van der Waals interactions introduce a strong coupling between the dihedral angle and bond-angle degrees of freedom.

Figure 2 shows a comparison of the calculated and experimental rms fluctuations on a residue-by-residue basis for reduced cytochrome c (101). The experimental values were corrected for an estimated disorder contribution by subtracting from all of them $< \Delta r^2 >_{dis} = 0.25$ Å^2, obtained from the average calculated results for the protein interior. There is generally good agreement between the experimental and theoretical values. This correlation confirms the reliablity of both the simulation results and the temperature factors as detailed measures of the internal mobility of proteins.

Since most of the molecular dynamics simulations have been done for a protein in vacuum, it is expected that, particularly for the exterior residues, the results will be in error owing to the absence of solvent and, with regard to X-ray temperature factors, the absence of the crystal environment. For cytochrome c, the most prominent differences between theoretical and experimental mean displacements (Figure 2) involve the residues calculated to have very large fluctuations; these are all charged sidechains (particularly lysines) that protrude from the protein and so are not correctly treated in the vacuum simulation. This result is confirmed by molecular-dynamics simulations of PTI in a Lennard-Jones solvent and in a crystal environment. The simulations show that the motion of the outside residues is significantly perturbed by the surrounding medium (107, 110); in particular, the interaction between charged sidechains of a given protein and its crystal neighbors can produce a reduction in the rms values (66, 110). Such results for the external residues contrast with those for the protein interior, where the environmental effects on the amplitude of fluctuations are found to be small. The dominant medium effect on the equilibrium properties of the PTI molecule is that the average structure in the solvent or crystal field is significantly closer to the X-ray structure than is the vacuum result (e.g. for the C^α-atoms, the vacuum simulation has an rms deviation from the X-ray structure equal to 2.2 Å, while those for the solvent and crystal simulations are 1.35 Å and 1.52 Å, respectively).

Recently, a crystal simulation of PTI, including the water molecules, was completed (111). Although the simulation is too short (12 ps) for definitive conclusions, the magnitude of the fluctuations corresponded to those found in the earlier simulations of PTI, while the dynamic average structure was somewhat closer to the X-ray result (C^α-atom rms deviation is 1.06 Å). That the integrity of the dynamic average structure is necessary to obtain the

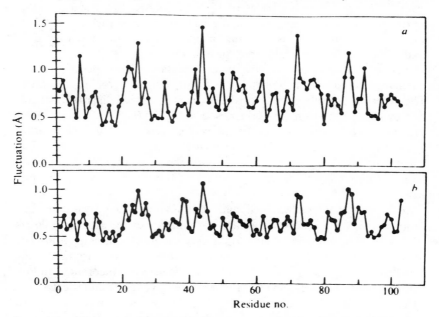

Figure 2 Calculated and experimental rms fluctuations of ferrocytochrome c; residue averages are shown as a function of residue number: (*a*) molecular-dynamics simulation; (*b*) X-ray temperature factor estimation corrected for mean disorder contribution.

correct rms fluctuations was noted in a simulation of rubredoxin (112);the agreement between calculated and experimental (113) temperature factors was poor, apparently because of a significant perturbation in the structure during the simulation.

Of interest also are the results from the dynamic simulation concerning deviations of the atomic motions from the isotropic, harmonic behavior assumed in most X-ray analyses of proteins. The motions of many of the atoms were found in the simulations to be highly anisotropic and somewhat anharmonic. The rms fluctuation of an atom in its direction of largest displacement is typically twice that in its direction of smallest displacement; larger ratios are not uncommon (106, 110, 114, 115). It is sometimes possible to rationalize these directional preferences in terms of local bonding, e.g. torsional oscillation of a small group around a single bond (115). In most cases, however, the directional preferences appear to be determined by larger-scale collective motions involving the atom and its neighbors (115–118). The atom fluctuations are generally also anharmonic; that is, the potentials of mean force for the atom displacements deviate from the simple parabolic forms that would obtain at sufficiently low temperature (106, 110, 119). The most markedly anharmonic atoms are those having multiple minima in their potentials of mean force. The shape of the PTI potential surface in the region of the native structure indicates that the anharmonicity is primarily associated with the softest collective modes of displacement in the protein (120).

Time-Dependence: Local and Collective Effects

Analyses of the time development of the atomic fluctuations were made for PTI (117, 118) and cytochrome c (116). The atomic fluctuations that contribute to the temperature factor (thermal ellipsoid) can be separated into local oscillations superposed on motions with a more collective character. The former have a subpicosecond time scale; the latter, which can involve only a few neighboring atoms, a residue, or groups of many atoms in a given region of the protein, have time scales ranging from 1–10 ps or longer ($\nu \cong 3$ to 30 cm^{-1}). By following the time development of the atomic fluctuations of PTI from 0.2–25 ps, it was shown that the high-frequency oscillations, which contribute about 40% of the average rms fluctuations of mainchain atoms, tend to be uniform over the structure. It is the longer-time-scale, more collective motions that introduce the variations in the fluctuation magnitudes that characterize different parts of the protein structure (117, 118). The correlations of anisotropy with local bonding are often destroyed by these large amplitude collective motions (115, 116).

The time dependence of atom and group motions in proteins can be characterized more fully by calculating appropriate time correlation function (72, 117, 118). The time correlation function of a fluctuating quantity describes the average manner in which a typical fluctuation decays (30, 31). For the positional fluctuations of individual atoms in the protein interior, the time correlation function has a partial loss of amplitude within the first 0.2 ps, followed by much slower decay on a time scale of several ps; the slow component has significant oscillations in many cases (72, 116–118). The decay times of the correlation functions are increased by including external solvent in the dynamic simulation; this effect is most pronounced for atoms at the protein surface (110, 117, 118). An analysis of the relaxation times for the atoms in PTI plus solvent yields a wide range (0.45–10 ps); in vacuum, the times shift to somewhat shorter values (0.2–6 ps).

Although there is no direct experimental measure of the time scale of the atomic fluctuations, it has been shown that NMR-relaxation parameters (T_1, T_2, and NOE values) are sensitive to picosecond motions. Of particular relevance are ^{13}C-NMR data since for protonated carbons, the C–H bond reorientation provides the dominant relaxation mechanism. Preliminary comparisons of ^{13}C-relaxation data for PTI suggest that the α-carbon mobility has a small effect on the relaxation parameters and that increasing effects are expected as the observed carbon is further out along a sidechain (121, 122). The effect of internal motions on other NMR parameters, such as chemical shifts (123) and vicinal-coupling constants, was also examined by molecular-dynamics simulations.

Biological Function

Although many of the individual atom fluctuations observed in the simulations or obtained from temperature factors may in themselves not be important for protein function, they contain information that is of considerable significance. The calculated fluctuations are such that the conformational space available to a protein at room temperature includes the range of local structural changes observed on substrate or inhibitor binding for many enzymes. There may be a correlated directional character to the active-site fluctuations that play a role in catalysis. Further, the small amplitude fluctuations are essential to all other motions in proteins; they serve as the "lubricant" which makes possible larger-scale displacements, such as domain motions (see Table I), on a physiological time scale. It may be possible to extrapolate from the short time fluctuations to larger-scale protein motions. This is suggested by the approximate correspondence between the rms fluctuations of hydrogen bond lengths in a dynamical simulation of PTI (124) and the relative exchange rate of the hydrogens as measured by NMR (42). Changes in the fluctuations induced by perturbations, e.g. ligand binding, are likely to be important as well, e.g. the entropy differences, for the study of which molecular-dynamics techniques were developed (125), may make a significant contribution to the binding free energy (15).

The collective modes are likely to be of particular significance in the biological function; they may be involved in the displacements of sidechains, loops or other structural units required for the transition from an inactive to the active configuration of a globular protein and in the correlated fluctuations that play a direct role in enzyme catalysis. Further, the extended nature of these motions makes them more sensible to the environment, e.g., differences in the simulation results between vacuum and solution results for PTI (117, 118). Because they involve sizable portions of the protein surface, the collective motions may be involved in transmitting external solvent effects to the protein interior (45). They might also be expected to be quenched at low temperature by freezing of the solvent. Their contribution to the mean-square fluctuations could explain the transition observed near 200°K in the temperature dependence of the fluctuations in proteins like in myoglobin (126).

SIDECHAIN MOTIONS

The motions of aromatic sidechains serve as a convenient probe of protein dynamics. The sidechain motions span a time range from picoseconds, during which local oscillations occur, to milliseconds or longer required for 180° rotations. To cover this range of motions requires use of a variety of approaches that complement each other in the analysis of protein dynamics. Further, the results obtained are typical of a class of motional phenomena that play a significant role.

Tyrosines in PTI

The torsional librations of buried tyrosines in PTI were studied in some detail (72). We focus on a particular aromatic sidechain, Tyr 21, whose ring is surrounded by and has a significant nonbonded interaction with atoms of its own backbone and of surrounding residues that are more distant along the polypeptide chain. Figure 3 shows a potential energy contour map for the sidechain dihedral angles χ_1 and χ_2 of Tyr 21 in the free dipeptide (*top*) and in the protein (*bottom*) (57). The minimum energy conformations are very similar in the two cases; this appears to be true for most interior residues of proteins. Where the plots differ is that the sidechain is much more rigidly fixed in position by its nonbonded neighbors in the protein than it is by interactions with the backbone of the chain in the dipeptide.

Figure 4 (*top*) shows the torsional fluctuations of Tyr 21 observed during a PTI simulation (72); the quantity plotted is $\Delta\phi = \phi - <\phi>$ where $<\phi>$ is the time average of the ring torsional angle. Figure 4 (*bottom*) shows corresponding torsional fluctuation history for the ring in an isolated tyrosine fragment simulation.

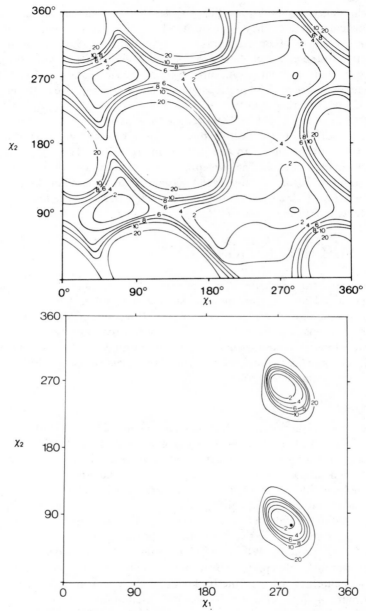

Figure 3 $(\kappa_1, \kappa_2))$ maps for Tyr-21 in PTI: (*top*) free peptide; (*bottom*) peptide in protein; the black dot corresponds to the X-ray value for (κ_1, κ_2) in the protein; energy contours in kcal/mol.

The torsional motion of the ring is less regular when it is surrounded by the protein matrix than in the separated fragment. In PTI, the rms fluctuation of the Tyr 21 torsion angle is 12°, while that for the tyrosine fragment is 15°. This relatively small difference in amplitudes as compared with the forms of the rigid rotation potentials (Figure 3) indicates that protein relaxation involving correlated fluctuations must play an important role in

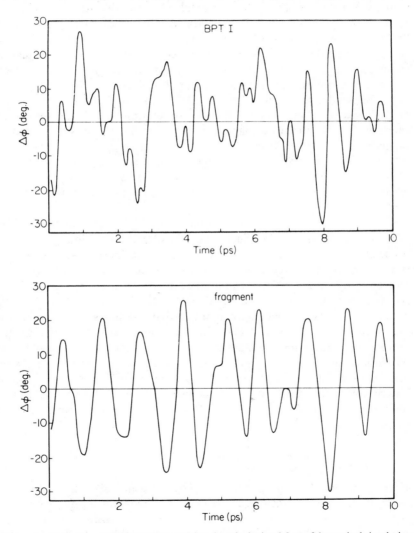

Figure 4 Evolution of the Tyr-21 ring torsional angle during 9.8 ps of dynamical simulation: (*top*) in the protein; (*bottom*) in the isolated tyrosine fragment.

the ring oscillations. The short time, local motion in the protein is consistent with a torsional Langevin equation that contains a harmonic restoring force (see Equation 3). The frictional random force terms are similar to those expected for ring rotation in an organic solvent; this is consistent with the hydrophobic environments of the rings in the protein. The time correlation functions for the torsional fluctuations decay to small values in a short time (~ 0.2 ps). However, the quantities involved in the relaxation times (110, 127) measured in fluorescence depolarization (trigonometric functions of the angles) decay much more slowly. For the tyrosine rings in PTI there is rapid partial decay in less than a picosecond to a plateau value equal to about 75% of the initial value; this behavior was recently confirmed by fluorescent depolarization measurements (128). Corresponding calculations (130) for the fluorescent depolarization of the tryptophan residues in lysozyme based on a molecular dynamics simulation (106) indicate a wide range of variation in the depolarization behavior. Since there are six tryptophans in a variety of environments, their behavior is expected to correspond to that which occurs more generally in proteins (129). Certain interior tryptophans have almost no decay over the time scale of the simulation while one in the active site (Trp 62) has its anisotropy reduced to 0.6 after 5 ps.

Tyrosine and phenylalanine ring rotations by 180° were studied by NMR in proteins (40, 43, 44, 56). Such ring "flips" occur very infrequently because of the large energy barrier due to steric hindrance (57–59). The long time intervals separating flips preclude systematic study by conventional molecular-dynamics methods. A modified molecular-dynamics method was recently developed to handle such local activated processes (59). This method is similar to adiabatic mapping in that one starts with an assumed "reaction coordinate" that defines the fundamental structural changes involved. It differs from the adiabatic method in that it involves consideration of all thermally accessible configurations and not just the minimum energy one for each value of the reaction coordinate. Also it provides a detailed description of the structural and dynamical features of the process. In this method, one calculates separately the factors in the rate constant expression (82, 131):

$$k = \frac{1}{2}\kappa < |\dot{\xi}| > [\rho(\xi^\dagger)/\int_i \rho(\xi)d\xi]. \qquad 5.$$

Here, ξ is the reaction coordinate, $\dot{\xi} = d\xi/dt$, and $\xi^\dagger$ is the value of ξ in the transition state region for the process. The factor in square brackets is the probability that the system will be in the transition state region, relative to the probability that it is in the initial stable state. This quantity corresponds roughly to the term $\exp(-\Delta G^\dagger/RT)$ in more familiar expressions for rate constants; it can be calculated by carrying out a sequence of simulations in which the system is constrained to stay near particular values of ξ. The remaining factors can be evaluated by analysis of trajectories initiated in the transition state region (59, 85, 86). The transmission coefficient κ is equal to one in ideal transition state theory (equilibrium populations maintained in the stable states and uninterrupted crossings through the transition state region); for real systems κ is less than one.

Application of this modified molecular dynamics method to the flipping of a tyrosine ring in PTI shows that the rotations themselves required only 0.5–1.0 ps (85, 86). At the microscopic level, the processes responsible for flipping are the same as those responsible for the smaller amplitude librations. The ring goes over the barrier not as the result of a particularly energetic collision with some cage atom, but as the result of a transient decrease in frequency and intensity of collisions that would drive the ring away from the barrier. These alterations of the collision frequency are caused by small, transient packing defects (86). The packing defects help to initiate ring rotation, but they are much too small to allow free rotation of the ring by a simple vacancy or free-volume mechanism (86, 132). The ring tends to be tightly encaged even in the transition state orientation. Collisions with cage atoms in the transition state produce frictional forces similar to those that occur in the stable state librations; these frictional effects reduce the transition rate to about 20% of the ideal transition-state theory value (59). As to the free energy of activation, the calculations suggest that the activation enthalpy contribution is similar to that found by adiabatic mapping techniques (57, 58) and that the activation entropy is small.

Although no enzyme has yet been studied by the techniques applied to the tyrosine ring flips, the methodology is applicable to the activated processes central to most enzymatic reactions. Further, many of the qualitative features found for the tyrosines (e.g. lowering of the potential of mean force by cage relaxation, alteration of the rate by frictional effects) should be present in general.

Ligand-Protein Interaction in Myoglobin

A biological problem where sidechain fluctuations are important concerns the manner in which ligands like carbon monoxide and oxygen are able to get from the solution through the protein matrix to the heme group in myoglobin and hemoglobin and then out again. The high-resolution X-ray structure of myoglobin (8, 9, 133, 134) does not reveal any path by which ligands such as O_2 or CO can move between the heme-binding site and the outside of the protein. Since this holds true both for the unliganded and liganded protein, i.e. myoglobin (133) and oxymyoglobin (134), structural fluctuation must be involved in the entrance and exit of the ligands. Empirical energy function calculations (96) showed that the rigid protein would have barriers on the order of 100 kcal/mol; such high barriers would make the transitions infinitely long on a biological time scale. Figure 5, panel I gives the nonbonded potential contour lines seen by a test particle representing an O_2 molecule in a plane (xy) parallel to the heme and displaced 3.2 Å from it in the direction of the distal histidine; the coordinate system in this and related figures has the iron at the origin and the z-axis normal to the heme plane. The low potential-energy minimum corresponds to the observed position of the distal O atom of an O_2 molecule forming a bent Fe–O–O bond (134). The shortest path for a ligand from the heme pocket to the exterior (the low energy region in the upper left of the figure) is between His E7 and Val E11. However, this path is not open in the X-ray geometry because the energy barriers due to the surrounding residues indicated in the figure are greater than 90 kcal/mol.

To analyse pathways available in the thermally fluctuating protein, ligand trajectories were calculated with a test molecule of reduced effective diameter to compensate for the use of the rigid protein structure (96). A trajectory was determined by releasing the test molecule with substantial kinetic energy (15 kcal/mol) in the heme pocket and following its classical motion for a suitable length of time. A total of 80 such trajectories were computed; a given trajectory was terminated after 3.75 ps if the test molecule had not escaped from the protein. Slightly more than half the test molecules failed to escape from the protein in the allowed time; 25 molecules remained trapped near the heme-binding site, while another 21 were trapped in two cavities accessible from the heme pocket. Most of the molecules that escaped did so between the distal histidine (E7) and the sidechains of Thr E10 and Val E11 (see Figure 5, panel I) A secondary pathway was also found; this involves a more complicated motion along an extension of the heme pocket into a space between Leu B10, Leu E4, and Phe B14, followed by squeezing out between Leu E4 and Phe B14. Figure 6 shows a typical model trajectory following this path. Additional, more complicated pathways also exist, as indicated by the range of motions observed in the trial trajectories.

In the rigid X-ray structure, the two major pathways have very high barriers for a thermalized ligand of normal size. Thus, it was necessary to study the energetics of barrier relaxation to determine whether either of the pathways had acceptable activation enthalpies. Local dihedral rotations of key sidechains, analogous to the tyrosine sidechain oscillations described above, were investigated; it was found that the bottleneck on the primary pathway could be relieved at the expense of modest strain in the protein by rigid rotations of the sidechains of His E7, Val E11, and Thr E10. The reorientation of these three sidechains and the resultant opening of the

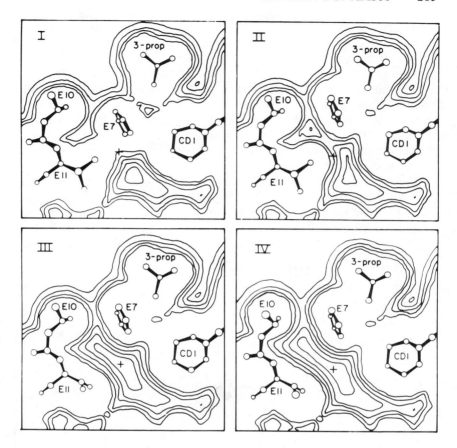

Figure 5 Myoglobin-ligand interaction contour maps in the heme (x,y) plane at $z = 3.2$ Å (the iron is at the origin) showing protein relaxation; a cross marks the iron atom projection onto the plane. Distances are in Å and contours in kcal; the values shown correspond to 90, 45, 10, 0, and −3 kcal/mol relative to the ligand at infinity. The highest contours are closest to the atoms whose projections onto the plane of the figure are denoted by circles. *Panel I:* X-ray structure; *panels II–IV:* sidechain rotations discussed in the text.

pathway to the exterior is illustrated schematically in Figure 5; Panel I shows the X-ray structure; in Panel II the distal histidine (E7) was rotated to $\chi_1 = 220°$ at an energy cost of 3 kcal/mol; in Panel III, Val E11 was also rotated to $\chi_1 = 60°$ (~5 kcal/mol); and Panel IV has the additional rotation of Thr E10 to $\chi_1 \cong 305°$ (<1 kcal/mol). In this manner a direct path to the exterior was created with a barrier of ~5 kcal/mol at an energy cost to the protein of ~8.5 kcal/mol, as compared with the X-ray structure value of nearly 100 kcal/mol. On the secondary path, however, no simple torsional motions reduced the barrier due to Leu E4 and Phe B14, since the necessary rotations led to larger strain energies. A test sphere was fixed at each of the bottlenecks and the protein was allowed to relax by energy minimization (adiabatic limit), in the presence of the ligand (57, 96).

Approximate values for the relaxed barrier heights were 13 kcal/mol and 6 kcal/mol for the two primary path positions and 18 kcal/mol for the secondary path position. These barriers are on the order of those estimated in the photolysis, rebinding studies for CO myoglobin by Frauenfelder et al (45, 135, 136). Further, a path suggested by the energy calculations was found to correspond to a high mobility region in the protein as determined by X-ray temperature factors (102).

The type of ligand motion expected for such a several-barrier problem can be determined from the trajectory studies mentioned earlier. What happens is that the ligand spends a long time in a given well, moving around in and undergoing collisions with the protein walls of the well (see Figure 6). When there occurs a protein fluctuation sufficient to significantly lower

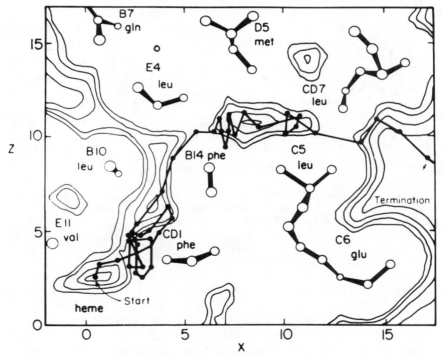

Figure 6   Diabatic-ligand trajectory following a secondary pathway (see text); a projection of the trajectory on the plane of the figure [(x,z) plane at $y = 0.5$ Å)] is shown with the dots at 0.15 ps intervals. The start of the trajectory at the heme iron and the termination point exterior to the protein are indicated by arrows.

the barrier, or the ligand gains sufficient excess energy from collisions with the protein, or more likely both at the same time, the ligand moves rapidly over the barrier and into the next well where the process is repeated. In a completely realistic trajectory involving a fluctuating protein and ligand-protein energy exchange, the time spent in the wells would be much longer than that found in the diabatic model calculations (Figure 6). Further, from the complexity of the range of pathways in the protein interior, it is likely that the motion of the ligand will have a diffusive character.

The analysis of myoglobin suggests that the native structure of a protein is often such that the small molecules that interact with the protein cannot enter or leave if the atoms are constrained to their average positions. Consequently, sidechain and other fluctuations may be required for ligand binding by proteins and for the entrance of substrates and exit of products from enzymes. Some analyses of the effects of such "gated" accessibility on the observed kinetics were made (137–139).

Exterior Sidechain and Loop Motions

In several enzymes, a displacement of surface sidechains or entire loops on substrate binding occurs. In carboxypepdidase A (140), for example, when the substrate binds, the structural changes include a large displacement of the sidechain of Tyr 248, which moves through more than 10 Å toward the active site. Another example is provided by an external loop in triophosphate isomerase, which was shown by X-ray diffraction to fold over the substrate when it is bound (141). If surface residues are involved, as is often the case, the motion is best treated by stochastic dynamics.

To study the motion of aliphatic sidechains in solution (76), the end of the chain attached to the macromolecule is held fixed and the Langevin equations of motion (Equation 3) for the atoms of the chain are solved simultaneously for periods of up to a microsecond. The methyl and methylene groups of the chain are treated as single extended atoms with a friction coefficient corresponding to methane in water and a generalized empirical potential energy function is used to represent the intramolecular interactions (nonbonded and torsional) in the usual way, except that they are slightly modified to take into account the presence of solvent; that is, a potential of a mean force replaces the isolated molecule potential function (142). It is found that the motion with respect to a given torsion angle separates into two time scales (76). The shorter time motion, on the order of tenths of picoseconds, corresponds to torsional oscillations within a potential well, and the longer, on the order of two hundred picoseconds, corresponds to transitions from one potential well to another; the torsional barrier used in the potential function is ~ 2.8 kcal/mol. Thus, analogous to the above description for an oxygen molecule moving through myoglobin, the sidechain spends most of the time oscillating about a single conformation (i.e. with each dihedral angle remaining in a given well) and only rarely makes a transition from one conformation to another. To test the validity of this type of calculation, comparisons of the stochastic trajectory results with NMR relaxation measurements (e.g. ^{13}C NMR) were made (122).

RIGID BODY MOTIONS

A type of motion that plays an important role in proteins is referred to as a rigid-body motion (Table I). It involves the displacement of one part of a protein relative to another such that each moving portion can be approximated as a rigid body. However, smaller fluctuations must accompany the rigid-body motions to reduce the required energy and permit them to proceed at a sufficiently rapid rate.

Hinge Bending

Many enzymes (23–25) and other protein molecules (e.g. immunoglobulins) consist of two or more distinct domains connected by a few strands of polypeptide chain that may be viewed as "hinges." In lysozyme, for example, it was noted in the X-ray structure (143) that when an active-site inhibitor is bound, the cleft closes down somewhat as a result of relative displacements of the two globular domains that surround the cleft. Other classes of proteins (kinases, dehydrogenases, citrate synthase) have considerably larger displacements of the two lobes on substrate binding than does lysozyme (23–25).

   In the theoretical analysis of lysozyme (88), the stiffness of the hinge was evaluated by the use of an empirical energy function (66, 88). An angle-bending potential was obtained by rigidly rotating one of the globular domains relative to a bending axis which passes through the hinge and calculating the changes in the protein conformational energy. This procedure overestimates the bending potential, since no allowance is made for the relaxation of the unfavorable contacts between atoms generated by the rotation. To take account of the relaxation, an adiabatic potential was calculated by holding the bending angle fixed at various values and permitting the positions of atoms in the hinge and adjacent regions of the two globular domains to adjust themselves so as to minimize the total potential energy. As in a previous adiabatic ring rotation calculation (57), only small (<0.3 Å) atomic displacements occurred in the relaxation process. Localized motions involving bond angle and local dihedral angle deformations occur. The frequencies associated with them (>100 cm^{-1}) are much greater than the hinge-bending frequency (≈ 5 cm^{-1}), so that the use of the adiabatic-bending potential is appropriate.

The bending potentials were found to be approximately parabolic, with the restoring force constant for the adiabatic potential about an order of magnitude smaller than that for the rigid potential (see Figure 7). However, even in the adiabatic case, the effective force constant is about 20 times as large as the bond-angle bending force constant of an α carbon (i.e. $N-C_\alpha-C$); the dominant contributions to the force constant come from repulsive nonbonded interactions involving on the order of fifty contacts. If the adiabatic potential is used and the relative motion is treated as an angular harmonic oscillator composed of two rigid spheres, a vibrational frequency of about 5 cm^{-1} is obtained. This is a consequence of the fact that, although the force constant is large, the moments of inertia of the two lobes are also large.

Although fluctuations in the interior of the protein, such as those considered in myoglobin, may be insensitive to the solvent (because the protein matrix acts as its own solvent), the domain motion in lysozyme involves two lobes that are surrounded by the solvent. To take account of the solvent effect in the simplest possible way, the Langevin equation (Equation 3) for a damped harmonic oscillator was used. The friction coefficient for the solvent damping term was evaluated by modeling the two globular domains as spheres (144). From the adiabatic estimate of the hinge potential and the magnitude of the solvent damping, it was found that the relative motion of the two globular domains in lysozyme is overdamped; i.e. in the absence of

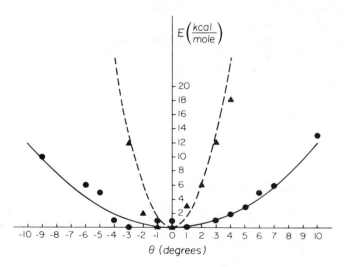

Figure 7   Change of conformational energy produced by opening ($\theta<0$) and closing ($\theta>0$) the lysozyme cleft; calculated values are for the rigid bending potential (triangles) and for the adiabatic-bending potential (circles); the origins for the two calculations are superposed.

driving forces the domains would relax to their equilibrium positions without oscillating. The decay time for this relaxation was estimated to be about 2×10^{-11}s. Actually, the lysozyme molecule experiences a randomly fluctuating driving force owing to collisions with the solvent molecules, so that the distance between the globular domains fluctuates in a Brownian manner over a range limited by the bending potential; a typical fluctuation opens the binding cleft by 1 Å and lasts for 20 ps.

The methodology developed in the lysozyme study is now being applied to a number of other proteins; they include antibody molecules (145), L-arabinose binding protein (146), and liver alcohol dehydrogenase (147). For the L-arabinose binding protein, calculations and experiment both suggest that the binding site is open in the unliganded protein but is induced to close by a hinge-bending motion upon ligation (146, 148). In the case of liver alcohol dehydrogenase, the open structure is stable in the crystal for the apoenzyme (149). Adiabatic energy-minimization calculations (147) suggest that the apoenzyme is highly flexible as far as its hinge-bending mode (rotation of the catalytic relative to the coenzyme binding domain) is concerned and that normal thermal fluctuations would lead to a closed structure (rotation of $\sim 10°$) similar to that found in the holoenzyme (150).

Since the hinge-bending motion in lysozyme and in other enzymes involves the active-site cleft, it is likely to play a role in the enzymatic activity of these systems. In addition to the possible difference in the binding equilibrium and solvent environment in the open and closed state, the motion itself could result in a coupling between the entrance and exit of the substrate and the opening and closing of the cleft (137–139). The interdomain mobility in immunoglobins may be involved in adapting the structure to bind different macromolecular antigens and, more generally, it may play a role in the cross-linking and other interactions required for antibody function. In the coat protein of tomato bushy stunt virus, a two-domain structure with a hinge peptide was identified from the X-ray structure (29, 151) and rotations about the hinge were shown to be involved in establishing different subunit interactions for copies of the same protein involved in the assembly of the complete viral protein shell.

Quaternary Structural Change

A classic case where large-scale motion plays an essential role is the allosteric transition in the hemoglobin tetramer (152–154). It is clear from the X-ray data that the subunits move relative to each other (quaternary change) and that more localized atomic displacements occur within each subunit (tertiary change). The coupling between these two types of structural changes is an essential factor in the cooperative mechanism of hemoglobin. As a first step in unraveling the nature of the motions involved, a reaction path for the tertiary structure change induced by ligand binding within a subunit was worked out by the use of empirical energy calculations (155, 156). The results are in good agreement with limited structural data available for intermediates in the ligation reaction (157). The calculations show how the perturbation introduced in the heme by the binding of ligand leads to displacements in the protein atoms, so that alterations appear in surface regions in contact with other subunits. This provides a basis for the coupling between tertiary and quaternary structural change, although the details of the motions leading from the unliganded to the liganded quaternary structure have yet to be worked out.

As is clear from the above, most of the information available on rigid motions comes from high-resolution crystal structures of proteins. Low angle X-ray scattering analyses of solutions provided evidence for radius of gyration changes that are in accord with the crystal results where available (158) or provide evidence for structural changes in cases where only one structure is known (148, 159). However, there is almost no experimental evidence on the time scale of the rigid-body motions. Fluorescence depolarization studies of labeled antibody molecules show that the time scale for internal motions is consistent with the diffusional displacements of flexibly hinged domains (145, 160). It would be of great interest to have corresponding data on the domain motions of enzymes.

α-HELIX MOTION: HARMONIC AND SIMPLIFIED MODEL DYNAMICS

Early evidence for motion in the interior of proteins or their fragments comes from analyzing vibrational spectroscopic studies. It is generally assumed in interpreting such data that a harmonic potential and the resulting normal-mode description of the motions is adequate. This approximation is most likely correct for the tightly bonded secondary structural elements, like α-helices and β-sheets. The fluctuations of a finite α-helix (hexadecaglycine) were determined from the normal modes of the system (63). At 300°K, the rms fluctuations of mainchain dihedral angles (ϕ and ψ) about their equilibrium values are equal to $\sim 12°$ in the middle of the helix and somewhat larger near the ends. The dihedral angle fluctuations are significantly correlated over two neighboring residues; these correlations tend to localize the fluctuations (63, 71, 79, 80). Fluctuations in the lengths between adjacent residues (defined as the projection onto the helix axis of the vector connecting the centers of mass of adjacent residues) ranged from about 0.15 Å in the middle of the helix to about 0.25 Å at the ends. These length fluctuations are negatively correlated for residue pairs ($i-1,i$) and ($i,i+1$) so as to preserve the overall length of the helix; positive correlations are observed for the pairs (4, 5), (8, 9) and (8, 9), (12, 13), suggesting that the motion of residue 8 is coupled to the motions of residues 4 and 12 to retain optimal hydrogen bonding.

Recently a full molecular-dynamics calculation was performed (81) for a decaglycine helix as a function of temperature between 5°–300°K and the results compared with those obtained in the harmonic approximation (63). For the mean-square positional fluctuations, $<\Delta r^2>$, of the atoms, the harmonic approximation is valid in the classical limit below 100°K, but there are significant deviations above that temperature; e.g. at 300°K, the average value of $<\Delta r^2>$ obtained for the α-carbons from the full dynamics is more than twice that found in the harmonic model. Quantum effects on the fluctuations are found to be significant only below 50°K. The temperature dependence of the fluctuations in the simulations is similar to that observed for α-helices in myoglobin between 80–300°K by X-ray diffraction (161).

As an approach to the helix-coil transition in α-helices, a simplified model for the polypeptide chain was introduced to permit a dynamic simulation on the submicrosecond time scale appropriate for this phenomenon (78): each residue is represented by a single interaction center ("atom") located at the centroid of the corresponding sidechain and the residues are linked by virtual bonds (97), as described earlier in the section on simplified model dynamics. The diffusional motion of the chain "atoms" expected in water was simulated by using a stochastic dynamics algorithm based on the Langevin equation with a generalized force term, Equation 3. Starting from an all-helical conformation, the dynamics of several residues at the end of a 15-residue chain were monitored in several independent 12.5-ns simulations at 298°K. The mobility of the terminal residue was quite large, with a rate constant $\approx 10^9 \ s^{-1}$ for the transitions between coil and helix states. This mobility decreased for residues further into the chain; unwinding of an interior residue required simultaneous displacements of residues in the coil, so that larger solvent frictional forces were involved. The coil region did not move as a rigid body, however; the torsional motions of the chain were correlated so as to minimize dissipative effects. Such concerted transitions are not consistent with the conventional idea that successive transitions occur independently. Analysis of the chain diffusion tensor showed that the frequent occurrence of the correlated transition results from the relatively small frictional forces associated with these motions (100).

PERSPECTIVE

Theoretical protein dynamics did not exist before 1977 when the first paper presenting a detailed molecular-dynamics simulation of a small protein was published (71). In the next five years more than 50 theoretical papers appeared. They explored dynamic phenomena in depth for a variety of proteins (protein inhibitors, transport and storage proteins, enzymes). The magnitudes and time scales of the motions were delineated and related to a variety of experimental measurements, including NMR, X-ray diffraction, fluorescent depolarization, infra-red spectroscopy and Raman scattering. It was shown how to extend dynamical methods from the subnanosecond time range accessible to standard molecular-dynamics simulations to much longer time scales for certain processes by the use of activated, harmonic and simplified model dynamics. Further, the effect of solvent was introduced by stochastic dynamic techniques or accounted for in full dynamic simulations including also the crystal environment. Concomitantly, a wealth of experimental information on the motions appeared. The interplay between theory and experiment provides a basis for the present vitality of the field of protein dynamics.

What is known and what remains to be done? On the subnanosecond time scale our basic knowledge of protein motions is essentially complete; that is, the types of motion that occur have been clearly presented, their characteristics evaluated and the important factors determining their properties delineated. Simulation methods have shown that the structural fluctuations in proteins are sizable; particularly large fluctuations are found where steric constraints due to molecular packing are small (e.g. in the exposed side chains and external loops), but substantial mobility is also found in the protein interior. Local atomic displacements in the interior of the protein are correlated in a manner that tends to minimize disturbances of the global structure of the protein. This leads to fluctuations larger than would be permitted in a rigid polypeptide matrix.

For motions on a longer time scale, our understanding is more limited. When the motion of interest can be described in terms of a reaction path (e.g. hinge-bending, local-activated event), methods exist for determining the nature and rate of the process. However, for the motions that are slow owing to their complexity and involve large-scale structural changes, extensions of the approaches described in this review are required. Harmonic and simplified model dynamics, as well as reaction-path calculations, can provide information on slower processes, such as opening fluctuations and helix-coil transitions, but a detailed treatment of protein folding is beyond the reach of present methods.

In the theory of protein dynamics there are two directions where active study and significant progress can be expected in the near future. One concerns the more detailed examination of processes of biological interest and the other, an improvement in approaches to longer-time dynamics. As to the latter, a variety of extensions of the methodology described in this review, as well as the availability of faster computers, may yield the necessary insights. As to the former, there are many biological problems to which current dynamical methods can be applied and for which a knowledge of the dynamics is essential for a complete understanding. Some of these are listed below.

For the transport protein hemoglobin, there is more evidence concerning the role of motion than for any other protein. The tertiary and quaternary structural changes that occur on ligand binding and their relation to the allosteric mechanism are well documented. An important role of the quaternary structural change is to transmit information over a longer distance than could take place by tertiary structural changes alone; the latter are generally damped out over rather short distances unless amplified by the displacement of secondary structural elements or domains. The detailed dynamics of the allosteric mechanism has yet to be investigated; in particular, the barriers along the reaction path from the deoxy to the oxy structure have not been analyzed, nor has the importance of the fluctuations for the activated processes involved been determined. For the related storage protein, myoglobin, fluctuations in the globin are essential to the binding process; that is, the protein matrix in the X-ray structure is so tightly packed that there is no sufficiently low energy path for the ligand to enter or leave the heme pocket. Only through structural fluctuations in certain bottleneck regions can the barriers be lowered sufficiently to obtain the observed rates of ligand binding and release. Although energy minimization was used to investigate the displacements involved and the resulting barrier magnitudes, activated dynamic studies are needed to analyze the activation entropies and rates of ligand motion across the barriers.

In many proteins and peptides, the transport of substances is through the molecule rather than via overall translation as in hemoglobin. The most obvious cases are membrane systems, in which fluctuations are likely to be of great importance in determining the kinetics of transport. For channels that open and close (e.g. gramicidin) as well as for active transport involving enzymes (e.g. ATPases), fluctuations, in some cases highly correlated ones, must be involved. At present, structural details and studies of the motions are lacking, but this is an area where dynamic analyses are likely to be made in the near future.

In electron-transport proteins, such as cytochrome c, protein flexibility is likely to play two roles in the electron transfer. Evidence now favors a vibronic-coupled tunneling mechanism for transfer between cytochrome c and other proteins, although outer-sphere mechanisms are not fully excluded. In the vibronic-coupled tunneling theory, processes which would be energetically forbidden for rigid proteins become allowed if the appropriate energies for conformational distortions are available. Experimental data indicate that the important fluctuations are characterized by an average frequency on the order of 250 cm^{-1}, close to that associated with the collective modes of proteins. Also, the transfer rate is a sensitive function of donor-acceptor distance and may be greatly increased by surface side-chain displacements that allow for the closer approach of the interacting proteins.

For proteins involved in binding, flexibility and fluctuations enter into both the thermodynamics and the kinetics of the reactions. For the rate of binding of two macromolecules (protein-antigen and antibody, protein-inhibitor and enzyme), as well as for smaller multisite ligands, structural fluctuations involving side chains, hydrogen-bonding groups, etc, can lead to lowering of the free energy barriers. Dividing the binding process into successive steps for which flexibility may be needed can increase the rate. The required fluctuations are likely to be sufficiently small and local that they will be fast relative to the binding and therefore not rate limiting.

The relative flexibility of the free and bound ligand, as well as changes in the binding protein, must be considered in the overall thermodynamics of the binding reaction. If the free species have considerable flexibility and fluctuations are involved in the binding step as described above, it is likely that the bound species will be less flexible and a significant entropic destabilization will result. Thus, for strong binding in cases where the rate is not important, relatively rigid species are desirable. This would reduce the conformational entropy decrease and could lead to a very favorable enthalpy of binding if there is high complementarity in the two binding sites. However, some flexibility and an increase in the conformational space available to the bound species has a stabilizing effect that partly compensates for the loss of translational and rotational entropy on binding. Conversely, the entropy loss of binding a flexible substrate or the rigidification of a protein on substrate binding can be used to modulate the binding constant even when strong, highly specific enthalpic interactions are present. The required balance between flexibility and rigidity will be determined by the function of the binding in each case. Dynamical techniques can be employed to determine the entropy differences for such systems. Further, the mechanism and rates of the binding processes, itself, are an ideal subject for dynamical analysis.

In the function of proteins as catalysts, there is the greatest possibility of contributions from motional phenomena. The role of flexibility per se has often been discussed, particularly from the viewpoint of structural changes induced by the binding of the substrate. In addition to cooperative effects caused by quaternary alterations, a variety of results can arise from the perturbation of the tertiary structure. One example is the ordered binding of several substrates (or effectors and substrates), with the first molecule to bind altering the local conformation so as to increase or decrease the subsequent binding of other molecules. The occurrence of large-scale changes, such as the closing of active-site clefts by substrate binding, as in certain kinases, has been interpreted in terms of catalytic specificity, alteration of the solvent environment of the substrate, and exclusion of water that could compete with the enzymatic reaction. In large enzymes with more than one catalytic site or in coupled enzyme systems, conformational freedom may be important in moving the substrate along its route from one site to the next. Many of these processes are ready for the application of dynamical methods, particularly in cases where structural data are available.

The flexibility of the substrate-binding site in enzymes can result in effects corresponding to those already considered in receptor binding. In the enzyme case there exists the often-discussed possibility of enhanced binding of a substrate with its geometry and electron distribution close to the transition state; for this to occur, conformation fluctuations are essential. Entropic effects also are likely to be of significance, both with respect to solvent release on substrate binding and possible changes in vibrational frequencies that alter the vibrational entropy of the bound system in the enzyme-substrate complex or in the transition state. There are also indications that the inactivity of enzyme precursors can result from the presence of conformational freedom in residues involved in the active site. The entropic cost of constraining them in the proper geometry for interacting with the substrate may be so high that the activity is significantly reduced relative to that of the normal enzyme where the same residues are held in place more rigidly. Such a control mechanism was suggested for the trypsin, trypsinogen system, and for other proteins. As to the time dependence of fluctuations and structural alternations, there are a variety of possibilities to be considered. In the binding of reactants and release of products, the time course of fluctuations in the enzyme could interact with the motion of the substrate. The opening and closing fluctuations of active-site clefts may be modified by interactions with the substrate as it enters or leaves the binding site.

Fluctuations could play an essential role in determining the effective barriers for the catalyzed reactions. If the substrate is relatively tightly bound, local fluctuation in the enzyme could couple to the substrate in such a way as to significantly reduce the barriers. If such coupling effects exist, specific structures could have developed through evolutionary pressure to introduce directionality and enhance the required fluctuations. Frictional effects that occur in the crossing of barriers in the interior of the protein could act to increase the transition state lifetime and so alter the reaction rates relative to those predicted by conventional rate theory. Energy released locally in substrate binding may be utilized directly for catalyzing its reaction, perhaps by inducing certain fluctuations. Whether such an effect occurs would depend on the rate of dissipation of the (mainly) vibrational energy and the existence of patterns of atoms and interactions to channel the energy appropriately. It will be of great interest to determine whether any of the rather speculative possibilities outlined here for the role of the energy and directionality of structural fluctuations in enzymatic reactions can be documented theoretically or experimentally for specific systems.

A wide range of biological problems involving proteins, not to mention nucleic acids and membrane lipids, are ready for study and exciting new results can be expected as dynamical methods are applied to them. In the coming years, we shall learn how to calculate meaningful rate constants for enzymatic reactions, ligand binding and many of the other biologically important processes mentioned above. The role of flexibility and fluctuations will be understood in much greater detail. It should become possible to determine the effects upon the dynamics of changes in solvent conditions and protein amino acid sequence. As the predictive powers of the theoretical approaches increase, applications will be made to practical problems arising in areas such as genetic engineering and industrial enzyme technology.

Literature Cited

1. Phillips, D. C. 1966. *Sci. Am.* 215:78
2. Dickerson, R. E., Geis, I. 1969. *Structure and Action of Proteins.* New York: Harper & Row
3. Fermi, G., Perutz, M. F. 1981. *Haemoglobin and Myoglobin.* Oxford: Clarendon
4. Richardson, J. A. 1981. *Adv. Protein Chem.* 34:167–339
5. Tanford, C. 1980. *The Hydrophobic Effect,* p. 142. New York: Wiley. 2nd ed.
6. Phillips, D. C. 1981. *Biomolecular Stereodynamics,* ed. R. H. Sarma, pp. 497–98. New York: Adenine
7. Marquart, M., Deisenhofer, J., Huber, R., Palm, W. 1980. *J. Mol. Biol.* 141:369–91
8. Perutz, M. F., Mathews, F. S. 1966. *J. Mol. Biol.* 21:199–202
9. Watson, H. C. 1969. *Prog. Stereochem.* 4:299
10. Williams, R. J. P. 1977. *Angew. Chem. Int. Ed. Engl.* 16:766–77
11. Blundell, T., Wood, S. 1981. *Ann. Rev. Biochem.* 51:123–54
12. Citri, N. 1973. *Adv. Enzymol. Relat. Areas Mol. Biol.* 37:397–648
13. Jencks, W. P. 1975. *Adv. Enzymol. Relat. Areas Mol. Biol.* 43:219–430
14. Koshland, D. E. Jr. 1976. *FEBS Lett.* 62:E47–E52 (Suppl.)
15. Sturtevant, J. M. 1977. *Proc. Natl. Acad. Sci. USA* 74:2236–41
16. Levich, V. G. 1966. *Adv. Electrochem. Electrochem. Eng.* 4:249–57
17. Hopfield, J. J. 1974. *Proc. Natl. Acad. Sci. USA* 71:3640–44
18. Jortner, J. 1976. *J. Chem. Phys.* 64:4860–67
19. Salemme, F. R. 1977. *Ann. Rev. Biochem.* 46:299–329
20. Mendelson, R. A., Morales, M. F., Botts, J. 1973. *Biochemistry* 12:2250–55
21. Harvey, S. C., Cheung, H. C. 1977. *Biochemistry* 16:5181–87
22. Highsmith, S., Kretzschmar, K. M., O'Konski, C. T., Morales, M. F. 1977. *Proc. Natl. Acad. Sci. USA* 74:4986–90
23. Anderson, C. M., Zucker, F. H., Steitz, T. A. 1979. *Science* 204:375–80
24. Johnson, L. N. 1983. In *Inclusion Compounds,* ed. J. L. Atwood, T. E. D. Davies, D. D. MacNicol. New York: Academic
25. Janin, J., Wodak, S. J. 1983. *Prog. Biophys. Mol. Biol.* In press
26. Givol, D. 1976. *Receptors and Recognition* (Ser. A), ed. P. Cuatrecasas, M. F. Greaves. New York: Halsted
27. DeLisi, C. 1976. In *Lectures Notes in Biomathematics,* Vol. 8. New York: Springer
28. Huber, R., Deisenhofer, J., Coleman, P. M., Matsushima, M., Palm, W. 1976. *Nature* 264:415–20
29. Harrison, S. C. 1978. *Trends Biochem. Sci.* 3:3–7

30. McQuarrie, D. A. 1976. *Statistical Mechanics* New York: Harper & Row. 641 pp.
31. Hansen, J. P., McDonald, I. R. 1976. *Theory of Simple Liquids.* New York: Academic
32. Chandler, D. 1974. *Acc. Chem. Res.* 7:246–51
33. Hynes, J. T. 1977. *Ann. Rev. Phys. Chem.* 28:301–21
34. Richards, F. M. 1977. *Ann. Rev. Biophys. Bioeng.* 6:151–76
35. McCammon, J. A., Karplus, M. 1980. *Ann. Rev. Phys. Chem.* 31:29–45
36. Karplus, M., McCammon, J. A. 1981. *CRC Crit. Rev. Biochem.* 9:293–349
37. Karplus, M. 1981. In *Structural Molecular Biology,* ed. D. B. Davis, W. Saenger, S. S. Danyluk, pp. 427–53. New York: Plenum
38. Karplus, M. 1982. *Ber. Bunsenges. Phys. Chem.* 86:386–95
39. McCammon, J. A., Karplus, M. 1983. *Acc. Chem. Res.* In press
39a. Levitt, M. 1982. *Ann Rev. Biophys. Bioeng.* 11:251–71
40. Campbell, I. D., Dobson, C. M., Williams, R. J. P. 1978. *Adv. Chem. Phys.* 39:55–107
41. Peticolas, W. L. 1978. *Methods Enzymol.* 61:425–58
42. Woodward, C. K., Hilton, B. D. 1979. *Ann. Rev. Biophys. Bioeng.* 8:99–127
43. Gurd, F. R. N., Rothgeb, T. M. 1979. *Adv. Protein Chem.* 33:73–165
44. Jardetzky, O. 1981. *Acc. Chem. Res.* 14:291–98
45. Debrunner, P. G., Frauenfelder, H. 1982. *Ann. Rev. Phys. Chem.* 33:283–99
46. Careri, G., Fasella, P., Gratton, E. 1979. *Ann. Rev. Biophys. Bioeng.* 8:69–97
47. Weber, G. 1975. *Adv. Protein Chem.* 29:1–83
48. Cooper, A. 1981. *Sci. Prog. Oxford* 66:473–97
49. Williams, R. J. P. 1979. *Biol. Rev.* 54:389–420
50. Williams, R. J. P. 1980. *Chem. Soc. Rev.* 9:325–64
51. O'Connor, M., ed. 1982. Mobility and Function in Proteins and Nucleic Acids. *Ciba Found. Symp. 93.* London: Pitman
52. Rossmann, M. G., Argos, P. 1981. *Ann. Rev. Biochem.* 50:497–532
53. Kim, P. S., Baldwin, R. L. 1982. *Ann. Rev. Biochem.* 51:459–89
54. Careri, G., Fasella, P., Gratton, E. 1975. *CRC Crit. Rev. Biochem.* 3: 141–64

55. Snyder, G. H., Rowan, R., Karplus, M., Sykes, B. D. 1975. *Biochemistry* 14: 3765–77
56. Wagner, G., DeMarco, A., Wüthrich, K. 1976. *Biophys. Struct. Mech.* 2: 139–58
57. Gelin, B. R., Karplus, M. 1975. *Proc. Natl. Acad. Sci. USA* 72:2002–06
58. Hetzel, R., Wüthrich, K., Deisenhofer, J., Huber, R. 1976. *Biophys. Struct. Mech.* 2:159–80
59. Northrup, S. H., Pear, M. R., Lee, C. Y., McCammon, J. A., Karplus, M. 1982. *Proc. Natl. Acad. Sci. USA* 79: 4035–39
60. Campbell, I. D., Dobson, C. M., Moore, G. R., Perkins, S. J., Williams, R. J. P. 1976. *FEBS Lett.* 70:96–98
61. *Structure and Motion in Molecular Liquids,* 1978. Faraday Disc. Chem. Soc., Vol. 66
62. Wilson, E. B., Decius, J. C., Cross, P. C. 1955. *Molecular Vibrations* New York: McGraw-Hill
63. Levy, R. M., Karplus, M. 1979. *Biopolymers* 18:2465–95
64. Warme, P. K., Scheraga, H. A. 1974. *Biochemistry* 13:757–67
65. Levitt, M. 1974. *J. Mol. Biol.* 82:393–420
66. Gelin, B., Karplus, M. 1979. *Biochemistry* 18:1256–68
67. Lifson, S. 1982. In *Structural Molecular Biology,* ed. D. G. Davies, W. Saenger, S. S. Danyluk. New York: Plenum
68. Pauling, L., Corey, R. R., Branson, H. B. 1951. *Proc. Natl. Acad. Sci. USA* 37:205–11
69. Ramachandran, G. N., Ramakrishnan, C., Sasisekharan, V. 1963. *J. Mol. Biol.* 7:95–99
70. Brooks, B., Bruccoleri, R. E., Olafson, B. D., States, D. J., Swaminathan, S., Karplus, M. 1983. *J. Comput. Chem.* In press
71. McCammon, J. A., Gelin, B. R., Karplus, M. 1977. *Nature* 267:585–90
72. McCammon, J. A., Wolynes, P. G., Karplus, M. 1979. *Biochemistry* 18: 927–42
73. van Gunsteren, W. F., Karplus, M. 1983. *Macromolecules.* In press
74. Chandrasekhar, S. 1943. *Rev. Mod. Phys.* 15:1–89
75. Ermak, D. L., McCammon, J. A. 1978. *J. Chem. Phys.* 69:1352–60
76. Levy, R. M., Karplus, M., McCammon, J. A. 1979. *Chem. Phys. Lett.* 65:4–11
77. Helfand, E., Wasserman, Z. R., Weber, T. A. 1980. *Macromolecules* 13:526–33
78. McCammon, J. A., Northrup, S. H.,

Karplus, M., Levy, R. M. 1980. *Biopolymers* 19:2033–45

79. Gō, M., Gō, N. 1976. *Biopolymers* 15:1119–27

80. Suezaki, Y., Gō, N. 1976. *Biopolymers* 15:2137–53

81. Levy, R. M., Perahia, D., Karplus, M. 1982. *Proc. Natl. Acad. Sci. USA* 79:1346–50

82. Chandler, D. 1978. *J. Chem. Phys.* 68:2959–70

83. Pechukas, P. 1976. In *Dynamics of Molecular Collisions,* Part B. ed. W. H. Miller. New York: Plenum

84. Keck, J. C. 1962. *Discuss. Faraday Soc.* 33:173–82

85. McCammon, J. A., Karplus, M. 1979. *Proc. Natl. Acad. Sci. USA* 76:3585–89

86. McCammon, J. A., Karplus, M. 1980. *Biopolymers* 19:1375–405

87. van Gunsteren, W. F., Berendsen, M. J. C. 1977. *Mol. Phys.* 34:1311–27

88. McCammon, J. A., Gelin, B. R., Karplus, M., Wolynes, P. G. 1976. *Nature* 262:325–26

89. McCammon, J. A., Wolynes, P. G. 1977. *J. Chem. Phys.* 66:1452–56

90. Karplus, M., Weaver, D. L. 1976. *Nature* 260:404–06

91. Karplus, M., Weaver, D. L. 1979. *Biopolymers* 18:1421–37

92. Miyazawa, T. 1967. In *Poly-α-Amino Acids,* ed. G. D. Fasman. New York: Marcel Dekker

93. Gō, N., Scheraga, H. A. 1976. *Macromolecules* 9:535–42

94. Bunker, D. L., Wang, F.-M. 1977. *J. Am. Chem. Soc.* 99:7457–59

95. Warshel, A., Karplus, M. 1974. *J. Am. Chem. Soc.* 96:5677–89

96. Case, D. A., Karplus, M. 1979. *J. Mol. Biol.* 132:343–68

97. Flory, P. J. 1969. *Statistical Mechanics of Chain Molecules.* New York: Wiley

98. Levitt, M. 1976. *J. Mol. Biol.* 104:59–107

99. Levitt, M. 1976. *Models for Protein Dynamics.* CECAM Workshop Report. Univ. Paris XI, Orsay, France

100. Pear, M. R., Northrup, S. H., McCammon, J. A., Karplus, M., Levy, R. M. 1981. *Biopolymers* 20:629–32

101. Northrup, S. H., Pear, M. R., McCammon, J. A., Karplus, M., Takano, T. 1980. *Nature* 287:659–60

102. Frauenfelder, H., Petsko, G. A., Tsernoglou, D. 1979. *Nature* 280:558–63

103. Sternberg, M. J. E., Grace, D. E. P., Phillips, D. C. 1979. *J. Mol. Biol.* 130:231–45

104. Artymiuk, P. J., Blake, C. C. F., Grace, D. E. P., Oatley, S. J., Phillips, D. C.,

Sternberg, M. J. E. 1979. *Nature* 280:563–68

105. Takano, T., Dickerson, R. E. 1980. *Proc. Natl. Acad. Sci. USA* 77:6371–75

106. Olafson, B., Ichiye, T., Karplus, M. 1983. *J. Mol. Biol.* Manuscript in preparation

107. van Gunsteren, W. F., Karplus, M. 1981. *Nature* 293:677–78

108. Kuryan, J., Swaminathan, S., Petsko, G., Karplus, M., Levy, R. M. 1983. *Biochemistry.* Manuscript in preparation

109. Aschaffenburg, R., Blake, C. C. F., Dickie, H. M., Gayen, S. K., Keegan, R., Sen, A. 1980. *Biochem. Biophys. Acta* 625:64–71

110. van Gunsteren, W. F., Karplus, M. 1982. *Biochemistry* 21:2259–74

111. van Gunsteren, W. F., Berendsen, M. J. C., Hermans, J., Hol, W. G. J., Postma, J. P. M. 1983. *Proc. Natl. Acad. Sci. USA* In press

112. Levitt, M. 1980. In *Protein Folding,* ed. R. Jaenicke, pp. 17–39. New York: Elsevier/North Holland

113. Watenpaugh, K. D., Margulis, T. N., Sieker, L. C., Jensen, L. H. 1978. *J. Mol. Biol.* 122:175–90

114. Karplus, M., McCammon, J. A. 1979. *Nature* 277:578

115. Northrup, S. H., Pear, M. R., Morgan, J. D., McCammon, J. A., Karplus, M. 1981. *J. Mol. Biol.* 153:1087–1109

116. Morgan, J. D., McCammon, J. A., Northrup, S. H. 1983. *Biopolymers.* In press

117. Karplus, M., Swaminathan, S., Ichiye, T., van Gunsteren, W. F. 1982. See Ref. 51

118. Swaminathan, S., Ichiye, T., van Gunsteren, W. F., Karplus, M. 1982. *Biochemistry* 21:5230–41

119. Mao, B., Pear, M. R., McCammon, J. A., Northrup, S. H. 1982. *Biopolymers.* 21:1979–89

120. Noguti, T., Gō, N. 1982. *Nature* 296:776–78

121. Levy, R. M., Karplus, M., McCammon, J. A. 1981. *J. Am. Chem. Soc.* 103:994–96

122. Levy, R. M., Karplus, M., Wolynes, P. G. 1981. *J. Am. Chem. Soc.* 103:5998–6011

123. Hoch, J. C., Dobson, C. M., Karplus, M. 1982. *Biochemistry* 21:1115–25

124. Levitt, M. 1981. *Nature* 294:379–80

125. Karplus, M., Kushick, J. N. 1981. *Macromolecules* 14:325–32

126. Parak, F., Frolov, E. N., Mössbauer, R. L., Goldanskii, V. I. 1981. *J. Mol. Biol.* 145:825–33

127. Levy, R. M., Szabo, A. 1982. *J. Am. Chem. Soc.* 104:2073–207
128. Kasprzak, A., Weber, G. 1982. *Biochemistry* 21: In press
129. Lakowicz, J. R., Maliwal, B., Cherek, H., Balter, A. 1983. *Biochemistry.* In press
130. Ichiye, T., Karplus, M. 1983. *Biochemistry.* Submitted for publication
131. Northrup, S. H., Hynes, J. T. 1980. *J. Chem. Phys.* 73:2700–14
132. Karplus, M., McCammon, J. A. 1981. *FEBS Lett.* 131:34–36
133. Takano, T. 1977. *J. Mol. Biol.* 110:569–84
134. Phillips, S. E. 1978. *Nature* 273:247–48
135. Austin, R. H., Beeson, K. W., Eisenstein, L., Frauenfelder, H., Gunsalus, I. C. 1975. *Biochemistry* 14:5355–73
136. Beece, D., Eisenstein, L., Frauenfelder, H., Good, D., Marden, M. C., et al. 1980. *Biochemistry* 19:5147–57
137. McCammon, J. A., Northrup, S. H. 1981. *Nature* 293:316–17
138. Northrup, S. H., Zarrin, F., McCammon, J. A. 1982. *J. Phys. Chem.* 86:2314–21
139. Szabo, A., Shoup, D., Northrup, S. H., McCammon, J. A. 1982. *J. Chem. Phys.* 77:4484–93
140. Hartsuck, J. A., Lipscomb, W. N. 1971. *Enzymes* 3:1–56
141. Banner, D. W., Bloomer, A. C., Petsko, G. A., Phillips, D. C., Pogson, D. I., et al. 1975. *Nature* 255:609–14
142. Pratt, L. R., Chandler, D. 1977. *J. Chem. Phys.* 67:3683–704
143. Imoto, T., Johnson, J. N., North, A. C. T., Phillips, D. C., Rupley, J. A. 1972. *Enzymes* 7:665
144. Wolynes, P. G., McCammon, J. A. 1977. *Macromolecules* 10:86–87
145. McCammon, J. A., Karplus, M. 1977. *Nature* 268:765–66
146. Mao, B., Pear, M. R., McCammon, J. A., Quiocho, F. A. 1982. *J. Biol. Chem.* 257:1131–33
147. Colonas, F., Perahia, D., Karplus, M., Ecklund, H., Bränden, C. I. 1983. *J. Mol. Biol.* In preparation
148. Newcomer, M. E., Lewis, B. A., Quiocho, F. A. 1981. *J. Biol. Chem.* 256:13218–22
149. Ecklund, M., Nordström, B., Zeppezauer, E., Söderlund, G., Ohlsson, I., et al. 1976. *J. Mol. Biol.* 102:27–59
150. Ecklund, H., Samama, J. P., Wallén, L., Bränden, C.-I., Åkeson, Å., Alwyn Jones, T. 1981. *J. Mol. Biol.* 146:561–87
151. Harrison, S. C. 1980. *Biophys. J.* 32:139–51
152. Perutz, M. F. 1979. *Ann. Rev. Biochem.* 48:327–86
153. Perutz, M. F. 1970. *Nature* 228:726–34
154. Baldwin, J., Chothia, C. 1979. *J. Mol. Biol.* 129:175–220
155. Gelin, B. R., Karplus, M. 1977. *Proc. Natl. Acad. Sci. USA* 74:801–05
156. Gelin, B. R., Lee, A. W.-M., Karplus, M. 1983. *J. Mol. Biol.* In press
157. Anderson, L. 1973. *J. Mol. Biol.* 79:495–506
158. McDonald, R. C., Steitz, T. A., Engelman, D. M. 1979. *Biochemistry* 18:338–42
159. Pickover, C. A., McKay, D. B., Engelman, D. M., Steitz, T. A. 1979. *J. Biol. Chem.* 254:11323–29
160. Hanson, D. C., Yguerabide, J., Schumaker, V. N. 1981. *Biochemistry* 20:6842–52
161. Hartmann, H., Parak, F., Steigemann, W., Petsko, G. A., Runge Ponti, D., Frauenfelder, H. 1982. *Proc. Natl. Acad. Sci. USA* 79:4967–71

Ann. Rev. Biochem. 1983. 52:301–54

CELLULAR ONCOGENES AND RETROVIRUSES

J. Michael Bishop

Department of Microbiology and Immunology, University of California, San Francisco, California 94143

CONTENTS

PERSPECTIVES AND SUMMARY 302
INTRODUCTION: CANCER GENES CONCEIVED 303
UNVEILING CELLULAR ONCOGENES: EXPERIMENTAL STRATEGIES
    AND DEFINITIONS... 304
    Transduction by Retroviruses 304
    Insertional Mutagenesis 306
    DNA-Mediated Gene Transfer (Transfection) 306
    Definitions.. 308
STRUCTURAL CHARACTERISTICS OF CELLULAR ONCOGENES.................. 308
PHYLOGENY OF CELLULAR ONCOGENES 309
EXPRESSION OF CELLULAR ONCOGENES 313
FUNCTIONS OF CELLULAR ONCOGENES.. 314
FAMILIES OF CELLULAR ONCOGENES 317
GENETIC LINKAGES OF CELLULAR ONCOGENES...................................... 319
TRANSDUCTION OF CELLULAR ONCOGENES BY RETROVIRUSES 320
    Does it Happen?.. 320
    Variations on the Theme 321
    The Architecture of Transduced Oncogenes 324
    The Mechanism of Transduction 325
    Forcing the Issue: Experimental Transduction of Cellular Oncogenes 328
    Retroviruses as Generalized Transducing Agents 330
THE ROLE OF CELLULAR ONCOGENES IN NORMAL CELLS...................... 331
CELLULAR ONCOGENES AND CANCER.. 334
    Viral and Cellular Oncogenes are Closely Related 334
    At Least Some Cellular Oncogenes Can Induce Neoplastic Transformation............... 335
    Implicating Cellular Oncogenes in Tumorigenesis 336
    Converting Proto-Oncogene to Oncogene... 338
    The Role of DNA Damage in Oncogenesis 340
    Accounting for Tissue-Specificity of Oncogenes 343
FITTING CELLULAR ONCOGENES INTO THE GENESIS OF NEOPLASIA 344
CONCLUSIONS AND PROSPECTS 346

0066-4154/83/0701-0301$02.00

"... death dwelled in the cell though the cell be looked in on at its most quick ..." (1)

PERSPECTIVES AND SUMMARY

The causes of cancer are legion, but all may play upon a common genetic substrate within the cell. The components of this substrate seemed far from view until the discovery that the oncogenes of retroviruses are copies of cellular genes, known as "proto-oncogenes" or "cellular oncogenes". Three experimental strategies have emerged to enlarge the repertoire of cellular oncogenes and to implicate these genes in carcinogenesis: (*a*) transduction by retroviruses continues to unveil new oncogenes derived from the cell; (*b*) insertional mutagenesis by retroviruses provides evidence that induction of cellular oncogenes participates in the genesis of tumors and, in the process, can finger novel oncogenes; and (*c*) DNA-mediated gene transfer can be used to detect and isolate active oncogenes from human tumor cells. On occasion, more than one of these strategies has hit upon the same cellular gene, evoking the image of a single family of cellular oncogenes.

Transduction by retroviruses has garnered the richest harvest of oncogenes—seventeen by present count, each with a distinct structure, and several with different mechanisms of action. Little is known of how transduction occurs: it exploits peculiarities of the retrovirus life cycle; it apparently occurs in several distinct steps, separated in time; it evidently employs nonhomologous recombination; it requires a spliced representation of the cellular gene as an intermediate form; and it is exceedingly rare. Cellular genes other than oncogenes are presumably subject to transduction by retroviruses as well, but the event would not be easy to perceive, and means by which to increase its frequency are not apparent.

Integration of retroviral DNA into the host chromosome can enhance the expression of adjacent cellular genes. In the rare event that the activated gene is a cellular oncogene, tumorigenesis may be initiated. A number of different tumors induced by retroviruses probably begin in this manner, promising further revelations of oncogenes.

Gene-transfer procedures have revealed active oncogenes in the DNA of a large variety of human tumors. The validity of this approach has been challenged on technical grounds, but the challenge now seems moot because the strategy has recovered genes whose structures or activities are manifestly abnormal in the tumors from which the genes were obtained. Several of the oncogenes isolated in this manner are derived from a family of cellular genes that is also subject to transduction by retroviruses. This happy conjunction should hasten elucidation of the mechanisms by which the genes act and their role in spontaneous oncogenesis.

The cellular oncogenes uncovered to date may be members of a superfamily with a single or very few founder genes. Whatever their origins, the genes can be traced from primitive metazoans to *Homo sapiens,* and they display remarkable conservation of both structure and function. These are cellular, not retroviral genes in another guise. Most are expressed in various tissues and at various times during the course of growth and development. Their function in normal cells is not known, although the prevailing speculation holds that they are active in differentiation.

Evidence is mounting that cellular oncogenes are involved in tumorigenesis, whatever its proximal cause. Some of these genes may serve to initiate the genesis of tumors, some to sustain the final neoplastic phenotype, some may serve either purpose in different contexts. Cellular oncogenes appear not to be tumorigenic in their native state; they must either be activated to abnormal levels of expression, or mutated so as to change some aspect of their function. Based on present evidence, either of these mechanisms may apply.

The recognition and isolation of cellular oncogenes has set the stage for fresh assaults on the mechanisms of carcinogenesis. The role of DNA damage can now be examined directly and the nature of the damage explored. The number of genetic functions required for the genesis of an individual tumor may become clear. The genes responsible for heritable predispositions to cancer can be sought. We hope to discern the variety and nature of the biochemical functions that can drive a cell to cancerous growth. The design of definitive strategies for the prevention or cure of human malignancy may have to await all of these accomplishments.

INTRODUCTION: CANCER GENES CONCEIVED

"Why, sometimes I've believed as many as six impossible things before breakfast." (2)

The genetic origins of cancer have been an article of faith for over half a century (3). Formal Mendelian analyses first gave substance to the faith: explicit genetic determinants of carcinogenesis were identified in several experimental systems, including fish, insects, and plants (4); the examination of human pedigrees engendered the concept of "cancer genes" whose inherited abnormalities predispose to specific forms of neoplasia (5–7). The discovery of vertically transmitted leukemia in inbred mice brought retroviruses on stage, leading eventually to the proposal that retroviral oncogenes reside in the germ lines of all species (8, 9). The proposal was faulty—the leukemogenic retroviruses carried by the murine germ line do not possess oncogenes. Paradoxically, however, the proposal proved to be only slightly wide of the mark, and it was from retroviruses that the seminal revelation sprang. Vertebrate species do indeed harbor genetic loci homologous to retrovirus oncogenes, but these loci are cellular, not viral genes, and they have assumed an importance that extends far beyond the confines of tumor virology.

Five years have passed since retroviruses were last reviewed in this series (10). At the time, all important aspects of these viruses could still be encompassed in a single, albeit condensed, essay. Now the scope of the subject is large enough to occupy a tome of intimidating bulk (11). Here I must be content to pursue the major promise with which the previous review concluded: the tumorigenic properties of retroviruses have taken center stage to provide exhilarating purchase on one of the preeminent problems in the biological sciences—the genesis of the neoplastic cell.

UNVEILING CELLULAR ONCOGENES: EXPERIMENTAL STRATEGIES AND DEFINITIONS

Transduction by Retroviruses

Many retroviruses bear genetic loci (oncogenes) whose activities are responsible for both the initiation and maintenance of neoplastic transformation induced by viral infection. At the time of their discovery, retroviral oncogenes were viewed as biological anomalies akin to the oncogenic determinants of other tumor viruses. We had no way to judge their possible significance or their potential relevance to tumorigenesis in human beings. Yet these genes were soon to provide our first glimpse of an apparatus that may lie at the core of all forms of carcinogenesis.

Oncogenes were first defined by genetic analysis (12, 13). The efforts required are arduous, however, and the oncogenic potentials of many retroviruses have resisted genetic dissection to this day. Physical descriptions of oncogenes proved easier to come by, acquiring great rigor with the advent of molecular cloning and the use of DNA-mediated gene transfer (or transfection) (14–16). As technical access improved, the ranks of oncogenes grew to unanticipated profusion. We can now account for seventeen distinctive retroviral oncogenes, and we have no reason to believe that the accretion of new genes is at an end. A detailed description of these genes is beyond the purview of the present essay; their structure and properties nevertheless pervade many of the arguments that follow and constitute most of what we know about the potential properties of cellular oncogenes. The reader is referred to a comprehensive treatise that has just appeared elsewhere (11). Table 1 summarizes a few important properties of the known retroviral oncogenes.

The evolutionary origins of retroviral oncogenes were every bit as obscure as the origins of other viral genes until the discovery that both avian (17) and mammalian (18) DNAs contain nucleotide sequences closely related to the oncogene src of Rous sarcoma virus. These findings prompted two suggestions that foreshadowed the present place of retroviruses in the study of cancer genes: (a) all vertebrates possess a highly conserved gene (cellular *src*) that is related to viral *src* and is therefore potentially oncogenic; and (*b*) viral *src* arose by the transduction of cellular *src* into a preexistent retrovirus. As additional retroviral oncogenes came to view, each in turn proved to have a cellular homologue from which the viral gene was apparently derived (11, 19). The sole exception to this rule is the Spleen Focus-Forming Virus of mice, whose oncogene appears to be a recombin-

Table 1 Oncogenes transduced by retroviruses

Viral oncogene[a]	Species of origin	Tumorigenicity	Protein products	
			Biochemical function	Subcellular location
v-*src*	Chicken	Sarsoma	PK(tyr)[b]	Plasma membrane
v-*fps*/v-*fes*	Chicken and cat	Sarcoma	PK(tyr)	Plasma membrane
v-*yes*	Chicken	Sarcoma	PK(tyr)	?[d]
v-*ros*	Chicken	Sarcoma	PK(tyr)	?
v-*ski*	Chicken		?	?
v-*myc*	Chicken	Carcinoma, sarcoma and myelocytoma	DNA binding	Nucleus
v-*erb*-A	Chicken	?	?	Cytoplasm
v-*erb*-B	Chicken	Erythroleukemia and sarcoma	?	Membranes
v-*myb*	Chicken	Myeloblastic leukemia	?	Nucleus
v-*rel*	Turkey	Lymphatic leukemia	?	?
v-*mos*	Mouse	Sarcoma	?	Cytoplasm
v-*abl*	Mouse and cat	B-cell lymphoma	PK(tyr)	Plasma membrane
v-*fos*	Mouse	Sarcoma	?	?
v-Ha-*ras*/ v-*bas*[c, e]	Rat and mouse	Sarcoma and erythroleukemia	Binds GTP; PK(thr)[f]	Plasma membrane
v-Ki-*ras*[e]	Rat	Sarcoma and erythroleukemia	Binds GTP; PK(thr)[f]	Plasma membrane
v-*fms*	Cat	Sarcoma	?	Membranes
v-*sis*	Woolly monkey and cat	Sarcoma	?	?

[a] The names of viral genes are treated here as generic terms, although in most instances several separate viral isolates are known and the proteins encoded by the homologous oncogenes in the different isolates may differ in size.

[b] PK denotes protein kinase. The amino acid subject to phosphorylation is given in parentheses.

[c] Names were assigned before adoption of standard nomenclature. Homologous genes transduced from different species are now given the same name. Examples include *abl* from mice and cats, *sis* from woolly monkeys and cats.

[d] Question marks signify uncertainty.

[e] Ha denotes Harvey strain, Ki denotes Kirsten strain.

[f] Phosphorylation has been observed only on the oncogene product itself, not on other substrates in *trans*.

ant form of the retrovirus envelope gene rather than the derivative of a cellular gene (20). It appears that retroviral oncogenes are but a sampling of a larger group of cellular genes with oncogenic potential—the "cancer genes" envisioned by medical geneticists (5–7). In this view, transduction by retroviruses has provided a rich diversity of genes whose properties may all speak to the ways and means by which tumorigenesis occurs, whatever its proximal cause.

Insertional Mutagenesis

Integration of viral DNA into the host genome to give a provirus is an inevitable consequence of infection by retroviruses (10, 21). Integration is potentially mutagenic on two counts: it may disrupt a vital region of the host genome, and it can bring powerful regulatory functions of the virus to bear on the expression of host genes. Virologists have long cherished the idea that insertional mutagenesis might account for some forms of viral oncogenesis, but the possibility lay dormant until B-cell lymphomas induced in chickens by avian leukosis viruses (ALVs) came under belated study. ALVs do not possess oncogenes (10, 22), so the tumorigenicity of these viruses requires more subtle explanations. Insertional mutagenesis provides at least part of the scheme. Almost all of the lymphomas induced by ALVs contain viral DNA integrated in a common domain of the host genome (23–25). Within this domain lies a cellular gene whose activity is apparently augmented as a consequence of the insertion of viral DNA (23, 24, 26, 27). By a stroke of great good fortune, the activated gene is already known to us as the cellular homolog (c-*myc*) of a retrovirus oncogene (v-*myc*) (26, 27). The inference that activation of c-*myc* initiates the process of oncogenesis in B-cells infected by ALV comes easily and has undeniable logical force. If other retroviruses without oncogenes act by similar means, the sites at which their proviruses integrate into cellular DNA may reveal previously unrecognized cellular genes whose activation can be implicated in tumorigenesis. Recent studies of several retroviruses with diverse oncogenic potentials indicate that this hope is likely to be realized (28, 29; D. Westaway, personal communication).

DNA-Mediated Gene Transfer (Transfection)

"If you hear hoof-beats, think of horses." (30)

The unveiling of potential cancer genes by transduction and insertional mutagenesis with retroviruses was largely serendipitous. A more deliberate assault on the problem was mounted by the use of transfection, in a quest for evidence that the DNA of tumor cells is abnormal (31). The quest has met with stunning success: DNA from a large variety of solid and hemopoietic tumors transforms certain lines of cultured cells to a neoplastic phenotype (31–45).

There has been abundant parlor talk among cancer biologists questioning the suitability of the transfection assay. The critisms focus on two main points: the assay utilizes established lines of cells with properties quite removed from those of normal cells, and the DNA for testing was at first obtained from established lines of tumor cells rather than from the tumors themselves. These concerns are obvious but now seem blunted: 1. Several different lines of cells have been reproducibly transformed, including cells that are stably diploid and display the multi-step pattern of chemically-induced transformation so familiar to students of oncogenesis (45). 2. DNAs from several types of primary tumors have now been shown to be effective in the assay (15, 16, 34, 39; M. Barbacid, G. Cooper, R. Weinberg, personal communication). We need fear no prevailing artifact due to the use of DNA from established cell lines. 3. The active oncogenes isolated from tumor DNA by combining the transfection assay with molecular cloning are manifestly abnormal; they transform cells with efficiencies that are orders of magnitude beyond the barely measurable transforming frequencies displayed by their cohorts isolated from normal cells (37a, 41, 43, 46). I prefer not to believe that this difference reflects wholly indiscriminate changes in the genomes of cancer cells. It seems far more likely that damaged genes with etiological importance have been isolated.

There are admitted constraints to the transfection assay: 1. Some active oncogenes may fail to transform the cells presently used in the assay; the precedence of retroviral oncogenes warns of severe host-specificity in some instances (11). 2. Exceptionally large genes may not perform well in the assay. 3. The success of transfection requires that the oncogene be a dominant effector. Recessive lesions have been implicated in some forms of tumorigenesis (47, 48), and I presume that these would remain hidden from the transfection assay. 4. The assay seems suitable only for the detection of genetic functions that maintain cells in a transformed phenotype. It seems unlikely that earlier steps in the process of tumorigenesis could be perceived (see below).

Given these cautions, we can only be surprised and pleased that transfection has proven so successful. A score of potentially novel oncogenes have been uncovered and are now enjoying the doting attention of molecular biologists. Some investigators anticipated that the cellular genes responsible for transformation by tumor DNA would prove to be a new variety of oncogene. They were wrong, at least in part. The ability to detect active oncogenes in tumor cell DNA permits the isolation of the genes by molecular cloning. Among the first harvests of this endeavor were genes previously known to retrovirologists as the cellular homologs (and presumed progenitors) of two retroviral oncogenes, the Harvey and Kirsten forms of v-*ras* (37a, 41, 43). These findings bring unity to the pursuit of cancer genes. We can now hope that all of the experimental approaches summarized here are addressed to the same cause: the isolation and characterization of cellular genes whose activities are authentic components in the pathway to malignancy.

Definitions

The rapid pace of discovery in the study of oncogenes has wreaked havoc on taxonomy. Retrovirologists reached an uneasy peace by adopting a nomenclature in which viral oncogenes are known as v-*onc*'s, the cellular progenitors of v-*onc*'s as cellular oncogenes (c-*onc*'s), and each of the viral and cellular genes by terms derived from the names of the viruses in question, e.g. v-*src*, c-*src*, v-*ras*, c-*ras* (49). Oncogenes identified by transfection have been known variously as "tumor genes," "cellular oncogenes," "transforming genes," or merely "oncogenes." The term "proto-oncogene" was used to denote either the cellular progenitors of retrovirus v-*onc*'s (19), or the cellular genes whose damage gives rise to the active oncogenes in tumor DNA (16, 50). The recognition that the oncogenes of retroviruses and tumor cells are of a kind will undoubtedly force the eventual adoption of a unifying nomenclature. Here I choose the conservative course of using the retrovirus terminology wherever appropriate, but I also extend the term "cellular oncogene" to include the active oncogenes in tumor DNAs, and I resurrect "proto-oncogene" to designate any cellular gene with the potential to become an oncogene.

STRUCTURAL CHARACTERISTICS OF CELLULAR ONCOGENES

Most of what we have learned of cellular oncogenes has come from the study of retroviral c-*onc*'s; they have been in hand far longer than the oncogenes of tumor DNA, and their viral counterparts provide convenient access to structure and function. According to early returns, however, the active oncogenes isolated from DNA of tumor cells will provably adhere to the principles enunciated with retrovirus c-*onc*'s.

Virtually all of the known retroviral c-*onc*'s have been obtained by molecular cloning from at least one species (generally the species that accommodated transduction of the cellular gene into a retrovirus genome; isolation of c-*onc*'s from human DNA has also received special attention). No c-*onc* has yet been described in full, however, principally because of ambiguities that arise from the means used to identify the genes. Retroviral c-*onc*'s are recognized by virtue of homology with v-*onc*'s. In no instance does this homology represent the entirety of the cellular gene; transduction of a c-*onc* generally excludes several cardinal features of the gene, including the regions where transcription from the gene starts and stops, introns, and, on occasion, portions of the coding region of the gene (see below). The boundaries of cellular oncogenes identified by transfection are even more vaguely defined, since the genes are carried on fragments of DNA whose arbitrary sizes are constrained only by the requirement that they include a sufficient portion of the cellular oncogene to assure its expression and functional competence. Even nucleotide sequencing can fail to resolve uncertainties regarding the transcriptional boundaries of the gene and the extent of the coding domain (51). We will be able to properly define c-*onc*'s only when we know where transcription and translation from the genes starts and stops.

The topographical diversity of c-*onc*'s is equivalent to anything observed with other cellular genes. They range in size from a few kilobasepairs (kbp) to over forty kbp (52–54). They may have no introns, within regions transduced by retroviruses at least (55–57), or more than a dozen (52–54). The size of individual introns can be minuscule (a few dozen bp) or gargantuan (thousands of bp), and both very large and very small introns are found within the same gene (52–54). Introns may demarcate functional domains within the protein encoded by the gene, but the evidence on this point is even less secure than for other, better studied genes and their products (58, 59).

Cellular oncogenes behave as classical Mendelian loci. They occupy constant positions within the genomes of particular species (60, 61), they are apparently present in all members of a given species (60, 61), and they segregate in a predictable fashion when breedings are analysed with the assistance of occasional structural polymorphisms that have been identified by restriction mapping (D. Spector, B. Vennstrom, personal communications).

PHYLOGENY OF CELLULAR ONCOGENES

"To assume that the current utility of a feature permits an inference about the reasons for its evolutionary origin is a lamentably common error. Current utility and historical origin are different subjects." (61a)

Evolutionary conservation is a hallmark of cellular oncogenes. The extent of conservation observed with early examples of transduced c-*onc*'s was sufficiently great to be viewed as evidence that the genes are vital to the species in which they are found (17–19). The techniques used at first to detect c-*onc*'s were not decisive, however, and doubts remained as to the reality of apparent similarities that stretched across extraordinary reaches of evolutionary time. These doubts have since been allayed by the use of molecular cloning to isolate and characterize homologous c-*onc*'s from creatures as disparate as *Drosophila* and humans. Although few of the isolated genes have as yet been analysed in detail, what we now know justifies the expectation that homologous c-*onc*'s transduced from different species will generally share common structural features, common functions, and common ancestors.

Similar tenets are emerging for cellular oncogenes identified in tumor DNA by transfection. Structural analysis of these genes has just begun, so we can speak with confidence only of those genes that had been recognized before as c-*onc*'s transduced by retroviruses (37a, 41, 43). But the activities of cellular oncogenes from the same types of tumors in rodents and humans can apparently be inactivated by cleavage with the same distinctive sets of restriction endonucleases (34–37, 39). These findings foretell structural homologies of a very high order.

The extent of evolutionary conservation varies from one cellular oncogene to another. Some are readily detectable only in closely related species; as a consequence, their evolutionary age is difficult to judge. Others appear to have taken recognizable form well before the emergence of vertebrates (62). What these differences connote, we cannot presently say. The tools used to trace the lineages of cellular oncogenes are blunt and rely upon the suspect assumption that conserved function demands preservation of nucleotide sequence. Perhaps all cellular oncogenes originated no later than the appearance of metazoan organisms. The issue is not trivial because arguments from evolution have informed most efforts to discern the roles of cellular oncogenes in normal cells (see below).

The discovery of c-*onc*'s in the insect *Drosophila* and the worm *Caenorhabditis elegans* (62) has received special attention, in part because we can now place the origins of the genes much deeper in the recesses of time than previously anticipated, but more importantly, because it may now be possible to bring sophisticated genetic analyses to bear on the function of oncogenes in both normal cells and heritable tumors (4, 63). Another view is possible, of course, that we now have pressed our techniques too far, that the homologies we detect reflect ancestral relationships without physiological meaning, that the ostensibly related oncogenes of insects and mammals are not counterparts in function. This view is not likely to hold sway. The homologies between *Drosophila* DNA and v-*src*, v-*abl* and v-*ras* encompass portions of the viral oncogenes that are critical to the functions of their protein products; initial results from nucleotide sequencing indicate that the c-*ras* genes of *Drosophila* and rodents are appreciably related; expression in the form of messenger RNAs has already been demonstrated for several oncogenes in *Drosophila*; and both embryonal cells and larval tissues of *Drosophila* contain a tyrosine-specific protein kinase similar to that encoded by v-*src* of Rous sarcoma virus and the c-*src* genes of birds and mammals (B. Shilo, M. Simon, personal communications).

The details of conservation among cellular oncogenes should attract the attention of evolutionary biologists. These genes represent both the largest sampling of low frequency eukaryotic genes presently available for study, and potential access to rules that govern the evolution of genes likely to possess regulatory functions. It is too early, however, to draw telling generalizations. Perhaps the most remarkable trend is an unruly variation of gene number from one species to another, substantially greater variation than has been observed for other unique sequence genes:

1. A single locus of c-*myc* was reported for chickens (64–66), whereas more than four loci related to c-*myc* exist in human DNA (67). The difficulty in comparisons of this sort is to establish satisfactory definitions of structural homology; it may prove possible to find further c-*myc* loci in chickens by reducing the stringency of the search. Despite these ambiguities, a structural lineage is apparent for c-*myc*; the locus identified in chickens (65, 66) and one of the human loci (67) share very similar topographies, whereas the multiplicity of other human loci related to c-*myc* may be only partial representations of the gene (67).

2. The family of c-*ras* genes has two arms, one transduced as v-*ras* in the Harvey strain of murine sarcoma virus, the other as v-*ras* in Kirsten murine sarcoma virus (68). Harvey and Kirsten c-*ras* are quite diverged (68–70), as if the two arms of the family evolved separately from an ancient gene duplication. Even *Drosophila* may possess distinct Harvey and Kirsten varieties of c-*ras* (B. Shilo, personal communication); if so, the ancestral duplication probably occurred before the phylogenetic radiation that produced insects. Harvey and Kirsten c-*ras* are distributed in three patterns among different species: as unique loci (possible examples include chickens and mice; see 68, 71); as duplications within the Harvey and Kirsten lineages (in rats and humans; see 68, 70, 72); and as substantially amplified gene families (in the rodent *Mus pohari* and the Chinese hamster; see 71). As a result of the duplications in rats and humans, the Harvey and Kirsten forms of *ras* are each represented by two loci, one with introns, the other without (68, 70, 72). The general topographies of the cognate rat and human genes appear to be similar. These interrelationships evoke the image of "processed genes"; the duplications of Harvey and Kirsten c-*ras* may have arisen when spliced versions of the original genes were transposed to new chromosomal locations (73). We cannot presently determine when the duplications might have occurred, or whether they have occurred only once during evolution. The Harvey locus that contains introns is the most

conserved of the c-*ras* loci: the number, size, and locations of exons and introns are analogous in mice, rats, and humans (68, 70–72); the nucleotide sequences of the exons diverge very little from mouse to human (71, 72); and the introns are appreciably conserved, albeit much less so than the exons (71). Harvey c-*ras* is amplified to approximately ten copies in *M. pohari* (but not in other species of rodents), Kirsten c-*ras* to at least eight copies in Chinese (but not Syrian) hamsters (71). These amplifications are unusual for unique sequence genes in two regards: magnitude, and occurrence relatively late in evolutionary divergence. The mechanism by which the amplifications arose is presently obscure. But *M. pohari* contains one locus of Harvey c-*ras* that is closely related to the conserved locus with introns found in rats, mice, and humans. It is tempting to regard this locus as the founder gene for the duplications of Harvey c-*ras* in rats and humans, and for the greater amplifications in *M. pohari*. We have no clues of this sort for the amplifications of Kirsten c-*ras*.

3. Less well-explored gene duplication is evident for c-*src* in chickens, two loci (74) and humans, two loci (R. Parker, personal communication). Two distinctive forms of c-*src* protein were observed in some human cells (75), but it is not known whether these proteins are encoded by different genes.

To what end are c-*onc*'s amplified? Do the supernumerary copies of the genes have a purpose, or were they abandoned to decay on a junkheap of needlessly duplicated genes? The evidence is mixed and inconclusive. On the one hand, all but one of the c-*myc* loci in human DNA appear to be incomplete and might be nonfunctional pseudogenes (67); one of the two copies of c-*src* in chicken and human DNAs may have abnormal structure (R. Parker, personal communication), and the ten-fold amplification of Harvey c-*ras* in *M. pohari* is not accompanied by augmented expression of the gene, as if the additional copies might be silent (71). On the other hand, both forms of Harvey c-*ras* in rats can transform cells in culture when linked to a retroviral promoter of transcription (70). The duplicated genes must still encode active proteins.

The fine details of evolutionary conservation have begun to emerge from analyses of nucleotide sequence. The mouse and human versions of c-*mos* are apparently single-copy genes without introns (55–57, 76). Within the coding regions of the mouse (57) and human (76) genes, 77% of the nucleotide sequences are identical; the deduced amino acid sequences of the encoded proteins are similarly related (75%). The genes c-*fps* and c-*fes* are avian and feline versions of the same genetic locus (77–79). The viral forms of *fps* and *fes* share about 70% of their nucleotide sequences and deduced amino acid sequences (80, 81). Conservation of this magnitude suggests that powerful forces have been at work to preserve essential functions. Paradoxically, there is no evidence for expression of c-*mos* in any species examined to date, whereas c-*fps*/c-*fes* is expressed in a variety of normal cells and tissues (see below).

Evolutionary arguments provided an early glimpse of the importance of cellular oncogenes, and continue to have exceptional value for interpreting the significance of these genes. From evolutionary arguments, we conclude that cellular oncogenes must be vital, and we take some assurance that the animal models most experimentalists prefer are legitimate alternatives to the often awkward use of human materials.

EXPRESSION OF CELLULAR ONCOGENES

The possibility that c-*onc*'s might be expressed in phenotypically normal cells first came into view with the discovery of RNA transcribed from c-*src* in uninfected fibroblasts and tissues of several avian species (82, 83). Since means to detect proteins encoded in cellular oncogenes have been slow to appear, measurement of transcription remains the only comprehensive strategy by which to search for expression of these genes. Virtually all c-*onc*'s are transcribed into RNA in normal (11, 82–88) and/or tumor (88–92) cells. There is only one exception at the moment: a continuing search has failed to find any evidence of transcription from c-*mos* (93, 94), a puzzling failure in view of the strong evolutionary conservation displayed by this gene (see above). Transcription from c-*fes* has also been difficult to find, but the apparently analogous gene (c-*fps*) of birds is transcribed in some tissues (87), and a protein thought to be encoded by c-*fes* has been found in cells derived from a number of mammals (95).

Transcription from c-*onc*'s follows a route already familiar from the study of other cellular genes (96). The initial transcript contains introns as well as exons and is then spliced into a series of smaller intermediates that are found only in the nucleus (85). Introns have been eliminated entirely from cytoplasmic RNAs derived from c-*onc*'s (85). Several of these cytoplasmic RNAs have now been rigorously identified as mRNAs: transcripts from c-*src* (83), c-*erb*-A, and c-*erb*-B (53) have been located in polyribosomes, and cytoplasmic RNAs representing the Kirsten variety of c-*ras* have been translated into authentic gene products in vitro (88).

As the study of transcription from c-*onc*'s broadened over the past few years, several general principles have emerged:

1. Transcription from c-*onc*'s occurs in a variety of tissues and in every vertebrate species that has been examined. Transcription of c-*src* and c-*myc* was also detected in *Drosophila* (M. Simon, S. Wadsworth, personal communication).

2. The amounts of RNA produced from c-*onc*'s in most cells are extremely small (about 1–10 copies per cell). Occasional exceptions have been found in both normal tissues (84–87) and tumor cells (89–92), but the significance of these variations and of the low constitutive levels of expression is not known (see below).

3. c-*onc*'s are not coordinately expressed as a group, nor with the genes of endogenous retroviruses, and the function of each gene may be required only in certain tissues (11, 84, 85, 87, 97).

4. Each c-*onc* gives rise to distinctive RNAs whose sizes are generally conserved in various types of cells and in different species (85, 97). The constancy of these RNAs among widely divergent species of vertebrates is further testimony to the selective pressures that have preserved the structure and function of c-*onc*'s.

5. Many of the mRNAs for c-*onc*'s are appreciably larger (three-fold or greater) than would be required to encode the presumed gene products (85, 97), as if the mRNAs had uncommonly large untranslated regions. The most dramatic example may be c-*erb*-B, whose mature mRNAs are 9 and 12 kb in length (53, 85) (although the protein(s) produced from these mRNAs has yet to be identified).

6. Most of the c-*onc*'s appear to produce a single cytoplasmic RNA, in accord with the expectation that each locus represents one gene (85). But avian embryos contain at least two mRNAs derived from c-*erb*-A and at least two derived from c-*erb*-B (53, 85). Each of these cellular loci apparently engendered a single viral gene (53, 98); it is therefore reasonable that each encodes a single protein. The multiplicity of c-*erb* mRNAs could be due to heterogeneity in the 3′ untranslated regions of the RNAs, as first described for the mRNAs of dihydrofolate reductase (99). In some instances, multiple RNAs representing a single cellular oncogene were reported, but no attempt was made to determine which were precursor forms and which were messengers (90, 91).

FUNCTIONS OF CELLULAR ONCOGENES

The biochemical functions of cellular oncogenes have not been easy to discern. Indeed, we would probably know nothing of the proteins encoded by these genes if it had not been for the existence of transduced c-*onc*'s. The experimental strategies developed for identification of proteins encoded by retroviral oncogenes can usually be deployed as well in a search for products of the corresponding c-*onc*'s. In particular, antisera prepared against a viral-transforming protein generally react with the homologous cellular protein. By these means, a handful of proteins encoded by c-*onc*'s have now been identified (Table 2). Identification of gene products,

Table 2 Proteins encoded by cellular oncogenes

| Oncogene | Protein[a] | Alleged function | |
		Cellular gene	Viral cognate
c-*src*	60K	tyr kinase[b]	tyr kinase[c]
c-*fps*	98K	tyr kinase	tyr kinase
c-*abl*	150K	?	tyr kinase
c-*ras* (Ha/Ki)	21K	thr kinase[c]	thr kinase[c]
c-*fes*	92K	?	tyr kinase

[a] Products of c-*onc*'s were identified by using antisera prepared originally against proteins encoded by the cognate v-*onc*'s. In two instances (*src* and *ras*), the proteins encoded by the viral and cellular genes have virtually identical sizes. Otherwise, the cellular proteins have molecular weights appreciably different from those of the viral proteins because the coding regions in the v-*onc*'s include substantial proteins of viral structural genes.

[b] Tyrosine-specific protein kinase activity was identified by phosphotransfer in immunoprecipitates (100, 102).

[c] The products of v-*ras* and c-*ras* have a selective kinase activity that phosphorylates threonine in the products of the oncogenes themselves, but in no other protein substrate tested to date.

however, does not guarantee discovery of function. Again, the procedure has been to argue by analogy from what is known of viral oncogenes. Once a function is found for a viral transforming protein, the same function is sought for the cellular cognate.

Characterization of the proteins encoded by cellular oncogenes proved difficult, principally because they have been available only in very small quantities. The product of vertebrate c-*src* is a 60,000 dalton phosphoprotein (pp60^{c-src}) with a tyrosine-specific protein kinase activity remarkably similar to that of the protein encoded by v-*src* (pp60^{v-src}) (100–110). Both the viral and cellular proteins are located on the plasma membrane (111) and appear to cluster in adhesion plaques (112, 113)—specialized regions of the plasma membrane that account for the adherence of cultured cells to glass or plastic. To date, only a single potential substrate has been identified for the enzymatic activity of pp60^{c-src} in vivo—an abundant cellular protein with a molecular weight of 36,000 (p36) (114–117). The function of p36 is not known, although the protein was found on the cytoplasmic aspect of the plasma membrane (118) and may also be attached to components of the cytoskeleton (119).

What part does pp60^{c-src} play in the physiology of the normal cell? Phosphorylation of tyrosine has been implicated in the cellular response to mitogenic growth factors (120–128), and the earliest events of this response take place in the plasma membrane (120, 126, 126a, 127). The enzymatic activity and subcellular location of pp60^{c-src} therefore suggest that the protein may participate in the control of cell growth and division. There is as yet only circumstantial evidence for this supposition: growth factors elicit the phosphorylation of tyrosine in p36, but the responsible kinase has not been identified (122–124, 128); and phosphorylation of pp60^{c-src} and enhancement of its protein kinase activity are among the immediate cellular responses to platelet-derived growth factor (R. Ralston, personal communication).

The two families of c-*ras* genes encode very similar proteins with molecular weights of about 21,000 (p21^{c-ras}) (129, 130). These proteins, like their viral homologs (p21^{v-ras}), bind guanine nucleotides with high affinity and carry out autophosphorylation on threonine (131, 132). No kinase activity for other protein substrates has been detected, however, and the function of the *ras* proteins therefore remains an enigma. The viral (133) and, presumably, the cellular versions of p21^{ras} are located on the plasma membrane, a provocative but so far unrevealing clue.

Relatively little is known of the other cellular oncogene products. A tenuous identification has been achieved for the product of c-*abl* (p150^{c-abl}): 1. It differs in size and composition from any of the proteins encoded by the several variants of v-*abl* (134), in part because the viral locus is a hybrid gene composed of portions of c-*abl* and portions of a viral structural gene (*gag*) (135, 136); there are nevertheless similarities between the phosphopeptide maps of the c-*abl* and v-*abl* proteins (137). 2. It was found in appreciable, but very small, amounts only in thymocytes and other lymphoid cells (134), whereas a wide variety of cells and tissues contain RNA transcribed from c-*abl* (91; D. Baltimore, personal communication). 3. Its phylogenetic distribution has not been reported. 4. Nothing is known of its function other than that tyrosine-specific kinase activity like that of the viral protein has not been found (137).

The products of c-*fps* and c-*fes* should represent homologous proteins, since the two genes are thought to be counterparts in birds and cats (77–79). The nominal sizes of the c-*fps* and c-*fes* proteins differ by 6000 Mr (Table 2), although direct comparisons have not been made. The presumed product of c-*fes* (p92^{c-fes}) was found in rats and related mammals, but not in other rodents or primates (95). Little else is known of the protein; the extent of its relationship to products of v-*fes* has not been critically assessed, no function has been reported, and there is no explanation for the failure to find RNA transcribed from c-*fes* in species known to produce p92^{c-fes}. The protein attributed to c-*fps* (p98^{c-fps}) is likely to be the authentic product of the gene because it contains numerous peptides identified previously in the transforming proteins encoded by variants of v-*fps* and displays tyrosine-specific protein kinase like that of the viral proteins (138).

If the functions of c-*onc*'s can be deduced from the properties of their viral homologs, then all but one of the identified c-*onc* products are likely to be tyrosine-specific protein kinases (see Table 2). The apparent frequency of this enzymatic activity among c-*onc*'s may be misleading on two counts, however. First, the current attributions of protein kinase activity to the cellular proteins rest largely on the rigor with which the functions of the viral proteins have been established. It could be argued that irrefutable evidence exists only for the products of v-*src* (139, 140) and v-*abl* (141). Second, with one exception (v-*abl*), the viral oncogenes that are thought to specify tyrosine protein kinases are found in sarcoma viruses and, hence, fail to adequately represent the diversity of tumorigenic mechanisms found among other retroviruses (see Table 1). For example, the *ras* transforming proteins (whose activities have been implicated in the genesis of several varieties of human carcinomas; see below) are not tyrosine kinases by any available evidence; the product of the carcinoma gene v-*myc* is said to be a DNA-binding protein (142) located in the nucleus (142, 143); and the products of v-*erb*-B and v-*fms* are membrane glycoproteins without detectable protein kinase activity (M. Privalsky, C. Sherr, personal communications). It would be premature to conclude that tyrosine phosphorylation is a common or even a typical function of cellular oncogenes.

FAMILIES OF CELLULAR ONCOGENES

Structural individuality has been the hallmark of retroviral oncogenes, and no single biochemical mechanism can explain their diverse actions. Many investigators were therefore unprepared for the mounting evidence that the rich variety of oncogenes may have sprung from a small number of ancient root stocks. The possibility of common origins for oncogenes first arose intuitively from the fact that the genes all share the dramatic property of tumorigenicity. Experimental observation has now supplanted intuition by making clear two forms of kinship among cellular oncogenes:

1. Single species can possess two or more cellular oncogenes representing the same immediate family. The origin and significance of these duplications have been explored above. The family of c-*ras* genes provides the most elaborate of available examples. Harvey and Kirsten varieties of *ras* appear only distantly related when their nucleotide sequences are compared (68, 144, 145), but the proteins encoded by these genes are antigenically similar (129, 130) and share 85% of their amino acid sequences (144, 145). The proteins can be distinguished serologically, however, by the use of monoclonal antisera (130). The difficulties inherent in identifying and explaining duplication of cellular oncogenes are compounded by unanticipated relationships between ostensibly distinct gene families. For example, c-*src* and c-*yes* are regarded as different genes, yet their nucleotide sequences are sufficiently related (146) to permit readily detectable cross-

reactions in molecular hybridizations carried out under relatively non-stringent conditions.

2. There may be a family or even a superfamily (147) of cellular oncogenes that share a single, ancient lineage. This possibility became apparent as soon as it was recognized that several different oncogenes encode tyrosine-specific protein kinases (Table 2), and that these enzymes may use a similar set of cellular proteins as natural substrates (148). Now comparisons of amino acid sequences from among the various oncogenes have given quantitative validity to these apparent relationships and, for good measure, have produced some surprises. When originally compared by molecular hybridization, the oncogenes *src*, *yes*, and *fps/fes* (avian and feline versions of the same gene) were thought not to be structurally related, even though they all encode protein kinases specific for tyrosine. But the amino acid sequences deduced for the proteins encoded by these genes are remarkably similar, particularly in the carboxy-terminal domain that carries the enzymatic activity of pp60^{v-src} (149). Using the amino acid sequence of pp60^{v-src} as reference point, the computed homologies are as follows: *yes*, 82% (146); *fps*, 40% (80); and *fes*, 45% (81). Moreover, appreciable homologies were found between the tyrosine-specific protein kinases of oncogenes and cellular protein kinases that phosphorylate serine and threonine (150); the homologies again predominate in regions with presumed import to function. Most surprisingly, the protein encoded by *mos* shares 25% of the amino acid sequence through the kinase domain of pp60^{v-src} (151), although there is for the moment no evidence that the *mos* protein has kinase activity.

Phosphotyrosine constitutes as little as 0.01% of the total phosphoaminoacids in the proteins of normal cells (107). Yet organisms may be equipped with no fewer than six tyrosine-specific protein kinases: c-*src*, c-*yes*, c-*fps*/c-*fes*, c-*ros*, c-*abl*, the receptor for epidermal growth factor (152, 153), and perhaps kinases activated by other growth factors (126, 126a, 127). Why is it necessary to have this profusion of enzymes to carry out a modification that is so scarce? Perhaps tyrosine phosphorylation is essential in many different developmental pathways, and it has proven desirable to assign each pathway an independently regulated gene that encodes the necessary kinase.

The relationships among the various protein kinases seem best interpreted as evidence for divergent evolution (147). It is not so surprising to find that all protein kinases may have a common ancestor. But the unexpected indication that *mos* has the same ancestry may prefigure a broad-reaching lineage for cellular oncogenes, no longer apparent in structure or biochemical function, but satisfying the prediction that all extant proteins have evolved from a very small number of archetypes (147).

If there is a superfamily of oncogenes, how large might it be? We know of 17 distinct retrovirus oncogenes, each with a corresponding c-*onc*. Transfection with tumor DNA has also uncovered new oncogenes, perhaps a half-dozen or more (15, 16). The number of oncogenes will probably continue to grow as efforts to identify novel isolates of retroviruses quicken, insertional mutagenesis becomes more widely deployed, and more entries emerge from transfection of tumor DNA. On the other hand, the number of cellular oncogenes may not be inordinately large. Most of the oncogenes first found in retroviruses have surfaced more than once in the same host species; several have turned up in more than one species; and at least two have been identified independently by transfection of tumor DNA (15). The reiterative emergence of cellular oncogenes in different viral isolates, and the discovery that some of the oncogenes isolated from tumor DNA had already been encountered in the study of retroviruses suggest that we may have the majority of cellular oncogenes already in view. The family might then comprise no more than 50–100 genes, a number that has been ventured before in speculations about human cancer genes (5, 6) and from efforts to enumerate the genes whose mutation leads to transformation of cells in culture (154–156).

GENETIC LINKAGES OF CELLULAR ONCOGENES

The remarkable prospect that cellular oncogenes may compose a single multigene family, and their potential role in spontaneous carcinogenesis, attach great interest to the location of these genes on chromosomes. Are any of the genes linked? Are their locations syntenic within the same species and/or homologous from one species to another and do their locations bear any relationship to chromosomal abnormalities found with increasing frequency in human tumors (157)? Answers to these questions are forthcoming from widespread efforts to map oncogenes to chromosomes in birds and mammals, but no certain picture has yet formed.

Established locations on human chromosomes include the following: c-*abl*, 9; c-*fes*, 15; c-*fos*, 2; c-*myc* (one locus), 8, band q24; Harvey c-*ras* (locus with introns), 11; Kirsten c-*ras* (locus with introns), 12; c-*mos*, 8; c-*myb*, 6; c-*sis*, 22; and c-*src*, 20. In mice, chromosome 2 carries c-*abl*, c-*fps/fes* and c-*src*; 4, c-*mos*. (References for the preceding include 158–161a; D. Baltimore, P. Leder, F. Ruddle, A. Sakeguchi, T. Shows, personal communications.) The data for chicken chromosomes are in conflict. Fractionation of chromosomes by rate-zonal centrifugation has located c-*src* on small macrochromosomes (162, 163), c-*myc* on one of the two or three largest chromosomes (64), and both c-*erb* loci on chromosomes of intermediate size, although it is not yet clear whether the two loci are linked or not (L. Sealy, M. Guyaux, personal communications). In contrast, hybridization in situ indicated that c-*src*, c-*myc*, c-*myb*, and the c-*erb* loci are all located on one or another of the large microchromosomes (10–12) of chickens (164–166; A. Tereba, personal communication). The discrepancies might arise from the fact that the cells used for chromosome fractionation are neoplastic and contain at least one chromosomal translocation, whereas the hybridizations in situ were performed with normal cells. On the other hand, fractionations have now been performed on chicken chromosomes from a different cell line, and the results with c-*src* and c-*myc* remain as before (G. Payne, L. Sealy, M. Guyaux, personal communications).

A few tentative conclusions can be ventured: 1. However common their origins, cellular oncogenes are now widely dispersed among the chromosomal complement of human beings. This is not surprising, since there are ample previous examples of related genes (or of genes whose functions are coordinately induced) that are located on different chromosomes (cf, for example, 163). 2. One provocative example of synteny is apparent: three genes encoding tyrosine-specific protein kinases (c-*src*, c-*fps/fes* and c-*abl*) are located on the same chromosome in mice (number 2). However, we have no evidence that the two genes might be more closely linked, and analogous synteny is not evident in human chromosomes. Indeed, the genes for three different tyrosine-specific protein kinases (c-*abl*, c-*fps/fes*, and c-*src*) are located on three different human chromosomes. 3. The human chromosomes carrying c-*abl*, c-*fps/fes* and c-*src* each share at least one other locus with the mouse chromosome carrying the same oncogene (A. Sakeguchi, personal communication): adenylate kinase in the case of c-*abl* (2 in mice, 20 in humans); sorbitol dehydrogenase and β-2 microglobulin in the case of c-*fps/fes* (2 in mice, 15 in humans); and inosine triphosphatase and adenosine deaminase in the case of c-*src* (2 in mice, 20 in humans). The dispersion of genes subsequent to the evolutionary divergence that produced rodents and humans may have spared at least some genetic linkages. On the other hand, the apparent congruences might be accidental, and similar homologies are not yet apparent for other cellular oncogenes.

TRANSDUCTION OF CELLULAR ONCOGENES BY RETROVIRUSES

"... all the evidence seems to concur in indicating that the Rous virus arises de novo in each tumor. There is no epidemiological evidence that cancer comes into the body from the outside ..." (166a)

Does It Happen?

Although cellular oncogenes were first envisioned as components of retrovirus genomes (8, 9), they are now clearly established to be cellular genes rather than viral genes in disguise. This conclusion rests on several points of evidence: (*a*) constancy of number and position for each cellular oncogene in every member of a species, in striking contrast to the diverse numbers and positions of the proviruses for endogenous retroviruses (60, 61, 167–169); (*b*) conservation of cellular oncogenes across immense reaches of evolutionary time—homologies among endogenous retroviruses can usually be found in only a limited number of closely related species (60, 168, 170, 171); (*c*) the presence of introns in most cellular oncogenes—a hallmark of eukaryotic genes and, again, a telling contrast with the organization of retrovirus genes; and (*d*) the fact that cellular oncogenes are neither within nor linked to proviruses of endogenous retroviruses (60, 61, 64).

What then are the ancestral relationships between cellular and retroviral oncogenes? How can we distinguish parent from progeny? First, by phylogenetic patterns: cellular oncogenes are widely distributed among metazoan organisms, whereas their retroviral counterparts appear only in occasional (or even single) species; moreover, the homology between viral oncogene and cellular homolog is greatest for the species in which the viral gene first appeared (17, 165, 172–175). Second, by the details of gene structure: retroviral oncogenes inevitably represent only a portion of the homologous cellular gene, disposed like a product of recombination within the genome of a retrovirus (see below and Figure 1). Third, by experimental recapitulation: transduction of cellular oncogenes into retrovirus genomes has now been witnessed directly both in cell culture (176–181) and in animals (182–188). The most sensible conclusion from all of these observations is that retrovirus oncogenes have arisen (and probably continue to arise) by transduction of cellular oncogenes. The only remaining formal alternative is that viral infection introduced oncogenes into the germ lines of ancient (presumably extinct) progenitors of contemporary metazoans. This intuitively unlikely proposal requires that the inserted genes be removed from the proviruses that brought them into the cell, acquire the means for independent expression, come under intense natural selection in order to survive until the present with only modest changes, and reemerge from the cellular genomes as viral genes on rare occasions during the course of vertebrate evolution. *Ipse res loquitur.*

Variations on the Theme

Three patterns of transduction have been observed: 1. Repeated emergence of the same cellular oncogene on separate occasions in the same species. This currently appears to be the most common pattern, but appearances may be misleading—independent isolates of virus can be difficult to identify on biological grounds alone, and the lineages of many contemporary stocks

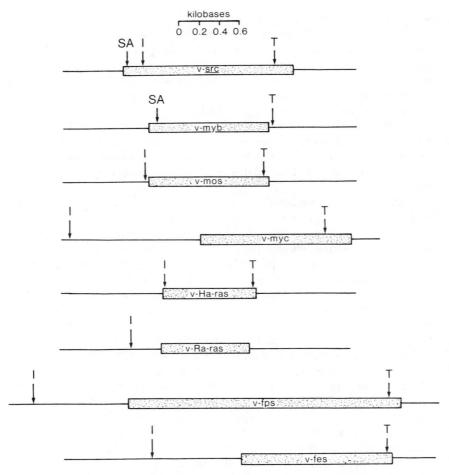

Figure 1   The topography of transduction by retroviruses. The drawing illustrates how viral and cellular domains are joined to create functional oncogenes in retroviruses. Portions derived from the genome of the transducing virus are indicated by solid lines, insertions transduced from cellular DNA by stippled blocks. The drawing is drawn approximately to scale. I denotes initiation codon; T, termination codon; and SA, splice acceptor site used to produce mRNA for the oncogene (not identified or not required in all instances). The v-*fps* and v-*fes* loci are cognates transduced from the chicken and cat, respectively.

The versions illustrated here are contained in the Fujinami avian sarcoma virus (80) and the Snyder-Theilen feline sarcoma virus (81). The two genes are aligned at their termination codons and are approximately homologous over the entirety of their shared cellular domains. The two versions of v-*ras* occur in the Harvey (Ha) and Rat (or Rasheed) (Ra) sarcoma viruses, and illustrate how the same gene can be transduced into different configurations with different outcomes for expression. The precise location of the termination codon in v-Ra-*ras* has not been established, but it must approximate that of the terminator for v-Ha-*ras*.

of retroviruses are opaque. 2. Transduction of related genes from the same species. The Harvey and Kirsten varieties of c-*ras* provide the preeminent example and pose a presently unanswerable question. Why should these oncogenes, among all others, be transduced from the rat genome in what may be a preferential manner? 3. Transduction of the same gene from different species by different retroviruses. This pattern should be prevalent if transduction proceeds with equal ease in different species. It was originally thought to be uncommon, but examples have materialized with increasing frequency, so that four can now be cited and others anticipated. Both rats and mice have yielded Harvey c-*ras* (68, 189, 190); mice and cats, c-*abl* (52, 191; P. Besmer, personal communication); woolly monkey and cats, c-*sis* (192, 193; personal communication); and chickens and cats, the gene known as either c-*fps* or c-*fes*, respectively (77, 87, 173).

The frequencies at which different cellular oncogenes are transduced apparently vary. For example, seven separate transductions of the *fps/fes* gene have been recorded, whereas only single transductions have been documented for genes such as *ros*, *erb*-A, *rel*, and *fms*. The Harvey c-*ras* seems to be another example of a relatively mobile oncogene because it can be repeatedly transduced (albeit at a very low frequency) from cells in culture (180).

The original transductions of Harvey and Kirsten c-*ras* followed a baroque path whose details still escape us. The resulting viruses have genomes constructed from three distinct components (69, 194): one derived from the virus that was used to initiate recovery of the sarcoma virus; a second derived from a defective endogenous retrovirus of rats; and the oncogene proper, derived true to form from a c-*ras* locus in the rat genome. We cannot explain why or how the endogenous rat virus participated in two separate transductions, carried out with different mouse leukemia viruses in different laboratories (189, 195). However, at least the Harvey c-*ras* can be transduced in a simpler manner because the gene has emerged from the genomes of both rats (as Rasheed/Rat sarcoma virus) and mice (as BALB murine sarcoma virus) accompanied only by portions of the transducing virus itself (196, 197).

Co-transduction of two or more separate cellular oncogenes seems most unlikely, but it appears to have occurred on at least one occasion. The v-*erb*-A and v-*erb*-B genes of avian erythroblastosis virus derived from distinct loci in the chicken genome (53) and remain separately expressed in the viral genome (98). If the c-*erb* loci are linked at all, it is across a distance of no less than 12 kbp (53). We cannot now determine whether the two genes were transduced simultaneously or sequentially. However, a recent viral isolate (AEV-H) contains only v-*erb*-B (K. Toyoshima, personal communication); transduction of one c-*erb* locus can apparently occur independently of the other, as would be required for sequential acquisition of the two genes.

The Architecture of Transduced Oncogenes

The sequencing of oncogenes has provided the beginnings of a molecular paleontology that may eventually describe the mechanism by which retroviruses transduce cellular genes (Figure 1). A number of useful generalizations can be made.

1. No predictable portion of a cellular oncogene is transduced. There is substantial latitude even in the repeated transduction of the same cellular gene. For example, the portions of c-*fps*/c-*fes* represented in various versions of the homologous viral oncogenes vary by almost two-fold (80, 81). It is essential only that a sufficient portion of the coding region be preserved from the cellular gene to engender an active protein.

2. The transduced genes are of necessity under the control of the retroviral promoter for transcription, since no other configuration would permit the oncogene to become part of the replicating viral genome (21). Any independent promoter within an oncogene (as has been suggested for Harvey v-*ras*; see 144, 198) must be overridden by the viral promoter if a complete viral genome is to be produced.

3. Most retroviral oncogenes are hybrid compositions that fuse the reading frame of a viral structural gene with the reading frame of the transduced cellular gene; v-*src* is the most prominent exception (11). As a consequence, the initiation codon or the termination codon for translation may not be derived from the cellular gene (see Figure 1). The hybrid structure is not essential to the tumorigenic properties of the gene, however; the domain contributed by the viral structural gene can be removed from molecular clones of the oncogene without damaging the capacity of the gene to transform cells (D. Baltimore, personal communication); and some cellular oncogenes have been transduced separately as hybrid and independent forms, both of which are oncogenic (178, 180, 199–201).

4. The reading frame of the cellular oncogene is reconstructed in the viral gene by the precise fusion of cellular exons (51). This reconstruction manifests a central feature of the mechanism by which transduction occurs (see below).

5. Some viral oncogenes are expressed from a spliced mRNA that unites a leader from the 5′ end of the viral genome to RNA transcribed from the cellular gene. The acceptor site for this splicing can be part of the transduced unit (51, 202), whereas the donor site resides within the genome of the transducing virus (98, 203, 204).

6. The points of recombination between the cellular gene and the genome of the transducing virus display very little in the way of constant features. The location of these recombinatory junctions within a given cellular gene may vary from one transducing event to another, or they can be reproduced with great precision in what are alleged to be two independent isolates of the same viral oncogene (81, 202). Little or no homology of nucleotide sequence exists between the cellular gene and the viral genome in the vicinity of the recombinatory junctions (51, 57, 202); the mechanism of the recombination therefore remains cryptic. A single rule of thumb may be developing, however: the leftward point of recombination can lie within either an intron or an exon of the cellular gene, whereas the rightward point may always fall within an exon (51, 54, 202). If these configurations prove to be consistent, they will have strong bearing on the mechanism of transduction (see below).

7. There appears to be little or no selection for the position on the genome of the transducing virus where the cellular gene is to be inserted. Transduced genes have been found within coding regions of several viral genes and within untranslated regions. With the sole exception of v-*src*, insertion of the oncogene has been accompanied by deletion of a substantial but varied portion of the transducing virus genome. The deletions can vary even from one transduction to another of the same cellular oncogene (81).

From these generalizations, only a single theme emerges. Any configuration of transduction that permits an appreciable (but very possibly truncated) portion of the coding region of a cellular oncogene to be expressed may give rise to a viral oncogene.

The Mechanism of Transduction

Retroviruses are adept at recombination: their genomes frequently exchange domains of unpredictable size during mixed infections (205), and their DNA forms enter the host genome with great facility (10, 21). They are furthermore diploid, allowing heterozygous combinations of different RNAs within single virions to provide topological linkage between viral genomes that could assist crossing-over (206, 207). These features underlay the first, prescient suggestions that retroviral oncogenes might be acquired by recombination with the host genome (208); they remain as linchpins for current models of how the acquisition (or transduction, as we now call it) transpires.

Several features of the transduction loom large to challenge any effort to explain how it happens: (*a*) its exceptional rarity; (*b*) its apparent independence of nucleotide sequence homology; (*c*) the elimination of introns from the products of transduction; (*d*) the lack of precision in the formation of recombinatory junctions between viral and cellular genetic domains; and (*e*) the monotony with which transduction seems to prefer oncogenes as opposed to other cellular genes. Two general proposals have taken the field. In the first, rearrangements of DNA mediated by transposable elements assemble a potpourri of cellular genes that together compose retroviral genomes as we now know them (209). This formulation—spawned by the original "protovirus hypothesis" of Temin (210)—is designed to account for the replicative genes of retroviruses as well as for their oncogenes; it does not constitute transduction as usually defined, since mobilization of oncogenes is not attributable to infection of cells by a pre-existent viral vehicle.

The alternative (and more generally invoked) model combines DNA rearrangements with recombination during retroviral replication to describe a pathway for authentic transduction of cellular oncogenes (Figure 2). The transducing vehicle is a preexistent retrovirus whose provirus integrates upstream from a cellular oncogene—a fortuitous and presumably infrequent event reflecting the general principles of retrovirus integration. Some workers have been quick to conclude that endogenous retroviruses, whose genomes are carried in the germ lines of normal animals, perform the transduction of c-*onc*'s. There is no evidence to support this view, and information derived from the nucleotide sequences of several retroviruses suggests that the view is not generally correct.

Rearrangement of DNA unites the provirus and oncogene into a hybrid transcriptional unit. If the rearrangement transposes the oncogene into a position within the provirus, transduction is formally complete; transcription and splicing would produce a possibly defective but otherwise conventional retrovirus genome now bearing at least a portion of the cellular oncogene. Precedents for transposition of this sort have not been described, but we know very little as yet about the gymnastics of which proviruses may be capable. Alternatively, in the generally preferred view, the rearrangement of DNA causes a deletion that fuses provirus and oncogene into a single genetic element, with the leftward portion of the provirus (including its transcriptional promoter) upstream and an unpredictable portion of the oncogene (including its site for termination of transcription) downstream. Transcription from the fused genetic element would begin on the retrovirus promoter, proceed through the c-*onc*, and generate a hybrid RNA from which the introns of the c-*onc* have been removed. The viral portion of the hybrid RNA permits the formation of heterodimers with the genome of a superinfecting (or co-resident) retrovirus and encapsidation of the RNA into virions (211). Recombination occurs when the heterozygous virions enter new host cells and begin reverse transcription from viral RNA. The crossing-over may have to occur without benefit of homology between the participating nucleic acids, but it otherwise can be described by any of the prevailing models for recombination among retrovirus genomes (205).

The product of recombination presents the cellular oncogene in its fully transduced form: stripped of introns, separated from its own signals for the initiation and termination of transcription, and included within the genome of a replicating retrovirus. The stipulated role of integration in these affairs provides an explanation for the rarity of transduction by retroviruses. The integration of retroviral DNA occurs at many positions on the host genome (21, 212) and may occur at random. Juxtaposition of a provirus to a particular cellular gene is therefore quite unlikely, and insertion of the provirus at a position that would conserve the necessary functions of the cellular gene (splicing and translation) and the integrity of its coding

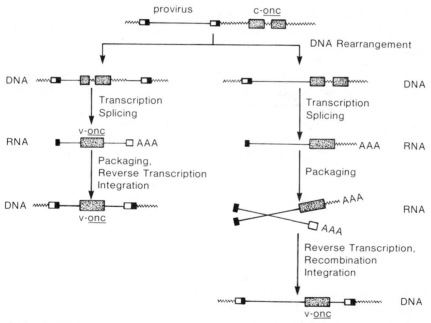

Figure 2 Models for transduction by retroviruses. The drawing illustrates how cellular proto-oncogenes might be transduced by pre-existent retroviruses. The scheme begins with an intact retroviral provirus fortuitously integrated upstream from a cellular proto-oncogene. The characteristic terminal redundancies of the provirus are illustrated by black and white boxes, exons of the cellular proto-oncogene by stippled boxes, viral nucleic acid by straight lines, and cellular nucleic acid by jagged lines. A postulated rearrangement of DNA begins transduction and could take either of two forms, as illustrated. Events thereafter proceed as outlined in the figure and described in detail by the text.

element is even less likely. The incalculable rarity of these events inspires gratitude for Nature's occasional benevolence to the experimentalist.

The transductive model accomodates a diverse array of supporting facts: 1. Established retroviral proviruses are subject to DNA rearrangements on at least rare occasions (213–215). 2. Experimental hybrids of covalently joined retroviral and cellular RNAs can be encapsidated and employed in subsequent nonhomologous recombination that transplants portions of the cellular RNA into the midst of a functional retrovirus genome (216). 3. Introns are removed from cellular genes carried within artificially constructed retroviral vectors (217, 218). 4. Nucleotide sequencing revealed retroviral oncogenes arise by the precise removal of introns from the cellular counterpart of the oncogene (51). 5. The leftward recombinatory junction between retrovirus genome and oncogene lies within an intron of the cellular oncogene in at least two examples (51, 202), a finding that can be readily explained only if the junction was formed through a DNA rearrangement rather than by means of a spliced intermediate. 6. The rightward recombinatory junction between retroviral genome and oncogene is within an exon of the cellular gene in at least one example (51), as expected if a spliced version of the cellular oncogene was used to form the junction. 7. No notable homology of nucleotide sequences has been found between viral and cellular domains at either the leftward or rightward recombinatory junctions (51, 57, 202). The recombinations that formed these junctions apparently occurred between nonhomologous nucleic acids. 8. Cellular oncogenes were deliberately mobilized in a transductive manner by infection of cells in culture or animals with retroviruses (see below).

Forcing the Issue: Experimental Transduction of Cellular Oncogenes

Virtually all of the retroviral oncogenes in hand arose from spontaneous events in the field or the laboratory. Could we plot the deliberate transduction of cellular oncogenes in the hope of expanding the repertoire? Many have tried; few have reported any success.

Given the evident rarity with which novel oncogenes have come on stage, transduction in cell culture seems a daunting undertaking. Two apparently successful protocols have been reported. Both are built on the intuitive expectation that oncogenes might be more readily mobilized from chemically transformed cells, but the success of the protocols remains unexplained. One of the two procedures reportedly produced retroviruses with an exceptional variety of oncogenic potentials (176, 179); none of these potentials has as yet been satisfactorily attributed to an oncogene. The other protocol gave a diametrically opposite result: repeated (albeit infrequent) transduction of the same genetic locus (Harvey c-*ras*), generally fused with a structural gene of the transducing virus (178, 180). The reproducibility with which c-*ras* emerges from this procedure and the variations in the precise architecture of the transduced element (180; see also Figure 1) lend credence to the claim that the gene is being newly transduced. But there is no apparent explanation for the repeated recovery of a single locus and of that locus alone from among all of the oncogenes known to reside in the rat genome.

A recent addition to the oncogene family (v-*ski*) appeared during the propagation of an avian retrovirus in cell culture (219). It seemed at first that the gene might have been transduced from the cultured cells, but a diligent search has now revealed the likelihood that the oncogene had been present in the viral stocks for some time before its recognition, and that the event of its transduction may be lost to history (E. Stavnezer, personal communication).

A partial deletion of v-*myc* was apparently repaired during growth of the mutant virus in cultured cells (220). Repair is thought to have occurred by homologous recombination with c-*myc*, and the reconstituted viral oncogene has novel biological properties that may be attributable to nucleotide sequences derived from c-*myc* (221). The recombination that is alleged to have occurred in this instance is inherently interesting, but its circumstances and presumed mechanism are atypical, and as a consequence, it does not provide a general analog for transduction by retroviruses.

Animal hosts offer a potentially favorable setting in which to accomplish and perceive transduction: the number of cells available to sustain viral replication is immense, enhancing the likelihood of transduction; tumorigenesis provides a powerful selective device for the detection of successful transduction; and the full representation of embryological lineages among the cells of a living animal permits the recognition of novel oncogenes.

Transduction of oncogenes was witnessed, but not immediately appreciated for what it was, when passage of murine leukemia viruses in rodents gave rise to the Harvey and Kirsten varieties of v-*ras* (189, 195) and to v-*abl* (191), and infection of cats with feline leukemia viruses produced sarcoma viruses (222). Administration of corticosteroid to mice appeared inexplicably to facilitate the recovery of *abl* (191). Otherwise, no useful adjuvants for transduction have been described. Surprisingly little use has been made of animals to attempt deliberate transduction. There exist only two contemporary examples where experimental infection led to the recovery of at least part of a cellular oncogene:

1. Large (but always incomplete) deletions in v-*src* can be reproducibly repaired by recombination with c-*src* in either chickens (184, 186, 187, 224, 225) or quail (185). The repair is presumed to occur by means of homologous recombination between the residual portion of *src* in the viral genome and c-*src* (187). If correct, this widely held presumption diminishes the value of the recombination as a model for nonhomologous transduction by retroviruses. The ambiguity was compounded by the fact that some of the deletion mutants retain only a small 3' domain from v-*src*, yet transduce c-*src* at appreciable frequencies (186). If the 3' domain of v-*src* is missing as well, however, transduction cannot be perceived (186, 187). Even the advantage conferred by homology has not permitted partial deletions in v-*src* to be repaired at perceptible frequencies in cell culture—a telling testimony to the favorable circumstances that enhance transduction by retroviruses in animal hosts.

2. The cellular oncogene *fps* has been recovered on a single occasion by infection of a chicken with an avian retrovirus (188). The genome of the transducing vehicle was devoid of appreciable homology with c-*fps*. It is not yet entirely certain that the emergence of c-*fps* in this single instance is attributable to the experimental infection, but structural details at least suggest that the recovered oncogene emerged from the chicken genome in the recent past (188).

Retroviruses as Generalized Transducing Agents

Is the seizure of cellular oncogenes by retroviruses a unique event, or might more prosaic genes also be transduced but go unrecognized among the vast number of progeny from retroviral infection? Neither the structure nor the expression of cellular oncogenes present any features that might make these genes uniquely accessible to the acquisitive habits of retroviruses. The problem is more likely to lie in the recognition of extraordinarily rare events. It is first of all possible that oncogenes survive and prosper in their transduced form because they offer a selective advantage to retroviruses, although it is difficult to construct a persuasive argument for such advantage. Second, the dramatic properties of oncogenes provide a dominant selective device by which the presence and activity of the genes can be detected with great sensitivity. Even a single cell harboring a newly emerged oncogene can engender a tumor bearing copious quantities of the gene safely ensconced in its new viral home. I can think of no other sort of gene with such advantages for recognition.

Has transduction of diverse genes by retroviruses occurred but gone undetected? We have very few clues: 1. Minuscule quantities of globin mRNA were reported in a complex with the genome of a murine retrovirus propagated on erythroleukemia cells (226), but the globin RNA gave no evidence of being recombined with the viral genome. 2. Nucleotide sequences derived from the chicken genome but of unknown function were found as transmissible constituents of a single stock of avian sarcoma virus (227). 3. The existence of processed genes raises the possibility that retroviruses have been shuttling genes from one point to another within (and among) eukaryotic cells from time immemorial (73). Processed genes can indisputably bear the markings of reverse transcription and retroviral integration. On the other hand, there is as yet no persuasive reason to believe that processed genes were ever part of a viral genome or ever negotiated the complete life cycle of a retrovirus (217).

Could retroviruses be adopted to use as generalized transducing agents by experimental ingenuity? At least three major considerations constrain this possibility: 1. The size of genes may govern their accessibility to transduction. According to available evidence, the haploid genome of retroviruses can accomodate no more than about 11 kb, including the admittedly small domains required for encapsidation, replication and expression of the genome; presently inapparent aspects of the transductive mechanism may also be constrained by the size of the target for transduction. 2. It may not be possible for all cellular genes to join a retrovirus provirus in the functional genetic hybrid postulated by prevailing models for transduction (see Figure 2). 3. The apparent rarity of transduction must be confronted, either by use of powerful selective procedures, or by the invention of means to focus integration to a specific domain of the host genome (an unlikely prospect, for the moment).

THE ROLE OF CELLULAR ONCOGENES IN NORMAL CELLS

We first encountered cellular oncogenes, and assigned to them a perjorative name, because of their kinship with oncogenic viral genes. But who could believe that the forces of natural selection have preserved this group of genes in response to their oncogenic potential? It seems more likely that these genes have survived the hazards of time because they are vital to the growth and development of the species in which they reside.

What are the work-a-day purposes of cellular oncogenes? The prevailing guess is that they are somehow involved in differentiation. This conjecture arose from several roots:

1. The heritable predispositions to cancer encountered among human beings are tissue specific; if there are "cancer genes" responsible for these predispositions, they act at fairly precise points in embryological lineages (5, 6).

2. The oncogenes of retroviruses display tissue specificity. Each induces tumors in only a certain organ(s) and transforms only certain cells in culture. The specificity of oncogenes is not merely a consequence of viral host range. Differentiated cells can contain and express an oncogene, yet resist its transforming power (228, 229). Apparently, the vulnerable cell contains substrates for the activity of the oncogene, the resistant cell does not.

3. Retroviral oncogenes meddle with differentiation; v-*src* induces the inappropriate expression of an embryonic gene (230) and suppresses the expression of differentiated properties (231), v-*erb* and v-*abl* arrest the progress of hemopoietic cells through their developmental pathways (erythroid and lymphoid, respectively) (232–234), and by some accounts, v-*myb* and v-*myc* can reverse the differentiation of monocytic hemopoietic cells (235, 236).

4. Cellular oncogenes identified by transfection display tissue-specificity. In some instances, the specificity was revealed by direct identification of the genes: Harvey c-*ras* from carcinoma of the bladder (37a, 41, 43); Kirsten c-*ras* from carcinoma of the lung and colon (37a); a poorly characterized member of the c-*ras* family from rat neuroblastoma (M. Wigler, personal communication); and a Kirsten variety of c-*ras* from rodent fibroblasts transformed by methylcholanthrene (R. Weinberg, personal communication). Otherwise, the specificity has been assessed by attempting to inactivate the oncogenes in different tumors with restriction endonucleases. By this means, distinctive oncogenes were implicated in each of five different forms of human leukemia (39). These findings in turn prompted the suggestion that the cellular oncogenes detected in the leukemias are normally stage-specific determinants of hemopoiesis (39). The validity of this formulation has been compromised in part, however, by the finding that the same oncogene may be active in two very different forms of hemopoietic tumors—a promyelocytic leukemia represented by the HL60 cell line, and the B-cell tumor represented by Burkitt's lymphoma (R. Weinberg, personal communication).

Are cellular oncogenes involved in differentiation? The pursuit of this question has been difficult and ineffectual. The underlying premise is that the expression of genes with importance to development will vary from one tissue to another, from one embryological lineage to another, from one time in embryogenesis to another. Variations in expression have indeed been found, some of which are provocative: 1. The expression of c-*myb* is prominent in hemopoietic tissues (84, 85) and primitive hemopoietic tumor cells (89), but negligible elsewhere. 2. c-*myc* is expressed in a wider variety of cells, although hemopoietic cells are again prominent (90). 3. The expression of both c-*myb* and c-*myc* diminishes abruptly and by a large degree in promyelocytic leukemia cells that have been caused to differentiate to more mature forms by chemical inducers (89, 90). 4. c-*erb*-A and c-*erb*-B are differentially transcribed, the former in bone marrow cells, the latter in embryonic fibroblasts (85, 97). 5. In neonatal mice, c-*fos* is vigorously expressed in bones, the target for neoplastic transformation by v-*fos* (86). 6. A pluripotent hemopoietic cell derived from mouse bone marrow contains exceptionally large quantities of the protein encoded by the Kirsten form of c-*ras* (132); Kirsten v-*ras* induces erythroleukemia as well as sarcomas (195, 237). 7. During the course of mouse embryogenesis, the expression of Harvey c-*ras* is sustained at high levels; the expression of c-*abl* and c-*fos* fluctuates appreciably and in distinctive patterns; and several cellular oncogenes are continually expressed in low amounts, whereas others are apparently silent (86). The tissues responsible for expression of the various oncogenes have not been identified. 8. Hybridization in situ has revealed that c-*erb* (the A and B loci have not been separately examined) is expressed at moderate levels in only occasional cells of the avian bone marrow (D. Stehelin, personal communication). It is presumed, but not yet proven, that these cells represent a distinct stage in hemopoiesis.

No coherent picture emerges from these findings. Granted, the expression of many cellular oncogenes varies as a correlate of differentiation. But the experimental strategy is compromised by several important caveats. It has relied too heavily upon tumor cells, in which expression of cellular oncogenes could be either a legitimate manifestation of cellular phenotype or an anomalous consequence of neoplastic transformation (gene amplification, presumably an anomaly, apparently contributes to the augmented expression of c-*myc* in HL60 leukemia cells; see 238, 239). It has failed to identify the individual cells that are responsible for expression within complex tissues. Most importantly, it cannot distinguish genes whose expression responds to development from genes whose activity helps direct development.

Virologists had hoped that the patterns of cellular oncogene expression would provide an explanation for the tissue-specificity of transformation by retroviruses. Viral oncogenes might affect only those cells in which homologous cellular oncogenes are expressed; alternatively, cells in which a given cellular oncogene is not normally active might be more vulnerable to the effects of the cognate viral gene. Neither expectation has been fulfilled; there has been no persuasive correlation between expression of cellular oncogenes and susceptibility to viral transformation (85, 87).

The discovery of cellular oncogenes in invertebrates has further obscured the relationship between these genes and organismal development. For example, it is difficult to conceive of genes whose functions serve both a highly specialized pathway of mammalian hemopoiesis and a developmental lineage in *Drosophila*. The paradox may be more apparent than real, however, and its resolution may reveal unexpected principles of developmental biology.

CELLULAR ONCOGENES AND CANCER

The study of viral oncogenes began with the vague hope that their mechanisms of action might mirror the biochemical abnormalities of neoplastic cells. This hope became a conviction when we learned that retroviral oncogenes are little more than copies of normal cellular genes. How likely is the possibility that cellular oncogenes mediate the effects of carcinogenic agents? Do these genes have oncogenic potential? Are they the long-sought "cancer genes"?

Viral and Cellular Oncogenes are Closely Related

The extraordinary resemblances between cognate viral and cellular oncogenes have exceeded the expectations of even the most sanguine retrovirologists. The corresponding nucleotide sequences of v-*mos* and c-*mos* (1157 nucleotides) differ at only 25 positions, the deduced amino acid sequences of the viral and cellular proteins by only 11 amino acids (57). The viral and cellular versions of *myb* display similarities of the same order: 15 differences among 1197 nucleotides, resulting in 11 substitutions in the amino acid sequences deduced for the viral oncogene product; an additional 10 amino acids are unique to the viral protein because translation from v-*myb* continues for a short distance into the envelope gene of the transducing virus (51). Retroviruses are subject to appreciable genetic variation during the course of propagation (205), perhaps because of the high frequency with which reverse transcriptase erroneously pairs nucleotides during polymerization (240). It therefore seems likely that the scattered differences between viral and cellular oncogenes have arisen as mutations in the viral genes subsequent to their transduction from cells.

Proteins encoded by viral oncogenes also bear striking resemblances to their cellular counterparts: 1. The viral and cellular forms of pp60^{src} are remarkably similar: they have the same molecular weight (101, 104, 105); display antigenic cross-reactivities (101, 104, 105); yield closely related peptide maps (109, 110); are both associated with the plasma membrane (111); are both phosphorylated and have similarly disposed phosphoamino acids, with phophoserine in the proximity of the amino-terminus (103) and phosphotyrosine within a carboxy-terminal domain (107, 110); and are both protein kinases specific for tyrosine (100, 104–108, 117). 2. The proteins encoded by the families of v-*ras* and c-*ras* also appear to be quite similar (129–132). They have molecular weights of about 21,000; they share at least some antigenic determinants; they produce related peptide maps; they possess the presently obscure ability to bind guanine nucleotides with high affinity; and they carry out autophosphorylation at threonine residues in vitro, whereas kinase activity for other protein substrates has not been detected. 3. The other identified products of cellular oncogenes have not been characterized in such detail, but their relationship to viral counterparts has been documented by one or more means (95, 134, 137, 138, 241): cross-reactive antigenicities (c-*abl*, c-*fps*, c-*fms* and c-*fes*), similar peptide maps (c-*abl* and c-*fps*), and shared function (tyrosine-specific protein kinase; c-*fps*).

The similarity between v-*src* and c-*src* was dramatized by the demonstration that partial deletions in v-*src* can be repaired when the deletion mutants recombine with c-*src* in chickens (see above). The recombinant gene can derive at least two-thirds of its substance from c-*src* and still remain active as an oncogene. Recombination between genes as similar as v-*src* and c-*src* is not easy to document with simple tools, so some observers have argued that the emergence of transforming virus from the infected chickens was due merely to unappreciated contamination of viral stocks (242). Doubts of this sort are no longer tenable; recombination with c-*src* has been accomplished with a deletion mutant purified by molecular cloning (L. Wang, personal communication), and the recombinant nature of the reconstituted v-*src* was demonstrated directly by analysis of nucleotide sequence (224, 225). There is now little reason to doubt that functional v-*src* can be formed largely (although perhaps not entirely; see below) of goods obtained from c-*src*.

At Least Some Cellular Oncogenes Can Induce Neoplastic Transformation

The experimentalist can recapitulate transduction of cellular oncogenes by the use of molecular cloning. In the simplest design, cellular oncogenes are isolated by cloning, then linked to a retrovirus transcriptional promoter to assure vigorous expression when the isolated genes are reintroduced into cells in culture. This exercise produced transforming genes with mouse c-*mos* (56, 242a), and with both rat (70) and human (46) c-*ras*. Loci of c-*ras* either with or without introns were used successfully. The cells transformed by c-*ras* produce large amounts of the 21,000 Mr protein encoded by the cellular oncogene and are tumorigenic in animals. The original impact of these findings has been blunted somewhat by the following.

1. The activated c-*mos* and c-*ras* genes have not yet been shown to be directly tumorigenic in animals, as are their viral counterparts.

2. Efforts to produce transforming genes by activation of human c-*mos*, chicken c-*src* and chicken c-*myc* have so far met with failure (G. Cooper, E. Matthews, R. Parker, G. Vande Woude, personal communications).

3. A battery of details have emerged to challenge the conclusion that proteins encoded by homologous cellular and viral oncogenes are alike in every regard (see below). As a consequence, we are left with the disappointing possibility that the effects of activated c-*mos* and c-*ras* do not represent prototypes for how cellular oncogenes might participate in tumorigenesis, but instead reflect special experimental circumstances, such as the use of established cell lines whose phenotypes are particularly vulnerable to neoplastic transformation, or the achievement of outlandish levels of oncogene expression. The ability of c-*mos* and c-*ras* to transform cells in culture has nevertheless provided powerful impetus for the idea that cellular oncogenes are agents of tumorigenesis.

Implicating Cellular Oncogenes in Tumorigenesis

Experiments with cell culture are imperfect substitutes for dissection of the tumor cell in its native form. How might we directly implicate cellular oncogenes in the genesis of tumors in animals? Three means have so far emerged:

1. Avian leukosis viruses are retroviruses without oncogenes (11, 22), yet they induce B-cell lymphomas in chickens at high frequencies (243). The first evidence of how these tumors might arise came with the discovery that viral DNA in the tumors is almost always integrated in the vicinity of a previously identified cellular oncogene, c-*myc*; as an apparent consequence of the insertion, transcription from c-*myc* is augmented by as much as 100-fold (26, 27). The insertions of viral DNA can take several forms: upstream of c-*myc*, in the same transcriptional orientation as the cellular gene (23, 24, 26, 27, 244); upstream of c-*myc*, but in the opposite transcriptional orientation (27); and downstream of c-*myc*, in the same transcriptional orientation (27). The mechanisms by which these various insertions augment transcription from c-*myc* must differ. The first configuration permits the promoter of the provirus to drive transcription from c-*myc* directly (26, 27), the other configurations may work indirectly by enhancing the activity of the c-*myc* promoter itself (27). Whatever the mechanism, it seems likely that the heightened expression of c-*myc* provokes and perhaps sustains the chain of events that engender B-cell tumors induced by leukosis viruses. A cellular oncogene has been caught *in flagrante delicto*. Similar findings are now emerging for tumors induced by other retroviruses without oncogenes. Examples include B-cell lymphomas caused by chicken syncytial virus (28), renal carcinomas induced by myeloblastosis-associated

virus (D. Westaway, personal communication), erythroblastosis caused by a variant of avian leukosis virus (H. J. Kung, H. Robinson, personal communications), and mammary carcinomas induced by mouse mammary tumor virus (29). In each instance, tumorigenesis apparently begins when integration of viral DNA induces the activity of a specific cellular gene or otherwise damages a particular domain within the host genome.

2. Cellular oncogenes identified and isolated by transfection are demonstrably abnormal; they induce neoplastic transformation without manipulation of their structure or intrinsic capacity for expression, whereas their counterparts isolated from normal tissue do not (15, 16, 50). It therefore is reasonable to argue that these genes are involved in the genesis of the tumors from which they have been isolated. The argument took on new implications when the active cellular oncogene isolated from cells of a human bladder carcinoma proved to be a Harvey c-*ras* gene (37a, 41, 43). Further identifications followed with exhilarating dispatch, and to date, all represent members of the c-*ras* family (see above). The reiterative appearance of c-*ras* is provocative but presently inexplicable. Tissue-specificity cannot suffice as an explanation; most of the tumors are epithelial in origin, but the transformed mouse cells are not. Harvey c-*ras* is expressed at high levels throughout the course of mouse embryogenesis (see above), as if the gene might be a universal growth function. Kirsten c-*ras* gives no such indications, yet the gene has been implicated in the genesis of two common forms of carcinomas.

There are other active oncogenes, such as those in a variety of human leukemias and in carcinoma of the breast, that remain unidentified and are unlikely to be homologs of any known retroviral oncogene (15, 16). But a remarkable unification has nevertheless been achieved: two disparate experimental strategies—one tracing the evolutionary origins of viral oncogenes, and the other isolating active oncogenes directly from tumor cells—have led us to overlapping sets of genes whose properties prefigure a role in carcinogenesis. These discoveries cannot reasonably be ascribed to coincidence. Rather, it is likely that at least some of the genetic means by which the cancer cell arises stand revealed.

3. The discovery that some forms of viral tumorigenesis may begin with the induction of cellular oncogenes unleashed a vigorous search for expression of these genes in the cells of human tumors. The first results of the search are in; they are provocative but indecisive. The strategy itself is deeply flawed: cohort normal cells for use as controls are often not available; the work to date has relied heavily upon established lines of cells whose properties may have changed drastically since the original explantation from tumors; the full repertoire of oncogenes is clearly not in hand, and each deficiency further compromises the search in human tumors;

it is not easy to distinguish expression that is etiological in nature from expression that is consequent to neoplastic transformation; and we now have reason to believe that the expression of active oncogenes in human tumors need not be appreciably augmented above normal levels (see below). Given these difficulties, it is not surprising that a coherent picture has yet to emerge. Some investigators report consistent elevation of c-*myc* expression in various hemopoietic tumors of B-cell origin (92), others do not (90). The search for c-*myc* expression in this setting represents an effort to test the hypothesis that all B-cell tumors, whether avian or human, may originate with the same genetic event. On a broader front, elevated expression of c-*myc* has been encountered in a wide variety of tumors, including sarcomas, carcinomas, leukemias and melanomas (90, 91). The elevations are sporadic, however, rather than consistent for any single form of tumor. In at least two instances (the cell lines HL60 and COLO320, representing promyelocytic leukemia and a neuroendocrine tumor arising in the colon, respectively), the augmented expression of c-*myc* is apparently due to many-fold amplification of the gene (238, 239; K. Alitalo, M. Schwab, personal communication). The significance of these afflictions of c-*myc* to the etiologies of the tumors are presently indeterminate; the active oncogene identified in HL60 cells by transfection is not c-*myc* (R. Weinberg, personal communication), and no active oncogene has yet been identified for COLO320 (K. Alitalo, M. Schwab, personal communication). Expression of c-*myb* was detected in leukemias derived from every major hemopoietic lineage, but only in cells representing relatively primitive stages in the lineages (89). The significance of these findings with regard to tumorigenesis is moot, however, since expression of c-*myb* is also prominent in normal hemopoietic tissues (84, 85). The expression of c-*sis* has proven to be relatively specific, occurring in sarcomas and glioblastomas, but not in a variety of other tumors (91). Expression of c-*abl* and c-*ras* was detected in diverse tumor cells (90, 91); the relative levels of expression were not assessed satisfactorily, however, and could represent nothing more than is found in a variety of normal cells. Perhaps the best that can be said of these efforts is that when more comprehensive surveys are complete, the results may point to embryological lineages in which to search for physiological functions of specific cellular oncogenes, and to particular genes whose structure and activity should be more closely examined in tumor cells.

Converting Proto-Oncogene to Oncogene

In their native form, cellular oncogenes appear to be incapable of transforming cells (15, 16, 32, 46, 56, 70). Why do these genes become oncogenic when transduced into the genomes of retroviruses, and what events give rise to the active oncogenes found in the DNA of some tumor cells? Efforts to answer these questions have invoked two contrasting mechanisms: 1. Quantitative (or "dosage"): excessive expression of an otherwise normal gene might convert cells to the neoplastic phenotype. Transduction by retroviruses places genes under powerful signals in the viral genome and promotes their expression in great abundance. Similarly, carcinogens that damage DNA might release potential oncogenes from the control of regulatory elements. 2. Qualitative: damage to the cellular gene could change its function. For example, the protein kinases encoded by a number of oncogenes might have distinctive substrate specificities that account for their tumorigenicity. Mutations in the coding domain of a cellular oncogene could arise during or subsequent to transduction into a retrovirus genome, or as damage inflicted on cellular DNA by a carcinogen.

Neither mechanism has yet received decisive support. Proponents for the dosage mechanism can point to the following facts. First, retroviral oncogenes such as v-*src* and v-*ras* are generally expressed in reasonable abundance, and reduction in the expression can cause the host cell to revert to an ostensibly normal phenotype (215, 245–247). Second, DNA from normal cells allegedly contains oncogenes capable of transforming cells in culture following activation by removal of *cis*-regulators (32) or insertion of retroviral enhancers for transcription (G. Cooper, personal communication). Third, the transformation of cells in culture by isolated c-*mos* and c-*ras* was achieved by augmenting the expression of these genes (46, 56, 70). Fourth, bladder carcinoma cells from which an active form of Harvey c-*ras* was isolated, and cells secondarily transformed by the isolated gene, appear to express the gene at levels above normal (38, 41, 43).

There is countervailing evidence, however, of equal or even greater weight. First, the prevailing view that v-*src* is inevitably expressed at high levels in cells transformed by Rous sarcoma virus may not be correct. Expression of v-*src* varies over a very wide range in transformed cells and can drop to levels that exceed the expression of c-*src* by no more than three to five-fold (246; H. Oppermann, personal communication). Second, oncogenes such as v-*mos* and v-*abl* are actually toxic to some cells when expressed at even moderate levels (248–250). Only minuscule quantities of the protein encoded by v-*mos* are required for neoplastic transformation (250). Third, efforts to transform cells by enhancing the expression of c-*src* and c-*myc* have failed (see above). Fourth, expression of at least two active oncogenes (Kirsten c-*ras* in carcinomas of the colon and lung, and an unidentified gene in B-cell lymphomas of chickens) is only marginally increased in the tumor cells where their activity is detectable by transfection (15, 37a). Fifth, close inspection has revealed subtle but potentially important differences between the proteins encoded by several cellular oncogenes and their viral counterparts: the cellular and viral versions of pp60^{src} are phosphorylated on different tyrosine residues (251) and have different amino acid sequences at their carboxytermini (H. Hanafusa, R. Swanstrom, personal communication); efforts to repair complete deletions of v-*src* by recombination with c-*src* have failed, although the failures may reflect impediments to the recombination rather than properties of c-*src* (186, 187); proteins encoded by the several forms of v-*ras* are phosphorylated, whereas the cellular counterparts are not because they lack the threonine residue on which phosphorylation of the viral protein occurs (129, 144); the proteins produced by the abnormal forms of Harvey and Kirsten c-*ras* found in some human tumors reveal their cellular origins by the absence of phosphorylation, but they also display subtle differences (mainly in electrophoretic mobilities) from the protein products of the normal c-*ras* genes (E. Scolnick, personal communication); and only the viral form of the *abl* protein is phosphorylated on tyrosine in vivo and displays phosphotransfer activity in vitro (137).

Our best hope for an incisive resolution of these ambiguities lies in the study of cellular oncogenes isolated from tumor cells. By locating and characterizing the lesions responsible for the transforming activity of these genes, we should be able to explain the conversion of proto-oncogene to oncogene. The answers may come quickly: recent results have revealed that the transforming activity of Harvey c-*ras* in carcinoma of the human bladder is due to a single mutation within the coding domain of the gene—a transversion of G to T, changing a glycine residue to valine in the deduced amino acid sequence of the gene product (251a, 251b). Similarly, a mutation at the same position may be responsible for the oncogenicity of the viral forms of *ras*—the residue is arginine rather than glycine in v-Ha-*ras*, and v-Ki-*ras* has serine in the analogous position (although the sequence of c-Ki-*ras* is not yet known). These findings lend considerable credence to the view that qualitative changes convert proto-oncogene to oncogene.

The Role of DNA Damage in Oncogenesis

It has become an article of faith among most investigators that damage to DNA initiates oncogenesis (252, 253), although occasional iconoclasts have argued effectively for reversible epigenetic events (254, 255). The identification and characterization of active oncogenes within human tumors provided the first direct evidence that the DNA of these tumors is in some way abnormal. But these exercises may not tell us how the original carcinogen wrought its mischief. The oncogenes identified by transfection are likely to represent the last step in the sequence of events that produces a malignant tumor (see below), whereas it is to the still mysterious first step that we must look for the lesions induced by proximal carcinogens (256).

The recent literature has burgeoned with imaginative proposals of how cellular oncogenes might be activated by damage to DNA (5, 257, 258). All of these proposals presume that dosage explains the effects of cellular oncogenes; all therefore seek to explain how the expression of oncogenes might be augmented and all will prove moot if the dosage hypothesis fails to survive the current rush of experimental data.

Point mutations induced by carcinogens could act in either *cis* or *trans*. Mutations within or adjacent to an oncogene might make the gene resistant to negative regulators. These would be dominant lesions of the sort a transfection assay might perceive if the oncogene were able to directly transform cells in culture. Alternatively, mutations in regulatory genes could destroy their activity and release oncogenes from their control. The resulting expression of the oncogene would be a recessive lesion, subject to reversal if the gene were transplanted into a normal cell, and hence not visible to a transfection assay.

Rearrangements of DNA, as opposed to point mutations, have gained considerable (but still hypothetical) currency as the potential driving force in carcinogenesis (253). The resulting damage to DNA could duplicate the anomalies of oncogene regulation outlined above for point mutations. Alternatively, rearrangements of DNA could bring previously silent oncogenes under the influence of positive regulators by attaching the genes to active promoters of transcription (257), or by approximating the genes to one of the mysterious regulatory elements presently known as enhancers or augmenters of transcription (259). The mechanism(s) by which enhancers work is not yet known, but their general effects are well described—they enhance or even evoke the activity of transcriptional promoters under circumstances (at positions in chromatin?) where the promoters would otherwise be sluggish or inactive (260–262). The long terminal repeats (LTRs) of retrovirus proviruses are known to be both transcriptional promoters (21, 263) and enhancers of transcription (21, 262); these activities account for the induction of c-*myc* expression during the genesis of B-cell tumors by infection with avian leukosis viruses (see above). Consequently, attempts were made to implicate the LTRs of endogenous retroviruses in the response to chemical carcinogens (264). The results have so far failed to make any coherent connection between the activity of the LTRs and neoplastic transformation induced by the carcinogens. It seems wiser to view the LTRs as merely one revealed form of a family of enhancers that are likely to be integral components of many if not all transcriptionally active zones in the chromatin of eukaryotic cells. Any of these enhancers might activate cellular oncogenes.

The most explicit attempt to describe how cellular oncogenes could be activated invokes chromosomal abnormalities as generators of increased oncogene dosage (257). Chromosomal duplications might elevate the dose of an oncogene and outstrip the effects of *trans*-active regulators on the gene. Translocations could arouse oncogenes by relocating them into transcriptionally active domains of chromatin. For example, the B-cell tumors of Burkitt's lymphoma generally display translocations involving chromosome 8 and a chromosome bearing an immunoglobulin gene that is active in the tumor: 2, kappa chains; 14, heavy chains; or 22, lambda chains (257, 265–269). Moreover, the genes for kappa chains were mapped to the region of the breakpoint for the translocations involving chromosome 2 (p12) (268), and heavy chain genes to the region of the breakpoint for translocations involving chromosome 14 (q32) (267). These findings have fueled speculation that the chromosomal translocations in Burkitt's tumors activate an oncogene on chromosome 8 by transposing it to the vicinity of transcribed immunoglobulin genes. To date, two cellular oncogenes have been mapped to human chromosome 8: c-*mos*, and one locus of c-*myc* (see above). c-*Mos* is an unlikely (but not impossible) offender in the genesis of B-cell tumors. The finding of c-*myc* on chromosome 8 is another matter, however, since the gene has been implicated in the induction of B-cell lymphomas by avian retroviruses (see above). This is a provocative conjunction of observations. Accordingly, investigators have sought and found evidence that c-*myc* is included in translocations between chromosomes 8 and 14 in Burkitt's lymphoma (269a, 269b). The relocated c-*myc* is immediately adjacent to the breakpoint for the translocation and, in some instances, is closely linked to a gene for heavy chain immunoglobulin (269b). Similarly, translocations between chromosomes 12 and 15 in mouse plasmacytomas producing IgA juxtapose c-*myc* (chromosome 15) to the constant region gene for alpha heavy chains (chromosome 12) (269b, 269c; K. Calame, personal communication). The significance of these very recent findings has yet to be fully evaluated, but the apparent analogy to the pathogenesis of B-cell lymphomas induced by retroviruses in chickens is arresting.

Other karyotypic abnormalities in human tumors can also be correlated with the chromosomal locations of cellular oncogenes. Illustrative examples include: c-*sis* on 22 (translocations in chronic myelogenous leukemia and certain B-cell tumors); c-*fes* on 15 (translocations in acute promyelocytic leukemia); c-*myb* on 6 (deletions in T cell leukemias and acute lymphatic leukemia, and translocations in certain ovarian adenocarcinomas); and, in a repeat performance, c-*myc* on 8 (translocations in acute myeloblastic leukemia). The translocations involving chromosomes 8, 15, and 22 are such that the hypothetical activator would be brought to the oncogene (269), rather than the converse postulated for Burkitt's lymphoma.

These are attractive ideas and provocative findings, but the issue remains in doubt. 1. The postulated role of chromosomal duplications and translocations relies too heavily on the now suspect dosage hypothesis (see above). For example, some early efforts to find augmented expression of c-*myc* in cells from Burkitt's lymphoma have failed (90). 2. In many instances, the agreement between chromosomal locations of oncogenes and karyotypic abnormalities in human tumors may be nothing more than accident, since both cellular oncogenes and the chromosomal changes are widely dispersed among the chromosomes; some congruence seems likely to occur by chance alone. Mapping of cellular oncogenes to their exact locations on chromosomes should prove helpful. 3. In several instances where oncogenes active in transfection have been isolated from human tumors, there is no evidence in the tumor cells for either amplification of the genes or major DNA rearrangements affecting the genes. 4. The abnormal chromosome in B-cell tumors that have been suitably studied bears the excluded rather than the active immunoglobulin allele, and the translocation of c-*myc* from chromosome 15 to chromosome 12 in mouse plasmacytomas also involves an excluded immunoglobulin allele (269d; K. Calame, personal communication). These findings require special pleading to implicate the translocations in gene activation.

Accounting for Tissue-Specificity of Oncogenes

From most vantage points, oncogenesis displays elements of specificity. Different forms of chemical carcinogens induce different and reproducible spectra of tumors in the same inbred animals (270), specificity in oncogenesis is one of the hallmarks of retroviral oncogenes (11), different types of tumors are disclosing different forms of active cellular oncogenes in transfection (15, 16, 50), and heritable predispositions to neoplasia in human families carry the risk of particular tumors (5, 6). All of these specificities may someday be understood in terms of the properties of cellular oncogenes.

The tissue specificity displayed by chemical carcinogenesis has generally been attributed to variation in the ability of different cells to convert chemicals to the state of proximal carcinogens. This facile explanation may not be uniformly applicable, however, and the problem seems ripe for investigation with tools that explore the activity and identity of cellular oncogenes.

The capabilities of retroviral oncogenes are manifest only in certain cells, and each oncogene has its own preferred set of target cells. The prevailing explanation for this specificity was outlined above: vulnerable cells contain the substrates on which oncogenes act, invulnerable cells do not.

Virologists suspect that this principle may be correct, but we know virtually nothing of the substrates in question nor of why they should vary from one type of cell to another.

Two contrasting hypotheses have been offered to explain the apparent specificity of active cellular oncogenes identified by transfection:

1. The specificity derives from vulnerability to damage; the oncogenes responsible for tumorigenesis in specific tissues represent genes that are susceptible to activation or structural mutation in those tissues (50). No potential explanation for selective susceptibility has yet been offered, but it is provocative that many suspected bladder carcinogens favor the transversion (G to T) implicated in the creation of a bladder oncogene (see above and 251a).
2. The specificity is analogous to that displayed by retroviral oncogenes; tumorigenesis occurs only in tissues that are susceptible to the effects of the active oncogene. I find the retroviral precedent a strong inducement to keep this possibility in contention, but we otherwise have no facts with which to conjure.

Heritable specificity in oncogenesis remains an enigma. It is tempting to frame hypotheses in terms of cellular oncogenes as we now know them, to attribute familial neoplasia to mutations that affect these genes. On the other hand, one attempt to demonstrate linkage between an inherited lesion that predisposes to carcinoma of the colon (Gardner's Syndrome) and the active cellular oncogene isolated from this form of tumor (Kirsten c-*ras*) has failed (R. White, personal communication). This is not surprising if the multistep nature of tumorigenicity is taken into account; inherited lesions are thought to represent the first of these steps (271), and active cellular oncogenes identified by transfection might represent the last (see below).

FITTING CELLULAR ONCOGENES INTO THE GENESIS IN NEOPLASIA

"False facts are highly injurious to the progress of science, for they often endure long; but false views, if supported by some evidence, do little harm, for everyone takes a salutory pleasure in proving their falseness." (271a)

It is generally agreed that two or more distinct events must occur to engender a malignant tumor (256, 271). The multistep progression to neoplasia has been viewed from several perspectives, including the interdependent roles of initiating agent and tumor promoters (256), the need for multiple mutations in a somatic cell before a tumor arises (271), and the incremental progression of the cell towards a frankly malignant phenotype (272). Here I frame tumor progression in terms of what we now know of cellular oncogenes, because the discovery and study of these genes may have given us access to at least two steps in the development of a tumor, the first and the last.

The first step in tumorigenesis has long been defined as initiation: the first lesion incurred; the response to carcinogen or the consequence of spontaneous mutagenesis; as most believe, damage to DNA of one sort or another. The last step I define as maintenance: the ostensibly abnormal biochemical activities whose sustained influence is required to maintain the neoplastic phenotype. It is likely that retroviruses have unveiled the genetic substance of both events: genes whose activities can initiate tumorigenesis, and genes whose activities maintain the neoplastic phenotype.

The revelation of maintenance functions came first, in the form of viral oncogenes which rapidly elicit and indefinitely sustain the neoplastic phenotype. In this view, cellular oncogenes identified by transfection have the same role; they are the final arbiters of malignancy, the effectors of the transformed phenotype. Viral and cellular oncogenes can even succeed one another as maintenance functions: tumors induced by Abelson murine leukemia virus can lose the viral oncogene by deletion and yet continue their malignant growth (273) because a cellular oncogene has come into play (G. Cooper, personal communication).

Insertional mutagenesis by retroviruses has provided our first glimpse of genes that might initiate tumorigenesis. When tumors are elicited by retroviruses without oncogenes, the induction of a cellular oncogene by integration of viral DNA is likely to be the first step towards malignant growth. The induced gene represents an initiation function. It now appears that the continued study of retroviruses without oncogenes will bring a number of such functions into view (see above).

The events that intervene between initiation and maintenance are a profound mystery. We know only that the initiated cell becomes increasingly wayward, that it progresses toward a tumorigenic phenotype (274, 275). We do not know how progression occurs, but it is provocative that evolving tumor cells are alleged to become hypermutable (276), a state that could hasten the advent of damage to the structure or regulation of a maintenance oncogene.

The production of B-cell lymphomas by infection with avian leukosis viruses illustrates all of these themes. Induction of c-*myc* activity by insertional mutagenesis may represent the initiation of tumorigenesis. As the initiated cells progress towards the malignant phenotype, they must survive immune surveillance, and they may be held in their path by the mitogenic effect of sustained antigenic stimulus (277, 278). Eventually, a maintenance function comes into play—a cellular oncogene, whose activity has been detected by transfection (15, 279), whose structure has been distinguished from that of c-*myc* (279), and whose function is presumed to maintain B-cells in neoplastic growth (15).

The implication of two distinct genes in the genesis of avian lymphomas provides experimental substance for multistep theories of oncogenesis. It also suggests how the dosage and qualitative theories of oncogenesis might be reconciled. Enhanced expression of an otherwise normal cellular oncogene (c-*myc*) initiates tumorigenesis, as it might if a chromosomal translocation were the first event in oncogenesis (see above). A subsequent qualitative change in a second cellular oncogene would provide a maintenance function without altered dosage, as appears to be the case in bladder carcinoma (see above). The active oncogene detected in the avian lymphomas by transfection is not expressed in exceptional amounts and may therefore be qualitatively different from its proto-oncogene (G. Cooper, personal communication).

The scheme for tumorigenesis by avian leukosis viruses remains as much fantasy as fact, but it brings vital questions into focus. How does the activity of c-*myc* initiate tumorigenesis? Might it be a mutator function, or might it activate other genes by epigenetic mechanisms? (The product of v-*myc* is said to be a DNA-binding protein, located in the nucleus, and bound to chromatin; see 142, 143, 280.) Why does the induction of c-*myc* figure in the production of B-cell tumors, when the viral form of the gene elicits primarily carcinomas? The discrepancy may be an illusion: variants of v-*myc* formed by recombination with c-*myc* rapidly induce lymphoid tumors rather than carcinomas (see 221). Must the action of c-*myc* be sustained, or does it evoke an irreversible train of events? How many steps intervene between the activation of c-*myc* and the final deployment of the cellular oncogene responsible for maintenance of the neoplastic phenotype? What is the nature of these steps, if they exist? Does the maintenance function come into play by dosage or by qualitative changes? And by what means does the maintenance gene sustain the neoplastic phenotype? [The active cellular oncogene isolated from B-cell lymphomas induced by avian retroviruses without oncogenes may encode a 6000-Mr protein whose chemical composition is reminiscent of polypeptide growth factors (see 15).] Pursuit of these questions promises a rich yield of insights into the course and mechanisms of tumorigenesis.

CONCLUSIONS AND PROSPECTS

Vertebrates and other metazoan organisms possess a family of proto-oncogenes whose altered activities may eventually explain both the initiation of tumorigenesis and the maintenance of cancerous growth. Transduction by retroviruses first brought these proto-oncogenes into view; insertional mutagenesis by retroviruses and DNA-mediated gene transfer have expanded the repertoire and provided direct evidence that at least some tumors contain active oncogenes.

The logic of evolution would not permit the survival of solely noxious genes. Powerful selective forces must have been at work to assure the conservation of proto-oncogenes throughout the diversification of metazoan phyla. Yet we know nothing of why these genes have been conserved, only that they are expressed in a variety of tissues and at various points during growth and development, that they are likely to represent a diverse set of biochemical functions, and that they may have all originated from one or a very few founder genes. Perhaps the proteins these genes encode are components of an interdigitating network that controls the growth of individual cells during the course of differentiation. We are badly in need of genetic tools to approach these issues, tools that may be forthcoming from the discovery of proto-oncogenes in *Drosophila* and nematodes.

Transduction by retroviruses is the only tangible means by which vertebrate genes have been mobilized and transferred from one animal to another without the intervention of an experimentalist. How does this transduction occur? What might its details tell us of the mechanisms of recombination in vertebrate organisms? What does it reflect of the potential plasticity of the eukaryotic genome? Can it transpose genetic loci other than cellular oncogenes? Has it figured in the course of evolution? How large is its role in natural as opposed to experimental carcinogenesis? These are ambitious questions, yet the means to answer most of them appear to be at hand.

The role of cellular oncogenes in carcinogenesis remains largely an inference. Direct tests of their tumorigenicity must be obtained, by novel means of gene-transfer, by incorporation into infectious viral vectors, and by the use of more appropriate cells and animal hosts as recipients. Transduction by retroviruses has uncovered at least 17 different proto-oncogenes. More may be in the offing. How many of these participate in tumorigenesis? The same small family of oncogenes (c-*ras*) has emerged repeatedly from the use of transfection to identify active oncogenes in tumor cells. Is this family of genes alone in its involvement in tumorigenesis, or will others emerge as well when the net has been cast more widely?

How is the net to be cast? It will be cast by the use of insertional mutagenesis to identify genes responsible for initiation of tumorigenesis, and by diversifying the techniques of gene-transfer used to search for active oncogenes. But we must also invent means to reveal recessive functions and lesions in regulatory loci that act on oncogenes. And the genetic substance of tumor progression should not remain an unmolested enigma.

Granted that at least some of the identified proto-oncogenes may figure in the development of malignancy. What is the nature of the conversion from proto-oncogene to oncogene? Does an increase in the amount of gene product suffice, or must the function of the gene be changed by damage to the coding domain? Is it unreasonable to expect that both answers may apply, perhaps each to a distinct set of carcinogens, oncogenes, or tumors? Or that dosage may serve to initiate tumorigenesis, qualitative changes to maintain neoplastic growth?

Once the nature of the lesions in active cellular oncogenes have been defined, we can begin the fascinating pursuit of how these lesions arise. Can they be traced to the direct action of proximal carcinogens or only to secondary events set in motion by the initiation of tumorigenesis? The long-elusive mechanisms by which carcinogens act may soon be laid bare.

We conclude where we began—with the hope that the mechanisms by which viral oncogenes transform cells can be deciphered and then parlayed into an understanding of how all forms of cancer arise. We have learned that tyrosine-specific protein kinases occupy an important place in the control of cell growth and in the induction of neoplastic transformation by some oncogenes. But we also have reasons to believe that protein phosphorylation is only one of several (perhaps many) threads in the tapestry we are trying to weave. So it is to the proteins encoded by oncogenes that our attentions must inevitably turn. The isolation of oncogenes, whether viral or cellular, does not solve the puzzle of how these genes work, and this puzzle is for me the end-game on which all else turns in cancer research.

> "We shall not cease from our exploration
> And the end of all our exploring
> Will be to arrive where we started
> And know the place for the first time." (281)

ACKNOWLEDGMENTS

I thank colleagues too numerous to cite for an invaluable blizzard of preprints, my friends in San Francisco for stimulating discussions, J. Marinos for patience and expert preparation of the manuscript, and the National Cancer Institute and American Cancer Society for support of research in my laboratory. This review was completed in September, 1982. KIB provided forbearance, but not lunch.

Literature Cited

1. Pynchon, T. 1966. *The Crying of Lot Forty Nine.* New York: Lippincott
2. Carroll, L. 1976. *Through the Looking Glass.* Vintage
3. Boveri, T. 1914. *Zur Frage der Erstehung Maligner Tumoren.* Jena: Fischer
4. Schwab, M. 1983. *Adv. Cancer Res.* In press
5. Knudson, A. G. 1979. *N. Engl. J. Med.* 301:606–7
6. Knudson, A. G. 1981. In *Genes, Chromosomes and Neoplasia,* ed. F. E. Arrighi, P. N. Rao, E. Stubblefield, pp. 453–62. New York: Raven
7. Comings, D. E. 1973. *Proc. Natl. Acad. Sci. USA* 70:3324–28
8. Huebner, R. J., Todaro, G. J. 1969. *Proc. Natl. Acad. Sci. USA* 64:1087–94
9. Todaro, G. J., Huebner, R. J. 1972. *Proc. Natl. Acad. Sci. USA* 69:1009–15
10. Bishop, J. M. 1978. *Ann. Rev. Biochem.* 47:35–88
11. Weiss, R., Teich, N., Varmus, H. E., Coffin, J., eds. 1982. *RNA Tumor Viruses.* Cold Spring Harbor, NY: Cold Spring Harbor
12. Eckart, W. 1972. *Ann. Rev. Biochem.* 41:503–40
13. Vogt, P. K. 1977. In *Comprehensive Virology,* ed. H. Fraenkel-Conrat, R. Wagner, 9:341–455. New York: Plenum
14. Graham, F. L. 1977. *Adv. Cancer Res.* 25:1–52
15. Cooper, G. M. 1982. *Science* 218:801–6
16. Weinberg, R. A. 1982. *Adv. Cancer Res.* 36:149–63
17. Stehelin, D., Varmus, H. E., Bishop, J. M., Vogt, P. K. 1976. *Nature* 260:170–73
18. Spector, D., Varmus, H. E., Bishop, J. M. 1978. *Proc. Natl. Acad. Sci. USA* 75:4102–6
19. Bishop, J. M. 1981. *Cell* 23:5–6
20. Linemeyer, D. L., Menke, J. G., Ruscetti, S. K., Evans, L. H., Scolnick, E. M. 1982. *J. Virol.* 43:223–33
21. Varmus, H. E. 1982. *Science* 216:812–27
22. Czernilofsky, A. P., Delorbe, W., Swanstrom, R., Varmus, H. E., Bishop, J. M., et al. 1980. *Nucleic Acids Res.* 8:2967–84
23. Neel, B. G., Hayward, W. S., Robinson, H. L., Fang, J., Astrin, S. M. 1981. *Cell* 23:323–34
24. Payne, G. S., Courtneidge, S. A., Crittenden, L. B., Fadley, A. M., Bishop, J. M., et al. 1981. *Cell* 23:311–22
25. Fung, Y.-K. T., Fadly, A. M., Crittenden, L. B., Kung, H.-J. 1981. *Proc. Natl. Acad. Sci. USA* 78:3418–22

26. Hayward, W. S., Neel, B. G., Astrin, S. M. 1981. *Nature* 290:475–79
27. Payne, G. S., Bishop, J. M., Varmus, H. E. 1982. *Nature* 295:209–17
28. Noori-Daloii, M. R., Swift, R. A., Kung, H.-J., Crittenden, L. B., Witter, R. L. 1981. *Nature* 294:574–76
29. Nusse, R., Varmus, H. E. 1982. *Cell* 31:99–109
30. Scolnick, E. M. 1978. Attribution in *The Gordon Conferences: An Historical Record.* Samizdat (Abstr.)
31. Shih, C., Shilo, B.-Z., Goldfarb, M. P., Dannenberg, A., Weinberg, R. A. 1979. *Proc. Natl. Acad. Sci. USA* 76:5714–18
32. Cooper, G. M., Okenquist, S., Silverman, L. 1980. *Nature* 284:418–21
33. Krontiris, T. G., Cooper, G. M. 1981. *Proc. Natl. Acad. Sci. USA* 78:1181–84
34. Lane, M. A., Sainten, A., Cooper, G. M. 1981. *Proc. Natl. Acad. Sci. USA* 78:5185–89
35. Perucho, M., Goldfarb, M., Shimizu, K., Lama, C., Fogh, J., et al. 1981. *Cell* 27:467–76
36. Shih, C., Padhy, L. C., Murray, M., Weinberg, R. A. 1981. *Nature* 290:261–63
37. Shilo, B.-Z., Weinberg, R. A. 1981. *Nature* 289:607–9
37a. Der, C. J., Krontiris, T. G., Cooper, G. M. 1982. *Proc. Natl. Acad. Sci. USA* 79:3637–40
38. Goldfarb, M., Shimizu, K., Perucho, M., Wigler, M. 1982. *Nature* 296:404–9
39. Lane, M.-A., Sainten, A., Cooper, G. M. 1982. *Cell* 28:873–80
40. Marshall, C. J., Hall, A., Weiss, R. A. 1982. *Nature* 299:171–73
41. Parada, L. F., Tabin, C. J., Shih, C., Weinberg, R. A. 1982. *Nature* 297:474–78
42. Pulciani, S., Santos, E., Lauver, A. V., Long, L. K., Robbins, K. C., Barbacid, M. 1982. *Proc. Natl. Acad. Sci. USA* 79:2845–49
43. Santos, E., Tronick, S. R., Aaronson, S. A., Pulciani, S., Barbacid, M. 1982. *Nature* 298:343–47
44. Shih, C., Weinberg, R. A. 1982. *Cell* 29:161–69
45. Smith, B. L., Anisowicz, A., Chodosh, L. A., Sager, R. 1982. *Proc. Natl. Acad. Sci. USA* 79:1964–68
46. Chang, E. H., Furth, M. E., Scolnick, E. M., Lowy, D. R. 1982. *Nature* 297:479–84
47. Ozer, H. L., Jha, K. K. 1977. *Adv. Cancer Res.* 25:53–94
48. Croce, C. M. 1980. *Biochim. Biophys. Acta* 605:411–30

49. Coffin, J. M., Varmus, H. E., Bishop, J. M., Essex, M., Hardy, W. D., et al. 1981. *J. Virol.* 40:953–57
50. Weinberg, R. A. 1982. *Cell* 30:3–4
51. Klempnauer, K.-H., Gonda, T. J., Bishop, J. M. 1982. *Cell.* In press
52. Goff, S. P., Gilboa, E., Witte, O. N., Baltimore, D. 1980. *Cell* 22:777–85
53. Vennstrom, B., Bishop, J. M. 1982. *Cell* 28:135–43
54. Chen, I. S. Y., Wilhelmsen, K. C., Temin, H. M. 1983. *J. Virol.* In press
55. Jones, M., Bosselman, R. A., Vanderhoeven, F. A., Berns, A., Fan, H., Verman, I. M. 1980. *Proc. Natl. Acad. Sci. USA* 77:2651–55
56. Oskarsson, M., McClements, W. L., Blair, D. G., Maizel, J. V., Vanderwoude, G. F. 1980. *Science* 207:1222–24
57. Van Beveren, C., van Straaten, F., Galleshaw, J. A., Verma, I. M. 1982. *Cell* 27:97–108
58. Gilbert, W. 1978. *Nature* 271:501–2
59. Crick, F. 1979. *Science* 204:264–71
60. Hughes, S. H., Vogt, P. K., Stubblefield, E. Robinson, H., Bishop, J. M., et al. 1979. *CSH Symp. Biol.* 44:1077–89
61. Hughes, S. H., Payvar, F., Spector, D., Schimke, R. T., Robinson, H. L., et al. 1979. *Cell* 18:347–59
61a. Gould, S. J. 1982. *Nat. Hist.* 91(9):6–13
62. Shilo, B.-Z., Weinberg, R. A. 1981. *Proc. Natl. Acad. Sci. USA* 78:6789–92
63. Gateff, E. 1978. *Science* 200:1448–58
64. Sheiness, D. K., Hughes, S. H., Varmus, H. E., Stubblefield, E., Bishop, J. M. 1980. *Virology* 105:415–24
65. Robins, T., Bister, K., Garon, C., Papas, T., Duesberg, P. 1982. *J. Virol.* 41:635–42
66. Vennstrom, B., Sheiness, D., Zabielski, J., Bishop, J. M. 1982. *J. Virol.* 42:773–79
67. Dalla-Favera, R. D., Gellman, E. P., Martinati, S., Franchini, G., Papas, T. S., et al. 1982. *Proc. Natl. Acad. Sci. USA* 79:6497–6501
68. Ellis, R. W., DeFeo, D., Shin, T. Y., Gonda, M. A., Young, H. A., et al. 1981. *Nature* 292:506–10
69. Chien, Y.-H., Lai, M., Shih, T. Y., Verma, I. M., Scolnick, E. M., et al. 1979. *J. Virol.* 31:752–59
70. DeFeo, D., Gonda, M. A., Young, H. A., Chang, E. H., Lowy, D. R., et al. 1981. *Proc. Natl. Acad. Sci. USA* 78:3328–32
71. Chattopadhyay, S. K., Chang, E. H., Lander, M. R., Ellis, R. W., Scolnick, E. M., Cowy, D. R. 1982. *Nature* 296:361–62
72. Chang, E. H., Gonda, M. A., Ellis, R. W., Scolnick, E. M., Lowy, D. R. 1982.
73. Marx, J. L. 1982. *Science* 216:969–70
74. Parker, R. C., Varmus, H. E., Bishop, J. M. 1981. *Proc. Natl. Acad. Sci. USA* 78:5842–46
75. Shealy, D. J., Erikson, R. L. 1981. *Nature* 293:666–68
76. Watson, R., Oskarsson, M., Vande Woude, G. F. 1982. *Proc. Natl. Acad. Sci. USA* 79:4078–82
77. Shibuya, M., Hanafusa, T., Hanafusa, H., Stephenson, J. R. 1980. *Proc. Natl. Acad. Sci. USA* 77:6536–40
78. Barbacid, M., Breitman, M. L., Lauver, A. V., Long, L. K., Vogt, P. K. 1981. *Virology* 110:411–19
79. Shibuya, M., Wang, L.-H., Hanafusa, H. 1982. *J. Virol.* 42:1007–16
80. Shibuya, M., Hanafusa, H. 1982. *Cell* 30:787–95
81. Hampte, A., Lapnevotte, J., Galibert, F., Fedele, L. A., Sher, C. J. 1982. *Cell* 30:775–85
82. Spector, D. H., Smith, K., Padyelt, T., McCombe, P., Roulland-Dussoix, D., et al. 1978. *Cell* 13:371–79
83. Spector, D. H., Baker, B., Varmus, H. E., Bishop, J. M. 1978. *Cell* 13:381–86
84. Chen, J. H. 1980. *J. Virol.* 36:162–70
85. Gonda, T. J., Sheiness, D. K., Bishop, J. M. 1982. *Mol. Cell. Biol.* 2:617–24
86. Muller, R., Slamon, D. J., Tremblay, J. M., Cline, M. J., Verma, I. M. 1982. *Nature* 299:640–44
87. Shibuya, M., Hanafusa, H., Balduzzi, P. C. 1982. *J. Virol.* 42:143–52
88. Ellis, R. W., DeFeo, D., Furth, M. E., Scolnick, E. M. 1983. *Mol. Cell. Biol.* 2:1339–45
89. Westin, E. H., Gallo, R. C., Arya, S. K., Eva, A., Souza, L. M., et al. 1982. *Proc. Natl. Acad. Sci. USA* 79:2194–98
90. Westin, E. H., Wongstaal, F., Gelmann, E. P., Favera, R. D., Papas, T. S., et al. 1982. *Proc. Natl. Acad. Sci. USA* 79:2490–94
91. Eva, A., Robbins, K. C., Andersen, P. R., Srinivasan, A., Tronick, S. R. 1982. *Nature* 295:116–19
92. Rovigatti, U. G., Rogler, C. E., Neel, B. G., Hayward, S. W., Astrin, S. M. 1982. *4th Ann. Bristol-Myers Symp. Tumor Cell Heterogeneity.* In press
93. Frankel, A. E., Fischinger, P. J. 1976. *Proc. Natl. Acad. Sci. USA* 73: 3705–9
94. Gattoni, S., Kirschmeier, P., Weinstein, I. B., Escobedo, J., Dina, D. 1982. *Mol. Cell. Biol.* 2:42–51
95. Barbacid, M., Beemon, K., Devare, S. G. 1980. *Proc. Natl. Acad. Sci. USA* 77:5158–62
96. Breathnach, R., Chambon, P. 1981. *Ann. Rev. Biochem.* 50:349–83

97. Bishop, J. M., Gonda, T., Hughes, S. H., Sheiness, D. K., Stubblefield, E. et al. 1980. *Mobilization and Reassembly of Genetic Information*, 12th Miami Winter Symp., pp. 261–73

98. Sheiness, D., Vennstrom, B., Bishop, J. M. 1981. *Cell* 23:291–300

99. Setzer, D. R., McGrogan, M., Nunberg, J. H., Schimke, R. T. 1980. *Cell* 22:361–70

100. Collett, M. S., Erikson, R. L. 1979. *Proc. Natl. Acad. Sci. USA* 75:2021–24

101. Collett, M. S., Brugge, J. S., Erikson, R. L. 1979. *Cell* 15:1363–70

102. Levinson, A. D., Oppermann, H., Levintow, L., Varmus, H. E., Bishop, J. M. 1978. *Cell* 15:561–72

103. Collett, M. S., Erikson, E., Purchio, A. F., Brugge, J. S., Erikson, R. L. 1979. *Proc. Natl. Acad. Sci. USA* 76:3159–63

104. Oppermann, H., Levinson, A. D., Varmus, H. E., Levintow, L., Bishop, J. M. 1979. *Proc. Natl. Acad. Sci. USA* 76:1804–8

105. Rohrschneider, L. R., Eisenman, R. N., Leitch, C. R. 1979. *Proc. Natl. Acad. Sci. USA* 76:4479–83

106. Collett, M. S., Purchio, A. F., Erikson, R. L. 1980. *Nature* 285:167–68

107. Hunter, T., Sefton, B. M. 980. *Proc. Natl. Acad. Sci. USA* 77:1311–15

108. Levinson, A. D., Oppermann, H., Varmus, H. E., Bishop, J. M. 1980. *J. Biol. Chem.* 255:11973–80

109. Sefton, B. M., Hunter, T., Beemon, K. 1980. *Proc. Natl. Acad. Sci. USA* 77:2059–63

110. Karess, R. E., Hanafusa, H. 1981. *Cell* 24:155–64

111. Courtneidge, S. A., Levinson, D., Bishop, J. M. 1980. *Proc. Natl. Acad. Sci. USA* 77:3783–87

112. Rohrschneider, L. R. 1980. *Proc. Natl. Acad, Sci. USA* 77:3514–18

113. Rohrschneider, L., Rosok, M., Shriver, K. 1982. *CSH Symp. Quant. Biol.* 46:953–65

114. Radke, K., Martin, G. S. 1979. *Proc. Natl. Acad. Sci. USA* 76:5212–16

115. Erikson, E., Erikson, R. L. 1980. *Cell* 21:829–36

116. Radke, K., Gilmore, T., Martin, G. S. 1980. *Cell* 21:821–28

117. Purchio, A. F., Erikson, E., Collett, M. S., Erikson, R. L. 1981. In *Cold Spring Harbor Conference on Cell Proliferation-Protein Phosphorylation*, ed. E. Krebs, O. Rosen, 8:1203–15. Cold Spring Harbor, NY: Cold Spring Harbor

118. Courtneidge, S. A., Ralston, R., Alitalo, K., Bishop, J. M. 1983. *Mol. Cell Biol.* In press

119. Cooper, J. A., Hunter, T. 1982. *J. Cell Biol.* 94:287–96

120. Ushiro, H., Cohen, S. 1980. *J. Biol. Chem.* 255:8363–65

121. Chinkers, M., Cohen, S. 1981. *Nature* 290:516–18

122. Cooper, J. A., Hunter, T. 1981. *J. Cell Biol.* 91:878–83

123. Erikson, E., Shealy, D. J., Erikson, R. L. 1981. *J. Biol. Chem.* 256:11381–84

124. Hunter, T., Cooper, J. A. 1981. *Cell* 24:741–52

125. Kudlow, J. E., Buss, J. E., Gill, G. N. 1981. *Nature* 290:519–20

126. Ek, B., Westermark, B., Wasteson, A., Heldin, C. H. 1982. *Nature* 295:419–20

126a. Nushimura, J., Huang, J. S., Deuel, T. F. 1982. *Proc. Natl. Acad. Sci. USA* 79:4303–7

127. Kasuga, M., Zick, Y., Blithe, D. L., Crettaz, M., Kahn, C. R. 1982. *Nature* 298:667–69

128. Cooper, J. A., Bowen-Pope, D. F., Raines, E., Ross, R., Hunter, T. 1983. *Cell* 31:263–73

129. Langbeheim, H., Shih, T. Y., Scolnick, E. M. 1980. *Virology* 106: 292–300

130. Furth, M. E., Davis, L. J., Fleurdelys, B., Scolnick, E. M. 1982. *J. Virol.* 43:294–304

131. Shih, T. Y., Papageorge, A. G., Stokes, P. E., Weeks, M. O., Scolnick, E. M. 1980. *Nature* 287:686–91

132. Scolnick, E. M., Weeks, M. O., Shih, T. Y., Ruscetti, S. K., Dexter, T. M. 1981. *Mol. Cell Biol.* 1:66–74

133. Willingham, M. C., Pastan, I., Shih, T. Y., Scolnick, E. M. 1980. *Cell* 19:1005–14

134. Witte, O. N., Rosenberg, N. E., Baltimore, D. 1979. *Nature* 281:396–98

135. Witte, O. N., Rosenberg, N., Paskind, M., Shields, A., Baltimore, D. 1978. *Proc. Natl. Acad. Sci. USA* 75:2488–92

136. Witte, O. N., Rosenberg, N., Baltimore, D. 1979. *J. Virol.* 31:776–79

137. Ponticelli, A. S., Whitlock, C. A., Rosenberg, N., Witte, O. N. 1982. *Cell* 29:953–60

138. Mathey-Prevot, B., Hanafusa, H., Kawai, S. 1982. *Cell* 28:897–906

139. Gilmer, T. M., Erikson, R. L. 1981. *Nature* 294:771–72

140. McGrath, J. P., Levinson, A. D. 1981. *Nature* 295:423–25

141. Wang, J. Y. J., Queen, C., Baltimore, D. 1982. *J. Biol. Chem.* 257:13181–84

142. Donner, P., Greiser-Wilke, I., Moelling, K. 1982. *Nature* 296:262–65

143. Abrams, H. D., Rohrschneider, L. R., Eisenman, R. N. 1982. *Cell* 29:427–39

144. Dhar, R., Ellis, R. W., Shih, T. Y.,

Onoszlan, S., Shapiro, B., et al. 1982. *Science* 217:934–37

145. Tsuchida, N., Ryder, T., Ohtsubo, E. 1982. *Science* 217:937–38

146. Kitamura, N., Kitamura, A., Toyoshima, K., Hirayama, Y., Yoshida, M. 1982. *Nature* 297:205–7

147. Doolittle R. F. 1981. *Science* 214:149–59

148. Cooper, J. A., Hunter, T. 1981. *Mol. Cell Biol.* 1:394–407

149. Levinson, A. D., Courtneidge, S. A., Bishop, J. M. 1981. *Proc. Natl. Acad. Sci. USA* 78:1624–28

150. Barker, W. C., Dayhoff, M. O. 1982. *Proc. Natl. Acad. Sci. USA* 79:2826–39

151. Van Beveren, C., Gallishaw, J. A., Jonas, V., Berns, A. J. M., Doolittle, R. F., et al. 1981. *Nature* 289:258–62

152. Buhrow, S. A., Cohen, S., Staros, J. V. 1982. *J. Biol. Chem.* 257:4019–22

153. Cohen, S., Ushino, H., Stoscheck, C., Chinkers, M. 1982. *J. Biol. Chem.* 257:1523–31

154. Barrett, J. C., Crawford, B. D., T'so, P. O. P. 1976. *J. Cell Biol.* 70:233–42

155. Parodi, S., Brambilla, G. 1977. *Mutat. Res.* 47:53–74

156. Barrett, J. C., T'so, P. O. P. 1978. *Proc. Natl. Acad. Sci. USA* 75:3297–301

157. Sandberg, A. A. 1979. *Chromosomes in Human Cancer and Leukemia.* New York: Elsevier North-Holland

158. Dalla-Favera, R., Franchini, G., Martinotti, S., Wong-Staal, F., Gallo, R. C., Croce, C. M. 1982. *Proc. Natl. Acad. Sci. USA* 79:4714–17

159. Dalla-Favera, R., Gallo, R. C., Giallongo, A., Croce, C. M. 1982. *Science* 218:686–89

160. Prakash, K., McBride, O. W., Swan, D. C., Devare, S. G., Tronick, S. R., Aaronson, S. A. 1982. *Proc. Natl. Acad. Sci. USA* 79:5210–14

161. Swan, D. C., McBride, O. W., Robbins, K. C., Keithley, D. A., Reddy, E. P., et al. 1982. *Proc. Natl. Acad. Sci. USA* 79:4691–95

161a. Heisterkamp, N., Groffen, J., Stephenson, J. R., Spurr, N. K., Goodfellow, P. N., et al. 1979. *Nature* 299:747–50

162. Padgett, T. G., Stubblefield, E., Varmus, H. E. 1977. *Cell* 10:649–57

163. Hughes, S. H., Stubblefield, E., Payrar, F., Engel, J. D., Dodyson, J. B., et al. 1979. *Proc. Natl. Acad. Sci. USA* 76:1348–52

164. Tereba, A., Lai, M. M. C., Murti, K. G. 1979. *Proc. Natl. Acad. Sci. USA* 76:6486–90

165. Wong, T. C., Tereba, A., Vogt, P. K.,

Lai, M. M. C. 1981. *Virology* 111:418–26

166. Tereba, A., Lai, M. M. C. 1982. *Virology* 116:654–57

166a. Boycott, A. E. 1928. *Proc. Roy. Soc. Med.* 22:55–69

167. Hughes, S. H., Toyoshima, K., Bishop, J. M., Varmus, H. E. 1980. *Virology* 108:189–207

168. Cohen, J. C., Majors, J. E., Varmus, H. E. 1979. *J. Virol.* 32:483–96

169. Varmus, H. E., Cohen, J. C., Hughes, S. H., Majors, J., Bishop, J. M. 1980. *Ann. NY Acad. Sci. Genetic Variation Viruses* 354:379–83

170. Callahan, R., Todaro, G. J. 1978. In *Origins of Inbred Mice*, ed. H. C. Morse, pp. 689–713. New York: Academic

171. Frisby, D. P., Weiss, R. A., Roussel, M., Stehelin, D. 1979. *Cell* 17:623–34

172. Frankel, A. E., Fischinger, P. J. 1977. *J. Virol.* 21:153–60

173. Frankel, A. E., Gilbert, J. H., Porzig, K. J., Scolnick, E. M., Aaronson, S. A. 1979. *J. Virol.* 30:821–27

174. Roussel, M., Saule, S., Lagroù, C., Rommens, C., Berg, H., et al. 1979. *Nature* 281:452–55

175. Sheiness, D., Bishop, J. M. 1979. *J. Virol.* 31:514–21

176. Rapp, U. R., Todaro, G. J. 1978. *Science* 201:821–23

177. Rasheed, S., Gardner, M. B., Huebner, R. J. 1978. *Proc. Natl. Acad. Sci. USA* 75:2972–76

178. Young, H. A., Shih, T. Y., Scolnick, E. M., Rasheed, S., Gardner, M. B. 1979. *Proc. Natl. Acad. Sci. USA* 76:3523–27

179. Rapp, U. R., Todaro, G. J. 1980. *Proc. Natl. Acad. Sci. USA* 77:624–28

180. Young, H. A., Rasheed, S., Sowder, R., Benton, C. V., Henderson, L. E. 1981. *J. Virol.* 38:286–93

181. Rasheed, S., Young, H. A. 1982. *Virology* 118:219–24

182. Hanafusa, H., Halpern, C. C., Buchhagen, D. L., Kawai, S. 1977. *J. Exp. Med.* 146:1735–47

183. Wang, L. H., Halpern, C. C., Nadel, M., Hanafusa, H. 1978. *Proc. Natl. Acad. Sci. USA* 75:5812–16

184. Karess, R. E., Hayward, W. S., Hanafusa, H. 1979. *Proc. Natl. Acad. Sci. USA* 76:3154–58

185. Wang, L.-H., Moscovici, C., Karess, R. E., Hanafusa, H. 1979. *J. Virol.* 32:546–56

186. Wang, L.-H., Snyder, P., Hanafusa, T., Hanafusa, H. 1980. *J. Virol.* 29:52–64

187. Hanafusa, H. 1981. *Harvey Lect.* 75:255–75

188. Neel, B. G., Wang, L. H., Mathey-Prevot, B., Hanafusa, T., Hanafusa, H.,

et al. 1982. *Proc. Natl. Acad. Sci. USA* 79:5088–92
189. Harvey, J. J. 1964. *Nature* 204:1104–5
190. Andersen, P. R., Devare, S. G., Tronick, S. R., Ellis, R. W., Aaronson, S. A., et al. 1981. *Cell* 26:129–40
191. Abelson, H. T., Rabstein, L. S. 1970. *Cancer Res.* 30:2213–22
192. Thelein, G. H., Gould, D., Fowler, M., Dungworth, D. L. 1971. *J. Natl. Cancer Inst.* 47:881–89
193. Wong-Staal, F., Favera, R. D., Gelmann, E. P., Manzari, V., Szala, S., et al. 1981. *Nature* 294:273–74
194. Ellis, R. W., DeFeo, D., Maryak, J. M., Young, H. A., Shih, T. Y., et al. 1980. *J. Virol.* 36:408–20
195. Kirsten, W. H., Mayer, L. A. 1961. *J. Natl. Cancer Inst.* 39:311–35
196. Andersen, P. R., Tronick, S. R., Aaronson, S. A. 1981. *J. Virol.* 40:431–39
197. Gonda, M. A., Young, H. A., Elser, J. E., Rashud, S., Talmadge, C. B., et al. 1982. *J. Virol.* 44:520–29
198. Gruss, P., Ellis, R. W., Shih, T. Y., Konig, M., Scolnick, E. M., Khoury, G. 1981. *Nature* 293:486–87
199. Bister, K., Ramsey, G., Hayman, M. J., Deusberg, P. H. 1980. *Proc. Natl. Acad. Sci. USA* 77:7142–46
200. Chiswell, D. J., Ramsay, G., Hayman, M. J. 1981. *J. Virol.* 40:301–4
201. Bister, K., Nunn, M., Moscovici, C., Perbal, B., Baluda, M. A., Deusberg, P. H. 1982. *Proc. Natl. Acad. Sci. USA* 79:3677–81
202. Swanstrom, R., Parker, R. C., Varmus, H. E., Bishop, J. M. 1983. *Proc. Natl. Acad. Sci. USA*. In press
203. Hackett, P. B., Swanstrom, R., Varmus, H. E., Bishop, J. M. 1982. *J. Virol.* 41:527–34
204. Swanstrom, R., Varmus, H. E., Bishop, J. M. 1982. *J. Virol.* 41:535–41
205. Coffin, J. M. 1979. *J. Gen. Virol.* 42:1–26
206. Weiss, R. A., Mason, W. S., Vogt, P. K. 1973. *Virology* 52:535–52
207. Hunter, E. 1978. *Curr. Top. Microbiol. Immunol.* 79:295–309
208. Weiss, R. A. 1973. In *Possible Episomes in Eukaryotes*, ed. L. G. Silvestri, pp. 130–41. London/Amsterdam: North-Holland
209. Temin, H. M. 1980. *Cell* 21:599–600
210. Temin, H. M. 1974. *Harvey Lect.* 69:173–96
211. Watanabe, S., Temin, H. M. 1982. *Proc. Natl. Acad. Sci. USA* 79:5986–90
212. Weinberg, R. 1980. *Ann. Rev. Biochem.* 49:197–226
213. Hsu, T. W., Taylor, J. M., Aldrich, C., Townsend, J. B., Seal, G., Mason, W. S.,

et al. 1981. *J. Virol.* 38:219–23
214. Deleted in proof
215. Varmus, H. E., Quintrell, N., Ortiz, S. 1981. *Cell* 25:23–36
216. Goldfarb, M. P., Weinberg, R. A. 1981. *J. Virol.* 38:136–50
217. Shimotohno, K., Temin, H. M. 1982. *Nature* 299:265–68
218. Sorge, J., Hughes, S. H. 1983. *J. Mol. Appl. Genet.* In press
219. Stavnezer, E., Gerhard, D. S., Binari, R. C., Balazs, I. 1981. *J. Virol.* 39:920–34
220. Ramsay, G. M., Enrietto, P. J., Graf, T., Hayman, M. J. 1982. *Proc. Natl. Acad. Sci. USA* 79:6885–89
221. Enrietto, P. J., et al. 1983. *Virology.* In press
222. Essex, M. 1975. *Adv. Cancer Res.* 21:175–248
223. Deleted in proof
224. Takeya, T., Feldman, R. A., Hanafusa, H. 1982. *J. Virol.* 44:1–11
225. Takeya, T., Hanafusa, H. 1982. *J. Virol.* 44:12–18
226. Ikawa, Y., Ross, J., Leder, P. 1974. *Proc. Natl. Acad. Sci. USA* 71:1154–58
227. Boccara, M., Pluquet, N., Coll, J., Rommens, C., Stehelin, D. 1982. *J. Virol.* 43:925–34
228. Graf, T., Beug, H., Hayman, M. J. 1980. *Proc. Natl. Acad. Sci. USA* 77:389–93
229. Durban, E. M., Boettiger, D. 1981. *Proc. Natl. Acad. Sci. USA* 78:3600–5
230. Groudine, M., Weintraub, H. 1980. *Proc. Natl. Acad. Sci. USA* 77:5351–54
231. Boettiger, D., Durban, E. M. 1979. *CSH Symp. Quant. Biol.* 44:1249–54
232. Graf, T., Ade, N., Beug, H. 1978. *Nature* 275:496–501
233. Beug, H., Palmieri, S., Freudenstein, C., Zentgraf, H., Graf, T. 1982. *Cell* 28:907–19
234. Baltimore, D., Rosenberg, N., Witte, W. N. 1979. In *Immunological Reviews*, ed. G. Moller, 48:3–22. Copenhagen: Munksgaard
235. Gazzolo, L., Moscovici, C., Moscovici, M. G., Samarut, J. 1979. *Cell* 16:627–38
236. Durban, E. M., Boettiger, D. 1981. *J. Virol.* 37:488–92
237. Hankins, W. D., Scolnick, E. M. 1981. *Cell* 26:91–100
238. Collins, S., Groudine, M. 1982. *Nature* 298:679–81
239. Favera, R. D., Wong-Staal, F., Gallo, R. C. 1982. *Nature* 299:61–63
240. Gopinathan, K., Weymouth, L., Kunkel, T., Loeb, L. 1979. *Nature* 278:857–58
241. Anderson, S. J., Furth, M., Wolff, L., Ruscetti, S. K., Sherr, C. J. 1982. *J. Virol.* 44:696–702

242. Lee, W.-H., Nunn, M., Duesberg, P. H. 1981. *J. Virol.* 39:758–76
242a. Blair, D. G., Oskarsson, M., Wood, T. G., McClements, W. L., Fischinger, P. J., et al. 1981. *Science* 212:941–43
243. Cooper, M. D., Payne, L. N., Dent, P. B., Burmester, B. R., Good, R. A. 1978. *J. Natl. Cancer Inst.* 41:373–89
244. Fung, Y.-K. T., Crittenden, L. B., Kung, H.-J. 1982. *J. Virol.* 44:742–46
245. Deng, C.-T., Stehelin, D., Bishop, J. M., Varmus, H. E. 1977. *Virology* 76:313–30
246. Bishop, J. M., Courtneidge, S. A., Levinson, A. D., Oppermann, H., Quintrell, N., et al. 1979. *CSH Symp. Quant. Biol.* 44:919–30
247. Porzig, K. J., Robbins, K. C., Aaronson, S. A. 1979. *Cell* 16:875–84
248. Ziegler, S. F., Whitlock, C. A., Goff, S. P., Gifford, A., Witte, O. N. 1981. *Cell* 27:477–86
249. Goff, S. P., Tabin, C. J., Wang, J. Y.-J., Weinberg, R., Baltimore, D. 1982. *J. Virol.* 41:271–85
250. Papkoff, J., Verma, I. M., Hunter, T. 1982. *Cell* 29:417–26
251. Smart, J. E., Opperman, H., Czernilovsky, R. P., Purchio, A. F., Erikson, R. L., Bishop, J. M. 1981. *Proc. Natl. Acad. Sci. USA* 78:6013–17
251a. Tabin, C. J., Bradley, S. M., Bargmann, C. I., Weinberg, R. A., Papageorge, A. G., et al. 1982. *Nature* 300:143–48
251b. Reddy, E. P., Reynolds, R. K., Santos, E., Barbacid, M. 1982. *Nature* 300:149–52
252. Ames, B. N. 1979. *Science* 204:587–93
253. Cairns, J. 1981. *Nature* 289:353–57
254. Rubin, H. 1980. *J. Natl. Cancer Inst.* 64:995–1000
255. Mintz, B., Fleischman, R. A. 1981. *Adv. Cancer Res* 34:211–78
256. Weinstein, I. B. 1981. *J. Supramol. Struct. Cell Biochem.* 17:99–120
257. Klein, G. 1981. *Nature* 294:313–18
258. Pall, M. L. 1981. *Proc. Natl. Acad. Sci. USA* 78:2465–68
259. Yaniv, M. 1982. *Nature* 297:17–18
260. Banerji, J., Rusconi, S., Schaffner, W. 1981. *Cell* 27:299–308
261. Capecchi, M. R. 1980. *Cell* 22:479–87
262. Levinson, B., Khoury, G., Vande Woude, G., Gruss, P. 1982. *Nature* 295:568–72
263. Temin, H. M. 1982. *Cell* 28:3–5
264. Kirschmeier, P., Gattoni-Celli, S., Dina, D., Weinstein, I. B. 1982. *Proc. Natl. Acad. Sci. USA* 79:2773–77
265. Cox, D. W., Markovic, V. D., Teschima, I. E. 1982. *Nature* 297:428–30
266. Lenoir, G. M., Preud'homme, J. L., Bernheim, A., Berger, R. 1982. *Nature* 298:474–76
267. Kirsch, I. R., Morton, C. C., Nakahara, K., Leder, P. 1982. *Science* 216:301–2
268. Malcolm, S., Barton, P., Murphy, C., Ferguson-Smith, M. A., Bentley, D. L., Rabbitts, T. H. 1982. *Proc. Natl. Acad. Sci. USA* 79:4957–61
269. Rowley, J. D. 1982. *Science* 216:749–51
269a. Dalla-Favera, R., Breyni, M., Erikson, J., Patterson, D., Gallo, R. C., Croce, C. M. 1982. *Proc. Natl. Acad. Sci. USA* 79: In press
269b. Taub, R., Kirsch, I., Morton, C., Lenoir, G., Swan, D., et al. 1982. *Proc. Natl. Acad. Sci. USA* 79: In press
269c. Calame, K., Kim, S., Lalley, P., Hill, R., Davis, M., et al. 1982. *Proc. Natl. Acad. Sci. USA* 79:6994–98
270. Becker, F. F., ed. 1975. *Cancer, A Comprehensive Treatise*, Vol. 1. New York/London: Plenum
271. Moolgavkar, S. H., Knudsen, A. G. 1981. *J. Nat. Cancer Inst.* 66:1037–52
271a. Darwin, C. 1859. *The Origin of Species by Means of Natural Selection*, New York: Modern Library
272. Farber, E., Cameron, B. 1980. *Adv. Cancer Res.* 31:125–226
273. Grunwald, D. J., Dale, B., Dudley, J., Lamph, W., Sugden, B., Ozanne, B., Risser, R. 1982. *J. Virol.* 43:92–103
274. Nowell, P. C. 1976. *Science* 194:23–28
275. Barrett, J. C., T'so, P. O. P. 1978. *Proc. Natl. Acad. Sci. USA* 75:3761–65
276. Citone, M. A., Fidler, I. J. 1981. *Proc. Natl. Acad. Sci. USA* 78:6949–52
277. McGrath, M. S., Weissman, I. L. 1979. *Cell* 17:65–76
278. Lee, J. C., Ihle, J. N. 1981. *Nature* 289:407–8
279. Cooper, G. M., Neiman, P. E. 1981. *Nature* 292:857–58
280. Bunte, T., Greiser-Wilke, I., Donner, P., Moelling, K. 1982. *EMBO J.* In press
281. Eliot, T. S. 1952. *Little Gidding: The Four Quartets.* New York: Harcourt Brace Jovanovich

Ann. Rev. Biochem. 1983. 52:355–77

LEUKOTRIENES

Sven Hammarström

Department of Physiological Chemistry, Karolinska Institutet,
S-10401 Stockholm, Sweden

CONTENTS

PERSPECTIVES AND SUMMARY 355
DISCOVERY AND STRUCTURE ELUCIDATION 356
    Leukotriene C_4 356
    Leukotriene B_4 359
BIOSYNTHESIS 360
    Formation of an Epoxide Intermediate 360
    Biosynthesis of Leukotriene A_4 362
    Leukotrienes Formed from Other Fatty Acids or by Oxygenation of Arachidonic Acid
        at Alternative Positions 363
OCCURRENCE 365
METABOLISM 366
    Modifications Involving the Peptide Part of Leukotriene C 366
    ω-Oxidation of Leukotriene B_4 367
DISTRIBUTION AND EXCRETION 369
ANALYTICAL METHODS 369
    High Performance Liquid Chromatography 369
    Mass Spectrometry 370
    Radioimmunoassay 370
BIOLOGICAL EFFECTS 371
    Respiratory Effects 371
    Microvascular Effects 372
    Effects on Leukocytes 373
    Gastrointestinal Effects 373
    Other Effects 373

PERSPECTIVES AND SUMMARY

The leukotrienes are a family of biologically active molecules, formed by leukocytes, mastocytoma cells, macrophages, and other tissues and cells in response to immunological and nonimmunological stimuli. They exhibit a number of biological effects such as contraction of bronchial smooth muscles, stimulation of vascular permeability, and attraction and activation of leukocytes. Compared to histamine, which causes constriction of

355

0066-4154/83/0701-0355$02.00

airways and edema formation, the leukotrienes are three to four orders of magnitude more potent and the effects have longer duration.

The leukotrienes were discovered in 1938 as a smooth muscle-contracting factor in lung perfusates. It was referred to as "slow reacting substance" (SRS) or "slow reacting substance of anaphylaxis" (SRS-A) until 1979 when its structure was reported. The term "leukotriene" was introduced at that time as a trivial name for the new type of compound. Leukotrienes C_4 and D_4 are glutathione and cysteinylglycine conjugates, respectively, of arachidonic acid. After hydrolytic release from phospholipids of the cell membrane, arachidonic acid is oxygenated by a lipoxygenase to 5-hydroperoxy-6,8,11,14-eicosatetraenoic acid. This product is further converted to leukotrienes by elimination of the 10-*pro-R* hydrogen and OH from the hydroperoxy group to give 5,6-oxido-7,9,11,14-eicosatetraenoic acid (leukotriene A_4). Nucleophilic opening of the epoxide at C-6 by the sulfhydryl group of glutathione gives leukotriene C_4, which is metabolized to leukotrienes D_4 and E_4 by sequential elimination of glutamic acid and glycine. The latter reactions are catalyzed by γ-glutamyl transpeptidase and a particulate dipeptidase from kidney. Alternatively, water may add at C-12 of leukotriene A_4, leading also to opening of the epoxide at C-6 with formation of 5,12-dihydroxy-6,8,10,14-eicosatetraenoic acid (leukotriene B_4). Leukotriene B_4 is metabolized by ω-hydroxylation to 20-hydroxy and 20-carboxy leukotriene B_4. Leukotrienes are also formed from eicosatrienoic acid (n-9) and eicosapentaenoic acid (n-3) after oxygenation at C-5 and from eicosatrienoic acid (n-6) and arachidonic acid after oxygenation at C-8 (eicosatrienoic acid) and C-12 or C-15 (arachidonic acid). Although they are formed from the same and additional fatty acids as prostaglandins and thromboxanes [reviewed in this series in (1)], the structures and the reactions involved in biosynthesis and catabolism of leukotrienes are completely separate from those required for prostaglandin formation and metabolism. The leukotrienes seem to provide a new system of biological regulators that are important in many diseases involving inflammatory or immediate hypersensitivity reactions.

DISCOVERY AND STRUCTURE ELUCIDATION

Leukotriene C_4

In 1938–1940 Feldberg, Kellaway, & Trethewie demonstrated that perfusates from dog lungs treated with cobra venom (2) and from sensitized guinea pig lungs challenged with antigen (3) induced characteristic contractions of guinea pig jejunum with a slow onset of the response. The active principle was called slow reacting substance [SRS (3)]; later slow reacting substance of anaphylaxis [SRS-A (4)]. Since SRS-A is formed by human asthmatic lung in response to the appropriate allergen, and because it is a potent stimulator of human bronchial smooth muscles (4), it was proposed that SRS-A mediates the symptoms of asthma (reviewed in 5). Because of this several laboratories investigated the chemical nature of SRS-A (e.g. 6–8). These studies suggested that SRS-A was a polar lipid possibly derived from arachidonic acid and possibly containing sulfur.

Structural work on SRS was performed on material isolated from murine mastocytoma cells, and the complete structure was reported in 1979 (10, 11). The cells were stimulated with L-cysteine and ionophore A23187 and the biologically active material was purified by ethanol precipitation of proteins, mild alkaline hydrolysis, XAD-8 and silicic acid column chromatographies, and reverse phase high performance liquid chromatography. The pure SRS thus obtained was characterized, except for stereochemical details, by spectroscopy and by chemical and enzymatic transformations with identification of reaction products. The structural work, summarized in Figure 1, is briefly described. Ultraviolet spectroscopy showed that the molecule had an unusual spectrum characteristic of a conjugated triene with an allylic sulfur substituent (12). Experiments with isotopically labeled arachidonic acid (^{3}H) and cysteines (^{14}C, ^{3}H, or ^{35}S) demonstrated that these isotopes were transformed into SRS. These results suggested that the molecule was a derivative of arachidonic acid containing three conjugated double bonds and cysteine, or a derivative of cysteine attached as an allylic thioether. This preliminary information directed the choice of degradation methods for the continued characterization of the compound. SRS isolated

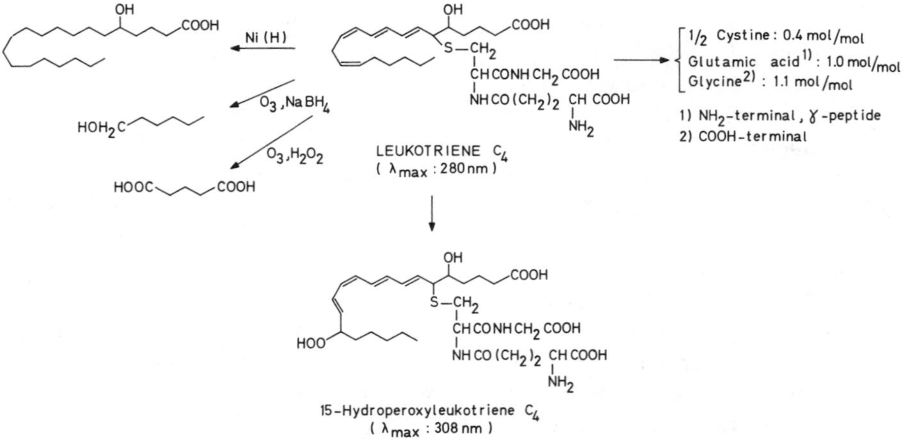

Figure 1   Reactions for structural work on leukotriene C_4.

after incubations with labeled arachidonic acid was treated with Raney nickel to remove sulfur. An ether-extractable, radioactive product was identified by gas-liquid chromatography-mass spectrometry as 5-hydroxy-eicosanoic acid. This indicated that SRS contained the complete carbon chain of arachidonic acid and a hydroxyl group at C-5 of the fatty acid. Oxidative ozonolysis yielded 1,5-pentanedioic acid probably due to oxidative cleavage of the bond between C-5 and C-6 in the molecule (cf below). Reductive ozonolysis gave tritium labeled 1-hexanol. This demonstrated that a double bond had been retained at the Δ^{14} position of the fatty acid and suggested the use of soybean lipoxygenase for further information on the double bond arrangements. This enzyme converts C_{20} fatty acids with cis double bonds at Δ^{11} and Δ^{14} to 15-hydroperoxy-11-cis-13-trans-eicosapolyenoic acids (13). After addition of enzyme to SRS a shift of the ultraviolet spectrum was observed, which suggested that the conjugated triene with an allylic sulfur substituent (λ_{max}: 280 nm; shoulders: 270 and 292 nm) had been converted to a conjugated tetraene with an allylic sulfur substituent (λ_{max}: 308 nm; shoulders: 295 and 323 nm). This result was very informative because it indicated the positions of the remaining three double bonds ($\Delta^{7,9,11}$) and of the sulfur substituent (C-6) as well as the geometry (cis) of the Δ^{11} and Δ^{14} double bonds. The cysteine-containing substituent was characterized by amino acid and sequence analyses. In addition to cystine, equimolar amounts of glutamic acid and glycine (1 mol/mol of triene) were obtained. Glutamic acid was the amino terminal and glycine the carboxyl terminal residue. The glutamyl-cysteine bond was resistant to Edman degradation but susceptible to hydrolysis by γ-glutamyl transpeptidase (see below: Metabolism). This indicated that the peptide bond involved the γ-carboxyl group of glutamic acid. The complete structure of the SRS from mastocytoma cells was therefore 5-hydroxy-6-S-γ-glutamylcysteinylglycine-7,9,11,14-eicosatetraenoic acid (10, 11). To facilitate future reference to this compound a trivial name was needed. The term leukotriene was chosen to indicate that SRS is formed by leukocytes (e.g. 8) and the presence of a conjugated triene in the molecule (10). A letter (C) was added to indicate the nature of the allylic substituent (glutathione), and a subscript was added to indicate the total number of double bonds in the molecule (14).

The structure of leukotriene C_4 was verified by total organic chemical synthesis (11, 15). This work also elucidated the stereochemistry of the molecule. Several isomers were synthesized. Their chemical, physical, and biological properties were compared with those of the natural compound (16). Since the geometry of the Δ^{11} and the Δ^{14} double bonds in leukotriene C_4 was cis (see above) 7,9-trans-11,14-cis- and the 7-trans-9,11,14-cis-5-hydroxy-6-S-glutathionyl-7,9,11,14-eicosatetraenoic acid were synthesized. Mixtures of $5(S),6(S)$-plus $5(R),6(R)$- or $5(S),6(R)$- plus $5(R),6(S)$-5-hydroxy-6-S-glutathionyl-7,9-*trans*-11,14-*cis*-eicosatetraenoic acids were initially obtained. The former two isomers were different from the natural substance. On the other hand, the latter mixture of isomers corresponded chromatographically to SRS. Syntheses of $5(S)$-hydroxy-6(R)-S-glutathionyl-7,9-*trans*-11,14-*cis*-eicosatetraenoic acid and its 9-*cis* isomer (15, 16) and comparisons with leukotriene C_4 isolated from mastocytoma cells showed that 9-*trans* isomer was identical to the natural compound.

In addition to leukotriene C_4, its 11-*trans* isomer is formed by mastocytoma cells (10, 17) This product was characterized by spectroscopy and chemical degradations and the structure was confirmed by chemical syntheses (17). Thiyl radicals formed from glutathione or cysteine in alkaline media induce isomerization of the Δ^{11} double bond in leukotriene C_4 (17). An analogous mechanism was proposed to explain the biosynthesis of 11-*trans* leukotriene C_4, namely enzymatic peroxide formation leading to thiyl radical formation and isomerization of leukotriene C_4 (18).

Leukotriene B_4

A dihydroxy metabolite of arachidonic acid, formed by rabbit peritoneal polymorphonuclear leukocytes, was described (19) at about the same time that the structure of leukotriene C_4 was reported (10). Mass spectrometric analyses of the native or hydrogenated metabolite indicated that it was a 5,12-dihydroxyeicosatetraenoic acid. Three double bonds (two *trans* and one *cis*) were conjugated. The remaining double bond probably had retained *cis* geometry. The positions of the double bonds and the stereochemistry of the hydroxyl groups were determined by oxidative ozonolysis of the methyl ester, *bis*-(−)menthoxycarbonyl derivative. The products obtained were identified as (−)menthoxycarbonyl derivatives of methyl-hydrogen-D-malate and methyl-hydrogen-2L-hydroxyadipate by mass spectrometry and by comparisons of retention times on gas-liquid chromatography with derivatives of enantiomeric reference compounds. The formation of 2L-hydroxyadipate showed that a double bond was present at Δ^6 and that the absolute configuration of the hydroxyl group at C-5 was (S). D-Malate was formed by cleavage of double bonds at Δ^{10} and Δ^{14}. Since the hydroxyl group was attached to C-12, its configuration was (R). The structure of the metabolite was therefore $5(S)$, $12(R)$-dihydroxy-6,8,10,14-eicosatetraenoic acid. The sequence of *cis* and *trans* double bonds in the triene (6-*cis*-8,10-*trans*) has subsequently been determined by total chemical synthesis and comparisons with the natural product (20). Since the 5,12-dihydroxy acid is metabolically related to leukotriene C_4 (see: Biosynthesis), it has been named leukotriene B_4 (14, 21).

Two stereoisomers of leukotriene B_4 (6-*trans* and 6-*trans*-12-*epi* leukotriene B_4) and two positional isomers (5,6-dihydroxy-7,9,11,14-eicosatetraenoic acids, epimeric at C-6) are formed in lower amounts by rabbit polymorphonuclear leukocytes (24). The characterization of the stereoisomers was analogous to that described above for leukotriene B_4.

BIOSYNTHESIS

Formation of an Epoxide Intermediate

Biosynthetic experiments were performed to study the mode of formation of leukotriene B_4. A related compound, 8,15-dihydroperoxyeicosatetraenoic acid, had been shown to be formed by double dioxygenation of arachidonic acid by soybean lipoxygenase (22, 23). An analogous mechanism for the biosynthesis of leukotriene B_4 was excluded because ^{18}O labeling indicated that only the oxygen of the hydroxyl group at C-5 originated in O_2. The oxygen of the hydroxyl group at C-12 was derived from water (25). This suggested that an intermediate was formed that could be hydrolyzed by water. Further evidence for this proposal was obtained by diluting mixtures of leukocytes and arachidonic acid with alcohols. This led to the formation of 12-*O*-alkyl derivatives of 6-*trans* and 6-*trans*-12-*epi* leukotriene B_4. Consequently, alcohols could replace water as agents reacting with the intermediate. Using HPLC for quantitative determination of 5(*S*)-hydroxy-12-methoxy-6,8,10,14(E,E,E,Z)-eicosatetraenoic acids, investigators demonstrated a transient accumulation of the intermediate. The highest concentration was observed after 45 sec incubation at 37°C. During the subsequent decline of 12-methoxy compounds a corresponding increase in the amounts of leukotriene B_4 was observed, which suggested that the intermediate gave rise both to the 12-methoxy compounds and to leukotriene B_4. The stability in a 1:1 mixture of incubation medium (pH 7.4) and acetone at 37°C was determined by the same method. The time for disappearance of 50% of the compound was 4 min. No degradation was observed after 4 min at or above pH 8.5. At pH 6 or lower, the rate of destruction was too rapid to be measured. Leukotriene B_4 was not formed during the decay of the intermediate in acetone/water mixtures. Instead, 6-*trans* and 6-*trans*-12-*epi* leukotriene B_4 and the two 5,6-dihydroxy-7,9,11,14-eicosatetraenoic acids mentioned above were formed. This indicated that leukotriene B_4 is formed by enzymatic hydrolysis of the intermediate and that the other dihydroxy compounds are formed by nonenzymatic hydrolysis. Based on the stability data and the nature of the degradation products it was proposed that the structure of the intermediate was 5,6-oxido-7,9,11,14-eicosatetraenoic acid (25). The same epoxide was suggested to be an intermediate in the biosynthesis of leukotriene C_4 and was called leukotriene A_4 (10). Experimental evidence supporting this proposal has subsequently been obtained (26): Using the methanol trapping procedure described above, investigators demonstrated transient formation of 5(S)-hydroxy-12-methoxy-6,8,10-14(E,E,E,Z)-eicosatetraenoic acids during leukotriene C_4 biosynthesis by mastocytoma cells. The time course for the formation and disappearance of these products was similar to that observed during leukotriene B_4 biosynthesis by polymorphonuclear leukocytes.

The stereochemical characterization of leukotriene C_4 (16) suggested that the stereochemistry of leukotriene A_4 (Figure 2) was *trans*-5(S),6(S)-

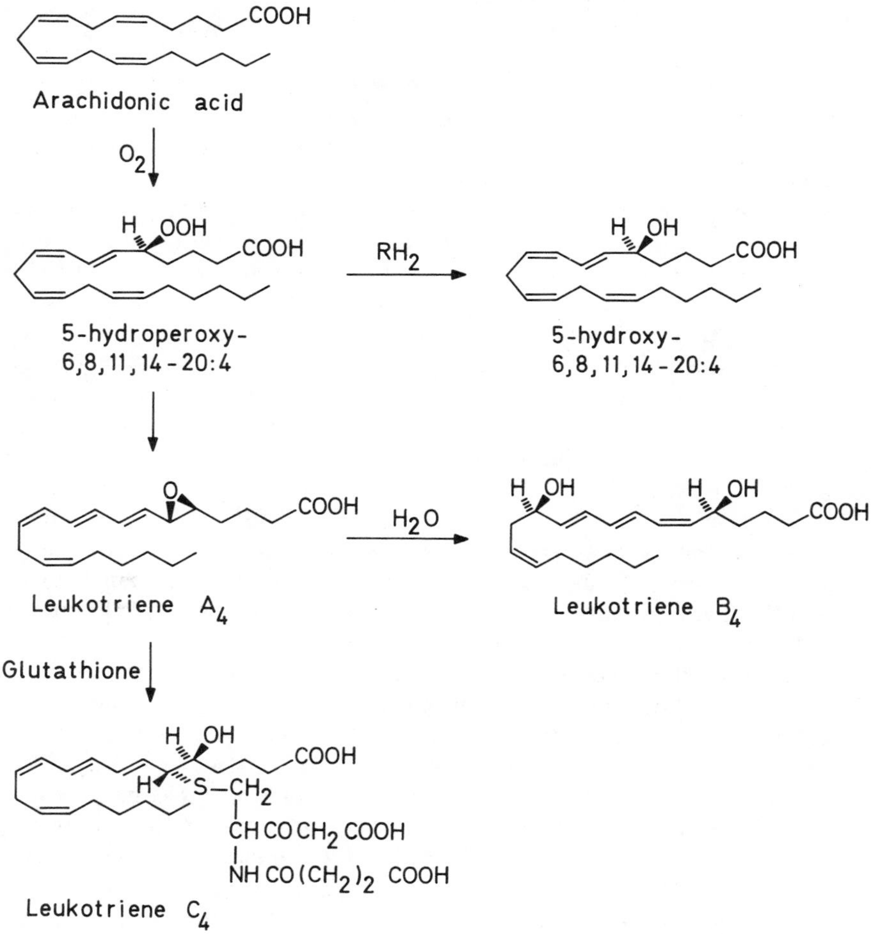

Figure 2   Biosynthesis of leukotrienes B_4 and C_4.

oxido-7,9-*trans*-11,14-*cis*-eicosatetraenoate. The methyl ester of this compound was synthesized as a chemical intermediate in the stereospecific synthesis of leukotriene C_4 (15). After hydrolysis of the ester group, the epoxide was enzymatically converted to a mixture of leukotriene C_4 and 11-*trans* leukotriene C_4 by murine mastocytoma cells (27) and to leukotriene B_4 by polymorphonuclear leukocytes (28). It was also used to facilitate the isolation of leukotriene A_4 from leukocytes (29). In the latter experiments the reaction mixture was made somewhat alkaline, treated with diazomethane and subsequently extracted at alkaline pH after addition of synthetic epoxide methyl ester. Comparisons of chromatographic properties on thin-layer chromatography and straight phase high performance liquid chromatography on aluminum oxide (or after acid hydrolysis by reverse phase HPLC) showed that the radioactive epoxide formed during the incubation was identical with the synthetic compound. These results provided evidence that leukotriene A_4 is a common intermediate in the biosynthesis of leukotrienes B_4 and C_4 and showed that the complete structure of leukotriene A_4 is *trans*-5(S), 6(S)-oxido-7,9-*trans*-11,14-*cis*-eicosatetraenoic acid (Figure 2). The transformations to leukotrienes B_4 and C_4 involve opening of the epoxide at the allylic side with formation of a 5(S)-hydroxy group. In the former reaction, OH from water adds at C-12 and the triene is isomerized to $\Delta^{6,8,10}$ (Z,E,E), and in the latter reaction sulfur of glutathione adds at C-6 without altering the position and stereochemistry of the triene. Indirect evidence that glutathione (rather than cysteine or a dipeptide) is added to leukotriene A_4 has also been provided by reports showing that experimental lowering of glutathione levels diminishes or prevents leukotriene C_4 biosynthesis in different cells (30–32).

Biosynthesis of Leukotriene A_4

A probable mechanism for the conversion of arachidonic acid to leukotriene A_4 involves the formation of a lipoxygenase product from arachidonic acid 5(S)-hydroperoxy-6-*trans*-8,11,14-*cis*-eicosatetraenoic acid (Figure 2) as an intermediate. The corresponding hydroxy acid (Figure 2), which is formed by reduction of the hydroperoxy acid, is a quantitatively important metabolite of arachidonic acid in murine mastocytoma cells and rabbit leukocytes (S. Hammarström, unpublished observation, 33), and has the same stereochemistry at C-5 [see correction to (33)] as leukotriene A_4. Studies on the mechanism of biosynthesis of leukotriene C_5 (see next section below) from eicosapentaenoic acid (34) have provided additional information regarding this pathway. Using the 10(S)- and the 10(R)-isomers of [1-^{14}C, 10-^{3}H]5,8,11,14,17-eicosapentaenoic acid, investigators showed that a hydrogen is stereospecifically eliminated from the 10-*pro*-(S) position[1]

---

[1] This position corresponds to 10-*pro*-R in arachidonic acid.

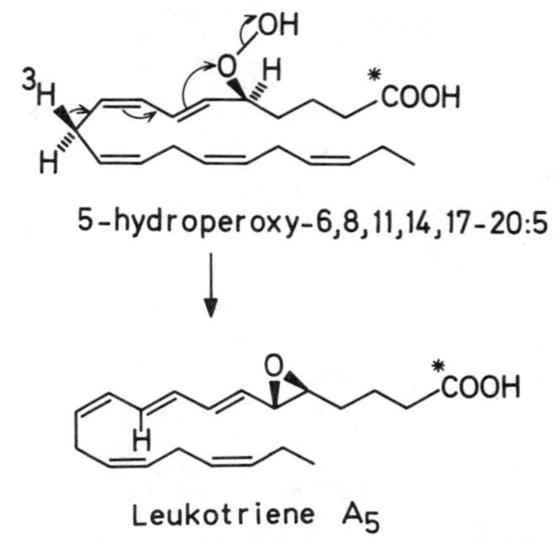

5-hydroperoxy-6,8,11,14,17-20:5

↓

Leukotriene A$_5$

Figure 3   Mechanism of leukotriene A$_5$ biosynthesis (* indicates the position of the ^{14}C label).

during the biosynthesis of leukotriene C$_5$. Since the enzymatic breaking of carbon-tritium bonds is slower than the breaking of a corresponding carbon-hydrogen bond (35), a kinetic isotope effect was expected in the immediate substrate for hydrogen elimination or in a stable derivative of this substrate. No enrichment of tritium was observed in eicosapentaenoic acid recovered after the incubations. Instead, a threefold increase in the ^{3}H/^{14}C ratio of 5-hydroxy-6,8,11,14,17-eicosapentaenoic acid formed from the 10(S)-tritium labeled precursor was observed. This demonstrated that oxygenation of eicosapentaenoic acid at C-5 occurs before hydrogen abstraction at C-10. The latter reaction is the rate-limiting step in the formation of a conjugated triene and an allylic epoxy group in leukotriene A$_5$ (see Figure 3).

Using a cell free system from basophilic leukemia cells, investigators observed parallel increases in the formation of 5-hydroxy-6,8,11,14-eicosatetraenoic acid and leukotrienes, as well as dependence on calcium ions for the conversion of arachidonic acid to 5-hydroxy eicosatetraenoic acid (36, 37). This agrees with earlier observations that the calcium ionophore A23187 stimulates the conversion of arachidonic acid to 5-hydroxy-6,8,11,14-eicosatetraenoic acid and leukotriene B$_4$ (38). Evidence for the conversion of [1-^{14}C]5-hydroperoxy-6,8,11,14-eicosatetraenoic acid to leukotriene C$_4$ by basophilic leukemia cells has also been published (39).

Leukotrienes Formed from Other Fatty Acids or by Oxygenation of Arachidonic Acid at Alternative Positions

The conversion of three other polyunsaturated fatty acids (besides arachidonic acid) to C-type leukotrienes has been demonstrated (40–42). The structures of these products (Figure 4) were determined using methods

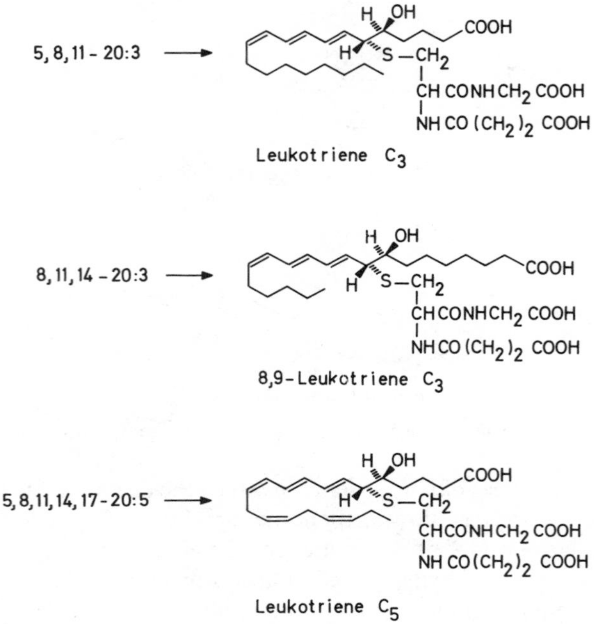

5,8,11 - 20:3 ⟶

Leukotriene C_3

8,11,14 - 20:3 ⟶

8,9-Leukotriene C_3

5,8,11,14,17 - 20:5 ⟶

Leukotriene C_5

Figure 4　Conversion of polyunsaturated fatty acids to leukotrienes.

similar to those described above for leukotriene C_4. Two compounds have structures analogous to leukotriene C_4 except for the number of non-conjugated double bonds: leukotriene C_3, which is formed from 5,8,11-eicosatrienoic acid, lacks a Δ^{14} double bond and leukotriene C_5 has an additional double bond at Δ^{17}. [Leukotriene C_5 is formed from 5,8,11,14,17-eicosapentaenoic acid (cf above).] The third product, 8,9-leukotriene C_3, is formed from 8,11,14-eicosatrienoic acid. The positions of the double bonds of this acid preclude oxygenation at C-5. The product formed is a positional isomer of leukotriene C_3 with the hydroxyl group at C-8, glutathione at C-9, and the triene at $\Delta^{10,12,14}$. The biosynthesis of this compound shows that leukotrienes can be formed after oxygenation of suitable precursor acids at positions other than C-5. This can also occur with arachidonic acid as substrate (43–48). Figure 5 shows conversions of 8-, 12-, and 15-hydroperoxy derivatives of 8,11,14-eicosatrienoic and arachidonic acid, respectively, to leukotrienes A and further transformations of the epoxides to leukotrienes B and C, namely 8,9-leukotriene C_3 (42), 5,12-leukotriene B_4 (a stereoisomer of 6-*trans* leukotriene B_4) plus 11,12-leukotriene B_4 (43), 8,15-leukotriene B_4 plus 14,15-leukotriene B_4 (44–47) and 14,15-leukotriene C_4 (48).

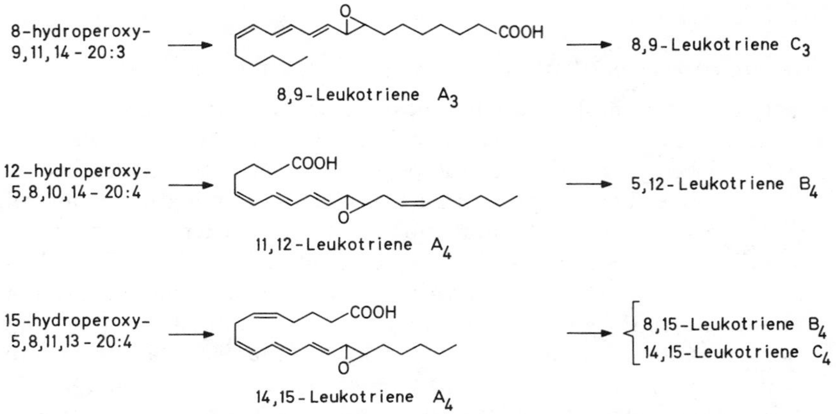

Figure 5   Leukotrienes formed from 8-, 12-, and 15-hydroperoxyeicosapolyenoic acids.

OCCURRENCE

The occurrence of leukotrienes formed after oxygenation at C-5 is summarized in Table 1. Leukotrienes formed after oxygenation at other positions have recently been identified in mastocytoma cells [oxygenation at C-8 (42)], polymorphonuclear leukocytes [oxygenation at C-12 (43) or C-15 (44, 46)], eosinophilic leukocytes [oxygenation at C-15 (46, 47)], and basophilic leukemia cells [conversion of 14,15-leukotriene A_4 to 14,15-leukotriene C_4 (48)].

Table 1 Occurrence of leukotrienes

Source	LTA_4	LTB_4	LTC_4	LTD_4	LTE_4	References
Body Fluids or Secretions						
Cystic fibrosis sputum		+	+			49
Rheumatoid synovial fluid		+				50, 51
Cells						
Leukocytes						
Basophilic leukemia	+		+	+	+	52–56
Eosinophilic		+	+	+		57
Polymorphonuclear	+	+	+			19, 25, 58, 59
Macrophages			+	+		60–63
Mastocytoma cells	+		+			10, 11, 26
Peritoneal cells (mixed)			+	+	+	64–67
Tissues						
Lung			+	+		66, 68
Paws				+	+	69
Spleen			+			70

METABOLISM

Modifications Involving the Peptide Part of Leukotriene C

Structural studies on slow reacting substance from rat basophilic leukemia cells indicated that it was related to leukotriene C_4 but was less polar (52). The difference was confined to the peptide part, which was identifed as cysteinylglycine. This suggested that the compound was a metabolite of leukotriene C_4 formed by elimination of the amino terminal glutamyl residue. Further evidence was obtained by treating leukotriene C_4 with γ-glutamyl transpeptidase. This enzyme catalyzes the initial step (removal of glutamic acid) in the metabolism of glutathione, glutathione disulfide, and various glutathione conjugates (71). The product obtained was identical to SRS from basophilic leukemia cells and was named leukotriene D_4 (52). Using inhibitors of γ-glutamyl transpeptidase it was shown that leukotriene C_4 is an intermediate in the formation of leukotriene D_4 (53, 72). Additional metabolic experiments using $[5,6,8,9,11,12\text{-}^3H_6]$ leukotriene C_3 (Figure 6) and guinea pig tissues showed that leukotriene D_3 was rapidly formed by lung homogenates (73). This reaction was inhibited by glutathione. Moreover, in the presence of glutathione, γ-glutamyl transpeptidase catalyzed the addition of a γ-glutamyl residue to the amino group of leukotriene D to form leukotriene C (Figure 6). This reverse reaction probably involves aminolysis by leukotriene D of a γ-glutamyl-enzyme intermediate (73). Kinetic evidence for the formation of such an intermediate has been reported before (74). The K_m value for leukotriene C_4 in the forward reaction catalyzed by γ-glutamyl transpeptidase is ap-

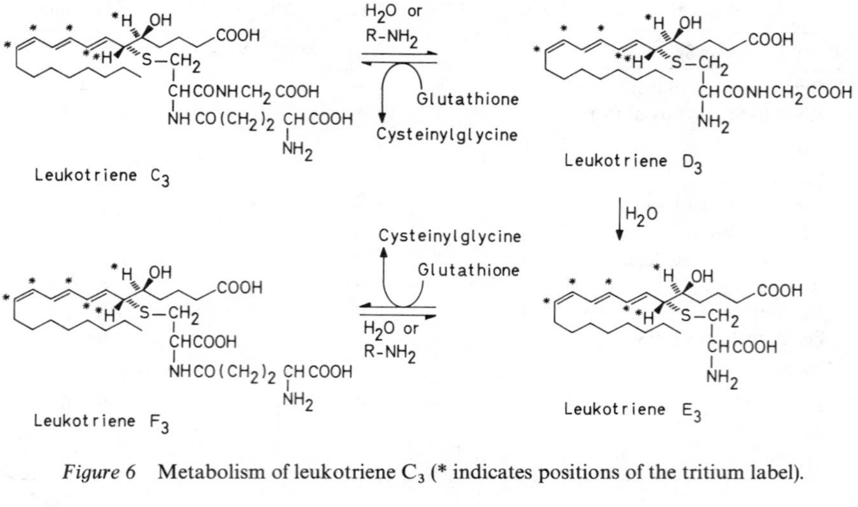

Figure 6   Metabolism of leukotriene C_3 (* indicates positions of the tritium label).

proximately the same as the K_m value for hydrolysis of glutathione [about 6 μM (75, 76)]. It is therefore likely that extracellular glutathione or glutathione disulfide will influence the metabolism of leukotrienes by γ-glutamyl transpeptidase. Leukotriene D_3 is degraded further by hydrolysis of the peptide bond (77). The product (leukotriene E_3, Figure 6) is formed rapidly by kidney and more slowly by lung and liver. The reaction is catalyzed by a membrane bound dipeptidase (76–78). Additional reports on in vitro conversions of leukotriene C to leukotriene E have also appeared (53, 55, 79). Leukotrienes D_4 and E_4 have been identified in various preparations of SRS-A (Table 1). Leukotriene E can be further converted to leukotriene F by addition of a γ-glutamyl residue to the amino group (76, 80). This reaction is analogous to the conversion of leukotriene D to leukotriene C and is catalyzed by γ-glutamyl transpeptidase in the presence of glutathione (Figure 6). Leukotriene F is not formed by direct elimination of glycine from leukotriene C_4 owing to the substrate specificity of the dipeptidase (77).

Similar transformations are observed in vivo (81, 82). Following injection into the right atrium of the heart of anesthetized monkeys [$^3H_6$]leukotriene C_3 was thus rapidly ($t_{1/2}$ about 30 sec) converted to leukotrienes D_3 and E_3 (81). After 2 min the radioactivity in blood corresponded mainly to leukotriene E_3 as judged by HPLC analyses. The initial rapid conversion to leukotriene D_3 probably occurred in the pulmonary circulation and the subsequent dipeptidase reaction in the vascular beds of other organs. In whole blood leukotriene C_3 is quite stable ($t_{1/2} > 2$ hr. unpublished observation). Leukotrienes C, D, and E are metabolized to more polar products prior to excretion (73, 82). The nature of the reactions involved is presently unknown. It has been suggested that leukotriene C_4 can be converted to leukotriene B_4 stereoisomers by peroxidase in eosinophilic leukocytes (83). There is however no evidence that these less polar metabolites are formed in vivo. A previously proposed inactivation of SRS-A by arylsulfatase has been reinvestigated (78, 84). No cleavage of the thioether bond of leukotrienes was observed and it appears that the inactivation was probably caused by a contaminating dipeptidase that converted leukotriene D to leukotriene E.

ω-Oxidation of Leukotriene B_4

Leukotriene B_4 is converted to 20-hydroxyleukotriene B_4 and 20-carboxyleukotriene B_4 by human leukocyte preparations (85). These reactions are analogous to ω-oxidation of prostaglandins (86) and fatty acids (87). Also, 20-Carboxyleukotriene B_4 was the major leukotriene formed following stimulation of human leukocytes with the chemotactic peptide N-formylmethionylleucylphenylalanine [fMLP (88)].

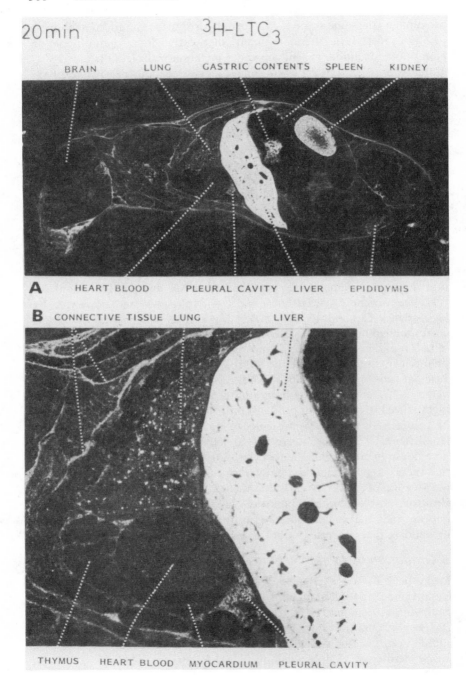

DISTRIBUTION AND EXCRETION

Studies on the distribution of tritium labeled leukotriene C_3 in mice by autoradiography (Figure 7) showed that the radioactivity was rapidly eliminated from blood by hepatic uptake and biliary excretion (82). A smaller uptake by the cortical parts of the kidneys was also observed. In addition tritium was taken up by the lungs, in pleural fluid, salivary glands, pancreas, ventral prostate, various fasciae, and in subcutaneous connective tissue. With the exception of the lungs, the physiological significance of the uptake of radioactivity by these tissues is unknown. It should stimulate further investigations regarding biological roles of leukotrienes in these organs. The data on excretion routes mentioned above were supported by measurements of tritium contents in feces and urine from guinea pigs following subcutaneous administration of $[^3H_6]$ leukotriene C_3 (73). The results showed that approximately 40% and 60% of the injected radioactivity were excreted in urine and feces, respectively.

Experiments using perfused liver and isolated hepatocytes from rats have shown that these preparations effectively take up leukotrienes C, D, and E (89, 90). Neither leukotriene B_4 nor the glutathione conjugate of paracetamol competed for the uptake, which was dependent on temperature and cellular ATP concentrations. The results suggested that liver cells contain a system for active transport into the cells of cysteine containing leukotrienes. Kinetic analyses indicated that the apparent K_m values were similar for leukotrienes C, D, and E (1 and 17 μM for high and low affinity sites, respectively). On the other hand the apparent V_{max} values were higher for leukotrienes D and E (0.1 and 1.35 nmol/min $\cdot 10^6$ cells) than for leukotriene C (0.07 and 1.0 nmol/min $\cdot 10^6$ cells; high and low affinity sites, respectively).

ANALYTICAL METHODS

High Performance Liquid Chromatography

This method has been used extensively for purification and analyses of leukotrienes by reverse phase techniques (10, 24, 38, 77, 91). In some instances (38, 91) prostaglandin B_2 has been used as an internal standard. Combinations of reverse phase and straight phase HPLC were used to separate leukotriene B_4 and 5(S),12(S)-dihydroxy-6,8,10,14-(E,Z,Z,E)-eicosatetraenoic acid [a stereoisomer of leukotriene B_4 formed by double dioxygenation of arachidonic acid in leukocytes (92)]. Straight phase

---

Figure 7   Distribution of $[5,6,8,9,11,12-^3H_6]$ leukotriene C_3 in the mouse as determined by autoradiography of a body section 20 min after i.v. administration (see 82).

HPLC has also been used to separate leukotriene B_4 from 20-hydroxy- and 20-carboxyleukotriene B_4 (85, 88) and for the separation of N-trifluoroacetyl, dimethyl ester derivatives of leukotriene D and E stereoisomers (93). Leukotriene A_4 has been analyzed intact by reverse phase HPLC (94) and as the methyl ester by straight phase HPLC on aluminum oxide (29). The former analyses were performed at pH 10 and the latter in the presence of triethylamine to minimize hydrolysis of the epoxide. It was reported that human serum albumin will prolong the half-life of leukotriene A_4 at pH 7.4 (94). Increased stability was also observed in ethanol-containing solutions.

Mass Spectrometry

Direct inlet, electron impact mass spectra have been published for the N-acetyl, O-trimethylsilyl dimethyl ester derivative of leukotriene D_4 (54, 68) and the corresponding trimethyl ester derivative of leukotriene C_4 (56). In the mass spectrum of the leukotriene D_4 derivative, ions corresponding to the molecular ion (M), M-15, M-31, and M-90 were found. Another ion of importance was one at m/e 203, owing to cleavage of the fatty acid between C-5 and C-6 with charge retention on the carboxyl-containing fragment. This ion indicates the presence of a hydroxyl group at C-5 of the fatty acid (10). The mass spectrum of the leukotriene C_4 derivative had ions at m/e 595 (interpreted as being formed by cleavage of the γ-peptide bond with elimination of the N-acetyl glutamyl residue), 578 (M-203 where 203 represents C-1 through C-5 of arachidonic acid), 563 (M-203-15), 507 (proposed to be formed from the ion at m/e 595 by further cleavage of the α-peptide bond and elimination of the glycyl methyl ester residue), and 203. Although the molecular ion was not observed, the spectrum is more informative than that of the leukotriene D_4 derivative, owing to the ions at m/e 595 and 507.

Fast atom bombardment mass spectrometry of underivatized leukotriene C_4 has also been reported (95). Both the positive and the negative ion spectra showed quasi-molecular ions (M-1 and M + 23(sodium)-2 for the latter spectrum). Cleavage of the bond between the sulfur atom and C-6 of the arachidonyl part gave ions consisting of the fatty acid and the peptide parts, respectively. The fast atom bombardment mass spectra thus give information on the fatty acid and the peptide parts separately as well as on the size of the intact molecule.

Radioimmunoassay

Reports describing protein conjugates of leukotrienes C_4 (96) and D_4 (97) as well as a radioimmunoassay for leukotrienes C, D, and E (97) have appeared. Leukotriene C_4 conjugates of bovine serum albumin or limpet hemocyanin were prepared with either 1,5-difluoro-2,4-dinitrobenzene or 6-N-maleimido hexanoyl chloride as coupling agents. The leukotriene D_4 protein conjugate was prepared using the anhydride obtained from N-trifluoroacetyl leukotriene D_4 monomethyl (glycine) ester and isobutyl chloroformate, and reacting it with bovine serum albumin. After removal of the protecting groups with K_2CO_3 and purification by Sephadex chromatography the conjugate (7 mol leukotriene D_4/mol albumin; 0.5 and 0.25 mg 21 days apart, i.m.) was injected into rabbits together with Freunds adjuvant. After 10 days serum from one rabbit bound 11-$trans$ [14,15-$^3H_2$] leukotriene C_4 as determined using goat antirabbit IgG as precipitating agent. At a dilution of 1:32, 23% of added 11-$trans$ leukotriene C_4 was bound by the serum. The association constant was 2.8×10^9 M^{-1} at 37°C. Specificity studies indicated that 11-$trans$ leukotriene D_4 was the most potent inhibitor followed (in decreasing order of effectiveness) by desamino leukotriene D_4, leukotriene C_4, 11-$trans$ leukotriene C_4, 11-$trans$ leukotriene E_4, leukotriene C_3, homocysteinyl leukotriene D_4, leukotriene D_4, and leukotriene E_4. The cross-reaction with leukotriene E_4 was 95% and greater for the other leukotrienes mentioned above. Less cross-reactivity was observed for 6-epi leukotriene D_4 (5%), 5,12-leukotriene C_4 (2%), and 11,12-leukotriene C_4 (2%), and there was no measurable cross reaction with 7-cis or 7-$trans$ hexahydroleukotriene B_4, 5-HETE, or arachidonic acid. A sensitive and specific radioimmunoassay for leukotriene C was obtained by coupling acetylated leukotriene C_4 to polyamino bovine serum albumin with a carbodiimide (98). The antiplasma obtained from immunized rabbits cross-reacted less than 0.5% with leukotriene F_4 and less than 0.1% with leukotrienes D_4 and E_4, and the lowest limit of detection was 0.05 pmol for leukotriene C_4.

BIOLOGICAL EFFECTS

The isolation, characterization and chemical syntheses of various leukotrienes has stimulated renewed interest in the pharmacological and physiological properties of these compounds. Close to a hundred original reports had been published by August 1982. A detailed coverage is not possible and the reader is referred to a comprehensive, recent review (99) for additional information.

Respiratory Effects

The effects of leukotrienes on respiratory functions have attracted special interest because of the proposed mediator role in asthma. Leukotrienes C_4, D_4, and E_4 are potent stimulators of airway smooth muscles from different species (100–103). In general, peripheral airways are more sensitive than central airways. The EC_{100} values (concentrations to increase the resting tension by 100%) for leukotriene C_4 were 0.6 and 50 nM in guinea pig lung strips and tracheas, respectively, and leukotriene C_4 was four orders of magnitude more potent than histamine on the former preparations (100). These contractile responses to leukotrienes were reversed by the SRS-A antagonist FPL 55712 (104). Airway smooth muscles from rats (102) and rabbits (100) did not respond to leukotrienes C_4 and D_4 whereas guinea pig (100–103), monkey (105), and human preparations (106–108) are very sensitive. Investigations on structure-activity relations for leukotriene C and D analogs on guinea pig airways (109–112) have indicated that the position of the peptide substituent (C-6) is important for maximal activity and that partial saturation of the triene (to the Δ^7-hexahydro compounds) reduced the smooth muscle stimulating activity (109). Free amino and carboxyl groups of the peptide were also essential but the arachidonyl carboxyl group could be derivatized without loss of activity (110). The stereochemistry at C-5 and C-6 is critical for respiratory and intestinal (see below) smooth muscle stimulating activity (110–112). Changing the number of amino acids (as in leukotrienes C_4, D_4, and E_4; cf Figure 6) or the total number of double bonds (leukotrienes C_3, C_4, and C_5, Figures 2, 4) did not influence the contractile activity on guinea pig lung strips (103). Moreover, 8,9-leukotriene C_3 (Figure 4) was as active as leukotriene C_4 (Figure 2) in the same preparation (103), which suggests that the peptide substituent can be moved to C-9 without loss of biological activity. Leukotrienes A_4 (104) and B_4 (85) were at least two orders of magnitude less potent than leukotrienes C_4, D_4, and E_4 on guinea pig lung strips. This indicates that the amino acid part of leukotrienes (cf Figure 2) is essential for smooth muscle stimulating activity. Contractions elicited by leukotriene B_4 also were not antagonized by the SRS-A antagonist (FPL 55712) but were abolished by indomethacin, which suggests that they were mediated by prostaglandin and/or thromboxane formation (113, 114).

In vivo, leukotriene C_4 (ED_{100}:0.12 nM) induced pronounced increases of insufflation pressures in anesthetized guinea pigs (100) or trans-pulmonary pressures of anesthetized monkeys (105) following systemic or aerosol administration. These effects were due to reductions in dynamic lung compliance rather than increases of pulmonary resistance (105). Leukotrienes C_4 and D_4 also impaired human respiratory functions in vivo (115).

Microvascular Effects

In addition to its respiratory effects, SRS-A was known to increase vascular permeability (5). Leukotrienes C_4 and D_4, in doses as low as 10^{-13} mol induced leakage of circulating Evans blue in guinea pig skin (100). Additional experiments were performed using the hamster cheek pouch model (116). When applied at 10^{-10} to 10^{-8} M concentrations, leukotrienes C_4 and D_4 induced transient constriction of arterioles and especially terminal arterioles, followed by leakage of circulating fluorescein isothiocyanate-conjugated dextran into the perivascular space around postcapillary venules (117). Histamine produced similar effects on capillary permeability at three to four orders of magnitude greater concentrations but had no vasoconstrictor activity. Leukotriene B_4 had no direct effects on vascular permeability but induced adhesion of leukocytes to the endothelial cells of postcapillary venules (117, 118) and subsequently leukocyte-dependent increases in vascular permeability.

Effects on Leukocytes

Besides causing adhesion of leukocytes to endothelial cells (117, 118; cf above) leukotriene B_4 stimulated directional (chemotaxis) and random movements (chemokinesis) of polymorphonuclear leukocytes (119–122). Leukotriene B_4 was more potent than 6-*trans* or 6-*trans*-12-*epi* leukotriene B_4 and leukotriene C_4 did not stimulate directional or random motion. A similar specificity pattern was observed for the ability of leukotriene B_4 to induce degranulation and release of lysosomal enzymes from human and rabbit polymorphonuclear leukocytes (123–127). The secretary actions of leukotriene B_4 depend partly on cytochalasin B and are enhanced by extracellular calcium (125, 127). Other reports have indicated that leukotriene B_4 induces calcium mobilization in rabbit neutrophils (129, 130) and acts as a calcium ionophore in liposomes (131). Leukotriene B_4 induced aggregation of polymorphonuclear leukocytes has also been reported (132).

Gastrointestinal Effects

Isolated guinea pig ileum has been the classical tissue for bioassay of SRS-A (5). Leukotrienes C_4, D_4, E_4, and F_4 induced contractions of this tissue (e.g. 10, 52, 77, 80) whereas leukotriene B_4 had no contractile effect (113, 114). The specificity differed somewhat from that of respiratory smooth muscle because leukotriene D_4 is more potent (52) and leukotriene E_4 less potent (77) than leukotriene C_4. Leukotrienes C_4 and D_4 also contracted guinea pig stomach in vitro and in vivo (133), which may explain similar effects on the stomach during anaphylaxis.

Other Effects

A number of other effects of leukotrienes C, D, and E have been reported such as constriction of coronary arteries and a negative inotropic effect on the heart (134, 135), effects on blood pressure (105, 136), stimulation of prostaglandin synthesis in macrophages (137, 138) and lung (139–142), excitation of cerebellar Purkinje neurons (143, 144), and inhibition of mitogen-induced lymphocyte transformation (145).

ACKNOWLEDGMENTS

The work from the author's laboratory was supported by grants from the Swedish Medical Research Council (project 03X-5914 and 03P-6396).

Literature Cited

1. Samuelsson, B., Goldyne, M. G., Granström, E., Hamberg, M., Hammarström, S., Malmsten, C. 1978. *Ann. Rev. Biochem.* 47:997–1029
2. Feldberg, W., Kellaway, C. H. 1938. *J. Physiol.* 94:187–226
3. Kellaway, C. H., Trethewie, E. F. 1940. *Q. J. Exp. Physiol. Cogn. Med. Sci.* 30:121–45
4. Brocklehurst, W. E. 1960. *J. Physiol.* 151:416–35
5. Orange, R. P., Austen, K. F. 1969. *Adv. Immunol.* 10:105–44
6. Strandberg, K., Uvnäs, B. 1971. *Acta Physiol. Scand.* 82:358–74
7. Orange, R. P., Murphy, R. C., Karnovsky, M. L., Austen, K. F. 1973. *J. Immunol.* 110:760–70
8. Jakschik, B. A., Falkenhein, S., Parker, C. W. 1977. *Proc. Natl. Acad. Sci. USA* 74:4577–81
9. Morris, H. R., Taylor, G. W., Piper, P. J., Sirois, P., Tippins, J. R. 1978. *FEBS Lett.* 87:203–6
10. Murphy, R. C., Hammarström, S., Samuelsson, B. 1979. *Proc. Natl. Acad. Sci. USA* 76:4275–79
11. Hammarström, S., Murphy, R. C., Samuelsson, B., Clark, D. A., Mioskowski, C., et al. 1979. *Biochem. Biophys. Res. Commun.* 91:1266–72
12. Koch, H. P. 1949. *J. Chem. Soc.*, pp. 387–94
13. Hamberg, M., Samuelsson, B. 1967. *J. Biol. Chem.* 242:5329–35
14. Samuelsson, B., Hammarström, S. 1980. *Prostaglandins* 19:645–48
15. Corey, E. J., Clark, D. A., Goto, G., Marfat, A., Mioskowski, C., et al. 1980. *J. Am. Chem. Soc.* 102:1436–39
16. Hammarström, S., Samuelsson, B., Clark, D. A., Goto, G., Marfat, A., et al. 1980. *Biochem. Biophys. Res. Commun.* 92:946–53
17. Clark, D. A., Goto, G., Marfat, A., Corey, E. J., Hammarström, S., et al. 1980. *Biochem. Biophys. Res. Commun.* 94:1133–39
18. Atrache, V., Sok, D.-E., Pai, J.-K., Sih, C. J. 1981. *Proc. Natl. Acad Sci. USA* 78:1523–26
19. Borgeat, P., Samuelsson, B. 1979. *J. Biol. Chem.* 254:2643–46
20. Corey, E. J., Marfat, A., Goto, G., Brion, F. 1980. *J. Am. Chem. Soc.* 102:7984–85
21. Bild, G. S., Ramadoss, C. S., Lim, S., Axelrod, B. 1977. *Biochem. Biophys. Res. Commun.* 74:949–54
22. Bild, G. S., Ramadoss, C. S., Axelrod, B. 1977. *Arch. Biochem. Biophys.* 184:36–41
23. Samuelsson, B., Borgeat, P., Hammarström, S., Murphy, R. C. 1979. *Prostaglandins* 17:785–87
24. Borgeat, P., Samuelsson, B. 1979. *J. Biol. Chem.* 254:7865–69
25. Borgeat, P., Samuelsson, B. 1979. *Proc. Natl. Acad. Sci. USA* 76:3213–17
26. Hammarström, S., Samuelsson, B. 1980. *FEBS Lett.* 122:83–86
27. Rådmark, O., Malmsten, C., Samuelsson, B. 1980. *Biochem. Biophys. Res. Commun.* 96:1679–87
28. Rådmark, O., Malmsten, C., Samuelsson, B., Clark, D. A., Goto, G., et al. 1980. *Biochem. Biophys. Res. Commun.* 92:954–61
29. Rådmark, O., Malmsten, C., Samuelsson, B., Goto, G., Marfat, A., et al. 1980. *J. Biol. Chem.* 255:11828–31
30. Parker, C. W., Fischman, C. M., Wedner, H. J. 1980. *Proc. Natl. Acad. Sci. USA* 77:6870–73
31. Rouzer, C. A., Scott, W. A., Griffith, O. W., Hamill, A. L., Cohn, Z. A. 1981. *Proc. Natl. Acad. Sci. USA* 78:2532–36
32. Rouzer, C. A., Scott, W. A., Griffith, O. W., Hamill, A. L., Cohn, Z. A. 1982. *Proc. Natl. Acad. Sci. USA* 79:1621–25
33. Borgeat, P., Hamberg, M., Samuelsson, B. 1976. *J. Biol. Chem.* 251:7816–20, Correction 1977. 252:8772
34. Hammarström, S. 1983. *J. Biol. Chem.* 258:1427–30
35. Northrop, D. B. 1981. *Ann. Rev. Biochem.* 50:103–31
36. Jakschik, B. A., Lee, L. H. 1980. *Nature* 287:51–52
37. Jakschik, B. A., Sun, F. F., Lee, L. H., Steinhoff, M. M. 1980. *Biochem. Biophys. Res. Commun.* 95:103–10
38. Borgeat, P., Samuelsson, B. 1979. *Proc. Natl. Acad. Sci. USA* 76:2148–52
39. Parker, C. W., Koch, D., Huber, M. M., Falkenhein, S. F. 1980. *Biochem. Biophys. Res. Commun.* 94:1037–43

40. Hammarström, S. 1980. *J. Biol. Chem.* 255:7093–94
41. Hammarström, S. 1981. *J. Biol. Chem.* 256:2275–79
42. Hammarström, S. 1981. *J. Biol. Chem.* 256:7712–14
43. Samuelsson, B. 1983. In *Adv. Prostaglandin, Thromboxane and Leukotriene Res.*, ed. B. Samuelsson, P. Ramwell, R. Paoletti. New York: Raven. In press
44. Lundberg, U., Rådmark, O., Malmsten, C., Samuelsson, B. 1981. *FEBS Lett.* 126:127–32
45. Jubiz, W., Rådmark, O., Lindgren, J. Å., Malmsten, C., Samuelsson, B. 1981. *Biochem. Biophys. Res. Commun.* 99:976–86
46. Maas, R. L., Brash, A. R., Oates, J. A. 1981. *Proc. Natl. Acad. Sci. USA* 78:5523–27
47. Turk, J., Maas, R. L., Brash, A. R., Roberts, L. J., Oates, J. A. 1982. *J. Biol. Chem.* 257:7068–76
48. Sok, D.-E., Han, C.-O., Sieh, W.-R., Zhou, B.-N., Sih, C. J. 1982. *Biochem. Biophys. Res. Commun.* 104:1363–70
49. Cromwell, O., Morris, H. R., Hodson, M. E., Walport, M. J., Taylor, G. W., et al. 1981. *Lancet* 2:164–65
50. Klickstein, L. B., Shapleigh, C., Goetzl, E. J. 1980. *J. Clin. Invest.* 66:1166–70
51. Davidson, E. M., Rae, S. A., Smith, M. J. H. 1982. *J. Pharm. Pharmacol.* 34:410
52. Örning, L., Hammarström, S., Samuelsson, B. 1980. *Proc. Natl. Acad. Sci. USA* 77:2014–17
53. Örning, L., Bernström, K., Hammarström, S. 1981. *Eur. J. Biochem.* 120:41–45
54. Morris, H. R., Taylor, G. W., Piper, P. J., Samhoun, M. N., Tippins, J. R. 1980. *Prostaglandins* 19:185–201
55. Parker, C. W., Falkenhein, S. F., Huber, M. M. 1980. *Prostaglandins* 20:863–86
56. Sok, D.-E., Pai, J.-K., Atrache, V., Sih, C. J. 1980. *Proc. Natl. Acad. Sci. USA* 77:6481–85
57. Jörg, A., Henderson, W. R., Murphy, R. C., Klebanoff, S. J. 1982. *J. Exp. Med.* 155:390–402
58. Hansson, G., Rådmark, O. 1980. *FEBS Lett.* 122:87–90
59. Siegel, M. I., McConnel, R. T., Bonser, R. W., Cuatrecasas, P. 1981. *Prostaglandins* 21:123–32
60. Doig, M. V., Ford-Hutchinson, A. W. 1980. *Prostaglandins* 20:1007–19
61. Rouzer, C. A., Scott, W. A., Cohn, Z. A., Blackburn, P., Manning, J. M. 1980. *Proc. Natl. Acad. Sci. USA* 77:4928–32
62. Hsueh, W., Sun, F. F. 1982. *Biochem. Biophys. Res. Commun.* 106:1085–91
63. Roubin, R., Mencia-Huerta, J. M., Benveniste, J. 1982. *Eur. J. Immunol.* 12:141–46
64. Bach, M. K., Brashler, J. R., Hammarström, S., Samuelsson, B. 1980. *J. Immunol.* 125:115–17
65. Bach, M. K., Brashler, J. R., Hammarström, S., Samuelsson, B. 1980. *Biochem. Biophys. Res. Commun.* 93:1121–26
66. Lewis, R. A., Austen, K. F., Drazen, J. M., Clark, D. A., Marfat, A., et al. 1980. *Proc. Natl. Acad. Sci. USA* 77:3710–14
67. Lewis, R. A., Drazen, J. M., Austen, K. F., Clark, D. A., Corey, E. J. 1980. *Biochem. Biophys. Res. Commun.* 96:271–77
68. Morris, H. R., Taylor, G. W., Piper, P. J., Tippins, J. R. 1980. *Nature* 285:104–6
69. Houglum, J., Pai, J.-K., Atrache, V., Sok, D.-E., Sih, C. J. 1980. *Proc. Natl. Acad. Sci. USA* 77:5688–92
70. Malik, K. U., Wong, P. Y.-K. 1981. *Biochem. Biophys. Res. Commun.* 103:511–20
71. Meister, A., Tate, S. S. 1976. *Ann. Rev. Biochem.* 45:559–604
72. Örning, L., Hammarström, S. 1980. *J. Biol. Chem.* 255:8023–26
73. Hammarström, S. 1981. *J. Biol. Chem.* 256:9573–78
74. Tate, S. S., Meister, A. 1974. *J. Biol. Chem.* 249:7593–602
75. Örning, L., Hammarström, S. 1982. *Biochem. Biophys. Res. Commun.* 106:1304–9
76. Anderson, M. E., Allison, R. D., Meister, A. 1982. *Proc. Natl. Acad. Sci. USA* 79:1088–91
77. Bernström, K., Hammarström, S. 1981. *J. Biol. Chem.* 256:9579–82
78. Sok, D.-E., Pai, J.-K., Atrache, V., Kang, Y.-C., Sih, C. J. 1981. *Biochem. Biophys. Res. Commun.* 101:222–29
79. Parker, C. W., Koch, D., Huber, M. M., Falkenhein, S. F. 1980. *Biochem. Biophys. Res. Commun.* 97:1038–46
80. Bernström, K., Hammarström, S. 1982. *Biochem. Biophys. Res. Commun.* 109:800–4
81. Hammarström, S., Bernström, K., Örning, L., Dahlén, S. E., Hedqvist, P., et al. 1981. *Biochem. Biophys. Res. Commun.* 101:1109–15
82. Appelgren, L. E., Hammarström, S. 1982. *J. Biol. Chem.* 257:531–35
83. Goetzl, E. J. 1982. *Biochem. Biophys. Res. Commun.* 106:270–75
84. Parker, C. W., Koch, D. A., Huber, M. M., Falkenhein, S. F. 1980. *Prostaglandins* 20:887–908
85. Hansson, G., Lindgren, J. Å., Dahlén, S.

E., Hedqvist, P., Samuelsson, B. 1981. *FEBS Lett.* 130:107–12
86. Powell, W. S. 1980. *Prostaglandins* 19:701–10
87. Björkhem, I., Danielsson, H. 1970. *Eur. J. Biochem.* 17:450–59
88. Jubiz, W., Rådmark, O., Malmsten, C., Hansson, G., Lindgren, J. Å., et al. 1982. *J. Biol. Chem.* 257:6106–10
89. Ormstad, K., Uehara, N., Orrenius, S., Örning, L., Hammarström, S. 1982. *Biochem. Biophys. Res. Commun.* 104:1434–40
90. Uehara, N., Ormstad, K., Örning, L., Hammarström, S. 1983. *Biochim. Biophys. Acta.* In press
91. Mathews, W. R., Rokach, J., Murphy, R. C. 1981. *Anal. Biochem.* 118:96–101
92. Lindgren, J. A., Hansson, G., Samuelsson, B. 1981. *FEBS Lett.* 128:329–35
93. McKay, S. W., Mallen, D. N. B., Shrubsall, P. R., Smith, J. M., Baker, S. R., et al. 1981. *J. Chromatogr.* 219:325–31
94. Wynalda, M. A., Morton, D. R., Kelly, R. C., Fitzpatrick, F. A. 1982. *Anal. Chem.* 54:1079–82
95. Murphy, R. C., Mathews, W. R., Rokach, J., Fenselau, C. 1982. *Prostaglandins* 23:201–6
96. Young, R. N., Kakushima, M., Rokach, J. 1982. *Prostaglandins* 23:603–13
97. Levine, L., Morgan, R., Lewis, R. A., Austen, K. F., Clark, D. A., et al. 1981. *Proc. Natl. Acad. Sci. USA* 78:7692–96
98. Lindgren, J. Å., Hammarström, S., Goetzl, E. J. 1983. *FEBS Lett.* 152:83–88
99. Hedqvist, P. 1983. See Ref. 43. In press
100. Hedqvist, P., Dahlén, S. E., Gustafsson, L., Hammarström, S., Samuelsson, B. 1980. *Acta Physiol. Scand.* 110:331–33
101. Drazen, J. M., Austen, K. F., Lewis, R. A., Clark, D. A., Goto, G., et al. 1980. *Proc. Natl. Acad. Sci. USA* 77:4354–58
102. Krell, R. D., Osborn, R., Vickery, L., Falcone, K., O'Donnell, M., et al. 1981. *Prostaglandins* 22:387–409
103. Dahlén, S. E., Hedqvist, P., Hammarström, S. 1983. *Eur. J. Pharmacol.* 86:207–15
104. Augstein, J., Farmer, J. B., Lee, T. B., Sheard, P., Tattersall, M. L. 1973. *Nature New Biol.* 245:215–17
105. Smedegård, G., Hedqvist, P., Dahlén, S. E., Revenäs, B., Hammarström, S., et al. *Nature* 295:327–29
106. Dahlén, S. E., Hedqvist, P., Hammarström, S., Samuelsson, B. 1980. *Nature* 288:484–86
107. Hanna, C. J., Bach, M. K., Pare, P. D.,

Schellenberg, R. R. 1981. *Nature* 290:343–44
108. Jones, T. R., Davis, C., Daniel, E. E. 1982. *Can. J. Physiol. Pharmacol.* 60:638–43
109. Drazen, J. M., Lewis, R. A., Austen, K. F., Toda, M., Brion, F., et al. 1981. *Proc. Natl. Acad. Sci. USA* 78:3195–98
110. Lewis, R. A., Drazen, J. M., Austen, K. F., Toda, M., Brion, F., et al. 1981. *Proc. Natl. Acad. Sci. USA* 78:4579–83
111. Baker, S. R., Boot, J. R., Jamieson, W. B., Osborne, D. J., Sweatman, W. J. F. 1981. *Biochem. Biophys. Res. Commun.* 103:1258–64
112. Tsai, B. S., Bernstein, P., Macia, R. A., Conaty, J., Krell, R. D. 1982. *Prostaglandins* 23:489–506
113. Sirois, P., Roy, S., Borgeat, P. 1981. *Prostagl. Med.* 6:153–59
114. Sirois, P., Roy, S., Borgeat, P., Picard, S., Corey, E. J. 1981. *Biochem. Biophys. Res. Commun.* 99:385–90
115. Holroyde, M. C., Altounyan, R. E. C., Cole, M., Dixon, M., Elliott, E. V. 1981. *Lancet* 2:17–18; 1981. *Agents Actions* 11:573–74
116. Svensjö, E., Arfors, K. E., Artursson, G., Rutili, G. 1978. *Upsala J. Med. Sci.* 83:71–79
117. Dahlén, S. E., Björk, J., Hedqvist, P., Arfors, K. E., Hammarström, S., et al. 1981. *Proc. Natl. Acad. Sci. USA* 78:3887–91
118. Björk, J., Hedqvist, P., Arfors, K. E. 1982. *Inflammation* 6:189–200
119. Ford-Hutchinson, A. W., Bray, M. A., Doig, M. V., Shipley, M. E., Smith, M. J. H. 1980. *Nature* 286:264–65
120. Malmsten, C., Palmblad, J., Udén, A. M., Rådmark, O., Engstedt, L., et al. 1980. *Acta Physiol. Scand.* 110:449–51
121. Goetzl, E. J., Pickett, W. C. 1981. *J. Exp. Med.* 153:482–87
122. Palmblad, J., Malmsten, C., Udén, A. M., Rådmark, O., Engstedt, L., et al. 1981. *Blood* 58:658–61
123. Bokoch, G. M., Reed, P. W. 1981. *J. Biol. Chem.* 256:5317–20
124. Rae, S. A., Smith, M. J. H. 1981. *J. Pharm. Pharmacol.* 33:616–17
125. Hafström, I., Palmblad, J., Malmsten, C. L., Rådmark, O., Samuelsson, B. 1981. *FEBS Lett.* 130:146–48
126. Feinmark, S. J., Lindgren, J. Å., Claesson, H. E., Malmsten, C., Samuelsson, B. 1981. *FEBS Lett.* 136:141–44
127. Showell, H. J., Naccache, P. H., Borgeat, P., Picard, S., Vallerand, P., et al. 1982. *J. Immunol.* 128:811–16
128. Deleted in proof

129. Molski, T. F. P., Naccache, P. H., Borgeat, P., Sha'afi, R. I. 1981. *Biochem. Biophys. Res. Commun.* 103:227–32
130. Sha'afi, R. I., Molski, T. F. P., Borgeat, P., Naccache, P. H. 1981. *Biochem. Biophys. Res. Commun.* 103:766–73
131. Serhan, C. N., Fridovich, J., Goetzl, E. J., Dunham, P. B., Weissmann, G. 1982. *J. Biol. Chem.* 257:4746–52
132. Cunningham, F. M., Carter, H. R., Smith, M. J. H., Ford-Hutchinson, A. W., Bray, M. A. 1981. *Agents Actions* 11:583–84
133. Francis, H. P., Goadby, P. 1981. *Br. J. Pharmacol.* 74:926P
134. Burke, J. A., Levi, R., Guo, Z.-G., Corey, E. J. 1982. *J. Pharmacol. Exp. Ther.* 221:235–41
135. Terashita, Z. I., Fukui, H., Hirata, M., Terao, S., Ohkawa, S., et al. 1981. *Eur. J. Pharmacol.* 73:357–61
136. Sirois, P., Kérouac, R., Roy, S., Borgeat, P., Picard, S., et al. 1981. *Prostagl. Med.* 7:363–73
137. Feuerstein, N., Foegh, M., Ramwell, P. 1981. *Br. J. Pharmacol.* 72:389–91
138. Feuerstein, N., Bash, J. A., Woody, J. N., Ramwell, P. 1981. *Biochem. Biophys. Res. Commun.* 100:1085–90
139. Folco, G., Hansson, G., Granström, E. 1981. *Biochem. Pharmacol.* 30:2491–93
140. Omini, C., Folco, G. C., Viganò, T., Rossoni, G., Brunelli, G., et al. 1981. *Pharmacol. Res. Commun.* 13:633–40
141. Schiantarelli, P., Bongrani, S., Folco, G. 1981. *Eur. J. Pharmacol.* 73:363–66
142. Piper, P. J., Samhoun, M. N. 1981. *Prostaglandins* 21:793–803
143. Palmer, M. R., Mathews, R., Murphy, R. C., Hoffer, B. J. 1980. *Neurosci. Lett.* 18:173–80
144. Palmer, M. R., Mathews, W. R., Hoffer, B. J., Murphy, R. C. 1981. *J. Pharmacol. Exp. Ther.* 219:91–96
145. Webb, D. R., Nowowiejski, I., Healy, C., Rogers, T. J. 1982. *Biochem. Biophys. Res. Commun.* 104:1617–22

Ann. Rev. Biochem. 1983. 52:379–409
Copyright © 1983 by Annual Reviews Inc. All rights reserved.

MECHANISM OF FREE ENERGY COUPLING IN ACTIVE TRANSPORT

Charles Tanford

Department of Physiology, Duke University Medical Center, Durham, N.C. 27710

CONTENTS

PERSPECTIVES AND SUMMARY 379
OVERALL REACTION AND THERMODYNAMICS 381
   Free Energy Bookkeeping 383
   Questions of Stoichiometry 385
   Closeness to Equilibrium 386
PATHWAY FOR ION TRANSLOCATION 386
   Alternating Access Model 386
   Access Channels 389
   How Can the Protein Change the Chemical Potential? 391
SEPARATE PATHWAY FOR ATP PROCESSING 395
   Pumps with Phosphoenzyme Intermediates 396
   F_0F_1 *Systems* 397
COUPLING BETWEEN SEPARATE PATHWAYS 398
   ATP-Linked Transport 398
   Counter-Transport Systems 400
CONFLUENT PATHWAYS 402
MONOMER OR OLIGOMER 403
CONCLUSION 404

PERSPECTIVES AND SUMMARY

This paper reviews current ideas about the mechanism of three kinds of processes: (*a*) uphill transport of ions across a membrane, in which a concentration gradient (chemical potential difference) is created and maintained by coupling to an exergonic chemical reaction, such as ATP hydrolysis; (*b*) the reverse of this process, the use of downhill transport of ions to drive an endergonic chemical reaction (ATP synthesis); and (*c*)

379

0066/4154/83/0701-0379$02.00

uphill transport of one ion, coupled to downhill transport of a second ion. All of the specific reactions to be considered, except one, are readily reversible in the laboratory, or even in vivo, so that the distinction between *a* and *b* is imposed by physiological demands on the transport protein and does not require an intrinsic difference in the reaction mechanism.

The problem of devising a mechanism for this kind of process (or designing a protein molecule to carry out the catalysis) is fundamentally different from the problem of devising a mechanism for catalysis of ordinary chemical transformations. In a chemical transformation, matter is exchanged between the interacting molecular species, and we can express the mechanism of catalysis in familiar chemical terms: close approach of interacting species to each other, breaking of one pattern of chemical bonds and replacement by a new pattern, elimination of H_2O, etc. The same kind of terminology may apply to part of an active transport process (e.g. when we consider how ATP is processed in ATP-linked transport), but the central problem in active transport is intrinsically different, because the transported species does not undergo chemical transformation. The interacting substrates exchange free energy but do not exchange matter; the transport protein functions as a thermodynamic engine, and not as a rate accelerator of a chemical reaction that could in principle occur (albeit slowly) in the absence of an enzyme. One consequence of the absence of chemical interchange is that close contact between interacting substrates is no longer an obvious advantage. For transport of ions other than H^+, an even stronger statement can be made: there is no plausible way that close contact can even be helpful in promoting free energy exchange. Close contact between Ca^{2+} and ATP, for example, would lead to formation of a thermodynamically stable complex, which would make it more difficult to promote Ca^{2+} to a state of high chemical potential. In the case of exchange transport between ions of like charge a mechanism involving a close contact verges on the absurd.

Free energy coupling in these systems appears likely to involve protein-mediated linkage between chemical events occurring some distance apart. The possibility that this may be a general feature of all active transport systems must be considered. An additional unique feature of the mechanism is the need to couple the chemical potential change of the transported ion to translocation from one side of the membrane to the other, but this is conceptually a less difficult problem than the problem of free energy transfer without direct contact.

This review attempts a unified approach to the mechanism of all proteins that catalyze reversible active transport of small inorganic cations. (Proteins that catalyze electron transport are excluded.) The only previous comprehensive discussion of all these systems together is by Mitchell (1, 2) but he endeavors to avoid distinction between ordinary enzymes and transport catalysts; i.e. he assumes close contact between interacting substrates and considers alternatives only in passing. This review begins with the alternative possibility, and the overall problem is dissected into component parts accordingly: ion translocation, treated as spatially distinct; ATP processing, treated as spatially distinct; and thermodynamic coupling between spatially distinct processes. At the end, the review returns to discussion of a Mitchell-type of mechanism, in which there is a single confluent pathway with transient direct interaction between transported ions and chemical free energy donor or acceptor.

There are numerous reviews of individual transport proteins. Recent reviews that include discussion of possible mechanisms for free energy coupling are available for the F_0F_1 class of ATP-linked proton transport proteins (3–9), ATP-driven Na^+K^+ pumps (10–14), the sarcoplasmic reticulum (SR) Ca^{2+} pump (15–18), plasma membrane Ca^{2+} pumps (18a), the gastric ATP-driven H^+, K^+ pump (19, 20), fungal ATP-driven proton pumps (21), the light-driven proton pump of bacteriorhodopsin (22, 23), the Na^+/Ca^{2+} exchange protein (24), and (H^+ or Na^+)/metabolite symports (25, 26). The erythrocyte anion exchange protein (a possible mechanistic prototype for other obligatory exchange processes) has also been reviewed (27). I have relied heavily on these reviews for information on individual systems.

There have been several attempts to approach active transport from a formal physical point of view, applying general principles of thermodynamics, statistical mechanics, and kinetics to the problem. Some of these may provide preliminary orientation for physical chemists unfamiliar with the problem, but none contain new ideas that are actually helpful in the formulation of a molecular mechanism. An extreme example is a book by Hill (28), entitled *Free Energy Transduction in Biology*, which asserts that it is not even possible to arrive at a step-by-step mechanism of the kind that biochemists normally seek. [For a simplified account, addressed to biochemists, see (29).] Jencks (30) provides a much more useful general analysis in a logical but nonmathematical paper that formulates explicit rules that any coupled vectorial process is obliged to obey. For ATP-coupled transport processes, a recent review of phosphoryl transfer reactions by Knowles (31) is helpful.

OVERALL REACTION AND THERMODYNAMICS

Table 1 lists some of the best-known transport systems with inorganic cations as transported species. They fall into four categories: (*a*) ATP-driven pumps with phosphoryl enzyme intermediates; (*b*) multisubunit

Table 1 Molecular parameters and thermodynamic characteristics at the physiological steady state

Protein	Free energy donor	Transported ions[a]	Approximate minimal M_r[b]	Distinct polypeptide chains	Free energy (kcal/mol)				References[e]
					Provided by donor[c]	Utilization[d]			
						Osmotic work	Electrical work	Excess	
Pumps without detectable E~P intermediate									
F_0F_1 (mitochondria)	ATP	$3H^+$	480,000	8	11	3.0	10.5	+2.5	32, 33
F_0F_1 (chloroplasts)	ATP	$3H^+$	480,000	8	14	14.5	0	+0.5	6, 34
F_0F_1 (bacteria)	ATP	$3H^+$	470,000	8	11	3.3[f]	9.6[f]	+1.9[f]	6
Pumps with established E~P intermediate									
H^+ pump (*Ascomyces*)	ATP	1 or $2H^+$	100,000	1	11.6		see note[g]		21, 35–38
H^+, K^+ pump (gastric)	ATP	$1H^+$, $1K^+$	100,000	1	14	11.0	0	−3	19, 39, 40
Na^+, K^+ pumps	ATP	$3Na^+$, $2K^+$	150,000	2[h]	14	7.8	1.6[i]	−4.6[i]	41
Ca^{2+} pump (sarc. retic.)	ATP	$2Ca^{2+}$	120,000	1[h]	14	13.5	0	−0.5	41
Ca^{2+} pump (plasma memb.)	ATP	1 or $2Ca^{2+}$	140,000	1	14		see text		42–44
Other									
Bacteriorhodopsin	light	$>2H^+$	26,000	1	50	see note[g]	see note[g]		45–47
Na^+/Ca^{2+} antiport	$3Na^+$	$1Ca^{2+}$	not known	not known	4.7[j]	6.5	−1.8	0	48, 49

[a] H^+ transport is treated as such, with the understanding that it may be impossible to distinguish between H^+ transport and OH^- transport in the opposite direction.

[b] Corresponding to one copy of the least abundant polypeptide chain. The values are not intended to be definitive, and no attempt has been made to evaluate the accuracy of individual determinations. References can be found in the reviews cited at the beginning of this paper.

[c] Systems involving ATP are formally treated in the direction of uphill transport. Excess ΔG is the same as ΔG for the overall reaction. If positive, it means that the reaction proceeds in the direction of ATP synthesis. ΔG for ATP hydrolysis in vertebrate cells is based on (50).

[d] ΔG for ion translocation is equal to $n(\Delta\mu_i + Z_i F\Delta\psi)$ where n is the number of ions moved per reaction cycle, μ_i is the conventional molar chemical potential, Z_i is the ionic charge, F is Faraday's constant, and $\Delta\psi$ is the trans-membrane potential against which electrical work is done.

[e] References refer to reaction stoichiometry and thermodynamic data.

[f] As with *E. coli*, aerobic conditions, external pH 7. The system works in the opposite direction when O_2 is in limited supply.

[g] As with *E. coli*, aerobic conditions, external pH 7. The system works in the opposite direction when O_2 is in limited supply.

[h] The "proteolipid" subunit sometimes mentioned in relation to these pumps is probably not a component of the pump protein per se.

[i] Data are for squid axon, but the large excess negative ΔG is typical for all Na^+, K^+ pumps.

[j] Osmotic component only. The net electrical component favors Ca^{2+} extrusion. All free energy data are for cardiac muscle (E. A. Johnson, personal communication).

ATP-linked proton transport systems without phosphoryl enzyme inter-mediates; (c) light-driven proton pumps (bacteriorhodopsin); and (d) obligatory exchange transport systems (antiports). The table indicates the probable stoichiometry for each system, molecular composition and minimal molecular weight, and (most important for this review) the thermodynamic parameters involved in the free energy coupling process. The molecular data are reasonably well established, apart from some remaining questions about the exact number of some of the polypeptides in each molecular unit of F_0F_1 (e.g. 51) and the possible presence of one or two functionally necessary polypeptides in addition to the eight that are definitely established (52, 53). The reaction stoichiometries and free energy parameters are subject to greater uncertainty, e.g. $4Na^+/Ca^{2+}$ remains a viable possibility for the Na^+/Ca^{2+} exchange protein (54). Reported parameters for the F_0F_1 systems vary over a wide range, but there is a limitation on what can be considered acceptable, because the net thermo-dynamic driving force must be in the direction in which the system is known to be actually operating (see below).

An immediately obvious feature of Table 1 is that the four categories of transport protein differ greatly in molecular composition. This can be regarded as a priori evidence against a common mechanism and grounds for questioning the wisdom of the unified approach of this review. However, all of the proteins are designed to solve the identical functional problem, and the problem can be subdivided into component parts in the same way for each of them. Given that we do not yet know for certain how any of the proteins copes with the task set for it, a unified approach is justified. Nature tends to conserve mechanisms it has found to be useful. With reference to the multiple polypeptide chains of the F_0F_1 class of proton transport proteins, there are well-established precedents for the evolution of enzymes from primitive multipolypeptide forms (in bacteria) to forms with one or two polypeptide chains in higher organisms. This evolutionary process can occur without significant alteration in the reaction mechanism. Fatty acid synthetase, which is reviewed in another chapter of this volume by S. J. Wakil et al is an excellent example (55).

Free Energy Bookkeeping

In order to define free energy coupling we must mentally separate the overall process into partial reactions, one energetically downhill and the other energetically uphill. We must also keep in mind that the proteins that catalyze the coupled process must be designed to *prevent* occurrence of either partial reaction without the other. (In terms of mechanism, this is perhaps the most difficult part of the problem. The problem of how uncoupled ATP hydrolysis is prevented is still essentially unsolved.) A simple way for the formal expression of both concepts is to introduce a slash into the equations for the partial reactions. Using the SR Ca^{2+} pump as an illustrative example, the overall reaction is:

$$ATP + 2Ca^{2+}(cyto) \rightleftharpoons ADP + P_i + 2Ca^{2+}(SR) \tag{1.}$$

where "cyto" refers to the cytoplasmic side of the SR membrane. The corresponding partial reactions may be written as:

$$ATP \rightleftharpoons ADP + P_i \tag{2.}$$

$$2Ca^{2+}(cyto) \rightleftharpoons 2Ca^{2+}(SR) \tag{3.}$$

and the corresponding free energy relation is $\Delta G_1 = \Delta G_2 + \Delta G_3$. Since ΔG is a thermodynamic variable, this relation is independent of the pathway, i.e. it in no way presupposes a spatial separation between the reaction pathways of reactions 2 and 3. The thermodynamic data in the Table are for the physiological steady state of a resting cell, where the efficiency of coupling is maximal (41, 56) and ΔG for the overall reaction is close to zero. It is evident that one cannot express that coupling takes place at all without explicit division of ΔG into the separate free energies of the partial reactions. This is an important point for the subsequent discussion of mechanism. If the phenomenon per se cannot be recognized without dividing the free energy into component parts, it is obvious that the mechanism cannot be discussed unless this division is maintained through all steps of the reaction cycle (56, 57).

Where ATP is involved, the data in Table 1 are formally presented in the direction of uphill transport, with ATP as free energy donor. The excess free energy in the table (equivalent to ΔG for the overall coupled reaction) is positive for the F_0F_1 systems, which means that these systems operate in the direction of ATP synthesis at the physiological steady state. Bacterial F_0F_1, however, works in the opposite direction when O_2 is deficient, and all but one of the listed systems can be made to operate in either direction in the laboratory by changing reactant concentrations or membrane potential (9, 24, 58, 59, 60). This reversibility is another important factor in the subsequent discussion of mechanism. The single exception is the light-driven proton pump of bacteriorhodopsin, for which the free energy donated by an absorbed photon far exceeds the requirements of the transport process. The availability of excess free energy allows a far greater range of mechanisms than is possible for reversible systems, and I shall therefore not discuss the large amount of experimental kinetic data that exists for bacteriorhodopsin, which has in any case not yet led to elucidation of the reaction mechanism (22, 23). There is, however, more structural information available for bacteriorhodopsin than for any other membrane protein; this information is often used (for want of any other likely prototype) when there is need for a speculative structural model for other transport proteins.

Questions of Stoichiometry

The number of transported ions per reaction cycle is a critical factor in thermodynamic bookkeeping, but many of the values in Table 1 are still controversial. If these values prove to be incorrect it could affect ideas about the reaction mechanism. For example, the stoichiometry for mitochondrial and bacterial F_0F_1 is often cited as $2H^+/ATP$ (9, 33). This would reduce the figures in Table 1 for ΔG utilized in ion transport by one third, and would make it impossible for F_0F_1 to operate in the direction of ATP synthesis. The only way to circumvent this difficulty would be to abandon the idea that F_0F_1 is an isolated proton pump at all under physiological conditions, as Williams (61) has long maintained. He postulated that F_0F_1 in vivo obtains H^+ ions by a direct pathway from the electron transport enzymes. These protons could have a higher $\tilde{\mu}_{H^+}$ than the protons in the aqueous solution, which would make a 2:1 ratio thermodynamically adequate. Attempts to find evidence for a direct pathway have not been successful (62, 63), but if a 2:1 ratio were to be experimentally established it would be irrefutable evidence for it (64). [Wilson and co-workers (65, 66) showed that ATP can be maintained at levels exceeding those permitted by a 3:1 stoichiometry in mitochondria, if the F_0F_1 system is constrained to operate at a very low rate. The F_0F_1 protein is presumably no longer the determining factor under these conditions.]

The stated stoichiometry (44) for plasma membrane Ca^{2+} pumps is $Ca^{2+}/ATP = 1$, but the response of the pump to changes in Ca^{2+} concentration is cooperative, suggesting a higher stoichiometry. The thermodynamic possibility of $Ca^{2+}/ATP = 2$ depends on whether the pump cycle involves counter-transport of other cations (or cotransport of anions), which is still a matter of dispute (67, 68): translocation of four uncompensated charges per reaction cycle against the membrane potential of muscle sarcolemma would add a prohibitive electrical component to the work of transport. Counter-transport also represents an unresolved issue for the SR Ca^{2+} pump (69–72). The uncertainty does not in this case affect the overall thermodynamics because $\Delta\psi$ across the SR membrane is zero and because the steady state concentrations of putative counterions, such as K^+, are the same on both sides of the membrane. The question is however obviously relevant to the problem of mechanism, e.g. K^+ counter-transport would establish a common relation with Na^+,K^+ and $H,^+K^+$ pumps, that would affect our thinking about all pumps with $E \sim P$ intermediates.

The gastric $H,^+K^+$ pump is electrically neutral, but the normal physiological pH gradient it needs to maintain does not permit $H^+/ATP > 1$. A stoichiometry of $H^+/ATP = 2$ has been reported under non-physiological conditions (73), and a possible explanation is that the reaction stoichiometry can vary, depending on experimental conditions. The fungal H^+ pump may likewise be capable of variable stoichiometry (37), and a recent review by Berman (74) suggests that even the SR Ca^{2+} pump may not have a rigidly fixed transport stoichiometry. In Table 1 and in most of the cited literature, stoichiometry for each transport protein is treated as fixed by kinetic constraints of the reaction cycle (as in Equation 1). This is not intended to preclude the possibility of alternate pathways that could affect the stoichiometry under special conditions, but existence of such pathways would have little or no effect on consideration of the mechanism of free energy coupling.

Closeness to Equilibrium

The close proximity of the physiological steady state to the true equilibrium state for many of the processes of Table 1 is remarkable. Lemasters (75, 76) suggested an even broader extension of this observation, with good evidence that the entire oxidative phosphorylation chain in mitochondria may be close to equilibrium when a cell is at rest. This conclusion would be incompatible with the often expressed idea that the cytochrome oxidase reaction is essentially irreversible (33). The basis for these discrepant opinions is again traceable to an unresolved question of stoichiometry, in this case the H^+/electron stoichiometry of cytochrome oxidase. Table 1 shows that $Na,^+K^+$ pumps do not approach as close to equilibrium as the other listed systems, and this is probably true for $Na,^+K^+$ pumps from all sources (41). No explanation for this difference exists.

PATHWAY FOR ION TRANSLOCATION

The unique feature of active transport is that it occurs without interchange of matter between the interacting species. In many systems the pathway for ion translocation through the transport protein may not at any point involve direct contact with the second substrate. Even if there is direct contact, the thermodynamic events affecting an ion along its translocation pathway (changes in chemical potential) need to be kept separate from the thermodynamic events affecting the free energy donor for the reasons given in the preceding section.

Alternating Access Model

This section considers the pathway for ion translocation from a structural point of view. It is reasonable to assume that all proteins that can carry ions across a membrane (actively or passively) possess a trans-membrane domain that resembles the trans-membrane domain of bacteriorhodopsin in being "inside-out" (77). The outer surface of the domain must be hydrophobic to keep the domain anchored in the phospholipid bilayer, and there must be a thread of polar amino acid residues running down or through the molecule to provide an *incipient* pathway for ion translocation. The incipient pathway cannot be a permanently open channel. In passive transport systems it can be a simple gated channel, with a "switch" enabling it to exist either in an open or closed state, but this is not a valid possibility for active transport because an open channel can catalyze only downhill movements of ions, regardless of what the energetics of opening or closing of the channel may be. An active transport translocation pathway must never be simultaneously accessible from both sides of the membrane because it must permit energetically uphill movement of the transported ions.

The only simple solution to this problem is an allosteric model, schematically illustrated (for a single transported ion) by Figure 1a. In this model the transport protein alternates in each reaction cycle between two different conformational states, in which protein-bound ions can exchange with like ions in the solution, first on one side of the membrane and then on the other side, but never simultaneously with both. Intermediate states, if they occur, have to be "occluded" states without ready access to either aqueous compartment. I shall call this the *alternating access model*.

Virtually all proposed models for the translocation pathway in active transport of ions (except those based on confluent pathways for ATP-linked H^+ transport) are variants of the alternating access model. The model was first proposed by Vidavar (78) and Jardetzky (79) and has been reproposed several times since (80, 81). It is a central element in Jencks' universal rules for active transport (30). It is the generally accepted model for the SR Ca^{2+} pump (17, 18) and for the ATP-driven Na^+K^+ pump (13). (The literature on Na^+K^+ pumps can be confusing on this subject, because it often refers to simultaneous accessibility from both sides of the membrane. The terminology here refers to the possibility of separate translocation devices for Na^+ and K^+, both of which alternate in the direction of access, but do so out of phase with each other.) The alternating access model was shown to be applicable to H^+ translocation in the only existing steady state kinetic analysis of H^+ transport in an F_0F_1 system (34). Alternating access is also an implicit feature of most proposals for the reaction mechanism of bacteriorhodopsin. The Schiff base link between the protein and retinal was originally proposed as the alternating proton binding site (23), but this specific suggestion is now thought to be invalid (82), because current data indicate that two or three protons are transported per reaction cycle (45–47). A speculative model that accommodates two protons has been proposed (83). (The abstract of this paper incorrectly implies that the model does not involve alternating access.) An alternating access model has been proposed for erythrocyte glucose transport (84) and would probably be consistent with many other transport processes that do not fall within the scope of this review.

It is questionable whether viable alternatives to the alternating access model exist. Separate primary uptake and discharge sites may be able to coexist throughout the transport cycle (see below), but passage of the bound ion from one site to the other must occur via intervening sites. A continuous pathway being disallowed, one of the intervening sites would have to be an

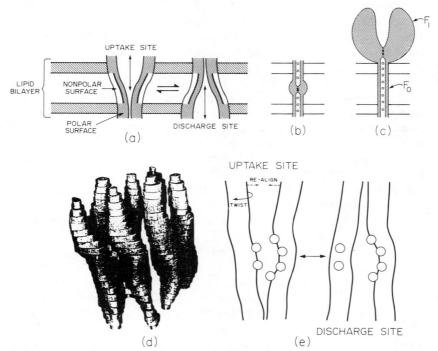

Figure 1   (*a*) Schematic representation of alternating access model for translocation of a single ion. Only the transmembrane portion of the protein is indicated. The width of the access openings is exaggerated. (*b*) Model with access channels, as postulated for bacteriorhodopsin (22). The small circles are transient sites for the ion. The notched area with double-headed arrow represents the alternating site, able to exist in two conformations, as in Figure 1*a*. Character of protein surface is not explicitly indicated. (*c*) Model for F_0F_1 class of proteins. Circles etc, defined as for Figure 1*b*. (*d*) Electron density representation of the structure of the membrane-embedded portion of bacteriorhodopsin (100). Each distorted cylinder represents a helical segment of polypeptide chain. (*e*) Hypothetical model for change in chemical potential of bound cation, occurring in synchrony with translocation (81). The circles are chelating groups, attached to transmembrane helices such as those of Figure 1*d*.

alternating access site, and alternation between its two conformational states would still constitute the core of the transport mechanism. The Stoeckenius model for bacteriorhodopsin (Figure 1b) is a model of this kind. Another possibility, a rotating carrier model (15, 85), was popular at one time. This is actually another version of the alternating access model, kinetically indistinguishable from the model of Figure 1a, but it is thermodynamically forbidden because it leads to mismatching of hydrophilic and hydrophobic surfaces in the transition from one access state to the other.

Access Channels

Figure 1a shows the binding sites in the two conformational states of the protein with unrestricted access to the adjacent aqueous media. This is not an essential feature of the model. Binding sites in one or both conformations may have only restricted access to the aqueous media via connecting access channels (Figures 1b, 1c) consisting of successive sites that the ion must transiently occupy on its way to the ultimate translocation site. If these hypothetical channel sites have only low affinity for the transported ion they would perhaps slow down the rate of diffusion of the ion to the translocation site, but would not affect thermodynamic considerations. If channel sites have binding affinities comparable to the affinities of the alternating sites to which they provide access, they would affect the thermodynamics of association between the transported ion and the protein, and could thereby play a more important role in the mechanism. For the sake of simplicity, this possibility is hereafter ignored. It would not affect the conceptual principles involved in free energy coupling and would only slightly complicate how one relates the process to the structure of the transport protein.

In the case of the SR Ca^{2+} pump, the sites for Ca^{2+} uptake from the cytoplasm have very high affinity for Ca^{2+}, and binding can be measured accurately under conditions where no transport occurs. The results show unambiguously that the number of binding sites is equal to the number of ions translocated in each reaction cycle (86, 87). There are no extra sites with comparable affinity, and there cannot be a thermodynamically significant access channel from the cytoplasmic side of the membrane. The same conclusion probably applies to the high affinity cation uptake sites of Na^+K^+ pumps, but the evidence is less secure. Both Ca^{2+} and Na^+K^+ pumps have low affinity sites in the conformational states from which the ions are discharged in uphill transport. The low affinity prevents meaningful binding measurements, so that there is no direct evidence regarding possible access channels leading to these sites.

Proton transport represents a special problem, for two reasons: 1.

Protons have a versatility in the possible mode of transit through a protein molecule that other cations cannot match, which greatly enlarges the range of plausible hypotheses. 2. The ubiquitous existence of proton binding sites unrelated to transport, with a broad range of pK_a values, makes it impossible to test hypotheses by means of direct binding measurements. This has led to an abundance of speculative and untestable proposals for access channels as part of the mechanism of bacteriorhodopsin, and the invention of a special term, *proton wire* (88). It has even been postulated that the polypeptide backbone helix can serve as proton conductance channel (89), which would imply that the thread of polar groups down the center of the bacteriorhodopsin structure might not serve a functional role. The most detailed proposals of this kind are the models of Nagle and co-workers (88, 90, 91). These models differ from all other transport models in that they treat alternating exposure as an integral property of the access channels per se. Proton conductance channels are viewed as asymmetric hydrogen-bonded chains, and alternation in access to the two sides of the membrane involves cooperative changes in hydrogen bonding patterns along the entire length of a channel.

For the F_0F_1 class of transport proteins, there is direct evidence for an access channel. For this class of proteins the component parts of the transport function exist on separate polypeptide chains. The F_0 and F_1 complexes are physically separable, and the F_0 segment by itself has been demonstrated to be an ionophore, conferring passive permeability to H^+ upon the membrane (92, 93). The ionophoric activity is ascribable to an oligomer of a low molecular weight polypeptide of F_0 (94), and can be reconstituted into lipid bilayers after purification. The complete transport protein must of course be designed to prevent passive H^+ exchange between the two sides of the membrane, and, as expected, the F_0F_1 complex has no ionophoric activity. In other words, F_0 plays the role of access channel, providing access (from the uptake side of the membrane in the direction of ATP synthesis) to a proton binding site on F_1. If a spatially separate alternating access model applies to F_0F_1, and if no structural changes occur when F_0 and F_1 are separated, the binding sites with alternating access have to be in F_1, and in a formal structural sense this would place them outside the physical domain of the lipid bilayer, as illustrated by Figure 1c. From a functional point of view this is a trivial distinction. One of the attributes of a membrane-incorporated protein can be to alter the local thickness of the permeability barrier separating the two sides of the membrane. In Figure 1a the protein is in effect viewed as shrinking the barrier; in Figure 1c it expands it.

It is conceivable that Figure 1c may be an accurate representation of the translocation pathway for ATP-driven pumps that work via a phosphoenzyme intermediate, as well as for the F_0F_1 type of protein, with the difference that the component parts would all be fused into a single polypeptide chain (or, at most, two chains). The Na^+,K^+ and SR Ca^{2+} pump molecules extend far from the membrane on the cytoplasmic side of the membrane, but not on the opposite side (95–97). The topology of this asymmetry is the same as for F_0F_1; i.e. F_0 faces the side of the membrane from which protons enter for ATP synthesis, which means that it would be functionally equivalent to the noncytoplasmic side of the membrane in Na^+,K^+ and Ca^{2+} pumps. As previously noted, there is no evidence from binding studies for or against the existence of restrictive access channels from that side. It has been claimed that intact Na^+,K^+ and Ca^{2+} pump molecules and proteolytic fragments derived from them possess ionophoric activity (98). For the intact molecules such activity would be contrary to Equation 3. The result would be reasonable for a proteolytic fragment, but critical examination of the actual experimental data (99) suggests that the claim of specific ionophoric activity is dubious.

MULTIPLE SITES   Most of the proteins in Table 1 transport several ions per reaction cycle. In the context of the models of Figure 1, this could reflect the presence of more than one binding site in a single alternating structural domain of the type of Figure 1a or the presence of several parallel alternating domains, one for each ion, as would be possible in a multihelical structure, such as illustrated in Figure 1d. Experimental data that might provide evidence concerning this question are not decisive. The two Ca^{2+} ions of the SR Ca^{2+} pump reaction and the three Na^+ ions of the Na^+,K^+ pump reaction are translocated simultaneously (13, 18), but the three H^+ ions of chloroplast F_0F_1 are translocated consecutively in two or three separate steps (34). Neither result necessarily favors one possibility over the other. The polypeptide chain composition of F_0F_1 is not yet completely established, but the available data favor three identical catalytic chains per F_1 unit, possibly suggesting that each of the three transported H^+ ions has a separate translocation unit. On the other hand, there appears to be only a single access channel for all three protons in the F_0 unit (4, 9, 94). The question of simultaneous or consecutive Na^+ and K^+ translocation in Na^+,K^+ pumps is discussed later.

How Can the Protein Change the Chemical Potential?

An active transport system, working in the direction of uphill transport, drives an otherwise unaltered solute from low to high chemical potential. This is an uncommon phenomenon, and there have been many speculations as to how the transport proteins accomplish it (101). Investigators concerned with transport of ions other than H^+ have usually addressed the problem in terms of an alternating exposure model, and have assumed that alternation of exposure is accompanied by alternation in binding affinity; i.e. the binding site is assumed to have a high binding constant for the transported ion on the uptake side of the membrane (in the direction of uphill transport), but a much lower affinity on the discharge side (30, 79–81). The possibility that sites in an access channel might play a significant thermodynamic role has usually not been explicitly considered, so that binding constants for uptake and discharge sites are in effect assumed to be the binding constants of the two states of the alternating site itself.

It is easy to show by rigorously correct thermodynamic and kinetic analysis that differing binding affinities are not formally essential for the design of an active transport process; theoreticians intent on cautioning their readers against uncritical acceptance of intuitive judgements (28, 30, 102) have stressed this. It is, however, easy to be overly formal and to ignore that a proposed mechanism must do more than demonstrate that it can account for the occurrence of active transport. The catalytic process must occur at an adequate rate, and must respond to changes in ion concentrations on the two sides of the membrane in a physiologically useful way, and must accomplish both goals with physically reasonable values of rate constants for individual steps of a reaction cycle. When these considerations are taken into account, a mechanism that does not involve different affinities becomes even theoretically improbable unless the chemical potential difference to be generated is very small (30, 103). Stated simply, it is in principle possible to drive uptake of an ion in spite of an unfavorable binding constant by means of strongly unidirectional equilibria in other parts of the reaction cycle, but in practice it requires unacceptably large rate constants to do so, or unacceptably small rate constants in one direction for a reaction that is known to be reversible.

Whether or not it must be so, all available data suggest that the binding constants at uptake and discharge sites in active transport are normally different; the differences have the effect of approximately compensating for the different ion concentrations on the two sides of the membrane with which the transport protein must cope in the course of its physiological function. Table 2 summarizes some of the available results. Additional evidence for conformationally linked affinity changes in F_0F_1 and related proteins of oxidative phosphorylation is given in many of the reviews cited initially in this paper. A more recent paper is by Hatefi et al (112). To be meaningful in the present context, each K_{eq} in Table 2 must apply to a binding process in which the protein is in the same conformational state when binding sites are ligand-free and when they are occupied by ligand. Experimental measurements have usually not tried to meet this condition. For this and other reasons the data in Table 2 should be regarded as

Table 2 Binding constants for transported ions[a]

Protein	Ion	Binding constant, K_{eq} (M^{-1})		References
		Uptake side	Discharge side	
SR Ca^{2+} pump	Ca^{2+}	$10^7 - 10^8$	300	86, 104–106
Na^+, K^+ pumps	Na^+	4×10^3	< 20	12, 107, 108
	K^+	2×10^4	< 10	12, 14, 107–109
Chloroplast F_0F_1	H^+	$> 10^8$	$< 10^6$	34
Na^+/Ca^{2+} exchange	Ca^{2+}	$2 \times 10^5 - 10^6$	400^b	48, 110

[a] Uptake and discharge sides are defined for direction of uphill transport of the ion for which data are given.
[b] A more recent report (111) suggests equal affinities for Ca^{2+} on the two sides of the membrane.

quantitatively subject to considerable uncertainty. The qualitative conclusion from the data is however not in doubt, except for Na^+/Ca^{2+} exchange.

Binding constants for ligand-protein association are determined by the reactive groups of the protein that enter into the coordination sphere of the bound ligand. In the case of Ca^{2+} there is extensive literature on this subject, both for model compounds and for proteins whose three-dimensional structure has been determined by x-ray crystallography (81, 113, 114). Six coordinating groups, including several COO^- groups, are required to form a high affinity site with $K_{eq} \sim 10^6 M^{-1}$ or greater. A smaller number of groups, with some coordination positions occupied by water molecules, suffices to generate a site with $K_{eq} \sim 10^3 M^{-1}$. The same principle undoubtedly applies to Na^+ and K^+ binding, but model compound data for these ions (115) are sparser than for Ca^{2+}, and not even speculative ideas have been advanced to suggest how affinities as high as those shown in Table 1 might be achieved (116). For H^+ ions, binding affinity depends primarily on the basic group to which the ion is bound and one intuitively tends to equate a change in pK_a with a transfer of the proton from one kind of basic group to another. However, as noted earlier, there has to be a place along the pathway where the direction of access changes while the proton remains attached to the same site (e.g. the Schiff base proton in one of the bacteriorhodopsin models), and a ΔpK_a at this point is necessary for any reasonably simple mechanism. Large changes in pK_a for a fixed group can in fact be brought about by changes in vicinal groups. A COOH group, for example, can have $pK_a \sim 2$ if there is a nearby positive charge. At the other extreme, lysozyme has a glutamic acid residue with $pK_a \sim 6.5$ (117) and β-lactoglobulin has a COOH group with an even higher pK_a (118). Papa (119) has discussed pK_a differences that might arise from vicinal metal ions in the electron transport proteins of mitochondrial or chloroplast membranes.

These considerations suggest that appropriate movement of amino acid side chains (and/or peptide groups) in and out of the ion binding sites must be one of the features of the conformational transition between the two states of an alternating access model. This kind of conformation-linked affinity change is a common phenomenon, exemplified by allosteric enzymes or by binding proteins such as hemoglobin (120). Early studies focused on oligomeric proteins, where the conformational difference is generated by alterations in contacts between subunits, but there are also examples where only a single polypeptide chain is involved, as in yeast hexokinase, where two states with very different affinities for glucose are generated by movement of separately folded domains towards and away from each other (121). In all these well-known examples the binding site faces the same solution in both conformations, whereas in transport proteins the change in binding affinity has to be synchronous with the change in accessibility from one side of the membrane to the other. A speculative model for how this might be brought about has been proposed (81), based on a hypothetical transmembrane protein domain resembling the transmembrane part of the structure of bacteriorhodopsin (100). This structure consists of seven roughly parallel polypeptide helices (Figure 1d), connected outside the membrane by short polypeptide segments that can serve as hinges or ball joints between the helices. If the binding sites are located in cavities between the helices, it is easy to imagine a concerted twisting and rocking movement (Figure 1e) that will achieve the desired result. A multihelical model of this type can accommodate simultaneous translocation-linked affinity changes for several ions, including the possibility that some of the bound ions (e.g. $Na^+ K^+$) move simultaneously in opposite directions.

Free energy is required to move ligating groups away from a tightly bound ion, and the equilibrium of the process envisaged in Figure 1e would by itself lie far to the left. This would therefore be a place in a reversible reaction cycle where free energy coupling would be necessary to achieve more nearly equal free energies for the two states.

EFFECT OF MEMBRANE POTENTIAL   There has been no discussion of how the foregoing considerations should be modified when the electrical term is the major component of the electrochemical potential difference ($\Delta\tilde{\mu}$) of the translocated ion, as it is for mitochondrial F_0F_1. Mitchell (122) has suggested that one of the functions of the F_0 access channel may be to convert the electrical contribution to $\Delta\tilde{\mu}_{H^+}$ to a purely osmotic chemical potential difference, in which case it would be appropriate to have RT $(\Delta lnK_{eq}) \simeq \Delta\tilde{\mu}_{H^+}$. My own suggestion (81) that this relation may have general applicability is probably incorrect. Availability of experimentally determined pK_a differences for mitochondrial F_0F_1, comparable to the data in Table 2 for chloroplast F_0F_1, would help to resolve the question.

SEPARATE PATHWAY FOR ATP PROCESSING

In systems with ATP as one of the participants, the pathway for ATP processing may be spatially separate from the pathway for ion translocation, the linkage between them being mediated by the protein. Where this occurs, the following premises may be made with reasonable confidence about the reaction mechanism. The premises are similar in spirit to the "rules" for coupled vectorial systems set forth by Jencks (30). They have the effect of keeping the relative populations of different kinetically important states of the reaction cycle at comparable levels, thereby facilitating response of the system (which includes ability to reverse the direction of the overall reaction) to changes in the concentration of substrates (30, 103):

1. The reaction pathway must be unable to catalyze the uncoupled hydrolysis of ATP. The simplest way is to have the protein exist in two conformational states, such that the overall process (which includes substrate binding and product release) is carried part way in one conformation and completed in the other.
2. At the points of the reaction cycle where ATP, ADP, or P_i are free to exchange between bound and free states (generally distinct points for each substance), the free energy difference between bound and aqueous species should be small under steady state conditions. This means that the overall reaction among bound species, ATP (when first bound) $\rightleftharpoons$ ADP (about to be released) + P_i (about to be released), must be accompanied by a large negative free energy change, -10 to -14 kcal/mol for the examples given in Table 1.
3. A general principle recently suggested by Knowles (31) for most enzyme-catalyzed phosphoryl transfer reaction is presumably applicable: the difference in free energy between ATP and its hydrolysis products at the site of bond rupture should be small.
4. It follows that there must be a step (or steps) in the reaction cycle where one (or more) of the bound substrate species undergoes a dramatic change in chemical potential, i.e. a transition between a "loosely" and a "tightly" bound state (high and low free energy). Only by this device can the large ΔG between bound substrates and products at the points of exchange with the aqueous medium be reduced to a small ΔG at the point of bond rupture. Reaction steps of this kind would by themselves be essentially irreversible; these are therefore the steps where thermodynamic coupling to the ion translocation pathway must take place.

Pumps with Phosphoenzyme Intermediates

In Na^+K^+ and Ca^+ pumps and in the gastric and fungal H^+ pumps, ATP processing occurs via an acyl phosphate intermediate, in which the phosphate is covalently linked to an aspartyl residue of the protein (19, 36, 37, 123). This intermediate exists in two very different states. One is a state of high free energy ($E \sim P$), thermodynamically roughly similar to the state of soluble aspartyl phosphate. The bound phosphate group in this state readily interacts with ADP to synthesize ATP. The second phosphoenzyme is a state of low free energy ($E'-P$), unprecedented for an acyl phosphate, and thermodynamically similar to a phosphate ester, e.g. seryl phosphate. The chemical potential for bound phosphate in $E \sim P$ is about the same as the chemical potential of ATP in the aqueous solution, whereas in $E'-P$ it is about the same as the chemical potential of aqueous P_i. The difference between them is of order 10–14 kcal/mol, as was first noted several years ago by Taniguchi & Post (58). Bond rupture in these proteins thus appears to take place at the level of free energy of undissociated ATP, and the requisite transition between high and low potential states is attributable entirely to the bound phosphate group. In the direction of ATP synthesis, employing the terminology often used in discussion of this subject, the substrate is "energized" entirely in the course of the conversion $E'-P \rightarrow E \sim P$ (17, 124).

There is still little comprehension of the underlying chemistry. It is not known what structural features of the protein prevent the spontaneous hydrolysis of $E \sim P$, which is of course an essential part of the mechanism. It is difficult to imagine how the acyl phosphate group can be readily accessible to ADP added from the adjacent solution (ATP synthesis is a relatively fast step in the reaction cycle), but inaccessible to H_2O. The structural features responsible for the existence of the low free energy $E-P$ state are also not understood. In the case of bound inorganic ions, discussed earlier, it is easy to visualize how different binding energies can be achieved, but no easy explanation is possible here. In the words of Jencks (30), "this phosphate molecule behaves as if it were surrounded by the physically impossible concentration of $10^8 M$ carboxylate groups." de Meis (18, 125) pointed out that an acyl phosphate link should be strengthened in a nonaqueous medium and suggested that the thermodynamic state of $E-P$ can be explained on the basis of a hydrophobic environment at the binding site. There is an absolute requirement for Mg^{2+} in the formation of $E'-P$, so that Mg^{2+} may be directly involved, but Mg^{2+} is a common cofactor for many other phosphoryl transfer enzymes as well, where no similar thermodynamic problem exists.

A phenomenon that has attracted some attention is a complex kinetic dependence of steady state ATP hydrolysis on ATP concentration (13, 18, 126). It is consistent with the possible existence of two ATP binding sites on each pump molecule, one of them serving a regulatory function (124, 126, 127). A simpler explanation is that there may be only a single ATP binding site, with different binding affinities in the two conformational states E and E' (126, 128). This explanation can account semiquantitatively for the experimental data, but whether it can do so completely has not been established (129).

F_0F_1 Systems

All available evidence indicates that ATP processing by the F_0F_1 proteins does not involve phosphoenzyme intermediates, but there has been little agreement on other aspects of the pathway. Some investigators have favored a confluent pathway in which the translocated H^+ ions interact directly with the chemical substrate (see below) and others have advocated a separate pathway with indirect free energy coupling (3). Boyer and coworkers (130–133) have been the most consistent advocates of a separate pathway, and their experimental data have led to the most detailed model for it. The model incorporates the four general precepts given at the beginning of this section, and in particular recognized that ATP bond cleavage per se occurs without much free energy change (134; see also 135, 136) several years before this was proposed as a general phenomenon for all phosphoryl transfer enzymes. Boyer emphasized that this leads to a need for tightly and loosely bound substrate species, but in F_0F_1 both ATP and its hydrolysis products appear to have states of differing binding affinity, in contrast to pumps with phosphoenzyme intermediates. Bond rupture or formation occurs at an intermediate free energy level, which means that free energy input in the direction of ATP synthesis (energization) has to be a two-stage process. In one stage bound ADP and/or P_i are promoted from their initial low free energy level to the intermediate level at which conversion to ATP takes place. In the second stage the bound ATP formed at this level is promoted to a loosely bound state at which its chemical potential becomes comparable to that of ATP in the adjacent solution, with which it is at this point freely exchangeable. The two stages of free energy transfer are postulated to occur as a single step in the overall reaction cycle, on the basis of evidence that the functional unit of the F_0F_1 is a dimer of two identical subunits that are always one stage apart in terms of substrate free energy (131, 132), so that free energy input simultaneously promotes bound ADP and P_i in one subunit and promotes bound ATP in the other subunit. (As noted earlier, the most recent data favor three identical subunits instead of two.)

The foregoing mechanism does not have the broad support accorded to the separate ATP processing pathway for pumps with phosphoenzyme intermediates. A complication that increases the difficulty of experimental resolution of the problem is the existence of tight binding of adenine nucleotides to noncatalytic sites on the F_1 moiety (4, 5, 53). The most recent paper on this subject is by Bruist & Hammes (137).

COUPLING BETWEEN SEPARATE PATHWAYS

ATP-Linked Transport

If there are separate pathways for ATP processing and ion translocation, the structural basis for coupling between them is virtually self-evident. The previous sections demonstrated that each separate process requires a conformational transition as part of its mechanism. Functional coupling is then automatically assured if the two conformational transitions are structurally coupled. The simplest possibility is that the protein molecule as a whole has *only* two conformational states, so that the change in accessibility of ion binding sites from one side of the membrane to the other could not occur without the simultaneous transition from E to E' that was discussed in relation to ATP processing.

One cannot specify a "pathway" for the structural link, because it is a property of the entire protein molecule. From an energetic point of view, however, specification of the structural connections between the two separate functional domains is not really relevant. It is well-established that long-range order in all proteins results from cooperative interactions in which all parts of the molecule are involved. Individual amino acid side chains can possess some local freedom of motion (138, 139), but it is usually not possible for a region of several residues to undergo a structural rearrangement independent of the rest of the molecule (140). Hemoglobin is the prototype example. It can exist in two conformational states, differing in oxygen affinity, and the transition between them is cooperative, such that all four oxygen binding sites change their affinity simultaneously. The separation between the heme groups that constitute the oxygen binding sites is of order 25Å, and there is no possibility of any significant direct interaction between them. Nevertheless, a conformational change in which the affinity is changed at one heme and not at the other does not occur. Changes in the pK_a values of acidic groups at the protein surface occur in synchrony with the change in oxygen affinity, and again are unable to occur independently.

The three-dimensional structure of hemoglobin has been known for a long time, and the numerous local interactions that are involved in the overall conformational transition have been analyzed in detail (141, 142).

Such detailed knowledge is, however, not necessary to accept conformational cooperativity as a valid element of a reaction mechanism. Action at a distance is a general phenomenon in protein chemistry and enzymology, and there is direct evidence for it for Na^+K^+ and Ca^{2+} pumps (143, 144). Mitchell (2) has criticized conformational coupling as a "black box" approach, but this criticism is not really warranted.

Conformational coupling, even with only two conformational states for the transport protein, permits two distinct mechanisms for energetic coupling. The more familiar one is the "rack" mechanism of Lumry (145), and this mechanism is often implicitly assumed when the idea of conformational coupling is invoked (130). In this mechanism the two protein conformations are viewed as having very different free energies, and coupling becomes a two-step process. In the first step the free energy released by the transition of bound substrate (in whichever form may be applicable) from "loosely bound" to "tightly bound" is used to drive the protein into its thermodynamically unfavorable conformational state. In a separate step, the protein returns to its more stable state, and the free energy released is used to raise the chemical potential of the bound ions. The same two steps, in the reverse direction, would be involved in ATP synthesis.

An equally plausible alternative is a mechanism in which free energy is exchanged directly between bound substrate and bound ions, in a single step (57). In this mechanism the two protein conformations are viewed as having about equal free energies, so that the protein becomes an energetically passive participant in the process. The conformational change that alters the free energy of bound ATP and/or ADP and/or P_i is viewed as occurring in synchrony with the conformational change that alters the chemical potential of the bound ions; the net free energy change in the combined process is close to zero.

With the assumption of only two significantly different conformational states (E and E'), available data favor a direct exchange mechanism for Na^+K^+ pumps and for the SR Ca^{2+} pump. $\Delta G°$ for the transition between unliganded E and E' in the SR Ca^{2+} pump has been estimated to be no larger than 4 kcal/mol (106, 146, 147). $\Delta G°$ for the same process in Na^+K^+ pumps can be estimated from various data (128, 148) to be 0 ± 2 kcal/mol. These free energies are far below the ~ 14 kcal/mol that has to be transferred between ATP and the transported ions, suggesting that free energy storage by the protein can play only a minor role in the mechanism.

Whether these pump proteins in fact possess only two distinct conformational states is more questionable. The existence of two distinct states is clearly established, most directly in the case of Na^+K^+ pumps (149, 150), and most published mechanisms invoke only two states. There is, however, no consensus on this point and multiconformation mechanisms have been proposed (86, 151, 152). Fluorescence changes are sometimes cited as evidence for additional conformational states (144, 153), but evidence from this source is dubious because fluorescence can change as a consequence of ligand binding without change in protein conformation (154). Some data indicate that occluded states for the transported ion occur as reaction intermediates [(150, 155–158), see however (159)], which would constitute a priori evidence for an additional conformational state. These unresolved questions do not affect conformational coupling as a conceptual mechanism, but do suggest that the process may involve two or more successive steps in the reaction mechanism.

An important experiment that provides impressive evidence for conformational coupling (largely in a single step) is the demonstration that the SR Ca^{2+} pump can synthesize ATP in the absence of a Ca^{2+} concentration gradient (17, 18, 160). All that is necessary is to have Ca^{2+} bound to the low affinity (high chemical potential) site of the protein, and it has been shown that this binding occurs to the E′–P state in which the bound phosphate is in its "tightly bound" state (161). The $Ca_2E′-P$ complex is spontaneously converted to the ADP-sensitive $Ca_2E \sim P$ complex, in which the bound phosphate is at high chemical potential (forming ATP when ADP is added) and the bound Ca^{2+} is at low potential. The interconversion takes place in a single step with no detectable intermediates other than possible transients of extremely short life time (162, 163).

Counter-Transport Systems

It makes no intrinsic difference whether free energy for uphill transport is provided by a chemical reaction or by downhill transport of another species. Ionic counter-transport systems, however, possess the unique possibility of using the same translocation pathway (Figure 1a) in both directions, i.e. the discharge state for ion X (low affinity for X) could be the uptake state for ion Y (high affinity for Y), and vice versa. If substantial changes in chemical potentials of the ions are involved, free energy coupling would in this case require a "rack" mechanism so that free energy released by downhill transport of one ion can be made available for uphill transport of the other ion when the conformational change in Figure 1a is reversed. The alternative possibility is separate translocation pathways for X and Y, in which case the coupling problem becomes indistinguishable from the problem of coupling ATP hydrolysis to ion transport. Separate translocation pathways could reside on separate but structurally coupled protein subunits or domains of a single subunit, or they could reside in separate crevices of a multihelical structure (Figure 1d).

Experimental studies generally ask the slightly different question of whether X and Y are transported simultaneously in opposite directions, or whether they are transported consecutively, with obligatory release of X before Y can be bound. Simultaneous transport requires dual translocation pathways. Consecutive transport is generally interpreted as indicating the existence of a single translocation unit, but does not in fact exclude the possibility of dual units because the two units could be structurally linked in a way that prevents simultaneous occupancy [see Figure 6 of (12)]. Kinetic data for the red blood cell anion exchange protein (164) are consistent with a consecutive mechanism, and have been interpreted in terms of a single unit. Kinetic data for Na^+/Ca^{2+} exchange, on the other hand, suggest simultaneous transport and dual translocation units (48). Indirect evidence for multiple translocators is provided by recent studies of mutant strains of bacteria (165, 166). Mutant strains deficient in Na^+/H^+ counter-transport function were also found to be deficient in Na^+-linked symports carrying amino acids or sugars into the cell. The simplest explanation is that these systems contain a common subunit for Na^+, which is combined with a separate subunit for translocation of H^+ or a metabolite. (In a eukaryotic organism the separate subunits could become separate domains of a single polypeptide chain.)

The same questions are relevant for translocation of Na^+ and K^+ in $Na,^+K^+$ pumps, although both ions are in this case pumped uphill. Most investigators currently favor a consecutive mechanism, and again consider this as implicit support for consecutive use of the same binding sites by Na^+ and K^+ (13, 155, 167), but support for the contrary view also exists (168, 169). Direct enumeration of binding sites of leaky vesicles (both sides of the membrane accessible) or with solubilized protein should in principle provide decisive evidence, but it is experimentally difficult to measure Na^+ and K^+ binding, and the measurement is complicated by the likelihood that the two ions can compete for each other's sites. Two recent studies (107, 109) indicate that five ion binding sites can be simultaneously occupied for each ATP processing site, which provides nominal support for separate translocation units, respectively specific for Na^+ and K^+. Matsui & Homaneda (108), on the other hand, reported that high affinity binding sites for Na^+ and K^+ cannot be occupied simultaneously, a result seemingly incompatible with these studies.

MORE COMPLEX MECHANISMS   It has been suggested (2, 170) that ATP-driven Na^+, K^+ and Ca^{2+} pumps might actually be ATP-driven H^+ pumps, intimately associated (intramolecularly) with antiports (H^+/Na^+, etc). This permits hypothetical consideration of a common confluent pathway for ATP processing in all ATP-driven pumps, but there is no evidence to support this idea. Racker (171) proposed that Mg^{2+} acts as intermediary in free energy coupling, a proposal based on calorimetric data that indicate absorption of 60–80 kcal/mol of heat when Mg^{2+} is bound to Na^+/K^+ and Ca^{2+} pumps when they work in the direction of ATP synthesis. The suggestion is thermodynamically incomplete because no attempt is made to reconcile the large ΔH value with the 14 kcal/mol of free energy involved in the coupling process, nor is there any indication of where in the reaction cycle the absorbed heat is again released to the surroundings. Data from other laboratories have not confirmed the high negative ΔH value itself (125, 172, 173).

CONFLUENT PATHWAYS

That metabolic free energy has to be the power source for generation of ion concentration gradients was recognized long ago, but Mitchell (174) was the first to recognize that active transport can be reversed and thereby can play a central role in bioenergetics. It is difficult today to appreciate why this idea was so slow to gain general acceptance. Mitchell (175) also made the valid point that the vectorial nature of transport processes is not their unique feature, as had been thought by many investigators. He suggests that all enzyme catalysis can be regarded as vectorial with reference to the enzyme molecule. A protein molecule traversing a membrane might therefore be inherently capable of translocating a substrate across the membrane, and no radically new principle need be involved. To support this idea, Mitchell suggested a mechanism for ATP-linked H^+ transport that is based on a single confluent pathway for ATP-processing and ion translocation. In the direction of ATP synthesis it involves direct association of protons (at high $\tilde{\mu}_{H^+}$) with P_i as a means for energizing a phosphate group (176). The catalytic pathway for processing the phosphate group (combination with ADP, formation and subsequent release of ATP) is seen as automatically carrying the bound H^+ ions along, with their ultimate release (at low $\tilde{\mu}_{H^+}$) on the opposite side of the membrane. The term "ligand conduction" was used to describe the mechanism. A major difficulty with this idea has been the lack of plausible candidates for the actual energized species (177, 178).

The ligand conduction mechanism requires two separate positions on the protein molecule for accessibility to H^+ ions from opposite sides of the mechanism, but the two access sites are topologically fixed and the transport protein does not have to undergo a conformational change as an essential part of the reaction mechanism. On the other hand, there is a new complication that does not apply to the models of Figure 1, this being a necessity for a phosphate loop through the transport protein. The model was designed explicitly for the F_0F_1 system, where the catalytic H^+ uptake site (in the direction of ATP synthesis) is located on F_1 at the end of an access channel (F_0) from one side of the membrane, as shown in Figure 1c. The phosphate group that is to be protonated must have a pathway to this same site, but from the opposite side of the membrane. There is no experimental evidence that a pathway to this uniquely defined position exists, and this makes the overall proposal highly speculative. Other model mechanisms that do not involve a conformational change of the type of Figure 1a have been proposed (61, 179, 180), and they share the difficulty of requiring a phosphate loop. The model of Williams (61) has an additional element of speculation because it proposes that protons travel directly between F_0F_1 and the redox proteins in the membrane, which requires another carrier for which there is no direct evidence (62, 63).

Well-designed steady state experimental studies of the chloroplast F_0F_1 system by Hammes and coworkers (34, 181) have demonstrated that ATP synthesis in this system is kinetically consistent with an alternating access model, and with structurally noncontiguous pathways for H^+ translocation and ATP processing. There is also direct evidence from chemical and physical studies for a transport-linked conformational change for this and other F_0F_1 systems (4, 5, 8, 182). These findings support the possibility that a model with separate pathways may apply to F_0F_1. Mitchell (2) has made the reciprocal suggestion that the ligand conduction mechanism can serve as a universal mechanism, adaptable even to Na^+K^+ and Ca^{2+} pumps. He speculated that these pumps might function as proton pumps coupled to antiports (see above) or that there might be direct interaction between Na^+ or Ca^{2+} and P_i that is energetically equivalent to P_i activation by covalent association with H^+. The latter idea seems farfetched because interaction between Ca^{2+} and P_i normally leads to stable complexes of low free energy. On balance, the evidence for a confluent pathway is not impressive.

MONOMER OR OLIGOMER

The state of association of transport proteins has been a subject of considerable debate (183), and some of it is relevant to the mechanism of free energy coupling. It has been frequently suggested, for example, that the contact area between subunits of an oligomeric protein is a more likely place for creation of a translocation channel than a polar passage through a single subunit (80, 183). The structure of bacteriorhodopsin (Figure 1d) shows however, that the contact area between helical segments of a single polypeptide chain can serve this purpose equally well. This protein appears to be fully functional in its monomeric state (184) although it exists in the native membrane as a crystalline sheet composed of trimeric building blocks (100).

A more cogent reason for an oligomeric state (applicable to ATP-linked transport) is a possible need for dual catalytic sites. Boyer's mechanism for ATP processing in F_0F_1, discussed earlier, is an example of such a need. A conceptually similar reason for a dimeric state has been proposed for the gastric ATP-driven $H,^+K^+$ pump (19). The alternating access model in its simplest form would require $\Delta pK_a \sim 7$ for the H^+ uptake and discharge sites; it was suggested that such a large ΔpK_a may require two successive stages, which could be coupled to each other by means of dual sites out of phase with each other, analogous to Boyer's model for ATP processing.

Mechanisms requiring dual catalytic sites have also been proposed for ATP-driven $Na,^+K^+$ and Ca^{2+} pumps (12, 185–188), but have received little support; there is even a four-subunit model for the SR Ca^{2+} pump (189). In the case of the $Na,^+K^+$ pump, support for a dimeric functional unit is perhaps indicated by attempts to solubilize the protein in benign detergents with retention of ATPase activity; active preparations obtained so far are in a dimeric state (190, 191). Earlier data suggesting that one can bind no more than 0.5 substrate or inhibitor molecules per catalytic chain of this protein in membrane-bound form (10, 192) are, however, not supported by the most recent results (193, 194). The SR Ca^{2+} pump protein can be solubilized in unquestionably monomeric form with retention of ATPase activity (95, 195, 196). It has been claimed that cooperativity of Ca^{2+} binding is lost when this is done and that there is only a single Ca^{2+} binding site per Ca^{2+} pump polypeptide, a dimer being required for the observed stoichiometry of $2Ca^{2+}/ATP$ (197). Data from other laboratories do not support this claim (198–201). A recent inactivation study provides persuasive evidence that the functional unit for ATPase activity is monomeric in both membrane-bound and soluble forms of the protein (201). The solubilized monomeric protein also appears to be able to carry out all the steps of the normal reaction cycle, including alternation between two states with differing binding affinity for Ca^{2+} (202). This strongly implies that the structural elements required for coupling between ion translocation and ATP hydrolysis are also contained within a single polypeptide chain.

CONCLUSION

Green (203) in 1974 emphasized the need for concepts and principles that can unify bioenergetics. The work reviewed in this paper demonstrates that a conceivable unitary mechanism for all kinds of active transport and chemiosmotic biosynthesis has in fact emerged. All systems of this kind may work by means of spatially separate pathways for free energy donor and acceptor, linked through the fabric of the transport protein molecule. Ion translocation may universally involve an alternating access pathway of the type illustrated by Figure 1. The possibility that ligand conduction can serve as a unifying principle seems more remote, although it cannot be rigorously excluded as a mechanism for ATP linked H^+ translocation.

The fact that a mechanism is conceivable does not mean that it is correct, and I have tried to indicate the areas of uncertainty where decisivie experiments need to be done or convincing theoretical explanations need to be provided. The putative difference of close to 14 kcal/mol between the chemical potentials of bound phosphate in the $E \sim P$ and $E'-P$ forms of phosphoenzyme intermediates needs an explanation, as does the resistance of $E \sim P$ to direct hydrolysis. Similar problems remain with respect to ATP processing by F_0F_1 proteins. All aspects of Na^+ and K^+ binding to $Na_+^+ K^+$ pumps are still questionable, and plausible structural models for binding sites with high affinity are not available. Definitive data on the minimal state of aggregation for a fully functional protein molecule are needed for all systems. The question of whether there are occluded binding sites for transported ions needs to be resolved, and the structural basis for occlusion needs to be explained if they exist. Detailed study of any of these problems could lead to quite different views of the probable reaction mechanism from any so far suggested. It have not reviewed the large volume of existing kinetic data for bacteriorhodopsin, both because no convincing mechanism has yet emerged and because the overall reaction is in this case irreversible, which allows a far wider range of mechanistic possibilities than for reversible pumps. Nevertheless, progress in understanding of the mechanism of bacteriorhodopsin may also provide ideas that could affect our thinking about reversible pumps.

Literature Cited

1. Mitchell, P. 1963. *Biochem. Soc. Symp.* 22:142–68
2. Mitchell, P. 1979. *Eur. J. Biochem.* 95:1–20
3. Boyer, P. D., Chance, B., Ernster, L., Mitchell, P., Racker, E., Slater, E. C. 1977. *Ann. Rev. Biochem.* 46:955–1026
4. Senior, A. E. 1979. In *Membrane Proteins in Energy Transduction*, ed. R. Capaldi. 2:233–78. New York: Dekker
5. Penefsky, H. S. 1979. *Adv. Enzymol.* 49:223–80
6. Fillingame, R. H. 1980. *Ann. Rev. Biochem.* 49:1079–113
7. Cross, R. L. 1981. *Ann. Rev. Biochem.* 50:681–714
8. Nelson, N. 1981. *Curr. Top. Bioenerget.* 11:1–33
9. Kagawa, Y. 1982. *Curr. Top Memb. Transp.* 16:195–213
10. Glynn, I. M., Karlish, S. J. D. 1975. *Ann. Rev. Physiol.* 37:12–53
11. Skou, J. C., Nørby, J. G., ed. 1979. *Na, K-ATPase* New York: Academic. 549 pp.
12. Robinson, J. D., Flashner, M. S. 1979. *Biochim. Biophys. Acta* 549:146–76
13. Cantley, L. C. 1981. *Curr. Top. Bioenerget.* 11:201–37
14. Jørgensen, P. L. 1982. *Biochim. Biophys. Acta.* 694:27–68
15. Tada, M., Yamamoto, T., Tonomura, Y. 1978. *Physiol. Revs.* 58:1–79
16. Hasselbach, W. 1979. *Top. Curr. Chem.* 78:1–56
17. de Meis, L., Vianna, A. L. 1979. *Ann. Rev. Biochem.* 48:275–92
18. de Meis, L. 1981. *The Sarcoplasmic Reticulum.* New York: Wiley. 163 pp.
18a. Schatzmann, H. J. 1982. In *Membrane Transport of Calcium*, ed. E. Carafoli, pp. 41–108. New York: Academic
19. Sachs, G., Wallmark, B., Saccomani, G., Rabon, E., Stewart, H. B., DiBona, D.

R., Berglindh, T. 1982. *Curr. Top Memb. Transp.* 16:135–60
20. Sachs, G. 1981. In *Chemiosmotic Proton Circuits in Biological Membranes*, ed. V. P. Skulachev, P. C. Hinkle, pp. 347–64. Reading, Mass: Addison-Wesley
21. Goffeau, A., Slayman, C. W. 1981. *Biochim. Biophys. Acta* 639:197–223
22. Stoeckenius, W. 1980. *Acc. Chem. Res.* 13:337–44
23. Stoeckenius, W., Lozier, R. H., Bogomolni. R. A. 1979. *Biochim. Biophys. Acta* 505:215–78
24. Mullins, L. J. 1981. *Ion Transport in Heart.* New York: Raven. 136 pp.
25. Christensen, H. N. 1979. *Adv. Enzymol.* 49:41–101
26. West, I. C. 1980. *Biochim. Biophys. Acta* 604:91–126
27. Knauf, P. A. 1979. *Curr. Top. Memb. Transp.* 12:249–363
28. Hill, T. L. 1977. *Free Energy Transduction in Biology.* New York: Academic. 229 pp.
29. Hill, T. L. 1977. *Trends Biochem. Sci.* 2:204–7
30. Jencks, W. P. 1980. *Adv. Enzymol.* 51:75–106
31. Knowles, J. R. 1980. *Ann. Rev. Biochem.* 49:877–919
32. Ferguson, S. J., Sorgato, M. C. 1982. *Ann. Rev. Biochem.* 52:185–217
33. Hinkle, P. C. 1981. See Ref. 20, pp. 49–58
34. Dewey, T. G., Hammes, G. G. 1981. *J. Biol. Chem.* 256:8941–46
35. Scarborough, G. A. 1980. *Biochemistry* 19:2925–31
36. Dame, J. B., Scarborough, G. A. 1980. *Biochemistry* 19:2931–37
37. Warncke, J., Slayman, C. L. 1980. *Biochim. Biophys. Acta* 591:224–33
38. Amory, A., Goffeau, A. 1982. *J. Biol. Chem.* 257:4723–30
39. Reenstra, W. W., Forte, J. G. 1981. *J. Memb. Biol.* 61:55–60
40. Stewart, B., Wallmark, B., Sachs, G. 1981. *J. Biol. Chem.* 256:2682–90
41. Tanford, C. 1981. *J. Gen. Physiol.* 77:223–29
42. Carafoli, E., Niggli, V., Penniston, J. T. 1980. *Ann. NY Acad. Sci.* 358:159–68
43. Caroni, P., Carafoli, E. 1981. *J. Biol. Chem.* 256:3263–70
44. Niggli, V., Adunyah, E. S., Penniston, J. T., Carafoli, E. 1981. *J. Biol. Chem.* 256:395–401
45. Govindjee, R., Ebrey, T. G., Croftes, A. R. 1980. *Biophys. J.* 30:231–42
46. Bogomolni, R. A., Baker, R. A., Lozier, R. H., Stoeckenius, W. 1980. *Biochemistry* 19:2152–59
47. Kuschmitz, D., Hess, B. 1981. *Biochemistry* 20:5950–57
48. Blaustein, M. P., Santiago, E. M. 1977. *Biophys. J.* 20:79–111
49. Bridge, J. H. B., Bassingthwaithe, J. B. 1983. *Science* 219:178–80
50. Veech, R. L., Lawson, J. W. R., Cornell, N. W., Krebs, H. A. 1979. *J. Biol. Chem.* 254:6538–47
51. Foster, D. L., Fillingame, R. H. 1982. *J. Biol. Chem.* 257:2009–15
52. Kagawa, Y., Sone, N., Hirata, H., Yoshida, M. 1979. *J. Bioenerg. Biomemb.* 11:39–78
53. Baird, B. A., Hammes, G. G. 1979. *Biochim. Biophys. Acta* 549:31–53
54. Mullins, L. J. 1979. *Am. J. Physiol.* 236:C103–10
55. Wakil, S. J., Stoops, J. K., Joshi, V. C. 1983. *Ann. Rev. Biochem.* 52:537–79
56. Tanford, C. 1982. *Proc. Natl. Acad. Sci. USA* 79:6527–31
57. Tanford, C. 1981. *Proc. Natl. Acad. Sci. USA* 78:270–73
58. Taniguchi, K., Post, R. L. 1975. *J. Biol. Chem.* 250:3010–18
59. Garrahan, P. J., Glynn, I. M. 1967. *J. Physiol.* 192:237–56
60. Hasselbach, W. 1978. *Biochim. Biophys. Acta* 515:23–53
61. Williams, R. J. P. 1978. *Biochim. Biophys. Acta* 505:1–44
62. McCarty, R. E. 1981. See Ref. 20, pp. 271–81
63. Storey, B. T., Lee, C-P., Wikstrom, M. 1981. *Trends Biochem. Sci.* 6:166–70
64. Williams, R. J. P. 1981. *Trends Biochem. Sci.* 6:R10 (Letter)
65. Holian, A., Wilson, D. F. 1980. *Biochemistry* 19:4213–21
66. Wilson, D. F., Forman, N. G. 1982. *Biochemistry* 21:1438–44
67. Niggli, V., Sigel, E., Carafoli, E. 1982. *J. Biol. Chem.* 257:2350–56
68. Rossi, J. P. F. C., Schatzmann, H. J. 1982. *J. Physiol.* 327:1–15
69. Chiesi, M., Inesi, G. 1980. *Biochemistry* 19:2912–18
70. Chiu, V. C. K., Haynes, D. H. 1980. *J. Memb. Biol.* 56:219–39
71. Ueno, T., Sekine, T. 1981. *J. Biochem. Tokyo* 89:1239–52
72. Garret, C., Brethes, D., Chevalier, J. 1981. *FEBS Lett.* 136:216–20
73. Rabon, E. C., McFall, T. L., Sachs, G. 1982. *J. Biol. Chem.* 257:6296–99
74. Berman, M. C. 1982. *Biochim. Biophys. Acta.* 694:95–121
75. Lemasters, J. J. 1980. *FEBS Lett.* 110:96–100
76. Lemasters, J. J., Billica, W. H. 1981. *J. Biol. Chem.* 256:2949–57
77. Engelman, D. M., Zaccai, G. 1980. *Proc. Natl. Acad. Sci. USA* 77:5894–98

78. Vidavar, G. A. 1966. *J. Theor. Biol.* 10:301–6
79. Jardetzky, O. 1966. *Nature* 211:969–70
80. Dutton, A., Rees, E. D., Singer, S. J. 1976. *Proc. Natl. Acad. Sci. USA* 73:1532–36
81. Tanford, C. 1982. *Proc. Natl. Acad. Sci. USA* 79:2882–84
82. Stoeckenius, W., Lozier, R. H., Bogomolni, R. A. 1981. See Ref. 20, pp. 283–309
83. Kalisky, O., Ottolenghi, M., Honig, B., Korenstein, R. 1981. *Biochemistry* 20:649–55
84. Gorga, F. R., Lienhard, G. E. 1981. *Biochemistry* 20:5108–13
85. Glynn, I. M. 1956. *J. Physiol.* 134:278–310
86. Inesi, G., Kurzmack, M., Coan, C., Lewis, D. E. 1980. *J. Biol. Chem.* 255:3025–31
87. Kalbitzer, H. H., Stehlick, D., Hasselbach, W. 1978. *Eur. J. Biochem.* 82:245–55
88. Nagle, J. F., Morowitz, H. J. 1978. *Proc. Natl. Acad. Sci. USA* 75:298–302
89. Krimm, S., Dwivedi, A. M. 1982. *Science* 216:407–8
90. Nagle, J. F., Mille, M., Morowitz, H. J. 1980. *J. Chem. Phys.* 72:3959–71
91. Nagle, J. F., Mille, M. 1981. *J. Chem. Phys.* 74:1367–72
92. Criddle, R. S., Packer, L., Shieh, P. 1977. *Proc. Natl. Acad. Sci. USA* 74:4306–10
93. Nelson, N., Eytan, E., Notsani, B-E., Sigrist, H., Sigrist-Nelson, K., Gitler, C. 1977. *Proc. Natl. Acad. Sci. USA* 74:2375–78
94. Sebald, W., Hoppe, J. 1981. *Curr. Top. Bioenerget.* 12:1–64
95. Møller, J. V., Andersen, J. P., le Maire, M. 1982. *Mol. Cell. Biochem.* 42:83–107
96. Saito, A., Wang, C-T., Fleischer, S. 1978. *J. Cell. Biol.* 79:601–16
97. Brady, G. W., Fein, D. B., Harder, M. E., Meissner, G. 1982. *Biophys. J.* 37:637–45
98. Shamoo, A. E., Murphy, T. J. 1979. *Curr. Top. Bioenerget.* 9:147–77
99. Shamoo, A. E. 1978. *J. Memb. Biol.* 43:227–42
100. Henderson, R., Unwin, P. N. T. 1975. *Nature* 257:28–32
101. Mitchell, P. 1959. *Biochem. Soc. Symp.* 16:73–93
102. Weber, G. 1972. *Proc. Natl. Acad. Sci. USA* 69:3000–3
103. Tanford, C. 1982. *Proc. Natl. Acad. Sci. USA* 79:6161–65
104. Ikemoto, N. 1975. *J. Biol. Chem.* 250:7219–24
105. de Meis, L., Tume, R. K. 1977. *Biochemistry* 16:4455–63
106. Tanford, C., Martin, D. W. 1982. *Z. Naturforsch. Teil C* 37:522–26
107. Yamaguchi, M., Tonomura, Y. 1980. *J. Biochem. Tokyo* 88:1365–75
108. Matsui, H., Homaneda, H. 1982. *J. Biochem. Tokyo* 92:193–217
109. Hastings, D., Skou, J. C. 1980. *Biochim. Biophys. Acta* 601:380–85
110. Caroni, P., Reinlib, L., Carafoli, E. 1980. *Proc. Natl. Acad. Sci. USA* 77:6354–58
111. Philipson, K. D., Nishimoto, A. Y. 1982. *J. Biol. Chem.* 257:5111–17
112. Hatefi, Y., Yagi, T., Phelps, D. C., Wong, S.-Y., Vik, S. B., Galante, Y. M. 1982. *Proc. Natl. Acad. Sci. USA* 79:1756–60
113. Kretsinger, R. H. 1976. *Ann. Rev. Biochem.* 45:239–66
114. Williams, R. J. P. 1977. In *Calcium Binding Proteins and Calcium Function*, ed. R. H. Wasserman et al., pp. 3–12. New York: North-Holland
115. Midgeley, D. 1975. *Chem. Soc. Rev.* 4:549–68
116. Grisham, C. M. 1979. *Adv. Inorg. Biochem.* 1:193–218
117. Rupley, J. A., Butler, L., Gerring, M., Hartdegen, F., Pecoraro, R. 1967. *Proc. Natl. Acad. Sci. USA* 57:1088–92
118. Tanford, C., Bunville, L. G., Nozaki, Y. 1959. *J. Am. Chem. Soc.* 81:4032–36
119. Papa, S. 1976. *Biochim. Biophys. Acta* 456:39–84
120. Monod, J., Wyman, J., Changeaux, J.-P. 1965. *J. Mol. Biol.* 12:88–118
121. Bennett, W. S. Jr., Steitz, T. A. 1978. *Proc. Natl. Acad. Sci. USA* 75:4848–52
122. Mitchell, P. 1976. *Biochem. Soc. Trans.* 4:399–430
123. Bastide, F., Meissner, G., Fleischer, S., Post, R. L. 1973. *J. Biol. Chem.* 248:8385–91
124. Boyer, P. D., Ariki, M. 1980. *Fed. Proc.* 39:2410–14
125. de Meis, L., Otero, A. de S., Martins, O. B., Alves, E. W., Inesi, G., Nakamoto, R. 1982. *J. Biol. Chem.* 257:4993–98
126. Neet, K. E., Green, N. M. 1977. *Arch. Biochem. Biophys.* 178:588–97
127. Cantley, L. C., Josephsen, L., Galles, J., Cantley, L. G. 1979. See Ref. 11, pp. 181–91
128. Moczydlowski, E. G., Fortes, P. A. G. 1981. *J. Biol. Chem.* 256:2357–66
129. Skou, J. C., Esmann, M. 1980. *Biochim. Biophys. Acta* 601:386–402
130. Boyer, P. D. 1977. See Ref. 3, pp. 957–66
131. Hutton, R. L., Boyer, P. D. 1979. *J. Biol. Chem.* 254:9990–93
132. Rosen, G., Gresser, M., Vinkler, C., Boyer, P. D. 1979. *J. Biol. Chem.* 254:654–61
133. Boyer, P. D. 1979. In *Membrane*

Bioenergetics, ed. C. P. Lee, G. Schatz, L. Ernster, pp. 461–79. Reading, Mass: Addison-Wesley

134. Boyer, P. D. 1974. In *Dynamics of Energy-Transducing Membranes*, ed. L. Ernster, R. W. Estabrook, E. C. Slater, pp. 289–301. Amsterdam: Elsevier

135. Slater, E. C. 1977. See Ref. 3, pp. 1015–26

136. Slater, E. C. 1974. See Ref. 134, pp. 1–20

137. Bruist, M. F., Hammes, G. G. 1981. *Biochemistry* 20:6298–305

138. Gurd, F. R. N., Rothgeb, T. M. 1979. *Adv. Protein. Chem.* 33:74–165

139. Karplus, M., McCammon, J. A. 1981. *CRC Crit. Rev. Biochem.* 9:293–349

140. Privalov, P. L. 1979. *Adv. Protein Chem.* 33:167–241

141. Baldwin, J., Chothia, C. 1979. *J. Mol. Biol.* 129:175–220

142. Pettigrew, D. W., Romeo, P. H., Tsapis, A., Thillet, J., Smith, M. L., Turner, B. W., Ackers, G. K. 1982. *Proc. Natl. Acad. Sci. USA* 79:1849–53

143. Carilli, C. T., Farley, R. A., Perlman, D. M., Cantley, L. C. 1982. *J. Biol. Chem.* 257:5601–6

144. Miki, K., Scott, T. L., Ikemoto, N. 1981. *J. Biol. Chem.* 256:9382–85

145. Lumry, R. 1974. *Ann. NY Acad. Sci.* 227:46–73

146. Loomis, C. R., Martin, D. W., McCaslin, D. R., Tanford, C. 1982. *Biochemistry* 21:151–56

147. Pick, U., Karlish, S. J. D. 1982. *J. Biol. Chem.* 257:6120–26

148. Beaugé, L. A., Glynn, I. M. 1980. *J. Physiol.* 299:367–83

149. Jørgensen, P. L., Peterson, J. 1975. *Biochim. Biophys. Acta* 401:399–415

150. Karlish, S. J. D. 1979. See Ref. 11, pp. 115–28

151. Yamada, S., Ikemoto, N. 1978. *J. Biol. Chem.* 253:6801–7

152. Esmann, M. 1982. *Biochim. Biophys. Acta* 688:260–70

153. Dupont, Y. 1978. *Biochem. Biophys. Res. Commun.* 82:893–900

154. Gorga, F. R., Lienhard, G. E. 1982. *Biochemistry* 21:1905–8

155. Post, R. L., Hegevary, C., Kume, S. 1972. *J. Biol. Chem.* 247:6530–40

156. Glynn, I. M., Richards, D. E. 1981. *J. Physiol.* 313:P31 (Abstr.)

157. Takisawa, H., Makinose, M. 1981. *Nature* 290:271–73

158. Dupont, Y. 1980. *Eur. J. Biochem.* 109:231–38

159. Takakuwa, Y., Kanazawa, T. 1981. *J. Biol. Chem.* 256:2691–95

160. Knowles, A. F., Racker, E. 1975. *J. Biol. Chem.* 250:1949–51

161. Suko, J., Plank, B., Preis, P., Kolossa, N., Hellman, G., Conca, W. 1981. *Eur. J. Biochem.* 119:225–36

162. Takakuwa, Y., Kanazawa, T. 1981. *J. Biol. Chem.* 256:2695–700

163. de Meis, L., Inesi, G. 1982. *J. Biol. Chem.* 257:1289–94

164. Gunn, R. B., Fröhlich, O. 1979. *J. Gen. Physiol.* 74:351–74

165. Guffanti, A. A., Cohn, D. E., Kaback, H. R., Krulwich, T. A. 1981. *Proc. Natl. Acad. Sci. USA* 78:1481–84

166. Zilberstein, D., Ophis, I. J., Pordana, E., Schuldiner, S. 1982. *J. Biol. Chem.* 257:3692–96

167. Sachs, J. R. 1980. *J. Physiol.* 302:219–40

168. Hoffman, P. G., Tosteson, D. C. 1971. *J. Gen. Physiol.* 58:438–66

169. Garrahan, P. G., Garay, R. P. 1976. *Curr. Top. Memb. Transp.* 8:29–97

170. Dame, J. B., Scarborough, G. A. 1981. *J. Biol. Chem.* 256:10724–30

171. Racker, E. 1977. See Ref. 114, pp. 155–63

172. Kanazawa, T. 1975. *J. Biol. Chem.* 250:113–19

173. Martin, D. W., Tanford, C. 1981. *Biochemistry* 20:4597–602

174. Mitchell, P. 1961. *Nature* 191:144–48

175. Mitchell, P. 1963. *Biochem. Soc. Symp.* 22:142–68

176. Mitchell, P. 1977. See Ref. 3, pp. 996–1005

177. Boyer, P. D. 1975. *FEBS Lett.* 50:91–94

178. Mitchell, P. 1975. *FEBS Lett.* 50:95–97

179. Kuzlov, I. A., Skulachev, V. P. 1982. *Curr. Top. Memb. Transp.* 16:285–301

180. Green, D. E., Vande Zande, H. D. 1981. *Proc. Natl. Acad. Sci. USA* 78:5344–47

181 Takabe, T., Hammes, G. G. 1981. *Biochemistry* 20:6859–64

182. Ryrie, I. J., Jagendorf, A. T. 1972. *J. Biol. Chem.* 247:4453–59

183. Klingenberg, M. 1981. *Nature* 290:449–54

184. Dencher, N. A., Heyn, M. P. 1979. *FEBS Lett.* 108:307–10

185. Stein, W. D. 1979. See Ref. 11, pp. 475–86

186. Repke, K. R. H., Dittrich, F. 1979. See Ref. 11, pp. 487–500

187. Ikemoto, N., Garcia, A. M., Kurobe, Y., Scott, T. L. 1981. *J. Biol. Chem.* 256:8593–601

188. Ikemoto, N., Miyao, A., Kurobe, Y. 1981. *J. Biol. Chem.* 256:10809–14

189. Hill, T. L., Inesi, G. 1982. *Proc. Natl. Acad. Sci. USA* 79:3978–82

190. Hastings, D. F., Reynolds, J. A. 1979. *Biochemistry* 18:817–21

191. Esmann, M., Skou, J. C., Christiansen, C. 1979. *Biochim. Biophys. Acta* 567:410–20

192. Jørgensen, P. L. 1980. *Physiol. Revs.* 60:864–917
193. Moczydlowski, E. G., Fortes, P. A. G. 1981. *J. Biol. Chem.* 256:2346–56
194. Peters, W. H. M., Swarts, H. G. P., dePont, J. J. H. H. M., Schuurmans-Stekhoven, F. M. A. H., Bonting, S. L. 1981. *Nature* 290:338–39
195. Dean, W. L., Tanford, C. 1978. *Biochemistry* 17:1683–90
196. Martin, D. W. 1983. *Biochemistry* 22: In press
197. Watanabe, T., Lewis, D., Nakamoto, R., Kurzmack, M., Fronticelli, C., Inesi, G. 1981. *Biochemistry* 20:6617–25

198. Meissner, G. 1973. *Biochim. Biophys. Acta* 298:907–26
199. Møller, J. V., Lind, K. E., Andersen, J. P. 1980. *J. Biol. Chem.* 255:1912–20
200. Murphy, A. J., Pepitone, M., Highsmith, S. 1982. *J. Biol. Chem.* 257:3551–54
201. Andersen, J. P., Møller, J. V., Jørgensen, P. L. 1982. *J. Biol. Chem.* 257:8300–7
202. Martin, D. W. 1983. *Ann. NY Acad. Sci.* 402:573–74
203. Green, D. E. 1974. *Ann. NY Acad. Sci.* 227:5

Ann. Rev. Biochem. 1983. 52:411–39

VITAMIN D: RECENT ADVANCES[1]

Hector F. DeLuca and Heinrich K. Schnoes

Department of Biochemistry, College of Agricultural and Life Sciences, University of Wisconsin-Madison, Madison, Wisconsin 53706

CONTENTS

PERSPECTIVES AND SUMMARY 412
METABOLISM OF VITAMIN D 412
   Pathways of Vitamin D Metabolism (Figure 1) 412
   Functional Significance of Known Pathways of Metabolism 418
MECHANISM OF ACTION OF 1,25-DIHYDROXYVITAMIN D_3 IN THE
   TARGET TISSUES 420
   Localization of 1,25-$(OH)_2D_3$ in Target Tissues 420
   Receptors for 1,25-$(OH)_2D_3$ 421
   Intestinal Calcium Transport Mechanism 424
   Intestinal Phosphate Transport 426
   The Mechanism of Action of 1,25-$(OH)_2D_3$ in Bone 426
REGULATION OF VITAMIN D METABOLISM 427
   1α-Hydroxylase 427
RECENT CHEMICAL WORK IN SUPPORT OF BIOCHEMICAL STUDIES ... 428
   Metabolite Preparation and Synthetic Methods 428
   Radiochemical Synthesis 429
   New 1,25-$(OH)_2D_3$ Analogs of Importance 430
VITAMIN D AND REPRODUCTION 430
   Studies in Rats 430
   Reproduction in Chickens 432
CONCLUDING REMARKS 432

[1] Abbreviations used: 25-OH-D_3, 25-hydroxyvitamin D_3; 1,25-$(OH)_2D_3$, 1α,25-dihydroxyvitamin D_3; 24,25-$(OH)_2D_3$, 24(R),25-dihydroxyvitamin D_3; 25,26-$(OH)_2D_3$, 25(S),26-dihydroxyvitamin D_3; 25-OH-D_3-26,23-lactone, 25-hydroxyvitamin D_3-26,23-lactone; 1α-OH-D_3, 1α-hydroxyvitamin D_3; 1α-OH-D_2, 1α-hydroxyvitamin D_2; 24,24-F_2-1,25-$(OH)_2D_3$, 24,24-difluoro-1α,25-dihydroxyvitamin D_3; 26,26,26,27,27,27-F_6-1,25-$(OH)_2D_3$, 26,26,26,27,27,27-hexafluoro-1α,25-dihydroxyvitamin D_3.

411

0066-4154/83/0701-0411$02.00

PERSPECTIVES AND SUMMARY

The field of vitamin D metabolism and mechanism of action has continued to be very active. Autoradiography has shown specific nuclear localization of $1,25\text{-}(OH)_2D_3$ in target organs prior to initiation of mechanism of action. Specific nuclear localization has also been demonstrated in a variety of other tissues not previously appreciated as targets of vitamin D action, suggesting the possibility that vitamin D carries out subtle functions previously unappreciated. A macromolecule believed to be a receptor that specifically binds $1,35\text{-}(OH)_2D_3$ has been found in the cells showing nuclear localization and in a number of tumor and cancer cell lines. Since $1,25\text{-}(OH)_2D_3$ has been found to cause differentiation of certain myeloid leukemia cells, a possible relationship between the vitamin D system and cancer has appeared. Substantial evidence exists that $1,25\text{-}(OH)_2D_3$ functions in a nuclear-mediated process, although some evidence exists that not all of the actions of $1,25\text{-}(OH)_2D_3$ are carried out through such a mechanism.

Substantial advances in our understanding of the metabolism of vitamin D have also been made. The presence of significant amounts of 1α-hydroxylase has been located in the placenta in addition to the kidney. Although there have been reports of extrarenal synthesis of $1,25\text{-}(OH)_2D_3$, these sites, if they produce $1,25\text{-}(OH)_2D_3$, produce it in insufficient amounts for function. The renal 1α-hydroxylase has been solubilized and shown to be a three-component system. The 25-hydroxylase in the liver has also been solubilized and shown to be a two-component mixed-function mono-oxygenase. New pathways of vitamin D metabolism include a 23-oxidation to form $23,25\text{-}(OH)_2D_3$ or a 23-hydroxylated form of $1,25\text{-}(OH)_2D_3$. $23,25\text{-}(OH)_2D_3$ is further oxidized to produce a $25\text{-}(OH)_2D_3$-26,23-lactone. Although these pathways are of significant magnitude, their roles remain unknown since the products have low biological activity. Important analogs of the vitamin D metabolites include $24,24\text{-}F_2\text{-}25\text{-}OH\text{-}D_3$ and the $26,26,26,27,27,27\text{-}F_6\text{-}25\text{-}OH\text{-}D_3$. These have been used to show that the 24-hydroxylation, the 26-hydroxylation, and the lactone formation do not play a significant role in the function of vitamin D. Their 1-hydroxy analogs have also been prepared and shown to be extremely biologically active, being somewhere around ten times more active than the native 1,25-dihydroxyvitamin D_3, illustrating that important analogs of the vitamin D system continue to be discovered.

METABOLISM OF VITAMIN D

Pathways of Vitamin D Metabolism (Figure 1)

Because vitamin D must be converted to metabolically active forms before it can function, research on vitamin D metabolism continues to be extremely active. Vitamin D_3 is produced in skin (1, 2), primarily the malpighian layer of the epidermis, by a nonenzymatic photolysis reaction. Previtamin D_3 is the expected intermediate, and side reaction products of this photolysis, namely lumisterol and tachysterol, have also been detected (3). Previtamin D_3 is not bound by plasma transport protein for vitamin D, whereas vitamin D_3 itself is transported from skin to the liver bound to the plasma transport protein (4). The production of vitamin D in skin from the previtamin appears to involve a purely photochemical reaction, not catalyzed by enzymes or proteins, as far as is currently known (1, 4). Pigmented skin appears to be less efficient in the conversion of 7-dehydrocholesterol to vitamin D_3 per unit of absorbed light (5). The photobiogenesis of vitamin D_3 in skin has been reviewed recently (4, 6, 7).

Vitamin D_3 is absorbed with the fats in the small intestine through the lacteal system into the chylomicrons. The requirement of bile salts has recently been reaffirmed (8, 9). Vitamin D_3 accumulates in the liver (10, 11), where it undergoes its first obligatory reaction, namely 25-hydroxylation. 25-Hydroxylation is the first metabolic reaction required for all subsequent metabolism of vitamin D (12). This occurs primarily in the endoplasmic

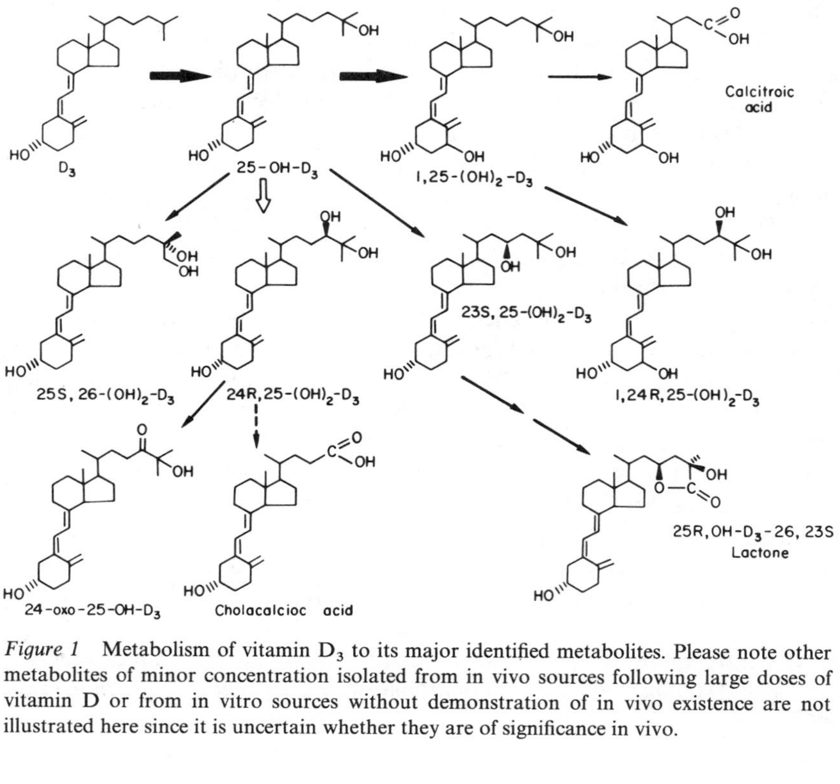

Figure 1 Metabolism of vitamin D_3 to its major identified metabolites. Please note other metabolites of minor concentration isolated from in vivo sources following large doses of vitamin D or from in vitro sources without demonstration of in vivo existence are not illustrated here since it is uncertain whether they are of significance in vivo.

reticulum by reaction requiring magnesium, NADPH, molecular oxygen, and a cytoplasmic factor (13–15). The cytoplasmic factor has been partially purified (16), but little else is known concerning its role. The specific cell types that carry out the 25-hydroxylation in liver are unknown. 25-Hydroxylation of vitamin D is also known to occur in the intestine and kidney and perhaps elsewhere (17, 18). Certainly, hepatectomy markedly reduces but does not eliminate the conversion of vitamin D to 25-hydroxyvitamin D (25-OH-D) (11). The K_m for vitamin D is 10^{-8} M and the microsomal system appears to be partially suppressed in animals given vitamin D (13, 19). The microsomal system has been solubilized and its components resolved into two enzymes, including a flavoprotein, presumably an NADPH-dependent cytochrome P-450 reductase, and a cytochrome P-450 (14). These components have not yet been purified, and thus their nature has not been elucidated. The system has, however, been reconstituted. There is also a mitochondrial vitamin D 25-hydroxylase having a K_m for vitamin D of 10^{-6} M (20, 21). This mitrochondrial system has been solubilized and shown to be a three-component mixed function monooxygenase involving an iron sulfur protein, a flavoprotein, and a cytochrome P-450 (21). This system, however, is not specific for vitamin D since it carries out other cholesterol hydroxylation reactions (22). It nevertheless will hydroxylate vitamin D in the 25-position when the substrate is present in high concentration, as for example with high doses of vitamin D_3 used in therapy. The 25-OH-D_3 rapidly leaves the liver bound to the plasma transport protein (10, 23).

The kidney is a major site of metabolism of 25-hydroxyvitamin D. The most important reaction occurring in the kidney is 1α-hydroxylation of 25-OH-D_3 to produce $1\alpha,25$-dihydroxyvitamin D_3 [$1,25$-$(OH)_2D_3$] (24, 25). This system has been solubilized and shown to be a three-component mixed-function monooxygenase requiring internally generated NADPH, molecular oxygen, and magnesium ions (26–28). The components are a flavoprotein known as renal ferredoxin reductase, an iron sulfur protein, called renal ferredoxin, and cytochrome P-450. Of the three components, only the renal ferredoxin from the chicken has been purified to homogeneity (28). It is a protein of M_r 11,800 and contains a two-iron, two-sulfur cluster (29). This system has been reconstituted and shown to be very similar to the beef adrenal steroidogenesis system. The first two components of the 1α-hydroxylase for vitamin D can be replaced by the corresponding beef adrenal proteins (26–28). As discussed below, 25-hydroxylation followed by 1α-hydroxylation are the two known activation reactions for vitamin D.

Recently, there has been much interest in possible extrarenal production of $1,25$-$(OH)_2D_3$. This was first demonstrated with anephric pregnant rats, which when injected with radiolabeled 25-OH-D$_3$ produced radiolabeled 1,25-(OH)$_2$D$_3$ (30, 31). This was followed by the demonstration of the 1α-hydroxylase in placental tissue, leaving no doubt that the placenta is a site of 1α-hydroxylation (32–34). More recently, there have been reports that bone tissue incubated in vitro with 25-OH-D$_3$ produces small amounts of 1,25-(OH)$_2$D$_3$ (35, 36). There seems little doubt that small amounts of 1,25-(OH)$_2$D$_3$ can be produced in vitro in various bone preparations. It is very likely that preparations from other tissue in vitro will also be shown to produce small amounts of 1,25-(OH)$_2$D$_3$ (37). The central issue, however, is whether this in vitro activity is expressed in vivo. Certainly nephrectomized rats injected with radiolabeled 25-OH-D$_3$ are not able to produce measurable amounts of 1,25-(OH)$_2$D$_3$ (24, 38). Because the early studies were carried out with radiolabeled 25-OH-D of low specific activity, these experiments have now been repeated using high specific activity labeled 25-OH-D$_3$. It can be shown that in anephric rats no 1,25-(OH)$_2$D$_3$ is produced from 25-OH-D$_3$ (L. Reeve, Y. Tanaka, H. F. DeLuca, submitted for publication), while under conditions of sham operation, or of ureturic ligation that produces uremia similar to that of the anephric state, considerable quantities of 1,25-(OH)$_2$D$_3$ are synthesized. Thus, it appears that except for the placenta, the extrarenal production of 1,25-(OH)$_2$D$_3$ is not significant in vivo, and the culture experiments must therefore be regarded as a biochemical curiosity. It is, however, of considerable interest that an anephric patient suffering from sarcoidosis had high blood levels of 1,25-(OH)$_2$D$_3$, which illustrated a large ectopic 1α-hydroxylase activity (39). Thus, the two in vivo exceptions to the concept of exclusive production of 1,25-(OH)$_2$D$_3$ by the kidney are the placenta and the abnormal or ectopic 1α-hydroxylation activity found in sarcoidosis. It has been reported that anephric subjects have measurable plasma levels of 1,25-(OH)$_2$D$_3$ (40). It is uncertain whether this is truly 1,25-(OH)$_2$D$_3$ in these subjects. If so, prolonged exposure of cells other than renal cells to high circulating levels of parathyroid hormone might result in induction of 1α-hydroxylase extrarenally. This result cannot be interpreted as a demonstration of significant extrarenal production of 1,25-(OH)$_2$D$_3$ under normal circumstances.

Quantitatively, the second most important reaction that occurs in kidney is 24R-hydroxylation. The 24R-hydroxylase is a cytochrome P-450-dependent reaction (41), which occurs not only in kidney but also in intestine, cartilage, and bone (42, 43). Nephrectomy does reduce the circulating levels of 24R,25-(OH)$_2$D$_3$; this illustrates that the kidney is an important contributor of this metabolite (44, 45). The ultimate disposition of this compound remains unknown, although a C-24 carboxylic acid, called cholecalcioic acid, has been isolated and identified from in vitro incubation mixtures (46). Whether this represents an important excretory product remains to be seen. The question of the relative importance of this hydroxylation from a functional point of view is discussed below.

The dihydroxylated vitamin, $25,26\text{-}(OH)_2D_3$, was isolated and identified from in vivo sources in 1970 (47). In addition to an early nonstereospecific synthesis (48), the two possible C-25-stereoisomers of this compound have also been synthesized, first by Redel and associates (49, 50) and later by Partridge et al (51) and Koizumi et al (52). Apparently, the natural form of $25,26\text{-}(OH)_2D_3$ in human plasma (51, 53) and in the plasma of experimental animals (54) is 25S. Surprisingly, this metabolite has recently been rediscovered (55) without reference to previous work—an unfortunately wasteful expenditure of research effort.

One of the newest pathways to be discovered is the functionalization of $25\text{-}OH\text{-}D_3$ at carbon 23, followed by oxidation at C-26 and lactonization to give $25\text{-}OH\text{-}D_3\text{-}26,23\text{-}lactone$ (56). This metabolite was isolated and identified from the plasma of chicks given either physiological or massive amounts of vitamin D. It has also been shown to be present in the plasma of rats (57), pigs (58), and man (59). Chemical synthesis of the lactone isomers has been accomplished (60–63). The stereochemistry of the natural product has been determined to be 23S and 25R (64). This finding has been confirmed (62). Although it was originally suggested that $25,26\text{-}(OH)_2D_3$ is a precursor of the lactone (65), work by Tanaka et al (66) clearly demonstrated that $23S,25\text{-}(OH)_2D_3$ is the intermediate in its production. The isolation and identification of $23S,25\text{-}(OH)_2D_3$ (67) together with its chemical synthesis (68) had been reported earlier, and a subsequent reidentification of the compound through the joint efforts of three laboratories led to the same result (69). Work in the Teijin Research Institute has supported the conclusion of Tanaka et al (66) that $23S,25\text{-}(OH)_2D_3$ is the intermediate in the biosynthesis of the lactone, and provided additional evidence that $23,25,26\text{-}(OH)_2D_3$ represents the next intermediate in the biosynthesis of this interesting metabolite (70). Both the $23S,25\text{-}(OH)_2D_3$ and the $25\text{-}OH\text{-}D_3\text{-}26,23\text{-}lactone$ are biologically inactive in the known systems responsive to vitamin D (H. F. DeLuca, C. M. Smith, J. K. Wichmann, unpublished results) suggesting that this entire lactonization pathway probably represents an inactivation route, although some evidence of biological activity has been presented (71).

Of considerable interest is the metabolic degradation of $1\alpha,25\text{-}(OH)_2D_3$, the hormonal form of the vitamin. Using $26,27\text{-}^{14}C$-labeled $1,25\text{-}(OH)_2D_3$, it was shown that this compound undergoes rapid and extensive side-chain oxidative cleavage (72). The product of this oxidative cleavage was isolated and identified as the C-23 carboxylic acid, calcitroic acid (73). This compound probably represents a major inactivation route since chemical synthesis (74) of calcitroic acid has been accomplished and the product shown to be biologically inactive (75). The intermediates in the process of side-chain oxidation have not yet been elucidated. $1\alpha,25\text{-}(OH)_2D_3$ also undergoes 23-oxidation in vitro to form 23-oxo-$1,25\text{-}(OH)_2D_3$, and 23-oxo-$1,25,26\text{-}(OH)_3D_3$ has also been isolated from intestinal preparations (76). This latter compound may undergo further conversion to form $1\alpha,25\text{-}(OH)_2D_3\text{-}26,23$-lactone. Thus, lactonization can take place on the $1,25\text{-}(OH)_2D_3$ molecule to form a 1α-hydroxylated form of vitamin D (77, 78). This pathway is, however, minor, if it exists at all in animals given physiologic doses of vitamin D (78).

Other minor metabolites that have been isolated and identified are: $1\alpha,25S,26\text{-}(OH)_3D_3$ (79, 80), 24-oxo-25-(OH)-D_3 (81, 82), $23,24,25\text{-}(OH)_3D_3$ (82), $24,25,26\text{-}(OH)_3D_3$ (82), 24-OH-D_3 (83), and 23-dehydro-25-OH-D_3 (82). Although these compounds can be isolated from in vitro incubations, or from animals given massive doses of vitamin D in vivo, they are probably of minor significance, and can be regarded as interesting analogs rather than significant in vivo metabolites. Figure 2 summarizes the structures of these minor known metabolites.

So far little is known concerning the pathways of excretion of the vitamin D metabolites. There is little doubt that the bile represents the major excretory route (84, 85). Only two compounds have been positively identified in the bile besides minor amounts of 25-OH-D_3, $1,25\text{-}(OH)_2D_3$ and of two dehydration products of $1,25\text{-}(OH)_2D_3$ (86). There is little doubt that a major biliary excretion product of $1,25\text{-}(OH)_2D_3$ is calcitroic acid (86), but most of the water-soluble bile products derived from D_3 remain to be identified. Furthermore, 25-OH-D_2-25-glucuronide has been isolated as a major metabolite of 25-OH-D_2 in birds (87). It must be recalled that birds discriminate against the vitamin D_2 series of compounds. The basis for the discrimination may well be rapid conversion to the 25-OH-D_2-25-glucuronide.

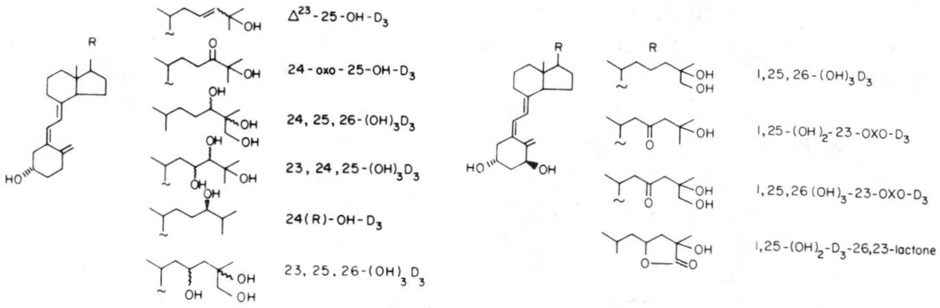

Figure 2   Minor metabolites of vitamin D_3.

5,6-*trans*-25-OH-D$_3$ has been isolated from the plasma of animals given radiolabeled vitamin D$_3$ (88) and suggested to be a bonified metabolite of vitamin D. This should be looked upon with some caution, because the report did not exclude the possibility of a nonenzymatic 5,6-*trans*-isomerization that can occur under a variety of circumstances.

Another series of compounds that have been claimed as metabolites of vitamin D are vitamin D sulfate and related compounds (89, 90). Vitamin D sulfate has been reported to be extremely biologically active, and to be the prime vitamin D active material in human breast milk and cows' milk (91–93). Chemical synthesis of vitamin D$_3$-3β-sulfate has been accomplished in two laboratories (94, 95) and both groups agree that vitamin D sulfate is essentially without biological activity. Furthermore, no vitamin D sulfate could be detected in milk using isotope dilution and isolation methods (96). Therefore, there is not yet significant evidence to establish the existence of vitamin D sulfates as bonified metabolites of vitamin D. Because of the interest in vitamin D sulfate, there have been efforts by two research groups to determine the vitamin D content of human and cows' milk (97–99). These studies do not support previous reports of high vitamin D activity in milk (91–93). Instead it appears that both human and cows' milk contain approximately 40 IU of vitamin D activity per liter (97–100), a level only slightly modified by giving larger doses of vitamin D to the mother (99). All of this vitamin D activity can be accounted for by vitamin D itself and by 25-OH-D. There is, therefore, no evidence to support the idea of a water-soluble form of vitamin D possessing high biological activity in milk.

Functional Significance of Known Pathways of Metabolism

Following the discovery of the pathways of vitamin D metabolism, there has been considerable interest in whether all or some of these pathways represent essential reactions for the function of vitamin D. The initial work with the rat showed quite clearly that 1α-hydroxylation is required for vitamin D to stimulate intestinal calcium and phosphorus transport in the small intestine and the mobilization of calcium from bone. There is no doubt that anephric rats do not respond by increasing their intestinal calcium transport, intestinal phosphate transport, or bone calcium mobilization when given physiologic amounts of 25-OH-D$_3$ (101–103). These results argue against the concept of a 1α-hydroxylase in bone cells (35, 36). There is a general concensus that 1α- and 25-hydroxylation are required for the function of vitamin D; and recent studies with certain D-analogs support this. For example, 25-fluoro-vitamin D$_3$ is considerably less active than vitamin D$_3$ itself (104). The small amount of activity observed with 25-fluoro-vitamin D$_3$ arises because the compound becomes 1α- and 24Rhydroxylated (105, 106). $1\alpha,24R$-hydroxylated vitamin D interacts very well with the intestinal cytosol receptor for $1,25\text{-}(OH)_2D_3$ (107). The $1\alpha,24$-dihydroxy-25-fluoro-vitamin D_3, therefore, is an excellent analog of $1,25$-$(OH)_2D_3$, which accounts for the biological activity seen in the fluoro-blocked compound (106). Also of considerable importance is the synthesis of the 1α-fluoro-vitamin D_3 (108). This compound has virtually no biological activity illustrating once again the importance of 1α-hydroxylation to the function of vitamin D. $1,25$-difluoro-vitamin D_3 also has no measurable biological activity (109). These results confirm the functional importance of 1α- and 25-hydroxylation.

A continuing effort is being made to establish a metabolic role for the 24R-hydroxylation of vitamin D compounds. It has been proposed that $24R,25\text{-}(OH)_2D_3$ plays an essential role in calcium homeostasis (110), in the mineralization of bone (111, 112), in the suppression of parathyroid hormone gland size (113), in the embryonic development of the chick (114), and in the growth and development of cartilage (115). These claims are based on experiments in which $1,25\text{-}(OH)_2D_3$ or combinations of $1,25$-$(OH)_2D_3$ and $24,25\text{-}(OH)_2D_3$ were administered in vivo. Other experiments used only in vitro techniques and thus might not have in vivo relevance. Perhaps the most definitive work on the importance of 24R-hydroxylation has come as a result of the chemical synthesis of $24,24\text{-}F_2\text{-}25\text{-}OH\text{-}D_3$ by two groups (116, 117). This compound cannot be converted to $24R,25\text{-}(OH)_2D_3$ (118, 119), and thus can be used to test the physiologic importance of 24R-hydroxylation. This compound proved to be as active as $25\text{-}OH\text{-}D_3$ in all known vitamin D responsive systems (118–120). It has been shown to produce normal mineralization of bone in growing animals and in young animals during early development (118, 119). A more recent extension of this investigation utilized female pups born to vitamin D-deficient rats (121, 122). These female pups were grown to maturity while receiving either $1,25\text{-}(OH)_2D_3$ administered by osmotic minipump parenterally, $25\text{-}OH\text{-}D_3$ delivered in the same manner, $24,24\text{-}F_2\text{-}25\text{-}OH\text{-}D_3$ given orally, or $25\text{-}OH\text{-}D_3$ given orally as their exclusive source of vitamin D. The female pups were grown to maturity, mated with normal males, and their pups were continued on the same regimen. These were compared with animals maintained on vitamin D-deficient diet throughout the entire period. The results of this experiment have shown that the rats supported on vitamin D compounds that cannot be converted to $24R,25\text{-}(OH)_2D_3$ are perfectly normal in every respect. Their growth is normal, blood calcium and phosphorus levels are normal, their tissues are normal, and furthermore, their bones mineralize normally (122). Histological examination has also revealed their mineralization processes to occur normally (M. Parfitt, R. Brommage, K. Jarnagin, C. Mathews, H. F. DeLuca, in preparation).

Thus, there is no evidence to support the idea that $24R,25\text{-}(OH)_2D_3$ has a significant in vivo role. If $24R,25\text{-}(OH)_2D_3$ is required for function, the function is so subtle as to escape notice through two generations of animals.

Recently, the possible role of $24R,25\text{-}(OH)_2D_3$ in the embryonic development in chicks has been examined using the 24,24-difluoro compound (123). Hens provided with $24,24\text{-}F_2\text{-}25\text{-}OH\text{-}D_3$ produce normal eggs that produce normal embryos (123). Thus, there is no solid evidence to support the idea that $24R,25\text{-}(OH)_2D_3$ plays an essential role in chick embryonic development. With respect to all these studies, it is important to note that a direct role for $25\text{-}OH\text{-}D_3$, without conversion to $1,25\text{-}(OH)_2D_3$, is possible. Thus, at this time, the only two metabolically active forms of vitamin D_3 must be regarded as $25\text{-}OH\text{-}D_3$ and $1,25\text{-}(OH)_2D_3$.

MECHANISM OF ACTION OF 1,25-DIHYDROXYVITAMIN D_3 IN THE TARGET TISSUES

Localization of $1,25\text{-}(OH)_2D_3$ in Target Tissues

This is perhaps the most active area of investigation and will continue to be so in the foreseeable future. Central to considering the mechanism is an understanding of the cellular site of localization of $1,25\text{-}(OH)_2D_3$ in the target tissues. Biochemical evidence using subcellular fractionation and tritiated $1,25\text{-}(OH)_2D_3$ of low specific activity had suggested localization in the nuclear-debris-chromatin fraction before initiation of target organ response (124–126). Chemical synthesis of radioactive $1,25\text{-}(OH)_2D_3$ having a specific activity of 160 Ci/mmol (127) permits frozen section autoradiography that eliminates the possibility of artifactual localization during cell fractionation. Furthermore, the label in the 26- and 27-positions eliminates recording of the major degradation product, namely calcitroic acid that would have lost the 26,27-tritium label. Using this labeled material, localization of $1,25\text{-}(OH)_2D_3$ in the nuclei of intestinal villus cells and crypt cells, but not submucosa, goblet cells, or smooth muscle of that organ, was demonstrated (128, 129). In addition, specific nuclear localization of $1,25\text{-}(OH)_2D_3$ was demonstrated in osteoblasts and lining cells of bone (130, 131) (but not chondrocytes of epiphyseal plate), distal renal tubular cells of the kidney (132, 133), the malpighian layer of skin (130), islet cells of the pancreas (134), certain cells of brain (135), the pituitary (130), and mammary gland (136), but not in skeletal muscle, heart muscle, liver, or spleen (130). Nuclear localization of $1,25\text{-}(OH)_2D_3$ was also detected in such tissues as the chorioallantoic membrane of eggs (137), and the endocrine cells of the stomach (130). In addition, therefore, to the known target organs of 1,25-(OH)$_2$D$_3$ (intestine, bone, kidney), specific nuclear localization was demonstrated in tissues not yet known to be target organs. This raises the possibility that 1,25-(OH)$_2$D$_3$ has more subtle actions not yet studied in detail in these other organs. In support of this, some studies have indicated 1,25-(OH)$_2$D$_3$ stimulation of prolactin secretion in GH$_4$ cells of the pituitary (138), 1,25-(OH)$_2$D$_3$-induced accumulation of 7-dehydrocholesterol in the malpighian layer of skin (139), and a function of 1,25-(OH)$_2$D$_3$ in the secretion of insulin in the islet cells of the pancreas (140). These results are all preliminary, but may suggest important functions of 1,25-(OH)$_2$D$_3$ not previously appreciated.

Of particular importance is the possible role of 1,25-(OH)$_2$D$_3$ in the parathyroid gland. Many reports have appeared regarding possible regulation of parathyroid secretion by 1,25-(OH)$_2$D$_3$. So far, this area has not been clarified and results have been published supporting either 1,25-(OH)$_2$D$_3$-dependent parathyroid hormone secretion, 1,25-(OH)$_2$D$_3$ suppression of parathyroid hormone secretion, or no action at all (141–145). The possible role of 1,25-(OH)$_2$D$_3$ in parathyroid gland cell physiology and function remains to be elucidated.

The presence of 1,25-(OH)$_2$D$_3$-dependent calcium-binding protein has now been demonstrated in a variety of tissues besides intestine and kidney. Of particular interest is its presence in skin (146), islet cells of the pancreas (147), and the parathyroid gland (148, 149). If it can be assumed that the calcium-binding protein is produced specifically in response 1,35-(OH)$_2$D$_3$, these studies also provide evidence for a role for 1,25-(OH)$_2$D$_3$ in these previously unappreciated target organs of 1,25-(OH)$_2$D$_3$ action.

Regardless of whether these sites illustrate new and unappreciated actions of 1,25-(OH)$_2$D$_3$, there is no doubt that nuclear localization of 1,25-(OH)$_2$D$_3$ in villus cells, distal renal tubule cells, and osteoblasts precedes onset of organ response. This strongly suggests that the mechanism of action of 1,25-(OH)$_2$D$_3$ in known target organs is nuclear mediated (150).

Receptors for 1,25-(OH)$_2$D$_3$

The first report of a specific receptor-like protein for 1,25-(OH)$_2$D$_3$ appeared in 1973 (151, 152) but was firmly established in 1976 (153). Since that time, many studies have been carried out demonstrating the existence in a variety of tissues of a macromolecule that specifically binds 1,25-(OH)$_2$D$_3$ with high affinity and low capacity (see 154). The first reports pertained to the small intestine of the chicken, the following to rat intestine (155, 156), rat embryonic bone (157) and rachitic chick bone (158). Thorough studies of the physical parameters of the receptor molecule (159, 160) revealed that the avian species have a receptor sedimenting at 3.7S, whereas the mammalian species have a receptor sedimenting at 3.2S. The

K_D for the receptor molecule ranges between 10^{-10} and 5×10^{-11} M (159, 160). The association rate and dissociation rate constants have been determined for the receptor from chicken, mouse, and human intestine (159–162). The selectivity of binding for 1,25-$(OH)_2D_3$ and its analogs have also been clearly demonstrated (163). The chick intestinal receptor aggregates and binds to other proteins at low salt concentrations (164). It does not aggregate under conditions of high salt and is a very labile molecule; the generation of monoclonal antibodies for this protein has been reported (165). The chick intestinal receptor has been purified to apparent homogeneity (166). Because this macromolecule binds to a large variety of substances under low salt conditions, studies involving so-called specific nuclear translocation of the receptor molecule must be viewed with considerable caution (167). Another problem that results from the behavior of the receptor in low and high salt media has surfaced in an attempt to determine the subcellular localization of the receptor prior to interaction with the ligand. Kream et al (153) first reported that under low salt conditions the receptor molecule is found with the nuclear fraction, and under high salt conditions is extracted into the cytoplasm. This finding has been further explored by Hunziker, Walters & Norman (168), who compared this behavior with that of the estrogen receptor and concluded that the nucleus may be the subcellular location of the receptor before interaction with the ligand. Unfortunately, it is not possible to carry out definitive experiments using biochemical measurements, and definitive work on location prior to interaction with the ligand must await the availability of specific antibodies to the receptor.

There has been some effort to learn about possible physiological regulation of receptor number. Thus, work from Feldman's laboratory (169) suggested that the receptor is found in highest concentration during the log phase growth period of cells and is found in lower concentration after the cells reach confluency. Glucocorticoids that inhibit intestinal calcium transport have been reported to suppress receptor levels by about 30% (170). Whether this small change in receptor number is sufficient to account for the glucocorticoid inhibition of intestinal calcium transport remains unknown.

Unfortunately, the regulation of the receptor level must be carried out in animals supplied with a source of vitamin D. The receptor, therefore, must exist both in the free and bound form. Scatchard plot analysis, used to measure receptor number, can only measure unoccupied receptor. To study the level of bound and unbound receptor, Hunziker et al (171) developed a kinetic analysis, but because of the complexity and a number of possible errors, this method unfortunately does not appear particularly accurate for these measurements (171). Coty's work (172) suggested the use of Mersalyl, a thiol blocking reagent to cause release of unlabeled $1,25\text{-}(OH)_2D_3$, followed by binding of labeled $1,25\text{-}(OH)_2D_3$. So far, however, this Mersalyl method has not been developed to a useable state for measuring occupied and unoccupied $1,25\text{-}(OH)_2D_3$ receptor. This is clearly needed before definitive results can be obtained on receptor level regulation.

Perhaps the most relevant of the recent receptor level regulation studies are those relating to a disease called vitamin D-dependency rickets Type II, and those on the development of intestinal calcium transport system in the neonatal suckling rat pup. Vitamin D-dependency rickets Type II is characterized by autosomal recessive inheritance, severe rickets with high circulating levels of $1,25\text{-}(OH)_2D_3$, and an alopecia (173, 174). Marx and collaborators (175) obtained cells from such patients and compared them with skin cells taken from normal subjects. Although skin cells from normal subjects contained the receptor to $1,25\text{-}(OH)_2D_3$, those from the affected children had no measurable receptor; this suggests that the disease likely results from a lack of the $1,25\text{-}(OH)_2D_3$ receptor (175). A study of reproduction and the role of vitamin D in the rat discovered that active intestinal calcium transport was absent during the suckling period and made its appearance only at the time of weaning, or when the animals began to eat solid food (176). Furthermore, the intestines of such animals showed a lack of responsiveness to administered vitamin D and no responsiveness to $1,25\text{-}(OH)_2D_3$ from exogenous sources (176). This contrasts with the bone calcium mobilization system that during this time is responsive to vitamin D and $1,25\text{-}(OH)_2D_3$ (176). The reason for the lack of responsiveness proved to be a lack of receptor for $1,25\text{-}(OH)_2D_3$. In the suckling rat, the receptor was absent from the small intestine during the suckling period, making its appearance at about 18 days postpartum (177). Adrenalectomy could delay the onset of receptor appearance and intestinal calcium transport, whereas injection of hydrocortisone could bring about a precocious appearance of the receptor (178). Mucosal cells explanted from suckling rats at 14 days postpartum and cultured for 24 hours do not produce a receptor. However, if hydrocortisone is included in the incubation medium, the receptor is produced in significant amounts (179). On the other hand, if intestinal cells are cultured from 16-day postpartum animals, the receptor will appear without addition of hydrocortisone, although hydrocortisone can further stimulate the receptor formation (179). Thus, it appears that the development of receptor for $1,25\text{-}(OH)_2D_3$ in the neonatal rat pup intestine is controlled both by hydrocortisone and by some intrinsic developmental programming. These two examples represent important and strong evidence that $1,25\text{-}(OH)_2D_3$ to elicit intestinal calcium transport response must interact with a receptor protein and make its appearance in the nucleus.

Intestinal Calcium Transport Mechanism

The mechanism of intestinal calcium transport is largely unknown. The response of intestinal calcium transport to $1,25\text{-}(OH)_2D_3$ in rats is biphasic in nature (180). There is a rapid response that peaks at 6 hr and decays to a low level at about 18 hr, only to have the response reappear, reach a maximum between 24–48 hr and remain high for several days. The initial (rapid) response can be reinduced by a second injection of $1,25\text{-}(OH)_2D_3$, but the second phase response is not reinduced by the second injection. The results argue, therefore, for two mechanisms of calcium transport response to $1,25\text{-}(OH)_2D_3$. One is by existing villus cells and is nuclear mediated and very rapid in nature. The second is a response of the crypt cells that are apparently programmed for calcium transport by $1,25\text{-}(OH)_2D_3$, and that retain this capacity as they progress along the villus. The existence of two mechanisms may explain the many controversial and apparently conflicting reports in the literature concerning the nature of the calcium transport response. It is entirely possible that under some circumstances the initial response has been studied, whereas in the longer adaptive studies, the longer-term response has been studied. Nevertheless, most of the mechanistic studies have been carried out on the initial response.

There has been considerable argument that the initial response to $1,25\text{-}(OH)_2D_3$ is not nuclear mediated, since its action is not blocked by the prior administration in vivo of actinomycin D of about 1 $\mu g/kg$ body weight (181–183). Unfortunately, this level of actinomycin D does not block all of the RNA formation; in fact, it is only blocked to about 30% (182). If larger amounts of actinomycin D are given, morbidity results, which makes conclusions impossible (183). Thus, the experiments involving actinomycin D and cycloheximide administration to intact animals are strongly suspect and cannot be used to determine whether transcription and translation is involved in the intestinal calcium transport response. Using intestinal organ cultures, the early work of Corradino (184) and subsequent work of Franceschi & DeLuca (185, 186) have provided strong evidence that the calcium transport response in chick embryonic intestine to $1,25\text{-}(OH)_2D_3$ is blocked by RNA and protein synthesis inhibitors in a reversible manner; this illustrates that transcription and translation are required to elicit a response in this organ. Thus, the strongest evidence suggests that a nuclear mechanism is involved in the action of $1,25\text{-}(OH)_2D_3$. Although it has been argued that the organ culture experiments are not representative of calcium transport in growing animals, the K_m for calcium is about 1 mM for both systems (185). The embryonic intestinal system lacks the component that brings about an active accumulation of calcium in an oxygen-dependent manner (185). Nevertheless, the embryonic intestinal system can be used to study the 1,25-$(OH)_2D_3$-dependent calcium uptake across the brush border membrane. In this system, it seems that 1,25-$(OH)_2D_3$ bound to the receptor somehow activates the transcription of specific genes that code for calcium transport protein(s).

Although nuclear binding of the receptor 1,25-$(OH)_2D_3$-complex has been reported (167), so far there is no good evidence on the nature of the interaction between the receptor and ligand with chromatin. The messenger RNA that is formed in response to 1,25-$(OH)_2D_3$ has not been studied except for that responsible for the formation of calcium-binding protein (187, 188). The nature of the gene products that make their appearance in response to 1,25-$(OH)_2D_3$ is largely unknown except for the calcium-binding protein of Wasserman and colleagues (189). This protein is no doubt a gene product that results from 1,25-$(OH)_2D_3$ interaction with the nucleus via the receptor (189, 190). In the chick, this protein weighs M_r 28,000 and in the rat about M_r 9000–12,000 (189). Its exact role in intestinal calcium transport remains unknown. It is now clear that this protein occurs in the cytoplasm, and available evidence suggests that it serves to protect the cell against the large fluxes of calcium that result from active calcium transport induced by 1,25-$(OH)_2D_3$ (187, 191). It is, therefore, apparent that there must be an additional protein or proteins that are induced by 1,25-$(OH)_2D_3$ that function in calcium transport. Although Lawson & Wilson (192) reported that actin is an induced protein, subsequent work has not confirmed this finding (193), and the search for other protein(s) made in response to 1,25-$(OH)_2D_3$ remains an active area of investigation.

There is little doubt that one of the sites of action of 1,25-$(OH)_2D_3$ is at the brush border membrane where calcium entry is stimulated (194). This has been clearly demonstrated by studies on isolated membrane vesicles from chick (195) and rat (196) intestinal brush borders. These vesicles show a clear increased ability to transport calcium into an osmotically active space in response to vitamin D. Goodman et al (197) and Rasmussen et al (198), however, believe that this calcium transport results from a change in lipid composition of the membrane. Thus, it has been demonstrated that the ratio of phosphatidylcholine to phosphatidylethanolamine changes in response to 1.25-$(OH)_2D_3$ (198), and there is a change in the ratio of cholesterol ester to free cholesterol (197). Exactly how the changes in lipid might stimulate or bring about increased calcium transport, or whether these changes are secondary to the transport process itself remains unclear, but represents an important phenomenon to be understood and integrated into the overall calcium transport mechanism.

There is some indication that 1,25-$(OH)_2D_3$ is involved at the basal-lateral membrane expulsion of calcium, but the evidence is not clear or strong (199, 200). There is evidence that sodium ions are required for the expulsion of calcium from the basal-lateral membrane although the exact nature of this requirement remains to be elucidated (201). Much remains to be learned concerning the intestinal calcium transport mechanism in response to $1,25\text{-}(OH)_2D_3$, and this subject will likely remain the most active area of investigation in the vitamin D field in the near future.

Intestinal Phosphate Transport

There is no doubt that $1,25\text{-}(OH)_2D_3$ stimulates intestinal phosphate transport (202–205). Sodium is required for the initial uptake process for phosphate (206), and the exact site of action of $1,25\text{-}(OH)_2D_3$ in this system remains to be determined. Thus far, no phosphate-binding protein has been discovered and the nature of this mechanism remains largely unknown.

The Mechanism of Action of $1,25\text{-}(OH)_2D_3$ in Bone

Less is known concerning the mechanism of action of $1,25\text{-}(OH)_2D_3$ in bone than in intestine. Besides nuclear localization and the presence of a receptor, as described above, little else of definitive nature is known. The rise in plasma calcium at the expense of bone in response to $1,25\text{-}(OH)_2D_3$ can be blocked by actinomycin D administration suggesting that protein and RNA synthesis is involved (207). The overall response for the mobilization of calcium from bone involving $1,25\text{-}(OH)_2D_3$ requires the presence of the parathyroid hormone (208, 209), and both hormones appear to act in concert on this system. The nature of the gene products made in response to $1,25\text{-}(OH)_2D_3$ remain unknown although there has been the report that calcium-binding protein appears in response to the vitamin D hormone (210).

Addition of $1,25\text{-}(OH)_2D_3$ to embryonic bone organ cultures causes an increase in osteoclastic-mediated bone resorption (211, 212). Exactly what this system means from a physiologic point of view remains obscure, since osteoclasts are probably not involved in the rapid adjustment of plasma calcium concentration (213). Some scientists do maintain that osteoclastic bone resorption is a significant contributor to calcium homeostatis (214), but definitive experiments are lacking. Osteoclastic-mediated bone resorption is intimately involved in the bone modeling and remodeling system (215). It is therefore likely that the embryonic organ culture responses represent the initial event in the bone modeling and remodeling systems. In the embryonic bone resorption system, $1,25\text{-}(OH)_2D_3$ and its analogs show a relative potency quite similar to their relative affinity for the intestinal receptor (216, 217).

It is possible that $1,25\text{-}(OH)_2D_3$ is involved in the differentiation of macrophages into osteoclasts. Very recently the work of Tanaka et al (218) and Abe et al (219) has shown that $1,25\text{-}(OH)_2D_3$ brings about a differentiation of myeloid leukemia cells to macrophages. This important observation suggests that 1,25-$(OH)_2D_3$ may be involved in the formation of osteoclasts, but perhaps not in the activity of osteoclasts thereafter. The action of 1,25-$(OH)_2D_3$ on the differentiation of myeloid leukemia cells, of course, suggests an important possible medical application of the 1-hydroxylated vitamin D compounds as useful drugs under some circumstances for the treatment of certain types of myeloid leukemia. It also suggests a possible role of the vitamin D hormone in the differentiation of cells that has been previously unappreciated.

REGULATION OF VITAMIN D METABOLISM

1α-Hydroxylase

It has been clearly demonstrated in vivo that 1,25-$(OH)_2D_3$ itself is an important regulator of the 25-OH-D-1α-hydroxylase of the kidney. At the same time, 1,25-$(OH)_2D_3$ appears to induce the 25-OH-D_3-24-hydroxylase (220, 221). Although 1,25-$(OH)_2D_3$ appears to be the most active in this system, other metabolites and analogs of vitamin D can also function in this capacity (222, 223). This regulatory system appears to involve nuclear activity since these responses are blocked by protein and RNA synthesis inhibitors (224). In almost every cell-culture system that has been devised to study the regulation of the 1α-hydroxylase, 1,25-$(OH)_2D_3$ additions cause the expected suppression of the 1α-hydroxylase and a stimulation of the 24-hydroxylase demonstrating that 1,25-$(OH)_2D_3$ acts directly on this system (223, 225).

Low calcium diets and hypocalcemia in intact animals results in a marked elevation of the 25-OH-D-1α-hydroxylase as demonstrated both in vivo (226) and in vitro (227, 228). Two groups have recently developed methods for measurement of mammalian 25-OH-D-1α-hydroxylase (227, 229). These have been used to demonstrate that parathyroid hormone regulates the 25-OH-D-1α-hydroxylase both in the D-deficient and vitamin D-treated animal (227). This confirms previous work carried out in the chick by Booth and colleagues (230). However, it has been very difficult to demonstrate a direct action of parathyroid hormone on cultured kidney cells from chickens or other species (231). Henry (232) reported that insulin is required for parathyroid hormone to bring about a significant increase in the 25-OH-D-1α-hydroxylase in chick kidney (232). Thus, there is little doubt that the parathyroid hormone is an important regulator of the 1α-hydroxylase in vivo, but so far experiments with cell cultures have been disappointing. The reason remains unknown.

Low phosphorus diets and hence hypophosphatemia markedly stimulate the 1α-hydroxylase measured both in vivo (233) and in vitro (227, 234).

However, the stimulation measured in vitro is much less than that seen with low calcium diets. Thus, it appears that hypophosphatemia, in addition to stimulating the 1α-hydroxylase, may affect vitamin D metabolism or function in the target tissues by some other mechanism.

The possible control of the 25-OH-D-1α-hydroxylase by other endocrine systems has also received some attention. In birds undergoing ovulation, estrogen together with progesterone and testosterone has been shown to markedly stimulate the 1α-hydroxylase (235–238). This phenomenon appears to be indirect, since addition of these sex hormones to cell cultures from chick fail to stimulate the 1α-hydroxylase (222, 223). Nevertheless, the in vivo effects of the sex hormones is so marked and clear that an understanding at the physiological and molecular level is needed.

Hypophysectomy results in a suppression of the plasma level of 1,25-$(OH)_2D_3$ that can be partially restored by growth hormone (239, 240). Some regulation is most probably exerted by the hypophysis on the 25-OH-D-1α-hydroxylase and growth hormone directly or indirectly appears to be one of the regulatory factory. There has also been the report that prolactin markedly stimulates the 25-OH-D-1α-hydroxylase (241), a phenomenon that has not been universally reproduced (242). The feeding of propyl-thiouracil, a suppressor of thyroid function, causes a suppression of 1,25-$(OH)_2D_3$ production, but increased intestinal calcium transport (243). This result suggests the existence of some other hormonal substance acting on intestinal calcium transport with consequent suppression of the 25-OH-D-1α-hydroxylase (243) due to elevated serum calcium levels. Nevertheless, the major controlling factors of the 25-OH-D-1α-hydroxylase remain hypocalcemia, hypophosphatemia, parathyroid hormone, and 1,25-$(OH)_2D_3$ itself.

RECENT CHEMICAL WORK IN SUPPORT OF BIOCHEMICAL STUDIES

With the development and the elucidation of the vitamin D metabolism picture, there has been increasing interest and activity in vitamin D synthetic chemistry. The preparation of metabolites, of structural analogs, and of radiolabeled derivatives has indeed been an important contributing factor in the rapid advances along the biochemical front. Much of the chemical work of the past decade has been reviewed previously (244–248), and the present summary will be restricted to more recent developments of possible interest to a biochemical audience.

Metabolite Preparation and Synthetic Methods

The development of efficient stereo-controlled syntheses of the various side chain-hydroxylated metabolites of vitamin D, has led to the preparation of both epimers of 23,25-$(OH)_2D_3$ (68) of 24,25-$(OH)_2D_3$ (249–252), and of 25,26-$(OH)_2D_3$ (50–54, 251); this work in turn established the configuration of the respective natural products as (23S), (24R), and (25S). Syntheses of the corresponding 1α-hydroxylated forms, i.e. of the 24R and S epimers of 1α,24,25-$(OH)_3D_3$ (253, 254), the corresponding 25-deoxy-analogs (255, 256), and the 25R and S epimers of 1α,25,26-$(OH)_3D_3$ have also been reported (257). An initial nonstereospecific preparation of the four possible isomers of the 25-OH-D_3 26,23-lactone (60) was followed by several more recent syntheses providing the four possible 23,25-stereoisomers in known configuration (61–64, 258), and established the stereochemistry of the natural product as 23S,25R (64).

There has been continuing interest in methods for the preparation of the biologically important 1α-hydroxylated metabolites or analogs. Recent efforts have focused on two approaches involving either the direct introduction of the C-1-hydroxy group into a preexisting vitamin D structure, or the total (or partial) synthesis of the carbon skeleton via coupling of 1α-hydroxylated ring A synthon to the ring C/D-portion. Exemplary of the first approach is the method of Paaren et al (259, 260), which involves selenium dioxide oxidation of 3,5-clyclovitamin D intermediates, and in a four-step process yields the desired 1α-hydroxyvitamin D compounds in about a 20% yield. Similar in concept and efficiency is the procedure of Salmond (261) involving the oxidation of vitamin D-selenide intermediates, and the method of Vanmaele et al (262) based on C-1-oxidation of triene-protected previtamin D intermediates. As recent examples of the total (or partial) synthesis approach, one may cite, in addition to the extensive investigations by Lythgoe's laboratory (245), the work of Okamura and colleagues (263) involving the thermal rearrangement of vinyl-allene intermediates for the construction of the ring A/triene system, and the elegant total synthesis of Baggiolini et al (264) in which d-carvone serves as starting material for the stereospecific construction of the 1α-hydroxy-ring A-unit.

Relevant to the problem of metabolite synthesis are also recent improvements in the preparation of 5,7-diene steroids (the provitamins) (265–267), new and efficient photochemical procedures for the conversion of provitamins to previtamins (268, 269), and an efficient photo-sensitized *trans → cis* isomerization process for the vitamin D chromophore (270).

Radiochemical Synthesis

Available synthetic methodology has been adapted to the convenient preparation of highly radiolabeled derivatives (see also 244). For example, both 25-OH-D_3 and 1α,25-$(OH)_2D_3$ have been prepared in tritium-labeled form (C-26,27) with specific activities up to 160 Ci/mmole (127, 271). The method, which involves preparation of the corresponding vitamin D-25-

carboxylic ester intermediates, and label introduction via a Grignard reaction as the last step, is highly efficient and suitable for the introduction of any isotope at C-26 and 27. A general method, based on cyclovitamin D chemistry, which provides for the conversion of essentially any vitamin D compound, to the corresponding C-6-tritium (or deuterium)-labeled form has also been reported recently (272). In utilizing highly tritium-enriched compounds for metabolism work, the tritium isotope effect must be borne in mind. For example, 26-hydroxylation is markedly suppressed by 26,27-^{3}H-substrate and C-24-tritium label has the same effect on the 24-hydroxylation (273).

New 1,25-(OH)$_2$D$_3$ Analogs of Importance

An important advance has been the synthesis of new analogs exhibiting very high bioactivity. One example is the synthesis of 24,24-difluoro-1α,25-dihydroxyvitamin D$_3$ [24,24-F$_2$-1,25-(OH)$_2$D$_3$] by two groups (274, 275), as a sequel to the synthesis of the 24-F$_2$-25-OH-D$_3$ prepared to elucidate the biological importance of 24R,25-(OH)$_2$D$_3$. The 24,24-F$_2$-1,25-(OH)$_2$D$_3$ proved to be approximately ten times more active than 1,25-(OH)$_2$D$_3$ in intestinal organ cultures in vitro (276) and in intestinal calcium transport, bone calcium mobilization and the mineralization of bone (274, 277). The reason for its increased activity remains unknown, although it is likely that the introduction of the fluoro groups at C-24 might block side-chain metabolism and hence increase the physiological half-life of the compound. Synthesis of a hexa-fluoro derivative, namely 26,26,26,27,27,27-F$_6$-1,25-(OH)$_2$D$_3$ has also been achieved (278, 279), and this compound also exhibits high potency (280, 281) and appears to be about ten times more active than 1,25-(OH)$_2$D$_3$. Because of these properties, both fluoro-analogs are likely to be extremely interesting and useful compounds, possibly as therapeutic successors to 1,25-(OH)$_2$D$_3$.

Another analog of interest, 1α-OH-D$_2$, has been known for some time (282); recent improvements in synthesis (260) should now make it available in quantity. This compound, though very similar to 1α-OH-D$_3$, appears to be less toxic in rats than 1α-OH-D$_3$. It has been suggested that this lower toxicity may result from a diminished ability to mobilize calcium from bone (283).

A summary of other known analogs is given in the review by DeLuca et al (244).

VITAMIN D AND REPRODUCTION

Studies in Rats

Plasma levels of 1,25-(OH)$_2$D$_3$ rise during the late stage of pregnancy and during lactation to very high values and then return to normal values after weaning. This has been found both in animals (284–286) and man (287, 288). As expected, intestinal calcium transport rises in direct correlation with plasma levels of 1,25-$(OH)_2D_3$ (289). It therefore appears that 1,25-$(OH)_2D_3$ is required for utilization of calcium during reproduction and lactation. To demonstrate this, experiments were carried out first to determine if vitamin D was at all required for reproduction and lactation and if so, which metabolite was required for this function. Surprisingly, vitamin D-deficient rats raised on low vitamin D diets from weaning to maturity were nevertheless able to reproduce, although not entirely normally (290, 291). There is no doubt that vitamin D is required for maximum fertility and reproductive efficiency (292, 293). It was, however, surprising that female rats and their offspring were able to survive the reproduction and lactation period without benefit of vitamin D (290). Thus, it might appear that placental transfer of calcium is not vitamin D dependent, although there are reports in sheep that it is (294). An examination of intestinal calcium transport of the vitamin D-deficient animals through pregnancy and lactation revealed that their intestinal calcium transport is increased during this period even though they had no vitamin D with which to increase it (289). This suggests the existence of some other humoral factor that might be responsible for stimulating intestinal calcium transport at this time (289). Prolactin injections into vitamin D-deficient male rats increase intestinal calcium transport and plasma calcium concentration; this suggests that prolactin may play a calcium-mobilizing role during pregnancy and lactation (295). Vitamin D-deficient mothers after weaning do not regain calcium in their bones, which remain markedly demineralized and osteoporotic in nature, whereas the D-replete animals are able to regain most of the calcium they lost during pregnancy and lactation (296). In any case, it appears that the hormones involved in the mobilization of calcium during the pregnancy and lactation period must be reexamined with the possibility of the existence of a new factor that plays the role in the mobilization of calcium that is quite different from vitamin D or the parathyroid hormone.

Of particular importance is the nature of the pups born to vitamin D-deficient mothers. They have approximately the same amount of calcium as the pups born to vitamin D-sufficient mothers (290). Their plasma calcium is somewhat lower than that of the pups from vitamin D-sufficient mothers at least for about seven days (297, 298). Furthermore, their bones, although showing somewhat greater appearance of osteoid and uncalcified epiphyseal plate, are not grossly abnormal (297, 298). Of considerable importance is that the intestine is able to absorb calcium even in the absence of vitamin D by some milk-supported process as discussed above (176). At about 7–10 days postpartum, vitamin D-deficient pups begin to show increased hypocalcemia and increased evidence of severe rickets (299). At weaning, severe hypocalcemia and a marked rachitic condition appears, despite the normal levels of calcium these animals receive in their diet. This is unlike weanling rats from normal mothers who do not develop rickets on this diet (300). On the other hand, although the intestine is unresponsive to the vitamin D hormone during the early stages of lactation (178), bone calcium mobilization apparently is vitamin D dependent since vitamin D-deficient pups are not able to control their plasma calcium when they are taken away from their dietary source of calcium (milk), whereas the vitamin D-sufficient pups are.

Reproduction in Chickens

Vitamin D-deficient birds lay decreasing numbers of eggs, which have thinner and thinner shells, until laying stops (301). There is no question that vitamin D is required for reproduction in birds. When it was attempted to learn which form of vitamin D was required for this function, $1,25\text{-}(OH)_2D_3$ was provided in increasing amounts to hens to learn that it can support the production of well-shelled eggs (302). However, when the eggs are subjected to incubation, although embryos develop, they fail to survive the third week of embryonic development, which results in a low hatchability and a high incidence of abnormal embryos (302). Thus, $1,25\text{-}(OH)_2D_3$ given to hens cannot itself support this reproductive function in the bird. This led to the addition of other forms of vitamin D to the diet in addition to the $1,25\text{-}(OH)_2D_3$. Large amounts of $24R,25\text{-}(OH)_2D_3$ added to the diet could restore hatchability and embryonic development to normal (114). On the other hand, small amounts of $25\text{-}OH\text{-}D_3$ could perform the same function (302). Furthermore, injection of a variety of forms of vitamin D into the egg itself taken from $1,25\text{-}(OH)_2D_3$-supported hens could also increase hatchability (302). Finally, $24,24\text{-}F_2\text{-}25\text{-}OH\text{-}D_3$ given to hens as their only source of vitamin D was totally able to support normal egg production, embryonic development, and hatchability (123). Since this compound cannot be 24-hydroxylated, it is evident that $24R,25\text{-}(OH)_2D_3$ is not itself required for embryonic development in the birds. It appears, therefore, that either $25\text{-}OH\text{-}D_3$ or some other of its metabolites is required by the bird to support normal reproductive function. It is entirely possible that this represents a transport problem, in which $1,25\text{-}(OH)_2D_3$ fails to get into the egg, whereas $25\text{-}OH\text{-}D_3$ is a readily transportable form of vitamin D. Continued investigation will undoubtedly elucidate this question.

CONCLUDING REMARKS

Vitamin D metabolism and mechanism of action remains an extremely active area of investigation. Several new metabolites of a minor nature have recently been isolated and identified. Advances have been made in our understanding of the molecular mechanism of action of the vitamin D hormone and especially its widespread distribution and possible new roles in unexpected target tissue. Furthermore, advances have also been made in our understanding of the role of vitamin D in a wide variety of physiologic processes. In all likelihood, the vitamin D field will continue to expand, opening many new areas of biochemical investigation into physiologic and medical processes.

ACKNOWLEDGEMENTS

Work in our laboratories was supported by program project grant AM-14881 from the National Institutes of Health, NASA grant NAG2-167 and contract NAS-9-15580, and the Harry Steenbock Research Fund of the Wisconsin Alumna Research Foundation.

Literature Cited

1. Esvelt, R. P., Schnoes, H. K., DeLuca, H. F. 1978. *Arch. Biochem. Biophys.* 188:282–86
2. Holick, M. F., Richtand, N. M., McNeill, S. C., Holick, S. A., Frommer, J. E., et al. 1979. *Biochemistry* 18:1003–8
3. Holick, M. F., MacLaughlin, J. A., Clark, M. B., Holick, S. A., Potts, J. T. Jr., et al. 1980. *Science* 210:203–5
4. Holick, M. F., Clark, M. B. 1978. *Fed. Proc.* 37:2567–74
5. Clemens, T. L., Henderson, S. L., Adams, J. S., Holick, M. F. 1982. *Lancet* 9:74–76
6. Holick, M. F. 1981. In *Hormones in Normal and Abnormal Human Tissues*, ed. K. Fotherby, S. B. Pal, pp. 223–50. New York: de Gruyter
7. Holick, M. F. 1981. *J. Invest. Dermatol.* 76:51–58
8. Dueland, S., Pedersen, J. I., Helgerud, P., Drevon, C. A. 1982. *J. Biol. Chem.* 257:146–52
9. Greaves, J. D., Schmidt, C. L. A. 1933. *J. Biol. Chem.* 102:101–12
10. Ponchon, G., DeLuca, H. F. 1969. *J. Clin. Invest.* 48:1273–79
11. Olson, E. B. Jr., Knutson, J. C., Bhattacharyya, M. H., DeLuca, H. F. 1976. *J. Clin. Invest.* 57:1213–20
12. DeLuca, H. F. 1983. In *Bone and Mineral Research*, ed. W. A. Peck, pp. 7–73. Amsterdam: Excerpta Medica
13. Madhok, T. C., DeLuca, H. F. 1979. *Biochem. J.* 184:491–99
14. Yoon, P. S., DeLuca, H. F. 1980. *Arch. Biochem. Biophys.* 203:529–41
15. Björkhem, I., Hansson, R., Holmberg, I., Wikvall, K. 1979. *Biochem. Biophys. Res. Commun.* 90:615–21
16. Miller, M. L., Ghazarian, J. G. 1980. *Biochem. Biophys. Res. Commun.* 96:1619–25
17. Tucker, G. III, Gagnon, R. E., Haussler, M. R. 1973. *Arch. Biochem. Biophys.* 155:47–57
18. Bhattacharyya, M. H., DeLuca, H. F. 1974. *Biochem. Biophys. Res. Commun.* 59:734–41
19. Bhattacharyya, M. H., DeLuca, H. F. 1973. *J. Biol. Chem.* 248:2969–73
20. Björkhem, I., Holmberg, I. 1978. *J. Biol. Chem.* 253:842–49
21. Björkhem, I., Holmberg, I., Oftebro, H., Pedersen, J. I. 1980. *J. Biol. Chem.* 255:5244–49
22. Pedersen, J. I., Björkhem, I., Gustafsson, J. 1979. *J. Biol. Chem.* 254:6464–69
23. Bouillon, R., Van Baelen, H., De Moor, P. 1980. *J. Steroid Biochem.* 13:1029–34
24. Fraser, D. R., Kodicek, E. 1970. *Nature* 228:764–66
25. Holick, M. F., Schnoes, H. K., DeLuca, H. F., Suda, T., Cousins, R. J. 1971. *Biochemistry* 10:2799–2804
26. Ghazarian, J. G., Jefcoate, C. R., Knutson, J. C., Orme-Johnson, W. H., DeLuca, H. F. 1974. *J. Biol. Chem.* 249:3026–33
27. Pedersen, J. I., Ghazarian, J. G., Orme-Johnson, N. R., DeLuca, H. F. 1976. *J. Biol. Chem.* 251:3933–41

28. Yoon, P. S., DeLuca, H. F. 1980. *Biochemistry* 19:2165–71
29. Yoon, P. S., Rawlings, J., Orme-Johnson, W. H., DeLuca, H. F. 1980. *Biochemistry* 19:2172–76
30. Weisman, Y., Vargas, A., Duckett, G., Reiter, E., Root, A. 1978. *Endocrinology* 103:1992–98
31. Gray, T. K., Lester, G. E., Lorenc, R. S. 1979. *Science* 204:1311–13
32. Tanaka, Y., Halloran, B., Schnoes, H. K., DeLuca, H. F. 1979. *Proc. Natl. Acad. Sci. USA* 76:5033–35
33. Weisman, Y., Harrell, A., Edelstein, S., David, M., Spirer, Z., Golander, A. 1979. *Nature* 281:317–20
34. Whitsett, J. A., Tsang, R. C., Norman, E. J., Adams, K. G. 1981. *J. Clin. Endocrinol. Metab.* 53:484–88
35. Howard, G. A., Turner, R. T., Sherrard, D. J., Baylink, D. J. 1981. *J. Biol. Chem.* 256:7738–40
36. Turner, R. T., Puzas, J. E., Forte, M. D., Lester, G. E., Gray, T. K., et al. 1980. *Proc. Natl. Acad. Sci. USA* 77:5720–24
37. Puzas, J. E., Turner, R. T., Forte, M. D., Kenny, A. D., Baylink, D. J. 1980. *Gen. Comp. Endocrinol.* 42:116–22
38. Gray, R., Boyle, I., DeLuca, H. F. 1971. *Science* 172:1232–34
39. Barbour, G. L., Coburn, J. W., Slatopolsky, E., Norman, A. W., Horst, R. L. 1981. *N. Engl. J. Med.* 305:440–43
40. Lambert, P. W., Stern, P., Avioli, R. C., Brockett, N. C., Turner, R. P., et al. 1982. *J. Clin. Invest.* 69:722–25
41. Kulkowski, J. A., Chan, T., Martinez, J., Ghazarian, J. G. 1979. *Biochem. Biophys. Res. Commun.* 90:50–57
42. Kumar, R., Schnoes, H. K., DeLuca, H. F. 1978. *J. Biol. Chem.* 253:3804–9
43. Garabedian, M., Lieberherr, M., Nguyen, T. M., Corvol, M. T., Dubois, M. B., Balsan, S. 1978. *Clin. Orthop. Relat. Res.* 135:241–48
44. Horst, R. L., Shepard, R. M., Jorgensen, N. A., DeLuca, H. F. 1979. *J. Lab. Clin. Med.* 93:277–85
45. Taylor, C. M., Mawer, E. B., Wallace, J. E., St. John, J., Cochran, M., et al. 1978. *Clin. Sci. Mol. Med.* 55:541–47
46. DeLuca, H. F., Schnoes, H. K. 1979. In *Vitamin D: Basic Research and its Clinical Application*, ed. A. W. Norman, K. Schaefer, D. v. Herrath, H. G. Grigoleit, J. W. Coburn, et al, pp. 445–58. Berlin: de Gruyter
47. Suda, T., DeLuca, H. F., Schnoes, H. K., Tanaka, Y., Holick, M. F. 1970. *Biochemistry* 9:4776–80
48. Lam, H.-Y., Schnoes, H. K., DeLuca, H. F. 1975. *Steroids* 25:247–56
49. Redel, J., Bell, P. A., Bazely, N., Calando, Y., Delbarre, F., Kodicek, E. 1974. *Steroids* 24:463–70
50. Cesario, M., Guilhem, J., Pascard, C., Redel, J. 1978. *Tetrahedron Lett.* 1978:1097–1101
51. Partridge, J. J., Shinney, S.-J., Chadha, N. K., Baggiolini, E. G., Blount, J. F., Uskokovic, J. J. 1981. *J. Am. Chem. Soc.* 103:1253–55
52. Koizumi, N., Morisaki, M., Ikekawa, N. 1978. *Tetrahedron Lett.* 1978:2899–902
53. Redel, J., Bazely, N., Mawer, E. B., Hann, J., Jones, F. S. 1979. *FEBS Lett.* 106:162–64
54. Redel, J., Bazely, N., Tanaka, Y., DeLuca, H. F. 1978. *FEBS Lett.* 94:228–230; 1980. (Erratum) 113:345
55. Napoli, J. L., Okita, R. T., Masters, B. S., Horst, R. L. 1981. *Biochemistry* 20:5865–71
56. Wichmann, J. K., DeLuca, H. F., Schnoes, H. K., Horst, R. L., Shepard, R. M., Jorgensen, N. A. 1979. *Biochemistry* 18:4775–80
57. Shepard, R. M., DeLuca, H. F. 1980. *Arch. Biochem. Biophys.* 202:43–53
58. Horst, R. L., Littledike, E. T. 1980. *Biochem. Biophys. Res. Commun.* 93:149–54
59. Littledike, E. T., Horst, R. L., Gray, R. W., Napoli, J. L. 1980. *Fed. Proc.* 39: 662 (Abstr.)
60. Wichmann, J. K., Paaren, H. E., Fivizzani, M. A., Schnoes, H. K., DeLuca, H. F. 1980. *Tetrahedron Lett.* 21:4667–70
61. Ikekawa, N., Hirano, Y., Ishiguro, M., Oshida, J.-I., Eguchi, T., Miyasaka, S. 1980. *Chem. Pharm. Bull.* 28:2852–54
62. Yamada, S., Nakayama, K., Takayama, H. 1981. *Chem. Pharm. Bull.* 29:2393–96
63. Morris, D. S., Williams, D. H., Norris, A. F. 1981. *J. Org. Chem.* 46:3422–28
64. Eguchi, T., Takatsuto, S., Ishiguro, M., Ikekawa, N., Tanaka, Y., DeLuca, H. F. 1981. *Proc. Natl. Acad. Sci. USA* 78:6579–83
65. Hollis, B. W., Roos, B. A., Lambert, P. W. 1980. *Biochem. Biophys. Res. Commun.* 95:520–28
66. Tanaka, Y., DeLuca, H. F., Schnoes, H. K., Ikekawa, N., Eguchi, T. 1981. *Proc. Natl. Acad. Sci. USA* 78:4805–8
67. Tanaka, Y., Wichmann, J. K., Schnoes, H. K., DeLuca, H. F. 1981. *Biochemistry* 20:3875–79
68. Ikekawa, N., Eguchi, T., Hirano, Y., Tanaka, Y., DeLuca, H. F., et al. 1981. *J. Chem. Soc. Chem. Commun.* 1981. 1157–59
69. Napoli, J. L., Pramanik, B. C., Partridge, J. J., Uskokovic, M. R.,

Horst, R. L. 1982. *J. Biol. Chem.* 257:9634–39

70. Ishizuka, S., Ishimoto, S., Norman, A. W. 1982. *FEBS Lett.* 138: 83–87
71. Ishizuki, S., Ishimoto, S., Norman, A. W. 1982. *FEBS Lett.* 139:267–70
72. Kumar, R., Harnden, D., DeLuca, H. F. 1976. *Biochemistry* 15:2420–23
73. Esvelt, R. P., Schnoes, H. K., DeLuca, H. F. 1979. *Biochemistry* 18:3977–83
74. Esvelt, R. P., Fivizzani, M. A., Paaren, H. E., Schnoes, H. K., DeLuca, H. F. 1981. *J. Org. Chem.* 46:456–58
75. Esvelt, R. P., DeLuca, H. F. 1981. *Arch. Biochem. Biophys.* 206:403–13
76. Ohnuma, N., Kruse, J., Popjak, G., Norman, A. W. 1982. *J. Biol. Chem.* 257:5097–5102
77. Ohnuma, N., Norman, A. W. 1982. *Arch. Biochem. Biophys.* 213:139–42
78. Tanaka, Y., Wichmann, J. K., Paaren, H. E., Schnoes, H. K., DeLuca, H. F. 1980. *Proc. Natl. Acad. Sci. USA* 77:6411–14
79. Tanaka, Y., Schnoes, H. K., Smith, C. M., DeLuca, H. F. 1981. *Arch. Biochem. Biophys.* 210:104–9
80. Reinhardt, T. A., Napoli, J., Praminik, B., Littledike, E. T., Beitz, D. C., et al. 1981. *Biochemistry* 20:6230–35
81. Takasaki, Y., Suda, T., Yamada, S., Takayama, H., Nishii, Y. 1981. *Biochemistry* 20:1681–86
82. Wichmann, J. K., Schnoes, H. K., DeLuca, H. F. 1981. *Biochemistry* 20:7385–91
83. Wichmann, J. K., Schnoes, H. K., DeLuca, H. F. 1981. *Biochemistry* 20:2350–53
84. Onisko, B. L., Esvelt, R. P., Schnoes, H. K., DeLuca, H. F. 1980. *Arch. Biochem. Biophys.* 205:175–79
85. DeLuca, H. F., Schnoes, H. K. 1976. *Ann. Rev. Biochem.* 45:631–66
86. Onisko, B. L., Esvelt, R. P., Schnoes, H. K., DeLuca, H. F. 1980. *Biochemistry* 19:4124–30
87. LeVan, L. W., Schnoes, H. K., DeLuca, H. F. 1981. *Biochemistry* 20:222–26
88. Kumar, R., Nagubandi, S., Jardine, I., Londowski, J. M., Bollman, S. 1981. *J. Biol. Chem.* 256:9389–92
89. Higaki, M., Takahashi, M., Suzuki, T., Sahashi, Y. 1965. *J. Vitaminol.* 11:261–65
90. Sahashi, Y., Suzuki, T., Higaki, M., Takahashi, M., Asano, T., et al. 1967. *J. Vitaminol.* 13:37–40
91. Le Boulch, N., Gulat-Marnay, C., Raoul, Y. 1974. *Int. J. Vitam. Nutr. Res.* 44:167–79
92. Lakdawala, D., Widdowson, E. M. 1977. *Lancet* 1:167–68

93. Sahashi, Y., Suzuki, T., Higaki, M., Asano, T. 1967. *J. Vitaminol.* 13:33–36
94. Reeve, L. E., DeLuca, H. F., Schnoes, H. K. 1981. *J. Biol. Chem.* 256:823–26
95. Nagubandi, S., Londowski, J. M., Bollman, S., Tietz, P., Kumar, R. 1981. *J. Biol. Chem.* 256:5536–39
96. Hollis, B. W., Roos, B. A., Draper, H. H., Lambert, P. W. 1981. *J. Nutr.* 111:384–90
97. Hollis, B. W., Roos, B. A., Draper, H. H., Lambert, P. W. 1981. *J. Nutr.* 111:1240–48
98. Reeve, L. E., Chesney, R. W., DeLuca, H. F. 1982. *Am. J. Clin. Nutr.* 36:122–26
99. Reeve, L. E., Jorgensen, N. A., DeLuca, H. F. 1982. *J. Nutr.* 112:667–72
100. Leerbeck, E., Sondergaard, H. 1980. *Br. J. Nutr.* 44:7–12
101. Boyle, I. T., Miravet, L., Gray, R. W., Holick, M. F., DeLuca, H. F. 1972. *Endocrinology* 90:605–8
102. Holick, M. F., Garabedian, M., DeLuca, H. F. 1972. *Science* 176:1146–47
103. Wong, R. G., Norman, A. W., Reddy, C. R., Coburn, J. W. 1972. *J. Clin. Invest.* 51:1287–91
104. Onisko, B. L., Schnoes, H. K., DeLuca, H. F., Glover, R. S. 1979. *Biochem. J.* 182:1–9
105. Napoli, J. L., Mellon, W. S., Fivizzani, M. A., Schnoes, H. K., DeLuca, H. F. 1979. *J. Biol. Chem.* 254:2017–22
106. Napoli, J. L., Mellon, W. S., Schnoes, H. K., DeLuca, H. F. 1979. *Arch. Biochem. Biophys.* 197:193–98
107. Smith, C. M., Tanaka, Y., DeLuca, H. F. 1982. *Proc. Soc. Exp. Biol. Med.* 170:53–58
108. Napoli, J. L., Fivizzani, M. A., Schnoes, H. K., DeLuca, H. F. 1979. *Biochemistry* 18:1641–46
109. Paaren, H. E., Fivizzani, M. A., Schnoes, H. K., DeLuca, H. F. 1981. *Arch. Biochem. Biophys.* 209:579–83
110. Norman, A. W., Henry, H. L., Malluche, H. H. 1980. *Life Sci.* 27:229–37
111. Rasmussen, H., Bordier, P. 1978. *Metab. Bone Dis. Relat. Res.* 1:7–13
112. Ornoy, A., Goodwin, D., Noff, D., Edelstein, S. 1978. *Nature* 276:517–19
113. Henry, H. L., Taylor, A. N., Norman, A. W. 1977. *J. Nutr.* 107:1918–26
114. Henry, H. L., Norman, A. W. 1978. *Science* 201:835–37
115. Corvol, M., Ulmann, A., Garabedian, M. 1980. *FEBS Lett.* 116:273–76
116. Kobayashi, Y., Taguchi, T., Terada, T., Oshida, J., Morisaki, M., Ikekawa, N. 1979. *Tetrahedron Lett.* 1978:2023–26
117. Yamada, S., Ohmori, M., Takayama,

H. 1979. *Tetrahedron Lett.* 1979:1859–62
118. Halloran, B. P., DeLuca, H. F., Barthell, E., Yamada, S., Ohmori, M., Takayama, H. 1981. *Endocrinology*, 108:2067–71
119. Okamoto, S., Tanaka, Y., DeLuca, H. F., Yamada, S., Takayama, H. 1981. *Arch. Biochem. Biophys.* 206:8–14
120. Tanaka, Y., DeLuca, H. F., Kobayashi, Y., Taguchi, T., Ikekawa, N., Morisaki, M. 1979. *J. Biol. Chem.* 254:7163–67
121. Jarnagin, K., Brommage, R., DeLuca, H. F., Yamada, S., Takayama, H. 1983. *Am. J. Physiol.* In press
122. Brommage, R., Jarnagin, K., DeLuca, H. F., Yamada, S., Takayama, H. 1983. *Am. J. Physiol.* In press
123. Ameenuddin, S., Sunde, M., DeLuca, H. F., Ikekawa, N., Kobayashi, Y. 1982. *Science* 217:451–52
124. Lawson, D. E. M., Wilson, P. W., Kodicek, E. 1969. *Biochem. J.* 115:269–77
125. Chen, T. C., Weber, J. C., DeLuca, H. F. 1970. *J. Biol. Chem.* 245:3776–80
126. Tsai, H. C., Wong, R. G., Norman, A. W. 1972. *J. Biol. Chem.* 247:5511–19
127. Napoli, J. L., Mellon, W. S., Fivizzani, M. A., Schnoes, H. K., DeLuca, H. F. 1980. *Biochemistry* 19:2515–21
128. Zile, M., Bunge, E. C., Barsness, L., Yamada, S., Schnoes, H. K., DeLuca, H. F. 1978. *Arch. Biochem. Biophys.* 186:15–24
129. Stumpf, W. E., Sar, M., Reid, F. A., Tanaka, Y., DeLuca, H. F. 1979. *Science* 206:1188–90
130. Stumpf, W. E., Sar, M., DeLuca, H. F. 1981. In *Hormonal Control of Calcium Metabolism*, ed. D. V. Cohn, R. V. Talmage, T. L. Matthews, pp. 222–29. Amsterdam/Oxford/Princeton: Excerpta Medica
131. Narbaitz, R., Stumpf, W. E., Sar, M., Huang, S., DeLuca, H. F. 1983. *Calcif. Tissue Int.* In press
132. Stumpf, W. E., Sar, M., Narbaitz, R., Reid, F. A., DeLuca, H. F., Tanaka, Y. 1980. *Proc. Natl. Acad. Sci. USA* 77:1149–53
133. Narbaitz, R., Stumpf, W. E., Sar, M., DeLuca, H. F. 1982. *Acta Anat.* 112:208–16
134. Clark, S. A., Stumpf, W. E., Sar, M., DeLuca, H. F., Tanaka, Y. 1980. *Cell Tissue Res.* 209:515–20
135. Stumpf, W. E., Sar, M., Clark, S. A., DeLuca, H. F. 1982. *Science* 215:1403–5
136. Narbaitz, R., Sar, M., Stumpf, W. E., Huang, S., DeLuca, H. F. 1981. *Horm. Res.* 15:263–69
137. Narbaitz, R., Stumpf, W. E., Sar, M.,

DeLuca, H. F., Tanaka, Y. 1980. *Gen. Comp. Endocrinol.* 42:283–89
138. Murdoch, G. H., Rosenfeld, M. G. 1981. *J. Biol. Chem.* 256:4050–55
139. Esvelt, R. P., DeLuca, H. F., Wichmann, J. K., Yoshizawa, S., Zurcher, J., et al. 1980. *Biochemistry* 19:6158–61
140. Norman, A. W., Frankel, B. J., Heldt, A. M., Grodsky, G. M. 1980. *Science* 209:823–25
141. Chertow, B. S., Baylink, D. J., Wergedal, J. E., Su, M. H. H., Norman, A. W. 1975. *J. Clin. Invest.* 56:668–78
142. Tanaka, Y., DeLuca, H. F., Ghazarian, J. G., Hargis, G. K., Williams, G. A. 1979. *Min. Electrolyte Metab.* 2:20–25
143. Golden, P., Greenwalt, A., Martin, K., Bellorin-Font, E., Mazey, R., et al. 1980. *Endocrinology* 107:602–7
144. Dietel, M., Dorn, G., Montz, R., Altenahr, E. 1979. *Endocrinology* 105:237–45
145. Canterbury, J. M., Lerman, S., Claflin, A. J., Henry, H., Norman, A., Reiss, E. 1978. *J. Clin. Invest.* 61:1375–83
146. Laouari, D., Pavlovitch, H., Deceneus, G., Balsan, S. 1980. *FEBS Lett.* 111:285–89
147. Christakos, S., Friedlander, E. J., Frandsen, B. R., Norman, A. W. 1979. *Endocrinology* 104:1495–99
148. Oldham, S. B., Fischer, J. A., Shen, L. H., Arnaud, C. D. 1974. *Biochemistry* 13:4790–97
149. Oldham, S. B., Mitnick, S. A., Coburn, J. W. 1980. *J. Biol. Chem.* 255:5789–94
150. DeLuca, H. F., Franceschi, R. T., Halloran, B. P., Massaro, E. R. 1982. *Fed. Proc.* 41:66–71
151. Brumbaugh, P. F., Haussler, M. R. 1973. *Biochem. Biophys. Res. Commun.* 51:74–80
152. Brumbaugh, P. F., Haussler, M. R. 1973. *Life Sci.* 13:1737–46
153. Kream, B. E., Reynolds, R. D., Knutson, J. C., Eisman, J. A., DeLuca, H. F. 1976. *Arch. Biochem. Biophys.* 176:779–87
154. Franceschi, R. T., Simpson, R. U., DeLuca, H. F. 1981. *Arch. Biochem. Biophys.* 210:1–13
155. Kream, B. E., DeLuca, H. F. 1977. *Biochem. Biophys. Res. Commun.* 76:735–38
156. Feldman, D., McCain, T. A., Hirst, M. A., Chen, T. L., Colston, K. W. 1979. *J. Biol. Chem.* 254:10378–84
157. Kream, B. E., Jose, M., Yamada, S., DeLuca, H. F. 1977. *Science* 197:1086–88
158. Mellon, W. S., DeLuca, H. F. 1980. *J. Biol. Chem.* 255:4081–86

159. Mellon, W. S., DeLuca, H. F. 1979. *Arch. Biochem. Biophys.* 197:90–95
160. Wecksler, W. R., Norman, A. W. 1980. *J. Biol. Chem.* 255:3571–74
161. Wecksler, W. R., Mason, R. S., Norman, A. W. 1979. *J. Clin. Endocrinol. Metab.* 48:715–17
162. Chen, T. L., Hirst, M. A., Feldman, D. 1979. *J. Biol. Chem.* 254:7491–94
163. Eisman, J. A., DeLuca, H. F. 1977. *Steroids* 30:245–57
164. Franceschi, R. T., DeLuca, H. F. 1979. *J. Biol. Chem.* 254:11629–35
165. Pike, J. W., Donaldson, C. A., Marion, S. L., Haussler, M. R. 1981. *Am. Soc. Bone Min. Res.*, 4th ed., p. S57 (Abstr.)
166. Simpson, R. U., DeLuca, H. F. 1982. *Proc. Natl. Acad. Sci. USA* 79:16–20
167. Colston, K., Feldman, D. 1980. *J. Biol. Chem.* 255:7510–13
168. Walters, M. R., Hunziker, W., Norman, A. W. 1980. *J. Biol. Chem.* 255:6799–6805
169. Chen, T. L., Feldman, D. 1981. *J. Biol. Chem.* 256:5561–68
170. Hirst, M., Feldman, D. 1982. *Biochem. Biophys. Res. Commun.* 105:1590–96
171. Hunziker, W., Walters, M. R., Norman, A. W. 1980. *J. Biol. Chem.* 255:9534–37
172. Coty, W. A. 1980. *J. Biol. Chem.* 255:8035–37
173. Bell, N. H., Hamstra, A. J., DeLuca, H. F. 1978. *N. Engl. J. Med.* 298:996–99
174. Rosen, J. F., Fleischman, A. R., Finberg, L., Hamstra, A., DeLuca, H. F. 1979. *J. Pediatr.* 94:729–35
175. Eil, C., Liberman, U. A., Rosen, J. F., Marx, S. J. 1981. *N. Engl. J. Med.* 304:1588–91
176. Halloran, B. P., DeLuca, H. F. 1980. *Am. J. Physiol.* 239:G473–79
177. Halloran, B. P., DeLuca, H. F. 1981. *J. Biol. Chem.* 256:7338–42
178. Massaro, E. R., Simpson, R. U., DeLuca, H. F. 1982. *Am. J. Physiol.* In press
179. Massaro, E. R., Simpson, R. U., DeLuca, H. F. 1982. *J. Biol. Chem.* 257:13736–39
180. Halloran, B. P., DeLuca, H. F. 1981. *Arch. Biochem. Biophys.* 208:477–86
181. Tanaka, Y., DeLuca, H. F., Omdahl, J., Holick, M. F. 1971. *Proc. Natl. Acad. Sci. USA* 68:1286–88
182. Bikle, D. D., Zolock, D. T., Morrissey, R. L., Herman, R. H. 1978. *J. Biol. Chem.* 253:484–88
183. Tsai, H. C., Midgett, R. J., Norman, A. W. 1973. *Arch. Biochem. Biophys.* 157:339–47
184. Corradino, R. A. 1973. *Nature* 243:41–43
185. Franceschi, R. T., DeLuca, H. F. 1981. *J. Biol. Chem.* 256:3840–47
186. Franceschi, R. T., DeLuca, H. F. 1981. *J. Biol. Chem.* 256:3848–52
187. Spencer, R., Charman, M., Wilson, P. W., Lawson, D. E. M. 1978. *Biochem. J.* 170:93–102
188. Thomasset, M., Desplan, C., Moukhtar, M., Mathieu, H. 1981. *FEBS Lett.* 134:178–80
189. Wasserman, R. H., Feher, J. J. 1977. In *Calcium Binding Proteins and Calcium Function*, ed. R. H. Wasserman, R. A. Corradino, E. Carafoli, R. H. Kretsinger, D. H. MacLennan, S. L. Siegel, pp. 292–302. New York: Elsevier
190. Bishop, C. W., Kendrick, N. C., DeLuca, H. F. 1983. *J. Biol. Chem.* In press
191. Taylor, A. N. 1981. *J. Histochem. Cytochem.* 29:65–73
192. Wilson, P. W., Lawson, D. E. M. 1978. *Biochem. J.* 173:627–31
193. Kendrick, N. C., Barr, C. R., Moriarity, D., DeLuca, H. F. 1981. *Biochemistry* 20:5288–94
194. DeLuca, H. F. 1978. In *Calcium Transport and Cell Function*, ed. A. Scarpa, E. Carafoli, pp. 356–76. New York: NY Acad. Sci.
195. Rasmussen, H., Fontaine, O., Max, E., Goodman, D. B. P. 1979. *J. Biol. Chem.* 254:2993–99
196. Bronner, F., Lipton, J., Pansu, D., Buckley, M., Singh, R., Miller, A. III. 1982. *Fed. Proc.* 41:61–65
197. Goodman, D. B. P., Haussler, M. R., Rasmussen, H. 1972. *Biochem. Biophys. Res. Commun.* 46:80–86
198. Rasmussen, H., Matsumoto, T., Fontaine, O., Goodman, D. B. P. 1982. *Fed. Proc.* 41:72–77
199. Schachter, D., Kowarski, S., Finkelstein, J. D., Wang, Ma, R. 1966. *Am. J. Physiol.* 211:1131–36
200. Ghijsen, W. E. J. M., Van Os, C. H. 1982. *Biochim. Biophys. Acta* 689:170–75
201. Martin, D. L., DeLuca, H. F. 1969. *Am. J. Physiol.* 216:1351–59
202. Harrison, H. E., Harrison, H. C. 1961. *Am. J. Physiol.* 201:1007–12
203. Chen, T. C., Castillo, L., Korycka-Dahl, M., DeLuca, H. F. 1974. *J. Nutr.* 104:1056–60
204. Kowarski, S., Schachter, D. 1969. *J. Biol. Chem.* 244:211–17
205. Wasserman, R. H., Taylor, A. N. 1973. *J. Nutr.* 103:586–99
206. Taylor, A. N. 1974. *J. Nutr.* 104:489–94
207. Tanaka, Y., DeLuca, H. F. 1971. *Arch. Biochem. Biophys.* 146:574–78
208. Rasmussen, H., DeLuca, H. F., Arnaud,

C., Hawker, C., von Stedingk, M. 1963. *J. Clin. Invest.* 42:1940–46

209. Garabedian, M., Tanaka, Y., Holick, M. F., DeLuca, H. F. 1974. *Endocrinology* 94:1022–27

210. Christakos, S., Norman, A. W. 1978. *Science* 202:70–71

211. Raisz, L. G., Trummel, C. L., Holick, M. F., DeLuca, H. F. 1972. *Science* 175:768–69

212. Stern, P. H., Trummel, C. L., Schnoes, H. K., DeLuca, H. F. 1975. *Endocrinology* 97:1552–58

213. Talmage, R. V. 1969. *Clin. Orthop. Relat. Res.* 67:210–24

214. Holtrop, M. F., Cox, K. A., Clark, M. B., Holick, M. F., Anast, C. S. 1981. *Endocrinology* 108:2293–97

215. Frost, H. M. 1966. *Bone Dynamics in Osteoporosis and Osteomalacia.* Henry Ford Hosp. Surg. Monogr. Ser., Springfield: Thomas

216. Stern, P. H., Tanaka, Y., DeLuca, H. F., Ikekawa, N., Kobayashi, Y. 1981. *Mol. Pharmacol.* 20:460–62

217. Stern, P. H., Mavreas, T., Trummel, C. L., Schnoes, H. K., DeLuca, H. F. 1977. *Mol. Pharmacol.* 12:879–86

218. Tanaka, H., Abe, E., Miyaura, C., Kuribayashi, T., Konno, K., et al. 1982. *Biochem. J.* 204:713–19

219. Abe, E., Miyaura, C., Sakagami, H., Takeda, M., Konno, K., et al. 1981. *Proc. Natl. Acad. Sci. USA* 78:4990–94

220. Tanaka, Y., Lorenc, R. S., DeLuca, H. F. 1975. *Arch. Biochem. Biophys.* 171:521–26

221. Colston, K. W., Evans, I. M. A., Spelsberg, T. C., MacIntyre, I. 1977. *Biochem. J.* 164:83–90

222. Henry, H. L. 1977. In *Vitamin D: Biochemical, Chemical and Clinical Aspects Related to Calcium Metabolism,* ed. A. W. Norman, K. Schaefer, J. W. Coburn, H. F. DeLuca, D. Fraser, et al, pp. 125–33. Berlin: de Gruyter

223. Trechsel, U., Bonjour, J.-P., Fleisch, H. 1979. *J. Clin. Invest.* 64:206–17

224. Larkins, R. G., MacAuley, S. J., MacIntyre, I. 1977. *Mol. Cell. Endocrinol.* 2:193–202

225. Juan, D., DeLuca, H. F. 1977. *Endocrinology* 101:1184–93

226. Boyle, I. T., Gray, R. W., DeLuca, H. F. 1971. *Proc. Natl. Acad. Sci. USA* 68:2131–34

227. Tanaka, Y., DeLuca, H. F. 1981. *Proc. Natl. Acad. Sci. USA* 78:196–99

228. Midgett, R. J., Spielvogel, A. M., Coburn, J. W., Norman, A. W. 1973. *J. Clin. Endocrinol. Metab.* 36:1153–61

229. Vieth, R., Fraser, D. 1979. *J. Biol. Chem.* 254:12455–50

230. Booth, B. E., Tsai, H. C., Morris, C.

1977. *J. Clin. Ivest.* 60:1314–20

231. Henry, H. L. 1979. *J. Biol. Chem.* 254:2722–29

232. Henry, H. L. 1981. *Endocrinology* 108:733–36

233. Tanaka, Y., DeLuca, H. F. 1973. *Arch. Biochem. Biophys.* 154:566–74

234. Baxter, L. A., DeLuca, H. F. 1976. *J. Biol. Chem.* 251:3158–61

235. Castillo, L., Tanaka, Y., DeLuca, H. F., Sunde, M. L. 1977. *Arch. Biochem. Biophys.* 179:211–17

236. Tanaka, Y., Castillo, L., Wineland, M. J., DeLuca, H. F. 1978. *Endocrinology* 103:2035–39

237. Tanaka, Y., Castillo, L., DeLuca, H. F. 1976. *Proc. Natl. Acad. Sci. USA* 73:2701–5

238. Baksi, S. N., Kenny, A. D. 1977. *Endocrinology* 101:1216–20

239. Pahuja, D. N., DeLuca, H. F. 1981. *Mol. Cell. Endocrinol.* 23:345–50

240. Spanos, E., Barrett, D., MacIntyre, I., Pike, J. W., Safilian, E. F., Haussler, M. R. 1978. *Nature* 273:246–47

241. Spanos, E., Colston, K. W., Evans, I. M. A., Galante, L. S., MacAuley, S. J., MacIntyre, I. 1976. *Nol. Cell. Endocrinol.* 5:163–67

242. Matsumoto, T., Horiuchi, N., Suda, T., Takahashi, H., Shimazawa, E., Ogata, E. 1979. *Metab. Clin. Exp.* 28:925–27

243. Pahuja, D. N., DeLuca, H. F. 1982. *Arch. Biochem. Biophys.* 213:293–98

244. DeLuca, H. F., Paaren, H. E., Schnoes, H. K. 1979. *Top. Curr. Chem.* 83:1–65

245. Lythgoe, B. 1980. *Chem. Soc. Rev.* 9:449–75

246. Bell, P. A. 1978. In *Vitamin D,* ed. D. E. M. Lawson. New York: Academic

247. Norman, A. W. 1979. *Vitamin D: The Calcium-Homeostatic Hormone.* New York: Academic

248. Yakhimovich, R. I. 1980. *Russ. Chem. Rev.* 49:371–83

249. Partridge, J. J., Toome, V., Uskokovic, M. R. 1976. *J. Am. Chem. Soc.* 98:3739–41

250. Seki, M., Koizumi, N., Ikekawa, N., Takeshita, T., Ishimoto, S. 1975. *Tetrahedron Lett.* 1975:15–18

251. Ishiguro, M., Koizumi, N., Yasuda, M., Ikekawa, N. 1981. *J. Chem. Soc. Chem. Commun.* 1981:115–17

252. Takayama, H., Ohmori, M., Yamada, S. 1980. *Tetrahedron Lett.* 1980:5027–28

253. Ikekawa, N., Morisaki, M., Koizumi, N., Kato, Y., Takeshita, T. 1975. *Chem. Pharm. Bull.* 23:695–97

254. Partridge, J. J., Shiuey, S.-J., Baggiolini, E. G., Hennessy, B., Uskokovic, M. R. 1977. See Ref. 222, pp. 47–55

255. Morisaki, M., Koizumi, N., Ikekawa,

N., Takeshita, T., Ishimoto, S. 1975. *J. Chem. Soc. Perkin Trans. I* 1975:1421–24

256. Ochi, K., Matsunaga, I., Shindo, M., Kaneko, C. 1978. *Chem. Pharm. Bull.* 26:2386–88

257. Partridge, J. J., Shiuey, S.-J., Chadka, N. K., Baggiolini, E. G., Hennessy, et al. 1981. *Helv. Chim. Acta* 64:2138–41

258. Eguchi, T., Tahatsuto, S., Hirano, Y., Ishiguro, M., Ikekawa, N. 1982. *Heterocycles* 17:359–75

259. Paaren, H. E., Hamer, D. E., Schnoes, H. K., DeLuca, H. F. 1978. *Proc. Natl. Acad. Sci. USA* 75:2080–81

260. Paaren, H. E., DeLuca, H. F., Schnoes, H. K. 1980. *J. Org. Chem.* 45:3253–58

261. Salmond, W. G. 1979. See Ref. 46, pp. 25–31

262. Vanmaele, L. J., De Clercq, P. J., Vandewalle, M. 1982. *Tetrahedron Lett.* 23:995–98

263. Condron, P. Jr., Hammond, M. L., Mourino, A., Okamura, W. H. 1980. *J. Am. Chem. Soc.* 102:6259–67

264. Baggiolini, E. G., Iacobelli, J. A., Hennessy, B. M., Uskokovic, M. R. 1982. *J. Am. Chem. Soc.* 104:2945–49

265. Salmond, W. G., Barta, M. A., Havens, J. L. 1978. *J. Org. Chem.* 54:2057–59

266. Confalone, P. N., Kulesha, I. D., Uskokovic, M. R. 1981. *J. Org. Chem.* 46:1030–32

267. Salmond, W. G., Barta, M. A., Cain, A. M., Sobola, M. C. 1977. *Tetrahedron Lett.* 1977:1683–86

268. Malatesta, V., Willis, C., Hachett, P. A. 1981. *J. Am. Chem. Soc.* 103:6781–83

269. Dauben, W. G., Phillips, R. B. 1982. *J. Am. Chem. Soc.* 104:355–56

270. Gielen, J. W. J., Koolstra, R. B., Jacobs, H. J. C., Havinga, E. 1980. *Rec. J. Roy. Neth. Chem. Soc.* 99:306–11

271. Napoli, J. L., Fivizzani, M. A., Hamstra, A. J., Schnoes, H. K., DeLuca, H. F. 1979. *Anal. Biochem.* 96:481–88

272. Paaren, H. E., Fivizzani, M. A., Schnoes, H. K., DeLuca, H. F. 1981. *Proc. Natl. Acad. Sci. USA* 78:6173–75

273. Tanaka, Y., DeLuca, H. F. 1982. *Arch. Biochem. Biophys.* 213:517–22

274. Tanaka, Y., DeLuca, H. F., Schnoes, H. K., Ikekawa, N., Kobayashi, Y. 1980. *Arch. Biochem. Biophys.* 199:473–78

275. Yamada, S., Ohmori, M., Takayama, H. 1979. *Chem. Pharm. Bull.* 27:3196–98

276. Corradino, R. A., Ikekawa, N., DeLuca, H. F. 1981. *Arch. Biochem. Biophys.* 208:273–77

277. Okamoto, S., Tanaka, Y., DeLuca, H. F., Kobayashi, Y., Ikekawa, N. 1983. *Am. J. Physiol.* In press

278. Kobayashi, Y., Taguchi, T., Kanuma, N., Ikekawa, N., Oshida, J.-I. 1980. *J. Chem. Soc. Chem. Commun.* 1980:459–60

279. Kobayashi, Y., Taguchi, T., Mitsuhashi, S., Eguchi, T., Ohshima, E., et al. 1982. *Chem. Pharm. Bull.* 30:4297–4302

280. Tanaka, Y., Pahuja, D. N., Wichmann, J. K., DeLuca, H. F., Kobayashi, Y., et al. 1982. *Arch. Biochem. Biophys.* 218:134–41

281. Tanaka, Y., Faber, D. B., DeLuca, H. F., Kobayashi, Y., Ikekawa, N. 1982. *Fed. Proc.* 41:7731 (Abstr.)

282. Lam, H.-Y., Schnoes, H. K., DeLuca, H. F. 1977. *Steroids* 30:671–77

283. Reeve, L. E., Schnoes, H. K., DeLuca, H. F. 1978. *Arch. Biochem. Biophys.* 186:164–67

284. Halloran, B. P., Barthell, E. N., DeLuca, H. F. 1979. *Proc. Natl. Acad. Sci. USA* 76:5549–53

285. Pike, J. W., Parker, J. B., Haussler, M. R., Boass, A., Toverud, S. U. 1979. *Science* 204:1427–29

286. Boass, A., Toverud, S. U., McCain, T. A., Pike, J. W., Haussler, M. R. 1977. *Nature* 267:630–31

287. Kumar, R., Cohen, W. R., Epstein, F. H. 1980. *N. Engl. J. Med.* 302:1143–45

288. Lund, B., Selnes, A. 1979. *Acta Endocrinol.* 92:330–35

289. Halloran, B. P., DeLuca, H. F. 1980. *Am. J. Physiol.* 239:E64–68

290. Halloran, B. P., DeLuca, H. F. 1979. *Science* 204:73–74

291. Boass, A., Toverud, S. U., Pike, J. W., Haussler, M. R. 1981. *Endocrinology* 109:900–4

292. Halloran, B. P., DeLuca, H. F. 1980. *J. Nutr.* 110:1573–80

293. Brommage, R., Neuman, W. F. 1981. *Calcif. Tissue Int.* 33:277–80

294. Ross, R. A., Care, D., Robinson, J. S., Pickard, D. W., Weatherley, A. J. 1980. *J. Endocrinol.* 87:P17–19

295. Pahuje, D. N., DeLuca, H. F. 1981. *Science* 214:1038–39

296. Halloran, B. P., DeLuca, H. F. 1980. *Endocrinology* 107:1923–29

297. Halloran, B. P., DeLuca, H. F. 1981. *Arch. Biochem. Biophys.* 209:7–14

298. Boass, A., Ramp, W. K., Toverud, S. U. 1981. *Endocrinology* 109:505–8

299. Miller, S. C., Halloran, B. P., DeLuca, H. F., Jee, W. S. S. 1982. *Calcif. Tissue Int.* In press

300. Steenbock, H., Herting, D. C. 1955. *J. Nutr.* 57:449–68

301. Bell, D. J., Freeman, B. M., eds. 1971. *Physiology and Biochemistry of the Domestic Fowl.* Vol. 3. London: Academic

302. Sunde, M. L., Turk, C. M., DeLuca, H. F. 1978. *Science* 200:1067–69

Ann. Rev. Biochem. 1983. 52:441–66

THE PATHWAY OF EUKARYOTIC mRNA FORMATION

Joseph R. Nevins

The Rockefeller University, New York, N.Y. 10021

CONTENTS

PERSPECTIVES AND SUMMARY 441
TRANSCRIPTIONAL UNIT DESIGN 443
    Elements of a Transcriptional Unit 443
    Mapping Transcriptional Units 443
    Promoters 444
    Transcriptional Unit Complexity 447
FORMATION OF THE PRIMARY TRANSCRIPT 447
    Initiation of Transcription 447
    Termination 449
    Poly(A) Site Selection 452
PROCESSING OF THE PRIMARY TRANSCRIPT 452
    Poly(A) Addition 452
    Splicing 454
    Capping and Methylation 458
NUCLEAR-CYTOPLASMIC TRANSPORT 459
CYTOPLASMIC STABILITY OF mRNA 460
    Role of Poly(A) in mRNA Stability 461
CONCLUDING REMARKS 462

PERSPECTIVES AND SUMMARY

The actual formation of a prokaryotic messenger RNA (mRNA) is a relatively simple process involving mainly the transcription of an appropriate gene by RNA polymerase. In most instances the primary transcript of the gene is the mRNA and functions as such even before the completion of the primary transcript. As much as any other difference

441

0066/4154/83/0701-0441$02.00

between a eukaryotic and a prokaryotic organism, the pathways for formation of messenger RNA diverge. In eukaryotic cells, the primary transcript must be processed in a myriad of ways before ending as a mature mRNA; furthermore, the final processed product must then be transported from one cellular compartment to another in order to carry out its function. Only then does the eukaryotic mRNA possess the properties of the still nascent *Escherichia coli* gene transcript.

Of central importance to the study of gene expression in both pro-karyotes and eukaryotes are the mechanisms by which the products of genes are regulated. The control of the expression of a particular eukaryotic gene is reflected as the concentration of functional mRNA in the cytoplasm. Thus, regulation can occur by modulating any of the many steps in mRNA biogenesis. This therefore is the reason for exploring the many details of RNA processing and metabolism: If one hopes to understand the regulation of eukaryotic gene expression, one must know all of the steps involved.

In recent years, a wealth of information has accumulated, documenting the events of eukaryotic mRNA biogenesis. The initial mapping of transcriptional units has led to the definition of promoters and sites for transcriptional initiation. The various events comprising RNA processing have been detailed: poly(A) addition, RNA splicing, and methylation. Finally, the process of nuclear-cytoplasmic transport has been studied, although it has yet to be fully understood. In several instances, specific events in mRNA formation have been shown to participate in the regulation of gene expression. Changes in transcriptional initiation, termination, and RNA splicing have been shown to fluctuate in specific situations, in an apparently regulated fashion. Even after the formation of a mRNA is complete, changes in the stability of the mRNA can alter the expression of the gene. It is thus becoming clear that the complex pathway of mRNA biogenesis itself may serve a regulatory role.

The purpose of this review is to describe the pathway involved in the formation of a eukaryotic mRNA by collating the pertinent facts concerning the various events of mRNA biogenesis. Many of the primary results were from the viral systems, most notably adenovirus and SV40. However, principally owing to advances in recombinant DNA technology, the more recent results draw from the variety of available gene systems. The emphasis of this review is on the events of transcription and RNA processing, i.e. those events of mRNA formation. The scope of the work to be discussed is limited, for the most part, to the higher eukaryotic systems. Finally, the involvement of the various steps of mRNA biogenesis in the regulation of gene expression is the underlying theme.

TRANSCRIPTIONAL UNIT DESIGN

Elements of a Transcriptional Unit

A consideration of the mechanism of formation of a eukaryotic mRNA must necessarily first consider the functional unit encoding the information contained in the mRNA, the transcriptional unit. Figure 1 shows a schematic of a "consensus" RNA polymerase II transcriptional unit. A transcriptional unit is that region of a genome containing the appropriate signals for the generation of a primary transcript, notably those for transcriptional initiation and termination. Thus, of interest to the discussion of the structure of a transcriptional unit is the nature of the promoter and the terminator for RNA polymerase. Although a considerable body of information has accumulated concerning the structure and function of eukaryotic promoters, there is yet no evidence for a termination signal for transcription by RNA polymerase II.

Mapping Transcriptional Units

Although we presently know a great deal concerning the nature of a promoter for RNA polymerase II, owing largely to recombinant DNA experiments and DNA sequencing, this knowledge only followed the initial descriptions and mapping of transcriptional units and in particular the start sites for transcription. The techniques employed to define and map RNA polymerase II transcriptional units were based on the same principle: nascent transcripts have in common a defined starting point. Two techniques were employed to define this point: UV transcription mapping (1) and RNA nascent chain analysis (2). Both procedures depend upon the ability to analyze briefly labeled RNA, particularly for the production and analysis of nascent chains, so as to avoid the problems of rapid degradation of processed portions of primary transcripts. As such, these experiments have largely utilized adenovirus transcription as the system, in which the

Components of Typical Transcription Unit

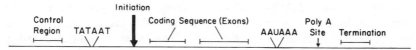

Figure 1   Schematic diagram depicting the essential components of a typical eukaryotic protein-coding (RNA polymerase II) transcriptional unit as discussed in text. The control region refers to those upstream sequences that are indispensible for the full in vivo expression of the gene.

output from a single transcriptional unit represents a significant fraction of the total transcription in the cell. A detailed discussion of the use of these techniques in mapping the late adenovirus transcriptional unit is presented in (3). These studies defined the boundaries of the late adenovirus transcriptional unit (2, 4–6) as well as several early adenovirus transcriptional units (6–10). the SV40 late transcriptional unit (11), and the mouse β-globin transcriptional unit (12).

The first detailed description of the actual initiation site for a polymerase II transcriptional unit came from studies by Ziff & Evans (13). Using both in vivo labeled material as well as isolated nuclei from late adenovirus infected cells, they localized the initiation site for late viral transcription to viral genome map coordinate 16.4 (16.4% of the distance from the left end of the genome). Then, by analyzing RNA fingerprints of nuclear RNA from this region and comparing to that of the 5' end of the late mRNA, they concluded that the sequence specifying the 5' cap-containing terminus of the late Ad-2 mRNAs was the site of initiation of transcription. In particular, they detected no transcription upstream of the mRNA cap site. Thus, there appeared to be no RNA processing event so as to produce the mature 5' terminus of the mRNA. Consistent with this has been the repeated failure to find nuclear mRNA precursors with 5' ends other than the mature mRNA 5' end (14–16). Of course, if the processing at the 5' end were rapid, such a precursor may never be found, and certainly not in steady state populations of nuclear RNA. The finding that the mRNA cap site was the transcriptional start site has had a considerable impact on subsequent studies, since it allowed the definition of sequences encoding the promoters for transcription. That is, in the absence of systems facilitating the definitive experiment described above, the principle that a cap site was the start site allowed the subsequent sequence analyses that demonstrated the sequence homologies important for promoter structure.

Promoters

A promoter for a prokaryotic gene is the site that functions to initiate transcription of a transcriptional unit (17). In the absence of such a sequence, transcription of the gene is greatly diminished. This site therefore involves the region of DNA for RNA polymerase binding so as to allow the subsequent initiation of transcription. It must also include other sequences, if such exist, that are indispensable for the normal in vivo transcription of the gene. Since the genetic analyses that were so powerful in the prokaryotic system cannot be employed, the definition of eukaryotic promoters has relied on two methods: comparative studies of sequence homologies and in vitro genetics to probe the function of the sequence homologies, as well as other important sequences. The most striking sequence homology found thus far is the TATA sequence (Goldberg-Hogness box) located 25–30 residues upstream from the start site of transcription (18). Virtually every gene analyzed to date, except the E2 gene and the IV_{a2} gene of adenovirus (19, 20) and the late genes of SV40 (21, 22) and polyoma (23), have some form of the TATA sequence. This sequence is indispensible for the in vitro transcription of polymerase II genes and appears to be the only significant sequence requirement for in vitro transcription (24, 25). The function of this sequence appears to involve the accurate positioning of the start of transcription. Deletions or mutations of the TATA sequence result in heterogeneous starts in vivo and deletion of sequences downstream from the TATA sequence results in transcription initiation approximately 25 nucleotides downstream from the TATA sequence (24–31). Although the TATA sequence appears to be related in sequence to the Pribnow box of prokaryotic promoters (32, 33), it clearly differs, in that deletion of the Pribnow box eliminates transcription of the prokaryotic gene (34); deletion of the TATA sequence does not abolish in vivo transcription of the eukaryotic gene.

Sequences essential for the transcription of polymerase II genes are located, in most cases, far upstream from the actual start of transcription (26–29, 35–38). In several cases the essential sequences are at a distance of 200–300 nucleotides from the transcription initiation site. The control region of the *Herpes simplex* virus thymidine kinase (tk) gene has been mapped in considerable detail through the combination of deletion mutants to yield selected point mutations (39). These mutants allowed the definition of two separate domains upstream of the tk gene, both of which are required for efficient expression. Furthermore, the position of the required sequences relative to each other and to the start site of transcription could be varied (39a). Possibly a more striking result is the finding that the control region appears to behave in a position-independent manner, since, as initially shown for the sea urchin histone H2A gene, it can be inverted in position and still retain function (35). Furthermore, the control regions from several viral promoters (SV40, polyoma, retroviruses) are able to enhance the expression, in a *cis* fashion, of heterologous genes (40–46). These DNA segments, termed "enhancer elements" and typified by the 72-base pair-repeated segment at the replication origin of SV40 DNA (21, 22), are able to confer a high level of expression to a gene that is normally functional only at a low level in the assay system. This effect is independent of position or orientation of the enhancer segment when inserted in the recipient plasmid. Whether the viral enhancers are distinct regulatory elements from normal promoters or rather just represent very strong promoters is not clear. It is of some interest that the sequences that exhibit enhancer activity are from the promoter regions of viral genes that must be expressed during the initial entry of the virus into the cell, although there is now apparently one example of a cellular derived sequence that exhibits the same properties (47). The mechanism by which these sequences promote the efficient transcription of the gene is not clear at this point. In view of the observation that the SV40 and polyoma control regions lie within a DNase-hypersensitive domain (50, 51), one possibility might be the establishment of a specific active chromatin structure (48, 49).

Transcription Unit Design

Simple

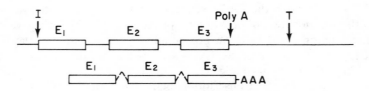

Complex

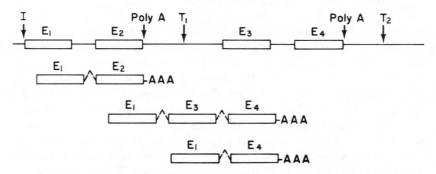

Figure 2   Features of transcriptional unit complexity:

A simple transcriptional unit contains a single transcription initiation site (I) and encodes a single mRNA 3′ terminus; therefore, there is a single site for poly(A) addition (poly A) and presumably a single site for transcription termination (T). Shown below the transcriptional unit is the structure of the resulting spliced mRNA.

A complex transcriptional unit contains a single transcription initiation site (I) and encodes multiple mRNA 3′ termini, and thus contains multiple sites for poly(A) addition (poly A). In such a transcriptional unit, there are potentially two transcriptional termination sites (T1 and T2). Shown below the transcriptional unit are three possible spliced mRNA products. The top mRNA utilizes the first poly(A) site while the middle and bottom mRNAs utilize the second poly(A) site.

Transcriptional Unit Complexity

Before leaving the discussion of transcriptional units, I will define their design as they relate to certain subsequent RNA processing events—namely, the distinction between simple transcriptional units, which encode a single mRNA 3' terminus, and complex transcriptional units, which encode multiple mRNA 3'-termini. As Figure 2 shows, a simple transcriptional unit would be typified by a single poly(A) addition site. An example of a simple transcriptional unit would be the one encoding the β-globin mRNA or the early SV40 transcriptional unit. A complex transcriptional unit would contain multiple poly(A) sites. The immunoglobulin μ heavy chain transcriptional unit (52, 53) and δ heavy chain transcriptional unit (54) as well as the late adenovirus transcriptional unit (3) are examples of complex transcriptional units; recently a description of the rat calcitonin gene (55) indicates that it is also a complex transcriptional unit.

But why include features such as poly(A) addition sites in the description of transcriptional units? As discussed in more detail in following sections, the selection of the late adenovirus poly(A) sites occurs during transcription and therefore, in essence, represents a transcriptional event. Furthermore, the use of distal poly(A) sites can be controlled through transcriptional termination.

The issue can be further complicated by the ability to achieve alternative splicing either in a simple transcriptional unit (early or late SV40) or in a complex transcriptional unit (late adenovirus). However, since RNA splicing is most surely a posttranscriptional event, this level of complexity must follow the discussion of transcriptional units.

FORMATION OF THE PRIMARY TRANSCRIPT

Initiation of Transcription

As detailed in previous sections, the transcription initiation site appears to be the mRNA cap site, i.e. the DNA sequence specifying the 5' terminus of the mRNA. The initial in vivo observation that the initiation site of transcription is the mRNA cap site (13) has now been extended, through the use of cell-free transcription systems. In fact, the first demonstration of accurate RNA polymerase II initiation in vitro made use of the adenovirus information and demonstrated that initiation occurs at the 5' cap site (56). This finding was repeated for the adenovirus late transcription utilizing another system (57) and was extended to many other genes (for review see 24, 25). These results once again suggest that the initiation site in vitro is the 5' cap sequence of the mRNA, although rapid processing of upstream sequences could not be ruled out in these in vitro studies. However, this issue was examined directly in an in vitro transcription of both the late adenovirus and β-globin transcriptional units (58). The cap structure in each of these transcripts was labeled with a β-P^{32} nucleotide (ATP) that served as the initiating nucleotide. Thus, capping must take place on an initiated terminus rather than a processed terminus. The same result was also obtained using isolated nuclei as the transcription system (60) or whole cells permeabilized to nucleotide triphosphates (61).

The transcription initiation event for a eukaryotic polymerase II gene can be very heterogeneous. Multiple mRNA 5' ends were initially observed for the late SV40 mRNAs (62) and the late polyoma RNAs (63) and have now been found for many other specific mRNAs (19, 20, 64–66). In general, the heterogeneous start sites derived from the mRNA sequences have been found to occur in the in vitro systems (67–69) with the exception of the extreme heterogeneity of start sites for the late SV40 and polyoma transcripts (67). In at least one instance alternative start sites may be regulatory. The mouse α-amylase gene is initiated at one site in the salivary gland but at another in the pancreas (70).

Progress has also been made toward defining the components of the in vitro systems that are required for accurate initiation. Roeder and colleagues (71) have defined at least four factors that together are essential for accurate transcription initiation. Since only one of the factors was purified to near homogeneity, the number of actual factors must be a minimal estimate. One factor appears to be responsible for the suppression of random, nonspecific transcription by RNA polymerase II without affecting the selective initiation at a bona fide promoter (in this case, the adenovirus major late). Furthermore, at least three of the factors and possibly the fourth are specific for polymerase II transcription and distinct from polymerase III stimulating factors. Finally, inhibition of host cell transcription by poliovirus infection appears to be the result of inactivation of one of these factors (72).

CONTROL OF INITIATION   Control of the initiation event has been directly demonstrated (through the actual measurement of nuclear transcription rates) in several systems under study, and certainly is implied in many other cases. Tissue specific gene expression (73) or developmentally regulated gene expression (74) have been shown to be due to transcriptional control. Negative control of transcription initiation, i.e. a shutoff of ongoing transcription, has been shown for early SV40 transcription (75–77) and for transcription from the E4 transcriptional unit of adenovirus (78). In both cases, the function responsible for the repression is known; T antigen shuts off early SV40 transcription (75–77) and the adenovirus DNA-binding protein shutsoff E4 transcription (79). However, the most detailed information concerning the mechanism is that for the T-antigen repression of early

SV40 transcription. T antigen has been shown to bind to the region of SV40 DNA that is essential for DNA replication and that includes the promoter for the early transcriptional unit (80–82). This region contains three specific binding sites for T antigen (83, 84). In vitro transcription assays have demonstrated that T-antigen binding at this region prevents early SV40 transcription, and that there is specificity to the repression, in that transcription from the adjacent late promoter is unaffected (85, 86).

Positive control of initiation, the activation of a previously silent transcriptional unit, has been shown for the early adenovirus genes (87), and the early herpes virus genes (88). The adenovirus E1A gene is the first to be expressed during a productive infection (78). The product of the E1A gene is then responsible for the activation of transcription from the five other early transcriptional units (89–91). A similar program of transcriptional activation operates during a herpes virus infection. The product of the *Herpes simplex* virus α-4 gene, one of the five viral genes initially expressed, is required for activation of the transcription of the early herpes genes (92–94).

Positive control of transcription has also been demonstrated for a variety of other genes, such as the heat shock genes (95), the metallothionein gene (96, 97), and the hormonally controlled genes (98–102), including the mouse mammary tumor virus genome (103). Recently, a DNA-binding site for the glucocorticoid hormone receptor at specific sites of the viral genome has been demonstrated (104, 105). One binding site is located 110–449 bp upstream from the transcription start site. Three other binding sites lie within the transcriptional unit. Furthermore, hormone receptor binding sites have been demonstrated for the ovalbumin gene using a fragment-specific competition assay (106).

Unfortunately, there has yet to be a clear demonstration of positive initiation control in an in vitro system. The most vivid illustration of this point is that the major late adenovirus promoter is probably the most active promoter as assayed in the in vitro systems prepared from uninfected cells (56, 57). There appears to be little or no enhancement, when an extract prepared from virus infected cells is used (68, 69). Yet, during an adenovirus infection, the late promoter is only active as a result of the action of the viral E1A gene product (89). This lack of regulation is probably reflected in the fact that the upstream control regions identified for a variety of genes are dispensable in in vitro transcription systems (24), indicating that the conditions of the present in vitro systems do not allow the recognition and utilization of potential positive regulatory sequences.

Termination

TRANSCRIPTION BEYOND POLY(A) ADDITION SITES   Although the 5′ terminus of the mRNA appears to correspond to the transcription initiation site, the same is not true for the 3' terminus of the mRNA. In each case where it has been examined, transcription does not terminate at the poly(A) site but rather at some distance downstream. This observation was first made for the late adenovirus transcriptional unit (108, 109), where five poly(A) sites are used in a single transcriptional unit. By measuring transcription rates along the length of the transcriptional unit it was concluded that transcription did not terminate at the poly(A) sites (108) but rather near the end of the genome. This same result has now also been found for several adenovirus simple transcriptional units that possess only a single site for poly(A) addition (8), for the SV40 late transcriptional unit (11), and now also for the mouse β-globin transcriptional unit (12). Therefore, the general rule in poly(A) containing mRNA biogenesis may be that the poly(A) site is not the transcription termination site, although more cases must be examined to establish the generality. In other cases, it is clear that transcription can proceed through poly(A) sites, but whether the formation of the poly(A) site that is utilized occurs by a cleavage or termination is not certain (52–54, 110–112).

If the mRNA 3' terminus is a processing site rather than a termination site, then what is the transcription termination site and what is the reason for transcription beyond the poly(A) site? As discussed previously, no clear example of termination has been established. Although termination can be mapped within a few hundred nucleotides (109, 113, 114) a clearly defined site has yet to be shown. It would appear that either the termination event is not precise or that terminated transcripts are rapidly degraded. As to the function for transcription beyond the poly(A) site, one has to assume at this point that there is a requirement for these 3' distal sequences in the formation of the site for cleavage at the poly(A) site. The simplest notion would be that poly(A) addition must accompany an RNA-chain cleavage, and thus at least one nucleotide beyond the poly(A) addition site must be transcribed in order to obtain a cleavage. A more likely situation would be that sequences beyond the actual poly(A) addition site participate in the formation of an RNase cleavage site, perhaps analogous to that for RNase III in *E. coli* (115). Thus these sequences that are located hundreds of nucleotides beyond the 3' terminus of the mature mRNA would be essential parts of the transcriptional unit. The answer to such a question will surely come from in vitro mutagenesis experiments, such as those already employed to define essential promoter sequences. The one clear example of a class of nonpolyadenylated mRNAs, the histone mRNAs, represent potential 3' ends generated by termination (116). However, even in this instance, it has not yet been clearly demonstrated that termination generates the 3' end of the RNA. Recent experiments employing in vitro deletion mutants suggested that sequences distal to the RNA 3' terminus are necessary for the formation of the 3' end (117). Whether these sequences are involved in termination or represent a portion of a cleavage site as discussed above is not known.

TERMINATION CONTROL   For a transcriptional unit encoding multiple proteins in a tandem array (i.e. the IgM heavy-chain and late adenovirus transcriptional units), a mechanism of control of transcription termination would be of importance. For instance, in the schematic of Figure 2, control of termination at T_1 would regulate the expression of the E3 and E4 exons. That such a mechanism can operate has been shown for the adenovirus late transcriptional unit. As originally defined, transcription of this genomic region, encompassing five poly(A) sites, during late viral infection terminates mainly at the end of the genome (108, 109). In sharp contrast, during early infection the same promoter is utilized, but transcripts terminate near the middle of the genome, thereby preventing the expression of the L4 and L5 regions (118–120). Obviously, such a mechanism could also regulate the expression of the immunoglobulin μ-membrane protein since these are the promoter distal exons of the transcriptional unit (52, 53) or the δ-protein exons which can be still further downstream in the same transcriptional unit (54). Another recent example of a complex transcriptional unit, that for the rat calcitonin mRNA, could also be so regulated (55).

Transcription termination could also regulate the expression of a simple transcriptional unit encoding only a single protein. In this instance, premature termination prior to the poly(A) site would prevent the formation of a mRNA. In principle, although certainly not in mechanism, such a situation would be analogous to attenuation in prokaryotic cells (121). Premature termination of transcription does appear to occur during adenovirus infection (112, 123) as well as SV40 infection (124, 125). Late in adenovirus infection as many as four out of five transcriptional starts from the major late promoter terminate early, before reaching any coding sequences (122). These transcripts are capped (126), but apparently never reach the cytoplasm (13). Transcripts from the late SV40 promoter can also terminate prematurely. In this instance, it appears that there is a specific stop site, 93–95 nucleotides from the major late start site of transcription (127). Furthermore, RNA secondary structure may play a role in the premature termination since treatment with proflavine enhances readthrough (128). In addition, it appears that a great many transcripts of cellular genes result as premature termination (129, 130), although in these cases it is not known if these transcriptional units are also producing mRNA. What has yet to be shown, however, is whether any of these events of premature termination are regulatory. That is, do they ever change in response to circumstances in the cell or are these just the fixed patterns that one observes?

Poly(A) Site Selection

Transcription of a complex transcriptional unit, as described previously, requires that a choice be made as to a particular poly(A) site. In the example of the late adenovirus transcriptional unit, this means that one out of five possible sites must be recognized and selected. As already described, the poly(A) site appears to be formed through an RNA-chain cleavage rather than by transcriptional termination (8, 11, 12, 108, 109). Furthermore, the selection of the poly(A) sites in the late adenovirus transcriptional unit appears to occur during transcription rather than as a posttranscriptional event (108). Thus, before the polymerase has completed the transcription of distal sequences in the transcriptional unit, a poly(A) site has already been chosen and poly(A) added; in essence, the poly(A) site selection is a transcriptional event. These results further suggest that the selection is not a random event, but rather is preprogrammed; since the selection occurs during transcription, one would expect a polar effect where the L1 site was selected most frequently, if the selection were random. Such is not the case; rather, the selection of the five poly(A) sites is roughly equal (108).

The possibility of regulating expression of a complex transcriptional unit through transcriptional termination was discussed earlier; i.e. a distal poly(A) site may not be utilized, owing to transcriptional termination occurring prior to the site. Variable selection of poly(A) sites is another possible mechanism by which regulation could occur. Again, in the example of Figure 2, selection of PA site 1 or 2 with termination occurring at T_2 would control the output of the transcriptional unit. During late adenovirus infection, the five poly(A) sites within the late transcriptional unit are utilized with a relative efficiency of about 1:2:3:2:2 (L1–L5) (108). When the late promoter is utilized during early infection, only the first three poly(A) sites are transcribed (118–120). Now the ratio of L1:L2:L3 is 3:1:1 (119); thus a sixfold change in the relative usage of L1:L2 has occurred. These results indicate that poly(A) site selection is controlled to alter gene expression.

PROCESSING OF THE PRIMARY TRANSCRIPT

Poly(A) Addition

ROLE OF THE AAUAAA HOMOLOGY   Poly(A)-containing mRNAs possess the homologous sequence AAUAAA, usually about 10–30 nucleotides upstream from the poly(A) (131). The sequence is highly conserved, which implies that it may play a role in poly(A) addition, possibly either in formation of a recognition site for cleavage or in recognition by the poly(A) polymerase. [There are now at least three exceptions to the AAUAAA homology. AUUAAA occurs in the chicken lysozyme mRNA (132) and in one of the mRNAs of the adenovirus E3 transcriptional unit (132a); AAUAUA is found in the minor form of the pancreatic alpha amylase mRNA (133).]

The role of the AAUAAA sequence in the polyadenylation step has been investigated by utilizing in vitro genetics. Deletion mutants of SV40 were constructed so as to contain modifications at and around the AAUAAA sequence that precedes by 12 nucleotides the poly(A) segment in the late RNAs (134). Deletion of the AAUAAA sequence abolished the formation of the late RNA, clearly indicating the involvement of the sequence in the formation of poly(A)$^+$ RNA. Deletions were also constructed around the AAUAAA, but the hexanucleotide sequence was left intact. Deletion of sequences 3′ to the AAUAAA [between the AAUAAA and the poly(A)] resulted in the production of RNAs in which poly(A) addition occurred downstream from the normal poly(A) site. In each instance, poly(A) addition took place at a site within 11–19 nucleotides from the AAUAAA. Furthermore, the production of RNAs utilizing downstream sites was nearly equally efficient as in the wild type. Therefore, it appears that there is no sequence requirement between the AAUAAA and the poly(A) addition site, but rather just a spatial requirement; in this respect, the function appears to resemble that of the TATAA sequence in positioning the transcription start site.

There is also heterogeneity in the site of poly(A) addition observed under normal in vivo circumstances. Poly(A) appears to be added to bovine prolactin mRNA at several sites within an 11-base pair region (135). The adenovirus E3 transcriptional unit is apparently a complex transcriptional unit possessing two poly(A) sites. The first site, at map position 82, is specified by an AUUAAA and poly(A) addition can occur at three possible sites, either 17, 26, or 29 nucleotides downstream (135a). Presumably, in both of these instances, this still represents a single poly(A) site, as there is only a single AAUAAA (an AUUAAA).

From the results obtained with SV40, it seems that the AAUAAA is essential for the formation of poly(A) containing RNA. However, the precise role that the sequence plays in poly(A) addition remains uncertain. The AAUAAA sequence could either specify the proper cleavage of the RNA chain, or it could serve as a recognition sequence for the poly(A) polymerase. The question is, in the absence of an AAUAAA, are proper 3′ ends generated with a subsequent failure of poly(A) addition or are the RNA chains simply left uncleaved and thus not a substrate for poly(A) addition? Recent experiments utilizing an adenovirus AAUAAA mutation suggested the former possibility, a role in cleavage. The AAUAAA in the E1A poly(A) site was converted to AAGAAA. Assay of the nuclear RNA by S1 mapping revealed that the E1A 3′ terminus produced by wild-type virus was found in only 10% of the transcripts in the mutant-infected cells (C. Montell, E. Fisher, M. Caruthers, A. J. Berk, in preparation). The majority of the transcripts apparently continued beyond this site and into the adjacent E1B transcriptional unit. The 10% of the transcripts that were cleaved at the normal site were found to contain poly(A). It would therefore appear that the AAUAAA sequence was required for the cleavage reaction but not for the subsequent poly(A) polymerization.

TEMPORAL ORDER OF POLY(A) ADDITION   The timing of the addition of poly(A) to the nuclear RNA has been demonstrated by kinetic labeling experiments and by sizing of newly formed polyadenylated RNAs. By analyzing the appearance of RNA sequences adjacent to poly(A) sites in adenovirus RNAs and total cellular RNA, it was concluded that poly(A) addition was an early and rapid event in mRNA biogenesis (108, 136, 137). Poly(A)$^+$ sequences could be detected within 90–120 seconds of the synthesis of the RNA. Furthermore, the very newly labeled (45 second pulse) and poly(A) containing adenovirus nuclear RNA was largely unspliced (108). Thus, poly(A) appears to be added very early and before splicing of the RNA. Although analysis of pulse-labeled poly(A)$^+$ nuclear RNA such as was done with adenovirus is not possible for most cellular genes, it is nevertheless likely that poly(A) addition precedes splicing. For most specific cellular RNAs that have been examined, poly(A)-containing precursors, larger than the corresponding mRNA, have been observed (24, 138). Thus, in these cases poly(A) addition can precede splicing, although it is not certain if this is the obligatory pathway.

Splicing

Following the initial observation that adenovirus mRNAs were constructed of spliced segments (139–141), it now is well documented that many mammalian mRNAs, if not the majority, are constructed of coding sequences separate in the genome. Splicing appears to occur exclusively in the nucleus (142, 143). In contrast to the rapid poly(A) addition process (108, 136, 137), the subsequent splicing of a precursor RNA to the mature mRNA is a relatively slow process. For the adenovirus fiber mRNA, 20 min are required after splicing is initiated until the mature mRNA can be seen in the nucleus (142). In many cases, intermediates in the splicing pathway have been found that indicate that the various introns in a transcript are not removed in one step (24, 138); in the splicing of the globin precursor, a single intron is apparently removed in steps (144). In at least one instance, it has been demonstrated that the splicing intermediates are obligatory precursors in the pathway to the final mRNA (142).

Analysis of the splicing of certain chimeric genes and deletion mutants has further elucidated sequence requirements, or rather the lack thereof, for splicing. SV40 mutants in which a large portion of the large T intron was removed, leaving as little as 12–13 nucleotides adjacent to the intron/exon borders was nevertheless spliced normally (145, 146). And, a plasmid construction in which the first exon of the SV40 T antigen was positioned upstream from the third exon of the mouse β-globin gene was found to produce a chimeric RNA that was spliced correctly (147).

However, the actual mechanisms by which splicing occurs, and in particular, how splice sites are selected, remain obscure. Clearly, in vitro systems are required such that the components can be fractionated and studied. Although considerable success has been achieved in the development of tRNA splicing systems from yeast (138), the equivalent for mRNA splicing lags far behind. There have been reports of splicing in mammalian extracts (148, 149), but these systems are rather inefficient and have not unambiguously demonstrated precise cleavage and ligation. However, a recent report employed primer extension analysis to demonstrate proper splicing of a fraction of the RNA synthesized in vitro from a globin DNA template (149a). Although once again the efficiency was very low, the assay demonstrated that the system could in fact carry out accurate splicing.

THE SPLICE SITE   An examination of sequences at splicing junctions revealed a strict conservation of nucleotides (24, 150). By analyzing the sequences at over 130 splice junctions (150), the consensus sequence of ^C_AAG/$\underline{GT}^A_G$AGT for the donor splice site and (^T_C)nN$^C_T\underline{AG}$/G for the acceptor splice site have been derived. Most notably, the first two (GT) and last two (AG) nucleotides of an intron (underlined nucleotides of the consensus sequence) are strictly conserved. Although these limited homologies certainly cannot explain selection of splice sites, they nevertheless most likely must be involved in the actual cleavage and ligation process. This was recently found through the analysis of the transcripts produced from defective globin genes. Mutations in the β-globin genes of certain β^0-thalessemic individuals (no production of β-globin) have been found to be point mutants resulting from single base changes. One of these β^0-mutations results in a G to A change at the 5' splice site of the second intron (151, 152). This results in the inactivation of this splice site and produces an mRNA that utilizes an alternate site found within the intron (153). Mutations resulting in β^+-thalessemias (low level of β-globin production) have also been found to affect splicing. Two of these have been localized to within the first intron of the β-globin gene and result in the creation of a splice acceptor sequence (154, 155) after the conversion of a G to an A. Analysis of mRNA production from the mutant genes revealed splicing abnormalities in which the mutated splice sites were utilized preferentially rather than the normal splice site (156, 157). Finally, in a case of α-thalessemia, mutation led to the deletion of five base pairs of DNA from the first intron immediately adjacent to exon 1. As a result the 5′ splice site was destroyed removing the GT conserved sequence (158). RNA produced from this gene failed to use the normal splice site (now deleted), but rather used an alternative site contained in the middle of exon 1 (159). Thus an mRNA not capable of coding for globin was produced.

Furthermore, a donor splice site in the adenovirus E1A transcriptional unit has been modified in vitro resulting in the conversion of a T to a G (160). This resulted in the failure to produce the E1A 12S mRNA normally utilizing the site, which indicates that the splice site had been inactivated. Therefore, from all of these results it appears that the hypothesis that the conserved intron sequences at the splice sites must be involved in the splicing reaction has been borne out.

What is still unanswered, however, is the mechanism for selective splicing. That is, when an RNA precursor contains multiple potential splicing sites such as the adenovirus, SV40, and polyoma RNAs (161), retrovirus RNAs (162), as well as the immunoglobulin μ-membrane precursor (52, 53), what is the mechanism by which the proper ligations are achieved? One proposal has suggested a scanning mechanism in which the first available splice site was utilized (163). Such a mechanism would explain how a simple transcript containing a multiple set of splice sites, but always generating a single mRNA (i.e. ovalbumin, globin) was always spliced correctly such that internal exons were not deleted. Potential sites not normally used, such as the site described above in the second β-globin intron, might be weak relative to others and thus bypassed. However, where alternative sites are involved, a scanning mechanism in either a 3′ or a 5′ direction surely cannot be operative based solely on sequence recognition. Such a mechanism would not allow a change in splicing site usage as occurs for the L1 region of adenovirus (see discussion below). However, the scanning mechanism need only in this case be modified by the action of specific factors that might block potential sites so as to achieve specificity. In such a way the splicing apparatus as it scans the RNA precursor would, in effect, not recognize a blocked site and thus proceed to the next.

ROLE OF POLY(A) IN SPLICING   As discussed previously, the addition of poly(A) is a rapid event, tightly linked to transcription, and poly(A) addition precedes the splicing event; this prompted the suggestion that poly(A) may be involved in splicing. This question was directly investigated with the drug cordycepin, known to block poly(A) addition as well as the accumulation of mRNA in the cytoplasm (164). In the presence of cordycepin, early adenovirus transcription continues, and processing of

several early nuclear RNAs was found to be normal (165). Poly(A)−
nuclear RNA from early regions E2 and E1B was found to be correctly
spliced. Thus, in the absence of poly(A), splicing can still take place and at
about the normal rate. And although poly(A) addition normally precedes
splicing, it is not required for splicing to take place. However, what is yet to
be determined is whether the proper 3′ end cleavage is necessary for splicing
to take place.

INVOLVEMENT OF SMALL NUCLEAR RNAS   Rogers & Wall (166) and Lerner et
al (167) originally proposed the hypothesis that certain small nuclear RNAs
may be involved in the splicing reaction. The proposal was based on two
observations: First, ribonucleoprotein particles containing the snRNAs
appeared to be associated with hnRNP; and second, there were significant
sequence homologies between the snRNAs and certain splice junctions
(168).

These proposals are, of course, only hypotheses consistent with certain
observations and can be proved only through the use of mutants or
fractionated in vitro systems. However, some evidence in favor of a role of
the snRNA in splicing has come from in vitro splicing experiments that used
antibody to snRNP. Nuclei from adenovirus-infected cells were shown to
synthesize and splice viral RNAs in vitro (169). Addition of anti-RNP
antisera (antibody reacting with U1 snRNP) or anti-Sm antisera (antibody
from patients with systemic lupus erthematosus that precipitates nuclear
RNP) to the nuclear system resulted in a failure to splice the adenovirus
RNAs (170). In contrast, addition of anti-Ro antisera (antibody against
cytoplasmic RNP) or anti-La antisera (antibody directed against a separate
class of nuclear RNP) had no effect on the splicing. These results thus
suggest that the U1-containing snRNPs are involved in some fashion in the
splicing reaction.

Finally, one other small RNA originally proposed to be involved in
splicing was the adenovirus VA-RNA, a small RNA polymerase III product
of the viral genome (171, 172). However, recent experiments argue against a
role for VA-RNA in splicing. An adenovirus VA-mutant was constructed
using a plasmid VA gene deleted in the control region. Analysis of late viral
gene expression revealed that late viral protein synthesis was reduced in the
VA-mutant infected cells as compared to WT-infected cells. However, late
viral mRNAs were found in wild-type amounts, which indicates that the
VA-RNA was not required for splicing of the late viral transcripts. Further
experiments revealed that the VA-RNA appears to be required for the
efficient translation of viral mRNA in the infected cells (173).

REGULATION   From the initial observation of splicing in adenovirus some
five years ago, it has become clear that a whole new dimension of regulation

of gene expression is possible. The ability to achieve alternative splicing obviously allows the cell to alter the output of the gene without affecting the transcription event by assembling various combinations of coding sequence. Recent experiments seem to show clearly that this possibility is a reality. It was obvious from the initial discovery that alternative splicing occurred to generate the adenovirus and SV40 RNAs (161) as well as the retrovirus mRNAs (162). However, only the controlled change of these alternatives indicates regulation. This occurrence was demonstrated during adenovirus infection. During late viral infection, the L1 region of the genome encodes three predominant mRNA species, each produced from a common 8 kb-RNA precursor. However, during early infection the precursor is spliced to yield a single mRNA distinct in structure from the three late species (119, 120, 174). By analyzing nuclear poly(A) RNA and pulse-labeled cytoplasmic RNA, it was possible to conclude that the change was due to splicing (119). Therefore, depending upon the environment in which the precursor is produced (early or late viral infection), quite different products result. These results thus suggest that factors acting in *trans* can affect the pathway of splicing of a particular precursor RNA. Certainly this represents the best possibility of identifying and isolating cellular or viral components involved in the splicing process.

Capping and Methylation

Most, if not all, RNA polymerase II products receive posttranscriptional modification in the form of specific methylations (175). The methylations are found in the $2'O$ ribose position of the penultimate nucleotide and at the N^6 position of adenylate residues throughout the RNA. In addition, the 5′ ends of all RNA polymerase II transcripts are blocked by the ^7mG cap (175, 176). It appears rather certain at this point that the cap structure functions in the initiation of translation through the interaction with specific proteins (for review see 175). Whether the cap plays any role during the processing of mRNA precursors is not yet clear. However, the timing of the addition of the cap and its subsequent fate during mRNA biogenesis are clear. The formation of a capped 5′ terminus is a very early event in mRNA biogenesis; there has yet to be a demonstration of uncapped polymerase II transcripts (130). Even nascent chains less than 100 nucleotides in length are found to be capped (126). Since even the prematurely terminated transcripts are capped and these molecules do not appear to be transported to the cytoplasm (13), the presence of a cap does not insure conservation of the RNA during mRNA biogenesis. In view of the rapid addition of the cap and poly(A) to mRNA precursors, it is tempting to speculate that the 5′ cap structure and the 3′ poly(A) segment are involved in the protection of these

RNAs from exonucleolytic degradation during the subsequent events of RNA processing.

Methylation of mRNAs also occurs at internal adenylate residues at the N^6 position (176). It has been suggested that these methylations are important in mRNA biogenesis since they are added at an early step in mRNA processing and are conserved during RNA processing (177). That is, m^6A residues are only added to sequences in a primary transcript of the late adenovirus transcriptional unit that are destined to become mRNA even when there are many potential sites of methylation.

Progress in the study of the importance of internal base methylations in mRNA biogenesis has been slow, owing largely to the infrequent nature of the methylation in a given RNA. Furthermore, inhibitors in general do not have the specificity to allow definitive experiments concerning the function of the m^6A residues. Experiments used the drug cycloleucine, a competitive inhibitor of the synthesis of S-adenosyl methionine, to reduce the frequency of m^6A in retrovirus RNAs during viral infection. Under these conditions, there appeared to be a reduced formation of the 3.5 kb subgenomic envelope mRNA species, an RNA formed by splicing of the full length RNA (178). There was also an increase in the synthesis of the protein products of the genome-length RNA. These results thus suggest that the m^6A residues play a positive role in splicing, although more direct experiments are needed to establish this point.

NUCLEAR-CYTOPLASMIC TRANSPORT

Once the mature fully processed eukaryotic mRNA is formed, transport to the cytoplasm must take place so as to allow function. Experiments performed both with adenovirus and with SV40 indicated that only fully processed molecules enter the cytoplasm (142, 143). There appears to be no cytoplasmic processing of RNA. Furthermore, portions of primary transcripts that are removed by RNA processing do not appear to be transported to the cytoplasm (108, 109). Neither are prematurely terminated transcripts from the adenovirus major late promoter transported to the cytoplasm (13). Thus, it is likely that a mechanism must exist for selective transport of RNA sequences from the nucleus to the cytoplasm. Can such a mechanism, that discriminates between functional and nonfunctional RNA, also be used to achieve regulation of transport of potentially functional mRNAs?

There is apparently at least one example of negative control of RNA transport. Late in adenovirus infection, most cellular mRNAs fail to appear in the cytoplasm, even though transcription of the genes continues as does apparently faithful RNA processing (179; Babich et al, in preparation).

Whether there is a specific shutoff of cellular mRNA transport or perhaps the viral mRNAs simply out-compete the cellular mRNAs for a transport factor is not known. Certainly before one can attempt to answer these questions, much more must be known about the normal mechanism of RNA transport.

Finally, it has long been thought that poly(A) may play a role in RNA transport; this is based upon the finding that cordycepin blocks accumulation of mRNA in the cytoplasm (164). However, experiments that used adenovirus-infected cells showed that RNA lacking poly(A) produced by cordycepin treatment can be transported at nearly normal efficiencies and enter into polyribosomes (180). Apparently the RNA then rapidly degrades upon entering the cytoplasm, resulting in a reduced accumulation. Thus, the role of poly(A) may be limited to the stabilization of mRNA in the cytoplasm (see below).

CYTOPLASMIC STABILITY OF mRNA

Of course the final concentration of functional mRNA available for translation into protein depends not only on the rate of delivery of the mRNA into the cytoplasm but also on the relative stability of the mRNA once in the cytoplasm. Clearly, in the face of a constant rate of formation, the level of expression of a particular gene could be regulated by altering the rate of decay of the mRNA. For instance, the specific stabilization of a cytoplasmic mRNA has as an example the casein mRNA (181). Upon exposure of breast explants to prolactin, the stability of the casein mRNA increases 25-fold, along with an increase in transcription of the gene. Furthermore, an adenovirus mRNA appears to accumulate during infection owing to a specific stabilization. The E1B 20S and 14S mRNAs decay with short half-lives during early infection. During late infection, however, the 20S remains unstable but the 14S accumulates owing to a long half-life (182).

In addition, an immediate shutdown of expression of a given gene after a turn off of transcription of the gene can only be accomplished if the mRNA is also rapidly degraded. That is, a gene may be rapidly turned off by simultaneously repressing transcription and destabilizing the preexisting mRNA. This phenomenon has been most clearly demonstrated for the developmental genes of Dictyostelium. Upon aggregation of single cells into a multicellular structure, a specific set of genes is activated (183, 184). The mRNAs encoded from these genes accumulate owing to a rather long half-life. When the aggregates are dispersed, transcription of the aggregation-stage genes is shutoff and the corresponding mRNAs are rapidly and selectively degraded (185). The stability of the preexisting common mRNAs that continue to be produced is unaffected. Thus the phenotype specified by these aggregation specific gene products very quickly disappears.

Role of Poly(A) in mRNA Stability

Several studies suggest that the poly(A) segment plays a role in determining mRNA stability. The most direct experiments involve the enzymatic removal of the poly(A), followed by assays for the functional longevity of the RNA. Globin mRNA, when injected into frog oocytes, retains activity for long periods of time. However, when the poly(A) tail is removed by enzymatic treatment, the mRNA rapidly loses the ability to direct the synthesis of globin after injection (186) owing to a rapid degradation of the mRNA (187). That this was not caused by the alteration of the mRNA itself was shown by restoring stability to the de-adenylated mRNA through in vitro poly(A) addition (188). Thus, the functional stability of globin mRNA upon injection into oocytes, or also upon injection into HeLa cells (189), depends directly on the presence of a poly(A) segment. Apparently a segment of only 30 adenylate residues is sufficient to confer stability on the mRNA (190). Furthermore, histone mRNA, a naturally occurring mRNA lacking poly(A), when injected into HeLa cells, was found to degrade rapidly (191). The RNA could be stabilized, however, by the prior addition of poly(A). Thus, a mRNA that normally turns over rapidly can be artificially stabilized by the presence of a 3' poly(A) segment.

A second approach has been to prevent the addition of poly(A) to mRNA inside the cell and then to investigate the fate of the mRNA. Cordycepin is known to prevent the addition of poly(A) to RNA in the nucleus (165). As described previously, under short-term conditions, cordycepin does not markedly affect transcription or processing of the RNA. However, RNA synthesized under these conditions did not accumulate in the cytoplasm, although almost normal amounts were transported (180). Thus, in the absence of poly(A), the RNA appears to enter the cytoplasm, but then rapidly turns over. Therefore, poly(A) apparently plays a role in determining the stability of an RNA in the cytoplasm.

What is the mechanism for differential stabilities of mRNAs, and more importantly, what causes a poly(A)-containing mRNA to decay rapidly in response to a specific stimulus? The adenovirus early mRNAs contain poly(A) and decay with a rapid half-life when synthesized early during a lytic infection (185, 186). However, when these same mRNAs are produced in an adenovirus transformed cell in which only a limited portion of the genome is expressed, they now are rather long lived (192; A. Babich, J. Nevins, in preparation). The rapid decay during an early lytic infection probably results from the action of the viral DNA-binding protein. In the absence of a functional DNA-binding protein during a lytic infection (through the use of a temperature-sensitive mutant), the same mRNAs that decay rapidly in wild-type infected cells now decay slowly (193). Thus, in some fashion, this protein appears to alter the cytoplasmic stability of the viral mRNAs. Preliminary evidence suggests that the protein affects the mRNA stability through a direct interaction, since the protein can be cross-linked to viral mRNA inside the cell (A. Babich, J. Nevins, in preparation).

CONCLUDING REMARKS

As outlined in this review, the pathway for the formation of a eukaryotic mRNA differs from that of an *E. coli* mRNA in virtually every manner. The DNA signals controlling the frequency of transcription initiation are quite different; those sequences in the eukaryotic gene apparently act in a position-independent manner. In contrast to prokaryotes, the termination of transcription may be an imprecise event, possibly reflected by the fact that termination does not produce the 3' end of a eukaryotic mRNA. However, as in prokaryotes, termination of transcription can be a regulatory event controlling the expression of distal parts of a transcriptional unit.

Processing of prokaryotic transcripts is limited to cleavages that serve to separate functions of a polycistronic RNA. The processing of a eukaryotic transcript represents a bewildering array of events including the generation of specific 3' termini, modification of both termini, internal methylations, and the splicing together of various portions of the primary transcript. After all of this is completed, the finished product is selectively transported to the cytoplasm. Possibly more amazing than the mere existence of this myriad of steps in mRNA manufacture is the finding that most, if not all, of these steps can be regulated to alter the level of expression of the gene.

What remains for the future is the business of determining the factors acting in this complex regulatory system. Virtually nothing is known of the details—promoter recognition and function, termination and antitermination, selective methylation, RNA splicing and the selection of splicing sites, RNA transport and the control thereof, and the mechanism of mRNA turnover and the basis for selective degradation. However, our knowledge of these processes and their action as regulatory events represents a considerable step toward our understanding of the mechanisms that regulate the expression of eukaryotic genes.

Literature Cited

1. Sauerbier, W. 1976. *Adv. Radiat. Biol.* 6:50–106
2. Bachenheimer, S. L., Darnell, J. E. 1975. *Proc. Natl. Acad. Sci. USA* 72:4445–49
3. Nevins, J. R., Chen-Kiang, S. 1981. *Adv. Virus Res.* 26:1–35
4. Goldberg, S., Weber, J., Darnell, J. E. 1977. *Cell* 10:617–21
5. Weber, J., Jelinek, W., Darnell, J. E. 1977. *Cell* 10:611–16
6. Evans, R. M., Fraser, N. W., Ziff, E., Weber, J., Wilson, M., Darnell, J. E. 1977. *Cell* 12:733–39
7. Wilson, M. C., Fraser, N. W., Darnell, J. E. 1979. *Virology* 94:175–84
8. Nevins, J. R., Blanchard, J.-M., Darnell, J. E. 1980. *J. Mol. Biol.* 144:377–86
9. Berk, A. J., Sharp, P. A. 1977. *Cell* 12:45–55
10. Fraser, N. W., Sehgal, P. B., Darnell, J. E. 1978. *Nature* 278:590–93
11. Ford, J. P., Hsu, M.-T. 1978. *J. Virol.* 28:795–801
12. Hofer, E., Darnell, J. E. 1981. *Cell* 23:585–93
13. Ziff, E. B., Evans, R. M. 1978. *Cell* 15:1463–76
14. Weaver, R. F., Weissmann, C. 1979. *Nucleic Acids Res.* 7:1175–93
15. Roop, D. R., Tsai, M.-J., O'Malley, B. W. 1980. *Cell* 19:63–68
16. Wasylyk, B., Kedinger, C., Corden, J., Brison, O., Chambon, P. 1980. *Nature* 285:367–73
17. Epstein, W., Beckwith, J. R. 1968. *Ann. Rev. Biochem.* 37:411–36
18. Goldberg, M. L. 1979. PhD thesis. Stanford Univ.
19. Baker, C. C., Herisse, J., Courtois, G., Galibert, F., Ziff, E. B. 1979. *Cell* 18:569–80
20. Baker, C. C., Ziff, E. B. 1981. *J. Mol. Biol.* 149:189–221
21. Reddy, V., Thimmappaya, B., Dhar, R., Subrumanian, K., Zain, S., et al. 1978. *Science* 200:494–500
22. Fiers, W., Contreras, R., Haegeman, G., Rogiers, R., Van de Voorde, A., et al. 1978. *Nature* 273:113–20
23. Soeda, E., Arrand, J. R., Smolar, N., Walsh, J. E., Griffin, B. E. 1980. *Nature* 283:445–53
24. Breathnach, R., Chambon, P. 1981. *Ann. Rev. Biochem.* 50:349–83
25. Shenk, T. 1981. *Curr. Top. Microbiol. Immunol.* 93:25–46
26. Benoist, C., Chambon, P. 1981. *Nature* 290:304–10
27. Dierks, P., van Ooyen, A., Mantei, N., Weissman, C. 1981. *Proc. Natl. Acad. Sci. USA* 78:1411–15
28. Grosschedl, R., Birnsteil, M. L. 1980. *Proc. Natl. Acad. Sci. USA* 77:1432–36
29. Gluzman, Y., Sambrook, J. F., Frisque, R. J. 1980. *Proc. Natl. Acad. Sci. USA* 77:3898–902
30. Ghosh, P. K., Lebowitz, P., Frisque, R. J., Gluzman, Y. 1981. *Proc. Natl. Acad. Sci. USA* 78:100–4
31. Kamen, R., Jat, P., Triesman, R., Favaloro, J. 1982. *J. Mol. Biol.* 159:189–224
32. Pribnow, D. 1975. *Proc. Natl. Acad. Sci. USA* 72:784–88
33. Schaller, H., Gray, C., Hermann, L. 1975. *Proc. Natl. Acad. Sci. USA* 72:737–41
34. Rosenberg, M., Court, D. 1979. *Ann. Rev. Genet.* 13:319–53
35. Grosschedl, R., Birnsteil, M. L. 1980. *Proc. Natl. Acad. Sci. USA* 77:7102–6
36. Grosveld, G. C., de Boer, E., Shewmaker, C. K., Flavell, R. A. 1982. *Nature* 295:120–25
37. Mellon, P., Parker, V., Gluzman, Y., Maniatis, T. 1982. *Cell* 27:279–88
38. McKnight, S. L., Gavis, E. R., Kingsbury, R., Axel, R. 1981. *Cell* 25:385–98
39. McKnight, S. L., Kingsbury, R. 1982. *Science* 217:316–24
39a. McKnight, S. L. 1982. *Cell* 31:355–65
40. Gruss, P., Dhar, R., Khoury, G. 1981. *Proc. Natl. Acad. Sci. USA* 78:943–47
41. Banerji, J., Rusconi, S., Schaffner, W. 1981. *Cell* 27:299–308
42. Moureau, P., Hen, R., Wasylyk, B., Everett, R., Gaub, M. P., Chambon, P. 1981. *Nucleic Acids Res.* 9:6047–68
43. Blair, D. G., McClements, W. L., Oskarsson, M. K., Fischinger, P. J., Vande Woude, G. F. 1980. *Proc. Natl. Acad. Sci. USA* 77:3504–8
44. de Villiers, J., Schaffner, W. 1981. *Nucleic Acids Res.* 9:6251–64
45. Capecchi, M. R. 1980. *Cell* 22:479–88
46. Tyndall, C., LaMantia, G., Thacker, C. M., Favalaro, J., Kamen, R. 1981. *Nucleic Acids Res.* 9:6231–50
47. Conrad, S. E., Botchan, M. R. 1982. *Mol. Cell. Biol.* 2:949–65
48. Weisbrod, S. 1982. *Nature* 297:289–95
49. Weintraub, H. 1979. *Nucleic Acids Res.* 7:781–92
50. Varshavsky, A. J., Sundin, O. H., Bohn, M. J. 1979. *Cell* 16:453–66
51. Herbomel, P., Saragosti, S., Blangy, D., Yaniv, M. 1981. *Cell* 25:651–58
52. Early, P., Rogers, J., Davis, M., Calame, K., Bond, M., Wall, R., Hood, L. 1980. *Cell* 20:313–19
53. Alt, F. W., Bothwell, A. L. M., Knapp,

M., Siden, E., Mather, E., et al. 1980. *Cell* 20:293–301

54. Maki, R., Roeder, W., Traunecker, A., Sidman, C., Wabl, M., et al. *Cell* 24:353–65

55. Amara, S. G., Jonas, V., Rosenfeld, M. G., Ong, E. S., Evans, R. M. 1982. *Nature* 298:240–45

56. Weil, P. A., Luse, D. S., Segall, J., Roeder, R. G. 1979. *Cell* 18:469–84

57. Manley, J. G., Fire, A., Cano, A., Sharp, P. A., Gefter, M. L. 1980. *Proc. Natl. Acad. Sci. USA* 77:3855–59

58. Hagenbuchle, O., Schibler, U. 1981. *Proc. Natl. Acad. Sci. USA* 78:2283–86

59. Deleted in proof.

60. Gidoni, D., Kahana, C., Canaani, D., Groner, Y. 1981. *Proc. Natl. Acad. Sci. USA* 78:2174–78

61. Contreras, R., Fiers, W. 1981. *Nucleic Acids Res.* 29:215–36

62. Ghosh, P. K., Reddy, V. B., Swinscoe, J., Lebowitz, P., Weissman, S. 1978. *J. Mol. Biol.* 126:813–46

63. Flavell, A. J., Cowie, A., Legon, S., Kamen, R. 1979. *Cell* 16:357–72

64. Haegeman, G., Fiers, W. 1980. *J. Virol.* 35:955–61

65. Cowie, A., Jat, P., Kamen, R. 1982. *J. Mol. Biol.* 159:225–55

66. Malek, L. T., Eschenfeldt, W. H., Munns, T. W., Rhoads, R. E. 1981. *Nucleic Acids Res.* 9:1657–73

67. Jat, P., Roberts, J., Cowie, A., Kamen, R. 1982. *Nucleic Acids Res.* 10:871–87

68. Fire, A., Baker, C. C., Manley, J. L., Ziff, E. B., Sharp, P. A. 1981. *J. Virol.* 40:703–19

69. Lee, D. C., Roeder, R. G. 1981. *Mol. Cell. Biol.* 1:635–51

70. Hagenbuchle, O., Bovey, R., Young, R. A. 1980. *Cell* 21:179–87

71. Matsui, T., Segall, J., Weil, P. A., Roeder, R. G. 1981. *J. Biol. Chem.* 255:11992–96

72. Crawford, N., Fire, A., Samuels, M., Sharp, P. A., Baltimore, D. 1981. *Cell* 27:555–61

73. Derman, E., Krauter, K., Walling, L., Weinberger, C., Ray, M., Darnell, J. E. 1981. *Cell* 23:731–39

74. Groudine, M., Peretz, M., Weintraub, H. 1981. *Mol. Cell. Biol.* 1:281–88

75. Tegtmeyer, P., Schwartz, M., Collins, J. K., Rundell, K. 1975. *J. Virol.* 16:168–78

76. Alwine, J. C., Reed, S. I., Stark, G. R. 1977. *J. Virol.* 24:22–27

77. Khoury, G., May, E. 1977. *J. Virol.* 23:167–76

78. Nevins, J. R., Ginsberg, H. S., Blanchard, J. M., Wilson, M. C., Darnell, J. E. 1979. *J. Virol.* 32:727–33

79. Nevins, J. R., Winkler, J. J. 1980. *Proc. Natl. Acad. Sci. USA* 77:1893–97

80. Jessel, D., Landau, T., Hudson, J., Lalor, T., Tenen, D., Livingston, D. M. 1976. *Cell* 8:535–45

81. Reed, S. I., Ferguson, J., Davis, R. W., Stark, G. R. 1975. *Proc. Natl. Acad. Sci. USA* 72:1605–9

82. Tjian, R. 1978. *Cell* 13:165–79

83. Tjian, R. 1978. *Cold Spring Harbor Symp. Quant. Biol.* 43:655–62

84. Myers, R., Rio, D., Robbins, A., Tjian, R. 1981. *Cell* 25:373–85

85. Rio, D., Robbins, A., Myers, R., Tjian, R. 1980. *Proc. Natl. Acad. Sci. USA* 77:5706–10

86. Hansen, U., Tenen, D., Livingston, D., Sharp, P. A. 1981. *Cell* 27:603–12

87. Persson, H., Philipson, L. 1982. *Curr. Top. Microbiol. Immun.* 97:157–203

88. Spear, P. G., Roizman, B. 1981. *Molecular Biology of Tumor Viruses*, ed. J. Tooze, pp. 615–745. Cold Spring Harbor, NY: Cold Spring Harbor Lab.

89. Nevins, J. R. 1981. *Cell* 26:213–20

90. Berk, A. J., Lee, F., Harrison, T., Williams, J., Sharp, P. A. 1979. *Cell* 17:935–44

91. Jones, N., Shenk, T. 1979. *Proc. Natl. Acad. Sci. USA* 76:3665–69

92. Dixon, R. A. F., Schaffer, P. A. 1980. *J. Virol.* 36:189–203

93. Preston, C. M. 1979. *J. Virol.* 29:275–84

94. Watson, R. J., Clements, J. B. 1980. *Nature* 285:329–30

95. Ashburner, M., Bonner, J. J. 1979. *Cell* 17:241–54

96. Durnam, D. M., Palmiter, R. D. 1981. *J. Biol. Chem.* 256:5712–16

97. Hager, L. J., Palmiter, R. D. 1981. *Nature* 291:340–42

98. Tsai, S., Roop, D. R., Tsai, M.-J., Stein, J. P., Means, A. R., O'Malley, B. W. 1979. *Biochemistry* 17:5753–80

99. LeMeur, M., Glanville, N., Mandel, J. L., Gerlinger, P., Palmiter, R., Chambon, P. 1981. *Cell* 23:561–71

100. McKnight, G. S., Palmiter, R. D. 1978. *J. Biol. Chem.* 254:9050–58

101. Swaneck, G. E., Nordstrom, J. L., Krenzalee, F., Tsai, M.-J., O'Malley, B. W. 1979. *Proc. Natl. Acad. Sci. USA* 76:1049–53

102. Wiskocil, R., Bensky, P., Dower, W., Goldberger, R. F., Gordon, J. I., Deeley, R. J. 1980. *Proc. Natl. Acad. Sci. USA* 77:4474–78

103. Ringold, G. M., Yamamoto, K. R., Bishop, S. M., Varmus, H. E. 1979. *Proc. Natl. Acad. Sci. USA* 74:2879–83

104. Payvar, F., Wrange, O., Carlstedt-Duke, J., Okret, S., Gustafsson, J. A., Yamamoto, K. R. 1981. *Proc. Natl. Acad. Sci. USA* 78:6628–32

105. Payvar, F., Firestone, G. L., Ross, S. R., Chandler, V. L., Wrange, O., et al. 1982. *Evolution of Hormone-Receptor Systems*, ed. R. A. Bradshaw, G. N. Gill. New York: Liss
106. Mulvihill, E. R., Le Pennec, J.-P., Chambon, P. 1982. *Cell* 28:621–32
107. Deleted in proof.
108. Nevins, J. R., Darnell, J. E. 1978. *Cell* 15:1477–93
109. Fraser, N. W., Nevins, J. R., Ziff, E. B., Darnell, J. E. 1979. *J. Mol. Biol.* 129:643–56
110. Acheson, N. H. 1978. *Proc. Natl. Acad. Sci. USA* 75:4754–58
111. Legon, S., Flavell, A. J., Cowie, A., Kamen, R. 1979. *Cell* 16:373–88
112. Setzer, D. R., McGrogan, M., Nunberg, J. H., Schimke, R. T. 1980. *Cell* 22:361–70
113. Fraser, N. W., Hsu, M.-T. 1980. *Virology* 103:514–16
114. Hofer, E., Hofer-Warbinek, R., Darnell, J. E. 1982. *Cell* 29:887–93
115. Apirion, D. 1983. *RNA Processing*, ed. D. Apirion. Boca Raton, Fla.: CRC Press. In press
116. Hentschel, C. C., Birnsteil, M. L. 1981. *Cell* 25:301–13
117. Birchmeier, C., Grosschedl, R., Birnsteil, M. L. 1982. *Cell* 28:739–45
118. Shaw, A. R., Ziff, E. B. 1980. *Cell* 22:905–16
119. Nevins, J. R., Wilson, M. D. 1981. *Nature* 290:113–18
120. Akusjarvi, G., Persson, H. 1981. *Nature* 292:420–26
121. Yanofsky, C. 1981. *Nature* 289:751–58
122. Evans, R., Weber, J., Ziff, E., Darnell, J. E. 1979. *Nature* 278:367–70
123. Fraser, N. W., Sehgal, P. B., Darnell, J. E. 1979. *Proc. Natl. Acad. Sci. USA* 2571–75
124. Laub, O., Bratosin, S., Horowitz, M., Aloni, Y. 1979. *Virology* 92:310–22
125. Laub, O., Jakobovits, E. B., Aloni, Y. 1980. *Proc. Natl. Acad. Sci. USA* 77:3297–301
126. Babich, A., Nevins, J. R., Darnell, J. E. 1980. *Nature* 287:246–48
127. Skolnik-David, H., Hay, N., Aloni, Y. 1982. *Proc. Natl. Acad. Sci. USA* 79:2743–47
128. Hay, N., Skolnik-David, H., Aloni, Y. 1982. *Cell* 29:183–93
129. Tamm, I., Kikuchi, T. 1979. *Proc. Natl. Acad. Sci. USA* 76:5750–54
130. Salditt-Georgieff, M., Harpold, M., Chen-Kiang, S., Darnell, J. E. 1980. *Cell* 19:69–78
131. Proudfoot, N. J., Brownlee, G. G. 1976. *Nature* 263:211–14
132. Jung, A., Sippel, A. E., Grez, M., Schutz, G. 1980. *Proc. Natl. Acad. Sci. USA* 77:5759–63
132a. Ahmed, C. M. I., Chanda, R. S., Stow, N. D., Zain, B. S. 1983. *Gene* In press
133. Tosi, M., Young, R. A., Hagenbuchle, O., Schibler, U. 1981. *Nucleic Acids Res.* 9:2313–23
134. Fitzgerald, M., Shenk, T. 1981. *Cell* 24:251–60
135. Sasavage, N., Smith, M., Gillam, S., Woychik, R. P., Rottman, F. M. 1982. *Proc. Natl. Acad. Sci. USA* 79:223–27
135a. Ahmed, C. M. I., Chanda, R., Stow, N., Zain, B. S. 1983. *Gene* In press
136. Salditt-Georgieff, M., Harpold, M., Sawicki, S., Nevins, J., Darnell, J. E. 1980. *J. Cell Biol.* 86:844–48
137. Weber, J., Blanchard, J.-M., Ginsberg, H., Darnell, J. E. 1980. *J. Virol.* 33:286–91
138. Abelson, J. 1979. *Ann. Rev. Biochem.* 48:1035–69
139. Berget, S., Moore, C., Sharp, P. A. 1977. *Proc. Natl. Acad. Sci. USA* 74:3171–75
140. Chow, L. T., Gelinas, R. E., Broker, T. R., Roberts, R. J. 1977. *Cell* 12:1–8
141. Klessig, D. F. 1977. *Cell* 12:9–21
142. Nevins, J. R. 1979. *J. Mol. Biol.* 130:493–506
143. Piper, P. W. 1979. *J. Mol. Biol.* 131:399–407
144. Kinniburgh, A. J., Ross, J. 1979. *Cell* 17:915–21
145. Thimmappaya, B., Shenk, T. 1979. *J. Virol.* 30:668–73
146. Volckaert, G., Feunteun, J., Crawford, L. V., Berg, P., Fiers, W. 1979. *J. Virol.* 30:674–82
147. Chu, G., Sharp, P. A. 1981. *Nature* 289:378–82
148. Goldenberg, C. J., Raskas, H. J. 1981. *Proc. Natl. Acad. Sci. USA* 78:5430–34
149. Weingartner, B., Keller, W. 1981. *Proc. Natl. Acad. Sci. USA* 78:4092–96
149a. Kole, R., Weissman, S. M. 1982. *Nucleic Acids Res.* 10:5429–45
150. Mount, S. M. 1982. *Nucleic Acids Res.* 10:459–72
151. Baird, M., Driscoll, C., Schreiner, H., Sciarratta, G. V., Sansone, G., et al. 1981. *Proc. Natl. Acad. Sci. USA* 78:4218–21
152. Orkin, S. H., Kazazian, H. H., Antonarkis, S. E., Goff, S. C., Boehm, C. D., et al. 1982. *Nature* 296:627–31
153. Treisman, R., Proudfoot, N. J., Shander, M., Maniatis, T. 1982. *Cell* 29:903–11
154. Spritz, R. A., Jagaddeeswaran, P., Choudary, P., Choudary, P. V., Biro, P. A., et al. 1981. *Proc. Natl. Acad. Sci. USA* 78:2455–59

155. Westaway, D., Williamson, R. 1981. *Nucleic Acids Res.* 9:1777–88
156. Busslinger, M., Moschonas, N., Flavell, R. A. 1981. *Cell* 27:289–98
157. Fukumaki, Y., Ghosh, P. K., Benz, E. J., Reddy, V. B., Lebowitz, P., et al. 1982. *Cell* 28:585–93
158. Orkin, S. H., Goff, S. C., Hechtman, R. L. 1981. *Proc. Natl. Acad. Sci. USA* 78:5041–45
159. Felber, B. K., Orkin, S. H., Hamer, D. H. 1982. *Cell* 29:895–902
160. Montell, C., Fisher, E. F., Caruthers, M. H., Berk, A. J. 1982. *Nature* 295:380–84
161. Ziff, E. B. 1981. *Nature* 287:491–99
162. Hayward, W. S., Neel, B. G. 1981. *Curr. Top. Microbiol. Immunol.* 91:217–76
163. Sharp, P. A. 1981. *Cell* 23:643–46
164. Darnell, J. E., Philipson, L., Wall, R., Adesnik, M. 1971. *Science* 174:507–10
165. Zeevi, M., Nevins, J. R., Darnell, J. E. 1981. *Cell* 26:39–46
166. Rogers, J., Wall, R. 1980. *Proc. Natl. Acad. Sci. USA* 77:1877–79
167. Lerner, M. R., Boyle, J. A., Mount, S. M., Wolin, S. L., Steitz, J. A. 1980. *Nature* 283:220–24
168. Busch, H., Reddy, R., Rothblum, L., Choi, Y. C. 1982. *Ann. Rev. Biochem.* 51:617–54
169. Yang, V. W., Flint, S. J. 1979. *J. Virol.* 32:394–407
170. Yang, V. W., Lerner, M. R., Steitz, J. A., Fling, S. J. 1981. *Proc. Natl. Acad. Sci. USA* 1371–75
171. Murray, V., Holliday, R. 1979. *FEBS Lett.* 106:5–7
172. Mathews, M. B. 1980. *Nature* 285:575–77
173. Thimmappaya, B., Weinberger, C., Schneider, R., Shenk, T. 1982. *Cell* 31:543–51
174. Chow, L. T., Broker, T. R., Lewis, J. B. 1979. *J. Mol. Biol.* 134:265–303
175. Banerjee, A. K. 1980. *Microbiol. Rev.* 44:175–205
176. Shatkin, A. J. 1976. *Cell* 9:645–53
177. Chen-Kiang, S., Nevins, J. R., Darnell, J. E. 1979. *J. Mol. Biol.* 135:733–52
178. Stoltzfus, C. M., Dane, R. W. 1982. *J. Virol.* 42:918–31
179. Beltz, G. A., Flint, S. J. 1979. *J. Mol. Biol.* 131:353–73
180. Zeevi, M., Nevins, J. R., Darnell, J. E. 1982. *Mol. Cell. Biol.* 2:517–25
181. Guyette, W. A., Matuski, R. J., Rosen, J. M. 1979. *Cell* 17:1013–23
182. Wilson, M. C., Darnell, J. E. 1981. *J. Mol. Biol.* 148:231–51
183. Blumberg, D. D., Lodish, H. F. 1980. *Dev. Biol.* 78:285–300
184. Blumberg, D. D., Lodish, H. F. 1981. *Dev. Biol.* 81:74–80
185. Chung, S., Landfear, S. M., Blumberg, D. D., Cohen, N. S., Lodish, H. F. 1981. *Cell* 24:785–97
186. Huez, G., Marbaix, G., Hubert, E., Leclercq, M., Nudel, U., et al. 1974. *Proc. Natl. Acad. Sci. USA* 71:3243–46
187. Marbaix, G., Huez, G., Burny, A., Cleuter, Y., Hubert, E., et al. 1975. *Proc. Natl. Acad. Sci. USA* 72:3065–67
188. Huez, G., Marbaix, G., Hubert, E., Cleuter, Y., Leclercq, M., et al. 1975. *Eur. J. Biochem.* 59:589–92
189. Huez, G., Bruck, C., Cleuter, Y. 1981. *Proc. Natl. Acad. Sci. USA* 78:908–11
190. Nudel, U., Soreq, H., Littauer, U. Z., Marbaix, G., Huez, G., et al. 1976. *Eur. J. Biochem.* 64:115–21
191. Huez, G., Marbaix, G., Gallwitz, D., Weinberg, E., Devos, R., et al. 1978. *Nature* 271:572–73
192. Wilson, M. C., Nevins, J. R., Blanchard, J.-M., Ginsberg, H. S., Darnell, J. E. 1980. *Cold Spring Harbor Symp. Quant. Biol.* 44:447–55
193. Babich, A., Nevins, J. R. 1981. *Cell* 26:371–79

Ann. Rev. Biochem. 1983. 52:467–506
Copyright © by Annual Reviews Inc. All rights reserved

THE GENE STRUCTURE AND REPLICATION OF INFLUENZA VIRUS

Robert A. Lamb[1] and Purnell W. Choppin

The Rockefeller University, New York, N.Y. 10021

CONTENTS

PERSPECTIVES AND SUMMARY 468
MORPHOLOGY OF THE INFLUENZA VIRUS PARTICLE
    (VIRION) 469
GENOME STRUCTURE 471
    The Segmented Genome 471
    Assigning Gene Functions to RNA Segments 472
    Sequences Common to all RNA Segments 474
THE SEQUENCES OF INFLUENZA VIRUS RNA SEGMENTS 474
    RNA Segments 1, 2, and 3: Transcriptase-Associated Proteins PB1, PB2, and PA ... 476
    RNA Segment 4: the Hemagglutinin (HA) 476
    RNA Segment 5: the Nucleocapsid Protein (NP) 481
    RNA Segment 6: the Neuraminidase (NA) 482
    RNA Segment 7: the Membrane Protein (M_1) and Nonstructural Protein (M_2) 484
    RNA Segment 8: Nonstructural Proteins NS_1 and NS_2 487
    Overlapping Coding Regions Using Different Reading Frames in Viruses 489
INFLUENZA VIRUS TRANSCRIPTION VIRUS AND REPLICATION 490
    Evidence for a Host-Cell Nuclear Requirement for Primary Transcription 490
    In Vitro Transcription of Influenza Virus mRNAs 491
    In Vivo Transcription of mRNAs 494
    Template A(−)cRNA Synthesis 495
    Virion RNA Synthesis 496
    The Control of RNA Synthesis 496
    Spliced mRNAs and Their Controlled Synthesis 497
    Subgenomic RNAs: A Replication Error Produces Defective Interfering Particles 498
    The Packaging of Eight RNA Segments 498
CONCLUDING REMARKS 499

[1] Present address: Department of Biochemistry, Molecular and Cell Biology, Northwestern University, Evanston, IL 60201.

0066/4154/83/0701-0467$02.00

PERSPECTIVES AND SUMMARY

Influenza viruses are biologically and biochemically unique. They continue to cause yearly epidemics with great health and economic impact, and in pandemic years, e.g. 1957 and 1968, hundreds of millions of cases occurred throughout the world with a high rate of excess mortality. Three immunological types of influenza viruses infect humans—A, B, and C. Influenza A viruses are also widespread in lower animals, particularly avian species, swine, and horses. Much of the challenge presented by influenza virus, biochemically and with respect to disease, derives from its segmented genome, consisting of eight segments of single-stranded RNA. This segmentation facilitates a high incidence of genetic reassortment of segments in infected cells. Such reassortment is apparently responsible for periodic antigenic shifts that occur with influenza A virus and lead to pandemics. The emergence of a new virus strain, to which the population lacks immunity, is thought to result from genetic reassortment between a human strain and an animal strain, yielding a virus that can replicate in humans, but has surface proteins acquired from the animal strain. Because resistance to infection depends on immunity to the viral surface proteins, particularly the hemagglutinin, and humans lack antibodies to these proteins on the new strain, a pandemic results. Furthermore, the previous influenza virus vaccines are not effective against such a new virus. In addition to major antigenic shifts due to genetic reassortment, the antigenicity of the surface proteins also changes (antigenic drift), owing to mutations in the genes coding for these surface proteins. Because of antigenic variation, vaccines must be changed frequently to be effective against new strains.

In addition to the important biological characteristics resulting from the segmented genome, influenza virus is unique among viruses in the site and mechanism by which the RNA genome is transcribed and replicated. Recently there has been a great increase in knowledge of the cellular and molecular biology of influenza. New technologies have facilitated this, such as molecular cloning, in vitro translation systems, and monoclonal antibodies. Many of the viral proteins have been characterized extensively with regard to structure and function, and previously unrecognized proteins have been identified. A striking example is provided by the hemagglutinin, biologically the most important viral protein. Its primary structure has been determined directly by amino acid sequencing and deduced for many strains from the nucleotide sequences of cloned DNA. The three-dimensional structure of the protein has been determined by X-ray crystallography, and antigenic sites have been identified. The amino acid changes in epidemiologically important natural strains and in variants selected by monoclonal antibodies are being located on the molecule, which will provide the basis for understanding at the molecular level the antigenic variation central to the problem of the disease. In addition, the active sites involved in receptor binding and in virus penetration are being characterized, and the mechanisms involved in these activities are being elucidated.

Exciting progress has also been made in defining the mechanisms of expression of the viral genome. A unique mechanism of transcription of viral messenger RNAs has been found, involving the use of cellular RNA polymerase II to synthesize cellular mRNAs, from which the viral mRNAs borrow the cap structure and several nucleotides using viral enzymatic activities. Viral proteins have been assigned to specific viral RNAs; in some instances two proteins are coded by a single viral RNA segment using overlapping genes read in different reading frames. The mRNA for one protein is a colinear transcript and for the other it is formed by splicing, the first evidence for splicing of an mRNA of an RNA virus. Thus, influenza virus provides a potentially valuable system for studying the mechanism of splicing of mRNA.

This article reviews recent advances in knowledge of virus structure, with emphasis on the viral genome, and the mechanisms of viral biosynthesis.

MORPHOLOGY OF THE INFLUENZA VIRUS PARTICLE (VIRION)

Laboratory-adapted strains of influenza virus are roughly spherical with a diameter of 80–120 nm; however, newly isolated strains are pleomorphic and contain long filamentous particles (1–3). The virions consist of an envelope composed of a membrane from which glycoproteins project like spikes. Within the envelope is the segmented helical ribonucleoprotein (nucleocapsid) of the virus. There are two types of surface glycoproteins— rod-shaped spikes, composed of the hemagglutinin (HA) protein and, in lesser abundance, mushroom-shaped spikes, the neuraminidase (NA) (4). The HA protein is involved in attachment of the virus to neuraminic acid-containing cellular receptors and in the penetration of the virus into the cell. The NA protein cleaves neuraminic acid from receptors on the cell surface, as well as from other sialoglycoproteins and glycolipids. The viral proteins are discussed below in sections dealing with the RNA segments and the proteins for which they code.

Influenza virus particles form by budding from the plasma membrane of infected cells (5, 6), with the lipid bilayer of the viral membrane being derived from the plasma membrane (7, 8). Just beneath the lipid bilayer is an electron dense layer composed of the viral membrane protein (M_1).

The internal ribonucleoprotein (RNP) structures containing the eight different segments of RNA have the appearance of flexible rods (9, 10) that may exhibit loops on one or both ends and alternating major and minor grooves; this suggests that they are formed by a strand folded back on itself and coiled into a double helix. As circular forms have also been observed, the termini of the RNA may not necessarily be at one end of the rod (11). The tightness of the coiling of the helix depends on the salt concentration (11). The RNPs can be separated into different size classes (10, 12), and each RNP appears to contain one segment of RNA.

The RNPs consist of four proteins and RNA. The nucleoprotein (NP) is the predominant protein, and three proteins (PB1, PB2, and PA) are found in lower quantities. This RNP complex has RNA-dependent RNA polymerase (transcriptase) activity (13–16). Figure 1 shows a schematic diagram of the virion.

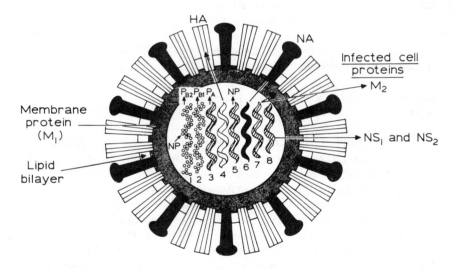

Figure 1 A schematic diagram of the structure of the influenza virion. Two types of glycoprotein spikes, which possess hemagglutinating (HA) and neuraminidase (NA) activities, are inserted through the lipid bilayer of the viral membrane. The distribution of the spikes are not meant to reflect the abundance of these proteins in the virion. On the inner surface of the lipid bilayer is the membrane protein (M_1). Within the envelope are the eight segments of single-stranded genome RNA contained in the form of helical ribonucleoproteins. Segments *1* and *2* show schematically the nucleocapsid (NP) proteins, which with small amounts of proteins PB1, PB2, and PA form a complex with RNA-dependent RNA polymerase (transcriptase) activity. The coding assignments of the eight RNA segments are also illustrated. RNA segments *7* and *8* each code for more than one protein; however M_2, NS_1 and NS_2 are nonstructural proteins found only in infected cells. The diagram is not drawn to scale.

GENOME STRUCTURE

The eight segments of single-stranded RNA code for seven virion structural proteins (PB1, PB2, PA, HA, NA, NP, and M_1) and three nonstructural proteins found only in infected cells (NS_1, NS_2, and M_2). The viral mRNAs are complementary to the genome RNAs. Because these mRNAs must be transcribed from virion RNAs by the virion-associated RNA transcriptase, naked RNA is not infectious. The complete nucleotide sequence of an influenza A virus has been established, and considerable information is known about the structure of the proteins including the X-ray crystallographic structure to 3-Å resolution of the HA and NA proteins.

The Segmented Genome

Early genetic studies showing a high rate of recombination between strains (17, 18) led to the suggestion that the influenza virus genome existed on discrete segments (19). Subsequently, biochemical evidence obtained by sucrose gradient analysis (20, 21) and polyacrylamide gel electrophoresis (22–24) confirmed this. Further genetic evidence for a segmented genome was provided by studies with temperature sensitive (ts) mutants (see 25) in which a high frequency of recombination was observed and complementation groups established. Recombination that results from the exchange of segments of influenza virus RNA is termed reassortment.

A major step in understanding the influenza virus genome was the electrophoretic separation of the virion RNA into eight segments on polyacrylamide gels containing 6 M urea (26–31). These, numbered 1–8 in decreasing order of electrophoretic mobility, are now known to range in size from 2341–890 nucleotides. The important demonstration that the eight RNAs are distinct was done by two-dimensional oligonucleotide finger printing (31). It was initially thought that each RNA segment coded for one protein, because the coding capacity of each segment, estimated from its chain length, corresponded to the sizes of the then known eight viral proteins (14, 28, 31–33). The segments from different virus strains had differing migration patterns on the acrylamide/urea gels (28), and this was exploited in assigning gene functions in recombination experiments (see below). However, when the RNA segments were completely denatured with glyoxal, their mobilities were identical among strains, except for those coding for HA and NA (34). Thus, the differences observed among other RNA segments on the acrylamide/urea gels are probably fortuitous and dependent on sequence differences and incomplete chain denaturation.

The RNAs of influenza B and C viruses have also been examined. Influenza B virus contains eight RNA segments, ranging in size from approximately 2500 to 1096 nucleotides; influenza C virus contains seven segments, ranging from 3000 to 1000 nucleotides (30, 34–38). The finding of only seven segments in influenza C virus may reflect the presence of only one glycoprotein that is subsequently cleaved into subunits (39, 40). Influenza C viruses adsorb to a different, non-neuraminic acid containing receptor and have a receptor-destroying activity that is not neuraminidase (41, 42).

Assigning Gene Functions to RNA Segments

The assignment of viral polypeptides to RNA segments has been made by the following three methods:

1. Identification was made possible by differences between strains in the electrophoretic mobility of RNA segments on polyacrylamide/urea gels, coupled with differences in specific proteins detected by immunological methods, or by differences in their electrophoretic mobility. Recombinants between two strains were made, and the mobilities of the RNA segments or polypeptides between the parents and the recombinants were compared. From analysis of the patterns obtained, gene assignments could be made (29, 33, 43–46).
2. Base sequence homologies were calculated between individual RNA segments of different strains, including the FPV strain (which is plaque selectable), ts mutants of FPV with known phenotypic defects, and recombinants between the wt strains and mutants. This permitted the parental origin of any segment to be obtained (47).
3. The most direct method is based on the ability to translate influenza virus mRNAs in vitro using cell extracts, but the inability to translate a mRNA/vRNA hybrid. Cytoplasmic poly(A) containing mRNAs were extracted from infected cells and hybridized to individual vRNA segments, and the total RNA was then translated in wheat germ extracts. Each vRNA segment specifically prevents the translation of the polypeptide(s) for which it codes (45–49).

The conclusions of such experiments were complicated by the RNA segments' failure to migrate on acrylamide/urea gels in the same order in all strains, e.g. segment 5 codes for NP in some strains and NA in others (29, 49a). Further, the P proteins do not migrate in the same order in different strains, e.g. the originally designated P2 of one strain is functionally equivalent to P3 of another strain (50). Thus the nomenclature of the P proteins is now based on their behavior during isoelectric focusing (two basic and one acidic), and the RNA segments are numbered according to migration order in glyoxal gels. The gene assignments are as follows: RNA segment 1 codes for PB2; 2 for PB1; 3 for PA; 4 for HA; 5 for NP; 6 for NA; 7 for M_1 and M_2; and 8 for NS_1 and NS_2. Table 1 shows these assignments.

Table 1 Influenza virus genome RNA segments and coding assignments

Segment	Length[a] (nucleotides)	mRNA length[b] (nucleotides)	Encoded[c] polypeptide	Nascent[d] polypeptide length (aa)	Approx. no.[e] molecules per virion	Remarks[f]
1	2,341	2,320	PB2	759	30–60	Host-cell RNA cap binding: component of RNA transcriptase
2	2,341	2,320	PB1	757	30–60	Initiation of transcription: possibly endonuclease activity: component of RNA transcriptase
3	2,233	2,211	PA	716	30–60	Elongation of mRNA chains?: component of RNA transcriptase
4	1,778	1,757	HA	566	500	Surface glycoprotein, trimer; major antigenic determinant
5	1,565	1,540	NP	498	1,000	Associated with RNA segments to form ribonucleoprotein: structural component of RNA transcriptase
6	1,413	1,392	NA	454	100	Surface glycoprotein: neuraminidase activity. Tetramer
7	1,027	1,005	M_1	252	3,000	Major protein component of virus: underlies lipid bilayer
		316	M_2	96		Spliced mRNA, nonstructural protein: function unknown
		276	?	?9		Spliced mRNA, peptide predicted by nucleotide sequence only
8	890	868	NS_1	230		Nonstructural protein: function unknown
		395	NS_2	121		Spliced mRNA, nonstructural protein: function unknown
Total	13,588					

[a] For A/PR/8/34 strain: see Table 2 for references.
[b] Deduced from RNA sequence, excluding poly A tract.
[c] Determined by biochemical and genetic approaches (see text).
[d] Determined by nucleotide sequence analysis and protein sequencing (see text).
[e] Adapted from (8).
[f] See text for references.

Sequences Common to all RNA Segments

The terminal nucleotide sequences of individual RNA segments are of interest because of their involvement in transcription, translation, or replication. Limited direct RNA sequencing of the 5' and 3' ends of the eight segments of influenza A viruses revealed common sequences of 13 nucleotides at the 5' end, and 12 nucleotides at the 3' end of each segment (51–53) [with some strains there is either a C or U residue at nucleotide 4 at the 3' end]. In addition, the conserved 3'- and 5'-terminal sequences show partial and inverted complementarity, which may be important for replication of the RNA (52, 53). The transcriptase may recognize the 3'-terminal sequences of the virion ($-$) strand and because of the inverted complementary sequences at the 5' end of the same strand, the polymerase could then bind to similar sequences at the 3' end of the complementary ($+$) strand RNA during replication. A third region common to all eight genome RNA segments is a uridine tract 15–21 residues from the 5' end, which is the signal for initiating polyadenylation during mRNA synthesis (54; see below). Influenza B and C viruses also contain common 3' and 5' ends that resemble those of the A viruses (51, 53).

THE SEQUENCES OF INFLUENZA VIRUS RNA SEGMENTS

Although the nucleotide sequence of the RNA segments could have been determined by successive ribonuclease digestions and homochromato-graphy (55), only limited progress was made (56), owing to the laborious-ness of the method. The development of molecular cloning of double-stranded DNA derived from single-stranded RNA, coupled with rapid DNA-sequencing methods (57, 58), greatly speeded the sequencing of the influenza virus genome. Purified virion RNA (vRNA) was the preferred strand from which to make cDNA copies, rather than mRNAs from infected cells (mRNAs are incomplete copies of the vRNA segment; see below). A variety of vectors have been used for cloning in bacteria, including the plasmid pBR322 and its high-copy derivative, pAT153, bacteriophage lambda, and the replicative form of derivatives of bacteriophage M13. Nucleotide sequences have been determined from cloned DNA copies of the RNA segments (see Table 2).

To investigate the biologically important genetic variation of the virus, sequencing of the RNA segments of many strains has begun, but this is an enormous undertaking. One approach that yields considerable information is to modify dideoxychain-termination sequencing so that reverse tran-scriptase primed with oligo dT_8dA can copy a polyadenylated vRNA segment (59–61). A disadvantage of the method is that the length of sequence obtained is related to the resolution of the polyacrylamide gel used to separate oligonucleotides.

The complete nucleotide sequences of all eight RNA segments of one strain of influenza A virus have been obtained (see Table 2), as well as for

Table 2 Compilation of complete nucleotide sequences of influenza virus RNA segments

RNA segment[a]	Encoded protein	Strain	Authors	Reference
1	PB2	A/PR/8/34	Fields & Winter (1982)	65
		A/WSN/33	Kaptein & Nayak (1982)	69
2	PB1	A/PR/8/34	Winter & Fields (1982)	66
		A/WSN/33	Sivasubramanian & Nayak (1982)	289
		A/NT/60/68	Bishop et al (1982)	67
3	PA	A/PR/8/34	Fields & Winter (1982)	65
		A/NT/60/68	Bishop et al (1982)	68
4	HA	A/PR/8/34 (H1)	Winter et al (1981)	279
		A/WSN/33 (H1)	Hiti et al (1981)	280
		A/Japan/305/57 (H2)	Gething et al (1980)	281
		A/Duck/Ukraine/63 (H3)	Fang et al (1981)	282
		A/Aichi/2/68 (H3)	Verhoeyen et al (1980)	283
		A/NT/60/68 (H3)	Both & Sleigh (1980)	284
		A/Memphis/102/72 (H3)	Sleigh et al (1980)	285
		A/Victoria/3/75 (H3)	Min Jou et al (1980)	286
		A/Bangkok/1/79 (H3)	Both & Sleigh (1981)	113
		A/FPV/Rostock/34 (H7)	Porter et al (1979)	287
		B/Lee/40	Krystal et al (1982)	116
5	NP	A/PR/8/34	Winter & Fields (1981)	142
		A/PR/8/34	Van Rompuy et al (1981)	143
		A/NP/60/68	Huddleston & Brownlee (1982)	144
6	NA	A/PR/8/34 (N1)	Fields et al (1981)	160
		A/WSN/33 (N1)	Hiti & Nayak (1982)	161
		A/NT/60/68 (N2)	Bentley & Brownlee (1982)	163
		A/Udorn/72 (N2)	Markoff & Lai (1982)	162
	NA and NB	B/Lee/40	Shaw et al (1982)	167
			Shaw et al (unpublished)	
7	M_1 and M_2	A/PR/8/34	Winter & Fields (1980)	180
		A/PR/8/34	Allen et al (1980)	181
		A/Udorn/72	Lamb & Lai (1981)	182
		A/FPV/Rostock/34	McCauley et al (1982)	288
		B/Lee/40	Briedis et al (1982)	186
8	NS_1 and NS_2	A/PR/8/34	Winter et al (1981)	207
		A/PR/8/34	Baez et al (1980)	205
		A/Udorn/72	Lamb & Lai (1980)	182
		A/Duck/Alberta/60/76	Baez et al (1981)	206
		A/FPV/Rostock/34	Porter et al (1980)	204
		B/Lee/40	Briedis & Lamb (1982)	209

[a] In migration order on polyacrylamide gels after glyoxal treatment (34).

several segments of other strains and for segments 4, 6, 7, and 8 of influenza B virus. The properties of each segment and its encoded polypeptide(s) are described below.

RNA Segments 1, 2, and 3: Transcriptase-Associated Proteins PB1, PB2, and PA

Early studies demonstrated one or two P proteins in virions (62, 63), and improved conditions for polyacrylamide gel electrophoresis revealed three proteins associated with virion RNA transcriptase activity (14, 16, 32) designated P1, P2, P3 in order of mobility on polyacrylamide gels (apparent M_r 95,000–82,000). Because of the differences in migration of these proteins of different strains, this nomenclature has been replaced by one based on isoelectric focusing and electrophoresis (64). The faster migrating basic P protein coded by segment 1 is designated PB2, the slower migrating basic protein coded by segment 2 is PB1, and the acidic protein coded by segment 3 is PA (see Table 1).

The sequences of segments 1, 2, and 3 (65–69), have revealed that 1 and 2 are 2341 nucleotides long and code for proteins of 759 and 757 amino acids, respectively. Because of the similar size of PB1 and PB2, the ability to separate them on polyacrylamide gels must be due to factors such as differential binding of sodium dodecylsulfate. RNA segment 3 is 2233 nucleotides long and codes for a protein of 716 amino acids. It has been suggested that, because the P proteins are involved in RNA synthesis and are so similar in size, they may have evolved from a single ancestral polymerase gene; however, no simple sequence homology has been detected (66).

Studies with ts mutants defective in RNA synthesis suggested that PB1 and PB2 are involved in mRNA synthesis and PA in vRNA synthesis (70, 71). Analysis of the in vitro transcription reaction indicated that PB2 is involved in binding to the cap-structure of host-cell mRNA (72–75) and PB1 in initiating transcription (75).

RNA Segment 4: the Hemagglutinin (HA)

HA is an integral membrane glycoprotein responsible for virus adsorption, so named because of the ability of the virus to agglutinate erythrocytes (76, 77) by attachment to specific neuraminic-acid-containing glycoproteins on erythrocytes (78). HA is the most important viral antigen against which neutralizing antibodies are directed (79), and antigenic variation in this protein is the major factor in influenza epidemics. In addition to mediating attachment of virus to the plasma membrane of the host cell, the HA is responsible for penetration of the virus into the cells (80, 81). Depending on the virus strain, host-cell type, and growth conditions, the HA molecule (M_r

$\sim$ 77,000) can be cleaved proteolytically into two polypeptide chains, HA1 ($M_r \sim$ 50,000) and HA2 ($M_r \sim$ 27,000) (63, 82–85), which are held together by disulfide bonds (86). Cleavage of HA does not affect its receptor-binding activity (83, 84) but activates the infectivity of the virus (80, 81), and is therefore important in the spread of infection in the organism and pathogenicity (87–89).

Examination of solubilized HA spikes by electron microscopy (4, 90), after their removal from the virion by detergent, suggested that they consisted of more than one subunit and were attached to the viral membrane by a hydrophobic base. The HA spike is now known to be a noncovalently linked trimer (91, 92). Correlation of the sizes of HA1 and HA2, released by protease or detergent with the observation that removal of the detergent from solubilized HA caused aggregation, while protease-released spikes were soluble, suggested that HA2 is inserted in the membrane (4, 62, 93). This was confirmed by protein-sequencing studies showing that HA1 contained the N-terminal portion of HA, and HA2 contained the C-terminal (94), hydrophobic region which is embedded in the viral membrane and removed by proteolysis.

THE STRUCTURE OF THE HEMAGGLUTININ GENE   The complete nucleotide sequence of RNA segment 4 derived by molecular cloning of cDNA copies has been obtained for the H1, H2, H3, and H7 subtypes of influenza A virus and for several variants of H1 and H3 strains (see Table 2). Extensive direct protein sequence data has been obtained for H2 and H3 strains (95, 96). The primary structures of the different HA types are very similar. For details of the amino acid changes among antigenically different HA types and for changes within a type, the reader is referred to reviews by Ward (95) and Webster et al (97). Figure 2 shows the primary structure of an HA molecule.

The RNA segment is 1765 nucleotides long; 20 nucleotides precede the AUG translation initiation codon, with the translated region extending 1698 nucleotides (specifying 566 amino acids) before a 3'-terminal untranslated region comprised of 35 nucleotides. The hydrophobic signal peptide is 16 amino acids; HA1 consists of 328 amino acids, and HA2, of 221 amino acids. With most strains, a single arginine residue is lost between HA1 and HA2 on proteolytic cleavage. After the initial action of a trypsin-like protease, an exopeptidase of the carboxypeptidase B type removes arginine from the cleavage site (98, 99). The N-terminus of HA2 is hydrophobic and highly conserved among strains, and is thought to be involved in virus penetration as described below. Near the C-terminus of HA2 is a stretch of 24 hydrophobic residues that span the lipid bilayer and anchor the HA in the lipid envelope. The precise location of a presumed signal interrupting the translocation of the HA across the membrane ["Stop-Transfer Signal" (100)] remains to be shown; however, it must be

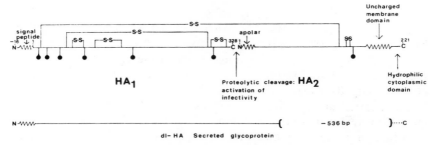

Figure 2 (Top) Schematic diagram of the A/Aichi/68 (H3) hemagglutinin polypeptide showing the *N*-terminal signal sequence, which is cleaved off during cotranslational membrane insertion, and the positions of the C-terminal uncharged membrane anchoring domain, and the site of proteolytic cleavage that yields polypeptide chains HA1 and HA2, and causes activation of infectivity. The disulfide bonds are shown and carbohydrate-attachment sites are illustrated by the line with the filled circle head. [Adapted from Wilson et al (92).]

(*Bottom*)   Schematic diagram of a hemagglutinin that has lost a large portion of the C-terminus by genetic manipulation using a simian virus 40-HA expression vector (indicated as a 536 base pair deletion in the DNA coding for HA). The synthesized HA molecule is a secreted glycoprotein [from Sveda et al (101)].

within 178 amino acids of the C-terminus of HA2, since in a eukaryotic expression system containing cloned HA DNA from which this region was deleted, the synthesis of a soluble HA molecule was observed (see Figure 2; 101).

The carbohydrate composition of HA has been examined by a number of workers (102–105). In the A/Aichi/68 strain there are seven possible glycosylation sites (asparagine-*x*-serine or asparagine-*x*-threonine). Analyses have shown four complex carbohydrate chains on HA1 and one on HA2, consisting of N-acetylglucosamine, mannose, galactose, and fucose, and two simple carbohydrate chains on HA1 consisting of N-acetylglucosamine and mannose. All chains are linked via N-glycosidic linkages to asparagine residues (92, 106). Terminal silalic acid residues are absent, owing to the action of the viral neuraminidase (107). There are four internal disulfide bonds in HA1, one in HA2, and one linking HA1 and HA2 (108, 109a).

THE THREE-DIMENSIONAL STRUCTURE OF THE HEMAGGLUTININ   Cleavage of HA with bromelain releases a structure (BHA) that is water soluble and has lost only the C-terminal hydrophobic region of HA2 (93, 94). BHA has been crystallized and the three-dimensional X-ray structure determined to 3-Å resolution (92; see Figure 3). The molecule is an elongated cylinder 135 Å long consisting of: (*a*) a long fibrous stem extending 76 Å from the membrane and containing two antiparallel α-helices that terminate near the membrane in a compact five-stranded antiparallel β-sheet globular fold; and (*b*) a distal globular region of antiparallel β-sheet structure. This

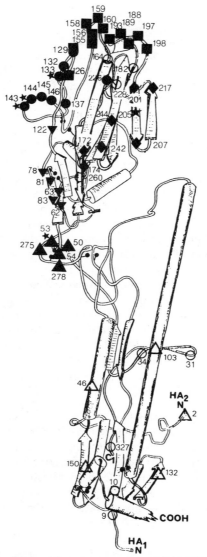

Figure 3 Schematic drawing of a monomer of the A/Aichi/68 hemagglutinin. Cylinders represent α-helices, flat-twisted arrows represent extended β-chains, and filled circles represent disulfide bonds. Amino acid substitutions observed between 1968–1979 are marked and grouped into five regions, A (filled circle), B (filled square), C (filled triangle), D (filled diamond), E (filled upside-down triangle). Conserved substitutions (open circle) and substitutions in HA2 (open triangle) are also shown [from Wiley et al. (109); J. J. Skehel, I. A. Wilson, D. C. Wiley, in preparation]. Stars indicate the positions of amino acid substitutions observed in the hemagglutinins of antigenic variants selected using monoclonal antibodies. Amino acid sequence data communicated by G. W. Both and J. J. Sleigh and from Webster & Laver (110). The antigenic sites described by these substitutions are also found (with the exception of C) for an H1 hemagglutinin from A/PR/8/34. From Caton et al (115a).

globular region is composed entirely of residues from HA1 and is connected to the HA2 fibrous stem by only two antiparallel chains from HA1. The C-terminus of HA1 is 21 Å from the N-terminus of HA2, indicating a substantial rearrangement in this region when HA is activated by cleavage. The N-terminus of HA1 is close to the membrane. The variable antigenic determinants are in the globular region. Four antigenic sites have been identified (109–111) from amino acid sequence changes in antigenic variants selected with monoclonal antibodies (112) and also in naturally occurring variants (92, 112–115). Mapping of additional changes in amino acid sequence and correlation with antigenic drift is continuing.

Although the X-ray structure has been determined only for the H3 HA, sequence comparison with the H1 and H2 subtypes indicates that 22% of HA_1 and 45% of HA_2 residues are absolutely conserved, including the six disulfide bridges and many structurally important residues such as proline (21 residues) and glycine (9 residues). Therefore it is expected that the basic structure of the HA of H1 and H2 subtypes is very similar to H3.

The influenza B virus HA has also been sequenced from a cDNA copy of RNA segment 4 (116). The gene is 1882 nucleotides long and codes for 584 amino acids. The major structural features of A viruses have been conserved in the B virus HA, including a hydrophobic signal peptide, hydrophobic N- and C-termini of HA2, and the HA1/HA2 cleavage site. Between influenza A and B viruses, 24% of the amino acids in HA1 and 39% of the amino acids in HA2 are conserved, suggesting a close evolutionary relationship. The largest area of conserved sequences is the N-terminus of HA2.

BIOSYNTHESIS OF THE HEMAGGLUTININ   The available evidence suggests that HA is translated from a monocistronic mRNA derived from RNA segment 4, and that its processing follows the pathway of many integral membrane proteins with cotranslational insertion into a vesicle after attachment of the growing polypeptide chain/ribosome complex (100). The hydrophobic N-terminal signal peptide is removed by proteolysis (59, 117) and the elongating chain is glycosylated in the rough endoplasmic reticulum vesicle (118–120) with further glycosylation occurring during transport of the vesicle to the plasma membrane (121, 122). A detailed description of the transport of HA has not been reported, but it is likely that it is mediated by clathrin coated vesicles as shown for the vesicular stomatitis virus (VSV) glycoprotein (see 123). In some cell types, influenza virus buds from the apical surface whereas VSV buds from the basolateral surface (124–126). The mechanism of the sorting of the glycoproteins between cell surfaces remains to be determined.

ACTIVATION OF INFECTIVITY AND IN VITRO FUSION   The most highly conserved sequence in the HA molecule is the apolar N-terminus of HA2, thought to be involved in virus penetration. As mentioned above, cleavage of HA to HA1 and HA2 is required for infectivity and fusion with cell membranes (80, 81, 127). Cleavage also results in a confirmational change in the HA molecule (81, 128). A hydrophobic sequence similar to that at the N-terminal of HA2, but longer, is present at the cleavage-activation site of the fusion protein (F) of Sendai virus (129–133). The importance of this sequence to the fusion process is indicated by the finding that synthetic peptides mimicking the appropriate sequence of either the HA2 N-terminus, or the Sendai virus F_1 protein N-terminus, inhibit virus-induced membrane fusion and virus penetration (132). It is difficult to visualize how the N-terminus of HA2 is directly involved in cell penetration unless a second confirmational change occurs after receptor binding, because it is located 100 Å from the distal tip and 35 Å from the membrane (92). Influenza virus HA can cause cell fusion in vitro at acid pH (5.0–5.55) (134–136), and it has been proposed that fusion activity at acid pH reflects the pH in endocytic vesicles, the presumed site of fusion of the viral membrane with the cell membrane in vivo.

RNA Segment 5: The Nucleocapsid Protein (NP)

The major protein of the nucleocapsid is coded for by RNA segment 5. The NP protein interacts not only with itself and RNA in the nucleocapsid complex, but also with the PB1, PB2, and PA proteins to form the transcriptase. In addition, biochemical evidence indicates that the RNAs can interact with the membrane protein (M_1) (137), which supports the early concept of nucleocapsids interacting with the viral membrane protein at the plasma membrane (138). NP is the type-specific antigen used to classify viruses as influenza A, B, and C viruses. Monoclonal antibody studies showed that NP contains at least three non-overlapping antigenic areas, and binding to one of these inhibited in vitro transcriptase activity (139). Whether this results from steric hinderance or direct involvement of this site in the function of NP is not clear. NP is phosphorylated at up to one serine residue per molecule (140, 141), but it is not clear what percentage of NP molecules are phosphorylated or what role the phosphate plays.

The sequence of RNA segment 5 has been obtained for two strains, A/PR/8/34 [(142, 143) with corrections by Min Jou, unpublished data, 1981] and A/NT/60/68 (144). Segment 5 is 1565 nucleotides long with a 5′-noncoding region of 45 nucleotides and a 3′-noncoding region of 26 nucleotides. The 1494 nucleotides code for 498 amino acids with a M_r of 56,101 for NP, which is very close to the observed value (9, 62, 63, 82). NP is rich in arginine with a net charge of + 14 at pH 6.5. There are no clusters of basic residues for the interaction with RNA, and therefore the RNA is probably associated with many regions of the molecule. Based on the total nucleotides of the viral genome (13,588) and approximately 500–900 molecules of NP per virion (8), there are approximately 20 nucleotides per protein subunit. The RNA in the nucleocapsid is susceptible to digestion with ribonuclease without disrupting the RNP structure (9, 12).

RNA Segment 6: the Neuraminidase (NA)

RNA segment 6 codes for NA, an integral membrane glycoprotein. The neuraminidase hydrolytically cleaves the glycosidic bond joining the keto group of N-acetylneuraminic acid (sialic acid) to D-galactose or D-galactosamine (145).

The NA spikes are mushroom shaped with a stalk and head (4, 146), and consist of a tetramer of $M_r \sim 220,000$. This structure has been confirmed by X-ray diffraction patterns (147). The individual polypeptide chains ($M_r \sim 56,000$) are held together by disulfide bonds (148). Incomplete NA molecules can be isolated by protease digestion of virus particles, and these "head" structures have lost their hydrophobic properties and the thin stalk, while retaining enzymatic and antigenic properties (61, 148–152).

The functional role of NA in virus replication is still not clear but appears to be related to the ability of the virus to free itself from neuraminic-acid-containing structures. During virus assembly this may promote release of budding virus particles from the host-cell membrane (153). Neuraminidase removal of sialic acid may also expose HA to cleavage (154); such a mechanism has been implicated in the neurovirulence of some virus strains (155). Neuraminidase activity may also enable the virus to elute from inhibitory sialoglycoproteins present in the respiratory tract, allowing it to find its way to the target epithelial cells. Antibodies to NA do not neutralize infectivity, but restrict multiple cycles of replication (156, 157) and may attenuate illness (158). The active site of the NA is thought to be relatively inaccessible to antibody, since inhibition of enzyme activity by antibodies is observed only when large substrates such as fetuin are employed, not with the small substrates sialyllactose (159), implying an indirect steric effect of antibody.

The nucleotide sequence of segment 6 has been obtained for the N1 and N2 subtypes (160–163). The PR/8/34 (N1) segment (160; see Figure 4) is 1413 nucleotides long with a 5'-noncoding region of 20 nucleotides and a 3'-noncoding region of 31 nucleotides; the 1362 nucleotides code for 453 amino acids with an M_r of 50,087. There are five possible glycosylation sites, and for the N2 Tokyo/67 strain the data suggest that the carbohydrate chains are of both the complex N-acetyl lactosamine and simple oligo-mannoside type attached via N-glycosidic linkages to asparagine residues (61).

The only hydrophobic region in the NA molecule sufficient to span a lipid

Figure 4 Schematic diagram of the neuraminidase polypeptide (N1 subtype). The position of the extended signal sequence, the only uncharged domain in the molecule capable of spanning a lipid bilayer is shown. The *N*-terminal methionine residue that initiates protein synthesis is not removed. Carbohydrate attachment sites are illustrated by the lines and filled circles.

bilayer is near the N-terminus and consists of 29 residues, 18 of which are hydrophobic and 11 neutral. On this basis, it was suggested that NA was inserted in the membrane by its N-terminus (160); this has been confirmed by protein-sequence studies on the intact and protease-cleaved "head" molecules (61). Thus, NA is similar in orientation to two other glyco-proteins, the intestinal brush border amino peptidase and isomaltase (164, 165), in having an "extended signal sequence" (100, 166) that is not cleaved and that both transfers the protein across the membrane and remains in the bilayer to anchor the protein. The N-terminal methionine of NA is not cleaved off (61); this is an unusual finding.

The nucleotide sequence of segment 6 of influenza B virus coding for NA has also been determined (167) and shows similar structural features to the NA of A viruses. An open reading frame starting at the second AUG codon from the 5′ end of the mRNA codes for a NA of 466 amino acids. In most eukaryotic mRNAs, including those of the other influenza virus genes sequenced, the first AUG is the initiation site for translation, although there are cases in which the first AUG codon is not used (reviewed in 168). What is unusual in the NA sequence of influenza B virus is that the first AUG codon, which is separated from the second by four nucleotides, is followed by an open reading frame of 100 amino acids overlapping the NA reading frame (167). Recent experiments using selected radioactive amino acids and hybrid-arrest translation have indicated that a polypeptide ($M_r \sim 11,000$), designated NB, is encoded in this reading frame. NB is synthesized both in infected cells and in vitro using wheat germ extracts (M. W. Shaw, R. A. Lamb, P. W. Choppin, unpublished results). The mRNAs for NA and NB are the same apparent size, and therefore it is possible that the transcript from RNA segment 6 is bicistronic. Comparisons of the nucleotide and amino acid sequences with those of NA of influenza A revealed seven regions of extensive homology within the central position of the molecule, including 12 conserved cysteine residues. Five other cysteine residues in the terminal portions were also conserved (167).

The biosynthesis of NA has proved difficult to examine, partly because in many strains NA migrates together on polyacrylamide gels with the nucleocapsid protein, and it is synthesized in smaller amounts than HA. However, the unglycosylated NA synthesized in vitro can be detected readily (45, 46). Association of partially glycosylated NA on rough endoplasmic reticulum, migration, and further glycosylation to smooth membranes and the plasma membrane has been observed (82, 118, 121, 169). However, a detailed analysis of the transport and processing should be most interesting in view of the N-terminal membrane insertion of NA.

RNA Segment 7: the Membrane Protein (M_1) and Nonstructural Protein (M_2)

Segment 7 codes for two proteins, M_1 and M_2, and possibly a small peptide (M_3). Three separate mRNAs are derived from segment 7, one of which is a colinear transcript with the RNA segment and two are spliced mRNAs containing interrupted regions (170).

The membrane (or matrix) protein (M_1) is the most abundant virion protein (62, 171) and the only protein in the virion in sufficient quantity to form a shell beneath the lipid bilayer (10, 172). The M_1 protein can interact with lipid in vitro (173–175). In addition to playing a structural role in the virion envelope, the M_1 proteins may recognize the viral glycoproteins and form a domain on the inner surfaces of the plasma membrane that subsequently provides a binding site for ribonucleoprotein during virus assembly (138). M_1 also appears to govern the sensitivity of the virus to the antiviral drug, amantidine (176, 177).

The complete sequence of segment 7 of influenza A virus has been obtained for three influenza A strains (see Table 2). The segment is 1027 nucleotides long with a 5'-noncoding region of 25 nucleotides and a 3'-noncoding region of 23 nucleotides. M_1 contains 252 amino acids (M_r = 27861) and the predicted amino acid sequence fits with amino acid composition and partial amino acid sequence data (60, 178). The protein is rich in arginine and methionine and is somewhat hydrophobic, which may account for its solubility in chloroform/methanol (179). There is one region of 37 amino acids in the middle of the molecule with 16 hydrophobic and only two charged amino acids. This region could be involved in hydrophobic interactions with either protein or lipid.

The coding region of the M_1 protein occupies only 75% of the length of segment 7. However, at the 5' end of the vRNA segment, there is a second reading frame, which could code for 97 amino acids, and which overlaps the first reading frame for the M_1 protein by 68 nucleotides (see Figure 5) (180–182). A protein derived from this reading frame (M_2: apparent $M_r \sim 15,000$) was identified in infected cells and shown to be coded by segment 7 in hybrid-arrest translation experiments, and by analysis of different recombinants (46, 183). Protein M_2 is translated from a distinct small mRNA (46). Nuclease S1 mapping analysis and nucleotide sequencing of the mRNAs showed that the M_2 mRNA contains an interrupted region of 689 nucleotides (170). The ~ 51 virus-specific nucleotides comprising the 5' end leader sequence of the M_2 mRNA are the same as those found at the 5' end of the colinear M_1 mRNA. Following the leader sequence, there is a 271 nucleotide body region that is 3'-coterminal with the M_1 mRNA. Because the 5' end sequences of the M_1 mRNA and M_2 mRNA are the same and

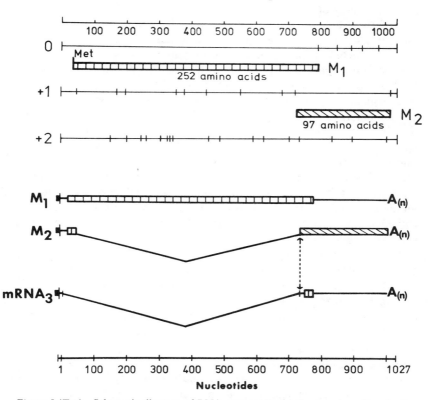

Figure 5 (Top) Schematic diagram of RNA segment 7 of influenza virus. Termination codons in all three reading frames are shown by vertical lines. The two open reading frames are shown as cross-hatched bars. [From Lamb & Lai (182).]

(Bottom) Model for the arrangement of the M_1, M_2-mRNAs and M-mRNA$_3$ and their coding regions. The thin lines at the 5'- and 3'-termini of the mRNAs represent the noncoding regions. In the region 740–1004 the M_2-mRNA is translated in a reading frame different from that used for M_1. No evidence has yet been obtained that M-mRNA$_3$ is translated. The V-shaped lines represent the interrupted regions. The filled-in bars before nucleotide 1 at the 5'-termini represent heterogeneous nucleotides derived from cellular mRNAs that are covalently linked to the viral sequences. [From Lamb et al (170).]

share the 5'-proximal initiation codon for protein synthesis, the first 9 amino acids are the same in the M_1 and M_2 proteins after which the sequences diverge. The ~ 271 nucleotide body region of the M_2 mRNA can be translated in the $+1$ reading frame, and the sequence indicates that M_1 and M_2 overlap by 14 amino acids (see Figure 5). No information about the function of M_2 is available, but it is found only in infected cells and is thus a nonstructural protein (46). A second interrupted mRNA (M mRNA$_3$) derived from segment 7 has also been found (170, 184). M mRNA$_3$ has a leader sequence of ~ 11 virus-specific nucleotides that are the same as the 5' end of the M_1 and M_2 mRNAs, followed by a body region of ~ 271 nucleotides that is the same as that of the M_2 mRNA. The coding potential of the M mRNA$_3$ is only nine amino acids, and these would be identical to the C-terminal region of the membrane protein M_1 (see Figure 5) (170). Such a peptide has not been identified.

The nucleotide sequences at both sides of the intervening regions of spliced eukaryotic mRNAs follow a distinct pattern, and a consensus sequence has emerged (184a). The sequences at the two donor sites and the common acceptor site on the colinear M_1 mRNA used to form the interrupted structure of the M_2 and M mRNA$_3$ exhibit such a pattern; thus these mRNAs are probably produced by splicing the colinear M_1 mRNA. Evidence that the M_1 mRNA can be spliced is provided by the finding that an identical splice junction is used when mRNAs are transcribed from a cloned DNA copy of segment 7 using a simian virus 40 promoter and cellular enzymes (185).

The nucleotide sequence of influenza B virus RNA segment 7 has been determined (186) and is similar in structure to that of the A viruses. Segment 7 is 1191 nucleotides long with an open reading frame of 744 nucleotides coding for an M_1 protein of 248 amino acids. A second coding region overlaps that of the M_1 protein by 86 amino acids in the $+2$ reading frame. It is likely that this region codes for a protein analogous to the influenza A virus M_2, but such a protein has not been identified. Comparison of the amino acid sequences of the M_1 proteins of the A and B viruses indicates that 63 amino acids are conserved. The internal hydrophobic domain, including three cysteine residues in the same positions, are also conserved, suggesting their functional significance (186).

The M_1 protein is synthesized on free ribosomes (169, 187); it lacks a signal peptide and its route to the plasma membrane is different from those of HA and NA. Although M_1 can be identified in association with various subcellular compartments (83, 118, 169, 187) it does not show a migration pattern, and therefore its association with these cell fractions may be artifactual.

RNA Segment 8: Nonstructural Proteins NS_1 and NS_2

The smallest RNA segment codes for two proteins that are found only in infected cells, the nonstructural proteins NS_1 ($M_r \sim 26,000$) and NS_2 ($M_r \sim 14,000$). NS_1 is formed in large amounts in infected cells (62, 83) and has been found associated with polysomes and in the nucleolus (83, 187–189). The function of NS_1 is unclear, but it has been suggested to be involved in the shut-off of host-cell protein synthesis and virion RNA synthesis (83, 187, 190, 191). NS_1 is a phosphoprotein with phosphate attached to one or two threonine residues per molecule (141, 192). Late in infection with some strains, NS_1 forms crystalline arrays (inclusions) (193, 194), which contain a mixture of cellular RNA species (195). The role of these inclusions, if any, is not known; they may be related to the abundance of NS_1 in the dying cell.

A polypeptide now designated NS_2 was observed early in infected cells (62, 188, 196–199). It was shown to be a virus-coded unique polypeptide on the basis of its peptide composition and strain-specific differences in its migration in polyacrylamide gels, leading to the suggestion that one virion RNA segment coded for two proteins (200). NS_2 is translated from a separate small mRNA (45, 49), and hybrid-arrest translation experiments and analysis of recombinants of defined genome composition showed it to be encoded by segment 8 (45, 49). On the basis of the estimated sizes of RNA segment 8, polypeptides NS_1 and NS_2, and hybrid-arrest translation data using DNA restriction fragments of cloned RNA segment 8, it was concluded that NS_1 and NS_2 must be encoded by overlapping reading frames (45, 49, 201). NS_2 is made late in infection and is located in the cytoplasm of infected cells (200, 202, 203), but its function is unknown.

The complete nucleotide sequence of segment 8 has been determined for several strains (202, 204–207). The RNA segment is 890 nucleotides long, with a 5'-noncoding region of 26 nucleotides, and an open reading frame coding for an NS_1 protein of 230–237 amino acids, depending on the virus strain. The predicted amino acid sequence agrees well with the amino acid composition of NS_1 (194) and the tryptic peptides (45, 200). A second open reading frame of 132 amino acids, found at the 5' end of segment 8, was predicted from S1 nuclease-mapping experiments, hybridization-blotting experiments, and hybrid-arrest translation experiments using cloned DNA of segment 8 (Figure 6) (201, 208). Nucleotide sequencing of the NS_2 mRNA showed that the mRNA contained an interrupted region of 473 nucleotides. The first ~ 56 virus-specific nucleotides at the 5' end of the NS_2 mRNA are the same as are found at the 5' end of the NS_1 mRNA. This leader sequence contains the initiation codon and coding information for nine amino acids that would be common to NS_1 and NS_2. The ~ 340-nucleotide body region of the NS$_2$ mRNA can be translated in the $+1$ reading frame, and the sequence indicates that NS$_1$ and NS$_2$ overlap by 70 amino acids that are translated from different reading frames (Figure 6). The nucleotides at the 5' and 3' junctions of the interrupted NS$_2$ mRNA are similar to those of intervening sequences of spliced eukaryotic mRNAs (184), indicating that the NS$_2$ mRNA is probably produced by splicing of the colinear NS$_1$

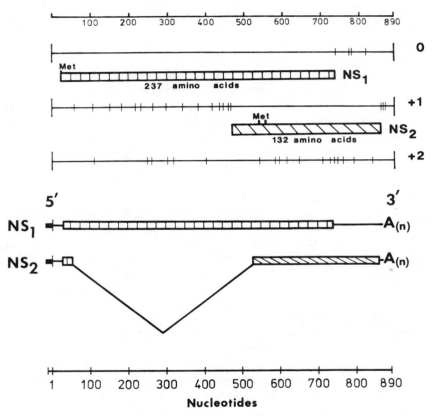

Figure 6 (Top)   Schematic representation of RNA segment 8 of influenza virus. Termination codons in all three reading frames are shown by vertical lines. The two open reading frames are shown by cross-hatched bars.

(Bottom)   Schematic representation for the arrangement of the NS$_1$- and NS$_2$-mRNAs. The thin lines at the 5'- and 3'-termini of the NS$_1$- and NS$_2$-mRNAs represent noncoding regions. The cross hatched bars represent coding regions. In the region 529–861, the NS$_2$-mRNA is translated in a reading frame different from that used for NS$_1$. The V-shaped thin line in the NS$_2$-mRNA represents the interrupted sequence in the NS$_2$-mRNA. The filled-in bars before nucleotide 1 at the 5'-termini represent heterogeneous nucleotides derived from cellular mRNAs. [Adapted from Lamb & Lai (202).]

mRNA. Evidence for splicing at these nucleotide sequences has been provided by finding that the identical splice junction is used when mRNAs are transcribed from cloned DNA of segment 8 using a simian virus 40 promoter and cellular enzymes (R. A. Lamb et al, manuscript in preparation).

The nucleotide sequence of influenza B virus RNA segment 8 has also been determined and is 1096 nucleotides long (209). It also codes for two proteins, NS_1 and NS_2 (38), with an overall structure similar to that of A viruses. The NS_1 protein has 281 amino acids, and NS_2 has the same N-terminus of ten amino acids as NS_1 before the splice junction in the mRNA, and then translation continues using the $+1$ reading frame. The NS_1 protein of influenza B virus is predicted to have a M_r of 32,026, but on polyacrylamide gels it migrates anomalously with an apparent M_r of $\sim 40,000$ (37, 38). NS_1 contains 122 amino acids, one more than the influenza A virus NS_1 and 52 amino acids overlap between NS_1 and NS_2.

Winter et al (207) speculated that NS_1 and NS_2 were originally colinear on the vRNA, but not overlapping. Readthrough of a terminator at the end of NS_1 then allowed NS_1 to become longer and use more of the available reading frame. Evidence is available that can be interpreted as supporting such a concept because: (a) the NS_1 protein is of varying lengths depending on the strain; (b) a mutant virus synthesizes a shortened NS_1 protein [ts 47; (210)] yet replicates normally at the nonpermissive temperature; and (c) comparison of nucleotide changes in different strains indicates that in the area of overlap between NS_1 and NS_2, the amino acid sequence of NS_2 is conserved at the expense of NS_1 (202, 205). These data could indicate that the C-terminal region of NS_1 is dispensable.

Examination of the nucleotide sequences of segment 8 of A/PR/8/34, A/Udorn/72, and A/FPV/Rostock/34 strains revealed that the negative (virion) strand had an initiation codon at position 98 from the 5' end and an open reading frame that extended for 167 or 216 amino acids (205). It has been speculated that such a reading frame, which statistically should only occur with a very low probability, might be translated (205). However, the virion RNA segment 8 from neither the A/duck/Alberta/60/76 (206) nor the B/Lee/40 virus (209) contains an open reading frame; therefore, it is unlikely to code for a necessary viral protein.

Overlapping Coding Regions Using Different Reading Frames in Viruses

The majority of the NS_2 polypeptide is translated from a reading frame different from that for NS_1; in the A viruses these overlap by 70 amino acids and in the B viruses by 52 amino acids (202, 209). The majority of the M_2 polypeptide is also translated from a reading frame different from that of

M_1, and these overlap by 14 amino acids (170). Previously, the use of overlapping reading frames has been found with bacteriophages ϕX174 and G4 (211, 212), and in bacteriophage Qβ (213, 214), in the large T and middle T antigens of polyoma virus (215), and in $VP_{2,3}$ and VP_1 of SV40 and polyoma virus (216, 217). Whether this efficient use of genomes occurs only in viruses with small genomes, or also in eukaryotic genomes is not yet known.

INFLUENZA VIRUS TRANSCRIPTION AND REPLICATION

Influenza virus can be distinguished from other nononcogenic RNA viruses in that its replication is dependent on an active host-cell nucleus both for viral transcription and for processing of the viral mRNAs. Two classes of genome transcript are involved in the replication of influenza virus. One class, A(+)cRNA, functions as mRNA, and the other class, A(−)cRNA, is used as template in genome replication (see Figure 7). The synthesis of the two classes of transcripts is mediated by the viral RNA transcriptase complex, which has a number of enzymatic activities. The initial transcription of viral RNA in the cell to form mRNAs [A(+)cRNA] is known as primary transcription, and this can be mimicked in vitro using disrupted virions under appropriate conditions. Secondary transcription is that of progeny RNA molecules and requires the production of template A(−)cRNA, which is dependent on viral protein synthesis.

Evidence for a Host-Cell Nuclear Requirement for Primary Transcription

The replication of influenza virus is inhibited by actinomycin D (which prevents DNA-dependent RNA transcription), by ultraviolet irradiation of cells before infection, and by other drugs that damage host-cell DNA (218–222). Inhibition of cellular DNA synthesis has no inhibitory effect (223, 224).

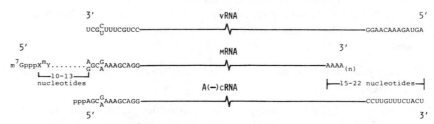

Figure 7 Schematic representation of influenza virus vRNA, mRNA, and A(−)cRNA. The conserved nucleotides at the ends of each RNA segment are shown. See text for details of the 5'- and 3'-terminal structure of the mRNA species.

A requirement for the host-cell nucleus was further indicated by the finding that influenza virus transcription was inhibited by α-amanitin, a potent inhibitor of DNA-dependent RNA polymerase II (225, 226). It was established that RNA polymerase II is required for viral primary transcription by the finding that viral transcription occurred in mutant cells containing an α-amanitin resistant polymerase II (227, 228). However, neither actinomycin D nor α-amanitin inhibited the in vitro activity of the virion-associated transcriptase (13, 229). This paradox has been resolved.

In Vitro Transcription of Influenza Virus mRNAs

Influenza virus transcriptase when assayed as initially described showed very low levels of activity (13, 230). The addition of guanosine or the dinucleotide GpC stimulated the reaction greatly, and it was shown that these molecules were specifically incorporated into the 5′ end of the newly synthesized RNA chain (231). The most efficient primer is the dinucleotide ApG and the only other dinucleotide that causes significant stimulation is GpC (232). The resulting RNA transcripts are polyadenylated and function as mRNAs in cell-free systems (232–234). These mRNAs are incomplete transcripts of the vRNA segments (232). No capping and methylating enzymes were detected in the virion (235); however, mRNAs isolated from infected cells do contain a 5′-terminal methylated type I cap structure (236). To explain this difference between in vitro and in vivo mRNAs, it was proposed that the 5′-cap found on in vivo mRNA is derived from a primer that is synthesized by the host RNA polymerase II (235).

Studies on an influenza virus coupled transcription-translation system using reticulocyte lysates indicated that in the absence of ApG as a primer, this system was much more active in transcription than an L-cell lysate (234); it was also found that globin mRNA and other cellular mRNAs stimulated the transcription. Translation of the globin mRNA-primed (but not the ApG-primed) viral RNA transcripts was inhibited by 7-methyl-guanosine 5′-phosphate in the presence of S-adenosylhomocysteine, which suggests that the globin mRNA-primed transcripts contained a 5′-terminal methylated cap structure. The priming by globin mRNA was, on a molar basis, about 1000-times more active than ApG. Direct evidence that the host-cell cap structure was transferred to the newly synthesized influenza virus mRNAs was obtained by using globin mRNA (237). Gel electrophoresis analysis of the globin mRNA-primed mRNA segments suggested that they were 10–15 nucleotides longer than ApG primed segments (237), indicating that globin nucleotides, in addition to the cap structure, were transferred to the newly synthesized mRNA. This was shown directly by partial sequencing of iodinated globin mRNA (238).

Removal of the m^7G of a cap eliminates priming, and by enzymatically recapping the mRNA, the priming is restored (237). The cap must contain methyl groups, since reovirus mRNAs with 5'-GpppG ends are not active (239), and both the 7-methyl on the terminal G and the 2'-O-methyl on the penultimate base strongly influence the priming of an mRNA (237, 239, 240). This is the first instance in which the 2'-O-methyl group of the cap has been shown to have a definite effect on a specific function of an mRNA. In cell-free translation systems, the absence of a 2'-O-methyl group from a cap, or the complete absence of a cap, only reduced the translational efficiency of the mRNA (see 241a); with influenza virus, uncapped RNAs can be translated readily (199, 234).

To investigate a possible sequence complementarity between the primer of transcription and the common 3' end of the vRNA templates, capped ribopolymers lacking a sequence complementarity to this region were used and found to be effective primers (240, 241). The observed stimulation of transcription of these fragments and the transfer of the cap and 10–15 nucleotides to the 5' end of the mRNA indicate that the priming could not result from hydrogen bonding between the primer and the 3' end of the vRNA. This finding suggests that the stimulation of transcription by the dinucleotides ApG and GpC (231, 232), which was the original basis for the experiments described here, is an artificial mechanism; this is further indicated by analysis of influenza virus mRNAs from infected cells.

Based on the above results, a mechanism for the priming of influenza viral RNA transcription by capped RNAs was proposed (16) and is illustrated in Figure 8. The capped RNA is cleaved by a virion-associated nuclease to generate a 5'-terminal fragment and this primer initiates transcription. Because hydrogen bonding does not have to be involved between the primer and the vRNA, but the structure of the 5'-terminal methylated cap structure is crucial to priming, it seemed likely that there must be a specific interaction between the capped RNA fragment and the transcriptase complex. This was shown to be the correct mechanism by uncoupling the cleavage and elongation steps from complete transcription, using either β-globin mRNA or alfalfa mosaic virus RNA 4 incubated with detergent-disrupted virions in the absence of ribonucleotide triphosphates, or with the addition of a single ribonucleotide triphosphate (16). It was found that the virion contains a cap-dependent endonuclease that cleaves capped RNAs preferentially at purine residues 10–13 nucleotides from the 5'-terminus. The primer does not have to be coupled directly to the nuclease reaction, because exogenously added fragments of the proper size and 3'-terminus can serve as primer directly. Elongation of the primers cleaved after an A or G residue occurs by incorporation of a G residue that is complementary to the second nucleotide from the 3' end of the vRNA segments. Thus, the virion transcriptase does not transcribe the 3'-U residue of the vRNA segment. However, after the specific cleavage of the primer, the transcriptase preferentially utilizes 3'-A-terminated fragments and an AGC sequence would be generated in the viral mRNA, complementary to the UCG sequence at the 3' end of the mRNA (16).

The cleavage of the capped primer RNA, initiation, elongation, and termination/polyadenylation of the transcript are all carried out by the transcriptase complex proteins (16), i.e. NP, PB1, PB2, and PA (14, 32, 72). To establish the function of specific proteins, transcriptase reactions with added primer were performed either in the absence of nucleoside 5'-triphosphates to allow cleavage to occur, or in the presence of a single ribonucleoside triphosphate to allow initiation to occur; then the reactions were u.v. irradiated to cross-link the proteins to nucleic acid. It was found that PB2 bound to the capped primer and PB1 to the initiating chain (72). It is not known whether PB1 is actually associated with the 3' end of the endonuclease-generated primer before the first nucleotide is linked to this fragment. If it is, then both PB1 and PB2 may constitute the endonuclease enzyme, with PB2 recognizing the 5' cap, and PB1 cleaving the phosphate bond (72). An interesting finding is that the molecules of PB2 that bind to

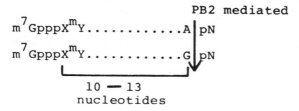

1. CLEAVAGE OF PRIMER

PB2 mediated

m^7GpppX^{m}Y . . . . .A|pN

m^7GpppX^{m}Y . . . . .G|pN

10 — 13
nucleotides

2. INITIATION ON PRIMER OF mRNA SYNTHESIS

PB1 mediated

UpCpGpUpUpUpUpCp . . . .

m^7GpppX^{m}Y . . . . .A G

3. ELONGATION OF CHAIN

PA mediated

UpCpGpUpUpUpUpCp . . . .

m^7GpppX^{m}Y . . . . .ApGpCpApApApApGp . . . .

Figure 8   Mechanism for the priming of influenza viral RNA transcription by capped RNAs. See text for description [adapted from Krug (257)].

the cap structure remain associated with it throughout transcription (72). Another approach to establishing that PB2 recognizes and binds a cap structure was the use of a photoreactive derivative of m^{7}GTP as an affinity label for the cap-binding protein (73, 74). By exclusion, PA is probably involved in elongating the nascent chain, but the elucidation of the precise role of all the proteins in the complex, including the polyadenylating function, may depend on solubilizing the proteins and reconstituting the transcriptive complex.

In Vivo Transcription of mRNAs

Each vRNA segment possesses a promoter for the transcription of mRNA (242, 243). The mRNAs isolated from infected cells are polyadenylated (244, 245), incomplete transcripts of the vRNA segments (51, 246, 247). Polyadenylation occurs at the signal U$_6$, 17–22 nucleotides from the 5' end of the segment (54). The mRNAs have a 5'-terminal type I cap structure (236) and 10–13 nucleotides obtained from host-cell mRNAs. Nucleotide sequence analysis of the 5' ends of individual cloned mRNA showed that different host-cell sequences are used (186, 202, 209, 248–250), and confirmed that the cleavage of the primer usually occurs after a purine residue; in most cases A or G is found opposite the 3'-terminal U of the vRNA. Direct sequencing of the 5' ends of a population of mRNAs derived from cloned DNA of an RNA segment indicated that there is a preference in primers used for GCA (170) or CA (250) at the 3' site of cleavage. Thus, although they show heterogeneity, the host-cell RNA species used as primers are a subset of capped RNAs containing the sequence GCA within 10–13 nucleotides. Although insufficient information is available to define precisely the nature of host cell sequences used as primers, in over 20 cloned viral mRNAs, the same host-cell sequence has not occurred (M. W. Shaw, R. A. Lamb, unpublished results), suggesting that highly reiterated cellular transcripts (e.g. *Alu* transcripts) are not utilized.

Influenza virus mRNAs in infected cells also contain on average three 6-methyl adenosine (m^{6}A) residues per chain (236), one of which is located within the nonviral host-cell sequences (251). However, the exact distribution of m^{6}A residues on each mRNA segment has not been determined, and because mRNAs from segments 7 and 8 can be spliced (170, 184, 202), it may be that only transcripts from these RNA segments contain internal m^{6}A residues.

The cellular site of influenza virus RNA transcription has been studied extensively. Assay of viral RNA polymerase activity in nucleus and cytoplasmic fractions is complicated because most of the cytoplasmic activity detected can be attributed to progeny nucleocapsids ready for assembly into virions. However, some nuclear activity was detected in early studies (252). Further analyses of the site of RNA transcription suggested that primary transcription probably occurs in the nucleus of infected cells (253, 254). Because of the leakiness of RNA from isolated nuclei, it could be argued that these results were inconclusive, as RNA was also detected in the cytoplasm. However, other evidence consistent with a nuclear site includes the presence of m^6A residues (236, 241) as cellular nuclear RNAs are methylated, and the processing of colinear transcripts of RNA segments 7 and 8 to form spliced mRNAs (170, 185, 202, 209). Evidence confirming that primary transcription occurs in the nucleus came from the use of nonaqueous procedures to isolate nuclei and the assay of rapidly pulse-labeled transcripts (255) and from the isolation of nuclear cages from infected cells with associated nascent viral transcripts (256).

It is now thought that influenza virus transcription is inhibited by α-amanitin because the cannibalized host-cell-capped RNAs are synthesized by DNA-dependent RNA polymerase II (257). However, because the addition of α-amanitin to cells just before infection effectively inhibits primary transcription (226, 227), influenza virus transcriptase is apparently primed only by newly synthesized capped RNA species, not preexisting RNAs in the nucleus.

Template $A(-)cRNA$ Synthesis

A second population of RNA transcripts [A(−)cRNA] complementary to the vRNA segments is found in infected cells and is thought to be the template for progeny vRNA synthesis. These are complete transcripts of the vRNA segment (51, 246, 247, 258), which do not contain polyA, and their 5'-terminus is pppA, with the A being complementary to the 3'-U of the vRNA strands (258). It seems likely that these A(−)cRNA molecules do not need a primer and are initiated de novo. Although the A(−)cRNAs can be translated in vitro (199), they are not associated with polysomes in infected cells (246).

The synthesis of A(−)cRNA is dependent on the continued production of functional viral proteins, whereas mRNA synthesis continues in the absence of protein synthesis (246). It is assumed that the virion transcriptase complex is involved in synthesis of the A(−)cRNAs and that a viral protein (e.g. nonstructural proteins NS_1, NS_2, or M_2) modifies transcription so that it is primer independent and allows the polymerase to proceed through the U_6 polyadenylation signal to the 5' end of the vRNA segment. However, the possibility has not been eliminated that mRNAs are synthesized as a complete transcript that is subsequently nucleolytically processed and polyadenylated. This is unlikely since in vivo the transcriptase makes only polyadenylated molecules in the absence of protein synthesis. The cellular site of synthesis of template A(−)cRNA is probably the nucleus (255).

Virion RNA Synthesis

The mechanism by which vRNA strands are replicated is not understood. The synthesis of vRNA segments is dependent on viral protein synthesis (246, 259). Virion RNA segments can be detected in both the nucleus and cytoplasm of infected cells (246, 253, 254, 260) but their site of synthesis has not been unequivocally established. The A(−)cRNA transcripts might remain in the nucleus to direct viron RNA synthesis or may be transported to the cytoplasm. However, newly synthesized vRNA segments would then have to be transported back to the nucleus for further mRNA synthesis.

The Control of RNA Synthesis

The synthesis of the three different classes of viral RNAs (mRNA, A(−)cRNA, and vRNA), and the relative abundance of specific mRNAs and vRNAs are controlled throughout infection. Messenger RNAs can be detected almost immediately after infection (246, 253, 254, 258) and their rate of synthesis peaks at about 2 hr. A(−)cRNA molecules can be detected after mRNAs; although these species accumulate for several hours (253), their maximal rate of synthesis is somewhat before that of mRNAs (246). The synthesis of vRNAs follows A(−)cRNA synthesis, and they accumulate for several hours (253), but it is not clear how early in infection vRNA synthesis declines (246, 259–261).

During primary transcription, similar amounts of each mRNA are synthesized (246) and estimations of functional mRNA abundance by translation of polypeptides both in vivo (32) and in vitro (262) have led to similar conclusions. Within the first hour of infection, the mRNAs coding for NP and NS_1 increase considerably, whereas those coding for PB1, PB2, and PA remain at relatively low amounts. Later in infection, after vRNA replication, the mRNAs coding for HA, NA, and M_1 greatly increase (246, 262). Throughout infection the relative amounts of each transcript of A(−)cRNA remain at approximately equimolar amounts (246, 261). However, the vRNA segments are produced in nonequivalent amounts. A correlation between the synthesis of the majority of the vRNAs and their corresponding mRNAs at early times during infection suggests that the controlled production of these mRNAs may be a direct consequence of the regulation of vRNA synthesis (261). Because both the vRNAs and mRNAs for NP and NS_1 are present in large amounts early in infection as compared to the other vRNAs, and the A(−)cRNAs are present in equimolar amounts, transcription from A(−)cRNA to vRNA is selective. In addition, transcription of vRNA to mRNA is also probably selective, because late in infection the amount of mRNAs for the three P proteins is much less than that of vRNA segments 1–3 (261).

The host cell also affects influenza virus RNA synthesis. Examples of this are the marked alterations in RNA synthesis in HeLa cells (263) and the greatly reduced amounts of mRNAs for NA and M_1 in L-cells infected with the A/FPV/34 strain (264). In addition there are differences between virus strains in different host-cell types. With the A/FPV/34 strain, the synthesis of the NS_1 mRNA and protein increases rapidly early in infection in most cell types with respect to M_1 mRNA and protein synthesis, and then synthesis of NS_1 declines. With the A/WSN/33 strain the increase in synthesis ahead of M_1 is not as apparent, and the decline of NS_1 cannot be observed with most cell types (227, 246, 265).

One unexplained feature of influenza virus replication is that various inhibitors of cellular DNA function [camptothecin, UV-irradiation, 5,6-dichloro-1-β-D-ribofuranosylbenzimidazole (DRB)] produce the same pattern of virus replication that is seen in the abortive infection of A/FPV/34 in L-cells, with greatly reduced amounts of HA, NA, and M_1 being synthesized (266, 267; R. A. Lamb et al, unpublished observations). It could be inferred that a host-cell product was necessary for the amplification of their mRNA/vRNA species.

Spliced mRNAs and Their Controlled Synthesis

As described above, NS_2 mRNA derived from segment 8, and M_2 mRNA and M mRNA$_3$ derived from segment 7 contain interrupted regions with 5'- and 3'-junctions similar to those of the consensus sequences at the junctions of spliced eukaryotic mRNAs (170, 202, 209). The relative abundance of these spliced mRNAs is low (5%–10% of their respective colinear transcripts), and a mechanism prevents all the colinear transcripts, NS_1 or M_1 mRNAs, in the nucleus from being processed. In addition, with the segment-7-derived mRNAs, a selection is made for one of two 5'-leader sequences by utilizing different 5'-splice sites. It was originally shown that NS_2 was synthesized late in infection, and that its synthesis depended on early viral protein synthesis (200) and this was confirmed (268). One interpretation of these findings is that influenza virus modified the host-cell-splicing mechanism. However, spliced mRNAs can be synthesized from cloned DNA copies of segments 7 and 8, using an SV40 promoter (185; R. A. Lamb et al, manuscript in preparation); thus viral protein synthesis is unlikely to be required. Therefore it is important to reexamine the time course of appearance of NS_2 and to establish whether primary transcript mRNAs, as well as secondary transcript mRNAs, are spliced. It is known that the relative amount of the NS_2 protein synthesized in different cell lines varies (45, 265), and that this reflects the relative abundance of the spliced mRNAs (R. A. Lamb, unpublished results).

Subgenomic RNAs: A Replication Error Produces Defective Interfering Particles

During replication of influenza viruses, subgenomic vRNAs are sometimes synthesized, and their formation is favored by high-multiplicity passaging (263, 269–271). The subgenomic RNAs are predominantly generated from RNA segments 1, 2, and 3, which code for PB1, PB2, and PA (272). These subgenomic RNAs have the same 5'- and 3'-terminal sequences as the vRNA segments, but contain large internal deletions (273–275). A mosaic subgenomic RNA containing sequences from RNA segments 1 and 3 has been found (65), which is thought to have been generated by the transcriptase "jumping" and reassociating with the same strand, or switching strands during vRNA synthesis and carrying with it the nascent RNA strand (65). However, no specific sequence has been recognized at the points where the transcriptase leaves and rejoins the RNA segments (65, 274, 275). Because the subgenomic RNAs have the signals for transcription and replication at both their 5' and 3' ends, they can be efficiently amplified. In mixed infections of particles with subgenomic RNAs and standard virus, the subgenomic RNAs become predominant and a decrease in the amount of the larger RNA segments is observed in the population of virions (263, 270, 276).

The generation of subgenomic RNAs within a virus population leads to the formation of defective-interfering (DI) virus particles. The DI particles have the following properties: (a) inability to propagate in the absence of helper virus; (b) ability to be complemented by and multiply in the presence of helper virus; (c) ability to decrease the yield of standard virus; and (d) the ability to increase their proportion in yield from cells coinfected with wild-type virus. In addition DI particles may play a role in the establishment of persistent viral infections. The generation of DI virus in cells is dependent not only on the multiplicity of infection but also on the host cell (263, 276, 277). This may reflect a function provided by the host cell for replication, as discussed above.

The Packaging of Eight RNA Segments

An influenza virion must have one copy of each of the eight RNA segments to be infectious. One of the intriguing questions about influenza virus replication is how this is accomplished. It is not now possible to conclude if this is an ordered process, with a complex mechanism existing for the selection of one of each of the eight RNA segments, or if it occurs by the random selection of RNA segments during the process of virus budding from the cell membrane. If the RNPs are incorporated randomly into virions from the intracellular pool, as has been proposed (19), the

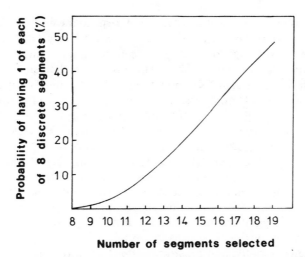

Figure 9   The probability of packaging eight discrete RNA segments of influenza virus into a virion on a random basis. The graph shows the probability (as a percentage) of a virion having one of each of the eight RNA segments vs the total number of randomly selected segments. For example, if 12 segments were randomly packaged, ten percent of the virions would have at least one copy of each of the eight segments. Calculated by D. S. Roos, The Rockefeller University, unpublished.

proportion of infective virions containing a copy of each segment in a population may be increased by incorporation of extra segments of RNA into the virions (see Figure 9). In support of such a concept is the finding that aggregates of influenza virions that either occur normally or are produced with nucleohistone have an enhanced infectivity (278), which presumably occurs by complementation of RNA segments. In support of an ordered selection process is the finding that the eight virion RNA segments are present in standard virions in approximately equimolar quantities (31) and this does not reflect the abundance of the vRNAs in infected cells (261). However, no evidence for linkage of RNPs during replication has been obtained.

CONCLUDING REMARKS

As indicated above, there have been striking advances in our understanding of influenza virus and its replication. In less than a decade we have gone from not knowing the correct number of viral proteins or genome RNA segments to establishing the complete nucleotide sequences of the entire genome and the three-dimensional structure of the most biologically important viral protein, and from the seeming paradox of the requirement of the nucleus for the replication of an RNA virus to the finding of a unique mechanism of transcription and of splicing of mRNAs transcribed from an RNA genome. Although progress has been great, many fascinating and important questions remain: What are the precise mechanisms involved in the transcription of viral mRNAs, including the capturing of a cap structure from cellular mRNA and splicing of some viral RNAs? How is the replication of the genome RNA strands carried out, and by what mechanism does the virus accomplish the extraordinary feat of getting eight independently replicated segments into the virus particle? What are the mechanisms by which genetic reassortment occurs under natural conditions, and why is there not genetic reassortment between influenza A and B viruses, whose genome segments are structurally and functionally analogous? Where are the epidemiologically most important antigenic sites on the viral hemagglutinin and neuraminidase? What are the precise events at the cell membrane during viral assembly (an important topic not covered in this review)? Finally, and most importantly, can this new cellular and molecular biological knowledge be used to prevent and treat more effectively the enormously important disease that recurs in annual epidemics and occasional devastating pandemics? The developments of the past decade suggest that the coming years will provide interesting answers to many of these questions.

ACKNOWLEDGMENTS

We are very grateful to Andrea Gifford for typing the manuscript, and to Dr. Don Wiley for making available Figure 3. Research by the authors was supported by Research Grants AI-05600 and AI-18746 from the National Institute of Allergy and Infectious Diseases, and PCM-80-13464 from the National Science Foundation. R. A. Lamb is an Irma T. Hirschl Awardee and an Established Investigator of the American Heart Association.

Literature Cited

1. Horne, R. W., Waterson, A. P., Farnham, A. E. 1960. *Virology* 11:79–98
2. Choppin, P. W., Murphy, J. S., Tamm, I. 1960. *J. Exp. Med.* 112:945–52
3. Choppin, P. W., Murphy, J. S., Stoeckenius, W. 1961. *Virology* 13:549–50
4. Laver, W. G., Valentine, R. C. 1969. *Virology* 38:105–19
5. Murphy, J. S., Bang, F. B. 1952. *J. Exp. Med.* 95:259–71
6. Morgan, C., Hsu, K. C., Rifkind, R. A., Knox, A. W., Rose, H. M. 1961. *J. Exp. Med.* 114:825–37
7. Klenk, H.-D., Choppin, P. W. 1970. *Virology* 40:939–47
8. Compans, R. W., Choppin, P. W. 1975. In *Comprehensive Virology*, ed. H. Fraenkel-Conrat, R. R. Wagner, 4:179–252. New York: Plenum
9. Pons, M. W., Schulze, I. T., Hirst, G. K. 1969. *Virology* 39:250–59
10. Compans, R. W., Content, J., Duesberg, P. H. 1972. *J. Virol.* 10:795–800
11. Heggeness, M. H., Smith, P. R., Ulmanen, I., Krug, R. M., Choppin, P. W. 1982. *Virology* 118:466–70

12. Duesberg, P. H. 1969. *J. Mol. Biol.* 42:485–99
13. Chow, N. L., Simpson, R. W. 1971. *Proc. Natl. Acad. Sci. USA* 68:752–56
14. Inglis, S. C., Carroll, A. R., Lamb, R. A., Mahy, B. W. J. 1976. *Virology* 74:489–503
15. Rochovansky, O. M. 1976. *Virology* 73:327–38
16. Plotch, S. J., Bouloy, M., Ulmanen, I., Krug, R. M. 1981. *Cell* 23:847–58
17. Burnett, F. M., Lind, P. E. 1954. *Nature* 173:627–30
18. Simpson, R. W., Hirst, G. K. 1961. *Virology* 15:436–51
19. Hirst, G. K. 1962. *Cold Spring Harbor Symp. Quant. Biol.* 27:303–9
20. Davies, P., Barry, R. D. 1966. *Nature* 211:384–87
21. Duesberg, P. H., Robinson, W. S. 1967. *J. Mol. Biol.* 25:383–405
22. Duesberg, P. H. 1968. *Proc. Natl. Acad. Sci. USA* 59:930–37
23. Pons, M. W., Hirst, G. K. 1968. *Virology* 34:385–88
24. Skehel, J. J. 1971. *Virology* 45:793–95
25. Scholtissek, C. 1979. *Adv. Genet.* 20:1–36
26. Bean, W. J. Jr., Simpson, R. W. 1976. *J. Virol.* 18:365–69
27. Pons, M. W. 1976. *Virology* 69:789–92
28. Palese, P., Schulman, J. L. 1976. *J. Virol.* 17:876–84
29. Palese, P., Schulman, J. L. 1976. *Proc. Natl. Acad. Sci. USA* 73:2142–46
30. Ritchey, M. B., Palese, P., Kilbourne, E. D. 1976. *J. Virol.* 18:738–44
31. McGeoch, D. J., Fellner, P., Newton, C. 1976. *Proc. Natl. Acad. Sci. USA* 73:3045–49
32. Lamb, R. A., Choppin, P. W. 1976. *Virology* 74:504–19
33. Ritchey, M. B., Palese, P., Schulman, J. L. 1976. *J. Virol.* 20:307–13
34. Desselberger, U., Palese, P. 1978. *Virology* 88:394–99
35. Cox, N. J., Kendal, A. P. 1976. *Virology* 74:239–41
36. Compans, R. W., Bishop, D. H. L., Meier-Ewert, H. 1977. *J. Virol.* 21:658–65
37. Racaniello, V. R., Palese, P. 1979. *J. Virol.* 29:361–73
38. Briedis, D. J., Lamb, R. A., Choppin, P. W. 1981. *Virology* 112:417–25
39. Meier-Ewert, H., Compans, R. W., Bishop, D. H. L., Herrler, G. 1978. In *Negative Strand Viruses and the Host Cell*, ed. B. W. J. Mahy, R. D. Barry, pp. 127–33. London: Academic
40. Meier-Ewert, H., Herrler, G., Nagele, A., Compans, R. W. 1980. In *Structure and Variation in Influenza Virus*, ed. G. Laver, G. Air, pp. 357–65. New York: Elsevier/North Holland
41. Hirst, G. K. 1950. *J. Exp. Med.* 91:177–84
42. Kendal, A. P. 1975. *Virology* 65:87–99
43. Schulman, J. L., Palese, P. 1976. *J. Virol.* 20:248–54
44. Palese, P., Ritchey, M. B., Schulman, J. L. 1977. *Virology* 76:114–21
45. Lamb, R. A., Choppin, P. W. 1979. *Proc. Natl. Acad. Sci. USA* 76:4908–12
46. Lamb, R. A., Choppin, P. W. 1981. *Virology* 112:729–37
47. Scholtissek, C., Harms, E., Rohde, W., Orlich, M., Rott, R. 1976. *Virology* 74:332–44
48. Inglis, S. C., McGeoch, D. J., Mahy, B. W. J. 1977. *Virology* 78:522–36
49. Inglis, S. C., Barrett, T., Brown, C. M., Almond, J. W. 1979. *Proc. Natl. Acad. Sci. USA* 76:3790–94
49a. Almond, J. W., McGeoch, D., Barry, R. D. 1977. *Virology* 81:62–73
50. Almond, J. W., Barry, R. D. 1979. *Virology* 92:407–15
51. Skehel, J. J., Hay, A. J. 1978. *Nucleic Acids Res.* 5:1207–19
52. Robertson, J. S. 1979. *Nucleic Acids Res.* 6:3745–57
53. Desselberger, U., Racaniello, V. R., Zazra, J. J., Palese, P. 1980. *Gene* 8:315–28
54. Robertson, J. S., Schubert, M., Lazzarini, R. A. 1981. *J. Virol.* 38:157–63
55. Sanger, F., Brownlee, G. G., Barrell, B. G. 1965. *J. Mol. Biol.* 13:373–98
56. Smith, J. C., Carey, N. H., Fellner, P., McGeoch, D., Barry, R. D. 1978. See Ref. 39, pp. 37–46
57. Maxam, A. M., Gilbert, W. 1977. *Proc. Natl. Acad. Sci. USA* 74:560–64
58. Sanger, F., Nicklen, S., Coulson, A. R. 1977. *Proc. Natl. Acad. Sci. USA* 74:5463–67
59. Air, G. M. 1979. *Virology* 97:468–72
60. Both, G. W., Air, G. M. 1979. *Eur. J. Biochem.* 96:363–72
61. Blok, J., Air, G. M., Laver, W. G., Ward, C. W., Lilley, G. G., et al. 1982. *Virology* 119:109–21
62. Compans, R. W., Klenk, H. D., Caliguiri, L. A., Choppin, P. W. 1970. *Virology* 42:880–89
63. Skehel, J. J. 1972. *Virology* 49:23–36
64. Horisberger, M. A. 1980. *Virology* 107:302–5
65. Fields, S., Winter, G. 1982. *Cell* 28:303–13
66. Winter, G., Fields, S. 1982. *Nucleic Acids Res.* 10:2135–43
67. Bishop, D. H. L., Huddleston, J. A., Brownlee, G. G. 1982. *Nucleic Acids Res.* 10:1335–43

68. Bishop, D. H. L., Jones, K. L., Huddleston, J. A., Brownlee, G. G. 1982. *Virology* 120:481–89
69. Kaptein, J. S., Nayak, D. P. 1982. *J. Virol.* 42:55–63
70. Palese, P., Ritchey, M. B., Schulman, J. L. 1977. *J. Virol.* 21:1187–95
71. Scholtissek, C., Bowles, A. L. 1975. *Virology* 67:576–87
72. Ulmanen, I., Broni, B. A., Krug, R. M. 1981. *Proc. Natl. Acad. Sci. USA* 78:7355–59
73. Blaas, D. B., Patzelt, E., Kuechler, E. 1982. *Virology* 116:339–48
74. Penn, C. R., Blaas, D., Kuechler, E., Mahy, B. W. J. 1982. *J. Gen. Virol.* 62:177–80
75. Ulmanen, I., Broni, B., Krug, R. M. 1983. *J. Virol.* 45:27–35
76. Hirst, G. K. 1941. *Science* 94:22–23
77. McClelland, L., Hare, R. 1941. *Can. J. Public Health* 32:530–38
78. Hirst, G. K. 1942. *J. Exp. Med.* 75:47–64
79. Laver, W. G., Kilbourne, E. D. 1966. *Virology* 30:493–501
80. Klenk, H.-D., Rott, R., Orlich, M., Blodorn, J. 1975. *Virology* 68:426–39
81. Lazarowitz, S. G., Choppin, P. W. 1975. *Virology* 68:440–54
82. Lazarowitz, S. G., Compans, R. W., Choppin, P. W. 1971. *Virology* 46:830–43
83. Lazarowitz, S. G., Compans, R. W., Choppin, P. W. 1973. *Virology* 52:199–212
84. Lazarowitz, S. G., Goldberg, A. R., Choppin, P. W. 1973. *Virology* 56:172–80
85. Klenk, H.-D., Scholtissek, C., Rott, R. 1972. *Virology* 49:723–34
86. Laver, W. G. 1971. *Virology* 45:275–88
87. Bosch, F., Orlich, M., Klenk, H. D., Rott, R. 1979. *Virology* 95:197–207
88. Rott, R. 1979. *Arch. Virol.* 59:285–98
89. Rott, R., Reinacher, M., Orlich, M., Klenk, H.-D. 1980. *Virology* 65:123–33
90. Griffith, I. P. 1975. In *Negative Strand Viruses*, ed. B. W. J. Mahy, R. D. Barry, 1:121–32. London: Academic
91. Wiley, D. C., Skehel, J. J., Waterfield, M. D. 1977. *Virology* 79:446–48
92. Wilson, I. A., Skehel, J. J., Wiley, D. C. 1981. *Nature* 289:366–73
93. Brand, C. M., Skehel, J. J. 1972. *Nature New Biol.* 238:145–47
94. Skehel, J. J., Waterfield, M. D. 1975. *Proc. Natl. Acad. Sci. USA* 72:93–97
95. Ward, C. W. 1981. *Curr. Top. Microbiol. Immunol.* 94/95:1–74
96. Waterfield, M. D., Espelie, K., Elder, K., Skehel, J. J. 1979. *Br. Med. Bull.* 35:57–63
97. Webster, R. G., Laver, W. G., Air, G. M., Schild, G. C. 1982. *Nature* 296:115–21
98. Dopheide, T. A., Ward, C. W. 1978. *Eur. J. Biochem.* 85:393–98
99. Garten, W., Bosch, F. X., Linder, D., Rott, R., Klenk, H.-D. 1981. *Virology* 115:361–74
100. Blobel, G. 1980. *Proc. Natl. Acad. Sci. USA* 77:1496–1500
101. Sveda, M. M., Markoff, L. J., Lai, C.-J. 1982. *Cell* 30:649–56
102. Schwarz, R. T., Schmidt, M. F. G., Anwer, U., Klenk, H.-D. 1977. *J. Virol.* 23:217–26
103. Keil, W., Klenk, H.-D., Schwarz, R. T. 1979. *J. Virol.* 31:253–56
104. Nakamura, K., Compans, R. W. 1978. *Virology* 86:432–42
105. Nakamura, K., Compans, R. W. 1979. *Virology* 95:8–23
106. Ward, C. W., Dopheide, T. A. A. 1981. *Biochem. J.* 193:953–62
107. Klenk, H.-D., Compans, R. W., Choppin, P. W. 1970. *Virology* 42:1158–62
108. Dopheide, T. A. A., Ward, C. W. 1980. *FEBS Lett.* 110:181–83
109. Wiley, D. C., Wilson, I. A., Skehel, J. J. 1981. *Nature* 298:373–78
109a. Waterfield, M. D., Gething, M.-J., Scrace, G., Skehel, J. J. 1980. See Ref. 40, pp. 11–20
110. Webster, R. G., Laver, W. G. 1980. *Virology* 104:139–48
111. Gerhard, W., Yewdell, J., Frankel, M. 1980. See Ref. 40, pp. 273–80
112. Laver, W. G., Air, G. M., Webster, R. G. 1981. *J. Mol. Biol.* 145:339–61
113. Both, G. W., Sleigh, M. J. 1981. *J. Virol.* 39:663–72
114. Yewdell, J. W., Webster, R. G., Gerhard, W. 1979. *Nature* 279:246–48
115. Laver, W. G., Gerhard, W., Webster, R. G., Frankel, M. E., Air, G. M. 1979. *Proc. Natl. Acad. Sci. USA* 76:1425–29
115a. Caton, A. J., Brownlee, G. G., Yewdell, J. W., Gerhard, W. 1982. *Cell* 31:417–27
116. Krystal, M., Elliott, R. M., Benz, E. W. Jr., Young, J. F., Palese, P. 1982. *Proc. Natl. Acad. Sci. USA* 79:4800–4
117. McCauley, J., Bye, J., Elder, K., Gething, M. J., Skehel, J. J., et al. 1979. *FEBS Lett.* 108:422–28
118. Klenk, H.-D., Wollert, W., Rott, R., Scholtissek, C. 1974. *Virology* 57:28–41
119. Elder, K. T., Bye, J. M., Skehel, J. J., Waterfield, M. D., Smith, A. E. 1979. *Virology* 95:343–50
120. McCauley, J., Skehel, J. J., Elder, K., Gething, M.-J., Smith, A., et al. 1980. See Ref. 40, pp. 97–104
121. Compans, R. W. 1973. *Virology* 55:541–45

122. Rott, R., Klenk, H.-D. 1977. *Cell Surf. Rev.* 2:47–81
123. Rothman, J. E., Fries, E., Dunphy, W. G., Urbani, L. J. 1982. *Cold Spring Harbor Symp. Quant. Biol.* 46:797–805
124. Rodriguez-Boulan, E., Sabatini, D. D. 1978. *Proc. Natl. Acad. Sci. USA* 75:5071–975
125. Roth, M. G., Fitzpatrick, J. P., Compans, R. W. 1979. *Proc. Natl. Acad. Sci. USA* 76:6430
126. Compans, R. W., Roth, M. G., Alonso, F. F. 1981. In *Genetic Variation among Influenza Viruses*, ed. D. P. Nayak, pp. 213–31. ICN-UCLA Symp. *Mol. Cell. Biol.* Vol. 21. New York: Academic
127. Huang, R. T. C., Rott, R., Klenk, H.-D. 1981. *Virology* 110:243–47
128. Skehel, J. J., Bayley, P. M., Brown, W. B., Martin, S. R., Waterfield, M. D., et al. 1982. *Proc. Natl. Acad. Sci. USA* 79:968–72
129. Scheid, A., Choppin, P. W. 1974. *Virology* 57:475–90
130. Gething, M. J., White, J. M., Waterfield, M. D. 1978. *Proc. Natl. Acad. Sci. USA* 75:2737–40
131. Scheid, A., Graves, M. C., Silver, S. M., Choppin, P. W. 1978. See Ref. 39, pp. 181–93
132. Richardson, C. D., Scheid, A., Choppin, P. W. 1980. *Virology* 105:205–22
133. Scheid, A., Choppin, P. W. 1977. *Virology* 80:54–66
134. Huang, R. T. C., Wahn, K., Klenk, H.-D., Rott, R. 1980. *Virology* 104:294–302
135. Maeda, T., Kawasaki, K., Ohnishi, S.-I. 1981. *Proc. Natl. Acad. Sci. USA* 78:4133–37
136. White, J., Kartenbeck, J., Helenius, A. 1982. *EMBO J.* 1:217–22
137. Rees, P. J., Dimmock, N. J. 1981. *J. Gen. Virol.* 53:125–32
138. Choppin, P. W., Compans, R. W., Scheid, A., McSharry, J. J., Lazarowitz, S. G. 1972. In *Membrane Research*, ed. C. F. Fox, pp. 163–79. New York: Academic
139. Van Wyke, K. L., Hinshaw, V. S., Bean, W. J., Webster, R. G. 1980. *J. Virol.* 35:24–30
140. Kamata, T., Watanabe, Y. 1977. *Nature* 267:460–62
141. Privalsky, M. L., Penhoet, E. E. 1981. *J. Biol. Chem.* 256:5368–76
142. Winter, G., Fields, F. 1981. *Virology* 114:423–28
143. Van Rompuy, L., Min Jou, W., Huylebroeck, D., Devos, R., Fiers, W. 1981. *Eur. J. Biochem.* 116:347–53
144. Huddleston, J. A., Brownlee, G. G. 1982. *Nucleic Acids Res.* 10:1029–38
145. Gottschalk, A. 1957. *Biochim. Biophys. Acta* 23:645–46
146. Wrigley, N. G., Skehel, J. J., Charlwood, P. A., Brand, C. M. 1973. *Virology* 51:525–29
147. Colman, P. M., Tulloch, P. A., Laver, W. G. 1980. See Ref. 40, pp. 351–56
148. Lazdins, I., Haslam, E. A., White, D. O. 1972. *Virology* 49:758–65
149. Mayron, L. W., Robert, B., Winzler, R. J., Rafelson, M. E. 1961. *Arch. Biochem. Biophys.* 92:475–83
150. Noll, H., Aoyagi, T., Orlando, J. 1962. *Virology* 18:154–57
151. Wilson, V. W., Rafelson, M. E. 1963. *Biochem. Prep.* 10:113–17
152. Wrigley, N. G., Laver, W. G., Downie, J. C. 1977. *J. Mol. Biol.* 109:405–21
153. Palese, P., Schulman, J. L. 1974. *Virology* 57:227–37
154. Schulman, J. L., Palese, P. 1977. *J. Virol.* 24:170–76
155. Sugiura, A., Ueda, M. 1980. *Virology* 101:440–49
156. Schulman, J. L., Khakpour, M., Kilbourne, E. D. 1968. *J. Virol.* 2:778–86
157. Webster, R. G., Laver, W. G. 1967. *J. Immunol.* 99:49–55
158. Rott, R., Becht, H., Orlich, M. 1974. *J. Gen. Virol.* 22:35–41
159. Rafelson, M. E., Schneir, M., Wilson, V. M. 1963. *Arch. Biochem. Biophys.* 103:424–30
160. Fields, S., Winter, G., Brownlee, G. G. 1981. *Nature* 290:213–17
161. Hiti, A. L., Nayak, D. P. 1982. *J. Virol.* 41:730–34
162. Markoff, L., Lai, C.-J. 1982. *Virology* 119:288–97
163. Bentley, D. R., Brownlee, G. G. 1982. *Nucleic Acids Res.* 10:5033–42
164. Maroux, S., Louvard, D. 1976. *Biochim. Biophys. Acta* 149:189–95
165. Brunner, J., Hauser, H., Braun, H., Wilson, K. J., Wacker, H., et al. 1979. *J. Biol. Chem.* 254:1821–28
166. Frank, G., Brunner, J., Hauser, H., Wacker, H., Semenza, G., et al. 1978. *FEBS Lett.* 96:183–88
167. Shaw, M. W., Lamb, R. A., Erickson, B. W., Briedis, D. J., Choppin, P. W. 1982. *Proc. Natl. Acad. Sci. USA* 79:6817–21
168. Kozak, M. 1981. In *Curr. Top. Microbiol. Immunol.* 93:81–123
169. Hay, A. 1974. *Virology* 60:398–418
170. Lamb, R. A., Lai, C.-J., Choppin, P. W. 1981. *Proc. Natl. Acad. Sci. USA* 78:4170–74
171. Schulze, I. T. 1970. *Virology* 42:890–904
172. Schulze, I. T. 1972. *Virology* 47:181–96
173. Bucher, D. J., Kharitonenkow, L. G., Zakomiridin, J. A., Grigoriev, V. B.,

Klimenko, S. M., et al. 1980. *J. Virol.* 36:586–90

174. Gregoriades, A. 1980. *J. Virol.* 36:470–79

175. Gregoriades, A., Frangione, B. 1981. *J. Virol.* 40:323–28

176. Lubeck, M. D., Schulman, J. L., Palese, P. 1978. *J. Virol.* 28:710–16

177. Hay, A. J., Kennedy, N. C. T., Skehel, J. J., Appleyard, G. 1979. *J. Gen. Virol.* 42:189–91

178. Robertson, B. H., Bhown, A. S., Compans, R. W., Bennett, J. C. 1979. *J. Virol.* 30:759–66

179. Gregoriades, A. 1973. *Virology* 54:369–83

180. Winter, G., Fields, S. 1980. *Nucleic Acids Res.* 8:1965–74

181. Allen, H., McCauley, J., Waterfield, M., Gething, M.-J. 1980. *Virology* 107:548–51

182. Lamb, R. A., Lai, C.-J. 1981. *Virology* 112:746–51

183. Palese, P., Elliott, R. M., Baez, M., Zazra, J. J., Young, J. F. 1981. See Ref. 126, pp. 127–40

184. Inglis, S. C., Brown, C. M. 1981. *Nucleic Acids Res.* 9:2727–40

184a. Lerner, M. R., Boyle, J. A., Mount, S. M., Wolin, S. L., Steitz, J. A. 1980. *Nature* 283:220–24

185. Lamb, R. A., Lai, C.-J. 1982. *Virology* 123:237–56

186. Briedis, D. J., Lamb, R. A., Choppin, P. W. 1982. *Virology* 116:581–88

187. Compans, R. W. 1973. *Virology* 51:56–70

188. Krug, R. M., Etkind, P. R. 1973. *Virology* 56:334–48

189. Krug, R. M., Soeiro, R. 1975. *Virology* 64:378–87

190. Wolstenholme, A. J., Barrett, T., Nichol, S. T., Mahy, B. W. J. 1980. *J. Virol.* 35:1–7

191. Koennecke, I., Boschek, C. B., Scholtissek, C. 1981. *Virology* 110:16–25

192. Almond, J. W., Felsenreich, V. 1982. *J. Gen. Virol.* 60:295–305

193. Morrongiello, M. P., Dales, S. 1977. *Intervirology* 8:281–93

194. Shaw, M. W., Compans, R. W. 1978. *J. Virol.* 25:605–15

195. Yoshida, T., Shaw, M., Young, J. F., Compans, R. W. 1981. *Virology* 110:87–97

196. Follett, E. A. C., Pringle, C. R., Wunner, W. H., Skehel, J. J. 1974. *J. Virol.* 13:394–99

197. Minor, P. D., Dimmock, N. J. 1975. *Virology* 67:114–23

198. Deleted in proof

199. Stephenson, J. R., Hay, A. J., Skehel, J. J.

1977. *J. Gen. Virol.* 36:237–48

200. Lamb, R. A., Etkind, P. R., Choppin, P. W. 1978. *Virology* 91:60–78

201. Lamb, R. A., Choppin, P. W., Chanock, R. M., Lai, C.-J. 1980. *Proc. Natl. Acad. Sci. USA* 77:1857–61

202. Lamb, R. A., Lai, C.-J. 1980. *Cell* 21:475–85

203. Mahy, B. W. J., Barrett, T., Briedis, D. J., Brownson, J. M., Wolstenholme, A. J. 1980. *Philos. Trans. R. Soc. London B Ser.* 288:349–57

204. Porter, A. G., Smith, J. C., Emtage, J. S. 1980. *Proc. Natl. Acad. Sci. USA* 77:5074–78

205. Baez, M., Taussig, R., Zazra, J. J., Young, J. F., Palese, P. 1980. *Nucleic Acids Res.* 8:5845–58

206. Baez, M., Zazra, J. J., Elliott, R. M., Young, J. F., Palese, P. 1981. *Virology* 113:397–402

207. Winter, G., Fields, S., Gait, M. J., Brownlee, G. G. 1981. *Nucleic Acids Res.* 9:237–45

208. Inglis, S. C., Gething, M.-J., Brown, C. M. 1980. *Nucleic Acids Res.* 8:3575–89

209. Briedis, D. J., Lamb, R. A. 1982. *J. Virol.* 42:186–93

210. Almond, J. W., McGeoch, D., Barry, R. D. 1979. *Virology* 92:416–27

211. Barrell, B. G., Air, G. M., Hutchinson, C. A. III. 1976. *Nature* 264:34–41

212. Shaw, D. C., Walker, J. E., Northrop, F. D., Barrell, B. G., Godson, G. N. 1978. *Nature* 272:510–15

213. Atkins, J. F., Steitz, J. A., Anderson, C. W., Model, P. 1979. *Cell* 18:247–56

214. Beremand, M. N., Blumenthal, T. 1979. *Cell* 18:257–66

215. Soeda, E., Arrand, J. R., Smolar, N., Griffin, B. E. 1979. *Cell* 17:357–70

216. Soeda, E., Arrand, J. R., Griffin, B. E. 1980. *J. Virol.* 33:619–30

217. Contreras, R., Rogiers, R., Van de Voorde, A., Fiers, W. 1977. *Cell* 12:529–38

218. Barry, R. D., Ives, D. R., Cruickshank, J. G. 1962. *Nature* 194:1139–40

219. Barry, R. D. 1964. *Virology* 24:563–69

220. White, D. O., Day, H. M., Batchelder, E. J., Dheyene, I. M., Wansbrough, A. J. 1965. *Virology* 280–302

221. Rott, R., Saber, S., Scholtissek, C. 1965. *Nature* 205:1187–90

222. Nayak, D. P., Rasmussen, A. F. 1966. *Virology* 30:673–83

223. Scholtissek, C., Rott, R. 1964. *Virology* 22:169–79

224. Brownson, J. M., Mahy, B. W. J. 1979. *J. Gen. Virol.* 42:579–88

225. Rott, R., Scholtissek, C. 1970. *Nature* 228:56

226. Mahy, B. W. J., Hastie, N. D.,

Armstrong, S. J. 1972. *Proc. Natl. Acad. Sci. USA* 69:1421–24
227. Lamb, R. A., Choppin, P. W. 1977. *J. Virol.* 23:816–19
228. Spooner, L., Barry, R. D. 1977. *Nature* 268:650–52
229. Penhoet, E., Miller, H., Doyle, M., Blatti, S. 1971. *Proc. Natl. Acad. Sci. USA* 68:1369–71
230. Bishop, D. H. L., Obijeski, J. F., Simpson, R. W. 1971. *J. Virol.* 8:74–80
231. McGeoch, D. J., Kitron, N. 1975. *J. Virol.* 15:686–95
232. Plotch, S. J., Krug, R. M. 1977. *J. Virol.* 21:24–34
233. Plotch, S. J., Krug, R. M. 1978. *J. Virol.* 25:579–86
234. Bouloy, M. B., Plotch, S. J., Krug, R. M. 1978. *Proc. Natl. Acad. Sci. USA* 75:4886–90
235. Plotch, S. J., Tomasz, J., Krug, R. M. 1978. *J. Virol.* 28:75–83
236. Krug, R. M., Morgan, M. A., Shatkin, A. J. 1976. *J. Virol.* 20:45–53
237. Plotch, S. J., Bouloy, M., Krug, R. M. 1979. *Proc. Natl. Acad. Sci. USA* 76:1618–22
238. Robertson, H. D., Dickson, E., Plotch, S. J., Krug, R. M. 1980. *Nucleic Acids Res.* 8:925–42
239. Bouloy, M., Morgan, M. A., Shatkin, A. J., Krug, R. M. 1979. *J. Virol.* 32:895–904
240. Bouloy, M., Plotch, S. J., Krug, R. M. 1980. *Proc. Natl. Acad. Sci. USA* 77:3952–56
241. Krug, R. M., Broni, B. A., LaFiandra, A. J., Morgan, M. A., Shatkin, A. J. 1980. *Proc. Natl. Acad. Sci. USA* 77:5874–78
241a. Shatkin, A. J. 1976. *Cell* 9:645–53
242. Abraham, G. 1979. *Virology* 97:177–82
243. Pons, M. W., Rochovansky, O. M. 1979. *Virology* 97:183–89
244. Etkind, P. R., Krug, R. M. 1974. *Virology* 62:38–45
245. Glass, S. E., McGeoch, D., Barry, R. D. 1975. *J. Virol.* 16:1435–43
246. Hay, A. J., Lomniczi, B., Bellamy, A. R., Skehel, J. J. 1977. *Virology* 83:337–55
247. Hay, A. J., Abraham, G., Skehel, J. J., Smith, J. C., Fellner, P. 1977. *Nucleic Acids Res.* 4:4197–4209
248. Caton, A. J., Robertson, J. S. 1980. *Nucleic Acids Res.* 8:2591–603
249. Dhar, R., Chanock, R. M., Lai, C.-J. 1980. *Cell* 21:495–500
250. Beaton, A., Krug, R. M., 1981. *Nucleic Acids Res.* 9:4223–36
251. Krug, R. M., Broni, B. A., Boyloy, M. 1979. *Cell* 18:329–34
252. Hastie, N. D., Mahy, B. W. J. 1973. *J. Virol.* 12:951–61
253. Barrett, T., Wolstenholme, A. J., Mahy, B. W. J. 1979. *Virology* 98:211–25
254. Mark, G. E., Taylor, J. M., Broni, B., Krug, R. M. 1979. *J. Virol.* 29:744–52
255. Herz, C., Stavnezer, E., Krug, R. M., Gurney, T. Jr. 1981. *Cell* 26:391–400
256. Jackson, D. A., Caton, A. J., McCready, S. J., Cook, P. R. 1982. *Nature* 296:366–68
257. Krug, R. M. 1981. In *Curr. Top. Microbiol. Immunol.* 93:125–49
258. Hay, A. J., Skehel, J. J., McCauley, J. 1982. *Virology* 116:517–22
259. Scholtissek, C., Rott, R. 1970. *Virology* 40:989–96
260. Taylor, J. M., Illmensee, R., Litwin, S., Herring, L., Broni, B., et al. 1977. *J. Virol.* 21:530–40
261. Smith, G. L., Hay, A. J. 1982. *Virology* 118:96–108
262. Inglis, S. C., Mahy, B. W. J. 1979. *Virology* 95:154–64
263. Choppin, P. W., Pons, M. W. 1970. *Virology* 42:603–10
264. Bosch, F. X., Hay, A. J., Skehel, J. J. 1978. See Ref. 39, pp. 465–74
265. Lamb, R. A., Choppin, P. W. 1978. See Ref. 39, pp. 229–38
266. Mahy, B. W. J., Carroll, A. R., Brownson, J. M. T., McGeoch, D. J. 1977. *Virology* 83:150–62
267. Minor, P. D., Dimmock, N. J. 1977. *Virology* 78:393–406
268. Mahy, B. W. J., Barrett, T., Briedis, D. J., Brownson, J. M., Wolstenholme, A. J. 1980. *Phil. Trans. R. Soc. London Ser. B* 288:349–57
269. Von Magnus, P. 1954. *Adv. Virus Res.* 2:59–78
270. Crumpton, W. M., Dimmock, N. J., Minor, P. D., Avery, R. J. 1978. *Virology* 90:370–73
271. Nayak, D. P., Tobita, K., Janda, J. M., Davis, A. R., De, B. K. 1978. *J. Virol.* 28:375–86
272. Davis, A. R., Nayak, D. P. 1979. *Proc. Natl. Acad. Sci. USA* 76:3092–96
273. Davis, A. R., Hiti, A. L., Nayak, D. P. 1980. *Proc. Natl. Acad. Sci. USA* 77:215–19
274. Winter, G., Fields, S., Ratti, G. 1981. *Nucleic Acids Res.* 9:6907–15
275. Nayak, D. P., Sivasubramanian, N., Davis, A. R., Cortini, R., Sung, J. 1982. *Proc. Natl. Acad. Sci. USA* 79:2216–20
276. Janda, J. M., Davis, A. R., Nayak, D. P., De, B. K. 1979. *Virology* 95:48–58
277. Choppin, P. W. 1969. *Virology* 39:130–34
278. Hirst, G. K., Pons, M. W. 1973. *Virology* 56:620–31
279. Winter, G., Fields, S., Brownlee, G. G. 1981. *Nature* 292:72–75

280. Hiti, A. L., Davis, A. R., Nayak, D. P. 1981. *Virology* 111:113–24
281. Gething, M. J., Skehel, J. J., Waterfield, M. D. 1980. *Nature* 287:301–6
282. Fang, R., Min Jou, W., Huylebroeck, D., Devos, R., Fiers, W. 1981. *Cell* 25:315–23
283. Verhoeyen, M., Fang, R., Min Jou, W., Devos, R., Huylebroeck, D., et al. 1980. *Nature* 771–76
284. Both, G. W., Sleigh, M. J. 1980. *Nucleic Acids Res.* 8:2561–75
285. Sleigh, M. J., Both, G. W., Brownlee, G.

G., Bender, V. J., Moss, B. A. 1980. See Ref. 40, pp. 69–79
286. Min Jou, W., Verhoeyen, M., Devos, R., Saman, E., Fang, R. 1980. *Cell* 19:683–96
287. Porter, A. G., Barber, C., Carey, N. H., Hallewell, R. A., Threlfall, G., et al. 1979. *Nature* 282:471–77
288. McCauley, J. W., Mahy, B. W. J., Inglis, S. C. 1982. *J. Gen. Virol.* 58:211–15
289. Sivasubramanian, N., Nayak, D. P. 1982. *J. Virol.* 44:321–29

Ann. Rev. Biochem. 1983. 52:507–35
Copyright © 1983 by Annual Reviews Inc. All rights reserved

RIBULOSE-1,5-BISPHOSPHATE CARBOXYLASE-OXYGENASE

Henry M. Miziorko

Department of Biochemistry, Medical College of Wisconsin, Milwaukee, Wisconsin 53226

George H. Lorimer

Central Research and Development Department, E. I. du Pont de Nemours & Co., Wilmington, Delaware 19898

CONTENTS

PERSPECTIVES AND SUMMARY 507
INTRODUCTION 509
   The Conundrum of Photorespiration 509
   The Genetics of RuBP Carboxylase and the Case for Subunit Heterogeneity 510
PHYSICOCHEMICAL PROPERTIES OF RuBP CARBOXYLASE-
   OXYGENASE 512
   Subunit Composition 512
   Amino Acid Sequences 513
   Quaternary Structure of RuBP Carboxylase 517
   Function of the Subunits 519
ACTIVATION OF THE ENZYME 521
   Historical Perspective 521
   Molecular Mechanism 521
   Physiological Aspects of Activation 524
CATALYSIS 526
   A Mechanism for the RuBP Carboxylase Reaction 526
   Additional Details Concerning Events at the Active Site 528
   A Mechanism for the RuBP Oxygenase Reaction 530

PERSPECTIVES AND SUMMARY

Ribulose-1,5-bisphosphate carboxylase (EC 4.1.1.39), which catalyzes the initial step in Calvin's reductive pentose phosphate cycle, is the major soluble leaf protein in plants. A decade has elapsed since it was discovered that this enzyme is also a monooxygenase, which catalyzes the primary

507

0066-4154/83/0701-0507$02.00

event in photorespiration. Improving the efficiency at which CO_2 competes with O_2 for reaction with ribulose bisphosphate would improve the rate of photosynthesis relative to that of photorespiration and could substantially increase productivity in a large variety of commercially important crops. This prediction has stimulated considerable study of the basic events that occur during the carboxylase and oxygenase reactions and investigations aimed at determining whether selective perturbation of these reactions is possible.

In addition to the conventional biochemical approaches to improve photosynthetic yield, the techniques of molecular biology have recently been applied to this problem. Recombinant DNA technology promises to facilitate the genetic manipulation that may lead to improved crop productivity. Initial results of the molecular biology-oriented studies have improved our concept of how the synthesis of the subunits of ribulose bisphosphate carboxylase occurs. There are encouraging indications that continuing investigations in this area will result in the elucidation of the mechanisms involved in processing, transporting, and assembling the newly synthesized peptides to produce a functional oligomeric enzyme.

It has become axiomatic that control of flux through a metabolic pathway is usually exerted at an irreversible enzymatic step that occurs early in that pathway. Ribulose bisphosphate carboxylase fulfills these criteria; thus, it is not surprising that enzyme activity can be modulated. Formation of a ternary enzyme-CO_2-divalent cation complex is required to convert the protein into a catalytically functional species. Carbamate formation involving the ε-amino group of lysine #201 occurs upon binding of activator CO_2, generating a form of the enzyme that exhibits tight, stoichiometric cation binding. The activator CO_2 is distinct from the CO_2 that is fixed in the reaction. In contrast, while two CO_2 molecules are required to support the reaction, only one divalent cation is necessary. The detailed role of the cation remains to be elucidated, although there is some evidence that it supports both activation and catalysis. If a dual function for cation is verified, it would suggest proximity of catalytic and activation sites. Portions of these sites have been mapped by use of active site–directed inhibitors and by trapping studies on the CO_2 activator. Peptides from these sites have been placed in the primary sequence of the enzyme's large subunit. Solution studies on the location and relative orientation of catalytic and activation sites may expedite a formulation of the three-dimensional structure of the enzyme; X-ray diffraction studies on this protein have not yet produced a high resolution structure.

Recent chemical studies have supported Calvin's mechanism for carboxylation of ribulose bisphosphate. It has been established that formation of an enediol from the sugar phosphate substrate occurs prior to attack on substrate CO_2. The data support the transient formation of a 2-carboxy-3-keto arabinitol bisphosphate intermediate, which is rapidly cleaved to generate two molecules of D-3-phosphoglycerate. The chemical events involved in the ribulose bisphosphate oxygenase reaction that are not strictly analogous to steps in the carboxylase reaction remain the subject of considerable speculation.

INTRODUCTION

In response to predictions of global food and energy shortages, increasing interest is evident in the enzyme responsible for annually fixing about 10^{11} tons of CO_2, ribulose-1,5-bisphosphate (RuBP) carboxylase-oxygenase. It is clearly the most abundant protein of all; some 40 million tons (or 20 lb for every man, woman, and child!) are needed globally to accomplish its prodigious task (1). It is also the source of a number of intriguing biochemical problems currently under investigation.

The Conundrum of Photorespiration

Photorespiration is defined as the uptake of O_2 *and* release of CO_2, associated with glycolate metabolism, which accompanies photosynthetic CO_2 fixation under aerobic conditions (2). An analogous process, chemo-respiration, accompanies aerobic CO_2 fixation by some chemosynthetic bacteria (2, 3). The function of chemo- or photorespiration is not at all apparent. Compounding this mystery is the fact that the enzyme that catalyzes the primary step in photorespiration, the oxygenation of RuBP, is the same enzyme that catalyzes the primary step in photosynthetic CO_2 fixation, the carboxylation of RuBP (4). Under natural conditions,[1] it is thought that the ratio of carboxylation to oxygenation in vivo is 3:1 to 4:1 (5). Consequently, the oxygenation of RuBP substantially decreases the overall rate of photosynthesis and hence of plant productivity. That one enzyme should catalyze the primary steps of two major, but diametrically opposing, metabolic pathways is a curious situation indeed. We know of no precedent for this. Since all RuBP carboxylases exhibit oxygenase activity, regardless of their taxonomic origin (including those in some anaerobic bacteria), it has been suggested (6) that oxygenation occurs because an obligatory intermediate in the carboxylation reaction is autoxidizable. While this speculation may be plausible, it cannot be regarded as proven until the mechanistic details of both reactions are uncovered.

---

[1] The term "natural conditions" refers to a gas phase containing 0.033% CO_2 and 21% O_2. At 25°C and normal pressure, a solution in equilibrium with this gas phase contains about 10 μM CO_2 and 250 μM O_2.

The Genetics of RuBP Carboxylase and the Case for Subunit Heterogeneity

THE LARGE SUBUNIT   The most common form of RuBP carboxylase is composed of eight large (catalytic) subunits and eight small subunits of unknown function (7, 8). A genetic analysis of the mode of inheritance of the large subunit in the genus *Nicotiana* provided the first evidence that the large subunit is encoded in the chloroplast DNA (9, 10). Direct evidence for this has been obtained by the application of molecular biological techniques (11–16). The large subunit gene is present as a single copy per chloroplast DNA molecule (12, 14). However, a chloroplast contains from 10- to 100-fold more DNA than can be accounted for by a single chloroplast DNA molecule [see Table 3 of (17)]. Thus, each chloroplast contains (depending on the species) 10–100 copies of the large subunit gene. In addition, since leaf cells may contain (again depending on the species) from 10–200 chloroplasts per cell (18), it follows that there may be several thousand copies of the gene for the large subunit per cell.

The basis of the genetic analysis of the large and small subunits in the genus *Nicotiana* was the variation in the isoelectric focusing patterns of carboxymethylated subunits (reviewed in 19, 20). This technique also revealed apparent charge heterogeneity in the composition of the large and small subunits. As discussed elsewhere (21–23), charge heterogeneity must be interpreted with considerable caution. Indeed, chemical analyses of the isoelectric variants of the large subunit failed to detect any differences between them, and Wildman's group (25) has concluded that the variants result from the modification of a single gene product. This conclusion is strengthened by findings (11–16) that only one gene for the large subunit appears to be present per chloroplast chromosome. A recent report indicates that charge heterogeneity arises artifactually during the alkylation procedure that precedes isoelectric focusing (24). In conclusion, the weight of evidence points to the large subunit being homogeneous, the product of a single gene.

THE SMALL SUBUNIT   The situation with regard to the small subunit is more complex. Genetic analysis in the genus *Nicotiana* first demonstrated that the small subunit was inherited biparentally; i.e. the gene was encoded within the nuclear DNA (26). This idea has been confirmed and considerably extended. Thus, it is now known that the mRNA for the precursor of the small subunit is transcribed in the nucleus (27, 28), and translated on cytoplasmic ribosomes (29–31) to yield a precursor of MW 4000–6000 larger than the mature polypeptide. The mRNA for the small subunit, prepared from cytoplasmic polysomes of pea, has been copied into DNA, cloned into an *Escherichia coli* plasmid, and the nucleotide sequence of the cDNA determined (32).

It is now apparent that a multigene family encodes the small subunit. In pea there are seven copies per haploid nuclear genome (184) and a similar number in soybean (34). Two cDNA clones from two different varieties of the pea *Pisum sativum* have been isolated and sequenced (32, 184). There are ten nucleotide differences that translate into eight differences in the amino acid sequence of the mature polypeptide without altering its charge (see Figure 2). While these differences might be due to the different varieties of pea, they might also reflect the presence of several species of small subunit mRNA (184).

Sequencing of the wheat and soybean genomic DNA has revealed the presence of one (wheat) and two (soybean) introns (34, 186). Interestingly, the first exon encodes the entire transit peptide and one or two amino acids of the mature polypeptide, a finding that supports Gilbert's (35) proposal that exons encode separate functional polypeptide domains.

Isoelectric focusing has revealed the presence of charge variants in the mature small subunit from pea (36). However, no heterogeneity was observed in the amino acid sequence (36), which was identical to that deduced from the nucleotide sequence of cDNA of mRNA (32). The charge variants were attributed to the partial loss of the N-terminal methionine followed by cyclization of the glutamine residue so exposed. This same mechanism has been invoked to account for the occurrence of two charge variants of the spinach subunit (37). Unequivocal proof of small subunit heterogeneity has been obtained for the spinach (37) and tobacco (25, 38) small subunits by amino acid sequencing and tryptic peptide fingerprint analysis. In the spinach subunit there is a tyrosine-proline substitution at residue 97 (37). The tyr-pro substitution does not account for the occurrence of charge variants which are probably artifacts. On the other hand, the occurrence of charge variants in the small subunit from *Nicotiana tabacum* is based on defined amino acid substitutions (38, 179) that support the amphidiploid origin of this species (20). Heterogeneity has been detected at three sites, tyr/ile at #6, asn/gly at #7, and his/arg at #48 (38, 179). The last change could account for the occurrence of charge variants in *N. tabacum*. These substitutions constitute strong evidence that the small subunit is encoded by more than one gene within the nuclear DNA. Assuming that each gene for the small subunit is equally transcribed and translated, the final assembled heteroenzyme would represent a population of different molecules. Two different small subunit genes would generate nine different heteroenzyme molecules, three different subunit genes would generate 45 different heteroenzyme molecules and so on (39). This raises the spectre that electrophoretically homogeneous preparations of purified enzyme might well represent a rather complex population of molecules. As Wildman has pointed out (20), electrically neutral substitutions would be difficult to detect. The occurrence of heterogeneity within the small subunit is especially worrying to those dealing with the kinetic properties of the enzyme. The kinetic constants determined with heterogeneous enzyme are not unique values but rather represent average values for the whole population of molecules.

PHYSICOCHEMICAL PROPERTIES OF RuBP CARBOXYLASE-OXYGENASE

Subunit Composition

RuBP carboxylase has been purified to homogeneity from a variety of plant, algal, and bacterial sources (40, 41). All of the plant and algal enzymes studied to date are of similar mass (560,000 daltons) and contain eight large (56,000 daltons) and eight small (14,000 daltons) subunits (41). There have been reports of heterogeneity (multiple bands on isoelectric focusing gels) in both large and small subunits (19, 42). However, in view of the genetic evidence outlined in the last section, and in the absence of any structural evidence to the contrary, e.g. amino acid sequence, peptide fingerprinting (25, 38), such reports must be interpreted with some caution. Bacterial RuBP carboxylase contrasts with the enzyme from higher organisms in that there is substantial variability in both the reported native molecular weights and the subunit composition of these proteins (41). The enzyme from *Rhodospirillum rubrum* is widely accepted to be dimeric, containing subunits that approach the large subunits of the plant enzyme in molecular weight (43, 44). Other bacterial enzymes have been reported with L_4, L_6, L_6S_6, and L_8 structures (for review see 41) and, in the case of *Rhodopseudonomas sphaeroides*, two forms of the enzyme with L_8S_8 and L_6 structures have been isolated (45). However, failure to detect the small subunit must be interpreted with caution. Absence of evidence does not constitute evidence of absence! For example, the RuBP carboxylases from *Anabaena cylindrica* (46) and *Thiobacillus intermedius* (47) were reported to lack the small subunit. Reinvestigation in other laboratories (48–50), however, demonstrated its presence. Both Codd & Stewart (51) and Andrews & Abel (52) have demonstrated the ease with which small subunits can be stripped from the native enzyme by mild acid treatment, leaving a core of large subunits intact. Studies in author Miziorko's laboratory that involve relatively mild partial denaturation of the spinach enzyme suggest that, even in the case of the eukaryotic protein, it is possible to generate oligomers of large subunits, which appear to be free of small subunits upon gel electrophoresis. Thus, care should be taken that the small subunit is not artifactually lost during enzyme purification.

A number of bacterial RuBP carboxylases have been reported to have an L_6S_6 structure (53–55). However, the recent thorough study of Andrews et al (52, 56) on the cyanobacterial enzyme from *Synechococcus* shows that these reports must be viewed with caution. When the molecular weight of the *Synechococcus* enzyme was determined by pore-gradient electrophoresis, gel-filtration chromatography, and density-gradient centrifugation, a value of M_r 430,000 was obtained (52). Since the protomer molecular weight was shown to be M_r 70,000, a hexameric L_6S_6 structure appeared likely. However, further investigations employing equilibrium sedimentation yielded a value of M_r 530,000, i.e. within the range for an L_8S_8 structure (56). Electron microscopic studies further revealed fourfold symmetry characteristic of an octomeric L_8S_8 structure. With the exception of the well-documented L_2 structure of the *R. rubrum* enzyme, we do not regard subunit compositions other than L_8S_8 as having been rigorously established. Speculation about evolutionary relationships that such data have prompted may require substantial revision.

Amino Acid Sequences

THE LARGE SUBUNIT   It has long been recognized that the large subunits of various RuBP carboxylases are similar in amino acid composition (57). Only recently has it become possible to ascertain whether the primary structure is also highly conserved. The genes coding for this subunit in maize (58), spinach (59), *Chlamydomonas* (60), and *Synechococcus* (33) have been cloned and sequenced, allowing the amino acid sequence to be deduced (Figure 1). Comparing the sequence of the evolutionarily most distant eukaryotic polypeptides (*Chlamydomonas* versus *Zea mais*) reveals greater than 85% homology in a total of 475 residues. This very high degree of homology is also apparent in the partial sequence of the barley enzyme's large subunit (62). Especially interesting is the complete sequence conservation between residues 169 and 220, and 321 and 340. These regions contain lysine residues at 175 and 335, which have been implicated in the catalytic process by affinity-labeling studies (63), as well as the lysine at 201, which bears the activator CO_2 (64, 65). Of additional interest is the observation (16) that region 161–192 contains only neutral or basic amino acids, a feature that may be of functional importance. When a negative charge was introduced by a gly → asp substitution at residue 171, the resulting protein was catalytically inactive (16).

A precursor to the large subunit, about 2000 daltons larger than the mature protein, has been detected in spinach (180) and pea (181). Treatment of the precursor with a chloroplast extract converted it to the mature form (180). Amino acid sequencing of the N-terminal region of the mature large subunit from barley (69), wheat and rape (P. G. Martin, personal communication) indicates that they begin with ala-gly-val-lys, i.e. [15]ala is the N-terminal residue of the mature protein. However, nucleotide

```
                    110         120         130         140         150         160         170         180         190         200
3  M S P Q T E T K A S V G F K A G V K D Y K L T Y Y T P E Y E T K D T D I L A A F R V T P Q L G V P P E E A G A A V A A E S S A G T W T T V W I D G L I S L D R Y K G H C Y H I L P V P D P D V I L Y
4  M V P Q T E T K A G A G F K A G V K D Y R L T Y Y T P D Y V V R D T D I L A A F R M T P Q L G V P P E E C G A A V A A E S S T G T W T T V W T D G L T S L D R Y K G R C Y D I E P V P G E D N Q Y I A Y
5  M - P K T Q S - A A - G Y K A G V K D Y K L T Y Y T P D Y T P K D T D L A A F P V S P Q P G V P A D E A G A A I A A E S S T G T W T T V W T D L L T D M D R Y K G K C Y H I E P V Q G E E N S Y F A F

                    110         120         130         140         150         160         170         180         190         200
1  V A Y P L D L F E E G S V T N M F F T S I V G N V F G F K A L R A L R L E D L R I P V A Y V K T F Q G P P H G I Q V E R D K L N K Y G R P L L G C T I K P K L G L S A K N Y G R A V Y E C L R G G L D F T
2                  M F F T S I V G N V F G F K A L
3  V A Y P I L L F E E G S V T N M F T S I V G N V F G F K A L R A L R L E D L R I P P A Y S K T F Q G P P R G M Q V E R D K L N K Y G R P L L G C T I K P K L G L S A K N Y G R A C Y E C L R G G L D F T
4  V A Y P L D L F E E G S V T N M F T S I V G N V F G F K A L R A L R L E D L R I P P A Y V K T F V G P P H G I Q V E R D K L N K Y G R G L L G C T I K P K L G L S A K N Y G R A V Y E C L R G G L D F T
5  I A Y P L D L F E E G S V T N I L T S I V G N V F G F K A I R S L R L E D I R F P V A L V K T F Q G P P H G I Q V E R D L L T K Y G R P L M G C T I K P K L G L S A K N Y G R A V Y E C L R G G L D F T

                    210         220         230         240         250         260         270         280         290         300
1  K D D E N V N S Q P F M R H R D R F L F C A E A L Y K A Q A E T G E I K G H Y L N A T A G T C E D M K R A V F A R E L G V P I V M H D Y L T G G F T A N T T L S H Y C R D I G L L L H I H R A M H A V
2        M R I R D R F L F C A E A L Y K A Q A E T G E I K G H Y L N A        M I K G A V F A R E L G V P        M H D Y L T G G F T A N T T L S H Y C R D B G L L L H I H R A M H A V
3  K D D E N V N S Q P F M R W R D R F V F C A E A I Y K S Q A E T G E I K G H Y L N A T A G T C D E M I L G A V F A R Q L G V P I V M H D Y L T G G F T A N T T L S H Y C R D N G L L L H I H R A M H A V
4  K D D E N V N S Q P F M R W R D R F L F V A E A I Y K A Q A E T G E V K G H Y L N A T A G T C E E M K R A V C A K E L G V P I I M H D Y L T G G F T A N T S L A I Y C R D N G L L L H I H R A M H A V
5  K D D E N I N S Q P F Q R W R D R F L F V A D A I H K S Q A E T G E V K G H Y L N V T A P T C E E M M K R A E F A K E L G M P I I M H D F L T A G F T A N T T L A K W C R D N G V L L H I H R A M H A V

                    310         320         330         340         350         360         370         380         390         400
1  I D R Q K N H G M H F R V L A K A L R L S G G D H I H S G T V V G K L E G E R D I T L G F V D L L R D D Y T E K D R S R I Y F T Q S W V S T P G V L P V A S G G I H V W H M P A L T E I F G D D S V L
2  I D R Q K B H G M H F R V L A K A L   M S G G D H I H S G T V V G K L E G E R E M T L G F V D L L R B B F I E K B
3  I D R Q K N H G M H F R V L A K A L R L S G G D H I H S G T V V G K L E G E R E I T L G F V D L L R D D F I R S R G I F F T Q D W V S   M P G V I P V A S G G I H V W H M P A L T E I L G D D S V L
4  I D R Q R N H G I H F R V L A K A L R M S G G D H L H S G T V V G K L E G E R E V T L G F V D L M R D D Y V E K D R S R G I Y F T Q D W C S M P G V M P V A S G G I H V W H M P A L V E I F G D D A C L
5  I D R Q R N H G I H F R V L A K C R L S G G D H L H S G T V V G K L E G D K A S T L G F V D L M R E D H I E R D R S R G V F F T Q D W A S M P G V L P V A S G G I H V W H M P A L V E I F G D D S V L

                    410         420         430         440         450         460         470
1  Q F G G G T L G H P W G N A P G A V A N R V A L E A C V Q A R N E G R D L A R E G N T I I R E A T K W S P E L A A A C E V W K E I K F E - F P A M D T V
2  Q F G G G T L G H P W G N A P G A A A N R V A L E A C V Q A R N E G R D L A R E                   A A C K
3  Q F G G G T L G H P W G N A H G A A A N R V A L E A C V Q A R N E G R D L A R E V Q - I I K A A C K W S P E L A A A C E I W K E I K F E F D G F K A M D T I
4  Q F G G G T L G H P W G N A P G A A A N R V A L E A C T Q A R N E G R D L A R E G G D V I R S A C K W S P E L A A A C E V W K E I K F E - F D T I D L L
5  Q F G G G T L G P P W G N A P G A T A N R V A L E A C V Q A R N E G R D L Y R E G G D I L R E A G K W S P E L A A A L D L W K E I K F E - F E T M D K L
```

A=Ala, B=Asx, C=Cys, D=Asp, E=Glu, F=Phe, G=Gly, H=His, I=Ile, K=Lys, L=Leu, M=Met, N=Asn, P=Pro, Q=Gln, R=Arg, S=Ser, T=Thr, V=Val, W=Trp, Y=Tyr.

Figure 1 The amino acid sequences of the large subunit of ribulose bisphosphate carboxylase from 1. *Spinacia oleracea* (59), 2. *Hordeum vulgaris* (62), 3. *Zea mais* (58), 4. *Chlamydomonas reinhardii* (60), and 5. *Synechococcus* PCC6301 (33).

sequencing (58–60) indicates that translation begins 14 residues before [15]ala. These observations support Langridge's (180) view that posttranslational processing converts the precursor of the large subunit to its mature form. The biological significance of this phenomenon remains to be established.

In sharp contrast to the homology that characterizes the primary structures of large subunits from the plant enzymes is the amino acid composition and sequence of the *R. rubrum* enzyme. Akazawa et al (57) had noted that the *R. rubrum* enzyme was much different in amino acid composition from the plant enzymes, which supports the idea that the prokaryotic and eukaryotic enzymes might not be very closely related in an evolutionary sense. Hartman and colleagues (66) have elucidated 70% of the primary structure of the *R. rubrum* enzyme by sequencing cyanogen bromide fragments. Only 28% of the residues are identical to those reported for the plant carboxylases. Despite the small degree of overall homology, two regions stand out as being highly conserved. These two regions contain the target sites for the affinity labels 2-bromoacetylaminopentitol-1,5-bisphosphate (68) and pyridoxal phosphate (67). The former reagent selectively alkylates methionine 335, which is adjacent to lysine 334. In the spinach enzyme, this lysine residue is derivatized by the affinity label 3-bromo-1,4-dihydroxy-2-butanone-1,4-bisphosphate (63). When this segment of the *R. rubrum* enzyme is compared with the corresponding segment of the spinach enzyme, eight residues out of 13 are identical. A more striking homology is seen at the lysine modified by pyridoxal phosphate, residue 175 of the spinach enzyme. This residue lies within a seven-residue segment that is identical in the two species. Thus, the suspicion (4) that the *R. rubrum* enzyme may have evolved independently has proven groundless.

THE SMALL SUBUNIT   The SΔQ analysis of relatedness performed by Takabe & Akazawa (70) on various eukaryotic and prokaryotic RuBP carboxylases indicated that the small subunit was less highly conserved than the large subunit. Recent sequencing work has borne this out; there was about 70% homology between the sequence of the mature spinach, soybeans, and pea small subunits. The primary structures of the small subunits from spinach, pea, and tobacco have been determined by classical protein-sequencing technique (36, 37, 179), while those from soybean and pea have been determined by cloning and DNA sequencing (32, 34). P. G. Martin (personal communication) has sequenced the first 40 amino acids from 46 angiosperm species (Figure 2). Highly conserved portions of the small subunit may provide clues to the function of the small subunit. The regions encompassing residues 3–5, 11–21, 38–45, 49–75, and 109–117 show much more homology than is apparent in the remainder of the molecule.

There is currently no information about the amino acid sequences of the small subunit from prokaryotic RuBP carboxylases.

The cloning procedure permits the amino acid sequence of the cytosolic precursor to the small subunit to be deduced. Upon processing to the mature small subunit, a protein of 123 amino acids is produced (29–31). It is interesting that no striking homologies are apparent when a comparison is made between the transit peptide sequences from higher plants (soybean, pea) and a green algae, *Chlamydomonas* (32, 34, 71). In addition, the transit peptide for the pea small subunit is 11 residues shorter than its counterpart in *Chlamydomonas*. These differences are reflected in the inability of intact pea chloroplasts to transport and process the small subunit precursor from *Chlamydomonas* (185). Significantly the transit peptides are characterized

```
                  10               20              30              40
    2 3 1 1 2 3 5 2 4 4 1 3 1 1 1 1 1 1 2 2 1 4 9 5 6 2 4 6 3 2 4 2 2 3 10 4 3 1 2 1
                        I N
1   M Q V W P P Y G K K K Y E T L S Y L P D L S Q E Q L L L E P D Y L L K D G W V P
2   M Q V W P P I G K K K F E T L S Y L P D L D D A Q L A K E V E Y L L R K G W I P
3   M Q V W P P L G L K K F E T L S Y L P P L T T E Q L L A E V N Y L L V K G W I P
4   M Q V W P P I G K K K F E T L S Y L P P L T R D Q L L K E V E Y L L R K G W V P
5   M Q V W P P I G K K K F E T L S Y L P P L T R D Q L L K E V E Y L L R K G W V P
6   M Q V W P I E G I K K F E T L S Y L P P L S T E A L L K Q V D Y L I R S K W V P

                  50               60              70              80
                          H
1   C L C F E T E R G F V Y R E N N K S P G Y Y D G R Y W T M W K L P M F G C T D A
2   C L E F E L E H G F V Y R E H N R S P - Y Y D G R Y W T M W K L P M F G C T D A
3   P L E F E V K D G F V Y R E H D K S P G Y Y D G R Y W T M W K L P M F G G T D P
4   C L E F E L L K G F V Y G E H N K S P R Y Y D G R Y W T M W K L P M F G T T D P
5   C L E F E L E K G F V Y R E H N K S P R Y Y D G R Y W T M W K L P M F G T T D A
6   C L E F S K V - G F V F R E H N S S P G Y Y D G R Y W T M W K L P M F G C T D A

                  90               100             110             120
1   T Q V L A E V G E A K K A Y P E A W I R I I G F D N V R Q V Q C I S F I A Y K P E G Y
2   S Q V L K E L Q E A K T A Y P N G F I R I I G F D N V R Q V Q C I S F I A Y K P P G F
                              Y
3   A Q V V N E V E E V K K A P P D A F V R F I G F D N K R E V Q C I S F I A Y K P A G Y
4   A Q V V K E L D E V V A A Y P E A F V P V I G F N N V R Q V Q C I S F I A H T P E S Y
5   S Q V L K E L D E V V A A Y P Q A F V P I I G F D N V R Q V Q C I S F I A H T P E S Y
6   T Q V L N E V E E V K K E Y P D A Y V R V I G F D N L R Q V Q C V S F I A F R P C E E S G K A
```

A = Ala, B = Asx, C = Cys, D = Asp, E = Glu, F = Phe, G = Gly, H = His, I = Ile, K = Lys, L = Leu, M = Met, N = Asn, P = Pro, Q = Gln, R = Arg, S = Ser, T = Thr, V = Val, W = Trp, Y = Tyr

Figure 2   The amino acid sequences of the mature small subunit of ribulose bisphosphate carboxylase from 1. *Nicotiana tabacum* (179), 2. *Glycine max* (34), 3. *S. oleracea* (37), 4. *Pisum sativum*, var, Feltham First (32, 36), 5. *P. sativum*, var. Progress No. 9 (184), and 6. *Triticum aestivum* (186). The values above the first 40 residues refer to the number of different amino acids detected at that position during P. G. Martin's (personal communication) sequencing of the first 40 amino acids of the small subunit from 46 taxonomically diverse Angiosperms; e.g., residue #3 is invariably valine, whereas ten different amino acids have been detected at Residue #35.

by a preponderance of basic amino acids. It has been suggested (71) that this feature is involved in the binding to and/or passage of the small subunit precursor through the negatively charged chloroplast envelope membrane (72).

Quaternary Structure of RuBP Carboxylase

Insight into the quaternary structure of RuBP carboxylase is largely based on electron microscopic and X-ray crystallographic studies on plant and bacterial enzymes which contain eight large and small subunits. Eisenberg's group (73) has studied the protein purified from tobacco and proposed a bilayer model (Figure 3) that indicates that four large subunits are stacked over another layer of four subunits. There is a channel between the large subunits that is easily detectable upon electron microscopic examination of

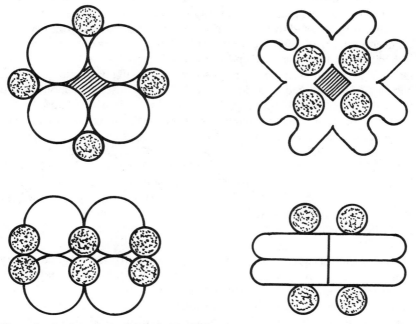

Figure 3 A schematic model depicting the quaternary structure of ribulose bisphosphate carboxylase from *N. tabacum* (*left*) and *Alcaligenes eutrophus* (*right*), based on X-ray diffraction and electron microscope studies. The top views show a planar array of four large subunits that stacks above the remaining four subunits, with formation of a well-defined central channel. The side view (*bottom*) indicates that the small subunits of the bacterial enzyme (*right*) are positioned above and below the planes of the large subunits. The positions of the small subunits of the tobacco enzyme (*left*) are speculative. The model depicts only one of several possibilities consistent with molecular symmetry and the mass distribution observed in electron micrographs of the tobacco enzyme.

negatively stained protein samples. The crystallographic data indicate a fourfold axis of symmetry, coincident with the center of the channel formed by eclipsed rings of large subunits. While the symmetry determined by crystallographic studies does not distinguish between a staggered or eclipsed arrangement of the layers of large subunits, the electron microscopy data favor the latter interpretation. The precise location of the small subunits of the tobacco enzyme is unclear; the tentative model shows them arranged in the same plane as the large subunits, situated peripherally in an array compatible with fourfold symmetry.

Bowien and colleagues (74) are studying the enzyme from *Alcaligenes eutrophus*, a chemoautotrophic hydrogen bacterium. This protein shows some similarity in quaternary structure to plant enzyme, since there are four large subunits involved in forming each of two stacked layers perpendicular to the fourfold rotational axis. The model for RuBP carboxylase from *A. eutrophus* (Figure 3) differs from Eisenberg's model in that the large subunits are suggested to be U-shaped rather than globular. Moreover, the small subunits are envisioned above and below the layers of large subunits, rather than around the edge and in the same plane as the large subunits. Data from immune electron microscopy, indicating very close contact between protein linked by antibodies prepared against small subunit, and the failure to observe small subunit specific antibodies bound to the periphery of the *A. eutrophus* enzyme's large subunits have been invoked to support the assignment of the small subunit locations. The crystal unit cell determined for the *A. eutrophus* enzyme has a 19% larger volume than that reported for the tobacco leaf enzyme and there are considerable differences in the dimensions and the ratios of the crystal cell axes. Packing of an enzyme molecule from one species into the crystal unit cell determined for enzyme from the other species is apparently not possible. This has prompted Bowien to argue in favor of different arrangements of the subunits.

Differences between the models for RuBP carboxylase quaternary structure are not entirely unexpected. Bowien's group grows crystals under conditions that generate activated RuBP carboxylase (i.e. 50 mM $NaHCO_3$; 10 mM $MgCl_2$), while the tobacco enzyme crystals are grown in the absence of activators (73). Circular dichroism experiments (75, 76) suggest that a conformational change occurs upon binding of bicarbonate and divalent cation to the protein. The ability of bicarbonate and cation to alter the reactivity of the enzyme toward cross-linking reagents (75) and reagents directed toward the RuBP binding site also argues for a conformational change upon activation (63). Thus, detailed crystal structures of activated and nonactivated enzyme may provide some interesting contrasts that will allow identification of the interactions that are critical for generating a catalytically competent RuBP carboxylase molecule.

In a novel approach to the question of quaternary structure Roy et al (77) have employed cross-linking reagents. Dimers of the small subunit were formed, which indicates that these subunits are closely paired.

Function of the Subunits

THE LARGE SUBUNIT   Assignment of a role to the large subunit is relatively straightforward. The observation that the *R. rubrum* enzyme, which contains only large subunits (43, 44), undergoes activation by CO_2 and Mg^{2+} via carbamate formation (78–80) and catalyzes both carboxylation and oxygenation reactions (81, 82) suggests that all the amino acids that mediate these processes reside on the large subunit. Considerable evidence has accumulated to establish this point for the L_8S_8 enzyme as well: 1. The activator CO_2 has been trapped on the large subunit of the plant enzyme and the ε-amino group of lysine 201 has been demonstrated to link the activator CO_2 to the protein (64, 65). In studies using cobalt as an exchange-inert probe of the activator cation binding site, metal ion has been shown to bind directly to amino acids of the large subunit of the spinach enzyme (83). 2. Reconstitution experiments with the spinach enzyme by Akazawa's group (84, 85) have established that the large subunit contains the catalytic site. 3. The work of Hartman's group (63) with active site–directed irreversible inhibitors has demonstrated that the large subunit contains the sugar phosphate substrate binding site. 4. The genetic analyses of a catalytically inactive point mutant of RuBP carboxylases in *Chlamydomonas* (15, 16) and of its revertants (86) confirms the catalytic role of the large subunit.

THE SMALL SUBUNIT   The function of the small subunit remains a mystery. It is not infrequently implied that it plays a regulatory role, but this proposition relies almost entirely upon analogy with other enzymes containing heterologous subunits. Unequivocal experimental evidence that clearly delineates a function for the small subunit has yet to be reported. The fact that the *R. rubrum* enzyme, which lacks small subunits, undergoes activation by CO_2 and catalyzes both oxygenation and carboxylation demonstrates that none of these processes is absolutely dependent on the presence of a small subunit.

One striking difference between the dimeric *R. rubrum* enzyme and that from higher plants concerns the $K_m (CO_2)$. The higher plant enzyme has an affinity for CO_2 that is about tenfold greater than the *R. rubrum* enzyme (20 μM vs 300 μM) (78, 89, 90). From this it might be supposed that the small subunit somehow modulates the affinity of the enzyme for CO_2. Yet this argument does not withstand close scrutiny, since, for example, L_8S_8 enzymes with equally poor affinities for CO_2 can be found: *Anabaena* K_m (CO_2) = 293 μM (91), and *Synechococcus* K_m (CO_2) = 250 μM (52).

In comparing the *R. rubrum* and higher plant enzymes, it was noted (78) that the second-order rate constant for the reaction of activator CO_2 with the enzyme was an order of magnitude smaller for the *R. rubrum* enzyme. The authors refrained from assigning a role to the small subunit, for this difference can just as well be attributed to differences in the large subunits (where the activation reaction occurs) as to the presence or absence of small subunits. Two molecular forms of the enzyme from *R. sphaeroides* have been isolated. Form I contains both large and small subunits, while form II apparently contains only large subunits (45). After analyzing the kinetics of activation of forms I and II, it was concluded that the small subunit accelerates the rate of activation (94). This conclusion is valid only if the large subunits of both forms are identical, i.e. the presence or absence of small subunits is the only difference between forms I and II. Judging from the lack of immunological cross reactivity (93) this does not appear to be the case.

RuBP carboxylases with small subunits are generally more susceptible to inhibition (competitive versus RuBP) by 6-phosphogluconate than those without small subunits (41, 45, 95). This could indicate differences in active-site topology or variations in protein structure that limit the enzyme's ability to assume a conformation that permits tight binding of the inhibitor. Since 6-phosphogluconate is bound at the catalytic site of the large subunit (88) and also stimulates the CO_2 and Mg^{2+}-dependent activation of the *R. rubrum* enzyme (78), the significance of the different sensitivities toward inhibition by 6-phosphogluconate vis-à-vis the function of the small subunit remains obscure.

In recent reconstitution experiments with the *Synechococcus* enzyme from which the small subunits had been partially stripped, it was observed (96) that the remaining catalytic activity was exactly proportional to the small subunit content. Upon addition of isolated small subunit to depleted preparations, reconstruction of catalytic activity to levels comparable with that of native protein occurred. From such results it was inferred that only those large subunits that have small subunits associated with them are catalytically competent.

In conclusion, while an unambiguous assignment of the role of the small subunit is not possible at this time, the hypothesis that it helps to maintain the protein in a stable conformation capable of activation to the catalytically competent enzyme species would accommodate all of the existing information.

ACTIVATION OF THE ENZYME

Historical Perspective

For many years following the discovery of RuBP carboxylase (97, 98) its kinetic properties (K_m for CO_2, V_{max}) were widely considered to be anomalous (99). Conceptual difficulties were encountered in understanding how the enzyme could possibly catalyze CO_2 fixation under natural conditions at the rates observed in vivo (100). The discovery of the oxygenase activity (101–104) only appeared to make matters worse. However, much of the kinetic data gathered during this period was compromised by the failure to appreciate that CO_2 is not only a substrate for carboxylation (105) and a competitive inhibitor of oxygenation (101, 106, 107) but also an activator for both carboxylation and oxygenation (108, 109). Some indication of the existence of an activated enzyme species had been provided by the experiments of Bahr & Jensen (61). An indication that CO_2 was somehow involved in stimulating RuBP carboxylase activity was provided by the early work of Pon et al (110). When this activation by CO_2 was taken into account, what emerged was an enzyme entirely capable of catalyzing both carboxylation and oxygenation at rates in vitro equivalent to those observed in vivo (111–114).

Molecular Mechanism

Figure 4 shows the molecular mechanism for the activation of RuBP carboxylase by CO_2 and divalent metal ions (Me^{2+}). It involves the formation of a carbamate with the ε-amino group of lysine 201 on the large catalytic subunit (64, 65) followed by the binding of Me^{2+} to form the catalytically competent ternary complex $E \cdot {}^ACO_2 \cdot Me^{2+}$. It is this ternary complex that catalyzes both carboxylation and oxygenation (108, 109). Carbamate formation, in general, is a readily reversible equilibrium process known to involve CO_2 (rather than carbonate or bicarbonate) and the uncharged amine as the reactive species (115, 116). The final position of equilibrium (activation state) of the enzyme is consequently dependent upon $[CO_2]$ and $[H^+]$. Additionally, since the activation equilibrium is displaced to the right by Me^{2+}, it is thought that Me^{2+} interacts directly with the carbamate (108). Kinetic (108) and spectroscopic (117) studies established that addition of ${}^ACO_2$ preceded the addition of Me^{2+}, that CO_2 is the active species in the activation reaction, and that the addition of ${}^ACO_2$ is the rate-determining step. From the response of the activation reaction to pH, it was concluded that ${}^ACO_2$ reacted with a group on the enzyme with a distinctly alkaline pK (108). Following the kinetic analyses of the activation reaction, the question arose as to whether the molecule of CO_2 which participates in activation is the same as that which becomes fixed during

$$E\text{-NH}_3^+ \xrightleftharpoons[\text{H}^+]{} E\text{-NH}_2 + {}^A\text{CO}_2 \xrightleftharpoons[\text{H}^+]{\text{slow}} E\text{-NH-ACOO}^- + \text{Me}^{2+} \xrightleftharpoons{\text{fast}} E\text{-NH-ACOO}^-\cdot\text{Me}^{2+}$$

$$\underset{\text{deactivated}}{\cdots} \cdots \cdots \underset{\text{activated}}{\cdots}$$

```
                           200                                                    210

Maize (58):          Leu Arg Gly Gly Leu Asp Phe Thr  Lys Asp Asp Glu Asn Val Asn Ser Gln Pro Phe Met Arg

Spinach (65):        Leu Arg Gly Gly Leu Asp Phe Thr  Lys Asp Asp Glu Asn Val Asn Ser Gln Pro Phe Met Arg

Chlamydomonas (60):  Leu Arg Gly Gly Leu Asp Phe Thr  Lys Asp Asp Glu Asn Val Asn Ser Gln Pro Phe Met Arg

Synechococcus (33):  Leu Arg Gly Gly Leu Asp Phe Thr  Lys Asp Asp Glu Asn Ile Asn Ser Gln Pro Phe Gln Arg

Rhodospirillum (182): Phe Trp Leu Gly Gly Asp Phe Ile  Lys Asn Asp Glu Pro Gln Gly Asn Gln Arg Phe Ala Pro
```

Figure 4 The mechanism of activation of RuBP carboxylase by CO_2 and Me^{2+} involves the formation of a carbamate on the ε-amino group of lysine #201. The structure of the activator CO_2 peptide, containing three acidic residues immediately adjacent to the lysine that becomes charged with ${}^A\text{CO}_2$, suggests that it is also the site for the binding of Me^{2+}. While coordination of Me^{2+} to the carbamate is an attractive idea, there is not yet any physical evidence to substantiate it. Note that the structure of the activator CO_2 peptide has been substantially conserved in passing from the nonsulfur purple bacterium *Rhodospirillum* through the cyanobacterium *Synechococcus*, and the green alga *Chlamydomonas* to higher plants.

carboxylation. Competitive binding (118) and kinetic turnover (119) experiments have clearly established that this is not the case (i.e. $^ACO_2 \neq {}^SCO_2$).

Enzyme-bound carbamates are generally quite labile and unlikely to withstand the rigorous handling associated with protein chemistry. Esterification (with diazomethane) to the corresponding methoxycarbonyl derivative offered a means to stabilize the labile C-N bond (120). The critical element of specificity was introduced with the observation (118, 121) that addition of the transition state analog, 2-carboxyarabinitol-1,5-bisphosphate (2CABP) to the activated ternary complex $E \cdot {}^ACO_2 \cdot Me$ leads to the formation of an extraordinarily stable quaternary complex of $E \cdot {}^ACO_2 \cdot Me \cdot 2CABP$ with a stoichiometry (protomer based) of 1:1:1:1. The usefulness of this complex lies in the fact that neither $^ACO_2$ nor Me^{2+} exchanges readily with unbound ligand (118, 121). Recent attempts to measure the rates of dissociation of $^ACO_2$ and Mn^{2+} from the quaternary complex indicate half-times in excess of one year (G. H. Lorimer, J. V. Schloss, unpublished data). Thus the binding of the transition state analog decreases the rates of dissociation of $^ACO_2$ and Me^{2+} by six to seven orders of magnitude. It was therefore possible to free the $E \cdot [^{14}C]^ACO_2 \cdot Mg \cdot 2CABP$ complex of nonspecifically bound $[^{14}C]CO_2$ by exchange with $[^{12}C]CO_2$ before performing the esterification step. In this manner the only radiolabeled CO_2 trapped on the enzyme is the $^ACO_2$. Identification of N-ε-methoxycarbonyl lysine as the radiolabeled residue followed (64). Sequencing of the resultant tryptic peptides revealed that $^ACO_2$ was bound to lysine 201 of the large subunit (65).

Activation by CO_2 and Me^{2+} is a property that appears to be common to all RuBP carboxylases regardless of their taxonomic origin. Indeed $^ACO_2$ is bound to a highly conserved region of the protein (residues 168–219) (Figure 1). Perfect homology in this region is seen between the *Chlamydomonas* (green alga) and spinach enzymes (60, 65). This highly conserved region also contains the lysine residue (#175) identified as being within the domain of the catalytic site (63). The dimeric form of the enzyme from *R. rubrum* is activated in much the same manner as is the spinach enzyme (78, 79). ^{13}C-NMR spectroscopy suggests that carbamate formation is also involved (80), although the residue involved has not been unequivocally defined. Recent nucleotide sequencing of the gene for the *R. rubrum* enzyme has defined a region with a strong resemblance to the activator peptide of the eukaryotic enzyme (Figure 4; 182). Interestingly, the putative site for the activation of the *R. rubrum* enzyme is separated from a lysine residue, identified by affinity labeling studies as being within the domain of the catalytic site (67, 68), by 25 residues (182). In the eukaryotic enzyme these two lysine residues are 26 residues apart.

While it is axiomatic that some structural change(s) must accompany activation, it may be so subtle as to defy detection. For example, carbamate formation on the α-amino groups of hemoglobin does not grossly alter the structure of the remainder of the molecule (122). In the case of RuBP carboxylase there is spectroscopic (123, 124), chemical (125), and physical (75, 76) evidence pointing to as yet undefined structural differences between activated and deactivated states. Recently Bowien & Gottschalk (126) demonstrated that the activation of the enzyme from the hydrogen bacterium *Alcaligenes* is accompanied by a very large decrease in the sedimentation coefficient from 17.5S to 14.1S. This change implies a substantial change in either the shape and/or the density of the enzyme.

The role of Me^{2+} in the activation reaction is yet to be unequivocally established. The currently accepted (but unproven) explanation (65, 127) is that Me^{2+} coordinates to and thus stabilizes the carbamate. Circumstantial evidence supports this view. For example, such a direct interaction represents the simplest explanation for the influence of $[Me^{2+}]$ and $[H^+]$ on the equilibrium position of the activation reaction (108) and for the requirement for CO_2 to be present for tight binding of one Mn^{2+} per protomer (117). In essence, carbamate formation converts a neutral or potentially positive site into an anionic one with which Me^{2+} could react. The ability of both 2CABP and 4CABP to anchor not only $^ACO_2$ but also Me^{2+} on the enzyme in a manner that decreases their rates of dissociation by six to seven orders of magnitude (G. H. Lorimer, J. V. Schloss, unpublished data) is also most simply explained by invoking direct carbamate-Me^{2+} interaction. This view is also supported by the acidic nature of the amino acids adjacent to lysine 201 (65). Their presence is likely to make the formation of a carbamate at lysine #201 thermodynamically unfavorable, since this introduces negative charge into an already anionic region (65). Coordination of Me^{2+} to the carbamate and possibly also to some of the adjacent acidic group(s) represents an attractive (if unproven) mechanism for neutralizing these negative charges. In some ways this proposed mechanism is similar to that thought to occur in prothrombin and other calcium-binding proteins (128). There, regions of high negative-charge density capable of binding Ca^{2+} are created, not by carbamate formation, but by the vitamin K–dependent carboxylation of glutamyl residues to yield γ-carboxylglutamyl residues (129).

Physiological Aspects of Activation

Although the molecular mechanism of activation is largely understood, the physiological role it plays in regulating the activity of the enzyme in vivo remains less clear. From the observation that the pool of ribulose bisphosphate did not completely disappear during light to dark transitions in *Chlorella* (130) [but see also (131)], it was inferred that the RuBP carboxylase activity had declined. Subsequently, several groups have reported changes in the activation state of the enzyme upon illumination and/or darkening of intact isolated chloroplasts (132–135) and of intact leaves (136, 137). A mutant of *Arabidopsis* that appears to be poorly activatable in vivo has been described (138). While most of these reports attribute physiological significance to these changes, the magnitude of these changes varies considerably. Indeed, Robinson et al (139) have wondered if the effect might not be too small, for in their experience with isolated protoplasts, the enzyme activity from darkened cells was quite enough to support the rates of light-dependent CO_2 fixation. Nevertheless, it is generally agreed that in illuminated cells the enzyme is substantially (and sometimes completely) activated. This creates a conceptual problem, for if one subjects the enzyme in vitro to the conditions thought to apply in vivo ($10\ \mu M\ CO_2$, $5-10\ mM\ Mg^{2+}$, pH about 8.0) it remains substantially in the inactivated state. This has prompted a number of laboratories to search for metabolites capable of stimulating the enzyme activity [see (140) for a summary]. The list of effective metabolites is long. However, before attributing physiological significance to the influence of these effectors, two constraining caveats must be mentioned.

First, the concentration of RuBP carboxylase active sites within the chloroplast is on the order of 4 mM, which clearly places some constraints upon mechanisms of regulation that might be considered. As has been pointed out elsewhere (140), in order for any putative effector to be physiologically relevant, it needs to be present in vivo in amounts that approach or exceed the in vivo concentration of RuBP carboxylase active sites. The pools of most of the metabolites reported to influence the activity of RuBP carboxylase in vitro are simply not large enough to satisfy this criterion (140).

The second caveat arises from the demonstration (88, 141) that the various effectors interact with the enzyme at a single site, the catalytic site for RuBP, and elicit their effect by stabilizing the activator carbamate. It follows, therefore, that an enzyme molecule cannot be simultaneously catalytically competent and activated by one of the effectors, since the latter involves occupancy of the substrate binding site. The physiological significance of the effects observed with the various sugar phosphates in vitro remains in doubt. The above caveat does not preclude the existence of a metabolite that promotes the activation reaction by interaction at a site distinct from the catalytic site. The view is commonly expressed that deactivation of the enzyme upon the darkening of photosynthetic organisms would be biologically advantageous. However, given the evidence for the light-dependent regulation of other Calvin cycle enzyme, e.g. fructose bisphosphatase (142) and phosphoribulokinase (142, 143), which would clearly stop CO_2 fixation in the dark, is not the additional deactivation of RuBP carboxylase superfluous?

CATALYSIS

RuBP carboxylase catalyzes the divalent cation-dependent carboxylation of D-ribulose 1,5-bisphosphate to yield two molecules of D-3-phosphoglycerate, as shown in Reaction 1:

$$
\begin{array}{c}
\text{O} \\
\parallel \\
\text{C} \\
\parallel \\
\text{O}
\end{array}
+
\begin{array}{c}
\text{CH}_2\text{-O-PO}_3^{2-} \\
| \\
\text{C=O} \\
| \\
\text{H-C-OH} \\
| \\
\text{H-C-OH} \\
| \\
\text{CH}_2\text{-O-PO}_3^{2-}
\end{array}
+ \text{H}_2\text{O} \xrightarrow{\text{Me}^{2+}} 2
\begin{array}{c}
\text{CO}_2^- \\
| \\
\text{H-C-OH} \\
| \\
\text{CH}_2\text{O-PO}_3^{2-}
\end{array}
+ 2\text{H}^+ \qquad 1.
$$

The enzyme also catalyzes an oxygenase reaction (Reaction 2), which is the primary event in the photorespiratory carbon oxidation cycle:

$$
\text{O}_2 +
\begin{array}{c}
\text{CH}_2\text{-O-PO}_3^{2-} \\
| \\
\text{C=O} \\
| \\
\text{H-C-OH} \\
| \\
\text{H-C-OH} \\
| \\
\text{CH}_2\text{-O-PO}_3^{2-}
\end{array}
\xrightarrow{\text{Me}^{2+}}
\begin{array}{c}
\text{CH}_2\text{-O-PO}_3^{2-} \\
| \\
\text{CO}_2^-
\end{array}
+
\begin{array}{c}
\text{CO}_2^- \\
| \\
\text{H-C-OH} \\
| \\
\text{CH}_2\text{-O-PO}_3^{2-}
\end{array}
+ 2\text{H}^+ \qquad 2.
$$

A great deal of effort has been expended in investigating whether these two reactions occur at the same active site. The observations that carboxylase and oxygenase activities are activated (109) and inhibited (68, 144) in parallel coupled with the report that the activities have co-evolved (145) strongly suggest that both reactions are catalyzed by the same enzyme.

A Mechanism for the RuBP Carboxylase Reaction

Figure 5 outlines a detailed mechanism that accounts for the carboxylation of RuBP. Key elements of this scheme, namely an enediol form of RuBP and a transient six-carbon intermediate, were suggested by Calvin in 1954 (146) and still appear to be substantially correct. According to the detailed mechanism, abstraction of a proton at C-3 of RuBP generates an enediol that is involved in the actual nucleophilic attack on CO_2. The resulting 2-carboxy-3-keto species is rapidly attacked at the C-3 position by a water molecule, yielding a D-phosphoglycerate molecule from the bottom half (as

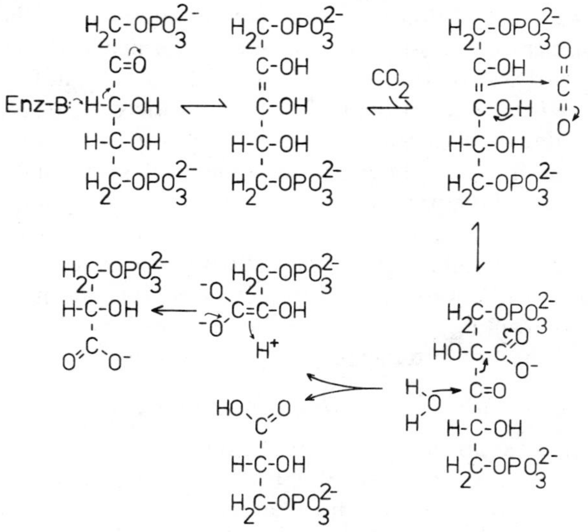

Figure 5   A mechanism for carboxylation of RuBP.

written) of the intermediate and the aci-acid form of phosphoglycerate from the top half. A stereochemically directed addition of a proton to the aci-acid yields the second molecule of product.

The experimental evidence that supports this sequence of events is substantial. Proton abstraction from C-3 of RuBP has been demonstrated to occur as a slow step in the overall reaction (147); the proton released is recovered in the aqueous medium. However, attempts to form RuBP that is tritiated at C-3 by incubating this substrate with enzyme in ^3H_2O (in the absence of CO_2) were initially unsuccessful. Recently, Knowles and colleagues (148–150) used CO_2 activated enzyme to convincingly demonstrate exchange between solvent protons and the C-3 proton of RuBP. These studies verify the intermediacy of the enediol form of RuBP in the enzymatic reaction. The observation that the oxygen atoms at C-2 and C-3 of RuBP are retained during the carboxylation reaction (151, 152) is also compatible with participation of an enediol intermediate and rules out mechanisms that involve covalent adducts between RuBP and enzyme (153). Generation of an enediol is required for nucleophilic attack on carbon dioxide (as opposed to the hydrated species, bicarbonate) which is the actual substrate species required for the carboxylation reaction (105). Attack of the enediol on CO_2 results in formation of a 2-carboxy-3-keto pentitol intermediate. Recent borohydride trapping experiments (154) strongly support the existence of this intermediate. Attack of a water molecule at C-3 of the six-carbon intermediate precedes the cleavage between C-2 and C-3 that has been demonstrated by Mullhofer & Rose (155). Upon cleavage, D-3-phosphoglycerate is formed from the bottom half of the six-carbon intermediate and a C-2 carbanion form of phosphoglycerate is produced from the top half. As Pierce et al (156) point out, since quenching of the carbanion by a solvent-derived proton would result in production of L-phosphoglycerate, some mechanism that controls the stereochemistry of the products must be invoked. One hypothesis involves formation of the aci-acid derivative of the top phosphoglycerate; such a species could be stabilized if complexed with a positively charged amino acid residue or with the divalent cation activator (156). Addition of a solvent proton to the front face of the double bond of the aci-acid would produce the observed reaction product, D-phosphoglycerate.

Additional Details Concerning Events at the Active Site

MODEL COMPLEX STUDIES   Recently a substantial amount of data has been generated in experiments that exploit transition state analogs that mimic 2-carboxy-3-keto-arabinitol bisphosphate or its gem-diol form, the six carbon intermediate formed during the carboxylation reaction. These results have enabled a considerable refinement of our understanding of enzyme activation and catalysis. 2-Carboxyarabinitol bisphosphate, the most widely used analog, and 4-carboxyarabinitol bisphosphate are slow, tight-binding inhibitors of the enzyme (117, 154, 156, 157). The slow inhibition by 2-carboxyarabinitol bisphosphate has been documented by both physical (117) and kinetic (156, 157) approaches. After an initial phase of rapid reversible interaction of analog with enzyme, there appears to be a slow conformational change in the protein (157) after which the inhibitor is functionally irreversibly bound (156, 157). Lane and colleagues (158) were the first to exploit extensively these inhibitors in studies on binding stoichiometries. Analog-containing complexes were useful in demonstrating the existence of a distinct CO_2 activator (118), in investigating its binding domain (64, 65, 121), and in examining the issue of allosteric effectors of the enzyme (88, 141). Stereochemical constraints involved in the mechanism of carboxylation have been proposed on the basis of analog experiments (156) and the existence of a formally hydrated gem-diol form of the 2-carboxy-3-keto arabinitol bisphosphate reaction intermediate has been predicted (154) on the basis of studies on the slow, tight-binding inhibitors. In addition, model complexes have clearly indicated that only one cation is bound to each enzyme site that is fully occupied with activator CO_2 as well as substrate CO_2 and RuBP; this observation applies to the dimeric R. rubrum enzyme as well as the octameric plant enzymes (83) and sets some constraints on the possible roles for cation and on the interaction between sites of activation and catalysis (121).

ROLES FOR THE DIVALENT CATION ACTIVATOR   The discovery of divalent cation involvement in the enzyme activation process prompted some to discount the involvement of divalent metal ion in catalysis (159). Subsequent kinetic investigation of the ability of various cations (e.g. Mg^{2+} vs Mn^{2+} or Co^{2+}) to support enzymatic activity indicated that changes in the partitioning between carboxylase and oxygenase activities occur, depending on the cation employed (160–162). On the basis of these data, an argument for cation involvement in catalysis has been offered (4). The strongest argument in support of such a role is based on ^{13}C-NMR data (117) that suggest that enzyme-bound Mn^{2+} is in close proximity to a CO_2 species (the bicarbonate resonance was observed in these experiments) that rapidly exchanges between free and enzyme-bound forms. The activator CO_2 exists as a carbamate on the enzyme, and exchanges very slowly on an NMR time scale ($k_{off} \sim 1 \times 10^{-2}$ s^{-1}). Thus, the rapidly exchanging CO_2 ($k_{off} \sim 1 \times 10^4$ s^{-1}) that binds closely to cation is identified as substrate CO_2, and, on the basis of proximity to the enzyme's substrate, cation is available to function in catalysis. Such an assignment is attractive from a chemical viewpoint and is compatible with results of recent ESR studies (187). There are several steps in the carboxylation mechanism where a divalent cation could serve as an electron sink, stabilizing an intermediate that occurs during the catalytic process. For example, the divalent cation bound in close proximity to substrate CO_2 would be able to stabilize the aci-acid form of 3-phosphoglycerate that has been postulated to be formed after cleavage of 2-carboxy-3-keto arabinitol bisphosphate (156).

The postulate that cation has a role in catalysis as well as in activation has an interesting corollary, in light of the stoichiometries established in model complex studies. If cation supports activation by coordinating directly to a carbamate and functions in catalysis while binding in close proximity to substrate CO_2, then, since only one cation binds per active site (83, 117, 121), the sites for activation and catalysis are in close proximity. With the primary sequence of the large subunit known (58), and several peptides involved in activation (65) and catalysis (cf below) sequenced and positioned within the large subunit backbone, a correct juxtaposition of these regions may expedite the long-range process of generating a three-dimensional model of the enzyme.

IDENTIFICATION OF ACTIVE-SITE AMINO ACIDS   While substantial information exists concerning the mechanistic aspects of RuBP carboxylation, relatively little progress has been made in identifying the amino acids that function in catalysis. A rigorous kinetic analysis, aimed at defining the pK's of the groups responsible for the acid-base chemistry is in progress (183).

Considerable efforts have been made to identify critical amino acid residues by the use of group-specific reagents. Recent work has suggested the existence of essential tyrosine and histidine residues, based upon inactivation of the enzyme with tetranitromethane (163) and diethylpyrocarbamate (164). Earlier work with phenylglyoxal (165, 166) and 2,3-butanedione (167) implied that arginine residues have a structural or functional role. Experiments with sulfhydryl reagents suggested cysteine involvement at the active site (168) and led to speculation that an essential sulfhydryl reacts with the RuBP carbonyl group to form a thiohemiketal (153). Such a mechanism has been ruled out by recent investigations on retention of substrate oxygen atoms (151, 152). The usefulness of group-specific reagents is limited by their inherent nonselectivity. Often multiple residues per site have been labeled (169), precluding any distinction between a structural or catalytically functional role for the residues. Arguments based upon the stoichiometry of group-specific reagent incorporation are less persuasive than the demonstration that a single specific tryptic peptide is the target of the reagent. It is premature to assign *functionality* to a residue as a result of its modification by group-specific or active site–directed reagents.

Active site–directed affinity labeling has yielded results that can be interpreted in a more straight-forward fashion. The inherent specificity of affinity labels has indicated the importance of lysine-175 of the large subunit of the spinach enzyme, which reacts with 3-bromo-1,4-dihydroxy-2-butanone-1,4-bisphosphate, N-bromoacetylethanolamine phosphate, and pyridoxal phosphate (170). As mentioned earlier, the importance of this amino acid is underscored by the observation that in the *R. rubrum* enzyme a lysyl residue in a homologous peptide also reacts with pyridoxal phosphate. 3-Bromo-1,4-dihydroxy-2-butanone-1,4-bisphosphate also modifies lys-334 of the spinach enzyme (169). Recently, Hartman's group reported that 2-bromoacetylaminopentitol-1,5-bisphosphate (68) and 2-N-chloroamino-2-deoxypentitol-1,5-bisphosphate (171) specifically modify a methionine residue of the *R. rubrum* enzyme, providing the first evidence for the presence of this amino acid at the active site of a RuBP carboxylase.

A Mechanism for the RuBP Oxygenase Reaction

Progress in elucidating the details of this reaction has largely been limited to those steps common to both carboxylation and oxygenation of RuBP. It has, however, been established with $^{18}O_2$ that one atom of ^{18}O is incorporated into the carboxylate group of phosphoglycolate; the other ^{18}O is apparently lost to the medium (104). The oxygen atom at C-2 of RuBP is retained during both the oxygenase reaction and the carboxylase reaction (172). In addition, cleavage occurs between C-2 and C-3 in the oxygenase reaction (172), as well as in the carboxylase reaction. A mechanism analogous to that proposed for carboxylation has been offered to explain the oxygenation of RuBP (Figure 6).

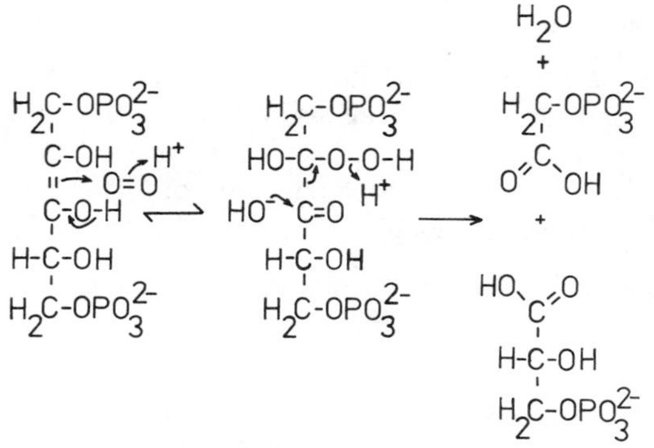

Figure 6 Formation and cleavage of a hydroperoxide intermediate in the RuBP oxygenase reaction.

Formation of an enediol form of RuBP precedes attack on molecular oxygen with the formation of a hydroperoxide. The detailed chemistry of this step is unclear. It has been proposed (4) that a radical pair is formed from singlet state RuBP and a triplet state oxygen molecule. In order to allow for spin inversion and subsequent formation of singlet state products to occur, stabilization of the radical pair would appear to be necessary. The existence of cage effects, as might be ascribed to a highly structured array of solvent molecules surrounding the radical pair, or some similar device, would make this mechanism more reasonable. In the case of RuBP carboxylase, the existence of highly immobilized solvent molecules at the active site has been postulated on the basis of physical experiments (117, 121). These observations, as well as the inability of the divalent metal ion in the quaternary $E \cdot {}^A CO_2 \cdot Me^{2+} \cdot CABP$ complex to exchange with unbound metal ion, might be construed as evidence for the existence of such a cage. The problem of spin inversion could also be resolved by invoking the participation of a transition metal ion or organic cofactor that may function to delocalize spin. Despite an early report that RuBP carboxylase contains copper (173), recent investigations (104, 174–176) have failed to demonstrate more than trace amounts of transition metal ions. Thus, by default, the stabilized radical pair mechanism, which appears to occur in other enzymatic reactions (177), remains the most plausible explanation for hydroperoxide formation. Attack on C-3 of this intermediate by a hydroxyl ion results in decomposition of the hydroperoxide to form phosphoglycolate and 3-phosphoglycerate, as well as in regeneration of a hydroxyl ion. Strong chemical precedent (178) exists for this element of the oxygenase mechanism, which is compatible with the observation that only the carboxyl group of phosphoglycolate contains oxygen from the substrate O_2 (104).

ACKNOWLEDGMENTS

The authors are grateful for the information transmitted in advance of publication by many of our colleagues. We regret that space constraints precluded the mention of some of these contributions. Author Miziorko appreciates the support of the US Department of Agriculture Competitive Research Grant Program (79-CRCR-1-0339) and the NIH (RCDA AM-00645).

Literature Cited

1. Ellis, R. J. 1979. *Trends Biochem. Sci.* 4:241–44
2. Lorimer, G. H., Andrews, T. J. 1981. In *The Biochemistry of Plants*, ed. M. D. Hatch, N. K. Boardman, 8:330–74. New York: Academic
3. Bowien, B., Schlegel, H. G. 1981. *Ann. Rev. Microbiol.* 35:405–52
4. Lorimer, G. H. 1981. *Ann. Rev. Plant Physiol.* 32:349–83
5. Farquhar, G. D., von Caemmerer, S., Berry, J. A. 1980. *Planta* 149:78–90
6. Andrews, T. J., Lorimer, G. H. 1978. *FEBS Lett.* 90:1–9
7. Rutner, A. C., Lane, M. D. 1967. *Biochem. Biophys. Res. Commun.* 28:531–37
8. Lane, M. D., Miziorko, H. M. 1978. In *Photosynthetic Carbon Assimilation*, ed. H. W. Siegelman, G. Hind, pp. 19–42. New York: Plenum
9. Sakano, K., Kung, S. D., Wildman, S. G. 1974. *Mol. Gen. Genet.* 130:91–97
10. Chan, P. H., Wildman, S. G. 1972. *Biochim. Biophys. Acta* 277:677–80
11. Coen, D. M., Bedbrook, J. R., Bogorad, L., Rich, A. 1977. *Proc. Natl. Acad. Sci. USA* 74:5487–91
12. Bedbrook, J. R., Coen, D. M., Beaton, A. R., Bogorad, L., Rich, A. 1979. *J. Biol. Chem.* 254:905–10
13. Bottomley, W., Whitfeld, P. R. 1979. *Eur. J. Biochem.* 93:31–39
14. Rochaix, J. P., Malone, P. 1979. In *Chloroplast Development*, ed. G. Akoyunoglou, J. H. Argyroudi-Akoyunoglou, pp. 581–86. Amsterdam: Elsevier
15. Spreitzer, R. J., Mets, L. T. 1980. *Nature* 285:114–15
16. Dron, M., Rahire, M., Rochaix, J. D., Mets, L. 1983. *Plasmid*. In press
17. Edelman, M. 1980. In *The Biochemistry of Plants*, ed. A. Marcus, 6:249–301. New York: Academic
18. Possingham, J. V. 1980. *Ann. Rev. Plant Physiol.* 31:113–29
19. Kung, S. D. 1976. *Science* 191:429–34
20. Wildman, S. G. 1979. *Arch. Biochem. Biophys.* 196:598–610
21. Melchers, G., Sacristan, M. D., Holder, A. A. 1978. *Carlsberg Res. Commun.* 43:203–18
22. Poulsen, C., Porath, D., Sacristan, M. D., Merlchers, G. 1980. *Carlsberg Res. Commun.* 45:249–67
23. Bedbrook, J. R., Kolodner, R. 1979. *Ann. Rev. Plant Physiol.* 30:593–620
24. O'Connell, P. B. H., Brady, C. J. 1982. *Biochim. Biophys. Acta* 670:355–61
25. Gray, J. C., Kung, S. D., Wildman, S. G. 1978. *Arch. Biochem. Biophys.* 185:272–81
26. Kawashima, N., Wildman, S. G. 1972. *Biochim. Biophys. Acta* 262:42–49
27. Smith, S. M., Ellis, R. J. 1981. *J. Mol. Appl. Gen.* 1:127–37
28. Gallagher, T. F., Ellis, R. J. 1982. *EMBO J.* In press
29. Dobberstein, B., Blobel, G., Chua, N. H. 1977. *Proc. Natl. Acad. Sci. USA* 74:1082–85
30. Highfield, P. E., Ellis, R. J. 1978. *Nature* 271:420–24
31. Cashmore, A. R., Broadhurst, M. K., Grey, R. E. 1978. *Proc. Natl. Acad. Sci. USA* 75:655–59
32. Bedbrook, J. R., Smith, S. M., Ellis, R. J. 1980. *Nature* 287:692–97
33. Reichelt, B. Y., DeLaney, S. F. 1983. *DNA*. In press
34. Berry-Lowe, S. L., McKnight, J. D., Shah, D. M., Meagher, R. B. 1982. *J. Appl. Mol. Gen.* In press
35. Gilbert, W. 1979. In *Eucaryotic Gene Regulation Symposium on Molecular and Cellular Biology*, ed. R. Arch, T. Maniatis, C. P. Fox, 14:1–14. New York: Academic
36. Takruri, I. A. H., Boulter, D., Ellis, R. J. 1981. *Phytochemistry* 20:413–15
37. Martin, P. G. 1979. *Aust. J. Plant Physiol.* 6:401–8
38. Strobaek, S., Gibbons, G. C., Haslett, B.

G., Boulter, D., Wildman, S. G. 1976. *Carlsberg Res. Commun.* 41:335–43
39. Hirai, A. 1977. *Proc. Natl. Acad. Sci. USA* 74:3443–45
40. Siegelman, H. W., Hind, G., eds. 1978. *Photosynthetic Carbon Assimilation.* New York: Plenum. 445 pp.
41. McFadden, B. A. 1980. *Acc. Chem. Res.* 13:394–99
42. Johal, S., Bowman, L. H., Chollet, R. 1982. *Biochim. Biophys. Acta* 704:374–78
43. Tabita, F. R., McFadden, B. A. 1974. *J. Biol. Chem.* 249:3459–64
44. Schloss, J. V., Phares, E. F., Long, M. W., Norton, I. L., Stringer, C. D., Hartman, F. C. 1979. *J. Bacteriol.* 137:490–501
45. Gibson, J. L., Tabita, F. R. 1977. *J. Biol. Chem.* 252:943–49
46. Tabita, F. R., Stevens, S. E., Gibson, J. L. 1976. *J. Bacteriol.* 125:531–39
47. Purohit, K., McFadden, B. A., Cohen, A. L. 1976. *J. Bacteriol.* 127:505–15
48. Bowman, L. H., Cholett, R. C. 1980. *J. Bacteriol.* 141:652–57
49. Takabe, T. 1977. *Agric. Biol. Chem.* 41:2255–60
50. Okabe, K., Codd, G. A. 1980. *Plant Cell Physiol* 21:1117–27
51. Codd, G. A., Stewart, W. D. P. 1977. *Arch. Microbiol.* 113:105–10
52. Andrews, T. J., Abel, K. M. 1981. *J. Biol. Chem.* 256:8445–51
53. Lawlis, V. B., Gordon, G. L. R., McFadden, B. A. 1979. *J. Bacteriol.* 139:287–98
54. Taylor, S., Dalton, H., Dow, C. 1980. *FEMS Microbiol. Lett.* 8:157–60
55. Taylor, S., Dow, C. 1980. *J. Gen. Microbiol.* 116:81–35
56. Andrews, T. J., Abel, K. M., Menzel, D., Badger, M. R. 1982. *Arch. Microbiol.* 130:344–48
57. Akazawa, T., Takabe, T., Asami, S., Kobayashi, H. 1978. See Ref. 40, pp. 209–25
58. McIntosh, L., Poulsen, C., Bogorad, L. 1980. *Nature* 288:556–60
59. Zurawski, G., Perrot, R., Bottomley, W., Whitfeld, P. R. 1981. *Nucleic Acids Res.* 9:3251–70
60. Dron, M., Rahiv, M., Rochaix, J. D. 1982. *J. Mol. Biol.* 162: 775–93
61. Bahr, J. T., Jensen, R. G. 1974. *Arch. Biochem. Biophys.* 164:408–13
62. Poulsen, C., Martin, B., Svendsen, I. 1979. *Carlsberg Res. Commun.* 44:191–99
63. Hartman, F. C., Norton, E. L., Stringer, C. D., Schloss, J. V. 1978. See Ref. 40, pp. 245–69
64. Lorimer, G. H., Miziorko, H. M. 1980. *Biochemistry* 19:5321–28
65. Lorimer, G. H. 1981. *Biochemistry* 20:1236–40
66. Hartman, F. C., Stringer, C. D., Omnaas, J., Donnelly, M. I., Fraij, B. 1982. *Arch. Biochem. Biophys.* 219:422–37
67. Herndon, C. S., Norton, I. L., Hartman, F. C. 1982. *Biochemistry* 21:1380–85
68. Fraij, B., Hartman, F. C. 1982. *J. Biol. Chem.* 257:3501–5
69. Poulsen, C. 1981. *Carlsberg Res. Commun.* 46:259–78
70. Takabe, T., Akazawa, T. 1975. *Plant Cell Physiol.* 16:1049–60
71. Schmidt, G. W., Devilleus-Thieny, A., Desnisseaux, H., Blobel, G., Chua, N. H. 1979. *J. Cell Biol.* 83: 615–22
72. Neuburger, M., Joyard, J., Douce, R. 1977. *Plant Physiol.* 59:1178–81
73. Eisenberg, D., Baker, T. S., Suh, S. W., Smith, W. W. 1978. See Ref. 40, pp. 271–81
74. Bowien, B., Mayer, F., Spiess, E., Pahler, A., English, U., Saenger, W. 1980. *Eur. J. Biochem.* 106:405–10
75. Grebanier, A. E., Champagne, D., Roy, H. 1978. *Biochemistry* 17:5150–55
76. Tomimatsu, Y., Donovan, J. W. 1981. *Plant Physiol.* 68:808–13
77. Roy, H., Valeri, A., Pope, D. H., Rueckert, L., Costa, K. A. 1978. *Biochemistry* 17:665–68
78. Christeller, J. T., Laing, W. A. 1978. *Biochem. J.* 173:467–73
79. Whitman, W. B., Martin, M. W., Tabita, F. R. 1979. *J. Biol. Chem.* 254:10184–89
80. O'Leary, M. H., Jaworski, R. J., Hartman, F. C. 1979. *Proc. Natl. Acad. Sci. USA* 76:673–75
81. Ryan, F. J., Jolly, S. O., Tolbert, N. E. 1974. *Biochem. Biophys. Res. Commun.* 59:1233–41
82. McFadden, B. A. 1974. *Biochem. Biophys. Res. Commun.* 60:312–17
83. Miziorko, H. M., Behnke, C. E., Houkom, E. C. 1982. *Biochemistry* 21:6669–74
84. Nishimura, M., Takabe, T., Sugiyama, T., Akazawa, T. 1973. *J. Biochem.* 74:945–54
85. Nishimura, M., Akazawa, T. 1974. *J. Biochem.* 76:169–76
86. Spreitzer, R. J., Jordon, D. B., Ogren, W. L. 1982. *FEBS Lett.* 148:117–21
87. Deleted in proof
88. Badger, M. R., Lorimer, G. H. 1981. *Biochemistry* 20:2219–25
89. Yeoh, H. H., Badger, M. R., Watson, L. 1980. *Plant Physiol.* 66:1110–12
90. Yeoh, H. H., Badger, M. R., Watson, L. 1981. *Plant Physiol.* 67:1151–55

91. Badger, M. R. 1980. *Arch. Biochem. Biophys.* 201:247–54
92. Deleted in proof
93. Gibson, J. L., Tabita, F. R. 1977. *J. Bacteriol.* 131:1020–22
94. Gibson, J. L., Tabita, F. R. 1979. *J. Bacteriol.* 140:1023–27
95. Tabita, F. R. McFadden, B. A. 1972. *Biochem. Biophys. Res. Commun.* 48:1153–59
96. Andrews, T. J., Ballment, B. 1982. *Proc. Int. Congr. Biochem.*, Perth, Australia. Abstr. 005–190
97. Quayle, J. R., Fuller, R. F., Benson, A. A., Calvin, M. 1954. *J. Am. Chem. Soc.* 76:3610–11
98. Weissbach, A., Smyrniotis, P. Z., Horecker, B. L. 1954. *J. Am. Chem. Soc.* 76:3611–12
99. Walker, D. A. 1973. *New Phytol.* 72:209–35
100. Peterkofsky, A., Racker, E. 1961. *Plant Physiol.* 36:409–14
101. Bowes, G., Ogren, W. L. 1972. *J. Biol. Chem.* 247:2171–76
102. Bowes, G., Ogren, W. L., Hageman, R. H. 1971. *Biochem. Biophys. Res. Commun.* 45:716–22
103. Andrews, T. J., Lorimer, G. H., Tolbert, N. E. 1973. *Biochemistry* 12:11–18
104. Lorimer, G. H., Andrews, T. J., Tolbert, N. E. 1973. *Biochemistry* 12:18–23
105. Cooper, T. G., Filmer, D., Wishnick, M., Lane, M. D. 1969. *J. Biol. Chem.* 244:1081–83
106. Laing, W. A., Ogren, W. L., Hageman, R. H. 1974. *Plant Physiol.* 54:678–85
107. Badger, M. R., Andrews, T. J. 1974. *Biochem. Biophys. Res. Commun.* 60:204–10
108. Lorimer, G. H., Badger, M. R, Andrews, T. J. 1976. *Biochemistry* 15:529–36
109. Badger, M. R., Lorimer, G. H. 1976. *Arch. Biochem. Biophys.* 175:723–29
110. Pon, N. G., Robin, B. R., Calvin, M. 1963. *Biochem. Z.* 338:7–19
111. Delaney, M. E., Walker, D. A. 1978. *Biochem. J.* 171:477–82
112. Seeman, J. R., Tepperman, J. M., Berry, J. A. 1981. *Carnegie Inst. Washington Yearb.* 80:67–72
113. Bauwe, H., Apel, P., Peisker, M. 1980. *Photosynthetica.* 14:550–56
114. von Caemmerer, S., Farquhar, G. P. 1981. *Planta* 153:376–87
115. Faurholt, C. 1925. *J. Chim. Phys. Phys. Chim. Biol.* 22:1–44
116. Ewing, S. P., Lockshon, D., Jencks, W. P. 1980. *J. Am. Chem. Soc.* 102:3072–84
117. Miziorko, H. M., Mildvan, A. S. 1974. *J. Biol. Chem.* 249:2743–50
118. Miziorko, H. M. 1979. *J. Biol. Chem.* 254:270–72
119. Lorimer, G. H. 1979. *J. Biol. Chem.* 254:5599–601
120. Katchalski, E., Klibanski, C. B., Berger, A. 1951. *J. Am. Chem. Soc.* 73:1829–31
121. Miziorko, H. M., Sealy, R. C. 1980. *Biochemistry* 19:1167–72
122. Arnone, A., Rogers, P. H., Briley, P. D. 1980. In *Biophysics and Physiology of Carbon Dioxide*, ed. C. Bauer, G. Gros, H. Bartels, pp. 67–74. Berlin: Springer
123. Wildner, G. F. 1976. *Z. Naturforsch. Teil C* 31:267–71
124. Vater, J., Salnikow, J., Kleinkauf, H. 1977. *Biochem. Biophys. Res. Commun.* 74:1618–25
125. Bhagwat, A. S., Ramakrishna, J. 1981. *Biochem. Biophys. Acta* 662:181–89
126. Bowien, B., Gottschalk, E. M. 1982. *J. Biol. Chem.* 257:11845–47
127. Lorimer, G. H., Miziorko, H. M. 1981. In *Photosynthesis IV Regulation of Carbon Metabolism*, ed. G. Akoyunoglou, pp. 3–16. Philadelphia: Balaban Int. Sci. Serv.
128. Magnusson, S., Peterson, T. E., Sottrup-Jensen, L., Claeys, H. 1975. In *Proteases and Biological Control*, ed. E. Reich, D. B. Rifkin, E. Shaw, pp. 123–49. Long Island, NY: Cold Spring Harbor
129. Suttie, J. W. ed. 1980. *Vitamin K. Metabolism and Vitamin K-Dependent Proteins*, pp. 592. Baltimore: University Park
130. Pederson, T. A., Kirk, M., Bassham, J. A. 1966. *Physiol. Plant.* 19:219–31
131. Strotmann, H., Heldt, H. W. 1969. In *Progress in Photosynthesis*, ed. H. Metzner, 3:1131–40. Tubingen: Int. Union Biol. Sci.
132. Bahr, J. T., Jensen, R. G. 1978. *Arch. Biochem. Biophys.* 185:39–48
133. Heldt, H. W., Chow, C. J., Lorimer, G. H. 1978. *FEBS Lett.* 92:234–40
134. Sicher, R. C., Hatch, A. L., Stumpf, D. K., Jensen, R. G. 1981. *Plant Physiol.* 68:252–55
135. Stumpf, D. K., Jensen, R. G. 1982. *Plant Physiol.* 69:1263–67
136. Machler, F., Nosberger, J. 1980. *J. Exp. Bot.* 31:1485–91
137. Perchorowicz, J. T., Raynes, D. A., Jensen, R. G. 1981. *Proc. Natl. Acad. Sci. USA* 78:2985–89
138. Somerville, C. R., Portis, A. R., Ogren, W. L. 1982. *Plant Physiol.* 70:381–87
139. Robinson, S. P., McNeil, P. H., Walker, D. A. 1979. *FEBS Lett.* 97:296–300
140. Lorimer, G. H., Badger, M. R., Heldt, H. W. 1978. See Ref. 40, pp. 283–306
141. McCurry, S. D., Pierce, J., Tolbert, N. E., Orme-Johnson, W. H. 1981. *J. Biol. Chem.* 256:6623–28

142. Buchanan, B. B. 1980. *Ann. Rev. Plant Physiol.* 31:341–74
143. Heldt, H. W., Laing, W. A., Lorimer, G. H., Stitt, M., Wirtz, W. 1981. See Ref. 127, pp. 213–26
144. Brown, H. M., Rejda, J. M., Chollet, R. 1980. *Biochim. Biophys. Acta* 614:545–52
145. Jordan, D. B., Ogren, W. L. 1981. *Nature* 291:513–15
146. Calvin, M. 1954. *Fed. Proc.* 13:697–711
147. Fiedler, F., Mullhofer, G., Trebst, A., Rose, I. A. 1967. *Eur. J. Biochem.* 1:395–99
148. Saver, B. G., Knowles, J. R. 1982. *Biochemistry* 21:5398–403
149. Sue, J. M., Knowles, J. R. 1982. *Biochemistry* 21:5404–9
150. Sue, J. M., Knowles, J. R. 1982. *Biochemistry* 21: 5410–14
151. Sue, J. M., Knowles, J. R. 1978. *Biochemistry* 17:4041–44
152. Lorimer, G. H. 1978. *Eur. J. Biochem.* 89:43–50
153. Rabin, B. R., Trown, P. W. 1964. *Nature* 202:1290–93
154. Schloss, J. V., Lorimer, G. H. 1982. *J. Biol. Chem.* 257:4691–94
155. Mullhofer, G., Rose, I. A. 1965. *J. Biol. Chem.* 240:1341–46
156. Pierce, J., Tolbert, N. E., Barker, R. 1980. *Biochemistry* 19:934–42
157. Siegel, M. I., Lane, M. D. 1972. *Biochem. Biophys. Res. Commun.* 48:508–16
158. Wishnick, M., Lane, M. D., Scrutton, M. C. 1970. *J. Biol. Chem.* 245:4939–47
159. Laing, W. A., Christeller, J. T. 1976. *Biochem. J.* 159:563–70
160. Christeller, J. T., Laing, W. A. 1979. *Biochem. J.* 183:747–50
161. Robison, P. D., Martin, M. N., Tabita, F. R. 1979. *Biochemistry* 18:4453–58
162. Wildner, G. F., Henkel, J. 1979. *Planta* 146:223–28
163. Robison, P. D., Tabita, F. R. 1979. *Biochem. Biophys. Res. Commun.* 88:85–91
164. Saluja, A. K., McFadden, B. A. 1980. *Biochem. Biophys. Res. Commun.* 94:1091–97
165. Schloss, J. V., Norton, I. L., Stringer, C. D., Hartman, F. C. 1978. *Biochemistry* 17:5626–31
166. Chollet, R. 1981. *Biochem. Biophys. Acta* 658:177–90
167. Lawlis, V. B., McFadden, B. A. 1978.

168. Rabin, B. R., Trown, P. W. 1964. *Proc. Natl. Acad. Sci. USA* 51:497–501
169. Hartman, F. C., Norton, I. L., Stringer, C. D., Schloss, J. V. 1978. See Ref. 40, pp. 245–69
170. Hartman, F. C., Fraij, B., Norton, I. L., Stringer, C. D. 1981. See Ref. 143, pp. 17–29
171. Christeller, J. T., Hartman, F. C. 1982. *FEBS Lett.* 142:162–66
172. Pierce, J., Tolbert, N. E., Barker, R. 1980. *J. Biol. Chem.* 255:509–11
173. Wishnick, M., Lane, M. D., Scrutton, M. C., Mildvan, A. S. 1969. *J. Biol. Chem.* 244:5761–63
174. Chollet, R., Anderson, L. L., Hovespian, L. C. 1975. *Biochem. Biophys. Res. Commun.* 64:97–107
175. Johal, S., Bourque, D. P., Smith, W. W., Suh, S. W., Eisenberg, D. 1980. *J. Biol. Chem.* 255:8873–80
176. Kosman, D. I., Ettinger, M. J., Bereman, R. D., Giordano, R. S. 1977. *Biochemistry* 16:1597–601
177. McCapra, F. 1976. *Acc. Chem. Res.* 9:201–8
178. Gleason, W. B., Barker, R. 1971. *Can. J. Chem.* 49:1425–32
179. Muller, K. D., Salnikow, J., Vater, J. 1982. *Biochim. Biophys. Acta.* In press
180. Langridge, P. 1981. *FEBS Lett.* 123:85–89
181. Roy, H., Bloom, M., Milos, P., Monroe, M. 1982. *J. Cell. Biol.* 94:20–27
182. Somerville, C. R., Fitchen, J., Somerville, S. C., McIntosh, L., Nargang, F. 1983. In *Advances in Gene Technology: Molecular Genetics of Plants and Animals*, ed. F. Ahmad, K. Downey, J. Schultz, R. Voellmy. New York: Academic. In press
183. Schloss, J. V. 1983. *Fed. Proc.* In press
184. Coruzzi, G., Broglie, R., Cashmore, A., Chua, N. H. 1983. *J. Biol. Chem.* In press
185. Chua, N. H., Schmidt, G. W. 1978. See Ref. 40, pp. 325–47
186. Coruzzi, G., Broglie, R., Lamppa, G., Chua, N. H. 1983. In *Structure and Function of Plant Genomes*, ed. O. Ciferri, L. Duve. New York: Plenum. In press
187. Miziorko, H. M., Sealy, R. C. 1983. *Fed. Proc.* In press

Ann. Rev. Biochem. 1983. 52:537–79

FATTY ACID SYNTHESIS AND ITS REGULATION[1]

Salih J. Wakil, James K. Stoops, and Vasudev C. Joshi

Verna and Marrs McLean Department of Biochemistry, Baylor College of Medicine, Houston, Texas 77030

CONTENTS

PERSPECTIVES AND SUMMARY 538
ACETYL-CoA CARBOXYLASE.. 540
    Animal Acetyl-CoA Carboxylase 541
    Yeast Acetyl-CoA Carboxylase 543
FATTY ACID SYNTHETASE 543
    Fatty Acid Synthetases of Eukaryotes 544
STRUCTURAL ORGANIZATION OF FATTY ACID
           SYNTHETASE 545
    Animal Fatty Acid Synthetase 545
    Yeast Fatty Acid Synthetase 552
FUNCTIONAL ORGANIZATION OF FATTY ACID SYNTHETASE.................... 556
    Acetyl and Malonyl Transacylases 556
    β-Ketoacyl Synthetase (Condensing Enzyme) 558
    β-Ketoacyl and Enoyl Reductases 562
    β-Hydroxyacyl Dehydratase 565
    Palmitoyl Thioesterase 565
MECHANISM OF ACTION OF FATTY ACID SYNTHETASE............................ 565
    Animal Synthetase 566
    Yeast Synthetase 567
REGULATION OF FATTY ACID SYNTHESIS 569
    Acetyl-CoA Carboxylase 569
    Fatty Acid Synthetase 573
SYNTHESIS OF BRANCHED AND MEDIUM-CHAIN FATTY
           ACIDS 575

[1] The following abbreviations are used: FAS, fatty acid synthetase; BCCP, biotin carboxyl carrier protein; BC, biotin carboxylase; CT, carboxyl transferase; ACP, acyl carrier protein; SDS, sodium dodecylsulfate; SDS-PAGE, sodium dodecylsulfate-polyacrylamide gel electrophoresis; KS, β-ketoacyl synthetase.

537

0066-4154/83/0701-0537$02.00

PERSPECTIVES AND SUMMARY

The biosynthesis of long-chain fatty acids occurs in two distinct steps. The first step is the conversion of acetyl-CoA to malonyl-CoA, a reaction catalyzed by a biotin-containing multienzyme system named acetyl-CoA carboxylase (Reaction 1). The second step (Reaction 2) is the conversion of acetyl-CoA and malonyl-CoA to palmitate in the presence of NADPH, a reaction catalyzed by the fatty acid synthetase (FAS).

$$CH_3COS–CoA + CO_2 + ATP \rightleftharpoons HOOCCH_2COS–CoA + ADP + P_i. \qquad 1.$$

$$CH_3COS–CoA + 7\ HOOCCH_2COS–CoA + 14\ NADPH$$
$$+ 14\ H^+ \longrightarrow CH_3CH_2(CH_2CH_2)_6CH_2COOH$$
$$+ 14\ NADP^+ + 8\ CoA–SH + 6\ H_2O. \qquad 2.$$

Each of these reactions consists of several partial reactions catalyzed by a specific enzyme activity. These partial activities are loosely associated with each other in prokaryotic and plant cells and can readily be separated by conventional procedures; they are tightly bound complexes, however, in eukaryotic cells.

It was assumed until recently that these complexes consisted of noncovalently bound enzymes that functioned as one unit. Research efforts were concentrated on dissociation of these complexes and isolation of their component enzymes. Several workers reported the isolation of some activities; others had difficulty substantiating these claims (1, 2).

A new conceptual understanding of the structure of FAS has now come from the discovery that yeast (3, 4) and animal synthetases (1, 2) are multifunctional enzymes consisting of polypeptide subunits ($M_r > 200,000$) that contain several enzyme activities and are arranged in structural units. These findings demonstrate unequivocally the existence of a new class of enzymes—the multifunctional enzymes. Subsequent studies on acetyl-CoA carboxylase (5) show it also to be a multifunctional enzyme, consisting of subunits with M_r 230,000. These enzymes have two characteristic properties: structurally, they consist of a single type of polypeptide chain; functionally, they have multiple catalytic activities (6). This implies that the active centers of the protein are generated by the folding of contiguous stretches of the polypeptide chain to yield autonomous globular structures or domains, each having a specific but different catalytic activity. This class of enzymes occurs in systems that catalyze sequential reactions in a metabolic pathway, e.g. palmitate synthesis. It is of interest to note that animal FAS is the largest known multifunctional protein, having the most catalytic domains.

There has been considerable study recently of the structure and function of acetyl-CoA carboxylase and FAS. The acetyl-CoA carboxylase of animal tissue consists of a subunit protein of M_r 225,000 and contains the catalytic domains of biotin carboxylase, transcarboxylase, biotin carboxyl carrier protein, and the regulatory allosteric site. The FAS of animal tissue is a homodimer with a protein subunit of M_r 263,000. The subunit contains the acyl carrier protein (ACP) site, with its 4'-phosphopantetheine group, and the seven catalytic activities (7–9). All of the component activities can be differentiated and assayed independently using appropriate substrates (10, 11). Considerable progress is being made regarding the relative location and functional organization of these activities on the synthetase subunit. A tentative peptide map and activity distribution have been constructed. Though each subunit contains all of the catalytic sites required for palmitate synthesis, studies show that the dimer is the only active form of this multifunctional protein; the monomer lacks β-ketoacyl synthetase activity. Investigation of this phenomenon using the bifunctional reagent 1,3-dibromo-2-propanone indicates that β-ketoacyl synthetase activity requires juxtapositioning of the cysteine-SH of the β-ketoacyl synthetase of one subunit and the cysteamine-SH of the ACP site of the other subunit (12). The two subunits are arranged head-to-tail, creating two centers for β-ketoacyl synthetase and, thus, for palmitate synthesis.

A similar thiol arrangement appears to function in the β-ketoacyl synthetase component of yeast FAS (13). The latter enzyme consists of two nonidentical subunits: α (M_r 213,000), which contains three activities, and β (M_r 203,000), which contains five different activities (3, 4, 10, 14). The active enzyme is an $\alpha_6\beta_6$-complex organized in an ellipsoid structure of "arches" and "plates". The α-subunit contains both the β-ketoacyl synthetase with its active cysteine-SH and the ACP with its cysteamine-SH. However, for a functional condensing enzyme, the active cysteine-SH of one α-subunit must be vicinal to the cysteamine-SH of an adjacent α-subunit. Thus, the assembly of two complementary half α-subunits and a β-subunit would constitute a single palmitate-synthesizing unit (7, 13). An $\alpha_6\beta_6$-structure, therefore, contains six such palmitate synthesizing units.

The synthesis of FAS and carboxylase is under long-term control and is regulated at both the transcriptional and the translational levels, with insulin playing a central role in the process. Fatty acid synthesis is also under allosteric or short-term control, primarily through the acetyl-CoA carboxylase. The allosteric regulation of the carboxylase by citrate and by fatty acyl-CoA is well established (15, 16). Citrate or isocitrate activates the carboxylase by polymerization of the inactive protomer to an active polymer. Recent evidence incidates that the carboxylase is also regulated by covalent modification. Phosphorylation of the enzyme both with cAMP-dependent and with cAMP-independent kinases has been reported (17, 18). This development makes it possible to relate fatty acid synthesis to carbohydrate metabolism and to understand the roles of glucagon, catecholamines, and insulin in these processes.

Workers in the field are investigating not only the structure, organization, and function of these highly complex enzymes, but also their regulation, at both the enzymatic and the protein synthesis levels. The latter involves isolation of the genes coding for these proteins, determination of their structure, and understanding of their expression into active protein. In this chapter we review the recent developments in the field of fatty acid synthesis with special emphasis on the mechanism of action of eukaryotic FAS. [For earlier reviews see (1, 15, 19, 20).]

ACETYL-CoA CARBOXYLASE

In 1958 Wakil & Gibson (21) discovered acetyl-CoA carboxylase. This was the first functional biotin enzyme to be recognized because of its high biotin content and the ability of avidin to inhibit the carboxylation of acetyl-CoA by specifically binding the enzyme-bound biotin. Other biotin enzymes involved in carboxylation and transcarboxylation reactions were later isolated, and the role of biotin in these reactions was delineated (22, 23).

The molecular mechanism involved in the carboxylation of acetyl-CoA has been studied extensively with the bacterial acetyl-CoA carboxylase. [For earlier reviews see (19, 24).] Briefly, the carboxylation of acetyl-CoA in *Escherichia coli* requires the presence of biotin carboxyl carrier protein (BCCP, a dimer of apparently identical subunits of M_r 22,500), biotin carboxylase (BC, a dimer of identical subunits of M_r 51,000), and carboxyl transferase (CT, having an $\alpha_2\beta_2$-tetrameric structure composed of M_r 30,000 and 35,000 subunits). The reaction proceeds in two steps: first, the carboxylation of the biotin moiety of the BCCP (Reaction 3), and second, the transfer of the carboxyl group to the acceptor acetyl-CoA (Reaction 4):

$$\text{BCCP} + \text{HCO}_3^- + \text{ATP} \xrightleftharpoons{\text{BC, Me}^{+2}} \text{BCCP} \sim \text{COO}^- + \text{ADP} + \text{P}_i. \qquad 3.$$

$$\text{BCCP} \sim \text{COO}^- + \text{CH}_3\text{COSCoA} \xrightleftharpoons{\text{CT}} \text{BCCP} + \overset{\text{COO}^-}{\underset{|}{\text{CH}_2}}\text{COSCoA}. \qquad 4.$$

The carboxylation of the biotin prosthetic group of the BCCP subunit occurs at the 1'-N-position of the ureido ring of biotin, yielding the 1'-N-carboxybiotinyl derivative (Reaction 3). Evidence of the transcarboxylation reaction (Reaction 4) is based on the observation that an adivinsensitive exchange reaction between malonyl-CoA and [^{14}C]acetyl-CoA is catalyzed only in the presence of BCCP $\sim$ COO$^-$ and transcarboxylase (24). Isotope studies indicate that the transcarboxylation reaction occurs in a concerted fashion (25–27), formation of the methylene-carboxyl bond being concomitant with abstraction of a methylene hydrogen, either by the ureido oxygen of carboxybiotin or by a basic residue of the enzyme.

Further insight into the mechanism of carboxylation-decarboxylation comes from molecular structure analyses (28) of free biotin and of the model compound N-1'-methoxycarbonyl biotin methyl ester. The results reveal that the ureido-carbonyl bond in the model compound has double bond (keto) character more than the corresponding bond in free biotin, which has single bond (enolate) character. It seems that upon decarboxylation the ureido-carbon bond (C-O$^-$) becomes more polarized, thereby facilitating the deprotonation of N-1' and increasing its nucleophilicity. As a result, the coenzyme can interact with polar species such as bicarbonate. Conversely, on carboxylation, the carbonyl bond is depolarized (C=O), allowing the carboxylated coenzyme to interact with and carboxylate nonpolar species such as the methyl group of acetyl-CoA. Thus, the carboxylation-decarboxylation of biotin serves the dual function both of turning on and off the polarization of the ureido-carbonyl bond and of modulating the nucleophilicity of N-1'.

The results of kinetic (24, 29) and protein subunit studies (30, 31) suggest that the half-reactions in which biotin participates occur at physically distinct catalytic sites. The process of biotin translocation, necessary for its participation in both half-reactions, was originally envisioned as a movement of the prosthetic group some 28 Å on the "swinging arm" of the 14 Å-side chain of biocytin connecting the bicyclic ring of the biotin to the enzyme backbone (24). Nuclear magnetic and electron spin resonance studies, however, indicate that the two catalytic sites are approximately 7 Å apart (32–34). X-ray diffraction studies of biotin and its vitamers suggest that the translocation involves *gauche* $\rightleftharpoons$ *trans* rotations about the two carbons of the valeryl side chain nearest the biotin bicyclic ring, resulting in movement of the carbon atom of the 1'-N-carboxy group approximately 7 Å (35).

Animal Acetyl-CoA Carboxylase

Homogeneous preparations of acetyl-CoA carboxylase have been isolated from chicken liver (36, 37), rat liver (5) and mammary gland (8, 38), and rabbit mammary gland (39, 40). Unlike the bacterial and plant carboxylases, the animal enzymes are isolated as tightly bound multienzyme complexes. The active enzyme (41) is a polymer (M_r 4–8 million) that can be dissociated into inactive protomers ($M_r \sim 400{,}000$). Factors affecting the dissociation and polymerization of the animal carboxylase have been reviewed (16).

There was little agreement in earlier studies on the subunit structure of the protomer form of animal acetyl-CoA carboxylase (41, 42). One reason for the disagreement stems from the enzyme's susceptibility to proteolysis, which results in proteolytic cleavage artifacts. Thus, the carboxylase subunit is reported to contain two polypeptides of M_r 117,000 and 129,000 (41). Biotin is associated with the 117,000 dalton peptide and the protomer contains two copies of each polypeptide. However, work in Numa's laboratory (5, 44) indicates that the carboxylase of rat liver consists of identical subunits of M_r 230,000 each containing one mole of biotin. Recent studies support Numa's conclusion and indicate that the native acetyl-CoA carboxylase from rat liver (43), chicken liver (36, 37, 45), and rat (18) and rabbit (39, 40) mammary gland is composed of subunits of M_r 230,000– 260,000. Hence the acetyl-CoA carboxylase protomer is the M_r 230,000 subunit containing the functions of biotin carboxylase, biotin carboxyl carrier protein, and transcarboxylase, as well as the regulatory allosteric site. The subunit is therefore a multifunctional protein.

Rapid purification procedures, polyethylene glycol precipitation, and monomer avidin-sepharose affinity chromatography overcome the problem of proteolysis encountered during carboxylase isolation. The acetyl-CoA carboxylase preparations obtained by various procedures have different kinetic parameters and are affected differently by allosteric regulators (see below). This may reflect the extent of their modification before or during isolation. For instance, the rat liver carboxylase obtained by the polyethylene glycol precipitation and avidin affinity procedure (43) has a specific activity one-fifth that of earlier preparations and contains six mol of alkali labile phosphate per mol of subunit. Moreover, the K_m for acetyl-CoA is 80 μM, which is eightfold higher than the value obtained with the enzyme prepared by the conventional procedure. In contrast, homogeneous chicken liver carboxylase preparations, isolated by the avidin affinity chromatography or conventional procedures, give similar V_m and K_m values for acetyl-CoA (45).

Kinetic studies of acetyl-CoA carboxylase are consistent with the overall process depicted by Reactions 3 and 4. Recent kinetic analyses of pyruvate carboxylase (46, 47) and acetyl-CoA carboxylase (45) indicate that a quaternary complex consisting of carboxylated biotin, ADP, P_i and acyl-CoA is formed. The results support the ordered addition of ATP, HCO_3^-, and acetyl-CoA to the citrate-activated carboxylase but not the double displacement mechanism previously proposed (16). Although the discrepancy between the two kinetic schemes for chicken liver acetyl-CoA carboxylase is not understood at present, it is suggested that the kinetics may be altered by proteolytic modification of the enzyme during its isolation (45).

Yeast Acetyl-CoA Carboxylase

Acetyl-CoA carboxylase has been purified to homogeneity from two different species of yeasts, *Saccharomyces cerevisiae* (48) and *Candida lypolytica* (49). The *S. cerevisiae* and *C. lypolytica* enzymes have a tetrameric structure (α_4) with subunit M_r of 190,000 and 230,000, respectively (30) and one mol of covalently bound *d*-biotin per subunit. In the *S. cerevisiae* enzyme, all four C-terminals have the sequence leu-lys, and all the terminal amino groups are masked (30, 48), suggesting the presence of four identical subunits comprising the active carboxylase. These findings indicate that yeast acetyl-CoA carboxylase, like the animal enzyme, is a multifunctional polypeptide (50).

FATTY ACID SYNTHETASE

The synthesis of long-chain fatty acids from acetyl-CoA and malonyl-CoA involves numerous sequential reactions and acyl intermediates. The nature of these reactions and the intermediates involved is known primarily from studies of fatty acid synthesis in cell-free extracts of *E. coli* (19, 51). A protein ($M_r \sim 10,000$) with a 4'-phosphopantetheine prosthetic group was identified as the coenzyme ACP that binds all acyl intermediates as thioester derivatives. The individual enzymes were then isolated and utilized in the reconstitution of the FAS system (19, 51). The following are the enzymes and reactions involved in the synthesis of palmitate, the major product:

Acetyl transacylase

$$CH_3COS-CoA + ACP-SH \rightleftharpoons CH_3COS-ACP + CoA-SH. \qquad 5.$$

Malonyl transacylase

$$\overset{\displaystyle COOH}{\underset{\displaystyle CH_2COS-CoA}{|}} + ACP-SH \rightleftharpoons \overset{\displaystyle COOH}{\underset{\displaystyle CH_2COS-ACP}{|}} + CoA-SH. \qquad 6.$$

β-Ketoacyl–ACP synthetase (condensing enzyme)

$$CH_3COS-ACP + Enz-SH \rightleftharpoons CH_3COS-Enz + ACP-SH. \qquad 7a.$$

$$CH_3COS-Enz + \overset{\displaystyle COOH}{\underset{\displaystyle CH_2COS-ACP}{|}} \longrightarrow CH_3COCH_2COS-ACP$$
$$+ CO_2 + Enz-SH. \qquad 7b.$$

β-Ketoacyl–ACP reductase

$$CH_3COCH_2COS-ACP + NADPH$$
$$+ H^+ \rightleftharpoons D-CH_3CHOHCH_2COS-ACP$$
$$+ NADP^+. \hspace{4cm} 8.$$

β-Hydroxyacyl–ACP dehydratase

$$CH_3CHOHCH_2COS-ACP \rightleftharpoons trans-CH_3CH = CHCOS-ACP$$
$$+ H_2O. \hspace{3cm} 9.$$

Enoyl–ACP reductase

$$CH_3CH = CHCOS-ACP + NADPH$$
$$+ H^+ \longrightarrow CH_3CH_2CH_2COS-ACP + NADP^+. \hspace{2cm} 10.$$

Thioesterase

$$CH_3(CH_2)_{14}COS-ACP + H_2O \longrightarrow CH_3(CH_2)_{14}COOH$$
$$+ ACP-SH. \hspace{4cm} 11.$$

Studies of the individual enzymes from *E. coli* have been reviewed extensively (19, 51) and are not considered here. The plant synthetases also dissociate readily into their component enzymes and are thus similar to the bacterial enzymes (52, 53), and are not considered further. Instead, this review covers recent developments concerning the FAS of eukaryotic cells, with special emphasis on the synthetases both of animal tissues and of yeast.

Fatty Acid Synthetases of Eukaryotes

The FAS of eukaryotic cells (plant cells excluded) are complexes of multifunctional proteins and can be divided into two groups, exemplified by the synthetases isolated from animal tissues and yeast. The animal synthetases are multifunctional proteins of $M_r \sim 500,000$ consisting of two identical subunits (M_r 250,000); they are therefore α_2 structures. The yeast synthetase is a complex of two nonidentical subunits, α (M_r 213,000) and β (M_r 203,000), with an apparent M_r of 2.4×10^6; it is an $\alpha_6\beta_6$-structure. The yeast synthetase requires FMN for activity and yields palmitoyl- or stearoyl-CoA derivatives as the products of synthesis. In contrast, the animal synthetases do not utilize FMN as a cofactor and yield palmitate and stearate as the products.

Until the early 1970s, the working hypothesis was that animal and yeast synthetases were multienzyme complexes, and research effort was concentrated on dissociating these enzyme complexes and isolating their component enzymes. Evidence that these synthetases were multienzyme aggregates seemed beyond question at the time. The yeast enzyme was reported to yield equal amounts of seven different N-terminal amino acids, and at least six different proteins were separated on starch gels of the urea-treated enzyme (54). In addition, the isolation from guanidinium chloride-treated synthetase of a M_r 16,000 peptide containing the 4'-phosphopante-theine was reported (55). Porter et al (56) reported similar results for the enzyme isolated from pigeon liver. Having found five different N-terminal amino acids and eight protein bands on phenol-acetic acid urea gels, they proposed that the synthetase was a multienzyme complex consisting of eight proteins (56). The multiprotein concept for the animal enzymes was further supported by the isolation of a protein (M_r 10,000) containing the 4'-phosphopantetheine prosthetic group from the FAS of pigeon (57), dog (58), chicken (59, 60), human (60), and rat liver (60). The separation of active enzyme components from the chicken liver enzyme was also reported (59).

Contrary to the above conclusion, independent investigations in Schweizer's laboratory (3, 4) of yeast synthetase and in our laboratory (2) of animal and yeast synthetases have revealed that these enzymes are comprised of multifunctional subunits. Schweizer's genetic studies of yeast synthetase indicate that the enzyme is encoded by two unlinked genes; he concludes that the yeast synthetase consists of two multifunctional proteins. The presence of multiple proteins in the yeast synthetase preparations is attributed to nonspecific proteolysis of the two multifunctional polypeptides.

Our laboratory has demonstrated the multifunctional enzyme nature of FAS isolated from animal tissues and yeast. We found (2) more extensive formation of low-molecular-weight peptides at high protein concentration (10 mg/ml) in SDS than at lower protein concentration (1 mg/ml). Such a result cannot be explained on the basis of incomplete dissociation in SDS and was therefore attributed to proteolysis of the synthetase subunit. Stoops et al (2) demonstrated that proteolysis does occur during preparation of the synthetase and is prominent in the presence of SDS as well as other denaturing agents (guanidinium chloride, urea). [For review of earlier reports on the multifunctional nature of FAS, see (1).]

STRUCTURAL ORGANIZATION OF FATTY ACID SYNTHETASE

Animal Fatty Acid Synthetase

The FAS of rat liver, adipose (61) and lactating mammary gland (62), chicken liver (63, 64), rabbit mammary gland (65), and uropygial gland (8) have many common features.

One of the first questions addressed was whether the two subunits are identical. Besides having the same size, the polypeptide subunits have the same charge and shape (61). Also, no free N-terminal amino acid residues have been detected, suggesting that the synthetase subunit polypeptides (animal as well as yeast) have blocked N-terminal residues (2). Recognition that the two polypeptide chains may be identical prompted us to reinvestigate the 4'-phosphopantetheine content of the enzyme. Application of several different analytical procedures to preparations of chicken and rat synthetases yields values of 1.4–1.8 mol of 4'-phosphopantetheine per mol of enzyme or about one mol per subunit (61, 63), with similar values subsequently found for synthetases of the uropygial gland of goose (8) and of the higher bacterium, *Mycobacterium smegmatis* (66). These results contrast with the value of one prosthetic group per mol of enzyme reported earlier for the synthetases of pigeon (67), rat (68) and dog liver (69), and rat lactating mammary gland (70). The finding of one prosthetic group per subunit strongly suggests that the two polypeptide subunits are identical, a conclusion upheld by further studies (see below).

Electron microscope studies of animal FAS further support the concept that synthetase is a multifunctional enzyme. Studies of negatively stained rat liver FAS (61) indicate that the enzyme subunit consists of a linear structure 200 Å long, containing at least four lobes 50 Å in diameter (Figure 1). Its mass, estimated to be about 200,000 daltons, is consistent with the estimated M_r of the subunit. A pseudotetrahedral set of images was also observed, with dimensions and forms indicating that the pseudotetrahedrally shaped mass may also be comprised of four major lobes, each about 50 Å in diameter, and therefore of the same mass as that of the linear structure (Figure 1). Since FAS is a dimer, and the numbers of linear and pseudotetrahedral structures observed are not equal, it is concluded that the linear structure may fold into the pseudotetrahedron, and that the dimer is comprised of chemically identical chains. This conclusion is compatible with the chemical and enzymological properties of animal FAS.

The isolation and characterization of synthetase mRNA and its translation product further support the concept that animal synthetase is a multifunctional enzyme (71–74). The FAS mRNA has recently been isolated from goose uropygial gland (71) and rat mammary gland (72). Poly(A)$^+$ RNA from these tissues is isolated by affinity chromatography on oligo (dT)-cellulose column. Translation of poly(A)$^+$ RNA in rabbit reticulocyte lysate produces, in addition to other proteins, a high mol wt polypeptide of M_r 250,000, which was recognized by antibodies to FAS, and competed with native synthetase for binding to antibody. Sucrose density gradient centrifugation of the poly(A)$^+$ RNA results in further purification of mRNA and gives sedimentation values of 35S and 37S for goose and rat mRNA, respectively. Gel electrophoresis of FAS mRNA after denaturation with methyl glyoxal or methylmercury hydroxide demonstrates that the goose and rat synthetase mRNA have M_r of 2.95×10^6 and 3.5×10^6, respectively. This is about 2000–3000 nucleotides greater than the minimum required to code for the entire FAS subunit.

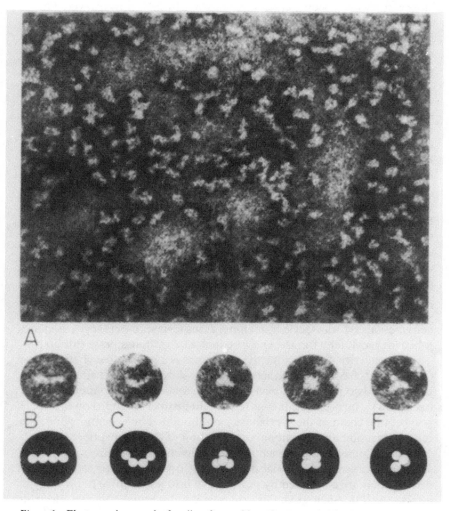

Figure 1   Electron micrograph of rat liver fatty acid synthetase. *A.* An electron micrograph of synthetase negatively stained with methylamine tungstate × 300,000. *B* and *C* are selected images of the linear form of the subunit × 400,000. *D, E,* and *F* are selected images of the pseudotetrahedral form of the subunit × 400,000. Below each subunit image is a photograph of a model of the subunit that has been folded and oriented so as to present correlative view with the image (61).

Goodridge's group (75) has recently cloned gene sequences for goose FAS. Partially purified FAS mRNA is isolated by extracting total poly(A)$^+$ RNA from goose uropygial gland by the rapid pH-5 precipitation technique, followed by fractionation on isokinetic sucrose gradients. Seven clones, containing sequences complementary to FAS mRNA, were identified by colony hybridization with ^{32}P-labeled cDNA transcribed from a partially purified FAS mRNA; the identity of the clones was confirmed by hybrid-selected translation. Two plasmids containing FAS sequences— pFAS1 and pFAS3—have inserts of 1400- and 1700-base pairs, respectively. A precise size determination of the FAS mRNA was made by electrophoresis of total uropygial gland RNA on denaturing agarose gel, followed by identification of the FAS mRNA band by Northern analysis using ^{32}P-labeled pFAS3 DNA. A value of 16 kilobase, considerably greater than previous estimates (71, 72), was obtained. In every case, however, the mRNAs are large enough to code for the synthetase subunit of M_r 250,000 or about 2300 amino acids (61). There are substantial noncoding regions in the FAS mRNA, which is not uncommon among mRNA molecules of animal tissues. These results, therefore, confirm the multifunctional nature of the synthetase and indicate that the synthetase subunit must arise as a single polypeptide chain synthesized from one contiguous mRNA.

The evidence mentioned above strongly indicates that the synthetase subunits are identical and that each polypeptide contains the sites of the partial reactions required for palmitate synthesis. Dissociation of the native enzyme to monomers results in the retention of six of the enzymatic activities: acetyl transacylase, malonyl transacylase, β-ketoacyl reductase, β-hydroxyacyl dehydratase, enoyl reductase and thioesterase (61, 76–78). The one activity absent from the monomer is the β-ketoacyl synthetase (condensing enzyme), whose active center is shown to be dependent on the presence of two juxtapositioned thiols (13), each derived from one subunit (see below). The presence, therefore, of eight distinct functions on the single polypeptide chain of the synthetase has stimulated investigations into the relative location and functional organization of the polypeptide.

There is increasing evidence that multifunctional proteins are actually arranged as a series of globular domains, the sites of catalytic or regulatory activity, that are connected by polypeptide bridges sensitive to proteolytic attack (6, 79). Such domains are readily discernible in the electron micrograph of the rat liver FAS (Figure 1). It has thus been possible to isolate active fragments from a number of multifunctional proteins such as DNA polymerase I, immunoglobulins, and the "CAD" protein of pyrimidine biosynthesis, among others (80–83). Indications are that FAS is organized in a similar fashion and is amenable to controlled analysis

employing proteolytic dissection. Several laboratories have treated synthetase with trypsin, elastase, or subtilisin to separate the thioesterase
activity from the remainder of the system (84–89). More recently, chicken
liver FAS was proteolyzed by a variety of proteases (chymotrypsin, elastase,
trypsin, myxobacterprotease, subtilisin A and B, and kallikrein), utilized
either individually or in combination (90; J. S. Mattick, Y. Tsukamoto, J.
Nickless, S. J. Wakil, manuscript in preparation). The proteolytic profiles
were analyzed with respect both to the kinetics and to the size of the peptide
fragments. The data strongly uphold the contention that the polypeptide
subunits of chicken liver FAS are identical. Also, a reasonably detailed map
of the synthetase subunit has been constructed (Figure 2). Analyses of all the
fragment patterns and summations of their M_r consistently give a value of
263,000 for the M_r of the intact synthetase subunit. The synthetase subunit

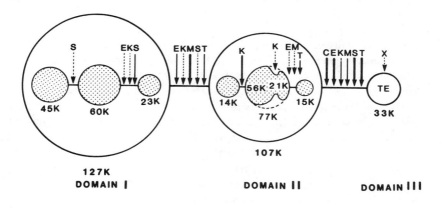

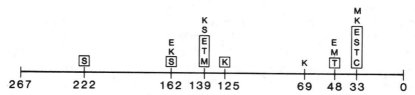

Figure 2   Proteolytic map of the chicken fatty acid synthetase. This map was constructed after
studies of the patterns of proteolytic digestions of the synthetase by various proteases. The
heavy arrows indicate a primary cleavage site; the light arrows indicate a secondary cleavage
site, and the dotted arrows indicate a tertiary cleavage site after longer incubation time.
Abbreviations for proteases are: T, trypsin; E, elastase; M, *Myxobacter* protease; S, subtilisin A
or B; C, α-chymotrypsin; K, kallikrein; X, all of these proteases. The lower figure shows
the actual distances (in daltons × 10^{-3}) of the protease cleavage sites from the thioesterase
C-terminus of the monomer. Proteases in boxes are capable of inflicting primary cleavage at
the sites indicated.

can be divided into a terminal 33,000 thioesterase (domain III) and a large peptide of M_r 230,000, which contains all of the "core" activities of the FAS sequence (Figure 2). α-Chymotrypsin specifically cleaves the synthetase at this site and separates the thioesterase from the multifunctional complex. Other proteases also attack this site, as well as hydrolyze the synthetase subunit at other sites (Figure 2). Since the intact synthetase is known to have a blocked N-terminus (2), N-terminal sequence analysis should identify which of the chymotryptic fragments has the free amino terminus, thereby specifying the orientation of the protein. The results of such determination show that the thioesterase (M_r 33,000) has a free N-terminus, whereas the 230,000-dalton fragment, like the intact synthetase, has a blocked N-terminus (J. S. Mattick, J. Nickless, M. Mizugaki, C. Y. Yang, S. Uchiyama, S. J. Wakil, manuscript in preparation). The sequence of the thioesterase at the N-terminus is found to be H_2N-Lys-Thr-Gly-Pro-Gly-Glu-Pro-Pro-.

The thiosterase itself is released from the synthetase complex as an intact and catalytically active fragment. It does, however, undergo a slow degradation into fragments of M_r 18,000 and 15,000 upon prolonged incubation with any of the proteases. This breakdown, which is accompanied by loss of catalytic activity, is evident at later stages in the cleavage patterns.

The 230,000-dalton "core" region may be segregated into two principal domains, designated I (M_r 127,000) and II (M_r 107,000), as defined by *Myxobacter* protease/tryptic cleavage (Figure 2). Domain II is located adjacent to the thioesterase (domain III) and contains the primary kallikrein site, 92,000 daltons from the thioesterase junction. Domain II also contains a secondary tryptic site, which leads to the release of a terminal 15,000-dalton segment identified as the ACP.

Recognition that thioesterase is located at the C-terminal end of the synthetase polypeptide makes it possible to establish the mapping of the functional centers on the synthetase subunit (Figures 2, 3). In these studies (J. S. Mattick, Y. Tsukamoto, H. Wong, S. J. Wakil, manuscript in preparation), the known properties of active sites of the component activities of the synthetase were employed whenever possible. For instance, radioactive pantetheine labeled the ACP site, which was then followed throughout the course of proteolysis. Similar approaches were followed using labeled substrates, specific inhibitors, or antibodies, either monoclonal or developed against homogeneous components or domains. A model for the synthetase polypeptide is proposed (Figure 3) that incorporates the results of such studies and illustrates the relative sizes of the domains and associated activities. As can be seen from this model, the β-ketoacyl synthetase (KS) and the ACP sites are located in separate domains of the synthetase subunit and are far removed from each other. Domain I contains the acetyl- and malonyl-transacylases as well as the condensing enzyme site, thus making this domain the substrate entry and chain elongation domain. On the other hand, domain II contains the β-ketoacyl reductase, the dehydratase, and the enoyl reductase activities, thus functioning as the processing domain for the reduction of the carbonyl carbon to the methylene analog by NADPH. The ACP and its 4′-phosphopantetheine arm is located next to the reduction domain (domain II) and connects it to the chain termination or palmitate release domain (thioesterase, domain III). The thioesterase was also shown to be the terminal domain of the rat and rabbit mammary gland FAS (84, 85). Elastase split the rabbit synthetase subunit into the 35,000-thioesterase component and 220,000-core polypeptide (84).

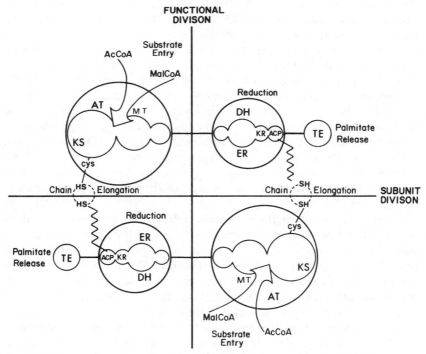

Figure 3   Proposed functional map of the chicken fatty acid synthetase. The sketch was based on the map shown in Figure 2 and the results were obtained, where possible, from assays of catalytic activities and the binding of substrates, specific inhibitors or antibodies. Two subunits are drawn in head-to-tail arrangement (subunit division) so that two sites of palmitate synthesis are constructed (functional division). The abbreviations for partial activities used are: AT, acetyl transacylase; MT, malonyl transacylase; KS, β-ketoacyl synthetase; KR, β-ketoacyl reductase; DH, dehydratase; ER, enoyl reductase; TE, thioesterase; and ACP, acyl carrier protein. The wavy line represents the 4′-phosphopantetheine prosthetic group.

Yeast Fatty Acid Synthetase

The yeast FAS catalyzes basically the same chemical reactions as the animal FAS, leading to the formation of long-chain fatty acids from acetyl-CoA, malonyl-CoA and NADPH. There are, however, some differences in both the architecture of the enzyme and in the mechanism of some of the reactions involved. For instance, the enoyl reductase requires FMN as a cofactor, and the product of synthesis is palmitoyl-CoA instead of free palmitate. Earlier it was proposed that yeast synthetase was a multienzyme complex consisting of seven individual enzymes and the ACP coenzyme held together by noncovalent interactions (23, 91). Recently, however, Schweizer et al (3, 4) have shown that the previous studies were misled by partial proteolysis of the synthetase during isolation of the enzyme; when the proteolysis is reduced, SDS-PAGE studies indicate that the enzyme consists of two subunits of M_r 185,000 and 180,000 respectively. We have reached a similar conclusion regarding the variable degradation of the enzyme by proteolysis and the need for protease inhibitors during enzyme isolation (92). We have also isolated the yeast synthetase from a yeast mutant that has reduced protease levels. This enzyme, prepared in the absence of inhibitors, showed the same SDS-PAGE patterns as the enzyme isolated in the presence of protease inhibitor. Tris-glycine-SDS-PAGE of the yeast synthetase gives two protein bands (designated α and β), which are present in equal amounts and have an estimated M_r of 213,000 and 203,000, respectively. These values are relatively higher than those reported by Schweizer et al (3, 4) and may reflect the different protein standards used. In any case, from these values it is concluded that the native yeast synthetase is an $\alpha_6\beta_6$-complex.

Estimation of the pantetheine content of the yeast synthetase gives values of 3.8–5.0 mol per mol of synthetase (M_r 2.3×10^6) or about one mol of pantetheine per two subunits (92, 93). Radioautography of the SDS-PAGE of the [^{14}C]pantetheine-labeled synthetase reveals that the α-subunit bears the prosthetic group, thus identifying the ACP site with α (3, 92). The peptides containing 4′-phosphopantetheine have been isolated from yeast (94) and rat liver (68) synthetases, and the sequence of amino acids around the serine-phosphopantetheine diester has been determined. As shown in Figure 4, there is considerable homology between amino acid sequences of corresponding peptides of the ACP of yeast, rat, and *E. coli* (95). Of the 18 amino acids around the 4′-phosphopantetheine-carrying serine of yeast and *E. coli* ACP, there are five identical amino acids (boxes) in analogous positions and eight pairs of amino acids (arrows) whose codons vary by a single point mutation (96). Homology between the animal and *E. coli* sequences is also good, despite the restricted number of amino acids sequenced in the animal enzyme. These sequence homologies support the

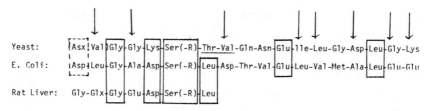

Yeast: [Asx¦Val])Gly]-Gly-[Lys]-Ser(-R)]-Thr-Val-Gln-Asn-Glu-Ile-Leu-Gly-Asp-Leu-Gly-Lys

E. Coli: ¦Asp-Leu-Gly-Ala-Asp]-Ser(-R)]-Leu-Asp-Thr-Val-Glu-Leu-Val-Met-Ala-Leu-Glu-Glu

Rat Liver: Gly-Glx-Gly-Glu-Asp]-Ser(-R)]-Leu

Figure 4 Sequence homologies of the acyl carrier proteins from yeast, *E. coli* and rat liver fatty acid synthetases in the neighborhood of the prosthetic group, 4'-phosphopantetheine (R) (94, 96).

thesis that the ACP of the three organisms are derived from the same ancestral gene with some mutation occurring at various noncritical positions.

Electron microscope studies of negatively stained yeast synthetase (Figure 5) have led us to propose a model for the enzyme (92): an ovate structure containing, on its short axis, plate-like protein structures around which six arch-like proteins are distributed three on either side (Figure 6). The claim that the complexes are made of two distinct structures, "arches"

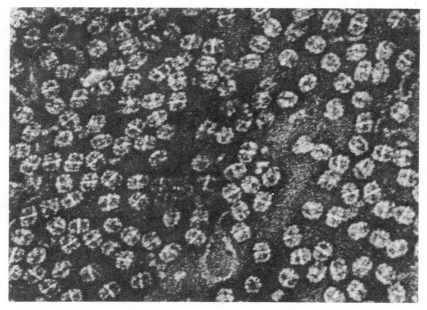

Figure 5 Electron micrograph of yeast fatty acid synthetase negatively stained with methylamine tungstate × 200,000. The protein was cross-linked with glutaraldehyde before exposure to the stain.

and "plates," is supported by the observation of images that contain these isolated structures (Figure 5). Measurements of the two distinct subunits that comprise the arch and plate structures reveal an approximate M_r of 200,000 each, agreeing with values obtained for the subunits by physico-chemical methods (92). The structural organization of yeast synthetase becomes apparent when stereoscopic images of particles are studied. These particles appear to contain three arches on each side of the plate, each beginning on one side of the plate and terminating on the opposite side of an adjacent plate subunit. This arrangement of arches suggests that the plates are alternately reversed (black and white surfaces shown in Figure 6) in their orientation with respect to their unique axis. This conclusion is compatible with and supportive of results obtained from our studies of the mechanism of action of the β-ketoacyl synthetase component of FAS (see below). Wieland et al (97) proposed a similar model, except that α and β form a V-shaped complex with the peptides protruding upwards and downwards from the plate proteins.

The α- and β-subunits of yeast FAS can be separated by first modifying the protein either with 3,4,5,6-tetrahydrophthalic anhydride followed by ion exchange chromatography (10) or with cyclic anhydrides like citraconic and dimethylmaleic anhydrides followed by sucrose gradient

Figure 6   Model of the yeast fatty acid synthetase. The arch-like structures are the β-subunits; the plate-like structures in the center are the α-subunits. The black and white faces of the α-subunits represent the alternate arrangement of these structures in the complex $\alpha_6\beta_6$.

centrifugation (14). When, after mild acid treatment of the separated subunits, the acylating agent is removed, the subunits slowly regain some of the activities. Assays of such fractions show that α contains, in addition to the phosphopantetheine site, the active thiol of the condensing site (β-ketoacyl synthetase) and the β-ketoacyl reductase site, while the β-subunit contains the transacylases, the dehydratase, and the enoyl reductase sites. These findings confirm earlier conclusions by Schweizer et al (4, Figure 7). Antibodies against α- and β-polypeptides have been prepared and utilized in identifying the α- and β-subunits in the model based on electron microscope studies (97). The results show that the α-subunits are the plates and the β-subunits are the arches, a conclusion fully supported by the cross-linking studies of Stoops & Wakil (10).

Recently, Kuziora et al (98; M. A. Kuziora, M. G. Douglas, S. J. Wakil, *J. Biol. Chem.*, submitted for publication) has used yeast transformation technique to isolate DNA clones of the genes coding for the α- and β-subunits of the synthetase through complementation of fatty acid auxotrophs of *S. cerevisiae*. Plasmids YEpFAS1 and YEpFAS2 were selected from a bank of yeast DNA sequences in the vector YEp13 by their ability to complement mutations in the *fas1* or *fas2* locus, respectively. Although both plasmids code for a functional protein in yeast cells, only YEpFAS2 produces in *E. coli* maxicells peptides that are antigenically reactive with antiyeast FAS antibody. The sizes of these peptides range from 100,000 to 160,000 daltons. Plasmid YEpFAS1 produces immunologically non-reactive peptides between 21,500 and 48,000 daltons. The inability to detect full-length α- or β-subunits in *E. coli* maxicells is attributed to difficulties encountered in expression of a eukaryotic gene by a prokaryotic organism, e.g. as lack of a functional promoter, or proteolysis of synthesized proteins.

Confirming evidence that plasmids YEpFAS1 and YEpFAS2 contain DNA sequences coding for the subunits of yeast synthetase is provided by the homology to plasmids 33F1 and 102B5, isolated from a yeast genomic bank (J. H. Chalmers, Jr., M. A. Kuziora, R. A. Hitzeman, S. J. Wakil, *J. Biol. Chem.*, submitted for publication). By employing the antibody

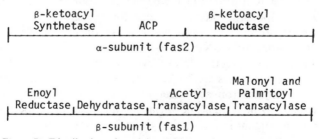

Figure 7   Distribution of partial activities on the α- and β-subunits (4).

selection method, approximately 5000 clones were screened containing randomly sheared yeast DNA inserted into the CoIE1 vector and two clones (33F1 and 102B5) were successfully identified that expressed FAS related antigen (99). The Southern blotting technique was used to show that 33F1 hybridized to a ^{32}P-labeled nick translated probe from YEpFAS1 and that 102B5 hybridized to a ^{32}P-probe from YEpFAS2, thus identifying these clones as containing DNA from the *fas1* and *fas2* loci, respectively. Restriction endonuclease mapping of the two clones further identifies the regions of homology between the two sets of plasmids.

FUNCTIONAL ORGANIZATION OF FATTY ACID SYNTHETASE

Palmitate synthesis is a cyclical process that requires an orderly involvement of seven different enzymes, five of which participate sequentially eight times each during the synthesis of one molecule of palmitate. Thus, the synthesis of palmitate from acetyl-CoA and malonyl-CoA involves at least 30 acyl intermediates that are covalently bound to the enzyme. The properties of the component enzymes and their functional relationships are summarized below.

Acetyl and Malonyl Transacylases

Acetyl and malonyl transacylases prime the FAS with the carbon atoms required for fatty acid synthesis (Reactions 5 and 6). The malonyl transacylase is specific for the malonyl group; the acetyl transacylase, however, though it normally uses acetyl-CoA, is species dependent and manifests some variability toward acyl-CoA substrates. Avian FAS shows preference for acetyl-CoA as a primer, while bovine and rodent synthetases utilize butyryl-CoA as well as acetyl-CoA (100, 101). These synthetases may also be sources of butyrate, for butyryl-S-enzyme is formed as an intermediate and in the presence of relatively low levels of malonyl-CoA, may either transfer to CoA to form butyryl-CoA or react with H_2O to form butyrate (100). It is not yet clear whether this built-in variation of the synthetase is important physiologically for generating or consuming butyrate or for both. The transacylases active on butyryl-CoA can also transacylate acetoacetyl-CoA or crotonyl-CoA and make them available as primers of FAS after reduction to butyryl-derivatives on the synthetase. In so doing, these synthetases are 20- to 50-times more efficient as primers in synthesizing palmitate from acetoacetyl-CoA or crotonyl-CoA than are the avian or yeast enzymes (102). The variation in transacylase specificity may be especially important in fatty acid synthesis in bovine tissue.

The transacylases of all synthetases have an active serine residue to which

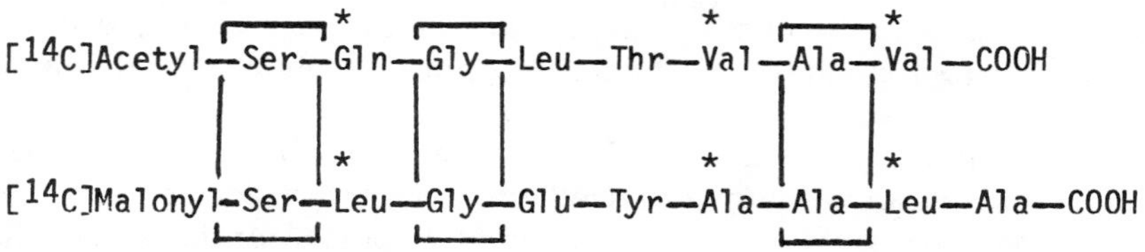

```
                    *                   *         *
[14C]Acetyl—Ser—Gln—Gly—Leu—Thr—Val—Ala—Val—COOH
          |       |*    |       |        |       |*
[14C]Malonyl-Ser—Leu—Gly—Glu—Tyr—Ala—Ala—Leu—Ala—COOH
          |___|   |___|                |___|
```

Figure 8 Comparison of active-site peptides from malonyl/palmitoyl transacylases and acetyl transacylase (96).

the acetyl and malonyl groups are bound in an O-ester linkage. Lynen and co-workers (103, 104) have isolated the serine-containing peptide of both acetyl and malonyl transacylases from yeast synthetase. As Figure 8 shows, the amino acids of octapeptides from the active sites of acetyl and malonyl transacylases have three pairs of identical amino acids and three pairs of amino acids whose codon differ by only one base. Though these sequences are too short to justify any conclusions, a common origin for the two transacylases has been postulated (96). The palmitoyl transferase of yeast FAS appears to share the same active serine site as the malonyl transacylase (Figure 9), for it not only contains the same sequences of amino acids neighboring the active serine residue but also appears to be coded by the same gene (105). Moreover, binding studies (106) of radiolabeled malonyl and palmitoyl groups to the synthetase show that malonyl binding excludes palmitoyl binding and vice versa.

The acetyl transacylase was isolated from yeast FAS after proteolysis with elastase (96). The protein fragment (M_r 56,000) contains none of the other component activities. The K_m values for the acetyl-CoA and acetyl-pantetheine are ten times greater when measured with the isolated enzyme than when measured with transacylase of the native FAS, suggesting that release of the enzyme from the synthetase results in some changes in protein conformation.

It was proposed earlier that the loading of an acetyl or malonyl group to the animal FAS was an ordered process; if the enzyme is charged with an acetyl group, it underwent a conformational change so that the malonyl

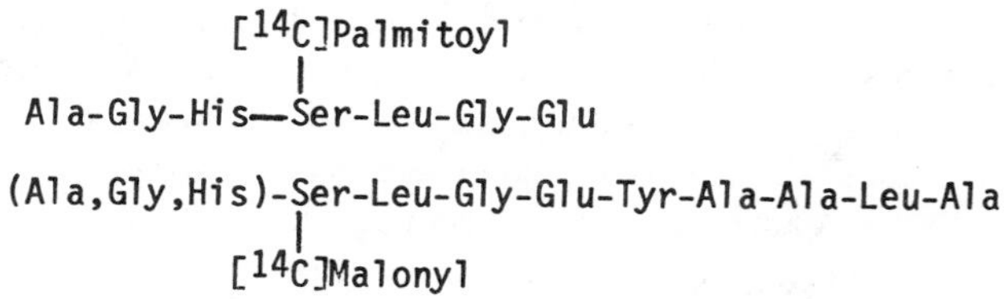

```
              [14C]Palmitoyl
                   |
       Ala-Gly-His—Ser-Leu-Gly-Glu

       (Ala,Gly,His)-Ser-Leu-Gly-Glu-Tyr-Ala-Ala-Leu-Ala
                         |
                   [14C]Malonyl
```

Figure 9 Sequence of active-site peptides from malonyl transacylase and palmitoyl transferase (96).

group is loaded preferentially in order to favor the condensation and chain elongation to palmitate (107). This does not appear to be the case. Recently it has been found that removal of CoA by a scavenging system, e.g. phosphotransacetylase, acetyl-CoA synthetase, or ATP citrate lyase, results in cessation of fatty acid synthesis (108). However, the system is reported to be reactivated by the addition of either CoA or thioesterase II (see below; 108). Based on these observations, CoA is proposed to be required for the formation of palmitoyl-CoA before its hydrolysis to free palmitate by the synthetase-bound thioesterase. Smith and co-workers (109, 110) and Poulose & Kolattukudy (111) have recently confirmed the CoA requirement for FAS, although they were unable to document the proposed requirement for thioesterase. Stern et al (109), however, have proposed that free CoA participates in a continuous exchange of acetyl and malonyl moieties between CoA thioesters and the enzyme. The acetyl and malonyl transacylases operate independently of each other; their respective acyl substrates compete for the pantetheine residue of the synthetase. In the absence of free CoA, the synthetase may be loaded with either two acetyl or malonyl groups, thus rendering it inactive, owing to its incorrect condensing partner. This conclusion is based on the premise that the presence of a CoA scavenging system blocks the unloading of the acetyl and malonyl groups from the synthetase and thus prevents the reloading of the appropriate substrate. Inhibition of the synthetase is then relieved when CoA or pantetheine is added. Thus, according to this hypothesis, the uptake of the acetyl and malonyl groups by the synthetase is a random process rather than a sequential one. Even though this proposal seems reasonable, kinetic studies are needed both to distinguish between the two processes and to shed more light on the role of free CoA in fatty acid synthesis.

β-Ketoacyl Synthetase (Condensing Enzyme)

The coupling of the acyl and malonyl groups to form β-ketoacyl derivatives is catalyzed by the β-ketoacyl synthetase component of the FAS system. Information concerning this reaction was obtained initially from studies of the condensation of acetyl-ACP and malonyl ACP (Reactions 7a and b) by the E. coli enzyme (112, 113). These studies show that the β-ketoacyl synthetase (DM) contains an active cysteine-SH, which forms an acyl-S-enzyme intermediate before coupling with malonyl-S-ACP to yield β-ketoacyl-ACP and CO_2. This same general mechanism operates in both the animal and yeast FAS. An active cysteine-SH is identified in the animal (13) and yeast (114, 115) enzymes, which upon alkylation with iodoacetamide inhibits the condensing activity only. Acetyl-CoA (but not malonyl-CoA) protects the enzyme against inhibition, suggesting formation of an acetyl-S-enzyme. The binding site of the acetyl group in the condensation site has been identified by isolating a carboxyamidomethyl peptide after radioactive iodoacetamide treatment and subsequent proteolysis. This peptide has the sequence: H_2N-Thr-Pro-Val-Gly-Ala-Cys-COOH (116). A similar acetate-containing peptide with identical properties has been isolated by high-voltage electrophoresis and paper chromatography (116).

The reaction of iodoacetamide with the cysteine-SH of the yeast synthetase is independent of pH between pH 5 and 9, indicating that the pK of this cysteine-SH residue is less than five and that cysteine-S$^-$ is the predominant form in the β-ketoacyl synthetase site (114). Such a perturbation has been measured previously for an active-site cysteine-SH residue and could be attributed to the presence of the positive charge of a nearby histidine residue (117). Attempts to identify such a residue in yeast synthetase have been unsuccessful (see below). The ionized state of the active thiol (cysteine-S$^-$) at pH 5 and above would favor acyl formation. It would also explain the high reactivity of this thiol toward iodoacetamide at pH 5 as compared to other cysteine residues that presumably are in the protonated state and therefore less reactive under the conditions used. Iodoacetamide reacts with the active cysteine-SH of yeast synthetase with a second-order rate constant of 185 M^{-1} min^{-1} (pH 7.5), comparable to the values reported for its reaction with ionized thiols (115). This further supports the contention that the pK of the active-site cysteine-SH is perturbed. Available evidence indicates that, in the eukaryotic synthetases, the acetyl and other acyl groups form a thioester with the cysteine-SH of the condensing enzyme, and that the malonyl group forms a thioester with the pantetheine-SH before condensation to form the β-ketoacyl derivative (for review see 1). Stoops & Wakil have studied the structure and mechanism of the β-ketoacyl synthetase of animal (13, 118) and yeast (12, 115) FAS. These studies provide solid experimental evidence supporting the previous assignment of the attachment sites of the acyl groups for the animal synthetase and reveal a novel yet common arrangement of the β-ketoacyl synthetase site for both animal and yeast FAS.

Centrifugation, used to determine the state of aggregation of the active animal synthetase during catalysis (61), indicates that the dimer is the active form of the synthetase in the pH range of 6.5–7.5. This conclusion is based on the values of the sedimentation coefficients (15.0 S–16.5 S) measured in the presence and absence of substrates. These values are shown by sedimentation equilibrium experiments to correspond to the dimeric form of the enzyme; the monomeric form (9.1 S) is inactive under these conditions.

The reasons for requiring the dimer form of the enzyme have become evident from Stoops & Wakil's studies (13, 118) of the role of active thiols in fatty acid synthesis. The FAS of chicken liver, for instance, is completely inhibited in the presence of 0.5-mM iodoacetamide, with a first-order rate constant estimated at 0.033 min^{-1}. When [^{14}C]iodoacetamide is reacted with the synthetase, over 80% of the ^{14}C-label is recovered after hydrolysis as [^{14}C]S-carboxymethyl-cysteine, but none as [^{14}C]S-carboxymethylcysteamine, indicating that inhibition of the enzyme is due to alkylation of the active cysteine-SH (13). This thiol was identified as the active cysteine-SH of the β-ketoacyl synthetase component of FAS because it was the only enzyme activity lost. Preincubation of the synthetase with acetyl-CoA, but not with malonyl-CoA, protects the enzyme from inhibition by iodoacetamide, suggesting that this thiol is the site of binding the acetyl group to the β-ketoacyl synthetase site, and that the malonyl-group binding site is not the cysteine-SH.

In contrast to the slow inhibition of the chicken FAS by iodoacetamide, the bifunctional reagent 1,3-dibromo-2-propanone inhibits the enzyme rapidly (within 30 sec) and completely (13). The loss of synthetase activity is due to inhibition of only the β-ketoacyl synthetase activity. Preincubation of the synthetase with acetyl-CoA protects the enzyme against inhibition by dibromopropanone, while malonyl-CoA does not. These results are similar to those found for iodoacetamide inhibition and clearly show that dibromopropanone competes with acetyl-CoA for the same thiol in the β-ketoacyl synthetase site.

When the dibromopropanone-inhibited synthetase is analyzed on SDS-PAGE, the synthetase subunit of M_r 220,000 is nearly absent while oligomers of higher M_r 400,000–500,000 concomitantly appear. Preincubation of the synthetase with acetyl-CoA or malonyl-CoA prevents cross-linking of the subunits by dibromopropanone. A similar result is also obtained when the synthetase is treated with iodoacetamide before its reaction with the dibromopropanone. The stoichiometry of inhibition indicates that binding of about 1.8 mol of dibromopropanone per mol of enzyme is required for complete inactivation of the FAS (13). Altogether, these results indicate that the dibromopropanone is reacting as a bifunctional reagent cross-linking the two subunits that comprise the enzymically active FAS dimer.

Finally, when [^{14}C]dibromopropanone is used as the cross-linking reagent, the cross-linked oligomers separated by SDS-PAGE contain over 85% of the protein-bound radioactivity. Oxidation of the ^{14}C-labeled oligomers with performic acid followed by HCl hydrolysis yields ^{14}C-labeled sulfones that have been identified by ion exchange chromatography (118). The ^{14}C-labeled sulfones of the hydrolysate co-chromatograph with standard S-carboxymethyl cysteine and S-carboxymethyl cysteamine sulfones and are present in equal amounts. These results indicate that dibromopropanone cross-links the two synthetase subunits by reacting with a cysteine-SH of one subunit and the cysteamine-SH of the other. This

observation has been confirmed using Ellman's reagent, which inhibits the
β-ketoacyl synthetase activity by forming the mixed disulfide consisting of
the cysteine-SH of one subunit and the pantetheine-SH of the adjacent
subunit (119).

The requirement for vicinal pantetheine and cysteine residues located on
separate subunits for the β-ketoacyl synthetase reaction explains the loss of
this partial activity and FAS activity on dissociation of the homodimer.
This stringent requirement for activity also explains on a molecular basis,
the cold inactivation of the enzyme (118). The pigeon (77) and chicken (118)
liver enzymes lose over 90% of their activity if incubated at 0°C for 12 hr.
Ultracentrifugal analysis and the lack of dependency of activity loss on
protein concentration demonstrates that this loss is not due to dissociation
of the dimer. Full reactivation can occur within 2 hr if the temperature of the
enzyme solution is raised to 25°C, or within five min if the enzyme is
incubated at 25°C in the presence of NADPH. Acetyl-CoA and malonyl-
CoA or NADH do not increase the rate of reactivation. Apparently the
proper positioning of the phosphate residue of NADPH plays an important
role in the reactivation of the enzyme.

The molecular basis for cold inactivation of the enzyme has been
elucidated with [^{14}C]dibromopropanone (118). At 0°C, it inhibits the
synthetase but does not cross-link the subunits. When the ^{14}C-labeled
enzyme is subjected to performic acid oxidation and HCl hydrolysis, 85% of
the radioactivity is associated with carboxymethyl cysteine sulfone while
less than 7% is associated with carboxymethyl cysteamine sulfone. As
shown above, radioactivity is equally distributed between these two
products when the active enzyme is similarly treated. It would appear, then,
that cold inactivation results from a conformational change that eliminates
the vicinal arrangement of the cysteine and pantetheine residues. In this
inactive conformation, dibromopropanone may no longer cross-link the
two residues, even though it reacts with the cysteine residue.

As discussed above, yeast FAS is very sensitive to inhibition by iodoacetamide, which reacts specifically with the active cysteine-SH of the β-
ketoacyl synthetase-component activity. The loss of FAS activity relative to
the amount of carboxamidomethyl groups bound to the enzyme has been
studied employing [^{14}C]iodoacetamide (115). Binding four to five mol of
carboxamidomethyl groups per mol of enzyme results in complete
inhibition of the synthetase. This value is somewhat lower than that
expected for six β-ketoacyl synthetase sites present in the $\alpha_6\beta_6$-structure of
yeast synthetase and may be explained on the basis that there are usually
four to five prosthetic groups (4'-phosphopantetheine) per mol of synthetase (92, 93). Thus, one or two sites of β-ketoacyl synthetase are
nonfunctional because they lack the prosthetic group. The presence of six

condensing sites per complex is supported by the finding that six carboxamidomethyl residues bind to the enzyme after complete reaction with [^{14}C]iodoacetamide (115). This result is in disagreement with the value of three reported earlier and is inconsistent with the proposal of half-site reactivity (114). Instead it supports the concept of full-site activity in the $\alpha_6\beta_6$-yeast structure and is in agreement with the results obtained from the dibromopropanone studies discussed below.

1,3-Dibromo-2-propanone inhibits yeast synthetase by reacting rapidly ($t_{1/2} = 7$ sec) with two juxtapositioned active sulfhydryl groups (12). SDS-PAGE of the dibromopropanone-inhibited synthetase shows the β-subunit to be intact and the α-subunit nearly absent, with a concomitant appearance of oligomers with an estimated M_r of 400,000 to 1.2×10^6. These results indicate that the α-subunits are cross-linked by the bifunctional reagent. Since the active centers of the dibromopropanone are 5Å apart, it is concluded that the α-subunits are closely packed and the reacting thiols of the adjacent α-subunits are juxtapositioned within 5Å of each other. Furthermore, since the plate-like structures in our model (Figure 6) are the only components arranged close enough to satisfy this requirement, it is proposed that the α-subunits are the plates and the β-subunits are the arches.

The identification of the residues with which dibromopropane reacts was analyzed in a similar manner to that described for the chicken enzyme. The results have led Stoops & Wakil (115) to propose sites of action of the dibromopropanone to be the active cysteine-SH of the β-ketoacyl synthetase of one α-subunit and the pantetheine-SH of the ACP moiety of an adjacent α-subunit. The active center of the β-ketoacyl synthetase would then consist of an acyl group attached to the cysteine-SH of one α-subunit (plate) and a malonyl group attached to the pantetheine-SH of an adjacent α-subunit (Figure 10). This arrangement appears to be necessary for the coupling of the acyl and β-carbon of the malonyl group, yielding CO_2 and the β-ketoacyl product. It may also explain the stringent requirements for the $\alpha_6\beta_6$-structure as the only active form of the enzyme (92). There exist six sites for β-ketoacyl synthetase in an $\alpha_6\beta_6$-structure, all of which may function simultaneously (Figure 10). This arrangement is a novel feature of our mechanism for the condensation reaction and for the synthesis of fatty acids by yeast FAS.

β-Ketoacyl and Enoyl Reductases

The two reductases exhibit high specificity for NADPH over NADH for all FAS, regardless of its source. The specific requirement of NADPH for these reactions suggests the possible involvement of an active arginine in its

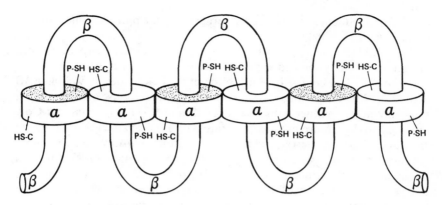

Figure 10 A linear drawing of the model shown in Figure 6 depicting the six sites of palmitate synthesis and the complementary arrangement of the 4'-phosphopantetheine-SH (P-SH) and the active cysteine-SH (HS-C) at the β-ketoacyl synthetase centers in each of these sites.

binding to the FAS. Poulose & Kolattukudy (120) showed that phenylglyoxal and 2,3-butanedione inactivates FAS of uropygial gland of goose by inhibiting the β-ketoacyl and enoyl reductases. $NADP^+$ and its 2'-phosphate-containing analogs partially protect the activities against inhibition by these reagents. The residue modified by phenylglyoxal is identified as arginine since acid hydrolysates of [^{14}C]phenylglyoxal-treated enzyme yielded the arginine adduct of phenylglyoxal. Binding analyses show that four mol of arginine per subunit are modified in order to obtain complete inactivation. Modification of the first two arginine residues has no effect on activity of either reductase or FAS, while modification of the next two results in complete inhibition of the reductases; presumably one arginine is modified at the β-keto reductase site and the other at the enoyl reductase site. The implication, therefore, is that the synthetase dimer consists of two identical subunits, each of which has the ketoacyl and enoyl reductase domains (120).

Recently it was shown that pyridoxal phosphate inhibits the enoyl reductase activity, presumably by forming a Schiff base with an active ε-amino group of a lysine residue at the active site (121). NADPH protects the enzyme against this inhibition by binding to this lysine. The formation of a Schiff base with the ε-amino group of a lysine is identified with labeled pyridoxal after reduction with NaB^3H_4. Binding analyses show that two lysine residues are modified per subunit. Reaction of one of these lysines has no effect on reductase activity; presumably this reactive lysine plays no role in catalysis. The second lysine is present on each of the synthetase subunits at the enoyl reductase site where it interacts with the pyrophosphate bridge of the NADPH.

Since $NADP^+$ protects both the β-ketoacyl and enoyl reductases from inhibition by the lysine and arginine directed inhibitors, it is not surprising that $NADP^+$ is a competitive inhibitor of both activities as well as FAS activity (122). Surprisingly, the condensing activity is inhibited in a noncompetitive manner with a value of K_i of 2.5 μM when malonyl-CoA and hexanoyl-CoA are used as substrates (122). This value is similar to the K_i value obtained for the reduction of crotonyl-CoA. When the enoyl reductase domain is inactivated by pyridoxal phosphate then $NADP^+$ no longer inhibits the condensing activity. From these results, it was proposed that the binding of $NADP^+$ to the enoyl reductase site results in a conformational change which causes the inactivation of β-ketoacyl synthetase, and that the functional interaction of the two domains may play a role in regulating fatty acid synthesis (122).

Removal of FMN from the yeast synthetase results in the loss of only the enoyl reductase activity (123). The flavin is associated with the β-subunit and six mol of flavin are bound per mol of synthetase, consistent with the $\alpha_6\beta_6$-structure of this enzyme. The enoyl reductase acts as a trans-hydrogenase by shuttling the hydride ion from NADPH to the double bonds of the fatty acyl group through FMN (124). No flavin free-radical intermediates are detected; apparently the reduction involves a two electron transfer (124).

The β-ketoacyl reductase component of the chicken liver FAS was recently isolated after sequential proteolysis of the complex with trypsin and subtilisin (H. Wong, S. J. Wakil, unpublished results). This enzyme proved to be more difficult to separate from the complex than the thioesterase component; it seems to bind tightly to other components, despite cleavages of the peptide bonds. Fractionation of the digest by ammonium sulfate and chromatography on a Procion Red affinity column permitted the isolation of a polypeptide (M_r 94,000) containing the β-ketoacyl reductase activity but no other partial activities normally associated with the synthetase. The specific activity of the β-ketoacyl reductase is increased two to three times in this fraction, an increase that is within the expected range based on the ratio of its M_r to that of the synthetase subunit and that indicates one β-ketoacyl reductase site per subunit. The kinetic parameters of the polypeptide towards NADPH and N-acetyl-S-acetoacetylcysteamine are essentially the same as those of the β-ketoacyl and reductase component of the intact synthetase. However, for reasons still unknown the purified fragment does not catalyze reduction of acetoacetyl-CoA, a substrate that is readily reduced by the intact synthetase.

The binding of NADP$^+$ to the isolated β-ketoacyl reductase fragment was also investigated. Fluorescence measurements with etheno-NADP$^+$ indicate the presence of about one NADP$^+$ binding site per 90,000 daltons, which is in agreement with the results obtained from fluorescence measurements with NADPH and the binding of a photoaffinity analog of NADP$^+$ ([3-^{3}H]arylazido-β-alanyl-NADP$^+$). Phenylglyoxal inhibits the β-ketoacyl reductase activity of the fragment, suggesting the involvement of an essential arginine at or near the active site.

β-Hydroxacyl Dehydratase

Little is known about the mechanism of this component activity. In the proteolytic and functional map it has been tentatively located at the "domain II" of the synthetase (Figures 2, 3).

Palmitoyl Thioesterase

The thioesterase domain is readily split from the animal synthetases by a number of proteases as described above. The enzyme has been purified from the synthetases derived from chicken liver (86), rat liver and mammary gland (125), and uropygial gland of goose (89). These enzymes have nearly the same mol wt (M_r 33,000) and exhibit the same chain length specificity towards long-chain acyl-CoA derivatives. Highest activity was shown with palmitoyl-CoA and, to a lesser extent, with stearoyl-CoA and myristoyl-CoA (86, 89). The thioesterases have an active serine residue that reacts readily with diisopropylfluorophosphate or phenylmethanesulfonyl fluoride, resulting in inhibition of the activity (86, 89, 126, 127). Studies of the binding of labeled diisopropylfluorophosphate to the synthetase indicate that there are two thioesterases per mol of native synthetase or one thioesterase domain per FAS subunit. The peptide containing the diisopropyl derivative of the active serine residue, isolated from the uropygial gland FAS, has the sequence Ser-Phe-Gly-Ala-Cys-Val-Ala-Phe (126). This sequence has a high degree of homology to those of proteases with active serine residues—human plasmin and bovine trypsin and carboxylesterase —suggesting evolutionary linkages among these hydrolases (126).

MECHANISM OF ACTION OF FATTY ACID SYNTHETASE

The recent observations concerning structure and function of the synthetase subunits necessitate a reevaluation of the mechanism of action of the FAS of both animal tissues and yeast. The proposed mechanisms for the animal and yeast synthetases are similar and are consistent with the known data, including those of electron microscopy.

Animal Synthetase

Since the subunits of the animal FAS are identical, with each subunit containing the same catalytic domains, including the active cysteine-SH of the β-ketoacyl synthetase and 4'-phosphopantetheine-SH of the ACP site, it is proposed that, in the dimer state, the two subunits are arranged in head-to-tail fashion (7, 115), as shown in Figure 11. This arrangement predicts the presence of two centers of β-ketoacyl synthetase (7, 115) and, therefore, two centers of palmitate synthesis. Recent studies using the core complex of the 230,000-dalton peptide allow the estimation of the stoichiometry of NADPH oxidation and fatty acids synthesized relative to the pantetheine content of the dimer. The results show that, in the absence of thioesterase, the core enzyme continues the chain elongation and reduction processes until fatty acids of C_{20} and C_{22} are synthesized as a limit (N. Singh, J. K. Stoops, S. J. Wakil, unpublished data). Palmitate and stearate, normally synthesized, are formed as minor products under these conditions,

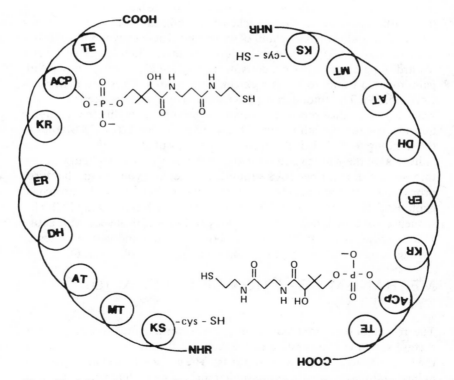

Figure 11 Schematic representation of the two multifunctional polypeptides and their head-to-tail association to form the enzymatically active homodimer. The abbreviations used (Figure 3) represent the activity domains on each of the polypeptides.

incidating that these fatty acyl groups are still attached to the pantetheine-SH and are further elongated to the C_{20}- and C_{22}-acids. The chain-terminating process in the native synthetase, therefore, is dictated by the thioesterase, which has the highest activity for palmitoyl and stearoyl thioesters. The data show that one mol of long-chain fatty acid is synthesized per mol of pantetheine associated with the core dimer; therefore, the two centers of palmitate synthesis are active simultaneously.

Libertini & Smith (128) studied fatty acid synthesis with rat mammary gland synthetase modified either by removal of thioesterase or by inhibiting the thioesterase with phenylmethanesulfonyl fluoride. The modified enzyme synthesized C_{16}–C_{22}-fatty acids covalently bound to the protein and a single enzyme-bound long-chain acyl thioester was formed by each molecule of modified synthetase dimer. Since they do not determine the pantetheine content of their preparation, it is not possible to answer from their studies the question of half-site or full-site reactivity.

Our studies predict that two centers may function independently of each other and may engage catalytic domains on the two subunits. In this arrangement (Figures 3, 11), each center has the entire complement of enzymes. Based on these results, the following mechanism for palmitate synthesis (Figure 12) is proposed (7): The active FAS in the dimer form interacts with the substrates acetyl-CoA and malonyl-CoA. The active cysteine-SH of the condensing site is charged with the acetyl group and the cysteamine-SH of the acyl carrier site is charged with the malonyl group via their respective transacylases (steps 1 and 2, Figure 12). The acetyl group of one subunit is coupled to the β-carbon of the malonyl group of the second subunit with a simultaneous release of CO_2 and the formation of acetoacetyl product. The cysteine-SH of the condensing enzyme is reset in the free thiol form. The acetoacetyl-S-pantetheine derivative is then processed as outlined in Figure 12 ultimately yielding palmitic acid. The essence of this mechanism is the involvement of the two subunits in the condensation reaction, where the acyl group "seesaws" between the cysteine-SH and cystamine-SH of the two subunits with each addition of a C_2-unit. In addition each active synthetase dimer has its own complement of enzymes and perhaps functions independently of the other. The multifunctional nature of the subunit and its organization and structural arrangement into the dimer form has produced a highly efficient enzyme capable of carrying out sequentially and repetitively a total of 37 reactions in the synthesis of a molecule of palmitate from acetyl-CoA and malonyl-CoA.

Yeast Synthetase

As stated earlier, the $\alpha_6\beta_6$-structure is the oligomer active in palmitate synthesis. In this structure, a palmitate-synthesizing center consists of

568

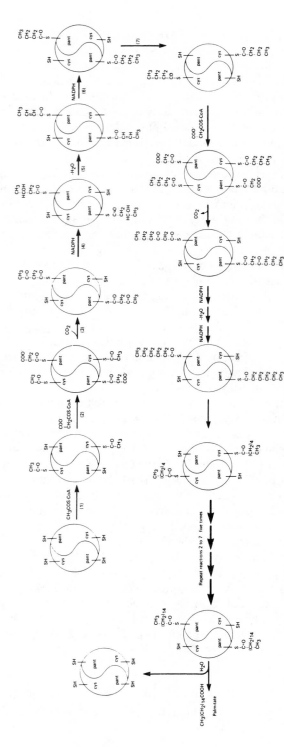

Figure 12 Scheme illustrating proposed mechanism for palmitate synthesis. The circles represent the multifunctional subunits of the fatty acid synthetase in its homodimer form. The cys-SH represents the active cysteine thiol of the β-ketoacyl synthetase site and pant-SH represents the pantetheine thiol of the acyl carrier protein site. Other catalytic domains are not shown and are presumed to be present in both subunits (cf Figures 3, 11).

complementary halves of two α-subunits and the arch β-subunit (Figure 10). In an $\alpha_6\beta_6$-structure, therefore, there are six sites for fatty acid synthesis, all of which function simultaneously (115). Condensation occurs by engaging the acyl-S-cysteine of one α-subunit and the malonyl-S-pantetheine of the second α-subunit; chain elongation occurs by transferring the acyl group back and forth between the pantetheine-SH and the cysteine-SH of the two complementary halves of the α-subunits. In essence this mechanism is analogous to the one proposed for animal synthetase (Figure 12). The condensation reaction yields the β-ketoacyl-S-pantetheine derivative, which is then reduced by NADPH to the β-hydroxy homolog at the β-keto reductase site of an α-subunit. Dehydration of the β-hydroxyacyl derivative by the dehyratase of the arch β-subunit yields the α-β unsaturated acyl homolog, which is then reduced by NADPH through the FMN of the enoyl reductase of the β-subunit to the saturated acyl derivative. The latter is then transferred from the pantetheine-SH to the active cysteine-SH of the β-ketoacyl synthetase of the α-subunit, where the acyl group was bound before condensation. The free pantetheine-SH is then recharged with another malonyl group and the sequence commences again. The sequential reactions are repeated until the acyl chain is elongated to 16 or 18 carbons, which are then transferred to CoA-SH by the palmitoyl transferase located on the arch β-subunit.

REGULATION OF FATTY ACID SYNTHESIS

Overall regulation of lipogenesis has recently been reviewed (129); we discuss here regulation of acetyl-CoA carboxylase and FAS with emphasis on recent findings. The activities of the enzymes involved in fatty acid synthesis appear to be controlled in two ways: (*a*) short-term or acute control, which involves allosteric or metabolic regulation, and covalent modification of enzymes; and (*b*) long-term control, involving changes in the amounts of the enzymes brought about by changes in the rates of synthesis and degradation.

Acetyl-CoA Carboxylase

METABOLITE CONTROL   Acetyl-CoA carboxylase catalyzes the committed step in the de novo synthesis of fatty acids and provides the earliest unique point at which control can be exerted. In animal tissues the precursor of cytoplasmic acetyl-CoA utilized for fatty acid synthesis is mitochondrical citrate (51, 130) whereas, in *E. coli*, acetyl-CoA is derived directly from pyruvate. Animal acetyl-CoA carboxylase is regulated by citrate and by long-chain acyl-CoAs. The interaction of citrate and other tricarboxylic acids with acetyl-CoA carboxylase has been discussed in previous reviews (16, 24, 130).

The crucial role of citrate in the regulation of fatty acid synthesis in avian hepatocytes (15, 131–135) can be summarized as follows: glucagon and dibutyryl cyclic AMP, which inhibit fatty acid synthesis, markedly decrease cytosolic citrate concentration; they also inhibit glycolysis at the level of phosphofructokinase, resulting in decreased glycolytic flux into pyruvate, which in turn decreases the mitochondrial synthesis of oxaloacetate and citrate. The decrease in citrate concentration results in a decrease in acetyl-CoA carboxylase activity.

Fatty acyl-CoAs, on the other hand, regulate fatty acid synthesis by inhibiting acetyl-CoA carboxylase as a result of its depolymerization. Both the binding of the acyl-CoA and depolymerization of the carboxylase is competitive with citrate and is reversed by albumin (136, 137). One acyl-CoA binds per carboxylase subunit with a dissociation constant of 5 nM (138, 139). In the presence of citrate, phosphatidylcholine (138) or 6-O-methylglucose polysaccharide (139) reverse the inhibition. Palmitoyl-CoA, stearoyl-CoA, and arachidyl-CoA are most effective in inhibiting the carboxylase (140). Palmitoyl-dephospho-CoA is a poor inhibitor with a 40-fold higher K_i than palmitoyl-CoA (140). These results indicate that there is a specific binding site for the acyl-CoA dependent on the chain length of the fatty acid and the 3'-phosphate of CoA. The regulation of the carboxylase by the long-chain acyl-CoA is reciprocal to that of citrate and may reflect the physiological role of these metabolites in regulating fatty acid synthesis.

Other effectors such as guanine nucleotides (141) and CoA (142, 143) have been implicated in the regulation of acetyl-CoA carboxylase. Guanine nucleotides stimulate carboxylase activity in dialyzed rat liver supernatant preparations at physiological concentration, yet have no effect on purified enzyme and the effect is therefore of uncertain significance. Physiological concentrations of CoA can activate purified acetyl-CoA carboxylase (142, 143). Activation of the partially purified enzyme by CoA is accompanied by a change in the value of K_m for acetyl-CoA from 0.2 mM to about 4 μM and by polymerization of the enzyme independent of citrate. There is one CoA binding site per subunit of acetyl-CoA carboxylase with a $K_d = 170 \ \mu$M. This binding is inhibited by palmitoyl-CoA; however, this inhibitory effect is reversed by CoA, together with bovine serum albumin. Changes in the cytosolic concentrations of CoA have not been related to changes in activity of acetyl-CoA carboxylase in situ; the physiological significance of CoA in the regulation of acetyl-CoA carboxylase remains to be established.

COVALENT MODIFICATION   Kim (144) first reported the regulation of acetyl-CoA carboxylase by phosphorylation-dephosphorylation of the protein. The hypothesis was attractive because it not only explained the short-term regulation of fatty acid synthesis but also tied such regulation to that of carbohydrate regulation through the cyclic nucleotides. However, the initial observation was made with partially purified rat liver acetyl-CoA carboxylase preparations and many investigators were unable to confirm it with highly purified enzyme (16, 145, 146).

More recently, this problem was reinvestigated employing highly purified kinases and phosphatases. Lent & Kim (17) found that a cyclic AMP-independent kinase from rat liver interacts strongly with acetyl-CoA carboxylase and copurifies with it in the early stages of isolation. On incubation with a homogeneous preparation of the kinase, acetyl-CoA carboxylase is completely inactivated when one mol of phosphate from ATP is incorporated per mol of carboxylase subunit, and is reactivated by incubation with phosphorylase phosphatase. The purified kinase phosphorylates acetyl-CoA carboxylases of rat liver and mammary gland and goose uropygial gland. Its K_m value for rat liver acetyl-CoA carboxylase (90 nM) is 10–50 times lower than for the phosphorylation of histones and protamines, indicating preference for the carboxylase. Coenzyme A may be an important factor in this inactivation, since in its absence, the ability of the kinase to inactivate the carboxylase is considerably reduced, whereas 100 μM CoA markedly enhances kinase action. The role of CoA in this inactivation is thus opposite to its activation of the carboxylase discussed above (142, 143).

Another kinase, a cyclic AMP-dependent protein kinase purified from rabbit skeletal muscle, inactivates rat mammary gland acetyl-CoA carboxylase (18). Both phosphorylation and inactivation of the carboxylase are blocked by the heat-stable protein inhibitor of cyclic AMP-dependent protein kinase and can be reversed by incubation with purified protein phosphatase-1 from rabbit skeletal muscle. Phosphorylation of the carboxylase does not change the apparent K_m values for its substrates, but reduces its catalytic efficiency by half and causes a twofold decrease in its affinity for citrate. The kinase incorporates more than one mol of phosphate into the carboxylase, but the 60% decrease in carboxylase activity is correlated only with the phosphorylation of a serine residue located in a tryptic peptide (147). This regulation may be physiologically significant because the phosphorylation of this same tryptic peptide is increased sixfold by adrenaline in the fat cell. This observation may explain the role of cyclic AMP as an intermediate in the inactivation of acetyl-CoA carboxylase by adrenaline in adipose tissue. Hardie & Guy (18) have reported that the activation of a low specific activity form of rabbit mammary gland acetyl-CoA carboxylase by protein phosphatase-1 is due to dephosphorylation at site(s) phosphorylated neither by endogeneous cyclic AMP-dependent protein kinase nor by cyclic AMP-independent protein kinase-2. These results indicate at least one other protein kinase in lactating rabbit mammary gland that phosphorylates and inactivates acetyl-CoA carboxylase.

In addition to β-adrenergic-mediated inactivation of acetyl-CoA carboxylase through the action of cyclic AMP-dependent protein kinase (18), the inhibitory effect of catecholamines on carboxylase in rat hepatocytes apparently is mediated through α-adrenergic receptor and calcium-dependent protein kinase (148). The twofold inactivation of acetyl-CoA carboxylase in hepatocytes by norepinephrine and phenylephrine is correlated with phosphate incorporation into the enzyme and is blocked by α-adrenergic blockers, phentolamine, and phenoxybenzamine. This type of regulation is reminiscent of the regulation of glycogen phosphorylase by the cyclic AMP-independent, but calcium-dependent, protein kinase. It is not known whether the three different protein kinases that inactivate acetyl-CoA carboxylase catalyze phosphorylation of three distinct and different sites.

The acetyl-CoA carboxylase phosphatase has been purified from rat epididymal fat pad; it dephosphorylates the carboxylase, thereby increasing its activity twofold (149). Like the cyclic AMP-independent acetyl-CoA carboxylase kinase, it occurs in a complex with the carboxylase. The phosphatase has a broad substrate specificity and acts on glycogen synthetase, phosphorylase a, HMG-CoA reductase, phosphoprotamine, and p-nitrophenylphosphate, in addition to acetyl-CoA carboxylase from adipose tissue and liver. Its K_m for carboxylase (1.5 μM) is 25 times lower than that for glycogen phosphorylase a. The phosphatase surprisingly requires no metal ion for activity and is not inhibited by the inhibitor protein for the rat liver phosphorylase phosphatase. The phosphatase appears to be specifically involved in the regulation of acetyl-CoA carboxylase since it exists as a complex with the carboxylase. The observations that endogenous phosphatase removes only 35% of the phosphate from carboxylase and that further treatment with exogenous phosphatase removes additional phosphate groups with corresponding increase in carboxylase activity suggest that more than one phosphatase might function, in vivo, for the removal of different types of phosphates in the overall regulation of acetyl-CoA carboxylase.

Exposure of rat fat cells or hepatocytes to insulin for a few minutes results in a twofold increase in the activity of acetyl-CoA carboxylase assayed immediately upon breaking open the cells in the absence of added citrate (150). The mechanism by which insulin stimulates acetyl-CoA carboxylase activity is not clear but could involve dephosphorylation of the less active enzyme (151). However, tryptic peptide analysis of the adipose tissue enzyme suggests that insulin causes increased phosphorylation of a site other than those phosphorylated by cyclic AMP-dependent protein kinase (152). A plasma membrane fraction from the rat adipose tissue catalyzes cyclic AMP-independent, but ATP-dependent, activation of partially purified adipose acetyl-CoA carboxylase, assayed in the absence of citrate. The protein kinase in the plasma membrane does not appear to require calcium for activity. Although insulin is known to stimulate phosphory-lation of several proteins in adipocytes, further work is required to support the thesis that insulin stimulates the plasma membrane protein kinase, which in turn activates acetyl-CoA carboxylase.

ADAPTIVE CHANGES IN ENZYME CONTENT

Animal   Long-term regulation of acetyl-CoA carboxylase is affected by diet, insulin, and thyroxine. It also occurs during cell differentiation and development. In all cases the changes in activity can be related to variations in the rate of synthesis of the enzyme (for review see 19).

Yeast   Growth of *S. cerevisiae* (153) and *C. lipolytica* (154) in media containing saturated, monounsaturated, or polyunsaturated fatty acids resulted in an 80% decrease in the amount of acetyl-CoA carboxylase. This decrease is caused by the reduction of mRNA coding for the enzyme (155). Numa et al (156–160) isolated mutant strains of *S. cerevisiae* defective in acyl-CoA synthetase and showed that yeast cells possess two distinct acyl-CoA synthetases. Acyl-CoA synthetase I produces long-chain acyl-CoA, which is utilized exclusively for the synthesis of cellular lipids. Acyl-CoA synthetase II provides long-chain acyl-CoA, which is specifically degraded via β-oxidation. The activation of exogenous fatty acids is catalyzed by acyl-CoA synthetase I only and is involved in the repression of acetyl-CoA carboxylase. Thus, long-chain acyl-CoA has a dual role in the regulation of acetyl-CoA carboxylase, acting not only as a putative co-repressor, but also as a specific allosteric inhibitor of the enzyme.

Fatty Acid Synthetase

METABOLIC CONTROL   Little is known about short-term regulation of FAS of animal tissues. Several hypotheses have been advanced during the past decade, including the role of phosphorylated sugars, especially fructose 1,6-diphosphate, in the stimulation of the synthetase and the inhibitory effect of palmitoyl-CoA on the yeast and animal synthetases (for review see 1). As discussed above, NADPH reactivates a cold-inactivated chicken liver FAS by restoring the β-ketoacyl synthetase activity to the synthetase dimer (118). The activation is relatively fast (< 5 min) and involves conforma-tional changes of the synthetase. $NADP^+$ is also an inhibitor of FAS and may play a role in its regulation (122).

ADAPTIVE CHANGES OF ENZYME CONTENT   Evidence indicates that the synthesis of FAS is a well-regulated process, affected by diet and hormones.

It is also regulated during neonatal development and differentiation (for review see 1, 19).

The increase in synthetase activity on refeeding or insulin administration has been directly related to increased amounts of synthetase as determined by immunotitration of the enzyme with antisynthetase antibodies (1, 19). Recently Goodridge's group (75) measured directly the synthetase mRNA levels in goose liver using DNA probes isolated from cloned plasmids containing cDNA sequences prepared from synthetase mRNA. On alimentation of unfed neonatal goslings, the rate of hepatic FAS synthesis increased 42-fold. Concomitantly, hepatic FAS mRNA, as estimated by the dot-blot hybridization analysis, increased 70-fold.

The murine 3T3-L1 cell line developed by Green et al (161, 162) has provided an excellent model system for studying the molecular mechanisms involved in the hormonal and metabolic regulation of major lipogenic enzymes. During adipocyte conversion, 3T3-L1 cells exhibit a coordinate rise in the enzymes of fatty acid (163–167) and triacyl-glycerol syntheses (168). The differentiation-dependent increase in the activities of acetyl-CoA carboxylase and FAS results from an increase in the amount of both enzymes caused by an enhanced rate of synthesis (163–167). Kasturi & Joshi (169) have demonstrated that 3T3-L1 cell differentiation and induction of FAS activity are both inhibited by anti-insulin antibodies, these inhibitions being relieved by the addition of exogenous insulin. These results indicate that insulin supports phenotypic expression as well as the induction of lipogenesis and FAS activity during adipose conversion of 3T3-L1 cells.

To further define factors involved in the regulation of synthetase, Joshi et al (170–173) characterized an avian liver explant culture system that is maintained in a serum-free chemically defined culture medium. In this system, physiological concentrations of insulin induce FAS fivefold. Triiodothyronine or hydrocortisone potentiate insulin induction of synthetase individually twofold and together fourfold. These hormones in the absence of insulin do not induce synthetase. Furthermore, studies of avian hepatocytes cultured in serum-free medium showed that both insulin and triiodothyronine (174) were required for maximal induction of FAS.

Recently, Joshi & Kasturi (175) observed that antimicrotubular agents such as colchicine, colcemid, and vinblastin inhibit insulin induction of FAS and stearoyl-CoA desaturase in cultured liver explants. Colchicine at 0.15–0.2 μM in the culture medium caused 50% inhibition of insulin induction of the two lipogenic enzymes, and at 1 μM, produced 90% inhibition. Colchicine has no effect on general protein synthesis, but specifically inhibits insulin-induced synthesis of synthetase and Δ^9 desaturase. Malic enzyme, which is specially induced by triiodothyronine but not by insulin

(170), is not inhibited by colchicine, suggesting that colchicine interferes with the induction of the synthetase by insulin. This action may be related to the depolymerization of microtubules by colchicine, since lumicolchicine, an inactive isomer, does not inhibit insulin induction of lipogenic enzymes.

SYNTHESIS OF BRANCHED AND MEDIUM-CHAIN FATTY ACIDS

Fatty acid synthesis catalyzed by animal FAS terminates with the release of free fatty acids, rather than transacylation to CoA as in yeast and in mycobacteria. The product of the former reaction is mostly palmitate and stearate and is determined by the specificity of the thioesterase and the activity of the β-ketoacyl synthetase. In the uropygial gland of goose, the synthesis of the branched fatty acids 2,4,6,8-tetramethyldecanoic acid and 2,4,6,8-tetramethylundecanoic acid has been described (176, 177). These acids are synthesized by FAS, utilizing methylmalonyl-CoA instead of malonyl-CoA. The rate of their synthesis is 1/100 that of palmitate synthesis (178). The methylmalonyl-CoA utilized in this reaction is produced by the carboxylation of propionyl-CoA. A cytosolic decarboxylase eliminates malonyl-CoA, thereby favoring the formation of branched-chain fatty acids from methylmalonyl-CoA (179, 180).

Another system in which the products of FAS are modified is described in the lactating mammary gland. In this system medium-chain as well as long-chain fatty acids are synthesized (181). This modification is brought about by a thioesterase termed thioesterase II to distinguish it from the thioesterase component of the FAS (181–186). This enzyme has been purified to homogeneity, has an M_r of 33,000, and hydrolyzes fatty acyl-CoA derivatives of chain length C_8 to C_{16} (186). Addition of this thioesterase to FAS results in early termination of fatty acid synthesis and production of C_8 to C_{12} fatty acids. This thioesterase is predominant in mammary tissues, where it plays a decisive role in determining the fatty acid content of milk.

ACKNOWLEDGMENTS

This work was supported in part by grants from the National Institutes of Health (GM 19091, AM 21286, and HL 17269), the National Science Foundation (PCM 77-00969), and the Robert A. Welch Foundation (Q-587). We wish to dedicate this chapter to the memory of Dr. Vasudev C. Joshi, who passed away on October 16, 1982.

Literature Cited

1. Stoops, J. K., Arslanian, M. J., Chalmers, J. H. Jr., Joshi, V. C., Wakil, S. J. 1977. *Bioorg. Chem.* 1:339–70
2. Stoops, J. K., Arslanian, M. J., Oh, Y. H., Aune, K. C., Vanaman, T. C., Wakil, S. J. 1975. *Proc. Natl. Acad. Sci. USA* 72:1940–44
3. Schweizer, E., Kniep, B., Castorph, H., Holzner, U. 1973. *Eur. J. Biochem.* 39:353–62
4. Schweizer, E., Dietlein, G., Gimmler, G., Knobling, A., Tahedl, H. W., Schweizer, M. 1975. *Proc. FEBS Meet.* 40:85–97
5. Tanabe, T., Wada, K., Okazaki, T., Numa, S. 1975. *Eur. J. Biochem.* 57:15–24
6. Kirschner, K., Bisswanger, H. 1976. *Ann. Rev. Biochem.* 45:143–66
7. Wakil, S. J., Stoops, J. K., Mattick, J. S. 1981. *Cardiovasc. Res. Cent. Bull.* (Houston) 20:1–23
8. Buckner, J. S., Kolattakudy, P. E. 1976. *Biochemistry* 15:1948–57
9. Smith, S., Stern, A. 1979. *Arch. Biochem. Biophys.* 197:379–87
10. Stoops, J. K., Wakil, S. J. 1978. *Biochem. Biophys. Res. Commun.* 84:225–31
11. Lynen, F. 1973. *Methods Enzymol.* 14:17–33
12. Stoops, J. K., Wakil, S. J. 1980. *Proc. Natl. Acad. Sci. USA* 77:4544–48
13. Stoops, J. K., Wakil, S. J. 1981. *J. Biol. Chem.* 256:5128–33
14. Wieland, F., Renner, L., Verfurth, C., Lynen, F. 1979. *Eur. J. Biochem.* 94:189–97
15. Lane, M. D., Watkins, P. A., Meredith, M. J. 1979. *CRC Crit. Rev. Biochem.* 7:121–41
16. Lane, M. D., Moss, J., Polakis, S. E. 1974. *Curr. Top. Cell. Regul.* 8:129–95
17. Lent, B., Kim, K.-H. 1982. *J. Biol. Chem.* 257:1897–901
18. Hardie, D. G., Guy, P. S. 1980. *Eur. J. Biochem.* 110:167–77
19. Valope, J. J., Vagelos, P. R. 1976. *Physiol. Rev.* 56:339–417
20. Bloch, K., Vance, D. 1977. *Ann. Rev. Biochem.* 46:263–98
21. Wakil, S. J., Titchener, E. B., Gibson, D. M. 1958. *Biochim. Biophys. Acta* 29:225–26
22. Kaziro, Y., Ochoa, S. 1964. *Adv. Enzymol.* 26:283–378
23. Lynen, F. 1967. *Biochem. J.* 102:381–400
24. Moss, J., Lane, M. D. 1971. *Adv. Enzymol.* 35:321–42
25. Retey, S., Lynen, F. 1965. *Biochem. Z.* 342:256–71
26. Rose, I.-A., O'Connell, E. J., Solomon, F. 1976. *J. Biol. Chem.* 251:902–4
27. Mildvan, A. S., Scrutton, M. D., Utter, M. F. 1966. *J. Biol. Chem.* 241:3488–98
28. Stallings, W. C., Monti, C. T., Lane, M. D., DeTitta, G. T. 1980. *Proc. Natl. Acad. Sci. USA* 77:1260–64
29. Northrop, D. B. 1969. *J. Biol. Chem.* 244:5808–19
30. Lynen, F. 1979. *CRC Crit. Rev. Biochem.* 7:103–19
31. Wood, H. G., Zwolinski, G. K. 1976. *CRC Crit. Rev. Biochem.* 5:47–122
32. Fung, C. H., Mildvan, A. S., Leigh, J. S. 1974. *Biochemistry* 13:1160–69
33. Fung, C. H., Feldmann, R. J., Mildvan, A. S. 1976. *Biochemistry* 15:75–84
34. Fung, C. H., Gupta, R. K., Mildvan, A. S. 1976. *Biochemistry* 15:85–92
35. DeTitta, G. T., Parthasarathy, R., Blessing, R. H., Stalling, W. 1980. *Proc. Natl. Acad. Sci. USA* 77:333–37
36. Mackall, J., Lane, M. D., Leonard, K. R., Pendergast, M., Kleinschmidt, A. K. 1978. *J. Mol. Biol.* 123:595–606
37. Fischer, P. W. F., Goodridge, A. G. 1978. *Arch. Biochem. Biophys.* 190:332–44
38. Ahmad, F., Ahmad, P. M., Pieretti, L., Watters, G. T. 1978. *J. Biol. Chem.* 253:1733–37
39. Hardie, D. G., Cohen, P. 1978. *FEBS Lett.* 91:1–7
40. Manning, R., Dils, R. D., Mayer, R. J. 1976. *Biochem. J.* 153:463–68
41. Guchhait, R. B., Zwergel, E. E., Lane, M. D. 1974. *J. Biol. Chem.* 249:4776–80
42. Inoue, H. D., Lowenstein, J. M. 1972. *J. Biol. Chem.* 247:4825–32
43. Song, C. S., Kim, K.-H. 1981. *J. Biol. Chem.* 256:7786–88
44. Nakanishi, S., Tanabe, T., Harikawa, S., Numa, S. 1976. *Proc. Natl. Acad. Sci. USA* 73:2304–7
45. Beaty, N. B., Lane, M. D. 1982. *J. Biol. Chem.* 257:924–29
46. Warren, G. G., Tipton, K. F. 1974. *Biochem. J.* 139:311–20
47. Easterbrook-Smith, S. B., Wallace, J. C., Keech, D. B. 1978. *Biochemistry* 169:225–28
48. Sumper, M., Riepertinger, C. 1972. *Eur. J. Biochem.* 29:237–48
49. Mishina, M., Kamiryo, T., Tanaka, A., Fukui, S., Numa, S. 1976. *Eur. J. Biochem.* 71:295–300
50. Obermayer, M., Lynen, F. 1976. *Trends Biochem. Sci.* 1:169–71
51. Wakil, S. J. 1970. In *Lipid Metabolism*, ed. S. J. Wakil, pp. 1–48. New York: Academic

52. Shimakata, T., Stumpf, P. K. 1982. *Arch. Biochem. Biophys.* 217:144–54
53. Shimakata, T., Stumpf, P. K. 1982. *Plant Physiol.* 69:1257–62
54. Lynen, F. 1964. In *New Perspectives in Biology,* ed. M. Sela, 4: 132–46. Amsterdam: Elsevier
55. Willecke, K., Ritter, E., Lynen, F. 1969. *Eur. J. Biochem.* 8:503–9
56. Yang, C. P., Butterworth, P. H. W., Bock, R. M., Porter, J. W. 1967. *J. Biol. Chem.* 242:3501–7
57. Lornitzo, F. A., Qureshi, A. A., Porter, J. W. 1974. *J. Biol. Chem.* 249:1654–56
58. Roncari, D. A. K. 1974. *J. Biol. Chem.* 249:7035–37
59. Bratcher, S. C., Hsu, R. Y. 1975. *Biochem. Biophys. Acta* 410:229–35
60. Qureshi, A. A., Lornitzo, F. A., Hsu, R. Y., Porter, J. W. 1976. *Arch. Biochem. Biophys.* 177:379–93
61. Stoops, J. K., Ross, P. R., Arslanian, M. J., Aune, K. C., Wakil, S. J., Oliver, R. M. 1979. *J. Biol. Chem.* 254:7418–26
62. Lin, C. Y., Smith, S. 1978. *J. Biol. Chem.* 253:1954–62
63. Arslanian, M. J., Stoops, J. K., Oh, Y. H., Wakil, S. J. 1976. *J. Biol. Chem.* 251:3194–96
64. Stoops, J. K., Arslanian, M. J., Aune, K. C., Wakil, S. J. 1978. *Arch. Biochem. Biophys.* 188:348–59
65. Paskin, N., Mayer, J. R. 1976. *Biochem. J.* 159:181–84
66. Wood, W. I., Peterson, D. O., Block, K. 1978. *J. Biol. Chem.* 253:2650–56
67. Jacob, E. J., Butterworth, P. H. W., Porter, J. W. 1968. *Arch. Biochem. Biophys.* 124:392–400
68. Roncari, D. A. K., Bradshaw, R. A., Vagelos, P. R. 1972. *J. Biol. Chem.* 247:6234–42
69. Roncari, D. A. K. 1974. *Can. J. Biochem.* 52:221–30
70. Smith, S., Abraham, S. 1970. *J. Biol. Chem.* 245:3209–17
71. Zehner, Z. E., Mattick, J. S., Stuart, R., Wakil, S. J. 1980. *J. Biol. Chem.* 255:9519–22
72. Mattick, J. S., Zehner, Z. E., Calabro, M. A., Wakil, S. J. 1981. *Eur. J. Biochem.* 114:654–51
73. Flick, P. K., Chen, J., Alberts, A. W., Vagelos, P. R. 1978. *Proc. Natl. Acad. Sci. USA* 75:730–34
74. Nepokroeff, C. M., Porter, J. W. 1978. *J. Biol. Chem.* 253:2279–83
75. Morris, S. M. Jr., Nilson, J. H., Jenik, R. A., Winberry, L. K., McDevitt, M. A., Goodridge, A. G. 1982. *J. Biol. Chem.* 257:3225–29
76. Butterworth, P. H. W., Yang, P. C., Bock, R. M., Porter, J. W. 1967. *J. Biol.*

Chem. 232:3508–16
77. Muesing, R. A., Lornitzo, F. A., Kumar, S., Porter, J. W. 1975. *J. Biol. Chem.* 250:1814–23
78. Yung, S., Hsu, R. Y. 1972. *J. Biol. Chem.* 247:2689–98
79. Wetlauter, D. B. 1973. *Proc. Natl. Acad. Sci. USA* 70:697–701
80. Porter, R. R. 1959. *Biochem. J.* 73:119–26
81. Setlow, P., Brutlag, D., Kornberg, A. 1972. *J. Biol. Chem.* 247:224–32
82 Pabo, C. O., Sauer, R. T., Sturtevant, J. M., Ptashne, M. 1979. *Proc. Natl. Acad. Sci.* 76:1608–12
83. Davidson, J. N., Rumsby, P. C., Tamaren, J. 1981. *J. Biol. Chem.* 256:5220–25
84. Smith, S., Stern, A. 1979. *Arch. Biochem. Biophys.* 197:379–87
85. Guy, P., Law, S., Hardie, G. 1978. *FEBS Lett.* 94:33–37
86. Crisp, D., Wakil, S. J. 1982. *J. Prot. Chem.* 1:241–55
87. Agradi, E., Libertini, L., Smith, S. 1976. *Biochem. Biophys. Res. Commun.* 68:894–900
88. Crisp, D. 1976. *Fed. Proc.* 35:1500
89. Bedord, C. J., Kolattukudy, P. E., Rogers, L. 1978. *Arch. Biochem. Biophys.* 186:139–51
90. Tsukamoto, Y. 1982. *Fed. Proc.* 41:1026
91. Lynen, F. 1967. *Organizational Biosynthesis,* ed. H. J. Vogel, J. O. Lampen, V. Bryson, pp. 243–66. New York/London: Academic
92. Stoops, J. K., Awad, E. S., Arslanian, M. J., Gunsberg, S., Wakil, S. J., Oliver, R. M. 1978. *J. Biol. Chem.* 253:4464–75
93. Schweizer, E., Piccinini, F., Duba, C., Gunther, S., Ritter, E., Lynen, F. 1970. *Eur. J. Biochem.* 15:483–99
94. Schreckenbach, T., Wobser, H., Lynen, F. 1977. *Eur. J. Biochem.* 80:13–23
95. Vanaman, T. C., Wakil, S. J., Hill, R. L. 1968. *J. Biol. Chem.* 243:6234–42
96. Lynen, F. 1980. *Eur. J. Biochem.* 112:431–42
97. Wieland, F., Siess, E. A., Renner, L., Verfurth, C., Lynen, F. 1978. *Proc. Natl. Acad. Sci. USA* 75:5792–96
98. Wakil, S. J., Kuziora, M. A. 1983. *Proc. Inst. Conf. on Manipulation and Expression of Genes in Eukaryotes,* ed. A. W. Nagley, W. J. Peacock, J. A. Pateman, pp. 131–40. Sydney: Academic
99. Chalmers, J. H. Jr., Hitzeman, R. A. 1980. *Fed. Proc.* 39:1829
100. Abdinejad, A., Fisher, A. M., Kumar, S. 1981. *Arch. Biochem. Biophys.* 208:135–45
101. Aprahamian, S. A., Arslanian, M. J.,

Wakil, S. J. 1982. *Comp. Biochem. Physiol. B* 71:577–82

102. Dodds, P. F., Guzman, M. G. F., Chalberg, S. C., Anderson, G. J., Kunar, S. 1980. *J. Biol. Chem.* 256:6282–90

103. Engeser, H., Hubner, K., Straub, J., Lynen, F. 1979. *Eur. J. Biochem.* 101:413–22

104. Ziegenhorn, J., Niedermeier, R., Nussler, C., Lynen, F. 1972. *Eur. J. Biochem.* 30:285–300

105. Knobling, A., Schiffmann, D., Sickinger, H. S., Schweizer, E. 1975. *Eur. J. Biochem.* 56:359–67

106. Engeser, H., Hubner, K., Straub, J., Lynen, F. 1979. *Eur. J. Biochem.* 101:407–12

107. Kumar, S., Phillips, G. T., Porter, J. W. 1972. *Int. J. Biochem.* 3:15–32

108. Linn, T. C., Srere, P. A. 1980. *J. Biol. Chem.* 255:10676–80

109. Stern, A., Sedgwick, B., Smith, S. 1982. *J. Biol. Chem.* 257:799–803

110. Smith, S. 1982. *Arch. Biochem. Biophys.* In press

111. Poulose, A. J., Kolattukudy, P. E. 1982. *Int. J. Biochem.* 14:445–48

112. Toomey, R. E., Wakil, S. J. 1966. *J. Biol. Chem.* 241:1159–65

113. D'Agnolo, G., Rosenfeld, I. S., Vagelos, P. R. 1975. *J. Biol. Chem.* 250:5283–88

114. Oesterhelt, D., Bauer, H., Kresze, G., Steber, L., Lynen, F. 1977. *Eur. J. Biochem.* 79:173–80

115. Stoops, J. K., Wakil, S. J. 1981. *J. Biol. Chem.* 256:8364–70

116. Kresze, G., Steber, L., Oesterhelt, D., Lynen, F. 1977. *Eur. J. Biochem.* 79:181–90

117. Lewis, S. D., Johnson, F. A., Shafer, J. A. 1976. *Biochemistry* 15:5009–17

118. Stoops, J. K., Wakil, S. J. 1982. *J. Biol. Chem.* 257:3230–35

119. Stoops, J. K., Wakil, S. J. 1982. *Biochem. Biophys. Res. Commun.* 104:1018–24

120. Poulose, A. J., Kolattukudy, P. E. 1980. *Arch. Biochem. Biophys.* 199:457–64

121. Poulose, A. J., Kolattukudy, P. E. 1980. *Arch. Biochem. Biophys.* 201:313–21

122. Poulose, A. J., Kolattukudy, P. E. 1981. *J. Biol. Chem.* 256:8379–83

123. Lynen, F. 1961. *Fed. Proc.* 20:941–50

124. Fox, J. L., Lynen, F. 1980. *Eur. J. Biochem.* 109:417–24

125. Smith, S., Agradi, E., Libertini, L., Dileepan, K. N. 1976. *Proc. Natl. Acad. Sci. USA* 73:1184–88

126. Poulose, A. J., Rogers, L., Kolattukudy, P. E. 1981. *Biochem. Biophys. Res. Commun.* 103:377–82

127. Dileepan, K. N., Lin, C. Y., Smith, S. 1978. *Biochem. J.* 175:199–206

128. Libertini, L. J., Smith, S. 1979. *Arch.*

Biochem. Biophys. 192:47–60

129. Saggerson, E. D. 1980. *Biochemistry of Cellular Regulation*, 2:207–56. Boca Raton, Fla.: CRC Press

130. Lane, M. D., Moss, J. 1971. *Metabolic Regulation*, p. 23. New York: Academic

131. Watkins, P. A., Tarlow, D. M., Lane, M. D. 1977. *Proc. Natl. Acad. Sci. USA* 74:1497–501

132. Meredith, M. J., Lane, M. D. 1978. *J. Biol. Chem.* 253:3381–83

133. Clarke, S. D., Watkins, P. A., Lane, M. D. 1979. *J. Lipid Res.* 20:974–85

134. Mooney, R. A., Lane, M. D. 1982. *Eur. J. Biochem.* 121:281–87

135. Lane, M. D., Mooney, R. A. 1981. *Curr. Top. Cell. Regul.* 18:221–41

136. Goodridge, A. G. 1972. *J. Biol. Chem.* 247:6946–52

137. Goodridge, A. G. 1973. *J. Biol. Chem.* 248:4318–26

138. Ogiwara, H., Tanabe, T., Nikawa, J., Numa, S. 1978. *Eur. J. Biochem.* 89:33–41

139. Sreekrishna, K., Gunsberg, S., Wakil, S. J., Joshi, V. C. 1980. *J. Biol. Chem.* 255:3348–51

140. Nikawa, J., Tanabe, T., Ogiwara, H., Shiba, T., Numa, S. 1979. *FEBS Lett.* 102:223–26

141. Witters, L. A., Friedman, S. A., Tipper, J. P., Bacon, G. W. 1981. *J. Biol. Chem.* 256:8573–78

142. Yeh, L.-A., Kim, K.-H. 1980. *Proc. Natl. Acad. Sci. USA* 77:3351–55

143. Yeh, L.-A., Song, C.-S., Kim, K.-H. 1981. *J. Biol. Chem.* 256:2289–96

144. Kim, K.-H. 1979. *Mol. Cell. Biochem.* 28:27–43

145. Pekala, P. H., Meredith, M. J., Tarlow, D. M., Lane, M. D. 1978. *J. Biol. Chem.* 253:5267–69

146. Moss, J., Lane, M. D. 1972. *J. Biol. Chem.* 247:4944–51

147. Brownsey, R. W., Hardie, D. G. 1980. *FEBS Lett.* 120:67–70

148. Ly, S., Kim, K.-H. 1981. *J. Biol. Chem.* 256:11585–90

149. Krakower, G. R., Kim, K.-H. 1981. *J. Biol. Chem.* 256:2408–13

150. Halestrap, A. P., Denton, D. M. 1974. *Biochem. J.* 142:365–77

151. Witters, L. A., Moriarity, D., Martin, D. B. 1979. *J. Biol. Chem.* 254:6644–49

152. Brownsey, R. W., Belsham, G. J., Denton, R. M. 1981. *FEBS Lett.* 124:145–50

153. Kamiryo, T., Numa, S. 1973. *FEBS Lett.* 38:29–32

154. Mishina, M., Kamiryo, T., Tanaka, A., Fukui, S., Numa, S. 1976. *Eur. J. Biochem.* 71:304–8

155. Horikawa, S., Kamiryo, T., Nakanishi,

S., Numa, S. 1980. *Eur. J. Biochem.* 104:191–98

156. Kamiryo, T., Parthasarathy, S., Numa, S. 1976. *Proc. Natl. Acad. Sci. USA* 73:386–90
157. Kamiryo, T., Mishina, M., Tashiro, S., Numa, S. 1977. *Proc. Natl. Acad. Sci. USA* 74:4947–50
158. Mishina, M., Kamiryo, T., Tashiro, S., Numa, S. 1978. *Eur. J. Biochem.* 82:347–54
159. Hosaka, K., Mishina, M., Tanaka, T., Kamiryo, T., Numa, S. 1979. *Eur. J. Biochem.* 93:197–203
160. Kamiryo, T., Nishikawa, Y., Mishina, M., Terao, M., Numa, S. 1979. *Proc. Natl. Acad. Sci. USA* 76:4390–94
161. Green, H., Meuth, M. 1974. *Cell* 3:127–33
162. Green, H., Kehinde, O. 1976. *Cell* 7:105–13
163. Russell, T. R., Ho, R. 1976. *Proc. Natl. Acad. Sci. USA* 73:4516–20
164. Mackall, J. C., Student, A. K., Polakis, S. E., Lane, M. D. 1976. *J. Biol. Chem.* 251:6462–64
165. Ahmad, P. M., Russell, T. R., Ahmad, F. 1979. *Biochem. J.* 182: 509–14
166. Student, A. K., Hsu, R. Y., Lane, M. D. 1980. *J. Biol. Chem.* 255:4745–57
167. Weiss, G. H., Rosen, O. M., Rubin, C. S. 1980. *J. Biol. Chem.* 255:4745–50
168. Coleman, R. A., Reed, B. C., Mackall, J. C., Student, A. K., Lane, M. D., Bell, R. M. 1978. *J. Biol. Chem.* 253:7526–61
169. Kasturi, R., Joshi, V. C. 1982. *J. Biol. Chem.* 257:12224–30

170. Joshi, V. C. 1981. *Methods Enzymol.* 72:743–47
171. Joshi, V. C., Aranda, L. P. 1979. *J. Biol. Chem.* 254:11783–86
172. Joshi, V. C., Aranda, L. P. 1979. *J. Biol. Chem.* 254:11779–82
173. Joshi, V. C., Sidbury, J. B. Jr. 1977. *Arch. Biochem. Biophys.* 182:214–20
174. Fischer, P. W. F., Goodridge, A. G. 1978. *Arch. Biochem. Biophys.* 190:332–44
175. Joshi, V. C., Kasturi, R. 1982. *Fed. Proc.* 41:969
176. Buckner, J. S., Kolattukudy, P. E. 1975. *Biochemistry* 14:1768–73
177. Buckner, J. S., Kolattukudy, P. E. 1975. *Biochemistry* 14:1774–82
178. Buckner, J. S., Kolattukudy, P. E., Rogers, L. 1978. *Arch. Biochem. Biophys.* 186:152–63
179. Kim, Y. S., Kolattukudy, P. E. 1978. *Arch. Biochem. Biophys.* 190:585–97
180. Kim, Y. S., Kolattukudy, P. E. 1978. *Arch. Biochem. Biophys.* 190:234–46
181. Smith, S., Abraham, S. 1975. *Adv. Lipid Res.* 13:195–239
182. Carey, E. M., Dils, R. 1973. *Biochim. Biophys. Acta* 306:156–67
183. Strong, C. R., Carey, E. M., Dils, R. 1973. *Biochem. J.* 132:121–23
184. Knudsen, J., Clark, S., Dils, R. 1976. *Biochem. J.* 160:683–91
185. Libertini, L., Lin, C. Y., Smith, S. 1976. *Fed. Proc.* 35:1671
186. Libertini, L. J., Smith, S. 1978. *J. Biol. Chem.* 253:1393–1401

Ann. Rev. Biochem. 1983. 53:581–615

PROKARYOTIC DNA REPLICATION SYSTEMS[1]

Nancy G. Nossal

Laboratory of Biochemical Pharmacology, National Institute of Arthritis, Diabetes, and Digestive and Kidney Diseases, National Institutes of Health, Building 4, Room 116, Bethesda, Maryland 20205

CONTENTS

PERSPECTIVES AND SUMMARY 581
BACTERIOPHAGE T7 583
   Elongation and Strand-Displacement Synthesis 584
   Primer Synthesis 586
   Initiation at the T7 Origin 587
BACTERIOPHAGE T4 588
   Primer Synthesis 589
   Elongation and Strand-Displacement Synthesis 590
ESCHERICHIA COLI PROTEINS 592
   Primer Synthesis by dnaG Primase 592
   Primer Synthesis by E. coli RNA Polymerase 595
   DNA Polymerase III Holoenzyme 596
   Strand Displacement Synthesis 598
   Initiation at the E. coli Chromosomal Origin 601
PLASMID REPLICATION 603
   ColEl Initiation and Replication 603
   Replication of Other Plasmids 606
   Plasmid Primases Involved in Conjugal Transfer 606
BACTERIOPHAGE LAMBDA 606
BACTERIOPHAGE Φ29 608
CONCLUSION 608

PERSPECTIVES AND SUMMARY

Replication of duplex helical DNA involves a complex, highly coordinated series of reactions in which new DNA chains are initiated and elongated on each strand of the duplex. Genetic studies of *Escherichia coli* and its phage

---

[1] The US Government has the right to retain a nonexclusive royalty-free license in and to any copyright covering this paper.

581

have identified a large number of gene products that participate in DNA replication and in some cases have identified the position of the replication origin(s). These genetic studies have provided the framework necessary for the development of multienzyme replication systems used to determine the sequences, proteins, and enzymatic reactions required for DNA replication.

The steps used to replicate duplex DNA are indicated schematically in Figure 1. The figure has been drawn to emphasize the similarity between replication on duplex eyes and on rolling circle intermediates. The figure is a compilation of incomplete information from many systems, and some of these steps have already been shown to differ in detail in the individual systems studied. Replication eyes are thought to be formed by the recognition and activation (unwinding) of a chromosomal origin site by specific proteins and the synthesis of a RNA primer for the leading strand (I*A*, Figure 1). Elongation by DNA synthesis of the leading strand proceeds with simultaneous unwinding of the duplex ahead of the replication fork (I*B*). Alternatively, an initiation site for leading strand synthesis can be created by a site-specific nick in one strand of the duplex (II*A*). If the duplex DNA is circular, elongation from such a nick leads to a rolling circle intermediate, in which the 5' end of the displaced strand is at least sometimes linked to the growing 3' terminus (II*B*). Elongation of the leading strand can be facilitated actively by a helicase such as the *rep* protein that unwinds in the 3' to 5' direction (II*A* and II*B*). Helicase direction is indicated, by convention, as the direction of movement of the

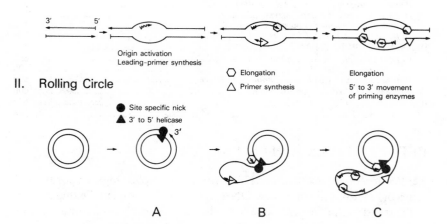

I.   Replication Eye

Origin activation
Leading-primer synthesis

○ Elongation
△ Primer synthesis

Elongation
5' to 3' movement
of priming enzymes

II.   Rolling Circle

● Site specific nick
▲ 3' to 5' helicase

A   B   C

Figure 1   Duplex DNA replication via replication eye, or rolling circle intermediates. Individual steps are described under the section "Perspectives and Summary."

helicase on the strand that is not displaced. On the lagging strand, synthesis proceeds by the synthesis (I and II*B*) and elongation (I and II*C*) of RNA primers to form discontinuous fragments, which, following removal of the primer and repair of the gap, are ligated to form a continuous strand. The priming systems that synthesize these short RNA primers range from the single bacteriophage T7 gene 4 protein to at least seven proteins in the *E. coli* primosome. In those systems in which it has been tested, the priming enzymes move 5′ to 3′ along the lagging strand template and have a nucleoside triphosphate-dependent 5′ to 3′ helicase activity to unwind the duplex DNA template ahead of the leading strand (I and II*C*).

This review is restricted to prokaryotic DNA replication enzymes that have been studied as part of multiprotein replication systems. It emphasizes recent developments since earlier reviews by Wickner (1) and Kornberg (2, 2a). Specific types of replication proteins have been discussed in recent reviews on DNA-binding proteins (3–5), helicases (6), DNA topoisomerases (7), and DNA polymerases (8–11). Related reviews have covered the initiation of DNA replication (12, 13), discontinuous DNA replication (14), the sequence and structure of replication origins (15), and viral (16) and eukaryotic (17) DNA replication.

BACTERIOPHAGE T7

The DNA of bacteriophage T7 is a linear duplex of about 40,000 base pairs (bp) with a terminal redundancy of 160 bp. The sequence of the entire molecule has been determined [(18); J. J. Dunn, cited in (19)]. The molecular biology (20) and DNA replication (19) of T7 have been reviewed recently. Initially, T7 DNA replicates as a unit-length linear molecule using a bidirectional origin. This origin was located at about 17% of the distance from the left end by electron microscopic studies (21, 22) and more recently placed between 14.75% and 15.0% by the analysis of deletion mutants (23) and cloned T7 fragments (24). If the primary origin is deleted, the phage replicates less efficiently using secondary origins, one located at approximately 4% (23). At later stages the replication intermediates are long linear concatemers, which are then processed to unit length.

Six T7 proteins encoded by genes 1, 2, 3, 4, 5, and 6 are required for DNA replication in vivo. Gene 5 encodes one subunit of T7 DNA polymerase. The gene 4 product is a multifunctional enzyme with RNA primer synthesis and helicase activities. Genes 3 and 6 encode nucleases that provide precursors by rapidly degrading host DNA, and also are involved in processing concatemers. Gene 6 mutants accumulate RNA-terminated Okazaki fragments (25). T7 RNA polymerase, encoded by gene 1, is required for the transcription of the other replication proteins and also participates directly in the initiation of replication, as described below. The gene 2 protein inhibits *E. coli* RNA polymerase (26–28); for reasons that are not yet clear, both phage mutants in this gene and *E. coli* mutants with RNA polymerase resistant to the gene 2 protein (29) are blocked at concatemer formation. Several other T7 proteins including DNA ligase and DNA-binding protein participate in DNA replication but can usually be replaced by host proteins (reviewed in 19).

Elongation and Strand-Displacement Synthesis

In vitro, the elongation of T7 DNA strands is catalyzed by T7 DNA polymerase, and stimulated by the gene 4 protein (Table 1). T7 DNA polymerase is a 1:1 complex of the T7 gene 5 protein (M_r 84,000) and the host thioredoxin (M_r 12,000) (30–33; Table 1). The polymerase complex has 3′ to 5′ exonuclease activities that degrade single- and double-stranded DNA to 5′ mononucleotides (34, 35). The gene 5 subunit retains the 3′ to 5′ exonuclease activity on single-stranded DNA, but has less than 2% of the polymerase or double-stranded DNA exonuclease of the T7 polymerase complex (33–35). Active polymerase can be reconstituted by mixing gene 5 protein with *E. coli* thioredoxin (31, 34, 35) or with any of the three spinach chloroplast thioredoxins (36). Thioredoxin from the T7-resistant *E. coli* mutant, tsnC7007, which differs from the wild type in a single amino acid, binds to the gene 5 protein, but does not reconstitute active polymerase (37).

T7 DNA polymerase has been isolated in two forms differing in their ability to be stimulated by the T7 gene 4 protein on nicked duplex DNA templates (19, 38–40). Early purification procedures gave enzyme now called Form I, which by itself catalyzes a limited strand-displacement reaction on nicked duplex templates (19, 39, 40) and is strongly stimulated by the T7 gene 4 protein (41–44). In contrast, Fischer & Hinkle (38) found that T7 DNA polymerase purified in the presence of EDTA (Form II) would not carry out strand displacement at nicks even with gene 4 protein. Form II has more exonuclease activity than Form I and can be converted to an enzyme resembling Form I by dialysis in the absence of EDTA. Recently R. L. Lechner & C. C. Richardson [(40); cited in (19)] made the important observation that Form II polymerase and gene 4 protein can carry out strand-displacement synthesis if the duplex template contains a single-stranded branch that presumably serves as a binding site for the gene 4 protein. The inability of Form II polymerase to strand-displace at nicks and at the junction of Okazaki fragments (see below) makes it better suited for synthesis at a replication eye (Figure 1).

The T7 gene 4 protein is a single-stranded DNA dependent nucleotidase with a preference for dATP and dTTP (41, 45), a helicase, and, as discussed below, a primase. Kolodner & Richardson (41) showed that β-γ-methylene

Table 1 Prokaryotic DNA replication proteins[a]

	Phage			Plasmids		Phage			General priming
	T7	T4	λ (λ dV)	oriC	ColE1	ΦX174	G4	fd, M13	
Chromosome initiation									
Replication bubble	T7 RNA pol[b] T7 gene 4[b] E. coli SSB[b] T7 DNA pol[b]	c	E. coli RNA pol[d] λ O[d] λ P[d] E. coli gyrase[d] Other factors[d]	E. coli proteins:[e] RNA pol dnaA dnaC dnaB SSB Other factors	E. coli proteins: RNA pol RNase H DNA pol I				
Rolling circle initiation									
Endonuclease						ΦX A	G4A	fd gene 2	
Helicase								E. coli rep	
Binding protein								E. coli SSB	
Primer synthesis (ss DNA)									
Primase	T7 gene 4	T4 gene 61				dnaG	dnaG	RNA pol	dnaG
Other proteins		T4 gene 41		f	f	n′ (Y) n, n″ (Z) i (X) dnaB dnaC SSB	SSB	E. coli proteins SSB	dnaB
Chain elongation									
Polymerase	T7 DNA pol	T4 DNA pol	f	f	f	DNA pol III core			
Polymerase accessory proteins		T4 gene 44/62 T4 gene 45				DNA pol III holoenzyme subunits (see Table 2)			
DNA-binding proteins	T7 gene 2.5 E. coli SSB	T4 gene 32				SSB	SSB	SSB	
Helicase	T7 gene 4	T4 gene 41				dnaB		rep	dnaB

[a] Requirements demonstrated in vitro, except as indicated. Further references are given in the text.
[b] Initiation in vitro is completely dependent on T7 RNA polymerase and T7 DNA polymerase and is stimulated by T4 gene 4 protein and E. coli SSB (58).
[c] Initiation has not been demonstrated in vitro. Rifampicin inhibition under certain conditions in vivo implicates E. coli RNA polymerase (65).
[d] Initiation achieved in extracts (295, 303, 304). Requirement for RNA polymerase and gyrase inferred from inhibition by rifampicin and novobiocin. Replication is inhibited by antisera to dnaG, dnaB, SSB, n, n′, and i proteins (295).
[e] Initiation in extracts is dependent on dnaA protein and is inhibited by rifampicin, nalidixic acid, and antibodies to dnaB, dnaC, and SSB proteins (144, 244, 245).
[f] Dependence on genes encoding primosome proteins and DNA pol III holoenzyme demonstrated in vivo. Reviewed in (1, 2, 12, 246, 247, 281).

dTTP, a strong inhibitor of the nucleotidase, decreased the gene 4 stimulation of strand-displacment synthesis. They suggested therefore that gene 4 protein was a nucleoside triphosphate-dependent helicase. Although gene 4 protein does not unwind a fragment completely annealed to a complementary single strand, it does unwind fragments with noncomplementary 3' single-stranded tails at least 6–7 nucleotides in length [S. W. Matson, C. C. Richardson, cited in (19)] (40). This unwinding requires the same nucleotides hydrolyzed by the gene 4 protein and is also inhibited by β-γ-methylene dTTP. It is likely that gene 4 protein is a 5' to 3' helicase, since it moves in that direction during primer synthesis [(46); Figure 1, IC]. Gene 4 protein purified by different procedures is a mixture of variable proportions of single polypeptides of 58,000 and 66,000 (38, 42, 44), with similar peptide maps (38, 44), which apparently are initiated at two AUGs separated by 63 codons (18). There is no evidence that the two proteins differ in their enzymatic activities (38, 44).

Both the T7 and *E. coli* single-stranded DNA-binding proteins stimulate strand-displacement synthesis by T7 DNA polymerase, with or without the gene 4 protein (38, 44, 47). Both proteins also increase the rate of copying of single-stranded templates (48). *E. coli* mutants in DNA-binding protein support the growth of wild-type T7, but not T7 mutants in gene 2.5, which encodes the T7 DNA-binding protein (49).

Primer Synthesis

Gene 4 protein catalyzes the synthesis of oligoribonucleotides, on single-stranded templates (44–46, 50–52). Both in vivo (53) and in vitro (46) this primase recognizes a 5-base template sequence, 3'-CTGN$_1$N$_2$-5'. On ΦX174 DNA the predominant recognition sites are 3'-CTGGT-5' and 3'-CTGGG-5', yielding the tetranucleotide primers pppACCA and pppACCC (46, 50, 52). Both *E. coli* and T7 single-stranded DNA-binding proteins stimulate primer synthesis (44, 51). Gene 4 protein also increases the utilization of short (< 6) ribo- or deoxyoligonucleotides as primer by T7 DNA polymerase (45).

The primers made by gene 4 are elongated by T7 DNA polymerase and DNA-binding protein to discontinuous fragments 1000–6000 nucleotides in length (52). With Form II DNA polymerase, synthesis of one fragment stops when it reaches the next so that following primer removal by the T7 gene 6 protein or *E. coli* pol I 5' to 3' exonuclease (54), the gaps can be filled in and the fragments ligated (19). Form I polymerase begins strand displacement when it reaches the next fragment, so that adjacent fragments cannot be ligated (39).

Initiation at the T7 Origin

The nucleotide sequence (55) of the primary T7 origin includes two rightward T7 RNA polymerase promoters (called 1.1A and 1.1B), followed by a 61 bp AT-rich noncoding region containing a rightward gene 4 protein primer site. There is a leftward gene 4 protein priming site 80 nucleotides to the right of the primary origin (55). The minor origin about 4% from the left of T7 also is near a T7 RNA polymerase promoter site followed by an AT-rich region (18). At least one of the promoters at the primary origin appears to be required, since inactivation of both promoters by introduction of a synthetic octomer, prevents initiation at the primary origin [S. Tabor, C. C. Richardson, cited in (19)] (40).

The proposal that T7 RNA polymerase participates in the initiation of DNA replication is based on both the location of the promoters in the origin region (55) and the observation that heat inactivation of T7 RNA polymerase inhibits DNA replication (56). Hinkle and co-workers (38, 57) first showed that purified T7 RNA polymerase stimulated T7 DNA synthesis with T7 DNA polymerase (Form II), gene 4 protein, and T7 DNA-binding protein. In their hands, DNA synthesis was almost completely dependent on both T7 RNA polymerase and gene 4 protein. Replication eyes in the product were not confined to the primary origin region, although they argued that the eyes were more common near known T7 RNA polymerase promoters (57). In contrast, using a similar system, Romano et al (58) recently reported that the formation of replication eyes (not total incorporation) was completely dependent on T7 DNA and RNA polymerases, but decreased only twofold in the absence of either the T7 gene 4 or T7 DNA-binding proteins. All of the replication eyes were located rightward of the origin between 16% to 24% from the T7 left, suggesting unidirectional replication from the primary origin.

Plasmids containing the T7 primary origin region are replicated by T7 extracts in vitro (24, 58, 59, 60). Romano et al (58) have shown preferential synthesis by T7 RNA polymerase, DNA polymerase, and gene 4 protein of a linearized plasmid containing the T7 origin cloned into pBR322 compared to a plasmid containing a late T7 promoter cloned into another site on the same vector. Analysis of the products by digestion with restriction enzymes suggests that replication proceeds rightward from a site near the primary origin [C. W. Fuller, C. C. Richardson, cited in (19)] (40). When supercoiled, both T7 promoter plasmids were copied at the same rate, which was higher than that of the vector (58). The role of T7 RNA polymerase may be to make a primer at the origin, or to open the helix to allow primer synthesis by the gene 4 protein (transcriptional activation). There is evidence for primer synthesis by RNA polymerase in vitro since replication eyes are present in the absence of the gene 4 protein (58). Furthermore, the 5' termini of the RNA attached to DNA chains correspond to those resulting from transcription at the 1.1A and 1.1B promoters [C. W. Fuller, C. C. Richardson, cited in (19)] (40). At this time it is not clear whether leading strand-rightward synthesis is ever initiated at the gene 4 priming site, nor what priming sites or possible additional proteins are used to establish lagging strand synthesis to the right and both leading and lagging strand synthesis leftward of the bidirectional origin.

BACTERIOPHAGE T4

The DNA of phage T4 is a 165-kb linear duplex that is both circularly permuted and terminally redundant (61). Replication normally proceeds through large, highly branched intermediates that are eventually cut into unit-length products. It has been proposed that recombination between replication intermediates is important in maintaining the normal high rate of DNA synthesis, since the formation of the large, branched intermediates is dependent on gene products required for recombination. In addition, most recombination-deficient mutants arrest DNA synthesis prematurely [reviewed in (62)]. Mosig and colleagues have proposed that: (*a*) T4 replication begins by origin-specific bidirectional replication intiated by *E. coli* RNA polymerase; (*b*) the formation of many additional forks then results from duplex invasion by single-stranded DNA termini generated in the early replication products (63, 64); and (*c*) origin-specific replication stops when the host RNA polymerase is modified by T4 proteins (65). In support of this model, they showed that rifampicin-inhibited DNA replication occurs after infection by a T4 recombination-deficient, DNA arrest mutant in genes 46 and 47, provided that the conversion of *E. coli* RNA polymerase to its T4-modified form is blocked by mutations in T4 genes 33 and 55 (65).

The number and position of the origin(s) of T4 replication are still controversial. Morris & Bittner (66) observed four possible origins by hybridization of early DNA to restriction fragments of T4 DNA, but noted that the pattern of origins depended on the conditions of infection and the choice of radioactive label ([^{32}P]-labeled inorganic phosphate or [^{3}H]thymidine). Mosig et al (66a) found a major origin located between genes 56 and *dda* (13–18 kb on the T4 map), and in addition a second origin at 111–115 kb that was used after heavy radiation damage to the host or with certain T4 deletion mutants. On the other hand, King & Huang (66b) reported one origin between genes 62 and 46 (29–35 kb) and a second between genes rI and *e* (62–64 kb), while Halpern et al (67) reported that the earliest DNA hybridized preferentially to DNA from plasmid clones with DNA from genes 50 to 5 (76–81 kb), with a less prominent peak from genes 10 to 29 (111–118 kb). However, all regions of the T4 genome were not represented in the clones tested in the latter study. An in vitro system specific for T4 origin initiation has not yet been developed. Alberts and colleages (68) have proposed that the T4 DNA topoisomerase, a multi-subunit nicking-closing enzyme encoded by three DNA delay genes, 39, 52, and 60 (68–70), participates in origin initiation.

Seven T4 proteins, the gene 41 and 61 priming proteins, gene 43 DNA polymerase, the gene 45 and complex of gene 44 and gene 62 protein polymerase accessory proteins, and the gene 32 helix-destabilizing protein (Table 1), catalyze efficient DNA synthesis in vitro on single-stranded or nicked duplex templates (71–75). The speed (72, 76, 77) and accuracy (78, 79) of this synthesis is comparable to that in vivo. Mutants in gene 61 are delayed in DNA synthesis in vivo. There is little or no DNA synthesis with mutants in the other six genes (80–82). Although there is persuasive evidence that the proteins required for dNTP synthesis are closely related to the DNA replication enzymes in vivo (83–86a and references therein), they have not yet been found to stimulate synthesis in vitro (72).

Primer Synthesis

The primers synthesized by the T4 gene 41 and 61 proteins on single-stranded DNA templates are pentanucleotides with the sequence $pppACN_3$ (87–89). This agrees with the finding of a terminal AC on at least 40% of the pentanucleotide primers made in vivo (90). It is not known whether primer synthesis requires a sequence other than that com-plementary to the primer. Since neither the 41 nor the 61 protein separately catalyzes phosphodiester bond formation, primase function has been ascribed to a complex of these two enzymes (76, 91, 92). 41 Protein has a single-stranded DNA-stimulated nucleotidase activity active on rGTP, rATP, dGTP, and dATP (93–95) and a helicase activity dependent on the same nucleotides (77). In each of its activities 41 protein shows a sigmoidal dependence on protein concentration consistent with an oligomerization of the protein. The activation of 41 protein by prior incubation with GTP, first described by Liu & Alberts (95), leads to a linear dependence on 41 protein concentration in the primer synthesis reaction (92). Under these conditions the stoichiometry suggests an oligomer of 41 protein functioning with a monomer of 61 protein (92). Primer synthesis is neither stimulated or inhibited by wild-type 32 protein (92, 96). It is inhibited, however, by a modified 32* protein (96) formed by proteolytic removal of 48 amino acids at the carboxyl end (97).

Elongation and Strand-Displacement Synthesis

T4 DNA polymerase, a monomer of 110,000, has a 3' to 5' exonuclease that may have a role in removing incorrect nucleotides (reviewed in 98, 98a). T4 polymerase alone is able to copy primed single-stranded templates, but incorporates only a few nucleotides with each binding to the template primer at physiological salt concentrations (99, 100). The T4 gene 45 and 44/62 polymerase accessory proteins increase the rate and processivity of this synthesis in an ATP-dependent reaction (99, 101–104). The gene 44 and 62 products copurify as a complex with an ATPase (dATPase) activity, stimulated by single-stranded DNA and the 45 protein (105). 45 Protein, recently sequenced (106), is required in vivo for both DNA replication and late RNA synthesis (80, 81, 107). With oligonucleotide-primed fd DNA (103) or restriction fragment-primed ΦX174 DNA (104), the accessory proteins increased the rate of synthesis through hairpin pause sites in the templates. Polymerase pause sites persisting with the 44/62 and 45 proteins were decreased by the gene 32 helix-destabilizing protein. The accessory proteins also greatly increased the rate of dNTP to dNMP conversion by the polymerase (104). Since dNMP formation correlated with the strength of hairpin barriers in the template primer and was decreased by the helix-destabilizing protein, it was interpreted as an idling reaction catalyzed by polymerase bound by the accessory proteins at pause sites in the template (104). While these studies (99–104) support the proposal that the accessory proteins "clamp" the polymerase to the DNA, there is no evidence yet that they actually travel with the polymerase. It has been suggested that ATP hydrolysis may be required for the assembly but not the movement of the complex of clamping proteins on single-stranded templates (76, 99, 103). This model is based on the findings that: (a) primer extension by these proteins on fd DNA was somewhat resistant to ATP-γ-S added after the reaction had started (103); and (b) within a short time frame (45 seconds) the number, but not the length, of oligo dT primers extended on poly dA by the polymerase, 44/62, 45, and 32 proteins, increased with ATP concentration (99). In the absence of ATP hydrolysis, the lifetime of the proposed complex must be short and influenced by the structure of the template. After addition of ATP-γ-S, the elongation of a fragment annealed to ΦX174 DNA was equivalent to less than 30-second synthesis at 31°C (N. G. Nossal, unpublished), and there was no synthesis on a nicked duplex template (76).

The accessory proteins and 32 protein also stimulate 3' to 5' hydrolysis of duplex templates by the polymerase (76, 108, 109), but inhibit hydrolysis of single-stranded DNA (108). A single-stranded DNA region more than three nucleotides long extending beyond the 3' OH of the DNA to be degraded is required for exonuclease activation by the 44/62 and 45 proteins and may serve as a binding site for one or more of these proteins (108).

T4 DNA polymerase alone is unable to catalyze strand-displacement synthesis at nicks in duplex DNA. However, T4 polymerase can carry out some strand displacement on a circular duplex molecule with a non-complementary single-strand branch (R. L. Lechner, C. C. Richardson, personal communication). Efficient synthesis on nicked molecules requires both the 44/62 and 45 polymerase accessory proteins and gene 32 protein (71–74). 32* protein can replace 32 protein (96). The rate of this synthesis is greatly increased by the gene 41 priming protein (72, 76, 77). 41 Protein has recently been shown to be a helicase that unwinds short fragments (50–400 bases) annealed to complementary single-stranded DNA beginning at their 3' termini, provided there is a single-stranded leader adjacent to the duplex region (77). This direction of unwinding is consistent with the idea, originally proposed by Alberts et al [(76); Figure 1, IC] that 41 protein destabilizes the helix ahead of the replication fork as it moves 5' to 3' on the displaced strand toward the next site for primer synthesis. 41 Helicase activity is stimulated by the 61 protein and the 44/62 plus 45 proteins, but strongly inhibited by 32 protein (77). Moreover, 41-protein stimulation of strand-displacement synthesis is greater when the 32 protein is decreased to concentrations that are limiting in the absence of 41 protein. Using circular duplex molecules with a single nick, Venkatesan et al (77) found a pronounced lag before stimulation by 41 protein that they attributed to the need for a prior synthesis of a single-stranded displaced strand relatively free of 32 protein to which 41 protein can attach. They have proposed that 41 protein thus precedes 32 protein on the lagging strand.

32 Protein and polymerase also play an important role in helix destabilization. The rate of fork movement on nicked templates increases with 32-protein concentration, in a range where there is sufficient 32 protein to bind all single-stranded DNA present (76). The defective polymerase of the T4 antimutator mutant, ts CB120, copies single-stranded homo-polymers at rates comparable to the wild-type enzyme, but is more in-hibited by hairpin pause sites in single-stranded DNA templates, and is unable to carry out strand-displacement synthesis at 30°C even with the 32, 44/62, 45, and 41 proteins [(110); N. G. Nossal, unpublished]. Gauss et al (111) recently showed that the CB120 mutant, as well as the T4 *dda* mutant, which codes for a helicase (112, 113), are unable to grow in the *E. coli optA* mutant originally isolated as unable to grow T7 gene 1.2 mutants (114). They have proposed that *optA* may control a helicase that becomes essential in the absence of the T4 *dda* helicase or the un-winding activity of the wild-type T4 polymerase (111). The *dda* helicase

has recently been shown to stimulate strand-displacement synthesis by the T4 43, 44/62, 45, and 32 proteins (114a). Moreover, this helicase allows the replication enzymes to displace RNA polymerase molecules that are bound to the DNA template.

ESCHERICHIA COLI PROTEINS

Primer Synthesis by dnaG *Primase*

Three different systems involving subsets of *E. coli* proteins have been described in which the primer is made by the *dnaG* protein (Table 1). The systems differ in the single-stranded DNA replicated and in the prepriming proteins and mechanism used to create a site recognized by the *dnaG* primase.

dnaG SSB PRIMING SYSTEM   The simplest of the *dnaG* systems employs only the primase and the *E. coli* single-stranded DNA-binding protein (SSB) and works with the closely related phage G4, ST-1, α-3, and ΦK DNA (115–118). *dnaG* Protein binds specifically to these DNAs coated with SSB (117). Two 40-base stretches of sequence that include a potential hairpin near the start of primer synthesis and a larger potential hairpin 35 to 40 bases 5′ downstream have been conserved in the four phage DNAs that can be primed by *dnaG* protein and SSB (119). Sims & Benz (120) showed that in the presence of SSB, a 115-base sequence surrounding the origin of phage ΦK is protected from micrococcal nuclease digestion by the *dnaG* protein. They concluded that both hairpins are essential parts of the *dnaG* protein-binding site, since removal of the downstream hairpin abolishes *dnaG* protein protection of the primer hairpin.

There is now general agreement that dNTP can be polymerized by *dnaG* protein, although at a lower rate than rNTP (117, 118, 121). The oligonucleotides made with dNTP or a combination of dNTP and rNTP are shorter and more heterogeneous than those made only with rNTP (118, 121). In the coupled reaction, the length of primers containing rNMP in chains extended by DNA pol III decreased with increasing dNTP concentrations (121), and oligonucleotides as short as di- or trinucleotides served as primers (118, 121). Approximately one 26- to 29-ribonucleotide-long primer is made with G4 or a α-3 DNA (118, 122, 123) at the site corresponding to the origin in vivo for each 60,000-dalton *dnaG* protein present.

dnaG GENERAL PRIMING SYSTEM   Primer synthesis on single-stranded DNA not coated with SSB takes place with only the *dnaB* protein in addition to the *dnaG* primase (124, 125). This primer synthesis in the absence of SSB was not observed in earlier studies with less-purified proteins (126, 127). As expected in view of the interactions between the *dnaB* and *dnaC* proteins described below, *dnaC* protein stimulates primer synthesis by the *dnaB* and *dnaG* proteins (R. McMacken, personal communication). With the *dnaB* and *dnaG* proteins, ribo-oligonucleotides of 10–60 nucleotides are synthesized at multiple sites on all natural single-stranded DNAs tested, as well as on poly dT (125). Deoxynucleotide primers are also formed on ΦX174 DNA if rATP, rGTP, rATP-γ-S, or rGTP-γ-S are present in addition to the four dNTP (124). This requirement for the ribonucleotides seems to be related to primer initiation since only rATP or rATP-γ-S permit oligo dA synthesis on poly dT. The formation of primers with only rATP-γ-S and four dNTP suggests that the *dnaB* DNA-dependent nucleotidase (active on rNTP but not dNTP) (128–131) is not required in the general priming reaction. *dnaB* Protein bound to uncoated ΦX174 DNA (with ATP) is freely exchangeable, in contrast to *dnaB* protein bound stably with the other primosome proteins (132). Thus it has been inferred that in the general priming reaction *dnaB* moves on and off the DNA (distributively), rather than moving processively along the DNA as proposed for the primosome reaction (see below).

Kornberg and coworkers (133) have suggested that the *dnaB* protein "engineers" the single-stranded DNA into a configuration recognized by primase, analogous to the complex hairpin structure required for primer synthesis by *dnaG* protein on SSB-covered G4 DNA. They showed that *dnaB* protein forms a ternary complex with nucleoside triphosphates and uncoated single-stranded DNA, and that complexes with nonhydrolyzable nucleotides are more stable (133, 134). The enhanced fluorescence of ethidium bromide bound to poly dT in the presence of *dnaB* protein and ATP (133) supports a *dnaB* protein-induced conformational change in the DNA. However, the binding of *dnaB* protein to DNA is not sufficient to allow primase to bind stably, since [³H]primase was not bound to ΦX174 DNA in a filterable complex with *dnaB* protein and App(NH)p or ATP (133). While primase was bound to DNA in the presence of *dnaB* protein and CTP, UTP, GTP, and ATP or App(NH)p, *dnaB* protein was not shown to be required for this binding. Both the primase and *dnaC* proteins increased the binding of *dnaB* protein (with ATP) to DNA (133). Thus, while *dnaB* protein binds and enables *dnaG* protein to prime on uncoated single-stranded DNA, the precise function of the *dnaB* protein is still not clear.

E. coli PRIMOSOME-SPECIFIC PRIMING SYSTEM   The first *E. coli* priming system studied has turned out to be the most complex. At least six proteins are required for a prepriming step that allows the *dnaG* primase to use SSB-covered ΦX174 single-stranded DNA. The six proteins include the *dnaB* and *dnaC* proteins and proteins referred to as factors X, Y, and Z by Wickner & Hurwitz (reviewed in 135) or protein n, n', n", and i by Kornberg and coworkers [(132, 136, 137); Table 1]. Proteins Y and n' have similar characteristics. The relationship of the other proteins is less clear. Functionally, factor X may be equivalent to protein i, and factor Z may be equivalent to proteins n + n" (138). Proteins n and n" have just recently been separated from each other (137). On the basis of studies cited below, Kornberg and colleagues have proposed that these six proteins interact to form an entity called a primosome that travels 5' to 3' along the DNA between primer sites [(134, 139); see Figure 1, IC].

There has been significant progress in determining the steps in assembling the prepriming complex. Briefly, protein n'(Y) binds to SSB-covered ΦX174 DNA (132, 135, 137, 140), which contains a protein n' recognition site (described below), and stimulates the binding of protein n (137). At higher protein concentrations, protein n (or factor Z) binds without protein n'(Y) (135, 137). In the next stage, *dnaB* and *dnaC* proteins incubated with ATP and Mg^{2+} form a protein complex (141–143). *dnaB* protein is transferred from the *dnaB-dnaC*-protein complex to ΦX174 DNA bound to factors Y and Z (135, 142) or proteins n', n, and n" (132, 137, 143), in a reaction that requires ATP and factor X or protein i. This prepriming DNA-protein complex can then be used by the *dnaG* primase. The *dnaB* protein (132, 134, 135, 142, 143) and factors Y plus Z (135, 142) or proteins n' plus n (132, 137) are tightly bound in the active prepriming complex isolated by gel filtration. The *dnaB*, n', and n proteins have also been found in gel-purified synthetic ΦX174 RF I formed by the addition of rNTP, dNTP, primase, pol III, pol I, and ligase (132, 137, 139). *dnaC* Protein (135, 142, 143) and factor X (135, 142) (assayed enzymatically) are not found on the prepriming complex. Although there is [³H]*dnaC* protein on gel-purified synthetic ΦX174 RF I (143), *dnaC*, i, and n" proteins must be added to gel-purified RF I for further RF synthesis (137, 139). However, protein i suffices with sucrose gradient-purified ΦX174 RF I (139). Primase may be stabilized on the DNA during primer synthesis and elongation. It is not found in a filtered complex formed by incubating the prepriming proteins, primase, and ATP with SSB-covered ΦX174 DNA (135), but is present in RF I made in a coupled reaction (139). The preparation of highly purified prepriming proteins, *dnaB* (145, 146), *dnaC* (144), n (137), n' (147), and i (136) will facilitate studies of their functions. The *dnaG* gene has been cloned (147a, 147b) and sequenced (147c). Primase synthesis requires sequences far upstream from the *dnaG* gene, and appears to be regulated in part by transcription attenuation (147a).

Protein n'(Y) recognizes a site on single-stranded DNA that is required for primer synthesis and is thought to act as an entry site for the primosome

(148). Wickner & Hurwitz (149) showed that factor Y was an ATPase (dATPase) that was stimulated (under some assay conditions) by ΦX174 DNA more than by several other single-stranded DNAs. The sequence on ΦX174 DNA that activates this ATPase has subsequently been shown to be a 55-nucleotide exonuclease VII-resistant fragment containing a potential 44-nucleotide hairpin (150), located at a site analogous to the origin used by dnaG protein on single-stranded G4 DNA. Recently two other sites that activate the protein n′(Y) ATPase were found on pBR322 (151, 152). When cloned into fl phage, these sites allow replication dependent on the primosome proteins (153). The three protein n′(Y) sites share only the sequence AAGCGG, which is not sufficient to define a factor Y site. This lends further support to the idea that protein n′(Y) may recognize a three-dimensional DNA structure (151). The HhaI-6 fragment of single-stranded ΦX174 DNA includes the protein n′ site and is the only ΦX174 HhaI fragment that supports primer-dependent DNA synthesis (148). Arai et al (134, 148) presented evidence that chains on HhaI-6 begin 3′ to the protein n′ site and that on intact ΦX174 DNA, priming sites are used in an order that suggests that the primosome enters at the protein n′ site and travels 5′ to 3′, i.e. in the opposite direction from DNA chain elongation. Movement of the primosome proteins has not been shown directly. It is inferred from the observations that: (a) although a protein n′ recognition sequence is required to use the specific priming system, primers are made at locations throughout the ΦX174 DNA (134); and (b) the primosome proteins n, n′, and dnaB, as well as primase, are present after synthesis and ligation of ΦX174 RF II (132, 137, 139). It has been proposed that the protein n′ nucleotidase provides the energy for this movement, because the synthesis of primers initiated by ATP-γ-S is stimulated by dATP (134), which can be used by the protein n′ but not the dnaB protein nucleotidase.

Primer Synthesis by E. coli RNA Polymerase

E. coli RNA polymerase synthesizes an RNA primer of about 30 nucleotides at a unique site on SSB-covered viral DNA of the small filamentous phage fd, fl, and M13 (154, 155). A 125-nucleotide stretch of SSB-covered fd viral DNA, protected from nuclease by RNA polymerase, contains four possible hairpins (reviewed in 156). Analogous regions in M13 and fl have been identified (157, 158). M13 phage deleted in the entire region specifying the primer are replicated, although much less well than the wild type, suggesting the possibility of secondary initiation sites (159).

In vitro, RNA polymerase can make primers on ΦX174 DNA as well as M13 and fd DNA, but it is clear that ΦX174 DNA is not replicated by a rifampicin-sensitive system in vivo. Crude extracts of E. coli catalyze rifampicin-sensitive replication on fd but not ΦX174 DNA (reviewed in 1,

2). Vicuna et al (160, 161) have presented evidence that two host factors (called discriminatory factors α and β), in addition to RNase H and SSB, are necessary to make the RNA polymerase reaction specific for fd DNA. In contrast, Kaguni & Kornberg (162) reported that the σ-subunit of RNA polymerase holoenzyme is sufficient to confer specificity for SSB-covered M13 versus ΦX174 DNA.

DNA Polymerase III Holoenzyme

The major replicative DNA polymerase in *E. coli* is a complex enzyme that has been viewed as either a holoenzyme with many subunits or as an activity that requires at least four separate proteins (reviewed in 1, 2, 2a, 10). Pol III can be purified as a holoenzyme of at least seven subunits (α to τ) by using the elongation of RNA-primed G4 DNA as an assay and selecting those procedures that maintain the integrity of the complex (10, 163–166; Table 2). The association of subunits in pol III holoenzyme is sufficiently weak that the holoenzyme can be divided by chromatography on phosphocellulose and hydroxylapatite into smaller complexes whose subunit composition and properties are summarized in Table 2. Each of the core subunits (α, ε, and θ) remains with the others during acrylamide gel electrophoresis under nondenaturing conditions (166). Alternatively, pol III core has been purified directly using synthesis on gapped DNA (167–170) or complementation of *dnaE* mutant extracts (171) as an assay. Wickner & Hurwitz (172) showed that the combination of pol III core, the *dnaZ* gene product and two elongation factors (EF I, EF III), was

Table 2   Forms of *E. coli* DNA polymerase III[a]

Subunit	α	ε	θ	τ	γ	δ	β[b]	Identifying reactions
Alternate names					*dnaZ*[c]	EF III[c]	EF I[c], Copol III[d]	
Size (kd, denatured)	140	25	10	83	52	32	37	
Gene	*dnaE*[e]				*dnaZ*[f]	*dnaX*[g]	*dnaN*[h]	
Polymerase designation								
Pol III (core)[i]	+	(+)[j]	(+)[j]					Synthesis on short gaps; 3' to 5', 5' to 3' exonuclease
Pol III'[k]	+	+	+	+				Limited synthesis on long gaps with spermidine
Pol III*[l]	+	+	+	+	+	+		Stimulated by *SSB* and β
Holoenzyme[m]	+	+	+	+	+	+	+	Processive synthesis on primed SS phage DNA

a Reviewed in (1, 2, 10, 170).
b (181).
c (135, 169, 170, 172).
d (2, 10).
e (173).
f (172, 175).

g (176).
h (174).
i (166–171, 179).
j not identified in all preparations (168, 170, 171).

k (164, 165, 180).
l (163, 164, 166, 180).
m (10, 163, 164, 166, 179, 182).

equivalent to pol III holoenzyme in elongating primed ΦX174 DNA. The α subunit is encoded by *dnaE* (173), β-subunit (EF I, copol III) by *dnaN* (174), γ by *dnaZ* (175). The δ-subunit, which appears to correspond to EF III (138), has been reported to be the *dnaX* gene product (176). The observation that mutations in *dnaQ* increase the temperature sensitivity of *dnaE* and *dnaZ* mutants (177) may indicate that the *dnaQ* product is, or interacts with, a subunit of the holoenzyme.

A variety of assays have been used to try to determine the function of the subunits. The 140-kd α-subunit has polymerizing activity by itself, as shown by in situ assay of the separated subunits after gel electrophoresis (178). No functional role has been shown for the ε or θ subunits of pol III core. The addition of the τ-subunit makes the pol III core more processive (179, 180) and allows some synthesis on the approximately 2000 nucleotide-long gaps present in randomly primed fd DNA in the presence of spermidine (Table 2; 164). Although τ-subunit is part of holoenzyme as judged by its precipitation in a 1:1 ratio with α-subunit by antibody to β-subunit (181), τ-subunit is not required to reconstruct holoenzyme-like activity on single-stranded phage DNA templates (165). The molecular weight of pol III' (M_r 400,000) is more than twice that of pol III core (M_r 160,000) (164); this finding led to the interesting suggestion that this and higher forms of pol III may have two active sites that replicate the two strands of the fork simultaneously [P. Burgers, A. Kornberg (2a), cited in (164)].

The addition of the γ- plus δ-subunits and possibly other subunits to form pol III* increases the processivity of the enzyme (180) and enables it to be stimulated by SSB (164; Table 2). These subunits are also necessary for subsequent addition of the β-subunit (copol III) to allow synthesis on primed G4 or ΦX174 DNA (163). The β-subunit, purified to homogeneity, is active as a dimer (181).

There is a good agreement that an isolatable initiation complex is made by incubating a primed DNA template, ATP (dATP), and either holoenzyme (182, 183) or pol III* plus β-subunit (181, 183, 184) or pol III plus EF III (δ), *dnaZ* protein (γ), and EF I (β) (135, 185, 186), and that EF I (185) or β-subunit (182, 184) is retained on the complex. The role of the EF III and *dnaZ* proteins remains controversial. Wickner has proposed that the EF III and *dnaZ* protein transfer EF I to the primed template, which can then bind to pol III core (185). Her model is based on the findings (185) that: (*a*) EF III and *dnaZ* protein (with ATP) are required to bind EF I to poly dA · oligo dT; and (*b*) all three proteins (EF I, EF III, and *dnaZ* protein) are needed to bind pol III core, but the EF III and *dnaZ* protein are not bound to the DNA. On the other hand, Johanson & McHenry (184, 187) and Burgers & Kornberg (183) have argued that γ-subunit (*dnaZ* protein) and δ-subunit (EF III), as well as β-subunit (EF I), must remain on the primed

G4 DNA initiation complex, since both groups found that the proteins present in the gel-purified complex can recycle and replicate added primed G4 DNA. Whether these differences are due to the template used (poly dA · oligo dT or SSB-covered primed G4 DNA), the strength of the interaction between the subunits in holoenzyme compared to that in the complex formed from purified subunits, or to additional subunits in the holoenzyme, remains to be determined.

Strand-Displacement Synthesis

ΦX174 RF → SS AND RF → RF REPLICATION   The replication of ΦX174 RF I has served as a model system for duplex DNA replication requiring strand-displacement synthesis with *E. coli* proteins (leading-strand synthesis) (Figure 1, pathway II). There is now general agreement that the replication of ΦX174 RF → ΦX174 RF occurs by a two-step procedure in which single-stranded DNA is made and then copied (138, 188–190). In the first stage, the phage-encoded ΦX174 A protein nicks supercoiled ΦX174 RF DNA and becomes covalently attached to the 5′ end of the nicked molecule (191, 192; Figure 1, II*A*). Supercoiled G4, ST-1, ΦK, and α-3 RF (138, 192, 193) as well as recombinant molecules with the ΦX174 plus strand origin cloned in pBR322 (194) are cut within an identical plus strand origin by the ΦX174 A and G4 A proteins (195, 196) and thus can be replicated in vitro by ΦX174 A protein-dependent systems developed for ΦX174 RF DNA. In the absence of gyrase, ΦX174 A protein cuts gel-purified synthetic ΦX174 RF bound to primosome proteins (139), but does not cut protein-free relaxed ΦX174 RF DNA (191). [For reasons that are unclear, gyrase was required with the synthetic ΦX174 RF protein complex that was not gel purified (197).]

Strand-displacement synthesis of a new plus strand occurs as nucleotides are added to the 3′ OH of the nick by pol III, and the DNA is unwound by the *E. coli rep* protein helicase and SSB (138, 188–190, 194, 198; Figure 1, II*B*). The role of *rep* protein is described below. Electron micrographs suggest that the ΦX174 A protein at the 5′ end remains associated with the replication fork, since duplex replication intermediates have looped out single-stranded circles (199) (Figure 1, II*B*). Upon synthesis of a new copy of the plus strand origin, the ΦX174 A protein apparently cuts the displaced single strand at the origin, ligates the two ends to form a new single-stranded circular plus strand, and is transferred to the new 5′ end of the displaced strand of the rolling circle intermediate (199). ΦX174 A protein has been shown to cut circular viral SS ΦX174 DNA at the plus strand origin and become linked to the 5′ end of the linear product (200, 201). The first (5′) 16 nucleotides of a 30-nucleotide sequence conserved in the plus strand origin of ΦX174, G4, and related phage are sufficient for ΦX174 A protein cutting of single-stranded DNA (201a), but do not allow ΦX174 A protein catalyzed nicking of supercoiled double-stranded DNA (201b). Although ligation of the two ends of single-stranded DNA by ΦX174 A protein has not been shown directly, ligation is catalyzed by the smaller, related ΦX174 A* protein under some conditions (202, 203). Ligation by ΦX174 A protein at a ΦX174 A protein recognition sequence is inferred from the formation of circular single-stranded DNA products of the appropriate size during unwinding of RF I DNA molecules containing one (198) or two (194a) ΦX174 A protein recognition sequences, in a reaction dependent on the ΦX174 A, rep, and SSB proteins.

In the second stage, the plus strands (ΦX174 viral DNA) are converted to new RF by synthesis of the complementary strand beginning with primers made by the primosome proteins (138, 188, 189; Figure 1, IIC). At least in vitro, the plus strand is made continuously, and only the synthesis of the complementary minus strand involves RNA-primed fragments (188) and is dependent on the dnaB, dnaC, and dnaG gene products (138). In addition to SS and RF DNA, a major product of the reaction catalyzed by pol III, ΦX174 A, rep SSB, and the primosome proteins is long, mostly double-stranded DNA, in some cases attached to duplex circular DNA (188, 189). It is likely that these arise when ΦX174 A protein fails to cut the newly synthesized plus strand so that the displaced strand becomes very long and then serves as a template for primosome-initiated DNA synthesis.

DNA pol III holoenzyme, rep, SSB, as well as the primosome proteins are bound to the intermediate that is purified by gel filtration of an RF → SS → RF reaction mixture, as judged by the significant, although not optimal, synthesis without further addition of these proteins (188). The tight binding of all of these enzymes has led to the suggestion that they may be organized in a multienzyme "replisome," but there is as yet no physical evidence for such a structure.

In vivo, the encapsulation of ΦX174 SS DNA into phage particles is coupled to SS DNA replication. A host factor(s) (different from rep, SSB, and pol III holoenzyme) that stimulates DNA synthesis in extracts of ΦX174-infected cells has been isolated, and is required for the formation of DNA protein complexes precipitable by antisera to the phage (204).

fd RF → SS SYNTHESIS   The synthesis of fd viral DNA, like ΦX174 DNA, begins with a site-specific cleavage of the viral strand in supercoiled RF I by a phage-encoded enzyme. The purified fd gene 2 protein, a monomer of 46,000 daltons (205), specifically cuts fd, fl, and M13 RF I DNA to give nicked RF II and a relaxed, covalently closed RF IV, which results from sealing the RF II (206). Thus the protein acts as a nicking-closing enzyme as well as a nuclease in the presence of Mg^{2+} and no other cofactors. The site of cleavage, between nucleotides 5781 and 5782 (207), lies within the region shown to be the origin of plus strand DNA replication in vivo (208).

In contrast to the ΦX174 A protein, the fd gene 2 protein does not become covalently attached to the 5' end of the cut viral strand (206). However, there is evidence that the fd gene 2 protein becomes at least weakly bound to the complementary strand across from the nick. Geider et al (209) showed that fd DNA nicked by the gene 2 protein is resistant to cutting opposite the nick by Bal31 nuclease prior to denaturation of the gene 2 protein, and that after denaturation resistance is restored upon addition of fresh gene 2 protein. Gene 2 protein cuts and joins the single strand in the rolling circle intermediate to form the viral circles. The nick made by gene 2 protein serves as a starting point for rolling circle DNA synthesis by either T7 DNA polymerase and gene 4 protein (210) or T4 DNA polymerase and gene 44/62, 45, and 32 proteins (211). Unit-length linear or circular single-stranded fd DNA is formed only when gene 2 protein is added in addition to the T7 or T4 proteins. There is no circularization if the gene 2 protein is added after synthesis of the rolling circle; this suggests that it recognizes the topology at the fork in addition to a specific fd DNA sequence (211). Removal of the 5' P from fd RF II made by gene 2 protein does not interfere with its use as a template for rolling circle DNA replication, but does prevent the circularization of the single-stranded product (209).

The fd RF I nicked by gene 2 protein is elongated by E. coli DNA polymerase III holoenzyme, SSB, and rep protein (209, 212). The characteristics of this sytem are very similar to that of ΦX174 RF $\rightarrow$ SS replication. Electron microscopy of the rolling circle DNA intermediates shows extended tails in contrast to the looped intermediates with ΦX174 DNA spread under the same conditions (209). However, these results do not exclude a looped intermediate if the interaction between 5' end and the enzymes at the fork is such that it is disrupted during the preparation of samples for microscopy. Active gene 2 protein is required during rep protein-catalyzed unwinding of fd RF II and thus has a role beyond the formation of a site-specific nick (209). Gene 2 protein stimulates DNA synthesis and presumably DNA unwinding on deproteinized fd RF II that had been cut by gene 2 protein, but not on randomly nicked RF II (212).

rep PROTEIN   The E. coli rep protein is required in vivo for ΦX174 replication and for the maximum rate of E. coli DNA synthesis at the replication fork (reviewed in 6). Rep protein is an ATP-dependent DNA helicase (unwinding enzyme) (193, 198, 213) that destabilizes the DNA beginning at the 5' terminus of the strand unwound (214, 215). As described above, in the presence of SSB rep protein unwinds RF I molecules that have been nicked by the ΦX174 A or fd gene 2 proteins. *Rep* protein also unwinds molecules containing a single-stranded gap adjacent to the duplex to be unwound (193). While unwinding of the partial duplexes also requires some SSB, this reaction is inhibited by concentrations of SSB sufficient to cover the SS DNA, presumably because SSB prevents *rep* protein from binding to the DNA (215). Experiments to test whether *rep* protein plus SSB allow strand displacement by pol III holoenzyme synthesis beginning on gapped templates have not been reported. Approximately two molecules of ATP (dATP) are hydrolyzed for each base pair separated in the ΦX174 A protein—ΦX174 RF complex when the helicase activity is uncoupled or coupled to DNA replication (213, 214, 216).

Purified *rep* protein is a monomer of 65,000 (217) with a nucleotidase activity dependent on SS DNA (216, 218). At a much higher concentration, the ΦX174 A protein—ΦX174 RF complex also acts as an effector of the *rep* protein nucleotidase (216). With this protein-DNA complex the nucleotidase reaction is distinguished by being more specific for ATP and dATP and by being stimulated rather than inhibited by SSB. Recent studies of the stability of the ternary complex of *rep* protein, SS DNA, and nucleotides are consistent with the conclusion that nonhydrolyzable ATP analogs increase, while ATP hydrolysis decreases, the affinity of *rep* proteins for SS DNA (218). In contrast, ATP hydrolysis does not increase the dissociation of *rep* protein bound to the ΦX174 A protein—ΦX174 RF complex (213). From these results it is inferred that the one molecule of *rep* protein at each fork remains tightly bound, perhaps because of its interaction with the ΦX174 A protein, and unwinds processively during DNA replication.

OTHER *E. coli* HELICASES   There are three *E. coli* helicases, I, II (219), and III (215), which unwind beginning at a 3′ terminus in addition to *rep* protein, which unwinds in the opposite direction (reviewed in 6). One would expect both types of helicase to promote fork movement by moving 5′ → 3′ on the lagging strand and 3′ to 5′ on the leading strand templates, respectively (Figure 1, II*C*). The role of helicases I and III in DNA replication is unknown. Antibody to *E. coli* helicase II inhibits *E. coli*, bacteriophage λ, and ColE1 DNA replication in extracts (220). Helicase II has been reported to promote replication of forked molecules by pol III holoenzyme and SSB in vitro [B. Kuhn, cited in (6)]. The *uvrD* gene encodes a DNA-dependent ATPase whose size and chromatographic properties resemble helicase II (221).

Initiation at the E. coli Chromosomal Origin

E. coli replicates bidirectionally from a single origin at 83.5 minutes (reviewed in 12). Initiation of replication is dependent on RNA polymerase and the products of *dnaA*, *dnaI*, and *dnaP* genes, as well as on the *dnaB* and *dnaC* primosome proteins. The origin, defined operationally as the sequence allowing ColE1-type vectors to replicate in pol A hosts, has been cloned and sequenced (222–224). A sequence of 232 bp to 245 bp has been shown to be essential for unidirectional replication of these vectors (222, 225). Additional sequence to the right of the minimum origin is required for bidirectional replication in vivo (225). The essential nucleotides within this region are being determined by comparing origin sequences in related Enterobacteriaceae (226, 227) and by site-specific mutagenesis of sequences within the cloned origin (228, 229). These studies [summarized in (230)] have identified highly conserved nucleotides required for the *ori*[+] phenotype, as well as other positions where base substitutions are tolerated but short insertions or deletions destroy *ori*[+] activity. Hirota and colleagues (229, 231) suggested that the origin is composed of specific sequences that are recognized by DNA initiation proteins and are arranged in the correct configuration with respect to each other by spacer regions of defined length.

The nucleotide where DNA synthesis begins in the L-strand of *ori* has not been identifed, nor is it clear whether a primer is made by RNA polymerase or the *dnaG* primase. Two RNA-DNA transition sites have been mapped on the opposite strand (232). Each strand of the origin region contains a promoter for transcription leading from inside to outside *ori*, which could be involved in initiation (233, 234). Other promoters have been identified in the region surrounding *ori* (234–237). Each strand of *ori* also contains a binding site for the protein B′ (238), a DNA-binding protein (denatured M.W. 68,000) with a preference for the origin region (239) whose function is unknown. It is not clear whether B′ is related to either of the two proteins reported to enhance the binding of DNA containing *ori* to isolated membrane preparations (240, 241). Plasmids containing *oriC* are incompatible with each other (241a). Thus *oriC* plasmids are excluded by other plasmids containing either *incA*, a region of 276 nucleotides including those required for autonomous replication of *oriC*, or *incB*, a region of 770 nucleotides just rightward of *oriC* (241a).

A cellophane disk system and a soluble system specific for *ori* have recently been described. In the disk system, *dnaC* protein stimulates extracts from *dnaC* mutants (242), while extracts of *dnaA* mutants are stimulated by extracts with *dnaA* extragenic suppressors, whose mode of suppression is unknown (243). In each case, the product hybridizes preferentially to *ori* DNA. There is no evidence that new chains are initiated in this system, which uses endogeneous template.

A soluble system specific for plasmids containing *ori* has been described that is dependent on RNA polymerase and the *dnaA*, *dnaB*, SSB (244), and *dnaC* (144) proteins. DNA synthesis is inhibited by nalidixic acid and coumermycin, but is not inhibited by chloramphenicol or puromycin. This system is distinguished from all others described previously in its complete dependence on hydrophilic polymers, like polyethylene glycol, which presumably act to concentrate the reactants. Replication proceeds bidirectionally from a position at or close to *ori* (245), even in a plasmid whose origin region was too short for bidirectional replication (of another vector) in vivo (225). This discrepancy could be due to different sequences to the right of *ori* in the plasmids used in the two studies or to other differences in the experimental systems.

PLASMID REPLICATION

ColE1 Initiation and Replication

The duplex circular DNA of the ColE1 plasmid can be replicated in cell-free extracts of *E. coli* that do not carry the plasmid, showing that plasmid-encoded proteins are not required. Genetic studies have shown that RNA polymerase, DNA polymerase I and III, as well as the primosome proteins (*dnaB*, *dnaC*, and *dnaG*) are required and that replication proceeds unidirectionally from a specific site (reviewed in 12, 246–248).

Replication is initiated on the H-strand leading to a D-loop intermediate containing a newly synthesized L-strand of about 6S (400b) (249). Itoh & Tomizawa (250, 250a) reported that initiation on the H-strand occurs in vitro with RNA polymerase, DNA polymerase I, and RNase H, with the DNA origin (RNA-DNA junction) occurring at the same three contiguous nucleotides used in the crude cell extract. The Klenow fragment of pol I, which lacks its 5′ to 3′ exonuclease, cannot replace DNA pol I in this reaction. RNase H increases the number of DNA chains initiated and specifically stimulates synthesis at the normal RNA-DNA junction. Gyrase is not required with supercoiled templates but does increase the length of the DNA products without increasing the number of chains initiated (250). Hillenbrand & Staudenbauer (251) have also developed a system for the initiation of ColE1 DNA replication consisting of DNA pol I, a "priming fraction" copurifying with RNA polymerase and DNA gyrase, and a "discriminating fraction" containing RNase H as well as other polypeptides whose size is about 20,000. With the "discriminatory fraction" there is origin-specific synthesis of the amplifiable plasmids ColE1, pBR322, RSF1030, and CloDF13, as judged by the position of D-loops in the products, and little synthesis of the nonamplifiable plasmids pSC101, pKN182, pSE014, and pSC138. In its absence there is random initiation of both sets of plasmids. This system is almost completely inhibited by novobiocin, perhaps because of residual relaxing activity in one or more of the protein fractions (251). In extracts, ColE1 replication is sensitive to gyrase inhibitors (252) and mutations in *gyrB* (253).

Itoh & Tomizawa (254) constructed a small plasmid, pNT7, containing 812 bp surrounding the ColE1 origin. Two transcripts are made from sites on opposite strands of this plasmid in a region 450–555 nucleotides upstream from the origin. This region is required for both replication and maintenance of the plasmid. RNase H cuts the transcript read from the H-strand (RNA II) at the replication origin to make a 555-nucleotide primer. This transcript is then extended by DNA polymerase I and the primer subsequently removed by RNase H. The second transcript, a 108 b RNA (RNA I) read from the L-strand, inhibits the synthesis of pNT7 DNA. RNA I prevents RNase H from cutting the primer transcript, apparently by interfering with the hybridization of the primer transcript to the DNA (255). RNA I from one plasmid inhibits the synthesis of that plasmid and other incompatible plasmids, but does not inhibit the synthesis of closely related compatible plasmids. Mutants of pNT7 selected as no longer incompatible with the related plasmid pMB9 were found to have a higher copy number than pNT7 and to have single base changes at or near the center of loops in three neighboring potential palindromes in the region that specifies both RNA I on the L-strand, and part of the primer transcript on the H-strand (256). Using plasmid DNA and RNA I purified from several of these mutants, these single base changes were shown to affect both the ability of RNA I to inhibit cleavage of the primer transcript by RNase H, as well as the sensitivity of the RNase H processing of a primer transcript to inhibition by RNA I.

Tomizawa & Itoh (256) have proposed that a palindromic structure formed by the interaction of two regions of the nascent primer transcript (RNA II) is essential for proper hybridization of RNA II to the DNA and thus for cutting by RNase H. RNA I is proposed to interfere with this process by annealing with one of the two regions of RNA II that is necessary to form this palindromic structure. The three neighboring palindromes in RNA I may facilitate its annealing with the complementary region of RNA II. Their conclusion that the secondary structure of RNA II in regions upstream from the DNA origin influences primer formation by RNase H is supported by their recent findings (256a) that primer formation is inhibited by the random substitution of IMP for less than 10% of the GMP in these regions, and that the addition of RNA I enhances pausing of transcription of RNA II at a site far downstream from the region where RNA I hybridizes to the RNA II nascent transcript. A mechanism of replication similar to that of ColE1 has been shown for the plasmids RSF1030, CloDF13, pBR322, and p15A (246, 255, 256b, 256c). Although there are differences in the sequences of these plasmids in the region determining the primer transcript, in each plasmid there is a potential set of three neighboring palindromes in the region specifying RNA I (256b, 256c).

Considerable progress has been made in showing which host proteins are required for later steps in ColE1 replication. Elongation of the 6S L-strand in extracts is dependent on the *dnaE* and *dnaZ* gene products, and these defects can be complemented by pol III holoenzyme (246, 257). L-strand elongation is also blocked by antisera to *dnaB* protein, but not by mutations in the *dnaG* primase (258). This has led to the proposal (258) that the *dnaB* protein may facilitate leading-strand synthesis in addition to its role in the synthesis of primers, analogous to the helicase function of the phage T7 gene 4 and T4 gene 41 priming proteins.

Discontinuous synthesis of the new H-strand is dependent on the *dnaB*, *dnaC*, and *dnaG* proteins. This has been shown both for plasmid synthesis in extracts (247, 259, 260) and by the requirements for DNA synthesis using the isolated L-strand as templates (261). Although elongation of the new L-strand on duplex supercoiled ColE templates does not require that *dnaG* protein as noted above, synthesis of the L-strand by copying the isolated H single strand is dependent on the *dnaB*, *dnaC*, and *dnaG* protein (261). A site for rifampicin-resistant initiation (*rri* site) has been identified on each strand of ColE1 by cloning fragments in M13 (262) and by restriction enzyme analysis of the DNA products of rifampicin-resistant synthesis in vitro (261). Protein n′(Y) ATPase sites were found in analogous positions of the related plasmid pBR322 (152). The *rri* sites on each strand are located ahead of the RNA-DNA junction for the 6S initiation intermediate for the new L-strand. The L-strand *rri* site presumably acts to assemble the primosome for discontinuous synthesis of the lagging H-strand, by the reactions described for ΦX174 SS → RF synthesis. It has been proposed that the H-strand *rri* site is where continuous L-strand synthesis by pol I stops and discontinuous synthesis of primosome-initiated synthesis by pol III begins (152), or alternatively, that it is required for the initiation of L-strand synthesis on the transferred H-strand during mobilization of the plasmid DNA (261).

There is conflicting evidence on the role of RNase III in the replication of plasmids related to ColE1. While Conrad & Campbell (260, 263) reported that plasmid RSF1030 synthesis was defective in preparations from *E. coli* with mutations in RNase III, Tomizawa & Itoh (256) found normal synthesis of early replicative intermediates in extracts from three RNase III mutants, and Hillenbrand & Staudenbauer (251) found no RNase III in their purified "priming" and "discriminatory" fractions. In vivo, ColE1 DNA replication has been reported to be unaffected by RNase III mutations [(264); G. Selzer, cited in (256)].

Replication of Other Plasmids

The plasmid R6K is replicated in vitro by a system that is dependent on a plasmid-encoded protein called π (265–268; reviewed in 269). Stalker et al (269a) determined the sequence of a 1565-bp autonomously replicating segment of R6K that contains both the structural gene for π and a functional replication origin (269b). A system to replicate miniplasmids derived from R1 has recently been developed (270), which requires simultaneous DNA-directed protein synthesis. Replication of the broad host range plasmid RSF1010 in vitro is dependent on one or more factors not expressed in extracts from the plasmid-free host, and is not inhibited by rifampicin or antisera to *E. coli* RNA polymerase (271).

Plasmid Primases Involved in Conjugal Transfer

During conjugal transfer of the IncIα plasmids, a single strand of the DNA enters the recipient (272) and is copied by a DNA pol III-dependent system (273). The observation that this strand can be copied in *dnaG* ts recipients at nonpermissive temperatures in the presence of rifampicin led to the suggestion that these plasmids might encode their own primase (274). Such primases have now been identified in plasmids from incompatibility groups B, Iα, Iγ, M, and P (275, 276). Hybridization studies indicate that there are at least two genetically distinct groups of plasmid primases (277). Derivatives of these plasmids that are derepressed for functions required for DNA transfer produce much higher levels of primase (275, 276), and in the case of the IncIα (278), IncIγ (279), and R16(B) (277), have been shown to suppress *dnaG* ts mutations. The ColI (IncIα) plasmid gene responsible for this suppression, called *sog* (suppressor of G), has been cloned and shown to encode two proteins read from different initiation sites in the same frame; the larger has primase activity (278, 280).

The primases from R64 and RP4 plasmids have been purified extensively (275, 276) and shown to resemble those of T4 and T7 phage in their ability to use a large variety of SS DNA templates. Maximum DNA synthesis primed by the R64 primase occurs with all four rNTP. The primers begin with C [E. Lanka, cited in (278)], and very little synthesis occurs in the absence of ATP or CTP (275). The R64 and RP4 plasmid primases are immunologically distinct from each other and from the *E. coli dnaG* primase (276).

BACTERIOPHAGE LAMBDA

The linear duplex DNA of bacteriophage λ has complementary single-stranded tails at each 5′ end that anneal to form circular DNA in vivo. Replication of these circular molecules begins unidirectionally from a defined origin. After one or a few rounds of ring-to-ring replication, there is a switch to rolling circle replication. λ DNA replication requires the phage encoded O and P proteins, *E. coli* RNA polymerase, and the host priming and elongation proteins *dnaG*, *dnaB*, *dnaZ*, and *dnaE*, as well as products of the host *dnaJ*, *dnaK*, *grpD*, and *grpE* genes, but is not dependent on the *dnaA*, *dnaC*, and *dnaI* gene products (reviewed in 186, 281–283).

The essential sites within the λ origin include a rightward promoter (pR) followed to the right by genes for the O and P proteins. Within the O gene is a short region called *ori*, defined by a series of *cis*-acting mutations that fail to replicate even with O and P provided in *trans* (284, 285). Normally initiation requires transcription from pR through *ori*, even if O and P proteins are supplied (reviewed in 286). It is not clear if RNA polymerase opens the duplex to make it accessible to the initiating proteins or provides a primer to be elongated. This transcription need not go through *ori*. Recent studies have shown that in λ *ri^c* mutants transcription originating about 100 b rightward of *ori* and proceeding away from the origin allows DNA replication in the absence of pR (286). Between pR and the O gene is a region called "*ice*" which Hobom and co-workers (287–289) have shown permits certain ColE1 type vectors containing λ *ori* to replicate at high temperatures in pol A ts hosts.

Shift-up experiments with ts mutants in λ O and P suggest that O protein is required continuously, while P protein is required during early (ring-to-ring) replication of λ (290). Studies with hybrid phage show that the N terminus of O protein interacts with *ori*, while the C terminus interacts with P protein (285, 291, 292). Functional interactions between the λ P protein and host *dnaB*, *dnaK*, *dnaJ*, *grpE*, and *grpD* proteins have been inferred from the ability of point mutants in λ P to bypass mutations in the host genes (reviewed in 282).

The λ O protein has been purified from cells with plasmids containing the O gene (293–296). O Protein (denatured M.W. 32,000) binds specifically to a 164-base fragment at "*ori*" in the absence of other host or phage proteins. In a series of elegant "footprinting" experiments, Tsurimoto & Matsubara (297, 298) showed that O protein binds to four tandemly arranged 19 bp repeats. At low concentration the O protein binds preferentially to the inner two repeats. From the size of O protein they have inferred that a dimer binds to each repeat.

P Protein (denatured M_r 25,000) may be associated with *dnaB* protein since P protein remains associated with the *dnaB* protein during initial purification steps (186, 295, 299, 300) and inhibits the *dnaB* ATPase (299) and *dnaB* complementation activity for ΦX174 SS → RF replication (186, 295, 299). Formation of a *dnaB—dnaC* protein complex by incubation with ATP protects *dnaB* protein from inhibition by λ P (186). No enzymatic activity other than these interactions with *dnaB* has yet been associated with the purified λ P protein. An altered *dnaB* protein has been purified from a *dnaB* mutant (*groP*), which is permissive for a λ mutant in P (301, 302).

The recent development of in vitro systems capable of replicating plasmids containing all of the sequences required for λ replication offers an opportunity to determine the biochemical roles of the O and P proteins (295, 303, 304). Replication is dependent on rNTP and an ATP-generating system as well as both the O and P proteins. It is inhibited by rifampicin and novobiocin but not by chloramphenicol. This replication is also inhibited by antibodies to the *dnaG*, *dnaB*, SSB, n, n', and i proteins but not by antibody to *dnaC* protein (295). The AT-rich region of λ *ori* was required for O-protein and P-protein-dependent replication of plasmids derived from pBR322, while the *ice* region could be deleted (304).

BACTERIOPHAGE Φ29

The DNA of the *Bacillus subtilis* bacteriophage Φ29 is a linear duplex of about 18 kb. The phage encoded protein p3 is covalently linked to the 5′ end of each DNA strand by a phosphodiester bond between a serine-OH and the terminal 5′ dAMP. There have been two recent reports (305, 306) that Φ29 DNA can be replicated in partially purified extracts of Φ29-infected *B. subtilis* by a reaction dependent on Φ29 DNA linked to intact p3 protein in which a new protein p3-AMP complex is first formed from ATP and then elongated.

CONCLUSION

The prokaryotic DNA replication systems have in common a DNA polymerase that is highly processive either as a holoenzyme or when assisted by accessory proteins, single-stranded DNA-binding proteins, helicases to remove secondary structure in single-stranded regions and unwind the duplex ahead of the fork, and enzymes making short RNA primers to initiate discontinuous synthesis on the lagging strand (Figure 1; Table 1). Helicase activity in the priming enzymes helps coordinate leading- and lagging-strand synthesis. These systems have evolved in different ways to meet the varying needs of replication of bacteriophage, plasmids, and stable chromosomes. The most complex system, and presumably therefore the system that can be most finely regulated, is used by the bacterial chromosome whose replication must be tightly coupled to cell division. I anticipate that eukaryotic replication will be carried out by still more complex variations of the same basic system.

Within the past few years many of the enzymes responsible for priming and chain elongation have been highly purified. Thus we can now determine at the molecular level the biochemical mechanisms used in these reactions. The characteristic redundancy of these systems, insuring replication by allowing one protein to substitute for another, suggests that additional priming and elongation proteins may be found. Finally, the recent development of initiation systems is already clarifying the reactions involved in this important step in replication.

ACKNOWLEDGEMENTS

I would like to thank the many authors who contributed to this review by generously sending preprints of their manuscripts. I would also like to express my gratitude to Dr. Deborah Hinton for critically reading the review, and to Mrs. Helen Jenerick for typing the manuscript.

Literature Cited

1. Wickner, S. H. 1978. *Ann. Rev. Biochem.* 47:1163–91
2. Kornberg, A. 1980. *DNA Replication,* San Francisco: Freeman
2a. Kornberg, A. 1982. *1982 Supplement to DNA Replication,* San Francisco: Freeman
3. Kowalczykowski, S. C., Bear, D. G., Von Hippel, P. H. 1981. *Enzymes* 14:373–444
4. Williams, K. R., Konigsberg, W. H. 1981. *Gene Amplication and Analysis,* ed. J. C. Chirikjian, T. Papas, 2:475–508. New York: Elsevier
5. Coleman, J. E., Oakley, J. L. 1980. *CRC Crit. Rev. Biochem.* 7:247–89
6. Geider, K., Hoffmann-Berling, H. 1981. *Ann. Rev. Biochem.* 50:233–60
7. Gellert, M. 1981. *Ann. Rev. Biochem.* 50:879–910
8. Lehman, I. R. 1981. *Enzymes* 14:16–38
9. Lehman, I. R. 1981. *Enzymes* 14:51–66
10. McHenry, C., Kornberg, A. 1981. *Enzymes* 14:39–50
11. Fujimura, R. K., Das, S. K., Allison, D. P., Roop, B. C. 1981. *Prog. Nucleic Acid Res. Mol. Biol.* 26:49–62
12. Tomizawa, J., Selzer, G. 1979. *Ann. Rev. Biochem.* 48:999–1034
13. Kolter, R., Helinski, D. R. 1979. *Ann. Rev. Genet.* 18:355–91
14. Ogawa, T., Okazaki, T. 1980. *Ann. Rev. Biochem.* 49:421–57
15. Hobom, G. 1981. *Curr. Top. Microbiol. Immunol.* 94:93–142
16. Mitra, S. 1980. *Ann. Rev. Genet.* 14:347–97
17. Challberg, M. D., Kelly, T. J. 1982. *Ann. Rev. Biochem.* 51:901–34
18. Dunn, J. J., Studier, F. W. 1981. *J. Mol. Biol.* 148:303–30
19. Richardson, C. C. 1983. *Developments in Molecular Virology,* ed. Y. Becker, Vol. 2. The Hague, The Netherlands: Martinus Nijhoff B. V. In press
20. Kruger, D. H., Schroeder, C. 1981. *Microbiol. Rev.* 45:9–51
21. Wolfson, J., Dressler, D., Magazin, M. 1972. *Proc. Natl. Acad. Sci. USA* 69:499–504
22. Wolfson, J., Dressler, D., Magazin, M. 1972. *Proc. Natl. Acad. Sci. USA* 69:998–1002
23. Tamanoi, F., Saito, H., Richardson, C. C. 1980. *Proc. Natl. Acad. Sci. USA* 77:2656–60
24. Panayotatos, N., Wells, R. D. 1979. *J. Biol. Chem.* 254:5555–61
25. Shinozaki, K., Okazaki, T. 1977. *Mol. Gen. Genet.* 154:263–67
26. Hesselbach, B. A., Nakada, D. 1977. *J. Virol.* 24:736–45
27. Hesselbach, B. A., Nakada, D. 1977. *J. Virol.* 24:746–60
28. LeClerc, J. E., Richardson, C. C. 1979. *Proc. Natl. Acad. Sci. USA* 76:4852–56
29. De Wyngaert, M. A., Hinkle, D. C. 1980. *J. Virol.* 33:780–88
30. Modrich, P., Richardson, C. C. 1975. *J. Biol. Chem.* 250:5508–14
31. Modrich, P., Richardson, C. C. 1975. *J. Biol. Chem.* 250:5515–22
32. Mark, D. F., Richardson, C. C. 1976. *Proc. Natl. Acad. Sci. USA* 73:780–84
33. Hori, K., Mark, D. F., Richardson, C. C. 1979. *J. Biol. Chem.* 254:11591–97
34. Hori, K., Mark, D. F., Richardson, C. C. 1979. *J. Biol. Chem.* 254:11598–604

35. Adler, S., Modrich, P. 1979. *J. Biol. Chem.* 254:11605–14
36. Harth, G., Geider, K., Schürmann, P., Tsugita, A. 1981. *FEBS Lett.* 136:37–40
37. Holmgren, A., Kallis, G.-B., Nordstrom, B. 1981. *J. Biol. Chem.* 256:3118–24
38. Fischer, H., Hinkle, D. C. 1980. *J. Biol. Chem.* 255:7956–64
39. Tamanoi, F., Engler, M. J., Lechner, R., Orr-Weaver, T., Romano, L. J., et al. 1980. *Mechanistic Studies of DNA Replication and Genetic Recombination*, ed. B. Alberts, pp. 411–28. New York: Academic
40. Fuller, C. W., Beauchamp, B. B., Engler, M. J., Lechner, R. L., Matson, S. W., et al. 1983. *Cold Spring Harbor Symp. Quant. Biol.* 47:669–79
41. Kolodner, R., Richardson, C. C. 1977. *Proc. Natl. Acad. Sci. USA* 74:1525–29
42. Kolodner, R., Masamune, Y., LeClerc, J. E., Richardson, C. C. 1978. *J. Biol. Chem.* 253:566–73
43. Kolodner, R., Richardson, C. C. 1978. *J. Biol. Chem.* 253:574–84
44. Scherzinger, E., Lanka, E., Morelli, G., Seiffer, D., Yuki, A. 1977. *Eur. J. Biochem.* 72:543–58
45. Hillenbrand, G., Morelli, G., Lanka, E., Scherzinger, E. 1978. *Cold Spring Harbor Symp. Quant. Biol.* 43:449–59
46. Tabor, S., Richardson, C. C. 1981. *Proc. Natl. Acad. Sci. USA* 78:205–9
47. Hinkle, D. C., Richardson, C. C. 1975. *J. Biol. Chem.* 250:5523–29
48. Reuben, R. C., Gefter, M. L. 1974. *J. Biol. Chem.* 249:3843–50
49. Araki, H., Ogawa, H. 1981. *Mol. Gen. Genet.* 183:66–73
50. Scherzinger, E., Lanka, E., Hillenbrand, G. 1977. *Nucleic Acids Res.* 4:4151–63
51. Romano, L. J., Richardson, C. C. 1979. *J. Biol. Chem.* 254:10476–82
52. Romano, L. J., Richardson, C. C. 1979. *J. Biol. Chem.* 254:10483–89
53. Fujiyama, A., Kohara, Y., Okazaki, T. 1981. *Proc. Natl. Acad. Sci. USA* 78:903–7
54. Richardson, C. C., Romano, L. J., Kolodner, R., LeClerc, J. E., Tamanoi, F., et al. 1979. *Cold Spring Harbor Symp. Quant. Biol.* 43:427–40
55. Saito, H., Tabor, S., Tamanoi, F., Richardson, C. C. 1980. *Proc. Natl. Acad. Sci. USA* 77:3917–21
56. Hinkle, D. C. 1980. *J. Virol.* 34:136–41
57. Wever, G. H., Fischer, H., Hinkle, D. C. 1980. *J. Biol. Chem.* 255:7965–72
58. Romano, L. J., Tamanoi, F., Richardson, C. C. 1981. *Proc. Natl. Acad. Sci. USA* 78:4107–11
59. Scherzinger, E., Lauppe, H. F., Voll, N.,

Wanke, M. 1980. *Nucleic Acids Res.* 8:1287–305
60. Campbell, J. L., Tamanoi, F., Richardson, C. C., Studier, F. W. 1978. *Cold Spring Harbor Symp. Quant. Biol.* 43:441–48
61. Wood, W. B., Revel, H. R. 1976. *Bacteriol. Rev.* 40:847–68
62. Broker, T. R., Doermann, A. H. 1975. *Ann. Rev. Genet.* 9:213–44
63. Mosig, G., Benedick, S., Ghosal, D., Luder, A., Dannenberg, R., Bock, S. 1980. See Ref. 39, pp. 527–43
64. Mosig, G., Luder, A., Rowen, L., Macdonald, P., Bock, S. 1981. *The Initiation of DNA Replication*, ed. D. Ray, pp. 277–95. New York: Academic
65. Luder, A., Mosig, G. 1982. *Proc. Natl. Acad. Sci. USA* 79:1101–05
66. Morris, C. F., Bittner, M. 1981. *J. Supramol. Struct. (Suppl.)* 5:334
66a. Macdonald, P. M., Seaby, R. M., Brown, W., Mosig, G. 1983. *Microbiology 1983*, ed. D. Schlesinger. In press
66b. King, G. J., Huang, W. M. 1982. *Proc. Natl. Acad. Sci. USA* 79:7248–52
67. Halpern, M. E., Mattson, T., Kozinski, A. 1979. *Proc. Natl. Acad. Sci. USA* 76:6137–41
68. Liu, L. F., Liu, C.-C., Alberts, B. M. 1979. *Nature* 281:456–61
69. Stettler, G. L., King, G. J., Huang, W. M. 1979. *Proc. Natl. Acad. Sci. USA* 76:3737–41
70. Seasholtz, A. F., Greenberg, G. R. 1983. *J. Biol. Chem.* 258:1221–28
71. Silver, L. L., Nossal, N. G. 1978. *Cold Spring Harbor Symp. Quant. Biol.* 43:489–94
72. Liu, C.-C., Burke, R. L., Hibner, U., Barry, J., Alberts, B. M. 1978. *Cold Spring Harbor Symp. Quant. Biol.* 43:469–87
73. Nossal, N. G., Peterlin, B. M. 1979. *J. Biol. Chem.* 254:6032–37
74. Bittner, M., Burke, R. L., Alberts, B. M. 1979. *J. Biol. Chem.* 254:9565–72
75. Sinha, N. K., Morris, C. F., Alberts, B. M. 1980. *J. Biol. Chem.* 255:4290–4303
76. Alberts, B. M., Barry, J., Bedinger, P., Burke, R. L., Hibner, U., et al. 1980. See Ref. 39, pp. 449–71
77. Venkatesan, M., Silver, L. L., Nossal, N. G. 1982. *J. Biol. Chem.* 257:12426–34
78. Hibner, U., Alberts, B. M. 1980. *Nature* 285:300–5
79. Sinha, N. K., Haimes, M. D. 1981. *J. Biol. Chem.* 256:10671–83
80. Epstein, R. H., Bolle, A., Steinberg, C. M., Kellenberger, E., Boy de la Tour, E., et al. 1964. *Cold Spring Harbor Symp. Quant. Biol.* 28:375–94

81. Warner, H. R., Hobbs, M. D. 1967. *Virology* 33:376–84
82. Yegian, C. D., Mueller, M., Selzer, G., Russo, V., Stahl, F. W. 1971. *Virology* 46:900–19
83. Chiu, C.-S., Cox, S. M., Greenberg, G. R. 1980. *J. Biol. Chem.* 255:2747–51
84. Allen, J. R., Reddy, G. P. V., Lasser, G. W., Mathews, C. K. 1980. *J. Biol. Chem.* 255:7583–88
85. Mathews, C. K., North, T. W., Reddy, G. P. V. 1979. *Adv. Enzyme Regul.* 17:133–56
86. Chao, J., Leacher, M., Karam, J. 1977. *J. Virol.* 24:557–63
86a. Vhiu, C.-S., Cook, K. S., Greenberg, G. R. 1982. *J. Biol. Chem.* 257:15087–97
87. Nossal, N. G. 1980. *J. Biol. Chem.* 255:2176–82
88. Liu, C.-C., Alberts, B. M. 1980. *Proc. Natl. Acad. Sci. USA* 77:5698–5702
89. Liu, C.-C., Alberts, B. M. 1981. *J. Biol. Chem.* 256:2821–29
90. Kurosawa, Y., Okazaki, T. 1979. *J. Mol. Biol.* 135:841–61
91. Silver, L. L., Venkatesan, M., Nossal, N. G. 1980. See Ref. 39, pp. 475–84
92. Silver, L. L., Nossal, N. G. 1982. *J. Biol. Chem.* 257:11696–705
93. Nossal, N. G. 1979. *J. Biol. Chem.* 254:6026–31
94. Morris, C. F., Moran, L. A., Alberts, B. M. 1979. *J. Biol. Chem.* 254:6797–802
95. Liu, C.-C., Alberts, B. M. 1981. *J. Biol. Chem.* 256:2813–20
96. Burke, R. L., Alberts, B. M., Hosoda, J. 1980. *J. Biol. Chem.* 255:11484–93
97. Williams, K. R., Lo Presti, M. B., Setoguchi, M., Konigsberg, W. K. 1980. *Proc. Natl. Acad. Sci. USA* 77:4614–17
98. Loeb, L. A., Kunkel, T. A. 1982. *Ann. Rev. Biochem.* 51:429–58
98a. Fersht, A. R. 1981. *Proc. R. Soc. London Ser. B* 212:351–79
99. Newport, J. W., Kowalczykowski, S. C., Lonberg, N., Paul, L. S., von Hippel, P. H. 1980. See Ref. 39, pp. 485–506
100. Das, S. K., Fujimura, R. K. 1979. *J. Biol. Chem.* 254:1227–32
101. Mace, D. C. 1975. In vitro interactions of some T4 replication proteins. PhD thesis. Princeton Univ., Princeton, N.J.
102. Piperno, J. R., Alberts, B. M. 1978. *J. Biol. Chem.* 253:5174–79
103. Huang, C. C., Hearst, J. E., Alberts, B. M. 1981. *J. Biol. Chem.* 256:4087–94
104. Roth, A. C., Nossal, N. G., Englund, P. T. 1982. *J. Biol. Chem.* 257:1267–73
105. Piperno, J. R., Kallen, R. G., Alberts, B. M. 1978. *J. Biol. Chem.* 253:5180–85
106. Spicer, E., Noble, J., Nossal, N. G., Konigsberg, W. H., Williams, K. R. 1982. *J. Biol. Chem.* 257:8972–79
107. Wu, R., Geiduschek, E. D., Cascino, A. 1975. *J. Mol. Biol.* 96:539–64
108. Venkatesan, M., Nossal, N. G. 1982. *J. Biol. Chem.* 257:12435–43
109. Bedinger, P., Alberts, B. M. 1982. *J. Biol. Chem.* 258: In press
110. Gillin, F. D., Nossal, N. G. 1976. *J. Biol. Chem.* 251:5219–24
111. Gauss, P., Doherty, D. H., Gold, L. 1983. *Proc. Natl. Acad. Sci. USA* 80:1669–73
112. Behme, M. T., Ebisuzaki, K. 1975. *J. Virol.* 15:50–54
113. Krell, H., Durwald, H., Hoffman-Berling, H. 1979. *Eur. J. Biochem.* 93:387–95
114. Saito, H., Richardson, C. C. 1981. *J. Virol.* 37:343–51
114a. Alberts, B. M., Barry, J., Bedinger, P., Formosa, T., Jongeneel, C. V., Kreuzer, K. 1983. *Cold Spring Harbor Symp. Quant. Biol.* 47: In press
115. Bouché, J. P., Zechel, K., Kornberg, A. 1975. *J. Biol. Chem.* 250:5995–6001
116. Zechel, K., Bouché, J. P., Kornberg, A. 1975. *J. Biol. Chem.* 250:4684–89
117. Wickner, S. 1977. *Proc. Natl. Acad. Sci. USA* 74:2815–19
118. Benz, E. W., Reinberg, D., Vicuna, R., Hurwitz, J. 1980. *J. Biol. Chem.* 255:1096–1106
119. Sims, J., Capon, D., Dressler, D. 1980. *J. Biol. Chem.* 254:12615–28
120. Sims, J., Benz, E. W. 1980. *Proc. Natl. Acad. Sci. USA* 77:900–4
121. Rowen, L., Kornberg, A. 1978. *J. Biol. Chem.* 253:770–74
122. Rowen, L., Kornberg, A. 1978. *J. Biol. Chem.* 253:758–64
123. Bouché, J. P., Rowen, L., Kornberg, A. 1978. *J. Biol. Chem.* 253:765–69
124. Arai, K., Kornberg, A. 1981. *J. Biol. Chem.* 256:5267–72
125. Arai, K., Kornberg, A. 1979. *Proc. Natl. Acad. Sci. USA* 76:4308–12
126. Wickner, S., Hurwitz, J. 1974. *Proc. Natl. Acad. Sci. USA* 71:4120–24
127. Schekman, R., Weiner, J. H., Weiner, A., Kornberg, A. 1975. *J. Biol. Chem.* 250:5859–65
128. Wickner, S., Wright, M., Hurwitz, J. 1974. *Proc. Natl. Acad. Sci. USA* 71:783–87
129. Reha-Krantz, L., Hurwitz, J. 1978. *J. Biol. Chem.* 253:4051–57
130. Arai, K., Kornberg, H. 1981. *J. Biol. Chem.* 256:5253–59
131. Ueda, K., McMacken, R., Kornberg, A. 1978. *J. Biol. Chem.* 253:261–69
132. Arai, K., Low, R., Kobori, J., Shlomai, J., Kornberg, A. 1981. *J. Biol. Chem.* 256:5273–80

133. Arai, K., Kornberg, A. 1981. *J. Biol. Chem.* 256:5260–66
134. Arai, K., Low, R. L., Kornberg, A. 1981. *Proc. Natl. Acad. Sci. USA* 78:707–11
135. Wickner, S. 1978. *The Single-Stranded DNA Phages*, ed. D. T. Denhardt, D. Dressler, D. S. Ray, pp. 255–71. Cold Spring Harbor, NY: Cold Spring Harbor Lab.
136. Arai, K., McMacken, R., Yasuda, S., Kornberg, A. 1981. *J. Biol. Chem.* 256:5281–86
137. Low, R. L., Shlomai, J., Kornberg, A. 1982. *J. Biol. Chem.* 257:6242–50
138. Reinberg, D., Zipursky, S. L., Hurwitz, J. 1981. *J. Biol. Chem.* 256:13143–51
139. Low, R. L., Arai, K., Kornberg, A. 1981. *Proc. Natl. Acad. Sci. USA* 78:1436–40
140. Shlomai, J., Kornberg, A. 1980. *J. Biol. Chem.* 255:6794–98
141. Wickner, S., Hurwitz, J. 1975. *Proc. Natl. Acad. Sci. USA* 72:921–25
142. Wickner, S., Hurwitz, J. 1975. *DNA Synthesis and Its Regulation*, ed. M. Goulian, P. Hanawalt, C. F. Fox, pp. 227–38. Menlo Park, CA: Benjamin Cummings
143. Kobori, J., Kornberg, A. 1982. *J. Biol. Chem.* 257:13770–75
144. Kobori, J., Kornberg, A. 1982. *J. Biol. Chem.* 257:13763–69
145. Reha-Krantz, L., Hurwitz, J. 1978. *J. Biol. Chem.* 253:4043–50
146. Arai, K., Yasuda, S., Kornberg, A. 1981. *J. Biol. Chem.* 256:5247–52
147. Shlomai, J., Kornberg, A. 1980. *J. Biol. Chem.* 255:6789–93
147a. Wold, M. S., McMacken, R. 1982. *Proc. Natl. Acad. Sci. USA* 79:4907–11
147b. Rowen, L., Kobori, J., Scherer, S. 1982. *Mol. Gen. Genet.* 187:501–9
147c. Smiley, B. L., Lupski, J. R., Svec, P. S., McMacken, R., Godson, G. N. 1982. *Proc. Natl. Acad. Sci. USA* 79:4550–54
148. Arai, K., Kornberg, A. 1981. *Proc. Natl. Acad. Sci. USA* 78:69–73
149. Wickner, S., Hurwitz, J. 1975. *Proc. Natl. Acad. Sci. USA* 72:3342–46
150. Shlomai, J., Kornberg, A. 1980. *Proc. Natl. Acad. Sci. USA* 77:799–803
151. Zipursky, S. L., Marians, K. J. 1980. *Proc. Natl. Acad. Sci. USA* 77:6521–25
152. Marians, K. J., Soeller, W., Zipursky, S. L. 1982. *J. Biol. Chem.* 257:5656–62
153. Zipursky, S. L., Marians, K. J. 1981. *Proc. Natl. Acad. Sci. USA* 78:6111–15
154. Geider, K., Beck, E., Schaller, H. 1978. *Proc. Natl. Acad. Sci. USA* 75:645–49
155. Meyer, T. F., Geider, K. 1980. See Ref. 39, pp. 579–88
156. Schaller, H. 1978. *Cold Spring Harbor Symp. Quant. Biol.* 43:401–08
157. Suggs, S. V., Ray, D. 1978. *Cold Spring Harbor Symp. Quant. Biol.* 43:379–88
158. Horiuchi, K., Ravetch, J. V., Zinder, N. D. 1978. *Cold Spring Harbor Symp. Quant. Biol.* 43:389–99
159. Ray, D. S., Cleary, J. M., Hines, J. C., Kim, M. H., Strathearm, M., et al. 1981. See Ref. 64, pp. 169–93
160. Vicuna, R., Hurwitz, J., Wallace, S., Giaard, M. 1977. *J. Biol. Chem.* 252:2524–33
161. Vicuna, R., Ikeda, J., Hurwitz, J. 1977. *J. Biol. Chem.* 252:2534–44
162. Kaguni, J. M., Kornberg, A. 1982. *J. Biol. Chem.* 257:5437–43
163. McHenry, C., Kornberg, A. 1977. *J. Biol. Chem.* 252:6478–84
164. McHenry, C. S. 1982. *J. Biol. Chem.* 257:2657–63
165. McHenry, C. S. 1980. See Ref. 39, pp. 569–88
166. McHenry, C. S., Crow, W. 1979. *J. Biol. Chem.* 254:1748–53
167. Kornberg, T., Gefter, M. 1972. *J. Biol. Chem.* 247:5369–75
168. Livingston, D., Hinkle, D., Richardson, C. 1975. *J. Biol. Chem.* 250:461–69
169. Hurwitz, J., Wickner, S. 1974. *Proc. Natl. Acad. Sci. USA* 71:6–10
170. Sumida-Yasumoto, C., Ikeda, J. E., Benz, E., Marians, K. J., Vicuna, R., et al. 1978. *Cold Spring Harbor Symp. Quant. Biol.* 43:311–29
171. Otto, B., Bonhoeffer, F., Schaller, H. 1973. *Eur. J. Biochem.* 34:440–47
172. Wickner, S., Hurwitz, J. 1976. *Proc. Natl. Acad. Sci. USA* 73:1053–57
173. Welch, M. M., McHenry, C. S. 1982. *J. Bacteriol.* 152:351–56
174. Burgers, P. M. J., Kornberg, A., Sakakibara, Y. 1981. *Proc. Natl. Acad. Sci. USA* 78:5391–95
175. Hübscher, U., Kornberg, A. 1980. *J. Biol. Chem.* 255:11698–703
176. Hübscher, U., Kornberg, A. 1979. *Proc. Natl. Acad. Sci. USA* 76:6284–88
177. Horiuchi, T., Maki, H., Sekiguchi, M. 1981. *Mol. Gen. Genet.* 181:24–28
178. Spanos, A., Sedgwick, S. G., Yarronton, G. T., Hübscher, U., Banks, G. R. 1981. *Nucleic Acids Res.* 9:1825–39
179. Fay, P. J., Johanson, K. O., McHenry, C. S., Bambara, R. A. 1981. *J. Biol. Chem.* 256:976–83
180. Fay, P. J., Johanson, K. O., McHenry, C. S., Bambara, R. A. 1982. *J. Biol. Chem.* 257:5692–99
181. Johanson, K. O., McHenry, C. S. 1980. *J. Biol. Chem.* 255:10984–90
182. Burgers, P. M. J., Kornberg, A. 1982. *J. Biol. Chem.* 257:11468–73
183. Burgers, P. M. J., Kornberg, A. 1982. *J. Biol. Chem.* 257:11474–78

184. Johanson, K. O., McHenry, C. S. 1982. *J. Biol. Chem.* 257:12310–15
185. Wickner, S. 1976. *Proc. Natl. Acad. Sci. USA* 73:3511–15
186. Wickner, S. H. 1978. *Cold Spring Harbor Symp. Quant. Biol.* 43:303–10
187. Johanson, K. O., McHenry, C. S. 1981. See Ref. 64, pp. 425–36
188. Arai, N., Polder, L., Arai, K., Kornberg, A. 1981. *J. Biol. Chem.* 256:5239–46
189. Arai, K., Arai, N., Shlomai, J., Kornberg, A. 1980. *Proc. Natl. Acad. Sci. USA* 77:3322–26
190. Eisenberg, S., Scott, J. F., Kornberg, A. 1976. *Proc. Natl. Acad. Sci. USA* 73:3151–55
191. Ikeda, J., Yudelevich, A., Shimamoto, N., Hurwitz, J. 1979. *J. Biol. Chem.* 254:9416–28
192. Eisenberg, S., Kornberg, A. 1979. *J. Biol. Chem.* 254:5328–32
193. Duguet, M., Yarranton, G., Gefter, M. 1978. *Cold Spring Harbor Symp. Quant. Biol.* 43:335–43
194. Zipursky, S. L., Reinberg, D., Hurwitz, J. 1980. *Proc. Natl. Acad. Sci. USA* 77:5182–86
194a. Reinberg, D., Zipursky, S. L., Weisbeek, P., Brown, D., Hurwitz, J. 1983. *J. Biol. Chem.* 258:529–37
195. Van Mansfeld, A. D. M., Langeveld, S. A., Weisbeek, P. J., Baas, P. D., Van Arkel, G. A., Jansz, H. S. 1978. *Cold Spring Harbor Symp. Quant. Biol.* 43:331–34
196. Weisbeek, P., Van Mansfeld, F., Kuhlemeier, C., Arkel, G., Langeveld, S. 1981. *Eur. J. Biochem.* 114:501–7
197. Shlomai, J., Polder, L., Arai, K., Kornberg, A. 1981. *J. Biol. Chem.* 256:5233–38
198. Scott, J. F., Eisenberg, S., Bertsch, L. L., Kornberg, A. 1977. *Proc. Natl. Acad. Sci. USA* 74:193–97
199. Eisenberg, S., Griffith, J., Kornberg, A. 1977. *Proc. Natl. Acad. Sci. USA* 74:3198–3202
200. Eisenberg, S. 1980. *J. Virol.* 35:409–13
201. Langeveld, S. A., Van Mansfeld, A. D. M., de Winter, J., Weisbeek, P. J. 1979. *Nucleic Acids Res.* 7:2177–88
201a. van Mansfeld, A. D. M., Langeveld, S. A., Baas, P. D., Jansz, H. S., van der Marel, G. A., Veeneman, G. H., van Boom, J. H. 1980. *Nature* 288:561–66
201b. Heidekamp, F., Baas, P. D., van Boom, J. H., Veeneman, G. H., Zipursky, S. L., Jansz, H. S. 1981. *Nucleic Acids Res.* 9:3335–54
202. Van der Ende, A., Langeveld, S. A., Teertstra, R., Van Arkel, G. A., Weisbeek, P. J. 1981. *Nucleic Acids Res.* 9:2037–53
203. Eisenberg, S., Finer, M. 1980. *Nucleic Acids Res.* 8:5305–15
204. Wolfson, R., Eisenberg, S. 1982. *Proc. Natl. Acad. Sci. USA* 79:5768–72
205. Meyer, T., Geider, K. 1979. *J. Biol. Chem.* 254:12636–41
206. Myer, T., Geider, K. 1979. *J. Biol. Chem.* 254:12642–46
207. Meyer, T. F., Geider, K., Kurz, C., Schaller, H. 1979. *Nature* 278:365–67
208. Horiuchi, K., Zinder, N. D. 1976. *Proc. Natl. Acad. Sci. USA* 73:2341–45
209. Geider, K., Bäumel, I., Meyer, T. F. 1982. *J. Biol. Chem.* 257:6488–93
210. Harth, G., Baümel, I., Meyer, T. F., Geider, K. 1981. *Eur. J. Biochem.* 119:663–68
211. Meyer, T. F., Baümel, I., Geider, K., Bedinger, P. 1981. *J. Biol. Chem.* 256:5810–13
212. Meyer, T. F., Geider, K. 1982. *Nature* 296:828–32
213. Arai, N., Kornberg, A. 1981. *J. Biol. Chem.* 256:5294–98
214. Yarranton, G. T., Gefter, M. L. 1979. *Proc. Natl. Acad. Sci. USA* 76:1658–62
215. Yarranton, G. T., Das, R. H., Gefter, M. L. 1979. *J. Biol. Chem.* 254:12002–06
216. Kornberg, A., Scott, J. F., Bertsch, L. L. 1978. *J. Biol. Chem.* 253:3298–3304
217. Scott, J. F., Kornberg, A. 1978. *J. Biol. Chem.* 253:3292–97
218. Arai, N., Arai, K., Kornberg, A. 1981. *J. Biol. Chem.* 256:5287–93
219. Kuhn, B., Abdel-Monem, M., Krell, H., Hoffmann-Berling, H. 1979. *J. Biol. Chem.* 254:11343–50
220. Klinkert, M., Klein, A., Abdel-Monem, M. 1980. *J. Biol. Chem.* 255:9746–52
221. Oeda, K., Horiuchi, T., Sekiguchi, M. 1982. *Nature* 298:98–100
222. Oka, A., Sugimoto, K., Takanami, M., Hirota, Y. 1980. *Mol. Gen. Genet.* 178:9–20
223. Sugimoto, K., Oka, A., Sugisaki, H., Takanami, M., Nishimura, A., et al. 1979. *Proc. Natl. Acad. Sci. USA* 76:575–79
224. Meijer, M., Beck, E., Hansen, F. G., Bergman, H. E. N., Messer, W., et al. 1979. *Proc. Natl. Acad. Sci. USA* 76:580–84
225. Meijer, M., Messer, W. 1980. *J. Bacteriol.* 143:1049–53
226. Zyskind, J. W., Smith, D. W. 1980. *Proc. Natl. Acad. Sci. USA* 77:2460–64
227. Zyskind, J. W., Harding, N. E., Takeda, Y., Cleary, J. M., Smith, D. W. 1981. See Ref. 64, pp. 13–25
228. Asada, K., Sugimoto, K., Oka, A., Takanami, M., Hirota, Y. 1982. *Nucleic Acids Res.* 10:3745–54

229. Hirota, Y., Oka, A., Sugimoto, K., Asada, K., Sasaki, H., Takanami, M. 1981. See Ref. 64, pp. 1–12
230. Zyskind, J. W., Smith, D. W., Hirota, Y., Mituru, T. 1981. See Ref. 64, pp. 26–28
231. Hirota, Y., Yamada, H., Nishimura, A., Sugimoto, A., Takanami, M. 1981. Prog. Nucleic Acid Res. Mol. Biol. 26:33–48
232. Okazaki, T., Hirose, S., Fujiyama, A., Kohara, Y. 1980. See Ref. 39, pp. 429–47
233. Lother, H., Messer, W. 1981. Nature 294:376–78
234. Lother, H., Buhk, H., Morelli, G., Heimann, B., Chakraborty, T., Messer, W. 1981. See Ref. 64, pp. 57–77
235. Morelli, G., Buhk, H. J., Fisseau, C., Lother, H., Yoshinaga, K. Messer, W. 1981. Mol. Gen. Genet. 184:255–59
236. Morita, M., Sugimoto, K., Oka, A., Takanami, M., Hirota, Y. 1981. See Ref. 64, pp. 29–36
237. Hansen, F. G., Koefoed, S., Von Meyenburg, K., Atlung, T. 1981. See Ref. 64, pp. 37–56
238. Jacq, A., Lother, H., Messer, W., Kohiyama, M. 1980. See Ref. 39, pp. 189–97
239. Jacq, A., Kohiyama, M. 1980. Eur. J. Biochem. 105:25–31
240. Hendrickson, W., Yamaki, H., Murchie, J., King, M., Boyd, D., Schaechter, M. 1981. See Ref. 64, pp. 79–90
241. Nagai, K., Hendrickson, W., Balakrishnan, R., Yamaki, H., Boyd, D., Schaechter, M. 1980. Proc. Natl. Acad. Sci. USA 77:262–66
241a. Yamaguchi, K., Yamaguchi, M., Tomizawa, J. 1982. Proc. Natl. Acad. Sci. USA 79:5347–51
242. Projan, S. J., Wechsler, J. A. 1981. Mol. Gen. Genet. 183:74–77
243. Projan, S. J., Wechsler, J. A. 1981. Mol. Gen. Genet. 182:263–67
244. Fuller, R. S., Kaguni, J. M., Kornberg, A. 1981. Proc. Natl. Acad. Sci. USA 78:7370–74
245. Kaguni, J. M., Fuller, R. S., Kornberg, A. 1982. Nature 296:623–27
246. Staudenbauer, W. L. 1978. Curr. Top. Microbiol. Immunol. 83:93–156
247. Tomizawa, J. 1978. DNA Synthesis: Present and Future, ed. I. Molineux, M. Kohiyama, pp. 797–826. New York: Plenum
248. Veltkamp, E., Stuitje, A. R. 1981. Plasmid 5:76–99
249. Tomizawa, J.-I. 1975. Nature 257:253–54
250. Itoh, T., Tomizawa, J. 1978. Cold Spring Harbor Symp. Quant. Biol. 43:409–17
250a. Itoh, T., Tomizawa, J. 1982. Nucleic Acids Res. 10:5949–65
251. Hillenbrand, G., Staudenbauer, W. L. 1982. Nucleic Acids Res. 10:833–53
252. Gellert, M., O'Dea, M. H., Itoh, T., Tomizawa, J. 1976. Proc. Natl. Acad. Sci. USA 73:4474–78
253. Orr, E., Staudenbauer, W. L. 1981. Mol. Gen. Genet. 181:52–56
254. Itoh, T., Tomizawa, J. 1980. Proc. Natl. Acad. Sci. USA 77:2450–54
255. Tomizawa, J., Itoh, T., Selzer, G., Som, T. 1981. Proc. Natl. Acad. Sci. USA 78:1421–25
256. Tomizawa, J.-I., Itoh, T. 1981. Proc. Natl. Acad. Sci. USA 78:6096–6100
256a. Tomizawa, J., Itoh, T. 1982. Cell 31:575–83
256b. Som, T., Tomizawa, J. 1982. Mol. Gen. Genet. 187:375–83
256c. Selzer, G., Som, T., Itoh, T., Tomizawa, J. 1983. Cell 32. In press
257. Staudenbauer, W. L. 1978. See Ref. 247, pp. 827–38
258. Staudenbauer, W., Scherzinger, E., Lanka, E. 1979. Mol. Gen. Genet. 177:113–20
259. Staudenbauer, W., Lanka, E., Schuster, H. 1978. Mol. Gen. Genet. 162:243–49
260. Conrad, S. E., Campbell, J. L. 1979. Nucleic Acids Res. 6:3289–3303
261. Böldicke, T. W., Hillenbrand, G., Lanka, E., Staudenbauer, W. L. 1981. Nucleic Acids Res. 9:5215–31
262. Nomura, N., Ray, D. 1981. See Ref. 64, pp. 157–68
263. Conrad, S. E., Campbell, J. L. 1979. Cell 18:61–71
264. Ely, S., Staudenbauer, W. L. 1981. Mol. Gen. Genet. 181:29–35
265. Inuzuka, M., Helinski, D. R. 1978. Biochemistry 17:2567–73
266. Inuzuka, M., Helinski, D. R. 1978. Proc. Natl. Acad. Sci. USA 75:5381–85
267. Inuzaka, N., Inuzuka, M., Helinski, D. 1980. J. Biol. Chem. 255:11071–74
268. Stalker, D. M., Shafferman, A., Tolun, A., Kolter, R., Yang, S., Helinski, D. R. 1981. See Ref. 64, pp. 113–24
269. Kolter, R. 1981. Plasmid 5:2–9
269a. Stalker, D. M., Kolter, R., Helinski, D. R. 1982. J. Mol. Biol. 161:31–43
269b. Kolter, R., Helinski, D. R. 1982. J. Mol. Biol. 161:45–56
270. Diaz, R., Nordström, K., Staudenbauer, W. L. 1981. Nature 289:326–28
271. Diaz, R., Staudenbauer, W. L. 1982. Nucleic Acids Res. 10:4687–4702
272. Vapnek, D., Lipman, M. B., Rupp, W. D. 1971. J. Bacteriol. 108:508–14
273. Wilkins, B. M., Hollom, S. E. 1974. Mol. Gen. Genet. 134:143–56
274. Boulnois, G. J., Wilkins, B. M. 1979. Mol. Gen. Genet. 175:275–79

275. Lanka, E., Scherzinger, E., Günther, E., Schuster, H. 1979. *Proc. Natl. Acad. Sci. USA* 76:3632–36

276. Lanka, E., Barth, P. T. 1981. *J. Bacteriol.* 148:769–81

277. Dalrymple, B. P., Boulnois, G. J., Wilkins, B. M., Orr, E., Williams, P. H. 1982. *J. Bacteriol.* 151:1–7

278. Wilkins, B. M., Boulnois, G. J., Lanka, E. 1981. *Nature* 290:217–21

279. Sasakawa, C., Yoshikawa, M. 1978. *J. Bacteriol.* 133:485–91

280. Boulnois, G. J., Wilkins, B. M., Lanka, E. 1982. *Nucleic Acids Res.* 10:855–69

281. Skalka, A. 1977. *Curr. Top. Microbiol. Immunol.* 78:201–37

282. Georgopoulos, C., Tilly, K., Yochem, J., Feiss, M. 1980. See Ref. 39, pp. 609–17

283. Matsubara, K. 1981. *Plasmid* 5:32–52

284. Furth, M. E., Blattner, F. R., McLeester, C., Dove, W. J. 1977. *Science* 198:1046–51

285. Moore, D. D., Denniston-Thompson, K., Kruger, K. E., Furth, M. E., Williams, B. G., et al. 1979. *Cold Spring Harbor Symp. Quant. Biol.* 43:155–63

286. Furth, M., Dove, W. F., Meyer, B. J. 1982. *J. Mol. Biol.* 154:65–83

287. Lusky, M., Hobom, G. 1979. *Gene* 6:137–72

288. Lusky, M., Hobom, G. 1979. *Gene* 6:173–97

289. Hobom, G., Kröger, M., Rak, B., Lusky, M. 1981. See Ref. 64, pp. 245–62

290. Klinkert, J., Kline, A. 1978. *J. Virol.* 25:730–37

291. Furth, M. E., McLeester, C., Dove, W. F. 1978. *J. Mol. Biol.* 126:195–225

292. Tomizawa, J.-I. 1971. *The Bacteriophage Lambda*, ed. A. D. Hershey, pp. 549–52. Cold Spring Harbor, NY: Cold Spring Harbor Lab.

293. Tsurimoto, T., Matsubara, K. 1981. *Mol. Gen. Genet.* 181:325–31

294. Kuypers, B., Reiser, W., Klein, A. 1980. *Gene* 10:195–203

295. Wold, M. S., Mallory, J. B., Roberts, J. D., Lebowitz, J. M., McMacken, R. L. 1982. *Proc. Natl. Acad. Sci. USA* 79:6176–80

296. Tsurimoto, T., Hase, T., Matsubara, H., Matsubara, K. 1982. *Mol. Gen. Genet.* 187:79–86

297. Tsurimoto, T., Matsubara, K. 1981. See Ref. 64, pp. 263–75

298. Tsurimoto, T., Matsubara, K. 1981. *Nucleic Acids Res.* 9:1789–99

299. Klein, A., Linka, E., Schuster, H. 1980. *Eur. J. Biochem.* 105:1–6

300. Klein, A., Reiser, W., Anderl, A. 1980. See Ref. 39, pp. 619–28

301. Günther, E., Lanka, E., Mikolajczyk, M., Schuster, H. 1981. *J. Biol. Chem.* 256:10712–16

302. Günther, E., Mikolajczyk, M., Schuster, H. 1981. *J. Biol. Chem.* 256:11970–73

303. Anderl, A., Klein, A. 1982. *Nucleic Acids Res.* 10:1733–40

304. Tsarimoto, T., Matsubara, K. 1982. *Proc. Natl. Acad. Sci. USA* 79:7639–43

305. Watabe, K., Shih, M.-F., Sugino, A., Ito, J. 1982. *Proc. Natl. Acad. Sci. USA* 79:5245–48

306. Peñalva, M. A., Salas, M. 1982. *Proc. Natl. Acad. Sci. USA* 79:5522–26

Ann. Rev. Biochem. 1983. 52:617–53

GLUCONEOGENESIS AND RELATED ASPECTS OF GLYCOLYSIS

H. G. Hers and L. Hue[1]

Laboratoire de Chimie Physiologique, Université de Louvain and International Institute of Cellular and Molecular Pathology, UCL 75.39 Avenue Hippocrate 75, B-1200 Brussels, Belgium

CONTENTS

PERSPECTIVES AND SUMMARY 618
GENERAL PROPERTIES OF GLUCONEOGENESIS AND GLYCOLYSIS
    IN THE LIVER.. 619
    Gluconeogenesis 619
    Glycolysis... 619
    Control Steps 620
ENZYME PROPERTIES... 620
    Pyruvate Carboxylase and Other Mitochondrial Involvement 620
    Phosphoenolpyruvate Carboxykinase... 623
    Pyruvate Kinase 625
    Fructose 1,6-Bisphosphatase... 626
    Fructose 2,6-Bisphosphatase... 629
    6-Phosphofructokinases.. 629
    Glucose 6-Phosphatase 633
    Glucokinase 634
THE CONTROL OF METABOLIC FLUX BY HORMONES AND OTHER
    EFFECTORS 634
    Cyclic AMP, Glucagon, and β-Adrenergic Agents 634
    Fructose 2,6-Bisphosphate 639
    α-Adrenergic Agents, Vasopressin, and Angiotensin 641
    Glucocorticoids... 643
    Glucose 643
THE FUTILE CYCLES.. 644
    Glucose/Glucose 6-Phosphate Cycle 644
    Fructose 6-Phosphate/Fructose 1,6-Bisphosphate Cycle 646
    Pyruvate/Phosphoenolpyruvate Cycle 647

[1] Present address: Hormone and Metabolic Research Unit, UCL 75.29 Avenue Hippocrate 75, B-1200 Brussels, Belgium.

0066-4154/83/0701-0617$02.00

PERSPECTIVES AND SUMMARY

Glycolysis is the metabolic sequence that converts glucose to pyruvate and lactate (ethanol and CO_2 in yeast) and provides energy under both aerobic and anaerobic conditions. In the liver and other lipogenic tissues, pyruvate formed by glycolysis is in great part used for the synthesis of fatty acids (1). Gluconeogenesis is the reverse of glycolysis; it consumes energy and provides glucose to fasted animals, as well as glucose 6-phosphate to microorganisms grown in the absence of glucose. It is thus essential for survival, because all cells require glucose derivatives for the synthesis of glycolipids, glycoproteins and structural polysaccharides, which are normal constituents of their membranes. Furthermore, glucose is the main fuel of the brain and is also the only energy source for the mammalian erythrocytes and avian retina (1). Because a short period of hypoglycemia may cause irreversible damage to the brain, a major role of the liver is to maintain a constant level of glycemia. This function is ensured by glycogen synthesis and breakdown, which are controlled by various hormones and also by the concentration of glucose itself (2, 3). The glycogen stores of the liver are, however, exhausted after a few hours of fasting, and the supply of glucose relies then entirely on its biosynthesis from nonglucidic stores, essentially amino acids, but also lactate and glycerol. Under these conditions, glycolysis is arrested in the liver and greatly diminished in other tissues, since fatty acids and ketone bodies may serve as alternate fuels and allow for a glucose sparing effect. Another role of gluconeogenesis is the disposal of lactate produced by glycolyzing tissues, such as erythrocytes and muscle, particularly during intense exercise.

Glycolysis and gluconeogenesis have most of their enzymes in common. These enzymes catalyze reversible reactions, the rate of which is controlled essentially by the concentration of substrates and products. Only at three levels are different enzymes used in glycolysis and gluconeogenesis, and they are the points of regulation discussed in this review. The activity of these enzymes is controlled in animals by several hormones and, in all organisms, by a series of metabolites; this control operates by various means, which include covalent modification of enzymes, formation of positive or negative effectors, and synthesis and degradation of protein. During the recent years, important progress has been made along these various lines; new substrates of cyclic AMP-dependent protein kinase have been discovered, but mechanisms not involving cyclic AMP have also been intensively investigated. Another discovery has been that of fructose 2,6-bisphosphate, a potent regulator of both phosphofructokinase and fructose 1,6-bisphosphatase.

Whereas glycolysis occurs in all living cells, gluconeogenesis operates only in liver and kidney, in certain plants and in microorganisms grown in the absence of glucose. In this review, the major interest is on the antagonistic control of glycolysis and gluconeogenesis in the liver, although problems encountered in other cells, mostly yeast, are also alluded to. The control of gluconeogenesis in animals has been the subject of extensive reviews (4–9), which provide additional information and references to early work. Considering, however, the progress made during the last two to three years, it seems appropriate to reconsider the physiological meaning of the various control mechanisms that were proposed. That is the purpose of this review.

GENERAL PROPERTIES OF GLUCONEOGENESIS AND GLYCOLYSIS IN THE LIVER

Gluconeogenesis

In perfused rat livers or in isolated rat hepatocytes, the highest rate of gluconeogenesis (2–2.5 μmol of glucose per min per g of liver) are obtained with fructose or dihydroxyacetone. The fact that lactate and pyruvate are about half as effective indicates that the limiting step in gluconeogenesis from lactate and pyruvate lies before the triose phosphates. Glycerol and alanine are relatively poor precursors, allowing rates about 50% smaller than does lactate. Saturation of the system with lactate or pyruvate occurs at concentrations well above the physiological range (half maximal rate at 2 mM lactate or 1 mM pyruvate versus physiological concentrations of 1 and 0.1 mM), rendering gluconeogenesis greatly dependent upon substrate concentration (10–12). In pigeon or chicken livers, pyruvate and alanine are very poor substrates (12–13).

Short-term control of gluconeogenesis can be exerted by hormones such as glucagon, insulin, and catecholamines. The direction of the catecholamine effect depends on the nutritional state. Gluconeogenesis is also controlled by its end product, glucose, and is increased during starvation and diabetes. Glucocorticoids appear to exert a permissive action.

Glycolysis

Liver does not glycolyze at high rates, except when in anoxia or when given a large excess of glucose (14, 15). The net glycolytic rate in the presence of physiological concentrations of glucose is close to 0.5 μmol of lactate min^{-1} g^{-1} and can be increased fourfold when glucose concentration is between 20 and 40 mM (14, 16). Glucose by itself is therefore a regulator of liver glycolysis, the other being anoxia, and it has been proposed that the major function of liver glycolysis is not to provide ATP but to allow the transformation of carbohydrate into fat (1, 14). The maximal rates of fatty acid and triglyceride synthesis [1–1.7 μmol of C_2 units/min per g (14, 15)] are indeed compatible with the glycolytic rate.

Control Steps

Identification of control steps in a metabolic pathway can be obtained by studying the changes in patterns of metabolite concentration during active gluconeogenesis and glycolysis. Two crossovers of metabolites have been observed, one at the level of fructose 6-phosphate/fructose 1,6-bisphosphate, the other at the level of pyruvate/P-enolpyruvate (17, 18, 19). These two steps should be controlled synchronously by a concerted mechanism, in order to prevent excessive accumulation of intermediary metabolites.

ENZYME PROPERTIES

Pyruvate Carboxylase and Other Mitochondrial Involvement

As Halestrap (20) pointed out, liver mitochondria have a relatively low respiratory-chain activity per mg of protein, and this is to be put in relation with their participation in several major metabolic pathways, such as gluconeogenesis and urea synthesis. Indeed, the first step of gluconeogenesis from pyruvate is catalyzed by pyruvate carboxylase, an exclusively mitochondrial enzyme (21, 22), whereas the second reaction controlled by P-enolpyruvate carboxykinase (see next section), is both cytosolic and mitochondrial with large variations from species to species. Transport of pyruvate and oxaloacetate or, in some species, P-enolpyruvate, in and out of the mitochondria are therefore included in the gluconeogenic pathway. Another role of the mitochondria is in the exchange of reducing equivalents required or produced by gluconeogenesis, depending on the substrate. Finally, pyruvate dehydrogenase is another exclusively mitochondrial enzyme that, although it is not on the path of gluconeogenesis, could play a role in its control by removing its substrate. Because its activity is not modified in conditions that affect gluconeogenesis (23, 24), its properties are not discussed in this review. Pyruvate dehydrogenase activity is, however, increased by vasopressin, phenylephrine, and angiotensin in livers of fed, but not of fasted, rats (25, 26). This effect may be related to the stimulatory action of these agents on glycolysis under the same conditions.

GENERAL PROPERTIES AND ROLE OF PYRUVATE CARBOXYLASE   Pyruvate carboxylase is a tetrameric protein of $\sim M_r$ 500,000, containing four biotinyl groups (27). The apparent K_m of the rat liver enzyme for pyruvate is in the physiological range of substrate concentration, which explains that the rate of gluconeogenesis from pyruvate is greatly dependent upon the substrate concentration. The activity of the enzyme is greatly stimulated by acetyl-CoA, favored by a high ATP/ADP ratio, and inhibited by glutamate (28). The maximal activity of the rat liver enzyme is close to 7 μmol $\cdot$ min^{-1} g^{-1} and is little influenced by diabetes, fasting, and refeeding (12). This lack of nutritional effect may be related to the need of the enzyme not only for gluconeogenesis, which predominates during fasting, but also for fatty acid synthesis (29), which operates only in the fed state. This is because pyruvate plays a catalytic role in the transfer of acetyl-CoA as citrate from the mitochondria to the cytosol associated with a stoichiometric generation of NADPH from NADH in the cytosol. This occurs by the action of pyruvate carboxylase, citrate synthase, citrate lyase, cytosolic malic dehydrogenase, malic enzyme, and pyruvate and citrate carriers. These considerations, together with the fact that pyruvate carboxylase is not on the path of gluconeogenesis from a number of amino acids, suggest that the enzyme is not a major site of hormonal control. However, the stimulation by glucagon of gluconeogenesis from pyruvate is accompanied by a decreased concentration of this substrate (18, 30), in great part explained by the inactivation of pyruvate kinase. The decrease in substrate concentration needs therefore to be at least compensated by a stimulation of pyruvate carboxylase; the possible mechanism is discussed in the following paragraphs.

HORMONAL EFFECTS ON MITOCHONDRIA   Haynes and co-workers first reported that mitochondria isolated from rats treated with glucagon, epinephrine, or cortisol (31), or from isolated hepatocytes incubated in the presence of glucagon, epinephrine, or cyclic AMP (32) have an increased rate of pyruvate metabolism, including a faster rate of pyruvate carboxylation. It then became progressively apparent that the faster metabolization of pyruvate was only one among many effects of glucagon on isolated mitochondria (reviewed in 20, 33, 34). The most striking effect was an increase in respiratory-chain activity (35), itself responsible for the proton-moted pyruvate transport. Other effects such as an increase in the activity of succinate dehydrogenase and glutaminase are unrelated to the respiratory chain activation. Similar effects were also obtained with α-adrenergic agents, being then independent of cyclic AMP. Recently several of the mitochondrial effects, including the faster rate of pyruvate carboxylation (36), but not all of them (37), were reported to be markedly smaller when mannitol rather than sucrose was used for the isolation of the mitochondrial fraction, or when agents known to decrease the fragility of the mitochondria were added to the preparation. These observations have led to the proposal that the effect of glucagon is to stabilize the mitochondria against inactivation during the process of their isolation, rather than to truly activate them (36). According to this view, the faster rate of pyruvate carboxylation observed with isolated mitochondria after glucagon would, therefore, not be representative of the in vivo situation. However, some of the glucagon-induced changes were found both in isolated mitochondria and in the liver of intact rats or in isolated hepatocytes and, therefore, do not suffer the same criticism. They include an increase in respiration, ATP content, glutamate consumption, and malate accumulation (33, 38–41). A high ATP and low glutamate concentration would favor pyruvate carboxylation.

It is not possible presently to define the molecular mechanism by which glucagon, cyclic AMP, α-adrenergic agents, and cortisol induce the mitochondrial changes. It however remains that these changes exist and may contribute to the overall stimulatory effect of glucagon and α-adrenergic agents on gluconeogenesis from pyruvate, lactate, and alanine.

CONTROL BY ACETYL-COA   Another important factor in the control of pyruvate carboxylase activity is the concentration of acetyl-CoA, which tends to increase after a glucagon treatment. The intramitochondrial concentration (0.6 mM) of acetyl-CoA is far above the K_a of the enzyme for its activator (15 μM). As discussed in detail elsewhere (28, 42, 43), a series of assumptions can, however, be made concerning the binding of acetyl-CoA to several mitochondrial proteins, the presence of competitive inhibitors, and the actual value of K_a at the intramitochondrial concentration of pyruvate carboxylase, which is many times greater than in the in vitro system. All these factors may explain that a relationship could be established between the concentration of acetyl-CoA and the rate of pyruvate carboxylation by isolated liver mitochondria with an apparent K_a of 0.17 (22) or 0.3 mM (28). Furthermore, the addition of fatty acids to perfused rat livers leads to both an increase in acetyl-CoA concentration and in gluconeogenic rate from lactate or pyruvate, although not from glutamine or fructose (11, 44–46). Conversely, the addition of 4-pentenoate, an inhibitor of fatty acid oxidation, causes a parallel decrease in acetyl-CoA and gluconeogenesis (47).

TRANSPORT OF METABOLITES AND EXCHANGE OF REDUCING EQUIVALENTS   The transport of pyruvate across the mitochondrial membrane is specifically inhibited by α-cyano-4-hydroxycinnamate (48). The finding that even a small amount of this compound inhibits gluconeogenesis in isolated hepatocytes indicates that pyruvate transport is limiting (49). This observation, however, was not confirmed by others (34).

Mitochondria provide reducing equivalents needed for the conversion of phosphoglycerate to triose phosphates when pyruvate is converted into glucose; the $NAD^+/NADH$ ratio is indeed much greater in the cytosol than in the mitochondria (500–700 vs 5–8) (50). In the mitochondria, oxaloacetate is reduced to malate, which is then transferred to the cytosol, where it regenerates oxaloacetate and NADH. When lactate is the source of carbon, it provides both NADH and pyruvate, and no additional reducing equivalent is needed. However, because the transport of oxaloacetate is a slow process, this metabolite needs first to be converted in the mitochondria to aspartate, which can cross the membrane and give rise in the cytosol to oxaloacetate and an amino group. Transaminases are thus involved and this explains that gluconeogenesis from lactate in rat liver is inhibited by aminooxyacetate, a transaminase inhibitor (51). This inhibition does not occur with the pigeon liver (52) in which gluconeogenesis is intra-mitochondrial up to the stage of P-enolpyruvate, a metabolite that can easily be transported in the cytosol. The fact that no reducing equivalent is conveyed to the cytosol by this mechanism may explain that in the pigeon liver gluconeogenesis from pyruvate and from amino acids is much slower than from lactate (12, 13, 52).

When reduced substrates, like sorbitol or glycerol are converted into glucose, reducing equivalents are formed and need to be transferred into the mitochondria. Because sorbitol is converted to glucose at the same rate as fructose, it appears that the reoxidation of NADH is not limiting. In contrast, gluconeogenesis from sorbitol and glycerol is impaired in hepatocytes incubated in the absence of calcium (53, 54). This is in agreement with the belief that cytosolic Ca^{2+} stimulates transfer of reducing equivalents from cytosol to mitochondria by enhancing the glycerol 3-phosphate shuttle (55).

Phosphoenolpyruvate Carboxykinase

GENERAL PROPERTIES P-enolpyruvate carboxykinase catalyzes the first step of gluconeogenesis, which is common to pyruvate and a number of amino acids. Its intracellular distribution varies between species: in rat and mouse liver, it is almost exclusively cytosolic, and in pigeon and rabbit liver, it is almost totally mitochondrial, whereas in guinea pig and man it is equally distributed between the cytosol and mitochondria (12, 56). Both the mitochondrial (57) and the cytosolic (58) enzymes are 74,000-dalton monomers. However, they differ by their kinetic and immunochemical properties (58–61). The apparent K_m of the rat cytosolic enzyme for oxaloacetate is in the micromolar range (61–62) and about the same as the cytosolic concentration of oxaloacetate for which values ranging from 5 to 50 μM have been reported (40, 63, 64). This concentration can be slightly increased by glucagon (19), but this effect has not always been observed (40).

SHORT-TERM REGULATION   Following their discovery that the administration of tryptophan to rats causes an inhibition of gluconeogenesis, Lardy and co-workers (reviewed in 65) have emphasized the role of Fe^{2+} in the regulation of P-enolpyruvate carboxykinase. They showed that, in the cell, the requirement of P-enolpyruvate carboxykinase for a divalent metal ion is probably fulfilled by Fe^{2+}, the action of which requires the presence of a protein called the ferroactivator. This ferroactivator is present in liver, kidney, and erythrocytes, but not in heart and skeletal muscle; its concentration is doubled in livers of starved and diabetic rats. Several ferroactivators have been isolated from liver and erythrocytes; one of them is a hemoprotein, very similar, if not identical, to catalase (66), an enzyme that is essentially peroxisomal in the liver. The effect of the ferroactivator is mimicked by 3-aminopicolinate, a ferrochelator, which can stimulate P-enolpyruvate carboxykinase in the presence of Fe^{2+}. The same compound causes a sustained hyperglycemia in the intact rat and stimulates glucose formation from [^{14}C]lactate, although not from glycerol or xylitol. More recently, it has been observed that mitochondria can release Fe^{2+} when incubated in the presence of physiological concentrations of Ca^{2+} (67). From these and other observations it has been proposed that a change in the concentration of cytosolic Ca^{2+}, as could be induced by α-adrenergic agents, stimulates P-enolpyruvate carboxykinase activity through the liberation of Fe^{2+} from mitochondrial stores (65, 67). In this view, the concentration of Fe^{2+} rather than that of ferroactivator would be the limiting factor in the activation of P-enolpyruvate carboxykinase.

The physiological meaning of this elegant mechanism should, however, be considered with caution. The lack of correlation between the presence of ferroactivator and of P-enolpyruvate carboxykinase in tissues and the peroxisomal location of catalase do not favor a role of that protein in the control of gluconeogenesis. On the other hand, according to Reynolds (68), the effect of ferrochelators on P-enolpyruvate carboxykinase is to protect the enzyme against the inactivation by Fe^{2+} rather than to activate it. Finally, two types of experimental approach indicate that P-enolpyruvate carboxykinase is not the target of short-term hormonal action. First, although glucose formation from lactate is inhibited by micromolar concentrations of mercaptopicolinate (69), a noncompetitive inhibitor of P-enolpyruvate carboxykinase [$K_i = 3$ μM (70)], a further analysis of these data allowed the conclusion (71) that this enzyme exerts only little control on gluconeogenesis from lactate. Secondly, in hepatocytes cultured in the absence of glucagon, P-enolpyruvate carboxykinase activity was greatly decreased and, most likely, was present in a limiting amount; however, gluconeogenesis was completely insensitive to stimulation by glucagon, which did nevertheless inhibit glycolysis (72).

LONG-TERM CONTROL   By contrast, the long-term control of P-enolpyruvate carboxykinase activity is well established. Up to fivefold differences in liver cytosolic enzyme activity and rate of synthesis have been reported to occur as adaptive changes to dietary and hormonal stimuli. Starvation, diabetes, glucagon, cyclic AMP, adrenaline, and gluco-corticoids are known to increase the hepatic content of the enzyme (up to 10 U/g), whereas insulin and refeeding have the reverse effect (56). This induction of P-enolpyruvate carboxykinase is associated with an increased synthesis of mRNA as measured by hybridization with a cDNA probe (73–75).

Pyruvate Kinase

GENERAL PROPERTIES   At least three different isoenzymes of pyruvate kinase have been identified in mammalian tissues. Muscle and brain contain the type M isoenzyme; the type L is the major isoenzyme in liver but a minor one in kidney. Pyruvate kinase type A (also called K or M_2) is present in most other tissues including kidney, adipose tissue, and lung. These enzymes are tetrameric proteins of $M_r \sim 250,000$ (76). The three classes can be distinguished by their kinetic and immunochemical properties (6, 77). Type A and L pyruvate kinase, but not type M, are allosterically regulated; the most active conformation is favored by P-enolpyruvate, fructose 1,6-bisphosphate, and low pH. The less active conformation, which is evidenced by sigmoidal kinetics with respect to P-enolpyruvate, is favored by ATP, alkaline pH, alanine, and several other amino acids. The enzyme has an absolute requirement for monovalent cations (76). The kinetics characteristics of the liver enzyme are such that in the presence of physiological concentrations of P-enolpyruvate, the activity is low and can be stimulated greatly by micromolar concentrations of fructose 1,6-bisphosphate (76, 77).

PHOSPHORYLATION AND DEPHOSPHORYLATION   Purified liver pyruvate kinase is phosphorylated by cyclic AMP-dependent protein kinase (78). Phosphorylation results in the incorporation of two (79) to four (78, 80) phosphates/tetramer and causes changes in the kinetic properties of the enzyme. After phosphorylation, the apparent affinity for P-enolpyruvate and for fructose 1,6-bisphosphate decreases, whereas the affinity for allosteric inhibitors, such as ATP and alanine, is enhanced; V_{max} remains unaffected (78). Dephosphorylation causes a reversal of these changes (81), and it has been shown that the enzyme responsible for reactivation is different from phosphorylase phosphatase (82). Thus phosphorylation of pyruvate kinase brings about changes in the kinetic properties such that under physiological conditions the activity is greatly diminished. Types A and M pyruvate kinase are not substrates for the cyclic AMP-dependent protein kinase (78–83). However the type M_2 from chicken liver can be phosphorylated by a cyclic AMP-independent protein kinase; this phosphorylation results in a decrease in V_{max} (84, 85).

Ligands of type L pyruvate kinase interfere with the phosphorylation and inactivation of the enzyme. As a rule, effectors that stimulate the activity are also those that prevent its inactivation and vice-versa (86–88). The more active conformation is thus a poor substrate for protein kinase. Similarly, phosphorylation and inactivation of type M_2-pyruvate kinase from chicken liver by cyclic AMP-independent protein kinase are inhibited by fructose 1,6-bisphosphate but stimulated by alanine (89).

HORMONAL EFFECTS   Treatment of perfused rat livers or hepatocytes with glucagon causes a dose-dependent change in the kinetic properties (78, 90–93) and phosphorylation state (94–96) of pyruvate kinase, similar to those observed with the purified enzyme; this can be correlated with the stimulation of gluconeogenesis (91). Furthermore, the proportion of the enzyme in the inactive form is greater during starvation and diabetes than under control conditions (86, 97). No effect of glucagon on pyruvate kinase activity has however been observed in chicken hepatocytes, although the hormone stimulates gluconeogenesis (98).

Treatment of hepatocytes with α-adrenergic agonists and vasopressin causes little (88, 99, 100), if any (49, 101, 102), change in the activity of pyruvate kinase, although they stimulate phosphorylation of the enzyme (100, 103, 104). Attempts to phosphorylate pyruvate kinase in a liver extract in the absence of cyclic AMP (86) or upon addition of Ca^{2+} (105) have been so far unsuccessful.

LONG-TERM REGULATION   Rat liver contains about 50 units of pyruvate kinase per g (106), which is at least fivefold more than the content in pyruvate carboxylase and P-enolpyruvate carboxykinase. Pyruvate kinase content is increased up to fivefold by feeding a high carbohydrate diet and is decreased threefold during starvation (107–109). Changes in pyruvate kinase content in response to dietary stimuli are related to alterations in the amount of functional mRNA coding for the enzyme (110).

Fructose 1,6-Bisphosphatase

GENERAL PROPERTIES OF THE LIVER ENZYME   Liver fructose 1,6-bisphosphatase (reviewed in 6–8, 111–114) is a tetrameric protein with M_r $\sim 140,000$. Its activity is magnesium dependent and is increased several fold in the presence of histidine, imidazole, or EDTA, which chelate an inhibitory metal, most likely Zn^{2+} (115, 116). Its concentration in the liver of a fed rat is ~ 15 U per g, i.e. five times that of phosphofructokinase (106). The major factors to be taken into consideration in the physiological control of its activity are the concentration of its substrate and of its two inhibitors, AMP and fructose 2,6-bisphosphate. Except for histidine, no naturally occurring positive effector of the enzyme is known.

The kinetics of fructose 1,6-bisphosphatase is Michaëlian, with a slight inhibition by excess substrate at concentrations above those usually found in the liver. The K_m is 3–5 μM, but could reach 20 μM in the presence of physiological concentrations of monovalent cations (117). The concentration of fructose 1,6-bisphosphate in the liver ranges between 5 and 50 μM; however, because most of it appears to be bound to proteins (42), the concentration of free substrate could lie well below K_m and the reaction rate be first order. P_i, the product of the reaction, is an inhibitor with a K_i in the physiological range of concentration.

INHIBITION BY AMP AND BY FRUCTOSE 2,6-BISPHOSPHATE   AMP (118) and fructose 2,6-bisphosphate (119, 120) are two potent inhibitors of fructose 1,6-bisphosphatase (K_i = 25 and 2.5 μM respectively), and their action is remarkably synergistic. The inhibition by AMP is cooperative (n = 2.35), noncompetitive (118), and less important at low than at high substrate concentration (121). The effect of fructose 2,6-bisphosphate is also slightly cooperative (n = 1.4). The action of the two inhibitors decreases with increasing temperature or pH (118, 119), whereas the affinity for the substrate is not affected by these modifications (122). Furthermore, fructose 2,6-bisphosphate but not the substrate protects the AMP allosteric site against acetylation (121) and proteolysis (123). Limited proteolysis removes inhibition by AMP (118) and decreases inhibition by fructose 2,6-bisphosphate [(123); see however (124)], but has little effect on the activity measured at pH 7.4 in the absence of inhibitors. All these properties indicate that, despite its obvious structural similarity with the substrate, fructose 2,6-bisphosphate does not act primarily by competing for the active site.

A remarkable effect of fructose 2,6-bisphosphate, which is not observed with AMP alone but which is reinforced by it, is to change the saturation curve of the enzyme for its substrate from hyperbolic to sigmoidal, again suggesting an allosteric type of interaction (119). Consequently, the inhibition by fructose 2,6-bisphosphate is much more important at low than at high substrate concentrations; this has led some investigators (120) to believe that the inhibition would be of a strictly competitive nature. This conclusion has been criticized on a technical basis (125). However, the inhibition by fructose 2,6-bisphosphate is competitive with P_i, when the reaction is investigated in the reverse direction (126). Furthermore, binding of fructose 2,6-bisphosphate to the active site is also suggested by the protection that, similarly to the substrate, it exerts on this site under various experimental conditions (121, 123, 124). This protection could, however, also result from a conformational change of the protein. It has been proposed that fructose 2,6-bisphosphate binds to a specific allosteric site, decreasing then the occupancy of the active site by the substrate (122).

FRUCTOSE 1,6-BISPHOSPHATASES FROM OTHER ORIGINS   Except for a greater sensitivity of the muscle enzyme towards AMP inhibition ($K_i = 10^{-7}$ M), the kinetic behavior of muscle and kidney fructose 1,6-bisphosphatases is qualitatively similar to that of the liver enzyme. Fructose 1,6-bisphosphatase from *Saccharomyces cerevisiae* is a dimer of M_r 115,000 (127). It is inhibited by both AMP and fructose 2,6-bisphosphate. However, the latter inhibitor does not induce cooperativity for the substrate and it affects both K_m and V_{max} (125). Higher plants contain two types of fructose 1,6-bisphosphatases; one is cytosolic and participates in sucrose formation, the other is chloroplastic and is involved in photosynthesis. The cytosolic enzyme is inhibited by both AMP and fructose 2,6-bisphosphate in a synergistic manner and both inhibitors are competitive with the substrate (115, 116). The chloroplastic enzyme is activated in the light by an interconversion catalyzed by thioredoxine (130); it is only weakly and nonsynergistically inhibited by high concentrations of both fructose 2,6-bisphosphate and AMP (124, 128, 129).

COVALENT MODIFICATIONS   Fructose 1,6-bisphosphatase from rat liver (131), but not from mouse or rabbit liver or pig kidney (132), is a substrate of cyclic AMP-dependent protein kinase; however, the phosphorylation of the enzyme induces minimal changes in kinetic properties. Phosphorylation of the rat liver enzyme has also been observed in isolated hepatocytes incubated in the presence of glucagon (133).

It has been known for many years that the proteolytic digestion of liver fructose 1,6-bisphosphatase causes the loss of its neutral activity associated with a severalfold increase of activity at alkaline pH (111, 112). Because the sensitivity to AMP inhibition is also greatly affected, it has been hypothesized that limited proteolysis by lysosomal enzymes may play a role in the physiological regulation of fructose 1,6-bisphosphatase, particularly during fasting (112, 134, 135). In *S. cerevisiae*, the proteolysis of fructose 1,6-bisphosphatase appears to play a role in the adaptation to the presence of glucose in the incubation medium (136). Under certain culture conditions (137, 138), but not when the cells are grown on a minimal medium (136), the inactivation of the enzyme proceeds in two steps. The first one occurs within 1 min and is reversible if glucose is removed from the medium. It involves an approximately 50% loss of activity, accompanied by various changes in kinetic properties (138, 139), and by a phosphorylation of the enzyme on a serine residue (140, 141). There is simultaneously a rise in fructose 2,6-bisphosphate (142), and, paradoxically, in cyclic AMP concentration (143–

145). The second step is a slow proteolysis, which is terminated in 1–2 hr. There is presently no proof that the two steps are connected. On the contrary, the fact that the proteolytic system, activated by glucose, maintains its activity on fructose 1,6-bisphosphatase, which was re-synthesized after transfer of the cells in a medium containing ethanol (136), argues against this assumption.

Fructose 2,6-Bisphosphatase

An enzyme that catalyzes the stoichiometric conversion of fructose 2,6-bisphosphate into fructose 6-phosphate and P_i and that can therefore be called a fructose 2,6-bisphosphatase has been purified from rat liver (146). Its concentration in the fresh tissue is ~ 60 mU per g. Contrary to fructose 1,6-bisphosphatase, this enzyme is not entirely magnesium dependent, is not inhibited by AMP and is not stimulated by EDTA. Its K_m is below micromolar [(146); see however (147, 148)], i.e. far below the usual concentration of fructose 2,6-bisphosphate in the liver (1–20 μM). One of its major regulatory properties is that it is strongly inhibited by fructose 6-phosphate, which decreases V_{max} and increases K_m. The inhibition by fructose 6-phosphate is counteracted by several phosphoric esters, including glycerol 3-phosphate, at concentrations normally present in the liver. An unexpected property of fructose 2,6-bisphosphatase is that it is stimulated by nucleotide triphosphates (146).

Like phosphofructokinase 2, from which it could not be separated (see next section), fructose 2,6-bisphosphatase has a M_r of $\sim 100,000$ and is a substrate for cyclic AMP-dependent protein kinase (146–149), which incorporates one phosphate into a 49,000-dalton peptide (148, 149). Phosphorylation of the enzyme causes an up to fourfold increase in activity (146, 147), which is essentially due to an increase in V_{max} (146). This agrees with the little possibility offered by the very low K_m of the enzyme for a regulation by an increase in affinity. The activation of the enzyme by glucagon has also been observed in isolated hepatocytes (146, 147, 149, 150).

6-Phosphofructokinases

The name phosphofructokinase is currently used to designate the enzyme of glycolysis that converts fructose 6-phosphate to fructose 1,6-bisphosphate by transphosphorylation from ATP. Another phosphofructokinase that forms fructose 2,6-bisphosphate by a similar reaction has now been discovered (151). It has been called phosphofructokinase-2, the classical phosphofructokinase becoming then phosphofructokinase-1. This nomenclature accounts for the carbon of fructose 6-phosphate, which is phosphorylated, and for the order of discovery of the two enzymes. A third enzyme, present in some microorganisms and in higher plants, uses pyrophosphate as a phosphoryl donor (152) and is called PP_i-phosphofructokinase.

PHOSPHOFRUCTOKINASE 1 Numerous isoenzymes of phospho-fructokinase-1 (reviewed in 153–157) have been isolated from various types of cell. Their general physicochemical and catalytic properties are usually similar. They are tetramers with M_r 340,000 and most of them show a strong tendency to aggregate.

Effectors Phosphofructokinase-1 is the prototype of a multimodulated enzyme. The regulators are the two substrates ATP and fructose 6-phosphate and a relatively large series of negative [citrate, H^+, and, as recently described (158), glycerol 3-phosphate] or positive [fructose 2,6-bisphosphate, fructose 1,6-bisphosphate, glucose 1,6-bisphosphate, AMP, NH_4^+, P_i, OH^-, and as recently reported (159), 6-phosphogluconate] effectors that reinforce the action of the substrates. ATP acts as a negative allosteric effector, which induces cooperativity for fructose 6-phosphate, whereas fructose 6-phosphate relieves the inhibition by ATP. The two major positive effectors of liver phosphofructokinase-1, relative to their concentration in the cell, are fructose 2,6-bisphosphate and AMP (160, 161). The saturation curves for both effectors are, like that for fructose 6-phosphate, typically sigmoidal and tend to become hyperbolic in the presence of increasing concentrations of any other positive effector, including fructose 6-phosphate.

When comparing the efficiency of various effectors on the activity of phosphofructokinase, it is therefore important to do so in strictly comparable conditions. It then appears that the efficiency of fructose 2,6-bisphosphate is 2–3.5 orders of magnitude greater than that of its isomers, fructose 1,6-bisphosphate and glucose 1,6-bisphosphate (160–162). Considering that in the liver the concentration of fructose 1,6-bisphosphate and fructose 2,6-bisphosphate are very similar and usually change in parallel, it appears that fructose 1,6-bisphosphate cannot be a physiological stimulator of phosphofructokinase, at least in the liver, and possibly in other cells too. Furthermore, fructose 1,6-bisphosphate at high concentration counteracts the positive effect of fructose 2,6-bisphosphate on the erythrocyte (163) and yeast (164) enzymes and can therefore be considered an inhibitor. These observations are important with regard to the longstanding belief that the activity of phosphofructokinase is paradoxically stimulated by its product, fructose 1,6-bisphosphate, and that such an effect would be important in the generation of glycolytic oscillations (165).

Stable modifications Several groups of investigators have observed under various experimental conditions changes in the kinetic properties of phosphofructokinase from various origins. These changes resisted gel filtration or partial purification and were therefore believed to result from a stable modification of the enzyme. They include the inactivation of rat liver phosphofructokinase by glucagon (166, 167), and of the chicken liver enzyme by epinephrine (168), and the activation of the heart muscle enzyme by epinephrine (169), as well as of the phosphofructokinase of quiescent 3T3 cells incubated in the presence of serum, insulin, or epidermal growth factor (170). However, a further investigation of the effect of glucagon showed that inactivation resulted from the destruction of a low-molecular-weight stimulator of the enzyme (171, 172), identified as fructose 2,6-bisphosphate (173). This conclusion was made possible thanks to an extensive gel filtration (at least 20 vol of gel for 1 vol of liver extract) performed in the presence of 0.1 mM-fructose 6-phosphate to stabilize the enzyme. The potential role of fructose 2,6-bisphosphate or of other low-molecular effectors in the other observations (168–170) remains an open problem.

Furthermore, it has been reported that the liver contains a polypeptide that stabilizes phosphofructokinase against heat denaturation, its concentration is reportedly greatly decreased in the liver by fasting and diabetes (174–175). The low phosphofructokinase content of the liver in the two latter conditions has been attributed to a faster rate of degradation, owing to the loss of this stabilizing factor (176). However, the procedure used to assay this factor does not discriminate against fructose 2,6-bisphosphate, the existence of which was unknown when this work was performed. Because fructose 2,6-bisphosphate is a potent stabilizer of phospho-fructokinase-1 (158, 161, 177), and because its concentration is greatly decreased in diabetes and prolonged fasting, it may also play a role in the long term adaptation of the enzyme. The existence and the role of a stabilizing polypeptide needs to be reconsidered (see 177a).

A directly related problem is to know whether phosphorylation of phosphofructokinase-1 (167, 178–182) can modify its catalytic properties. Claus et al (182) have shown that liver phosphofructokinase can be phosphorylated by cyclic AMP-dependent protein kinase, but without change in its catalytic properties. Similar observations were also made in the case of the muscle enzyme, in which only minor kinetic changes were observed (183). The phosphorylation rate of the liver enzyme is increased by fructose 2,6-bisphosphate, as is also its stability during the extraction procedure (158). These effects have been proposed (158) as an explanation for the observation made by Söling and co-workers (178, 179) that phosphorylation of the liver enzyme occurs in the presence of glucose and is associated with an increased maximal velocity. This interpretation accounts for the fact that both glucose and glucagon enhance the incorporation of ^{32}P into phosphofructokinase-1 in isolated hepatocytes, but have opposite effects on the level of fructose 2,6-bisphosphate and on the glycolytic flux. Finally, the claim (184) that a high phosphate-containing form of liver phosphofructokinase has a lower affinity for fructose 2,6-bisphosphate than a low phosphate-containing form has not been confirmed (158).

PHOSPHOFRUCTOKINASE 2   An enzyme partially purified from rat liver catalyzes the transfer of the γ-phosphoryl group of ATP onto carbon 2 of fructose 6-phosphate, causing the stoichiometric formation of fructose 2,6-bisphosphate. Rat liver contains approximately 20 mU of this 6-phosphofructo 2-kinase per g. The kinetics for fructose 6-phosphate is hyperbolic and the affinity for this substrate is increased by P_i and by AMP. In the presence of physiological concentrations of P_i, the K_m for fructose 6-phosphate is 50 μM, which is the physiological concentration in the liver cell (151). Like phosphofructokinase-1, phosphofructokinase-2 is inhibited by citrate, P-enolpyruvate (151), and glycerol 3-phosphate (158), although not by an excess ATP $[K_m = 0.2–0.4\,\text{mM}\ (151, 185)]$. The M_r of the enzyme, as estimated by gel filtration, is close to 100,000 (185, 186, 187). Indirect evidence for the existence of a 6-phosphofructo 2-kinase can also be found in several publications (185, 188, 189).

Similar to fructose 2,6-bisphosphatase, phosphofructokinase-2 is a substrate for cyclic AMP-dependent protein kinase (149, 186, 187, 190), which incorporates one phosphate into a 49,000 dalton peptide (149, 186). After this treatment, the K_m for fructose 6-phosphate is increased and the enzyme becomes less sensitive to its positive effectors and more sensitive to its inhibitors. The result of the phosphorylation is therefore a marked decrease in the activity of phosphofructokinase-2, which is most apparent when its activity is measured at pH 6.6 and in the presence of 5 mM P_i (190). Under these conditions the phosphoenzyme retains about 20% of the activity of the dephospho form. This is also the residual activity observed in isolated hepatocytes incubated in the presence of saturating amounts of glucagon [(151, 186, 189, 190); see however (150)].

The failure to separate phosphofructokinase-2 from fructose 2,6-bisphosphatase by various procedures based on the charge, size and hydrophobicity of the proteins not only indicates that the two enzymes are very similar in their physicochemical properties, but also suggests that they could be parts of a single multifunctional enzyme (146, 147, 149, 190). The latter would then be the first in this group to be modified by phosphorylation and dephosphorylation.

PP$_i$-PHOSPHOFRUCTOKINASE   PP$_i$: fructose 6-phosphate 1-phosphotransferase is widespread in higher plants. The activity of the mung bean enzyme is greatly increased in the presence of fructose 2,6-bisphosphate by an effect on both V_{max} and K_m (191). This effect has been confirmed for the enzymes from other plants (125, 128, 192).

PP_i-phosphofructokinase catalyzes an easily reversible reaction, which could therefore play a role in gluconeogenesis as well as in glycolysis. A glycolytic role is hindered by the presumably low concentration of PP_i in the cell. The kinetic properties of PP_i-phosphofructokinase purified from potato tubers have been investigated both in the forward and in the reverse direction (192). These properties greatly favor the glycolytic function of the enzyme at low fructose 2,6-bisphosphate concentrations. The role of PP_i-phosphofructokinase in plant glycolysis is unknown as are also the conditions that affect the concentrations of fructose 2,6-bisphosphate in plants. Illumination of wheat leaves had no effect on this concentration (129).

Glucose 6-Phosphatase

It is a current belief that, because it occupies a strategic position in the maintenance of glucose homeostasis, glucose 6-phosphatase should be tightly regulated (reviewed in 193, 194). However, the contrary is true, since, no mechanism other than substrate concentration is known to control the activity of the enzyme. Because the concentration of glucose 6-phosphate in the liver is well below K_m (3 mM), the reaction rate is usually first order and the major regulator is glucose 6-phosphate itself. There is a slight inhibition by the two reaction products, glucose and P_i, but this effect is too small to play a role under physiological conditions. The recent proposal that the enzyme could be phosphorylated and simultaneously activated by cyclic AMP-dependent protein kinase (195) could not be confirmed (196). The liver of fed rat contains approximately 14 U of glucose 6-phosphatase per g, and this amount is doubled by starvation or diabetes.

Glucose 6-phosphatase is tightly bound to the endoplasmic reticulum of the liver cells, without the significance of the specific location being apparent. Arion et al (197, 198) proposed that three components of the endoplasmic reticulum participate in the process of glucose 6-phosphate hydrolysis: (a) a glucose 6-phosphate specific transporter that mediates penetration of the hexose phosphate into the microsomal cisternae and is rate limiting for the hydrolysis of glucose 6-phosphate by intact microsomes; (b) a relatively nonspecific phosphohydrolase localized on the luminal site of the reticulum network; and (c) a second translocase controlling the permeability of microsomes to P_i. Whereas the three components would be required for the hydrolysis of glucose 6-phosphate by native microsomes, only the second is necessary after disruption of the membranes by various means. A 54,000-dalton polypeptide, which appears to be implicated in the transport of glucose 6-phosphate, has been isolated (199). The model is supported mostly by the differences in substrate affinity and specificity induced by the disruption of the membranes as well as the effect of inhibitors and a series of kinetic studies. It has received additional support from the discovery that type Ib glycogenosis could be due to a specific defect of the transporter system. This congenital disorder is characterized by a normal activity of glucose 6-phosphatase in frozen livers, but also by its absence in fresh tissue (200, 201).

Glucokinase

Rat liver contains a cytosolic high K_m hexokinase (reviewed in 202, 203), the substrate specificity of which is similar to that of other hexokinases. Whereas the K_m for glucose is in the range of 10 mM, i.e. around physiological, the K_m for other sugars like mannose and fructose is so high that the role of the enzyme is limited to the phosphorylation of glucose. The name glucokinase is therefore appropriate for that enzyme, also called hexokinase IV. The K_i for glucose 6-phosphate is also very high (60 mM) so that there is no inhibition by the reaction product under physiological conditions. The concentration of glucokinase in rat liver (3 U per gram) is decreased by about 50% by prolonged fasting and also by diabetes. In bird livers, glucokinase is membrane bound and can be solubilized by vigorous mechanical disruption of the tissue (204).

Glucokinase is a monomeric protein with $M_r \sim 50,000$, which is only one-half that of most other hexokinases. However, the saturation curve for glucose is slightly sigmoidal (nH = 1.6 at saturating ATP and somewhat smaller at lower ATP) (205, 206), and a mnemonical mechanism has been proposed (203, 206) to explain the observed cooperativity.

The phosphorylation of glucose in isolated hepatocytes, as measured by the release of ^3H_2O in the presence of [2-^{3}H]glucose, is entirely accounted for by the activity of glucokinase. Cooperativity for glucose was also observed in the cellular system and, remarkably, it was decreased when the cells were incubated in the presence of a K^+-rich medium (207). The mechanism of this ionic effect is unknown but quite intriguing. The consequence of the sigmoidal shape of the saturation curve is to render the reaction most sensitive to a small change in glucose concentration in the range of 5–10 mM. No other short-term regulation mechanism of liver glucokinase is known; the control is therefore essentially by substrate concentration.

THE CONTROL OF METABOLIC FLUX BY HORMONES AND OTHER EFFECTORS

Cyclic AMP, Glucagon, and β-Adrenergic Agents

CHARACTERISTICS OF THE HORMONAL EFFECT   The property of glucagon to stimulate gluconeogenesis from lactate in isolated liver preparations was first reported by Schimassek & Mitzkat in 1963 (17). It has since been confirmed by numerous other groups who also observed a simultaneous inhibition of glycolysis. Glucagon increases the apparent rate of gluco-neogenesis by 30–70% when livers from fasted rats are used, and as much as two- to fourfold with the livers from fed animals (4, 11, 19, 208, 209, 210). In the latter case, however, because of intense glycogenolysis, gluconeogenesis cannot be measured by the net formation of glucose, but only by the incorporation of a radioactive precursor into glucose. This incorporation results from futile recycling at various levels and does not represent true gluconeogenesis because the net flux of metabolites is initially glycolytic and becomes gluconeogenic in the presence of the hormone. Glucagon stimulates gluconeogenesis not only from lactate and pyruvate, but also from substrates like glutamate (39), glutamine (211), and propionate (212), which enter the pathway at the level of oxaloacetate, or like fructose (213), glyceraldehyde, and dihydroxyacetone (214), which are converted to triose phosphates. There is, however, no effect on gluconeogenesis from glycerol (214–215), or from high (5–10 mM) concentrations of fructose (11, 18).

All the above effects of glucagon are obtained at the same doses of hormones that stimulate glycogenolysis and can be mimicked by cyclic AMP. There is therefore a general belief that these effects are mediated by the action of cyclic AMP-dependent protein kinase and the phosporylation of regulatory enzymes. As reported in the preceding sections, five enzymes related to glycolysis or gluconeogenesis are substrates for cyclic AMP-dependent protein kinase, namely phosphofructokinase-1, fructose 1,6-bisphosphatase, phosphofructokinase-2, fructose 2,6-bisphosphatase, and pyruvate kinase. Only the latter three enzymes qualify as direct targets of the hormonal regulation, since phosphorylation greatly affects their kinetic properties. By contrast, phosphofructokinase-1 and fructose 1,6-bisphos-phatase are indirectly modulated by the concentration of fructose 2,6-bisphosphate. The regulation appears therefore to occur at the level of the fructose 6-phosphate/fructose 1,6-bisphosphate and of the pyruvate/P-enolpyruvate interconversions, where the cross-over points indicated by the changes in metabolite concentration are also observed.

Adrenalin also increases cyclic AMP concentration in the liver, although much less than glucagon (4); phenylephrine, a synthetic α-adrenergic agonist, has a similar effect on hepatocytes incubated in the absence of external calcium (99, 101). The stimulation of gluconeogenesis by adrenalin is only 30–50% of that by glucagon (91, 99, 102, 209) and is predominantly α-mediated (216).

RELATIONSHIP BETWEEN THE TWO POINTS OF REGULATION Since there are two points of action of glucagon on gluconeogenesis, both mediated by cyclic AMP-dependent protein kinase, it is of interest to compare them and to see how they are connected. One difference between the two systems is that, in isolated hepatocytes, the changes in activity of phosphofructokinase-2 and fructose 2,6-bisphosphatase, and in the concentration of fructose 2,6-bisphosphate are notably slower (lasting 5–10 min) than the inactivation of pyruvate kinase, as well as the activation of phosphorylase (obtained in 1–2 min) (217). The physiological significance of this difference is not apparent except maybe during a very fast change in hormone concentration. All systems were, however, equally sensitive to the hormone [(half maximal effect at ~ 0.3 nM glucagon (217)]. In contrast, other investigators have reported 10- (147) to 1000-fold (150, 218, 219) greater sensitivity to glucagon for the fructose 2,6-bisphosphate system that was proposed to be affected by glucagon through a mechanism independent of cyclic AMP (218). The reason for these discrepancies is not clear and the latter proposal appears unrealistic.

The two regulated steps of gluconeogenesis may also be connected by metabolites, which tend to reinforce the hormonal effect. Indeed, the inactivation of pyruvate kinase causes a several-fold increase in the concentration of P-enolpyruvate, which further inhibits phosphofructokinase-2 and stimulates fructose 2,6-bisphosphatase; through this feedforward mechanism, it can decrease the concentration of fructose 2,6-bisphosphate and open the second half of the pathway. Conversely, although fructose 2,6-bisphosphate by itself is without action on pyruvate kinase (125, 177), it causes an increase in the concentration of fructose 1,6-bisphosphate, which is a potent stimulator of liver pyruvate kinase, as well as an inhibitor of its inactivation by cyclic AMP-dependent protein kinase. This again is a potential feedforward effect to be taken into consideration in both the effect of glucose to stimulate glycolysis and that of glucagon to inhibit it. Experimental evidence for such an effect of fructose 1,6-bisphosphate has been obtained in conditions in which the concentration of this metabolite in hepatocytes was greatly increased by the presence of dihydroxyacetone (88), whereas in experiments performed with glucose as substrate, the results were inconclusive (91).

RELATIONSHIP WITH THE GLYCOGENOLYTIC EFFECT   Glycogen breakdown in the liver is increased by glucagon and other hormones. It is always accompanied by an important rise in the concentration of hexose 6-phosphates, which causes a proportional increase in glucose 6-phosphate hydrolysis. Because of the property of fructose 6-phosphate to activate phosphofructokinase-1 both by its substrate cooperative effect and by its ability to cause the formation of fructose 2,6-bisphosphate, glycogenolysis is expected to be followed by an increased rate of glycolysis; this is the sequence of events that actually occurs when glycogen breakdown is initiated by cyclic AMP-independent agents. In contrast, the property of glucagon to prevent the accumulation of fructose 2,6-bisphosphate tends to counteract this glycolytic action of fructose 6-phosphate and to orient the metabolic flux in the direction of glucose. This latter effect is not necessarily complete and phosphofructokinase can remain active in the liver of fed rats treated with glucagon, causing then a futile recycling of metabolites between fructose 6-phosphate and fructose 1,6-bisphosphate (171, 172).

CONDITIONS IN WHICH GLUCAGON HAS NO EFFECT   Glucagon is without effect on gluconeogenesis in isolated liver preparations incubated in the presence of fructose at concentrations greater than 5 mM (11, 18) or, at least in the fasted state, of glycerol (214–215). This can be related to the great decreases in the concentrations of ATP, P_i (220), and fructose 2,6-bisphosphate (see next section), observed under these conditions independently of the presence of glucagon. Because P_i and fructose 2,6-bisphosphate are inhibitors of fructose 1,6-bisphosphatase, this enzyme can now operate at a fast rate. It therefore appears that the presence of fructose 2,6-bisphosphate is important for the expression of the glucagon effect on gluconeogenesis from substrates that enter the pathway at the level of triose phosphates.

ROLE OF FATTY ACIDS   After a stimulatory effect of glucagon on hepatic lipolysis had been reported (221), it was proposed that stimulation of gluconeogenesis by glucagon could be explained by its effect on lipolysis and generation of both NADH and acetyl-CoA (19, 30, 44). There is now general agreement to discard this proposal because of the following arguments: (a) the stimulation of gluconeogenesis by glucagon persists even in the presence of an inhibitor of lipolysis (222); (b) the effect of glucagon and fatty acids are additive, at least under certain conditions (11, 223); (c) glucagon but not oleate causes an increase in P-enolpyruvate and a decrease in glutamate and 2-oxoglutarate concentrations (40, 46); and (d) glucagon, but not oleate, stimulates gluconeogenesis from substrates entering the gluconeogenic pathway at the level of oxaloacetate and triose phosphates (46). It remains, however, that fatty acids do stimulate gluconeogenesis from pyruvate, and that this effect may be of significance in diabetes and during starvation, when fatty acid oxydation is increased.

THE ANTAGONISM BY INSULIN   The only clearly reproducible short-term effect of insulin in the liver is to counteract the action of small doses of glucagon (4). This was demonstrated for the formation of cyclic AMP (224) and for the inactivation of pyruvate kinase (90, 91) and of phosphofructokinase (166); the latter inactivation is now known to be mediated by the disappearance of fructose 2,6-bisphosphate. Fasting and diabetes are two conditions in which the concentration of cyclic AMP in the liver is increased (4, 224, 225); this offers an explanation for the presence of pyruvate kinase in its inactive form (86, 97, 226) and a low content in fructose 2,6-bisphosphate (see next section).

CONCLUSION   It appears that the effect of glucagon to stimulate gluconeogenesis and to inhibit glycolysis in the liver is adequately explained by the property of cyclic AMP-dependent protein kinase to phosphorylate phosphofructokinase-2, fructose 2,6-bisphosphatase and pyruvate kinase and by the various interconnections that exist between these effects. In the fed state, the glycogenolytic action of the hormone, because it increases the concentration of hexose 6-phosphates, favors glycolysis, and results in metabolite recycling. Figure 1 summarizes these cyclic AMP-mediated effects of glucagon. The effect of glucagon to stimulate pyruvate carboxy-

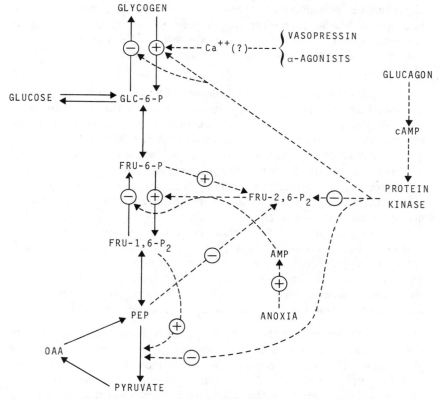

Figure 1 Control of glycolysis and gluconeogenesis in the liver. Abbreviation: PEP, phosphoenolpyruvate.

lation might also play a role in the stimulation of gluconeogenesis from pyruvate. It has not been discussed again in this section, because it has not been explained in terms of cyclic AMP- dependent protein kinase.

Fructose 2,6-Bisphosphate

DISTRIBUTION   Fructose 2,6-bisphosphate was discovered in 1980 as a low-molecular-weight stimulator of liver phosphofructokinase, which was formed in the liver after a load of glucose, and was destroyed after glucagon treatment (172, 173; reviewed in 125, 227–229). Its concentration in the liver of a well-fed rat is in the range of 10–20 μM, and reaches very low values during prolonged fasting and diabetes (172, 230–232), but remains elevated in the liver of fasted genetically obese mice (233). In hepatocytes from fed rats incubated in the presence of 20-mM glyceraldehyde, fructose, lactate, glycerol, pyruvate, or alanine, the concentration of fructose 2,6-bisphosphate was only 10%–50% that in the control (231). In hepatocytes from fasted rats, dihydroxyacetone (2–20 mM) and small concentrations (2 mM or below) of fructose cause an accumulation of fructose 2,6-bisphosphate (217).

Fructose 2,6-bisphosphate is present in a series of gluconeogenic (kidney, yeast and other fungi, higher plants) or nongluconeogenic (skeletal muscle, heart, brain, adipose tissues, lung, pancreatic islets) cells (129, 142, 191, 231, 232, 234). It is not measurable in *S. cerevisiae* grown in the absence of glucose but appears within 1 min after the addition of glucose to the medium (142). Up to now it has not been detected in bacteria. Fructose 2,6-bisphosphate was not found in isolated chloroplasts and therefore does not seem to play a role in photosynthesis (125, 128).

BIOLOGICAL EFFECTS   Fructose 2,6-bisphosphate has the property to stimulate all mammalian phosphofructokinases-1 that have been tested (160, 161, 235, 236), including those extracted from cells that, like erythrocytes (163), do not contain fructose 2,6-bisphosphate. It is also active on the enzyme from *Xenopus* oocytes, yeast (164, 237), and other fungi, and on the plastid phosphofructokinase from higher plants, although not on phosphofructokinases from plant cytosol (238) or *Escherichia coli*. It also greatly stimulates PP_i-phosphofructokinase from higher plants but not the microbial enzyme (128, 191, 192) and is a potent inhibitor of fructose 1,6-bisphosphatase, not only from the liver, but also from skeletal muscle, yeast, and higher plants (119–126).

BIOSYNTHESIS AND DEGRADATION   In the liver, fructose 2,6-bisphosphate is formed by phosphofructokinase-2 and destroyed by fructose 2,6-bisphosphatase. The ability of cyclic AMP-dependent protein kinase to phosphorylate these two enzymes and to affect their kinetic properties explains the disappearance of fructose 2,6-bisphosphate from the liver after a glucagon treatment. The dependency of the activity of phospho-fructokinase-2 upon fructose 6-phosphate concentration in the physiological range, as well as the inhibition of fructose 2,6-bisphosphatase by the same ester, renders the synthesis and degradation of fructose 2,6-bis-phosphate greatly dependent upon the availability of hexose 6-phosphates. It is on this basis that one can explain the high concentration of fructose 2,6-bisphosphate in the liver of well-fed animals, as well as the action of a glucose load (189). Glycerol 3-phosphate has an effect antagonistic to that of fructose 6-phosphate (146, 158). This may explain the low concentration of fructose 2,6-bisphosphate in the presence of glycerol or of ethanol (158, 231).

ROLE IN THE CONTROL OF GLYCOLYSIS AND OF GLUCONEOGENESIS   The role of fructose 2,6-bisphosphate in liver glycolysis is illustrated by the conclusion reached by Reinhart & Lardy (239) soon before the discovery of fructose 2,6-bisphosphate, that at the hepatic concentration of substrate, and of positive and negative effectors known at that time, phosphofructokinase-1 would always be inactive. Since fructose 2,6-bisphosphate is formed from fructose 6-phosphate when present in sufficient concentrations, fructose 2,6-bisphosphate may be considered an amplificator of its precursor fructose 6-phosphate, having a similar effect on the activity of phospho-fructokinase although at 100- to 1000-times smaller concentrations.

The major role of fructose 2,6-bisphosphate is obviously the control of the fructose 6-phosphate/fructose 1,6-bisphosphate cycle (Figure 1), the significance of which is discussed in another section. It is remarkable that the two enzymes of this cycle are substrates of cyclic AMP-dependent protein kinase, and surprising that evolution has not selected a mechanism of control based on a change of their catalytic properties associated with their phosphorylation. One advantage of the rather complex system based on the formation and destruction of fructose 2,6-bisphosphate is presumably that it integrates more information than does cyclic AMP by itself. This includes the concentration of various metabolites like fructose 6-phosphate, glycerol 3-phosphate and P-enolpyruvate. It may also be hypothesized that fructose 2,6-bisphosphate has other biological effects not yet discovered.

In yeast, one major role of fructose 2,6-bisphosphate appears to be the arrest of gluconeogenesis as soon as glucose is supplied to the cell and before the slower mechanism of proteolytic degradation of fructose 1,6-bisphosphatase has reached completion. A role of fructose 2,6-bisphos-phate in the latter process has been hypothesized (125, 142). Fungi and higher plants are a promising field for future research on fructose 2,6-bisphosphate.

The concentration of fructose 2,6-bisphosphate in isolated hepatocytes is markedly decreased by anoxia (16), but the mechanism of this effect is poorly understood. It therefore appears that, at least in the liver, the Pasteur effect is not mediated by an effect of fructose 2,6-bisphosphate, but most likely by a change in the concentration of other effectors of phosphofructokinase, mostly a decrease in ATP and an increase in AMP and P_i concentrations. In contrast, it has been reported that anoxia increased the fructose 2,6-bisphosphate content of yeast (240).

FRUCTOSE 2,6-BISPHOSPHATE VS CYCLIC AMP   As emphasized in other reviews (125, 228), fructose 2,6-bisphosphate and cyclic AMP have in common that they are regulatory molecules with a long evolutionary history. Both are "signals" in the sense defined by Stryer (241), because they are formed from ubiquitous molecules, but are not part of a major metabolic pathway. Whereas cyclic AMP has been called a "hunger signal," which signifies that glucose is lacking, fructose 2,6-bisphosphate is formed in plethoric conditions and signifies that glucose can be utilized and that gluconeogenesis can be stopped.

With regard to the control of carbohydrate metabolism in the liver, fructose 2,6-bisphosphate acts as an integrator that concentrates information from both hormonal and metabolic origin on the fructose 6-phosphate/fructose 1,6-bisphosphate interconversion. In contrast, cyclic AMP spreads out hormonal information to three points of action: glycogen synthesis and degradation, fructose 2,6-bisphosphate, and pyruvate kinase (Figure 1). The effect of both signals on fat metabolism appears to be secondary to their glycolytic action (125).

α-Adrenergic Agents, Vasopressin, and Angiotensin

STIMULATION OF GLYCOGENOLYSIS AND ROLE OF CALCIUM   α-adrenergic agents, as well as vasopressin and angiotensin, have several connected effects on liver carbohydrate metabolism. The predominant effect is an activation of glycogen phosphorylase and stimulation of glycogen breakdown (242); it appears to be mediated by an increased cytosolic concentration of free Ca^{2+} and the binding of this cation to calmodulin, which is a subunit of phosphorylase kinase, at least in the case of the muscle enzyme (reviewed in 243, 244). The major arguments in favor of a Ca^{2+}-mediated mechanism are: (a) the absence of modification of the cyclic AMP content of the cell; (b) various alterations in calcium fluxes; (c) the possibility to mimic the effect by a calcium ionophore; and (d) the suppression of the vasopressin effects on both glycogenolysis and gluconeogenesis by removal of Ca^{2+} from the incubation medium. Under the latter condition, however, phenylephrin acts as a β-adrenergic agent, since it causes a slight increase in cyclic AMP and activates protein kinase. The calcium movement, induced by the α-adrenergic agents, appears to originate mainly from the mitochondrial stores, but the mechanism by which the hormonal message could be transmitted from the plasma membrane to the mitochondria is completely unknown.

STIMULATION OF GLYCOLYSIS   In hepatocytes obtained from fed rats, the stimulation of glycogenolysis causes an increase in the concentration of hexose 6-phosphates and, secondarily, of fructose 2,6-bisphosphate (189, 245). This is because α-adrenergic agents do not promote under normal conditions the formation of cyclic AMP, which would otherwise prevent the formation of fructose 2,6-bisphosphate. Glycolysis is therefore increased, as evidenced by an increase in the net formation of lactate (245–247). Paradoxically, there is simultaneously a slight (30%–50%) increase in the formation of labeled glucose from radioactive precursors (99, 101, 215), which indicates a relatively large recycling of metabolites, and originated the widespread belief that α-adrenergic agents stimulate gluconeogenesis rather than glycolysis. A slight inactivation of pyruvate kinase has been observed under these conditions by some groups of investigators (88, 99, 100), although not by others (49, 101, 102). This inactivation is, in contrast, a constant finding when the hepatocytes are incubated in the absence of calcium, as expected from the activation of cyclic AMP-dependent protein kinase; under these conditions, glycolysis is inhibited (101). It was recently reported that α-adrenergic agents prevent the stimulation of [^{14}C]lactate conversion to glucose by glucagon and exogenously added cyclic AMP (248).

STIMULATION OF GLUCONEOGENESIS   In hepatocytes obtained from fasted rats, because of the near absence of glycogen, glycolysis is not stimulated, except at relatively high concentrations of glucose (10 mM or higher) (245), which are not physiological in the fasting state. The predominant effect of catecholamines (209, 216, 249), vasopressin (250) and angiotensin (251) is then a 30%–50% increase in the rate of gluconeogenesis, which has been detected not only by isotopic studies, but also by an increase in the net formation of glucose. The mechanism of this effect is still obscure. Changes in metabolite concentration during stimulation of gluconeogenesis by phenylephrine were measured only in the. presence of an inhibitor of phosphodiesterase (99); they are then similar to those observed in the presence of glucagon, but could result from a cyclic AMP-dependent mechanism. To the best of our knowledge, there are no reports of a similar study performed with phenylephrine or vasopressin alone.

PHYSIOLOGICAL IMPLICATION   Exton et al (4) have emphasized the unlikelihood that hepatic glucose production is regulated by blood levels of catecholamines. This is because the effective concentrations of epinephrine and of norepinephrine for stimulation of glucose output are considerably higher than those currently measured in blood plasma. Evidence for a control exerted by the sympathetic nervous activity within the liver has been recently presented (252).

Glucocorticoids

Adrenalectomized rats do not increase their rate of gluconeogenesis during starvation or diabetes as much as normal animals (253). Under these conditions, their liver contains more pyruvate and less phosphoenolpyruvate than the controls and this can be put in relation with a lower content in P-enolpyruvate carboxykinase. Furthermore, gluconeogenesis is less sensitive to stimulation by glucagon in adrenalectomized animals. Since the accumulation of cyclic AMP and activation of protein kinase are normal, this effect appears to lie beyond the activity of protein kinase. Accordingly, phosphorylation and inactivation of pyruvate kinase in hepatocytes from adrenalectomized rats is markedly less stimulated by glucagon than in control cells (254).

Glucose

Although the effect of glucose to inhibit hepatic gluconeogenesis is controversial, it appears now that this controversy could be resolved in simple terms. From all of what is known of the control of the gluconeogenic enzymes, one concludes that an inhibition of gluconeogenesis by glucose, when it occurs, would be mediated by the formation of fructose 2,6-bisphosphate, itself secondary to an increase in the concentration of fructose 6-phosphate. These glucose-induced changes in the concentration of phosphoric esters occur in hepatocytes isolated from fasted rats (172), but not necessarily in vivo. This is because in isolated liver preparations, the synthesis of glycogen is relatively deficient and hexose phosphates can be formed from glucose faster than they are used. Accordingly, a 40%–50% inhibition by glucose of gluconeogenesis from alanine (255, 256), or low concentrations of pyruvate (91) or lactate (257), has been observed with such preparations; this was, however, not the case in the presence of high concentrations of lactate (10–30 mM) (10, 256), which are known to decrease fructose 2,6-bisphosphate levels in the hepatocytes. In vivo, glycogen metabolism is more sensitive to glucose, the administration of which causes a rapid inactivation of glycogen phosphorylase and subsequent activation of glycogen synthase (2, 3). Under these conditions, the concentration of hexose 6-phosphates remains low and no formation of fructose 2,6-bisphosphate is expected. The effect of glucose is then to favor glycogen synthesis from gluconeogenic precursors at the expense of glucose formation, with no change in gluconeogenic flux (258). It may however be that at a higher glucose concentration, or when glycogen synthesis starts to slow down, glucose 6-phosphate is formed more rapidly than it is converted to glycogen; its concentration and that of fructose 2,6-bisphosphate would then increase, thus inhibiting gluconeogenesis (259).

Through these mechanisms, the first effect of glucose on the control of carbohydrate metabolism in the liver is at the level of phosphorylase a and glycogen metabolism (2, 3). This indicates that the temporary disposal of glucose as glycogen is a primary function of the liver, which is initiated as soon as the level of glycemia rises. Gluconeogenesis is another essential function inhibited by glucose only when the glycogen store has been replenished. Glycolysis can then occur and the excess of carbohydrates is converted to fat.

THE FUTILE CYCLES

A futile cycle is a metabolic conversion, the net balance of which is the hydrolysis of ATP into ADP and P_i. The simplest futile cycles are made of two nonequilibrium reactions, e.g. a kinase and a phosphatase for which a product of one reaction is a substrate of the other. When they operate at the same rate there is no net flux of metabolites, but an apparently futile loss of energy as heat. There are three potential futile cycles in the gluconeogenic pathway, two of which are of this simple kinase/phosphatase type. These cycles, and the complex methodology involved in their measurement, have been reviewed (8, 260, 261).

Futile recycling can be avoided if both enzymes are controlled by a on/off mechanism. A typical example of this rigorous type of regulation is at the level of glycogen synthesis and degradation in the liver of fed rats, since the system is arranged in such a way that glycogen synthase is inactive when phosphorylase is active and vice versa (2, 3). None of the cycles of gluconeogenesis correspond to this model.

Glucose/Glucose 6-Phosphate Cycle

PROPERTIES   Because there is apparently no on/off mechanism of control of glucokinase and glucose 6-phosphatase, these two enzymes are always simultaneously in operation in the adult liver. Both of them are characterized by a K_m that exceeds the usual concentration of either glucose or glucose 6-phosphate and their activity is essentially controlled by substrate concentration. Consequently, increasing the concentration of the substrates of one reaction would accelerate its conversion to product, a higher level of which should then favor the reverse reaction, making recycling more intense. This however does not occur for the following reasons: 1. When glucose concentration is high, it not only accelerates glucokinase, but it also inactivates glycogen phosphorylase and activates glycogen synthase, two effects that cooperate to maintain a low level of glucose 6-phosphate (2, 3). Furthermore, an increase in hexose 6-phosphates would secondarily induce the formation of fructose 2,6-bisphosphate, resulting in a higher rate of glycolysis. 2. When the concentration of glucose 6-phosphate is high, glucose is formed at a proportional rate but, because it diffuses out of the liver and is diluted in the body fluids, its concentration is only moderately increased.

There is solid experimental evidence that recycling between glucose and glucose 6-phosphate actually occurs in normal livers at a rate close to 1 μmol/min per g. It is less intense in the fasted condition, because of a low concentration of both glucose and glucose 6-phosphate, and because of the marked decrease in glucokinase content during fasting.

ROLE   Although this recycling consumes energy, it has several advantages that explain its resistance to evolutionary pressure:

1. It can be estimated that at the normal level of glycemia, glucokinase and glucose 6-phosphatase operate at the same rate, and that the net flux of metabolites is zero. Therefore, even a small positive or negative change in the concentration of either glucose or glucose 6-phosphate allows a net uptake or output of glucose and the relative change in metabolite flux is enormous, even unlimited.

2. Because glucose 6-phosphate is the substrate or product of several major metabolic pathways, including glycolysis, gluconeogenesis, glycogen synthesis and degradation, and the pentose phosphate pathway, an on/off control of glucokinase and glucose 6-phosphatase would need to be in phase with the activity of each of these conversions and would be exceedingly complex. The advantage of this system in the case of glycogen metabolism has been previously emphasized (262). A similar situation exists for glycolysis and gluconeogenesis, which are controlled at the level of phosphofructokinase-1, fructose 1,6-bisphosphatase, and pyruvate kinase, but not primarily at the level of glucokinase and glucose-6-phosphatase.

3. The recycling system may be considered a substitute for the inhibition of hexokinase by glucose 6-phosphate, which is currently considered a prerequisite for the control of glycolysis at the level of phosphofructokinase in the case of the Pasteur effect (263). In the liver, glucose 6-phosphate does not inhibit glucokinase but inhibits glucose uptake, which is the difference between the activity of glucokinase and that of glucose 6-phosphatase.

Fructose 6-Phosphate/Fructose 1,6-Bisphosphate Cycle

PROPERTIES   The recycling between fructose 6-phosphate and fructose 1,6-bisphosphate is controlled by a mechanism that is a compromise between the two extreme systems described above, and could be called an incomplete on/off mechanism. The system oscillates between two extreme conditions. During fasting, because of the low concentration of fructose 2,6-bisphosphate, phosphofructokinase is inactive and the flux of metabolites is unidirectionally gluconeogenic (172, 173). In contrast, in the fed state, fructose 2,6-bisphosphate is present and activates phosphofructokinase as well as it inhibits fructose 1,6-bisphosphatase, at least in a first stage (on/off mechanism). Under these conditions, futile recycling would be avoided. The experimental evidence (172, 173, 264) indicates, however, that in the fed state, as much as 30% of fructose 1,6-bisphosphate formed by phosphofructokinase is converted back to glucose. There is then an up to tenfold increase in the concentration of fructose 1,6-bisphosphate, which is the equivalent, although on a much smaller scale, of the enormous accumulation of fructose 1-phosphate that follows the administration of large doses of fructose. The latter accumulation is explained by the greater capacity of fructokinase relative to the enzymatic system able to utilize fructose 1-phosphate, and is most damaging to the cell (265). Because of the competitive aspect of the inhibition of fructose 1,6-bisphosphatase by fructose 2,6-bisphosphate, the increased concentration of substrate reactivates the enzyme and allows a substantial part of the metabolites to be converted back to fructose 6-phosphate. Because the saturation curve for fructose 1,6-bisphosphate is sigmoidal in the presence of fructose 2,6-bisphosphate, recycling occurs only at relatively high concentrations of substrate. It must be recognized, however, that one would have expected a more complete inhibition of fructose 1,6-bisphosphatase by fructose 2,6-bisphosphate; therefore, the recycling observed suggests that the inhibition may be less complete in vivo than expected from in vitro observations. One major uncertainty is the concentration of free AMP in the cell (42).

ROLE   From the above considerations it appears that the futile recycling of metabolites between fructose 6-phosphate and fructose 1,6-bisphosphate in the liver results from an overflow of fructose 1,6-bisphosphate when this compound is formed by phosphofructokinase in excess over the glycolytic capacity of the cell and has none of the advantages described for the glucose/glucose 6-phosphate cycle. Its *raison d'être*, if any, appears to prevent a deleterious accumulation of fructose 1,6-bisphosphate, as well as that of lactic acid, in excess of the lipogenic capacity of the liver (8, 261). It can therefore be considered a brake put on glycolysis, rather than a very fine system of amplification. It could resist evolutionary pressure, because it operates only when the amount of oxidizable substrate exceeds the energy needs of the cell.

A more economical mechanism would have been a feedback inhibition of phosphofructokinase by its product. Such an inhibition occurs only at millimolar concentrations of fructose 1,6-bisphosphate never observed in the liver, but well-observed in yeast. Again, the very high concentration of fructose 1,6-bisphosphate recorded in these cells indicates that phosphofructokinase activity limits the rate of glycolysis only at the lowest range of metabolic flux. Marginal, if any, recycling of metabolites was observed in yeast under various experimental conditions (266).

Pyruvate-Phosphoenolpyruvate Cycle

PROPERTIES   This cycle is made of three reactions. One, catalyzed by pyruvate kinase, forms 1 ATP; the two others, controlled by pyruvate carboxylase and phosphoenolpyruvate carboxykinase, each consume 1 ATP. The net balance is the hydrolysis of 1 ATP into ADP and P_i. There is good evidence that recycling is continuously operating and is more intense in the fed than in the fasted state (267). A quantitative estimate of the recycling is, however, difficult.

One important particularity of the cycle is the compartmentation of its constitutive enzymes. Pyruvate kinase is cytosolic and is inactivated by cyclic AMP-dependent protein kinase. Pyruvate carboxylase is mitochondrial, and thus not accessible to the control by cyclic AMP-dependent protein kinase. P-enolpyruvate carboxykinase is both cytosolic and mitochondrial, with a variable distribution from species to species. A purely cytosolic or purely mitochondrial control would, therefore, be of limited efficiency. The activity of this enzyme is mostly regulated by protein synthesis. One concludes therefore that the two gluconeogenic enzymes of the cycle are apparently not submitted to short-term hormonal regulation, at least of the on/off type, rendering futile recycling inescapable under glycolytic conditions. The long-term decrease in the amount of P-enolpyruvate carboxykinase in the fed state and of pyruvate kinase during fasting is a means of decreasing recycling severalfold. When gluconeogenesis is predominant, the intensity of cycling is controlled by cyclic AMP-dependent inactivation of pyruvate kinase. This inactivation is however incomplete even under maximal hormonal stimulation (91).

ROLE   The pyruvate/P-enolpyruvate cycle is inherent to the subcellular location of its two gluconeogenic enzymes. Like the fructose 6-phosphate/fructose 1,6-bisphosphate cycle, it does not offer the great amplification of control that has been proposed (268) to justify the energy loss involved in recycling and that, in the case of the glucose/glucose 6-

phosphate cycle, compensates for the poverty of other control mechanisms. Indeed, like phosphofructokinase, pyruvate kinase is a highly regulated allosteric enzyme for which an additional control appears unnecessary; because the activity of the two enzymes can be shut off, the amplification by recycling is small.

There is an interesting analogy to be drawn between the role of P-enolpyruvate in gluconeogenesis and that, described above, of fructose 1,6-bisphosphate in glycolysis. Indeed, the high concentration of these two metabolites occurring under opposite nutritional conditions is self-limiting, both because of the feedforward stimulation that they exert on the second half of the pathway (see section on glucagon) and because they favor their own recycling. In the case of P-enolpyruvate, this occurs by the substrate effect on the activity of pyruvate kinase, as well as by the prevention of the cyclic AMP-dependent inactivation of the same enzyme.

ACKNOWLEDGMENTS

We are grateful to Dr. E. Van Schaftingen for critical reading of the manuscript and to the Fonds de la Recherche Scientifique Médicale and the US Public Health Service (Grant AM 9235) for their support. L. Hue is Maître de Recherches of the Fonds National de la Recherche Scientifique.

Literature Cited

1. Krebs, H. A. 1972. *Essays Biochem.* 8:1–34
2. Hers, H. G. 1976. *Ann. Rev. Biochem.* 45:167–89
3. Stalmans, W. 1976. *Curr. Top. Cell. Regul.* 11:51–97
4. Exton, J. H., Mallette, L. E., Jefferson, L. S., Wong, E. H. A., Friedmann, N., et al. 1970. *Recent Prog. Horm. Res.* 26:411–61
5. Hanson, R. W., Mehlman, M. A. 1976. *Gluconeogenesis. Its Regulation in Mammalian Species.* New York: Wiley. 592 pp.
6. Pilkis, S. J., Park, C. R., Claus, T. H. 1978. *Vitam. Horm. NY* 36:383–460
7. Claus, T. H., Pilkis, S. J. 1981. In *Biochemical Action of Hormones,* ed. G. Litwack, 8:209–71
8. Hue, L. 1981. *Adv. Enzymol.* 52:247–331
9. Hue, L., van de Werve, G. 1981. *Short-term Regulation of Liver Metabolism.* Amsterdam: Elsevier/North Holland. 464 pp.
10. Exton, J. H., Park, C. R. 1967. *J. Biol. Chem.* 242:2622–36
11. Ross, B. D., Hems, R., Krebs, H. A. 1967. *Biochem. J.* 102:942–51
12. Söling, H. D., Kleineke, J. 1976. See Ref. 5, pp. 369–462
13. Dickson, A. J., Langslow, D. R. 1978. *Mol. Cell. Biochem.* 22:167–81
14. Woods, H. F., Krebs, H. A. 1971. *Biochem. J.* 125:129–39
15. Brunengraber, H., Boutry, M., Lowenstein, J. M. 1973. *J. Biol. Chem.* 248:2656–69
16. Hue, L. 1982. *Biochem. J.* 206:359–65
17. Schimassek, H., Mitzkat, H. J. 1963. *Biochem. Z.* 337:510–18
18. Exton, J. H., Park, C. R. 1969. *J. Biol. Chem.* 244:1424–33
19. Williamson, J. R., Browning, E. T., Thurman, R. G., Scholz, R. 1969. *J. Biol. Chem.* 244:5055–64
20. Halestrap, A. P. 1981. See Ref. 9, pp. 389–409
21. Böttger, I., Wieland, O., Brdiczka, D., Pette, D. 1969. *Eur. J. Biochem.* 8:113–19
22. Walter, P. 1976. See Ref. 5, pp. 239–65
23. Patzelt, C., Löffler, G., Wieland, O. H. 1973. *Eur. J. Biochem.* 33:117–22
24. Claus, T. H., Pilkis, S. J. 1977. *Arch. Biochem. Biophys.* 182:52–63
25. Hems, D. A., McCormack, J. G.,

Denton, R. M. 1978. *Biochem. J.* 176:627–29

26. Denton, R. M., McCormack, J. G., Oviasu, O. A. 1981. See Ref. 9, pp. 159–74

27. Utter, M. F., Barden, R. E., Taylor, B. L. 1975. *Adv. Enzymol.* 42:1–72

28. Barritt, G. J., Zander, G. L., Utter, M. F. 1976. See Ref. 5, pp. 3–46

29. Utter, M. F., Scrutton, M. C. 1969. *Curr. Top. Cell. Regul.* 1:253–96

30. Williamson, J. R. 1966. *Biochem. J.* 101:11C–14C

31. Adam, P. A., Haynes, R. C. Jr. 1969. *J. Biol. Chem.* 244:6444–50

32. Garrison, J. C., Haynes, R. C. 1975. *J. Biol. Chem.* 250:2769–77

33. Siess, E. A., Wieland, O. 1980. *Eur. J. Biochem.* 110:203–10

34. Martin, A. D., Titheradge, M. A. 1983. *Biochem. Soc. Trans.* 11:78–81

35. Yamazaki, R. K. 1975. *J. Biol. Chem.* 250:7924–30

36. Siess, E. A., Fahimi, F. M., Wieland, O. H. 1981. *Hoppe-Seylers Z. Physiol. Chem.* 362:1643–51

37. Verhoeven, A. J., Hensgens, H. E. S. J., Meijer, A. J., Tager, J. M. 1982. *FEBS Lett.* 140:270–72

38. Ui, M., Claus, T. H., Exton, J. H., Park, C. R. 1973. *J. Biol. Chem.* 248:5344–49

39. Ui, M., Exton, J. H., Park, C. R. 1973. *J. Biol. Chem.* 248:5350–59

40. Siess, E. A., Brocks, D. G., Lattke, H. K., Wieland, O. H. 1977. *Biochem. J.* 166:225–35

41. Cook, G. A., Nielsen, R. C., Hawkins, R. A., Mehlman, M. A., Lakshmanan, M. R., Veech, R. L. 1977. *J. Biol. Chem.* 252:4421–24

42. Sols, A., Marco, R. 1970. *Curr. Top. Cell. Regul.* 2:227–73

43. Scrutton, M. C., Griffiths, J. R. 1981. See Ref. 9, pp. 175–98

44. Struck, E., Ashmore, J., Wieland, O. H. 1965. *Biochem. Z.* 343:107–10

45. Williamson, J. R., Browning, E. T., Scholz, R. 1969. *J. Biol. Chem.* 244:4607–16

46. Brocks, D. G., Siess, E. A., Wieland, O. H. 1980. *Eur. J. Biochem.* 113:39–43

47. Williamson, J. R., Rostand, S. G., Peterson, M. J. 1970. *J. Biol. Chem.* 245:3242–51

48. Halestrap, A. P., Denton, R. M. 1974. *Biochem. J.* 138:313–16

49. Thomas, A. P., Halestrap, A. P. 1981. *Biochem. J.* 198:551–64

50. Williamson, D. H., Lund, P., Krebs, H. A. 1967. *Biochem. J.* 103:514–27

51. Rognstad, R., Clark, D. G. 1974. *Arch. Biochem. Biophys.* 161:638–46

52. Söling, H. D., Kleineke, J., Willms, B.,

Janson, G., Kuhn, A. 1973. *Eur. J. Biochem.* 37:233–43

53. Kneer, N. M., Wagner, M. J., Lardy, H. A. 1979. *J. Biol. Chem.* 254:12160–68

54. Yip, B. P., Lardy, H. A. 1981. *Arch. Biochem. Biophys.* 212:370–77

55. Ochs, R. S., Lardy, H. A. 1981. *FEBS Lett.* 131:119–21

56. Tilghman, S. M., Hanson, R. W., Ballard, F. J. 1976. See Ref. 5, pp. 47–91

57. Chang, H. C., Lane, M. D. 1966. *J. Biol. Chem.* 241:2413–20

58. Ballard, F. J., Hanson, R. W. 1969. *J. Biol. Chem.* 244:5625–30

59. Nordlie, R. C., Lardy, H. A. 1963. *J. Biol. Chem.* 238:2259–63

60. Holten, D. D., Nordlie, R. C. 1965. *Biochemistry* 4:723–31

61. Ballard, F. J. 1970. *Biochem. J.* 120:809–14

62. Walsh, D. A., Chen, L. J. 1971. *Biochem. Biophys. Res. Commun.* 45:669–75

63. Tischler, M. E., Friedrichs, D., Coll, K., Williamson, J. R. 1977. *Arch. Biochem. Biophys.* 184:222–36

64. Siess, E. A., Brocks, D. G., Wieland, O. H. 1982. *Metabolic Compartmentation*, ed. H. Sies, pp. 235–57. London: Academic. 561 pp.

65. Lardy, H. A., Merryfield, M. L. 1981. *Curr. Top. Cell. Regul.* 18:243–53

66. Merryfield, M. L., Kramp, D. C., Lardy, H. A. 1982. *J. Biol. Chem.* 257:4646–54

67. Merryfield, M. L., Lardy, H. A. 1982. *J. Biol. Chem.* 257:3628–35

68. Reynolds, C. H. 1980. *Biochem. J.* 185:451–54

69. Rognstad, R. 1979. *J. Biol. Chem.* 254:1875–79

70. Jomain-Baum, M., Schramm, V. L., Hanson, R. W. 1976. *J. Biol. Chem.* 251:37–44

71. Groen, A. K., Van der Meer, R., Westerhoff, H. V., Wanders, R. J. A., Akerboom, T. P. M., Tager, J. M. 1982. See Ref. 64, pp. 9–37

72. Probst, I., Schwartz, P., Jungermann, K. 1982. *Eur. J. Biochem.* 126:271–78

73. Beale, E. G., Hartley, J. L., Granner, D. K. 1982. *J. Biol. Chem.* 257:2022–28

74. Cimbala, M. A., Lamers, W. H., Nelson, K., Monahan, J. E., Yoo-Warren, H., Hanson, R. W. 1982. *J. Biol. Chem.* 257:7629–36

75. Lamers, W. H., Hanson, R. W., Meisner, H. M. 1982. *Proc. Natl. Acad. Sci. USA* 79:5137–41

76. Kayne, F. J. 1973. *Enzymes* 8:353–82

77. Seubert, W., Schoner, W. 1971. *Curr. Top. Cell. Regul.* 3:237–67

78. Engström, L. 1978. *Curr. Top. Cell. Regul.* 13:29–51

79. Blair, J. B., Cimbala, M. A., James, M. E. 1982. *J. Biol. Chem.* 257:7595–602
80. El-Maghrabi, M. R., Haston, W. S., Flockhart, D. A., Claus, T. H., Pilkis, S. J. 1980. *J. Biol. Chem.* 255:668–75
81. Titanji, V. P. K. 1977. *Biochim. Biophys. Acta* 481:140–51
82. Jett, M.-F., Hue, L., Hers, H. G. 1981. *FEBS Lett.* 132:183–86
83. Humble, E., Berglund, L., Titanji, V., Ljungström, O., Edlund, B., et al. 1975. *Biochem. Biophys. Res. Commun.* 66:614–21
84. Eigenbrodt, E., Schoner, W. 1977. *Hoppe-Seylers Z. Physiol. Chem.* 358:1033–46
85. Eigenbrodt, E., Abdel-Fattah Mostafa, M., Schoner, W. 1977. *Hoppe-Seylers Z. Physiol. Chem.* 358:1047–55
86. Felíu, J. E., Hue, L., Hers, H. G. 1977. *Eur. J. Biochem.* 81:609–17
87. Pilkis, S. J., Pilkis, J., Claus, T. H. 1978. *Biochem. Biophys. Res. Commun.* 81:139–46
88. Claus, T. H., El-Maghrabi, M. R., Pilkis, S. J. 1979. *J. Biol. Chem.* 254:7855–64
89. Eigenbrodt, E., Schoner, W. 1977. *Hoppe-Seylers Z. Physiol. Chem.* 358:1057–67
90. Blair, J. B., Cimbala, M. A., Foster, J. L., Morgan, R. A. 1976. *J. Biol. Chem.* 251:3756–62
91. Felíu, J. E., Hue, L., Hers, H. G. 1976. *Proc. Natl. Acad. Sci. USA* 73:2762–66
92. Van Berkel, T. J. C., Kruijt, J. K., Koster, J. F., Hülsmann, W. C. 1976. *Biochem. Biophys. Res. Commun.* 72:917–25
93. Riou, J. P., Claus, T. H., Pilkis, S. J. 1976. *Biochem. Biophys. Res. Commun.* 73:591–600
94. Garrison, J. C., Borland, M. K. 1978. *J. Biol. Chem.* 253:7091–7100
95. Ljungström, O., Ekman, P. 1977. *Biochem. Biophys. Res. Commun.* 78:1147–55
96. Riou, J. P., Claus, T. H., Pilkis, S. J. 1978. *J. Biol. Chem.* 253:656–59
97. Van Berkel, T. J. C., Kruijt, J. K., Koster, J. F. 1977. *Eur. J. Biochem.* 81:423–32
98. Ochs, R. S., Harris, R. A. 1978. *Arch. Biochem. Biophys.* 190:193–201
99. Chan, T. M., Exton, J. H. 1978. *J. Biol. Chem.* 253:6393–400
100. Steiner, K. E., Chan, T. M., Claus, T. H., Exton, J. H., Pilkis, S. J. 1980. *Biochem. Biophys. Acta* 632:366–74
101. Hue, L., Felíu, J. E., Hers, H. G. 1978. *Biochem. J.* 176:791–97
102. Blair, J. B., James, M. E., Foster, J. L. 1979. *J. Biol. Chem.* 254:7585–90
103. Garrison, J. C., Borland, M. K. 1979. *J. Biol. Chem.* 254:1129–33
104. Nagano, M., Ishibashi, H., McCully, V., Cottam, G. L. 1980. *Arch. Biochem. Biophys.* 203:271–81
105. Van den Berg, G. B., Van Berkel, T. J. C., Koster, J. F. 1980. *Eur. J. Biochem.* 113:131–40
106. Scrutton, M. C., Utter, M. F. 1968. *Ann. Rev. Biochem.* 37:249–302
107. Krebs, H. A., Eggleston, L. V. 1965. *Biochem. J.* 94:3C–4C
108. Hopkirk, T. J., Bloxham, D. P. 1979. *Biochem. J.* 182:383–97
109. James, M. E., Blair, J. B. 1982. *Biochem. J.* 204:329–38
110. Cladaras, C., Cottam, G. L. 1980. *J. Biol. Chem.* 255:11499–503
111. Pontremoli, S., Horecker, B. L. 1971. *Enzymes* 4:611–46
112. Horecker, B. L., Melloni, E., Pontremoli, S. 1975. *Adv. Enzymol.* 42:193–226
113. Benkovic, S. J., deMaine, M. M. 1982. *Adv. Enzymol.* 53:45–82
114. Tejwani, G. A. 1982. *Adv. Enzymol.* 54:121–94
115. Nimmo, H. G., Tipton, K. F. 1975. *Biochem. J.* 145:323–34
116. Tejwani, G. A., Pedrosa, F. O., Pontremoli, S., Horecker, B. L. 1976. *Proc. Natl. Acad. Sci. USA* 73:2692–95
117. Nakashima, K., Tuboi, S. 1976. *J. Biol. Chem.* 251:4315–21
118. Taketa, K., Pogell, B. M. 1965. *J. Biol. Chem.* 240:651–62
119. Van Schaftingen, E., Hers, H. G. 1981. *Proc. Natl. Acad. Sci. USA* 78:2861–63
120. Pilkis, S. J., El-Maghrabi, M. R., Pilkis, J., Claus, T. H. 1981. *J. Biol. Chem.* 256:3619–22
121. Pilkis, S. J., El-Maghrabi, M. R., McGrane, M. M., Pilkis, J., Claus, T. H. 1981. *J. Biol. Chem.* 256:11489–95
122. François, J., Van Schaftingen, E., Hers, H. G. 1983. *Eur. J. Biochem.* In press
123. Pontremoli, S., Melloni, E., Michetti, M., Salanino, F., Sparatore, B., Horecker, B. L. 1982. *Arch. Biochem. Biophys.* 219:609–13
124. Gottschalk, M. E., Chatterjee, T., Edelstein, I., Marcus, F. 1982. *J. Biol. Chem.* 257:8016–20
125. Hers, H. G., Van Schaftingen, E. 1982. *Biochem. J.* 206:1–12
126. Ganson, N. J., Fromm, H. J. 1982. *Biochem. Biophys. Res. Commun.* 108:233–39
127. Funayama, S., Molano, J., Gancedo, C. 1979. *Arch. Biochem. Biophys.* 197:170–77
128. Cséke, C., Weeden, N. F., Buchanan, B. B., Uyeda, K. 1982. *Proc. Natl. Acad. Sci. USA* 79:4322–26

129. Stitt, M., Mieskes, G., Söling, H. D., Heldt, H. W. 1982. *FEBS Lett.* 145:217–22
130. Buchanan, B. B. 1980. *Ann. Rev. Plant Physiol.* 31:341–74
131. Riou, J. P., Claus, T. H., Flockhart, D. A., Corbin, J. D., Pilkis, S. J. 1977. *Proc. Natl. Acad. Sci. USA* 74:4615–19
132. Hosey, M. M., Marcus, F. 1981. *Proc. Natl. Acad. Sci. USA* 78:91–94
133. Claus, T. H., Schlumpf, J., El-Maghrabi, M. R., McGrane, M., Pilkis, S. J. 1981. *Biochem. Biophys. Res. Commun.* 100:716–23
134. Melloni, E., Pontremoli, S., Salamino, F., Sparatore, B., Michetti, M., Horecker, B. L. 1981. *Proc. Natl. Acad. Sci. USA* 78:1499–1502
135. Pontremoli, S., Melloni, E., Horecker, B. L. 1981. In *Metabolic Interconversion of Enzymes 1980*, ed. H. Holzer, pp. 186–98. Berlin/Heidelberg/New York: Springer. 397 pp.
136. Funayama, S., Gancedo, J. M., Gancedo, C. 1980. *Eur. J. Biochem.* 109:61–66
137. Lenz, A. G., Holzer, H. 1980. *FEBS Lett.* 109:271–74
138. Tortora, P., Birtel, M., Lenz, A. G., Holzer, H. 1981. *Biochem. Biophys. Res. Commun.* 100:688–95
139. Gancedo, J. M., Mazon, M. J., Gancedo, C. 1982. *Arch. Biochem. Biophys.* 218:478–82
140. Müller, D., Holzer, H. 1981. *Biochem. Biophys. Res. Commun.* 103:926–33
141. Mazon, M. J., Gancedo, J. M., Gancedo, C. 1982. *J. Biol. Chem.* 257:1128–30
142. Lederer, B., Vissers, S., Van Schaftingen, E., Hers, H. G. 1981. *Biochem. Biophys. Res. Commun.* 103:1281–87
143. Mazon, M. J., Gancedo, J. M., Gancedo, C. 1982. *Eur. J. Biochem.* 127:605–8
144. Londesborough, J. 1982. *FEBS Lett.* 144:269–72
145. Purwin, C., Leidig, F., Holzer, H. 1982. *Biochem. Biophys. Res. Commun.* 107:1482–89
146. Van Schaftingen, E., Davies, D. R., Hers, H. G. 1982. *Eur. J. Biochem.* 124:143–49
147. El-Maghrabi, M. R., Claus, T. H., Pilkis, J., Fox, E., Pilkis, S. J. 1982. *J. Biol. Chem.* 257:7603–7
148. Furuya, E., Yokoyama, M., Uyeda, K. 1982. *Biochem. Biophys. Res. Commun.* 105:264–70
149. El-Maghrabi, M. R., Fox, E., Pilkis, J., Pilkis, S. J. 1982. *Biochem. Biophys. Res. Commun.* 106:794–802
150. Richards, C. S., Yokoyama, M., Furuya, E., Uyeda, K. 1982. *Biochem. Biophys. Res. Commun.* 104:1073–79
151. Van Schaftingen, E., Hers, H. G. 1981. *Biochem. Biophys. Res. Commun.* 101:1078–84
152. Reeves, R. J., South, D. J., Blytt, H. J., Warren, L. G. 1974. *J. Biol. Chem.* 249:7737–45
153. Bloxham, D. P., Lardy, H. A. 1973. *Enzymes* 8:239–78
154. Hofmann, E. 1976. *Rev. Physiol. Biochem. Pharmacol.* 75:1–68
155. Goldhammer, A. R., Paradies, H. H. 1979. *Curr. Top. Cell. Regul.* 15:109–41
156. Uyeda, K. 1979. *Adv. Enzymol.* 48:194–244
157. Sols, A., Castano, J. G., Aragon, J. J., Domenech, C., Lazo, P. A., Nieto, A. 1981. See Ref. 135, pp. 111–23
158. Claus, T. H., Schlumpf, J. R., El-Maghrabi, M. R., Pilkis, S. J. 1982. *J. Biol. Chem.* 257:7541–48
159. Sommercorn, J., Freedland, R. A. 1982. *J. Biol. Chem.* 257:9424–28
160. Van Schaftingen, E., Jett, M.-F., Hue, L., Hers, H. G. 1981. *Proc. Natl. Acad. Sci. USA* 78:3483–86
161. Uyeda, K., Furuya, E., Luby, L. J. 1981. *J. Biol. Chem.* 256:8394–99
162. Sener, A., Malaisse-Lagae, F., Malaisse, W. J. 1982. *Biochem. Biophys. Res. Commun.* 104:1033–40
163. Heylen, A., Van Schaftingen, E., Hers, H. G. 1982. *FEBS Lett.* 143:141–43
164. Bartrons, R., Van Schaftingen, E., Vissers, S., Hers, H. G. 1982. *FEBS Lett.* 143:137–40
165. Boiteux, A., Hess, B., Sel'kov, E. E. 1980. *Curr. Top. Cell. Regul.* 17:171–203
166. Castano, J. G., Nieto, A., Felíu, J. E. 1979. *J. Biol. Chem.* 254:5576–79
167. Kagimoto, T., Uyeda, K. 1979. *J. Biol. Chem.* 254:5584–87
168. Fister, P., Eigenbrodt, E., Schoner, W. 1982. *Mol. Cell. Endocrinol.* 26:19–30
169. Clark, M. G., Patten, G. S. 1981. *J. Biol. Chem.* 256:27–30
170. Schneider, J. A., Diamond, I., Rozengurt, E. 1978. *J. Biol. Chem.* 253:872–77
171. Van Schaftingen, E., Hue, L., Hers, H. G. 1980. *Biochem. J.* 192:263–71
172. Van Schaftingen, E., Hue, L., Hers, H. G. 1980. *Biochem. J.* 192:887–95
173. Van Schaftingen, E., Hue, L., Hers, H. G. 1980. *Biochem. J.* 192:897–901
174. Dunaway, G. A., Segal, H. L. 1974. *Biochem. Biophys. Res. Commun.* 56:689–96
175. Dunaway, G. A., Segal, H. L. 1976. *J. Biol. Chem.* 251:2323–29
176. Dunaway, G. A., Leung, G. L.-Y.,

Thrasher, J. R., Cooper, M. D. 1978. *J. Biol. Chem.* 253:7460–63
177. Söling, H. D., Kuduz, J., Brand, I. A. 1981. *FEBS Lett.* 130:309–13
177a. Van Schaftingen, E., Hers, H. G. 1983. *Biochem. Biophys. Res. Commun.* In press
178. Brand, I. A., Müller, M. K., Unger, C., Söling, H. D. 1976. *FEBS Lett.* 68:271–4
179. Söling, H. D., Brand, I. A. 1981. *Curr. Top. Cell. Regul.* 20:107–38
180. Brand, I. A., Söling, H. D. 1982. *Eur. J. Biochem.* 122:175–81
181. Kagimoto, T., Uyeda, K. 1980. *Arch. Biochem. Biophys.* 203:792–99
182. Claus, T. H., Schlumpf, J. R., El-Maghrabi, M. R., Pilkis, J., Pilkis, S. J. 1980. *Proc. Natl. Acad. Sci. USA* 77:6501–05
183. Foe, L. G., Kemp, R. G. 1982. *J. Biol. Chem.* 257:6368–72
184. Furuya, E., Uyeda, K. 1980. *J. Biol. Chem.* 255:11656–59
185. El-Maghrabi, M. R., Claus, T. H., Pilkis, J., Pilkis, S. J. 1981. *Biochem. Biophys. Res. Commun.* 101:1071–77
186. El-Maghrabi, M. R., Claus, T. H., Pilkis, J., Pilkis, S. J. 1982. *Proc. Natl. Acad. Sci. USA* 79:315–19
187. Furuya, E., Yokoyama, M., Uyeda, K. 1982. *Proc. Natl. Acad. Sci. USA* 79:325–29
188. Furuya, E., Uyeda, K. 1981. *J. Biol. Chem.* 256:7109–12
189. Hue, L., Blackmore, P. F., Exton, J. H. 1981. *J. Biol. Chem.* 256:8900–3
190. Van Schaftingen, E., Davies, D. R., Hers, H. G. 1981. *Biochem. Biophys. Res. Commun.* 103:362–68
191. Sabularse, D. C., Anderson, R. L. 1981. *Biochem. Biophys. Res. Commun.* 103:848–55
192. Van Schaftingen, E., Lederer, B., Bartrons, R., Hers, H. G. 1982. *Eur. J. Biochem.* 129:191–95
193. Nordlie, R. C. 1971. *Enzymes* 4:543–610
194. Nordlie, R. C. 1976. See Ref. 5, pp. 93–152
195. Begley, P. J., Craft, J. A. 1981. *Biochem. Biophys. Res. Commun.* 103:1029–34
196. Burchell, A., Burchell, B., Arion, W. J., Walls, H. E. 1982. *Biochem. Biophys. Res. Commun.* 107:1046–52
197. Arion, W. J., Wallin, B. K., Lange, A. J., Ballas, L. M. 1975. *Mol. Cell. Biochem.* 6:75–83
198. Arion, W. J., Lange, A. J., Walls, H. E., Ballas, L. M. 1980. *J. Biol. Chem.* 255:10396–406
199. Zoccoli, M. A., Hoopes, R. R., Karnovsky, M. L. 1982. *J. Biol. Chem.* 257:3919–24

200. Narisawa, K., Igarashi, Y., Otomo, H., Tada, K. 1978. *Biochem. Biophys. Res. Commun.* 83:1360–64
201. Lange, A. J., Arion, W. J., Beaudet, A. L. 1980. *J. Biol. Chem.* 255:8381–84
202. Weinhouse, S. 1976. *Curr. Top. Cell. Regul.* 11:1–50
203. Pollard-Knight, D., Cornish-Bowden, A. 1982. *Mol. Cell. Biochem.* 44:71–80
204. Wals, P. A., Katz, J. 1981. *Biochem. Biophys. Res. Commun.* 100:1543–48
205. Niemeyer, H., Cardenas, M. L., Rabajille, E., Ureta, T., Clark-Turri, L., Penaranda, J. 1975. *Enzyme* 20:321–33
206. Storer, A. C., Cornish-Bowden, A. 1976. *Biochem. J.* 159:7–14
207. Bontemps, F., Hue, L., Hers, H. G. 1978. *Biochem. J.* 174:603–11
208. Johnson, M. E. M., Das, N. M., Butcher, F. R., Fain, J. N. 1972. *J. Biol. Chem.* 247:3229–35
209. Garrison, J. C., Haynes, R. C. Jr. 1973. *J. Biol. Chem.* 248:5333–43
210. Claus, T. H., Pilkis, S. J., Park, C. R. 1975. *Biochim. Biophys. Acta* 404:110–23
211. Joseph, S. K., McGivan, J. D. 1978. *Biochim. Biophys. Acta* 543:16–28
212. Blair, J. B., Cook, D. E., Lardy, H. A. 1973. *J. Biol. Chem.* 248:3608–14
213. Veneziale, C. M. 1971. *Biochemistry* 10:3443–47
214. Veneziale, C. M. 1972. *Biochemistry* 11:3286–89
215. Pilkis, S. J., Riou, J. P., Claus, T. H. 1976. *J. Biol. Chem.* 251:7841–52
216. Tolbert, M. E. M., Butcher, F. R., Fain, J. N. 1973. *J. Biol. Chem.* 248:5682–92
217. Unpublished results from authors' laboratory
218. Richards, C. S., Furuya, E., Uyeda, K. 1981. *Biochem. Biophys. Res. Commun.* 100:1673–79
219. Richards, C. S., Uyeda, K. 1982. *J. Biol. Chem.* 257:8854–61
220. Iles, R. A., Griffiths, J. R. 1982. *Biosci. Rep.* 2:735–42
221. Claycomb, W. C., Kilsheimer, G. S. 1969. *Endocrinology* 84:1179–83
222. Fröhlich, J., Wieland, O. H. 1971. *Eur. J. Biochem.* 19:557–62
223. Exton, J. H., Corbin, J. G., Park, C. R. 1969. *J. Biol. Chem.* 244:4095–4102
224. Jefferson, L. S., Exton, J. H., Butcher, R. W., Sutherland, E. W., Park, C. R. 1968. *J. Biol. Chem.* 243:1031–38
225. Gumma, K. A., McLean, P., Greenbaum, A. L. 1971. *Essays Biochem.* 7:39–86
226. Kohl, E. A., Cottam, G. L. 1977. *Biochim. Biophys. Acta* 484:49–58
227. Pilkis, S. J., El-Maghrabi, M. R.,

McGrane, M., Pilkis, J., Fox, E., Claus, T. H. 1982. *Mol. Cell. Endocrinol.* 25:245–66

228. Hers, H. G., Hue, L., Van Schaftingen, E. 1982. *Trends Biochem. Sci.* 7:329–31
229. Uyeda, K., Furuya, E., Richard, C. S., Yokoyama, M. 1982. *Mol. Cell. Biochem.* 48:97–120
230. Neely, P., El-Maghrabi, M. R., Pilkis, S. J., Claus, T. H. 1981. *Diabetes* 30:1062–64
231. Hue, L., Blackmore, P. F., Shikama, H., Robinson-Steiner, A., Exton, J. H. 1982. *J. Biol. Chem.* 257:4308–13
232. Kuwajima, M., Uyeda, K. 1982. *Biochem. Biophys. Res. Commun.* 104:84–88
233. Hue, L., van de Werve, G. 1982. *FEBS Lett.* 145:263–66
234. Malaisse, W. J., Malaisse-Lagae, F., Sener, A. 1982. *Diabetes* 31:90–93
235. Malaisse, W. J., Malaisse-Lagae, F., Sener, A., Van Schaftingen, E., Hers, H. G. 1981. *FEBS Lett.* 125:217–19
236. Pilkis, S. J., El-Maghrabi, M. R., Pilkis, J., Claus, T. H., Cumming, D. A. 1981. *J. Biol. Chem.* 256:3171–74
237. Avigad, G. 1981. *Biochem. Biophys. Res. Commun.* 102:985–91
238. Miernyk, J. A., Dennis, D. T. 1982. *Biochem. Biophys. Res. Commun.* 105:793–98
239. Reinhart, G. D., Lardy, H. A. 1980. *Biochemistry* 19:1477–84
240. Furuya, E., Kotaniguchi, H., Hagihara, B. 1982. *Biochem. Biophys. Res. Commun.* 105:1519–23
241. Stryer, L. 1981. *Biochemistry.* San Francisco: Freeman. 949 pp.
242. Hems, D. A., Whitton, P. D. 1980. *Physiol. Rev.* 60:1–50
243. Exton, J. H. 1981. *Mol. Cell. Endocrinol.* 23:233–64
244. Williamson, J. R., Cooper, R. H., Hoek, J. B. 1981. *Biochim. Biophys. Acta* 639:243–95
245. Hue, L., Van Schaftingen, E., Blackmore, P. F. 1981. *Biochem. J.* 194:1023–26
246. Jakob, A., Diem, S. 1975. *Biochim. Biophys. Acta* 404:57–66

247. Williamson, D. H., Ilic, V., Tordoff, A. F. C., Ellington, E. V. 1980. *Biochem. J.* 186:621–24
248. Assimacopoulos-Jeannet, F. D., Blackmore, P. F., Exton, J. H. 1982. *J. Biol. Chem.* 257:3759–65
249. Tolbert, M. E. M., Fain, J. N. 1974. *J. Biol. Chem.* 249:1162–66
250. Hems, D. A., Whitton, P. D. 1973. *Biochem. J.* 136:705–9
251. Whitton, P. D., Rodrigues, L. M., Hems, D. A. 1978. *Biochem. J.* 176:893–98
252. Hartmann, H., Beckh, K., Jungermann, K. 1982. *Eur. J. Biochem.* 123:521–26
253. Exton, J. H. 1979. *Glucocorticoid Hormone Action*, ed. J. D. Baxter, G. G. Rousseau, pp. 535–60. Berlin: Springer. 638 pp.
254. Postle, A. D., Bloxham, D. P. 1982. *Eur. J. Biochem.* 124:103–8
255. Ruderman, N. B., Herrera, M. G. 1968. *Am. J. Physiol.* 214:1346–51
256. Solanki, K., Nyfeler, F., Mozer, U. K., Walter, P. 1980. *Biochem. J.* 192:377–80
257. Rognstad, R. 1982. *Arch. Biochem. Biophys.* 217:498–502
258. Shikama, H., Ui, M. 1978. *Am. J. Physiol.* 235:E354–60
259. McDaniel, H. G. 1975. *Am. J. Physiol.* 229:1569–75
260. Katz, J., Rognstad, R. 1976. *Curr. Top. Cell. Regul.* 10:237–89
261. Hue, L. 1982. See Ref. 64, pp. 71–97
262. Hue, L., Hers, H. G. 1974. *Biochem. Biophys. Res. Commun.* 58:540–48
263. Racker, E. 1974. *Mol. Cell. Biochem.* 5:17–23
264. Rognstad, R., Katz, J. 1980. *Arch. Biochem. Biophys.* 203:642–46
265. Van den Berghe, G. 1978. *Curr. Top. Cell. Regul.* 13:97–135
266. Banuelos, M., Fraenkel, D. G. 1982. *Mol. Cell. Biol.* 2:921–29
267. Friedmann, B., Goodman, E. H. Jr., Saunders, H. L., Kostos, V., Weinhouse, S. 1971. *Arch. Biochem. Biophys.* 143:566–78
268. Newsholme, E. A., Gevers, W. 1967. *Vitam. Horm. NY* 25:1–86

Ann. Rev. Biochem. 1983. 52:655–709

HUMAN PLASMA PROTEINASE INHIBITORS

J. Travis and G. S. Salvesen

Department of Biochemistry, University of Georgia,
Athens, Georgia 30602

CONTENTS

PERSPECTIVES AND SUMMARY 655
INTRODUCTION 657
MECHANISM OF ACTION OF PROTEINASE INHIBITORS 659
   The Standard Mechanism 660
   Deviations from the Standard Mechanism 661
ALPHA$_1$-PROTEINASE INHIBITOR 663
   Polymorphysm of α_1 Proteinase Inhibitor 665
   Mechanism Studies 667
   Physiological Role of α_1 Proteinase Inhibitor 672
ANTITHROMBIN III 674
   Effect of Heparin 674
   Mechanism Studies 675
   Physiological Role of Antithrombin III 676
ALPHA$_2$-ANTIPLASMIN 676
ALPHA$_1$-ANTICHYMOTRYPSIN 678
C$\bar{1}$-INHIBITOR 681
ALPHA$_2$-MACROGLOBULIN 684
   The Trapping Reaction 685
   The Covalent-Linking Reaction 690
   The Adherence Reaction 693
   Physiological Role of α_2-Macroglobulin 693
INTER-ALPHA-TRYPSIN INHIBITOR 695
BETA$_1$-ANTICOLLAGENASE 696
ALPHA-CYSTEINE PROTEINASE INHIBITOR 698
OTHER INHIBITORS 699
CONCLUDING REMARKS 700

PERSPECTIVES AND SUMMARY

The plasma proteinase inhibitors, after albumin and the immunoglobulins, constitute by weight the third largest group of functional proteins in human

655

0066-4154/83/0701-0655$02.00

plasma. Representing nearly 10% of the total protein in plasma, they control a variety of critical events associated with connective tissue turnover, coagulation, fibrinolysis, complement activation, and inflammatory reactions. In addition, at least one of the inhibitors, α_2-macroglobulin ($\alpha_2 M$), plays an important secondary role in backing up the primary function of the other inhibitors.

Many of the plasma inhibitors are reported to be congenitally deficient in specific individuals. When this occurs the development of specific disease states becomes manifest, and this can make it relatively simple, as in the case of α_1-proteinase inhibitor ($\alpha_1 PI$) deficiency, to determine a physiological function for the inhibitor. However, deficiency states have not been reported for all of the inhibitors, so this method of elucidating a physiological role is not always possible. The problem is complicated further by the many claims for multiple roles of an individual inhibitor in a variety of physiological processes. This is highly unlikely since patients with deficiencies in one inhibitor, for example $\alpha_1 PI$, tend to develop only one major disease, in this case familial emphysema, while showing little or none of the effects expected if the inhibitor had the claimed multiple functions.

The difficulty in determining the current role of proteinase inhibitors could be easily clarified if the rate of complex formation between inhibitor and suspected enzyme were available. This, together with a measurement of the distribution of the enzyme in question among all of the inhibitors in plasma, would substantially reduce false interpretations with regard to the specific role of a particular inhibitor. Several reports, for example, only indicate that specific inhibitors and enzymes complex slowly, progressively, moderately, or fast. This really says very little as to whether such a reaction could be physiologically significant. However, with available kinetic data, and by using the formula described by Bieth (1), it is possible to measure the in vivo half-time of inhibition, as follows:

$$t_{1/2\,ass} \simeq 1/(k_{ass})(I)$$

where k_{ass} is the apparent second-order rate constant for association of inhibitor and enzyme, and I is the normal inhibitor concentration in plasma. In Table 1 we have compiled the $t_{1/2\,ass}$ in vivo for several of the known plasma inhibitors with individual enzymes. This data has been obtained from the literature values for the apparent second order association rates measured at either 37° or room temperature, as well as the known plasma concentration of each inhibitor, as described herein. The results clearly show the relative contributions of each inhibitor in controlling the activity of a specific proteinase and also indicate that several of the purported functions of plasma inhibitors are too slow to be of biological significance. For most inhibitor reactions a $t_{1/2}$ of 100 msec is the maximum that should be allowed for effective control of a particular enzyme. This is based on the fact that while α_1PI can inhibit cathepsin G with a $t_{1/2}$ in this range, a second inhibitor, α_1-antichymotrypsin (α_1Achy) has an even faster $t_{1/2}$ of 5 msec. Teleologically, our interpretation is that α_1PI simply cannot inactivate cathepsin G rapidly enough to control the activity of this enzyme and that α_1Achy has thus been developed for this purpose. Some of the reactions that have been measured, such as those between $C\overline{1}$-inhibitor ($C\overline{1}$-Inh) and $C\overline{1}r$ or $C\overline{1}s$, probably do not represent true in vivo rates, because the enzyme(s) is in a multicomplex. This may also be true for kallikrein and Factor Xa. Alternatively, there may be other undiscovered factors in plasma or tissues that increase the rates of inhibition in the manner in which heparin activates antithrombin III (AT III). In any case, with these few exceptions all the enzymes listed can be controlled by inhibitors that interact with a $t_{1/2}$ of less than 100 msec.

The data given in Table 1 also show that the original function(s) given to at least two of the inhibitors, α_1PI, and α_1Achy, do not represent their true role in controlling physiological processes involving proteolytic enzymes. In fact, it is probable that several plasma inhibitors specifically control the activity of individual proteinases, and that many currently used common names must be revised as their functions become clear. Potential functions for plasma inhibitors should, therefore, not be stated unless the physical defect associated with a known deficiency, kinetic data indicating rapid complexing of specific enzymes, and the partitioning of the enzyme in question among all plasma proteins are available and compatible. Where deficiency states are not known, both of the latter two sets of data still aid markedly in determining whether the inhibitor can inactivate the enzyme in question rapidly enough to be of physiological significance.

INTRODUCTION

The existence of proteinase inhibitor activity in human plasma was apparently first noted by Fermi & Pernossi in 1894 (2). Since that time a host of investigations have been made to determine the various inhibitory activities in this tissue derivative, primarily by adding a larger number of proteinases of varying specificities and catalytic mechanisms to plasma and plasma fractions. These experiments were performed to determine the number, type, concentration, and mechanism of action of the various inhibitors, and with the hope of understanding their role in the control of proteolytic events within the body. We now recognize at least nine separate, well-characterized proteins in human plasma that share the capacity to inhibit the activity of various proteinases.

The plasma of certain animals contains homologous inhibitors, but as

Table 1   Half-time of association of proteinases with human plasma proteinase inhibitors[a]

Inhibitor[b]	Enzyme									
	Elastase	Cat. G	Trypsin	Chymotrypsin	Kallikrein	Thrombin	Plasmin	Clr	Cls	Factor Xa
Alpha$_1$-PI	0.61[c]	102[c]	3,600[c]	8[c]	+	8.3×10^5[c]	2.1×10^5[c]	–	–	1.7×10^5
Alpha$_1$-Achy	–	5[c]	–	27,000[c]	–	–	–	–	–	–
Alpha$_2$-M	7.2[c]	93[c]	19[c]	27[c]	6,125	5×10^5	+	–	–	3.5×10^5
Cl-Inh	–	–	–	–	8,333	?	+	2.0×10^5	4.7×10^4	?
Alpha$_2$-AP	–	–	617[c]	11,100[c]	+	–	29[c]	–	–	+
AT 111	–	–	+	–	+	847	+	–	–	2.3×10^5

[a] In milliseconds.
[b] (–), No interaction; (+), inactivation but no kinetic data; (?), no data available.
[c] Measured at room temperature.

the majority of attention was focused on the human proteins, we confine this review to the protein inhibitors of endopeptidases (proteinases) in human blood plasma. We group several of the inhibitors together, including $\alpha_1 PI$, AT III, $\alpha_1 Achy$, C̄I-Inh, and α_2-antiplasmin ($\alpha_2 AP$), as they appear to share several features, in particular their mechanism of action. However, they are not sufficiently homologous to be considered an inhibitor family as defined in a previous review (3). Therefore, we avoid this term and refer to them as the $\alpha_1 PI$ class, since this inhibitor is by far the most characterized of the group.

$\alpha_2 M$ inhibits members of each of the four proteinase catalytic classes— serine, cysteine, aspartic, and metallo proteinases (4)—but each of the other inhibitors is class specific. The $\alpha_1 PI$ group and the inter-α-trypsin inhibitor (IαI) inhibit only serine proteinases, α_1-cysteine proteinase inhibitor (αCPI) inhibits only cysteine proteinases, and the β_1-anticollagenase ($\beta_1 AC$) inhibits only collagenolytic enzymes of the metalloenzyme class. Significantly, no class specific plasma inhibitor of aspartic proteinases is known.

With the possible exception of $\beta_1 AC$, all the inhibitors react with several proteinases from such a wide variety of sources that it is difficult to assign a definite role for each. We hope to shed light on this problem by attempting to establish the physiologically relevant "target" proteinases for each of the inhibitors. We do so by considering current data on the effective reaction rates of proteinases with each inhibitor that leads to the formation of a stable enzyme-inhibitor complex.

MECHANISM OF ACTION OF PROTEINASE INHIBITORS

Little is known of the mechanism by which proteinase inhibitors of non-serine proteinases function ($\beta_1 AC$ and αCPI, for example), with the exception of $\alpha_2 M$. We thus concentrate on only the $\alpha_1 PI$ class of inhibitors; $\alpha_2 M$ and IαI are dealt with in separate sections.

Protein proteinase inhibitors act competitively by allowing their target enzymes to bind directly to a substrate-like region contained within the amino acid sequence of the inhibitor (3). This reaction between enzyme and inhibitor is essentially second order, and the resultant complex contains one molecule of each of the reactants, with the exception of those inhibitors that contain more than one inhibitory domain (the "multiheaded" inhibitors). However, if one considers the stoichiometry of reaction in terms of domains rather than entire molecules, all proteinase-proteinase inhibitor complexes are equimolar with respect to inhibitory domains (or more accurately, "reactive sites") and that the multiheaded inhibitors constitute a special case of the general rule.

The Standard Mechanism

The generally accepted mechanism for the inhibition of serine proteinases by protein inhibitors was reviewed recently (3). However, although these inhibitors share many features in common with the α_1 PI class, there are sufficient differences for the latter to be considered a separate mechanistic subclass. Before dealing with these deviations, however, we summarize briefly those features that help make a protein an efficient inhibitor.

Of primary importance, the protein must combine rapidly with its target proteinase to form a stable complex. Apparent second-order association rates (k_{ass}) are generally faster than 10^5 M^{-1} sec^{-1}, and dissociation rates (k_{diss}) are usually less than 10^{-5} sec^{-1}, giving an equilibrium constant of 10^{10} M or greater. The kinetic constants we choose to distinguish a good inhibitor from a poor one are, of course, arbitrary, but we do so primarily to provide a guide, and with the knowledge that many proteinase-inhibitor complexes are reported to have equilibrium constants much less than 10^{10} M (3).

Inhibition occurs as a consequence of binding of the active-site substrate-binding region of a proteinase to the corresponding substrate-like region (reactive site) on the surface of the inhibitor (5). At or near the center of the reactive site is an amino acid residue specifically recognized by the primary substrate binding site of the target proteinase (3) and termed the P_1 residue (6). Adjacent to P_1 in the direction of the carboxy terminus is the residue referred to as P'_1. It is the peptide bond joining these two residues, known as the reactive-site peptide bond, that is hydrolyzed during complex formation between an inhibitor and its target enzyme. The k_{cat}/K_m value for the hydrolysis of this peptide bond in inhibitors obeying the standard mechanism is very high, $10^4 - 10^6$ M^{-1} sec^{-1} (7, 8), and X-ray analyses reveal that the geometry of the reactive sites of certain inhibitors is such that they should act as good substrates (3, 9). Indeed, some inhibitors are easily hydrolyzed by certain proteinases (3, 8), yet they are extremely efficient inactivators of other, homologous proteinases. Structural studies may so far have failed to explain exactly what makes an inhibitor an inhibitor, rather than just a good substrate, but kinetic data have clarified the major effects of such inhibition. For example, although values of k_{cat}/K_m for the cleavage of several inhibitor reactive-site peptide bonds are high, the individual values of k_{cat} and K_m are several orders of magnitude lower than those for normal substrates (3) with the result that hydrolysis of the bond is extremely slow. In the case of those inhibitors obeying the standard mechanism, hydrolysis of the reactive-site peptide bond does not proceed to completion (3). Rather, an equilibrium near unity is established (3) between the form of the inhibitor with peptide bond intact ("virgin inhibitor") and the form with peptide bond cleaved ("modified inhibitor"). The following simplified kinetic scheme summarizes the reaction:

$$E + I \rightleftharpoons EI \rightleftharpoons C \rightleftharpoons EI^* \rightleftharpoons E + I^*$$

where E is the proteinase, I and I* are virgin and modified inhibitor, respectively, and C is the stable complex. The full kinetic scheme contains far more terms, relating to intermediates in the reaction, than those shown above, and we intend this scheme simply to be a guide. The stable complex C is formed from either virgin or modified inhibitor, both of which are thermodynamically strong inhibitors; however, the apparent rate of association of modified inhibitor with enzyme is generally slower than the corresponding rate for virgin inhibitor (3, 7). Significantly, the capacity for peptide bond hydrolysis is not essential for the establishment of a stable complex as those formed between anhydrotrypsin (with the active-site serine residue chemically modified to dehydroalanine) and certain inhibitors are of comparable stability, with similar geometric arrangement to those between trypsin and the same inhibitor (10–12). X-ray analysis reveals that the reactive-site peptide bond is intact within the complex, but that the carbonyl function is partially tetrahedral rather than trigonal, as it is in its uncomplexed state (13). The distortion of this bond is caused by attraction of the carbonyl oxygen to the enzyme oxyanion binding pocket (rather than to attack of the nucleophilic O^γ in the catalytic serine residue as once thought) and is envisaged as one of the attractions that help stabilize the enzyme-inhibitor complex. Other interactions, including hydrogen bonds, Van der Waal's forces, and possibly salt bridges, in the close contact area between the reactive site and the corresponding substrate-binding region of the proteinase all contribute to the characteristic stability of the complex (3).

Deviations from the Standard Mechanism

The members of the $\alpha_1 PI$ class are all high molecular weight single-chain glycoproteins with single reactive centers. Obeying at least a part of the standard mechanism, they rapidly form equimolar complexes with their target enzymes via reactive-site regions (Figure 1), and at least one of these inhibitors, $\alpha_1 PI$, shares significant homology with other plant and animal reactive sites (14). There is also some homology within the class of inhibitors listed at their reactive centers; this extends to other regions of at least two of the inhibitors, AT III and $\alpha_1 PI$ (15). Data on the reactive centers of $\alpha_2 AP$ and $C\bar{1}$-Inh, however, are still preliminary. Despite the similarities, there are two obvious points of deviation of the $\alpha_1 PI$ class from those operating via the standard mechanism: 1. Modified inhibitors are inactive and cannot recombine with a proteinase. 2. The strength of the interaction is so strong as to imply a covalently stabilized complex.

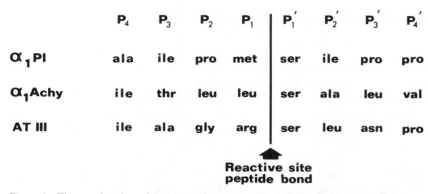

	P_4	P_3	P_2	P_1	P_1'	P_2'	P_3'	P_4'
α_1PI	ala	ile	pro	met	ser	ile	pro	pro
α_1Achy	ile	thr	leu	leu	ser	ala	leu	val
AT III	ile	ala	gly	arg	ser	leu	asn	pro

**Reactive site
peptide bond**

Figure 1   The reactive sites of members of the alpha$_1$-proteinase inhibitor class of human plasma proteinase inhibitors. Sequence data were taken from: α_1PI (14), α_1Achy (M. Morii, J. Travis, manuscript in preparation), AT III (16). The terminology of the reactive-site residue positions is taken from Schechter & Berger (6), with the reactive-site peptide bonds arrowed.

INACTIVE MODIFIED INHIBITOR   In contrast to those inhibitors obeying the standard mechanism, cleavage of the reactive-site peptide bond of members of the α_1PI class apparently results in a modified protein that is no longer inhibitory. Thus, thrombin, Factor IXa, and Factor Xa each produce a modified AT III that is unable to inhibit thrombin (16, 17); chymotrypsin and cathepsin G produce an inactive form of α_1Achy (18); and trypsin causes the formation of a modified form of α_1PI (19, 20). The rate of breakdown of complexes to form inactive inhibitors has also been determined, and, although it is quite slow ($k_3 = 6 \times 10^{-6}$ sec^{-1} to 3 $\times 10^{-8}$ sec^{-1} in the scheme given below) for various trypsin-α_1PI complexes (20, 21), peptide bond cleavage often goes to virtual completion (22).

Modified inhibitor can also be formed by direct turnover of the reactive-site peptide bond by non-serine proteinases such as papain (23), *Pseudomonas aeroginosa* elastase (24), and *Serratia marcescens* metallo-proteinase (25). This may be of some physiological relevance in depleting active α_1PI levels in tissues, for example, and is discussed later.

COVALENT COMPLEX FORMATION   Unique to the members of the α_1PI class is their ability to form complexes with their target enzymes that resist dissociation by denaturing agents such as sodium dodecylsulfate (NaDodSO$_4$) or urea. This implies that a covalent interaction must exist between proteinase and inhibitor (26–30). If a bond is formed it must exist between the O$^\gamma$ of the catalytic serine of the enzyme and the carbonyl carbon of the reactive-site peptide bond of the inhibitor; in other words, inhibition is presumed to result from the establishment of a stable acyl-enzyme derivative. Whether the native complex contains a fully formed covalent bond between enzyme and inhibitor or whether this bond is formed as a consequence of the denaturing conditions required for its detection remains to be determined; however, the establishment of a covalent link has been claimed as evidence for a functionally irreversible complex (26, 27). In opposition to this is the recent report that $\alpha_2 M$ can bind proteinases originally complexed to $\alpha_1 PI$, presumably through dissociation of the inhibitory complex. Since at least a part of the $\alpha_1 PI$ recovered in these experiments was active toward other serine proteinases, the inhibitory complex formed between this inhibitor and a given proteinase apparently can exist in a non-acylated form, and acylation is not required for proteinase inhibitor function (21).

The inhibition of proteinases by members of the $\alpha_1 PI$ class is thought to obey the following scheme, modified from that proposed for the inhibition of acetylcholinesterase by methane sulfonic acid derivatives (21):

$$E + I \underset{k_{-1}}{\overset{k_1}{\rightleftharpoons}} EI \overset{k_2}{\rightarrow} EI^* \overset{k_3}{\rightarrow} E + I^*$$

Obviously, it is difficult to know whether any of the stable inhibitor complex exists in an acyl form (peptide bond cleaved). Data showing the binding of anhydrotrypsin to $\alpha_2 AP$ indicates that this is not necessary (30); this is supported by the $\alpha_2 M$ reversibility results described above (31). The confusion over the nature of the stable complex between a proteinase and members of the $\alpha_1 PI$ class will not be resolved until X-ray crystallographic studies have been carried out. However, the acyl-intermediate form may not be required for inhibition by this class of inhibitors.

ALPHA₁-PROTEINASE INHIBITOR

Human $\alpha_1 PI$ has been the subject of intense investigations, primarily because of its role in controlling proteolytic events in tissues. Originally named $\alpha_1$antitrypsin (32) because of its ability to inactivate pancreatic trypsin, this protein has since been found to be far more effective in controlling the activity of a number of other serine proteolytic enzymes (33). Because of this and the confusion associated with the term "antitrypsin," we have chosen the name $\alpha_1 PI$ as a more appropriate term associated with the spectrum of proteinases with which it can form complexes (34).

The reason for the interest in $\alpha_1 PI$ at both the biochemical and clinical levels is that many individuals with circulating levels of this inhibitor that are less than 15% of normal are susceptible to the development of lung disease (familial emphysema) at an early age (35). Thus, this inhibitor could represent an important part of the defense mechanism of the lung towards proteolytic attack. In actuality, the development of emphysema is not restricted to individuals with low, circulating protein levels of α_1PI; therefore, intense investigations of the structure, mechanism of action, biosynthesis, and secretion of this inhibitor have been undertaken to define more clearly its important role in controlling proteolysis of the connective tissue of the lung.

The isolation of α_1PI was performed in numerous laboratories, which took advantage of the known properties of the inhibitor to develop novel, yet efficient techniques for its purification to homogeneity. The three most commonly used procedures involve, as fundamental steps, either the removal of albumin from plasma by affinity chromatography on dye-linked matrices (34, 36–40), affinity chromatography on concanavalin A sepharose to bind most of the glycoproteins in plasma but not albumin (37, 41–47), or disulfide interchange between the single cysteinyl residue in reduced α_1PI and matrix-bound mixed disulfide bonds of immunoglobulin kappa chains to specifically bind this inhibitor (47, 48). However, all methods still required at least one additional step to obtain homogeneous preparations. Even the use of matrix-bound antibodies to α_1PI, which should specifically adsorb this inhibitor from whole plasma, has not been completely successful since minor contaminants were also bound and desorbed (49).

Human α_1PI is a glycoprotein of M_r near 53,000 (34, 50). The inhibitor exists as a single polypeptide chain with no internal disulfide bonds and only a single cysteinyl residue normally intermolecularly disulfide-linked to either cysteine or glutathione (48). The reason for this unusual structure is not clear. However, in its native form the α_1PI may exist in a reduced state with its free thiol group serving in a protective role as a scavenger for oxidants that could inactivate the inhibitor.

Preparations of α_1PI have been reported to have either a blocked amino terminus (38, 51–53) or glutamic acid as the first residue (34, 43, 50, 54). A similar controversy was also found at the carboxy terminus, which was reported to be either lysine (43, 53, 55) or leucine (51, 53, 54). These controversies, however, were resolved when the primary structure of the inhibitor was elucidated, through both classical protein chemistry techniques (14, 56–59) and cloning and sequencing of cDNA for baboon and human α_1PI (60). The results indicate that the mature protein consists of 394 amino acid residues with an amino terminal glutamic acid, a carboxy terminal lysine, and a single cysteinyl residue. Comparison of this sequence with that of both AT III and chicken ovalbumin suggests 30% homology among these proteins (15, 59). However, the exact meaning of this is not yet established, especially in terms of whether ovalbumin was once a proteinase inhibitor.

Polymorphism of α_1 *Proteinase Inhibitor*

Human α_1PI occurs in at least twenty different forms in the general population. Hence a classification system based on the electrophoretic mobility of the inhibitor isoforms in individual samples has been developed (61, 62). This method of phenotyping, known as the Pi system, refers to the commonest genotype of α_1PI as Pi_M, with the phenotype being Pi_{MM}. However, even in such homozygous individuals there are at least five isoinhibitor forms, which occur as two major and three minor forms (48). The microheterogeneity of individual phenotypes was originally attributed to differences in sialic acid content of the attached carbohydrate side chains (63, 64). However, treatment of α_1PI$_{MM}$ with neuraminidase did not substantially alter the five-band pattern on isoelectric focusing gels, except to shift it toward higher isoelectric points (48, 50, 65). The two major components differed in sialic acid content by one residue, yet remained separate after neuraminidase treatment; this implied other structural differences between these two forms (48).

Several investigators examined the carbohydrate side-chain structure of α_1PI$_{MM}$ to determine whether changes in this moiety, which represents 12% of the molecular weight of the inhibitor, were still responsible for its heterogeneous pattern in a homozygous phenotype. It was first reported that the protein contained two distinct types of oligosaccharide chains attached as two sets to four positions in the protein (51). However, others disputed this and suggested that the inhibitor contained four attachment sites for three or four types of oligosaccharide chains (66). Later (67), two types of carbohydrate chains were discovered in α_1PI. These chains had a biantennary and triantennary structure as shown below:

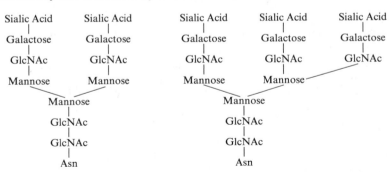

The original hypothesis of four oligosaccharide chains was then re-examined (68, 69), and three carbohydrate side chains in α_1PI at positions 46, 83, and 247, numbering from the amino terminus of the polypeptide (70,

71) were found. However, the heterogeneity associated with individual forms of $\alpha_1 PI$ present in Pi_{MM} individuals was clarified only by the finding that the isoproteins themselves contained variable amounts of each of the two forms of chains (47, 72). This was determined primarily by isolating individual isoinhibitors through their different affinities on concanavalin A sepharose; the biantennary glycoproteins or glycopeptides were bound and the triantennaries showed no affinity (42). The two major isoinhibitors in $\alpha_1 PI_{MM}$ were separated by this procedure; one was more tightly adsorbed to the lectin column than the other. Examination of the carbohydrate side-chain composition indicated that the form most tightly bound contained three biantennary side chains while the other had two biantennary and one triantennary. More recently (47), a third minor isoinhibitor from $\alpha_1 PI_{MM}$ was found to contain one biantennary and two triantennary chains. The two isoforms of side chains provide much of the explanation for the microheterogeneity of $\alpha_1 PI$ in a homozygous individual, with only minor differences attributable to desialylation of terminal oligosaccharide residues.

While the above explanation appears to satisfy the problem of micro-heterogeneity in the $\alpha_1 PI$ obtained from homozygotes, it does not necessarily explain the differences associated with the electrophoretic mobility of $\alpha_1 PI$ isoinhibitors obtained from individuals with varying phenotypes. The data indicate clearly at least one fundamental difference in the primary structure of the various inhibitors secreted by individual $\alpha_1 PI$ genotypes. Such studies were initiated to explain the low concentration of $\alpha_1 PI$ in the plasma of individuals first described as also developing lung disease (35). This group secretes a form of $\alpha_1 PI$ whose electrophoretic pattern, referred to as Pi_{ZZ}, indicates that the isoinhibitor forms have isoelectric points farthest away and higher than that of Pi_{MM}. The protein is present in concentrations less than 15 mg/100 ml in plasma (73), relative to the Pi_{MM} phenotypes whose plasma concentration is between 100–130 mg/100 ml (36, 48). However, the problem in Pi_{ZZ} individuals is one of secretion rather than synthesis, since inclusion bodies, consisting primarily of an insoluble, polymerized form of $\alpha_1 PI$, are localized within the rough endoplasmic reticulum of their hepatocytes (74, 75). For this reason attempts have been made to characterize both the plasma and liver inclusion body forms of $\alpha_1 PI_{ZZ}$ and compare them with that of plasma $\alpha_1 PI_{MM}$. It was suggested that plasma $\alpha_1 PI_{ZZ}$ differed primarily from $\alpha_1 PI_{MM}$ by having a substantially lower sialic acid content (51, 62, 63). However, recent data indicate that the carbohydrate content of both plasma $\alpha_1 PI_{MM}$ and $\alpha_1 PI_{ZZ}$ are virtually identical (48, 76), and that neuraminidase treatment yields changes in patterns of both sets of proteins. Concurrently, examinations of the comparative protein structures of $\alpha_1 PI$ variants indicated that a single amino acid replacement of glutamic acid in $\alpha_1 PI_{MM}$ for lysine in $\alpha_1 PI_{ZZ}$ at position 342 in the amino acid sequence had occurred; this change might affect subsequent posttranslational modifications, thereby reducing secretion of $\alpha_1 PI_{ZZ}$ (77–79). To test this possibility, an examination of the $\alpha_1 PI_{ZZ}$ protein in liver inclusion bodies was made. Liver $\alpha_1 PI_{ZZ}$ contained very low levels of galactose, no sialic acid, and decreased N-acetyl glucosamine, relative to either plasma $\alpha_1 PI_{MM}$ or $\alpha_1 PI_{ZZ}$ (80), but also a twofold increase in mannose content (81). Subsequent examination of the protein indicated that partial processing to form normal biantennary or triantennary side chains had begun and that all enzymes necessary for normal synthesis were present (82).

A reasonable explanation for the above results is that the change in primary structure from glutamic acid to lysine had effected a conformational change, thereby reducing the rate of processing and consequent retention of the $\alpha_1 PI_{ZZ}$ in the liver. It therefore appears that with the exception of the point mutation in the $\alpha_1 PI_{ZZ}$ protein, there are no other differences in this plasma protein from $\alpha_1 PI_{MM}$ and that interference with secretion occurs at the level of posttranslational modification. A similar explanation probably applies to other variants as well, including the Pi_S phenotype, for which a glutamic acid to valine substitution has been reported, although not at the same position as in the Pi_Z variant, residue 264 (82).

Mechanism Studies

Human $\alpha_1 PI$ has been shown to inactivate virtually all mammalian serine proteinases tested to date, including pancreatic and neutrophil elastase (83–85), pancreatic trypsin and chymotrypsin (83), neutrophil cathepsin G (33), thrombin (86), plasmin (87, 88), acrosin (89), tissue kallikrein (90), Factor Xa (91), Factor XIa (92), skin and synovial collagenases (93, 94), and urokinase (95). In addition, microbial serine proteinases have also been reported to be inactivated by $\alpha_1 PI$ (96–98). However, inactivation of an enzyme is not, in itself, a sufficient criterion for a physiological role, as outlined earlier. Rather, the rate of inactivation is of primary importance as this varies from an apparent k_{ass} of $6.5 \times 10^7 \, M^{-1} \, sec^{-1}$ for neutrophil elastase to 4.8×10^1 $M^{-1} \, sec^{-1}$ for thrombin (33), with many other proteinases reacting at rates between the two extremes. The data shown here and in Table 1 shows that a primary function of $\alpha_1 PI$ is to control the activity of neutrophil elastase since the k_{ass} is more than tenfold higher than that of other proteinases tested to date and is only 15-fold lower than diffusion controlled limits.

The interaction of $\alpha_1 PI$ with serine proteinases occurs through strong interactions presumably caused by proteolytic attack on the inhibitor, the latter acting initially as a substrate for the enzyme. The rationale for this type of initial interaction is that (*a*) diisopropylphosphoryl-trypsin does not form complexes with $\alpha_1 PI$; (*b*) anhydrochymotrypsin forms only a loose complex with $\alpha_1 PI$ (100), which dissociates upon treatment with sodium dodecylsulfate (NaDodSO$_4$), while anhydrotrypsin does not react at all (101); and (*c*) zymogens react slowly to form NaDodSO$_4$ stable complexes (102, 103), analogous to the inhibitory effect of diisopropylfluorophosphate on pancreatic zymogens (104). These data offer substantial evidence that the reactive-site serine residue of the proteinase is involved in some type of covalent bonding with the inhibitor and would normally stabilize the inhibitor-enzyme complex.

There are also substantial ionic and hydrophobic components involved in complex formation since modification of conditions for inhibition or chemical modification of the inhibitor prior to interaction with a specific proteinase have marked effects. For example, increases in both temperature and ionic strength enhance $\alpha_1 PI$ complex formation with porcine elastase (105, 106), while the addition of 20% dioxane completely blocks complex formation with this enzyme (107). Significantly, the addition of dioxane has little or no effect on the inhibition of trypsin or chymotrypsin and can be reversed for elastase by dilution.

Chemical studies have also been carried out to relate a loss of inhibitory activity with amino acid residues involved in complex formation. Alterations of lysyl and arginyl residues has received the most attention, primarily because the original name for this inhibitor, α_1-antitrypsin, implied that the reactive center of the inhibitor must contain either one of these basic residues in order to satisfy the specificity of the enzyme. Treatment of $\alpha_1 PI$ with maleic anhydride (108), acetic anhydride (109), or citraconic anhydride (110), all of which react with lysyl residues, caused a reversal of charge and inactivated the inhibitor towards trypsin, chymotrypsin, and elastase. Modification of lysyl residues with only charge neutralization, using trinitrobenzene sulfonic acid, also caused a loss of inhibitory activity towards elastase, chymotrypsin, and trypsin but introduced a bulky side chain derivative (111). It was concluded that $\alpha_1 PI$ had a lysyl residue at its reactive center similar to that found in many other plant and animal inhibitors of lower molecular weight (3). However, when the lysyl residues were modified by reductive methylation without changing the charge on the lysyl residues, and by introduction of less bulky methyl groups, no inhibitory activity was lost (112), leaving conflicting results as to the direct importance of lysyl residues in $\alpha_1 PI$ function.

Examination of other potential amino acids involved in the $\alpha_1 PI$ inhibitory mechanism have not, with the exception of methionine modification (to be discussed in detail separately), been investigated in depth. Reports of the importance of arginyl residues by inactivation of $\alpha_1 PI$ with phenylglyoxal hydrate (99) are not yet substantiated. It was clearly shown that the reagent 1,2 cyclohexanedione, which is specific for arginyl residues, has no effect on the inhibitory activity of $\alpha_1 PI$ (110). Recently (113), the elastase inhibitory activity of $\alpha_1 PI$, but not that of trypsin or chymotrypsin, was abolished by treatment with tetranitromethane or N-acetyl imidazole, reagents which modify tyrosyl residues. In this case a single residue was modified; however, its position in the primary structure is not yet identified.

Because $\alpha_1 PI$ has such a broad specificity towards serine proteinases, multiple inhibitory sites on the protein have been suggested (113–116). Nevertheless, the inhibition stoichiometry remains at 1:1 on a molar basis, although one group initially reported 2:1 binding of trypsin (117), which was later retracted (85). The existence of a single site or closely overlapping sites in $\alpha_1 PI$ was clearly demonstrated through competition studies with various proteinases (39, 99, 118, 119). The first demonstration of a probable single reactive center for $\alpha_1 PI$ was performed by examining individual complexes formed with pancreatic trypsin, chymotrypsin, and elastase (23, 26, 120). Complexes were dissociated with nucleophiles and the $\alpha_1 PI$ reisolated. In all cases an identical, new amino terminal sequence was detected, indicating peptide bond cleavage at the same site by all three proteinases. Indeed, even papain, a non-serine proteinase, cleaved $\alpha_1 PI$ at the same site, resulting in the inactivation of the inhibitor by catalytic amounts of enzyme (23). Others reported similar results, with the major differences that some nonspecific cleavage by enzymes at other sites on the inhibitor molecule also occurred (26, 54, 106, 114, 121). This result was primarily attributable to experiments carried out either under 1:1 stoichiometric conditions or in enzyme excess. Under such conditions some proteolysis at other available sites on the $\alpha_1 PI$ molecule increases dramatically, together with degradation of enzyme already in complex (106, 122, 123). Curiously, inhibitor already in complex does not seem to be susceptible to further proteolytic action (22).

During complex dissociation, some active enzyme and a modified form of $\alpha_1 PI$, referred to as $\alpha_1 PI^*$ (or modified $\alpha_1 PI$), are released (15, 120, 124, 125), and the molecular weight of $\alpha_1 PI^*$ reduced to less than 50,000. This indicates that cleavage of a small peptide occurred during complex dissolution. The positioning of this peptide in the primary sequence, whether it arises from the amino or carboxyl terminus of native $\alpha_1 PI$, has caused considerable confusion, primarily because the amino terminus of the native inhibitor is glutamic acid (34, 43, 50, 54), an amino acid that is frequently blocked by cyclization to pyrrolidone carboxylic acid (126). In addition, the small peptide fragment has a tenacious hydrophobic adhesion to the larger $\alpha_1 PI^*$ (58, 127). Consequently, it is not easily removed except by treatment with strong denaturing agents such as $NaDodSO_4$. Therefore, during sequence analysis of modified $\alpha_1 PI^*$, what was really being examined was possibly a tightly bound peptide adsorbed onto a protein with a cyclized amino terminus, the latter formed during the dissociation of complex. Thus, one would only see the sequence of the peptide and believe it to be due to $\alpha_1 PI^*$. This reasoning explains why several groups reported peptide bond cleavage during complex dissociation at the amino terminus of native $\alpha_1 PI$ (14, 114, 116, 128). In fact, the confusion was resolved when the low-molecular-weight peptide was isolated and sequenced (58, 127) and shown to have sequence homology with the carboxy terminal of $\alpha_1 PI$ (55). Final confirmation of the position of the cleavage site was obtained during determination of the primary structure of the inhibitor (59, 60, 70).

The amino acid sequence at the putative reactive center of $\alpha_1 PI$ was determined by isolation and analysis of peptides whose structures overlapped with that obtained by amino-terminal sequence analysis of dissociated complexes. The results indicated that a methionyl–seryl sequence was at the $P_1 - P_1'$ of the inhibitor (26), the methionyl residue having also been suggested on the basis of the sensitivity of $\alpha_1 PI$ to oxidizing agents (129). The presence of methionine at the reactive site of $\alpha_1 PI$ was not unusual since it has a polarizable side chain, which permits it to assume a positive charge in accommodating the active site of trypsin-like enzymes or to fit equally well into the hydrophobic centers of both chymotrypsin and elastase. Indeed, synthetic substrates containing sequences homologous or analogous with those near the reactive site of $\alpha_1 PI$ have been investigated for hydrolysis by trypsin, chymotrypsin, pancreatic elastase, neutrophil elastase, and cathepsin G (130–132). These compounds were all cleaved after the methionyl residue by the enzymes tested but at vastly different rates, with chymotrypsin by far the most effective and trypsin least effective. Since neutrophil elastase has a higher apparent k_{ass} for $\alpha_1 PI$ than bovine chymotrypsin ($10^7 \ M^{-1} \ sec^{-1}$ vs $10^6 \ M^{-1} \ sec^{-1}$) (33), other factors are clearly involved in complex formation. Significantly, chemical oxidation of the methionine-containing peptide substrates to the corresponding sulfoxide derivatives reduced markedly the rates of hydrolysis, indicating that the methionyl residue is still of primary importance in $\alpha_1 PI$ inhibitory function.

EFFECT OF OXIDANTS ON α_1PROTEINASE INHIBITOR   The presence of a methionyl residue in a critical position for $\alpha_1 PI$ function resulted in intense investigations into the effect of oxidizing agents on inhibitory activity. Using defined chemical oxidants, investigators demonstrated readily that the oxidized inhibitor had lost all of its initial combining activity with porcine elastase (14, 106, 129, 133, 134), and had a reduced rate of interaction with trypsin (133, 134). However, there was only a moderate change in its interaction with chymotrypsin, despite the poor hydrolysis of oxidized synthetic substrates analogous in structure to the $\alpha_1 PI$ reactive center (131). Oxidation occurred at two methionyl residues, P_1 and P_8 of the reactive-site sequence, confirming its expected susceptibility to oxidizing agents (133). Nearly all of this data was explained by measurement of the k_{ass} for both native and oxidized inhibitor with several serine proteinases (33). The results indicated significant reductions in k_{ass} after oxidation from $10^7 \, M^{-1} \sec^{-1}$ to $10^4 \, M^{-1} \sec^{-1}$ for neutrophil elastase, and from $10^5 \, M^{-1} \sec^{-1}$ to zero for porcine pancreatic elastase, for example. In fact, with the exception of chymotrypsin interactions, there was a reduced rate of inactivation of all enzymes tested with oxidized inhibitor.

Biological oxidants have also been tested as potential inactivators of $\alpha_1 PI$ and the data have been recently summarized (135). The most common mammalian system for generating oxidants is that which uses myeloperoxidase, H_2O_2 and Cl^-. This enzyme system is normally utilized in phagocytic processes in the neutrophil, where it is found in abundance. When tested with $\alpha_1 PI$ it was found to rapidly and catalytically inactivate the inhibitor (129, 135–138), yielding a modified inhibitor with similar properties to that obtained by chemical oxidation.

During phagocytosis neutrophils release some oxidants outside of the cell, since $\alpha_1 PI$ exposed to dialyzable components produced by these cells reduced markedly its elastase inhibitory capacity (139). When these studies were extended to blood monocytes and alveolar macrophages, reactive oxygen species were also released, which reduced elastase inhibition (140). Therefore, oxidants produced by phagocytizing cells may play a significant role in the inactivation of $\alpha_1 PI$ outside of the cell.

Because of the potential relationship between low levels of αPI and the development of pulmonary emphysema (73), experiments were performed to determine whether cigarette smoke could also inactive this inhibitor. Unfractionated cigarette smoke rapidly reduced the inhibitory activity of $\alpha_1 PI$ toward both pancreatic and neutrophil elastase (141–143); antioxidants prevented this inactivation. Furthermore, examination of the $\alpha_1 PI$ in the bronchial lavage fluids of the lungs of rat and man after exposure to cigarette smoke indicated a marked reduction in elastase inhibitory capacity relative to controls (144, 145) as well as the oxidation of four of the methionyl residues of the human inhibitor (146). Recently (147), this oxidation was also measured in vivo; more than 23% of the $\alpha_1 PI$ was reported oxidized in the serum of individuals after inhalation of cigarette smoke. These data concur with those obtained earlier in the examination of $\alpha_1 PI$ isolated from rheumatoid synovial fluid (148), which was also found to contain four methionine sulfoxide residues, two of which were at the P_1 and P_8 positions, and to be inactive as an inhibitor.

The effect of oxidation on the second order k_{ass} with neutrophil elastase is such as to reduce it by more than a factor of 10^3. If it is assumed that all of the $\alpha_1 PI$ in the lung or other tissues is totally oxidized by either chemical (cigarette smoke) or biological (myeloperoxidase) oxidants, as reported (144–146), the half-time of association, $t_{1/2}$, for the interaction with neutrophil elastase would be increased from 0.64 msec to 1.33 sec. Such a change would be exceedingly deleterious to the protection of lung tissue, particularly since it has been reported that lung elastin already competes effectively with active $\alpha_1 PI$ for binding sites on neutrophil elastase (149).

Physiological Role of α_1 Proteinase Inhibitor

The control of elastolytic activity secreted primarily by neutrophils appears to be the primary function of this protein. This is based on the following data: (a) $\alpha_1 PI$-deficient individuals develop pulmonary emphysema rapidly and at an early age (73); (b) the development of emphysema, in general, is associated with increased turnover of lung connective tissue proteins, particularly elastin (150); and (c) kinetic experiments show clearly that $\alpha_1 PI$ inactivates neutrophil elastase at a rate more than ten times faster than any other proteinase tested so far (33).

In individuals with normal serum protein levels of $\alpha_1 PI$, as long as there is no burden to the lung, there is sufficient active inhibitor to control proteolysis. However, upon inhalation of foreign particulate matter, cigarette smoke for example, a series of events is presumed to occur that alters the proteinase inhibitor–proteinase balance in the lung. First, the foreign material itself may contain oxidants that directly reduce active $\alpha_1 PI$ levels, both in the lungs and in the serum (141, 146, 147, 151). Second, activation and further recruitment of phagocytic cells (neutrophils, macrophages), followed by phagocytosis, would likely produce oxidants that could readily cause $\alpha_1 PI$ inactivation (129, 135, 138). Because these cells produce proteinases, together with oxidants, and both can be released, leak out during phagocytosis, or appear after cell death, a rapid change in active proteinase inhibitor-active proteinase levels would take place, even though protein $\alpha_1 PI$ levels remain normal.

Proteolytic enzymes can reportedly also enhance production of oxidants (152). Together with the production of proteinases and oxidants by phagocytosis, this would tend to set up a vicious cycle whereby proteolytic enzymes would not only degrade tissue abnormally, but also increase the production of oxidants to turn off inhibitor control of the proteinases, themselves.

An animal model is now developed that tests the above potential role of $\alpha_1 PI$ by maintaining very low active levels of this inhibitor in the blood of animals through a regimented infusion of oxidizing agents (134). The results of this relatively short-term experiment (3–27 weeks) clearly indicates the development of emphysema-like alterations in the lungs of the animals (143) and strongly supports the normal protective role of α_1PI. In any disease state in which there is primarily abnormal connective-tissue turnover, the major cause may be inactivation of α_1PI, presumably by phagocytic cells attracted to the tissue in question. For example, in diseases in which immune complex buildup is a primary result, one would expect the migration of neutrophils and/or macrophages to the disease site(s), primarily to degrade these complexes. The result would be the same as in the emphysema condition, with a change in the equilibrium between protective α_1PI and the proteinases released by the phagocytic cells. Hence, rheumatoid arthritis and glomerulonephritis, in which connective-tissue turnover is a primary effect, may be caused by the inability of sufficient active α_1PI to be maintained as a protective agent for the tissue in question.

Recently (153), an enzyme was noted that can reactivate oxidized α_1PI by reduction of the methionine sulfoxide residues. This enzyme, methionine sulfoxide reductase, has also been found in human neutrophils (N. Brot, personal communication). Whether it plays any role in maintaining active α_1PI levels in the lung and other tissues remains to be established. However, many systems can augment either the oxidation of α_1PI or the protection against this effect through the production of either oxidizing or reducing agents. This may explain why some individuals are more prone to the development of lung disease than others, as their levels of various enzymes such as glutathione reductase, glutathione peroxidase, superoxide dismutase, and methionine sulfoxide reductase, for example, are altered to either higher or lower values than normal.

Finally, we ask why a methionyl residue is at the reactive site of α_1PI, when a valyl residue would provide the host with a more potent and permanent inhibitor of neutrophil elastase (130–132), particularly one not susceptible to chemical or biological oxidants. There are at least three reasons: 1. The inhibitor, even with a methionyl residue, still inactivates neutrophil elastase at a rate faster than that reported for any other proteinase inhibitor:proteinase interaction. 2. Because the inhibitor is sensitive to oxidants, its inactivation at localized areas in tissue under normal conditions allows some tissue remodeling to occur, which would be very difficult with any other residue at the reactive center. 3. The presence of α_1PI near phagocytizing cells may be deleterious to the process. Oxidants and proteinases released by these cells may, through oxidative inactivation and proteolysis, provide an area of liquefaction for the initial processes required to expel the pustule containing the phagocytized material (59). A methionyl residue, therefore, provides a back-up control over the inhibitor when conditions occur where inhibition of proteolysis would be deleterious to the maintenance of homeostasis.

ANTITHROMBIN III

Human AT III is a serum glycoprotein (normal concentration $= 29\,\text{mg}/100$ ml) that plays a major role in controlling serine proteinases in the coagulation cascade scheme. In particular, it inactivates thrombin, an effect enhanced by the presence of heparin (154). Several groups (155–159) isolated the protein, using both ion-exchange and gel filtration chromatography, or affinity chromatography on heparin-bound sepharose. The purified protein is a single-chain molecule of M_r 58,000, contains 15% carbohydrate, and has six disulfide bonds (157, 160, 161). The amino acid sequence is nearly complete (162) and shows extensive homology with that of $\alpha_1 PI$, as well as ovalbumin (15, 59). The reason for the latter is still unknown. The protein appears to contain four glycosylation sites and their carbohydrate structures are nearly identical with those reported for the biantennary chains of $\alpha_1 PI$ (72, 163, 164).

In addition to thrombin (156), human AT III inactivates Factors IXa, Xa, and XIa (92, 165–168), plasmin (169), and plasma kallikrein (170), as well as trypsin and chymotrypsin (20). However, with the exception of thrombin, Factor Xa, and Factor XIa, no rate studies have been performed on the inhibition of any of these enzymes by AT III. In addition, $\alpha_1 PI$ was found recently to inactivate both Factors Xa and XIa at a more rapid rate than AT III, which suggests that Xa and XIa–AT III interactions are not physiologically relevant (91, 92). However, this does not indicate a potential role for $\alpha_1 PI$ in the coagulation process since both the apparent k_{ass} and $t_{1/2}$ of association (see Table 1) are far too slow to be of significance. It is also likely that all of the reactions of AT III with proteinases, with the exception of thrombin, are unimportant even with the accelerating effect heparin gives to such interactions.

Effect of Heparin

The inactivation of proteolytic enzymes by AT III proceeds only at a moderate rate in the absence of heparin but is significantly accelerated in its presence (k_{ass} increases from 10^3 to 10^5; 156). There is thus significant interest in both the chemistry of heparin and its effect on AT III in enhancing the rate of complex formation (171). The former has been adequately reviewed recently (172). Therefore, we focus on the role of heparin in altering the conformation of the AT III protein moiety.

The role of heparin in AT III function was viewed as an allosteric activator in which binding at one site of the protein effected a conformational change to accelerate the inhibition of thrombin and other serine proteinases (156). This simplistic role has not lost favor although the effect might also be on thrombin, since this enzyme can also bind heparin (173). This view is still unproved, however, and virtually all studies to date involved heparin and AT III.

The activation of AT III by heparin appears to involve specific interactions with or near a peptide sequence containing a reactive tryptophanyl residue (174–181) located near the amino terminus of the protein (residue 49; 181). This conclusion is based on difference spectroscopy studies of AT III in the presence and absence of heparin (174–178), circular dichroism experiments (178), and chemical modification of AT III by either oxidation (179) or alkylation with tryptophan specific reagents (180, 181). The addition of heparin to AT III is believed to cause a conformational change so that one tryptophan residue is buried (174–178). Chemical modification of AT III caused complete interruption of its binding to heparin (179–181), although the modified protein could still bind thrombin. Significantly, mild reduction of AT III to cleave only one of the three disulfide bridges also caused a marked decrease in heparin binding, while complex formation with thrombin was still retained (182). All of these data therefore imply that heparin and thrombin binding must occur at separate sites on the AT III molecule.

Mechanism Studies

The inhibition of serine proteinases, particularly thrombin, by AT III occurs by the same mechanism as that found for α_1PI, described previously. The interaction of the inhibitor and enzyme results in the formation of an extremely stable equimolar complex (27), the nature of which is not yet known. The AT III-thrombin complex was dissociated with strong nucleophiles, and a modified, inactive, two-chain form of AT III reisolated (183), indicating peptide bond cleavage during this reaction. Significantly, no peptide is released during this dissociation; disulfide bonds maintain a covalent link between the heavy and light chains formed. Subsequent amino-terminal sequence analyses of the two chains, as well as comparison with the partial structure of AT III, indicated that an arginyl-seryl peptide bond at positions 384–385 had been cleaved as a consequence of thrombin–AT III interaction (16, 162, 184). Recently (17), the same site was also cleaved by Factors IXa and Xa after dissociation of AT III complexes containing these enzymes, thereby supporting the postulation that this is the reactive-site region of AT III.

Curiously, hydrolysis of the reactive-site peptide bond by thrombin to give inactive AT III, without apparent complex formation, has also been described (185–188). The reason for this inactivation is unclear but may be:

(*a*) inactive AT III in preparations utilized in these experiments; (*b*) partially autolyzed thrombin used in these studies; or (*c*) proteolysis of the AT III-thrombin complex by excess thrombin. Any other explanation would be unclear in view of the known stability of the AT III-thrombin complex. However, inactivation of AT III by proteinases can occur. For example, neutrophil elastase, which is not inhibited by AT III, slowly inactivates the protein by proteolytic cleavage, although the site is not yet identified (189). Similarly, snake venom proteinases inactivate AT III and, in this case, peptide bond cleavage occurred between ala_{375} and ser_{376} of the inhibitor, only eight residues from the reactive center (190, 191). Whether such an inactivation occurs in vivo after venom injection remains to be determined.

Physiological Role of Antithrombin III

Unlike several other inhibitors discussed in this review, AT III has the primary function to modulate the activity of thrombin. In the presence of heparin this inhibitor can rapidly shut down the coagulation process by inactivating this single proteinase. In contrast, many other enzymes reported to be inactivated by AT III do so at a relatively slow rate, which suggests only a minor role, if any, for this inhibitor in controlling their function. The need for control of Factors IXa, Xa, and XIa in coagulation processes when AT III–heparin complexes are so effective against thrombin is unclear. Furthermore, many of these enzymes already exist in large complexes in vivo, and evidence suggests that at least one, Factor Xa, is only slowly complexed by AT III–heparin when the enzyme is bound to either platelets (192) or to the Factor Va–phospholipid complex (193).

Quite obviously, much more work is needed to offer further functions for AT III. In particular, attention should be given to heparin-like substitutes occurring in vivo, since the vascular endothelium catalyzes the interaction of AT III with thrombin at a rate paralleling that of AT III–heparin (194). Whether this also applies to other proteinase interactions with AT III remains to be investigated.

ALPHA$_2$-ANTIPLASMIN

In 1974 Mullertz (195) postulated the existence of a plasma inhibitor of plasmin, separate from $\alpha_2 M$ and $\alpha_1 PI$, on the basis of the rapid interaction of this enzyme with an unknown component in plasma. Several groups isolated the inhibitor and showed it to be distinct from the other known plasma inhibitors (196–198). The classical method of purification is by affinity chromatography on plasminogen-bound sepharose (199), with the resultant product characterized as a single-chain glycoprotein of M_r 65,000–70,000, containing 11% carbohydrate, and asparagine and leucine as the amino and carboxy terminal residues, respectively (196, 200, 201). The normal plasma concentration of $\alpha_2 AP$ is 6 mg/100 ml (202).

Synonyms for this inhibitor include antiplasmin (197), primary plasmin inhibitor (195), and $\alpha_2$plasmin inhibitor (196). All are sufficient, but we follow the International Committee on Thrombosis and Haemostasis (203), which uses the name $\alpha_2$antiplasmin.

As discussed in detail below, $\alpha_2 AP$ is an extremely efficient, fast-reacting inhibitor of plasmin. However, several reports indicate that it can also inactivate trypsin, plasma kallikrein, chymotrypsin, thrombin, Factor Xa, and Factor XIa (199, 204, 205). The inhibition of all of these enzymes, with the exception of trypsin is very slow (e.g. k_{ass} for chymotrypsin is 1.0 $\times 10^5 M^{-1} sec^{-1}$; 199), and it is unlikely that $\alpha_2 AP$ plays any role in controlling these activities. However, bovine trypsin reacts reasonably rapidly at 25° ($k_{ass} = 1.8 \times 10^6 M^{-1} sec^{-1}$) to form an equimolar, denaturation-resistant complex (206). In view of this rapid, tight binding, it is surprising that the inhibitor was not detected earlier as a trypsin inhibitor. We conclude that its inhibitory activity was masked by the higher concentration of the other inhibitors in plasma, including $\alpha_2 M$, $\alpha_1 PI$, and IαI. Nevertheless, on the basis of Table 1, this inhibitor may play some role in controlling tryptic activity, should this enzyme appear in tissues other than the blood.

Several other enzymes have been tested against $\alpha_2 AP$, but none were inactivated. These include C$\overline{1}$s, Factors IXa and Xa, neutrophil elastase, cathepsin G, and papain (205, 207, 208). In view of the reported reactive-site P_1 residue of $\alpha_2 AP$ as being leucine (201), we find it surprising that cathepsin G is not inhibited and that chymotrypsin is only slowly inhibited when compared to either plasmin or trypsin, which prefer positively charged side chains, such as lysine or arginine. The assignment of leucine as P_1 in $\alpha_2 AP$ was performed by digestion of a modified form of the inhibitor with carboxypeptidase Y (201), a procedure known to give leucine as a carboxy-terminal erroneously because of endopeptidase activity associated with preparations of this enzyme (209). Consequently, we are skeptical in regard to leucine as P_1, until an independent method of determination is described.

The interaction of plasmin with $\alpha_2 AP$ is very rapid with a k_{ass} of 1.8–3.8 $\times 10^7 M^{-1} sec^{-1}$ at 25°, depending on the type of plasmin used. The inactivation appears to follow the scheme outlined earlier, resulting in an equimolar, denaturation-resistant complex (30, 201, 210). The rapid rate of inhibition has been suggested to depend partly on the existence of unoccupied lysine binding sites on the plasmin A-chain (30, 211). The k_{ass} for this reaction is estimated to be 10-fold slower in the presence of 1–10 mM 6-amino-hexanoic acid (a lysine analog) than in its absence (205), and one form of plasmin lacking its lysine binding sites ("miniplasmin"), yet retaining catalytic activity, reacted 60-fold more slowly than intact plasmin (211). The importance of the lysine binding sites is emphasized by the finding that the rates of reaction of trypsin, which contains no plasmin-like binding sites, with α_2AP is unaffected by concentrations of 6-amino-hexanoic acid up to 0.1M.

Collen (212) discussed the central function played by α_2AP in the molecular mechanism of fibrinolysis and suggested that the control of clot dissolution in vivo requires a continuous replacement at the fibrin surface of plasminogen because of inactivation of plasmin by α_2AP at this site. The plasminogen would then be activated by fibrin-bound plasminogen activator molecules. Since the lysine binding sites on plasmin are involved in interactions with fibrin as well as α_2AP (213), both the substrate and inhibitor would also compete for binding sites on plasminogen, as reported (214). However, upon activation to plasmin the affinity for lysine increases dramatically (215), and those plasmin molecules actively involved in the degradation of fibrin clots are inhibited by α_2AP at a much slower rate than those free in solution (216). Consequently, those plasmin molecules bound to fibrin have a half-life of about ten sec (216), whereas the half-life of free plasmin in plasma is less than 0.1 sec. This difference enables the enzyme to carry out its function of clot dissolution without cleaving circulating fibrinogen or posing a danger to other plasma proteins. It is tempting to suggest that α_2AP has evolved sites on its surface to enable it to "distinguish" plasmin molecules directly involved in clot lysis from those molecules free in plasma.

The half-life of circulating α_2AP-plasmin complexes is about 0.52 days, compared with that of 2.6 days for unreacted α_2AP (217). There is no data to indicate any "transfer" of α_2AP-complexed plasmin to α_2M (see section on Physiological Role of α_2-Macroglobulin). The inhibitor is temporarily exhausted during thrombolytic therapy because of the formation of complexes with plasmin, and patients suffering from a deficiency of α_2AP are subjected to episodes of bleeding (218, 219). It is unlikely that the inhibitor has other physiological functions, except perhaps the control of one or more forms of plasminogen activator (199). However, there are as yet no kinetic or physiological data to support this contention.

ALPHA$_1$-ANTICHYMOTRYPSIN

α_1Achy is a plasma glycoprotein first isolated and characterized without knowledge of its function (220). Subsequently, it was shown to have inhibitory activity towards chymotrypsin (221), together with α_1PI and

$\alpha_2 M$. The inhibitor is a major acute phase protein, whose concentration increases rapidly and dramatically after a variety of events, including surgery (222), burn injuries (223, 224), Crohn's disease and ulcerative colitis (225), and some types of cancer (226). In fact, with the exception of C-reactive protein, whose concentration in normal plasma is usually undetectable, α_1Achy shows the most immediate response as an acute phase protein, doubling in concentration within eight hours of insult. Thus, this inhibitor might play an important role in controlling specific systems associated with inflammatory episodes.

Studies on the structure and function of α_1Achy, unlike those with α_1PI and $\alpha_2 M$, are quite limited, despite its marked increase in production under traumatic conditions relative to other inhibitors. However, because no clear function is associated with α_1Achy, and because of its relatively low concentration in plasma (25 mg/100 ml, normally) it has not received the attention given to other inhibitors.

Three laboratories isolated α_1Achy (56, 220, 227), using a variety of conditions. In all cases the product obtained stoichiometrically inhibited chymotrypsin to give a 1:1 complex that was denaturation resistant to dissociation. The molecular weight of the native protein was between 58,000 (227) and 68,000 (56, 220); the differences were presumably attributable to the methodology and the high carbohydrate content (about 26%) of this glycoprotein. An amino-terminal sequence with arginine as the first amino acid has been reported (56) and confirmed for the inhibitor as isolated from pleural fluid (227). However, this does not represent the true amino terminus, since isolation of α_1Achy by affinity chromatography techniques recently yielded a protein with an amino-terminal aspartic acid and a fourteen amino acid extension prior to the arginine residue originally reported (M. Morii, J. Travis, manuscript in preparation). Glycine has been found at the carboxy terminus (56). Significantly, some structural homology beyond the arginine residue has been found with the amino-terminal sequence of α_1PI (56), which suggests a common evolutionary origin for these two proteins as well as with AT III (15).

Human α_1Achy is a specific inhibitor of chymotrypsin-like proteinases, forming stable complexes with bovine and human chymotrypsin (33, 56, 119), human neutrophil cathepsin G (33, 119, 228), and dog and human mast cell "chymases" (119, 229). No inhibition of either human trypsin or neutrophil elastase has been found (119), and evidence indicates that the latter enzyme may actually turn over α_1Achy.

As with α_1PI, human α_1Achy rapidly forms complexes with chymotrypsin-like serine proteinases; the rate is by far the fastest with cathepsin G ($k_{ass} = 5.1 \times 10^7 \, M^{-1} \, sec^{-1}$), and at least 1000 times slower with bovine and human chymotrypsin (33). Such data confirm the original observations that showed by partition studies that $\alpha_1 PI$ and $\alpha_1 Achy$ interacted preferentially with human chymotrypsin and cathepsin G, respectively (119). Mast cell "chymase" has also been shown to form complexes but the k_{ass} for this interaction is not yet reported (119).

Because of the narrow specificity of $\alpha_1 Achy$ towards chymotrypsin-like enzymes, it would seem that the normal function of this inhibitor could be readily understood. Unfortunately, this is not the case, since a role for cathepsin G, with which it appears to interact most rapidly, has not yet been established. The enzyme has been found to degrade proteoglycan (230) and fibronectin (231). Therefore, the function of the inhibitor may involve control of connective tissue breakdown and cell–cell interactions, although this is not proven. A more likely role is in the regulation of angiotensin II production, where both cathepsin G and mast cell chymase can act as effective angiotensin converting enzymes (239). Since angiotensin I to angiotensin II conversion is an inflammatory reaction, the effect of increased synthesis of $\alpha_1 Achy$ would be to produce sufficient inhibitor to control smooth muscle contraction and aldosterone production. In this context, $\alpha_1 Achy$ has been reported more readily concentrated in bronchial secretions of patients with chronic bronchitis (232, 233), a condition involving bronchial inflammation. Whether the higher $\alpha_1 Achy$ levels are present to counteract angiotensin II formation in this tissue remains to be established. However, in the bronchial tree, large populations of mast cells exist that could serve as a source of "chymase" for angiotensin II production.

Chymotrypsin-like enzymes have been detected in a variety of tissues besides the pancreas, neutrophil, and mast cell. These include unstimulated lymphocytes (234), activated peritoneal macrophages (235), natural killer lymphocytes (236), and chromatin (237). Significantly, $\alpha_1 Achy$ can be isolated from serum through its ability to bind to DNA (238); this indicates some potential role in controlling the chymotrypsin-like enzyme found in chromatin. A far more important role, however, has been suggested in a study of incubation of natural killer cells with $\alpha_1 Achy$. The cells lost both their killing activity and the antibody-dependent cytolytic activity (236). Since the function of natural killer cells is usually associated with the lysis of tumor cells, it is not surprising that at least two tumor cell lines have been identified that produce $\alpha_1 Achy$ (239, 240), presumably as a defensive agent against the killer cells. In both cases, however, the protein isolated was inactive as an inhibitor, possibly because of its presence in a precursor form, and it remains to be established whether an active form of $\alpha_1 Achy$ is ever elicited as a protective agent for the tumor against natural killer lymphocytes.

C1̄-INHIBITOR

As its name implies, C1̄-Inh was identified originally as an inhibitor of the activated form of the first component of complement (241, 242), and hence is also referred to as C1̄-esterase inhibitor and C1̄-inactivator; however, we feel these names to be inappropriate as the first implies that only the esterase function of C1̄ is inhibited, while the second may lead to confusion with C3b-inactivator, which is a proteinase.

Human C1̄-Inh is a single chain glycoprotein of M_r 96,000–104,000, containing approximately 35% carbohydrate (243–245). The normal concentration in plasma is 17 mg/100 ml (244) and the protein is readily isolated by standard chromatographic techniques (243–245). The inhibitor has an amino-terminal asparagine (244; R. Harrison, personal communication), but no other structural studies have been reported to date.

Both C1̄r and C1̄s are inhibited by C1̄-Inh (246). In addition, plasma kallikrein (247), Factors XIa and XIIa (248), and plasmin (28) are inactivated by this inhibitor. Trypsin (249), porcine pancreatic elastase, and bovine chymotrypsin are not inhibited (28; G. Salvesen, J. Travis, manuscript in preparation); each of these enzymes produces a proteolytically modified form of the inhibitor unable to react with either C1̄s or plasma kallikrein.

C1̄r and C1̄s, as subcomponents of C1̄, possess limited proteolytic activity (249). C1r is activated to C1̄r by a process believed to involve a change in conformation promoted by C1-antibody-antigen complexes; this leads in turn to the activation of C1s by limited proteolysis. C1̄s activates C4 and C2 by a similar proteolytic cleavage, although with a different specificity (250). Thus, both enzymes play a critical role in complement activation.

Isolated C1̄r and C1̄s both react to form equimolar complexes with C1̄-Inh (28, 251), although under certain conditions C1̄r forms large complexes thought to have the composition $(C1̄r)_2–(C1̄-Inh)_2$ (252). Prior reaction of both C1̄r and C1̄s with diisopropylfluorophosphate prevents complex formation with C1̄-Inh, indicating that binding between the enzyme and inhibitor occur in the normal manner, via the active site of each proteinase (251). C1̄r reacts more slowly with C1̄-Inh ($k_{ass} = 2.8 \times 10^3 \, M^{-1} \, sec^{-1}$) relative to C1̄s ($k_{ass} = 1.2 \times 10^4 \, M^{-1} \, sec^{-1}$), and it appears that isolated C1̄r is far more sensitive to changes in ionic strength and Ca^{2+} concentration than is C1̄s for its reactions with physiological substrates as well as with C1̄-Inh (246, 253). Equilibrium constants (246) of the reversible reaction (K_i) between C1̄-Inh and C1̄r or C1̄s are almost identical ($10^{-7}M$); this implies that the C1̄r reversible complex is more stable than the C1̄s complex. However, there has been no attempt to measure a k_2 rate for these reactions. When these become available we expect revisions in the estimation of K_i.

The rate of reaction of $C\bar{1}s$ and, to a lesser extent $C\bar{1}r$, with C1-Inh increases in the presence of heparin (246). The mechanism of this potentiation is not known but is assumed to be closely related to the effect of heparin on the reactivity of AT III, particularly since $C\bar{1}$-Inh binds to heparin-sepharose (254). In contrast, heparin may have the reverse effect on the reaction of $C\bar{1}s$ with $C\bar{1}$-Inh, i.e. the rate is decreased (255), and $C\bar{1}s$ contained two or more reactive sites that served separate functions with respect to the hydrolysis of synthetic substrates and reaction with $C\bar{1}$-Inh. Unfortunately, a detailed examination of this report revealed the use of inappropriate substrates and dubious purification procedures. As suggested separately (256), it is possible that there are at least two separate species of $C\bar{1}s$, each with significantly different sensitivities to $C\bar{1}$-Inh and activities towards synthetic substrates. However, a systematic examination of this possibility has not yet been undertaken (257).

Plasma kallikrein reacts with $C\bar{1}$-Inh ($k_{ass} = 6.9 \times 10^4 \, M^{-1} sec^{-1}$) to form stable complexes, via the reactive site–containing chain of this enzyme. (242, 258). However, as Table 1 shows, $C\bar{1}$-Inh shares its role as a primary inhibitor of plasma kallikrein with $\alpha_2 M$; although the rate of inactivation by $C\bar{1}$-Inh is greater, the concentration of $\alpha_2 M$ in normal plasma is higher, thus offsetting its slower rate of inhibition. The in vivo rate of reaction of plasma kallikrein with $C\bar{1}$-Inh (as well as with other inhibitors) may be slower than estimated from data derived from purified systems, as a consequence of binding to circulating high-molecular-weight kininogen (258, 259). The $C\bar{1}$-Inh-plasma kallikrein complex is essentially irreversible (242, 259) and, although we expect that complex formation will follow the scheme outlined earlier, so far no reversible step has been detected (258), possibly because k_2 is very much larger than k_{-1}. In contrast to its reaction with $C\bar{1}r$ and $C\bar{1}s$, heparin appears to have no effect on the rate of reaction of $C\bar{1}$-Inh with plasma kallikrein (G. Salvesen, J. Travis, unpublished observations).

Plasmin forms a stable complex with $C\bar{1}$-Inh (28), but no rates of inhibition have been determined. This complex can be further degraded by excess plasmin to give an inactive form of the inhibitor with reduced molecular weight, as shown for other members of the $\alpha_1 PI$ class (28). Curiously, papain, ficin, and bromelain also produce a proteolytically modified inhibitor; however, in this case the protein retains its capacity to inhibit $C\bar{1}s$ and plasma kallikrein (G. Salvesen, J. Travis, manuscript in preparation). These data, therefore, suggest that $C\bar{1}$-Inh contains a single reactive site, possibly with a lysyl residue in the P_1 reactive site based on chemical modification studies (260), which accommodates $C\bar{1}s$ or plasma kallikrein (and also presumably $C\bar{1}r$). This deduction is supported by the finding that pretreatment of the inhibitor with plasmin prevents the subsequent inhibition of $C\bar{1}s$ (248).

$C\bar{1}$-Inh is the only known plasma inhibitor of $C\bar{1}r$ and $C\bar{1}s$ (252), and in this role it is presumed to be pivotal in controlling the activation of the classical pathway of complement. Under physiological conditions the reaction of $C\bar{1}$-Inh with antigen-antibody-$C\bar{1}$ leads to the dissociation of the $C\bar{1}$ complex (246, 251, 261, 262), and the proteins are released in the form $(C\bar{1}r_1)(C\bar{1}s_1)-(C\bar{1}Inh)_2$ (261). The $C1_q$, denuded of $C\bar{1}r$ and $C\bar{1}s$ in this way, may be recognized by cells possessing $C1_q$ receptors (250), presumably aiding endocytosis of the antibody coated antigen. Further speculation of the physiological role of $C\bar{1}$-Inh-mediated disruption of $C\bar{1}$ complexes is beyond the scope of this review.

Inhibition of plasma kallikrein by $C\bar{1}$-Inh has long been known (247), and recent reports suggest that the enzyme is controlled equally by $C\bar{1}$-Inh and α_2M in plasma (242, 258, 263). This has tremendous potential importance since plasma kallikrein is thought to have two main functions: generation of the pharmacologically active peptide, bradykinin, from high-molecular-weight kininogen (264) and amplification of the early phase of the intrinsic pathway of blood coagulation by a feedback mechanism to activate Factor XIIa. Despite the lack of a systematic evaluation of the control of this latter enzyme by plasma inhibitors, it has been suggested that $C\bar{1}$-Inh is its primary inhibitor (265). If this postulation proves correct then $C\bar{1}$-Inh plays the critical role in controlling the activation of the first two proteinases involved in intrinsic pathway activation during coagulation.

Whether $C\bar{1}$-Inh exerts its main physiological role by controlling the initiation of complement or the initiation of the intrinsic pathway, or both, may not be known for a long period. However, with the information available concerning an inherited deficiency of this protein (266) and reflected by the development in such individuals of hereditary angioneurotic edema, an autosomal dominant trait resulting in frequent outbreaks of edema of organs, skin, and mucosa of the upper respiratory tract, the main function of the inhibitor is apparently to control complement activation rather than intrinsic pathway activation or bradykinin generation. This is based on the existence of variants of the deficiency state, in which the antigenic concentration of circulating inhibitor is normal, yet the protein does not block the hemolytic activity of $C\bar{1}$. Furthermore, the varient proteins retain their ability to inhibit plasma kallikrein (247) and bind plasmin (267).

As Table 1 showed, a major concern in all hypotheses regarding the function of $C\bar{1}$-Inh is the slow rates of $C\bar{1}r$, $C\bar{1}s$, and plasma kallikrein inactivation. These appear to be far too inefficient interactions to be of any physiological significance. However, when $C\bar{1}r$ and $C\bar{1}s$ are present as part of the $C\bar{1}$ complex, inactivation rates may be far more rapid, presumably because of conformational changes in the two proteinases while they are bound to the other proteins in $C\bar{1}$, or possibly because $C\bar{1}$-Inh associates with unactivated C1 complexes (268). Such hypotheses remain to be tested.

ALPHA$_2$-MACROGLOBULIN

The nature of the reaction of $\alpha_2 M$ with proteinases is unique in at least one respect, the retention of proteolytic activity of bound proteinases against low-molecular-weight substrates (269). This contrasts to the almost total inhibition of activity of bound proteinases against high-molecular-weight substances such as hide powder, elastin, and fibrinogen (270–272). In this chemical sense $\alpha_2 M$ is so markedly different from all other known inhibitors that some scientists question whether it should be put in this class at all (3).

Although we have a very clear impression of the unique mechanism by which $\alpha_2 M$ functions, little is known of its physiological role, principally because it reacts with such a vast array of proteinases having different specificities, catalytic mechanisms, and tissue sources. Paradoxically, this lack of knowledge, coupled with a better understanding of its mechanism, led to the appearance of several reviews during the past few years (273–279). In keeping within the scope of this review we therefore focus primarily on the nature of the binding reactions of $\alpha_2 M$ and on the more recent suggestions for a physiological role.

Human $\alpha_2 M$ is a glycoprotein of M_r 725,000, containing 8%–11% carbohydrate (280, 281), and present in plasma at a concentration of 250 mg per 100 ml. It is relatively easy to isolate because of its large molecular weight and purification has been attained by gel filtration chromatography (282), dye-ligand affinity chromatography (283), and zinc-chelate adsorption chromatography (284). The purified protein is composed of a tetramer of identical subunits of $M_r \sim 185,000$, linked in pairs by disulfide bonds. Two pairs associate by noncovalent binding to form the native, tetrameric molecule (273, 274). The inhibitor has been largely sequenced (285) and the subunits found to contain approximately 1450 amino acids, with an amino-terminal serine and a carboxy-terminal alanine (286, 287).

The characteristics of the reaction of proteinases with $\alpha_2 M$ suggest a unique reaction mechanism. Barrett & Starkey (288) used these to formulate a proposal known as the "trap hypothesis." However, while this apparently explains the mechanism by which $\alpha_2 M$ inactivates proteinases, at least three types of binding reactions may occur during inhibitor-enzyme interactions, including: (*a*) a steric trapping reaction, specific for proteinases, (*b*) a covalent linking of proteinases and other molecules containing nucleophilic groups, and (c) a noncovalent, nonsteric adherence reaction with a number of other proteins and other molecules, unrelated to proteolytic activity (279, 289).

The Trapping Reaction

The postulates of the "trap hypothesis" have been reviewed elsewhere (273, 279). Nevertheless, they merit mention here as they are central to the current understanding of the mechanism of action of $\alpha_2 M$. The hypothesis suggests the following reaction between a proteinase and $\alpha_2 M$, and it should be emphasized that this is not a kinetic scheme but rather a summary of the mechanism:

$$\alpha_2 M + E \overset{1}{\rightarrow} \alpha_2 M - E \overset{2}{\rightarrow} \alpha_2 M^* - E \overset{3}{\rightarrow} \alpha_2 M^* - (E)$$

E is an active proteinase and $\alpha_2 M^*$ denotes peptide bond cleavage in the inhibitor. An active endopeptidase is pictured as recognizing a particular sequence in the $\alpha_2 M$ subunit and forms a loose complex (step 1). The target peptide bond is then hydrolyzed (step 2), causing a rapid conformational change which physically entraps the enzyme molecule within the bulk of the $\alpha_2 M^*$ molecule (step 3). Entrapment of the proteinase results in a sterically induced blockade of certain of its properties, symbolized here as (E). Thus, the proposed conformational alteration is the key to the action of $\alpha_2 M$, causing inhibition by hindering the access of high-molecular-weight proteins to the entrapped proteinase.

The most convincing evidence for the proposed conformational changes given above are those shown in electron micrographs (290), which indicate a more compact shape for individual inhibitor-enzyme complexes relative to native $\alpha_2 M$. In addition, conformational changes have also been shown to be produced by a number of proteinases, including trypsin, chymotrypsin, thrombin, and plasmin, after inactivation by $\alpha_2 M$, as evidenced by an increase in the rate of migration of complexes in nondenaturing polyacrylamide gels (282, 291), further penetration into gels of increasing polyacrylamide concentration (pore limit gels) (282), decrease in frictional ratio (292), and increase in fluorescence intensity (293). Small alkylamines cause a similar conformational change in $\alpha_2 M$ to both inactivate the inhibitor and also convert it into a more rapidly migrating form after electrophoresis (282). The proteinase-reacted form of $\alpha_2 M$ is referred to as fast-$\alpha_2 M$ (F-$\alpha_2 M$) to distinguish it from the normal, slower migrating form (S-$\alpha_2 M$) after gel electrophoresis (282). These conformational changes, however, although apparently significant, are still not distinct enough to be detected by the triplet probe depolarization technique (294).

All evidence indicates that $\alpha_2 M$ can inactivate the majority of known proteinases from all four catalytic classes (273, 288), including some showing very limited substrate specificity such as vertebrate collagenase (295) and plasma kallikrein (296). If one considers this range of proteinases bound by $\alpha_2 M$, one can conclude only that the ability to hydrolyze peptide bonds is the sole property that governs this interaction. This fact led to the proposal that a conformational change was triggered by proteolytic cleavage of one or more of the subunits of $\alpha_2 M$ (288). Evidence for such a cleavage, first reported by Harpel (297), is now abundant; proteinases of different classes cleave near the middle of the $\alpha_2 M$ subunit to give two polypeptides of apparent M_r 85,000 and 95,000 (282, 297, 298). This has given rise to the concept of a restricted sequence in the subunit chain that is particularly sensitive to proteolytic attack, referred to as the "bait region" in analogy to the waiting "trap" (299, 300). The bait region is believed to be encompassed within a 27-residue sequence, all proteinases inactivated by $\alpha_2 M$ having cleaved the protein in this specific region (Figure 2; 301), according to their known peptide bond specificity (300, 301, 322).

In view of the known primary structure of the bait region and considering the proposed nature of the trapping reaction we should not be surprised if some proteinases do not react at an appreciable rate with $\alpha_2 M$, notably those showing a highly selective specificity and those that may be small enough to escape from the confines of the trap, or that may be too large to become enclosed in it. We know of no proteinases small enough to escape from the trap, but those known not to be inhibited are either very large or show selective substrate specificity. These include $C\bar{1}r$, $C\bar{1}s$, Factor XIIa, urokinase, and renin (252, 302–304).

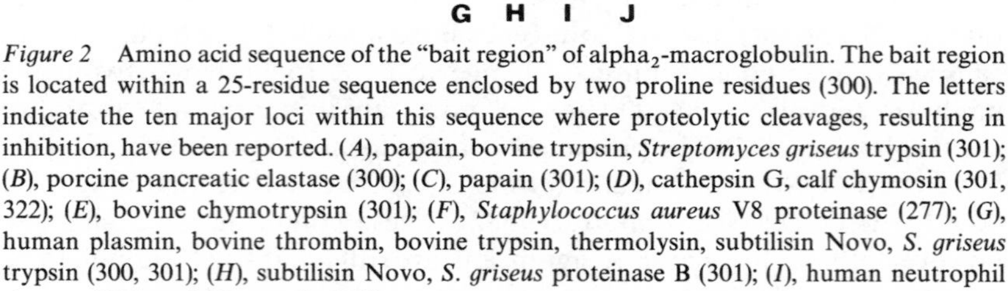

A B C D E F

NH₂—— pro-glu-gly-leu-arg-val-gly-phe-tyr-glu-ser-asp-val-met-

-gly-arg-gly-his-ala-arg-leu-val-his-val- glu-glu-pro ——COOH

G H I J

Figure 2 Amino acid sequence of the "bait region" of alpha₂-macroglobulin. The bait region is located within a 25-residue sequence enclosed by two proline residues (300). The letters indicate the ten major loci within this sequence where proteolytic cleavages, resulting in inhibition, have been reported. (*A*), papain, bovine trypsin, *Streptomyces griseus* trypsin (301); (*B*), porcine pancreatic elastase (300); (*C*), papain (301); (*D*), cathepsin G, calf chymosin (301, 322); (*E*), bovine chymotrypsin (301); (*F*), *Staphylococcus aureus* V8 proteinase (277); (*G*), human plasmin, bovine thrombin, bovine trypsin, thermolysin, subtilisin Novo, *S. griseus* trypsin (300, 301); (*H*), subtilisin Novo, *S. griseus* proteinase B (301); (*I*), human neutrophil elastase (322); and (*J*), papain (301).

PROPERTIES OF α_2M-BOUND PROTEINASES Several of the properties of proteinases bound to α_2M are consistent with a sterically induced blockade of proteolytic activity as suggested by the trap hypothesis. For example, proteinases bound to this inhibitor typically retain 80%–100% of their hydrolytic activity against low-molecular-weight substrates but only up to 10% against large proteins (279). These include both plasmin and neutrophil elastase which lose more than 99% of their activity against fibrinogen and tropoelastin, respectively, when bound to α_2M (272, 305). Complexes do, however, exhibit a significant activity against peptides or proteins in which there is no highly ordered secondary or tertiary structure. Trypsin bound to α_2M cleaves proinsulin to desalanyl insulin at a rate 1/10th that of free trypsin (306); this complex has also been shown to be capable of inactivating other peptide hormones through the same process (307). Unfortunately, no comparable rates were determined in the latter report. Casein and denatured azohemoglobin are also cleaved by α_2M-bound chymotrypsin and cathepsin B, respectively (288, 308). Much has been made of the physiological relevance of such residual activity, even to the point of claiming a role for the neutrophil elastase-α_2M complex in the development of lung damage through elastin degradation (309, 310). However, proteinases complexed to α_2M are not only markedly reduced in their enzymatic activity but also cleared rapidly from the circulation by cells of the reticuloendothelial system (311–313), so that one might only see limited degradation at extravascular sites, if at all.

The hydrolytic activity of proteinases bound to α_2M can be abolished by low-molecular-weight inhibitors but is still protected from those with higher molecular weights. The esterolytic activity of α_2M-bound trypsin, for example, is completely inhibited by diisopropylfluorophosphate and tosyl-L-lysyl chloromethyl ketone (314) but unaffected by α_1PI, even after long incubation periods (20, 21). Indeed, the "protection" of the activity of trypsin against soybean trypsin inhibitor, while in complex with α_2M, has been used to measure the active concentration of the α_2M in solution (315). This assay, while used routinely to measure active α_2M, does have the important drawback that a certain proportion of α_2M-proteinase complexes do react, albeit slowly, with the M_r 21,000–soybean inhibitor (316).

We know of no convincing reports that suggest that α_2M-bound proteinases are able to react with other inhibitors in plasma. It is therefore probable that the soybean inhibitor defines the limit of size or shape of molecules that can gain access to entrapped proteinases.

KINETICS AND BINDING RATIO As described earlier, the most important properties in the interaction of a proteinase with an inhibitor are the rate at which binding occurs and the stability of the resulting complex. Certainly, α_2M-complex formation efficiently fulfills both requirements. In fact, those complexes that received detailed study are so stable that no leakage of proteinases, as measured by transfer to other inhibitors, has ever been detected, even after prolonged incubation with α_1PI (20, 21), soybean trypsin inhibitor (317), or $C\bar{1}$-Inh (244). Once bound, a proteinase cannot be displaced from its complex with α_2M by another proteinase. Hence papain and cathepsin B blocked any binding of trypsin or chymotrypsin, respectively, when the former pair were first complexed to α_2M (288), and plasmin and thrombin were able to compete equally for an equivalent binding site in α_2M (318). These findings indicate clearly that proteinases share the same binding site within the trap and imply that this binding is essentially irreversible. (Recent reports suggest that certain chemically modified forms of trypsin make dissociable complexes with α_2M; we consider these later (319–321).)

If the binding of proteinases by α_2M is essentially irreversible we conclude that the free energy change of one or more of the trapping reactions is so large as to all but preclude the reverse reaction on thermodynamic grounds. The molecular manifestations of the thermodynamically stable "closed" trap (F-α_2M) have already been mentioned, and an investigation of the changes in subunit interactions after conversion of S-α_2M to F-α_2M reveals significant differences (282); subunits of F-α_2M are far less susceptible to dissociation by denaturation than those of S-α_2M. Furthermore, F-α_2M subunits are more closely associated as shown by chemical cross-linking studies (282). The stable structure of the F-α_2M tetramer probably also precludes dissociation of entrapped proteinase molecules through proteolysis from without or within since proteins with little conformational freedom tend to be resistant to proteolytic attack (312).

Given that the binding of a proteinase by α_2M is essentially irreversible, the only meaningful kinetic parameter, at least from a physiological viewpoint, is the rate of binding. Conventional estimates of such rates are complicated by the fact that α_2M-bound proteinase complexes retain variable amounts of proteolytic activity. Recently, however, several workers determined the apparent k_{ass} of various proteinases with α_2M, yielding values of $1.5 \times 10^7\,M^{-1}\,sec^{-1}$ for bovine trypsin at 25° (317), $4.0 \times 10^7\,M^{-1}\,sec^{-1}$ for neutrophil elastase and $3.1 \times 10^6\,M^{-1}\,sec^{-1}$ for cathepsin G (322), $2.0 \times 10^6\,M^{-1}\,sec^{-1}$ for bovine chymotrypsin (J. Bieth, personal communication), $4.8 \times 10^4\,M^{-1}\,sec^{-1}$ for plasma kallikrein (244), $5.0 \times 10^2\,M^{-1}\,sec^{-1}$ for thrombin (323), and $6.7 \times 10^2\,M^{-1}\,sec^{-1}$ for Factor Xa (91). Some of these rates are relatively rapid, not surprising if one considers that the diffusion controlled rate of interaction of a small molecule with one much larger is higher than if the molecules were of similar size, because of the combination of the large target area of one and the high mobility of the other. In addition, each molecule of $\alpha_2 M$ has four bait regions; therefore, the fraction of interactions that lead to a "productive" reaction is increased by a factor of four if the subunits are arranged symmetrically.

There are two possible gross rate–limiting steps in the trapping of a proteinase by $\alpha_2 M$, either the bait region cleavage or the conformational change. No kinetic data are available on the latter, but this should not be different for individual proteinases. In other words, once the trap is sprung by cleavage at the bait region, then chain folding—leading to compacting of the molecule—should proceed as an intramolecular series of reactions whose rate is determined solely by the structure of the inhibitor molecule. Whether this conformational change or bait region cleavage is rate limiting depends on the rate of cleavage of the scissile peptide bond. Consequently, the different rates of inhibition of proteinases by $\alpha_2 M$ probably directly reflects the rate of hydrolysis of their target peptide bonds in the bait region. In contrast to those inhibitors that follow the standard mechanism of proteinase inactivation (3), the peptide bond cleavage of $\alpha_2 M$ must proceed to completion since the active site of a bound enzyme is not blocked. Indeed, the scissile bond(s) in the bait region must be cleaved very rapidly by trypsin, neutrophil elastase, and cathepsin G, as their k_{ass} are extremely high. In this respect the bait region must act as a true substrate for proteinases, i.e. as a substrate with a high value of k_{cat}; this contrasts with the reactive-site bond in other inhibitors for which proteinases have a much smaller k_{cat} (3).

Between one and two molecules of proteinase are trapped by one molecule of $\alpha_2 M$. The actual binding ratio varies with experimental conditions (temperature, pH, and concentration of reactants) because the trapping reaction can be more or less efficient (279, 322, 324, 325). For example, the apparent binding ratio of trypsin rose 1.0–1.6 when the concentration of reactants was increased from $10^{-7} M$–$10^{-5} M$ (324). Similarly, the maximum binding ratio of trypsin was 2.0, when under the same conditions the ratio for chymotrypsin was 1.4 and that for plasmin 1.0 (325). The latter report showed that the binding ratio was directly related to the rate of trapping of each proteinase by $\alpha_2 M$, trypsin being much faster than plasmin, while chymotrypsin was intermediate. The most simple explanation is that $\alpha_2 M$ can bind two molecules of proteinase if the rate of conformational change is much slower than the initial binding step, whereas only one molecule is bound if the first step is much slower than the conformational change. Obviously, intermediate binding ratios are obtained if the rates are similar, and the scatter of values obtained illustrate two important features of the nature of the reaction of proteinases with $\alpha_2 M$: (*a*) two molecules of proteinase can be trapped, but only one is required to spring the trap; and (*b*) the data show the difference between the result of a trapping mechanism and one in which a definite number of actual binding sites become saturated (279).

The Covalent-Linking Reaction

The trap theory of proteinase inhibition by $\alpha_2 M$ is well supported by the characteristics of the reaction; however, the validity of this hypothesis has been challenged recently following the discovery that proteinases can become covalently attached to $\alpha_2 M$ (299, 321, 326–328). Since $\alpha_2 M$ complexes do retain the ability to hydrolyze low-molecular-weight substrates, the bond formation cannot involve the active site of the proteinase in question, in contrast to that which occurs between other inhibitors and their target enzymes. Furthermore, this linking reaction is not restricted to proteinases, but it does require activation by a proteinase to "prime" the site responsible for covalent bonding. However, once the active species of $\alpha_2 M$, referred to as "nascent" $\alpha_2 M$, is formed, it can react with any nucleophile in its immediate vicinity before it decays, presumably by reaction with water and with a half-life of 5–120 sec (289, 329, 330).

The extent of covalent linking is dependent on the proteinase, and varies, for example, from 17% for neutrophil elastase to 61% for bovine trypsin (299, 321). Certain small nucleophiles, such as primary amines and hydroxamic acids, decrease the extent of covalent bond formation by competing with nucleophilic groups on proteins and thus becoming irreversibly incorporated into the $\alpha_2 M$ complex (289, 328–330). This linking to $\alpha_2 M$ is via lysine side chains on the proteinase, as evidenced by: (*a*) a correlation between the lysine content of individual proteinases and their extent of cross-linking (289), (*b*) a correlation between the degree of remaining amine content of partially lysine derivatized trypsin molecules and their extent of covalent linking (289, 321, 328), and (*c*) sequence localization of the site of linking of trypsin to $\alpha_2 M$ (330).

The chemical nature of the covalent-linking site (electrophilic site) in nascent $\alpha_2 M$ has been deduced from studies of the reaction of the protein with methylamine. As mentioned earlier, small alkylamines cause a shift in conformation in S-$\alpha_2 M$ so that it is converted into F-$\alpha_2 M$, and methylamine is incorporated into a glutamyl residue, as Figure 3 shows; this forms a glutamylmethylamide with a stoichiometry indicating one such site per subunit (331). The site of incorporation of methylamine into $\alpha_2 M$ has also been recently shown to be that of the covalent linking of proteins into nascent $\alpha_2 M$ (289, 333). The unusual reactivity of the site was first postulated to be caused by an internal pyroglutamyl residue (332), but it is

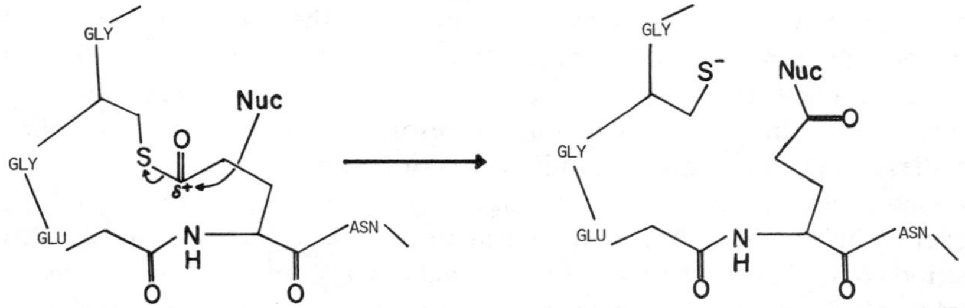

Figure 3   Postulated nucleophile reactivity of the covalent-linking site in alpha$_2$ macroglobulin. The diagram shows the thiol-ester structure proposed responsible for the nucleophilic reactivity of α_2M. It is envisaged that small nucleophiles, particularly amines such as methyalmine, can react directly with the electrophilic center, whereas larger ones, including proteins, must await formation of nascent α_2M by proteolytic activation to expose this site before they can react.

now accepted that an intrachain thiol ester gives rise to the unusual reactivity of this region of α_2M (289, 333, 334); however, although the characteristics of the covalent-linking site are suggestive of a thiol ester (289, 323, 333), such an assignment is still presumptive (334) and will remain so until more direct data can be obtained.

Each subunit of S-α_2M contains a peptide bond about two-thirds down stream from the amino terminus that tends to be cleaved under denaturing conditions (14, 335). This "autolytic" bond is adjacent to the covalent-linking site (322), and the sequence of events postulated to result in its cleavage are outlined in Figure 4 (289, 336).

The proposed thiol ester in S-α_2M is relatively stable and probably sterically shielded from reaction with large nucleophiles, but is revealed as a highly reactive group, free from steric hindrance, after the proteolytic cleavage (289, 329). Curiously, this highly reactive site is also shared by complement components C3 and C4 (336–340). However, unlike the presumed function of $\bar{\text{C}}$3 and $\bar{\text{C}}$4 sites (to enable fixation of these components to cell walls and the Fc regions of IgG molecules, respectively), the function of the α_2M electrophilic site remains a puzzle. It is not essential for inhibition, since proteinases that do not become linked to a great extent, such as neutrophil elastase, are still rapidly and efficiently inhibited. It is possible to almost completely out-compete the covalent linking of proteinases to α_2M by using specific potent nucleophiles in the presence of the enzyme being inhibited, but without affecting the inhibition in any major way (289, 328). The amount of covalent linking of trypsin to α_2M is extensively decreased when the lysine residues of trypsin are specifically blocked (289, 321, 328). One report suggests that the derivatized trypsin can be dissociated from its complex by addition of anhydrotrypsin (321), implying that a covalent link is required for the establishment of an irreversible complex. In contrast, others have shown that inhibition of lysine-modified trypsin is just as efficient as unmodifed trypsin (289, 328). These conflicting results will only be resolved through further detailed studies of this important and unique function of $\alpha_2 M$.

Two functions have been envisaged for the binding reaction: 1. The reaction helps to stabilize complexes against dissociation over a prolonged period (287, 321). 2. The thiol ester site acts as a "conformational trigger," holding $\alpha_2 M$ in its S-conformation, and thus promoting a physical change upon cleavage at this position (328, 334, 341). The first function, while intriguing, is not logical in view of the rapid uptake of $\alpha_2 M$-proteinase complexes by the reticuloendothelial system (312), nor does it explain how those proteinases with only one lysine residue, e.g. neutrophil elastase (85), are inhibited so well by $\alpha_2 M$ (299, 305).

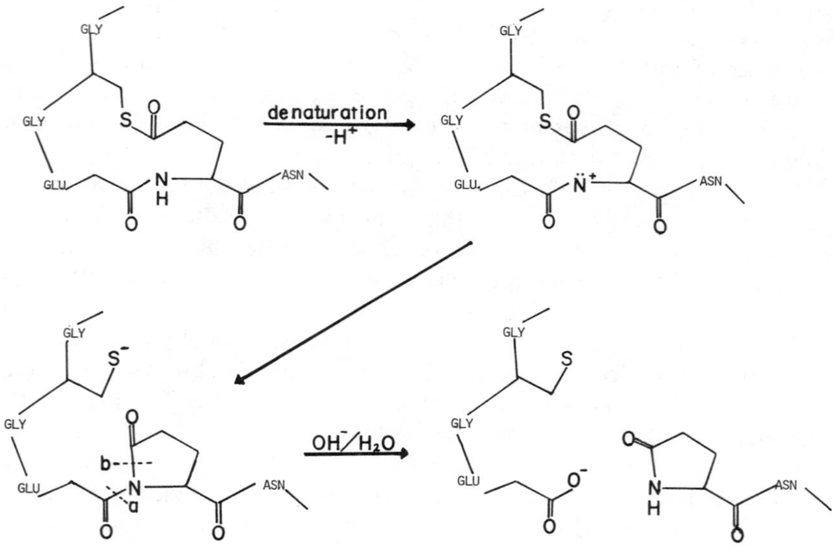

Figure 4 Proposed reaction sequence leading to spontaneous peptide-bond cleavage in denatured slow alpha$_2$ macroglobulin. Positions of the labile bond and activated thiol ester are taken from Howard et al (332). It is proposed that attack at the electrophilic center occurs under alkaline, denaturing conditions by peptide bond nitrogen atom to give peptide bond hydrolysis, with an internal pyroglutamic acid residue as an intermediate. Prior reaction of the electrophilic center, in S-$\alpha_2 M$ or nascent $\alpha_2 M$, would eliminate the potential for peptide bond cleavage by destroying the requirement for an activated center. Hence, the cleavage is not seen in methylamine-treated $\alpha_2 M$ (282) or in proteinase-reacted $\alpha_2 M$ (299). Peptides containing a pyroglutamic acid residue undergo hydrolysis in alkali at one of the two sites designated *a* and *b*. Hydrolysis along pathway *a*, leading to peptide bond cleavage, may take precedence in denatured S-$\alpha_2 M$, and this pathway is particularly favored at high temperatures (334).

The Adherence Reaction

α_2M has an affinity for various proteinases quite separate from its reaction with active proteinases. This "adherence" reactivity is unrelated to the trapping and covalent-linking reactions and is probably mediated by different parts of the molecule (279). Binding by this mechanism occurs largely through ionic and hydrophobic interactions and several basic proteins, e.g., carboxypeptidase A (288), aspartic acid amino transferase (342), myelin basic protein (343), histone H4 (344), anhydrotrypsin (319, 320), are bound by the acidic α_2M molecule. Unlike trypsin, α_2M-bound anhydrotrypsin is displaced from α_2M by other trypsin inhibitors, is not shielded sterically from reaction with antibodies to trypsin, and adheres to F-α_2M as well as S-α_2M, thereby exhibiting none of the properties of an entrapped proteinase (320).

Affinity of α_2M for liposomes (345) and cell membranes (346) is also probably mediated by hydrophobic interactions (279), and binding in this manner is responsible for the aggregation shown by F-α_2M and not by S-α_2M (324). Adherence to α_2M is reversible (279, 320), does not prevent the trapping of proteinases (288, 320) and is not known to be associated with any conformational change (239, 320, 342). No evidence indicated that the adherence reaction serves any physiological role, but the possibility is open to further experimentation.

Finally, we consider that the trap hypothesis best describes the nature of the inhibition of proteinases by α_2M, and we are skeptical of other recently suggested models (316, 321, 347) for the following reasons: 1. The covalent linking reaction cannot explain the nature of inhibition for reasons pointed out earlier and certainly fails to explain the mechanism of inhibition of those proteinases not covalently linked. We suggest that those interested in pursuing this further consider the reaction of proteinases other than trypsin. 2. A "conventional binding site" model (see section on α_1PI class mechanism) cannot be allowed as it cannot explain the obvious differences in the binding of trypsin versus anhydrotrypsin. Furthermore, in such a model it would be difficult to envisage a situation in which one (or even two) binding sites could accommodate all the proteinases we know to be inhibited by α_2M (279).

Physiological Role of α_2-Macroglobulin

In evolutionary terms it would appear that α_2M is an ancient protein, as homology with species as primitive as the hagfish and lamprey has been detected (348); these incidentally are the species possessing the earliest forms of immunoglobulins. Thus, whatever its function, it is of early and presumably fundamental derivation. No total deficiency of α_2M is yet detected, despite many searches, including one of over 100,000 sera (349). Consequently, it is difficult to assign a specific physiological role to the protein. However, a variety of suggestions have been made and we refer readers to reviews not directly related to the function of $\alpha_2 M$ to inhibit proteinases (277, 278, 350).

The role of $\alpha_2 M$ in the control of coagulation (351) and fibrinolysis (352) have been examined. The inhibitor does not inhibit either Factor XIIa or XIa (92, 302), but it does complex Factor Xa and thrombin, albeit at slow rates (Table 1; 91, 323). However, comparison with AT III indicates that it is a far less effective inhibitor during normal coagulation (351, 353). Therefore, if $\alpha_2 M$ aids in controlling the coagulation cascade, it is probably through inhibition of plasma kallikrein, a role it shares with $C\bar{1}$-Inh (244, 259). Since the discovery of $\alpha_2 AP$, several studies have been conducted to determine the relative effectiveness of $\alpha_2 M$ as an inhibitor of plasmin (354–356). The data indicate that it can only act in a minor role, if at all, in controlling normal fibrinolysis, the main regulator being $\alpha_2 AP$. It is only when the local or systemic concentrations of an inhibitor become lowered significantly, for example during pathological episodes such as disseminated intravascular coagulation, during which the majority of the AT III is exhausted (357), and during thrombolytic therapy, during which $\alpha_2 AP$ levels drop precipitously (354), that $\alpha_2 M$ exerts any significant control of the coagulation and fibrinolytic systems (352, 358).

The rapid clearance of $\alpha_2 M$-proteinase complexes by reticuloendothelial cells (312, 361) suggests that $\alpha_2 M$ controls proteolytic activity in the circulation by removing enzymes and that inhibition is largely incidental. The suggestion that proteinase inhibitors in plasma function to complex proteinases and "transfer" these to $\alpha_2 M$ for rapid clearance (312) is now in doubt, because of the finding that the clearance of $\alpha_2 M$ complexes in the mouse is separate from the clearance of other inhibitor complexes (362, 363). However, when such a transfer does occur, the free inhibitor released, in this case $\alpha_1 PI$, is active and could potentially be re-utilized (21). It is therefore not completely clear whether this "piggyback" mechanism is physiologically significant.

Clearance of $\alpha_2 M$-proteinase complexes from the circulation is almost certainly accomplished by their binding to cellular receptors, as shown by their internalization via receptor-mediated endocytosis by macrophages (313, 361, 366) and fibroblasts (365–368) in vitro. $F-\alpha_2 M$ molecules only are recognized by cellular receptors (364) and subsequently internalized (369–370). The uptake of $F-\alpha_2 M$ is independent of the proteinase involved in the complex, so the altered conformation of the molecule appears to be recognized by the receptor (361, 365); this is endorsed by the finding that methylamine-generated $F-\alpha_2 M$ is endocytosed in the same way as proteinase-generated F-α_2M (313, 371, 372). The mechanism of endocytosis of F-α_2M is discussed in more detail elsewhere (276, 277, 373).

By virtue of its broad specificity, α_2M may contribute as a defense against invading pathogens and parasites by inactivating proteinases elaborated by these organisms, should these reach the blood (273). The proteinases contained in venoms from some snakes are also inhibited by α_2M, although other proteinases from the same sources appear to inactivate the inhibitor (359, 360). One interesting speculation resulting from these findings is that, whereas the class-specific inhibitors of human plasma seem to be highly specialized for control of human proteinases, α_2M can eliminate proteinases of all classes even when they are of exogenous origin and there has been no opportunity, therefore, to develop specific protective mechanisms (273). In this way, α_2M should perhaps be considered part of the immune defense system.

INTER-ALPHA-TRYPSIN INHIBITOR

The protein referred to as IαI is a serine proteinase inhibitor first described and isolated from plasma in 1961 (374). However, its function was not established until 1965 when it was found to have inhibitory activity against trypsin (375). Since that time only limited studies have been performed to determine the function of IαI (108, 376), despite the excretion of extraordinarily high levels of a degraded but active form of the inhibitor by patients with various types of cancer (377) or inflammatory diseases (378, 379).

The isolation of IαI has been fraught with difficulties, primarily because of the instability of the purified protein and the large number of steps usually required (380). Some of this has been overcome recently by taking advantage of the zinc-binding capacity of the protein for its isolation (381). IαI (M_r 180,000) is a single-chain glycoprotein, containing 8.4% carbohydrate (108), and nearly 1.0 gram atoms of zinc. An amino-terminal arginine has been reported for the native protein (380), but no other structural analyses on this form have been undertaken (however, see below).

IαI reportedly inhibits human and bovine trypsins and chymotrypsins by forming 1:1 complexes (29, 382), although those with the human enzymes are not stable. It can also inactivate human acrosin (89), as well as human plasmin (383), although complexes with the latter are slow to form. No inhibition of pancreatic elastase (384), or neutrophil elastase and cathepsin G have been observed with IαI (M. Morii, J. Travis, unpublished results).

Unlike other proteinase inhibitors discussed in this review, the mechanism by which IαI forms complexes with proteinases is not known. However, arginine may be at the reactive site on the basis of the abolition of inhibitory activity after treatment with butanedione (385). Complexes of IαI and trypsin are unstable in acid or denaturing agents and enzyme activity is readily recovered. As pointed out the purified inhibitor is unstable; however, this does not affect its function as an inhibitor. Molecular-weight species of less than 50,000 and generated by an as yet unknown mechanism are obtained with no loss of inhibitory activity (380).

Hochstrasser et al have taken advantage of the results summarized above to isolate inhibitory proteins from both serum and urine by perchloric acid treatment, which cross-reacted immunologically with antisera to IαI, yet had molecular weights substantially less than the native protein [M_r 50,000 and 30,000 species (386–388)]. Furthermore, structural studies indicated that they were related to the Kunitz family of inhibitors (3) by the striking homology seen with bovine pancreatic trypsin inhibitor (389–391). The native inhibitor from which these fragments are derived probably also functions by mechanisms described for the Kunitz family (3). However, there is still puzzlement, since inhibitors that cross-react with antisera to IαI can be demonstrated in serum first depleted of IαI, and then treated with thermolysin (392). This type of inhibitor can apparently complex with IgG, thereby protecting it from precipitation as an antigen-antibody complex. The mechanism by which such complex formation occurs, the source of the inhibitor, and its biological significance is unknown.

As yet, no physiological role for IαI has been elucidated. This is primarily because: (a) no deficiency state has ever been noted for the inhibitor; and (b) with the exception of the pancreatic proteinases, no other enzymes have been obtained that are rapidly complexed by IαI. It is doubtful that IαI plays any role in controlling the pancreatic proteinases, even though structural studies indicate its relationship to pancreatic inhibitors (390, 391). Based on its rapid turnover under malignant or inflammatory conditions (377, 378), it is more likely involved in normally controlling enzymes released from tissues other than the pancreas. However, the composition of these enzymes, as well as their ultimate source, must be determined before a function can be designated for IαI.

BETA$_1$-ANTICOLLAGENASE

The inhibitory activity in serum toward mammalian collagenases appears to reside in three proteins, α_2M, β_1AC, and a third, as yet uncharacterized cationic protein (393). Wooley et al first detected the β_1AC (394) and determined the properties of the partially purified protein in 1975 (294, 395). However, more recently (396) a homogeneous preparation of this inhibitor was obtained by using affinity chromatography through disulfide-thiol interchange.

Human serum β_1AC is a single chain glycoprotein of M_r 30,000–33,000 (394–396), containing only traces of carbohydrate. The inhibitor contains two free thiol groups, at least one of which is involved in its normal function since alklylation irreversibly inactivates the protein (396).

β_1AC specifically inactivates collagenases from skin, rheumatoid synovial fluid, gastric mucosa, and granulocytes (394). Inhibition of trypsin, elastase, papain, thermolysin, or pepsin by this inhibitor has not been demonstrated (395), indicating the unique specificity of this protein.

The interaction of collagenases with β_1AC appears to occur through a thiol-disulfide interchange between inhibitor and enzyme (396) to give a mixed disulfide similar to that reported for a collagenase inhibitor-collagenase complex present in neutrophils (397). This inactive complex can be reactivated by oxidized glutathione or myeloperoxidase derived oxidants to yield active enzyme and inactive inhibitor. Whether this mechanism occurs in vivo remains to be elucidated.

Inhibitor-collagenase complexes are not dissociated in the presence of α_2M (396). Therefore, these complexes must be taken up rapidly in an intact form by phagocytic cells before they come into contact with either oxidized glutathione or phagocyte derived oxidants. Most phagocytic cells, however, produce copious quantities of oxidants, making it difficult to understand exactly how β_1AC-collagenase complexes could be removed from tissues before free, potentially damaging enzyme is released.

Since β_1AC is responsible only for minor collagenase inhibitory activity, in serum, about 5% (396), the remainder attributed primarily to α_2M, a role for the inhibitor is not intuitively obvious. Nevertheless, by virtue of its relatively low molecular weight as a plasma inhibitor it should be capable of penetrating into tissues and specifically complexing with collagenolytic enzymes. This has been suggested as the normal role for this inhibitor, especially in the control of neutral collagenase in the synovium (395). However, this is probably not the major contributing inhibitor in connective tissue degradation since other enzymes, such as neutrophil elastase and cathepsin G, present in higher concentrations in tissues undergoing rapid destruction, are also potent collagenases (398), controlled normally by α_1PI and α_1Achy (33). Curiously, β_1AC has only a limited effect on the collagenases released by various human tumors so that even a role in controlling tumor growth cannot be established clearly (399). In fact, when one considers the number of low-molecular-weight inhibitors of collagenases reported in tissues (400), the potential presence of specific serine proteinases having collagenolytic activity far in excess of classical collagenases in tissues, and the inability of $\beta_1 AC$ to control tumor collagenases, it is difficult to establish any important role for this inhibitor in the control of collagen turnover. Perhaps immunochemical localization of the inhibitor in normal and diseased tissues will shed light on its physiological function.

ALPHA-CYSTEINE PROTEINASE INHIBITOR

Originally described in 1976 as a plasma protein with a marked specificity for the inactivation of cysteine proteinases (401), αCPI has now been isolated to homogeneity and characterized by several groups (402–404). The protein is unusually heterogenous with respect to both charge and molecular weight, which helps explain why it was overlooked, despite its high plasma concentration (0.5 mg/ml) (403), until the recent affinity-based improvements in plasma fractionation were developed. The protein was originally named $\alpha_2$thiol proteinase inhibitor (402), but because of its habit of migrating as both α_1 and α_2 components in immunoelectrophoresis (403, 404), and following the current terminology of proteinases (4), we endorse the name "α-cysteine proteinase inhibitor" suggested elsewhere (405).

There are at least two forms of αCPI in plasma with molecular weights in the range 57,000–175,000 (402–404, 406, 407). Both forms have similar pH stabilities and inhibitory spectra and share immunological identity with each other (402, 407). The high-molecular-weight forms may result from multimer formation, mediated primarily by interchain disulfide bonds from the lowest-molecular-weight form, single chain (402–406). A single amino-terminal sequence of a mixture of the forms has been established (406) with isoleucine being the first residue. This result indicates that the different molecular forms are identical, at least in this region, and supports the conclusion that the high-molecular-weight forms are multimers.

All evidence suggests that αCPI is highly specific for cysteine proteinases; the activity of other types of enzymes with a cysteinyl residue involved in catalysis, such as Factor XIIIa, glyceraldehyde-3-phosphate dehydrogenase, and alcohol dehydrogenase, is unaffected by the inhibitor (402, 406). The inhibitor rapidly inactivates cathepsin H (405, 406) and cathepsin L (408), whereas cathepsin B reacts only slowly (402–406). It also complexes the nonmammalian proteinases papain and ficin rapidly (402, 405, 406) but only appears to interact slowly with bromelain. No inhibition of trypsin, chymotrypsin, or pancreatic elastase has been observed (406, 407). In all these experiments no kinetic investigations were made; thus, the meaning of the inhibition of any of the enzymes listed cannot be established.

Complex formation between αCPI and ficin or papain is detectable after electrophoresis in agarose gels (402, 405), and the resultant complexes are also stable to gel filtration (407), although only under nondenaturing conditions. In the presence of $NaDoDSO_4$, complexes dissociate (405), indicating that they are not stabilized by covalent interactions as they occur in the $\alpha_1 PI$ class of inhibitors. The number of inhibitory sites contained within each αCPI molecule has yet to be determined accurately, although one report suggested an equimolar binding ratio with papain (405). Two factors complicate such determinations. First, the high-molecular-weight forms of the protein may have multiple binding sites, since they seem to be composed of two or more single-chain αCPI molecules. Second, αCPI forms complexes with inactive cysteine proteinases [notably, active-site inactivated carboxymethyl papain (405, 406)], so that it becomes essential to use only fully active proteinases, and not active-site standardized enzymes.

The presence in plasma of an inhibitor specific for cysteine proteinases seems redundant when one considers that there are no known members of the circulating proteinases that use a cysteine catalytic mechanism. All of the well-characterized human cysteine proteinases are thought to function within cells (409), and there is no evidence that αCPI is present intracellularly. However, there is little doubt that these same intracellular proteinases are released into tissue spaces upon cell death and related events (410) where they are believed to cause significant and potentially harmful protein degradation (411). It is at this level that αCPI probably acts to prevent tissue damage. We already know that $\alpha_1 PI$ diffuses from the blood into tissue spaces where it aids in the regulation of neutrophil elastase activity, and we believe that αCPI, in particular the low-molecular-weight forms of this inhibitor, acts similarly. Support for this hypothesis can be found in the detection of a cysteine proteinase inhibitor of similar molecular weight in skin and muscle extracts from various mammalian species (401, 412), although it was not confirmed as to whether this activity was related to αCPI.

Should cysteinyl proteinases ever find their way into the blood they would be inhibited not only by αCPI but also by $\alpha_2 M$, which is also known to react with cathepsins B and H (413, 414) and which does not normally diffuse into extravascular sites because of its large size. In fact, it is probable that $\alpha_2 M$ is a much better inhibitor of cathepsin B than αCPI (413). Consequently, we speculate that the relatively high concentration of αCPI in plasma exists solely to help maintain a protective level against cysteine proteinases in intracellular tissue spaces.

OTHER INHIBITORS

In the past there have been numerous reports indicating the presence of new inhibitors in plasma. For the most part these have been found to be either components leaking out of other tissues into the blood or inhibitors already characterized and present in trace quantities in partially purified plasma protein fractions. In fact, with the exception of the αCPI and the β_1AC, there has been very little research performed to indicate new inhibitors in plasma.

Two authentic new inhibitors have, however, been recently isolated. The first, referred to as heparin cofactor II, was originally detected as an inhibitor of thrombin, requiring higher concentrations of heparin for activity, relative to AT III (415). This inhibitor has recently been purified (416) and its properties determined. It is a single-chain glycoprotein of M_r 65,000–70,000, which forms a covalent 1:1 molar complex with thrombin. This inactivation is accelerated nearly 1000-fold in the presence of optimal heparin concentrations. Significantly, inhibition of Factor Xa is very slow, even in the presence of heparin. A physiological role for the inhibitor has not yet been found, but it may play an important role in the control of various thrombin activities (416, 417).

The second recently isolated inhibitor complexes with protein C. The latter is a plasma serine proteinase that inactivates both Factor Va and Factor VIIIa by limited proteolysis (418) and therefore acts as an anticoagulant. The inhibitor was originally suggested to be present in order to explain a combined Factor V/VIII deficiency state (419). While this explanation may not be correct, an inhibitor to protein C has been detected and isolated, but the final product was inactive. The reason for this is unknown, but the purified component was characterized as a glycoprotein of M_r 96,000 and an amino-terminal sequence homologous with both a colostrum inhibitor and the bovine trypsin inhibitor, Kunitz type (3). Based on the ability of partially purified preparations of the inhibitor to form NaDodSO$_4$ stable complexes, it seems to be a member of the α_1PI class.

CONCLUDING REMARKS

In summary, the data now available strongly suggest the following specific inhibitor-target enzyme pairings: α_1PI-neutrophil elastase; AT III-thrombin; α_1Achy-cathepsin G; and α_2-AP-plasmin. It is also likely that C$\bar{1}$-Inh controls the complement proteinases C$\bar{1}$r and C$\bar{1}$s, although this remains to be firmly established, and that α_1Achy is also involved in the regulation of mast cell chymase. Human α_2M, by virtue of its ability to inactivate proteinases from all classes, may act as a rapid and efficient clearing agent for these enzymes when they appear free in the circulation. Currently, clear functional roles for IαI, β_1AC, and αCPI are not known, despite the narrow specificity of these inhibitors.

ACKNOWLEDGEMENTS

We gratefully acknowledge all of our colleagues who provided unpublished manuscripts, and the National Heart and Lung Institute and The Council for Tobacco Research-U.S.A. for their support.

Literatured Cited

1. Bieth, J. G. 1980. *Bull. Euro. Physiopath. Resp.* 16:183–95
2. Fermi, C., Pernossi, L. 1897. *Zgcar. Hyg.* 18:83
3. Laskowski, M. Jr., Kato, I. 1980. *Ann. Rev. Biochem.* 49:593–626
4. Barrett, A. J. 1980. *Protein Degradation in Health and Disease, Ciba Found. Symp.* 75, ed. D. Evered, J. Whelan, pp. 1–13. Amsterdam: Excerpta Medica
5. Ozawa, K., Laskowski, M. Jr. 1960. *J. Biol. Chem.* 241:3455–61
6. Schechter, I., Berger, A. 1967. *Biochem. Biophys. Res. Commun.* 27:157–62
7. Finkenstadt, W. R., Hamid, M. A., Mathis, J. P., Schrode, J., Sealock, R. W., et al. 1974. *Proc. 5th Bayer Symp. Proteinase Inhibitors,* ed. H. Fritz, H. Tscheche, L. J. Greene, E. Truscheit, pp. 389–411, Berlin/Heidelberg/New York: Springer
8. Estell, D. A., Wilson, K. A., Laskowski, M. Jr. 1980. *Biochemistry* 19:131–37
9. Robertus, J. D., Kraut, J., Alden, R. A., Birkhoff, J. J. 1972. *Biochemistry* 11:4293–304
10. Vincent, J. P., Lazdunski, M. 1972. *Biochemistry* 11:2967–77
11. Ako, H., Foster, R. J., Ryan, C. A. 1974. *Biochemistry* 13:132–39
12. Huber, R., Bode, W., Kukla, D., Kohl, U., Ryan, C. A. 1975. *Biophys. Struct. Mech.* 1:189–201
13. Huber, R., Bode, W. 1978. *Acc. Chem. Res.* 11:114–22
14. Johnson, D., Travis, J. 1978. *J. Biol. Chem.* 253:7142–44
15. Hunt, L. T., Dayhoff, M. O. 1982. *Biochem. Biophys. Res. Commun.* 95:864–71
16. Jornvall, H., Fish, W. W., Bjork, I. 1979. *FEBS Lett.* 106:358–62
17. Bjork, I., Jackson, C. M., Jornvall, H., Lavine, K. K., Nordling, K., Salsgiver, W. J. 1982. *J. Biol. Chem.* 257:2406–11
18. Laine, A., Davril, M., Hayem, A., Loucheux-Lefevbre, M. H. 1982. *Biochem. Biophys. Res. Commun.* 107:337–44
19. Johnson, D. A., Travis, J. 1976. *Biochem. Biophys. Res. Commun.* 72:33–39
20. Aubry, M., Bieth, J. 1972. *Clin Chim. Acta* 78:371–80
21. Beatty, K. G., Travis, J., Bieth, J. 1982. *Biochim. Biophys. Acta* 704:221–26
22. Travis, J., Johnson, D., Pannell, R. 1974. See Ref. 7, pp. 31–39
23. Johnson, D., Travis, J. 1977. *Biochem. J.* 163:639–42
24. Morihara, K., Tsuzuki, H., Oda, K. 1979. *Infect. Immun.* 24:188–93
25. Virca, G. D., Lyerly, D., Kreger, A., Travis, J. 1982. *Biochim. Biophys. Acta* 704:267–71
26. Moroi, M., Yamasaki, M. 1974. *Biochim. Biophys. Acta* 359:130–41
27. Owen, W. G. 1975. *Biochim. Biophys. Acta* 405:380–87
28. Harpel, P. C., Cooper, N. R. 1975. *J. Clin. Invest.* 55:593–604
29. Bieth, J., Aubrey, M., Travis, J. 1974. See Ref. 7, pp. 53–62
30. Moroi, M., Aoki, N. 1977. *Biochim. Biophys. Acta* 482:412–20
31. Kitz, R., Wilson, I. B. 1962. *J. Biol. Chem.* 237:3245–49
32. Schultze, H. E., Heide, K., Haupt, H. 1962. *Klin. Wochenschr.* 40:427–34
33. Beatty, K., Bieth, J., Travis, J. 1980. *J. Biol. Chem.* 255:3931–34
34. Pannell, R., Johnson, D., Travis, J. 1974. *Biochemistry* 13:5439–45
35. Eriksson, S. 1965. *Acta Med. Scand.* 177 (Suppl. 432): 1–85
36. Travis, J., Pannell, R. 1973. *Clin. Chim. Acta* 49:49–52
37. Bagdasarian, R., Colman, R. W. 1978. *Blood* 51:139–56
38. Plancot, M., Delacourte, A., Han, K., Dautrevaux, M., Biserte, G. 1977. *Int. J. Protein Res.* 10:113–19
39. Bloom, J. W., Hunter, M. J. 1978. *J. Biol. Chem.* 283:547–59
40. Birkenmeier, G., Kopperschlaeger, G. 1982. *J. Chromatogr.* 235:237–48
41. Murthy, R. J., Hercz, A. 1973. *FEBS Lett.* 32:243–46
42. Liener, I. E., Garrison, O., Pravda, Z. 1973. *Biochem. Biophys. Res. Commun.* 51:436–43
43. Morii, M., Odani, S., Koide, T., Ikenaka, T. 1978. *J. Biochem.* 83:269–77

44. Musiani, P., Massi, G., Piantelli, M. 1976. *Clin. Chim. Acta* 73:561–65
45. Musiani, P., Tomasi, D. 1976. *Biochemistry* 15:798–804
46. Hercz, A., Barton, M. 1977. *Can. J. Biochem.* 55:661–65
47. Vaughan, L., Lorier, M., Carrell, R. W. 1982. *Biochim. Biophys. Acta* 701:339–45
48. Jeppson, J. O., Laurell, C. B., Fagerhol, M. 1978. *Eur. J. Biochem.* 83:143–55
49. Sugiura, M., Hayakawa, S., Adache, T., Ito, Y., Hirano, K., Sawaki, S. 1981. *J. Biochem. Biophys. Methods* 5:243–49
50. Crawford, I. P. 1973. *Arch. Biochem. Biophys.* 156:215–22
51. Chan, S. K., Rees, D. C., Li, S. C., Li, Y. T. 1976. *J. Biol. Chem.* 251:471–76
52. Mahoney, W. C., Kurachi, K., Hermodson, M. A. 1980. *Eur. J. Biochem.* 105:545–52
53. Horng, W., Gan, J. C. 1982. *Tex. Rep. Biol. Med.* 32:489–504
54. Cohen, A. B., Geczy, D., James, H. L. 1978. *Biochemistry* 17:392–400
55. Travis, J., Johnson, D. 1978. *Biochem. Biophys. Res. Commun.* 84:219–24
56. Travis, J., Garner, D., Bowen, J. 1978. *Biochemistry* 17:5647–50
57. Shochat, D., Staples, S., Hargrove, K., Kozel, J. S., Chan, S. K. 1978. *J. Biol. Chem.* 253:5630–34
58. Carrell, R., Owen, M., Brennan, S., Vaughan, L. 1979. *Biochem. Biophys. Res. Commun.* 91:1032–37
59. Carrell, R. W., Jeppson, J. O., Laurell, C. B., Brennan, S. O., Owen, M. C., et al. 1982. *Nature* 298:329–34
60. Kurachi, K., Chandra, T., Degen, S. J. F., White, T. T., Marchiano, T. L., et al. 1981. *Proc. Natl. Acad. Sci. USA* 78:6826–30
61. Fagerhol, M. K., Laurell, C. B. 1970. *Prog. Med. Genet.* 7:96–111
62. Pierce, J., Jeppson, J. O., Laurell, C. B. 1976. *Anal. Biochem.* 74:227–41
63. Bell, O. F., Carrell, R. W. 1973. *Nature* 243:410–11
64. Cox, D. W. 1973. *Lancet* 2:844–45
65. Talamo, R. C., Alpert, E., Langley, C. E. 1975. *Pediatr. Res.* 9:123–26
66. Roll, D. E., Aguanno, J. T., Coffee, C. J., Glew, R. H. 1978. *J. Biol. Chem.* 253:6992–96
67. Hodges, L. C., Laine, R., Chan, S. K. 1979. *J. Biol. Chem.* 254:8208–12
68. Mega, T., Lujan, E., Yoshida, A. 1980. *J. Biol. Chem.* 255:4053–56
69. Mega, T., Lujan, E., Yoshida, A. 1980. *J. Biol. Chem.* 255:4057–61
70. Carrell, R. W., Jeppson, J. O., Vaughan, L., Brennan, S. O., Owen, M. C., Boswell, D. R. 1981. *FEBS Lett.* 135:301–3
71. Hodges, L. C., Chan, S. K. 1982. *Biochemistry* 21:2805–10
72. Vaughan, L., Carrell, R. 1981. *Biochem. Int.* 2:461–67
73. Laurell, C. B., Eriksson, S. 1963. *Scand. J. Clin. Lab. Invest.* 15:132–40
74. Lieberman, J., Mittman, C., Gordon, H. W. 1972. *Science* 175:63–65
75. Eriksson, S., Larsson, C. 1975. *N. Engl. J. Med.* 292:176–80
76. Hercz, A., Barton, M. 1977. *Eur. J. Biochem.* 74:603–10
77. Yoshida, A., Lieberman, J., Gridulus, L., Ewing, C. 1976. *Proc. Natl. Acad. Sci. USA* 73:1324–28
78. Jeppson, J. O. 1976. *FEBS Lett.* 65:195–97
79. Owen, M. C., Carrell, R. W. 1977. *FEBS Lett.* 79:245–47
80. Jeppson, J. O., Larsson, C., Eriksson, S. 1975. *N. Engl. J. Med.* 293:576–79
81. Hercz, A., Katona, E., Cutz, E., Wilson, J. R., Barton, M. 1978. *Science* 201:1229–32
82. Hercz, A., Hardaz, N. 1980. *Can. J. Biochem.* 58:644–48
83. Schwick, H. G., Heimburger, N., Haupt, H. 1966. *Z. Gesamte Inn. Med. Ihre Grenzgeb.* 21:1–3
84. Janoff, A. 1972. *Ann. Rev. Med.* 23:177–89
85. Baugh, R., Travis, J. 1976. *Biochemistry* 15:836–48
86. Matheson, N. R., Travis, J. 1976. *Biochem. J.* 159:495–502
87. Rimon, A., Shamash, Y., Shapiro, B. 1966. *J. Biol. Chem.* 241:5102–7
88. Hercz, A. 1974. *Eur. J. Biochem.* 49:287–92
89. Fritz, H., Heimburger, N., Meier, M., Arnhold, M., Zanneveld, L. J. D., Schumacher, G. F. 1972. *Hoppe-Seylers Z. Physiol. Chem.* 353:1953–56
90. Fritz, H., Brey, B., Schmal, A., Werle, E. 1969. *Hoppe-Seylers Z. Physiol. Chem.* 350:1551–57
91. Ellis, V., Scully, M., Macgregor, I., Kakkar, V. 1982. *Biochim. Biophys. Acta* 701:24–31
92. Scott, C. F., Schapira, M., James, H. L., Cohen, A. B., Colman, R. W. 1982. *J. Clin. Invest.* 69:844–52
93. Tokoro, Y., Eisen, A. Z., Jeffrey, J. T. 1972. *Biochim. Biophys. Acta* 258:289–302
94. Harris, E. D., Dibona, D. R., Krane, S. M. 1969. *J. Clin. Invest.* 48:2104–13
95. Clemmensen, I., Christensen, U. 1976. *Biochim. Biophys. Acta* 249:591–99
96. Bergvist, R. 1963. *Acta Chem. Scand.* 17:2239–46

97. Wicher, V., Dolovich, J. 1973. *Immunochemistry* 10:239–48
98. Sasaki, M., Yamamoto, H., Yamamoto, H., Iidas, S. 1975. *J. Biochem.* 75:171–77
99. Cohen, A. B. 1973. *J. Biol. Chem.* 7055–59
100. Cohen, A. B. 1974. *Fed. Proc.* 33:1311 (Abstr.)
101. Moroi, M., Yamasaki, M., Aoki, N. 1975. *J. Biochem.* 78:925–28
102. Largman, C., Brodrick, J. W., Geokas, M. C., Sischo, W. M., Johnson, J. H. 1979. *J. Biol. Chem.* 254:8516–23
103. Brodrick, J. W., Glaser, C. B., Largman, C., Geokas, M. C., Graceffo, M., et al. 1980. *Biochemistry* 19:4865–70
104. Robinson, N. C., Neurath, H., Walsh, K. A. 1973. *Biochemistry* 12:420–26
105. Saklatvala, J., Wood, G. C., White, D. D. 1976. *Biochemical J.* 157:339–51
106. Satoh, S., Kurecki, T., Kress, L. F., Laskowski, M. Sr. 1979. *Biochem. Biophys. Res. Commun.* 86:130–37
107. Cohen, A. B. 1975. *Biochim. Biophys. Acta* 391:193–200
108. Heimburger, N., Haupt, H., Schwick, H. 1971. In *Proc. 1st Int. Res. Conf. Proteinase Inhibitors*, ed. H. Fritz, H. Tscheche, pp. 1–12. Berlin: de Gruyter
109. Fretz, C., Gan, J. C. 1978. *Biochim. Biophys. Acta* 537:226–37
110. Johnson, D. A., Travis, J. 1975. *Protides Biol. Fluids Proc. Colloq.* 23:35
111. Busby, J. F., Gan, J. C. 1976. *Arch. Biochem. Biophys.* 177:552–60
112. Busby, J. F., Yu, S. D., Gan, J. C. 1977. *Arch. Biochem. Biophys.* 184:267–75
113. Feste, A., Gan, J. C. 1981. *J. Biol. Chem.* 256:6374–80
114. James, H. L., Cohen, A. B. 1978. *J. Clin. Invest.* 62:1344–53
115. Cohen, A. B. 1979. *Am. Rev. Respir. Dis.* 119:953–60
116. Martodam, R., Liener, I. E. 1981. *Biochim. Biophys. Acta* 667:328–40
117. Johnson, D. A., Pannell, R. N., Travis, J. 1974. *Biochem. Biophys. Res. Commun.* 57:584–90
118. Bundy, H. F., Mehl, J. W. 1959. *J. Biol. Chem.* 234:1124–28
119. Travis, J., Bowen, J., Baugh, R. 1978. *Biochemistry* 17:5651–56
120. Johnson, D. A., Travis, J. 1976. *Biochem. Biophys. Res. Commun.* 72:33–39
121. Lo, T. N., Cohen, A. B., James, H. L. 1976. *Biochim. Biophys. Acta* 453:344–46
122. Oda, K., Laskowski, M. Sr., Kress, L. P., Kowalski, D. 1977. *Biochem. Biophys. Res. Commun.* 76:1062–70
123. Baumstark, J. S., Lee, C. T., Luby, R. J.

1977. *Biochim. Biophys. Acta* 482:400–11
124. Hercz, A. 1973. *Can. J. Biochem.* 51:1447–50
125. James, H. L., Cohen, A. B., 1979. *Biochem. Biophys. Res. Commun.* 90:547–53
126. Blomback, B. 1967. *Methods Enzymol.* 44:398–411
127. Morii, M., Odani, S., Ikeneka, T. 1979. *J. Biochem.* 86:915–21
128. Kress, L. F., Kurecki, T., Chan, S. K., Laskowski, M. Sr. 1979. *J. Biol. Chem.* 254:5317–20
129. Carp, H., Janoff, A. 1978. *Am. Rev. Respir. Dis.* 118:617–21
130. DelMar, E. G., Brodrick, J. W., Geokas, M. C., Largman, C. 1979. *Biochem. Biophys. Res. Commun.* 88:346–50
131. Nakajima, K., Powers, J. C., Ashe, B. M., Zimmerman, M. 1979. *J. Biol. Chem.* 254:4027–32
132. McRae, B., Nakajima, K., Travis, J., Powers, J. C. 1980. *Biochemistry* 19:3973–80
133. Johnson, D., Travis, J. 1979. *J. Biol. Chem.* 254:4022–26
134. Cohen, A. B. 1979. *Am. Rev. Respir. Dis.* 119:953–60
135. Matheson, N. R., Wong, P. S., Travis, J. 1979. *Biochem. Biophys. Res. Commun.* 88:402–9
136. Matheson, N. R., Wong, P. S., Travis, J. 1980. *Biochemistry* 20:325–30
137. Matheson, N. R., Wong, P. S., Schuyler, M., Travis, J. 1980. *Biochemistry* 20:331–36
138. Clarke, R. A., Stone, P. J., Hag, A. E., James, D. C., Franzblau, C. 1981. *J. Biol. Chem.* 256:3348–53
139. Carp, H., Janoff, A. 1979. *J. Clin. Invest.* 63:793–97
140. Carp, H., Janoff, A. 1980. *J. Clin. Invest.* 66:987–95
141. Janoff, A., Carp, H. 1977. *Am. Rev. Respir. Dis.* 116:65–72
142. Ohlsson, K., Fryksmark, U., Tegner, H. 1980. *Eur. J. Clin. Invest.* 10:373–79
143. Abrams, W., Cohen, A. B., Damiano, V. V., Elirez, A., Kimbel, P., et al. 1981. *J. Clin. Invest.* 68:1132–39
144. Janoff, A., Carp, H., Lee, D. K., Drew, R. T. 1979. *Science* 206:1313–14
145. Gadek, J., Fells, G. A., Crystal, R. G. 1979. *Science* 206:1315–16
146. Carp, H., Miller, F., Hoidal, J. R., Janoff, A. 1982. *Proc. Natl. Acad. Sci. USA* 79:2041–45
147. Beatty, K., Robertie, P., Senior, R. M., Travis, J. 1982. *J. Lab. Clin. Med.* 100:186–92
148. Wong, P. S., Travis, J. 1980. *Biochem. Biophys. Res. Commun.* 96:1449–54

149. Reilly, C. F., Travis, J. 1980. *Biochim. Biophys. Acta* 621:147–57
150. Mittman, C., ed. 1972. *Pulmonary Emphysema and Proteolysis.* New York: Academic
151. Cohen, A. B., James, H. L. 1982. *Am. Rev. Respir. Dis.* 126:125–30
152. Kitagawa, S., Takaku, F., Sakamoto, S. 1980. *J. Clin. Invest.* 65:74–81
153. Abrams, W. R., Weinbaum, G., Weissbach, L., Weissbach, H., Brot, N. 1981. *Proc. Natl. Acad. Sci. USA* 78:7483–86
154. Abildgaard, U. 1968. *Scand. J. Haematol.* 5:440–53
155. Fagerhol, M. K., Abildgaard, U. 1970. *Scand. J. Haematol.* 7:10–17
156. Rosenberg, R. D., Damus, P. S. 1973. *J. Biol. Chem.* 248:6490–505
157. Miller-Anderson, M., Borg, H., Anderson, L. O. 1974. *Thromb. Res.* 5:439–52
158. Damus, P. S., Wallace, C. A. 1974. *Biochem. Biophys. Res. Commun.* 61:1147–53
159. Thaler, E., Schmer, G. 1975. *Br. J. Haematol.* 31:233–43
160. Kurachi, K., Schmer, G., Hermondson, M. A., Teller, D. C., Davie, E. W. 1976. *Biochemistry* 15:368–73
161. Nordenman, B., Nyson, C., Bjork, I. 1977. *Eur. J. Biochem.* 78:195–203
162. Peterson, T. E., Dudek-Wojciechowska, G., Sottrup-Jensen, L., Magnusson, S. 1979. In *Physiological Inhibitors of Blood Coagulation and Fibrinolysis,* ed. D. Collen, B. Wirman, M. Verstraete, pp. 43–54. Amsterdam: Elsevier
163. Mizuochi, T., Fujii, J., Kurachi, K., Kobata, A. 1980. *Arch. Biochem. Biophys.* 203:458–65
164. Franzen, L. E., Svensson, S., Larm, O. 1980. *J. Biol. Chem.* 255:5090–93
165. Yin, E. T., Wessler, S., Stoll, P. 1971. *J. Biol. Chem.* 246:3712–19
166. Damus, P. S., Hicks, M., Rosenberg, R. D. 1973. *Nature* 246:355–57
167. Kurachi, K., Fujikawa, K., Schmer, G., Davie, E. W. 1976. *Biochemistry* 15:373–77
168. Rosenberg, J. S., Mckenna, P., Rosenberg, R. D. 1975. *J. Biol. Chem.* 250:8883–88
169. Highsmith, R. F., Rosenberg, R. D. 1974. *J. Biol. Chem.* 249:4335–38
170. Lahire, B., Bagdasarian, A., Mitchell, B., Talamo, R. C., Colman, R. W., Rosenberg, R. D. 1976. *Arch. Biochem. Biophys.* 175:737–47
171. Abildgaard, U., Egeberg, O. 1968. *Scand. J. Haematol.* 5:155–62
172. Rosenberg, R. D. 1979. In *Chemistry and Physiology of the Human Plasma Proteins,* ed. D. H. Bing, pp. 353–68. New York: Pergamon
173. Li, E. H., Orton, C., Feinman, R. D. 1974. *Biochemistry* 13:5012–17
174. Villanueva, G. B., Danishefsky, I. 1977. *Biochem. Biophys. Res. Commun.* 74:803–9
175. Einarsson, R., Andersson, L. O. 1977. *Biochim. Biophys. Acta* 490:104–11
176. Nordenman, B., Danielsson, A., Bjork, I. 1978. *Eur. J. Biochem.* 90:1–6
177. Danielsson, A., Bjork, I. 1978. *Eur. J. Biochem.* 90:7–12
178. Nordenman, B., Bjork, I. 1978. *Biochemistry* 17:3339–44
179. Bjork, I., Nordling, K. 1979. *Eur. J. Biochem.* 102:497–502
180. Villanueva, G. B., Perret, V., Danishefsky, I. 1980. *Arch. Biochem. Biophys.* 203:453–57
181. Blackburn, M. N., Smith, R. L., Sibley, C. C., Johnson, V. A. 1981. *Ann. NY Acad. Sci.* 370:700–08
182. Longas, M. O., Ferguson, W. S., Finlay, T. H. 1980. *J. Biol. Chem.* 255:3436–44
183. Fish, W. W., Bjork, I. 1979. *Eur. J. Biochem.* 101:31–38
184. Longas, M. O., Finlay, T. H. 1980. *Biochem. J.* 189:481–89
185. Bjork, I., Danielsson, A., Fenton, J. W., Jornvall, J. 1981. *FEBS Lett.* 126:257–60
186. Fish, W. W., Bjork, I. 1981. In *Chemistry and Biology of Heparin,* ed. R. L. Lundblad, W. V. Brown, K. G. Mann, H. R. Roberts, pp. 335–43. Ann Arbor Mich.: Ann Arbor Science
187. Griffith, M. J., Lundblad, R. L. 1981. *Biochemistry* 20:105–10
188. Wallgren, P., Nordling, K., Bjork, I. 1981. *Eur. J. Biochem.* 116:493–96
189. Jochum, M., Lander, S., Heimburger, N., Fritz, H. 1981. *Z. Physiol. Chem.* 362:103–12
190. Kress, L. F., Catanese, J. 1980. *Biochim. Biophys. Acta* 615:178–86
191. Kress, L. F., Catanese, J. 1981. *Biochemistry* 20:7432–38
192. Marciniak, E. 1973. *Brit. J. Haematol.* 24:391–400
193. Miletich, J. P., Jackson, C. M., Majerus, P. W. 1978. *J. Biol. Chem.* 253:6908–17
194. Busch, P. G., Owen, W. G. 1981. *Thromb. Haemostasis* 46:38
195. Mullertz, S. 1974. *Biochem. J.* 143:273–83
196. Moroi, M., Aoki, N. 1976. *J. Biol. Chem.* 251:5956–65
197. Collen, D. 1976. *Eur. J. Biochem.* 69:209–16
198. Mullertz, S., Clemmensen, I. 1976. *Biochem. J.* 159:545–53

199. Wiman, B. 1981. *Methods Enzymol.* 80:395–408
200. Wiman, B., Collen, D. 1977. *Eur. J. Biochem.* 78:19–26
201. Wiman, B., Collen, D. 1979. *J. Biol. Chem.* 254:9291–97
202. Matsuda, M., Wakabayashi, K., Aoki, N., Morioka, Y. 1979. *Thromb. Res.* 17:527–32
203. Subgroup on Inhibitors, International Committee on Thrombosis and Haemostasis. 1978. *Thromb. Haemostasis* 39:524–26
204. Edy, J., Collen, D. 1977. *Biochim. Biophys. Acta* 484:423–32
205. Saito, H., Goldsmith, G. H., Moroi, M., Aoki, N. 1979. *Proc. Natl. Acad. Sci. USA* 76:2013–17
206. Wiman, B., Collen, D. 1978. *Eur. J. Biochem.* 84:573–78
207. Moroi, M., Aoki, N. 1977. *J. Biochem.* 82:969–72
208. Klingeman, H. G., Egbring, R., Holst, F., Gramse, M., Havemann, K. 1981. *Thromb. Res.* 24:479–83
209. Christensen, U., Clemmensen, I. 1977. *Biochem. J.* 163:389–91
210. Hayashi, R. 1977. *Methods Enzymol.* 47:84–93
211. Wiman, B., Boman, L., Collen, D. 1978. *Eur. J. Biochem.* 87:143–46
212. Collen, D. 1979. *Lancet* 1:1039–40
213. Wiman, B., Wallen, P. 1977. *Thromb. Res.* 10:213–22
214. Moroi, M., Aoki, N. 1979. *Thromb. Res.* 10:851–56
215. Wiman, B., Lijren, H. R., Collen, D. 1979. *Biochim. Biophys. Acta* 579:142–54
216. Wiman, B., Collen, D. 1979. See Ref. 162, pp. 247–54
217. Collen, D., Wiman, B. 1979. *Blood* 53:313–24
218. Aoki, N., Saito, H., Kamiya, T., Koie, K., Sakata, Y., Kobakuru, M. 1979. *J. Clin. Invest.* 63:877–84
219. Aoki, N., Sakuta, Y., Matsuda, M., Tatero, K. 1980. *Blood* 55:483–88
220. Schultze, H. E., Heide, K., Haupt, H. 1962. *Naturwissenschaften* 49:133
221. Heimburger, N., Haupt, H. 1965. *Clin. Chim. Acta* 12:116–18
222. Aronsen, K. F., Ekeland, G., Kindmark, C. O., Laurell, C. B. 1972. *Scand. J. Clin. Lab. Invest.* 29 (Suppl. 124): 127–36
223. Coombes, J., Shakespeare, P. G., Batstone, C. F. 1979. *Clin. Chim. Acta* 95:201–9
224. Daniels, J. C., Larson, D. L., Abston, S., Ritzmann, S. E., 1974. *J. Trauma* 14:153–62
225. Weeke, B., Jarnum, S. 1971. *Gut* 12:297–302
226. Kelly, U. L., Cooper, E. H., Alexander, C., Stone, J. 1978. *Biomedicine* 28:209–15
227. Laine, A., Hayem, A. 1981. *Biochim. Biophys. Acta* 668:429–38
228. Laine, A., Davril, M., Hayem, A. 1982. *Biochem. Biophys. Res. Commun.* 105:186–93
229. Reilly, C. F., Tewksbury, D., Schechter, N., Travis, J. 1982. *J. Biol. Chem.* 257:8619–22
230. Roughley, P. J., Barrett, A. J. 1977. *Biochem. J.* 167:629–37
231. Vartio, T., Seppa, H., Vaheri, A. 1981. *J. Biol. Chem.* 256:471–77
232. Ryley, H. C., Brogan, T. D. 1973. *J. Clin. Path.* 26:852–56
233. Stockley, R. A., Burnett, D. 1980. *Am. Rev. Respir. Dis.* 122:81–88
234. Hatcher, V. B., Oberman, M. S., Lazarus, G. S., Grayzel, A. I. 1978. *J. Immunol.* 120:665–70
235. Adams, D. O. 1980. *J. Immunol.* 124:286–92
236. Hudig, D., Haverty, T., Fulcher, C., Redelman, D., Mendelson, J. 1981. *J. Immunol.* 126:1564–74
237. Watson, D. K., Moudrianakis, E. N. 1982. *Biochemistry* 21:248–56
238. Katsunuma, T., Tsuda, M., Kusumi, T., Ohkubo, T., Mitoki, T., et al. 1980. *Biochem. Biophys. Res. Commun.* 93:552–57
239. Gaffer, S. A., Pringler, G. L., McIntyre, K. R., Braatz, J. 1980. *J. Biol. Chem.* 255:8334–39
240. Kondo, Y., Ohsawa, N. 1982. *Cancer Res.* 42:1549–54
241. Ratnoff, O. D., Lepow, I. H. 1957. *J. Exp. Med.* 106:327–43
242. Levy, L., Lepow, I. H. 1957. *Proc. Soc. Exp. Biol. Med.* 101:608–11
243. Pensky, J., Schwick, H. 1969. *Science* 163:698–99
244. Salvesen, G., Virca, G. D., Travis, J. 1982. *Kinin 81.* In press
245. Haupt, H., Heimburger, N., Kranz, T., Schwick, H. 1970. *Eur. J. Biochem.* 17:254–61
246. Sim, R., Arlaud, G., Colomb, M. 1980. *Biochim. Biophys. Acta* 612:433–49
247. Gigli, I., Mason, J., Colinan, R., Austen, F. 1970. *J. Immunol.* 104:574–81
248. Forbes, C., Pensky, J., Ratnoff, O. 1970. *J. Lab. Clin. Med.* 76:809–15
249. Reid, K. B. M., Porter, R. R. 1981. *Ann. Rev. Biochem.* 50:433–64
250. Sim, R. B., Reboul, A. 1981. *Methods Enzymol.* 80:26–42
251. Arlaud, G., Reboul, A., Sim, R., Colomb, M. 1979. *Biochim. Biophys. Acta* 576:151–62
252. Sim, R., Reboul, A., Arlaud, G., Villiers,

C., Colomb, M. 1979. *FEBS Lett.* 97:111–15

253. Gigli, I., Porter, R., Sim, R. 1976. *Biochem. J.* 157:541–48

254. Williams, C., Wickerhauser, M., Busby, T. F., Ingham, K. C. 1982. *Fed. Proc.* 41:2877

255. Takada, A., Takada, Y. 1980. *Thromb. Res.* 18:847–59

256. Takahashi, K., Nagasawa, S., Koyama, J. 1980. *Biochim. Biophys. Acta* 611:196–204

257. Sim, R., Reboul, A. 1981. *Methods Enzymol.* 80:43–54

258. Schapira, M., Scott, C., Colman, R. 1981. *Biochemistry* 20:2738–43

259. Schapira, M., Scott, C., James, A., Silver, L., Kueppers, F., James, H., Colman, R. 1982. *Biochemistry* 21:567–72

260. Minta, J. O., Aziz, E. 1981. *J. Immunol.* 126:250–55

261. Ziccardi, R., Cooper, N. 1979. *J. Immunol.* 123:788–92

262. Laurell, A. B., Johnson, U., Martensson, U., Sjoholm, A. G. 1978. *Acta Pathol. Microbiol. Scand. Sect. C* 86:299–306

263. Gallimore, M., Amundsen, E., Larsbraaten, M., Lyngaas, K., Fareid, E. 1979. *Thromb. Res.* 16:695–703

264. Muller-Esterl, W., Fritz, H. 1981. In *New Trends in Allergy*, ed. J. Ring, G. Bieng, pp. 81–90. Berlin/Heidelberg: Springer

265. Heimark, R. L., Kurachi, K., Fujikawa, K., Davie, E. W. 1980. *Nature* 286:456–60

266. Donaldson, V. H., Evans, R. R. 1963. *Am. J. Med.* 35:37–44

267. Donaldson, V. H., Harrison, R. A. 1982. *Blood* 60:121–29

268. Ziccardi, R. J. 1982. *J. Immunol.* 128:2505–8

269. Haverback, B. J., Dyce, B., Bundy, H. F., Wirtschafter, S. K., Edmondson, H. A. 1962. *J. Clin. Invest.* 41:972–80

270. Rinderknecht, H., Silverman, P., Geokas, M. C., Haverback, B. J. 1970. *Clin. Chim. Acta* 28:239–45

271. Bieth, J., Pichoir, M., Metais, P. 1970. *FEBS Lett.* 8:319–21

272. Harpel, P. C., Mosseson, M. W. 1973. *J. Clin. Invest.* 52:2175–84

273. Starkey, P. M., Barrett, A. J. 1977. In *Proteinases in Mammalian Cells and Tissues*, ed. A. J. Barrett, pp. 661–91. Amsterdam: Elsevier/North Holland Biomedical

274. Harpel, P. C., Rosenberg, R. D. 1976. *Prog. Hemostasis Thromb.* 3:145–89

275. Starkey, P. M. 1979. See Ref. 162, pp. 221–30

276. Van Leuven, F. 1982. *Trends Biol. Sci.* 7:185–87

277. Roberts, R. C. 1983. *Rev. Hematol.* In press

278. James, K. 1980. *Trends Biol. Sci.* 5:43–47

279. Barrett, A. J. 1981. *Methods Enzymol.* 80:737–54

280. Roberts, R. C., Riesen, W. A., Hall, P. K. 1974. See Ref. 7, pp. 63–71

281. Dunn, J. T., Spiro, R. G. 1967. *J. Biol. Chem.* 242:5556–63

282. Barrett, A. J., Brown, M. A., Sayers, C. A. 1979. *Biochem. J.* 181:401–18

283. Virca, G. D., Travis, J., Hall, P. K., Roberts, R. C. 1978. *Anal. Biochem.* 89:274–78

284. Kurecki, T., Kress, L. F., Laskowski, M. Sr. 1979. *Anal. Biochem.* 99:415–20

285. Sottrup-Jensen, L., Stepanik, T. M., Jones, C. M., Petersen, T. E., Magnusson, S. 1979. See Ref. 162, pp. 255–72

286. Swenson, R. P., Howard, J. B. 1979. *J. Biol. Chem.* 254:4452–56

287. Hall, P. K., Nelles, L. P., Travis, J., Roberts, R. C. 1981. *Biochem. Biophys. Res. Commun.* 100:8–16

288. Barrett, A. J., Starkey, P. M. 1973. *Biochem. J.* 133:709–24

289. Salvesen, G. S., Sayers, C. A., Barrett, A. J. 1981. *Biochem. J.* 195:453–61

290. Barrett, A. J., Starkey, P. M., Munn, E. A. 1974. See Ref. 7, pp. 72–77

291. Nelles, L. P., Hall, P. K., Roberts, R. C. 1980. *Biochim. Biophys. Acta* 623:46–56

292. Bjork, I., Fish, W. W. 1982. *Biochem. J.* 207:347–56

293. Richman, J. B. Y., Verpoorte, J. A. 1981. *Can. J. Biochem.* 59:519–23

294. Pochon, F., Amand, B., Lavalette, D., Bieth, J. 1978. *J. Biol. Chem.* 253:7496–99

295. Werb, Z., Burleigh, M. C., Barrett, A. J., Starkey, P. M. 1974. *Biochem. J.* 139:359–68

296. Harpel, P. C. 1970. *J. Exp. Med.* 132:329–52

297. Harpel, P. C. 1973. *J. Exp. Med.* 138:508–21

298. Hall, P. K., Roberts, R. C. 1978. *Biochem. J.* 171:27–38

299. Salvesen, G. S., Barrett, A. J. 1980. *Biochem. J.* 187:695–701

300. Sottrup-Jensen, L., Lonblad, P. B., Stepanik, T. M., Petersen, T. E., Magnusson, S., Jornvall, H. 1981. *FEBS Lett.* 127:167–73

301. Mortensen, S. B., Sottrup-Jensen, L., Hansen, H. F., Petersen, T. E., Magnusson, S. 1981. *FEBS Lett.* 135:295–300

302. Chan, J. Y. C., Burrows, C. E., Habal, F.

M., Movat, H. Z. 1977. *Biochem. Biophys. Res. Comm.* 74:150–58
303. Vahtera, E., Hamburg, U. 1978. *Biochem. J.* 171:767–70
304. Scharpe, S., Eid, M., Cooreman, W., Lauwers, A. 1976. *Biochem. J.* 153:505–7
305. Kueppers, F., Abrams, W. R., Weinbaum, G., Rosenbloom, J. 1981. *Arch. Biochem. Biophys.* 211:143–50
306. Largman, C., Johnson, J. H., Brodrick, J. W., Geokas, M. C. 1977. *Nature* 269:168–70
307. Karic, L., Glaser, C. B. 1981. *Int. J. Pept. Protein Res.* 18:416–19
308. Iwamoto, M., Abiko, Y. 1970. *Biochim. Biophys. Acta* 214:402–10
309. Galdston, M., Levytska, V., Liener, I. E., Twumasi, D. Y. 1979. *Am. Rev. Respir. Dis.* 119:435–41
310. Stone, P. J., Calore, J. D., Snider, G. L., Franzblau, C. 1979. *Am. Rev. Respir. Dis.* 120:577–87
311. Nilenh, J.-E., Ganrot, P. O. 1967. *Scand. J. Clin. Lab. Invest.* 20:113–21
312. Ohlsson, K., Laurell, C. B. 1976. *Clin. Sci. Mol. Med.* 51:87–92
313. Imber, M. J., Pizzo, S. V. 1981. *J. Biol. Chem.* 256:8134–39
314. Nagasawa, S., Sugihara, H., Han, B. H., Suzuki, T. 1970. *J. Biochem.* 67:809–19
315. Ganrot, P. O. 1966. *Clin. Chim. Acta* 14:493–501
316. Bieth, J. G., Tourbez-Perrin, M., Pochon, F. 1981. *J. Biol. Chem.* 256:7954–57
317. Barrett, A. J., Slavesen, G. S. 1979. See Ref. 162, pp. 247–54
318. Ganrot, P. O., Nilehn, J.-E. 1967. *Clin. Chim. Acta* 17:511–13
319. Tsuru, D., Kado, K., Fujiwara, K., Tomimatsu, M., Ogita, K. 1978. *J. Biochem.* 83:1345–53
320. Sayers, C. A., Barrett, A. J. 1980. *Biochem. J.* 189:255–61
321. Wu, K., Wang, D., Feinman, R. D. 1981. *J. Biol. Chem.* 256:10409–14
322. Virca, G. D. 1982. Characterization studies on human plasma α_2-macroglobulin. PhD thesis. Univ. Georgia, Athens
323. Downing, M. R., Bloom, J. W., Mann, K. G. 1978. *Biochemistry* 17:2649–53
324. Salvesen, G. S. 1981. Aspects of the structure of α_2-macroglobulin and its reaction with proteinases. PhD thesis. Cambridge Univ., Cambridge, UK
325. Howell, J. B., Beck, T., Bates, B., Hunter, M. J. 1983. *Arch. Biochem. Biophys.* 221: In press
326. Harpel, P. C. 1977. *J. Exp. Med.* 146:1033–40
327. Sottrup-Jensen, L., Petersen, T. E., Magnusson, S. 1981. *FEBS Lett.* 128:127–32
328. Van Leuven, F., Cassiman, J.-J., Van Den Berghe, H. 1981. *J. Biol. Chem.* 256:9023–27
329. Sottrup-Jensen, L., Petersen, T. E., Magnusson, S. 1981. *FEBS Lett.* 128:123–26
330. Sottrup-Jensen, L., Hansen, H. F. 1982. *Biochem. Biophys. Res. Commun.* 107:93–100
331. Swenson, R. P., Howard, J. B. 1979. *Proc. Natl. Acad. Sci. USA* 76:4313–16
332. Howard, J. B., Vermeulen, M., Swenson, R. P. 1980. *J. Biol. Chem.* 255:3820–28
333. Sottrup-Jensen, L., Petersen, T. E., Magnusson, S. 1980. *FEBS Lett.* 121:275–79
334. Howard, J. B. 1981. *Proc. Natl. Acad. Sci. USA* 78:2235–39
335. Harpel, P. C., Hayes, M. B., Hugli, T. E. 1979. *J. Biol. Chem.* 254:8669–78
336. Sim, R. B., Sim, E. 1981. *Biochem. J.* 193:129–41
337. Tack, B. F., Harrison, R. A., Janatova, J., Thomas, M. L., Prahl, J. W. 1980. *Proc. Natl. Acad. Sci. USA* 77:5764–68
338. Howard, J. B. 1980. *J. Biol. Chem.* 255:7082–84
339. Campbell, R. D., Dodds, A. W., Porter, R. R. 1980. *Biochem. J.* 189:67–80
340. Sim, R. B., Twose, T. M., Sim, E., Paterson, D. S. 1981. *Biochem. J.* 193:115–27
341. Swenson, R. P., Howard, J. B. 1980. *J. Biol. Chem.* 255:8087–91
342. Boyde, T. R. C. 1969. *Biochem. J.* 111:59–61
343. McPherson, T. A., Marchalonis, J. J., Lennon, U. 1970. *Immunology* 10:929–33
344. Stollar, B. D., Rezuke, W. 1978. *Arch. Biochem. Biophys.* 190:398–4
345. Black, D. V., Gregoriadis, G. 1976. *Biochem. Soc. Trans.* 4:253–56
346. Nachman, R. L., Harpel, P. C. 1976. *J. Biol. Chem.* 251:4514–20
347. Granelli-Piperno, A., Reich, E. 1978. *J. Exp. Med.* 148:223–34
348. Starkey, P. M., Barrett, A. J. 1982. *Biochem. J.* 205:91–95
349. Laurell, C. B., Jeppson, J. O. 1975. In *Plasma Proteins, Structure Function, and Genetic Control*, ed. F. W. Putnam, 1:229–64. New York: Academic. 2nd ed.
350. Hubbard, W. J. 1978. *Cell. Immunol.* 39:388–94
351. Abildgaard, U. 1979. See Ref. 162, pp. 239–41
352. Mullertz, S. 1979. See Ref. 162, pp. 243–45
353. Shapiro, S. S., Anderson, D. B. 1977. In

Chemistry and Biology of Thrombin, ed. R. L. Lundblad, J. W. Fenton II, K. G. Mann, pp. 361–74. Ann Arbor, MI: Ann Arbor Press

354. Edy, J., Deloch, F., Collen, D. 1976. *Thromb. Res.* 8:513–18

355. Highsmith, R. E., Weirich, C. J., Burnett, C. J. 1978. *Biochem. Biophys. Res. Commun.* 79:648–56

356. Harpel, P. C. 1981. *J. Clin. Invest.* 68:46–55

357. Damus, P. S., Gaeri, A., Wallace, C. A. 1975. *Thromb. Res.* 6:27–38

358. Osterud, B., Miller-Andersson, M., Abildgaard, U., Prydz, H. 1976. *Thromb. Haemostasis* 35:295–304

359. Kress, L. F., Kurecki, T. 1980. *Biochim. Biophys. Acta* 613:469–75

360. Kress, L. F., Catanese, J. J. 1981. *Toxicon* 19:501–7

361. Debanne, M. T., Bell, R., Dolovich, J. 1975. *Biochim. Biophys. Acta* 411:295–304

362. Fretz, J. C., Gan, J. C. 1980. *Int. J. Biochem.* 12:597–603

363. Fuchs, H. E., Shifman, M. A., Pizzo, S. V. 1982. *Biochim. Biophys. Acta* 716:151–57

364. Kaplan, J., Nielsen, M. L. 1979. *J. Biol. Chem.* 254:7323–28

365. Van Leuven, F., Cassiman, J.-J., Van Den Berghe, H. 1979. *J. Biol. Chem.* 254:5155–60

366. Mosher, D. F., Vaheri, A. 1980. *Biochim. Biophys. Acta* 627:113–22

367. Pastan, I. H., Willingham, M. C., Anderson, W. B., Gallo, M. G. 1977. *Cell* 12:609–17

368. Willingham, M. C., Maxfield, F. R., Pastan, I. H. 1979. *J. Cell Biol.* 82:614–25

369. Davies, P. J. A., Davies, D. R., Levitzki, A., Maxfield, F. R., Milhaud, P., Willingham, M. C., Pastan, I. H. 1980. *Nature* 283:162–67

370. Mortensen, S. B., Sottrup-Jensen, L., Hansen, H. F., Rider, D., Petersen, T. E., Magnusson, S. 1981. *FEBS Lett.* 129:314–17

371. Kaplan, J., Ray, F. A., Keogh, E. A. 1981. *J. Biol. Chem.* 256:7705–7

372. Marynen, P., Van Leuven, F., Cassiman, J. J., van den Berghe, H. 1981. *J. Immunol.* 127:1782–86

373. Pastan, I. H., Willingham, M. C. 1981. *Ann. Rev. Physiol.* 43:239–50

374. Steinbuch, M., Loeb, J. 1961. *Nature* 192:1196

375. Heide, K., Heimburger, N., Haupt, H. 1965. *Clin. Chim. Acta* 11:82–85

376. Haupt, H., Heimburger, N., Krantz, T., Schwick, H. G. 1970. *Eur. J. Biochem.* 17:254–62

377. Chawla, R. K., Wadsworth, A. A., Rudman, D. 1978. *J. Immun.* 121:1636–39

378. Hochstrasser, K., Niebel, J., Feuth, H., Lempart, K. 1977. *Klin. Wochenschr.* 55:337–42

379. Hochstrasser, K., Niebel, J., Lempart, K. 1977. *Klin. Wochenschr.* 55:343–45

380. Steinbuch, M. 1976. *Methods Enzymol.* 45:760–72

381. Salier, J. P., Martin, J. P., Lambin, P., McPhee, H., Hochstrasser, K. 1980. *Anal. Biochem.* 109:273–83

382. Aubry, M., Bieth, J. 1976. *Biochim. Biophys. Acta* 438:221–30

383. Lambin, P. 1978. *Thromb. Res.* 13:563–68

384. Meyer, J. F., Bieth, J., Metais, P. 1975. *Clin. Chim. Acta* 62:43–53

385. Fritz, H., Brey, B., Muller, M., Gebhardt, M. 1971. See Ref. 108, pp. 28–34

386. Hochstrasser, K., Bretzel, G., Feuth, H., Hilla, W., Lempart, K. 1976. *Hoppe-Seylers Z. Physiol. Chem.* 357:153–62

387. Hochstrasser, K., Wachter, E. 1979. *Hoppe-Seylers Z. Physiol. Chem.* 360:1285–96

388. Hochstrasser, K., Feuth, H., Steiner, O. 1973. *Hoppe-Seylers Z. Physiol. Chem.* 354:927–32

389. Wachter, E., Hochstrasser, K., Bretzel, G., Heindl, S. 1979. *Hoppe-Seylers Z. Physiol. Chem.* 360:1297–303

390. Wachter, E., Hochstrasser, K. 1979. *Hoppe-Seylers Z. Physiol. Chem.* 360:1305–11

391. Wachter, E., Hochstrasser, K. 1981. *Hoppe-Seylers Z. Physiol. Chem.* 362:1351–55

392. Hochstrasser, K., Schonberger, D. L., Lempart, K., Metzger, M. 1981. *Hoppe-Seylers Z. Physiol. Chem.* 362:1363–67

393. Borth, W., Menzel, E. J., Salzer, M., Stefren, C. 1981. *Clin. Chim. Acta* 177:219–55

394. Woolley, D. E., Roberts, D. R., Evanson, J. M. 1975. *Biochem. Biophys. Res. Commun.* 66:747–54

395. Woolley, D. E., Roberts, D. R., Evanson, J. M. 1976. *Nature* 261:325–27

396. McCartney, H. W., Tscheche, H. 1982. *Eur. J. Biochem.* In press

397. Tscheche, H., McCartney, H. W. 1981. *Eur. J. Biochem.* 120:183–90

398. Starkey, P. M., Barrett, A. J. 1977. *Biochim. Biophys. Acta* 483:386–97

399. Woolley, D. E., Tetlow, L. C., Mooney, C. J., Evanson, J. M. 1980. In *Proteinases and Tumor Invasion*, ed. P. Struli et al, pp. 97–115. New York: Raven

400. Kuettner, K., Soble, L., Croxen, R. L., Marczynska, B., Hiti, J., Harper, E. 1977. *Science* 196:653–54
401. Jarvinen, M. 1976. *Acta Chem. Scand. B* 30:933–40
402. Sasaki, M., Minakata, K., Yamamoto, H., Niwa, M., Kato, T., Ito, N. 1977. *Biochem. Biophys. Res. Commun.* 76:917–24
403. Ryley, H. C. 1979. *Biochem. Biophys. Res. Commun.* 89:871–78
404. Jarvinen, M. 1979. *FEBS Lett.* 108:461–64
405. Gounaris, A. D., Barrett, A. J. 1982. *Fed. Proc.* 41:4392 (Abstr.)
406. Brant, B. E. 1982. The purification and characterization of the human plasma cysteine proteinase inhibitor. MS thesis. Univ. Georgia, Athens
407. Sasaki, M., Taniguchi, K., Minakata, K. 1980. *J. Biochem.* 89:169–77
408. Pagano, M., Engler, R. 1982. *FEBS Lett.* 138:307–10
409. Barrett, A. J., McDonald, J. K. 1980. In *Mammalian Proteases: A Glossary and Bibliography*, Vol. 1, *Endopeptidases.* New York: Academic
410. Melloni, E., Pontremoli, S., Salamino, F., Sparatore, B., Michetti, M., Horecker, B. L. 1981. *Arch. Biochem. Biophys.* 208:175–83
411. Gordon, S. G., Cross, R. A. 1981. *J. Clin. Invest.* 67:1665–71
412. Schwartz, W. N., Bird, J. W. 1977. *Biochem. J.* 167:811–20
413. Starkey, P. M., Barrett, A. J. 1973. *Biochem. J.* 131:823–31
414. Schwartz, W. N., Barrett, A. J. 1980. *Biochem. J.* 191:487–97
415. Tollefsen, D. M., Blank, M. K. 1981. *J. Clin. Invest.* 68:589–96
416. Tollefsen, D. M., Majerus, D. W., Blank, M. K. 1982. *J. Biol. Chem.* 257:2162–69
417. Wunderwald, P., Schrenk, W. J., Port, H. 1982. *Thromb. Res.* 25:177–91
418. Kisiel, W., Canfield, W. M. 1982. *Fed. Proc.* 41:655 (Abstr.)
419. Marler, R. A., Griffin, J. H. 1980. *J. Clin. Invest.* 66:1186–89

Ann. Rev. Biochem. 1983. 52:711–60

GLUTATHIONE

Alton Meister and Mary E. Anderson

Department of Biochemistry, Cornell University Medical College, 1300 York Avenue, New York, New York 10021

CONTENTS

PERSPECTIVES AND SUMMARY 712
METABOLISM OF GLUTATHIONE—AN OVERVIEW 713
TRANSPORT OF GLUTATHIONE 715
TRANSPORT OF γ-GLUTAMYL AMINO ACIDS 719
FUNCTIONS OF THE γ-GLUTAMYL CYCLE 721
ENZYMES OF THE γ-GLUTAMYL CYCLE 725
   γ-Glutamylcysteine Synthetase 725
   Glutathione Synthetase 727
   γ-Glutamyl Transpeptidase 727
   γ-Glutamyl Cyclotransferase 729
   5-Oxoprolinase 730
   Dipeptidase 732
INTERCONVERSION OF GLUTATHIONE AND GLUTATHIONE
   DISULFIDE 733
   Glutathione Oxidation 733
   Glutathione Peroxidase 735
   Glutathione Transhydrogenases 735
   Glutathione Reductase 738
CONJUGATION OF GLUTATHIONE 739
   Exogenous Compounds; Glutathione S-transferases 739
   Endogenous Compounds 740
OTHER ASPECTS OF GLUTATHIONE FUNCTION 741
   Coenzyme Functions 741
   Radiation and Oxygen Toxicity 742
   Cancer 743
   Calcium Metabolism 744
GLUTATHIONE DEFICIENCY AND DEPLETION 744
   Effects of Inhibition of Glutathione Synthesis 744
   Mutant Microorganisms Deficient in Glutathione
      Synthesis 745
   Human Diseases Involving Defects of Glutathione
      Synthesis and Metabolism 746
ANALYTICAL PROCEDURES 748
CONCLUDING REMARKS 750

PERSPECTIVES AND SUMMARY

This ubiquitous tripeptide (L-γ-glutamyl-L-cysteinylglycine), usually the most prevalent intracellular thiol, is now known to function directly or indirectly in many important biological phenomena, including the synthesis of proteins and DNA, transport, enzyme activity, metabolism, and protection of cells. The multifunctional properties of glutathione are reflected by the growing interest in this small molecule on the part of investigators of such diverse subjects as enzyme mechanisms, biosynthesis of macromolecules, intermediary metabolism, drug metabolism, radiation, cancer, oxygen toxicity, transport, immune phenomena, endocrinology, environmental toxins, and aging.

This chapter is concerned with current progress in unraveling the biochemical bases of the physiological roles of this important compound. Detailed information is now available about glutathione synthesis and its metabolism by the reactions of the γ-glutamyl cycle, and its function in reductive processes that are essential for the synthesis (and the degradation) of proteins, formation of the deoxyribonucleotide precursors of DNA, regulation of enzymes, and protection of the cell against reactive oxygen compounds and free radicals. In addition, glutathione is a coenzyme for several reactions; it conjugates with foreign compounds (e.g. drugs) and with compounds formed in metabolism (e.g. estrogens, prostaglandins, leukotrienes), and thus participates in their metabolism.

An important recent finding is that cellular turnover of glutathione is associated with its transport, in the form of GSH, out of cells. The functions of such GSH transport include formation by membrane-bound γ-glutamyl transpeptidase of γ-glutamyl amino acids, which can be transported into certain cells, and thus serve as one mechanism of amino acid transport. Transported GSH probably also functions in reductive reactions that may involve the cell membrane and the immediate environment of the cell. In the mammal, such transported GSH may enter the blood plasma and be transferred to other cells. Glutathione thus appears to be a storage form and a transport form of cysteine.

Much of the new information about glutathione has arisen through studies with selective inhibitors of the enzymes involved in its metabolism. Thus, inhibition in vivo of γ-glutamyl transpeptidase, γ-glutamyl cyclotransferase, 5-oxoprolinase, and glutathione synthesis has been achieved, and the effects observed have contributed importantly to the understanding of glutathione metabolism and function. Studies on the inhibition of γ-glutamyl transpeptidase have elucidated the transport of GSH and the formation and transport of γ-glutamyl amino acids. Inhibition of glutathione synthesis by sulfoximine compounds that inactivate γ-glutamylcysteine synthetase has also contributed to such knowledge, as well as to information about the roles of glutathione in protection against both free radicals and reactive oxygen compounds, and in metabolism. These enzyme inhibitors, and other compounds that increase in vivo glutathione synthesis have opened the way to selective modulation of glutathione metabolism; this has made several new therapeutic approaches possible.[1]

METABOLISM OF GLUTATHIONE— AN OVERVIEW[2]

Glutathione is synthesized intracellularly (Figure 1) by the consecutive actions of γ-glutamylcysteine synthetase (Reaction 1) and GSH synthetase (Reaction 2). Reaction 1 is feedback inhibited by GSH. The breakdown of GSH (and also of GSSG and S-substituted GSH) is catalyzed by γ-glutamyl transpeptidase, which catalyzes transfer of the γ-glutamyl moiety to acceptors—amino acids, e.g. cystine, glutamine, and methionine, certain dipeptides, water, and GSH itself—(Reaction 3). GSH occurs mainly intracellularly and a major fraction of the transpeptidase is on the external surface of the cell membranes. GSH transported across cell membranes interacts with γ-glutamyl transpeptidase. γ-Glutamyl amino acids formed by γ-glutamyl transpeptidase are transported into cells; evidence for such formation and transport of γ-glutamyl amino acids is given below. Intracellular γ-glutamyl amino acids are substrates of γ-glutamyl cyclotransferase (Reaction 4), which converts these compounds into the corresponding amino acids and 5-oxo-L-proline. The ATP-dependent conversion of 5-oxo-L-proline to L-glutamate is catalyzed by the intracellular enzyme 5-oxo-prolinase (Reaction 5). The cysteinylglycine formed in the transpeptidase reaction is split by dipeptidase (Reaction 6). These six reactions constitute the γ-glutamyl cycle, which thus accounts for the synthesis and degradation of GSH. Two of the enzymes of the cycle also function in the

---

[1] More than 2000 current papers on glutathione have come to the authors' attention through a computer search and the kindness of many investigators, who have supplied us with manuscripts. Because of space limitations we cannot cite the full literature here; while we have included references to many recent developments, we have probably inadvertently overlooked some relevant papers. Readers should also consult the published proceedings of several recent meetings and earlier reviews (1–8).

[2] We follow current usage in abbreviating glutathione as GSH and glutathione disulfide as GSSG. In this field it is common to use the term glutathione (GSH) to include both GSH and GSSG, because the relative amounts of each form may not be known, and the analytical methods used may determine the sum of both forms. Although to some extent we follow this practice here, we indicate, where known, the predominant redox form. The term "total glutathione" has been used in the literature to indicate the sum of GSH and GSSG in GSH equivalents.

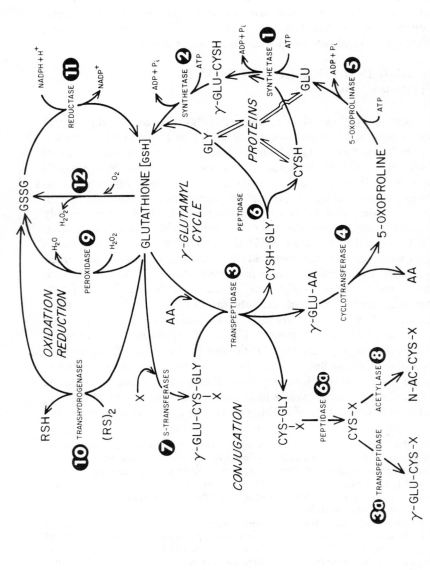

Figure 1   Overall summary of glutathione metabolism (see text): Reaction 1. γ-GLU-CYSH synthetase; Reaction 2. GSH synthetase; Reaction 3 and 3*a*. glutamyl transpeptidase; Reaction 4. γ-glutamyl cyclotransferase; Reaction 5. 5-oxoprolinase; Reactions 6 and 6*a*. dipeptidase; Reaction 7. GSH S-transferases; Reaction 8. *N*-acetylase; Reaction 9. GSH peroxidase; Reaction 10. transhydrogenases; Reaction 11. GSSG reductase; Reaction 12. oxidation of GSH by O₂; conversion of GSH to GSSG is also mediated by free radicals.

metabolism of S-substituted GSH derivatives, which may be formed nonenzymatically by reaction of GSH with certain electrophilic compounds or by GSH S-transferases (Reaction 7). The γ-glutamyl moiety of such conjugates is removed by the action of γ-glutamyl transpeptidase (Reaction 3), a reaction facilitated by γ-glutamyl amino acid formation. The resulting S-substituted cysteinylglycines are cleaved by dipeptidase (Reaction 6a) to yield the corresponding S-substituted cysteines, which may undergo N-acylation (Reaction 8) or an additional transpeptidation reaction to form the corresponding γ-glutamyl derivative (Reaction 3a); the latter reaction has been demonstrated with a substrate of exogenous origin (9), and with leukotriene E (10) (see p. 741).

Intracellular GSH is converted to GSSG by selenium-containing GSH peroxidase, which catalyzes the reduction of H_2O_2 and other peroxides (Reaction 9); there is evidence that certain GSH S-transferases can also catalyze such reactions. GSH is also converted to GSSG by transhydrogenation (Reaction 10); a number of reactions of this type are considered later. Reduction of GSSG to GSH is mediated by the widely distributed enzyme GSSG reductase which uses NADPH (Reaction 11). Extracellular conversion of GSH to GSSG has also been reported; the overall reaction requires O_2 and leads to formation of H_2O_2 (Reaction 12) (see p. 733). GSSG is also formed by reaction of GSH with free radicals.

TRANSPORT OF GLUTATHIONE

The intracellular level of GSH in mammalian cells is in the millimolar range (0.5–10 mM), whereas micromolar concentrations are typically found in blood plasma. Several lines of evidence (2) indicate that γ-glutamyl transpeptidase is accessible to external substrate, and that the enzyme is largely bound to the outer surface of cell membranes. Nevertheless, many findings indicate that intracellular GSH is the major substrate of transpeptidase. The finding of an enzyme and its substrate on opposite sides of a membrane led to the postulate that intracellular GSH is transported to the membrane-bound transpeptidase (11). Studies on a patient with γ-glutamyl transpeptidase deficiency who has marked glutathionuria and glutathionemia (12) led to the suggestion that transport of intracellular GSH to the plasma and glomerular filtrate in this patient reflects an aspect of the normal process that provides substrate to the membrane-bound enzyme (11, 13). Thus, in the absence of significant transpeptidase activity, substantial amounts of GSH appear extracellularly. This interpretation is supported by studies in which marked glutathionuria and glutathionemia were found in mice and rats that had been treated with transpeptidase inhibitors (14, 15, 16). Animals given such inhibitors exhibit glutathionuria during and for a short period after such treatment without microscopic evidence of tissue damage. With one inhibitor, urinary concentrations as high as 29 mM were observed (16). The kidney is a major source of such urinary GSH (14), but some undoubtedly arises also from the liver.

When the inhibitors of GSH synthesis (e.g. prothionine sulfoximine, buthionine sulfoximine) are given to mice and rats, plasma GSH levels decrease substantially (2, 14, 15). The rapid and marked increase in plasma GSH after inhibiting transpeptidase, and the considerable decrease seen soon after inhibition of GSH synthesis indicate active turnover of plasma GSH. The findings suggest that there is normally an appreciable flow of GSH from liver into plasma, and that cells that have high transpeptidase levels utilize plasma GSH; the products formed (including γ-glutamyl amino acids) enter the cells. The major organs involved in this inter-organ circulation of GSH are the liver and kidney, but undoubtedly other organs also participate. Studies on anephric animals treated with transpeptidase inhibitors show that about 67% of the plasma GSH is used by the kidney and the remainder by extrarenal transpeptidase [(15); see also (15a)]. These in vivo investigations indicate that GSH is normally translocated to membrane-bound transpeptidase as a discrete step in the γ-glutamyl cycle (14); in vitro studies support this. Thus, GSH export from cells to medium was found in human lymphoid cells (17, 18) and skin fibroblasts (19), and macrophages (20). Isolated perfused liver preparations show efflux of GSH to the perfusate (21).

Independent evidence for inter-organ GSH transport came from studies on the plasma levels of GSH in various blood vessels of the rat (22). Hepatic vein plasma has a much higher level than does arterial blood plasma or that obtained from the inferior vena cava. Renal vein plasma has about 20% of the level found in arterial plasma, indicating that kidney has a mechanism in addition to filtration for removal of plasma GSH (15a, 22, 23). Disappearance of plasma GSH on passage through the kidney decreased markedly when either of two types of transpeptidase inhibitors were given; this indicates that the nonfiltration mechanism that utilizes GSH involves the action of γ-glutamyl transpeptidase (22). Although it was previously believed that virtually all renal transpeptidase is localized on the brush border side of the cell, electron micrographic studies (24) provide evidence that there is transpeptidase on the basolateral side as well. Thus, transpeptidase acts on plasma GSH on both the brush border and the basolateral sides. About 80%–90% of renal tubular GSH arises from kidney cells; the remainder comes from the plasma (25, 26). The amount reaching the tubule from the plasma (estimated for a 30-g mouse from the glomerular filtration rate and the plasma GSH level) is about 0.8 μmole/hr. The turnover rate of renal GSH indicates a flow of about 4.1 μmole/hr from renal cells to tubule; most of this is utilized in the tubule. It has been tentatively estimated that about 2.4 μmole/hour of GSH is utilized on the basolateral side (26). Such basolateral utilization of GSH, described earlier as "extraction" (15a) and "uptake" (27), is mainly the result of extracellular conversion to other products, chiefly the amino acid constituents of GSH (28).

Liver GSH is transported in substantial amounts to hepatic vein plasma and to the bile; in the rat the GSH levels are, respectively, about 26 μM (22) and 1–6 mM (29–33). Rat bile also contains about 1 mM cyst(e)inylglycine and about 0.2 mM cyst(e)ine (32, 33). Studies on rat bile are often complicated by the high entrance of the pancreatic ducts into the bile duct, so that the bile collected contains pancreatic juice; this may lead to high GSSG values and low values of total glutathione. Proper cannulation yields "pure" bile, which has very low transpeptidase activity and 1–6 mM GSH; pancreatic juice, which has no detectable GSH or GSSG, has high levels of transpeptidase (33). The GSH level in bile decreases markedly after treatment of rats with buthionine sulfoximine, and it increases approaching intracellular levels after treatment with a transpeptidase inhibitor (33). The findings suggest that GSH transported from hepatocytes interacts with the transpeptidase of ductule cells, and that there is substantial reabsorption of metabolites by ductule epithelium. These findings on intact rats indicate a greater extent of GSH export from liver cells into bile than has been deduced by analysis of the excreted bile. In the intact rat, about 12 and 4 nmoles/g/min of GSH appear in the hepatic vein and bile, respectively.

Studies on isolated perfused rat liver indicate transport of GSH and GSSG into the total perfusate of 12 and 1 nmole/g/minute, respectively (21, 33a). It was concluded (33b, 33c), that GSSG and GSH-conjugates are released from the liver preferentially into bile by a process in which these compounds mutually compete for secretion. Other data [see Figures 1B and 3 in (34); see also (34a)] show, however, that substantial amounts of GSSG and GSH-conjugates enter the venous outflow. The quantitative aspects of the suggested competition indicate that the system has a much greater affinity for GSH-conjugates than for GSSG (33). The data on transport of GSH and GSSG from liver to bile obtained with the isolated perfused liver system differ substantially from those obtained on anesthetized intact rats. A major difference is that, in contrast to the intact rat, little or no GSH is found in the bile in the perfusion system.

Intracellular glutathione is normally over 99% GSH; GSH is the major transport form. Analyses of mouse blood plasma (35) and rat blood plasma, i.e. arterial, hepatic vein, renal vein, and vena cava (22, 36), and rat bile (31,

29) show that about 90% is in the GSH form. About 70% of mouse urinary glutathione was found as GSH after potent inhibition of transpeptidase (16). Human lymphoid cells export glutathione that is at least 90% GSH (17). These findings contrast markedly with earlier reports that rat blood plasma (15a, 21, 23, 37) and bile (34) contain glutathione that is predominantly GSSG (see p. 748). In earlier studies that suggested that GSH and GSSG are transported across cell membranes, the possibility that the GSH and GSSG found extracellularly was related to oxidative or other types of cell damage could not be excluded.

Under conditions of marked toxicity or oxidative stress, intracellular GSSG increases substantially (38), and there may be a mechanism for its export. It has been concluded that the export of glutathione from erythrocytes involves transport of GSSG (39–42); however, normal transport of GSH is difficult to exclude. GSSG transport was reported in erythrocytes whose GSSG level (normally, <0.1%) was artificially raised by a GSH-oxidizing agent; such transport did not occur in ATP-depleted cells. Evidence for an active transport system for GSSG has been obtained in other studies on erythrocytes (40, 41) and lens (43); a similar pathway may exist in liver (38, 44). Beutler (39) has suggested that active transport of GSSG may be an emergency mechanism to protect cells from toxic effects of GSSG. The question as to how much GSSG transport occurs normally is difficult to answer. Intracellular GSSG levels are extremely low, and accurate measurement of these small amounts in the presence of very large amounts of GSH constitutes a formidable technical problem.

The rate of transport of GSH from lymphoid cells has been found to be proportional to the intracellular GSH level (18). Little if any intact GSH is taken up by these cells. Lymphoid cells that have been depleted of GSH by treatment with buthionine sulfoximine, and which are therefore more sensitive to the effects of irradiation (18), are not protected by suspension in GSH-containing media (45). Repletion of intracellular GSH by suspending cells in media containing GSH takes place by a process involving enzymatic degradation of the external GSH, uptake of the products, and intracellular resynthesis of GSH (45).

Transport of GSH out of cells is a property of many cells (2, 46). In the mammal, this seems to function in the transfer of cysteine sulfur between cells. Transport of GSH may protect the cell membrane against oxidative damage by maintaining essential SH-groups. It may provide a way of reducing compounds in the immediate environment of the cell; this might protect the cell or facilitate transport of certain compounds, e.g. disulfides. Transport of GSH to membrane-bound transpeptidase leads to γ-glutamyl amino acid formation, a process that is part of a transport system and which serves also in the recovery of the amino acid constituents of GSH.

TRANSPORT OF γ-GLUTAMYL AMINO ACIDS

Indirect evidence for transport of γ-glutamyl amino acids (1, 2) has been supplemented by direct observation of γ-glutamyl amino acid uptake after their administration to mice. Uptake of L-γ-glutamyl-L-methionine sulfone, a poorly metabolized compound, by kidney was much greater than that of the corresponding free amino acid, and uptake was inhibited by γ-glutamyl amino acids but not by free amino acids (47). Appreciable amounts of L-γ-glutamyl-L-methionine sulfone were found in the kidney, and the data suggest that other tissues (liver, pancreas) may also transport γ-glutamyl amino acids. Under physiological conditions, γ-glutamyl amino acids might be hydrolyzed at the membrane and the amino acid products of such hydrolysis may be effectively transported.

However, in vivo studies show transport of intact γ-glutamyl amino acids (47) and there is increased formation of 5-oxoproline after suspension of rat kidney slices in media containing γ-glutamyl amino acids (48). This shows that γ-glutamyl amino acids are transported as such under these conditions, but does not exclude some membranous hydrolysis. Kidney and possibly other cells thus have a transport system for γ-glutamyl amino acids that is not shared by free amino acids. In the kidney, γ-glutamyl amino acid uptake takes place predominantly on the lumenal side rather than the basolateral side (28). As discussed later, there is good evidence that γ-glutamyl amino acids are formed by transpeptidation under physiological conditions.

Patients with transpeptidase deficiency exhibit glutathionuria and glutathionemia, as do animals given transpeptidase inhibitors (see above). When the urine of these patients and animals was treated with dithiothreitol and 2-vinylpyridine, chromatographic analysis showed, in addition to the vinylpyridine derivative of GSH, appreciable quantities of vinylpyridine derivatives of cysteine and γ-glutamylcysteine (16). The discovery of increased urinary excretion of GSH, γ-glutamylcysteine, and cysteine in both human and experimental animal γ-glutamyl transpeptidase deficiency suggests that the physiological function of transpeptidase is closely associated with the metabolism or transport (or both) of these compounds. Further studies showed that urinary γ-glutamylcyst(e)ine in animals treated with transpeptidase inhibitors is formed by the action of the residual γ-glutamyl transpeptidase (9). The patients are markedly deficient, but not altogether lacking in transpeptidase. Similarly, although the enzyme may be $\sim 90\%$ inhibited in the experimental animals, there is sufficient activity remaining to catalyze transpeptidation between GSH and cystine to form γ-glutamylcystine. It is relevant to note that cystine is an excellent acceptor substrate (K_m 30 μM) of transpeptidase (49–51).

Thus, transpeptidation between GSH and cystine can occur in the presence of substantial transpeptidase deficiency or inhibition. An alternative explanation for the finding of urinary γ-glutamylcysteine, i.e. that cleavage of the cys-gly bond of GSH (or of GSSG) occurs, was rendered unlikely by studies with [^{35}S]GSH, and by failure to find such activity after an extensive search (9). When there is marked deficiency of transpeptidase, γ-glutamylcystine is apparently formed more rapidly than it is transported. The high extracellular levels of GSH that accompany inhibition of transpeptidase inhibit transport of γ-glutamyl amino acids (52); under these conditions, γ-glutamylcystine appears in the urine.

Additional evidence for transport of γ-glutamyl amino acids has come from studies in which administration of γ-glutamylcystine to mice increased renal GSH levels. The level of GSH in the kidney is regulated by feedback inhibition of γ-glutamylcysteine synthetase by GSH (53). When either γ-glutamylcystine or γ-glutamylcysteine disulfide was administered to mice, unusually high levels of renal GSH were found, evidently because the feedback-regulated step was bypassed (54). Thus, GSH levels of 6–7 mM were found in animals given γ-glutamylcystine or γ-glutamylcysteine disulfide as compared to values of 4.5–5.2 mM for untreated animals or controls given glutamate, cysteine and cystinyl-bis-glycine. These observations indicate transport of γ-glutamyl amino acids, and also suggest an alternative pathway of GSH biosynthesis.

Administration of γ-glutamylcystine or of γ-glutamylcysteine disulfide to animals previously treated with buthionine sulfoximine, to lower markedly the intracellular levels of GSH, had little effect on renal GSH levels, but when such animals were given γ-glutamylcysteine, a substantial increase in renal GSH occurred. When γ-glutamylcystine labeled selectively with ^{35}S in either the external or internal S atom were given to normal mice, the specific radioactivity of the renal GSH 15 min after giving the internally labeled compound was much greater than that found after giving the externally labeled compound (55). This result, obtained both with large doses and with tracer doses of the labeled γ-glutamylcystines, indicates that the administered γ-glutamylcystine is reduced to cysteine and γ-glutamylcysteine; the latter is converted directly to GSH. The experiments in which the animals were pretreated with buthionine sulfoximine indicate an apparent requirement for intracellular GSH for reduction of the administered disulfides. The findings support a metabolic-transport scheme (Figure 2) in which GSH is transported to a membrane site containing transpeptidase where it interacts with cystine to form γ-glutamylcystine. γ-Glutamylcystine is transported and reduced by transhydrogenation involving GSH. Extracellular GSH can inhibit γ-glutamylcystine transport, and intracellular GSH inhibits utilization of cysteine by γ-glutamylcysteine syn-

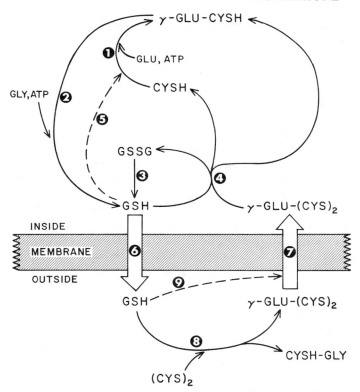

Figure 2   Metabolic-Transport Scheme (see text): 1. γ-glutamylcysteine synthetase; 2. GSH synthetase; 3. GSSG reductase; 4. transhydrogenation [GSH + γ-GLU-(CYS)$_2$]; 5. feedback inhibition by GSH of 1; 6. transport of GSH; 7. transport of γ-GLU-(CYS)$_2$; 8. γ-glutamyl transpeptidase; 9. inhibition by GSH of transport of γ-GLU-(CYS)$_2$.

thetase; these phenomena may function in regulation of GSH synthesis and transport.

FUNCTIONS OF THE γ-GLUTAMYL CYCLE

Since the cycle accounts for the biosynthesis of GSH, its functions are intimately associated with those of GSH, which include its roles in a variety of protective, metabolic, transport, and catalytic processes, in the regulation of GSH synthesis and utilization, and in the storage and transport of cysteine.

The hypothesis that the γ-glutamyl cycle functions in amino acid transport is based on the idea that γ-glutamyl amino acids are formed in or on the cell membrane through interaction of γ-glutamyl transpeptidase, intracellular GSH, and extracellular amino acids. That this mechanism constitutes one of the pathways that mediates transport of amino acids is strongly supported by the data reviewed above that demonstrate in vivo formation of γ-glutamyl amino acids and their transport. This pathway has not been proposed to function in all cells, or in a given cell for all amino acids (1, 2). Cystine and glutamine are the most likely candidates for this type of transport, but other amino acids may also participate. Amino acids are transported by "systems" (56). Although little is known about the mechanisms involved, there is evidence, based largely on competition studies, for several amino acid transport systems that overlap in specificity. Presently available data indicate that the γ-glutamyl amino acid pathway plays a role in amino acid transport; its relationship to other "systems" that have been postulated is not yet clear.

The arguments against the γ-glutamyl amino acid transport idea include: (a) equivalence between GSH turnover and amino acid transport has not been observed; (b) the transpeptidase functions in vivo solely as a hydrolase; (c) the distribution of transpeptidase activity in the renal tubule is not parallel to amino acid uptake; (d) blocks of the cycle are not always associated with defective amino acid transport; and (e) too much energy is needed.

The considerations reviewed above indicate that one would not expect equivalence between GSH turnover and amino acid transport, which some investigators have expected (57). With respect to the energy utilized, the export of GSH from cells, apparently an obligatory step in GSH turnover, followed by its extracellular or membranous breakdown, must require at least an amount of energy equivalent to the cleavage of two molecules of ATP to ADP. Cleavage of another molecule of ATP would be required if the process involves formation and utilization of 5-oxoproline. Possibly less energy than this would be needed if intracellular or membranous hydrolysis of γ-glutamyl amino acids occurs. The need for energy in this system is not unusually great when considered in relation to processes such as protein synthesis and urea formation (see 1).

Despite ample data indicating that transpeptidation is a physiologically significant function of γ-glutamyl transpeptidase, it has nevertheless been concluded (58, 59), that the sole function of this enzyme is to hydrolyze GSH and GSSG. This conclusion, which rests on invalid assumptions, has been dealt with in detail (60), but studies in two laboratories (58, 60) indicate the participation of about half of the GSH utilized by transpeptidase in transpeptidation. One study was based on an estimate (58). In another (60), the enzyme was incubated with GSH or GSSG and a mixture of amino acids approximating the amino acid composition of plasma, and the relative extents of transpeptidation and hydrolysis were determined by measuring the products formed. At pH 7 in the presence of 50 μM GSH and the amino acid mixture, about half of the GSH utilized participated in transpeptidation; L-cystine and L-glutamine are the most active acceptors and other amino acids are also active. Thus, a major fraction of the GSH transported from cells participates in transpeptidation. Since membrane-bound transpeptidase must always be in close contact with amino acids and GSH, γ-glutamyl amino acids must be formed continuously; although they can also serve as substrates of transpeptidase, there is direct evidence for their transport into cells. There is a high concentration of transpeptidase in the renal brush border (about 0.1 mM), which is of about the same order as that of many amino acid acceptors. It is doubtful that parameters derived from Michaelis-Menten kinetics can be validly applied here. Nevertheless, it is of interest that the apparent K_m for L-cystine is lower than the plasma level of this amino acid.

Studies on the distribution of transpeptidase in segments of renal tubules showed the enzyme to be present in considerable concentration in both the proximal convoluted and straight segments, as well as in other regions of the nephron and elsewhere in the kidney (61, 62, 63). It is therefore unlikely that amino acids would escape the action of transpeptidase in the proximal convoluted tubule. Other considerations that support the conclusion that the transpeptidase functions in transpeptidation include the finding of γ-glutamyl amino acids in blood plasma (64; D. Wellner, unpublished), urine (65, 66), and various tissues (67–71), and the formation of γ-glutamyl amino acids by isolated renal tubules (72).

Dramatic defects in amino acid transport have not always been observed in humans or animals blocked at various steps of the cycle; this is not unexpected because the inhibitions thus far achieved are far from complete. Furthermore, there are multiple amino acid transport systems with overlapping specificity that can therefore supplement each other. Multiple overlapping mechanisms seem to have evolved for amino acid transport. There are other examples (e.g. deoxyribonucleotide synthesis) (see p. 738) in which multiple pathways have developed for important functions. Generalized aminoaciduria occurs in patients with severe γ-glutamyl-cysteine synthetase deficiency (73); cystinuria and possibly other defects in amino acid transport occur in γ-glutamyl transpeptidase deficiency (1, 16).

Renal GSH levels decrease when there is an appreciable increase in transpeptidation. Rats treated with glycylglycine, an excellent acceptor substrate, showed a marked decline in renal GSH levels (74). Administration of moderate amounts of amino acids to mice also led to a decline in the level of kidney GSH (11).[3] Such effects were not found when

---

[3] Administration of α-aminoisobutyrate, which is not a substrate of transpeptidase, also led to lower renal GSH levels and to increased plasma amino acid levels. The latter might contribute to a decrease in renal glutathione, but the effect of α-aminoisobutyrate was only partially blocked by serine plus borate suggesting that α-aminoisobutyrate may also act by another mechanism (11).

mice were also treated with a mixture of serine and borate, a combination that effectively inhibits transpeptidase (75). Similar observations were made on isolated renal cells; suspension of the cells in media containing amino acids led to depletion of GSH (76). The GSH content of perfused kidneys decreases when the perfusate contains amino acids (77). Similar findings were made on the mammary gland (see below). The rate of turnover of liver GSH in intact animals increases after amino acid administration (78).

The rate at which the intracellular renal GSH level declines after giving an inhibitor of GSH synthesis is slowed by giving inhibitors of transpeptidase (11, 14). This finding, and those summarized above on the effects of amino acids on GSH levels and turnover indicate significant connections between amino acids, export of GSH, and transpeptidase. A full explanation of these phenomena will require, minimally, more detailed understanding of the membranous orientation of the transpeptidase, the mechanisms of GSH and γ-glutamyl amino acid transport, and associated cellular and membrane functions.

Several recent studies are relevant to the findings discussed above. An apparent association between amino acid transport and localization of transpeptidase in the intestine has been noted (79, 80); thus, the transpeptidase activity of villus tip cells is far greater than that of crypt cells, which are much less active in transport. The pattern of relative rates of uptake of various amino acids closely resembles the acceptor specificity pattern of the transpeptidase. Transpeptidase, 5-oxoprolinase, and γ-glutamylcysteine synthetase activities increase during lactogenesis in the rat mammary gland (81), and factors that affect transpeptidase activity in the gland induce parallel changes in amino acid transport (82, 83, 84). Plasma arterial-venous amino acid differences measured across the mammary gland were decreased by treatment with serine plus borate, AT-125, and L-γ-glutamyl-(o-carboxy)-phenylhydrazide; this effect was greatest on amino acids that are good acceptor substrates of transpeptidase. Incubation of isolated acini from lactating glands with high concentrations of amino acids led to decreased GSH levels, an effect prevented by serine plus borate. A correlation was found between length of lactation, increase of the amino acid arterial-venous difference, and transpeptidase activity. Similar correlations were noted in studies on the effects of milk accumulation and of hormones.

Sensitivity of certain cell lines to azaserine correlates with transpeptidase activity (85). Azaserine-resistant cells are less efficient in concentrating azaserine than the sensitive parental lines. Human Wilms' tumor cells are sensitive to azaserine, but azaserine-resistant strains derived from these cells have greatly reduced transpeptidase activity. The findings suggest that azaserine may be transported into certain cells as the γ-glutamyl derivative.

Kalra and colleagues (86–88) incorporated detergent-isolated renal transpeptidase into phospholipid vesicles containing entrapped GSH; the vesicles took up glutamate, but not proline, and uptake, which was inhibited by inhibitors of transpeptidase, was associated with internal γ-glutamylglutamate formation. Transpeptidase was also incorporated into human erythrocyte membranes, and such modified cells took up alanine and glutamate but not methionine; uptake was inhibited by transpeptidase inhibitors. This system seems to be an interesting and potentially valuable model for further study of transpeptidase-mediated transport.

Several modifications of the γ-glutamyl cycle have been considered (2). For example, a model was suggested in which γ-glutamylcysteine (or another γ-glutamyl amino acid formed by transpeptidation) might function in an exchange diffusion system or in an active transport system (1).

ENZYMES OF THE γ-GLUTAMYL CYCLE

γ-Glutamylcysteine Synthetase

This enzyme has been purified from several sources (89, 90), and many of its structural and catalytic properties have been examined, expecially in work on the rat kidney enzyme (91). The enzyme is inhibited nonallosterically by GSH under conditions similar to those that prevail in vivo; this indicates a physiologically significant feedback mechanism (53). GSH also inhibits the γ-glutamylcysteine synthetase activities of fetal liver and Novikoff hepatoma (92). Treatment of rats with sodium selenite increases the activity of liver γ-glutamylcysteine synthetase, an apparent effect of increased enzyme synthesis (93). Decreased γ-glutamylcysteine synthetase occurs in a human inborn error of metabolism (see p. 747) and also in the erythrocytes of certain GSH-deficient sheep (94, 95).

The acceptor amino acid specificity of the enzyme is rather broad, and it also interacts with several glutamate analogs including β-aminoglutarate (91), and α-aminomethylglutarate (96). The mechanism of the reaction involves formation of enzyme-bound γ-glutamyl phosphate and interaction of the latter with the amino group of cysteine (89, 97). Kinetic investigations are consistent with this possibility, but did not demonstrate γ-glutamyl phosphate as a discrete covalent complex (98). Methionine sulfoximine is an effective irreversible inhibitor of the enzyme and in the presence of MgATP is converted to methionine sulfoximine phosphate, which binds tightly to the enzyme (99). Of the four stereoisomers of methionine sulfoximine, only L-methionine-S-sulfoximine inhibits, the same stereoisomer that irreversibly inhibits glutamine synthetase and causes convulsions in animals (100). Chemically synthesized L-methionine sulfoximine phosphate inhibits γ-glutamylcysteine synthetase; the mechanism of inhibition is thus similar to that previously shown for glutamine synthetase. However, methionine sulfoximine phosphate binds less tightly to γ-glutamylcysteine synthetase than it does to glutamine synthetase, and under certain conditions the inhibitor can be released from the inhibited enzyme with restoration of catalytic activity. The findings are consistent with intermediate formation of enzyme-bound γ-glutamyl phosphate, since phosphorylation of methionine sulfoximine seems to reflect phosphorylation of glutamate in the normal reaction. α-Ethylmethionine sulfoximine induces convulsions in mice and inhibits glutamine synthetase irreversibly, but it does not inhibit γ-glutamylcysteine synthetase (101). The reciprocal goal of inhibiting γ-glutamylcysteine synthesis without perturbing glutamine synthesis was achieved by preparing inhibitors in which the methyl moiety of methionine sulfoximine was replaced by propyl and butyl moieties, i.e. prothionine sulfoximine (102) and buthionine sulfoximine (103). A number of other sulfoximine analogs of methionine sulfoximine were also prepared and studied (103). Buthionine sulfoximine inhibits the enzyme more effectively than prothionine sulfoximine and at least 100 times more effectively than methionine sulfoximine. The data suggest that the S-alkyl moiety of the sulfoximine binds at the enzyme site that normally accepts L-cysteine. α-Methylbuthionine sulfoximine is almost as effective as buthionine sulfoximine (103). The higher homologs, penta-, hexa- and heptathionine sulfoximine, also inhibit the enzyme (104). The hexa- and heptathionine derivatives are unexpectedly toxic for reasons not yet understood. Buthionine sulfoximine, like methionine sulfoximine (99) is phosphorylated at the active site of γ-glutamylcysteine synthetase (104).

The enzyme is also inhibited by cystamine (105, 106), the optical isomers of 3-amino-1-chloro-2-pentanone (107) (the α-chloroketone analogs of L- and D-α-aminobutyrate), and L-2-amino-4-oxo-5-chloropentanoate (108). A sulfhydryl group at the active site interacts with the chloroketone inhibitors and also with cystamine; these compounds bind to the enzyme as glutamate analogs. The possibility that a γ-glutamyl-S-enzyme intermediate may be formed in the reaction needs to be considered. There is evidence that divalent metal ions play a role in the binding of amino acid substrates.

γ-Glutamylcysteine synthetase binds covalently to cystamine-Sepharose, an interaction facilitated by ATP and inhibited by Mg^{2+} plus glutamate (109). A large fraction of the enzyme applied to such columns binds apparently by forming a disulfide bond between cysteamine-Sepharose and a sulfhydryl group at or near the active site. The enzyme may be released by treatment with dithiothreitol. The enzyme does not bind to columns of S-(S-methyl)cysteamine-Sepharose, whereas free S-(S-methyl)cysteamine is a potent inhibitor. A cysteamine-S disulfide moiety derived from the external cysteamine residue of cystamine-Sepharose seems to be the critical group recognized by the enzyme. Studies with a number of cystamine analogs supported this conclusion and led to the further conclusion that a disulfide (or diselenide) moiety and a single free amino group are required for inhibition (G. F. Seelig, A. Meister, see 110).

γ-Glutamylcysteine synthetase from rat kidney has a single disulfide bond and two free sulfhydryl groups per M_r 100,000. Only one SH-group of the native enzyme is titrable with DTNB, and such reaction does not affect enzyme activity (G. F. Seelig, A. Meister, see Ref. 110). The enzyme (M_r 100,000) can be dissociated by treatment with SDS and thiol into heavy ($M_r \sim 74,000$) and light ($M_r \sim 24,000$) subunits; after the enzyme is cross-linked with dimethylsuberimidate, a species of $M_r \sim 100,000$ is obtained (108). Dissociation of the native enzyme by treatment with dithiothreitol yields a heavy subunit that exhibits full enzymatic activity and inhibition by GSH. Reassociation of the subunits has also been achieved (G. F. Seelig, A. Meister, see Ref. 111); the function of the light subunit needs additional study.

Glutathione Synthetase

This enzyme has been purified from several sources (89). There is evidence that enzyme-bound γ-glutamylcysteinyl phosphate is formed in the reaction, whose mechanism is thus similar to those catalyzed by γ-glutamylcysteine and glutamine synthetases. The most highly purified preparation was obtained from rat kidney by a method involving ATP-affinity chromatography (112). The enzyme (M_r 118,000) has two apparently identical subunits. The rat kidney enzyme is about 20 times more active than the apparently homogeneous preparation of this activity from erythrocytes. The regions of the active site that bind glycine and the cysteine moiety of L-γ-glutamyl-L-cysteine are highly specific. On the other hand, the enzyme acts on substrates in which the L-γ-glutamyl moiety of the dipeptide substrate is replaced by D-γ-glutamyl, β-aminoglutaryl, mono-methyl (α-methyl, β-methyl, γ-methyl, N-methyl) glutamyl, glutaryl, and N-acetyl moieties.

γ-Glutamyl Transpeptidase

γ-Glutamyl transpeptidase (113, 114), an enzyme of major importance in GSH metabolism, initiates GSH degradation. It can catalyze three types of reactions: (a) transpeptidation, in which the γ-glutamyl moiety is transferred to an acceptor; (b) autotranspeptidation, in which the γ-glutamyl moiety is transferred to GSH to form γ-glutamyl-GSH; and (c) hydrolysis, in which the γ-glutamyl moiety is transferred to water. GSH, GSSG, S-substituted GSH, and other γ-glutamyl compounds are substrates. The L-isomers of cystine, methionine, and other amino acids, as well as many dipeptides, especially aminoacylglycines, are good acceptors.

Histochemical studies indicate substantial transpeptidase activity in the membranes of cells that exhibit secretory or absorptive functions such as the epithelial cells of proximal renal tubules, jejunum, biliary tract, epididymis, seminal vesicles, choroid plexus, ciliary body, retinal pigment membrane, bronchioles, thyroid follicles, canalicular region of hepatocytes, pancreatic acinar and ductule cells (61); it is present in microvillus membranes. Such activity has been found on human lymphoid cell membranes (115), endoplasmic reticulum and Golgi, and there is evidence for its presence in renal basolateral membranes (24).

Highly purified enzyme preparations have been obtained from kidney of several species and from other tissues; kidney exhibits the highest activity, followed by pancreas, epididymis, seminal vesicle, jejunal epithelial cells, liver, and spleeen (114). This activity has also been found in soluble form in urine, seminal fluid, and pancreatic juice (29); there is little present in rat bile. The blood plasma of rats and mice is devoid of activity, but the enzyme is detectable in human plasma, where its elevation is of some clinical value in detection of liver disease (116, 117). Several γ-glutamyl substrates suitable for convenient determination of its activity have been developed. The most widely used is L-γ-glutamyl-p-nitroanilide (118). Useful spectrophotometric methods based on S-derivatives of GSH are also available (1, 119), and the fluorogenic substrate, L-γ-glutamyl-7-amino-4-methylcoumarin also provides a sensitive assay (120).

The enzyme has a donor site that interacts with both L- and D-γ-glutamyl compounds. The acceptor site consists of subsites for the cysteinyl and glycine moieties of cysteinylglycine; L-amino acids bind to the former, which prefers amino acids such as the L-isomers of cystine, glutamine, and methionine. Branched chain amino acids and aromatic amino acids are bound with less affinity. Kinetic studies are consistent with a ping-pong mechanism involving a γ-glutamyl-enzyme (1). The various complexities of the catalytic reaction have led to a number of interesting kinetic studies. Recently, it was found that compounds such as maleate (1), hippurate (121), various bile acids and their conjugates (S. J. Gardell, S. S. Tate, see Ref. 122; W. Abbott, A. Meister, see Ref. 123), as well as phospholipids (124), can modulate relative extents of hydrolysis and transfer.

The enzyme may be solubilized by treatment with a detergent or organic solvent or with proteinases (113, 114, 125, 126). The purified preparations are glycoproteins and exhibit heterogeneity associated with the presence of isozymes containing different amounts of sialic acid. The proteinase-solubilized rat kidney enzyme (M_r 68,000) consists of two subunits (M_r 46,000, 22,000). The M_r 22,000–subunit contains the active site residue (probably on OH group) involved in formation of the γ-glutamyl-enzyme intermediate. Although there is relatively little species variation in size of the light subunit, the transpeptidases fall into two groups with the respect to the heavy subunit: (M_r 46,000–50,000) rat and rabbit; $M_r > 60,000$ (bovine, sheep, hog, and human). Treatment of the rat kidney enzyme with urea at neutral pH leads to digestion of the heavy chain (owing to latent proteinase activity of the light subunit (127)); however, if treatment with urea is carried out in the presence of acetic acid, the subunits are not significantly degraded and their reconstitution may be achieved (128) with return of activity.

The heavy subunit of the detergent-solubilized enzyme differs from that of the protease-treated enzyme; the former contains a short, relatively hydrophobic, protease-sensitive amino terminal segment that probably anchors the enzyme to the cell membrane (125, 126). The molecular weights of the corresponding heavy subunits are estimated to be 51,000 and 46,000. The light subunit, which does not appear to interact directly with the lipid bilayer, is bound noncovalently to the heavy subunit. The biosynthesis of the enzyme has been studied in kidney slices suspended in media containing L-[^{35}S]methionine (129). In studies involving immunoprecipitation with antitranspeptidase antibodies, a protein (M_r 78,000) was identified that appears to be a precursor of the two enzyme subunits. The evidence suggests that this protein is initially synthesized in renal cells and that its cleavage occurs prior to transport of the enzyme to the brush border membrane (130).

The enzyme is inhibited by L-serine plus borate, owing to formation of a tetrahedral borate complex, a transition state analog (75). Other inhibitors include γ-glutamylhydrazones of α-keto acids (131) and various γ-glutamylphenylhydrazides (14). The L- and D-isomers of γ-glutamyl-(o-carboxy)phenylhydrazide (14, 132) are effective competitive inhibitors. Glutamine antagonists such as 6-diazo-5-oxo-L-norleucine, L-azaserine, and L-(αS, 5S)-α-amino-3-chloro-4,5-dihydro-5-isoxazole acetic acid (AT-125) are effective irreversible inhibitors of the enzyme and serve as affinity labels of the γ-glutamyl site (16, 133, 133a, 134, 113, 114). Of these compounds, AT-125 is the most effective. As discussed above, when potent transpeptidase inhibitors are given to mice, glutathionuria occurs. Glutathionuria also results from administration of γ-glutamyl amino acids, presumably an effect of competitive inhibition (135).

γ-Glutamyl Cyclotransferase

This soluble enzyme, which catalyzes Reaction 4 (Figure 1), is widely distributed in mammalian tissues (1), and is highly active toward the L-γ-glutamyl derivatives of glutamine, alanine, cysteine, cystine, methionine, and several other amino acids. In general, its substrate specificity parallels

the amino acid acceptor specificity of γ-glutamyl transpeptidase, supporting the idea that the cyclotransferase and transpeptidase function in
sequence in metabolism. The enzyme also acts on di-γ-glutamyl amino
acids; indeed, it is much more active toward certain di-γ-glutamyl amino
acids than toward the corresponding γ-glutamyl amino acids. For example,
the enzyme acts on di-L-γ-glutamyl-L-proline, but has no detectable activity
toward L-γ-glutamyl-L-proline. Similar findings were made with the
corresponding derivatives of valine, tyrosine, and leucine. On the other
hand, the enzyme exhibits a high order of activity toward both the L-γ-
glutamyl and di-L-γ-glutamyl derivatives of glutamine and alanine.

The enzyme has been purified from human and sheep brain (136), hog
(137) and rat (138) liver, human erythrocytes (139) and rat kidney (140).
Several forms of the enzyme, separable by ion exchange chromatography
and electrophoresis, have been found. The enzyme is relatively unstable and
undergoes substantial changes in physical and catalytic properties during
preparation and storage. The most highly purified preparation, obtained
from rat kidney (140), contains several readily accessible SH groups, whose
modification is associated with appearance of multiple enzyme forms. The
enzyme was obtained, after 1000-fold purification, in apparently homogeneous form by a procedure involving treatment with dithiothreitol followed
by chromatography on thiol-Sepharose. It was also isolated in highly active
stable form after reduction and treatment with iodoacetamide. It appears
that most of the cyclotransferase activity of rat kidney is associated with a
single protein. Apparent multiple forms may be produced by mechanisms
involving intramolecular disulfide bond formation, perhaps associated
with conformational changes in the protein, and also by mixed disulfide
formation with low-molecular-weight thiols. Such changes might be
involved in regulation of the enzyme.

5-Oxoprolinase

This enzyme catalyzes the ATP-dependent cleavage of 5-oxoproline to
glutamate:

$$5\text{-Oxo-L-proline} + ATP + 2H_2O \underset{K^+(NH_4^+)}{\overset{Mg^{2+}}{\rightleftharpoons}} \text{L-glutamate} + ADP + P_i$$

The ATP requirement is mandated by the unusual stability of the 5-
oxoproline amide bond. 5-Oxoprolinase has been found in mammalian
tissues (141), plants (142), and microorganisms (143). 5-Oxo-L-proline is
formed by the action of γ-glutamyl cyclotransferase on γ-glutamyl amino
acids, by enzymatic cleavage of amino terminal 5-oxo prolyl residues of

peptides and proteins, and by the action of γ-glutamylamine cyclotransferase (144) on ε-(L-γ-glutamyl)-L-lysine derived from the degradation of proteins containing transglutaminase-generated cross-links. 5-Oxoproline may also arise from dietary sources and be formed by nonenzymatic cyclization of γ-glutamyl compounds such as glutamine. 5-Oxoprolinase is found in virtually all mammalian tissues except the erythrocyte and the lens.[4] The enzyme, which was isolated in apparently homogenous form from rat kidney in 50% yield after 1700-fold purification (M_r 325,000), is composed of two apparently identical subunits (146). [An earlier report (147) of M_r 420,000 is probably incorrect.] The enzyme contains 27 SH groups per monomer, six of which can be titrated in the native enzyme, and two of which are required for catalysis (146). At least one SH is at or close to the nucleoside triphosphate (NTP) binding site. The enzyme can bind 5-oxo-L-proline in the absence of NTP, and it can bind NTP in the absence of 5-oxo-L-proline. When ATP is replaced by other NTP's such as ITP or UTP, there is rapid cleavage of the NTP in the absence of 5-oxoproline (148, 149). When 5-oxoproline is added to enzyme catalyzing hydrolysis of NTP, there is a slow time-dependent decrease in the V_{max} and an increase in the K_m for NTP. These findings suggest that 5-oxoprolinase is a hysteretic enzyme, and that in the normal reaction, binding of both ATP and 5-oxoproline induces a conformational change that brings the substrates into a juxtaposition that facilitates the reaction.

The 2 hydrolytic activities of 5-oxoprolinase may be uncoupled by: (*a*) substitution of ATP by certain other NTP's; (*b*) replacement of 5-oxoproline by certain analogs; or (*c*) substitution for Mg^{2+} by Mn^{2+} (in excess), Ca^{2+}, or Co^{2+} (150, 146). Modification of the enzyme by treatment with *N*-ethylmaleimide, 5,5'-dithiobis-(2-nitrobenzoic acid), 5'-*p*-fluorosulfonylbenzoyl adenosine, or 5'-*p*-fluorosulfonylbenzoyl inosine leads to loss of all catalytic activities; inactivation is reduced by either ATP or ITP (146). The enzyme does not catalyze ATP-ADP exchange reactions. Studies with ^{18}O-labeled substrates show that: (*a*) all three oxygens of 5-oxoproline are recovered in glutamate; and (*b*) the two water molecules used contribute one oxygen atom to P_i and one to the γ-carboxyl of glutamate (151). The enzyme also catalyzes the intrinsically exergonic hydrolysis of α-hydroxyglutarate lactone, a reaction that is ATP-dependent. Although conversion of 5-oxo-L-proline to L-glutamate by 5-oxoprolinase is essentially irreversible, very slow reversal was shown by

---

[4] Evidence for the occurrence of 5-oxoprolinase in rabbit lens has been reported (145); since this activity was reported not to require ATP, thermodynamic considerations suggest that the observed reaction may not be due to 5-oxoprolinase and that another explanation for the findings should be sought.

measuring ATP formation in the presence of high glutamate concentrations. The most probable mechanism is one in which 5-oxoproline is phosphorylated by ATP on the amide carbonyl oxygen followed by hydrolysis of the resulting intermediate to yield γ-glutamyl phosphate, which is then hydrolyzed to glutamate and P_i.

The reaction has been studied with a variety of 5-oxoproline analogs, which fall into three categories (150, 152): (*a*) analogs that promote ATP cleavage but are not themselves hydrolyzed; (*b*) analogs whose hydrolysis is partially coupled to cleavage of ATP, i.e. molar formation of ADP exceeding that of imino acid; and (*c*) analogs that participate in fully coupled reactions. Mapping studies show that binding of imino acid substrate requires a 5-carbonyl (or $=NH$) moiety, an unsubstituted N_1 atom, and a C_2 atom of L-configuration; substantial modification of the 5-oxo-L-proline molecule in the region of C_3 and C_4 atoms is possible with retention of binding properties. Uncoupling phenomena seem to be associated with significant differences in orientation of analogs at the active site, which facilitate access of water to the nucleoside triphosphate.

The 5-oxo-L-proline analog in which the 4-methylene moiety is replaced by S (i.e. L-2-oxothiazolidine-4-carboxylate) is a good substrate (152, 153, 154). It has high affinity for the enzyme and is rapidly converted to L-cysteine, presumably through formation and nonenzymatic breakdown of S-carboxy-L-cystine. When L-2-oxothiazolidine-4-carboxylate is administered to mice, liver GSH increases suggesting that administration of this thiazolidine would be an effective cysteine delivery system. Recent studies show that L-2-oxothiazolidine-4-carboxylate is more effective than N-acetyl-L-cysteine in stimulating GSH synthesis. Protection against acetaminophen toxicity (153, 154) is mediated through increased synthesis of GSH. Thus, administration of buthionine sulfoximine inhibits the increase in hepatic GSH found after L-2-oxothiazolidine-4-carboxylate is given. This thiazolidine may be useful as a component of amino acid mixtures used in diets and in parenteral administration. Most such preparations do not now contain cysteine because of the toxicity of this amino acid (155–158) and because it is rapidly oxidized to the very insoluble cystine. The thiazolidine may also be useful for the growth of cells sensitive to the toxic effects of cysteine (158), and as a safener in agriculture to protect crop plants against herbicides.

Dipeptidase

Few studies have been carried out on the activities that cleave cysteinylglycine, the disulfide form of this dipeptide, and S-substituted cysteinylglycine derivatives. Dipeptides of this type are split by aminopeptidase M, an activity concentrated in the kidney and jejunal microvillus membranes

(159, 160, 161). This enzyme splits dipeptides and a variety of larger peptides. The finding of relatively little aminopeptidase M in rat epididymis led to the search for another activity that might be responsible for the cleavage of dipeptides released in the course of GSH metabolism (162). A membrane-bound dipeptidase was isolated from rat kidney that exhibits a specific activity toward L-alanylglycine, which is about 70-fold greater than that shown by aminopeptidase M. This dipeptidase is much more active than aminopeptidase M in catalyzing hydrolysis of substrates such as S-methylcysteinylglycine, cystinyl-bis-glycine and leukotriene D4 (10, 162). The enzyme is present in the microvillus membranes of rat kidney, jejunum, and epididymis. The kidney enzyme (M_r 100,000) has two apparently identical subunits; it contains about two gram atoms of Zn per subunit. Neither this dipeptidase nor aminopeptidase M exhibits much activity toward L-cysteinylglycine. This dipeptide, formed in transpeptidation of GSH, may be transported and hydrolyzed intracellularly. Several studies show that there is substantial cytosolic dipeptidase activity (e.g. 163). Extracellular oxidation of L-cysteinylglycine may also occur.

INTERCONVERSION OF GLUTATHIONE AND GLUTATHIONE DISULFIDE

Glutathione Oxidation

Elvehjem et al (164) found that mouse kidney homogenates catalyze O_2-dependent conversion of GSH to GSSG, that this reaction is inhibited by cyanide, and that kidney homogenates are much more active than those of liver. Orrenius et al (165, 166) confirmed these findings in isolated rat kidney and liver cells and obtained evidence for stepwise enzymatic breakdown of GSSG. Conversion of GSH to GSSG was ascribed to a "GSH oxidase". Tate et al (167) found such oxidase activity in kidney, epididymis, jejunum, choroid plexus and other tissues, and that this activity is membrane-bound and associated with renal and jejunal brush borders; its distribution follows a pattern similar to that of γ-glutamyl transpeptidase. Efforts to separate the two renal activities failed. Highly purified preparations of rat kidney transpeptidase were found to exhibit oxidase activity, as did each of the 12 isozymic forms of this enzyme (167, 168). Griffith & Tate (169) showed that the apparent GSH oxidase activity of transpeptidase is mediated by the cysteinylglycine formed in the transpeptidase-catalyzed degradation of GSH. This dipeptide is rapidly and nonenzymatically oxidized to form cystinyl-bis-glycine. The oxidation of GSH takes place by nonenzymatic transhydrogenation between GSH and cystinyl-bis-glycine (or between GSH and the mixed disulfide between cysteinylglycine and GSH). Such oxidation, which is inhibited by EDTA, is undoubtedly mediated by metal ions. In support of this mechanism, it was found that a covalent inhibitor of transpeptidase essentially abolishes both transpeptidase and oxidase activities of purified transpeptidase as well as of kidney homogenates. Notably, spontaneous oxidation of cysteinylglycine is more rapid than that of cysteine; in comparison, GSH reacts sluggishly with oxygen. This mechanism accounts for the oxidation of GSH observed with purified preparations of transpeptidase; thus, oxidation is *not* mediated by the transpeptidase, but rather by cysteinylglycine. Only a very low concentration of this dipeptide is needed to mediate GSH oxidation, and therefore there may or may not be a direct relationship between "oxidase" activity and the amount of transpeptidase. Indeed, a small amount of transpeptidase is sufficient to generate enough cysteinylglycine to catalyze rapid GSH oxidation; in the presence of large amounts of transpeptidase, oxidation of GSH may be greatly reduced because of substantial enzymatic degradation of GSH. These considerations are relevant to subsequent studies in which the finding of partial separations of transpeptidase and "GSH oxidase" activities were considered as evidence for existence of a separate renal GSH oxidase (170, 171, 172). The finding of GSH oxidation in kidney perfusion studies led to the conclusion that the oxidase is localized on the basolateral side of the renal cell (173), but this region apparently has transpeptidase (24). The reported resolution of a renal sulfhydryl oxidase from transpeptidase (170) is of doubtful validity because the fraction exhibiting "oxidase" activity was not shown to contain protein. Whether copper is involved in this oxidation, as suggested (171), could be determined if the enzyme had been isolated and analyzed.

Although GSH oxidation can be explained by nonenzymatic phenomena, kidney and other tissues may contain proteins capable of catalyzing this reaction; there is evidence consistent with this. Sulfhydryl oxidases have been isolated from milk (174, 175), kidney (176), and rat seminal vesicle secretions (177). Although even the most purified of these (176) exhibits some transpeptidase activity, these proteins may oxidize thiols directly. Trace metals and heme compounds can oxidize thiols rapidly, as might several metalloproteins as well. These considerations indicate that in all probability the cysteinylglycine-mediated mechanism (169) explains the oxidation of GSH catalyzed by purified transpeptidase as well as that observed in kidney homogenates (164) and with suspensions of kidney cells (165, 166). An aspect of this problem that needs more attention relates to the specificity for various thiols.

Since 10–15% of the total glutathione of rat and mouse blood plasma may be GSSG, some oxidation of GSH seems to occur in vivo, but the significance of this is not clear. The H_2O_2 formed could serve a protective function by destroying microorganisms. GSSG formation may promote

GSH translocation (14, 15) and provide a substrate for cystine formation (25).

Glutathione Peroxidase

This enzyme, present in a number of tissues, catalyzes GSH-dependent reduction of H_2O_2 (Figure 1, Reaction 9). The reduction of H_2O_2 in erythrocytes in the presence of GSH and GSH peroxidase is coupled with oxidation of glucose-6-phosphate and of 6-phosphogluconate, which provides NADPH for reduction of GSSG by GSSG reductase. This is a major pathway of H_2O_2 metabolism in many cells, and one which also catalyzes reduction of other peroxides (7, 178, 179, 180). It is thus important for the protection of membrane lipids against oxidation. Intermediates such as O_2^- and H_2O_2 are formed extensively in biological systems, and these produce reactive oxygen species that can lead to organic peroxide formation (181). Increased oxygen tension leads to increased formation of reactive oxygen species, and this may cause an increase in formation and release of GSSG. This subject has been extensively reviewed (179, 181, 182). Reactive oxygen species may produce other deleterious effects including those associated with mutagenesis. GSH peroxidase also plays a role in the biosynthesis of prostaglandins and in the regulation of prostacyclin formation (180).

GSH peroxidase of beef erythrocytes (M_r 85,000), consists of four apparently identical subunits, each of which contains one atom of selenium (180). It has long been known that Se is an essential nutrient. Se-deficient animals have markedly decreased GSH peroxidase activity. The reduced form of GSH peroxidase appears to contain selenocysteine selenol ($-$ SeH) at the active site (180, 183–185a).

Another type of GSH peroxidase activity found in rat liver (186, 187) is unaffected by Se deficiency and exhibits much less activity toward H_2O_2 as compared to organic hydroperoxides. Studies with a hemoglobin-free liver perfusion system led to evidence that the Se-independent GSH peroxidase (a GSH S-transferase) reduces organic hydroperoxides, but does not function in reduction of H_2O_2. In this system, both Se-deficient and control liver preparations released GSSG after infusion of tert-butyl hydroperoxide.

Glutathione Transhydrogenases

Many metabolic and physiological functions involve thiol-disulfide exchange, e.g. protein synthesis, protein degradation, activation and inactivation of enzymes, synthesis of the deoxyribose intermediates required for DNA synthesis, reduction of cystine. It has long been thought that many such reactions involve participation of GSH, since it is the major intracellular thiol, and because GSSG reductase is widely distributed. Early reports described a beef liver activity that catalyzes GSH-dependent reduction of homocystine (188), and a similar system in yeast that catalyzes reduction of cystine and related compounds (189). Later, evidence was obtained for several apparently separate enzymes of low specificity that catalyze thiol-disulfide interchange between GSH and low-molecular-weight disulfides, as well as between GSH and protein disulfides (190). Attention has also been directed toward systems that catalyze reduction and rearrangement of protein disulfide bonds. GSSG reductase, contrary to early reports, does not directly catalyze such reductions, which seem to be mediated by separate transhydrogenases. Transhydrogenase activities are widely distributed, and the task of sorting out and identifying the individual catalysts involved, which are undoubtedly of major physiological importance, is being pursued in several laboratories. Because many varied substrates participate in thiol-disulfide interchange reactions, there may be many transhydrogenases. On the other hand, the very large number of reactions involved in the assembly of proteins, regulation of enzymes, and reductive degradation of proteins, makes it necessary to consider that the number of individual transhydrogenases may be much smaller than the number of reactions catalyzed. Freedman (191) has considered the idea (among others) that there may be separate enzymes for each of four reaction types.

Protein disulfide isomerase activity was first observed in studies on reactivation of reduced pancreatic ribonuclease (192–198). Several investigators (199–205) purified an activity from liver that catalyzes reductive cleavage of the disulfides of insulin in the presence of GSH. Both protein disulfide isomerase and GSH-insulin transhydrogenase activities were found to copurify (206–208), although some differences in the activity ratios were noted (209, 210). A recent characterization of thiol-protein disulfide oxidoreductase from rat liver (198) was based on an earlier procedure for purification of this activity from bovine liver (211). Comparison of the physicochemical properties of these enzymes with those previously isolated from the same and other sources showed substantial similarity, and suggests that thiol protein disulfide oxidoreductase, protein disulfide isomerase, and GSH-insulin transhydrogenase may be identical.

The physiological functions associated with this protein are not yet clear. Evidence that the widely distributed GSH-insulin transhydrogenase activity functions in insulin degradation has been published (212–216). Thiol protein oxidoreductase of liver and plasmacytoma catalyzes formation of interchain disulfides of monomeric immunoglobulin M (217), and the liver enzyme catalyzes activation of choleragen by reductive cleavage of the disulfide bond linking A_1 and A_2 peptides of the toxin (218). The enzyme has been localized in the endoplasmic reticulum of all sources examined. The findings are consistent with a role in the synthesis of protein disulfide bonds (197), but the enzyme may also be involved in enzyme regulatory mechanisms (219) and in protein degradation (212–216).

Although GSH transhydrogenases probably have important functions that relate to the synthesis, structure, degradation, and function of proteins, other systems that affect the thiol-disulfide status of cells are also of considerable importance. These include the widely distributed thioredoxin system, which exhibits high reducing potential (220, 221). Formation of cellular disulfides is also an important process in which GSSG might play a role. Ziegler and collaborators (222–224) proposed that oxidation of cysteamine to cystamine, catalyzed by a membrane-bound monoxygenase, might serve as a source of cellular disulfides.

Reichard and collaborators (220, 225), showed that deoxyribonucleotides are formed by direct reduction of ribonucleotides, and isolated the proteins involved, i.e. ribonucleotide reductase, thioredoxin, and thioredoxin reductase. A mutant of *Escherichia coli* was later found that lacked thioredoxin, but which is active in NADPH-dependent ribonucleotide reduction (226). In this mutant, GSH and glutaredoxin (which has been purified to homogeneity from cells of the wild type), function in the reduction (226–229). Glutaredoxin is an acidic protein (M_r 12,000) that exhibits high and specific transhydrogenase activity. *E. coli* ribonucleotide reductase is composed of two nonidentical subunits, protein B1 (M_r 160,000), and protein B2 (M_r 78,000), neither of which is active separately (220). The subunits combine in the presence of Mg^{2+} to yield active enzyme, which can reduce all four ribonucleoside diphosphates to the corresponding deoxyribose compounds. Protein B1 contains two ribonucleotide binding sites and binding sites for four allosteric effectors (nucleoside triphosphates) as well as thiol groups which, together with protein B2, mediate stoichiometric reduction of substrate in the absence of GSH. Protein B2 contains two nonheme iron atoms and an organic-free radical on a tyrosine residue. The catalytic site of the reductase is formed from both subunits and involves the free radical moiety and a dithiol moiety of protein B1 (228). Chemically reduced glutaredoxin is enzymically active in conversion of CDP to dCDP. Holmgren et al (229) showed that, in the presence of glutaredoxin, the substrates for ribonucleotide reductase are GSH and each of the four ribonucleoside 5'-diphosphates. Glutaredoxin contains about 89 amino acid residues including two half-cystine residues that form a single disulfide bridge that is readily reduced to a dithiol by GSH, NADPH and GSSG reductase. Such reduction does not occur with NADPH and thioredoxin reductase. Glutaredoxin and thioredoxin are structurally unrelated proteins, and apparently are separate gene products.

Highly purified glutaredoxin has GSH-disulfide dehydrogenase activity toward 2-hydroxyethyl disulfide, but this accounts for less than 1% of this type of transhydrogenase activity of crude extracts of *E. coli*.

Glutaredoxin has also been purified to homogeneity from calf thymus (230, 231). This protein ($M_r \sim$ 11,000) is similar but not identical to the *E. coli* protein. It also exhibits GSH-disulfide transhydrogenase activity. However, thymus, in contrast to *E. coli*, appears to have only one cytoplasmic type of transhydrogenase, which is active with glutaredoxin, 2-hydroxyethyl disulfide, L-cystine, and several other substrates.

Interestingly, loss of the thioredoxin system in a mutant of *E. coli* does not lead to loss of ribonucleotide reduction (226), and GSH-deficient mutants grow at normal rates (see p. 745); thus, either pathway alone can mediate ribonucleotide reduction.

Glutathione Reductase

This widely distributed flavoprotein, which catalyzes Reaction 11 (Figure 1) was first observed in the 1930s and later purified from several sources (7). Reaction 11 is essentially irreversible; this reaction accounts for the very high GSH:GSSG ratios found in cells. Although early studies suggested that the enzyme can catalyze reduction of several mixed disulfides between GSH and other compounds including proteins, it now appears that only GSSG, mixed disulfides between GSH and γ-glutamylcysteine (16), and between GSH and coenzyme A (232) are significant substrates. New methods have been described for the isolation of the enzyme from liver [calf (233), mouse (234), rabbit (235), rat (236)], and erythrocytes [pig (237), human (238–240)]. Extensive amino acid sequence homology was found between nine tryptic peptides of pig heart lipoamide dehydrogenase and that of human erythrocyte GSSG reductase. The homology apparently extends throughout the molecules, which have both mechanistic and structural properties in common. Both contain a cystine residue that undergoes reduction and oxidation during the catalytic cycle, and the sequences adjacent to the active cystine residue are highly homologous (241). The three-dimensional structure of human erythrocyte glutathione reductase at 2-Å resolution has appeared (242). The amino acid sequence was fitted unambigiously to this map. This enzyme (M_r 104,800) has two identical subunits, each of which can be divided into four domains and a flexible segment of 18 amino acid residues at the *N*-terminus. The X-ray analysis indicates that the binding positions for NADPH and GSSG are at the opposite sides of one subunit. Transfer of reduction equivalents is mediated by the ring of FAD, which is located at the center of each subunit, and by the adjacent active cystine residue. Each subunit contributes to both of the sites at which GSSG is bound and reduced. Continuation of these studies should facilitate elucidation of the reaction mechanism and other aspects of this important intracellular catalyst.

CONJUGATION OF GLUTATHIONE

Exogenous Compounds; Glutathione S-Transferases

It was reported in 1879 that administration of bromobenzene or of chlorobenzene to dogs was followed by urinary excretion of compounds called mercapturic acids. Later work established that GSH reacts with a very large number and variety of foreign compounds to form GSH conjugates (3, 243–249). Compounds with an electrophilic center can readily conjugate with GSH; in some instances an electrophilic center is introduced by another reaction, for example, by action of microsomal oxygenase to form an epoxide that reacts with GSH. The interaction of foreign compounds with GSH may be spontaneous or catalyzed by GSH S-transferases. GSH conjugates are typically converted to mercapturic acids by a series of reactions initiated by γ-glutamyl transpeptidase in which the γ-glutamyl moiety of the conjugate is transferred to an acceptor; the resulting cysteinylglycine conjugate is converted by the action of dipeptidase to the corresponding cysteinyl conjugate which is N-acetylated to form an N-acetylcysteine conjugate (a mercapturic acid). The pathways of GSH conjugate metabolism and the mode of ultimate excretion vary with different compounds and the species; a voluminous literature is available (3, 247, 248).

GSH S-transferases occur in substantial quantities in liver and other mammalian tissues, e.g. erythrocytes (250), intestine (251–252); such enzymes also exist in plants and in insects. There are multiple forms of GSH S-transferase (253, 254), and there has been much interest in their purification and characterization, especially those of mammalian liver. The GSH S-transferases account for about 10% of the soluble protein of rat liver. They are identical with ligandins, first recognized as proteins that bind a variety of anionic compounds (255–258). It was suggested that ligandins function in the transport of these compounds from blood plasma into liver cells (255). Ligandins bind bilirubin, certain carcinogens, steroids, and azo dyes (259). The ligandin-GSH S-transferase family appears to have 3 detoxication functions: (a) catalysis, (b) binding of ligands which are not substrates; and (c) covalent bond formation with very reactive compounds leading to inactivation and destruction of the protein (259–264).

GSH-S-transferases, from liver, are dimeric enzymes that may contain four types of subunits, designated as Ya (M_r 22,000), Yb (M_r 23,500), Yb' (M_r 23,500) and Yc (M_r 25,000) (253, 254, 265, 266). These may combine to form six isozymes that were originally named A, B, C, D, E, and AA on the basis of chromatographic properties. Two groups of GSH S-transferases may be distinguished by reaction with either antibody to GSH S-transferase B (subunits Ya and Yc) or to GSH S-transferase C (subunits Yb and Yb'). For each group there are three possible combinations of the two subunits: YaYa, YaYc, YcYc or YbYb, YbYb', Yb'Yb'. Enzymes containing Ya or Yc exhibit high GSH peroxidase activity. A recent study of the subunit composition of rat liver GSH S-transferases, in which the poly (A) − RNA species for the S-transferase subunits were detected by a cDNA plasmid containing partial coding sequences for one of the rat liver subunits, has led to the conclusion that the minimum number of subunits for the liver isozymes is four (267).

The physiological functions of GSH S-transferases are undoubtedly diverse (see below); they may have a relationship to ethanol metabolism (268), heavy metals (269), reduction of peroxides (270, 187) and drug metabolism (249). These activities may be induced by drugs such as phenobarbital (271, 272).

Endogenous Compounds

Although most studies of GSH conjugation have dealt with detoxication of foreign compounds, there is increasing interest in conjugation reactions involving endogenous metabolites. For example, conversion of Δ^5-3-ketosteroids to the corresponding α,β-unsaturated Δ^4-3-ketosteroids is catalyzed by proteins of human and rat liver that appear to be identical with GSH S-transferases (273). Δ^5-3-Ketosteroid isomerase activity of human liver was found in several of the GSH S-transferases of this tissue. GSH S-transferases may have additional catalytic functions; there is evidence that form B of GSH S-transferase accounts for the nonselenium-containing GSH peroxidase activity. Estradiol-17-β also conjugates with GSH in rat liver preparations and in vivo (274–279). When the GSH conjugate of 2-hydroxy-estradiol-17-β was administered to rats, there was biliary excretion of the corresponding derivatives of cysteinylglycine, cysteine, and N-acetylcysteine. There are also reports on the formation of GSH conjugates of prostaglandins (280–282). GSH seems to function in several aspects of prostaglandin metabolism (283–285).

5-S-Glutathione-3,4-dihydroxyphenylalanine produced in melanocytes appears to be an intermediate in incorporation of cysteine sulfur into certain malanins (286). Interestingly, 5-S-cysteinyl-3,4-dihydroxy-phenylalanine is excreted in large amounts in the urine of some patients with widespread malignant melanoma (287).

Conjugation with GSH also functions in the metabolism of the leukotrienes (288, 289). In this pathway, leukotriene A, an epoxide derived from arachidonic acid, reacts with GSH to form the conjugate leukotriene

C. The γ-glutamyl moiety of this conjugate is removed by the action of γ-glutamyl transpeptidase, a reaction that proceeds more rapidly in the presence of amino acids (10), and leads to formation of leukotriene D, a slow reacting substance of anaphylaxis. Leukotriene D is further metabolized by the action of dipeptidase (10, 162) to the corresponding S-substituted cysteine, leukotriene E. A new leukotriene was recently discovered in studies in which γ-glutamyl transpeptidase was incubated with GSH and leukotriene E (10). The new compound, γ-glutamyl leukotriene E, was subsequently named leukotriene F (see review by S. Hammarström, this volume).

The apparent K_m of leukotriene C4 for the transpeptidase is about the same as that for GSH, which would probably compete in vivo with leukotriene C4 (10). The complexity of this system is further suggested by the possible conversion of leukotriene D4 by the transpeptidase, under physiological conditions, to the C4 derivative, to di- or poly-γ-glutamyl leukotriene derivatives, and to leukotriene F4. Notably, two enzymes of the γ-glutamyl cycle (transpeptidase and dipeptidase) are involved in the transformations of the leukotrienes. The enzyme that catalyzes formation of leukotriene C from the epoxide leukotriene A has not yet been identified.

OTHER ASPECTS OF GLUTATHIONE FUNCTION

Coenzyme Functions

The activities of many enzymes are influenced by GSH and by other thiols, and such effects may reflect significant physiological regulatory phenomena. There are several enzymatic reactions in which GSH participates as a coenzyme. The best known is the glyoxylase reaction in which the hemimercaptal formed by nonenzymatic interaction of methylglyoxal and GSH is converted to S-lactyl GSH by glyoxylase I, which is hydrolyzed (by glyoxylase II) to D-lactate and GSH. Many studies (e.g. 7, 290–297a) have been reported on this enzyme, whose physiological function needs study. The cis-trans isomerization of maleylacetoacetate to fumarylacetoacetate (298, 299), and the analogous reaction of maleylpyruvate (300, 301) also require GSH. The dehydrochlorination of 1,1,1-trichloro-2,2-bis(p-chlorophenyl)ethane (DDT) is catalyzed by a GSH-requiring enzyme found in houseflies; DDT apparently reacts with GSH to give an alkylated intermediate which breaks down to GSH, HCl, and the dehydrochlorinated product (302–305). The reactions catalyzed by prostaglandin endoperoxide D-isomerase and E-isomerase (284, 285, 306–309) also exhibit specific GSH involvement. Formaldehyde dehydrogenase catalyzes reversible conversion of formaldehyde, GSH, and NAD^+ to NADH, H^+, and S-formyl-GSH; the latter is cleaved by a hydrolase to formate and GSH.

Dehydrogenase and hydrolase activities have been found in many animal tissues and plants (310–320). The function of formaldehyde dehydrogenase apart from detoxication of formaldehyde is not yet clear.

Radiation and Oxyen Toxicity

Guzman-Barron (321) noted that thiols are easily oxidized, and postulated in 1944 that ionizing radiations would rapidly oxidize the thiol groups of cells. Later, it became apparent that radiation can produce a decrease in the cellular concentration of GSH and lead to formation of GSSG, and also that administration of various thiols can protect animals against the effects of radiation. Many observations show that the radiosensitivity of cells depends on the intracellular thiol level. The effects of radiation in air are generally greater than those found under hypoxic conditions, and the "oxygen effect", expressed as "O_2-enhancement ratio (OER)" is about three. The oxygen effect has been explained in terms of competition between thiol and O_2 for radicals produced by irradiation (322). Thus, ionizing radiation abstracts a hydrogen from a molecule to form a radical; thiols restore hydrogen to the radical, whereas reaction of radical with O_2 "fixes" damage by virtue of further chemical transformation. Although complex, the oxygen effect might be explained by the fact that in O_2, cells produce reactive oxygen compounds, which are normally destroyed by reaction with GSH. Under anoxic conditions, more GSH would be available to react with radiation-induced radicals. Thus, there would be less radiation damage in anoxia than in air. This predicts that the OER would decrease in the absence of GSH.

Recent relevant research includes studies on human lymphoid cells depleted of GSH by suspension in media containing buthionine sulfoximine (18), and experiments on fibroblasts from a patient (323, 324) with severe GSH synthetase deficiency (see p. 746). Lymphoid cell viability was markedly reduced by γ radiation of cells containing about 5% of the control level of GSH (18). The in vitro clonogenic survival of GSH-deficient fibroblasts and controls treated with X rays was studied under oxic and anoxic conditions (325); the OER was 1.5 (controls, 2.9). When irradiated in 95% air–5% CO_2 the GSH-deficient cells were slightly more radiosensitive than the controls. With 100% oxygen, there was a moderate effect on controls, but only a very low effect on the GSH-deficient cells. Misonidisole, a radiosensitizing agent, had a much greater effect on controls than on GSH-deficient cells; misonidisole seems to act, at least in part, by interacting with GSH (326; see also 327–331).

V79 fibroblasts depleted of GSH with buthionine sulfoximine gave OER values of 2.2; control 2.9 (332). These and other findings indicate the potential usefulness of buthionine sulfoximine as a drug. Since the rate of

GSH depletion with this agent depends on the rate of GSH turnover, it is important to monitor the GSH levels during such experiments and therapy.

The role of GSH in the prevention of oxygen toxicity is closely related to phenomena associated with radiation (333, 334). Increased oxygen tension leads to increased formation of H_2O_2, other reactive oxygen species, free radicals and GSSG. Rats fed a low protein diet show increased susceptibility (decreased survival time) to 98% oxygen; this effect was associated with deficiency of dietary sulfur amino acids (335). Rats on such a diet (in contrast to controls) did not develop increased levels of lung GSH in response to hyperoxia. The significant relationship between dietary cysteine and GSH synthesis has been examined (335a). The implications of these studies with regard to oxygen therapy are clear, and the observations also reflect a significant function of GSH in protection against reactive oxygen compounds.

Cancer

Potentially significant connections between GSH and carcinogenesis have attracted attention; see for example (336–343). Administration of certain carcinogens increases levels of GSH and of γ-glutamyl transpeptidase. Some tumors (e.g. skin, liver, colon) have unusually high levels of transpeptidase. It has been concluded that increased transpeptidase is a marker for very early, as well as late, putative liver preneoplastic and neoplastic cells (337). A recent study of transpeptidase in the developing mouse tooth, intervertebral disc, and hair follicle concluded that this enzyme is "a marker of cell differentiation, cell aging, and/or reduced cellular proliferation," and that it can be a marker of cell fetalization (338).

The reactive ultimate carcinogenic forms of chemical carcinogens are electrophiles, and thus good candidates for detoxication by reactions catalyzed by GSH S-transferases. A number of studies have been carried out on compounds that inhibit carcinogenesis by apparently increasing GSH S-transferases; elevations of other enzymes and of GSH levels have also been observed (344–351). For example, butylated hydroxyanisole, a food additive, has been found to protect against chemical carcinogenesis, and to increase the levels of liver microsomal epoxide hydratase, GSH S-transferase, and GSH.

A finding of great potential interest was reported by Novi (343), who treated rats bearing aflatoxin-induced liver tumors with large oral doses of GSH, and noted substantial regression of the tumors. Treatment was started three months after withdrawal of aflatoxin, at which time the tumors were presumably well established. Further investigations of this phenomenon will be of interest.

Calcium Metabolism

Orrenius and colleagues (352, 353) reported the formation of blebs in isolated hepatocyte membranes due to toxic compounds (e.g. bromobenzene) to be associated with GSH depletion and loss of extramitochondrial Ca^{2+}. Bleb formation is related to changes in Ca^{2+} level that affect the cytoskeleton and cell surface morphology. Evidence that the level of cellular thiols controls the Ca^{2+} pool suggests a significant link between GSH and Ca^{2+} homeostasis. This would be consistent with finding that Ca^{2+} efflux from perfused liver is stimulated by diamide (353a).

GLUTATHIONE DEFICIENCY AND DEPLETION

Effects of Inhibition of Glutathione Synthesis

Studies on the effects of GSH depletion should yield significant information about the physiological functions of this compound. The use of oxidants including hydroperoxides and Kosower's reagent, diamide (354), led to interesting findings. However, such compounds have certain disadvantages; their effects are short lived, are invariably associated with large increases in GSSG, and are nonspecific since these reagents oxidize other cellular components. Depletion of GSH has also been attempted with compounds that conjugate with GSH such as 1-chloro-2,4-dinitrobenzene (355) and diethylmaleate, which is converted to maleate in tissues (for review see 355a). These compounds, however, are also not specific for GSH; their effects may also be of short duration, since GSH synthesis increases rapidly after acute depletion.

An excellent method for depletion of intracellular GSH is administration of sulfoximine inhibitors of γ-glutamylcysteine synthetase (102, 103); buthionine sulfoximine is a potent and selective inhibitor of GSH synthesis that is highly effective in vitro and in vivo. When present in high concentrations this compound may also have other effects; it may be considered a γ-glutamyl amino acid analog, and may compete with γ-glutamyl amino acids for transport (47). Injection of mice with buthionine sulfoximine produces a rapid decrease in the GSH levels of the kidney, liver, plasma, pancreas, and muscle; other tissues were also affected to a lesser extent (15). When the sulfoximine was given orally for 15 days, the GSH levels of virtually all tissues were greatly decreased. In these studies, the rate at which tissue GSH levels decline after administration of the inhibitor provides a measure of GSH utilization, which is about equivalent to the export of GSH. As discussed above, the marked decrease in the plasma level of GSH is a reflection of its inter-organ transport. When buthionine sulfoximine was given to mice and rats, the level of GSH in liver and kidney declined rapidly to about 20% of the control, and further decrease occurred much more slowly; this may reflect existence of an intracellular pool of GSH, perhaps in the mitochondria, that turns over relatively slowly.

Studies of this type were also carried out on suspensions of human erythrocytes (356), macrophages grown in culture (20), and human lymphoid cells (17, 18, 45). Essentially exponential decline of the intracellular GSH levels was observed and substantial amounts of GSH were found extracellularly (18). Human lymphoid cells depleted of GSH exhibited increased sensitivity to irradiation. It was suggested earlier that GSH depletion by sulfoximine inhibitors renders tumor cells more susceptible to irradiation and to chemotherapeutic agents that are detoxified by reactions involving GSH (336). Tumor cells depleted of GSH by treatment with buthionine sulfoximine exhibit increased susceptibility to cytolysis by reactive oxygen intermediates (357). Treatment of mice infected with trypanosomes with buthionine sulfoximine led to prolonged survival of the animals (358); trypanosomes, which lack catalase and have high intracellular H_2O_2 levels, are evidently more susceptible to the effects of GSH depletion than are the host cells. Treatment of macrophages with buthionine sulfoximine led to decreased synthesis of leukotriene C and prostaglandin E_2 (359).

These studies have provided significant evidence for the proposed functions of GSH in destroying free radicals, and in the reduction of reactive oxygen intermediates. The findings are consistent with current thinking about the effects of radiation; they indicate that buthionine sulfoximine may be, as suggested (336), an important adjuvant in a wide range of chemotherapy and radiation therapy. Such an approach requires knowledge of the GSH status of the host cells and of the invading cell.

This approach may also be useful in elucidating additional functions of GSH, for example, the possibility that it is involved in thyroid hormone metabolism (360), melanin information (286, 287), the immune system (361), neurotransmission (362, 363), microfilament structure and function (364), and in other biological processes.

Mutant Microorganisms Deficient in Glutathione Synthesis

Mutants of *E. coli* K 12 deficient in the activities of γ-glutamylcysteine synthetase or of GSH synthetase were isolated by Apontoweil & Berends (365, 366). No GSH could be detected in two mutants; two others had 1% and 12% of the control level of GSH. The mutants, which grew normally, did not show increased sensitivity to X rays, but were more susceptible than their parent strains to a number of sulfhydryl reagents and antibiotics.

Fuchs & Warner (367) isolated a mutant of *E. coli* deficient in GSH synthetase that accumulates γ-glutamylcysteine at a level about equal to that of GSH in the parent line. The activity of the B1 subunit of ribonucleotide reductase in this mutant was also greatly reduced. This mutant grew at a normal rate, but was much more sensitive to diamide and methylglyoxal than the parent; it was more sensitive to X rays under anoxic conditions. Mutants deficient in GSSG reductase were also isolated (368, 369); they grew normally and contained normal amounts of GSH, indicating that this organism can reduce GSSG by another pathway, presumably involving thioredoxin. As noted above, GSH and the thioredoxin system can participate in alternate pathways of ribonucleotide reduction.

Saccharomyces cerevisiae, in contrast to *E. coli*, contains all of the γ-glutamyl cycle enzymes (370), and a mutant of this organism deficient in GSH synthetase exhibited decreased uptake of glycine (371). Further studies on this mutant, which might have some properties similar to those found in 5-oxoprolinuria (see below), would be of interest.

Human Diseases Involving Defects of Glutathione Synthesis and Metabolism

Several human diseases are associated with deficiencies of specific enzymes of GSH metabolism. Since this subject has recently been reviewed (372), only the salient features are considered. Severe GSH synthetase deficiency (also known as 5-oxoprolinuria and pyroglutamic aciduria) is characterized by massive urinary excretion of 5-oxoproline, increased blood and cerebrospinal fluid levels of 5-oxoproline, severe metabolic acidosis, tendency to hemolysis, and defective central nervous system function. Severe generalized deficiency of GSH synthetase leads to GSH deficiency. Since GSH regulates its biosynthesis by inhibiting γ-glutamylcysteine synthetase, when there is a marked reduction in the intracellular level of GSH, there is increased formation of γ-glutamylcysteine. This compound is converted to 5-oxoproline and cysteine by γ-glutamyl cyclotransferase, and the overproduction of 5-oxoproline exceeds the capacity of 5-oxoprolinase, so that a substantial amount (about (30%) of the 5-oxoproline formed is excreted. The biochemical and clinical aspects of this condition have been considered in detail (372–375). There are now 12 reported cases, most of whom have developed central nervous system damage even when maintained on constant bicarbonate therapy for chronic acidosis.

Several patients have been found with decreased erythrocyte GSH levels and well-compensated hemolytic disease associated with a milder form of GSH synthetase deficiency without substantial 5-oxoprolinuria. The genetic lesion appears to lead to synthesis of an unstable GSH synthetase molecule; the rate of replacement of this defective but active enzyme is sufficiently rapid so that most tissues can compensate for the defect, but such compensation is not possible in erythrocytes, which do not synthesize protein.

Patients with γ-glutamylcysteine synthetase deficiency experience hemolytic anemia, spinocerebellar degeneration, peripheral neuropathy, myopathy, and aminoaciduria. Patients with apparently generalized γ-glutamyl transpeptidase deficiency exhibit marked glutathionemia, urinary excretion of GSH, γ-glutamylcysteine and cysteine moieties, and may have defective renal amino acid transport.

5-Oxoprolinase deficiency has recently been reported in three individuals who present somewhat different clinical problems; they do not have obvious abnormalities of amino acid transport nor do they suffer acidosis probably because the accumulation of 5-oxoproline is not large. They excrete 29–71 mmoles of 5-oxoproline per day in their urine, substantially less than found in patients with severe GSH synthetase deficiency. Cultured skin fibroblasts of these patients contain about 2% of the control level of 5-oxoprolinase (375a).

5-Oxoprolinuria has also been observed in homocystinuria (375), in which 5-oxoproline formation appears to involve accumulation of homocysteine, a substrate of γ-glutamylcysteine synthetase (376). γ-Glutamylhomocysteine, a substrate of γ-glutamyl cyclotransferase, is converted to homocysteine and 5-oxoproline (375). Some patients with homocystinuria do not excrete 5-oxoproline; 5-oxoproline formation seems to be associated with a plasma level of homocysteine that is at least about 0.2 mM. Transient moderate 5-oxoprolinuria has been observed in a patient with an inherited defect of tyrosine metabolism in which a cysteine derivative [(2-L-cysteine-S-yl-1,4-dihydroxycyclohex-5-en-1-yl)-acetic acid] named hawkinsin appears in the urine; the mechanism by which 5-oxoproline is produced is not yet clear (377). 5-Oxoprolinuria may also arise from diets high in this compound (378), in end-stage renal disease (379), and in severe burns or allergic disease (379). The depletion of GSH that occurs in hereditary tyrosinemia (380, 381) may be related to accumulation of maleyl- or fumarylacetoacetate (whose cleavage is genetically blocked); these compounds may be produced in sufficient quantity to interact with and deplete GSH. Such phenomena may be associated with increased incidence of cancer in this condition. Therapy with thiol compounds, including GSH, has been tried (381).

Some patients with moderate degrees of erythrocyte GSH peroxidase deficiency have increased hemolysis, but the mechanisms involved require further study (382). Moderate erythrocyte GSSG reductase deficiency, correctable by administration of riboflavin, is not generally associated with symptoms (383). Severe GSSG reductase activity, unaffected by administration of riboflavin or by addition of FAD to the enzyme assays, has been observed in three siblings whose symptoms include hemolysis and cataracts; their parents exhibit intermediate levels of reductase (384).

ANALYTICAL PROCEDURES

Progress in almost any area of biochemistry depends importantly on reliable analytical methods. That there are many methods for the determination of glutathione suggests that such analyses are indeed not always satisfactory. In early work, total nonprotein thiols were determined, e.g. by iodometric titration or by reaction with 5,5′-dithiobis-(2-nitrobenzoic acid) (DTNB). o-Phthaldehyde (OPA) has been used (385, 386), but it reacts with many primary amines (387). It was proposed that the fluorophor formed on reaction of GSSG with OPA in alkali could serve as the basis of a method in which N-ethylmaleimide (NEM) was used to alkylate the GSH present. However, the procedure gives erroneously high values of GSSG (388).

Enzymatic determination of GSH has been accomplished using glyoxylase (389), and by a method using both glyoxylase and GSSG reductase (390) in which GSH is determined from A_{240nm} (lactoyl-GSH), and a second assay with the reductase (based on NADPH oxidation) measures GSSG. The requirement for GSH by maleylpyruvate isomerase (391) and by formaldehyde dehydrogenase (392) has been used in spectrophotometric assays. Methods using GSH S-transferase are also available (393, 394). In one, the sample is incubated with GSH S-transferase and o-dinitrobenzene and the nitrite released on conjugation is determined with N-(1-naphthyl)ethylene diamine (394).

In addition to methods involving stoichiometric reactions, there are recycling assays which offer higher sensitivity. Thus, the rate of formation of 2-nitro-5-thiobenzoic acid is measured in a system containing DTNB, GSSG reductase, phosphate buffer, EDTA, NADPH, and sample (35, 37, 395, 397). The GSH + GSSG present is determined by comparison of the result with an appropriate standard curve. It is of great importance to determine the standard curve under the same conditions, for example, in the presence of the protein precipitant used in preparation of the sample; other important details have been described (35, 36). The procedure can be made specific for GSSG by masking GSH with NEM (37, 398–400) or 2-vinylpyridine (35); NEM inhibits GSSG reductase, so it must be removed before assay.

Many column chromatographic procedures have been described (16, 396, 401–405); most are lengthy. HPLC is more rapid. Electrochemical detection has been used (406, 407); the electrodes thus far used are sensitive to other reducing substances (406). Precolumn derivatization followed by HPLC has been used with fluorescent reagents such as N-(9-

acridinyl)maleimide, but difficulties arise, owing to fluorescence of these reagents and their products (408, 409). Precolumn derivatization with DTNB has been used with HPLC (410), however, detection at 280 nm is troublesome because other compounds absorb. An HPLC method in which thiols are converted to S-carboxymethyl derivatives followed by reaction with 1-fluoro-2,4-dinitrobenzene to yield 2,4-dinitrophenyl derivatives has been described (411). Several postcolumn HPLC methods are available (412–414); one is based on reaction of GSH with N-chloro-5-dimethylaminonaphthalene-1-sulfonamide (412). In others, OPA is used (413, 414). A sensitive and selective precolumn derivatization procedure has been developed by Fahey et al (415) using Kosower's monobromobimane reagent (416). Experience with this method in the authors' laboratory has been satisfactory.

Several investigators have reported data suggesting occurrence in tissues of mixed disulfides between proteins and GSH or other thiols (e.g. 417–424). In evaluating these findings, it is important to consider such factors as freshness of the sample (e.g. 418), and the procedures used for homogenization and deproteinization. Samples must be worked up promptly. Rat blood plasma was found to contain 19–23 μM GSH and 1.5–2 μM GSSG in samples worked up within 3 min. Samples examined 7 min. later contained about 11 μM GSH and 3 μM GSSG; under these conditions, some of the GSH that disappeared was found in the form of mixed difulfides (36). Homogenization at pH values higher than about 2 (e.g. 420, 421) is undesirable because of the likelihood of thiol-disulfide interchange and thiol oxidation. The nature of the protein precipitant is of great importance. Perchloric acid, trichloroacetic acid, and metaphosphoric acid are generally unsuitable because GSH:GSSG ratios are not maintained. In the authors' experience, sulfosalicylic acid and picric acid are more satisfactory. Although acid homogenization decreases its rate, disulfide interchange may still occur. The possible effects of anesthesia also need to be considered (423, 424). Treatment of tissue extracts with metal hydrides and other reducing agents may liberate thiols from thiol ester linkage. The various manipulations used may also liberate noncovalently bound GSH and GSSG from proteins, e.g. GSH S-transferases. That the extent of artifactitious GSH binding can be substantial is indicated by rapid binding of added [^{35}S]GSH to plasma proteins (36). When kidney homogenates were treated with the internal standard norleucine, about 20% of the norleucine (and therefore a similar fraction of the free GSH and GSSG present in the homogenate) remained in the pellet. No evidence for significant amounts of protein-GSH mixed disulfides was found (O. W. Griffith, unpublished). Several enzymes do form mixed disulfides with GSH and other thiols, but the significance of such findings is not yet clear (e.g. 425); they may represent artifacts. We believe that GSH probably does form mixed disulfides with certain proteins under physiological conditions. Tissues may also contain thiol esters of GSH and may bind GSH and GSSG noncovalently. However, the published data on forms designated as mixed disulfides between proteins and low-molecular-weight thiols including GSH are difficult to interpret and should be given serious critical attention. Studies suggesting that mixed disulfides between proteins and low-molecular-weight thiols are increased after treatment with certain drugs, or after oxidative stress, while potentially interesting, require refinement.

CONCLUDING REMARKS

The development of selective enzyme inhibitors (e.g. the sulfoximine inhibitors of γ-glutamylcysteine synthetase), and their effective use in vivo has facilitated biochemical dissection of this complex system. Apart from their usefulness in research, these inhibitors and compounds that increase GSH synthesis have promise in therapy (425a). As one example, approaches involving use of toxic agents together with L-2-oxothiazolidine-4-carboxylate might be considered in situations in which the cell to be killed and the cell to be spared exhibit appropriate quantitative differences in GSH S-transferase and 5-oxoprolinase activities (154).

Despite much progress in defining the biochemistry that underlies the functions of GSH, many important questions remain. We need to learn more about the mechanism of GSH transport, the factors that regulate it and its significance in cell membrane function. Further work is required on the metabolic interactions between amino acids and GSH, and on the formation, transport, and utilization of γ-glutamyl amino acids. The voluminous literature on γ-glutamyl transpeptidase does not contain answers to many questions about its structure, membranous orientation, and induction in neoplastic cells.

There is good evidence that GSH has the important function of destroying reactive oxygen intermediates and free radicals that are constantly formed in metabolism, and in larger amounts after administration of certain drugs, oxygen, and X rays. However, many aspects of the interconversion of GSH and GSSG require study, e.g. the significance of extracellular GSH oxidation by O_2, and the nature and number of GSH transhydrogenases. Although GSH is involved in many intracellular reductive processes, there is good evidence that the thioredoxin system is also important in such reactions. Both GSH and the thioredoxin system may perform some of the same functions. Similarly, the existence of an oxidizing system that promotes disulfide formation requires further attention. The occurrence of mixed disulfides between GSH (and other thiols) and proteins needs careful investigation.

The several GSH S-transferases require further sorting out; more data are needed about the functions of these widely distributed enzymes in normal metabolism. Explorations of the biochemical functions of GSH in protein synthesis and degradation are needed, as are explorations of its probable functions in other processes such as the immune system and neurotransmission. Further biochemical study of the GSH-induced feeding response of hydra (426–428) might be productive. The metabolic functions of several of the enzymes that use GSH as a coenzyme, including glyoxylase and formaldehyde dehydrogenase need more study. The decrease in tissue GSH levels with age (429–430) is another area of potentially fruitful study.

Much less is known about GSH metabolism in bacteria and plants than in mammalian systems; certain phases of GSH metabolism in microorganisms have been reviewed (431, 431a). Observations on human diseases associated with GSH deficiency indicate that GSH is essential for normal health. However, a number of bacteria, including certain anaerobes, do not contain GSH (432), and at least one organism (*E. coli*) that normally contains large amounts of GSH does not seem to require it for growth. Interestingly, both microorganisms and plants (433, 434) seem to export GSH, at least in the few instances thus far studied. *E. coli* accumulates the novel compound glutathionyl-spermidine (435), and homoglutathione (γ-glutamyl-cysteinyl-β-alanine) has been found in mung beans (436). Thioredoxin, and possibly also GSH, function in photosynthesis (437). Plants contain GSH S-transferases (438). Clearly much potentially important research on GSH remains to be done.

Literature Cited

1. Meister, A., Tate, S. S. 1976. *Ann. Rev. Biochem.* 45:559–604
2. Meister, A. 1981. *Curr. Top. Cell. Regul.* 18:21–57
3. Arias, I. M., Jakoby, W. B., eds. 1976. *Glutathione: Metabolism and Function*, Kroc Found. Ser. 6:1–382. New York: Raven
4. Flohe, L., Benohr, H. Ch., Sies, H., Waller, H. D., Wendel, A., eds. 1974. *Glutathione, Proc. 16th Conf. German Soc. Biol. Chem. Tubingen, March 1973*, pp. 1–316. Stuttgart: Thieme
5. Sies, H., Wendel, A., eds. 1978. *Functions of Glutathione in Liver and Kidney, Proc. Life Sci.*, pp. 1–212. Berlin/Heidelberg/New York: Springer
6. Larsson, A., Orrenius, S., Holmgren, A., Mannennik, B., eds. 1983. *Functions of Glutathione* (Nobel Conf.). New York: Raven.
7. Meister, A. 1975. *Metabolism of Sulfur Compounds, Metabolic Pathways*, ed. D. M. Greenberg, 7:101–88. New York: Academic. 3rd ed.
8. Kosower, N. S., Kosower, E. M. 1978. *Int. Rev. Cytol.* 54:109–59
9. Griffith, O. W., Bridges, R. J., Meister, A. 1981. *Proc. Natl. Acad. Sci. USA* 78:2777–81
10. Anderson, M. E., Allison, R. D., Meister, A. 1982. *Proc. Natl. Acad. Sci. USA* 79:1088–91
11. Griffith, O. W., Bridges, R. J., Meister, A. 1978. *Proc. Natl. Acad. Sci. USA* 75:5405–8
12. Schulman, J. D., Goodman, S. I., Mace, J. W., Patrick, A. D., Tietze, F., Butler,

E. J. 1975. *Biochem. Biophys. Res. Commun.* 65:68–74
13. Meister, A. 1978. See Ref. 5, pp. 43–59
14. Griffith, O. W., Meister, A. 1979. *Proc. Natl. Acad. Sci. USA* 76:268–72
15. Griffith, O. W., Meister, A. 1979. *Proc. Natl. Acad. Sci. USA* 76:5606–10
15a. Haberle, D., Wahllander, A., Seis, H. 1979. *FEBS Lett.* 108:335–40
16. Griffith, O. W., Meister, A. 1980. *Proc. Natl. Acad. Sci. USA* 77:3384–87
17. Griffith, O. W., Novogrodsky, A., Meister, A. 1979. *Proc. Natl. Acad. Sci. USA* 76:2249–52
18. Dethmers, J. K., Meister, A. 1981. *Proc. Natl. Acad. Sci. USA* 78:7492–96
19. Bannai, S., Tsukedo, H. 1979. *J. Biol. Chem.* 254:3444–50
20. Rouzer, C. A., Scott, W. A., Griffith, O. W., Hamill, A. L., Cohn, Z. A. 1982. *J. Biol. Chem.* 257:2002–8
21. Bartoli, G. M., Sies, H. 1978. *FEBS Lett.* 86:89–91
22. Anderson, M. E., Bridges, R. J., Meister, A. 1980. *Biochem. Biophys. Res. Commun.* 96:848–53
23. Bartoli, G. M., Haberle, D., Sies, H. 1978. See Ref. 5, pp. 27–31
24. Spater, H. W., Poruchynsky, M. S., Quintana, N., Inoue, M., Novikoff, A. B. 1982. *Proc. Natl. Acad. Sci. USA* 79:3547–50
25. Griffith, O. W. 1981. *J. Biol. Chem.* 256:12,263–68
26. Meister, A. 1983. See Ref. 6, pp. 1–22
27. Rankin, B. B., Curthoys, N. P. 1982. *FEBS Lett.* 147:193–96
28. Bridges, R. J., Abbott, W. A., Meister, A. 1983. *Fed. Proc.* 42:2644
29. Refsvik, T., Norseth, T. 1975. *Acta Pharmacol. Toxicol.* 36:67–68
30. Sies, H., Koch, O. R., Martino, E., Boveris, A. 1979. *FEBS Lett.* 103:287–90
31. Eberle, D., Clarke, R., Kaplowitz, N. 1981. *J. Biol. Chem.* 256:2115–17
32. Abbott, W. A., Meister, A. 1982. *Fed. Proc.* 41:1430
33. Abbott, W. A., Meister, A. 1982. Unpublished
33a. Sies, H., Bartoli G. M., Burk, R. F., Waydhas, C. 1978. *Eur. J. Biochem.* 89:113–18
33b. Sies, H., Wahllander, A., Waydhas, C., Soboll, S., Haberle, D. 1980. *Adv. Enzyme Regul.* 18:303–20
33c. Akerboom, T. P. M., Bilzer, M., Sies, H. 1982. *FEBS Lett.* 140:73–76
34. Sies, H., Wahllander, A., Waydhas, C. 1978. See Ref. 5, pp. 120–26
34a. Wahllander, A., Sies, H. 1979. *Eur. J. Biochem.* 96:441–46
35. Griffith, O. W. 1980. *Anal. Biochem.*

106:207–12
36. Anderson, M. E., Meister, A. 1980. *J. Biol. Chem.* 255:9530–33
37. Tietze, F. 1969. *Anal. Biochem.* 27:502–22
38. Sies, H., Gerstenecker, C., Summer, K. H., Menzel, H., Flohe, L. 1974. See Ref. 4, pp. 261–76, and 1972. *FEBS Lett.* 27:171–75
39. Beutler, E. 1983. See Ref. 6, pp. 65–74
40. Kondo, T., Dale, G. L., Beutler, E. 1981. *Biochim. Biophys. Acta* 645:132–36
41. Prchal, J., Srivastava, S. K., Beutler, E. 1975. *Blood* 46:111–17
42. Lunn, G., Dale, G. L., Beutler, E. 1979. *Blood* 54:238–44
43. Srivastava, S. K., Beutler, E. 1968. *Proc. Soc. Exp. Biol. Med.* 127:512–14
44. Akerboom, T. P. M., Bilzer, M., Sies, H. 1982. *J. Biol. Chem.* 257:4248–52
45. Jensen, G., Meister, A. 1982. Unpublished
46. Meister, A., Griffith, O. W., Novogrodsky, A., Tate, S. S. 1980. *CIBA Found. Symp.* 72:135–61. Elsevier, North-Holland: Excerpta Medica
47. Griffith, O. W., Bridges, R. J., Meister, A. 1979. *Proc. Natl. Acad. Sci. USA* 76:6319–22
48. Bridges, R. J., Meister, A. 1982. Unpublished
49. Thompson, G. A., Meister, A. 1975. *Proc. Natl. Acad. Sci. USA* 72:1985–88
50. Thompson, G. A., Meister, A. 1976. *Biochem. Biophys. Res. Commun.* 71:32–36
51. Thompson, G. A., Meister, A. 1977. *J. Biol. Chem.* 252:6792–97
52. Bridges, R. J., Meister, A. 1983. Unpublished
53. Richman, P., Meister, A. 1975. *J. Biol. Chem.* 250:1422–26
54. Anderson, M. E., Meister, A. 1982. *Fed. Proc.* 41:1168 (Abstr.)
55. Anderson, M. E., Meister, A. 1983. *Proc. Natl. Acad. Sci. USA* 80:701–11
56. Christensen, H. N. 1979. *Adv. Enzymol.* 49:41–101
57. Robins, R. J., Davies, D. D. 1981. *Biochem. J.* 194:63–70
58. McIntyre, T. M., Curthoys, N. P. 1979. *J. Biol. Chem.* 254:6499–504
59. Elce, J. S., Broxmeyer, B. 1976. *Biochem. J.* 153:223–32
60. Allison, R. D., Meister, A. 1981. *J. Biol. Chem.* 256:2988–92
61. Meister, A., Tate, S. S., Ross, L. L. 1976. In *Membrane-Bound Enzymes*, ed. A. Martinosi, 3:315–47. New York: Plenum
62. Marathe, G. V., Nash, B., Haschemeyer,

R. H., Tate, S. S. 1979. *FEBS Lett.* 107:436–40

63. Heinle, H., Wendel, A., Schmidt, U. 1977. *FEBS Lett.* 73:220–24
64. Hagenfeldt, L., Arvidsson, A., Larsson, A. 1978. *Clin. Chim. Acta* 85:167–73
65. Buchanan, D. L., Haley, E. E., Markiw, R. T. 1962. *Biochemistry* 1:612–20
66. Peck, H., Pollitt, R. J. 1979. *Clin. Chem. Acta* 94:237–40
67. Kakimoto, Y., Nakajima, T., Kanazawa, A., Takesada, M., Sano, I. 1964. *Biochim. Biophys. Acta* 93:333–38
68. Reichelt, K. L. 1970. *J. Neurochem.* 17:19–25
69. Versteeg, D. H. G., Witter, A. 1970. *J. Neurochem.* 17:41–52
70. Kanazawa, A., Kakimoto, Y., Nakajima, T., Sano, I. 1965. *Biochim. Biophys. Acta* 111:90–95
71. Sano, I. 1970. *Int. Rev. Neurobiol.* 12:235–63
72. Wendel, A., Hahn, R., Guder, W. G. 1976. *Curr. Probl. Clin. Biochem.* 6:426–36
73. Richards, F. II, Cooper, M. R., Pearce, L. A., Cowan, R. J., Spurr, C. L. 1974. *Arch. Intern. Med.* 134:534–37
74. Palekar, A. G., Tate, S. S., Meister, A. 1975. *Biochem. Biophys. Res. Commun.* 62:651–57
75. Tate, S. S., Meister, A. 1978. *Proc. Natl. Acad. Sci. USA* 75:4806–9
76. Ormstad, K., Jones, D. P., Orrenius, S. 1980. *J. Biol. Chem.* 255:175–81
77. Ormstad, K., Lastbom, T., Orrenius, S. 1980. *FEBS Lett.* 112:55–59
78. Lauterburg, B. H., Mitchell, J. R. 1981. *J. Clin. Invest.* 67:1415–24
79. Garvey, T. Q. III, Hyman, P. E., Isselbacher, K. J. 1976. *Gastroenterology* 71:778–85
80. Cornell, J. S., Meister, A. 1976. *Proc. Natl. Acad. Sci. USA* 73:420–22
81. Puente, J., Castellon, E., Sapag-Hager, M. 1982. *Experientia* 38:531–32
82. Vina, J. R., Puertes, I. R., Vina, J. 1981. *Biochem. J.* 200:705–8
83. Vina, J., Puertes, I. R., Estrela, J. M., Vina, J. R., Galbis, J. L. 1981. *Biochem. J.* 194:99–102
84. Vina, J., Vina, J. R. 1983. See Ref. 6, pp. 23–30
85. Perantoni, A., Berman, J. J., Rice, J. M. 1979. *Exp. Cell Res.* 122:55–61
86. Sikka, S. C., Kalra, V. K. 1980. *J. Biol. Chem.* 255:4399–402
87. Kalra, V. K., Sikka, S. C., Sethi, G. S. 1981. *J. Biol Chem.* 256:5567–71
88. Sikka, S. C., Green, G. A., Chauhan, V. P. S., Kalra, V. K. 1982. *Biochemistry* 21:2356–66

89. Meister, A. 1974. *Enzymes* 10:671–97
90. Kumagai, H., Nakayama, R., Tochikura, T. 1982. *Agric. Biol. Chem.* 46:1301–9
91. Sekura, R., Meister, A. 1977. *J. Biol. Chem.* 252:2599–605
92. Wirth, P. J., Thorgeirsson, S. S. 1978. *Cancer Res.* 38:2861–65
93. Chung, A.-S., Maines, M. D. 1981. *Biochem. Pharmacol.* 30:3217–23
94. Board, P. G., Smith, J. E., Moore, K., Ou, D. 1980. *Biochim. Biophys. Acta* 613:534–41
95. Young, J. D., Tucker, E. M. 1983. See Ref. 6, pp. 371–82
96. Sekura, R., Hochreiter, M., Meister, A. 1976. *J. Biol. Chem.* 251:2263–70
97. Orlowski, M., Meister, A. 1971. *J. Biol. Chem.* 246:7095–7105
98. Yip, B., Rudolph, F. B. 1976. *J. Biol. Chem.* 251:3563–68
99. Richman, P. G., Orlowski, M., Meister, A. 1973. *J. Biol. Chem.* 248:6684–90
100. Meister, A. 1978. In *Enzyme-Activated Irreversible Inhibitors*, ed. N. Seiler, M. J. Jung, J. Koch-Weser, pp. 187–211. Amsterdam: Elsevier North-Holland Biomedical
101. Griffith, O. W., Meister, A. 1978. *J. Biol. Chem.* 253:2333–38
102. Griffith, O. W., Anderson, M. E., Meister, A. 1979. *J. Biol. Chem.* 254:1205–10
103. Griffith, O. W., Meister, A. 1979. *J. Biol. Chem.* 254:7558–60
104. Griffith, O. W. 1982. *J. Biol. Chem.* 257:13704–12
105. Lebo, R. V., Kredich, N. M. 1978. *J. Biol. Chem.* 253:2615–23
106. Griffith, O. W., Larsson, A., Meister, A. 1977. *Biochem. Biophys. Res. Commun.* 79:919–25
107. Beamer, R. L., Griffith, O. W., Gass, J. D., Anderson, M. E., Meister, A. 1980. *J. Biol. Chem.* 255:11732–36
108. Sekura, R., Meister, A. 1977. *J. Biol. Chem.* 252:2606–10
109. Seelig, G. F., Meister, A. 1982. *J. Biol. Chem.* 257:5092–96
110. Seelig, G. F., Meister, A. 1983. *J. Biol. Chem.* 258: in press
111. Seelig, G. F., Meister, A. 1983. *Fed. Proc.* 42:1688
112. Oppenheimer, L., Wellner, V. P., Griffith, O. W., Meister, A. 1979. *J. Biol. Chem.* 254:5184–90
113. Meister, A., Tate, S. S., Griffith, O. W. 1981. *Methods Enzymol.* 77:237–53
114. Tate, S. S., Meister, A. 1981. *Mol. Cell. Biochem.* 39:357–68
115. Marathe, G. V., Damle, N. S., Haschemeyer, R. H., Tate, S. S. 1980. *FEBS Lett.* 115:273–77

116. Goldberg, D. M. 1980. *CRC Crit. Rev. Clin. Lab. Sci.* 12:1–58
117. Rosalki, S. B. 1975. *Adv. Clin. Chem.* 17:53–107
118. Orlowski, M., Meister, A. 1963. *Biochim. Biophys. Acta* 73:679–81
119. Tate, S. S. 1980. In *Enzymatic Basis of Detoxication*, ed. W. B. Jakoby. 2:315–47. New York: Academic
120. Smith, G. D., Ding, J. L., Peters, T. J. 1979. *Anal. Biochem.* 100:136–39
121. Thompson, G. A., Meister, A. 1980. *J. Biol. Chem.* 255:2109–13
122. Gardell, S. J., Tate, S. S. 1983. *J. Biol. Chem.* 258: in press
123. Abbott, W., Meister, A. 1983. *J. Biol. Chem.* 258: in press
124. Butler, J. De B., Spielberg, S. P. 1979. *J. Biol. Chem.* 254:3152–55
125. Hughey, R. P., Coyle, P. J., Curthoys, N. P. 1979. *J. Biol. Chem.* 254:1124–28
126. Tsuji, A., Matsuda, Y., Katunuma, N. 1980. *J. Biochem. Tokyo* 87:1567–71
127. Gardell, S. J., Tate, S. S. 1979. *J. Biol. Chem.* 254:4942–45
128. Gardell, S. J., Tate, S. S. 1981. *J. Biol. Chem.* 256:4799–804
129. Nash, B., Tate, S. S. 1982. *J. Biol. Chem.* 257:585–88
130. Nash, B., Tate, S. S. 1983. *Fed. Proc.* 42:20115
131. Tate, S. S., Meister, A. 1974. *J. Biol. Chem.* 249:7593–602
132. Minato, S. 1979. *Arch. Biochem. Biophys.* 192:235–40
133. Allen, L., Meck, R., Yunis, A. 1980. *Res. Commun. Chem. Pathol. Pharmacol.* 27:175–82
133a. Reed, D. J., Ellis, W. W., Meck, R. A. 1980. *Biochem. Biophys. Res. Commun.* 94:1273–77
134. Tate, S. S., Meister, A. 1977. *Proc. Natl. Acad. Sci. USA* 74:931–35
135. Anderson, M. E., Meister, A. 1982. *Fed. Proc.* 41:5246
136. Orlowski, M., Richman, P. G., Meister, A. 1969. *Biochemistry* 8:1048–55
137. Adamson, E. D., Szewczuk, A., Connell, G. E. 1971. *Can. J. Biochem.* 49:218–26
138. Orlowski, M., Meister, A. 1973. *J. Biol. Chem.* 248:2836–44
139. Board, P. G., Moore, K. A., Smith, J. E. 1978. *Biochem. J.* 173:427–31
140. Taniguchi, N., Meister, A. 1978. *J. Biol. Chem.* 253:1799–806
141. Van Der Werf, P., Meister, A. 1975. *Adv. Enzymol.* 43:519–56
142. Mazelis, M., Creveling, R. K. 1978. *Plant Physiol.* 62:798–801
143. Van Der Werf, P., Meister, A. 1974. *Biochem. Biophys. Res. Commun.* 56:90–96
144. Fink, M. L., Folk, J. E. 1981. *Mol. Cell. Biochem.* 38:59–67
145. Reddy, V. N., Chakrapani, B., Rathbun, W. B., Hough, M. M. 1975. *Invest. Ophthalmol.* 14:228–32
146. Williamson, J. M., Meister, A. 1982. *J. Biol. Chem.* 257:9161–72
147. Wendel, A., Flugge, U.-I., Flohe, L. 1976. See Ref. 3, pp. 71–75
148. Griffith, O. W., Meister, A. 1976. *Biochem. Biophys. Res. Commun.* 70:759–65
149. Griffith, O. W., Meister, A. 1982. *J. Biol. Chem.* 257:4392–97
150. Van Der Werf, P., Griffith, O. W., Meister, A. 1975. *J. Biol. Chem.* 250:6686–92
151. Griffith, O. W., Meister, A. 1981. *J. Biol. Chem.* 256:9981–85
152. Williamson, J. M., Meister, A. 1982. *J. Biol. Chem.* 257:12039–42
153. Williamson, J. M., Meister, A. 1981. *Proc. Natl. Acad. Sci. USA* 78:936–39
154. Williamson, J. M., Boettcher, B., Meister, A. 1982. *Proc. Natl. Acad. Sci. USA* 79:6246–49
155. Olney, J. W., Ho, O. L., Rhee, V. 1971. *Exp. Brain Res.* 14:61–76
156. Karlsen, R. L., Grofova, I., Malthe-Sorenssen, D., Fonnum, F. 1981. *Brain Res.* 208:167–80
157. Birnbaum, S. M., Winitz, M., Greenstein, J. P. 1957. *Arch. Biochem. Biophys.* 72:428–36
158. Nishiuch, Y., Sasaki, M., Nakayasu, M., Oikawa, A. 1976. *In Vitro* 12:635–38
159. Hughey, R. P., Rankin, B. B., Elce, J. S., Curthoys, N. P. 1978. *Arch. Biochem. Biophys.* 186:211–17
160. Rankin, B. B., McIntyre, T. M., Curthoys, N. P. 1980. *Biochem. Biophys. Res. Commun.* 96:991–96
161. Okajima, K., Inoue, M., Morino, Y. 1981. *Biochim. Biophys. Acta* 675:379–85
162. Kozak, E. M., Tate, S. S. 1982. *J. Biol. Chem.* 257:6322–27
163. Das, M., Radhakrishnan, A. N. 1973. *Biochem. J.* 135:609–15
164. Ziegenhagen, A. J., Ames, S. R., Elvehjem, C. A. 1947. *J. Biol. Chem.* 167:129–33
165. Moldeus, P., Jones, D. P., Ormstad, K., Orrenius, S. 1978. *Biochem. Biophys. Res. Commun.* 83:195–200
166. Jones, D. P., Moldeus, P., Stead, A. H., Ormstad, K., Jornvall, H., Orrenius, S. 1979. *J. Biol. Chem.* 254:2787–92
167. Tate, S. S., Grau, E. M., Meister, A. 1979. *Proc. Natl. Acad. Sci. USA* 76:2715–19
168. Tate, S. S., Orlando, J. 1979. *J. Biol. Chem.* 254:5573–75

169. Griffith, O. W., Tate, S. S. 1980. *J. Biol. Chem.* 255:5011–14
170. Ashkar, S., Binkley, F., Jones, D. P. 1981. *FEBS Lett.* 124:166–68
171. Ormstad, K., Lastbom, T., Orrenius, S. 1981. *FEBS Lett.* 130:239–43
172. Lash, L. H., Jones, D. P. 1982. *Biochem. J.* 203:371–76
173. Ormstad, K., Orrenius, S. 1980. *Biochem. Biophys. Res. Commun.* 92:540–45
174. Janolino, V. G., Swaisgood, H. E. 1975. *J. Biol. Chem.* 250:2532–38
175. Sliwkowski, M. B., Sliwkowski, M. X., Swaisgood, H. E., Horton, H. R. 1981. *Arch. Biochem. Biophys.* 211:731–37
176. Schmelzer, C. H., Swaisgood, H. E., Horton, H. R. 1982. *Biochem. Biophys. Res. Commun.* 107:196–201
177. Ostrowski, M. C., Kistler, W. S. 1980. *Biochemistry* 19:2639–45
178. Flohe, L., Gunzler, W. A. 1974. See Ref. 4, pp. 132–45
179. Chance, B., Boveris, A., Nakase, Y., Sies, H. 1978. See Ref. 5, pp. 95–106
180. Flohe, L. 1979. *CIBA Found. Symp.* 65:95–122. New York: Excerpta Medica
181. Chance, B., Sies, H., Boveris, A. 1979. *Physiol. Rev.* 59:527–605
182. Sies, H., Brigelius, R., Akerboom, T. P. M. 1983. See Ref. 6, pp. 51–64
183. Zakowski, J. J., Forstrom, J. W., Condell, R. A., Tappel, A. L. 1978. *Biochem. Biophys. Res. Commun.* 84:248–53
184. Forstrom, J. W., Tappel, A. L. 1979. *J. Biol. Chem.* 254:2888–91
185. Forstrom, J. W., Zakowski, J. J., Tappel, A. L. 1978. *Biochemistry* 17:2639–44
185a. Sies, H., Wendel, A., Burk, R. F. 1982. In *Oxidases and Related Redox Systems*, pp. 169–89. Oxford/New York: Pergamon
186. Lawrence, R. A., Burk, R. F. 1976. *Biochem. Biophys. Res. Commun.* 71:952–58
187. Burk, R. F., Nishiki, K., Lawrence, R. A., Chance, B. 1978. *J. Biol. Chem.* 253:43–46
188. Racker, E. 1955. *J. Biol. Chem.* 217:867–74
189. Nagai, S., Black, S. 1968. *J. Biol. Chem.* 243:1942–47
190. Mannervik, B., Eriksson, S. A. 1974. See Ref. 4, pp. 120–32
191. Freedman, R. B. 1979. *FEBS Lett.* 97:201–10
192. Anfinsen, C. B., Haber, E. 1961. *J. Biol. Chem.* 236:1361–63
193. DeLorenzo, F., Fuchs, S., Anfinsen, C. B. 1966. *Biochemistry* 5:3961–65
194. DeLorenzo, F., Goldberger, R. F., Steers, E., Givol, D., Anfinsen, C. B. 1966. *J. Biol. Chem.* 241:1562–67
195. Goldberger, R. F., Epstein, C. J., Anfinsen, C. B. 1963. *J. Biol. Chem.* 238:628–35
196. Venetianer, P., Straub, F. B. 1964. *Biochim. Biophys. Acta* 89:189–90
197. Freedman, R. B., Brockway, B. E., Forster, S. J., Lambert, N., Mills, E. N. C., Roden, L. T. 1983. See Ref. 6, pp. 271–82
198. Morin, J. E., Axelsson, K., Dixon, J. E. 1983. See Ref. 6, pp. 283–94
199. Katzen, H. M., Tietze, F., Stetten, D. W. 1963. *J. Biol. Chem.* 238:1006–11
200. Katzen, H. M., Tietze, F. 1966. *J. Biol. Chem.* 241:3561–70
201. Tomizawa, H. H. 1962. *J. Biol. Chem.* 237:3393–96
202. Tomizawa, H. H., Varandani, P. T. 1965. *J. Biol. Chem.* 240:3191–94
203. Narahara, H. T., Williams, R. H. 1959. *J. Biol. Chem.* 234:71–77
204. Varandani, P. T., Plumley, H. 1968. *Biochim. Biophys. Acta* 151:273–75
205. Varandani, P. 1973. *Biochim. Biophys. Acta* 320:249–57
206. Ansorge, S., Bohley, P., Kirschke, H., Langner, J., Marquardt, I., et al. 1973. *FEBS Lett.* 37:238–40
207. Ansorge, S., Bohley, P., Kirschke, H., Langner, J., Wiederanders, B., Hanson, H. 1973. *Eur. J. Biochem.* 32:27–35
208. Chandler, M. L., Varandani, P. T. 1975. *Biochemistry* 14:2107–15
209. Hawkins, H. C., Freedman, R. B. 1976. *Biochem. J.* 159:385–93
210. Ibbetson, A. L., Freedman, R. B. 1976. *Biochem. J.* 159:377–84
211. Carmichael, D. F., Keefe, M., Pace, M., Dixon, J. E. 1979. *J. Biol. Chem.* 254:8386–90
212. Varandani, P. T. 1973. *Biochim. Biophys. Acta* 295:630–36
213. Varandani, P. T., Nafz, M. A. 1970. *Int. J. Biochem.* 1:313–21
214. Varandani, P. T. 1974. *Biochim. Biophys. Acta* 371:577–81
215. Varandani, P. T., Nafz, M. A., Chandler, M. L. 1975. *Biochemistry* 14:2115–20
216. Varandani, P. T., Nafz, M. A. 1976. *Diabetes* 25:173–79
217. Roth, R. A., Koshland, M. E. 1981. *Biochemistry* 20:6594–99
218. Moss, J., Stanley, S. J., Morin, J. E., Dixon, J. E. 1980. *J. Biol. Chem.* 255:11085–87
219. Mannervik, B., Axelsson, K. 1980. *Biochem. J.* 190:125–30
220. Thelander, L., Reichard, P. 1979. *Ann. Rev. Biochem.* 48:133–58

221. Holmgren, A. 1981. *Trends Biochem. Sci.* 6:26–29
222. Ziegler, D. M., Poulsen, L. L. 1977. *Trends Biochem. Sci.* 2:79–81
223. Ziegler, D. M., Duffel, M. W., Poulsen, L. L. 1980. *CIBA Found. Symp.* 72:191–204. New York: Excerpta Medica
224. Ziegler, D. M., Poulsen, L. L., York, B. M. 1983. See Ref. 6, pp. 295–304
225. Hammarsten, E., Reichard, P., Saluste, E. 1950. *J. Biol. Chem.* 183:105–9
226. Holmgren, A. 1981. *Curr. Top. Cell. Regul.* 19:47–76
227. Holmgren, A. 1978. *J. Biol. Chem.* 253:7424–30
228. Holmgren, A. 1979. *J. Biol. Chem.* 254:3664–71, 3672–78
229. Hoog, J.-O., Holmgren, A., D'Silva, C., Douglas, K. T., Seddon, A. P. 1982. *FEBS Lett.* 138:59–61
230. Luthman, M., Eriksson, S., Holmgren, A., Thelander, L. 1979. *Proc. Natl. Acad. Sci. USA* 76:2158–62
231. Luthman, M., Holmgren, A. 1982. *J. Biol. Chem.* 257:6686–89
232. Carlberg, I., Mannervik, B. 1977. *Biochim. Biophys. Acta* 484:268–74
233. Carlberg, I., Mannervik, B. 1981. *Anal. Biochem.* 116:531–36
234. Lopez-Barea, J., Lee, C.-Y. 1979. *Eur. J. Biochem.* 98:487–99
235. Zanetti, G. 1979. *Arch. Biochem. Biophys.* 198:241–46
236. Carlberg, I., Mannervik, B. 1975. *J. Biol. Chem.* 250:5475–80
237. Boggaram, V., Larson, K., Mannervik, B. 1978. *Biochim. Biophys. Acta* 527:337–47
238. Worthington, D. J., Rosemeyer, M. A. 1974. *Eur. J. Biochem.* 48:167–77
239. Worthington, D. J., Rosemeyer, M. A. 1975. *Eur. J. Biochem.* 60:459–66
240. Worthington, D. J., Rosemeyer, M. A. 1976. *Eur. J. Biochem.* 67:231–38
241. Williams, C. H. Jr., Arscott, L. D., Schulz, G. E. 1982. *Proc. Natl. Acad. Sci. USA* 79:2199–2201
242. Thieme, R., Pai, E. F., Schirmer, R. H., Schulz, G. E. 1981. *J. Mol. Biol.* 152:763–82
243. Boyland, E., Chasseaud, L. F. 1969. *Adv. Enzymol.* 23:173–219
244. Wood, J. L. 1970. In *Metabolic Conjugation and Metabolic Hydrolysis*, ed. W. H. Fishman, 2:261–99. New York: Academic
245. Chasseaud, L. F. 1973. In *Symposium on Glutathione*, ed. L. Flohe, pp. 90–108. Stuttgart: Thieme
246. Chasseaud, L. F. 1973. *Drug. Metab. Rev.* 2:185–219
247. Jakoby, W. B., ed. 1980. *Enzymatic Basis of Detoxication*, Vol. 2. New York: Academic
248. Jakoby, W. B., ed. 1981. *Methods Enzymol.* 77:1–476
249. Chasseaud, L. F. 1979. *Adv. Cancer Res.* 29:175–274
250. Marcus, C. J., Habig, W. H., Jakoby, W. B. 1978. *Arch. Biochem. Biophys.* 188:287–93
251. Pinkus, L. M., Ketley, J. N., Jakoby, W. B. 1977. *Biochem. Pharmacol.* 26:2359–63
252. Clifton, G., Kaplowitz, N. 1977. *Cancer Res.* 37:788–91
253. Mannervik, B., Gutenberg, C., Jensson, H., Warholm, M. 1983. See Ref. 6, pp. 75–88
254. Habig, W. H., Pabst, M. J., Jakoby, W. B. 1974. *J. Biol. Chem.* 249:7130–39
255. Reyes, H., Levi, A. J., Gatmaitan, Z., Arias, I. M. 1971. *J. Clin. Invest* 50:2242–52
256. Levi, A. J., Gatmaitan, Z., Arias, I. M. 1969. *J. Clin. Invest.* 48:2156–67
257. Ketley, J. N., Habig, W. H., Jakoby, W. B. 1975. *J. Biol. Chem.* 250:8670–73
258. Habig, W. H., Pabst, M. J., Fleischner, G., Gatmaitan, Z., Arias, I. M., Jakoby, W. B. 1974. *Proc. Natl. Acad. Sci. USA* 71:3879–82
259. Vander Jagt, D. L., Wilson, S. P., Dean, V. L., Simons, P. C. 1982. *J. Biol. Chem.* 257:1997–2001
260. Keen, J. H., Jakoby, W. B. 1978. *J. Biol. Chem.* 253:5654–57
261. Jakoby, W. B. 1978. *Adv. Enzymol.* 46:383–414
262. Mannervik, B., Guthenberg, C., Jakobson, I., Warholm, M. 1978. In *Conjugation Reactions in Drug Biotransformation*, ed. A. Aitio, pp. 101–10. Amsterdam: Elsevier/North-Holland Biomedical
263. Ketterer, B., Tipping, E. 1978. In *Conjugation Reactions in Drug Biotransformation*, ed. A. Aitio, pp. 91–100. Amsterdam: Elsevier/North-Holland Biomedical
264. Smith, G. J., Ohl, V. S., Litwack, G. 1977. *Cancer Res.* 37:8–14
265. Mannervik, B., Jensson, H. 1982. *J. Biol. Chem.* 257:9909–12
266. Maruyama, H., Inoue, M., Arias, I. M., Listowski, I. 1983. See Ref. 6, pp. 89–98
267. Tu, C.-P. D., Weiss, M. J., Reddy, C. C. 1982. *Biochem. Biophys. Res. Commun.* 108:461–67
268. Hetu, C., Yelle, L., Joly, J.-G. 1982. *Drug Metab. Dispos.* 10:246–50
269. Dierickx, P. J. 1982. *Enzyme* 27:25–32
270. Saneto, R. P., Awasthi, Y. C., Srivastava, S. K. 1982. *Biochem. J.* 205:213–17

271. Reyes, H., Levi, A. J., Gatmaitan, Z., Arias, I. M. 1969. *Proc. Natl. Acad. Sci. USA* 64:168–70
272. Down, W. H., Chasseaud, L. E. 1979. *Biochem. Pharm.* 28:3525–28
273. Benson, A. M., Talalay, P., Keen, J. H., Jakoby, W. B. 1977. *Proc. Natl. Acad. Sci. USA* 74:158–62
274. Kuss, E. 1967. *Hoppe-Seylers Z. Physiol. Chem.* 348:1707–8
275. Kuss, E. 1968. *Hoppe-Seylers Z. Physiol. Chem.* 349:1234–36
276. Kuss, E. 1969. *Hoppe-Seyler Z. Physiol. Chem.* 350:95–97
277. Kuss, E. 1971. *Hoppe-Seylers Z. Physiol. Chem.* 352:817–36
278. Jellinck, P. H., Lewis, J., Boston, F. 1967. *Steroids* 10:329–46
279. Elce, J. S., Harris, J. 1971. *Steroids* 18:583–91
280. Cagen, L. M., Fales, H. M., Pisano, J. J. 1976. *J. Biol. Chem.* 251:6550–54
281. Cagen, L. M., Pisano, J. J. 1979. *Biochim. Biophys. Acta* 573:547–51
282. Chaudhari, A., Anderson, M. W. Eling, T. E. 1978. *Biochim. Biophys. Acta* 531:56–64
283. Nugteren, D. H., Hazelhof, E. 1973. *Biochim. Biophys. Acta* 326:448–61
284. Ogino, N., Miyamoto, T., Yamamoto, S., Hayaishi, O. 1977. *J. Biol. Chem.* 252:890–95
285. Raz, A., Kenig-Wakshal, R., Schwartzman, M. 1977. *Biochim. Biophys. Acta* 488:322–29
286. Prota, G. 1980. In *Natural Sulfur Compounds*, ed. D. Cavallini, G. E. Gaull, V. Zappia, pp. 391–98. New York: Plenum
287. Agrup, G., Agrup, P., Andersson, T., Falck, B., Hansson, J.-A., et al. 1975. *Acta Derm. Venereol.* 55:337–41
288. Orning, L., Hammarström, S., Samuelsson, B. 1980. *Proc. Natl. Acad. Sci. USA* 77:2014–17
289. Hammarström, S., Samuelsson, B., Clark, D. A., Goto, G., Marfat, A., et al. 1980. *Biochem. Biophys. Res. Commun.* 92:946–53
290. Knox, W. E. 1960. *Enzymes (Part A)*, 2:253–94
291. Marmstal, E., Mannervik, B. 1979. *Biochim. Biophys. Acta* 566:362–70
292. Han, L.-P. B., Schimandle, C. M., Davison, L. M., Vander Jagt, D. L. 1977. *Biochemistry* 16:5478–84
293. Lyon, P. A., Vince, R. 1977. *J. Med. Chem.* 20:77–80
294. Douglas, K. T., Nadvi, I. N. 1979. *FEBS Lett.* 106:393–96
295. Ball, J. C., Vander Jagt, D. L. 1979. *Anal. Biochem.* 98:472–77
296. Uotila, L., Koivusalo, M. 1980. *Acta

297. Sellin, S., Aronsson, A. C., Eriksson, L. E. G., Larsen, K., Tibbelin, G., Mannervik, B. 1983. See Ref. 6, pp. 185–96
297a. Kozarich, J. W., Chari, R. V. J. 1982. *J. Am. Chem. Soc.* 104:2655–57
298. Edwards, S. W., Knox, W. E. 1956. *J. Biol. Chem.* 220:79–91
299. Saltzer, S. 1973. *J. Biol. Chem.* 248:215–22
300. Lack, L. 1961. *J. Biol. Chem.* 236:2835–40
301. Seltzer, S., Lin, M. 1979. *J. Am. Chem. Soc.* 101:3091–97
302. Lipke, H., Kearns, C. W. 1959. *J. Biol. Chem.* 234:2123–28
303. Goodchild, B., Smith, J. N. 1970. *Biochem. J.* 117:1005–9
304. Balabaskaran, S., Smith, J. N. 1970. *Biochem. J.* 117:989–96
305. Dinamarca, M. L., Levenbrook, L., Valdes, E. 1971. *Arch. Biochem. Biophys.* 147:374–83
306. Nugteren, D. H., Hazelhof, E. 1973. *Biochim. Biophys. Acta* 326:448–61
307. Cottee, F., Flower, R. J., Moncada, S., Salmon, J. A., Vane, J. R. 1977. *Prostaglandins* 14:413–23
308. Saeed, S. A., Cuthbert, J. 1977. *Prostaglandins* 13:565–75
309. Duchesne, M. J., Thaler-Dao, H., de Paulet, A. C. 1978. *Prostaglandins* 15:19–41
310. Uotila, L., Koivusalo, M. 1983. See Ref. 6, pp. 173–84
311. Uotila, L., Koivusalo, M. 1974. *J. Biol. Chem.* 249:7653–63
312. Uotila, L., Koivusalo, M. 1974. *J. Biol. Chem.* 249:7664–72
313. Uotila, L. 1973. *Biochemistry* 12:3944–51
314. Uotila, L. 1973. *Biochemistry* 12:3938–43
315. Uotila, L., Mannervik, B. 1980. *Biochim. Biophys. Acta* 616:153–57
316. Uotila, L., Mannervik, B. 1979. *Biochem. J.* 177:869–78
317. Uotila, L., Koivusalo, M. 1981. *Methods Enzymol.* 77:314–20
318. Uotila, L., Koivusalo, M. 1981. *Methods Enzymol.* 77:320–25
319. Uotila, L. 1981. *Methods Enzymol.* 77:424–30
320. Uotila, L. 1979. *J. Biol. Chem.* 254:7024–29
321. Guzman-Barron, E. S. 1951. *Adv. Enzymol.* 11:201–66
322. Alexander, P., Charlesby, A. 1955. In *Radiobiology Symp., Liege, 1954*, pp. 49–60, ed. Z. M. Bacq, P. Alexander. London: Butterworth
323. Wellner, V. P., Sekura, R., Meister, A.,

Larsson, A. 1974. *Proc. Natl. Acad. Sci. USA* 71:2505–9

324. Hagenfeldt, L., Larsson, A., Zetterstrom, R. 1974. *Acta Paediatr. Scand.* 63:1–8

325. Deschavanne, P. J., Midander, J. Edgren, M., Larsson, A., Malaise, E. P., Revesz, L. 1981. *Biomedicine* 35:35–37

326. Midander, J., Deschavanne, P.-J., Malaise, E.-P., Revesz, L. 1982. *Int. J. Radiat. Oncol. Biol. Phys.* 8:443–46

327. Guichard, M., Malaise, E.-P. 1982. *Int. J. Radiat. Oncol. Biol. Phys.* 8:465–68

328. Deschavanne, P. J., Malaise, E.-P., Revesz, L. 1981. *Br. J. Radiol.* 54:361–62

329. Edgren, M., Larsson, A., Nilsson, K., Revesz, L., Scott, O. C. A. 1980. *Int. J. Radiat. Biol.* 37:299–306

330. Edgren, M., Revesz, L., Larsson, A. 1981. *Int. J. Radiat. Biol.* 40:355–63

331. Revesz, L., Edgren, M., Larsson, A. 1979. In *Radiation Research, Proc. 6th Int. Congr. Radiat. Res.*, *May 13–19, Tokyo, Japan*, ed. S. Okada, M. Imamura, T. Terashima, H. Yamaguchi, pp. 862–66. San Diego: Academic

332. Guichard, M., Jensen, G., Meister, A., Malaise, E. 1983. *Proc. Radiation Research Society*, p. 10

333. Rink, H. 1974. See Ref. 4, pp. 206–16

334. Kosower, N. S., Kosower, E. M. 1974. See Ref. 4, pp. 216–27

335. Deneke, S. M., Gershoff, S. N., Fanburg, B. L. 1983. *J. Appl. Physiol.* 54:47–51

335a. Tateishi, N., Higashi, T., Naruse, A., Hikita, K., Sakamoto, Y. 1981. *J. Biochem.* 90:1603–10

336. Meister, A., Griffith, O. W. 1979. *Cancer Treat. Rep.* 63:1115–21

337. Cameron, R., Kellen, J., Kolin, A., Malkin, A., Farber, E. 1978. *Cancer Res.* 38:823–29

338. Richards, W. L., Astrup, E. G. 1982. *Cancer Res.* 42:4143–52

339. Boyd, S. C., Sasame, H. A., Boyd, M. R. 1979. *Science* 205:1010–12

340. Huberman, E., Montesano, R., Drevon, C., Kuroki, T., St. Vincent, L., et al. 1979. *Cancer Res.* 39:269–72

341. Fiala, S., Trout, E. C. Jr., Teague, C. A., Fiala, E. S. 1980. *Cancer Detect. Preven.* 3:471–85

342. Fiala, S., Mohindru, A., Kettering, W. G., Fiala, A. E., Morris, H. P. 1976. *J. Natl. Cancer Inst.* 57:591–98

343. Novi, A. M. 1981. *Science* 212:541–42

344. Benson, A. M., Batzinger, R. P., Ou, S.-Y. L., Bueding, E., Cha, Y.-N., Talalay, P. 1978. *Cancer Res.* 38:4486–95

345. Benson, A. M., Cha, Y.-N., Bueding, E., Heine, H. S., Talalay, P. 1979. *Cancer Res.* 39:2971–77

346. Lam, L. K. T., Sparnins, V. L., Hochalter, J. B., Wattenberg, L. W. 1981. *Cancer Res.* 41:3940–43

347. Lam. L. K. T., Sparnins, V. L., Wattenberg, L. W. 1982. *Cancer Res.* 42:1193–98

348. Sparnins, V. L., Chuan, J., Wattenberg, L. W. 1982. *Cancer Res.* 42:1205–7

349. Nakagawa, Y., Hiraga, K., Suga, T. 1981. *J. Pharm. Dyn.* 4:823–26

350. Batzinger, R. P., Ou, S.-Y. L., Bueding, E. 1978. *Cancer Res.* 38:4478–85

351. Wattenberg, L. W. 1979. In *Environmental Carcinogenesis*, ed. P. Emmelot, E. Kriek, pp. 241–63. Amsterdam: Elsevier/North-Holland Biomedical

352. Orrenius, S., Jewel, S. A., Belloma, G., Thor, H., Jones, D. P., Smith, M. T. 1983. See Ref. 6, pp. 259–70

353. Jewell, J. A., Bellomo, G., Thor, H., Orrenius, S., Smith, M. T. 1982. *Science* 217:1257–79

353a. Sies, H., Graf, P., Estrela, J. M. 1981. *Proc. Natl. Acad. Sci. USA* 78:3358–62

354. Kosower, E. M., Kosower, N. S. 1969. *Nature* 224:117–20

355. Novogrodsky, A., Nehring, R. E. Jr., Meister, A. 1979. *Proc. Natl. Acad. Sci. USA* 76:4932–35

355a. Plummer, J. L., Smith, B. R., Sies, H., Bend, J. R. 1981. Detoxication and Drug Metabolism: Conjugation and Related Systems, *Methods Enzymol.* 77:50–59

356. Griffith, O. W. 1981. *J. Biol. Chem.* 256:4900–4

357. Arrick, B. A., Nathan, C. F., Griffith, O. W., Cohn, Z. A. 1982. *J. Biol. Chem.* 257:1231–37

358. Arrick, B. A., Griffith, O. W., Cerami, A. 1981. *J. Exp. Med.* 153:720–25

359. Rouzer, C. A., Scott, W. A., Griffith, O. W., Hamill, A. L., Cohn, Z. A. 1982. *Proc. Natl. Acad. Sci. USA* 79:1621–25

360. Visser, T. J. 1980. *Trends Biochem. Sci.* 5:222–24

361. Folkers, K., Dahmen, J., Ohta, M., Stepien, H., Leban, J., et al. 1980. *Biochem. Biophys. Res. Commun.* 97:590–94

362. Meister, A. 1974. In *Brain Dysfunction in Metabolic Disorders*, ed. F. Plum, *Res. Publ. Assoc. Res. Nerv. Ment. Dis.* 53:273–91

363. Meister, A. 1978. *Adv. Neurol.* 21:289–302

364. Oliver, J. M., Spielberg, S. P., Pearson, C. B., Schulman, J. D. 1978. *J. Immunol.* 120:1181–86

365. Apontoweil, P., Berends, W. 1975. *Biochim. Biophys. Acta* 399:10–22

366. Apontoweil, P., Berends, W. 1975. *Mol. Gen. Genet.* 141:91–95
367. Fuchs, J. A., Warner, H. R. 1975. *J. Bacteriol.* 124:140–48
368. Fuchs, J. A., Haller, B., Tuggle, C. K. 1983. See Ref. 6, pp. 383–90
369. Davis, N. K., Greer, S., Jones-Mortimer, M. C., Perham, R. N. 1982. *J. Gen. Microbiol.* 128:1631–34
370. Mooz, E. D., Wigglesworth, L. 1976. *Biochem. Biophys. Res. Commun.* 68:1066–72
371. Mooz, E. D. 1979. *Biochem. Biophys. Res. Commun.* 90:1221–28
372. Meister, A. 1982. In *Metabolic Basis of Inherited Diseases*, ed. J. B. Stanbury, J. B. Wyngaarden, D. S. Frederickson, J. L. Goldstein, M. S. Brown, pp. 348–59. New York: McGraw-Hill. 5th ed.
373. Larsson, A. 1981. In *Transport and Inherited Disease*, ed. N. R. Belton, C. Toothill, pp. 277–306. London: MTP
374. Larsson, A., Hagenfeldt, L. 1983. See Ref. 6, pp. 315–22
375. Jellum, E., Marstein, S., Skullerud, K., Munthe, E. 1983. See Ref. 6, pp. 345–52
375a. Larsson, A., Mattsson, B., Wauters, E. A. K., van Gool, J. D., Duran, M., Wadman, S. K. 1983. See Ref. 6, pp. 323–34
376. Orlowski, M., Meister, A. 1971. *Biochemistry* 10:372–80
377. Wilcken, B., Hammond, J. W., Howard, N., Bohane, T., Hocart, C., Halpern, B. 1981. *N. Engl. J. Med.* 305:865–69
378. Palekar, A. G., Tate, S. S., Sullivan, J. F., Meister, A. 1975. *Biochem. Med.* 14:339–45
379. Tham, R., Nystrøm, L., Holmstedt, B. 1968. *Biochem. Pharmacol.* 17:1735–38
380. Stoner, E., Starkman, H., Wellner, D., Wellner, V. P., Sassa, S., et al. 1983. *J. Pediatrics.* In press
381. Lindblad, B. 1983. See Ref. 6, pp. 335–44
382. Valentine, W. N., Tanaka, K. R., Paglia, D. E. 1982. See Ref. 372, pp. 1606–28
383. Beutler, E. 1974. See Ref. 4, pp. 109–44
384. Loos, H., Roos, D., Weening, R., Houwerzijl, J. 1976. *Blood* 48:53–62
385. Cohn, V. H., Lyle, J. 1966. *Anal. Biochem.* 14:434–40
386. Hissin, P. J., Hilf, R. 1976. *Anal. Biochem.* 74:214–26
387. Benson, J. R., Hare, P. E. 1975. *Proc. Natl. Acad. Sci. USA* 72:619–22
388. Beutler, E., West. C. 1977. *Anal. Biochem.* 81:458–60
389. Woodward, G. E. 1935. *J. Biol. Chem.* 109:1–10
390. Bernt, E., Bergmeyer, H. U. 1974. *Methods Enzymatic Anal.* 4:1643–47
391. Lack, L., Smith, M. 1964. *Anal. Biochem.* 8:217–22
392. Koivusalo, M., Uotila, L. 1974. *Anal. Biochem.* 59:34–45
393. Crowley, C., Gillham, B., Thorn, M. B. 1975. *Biochem. Med.* 13:287–92
394. Asaoka, K., Takahashi, K. 1981. *J. Biochem.* 90:1237–42
395. Owens, C. W. I., Belcher, R. V. 1965. *Biochem. J.* 94:705–11
396. Tabor, C. W., Tabor, H. 1977. *Anal. Biochem.* 78:543–53
397. Grassetti, D. R., Murray, J. F. Jr. 1967. *Anal. Biochem.* 21:427–34
398. Wendell, P. L. 1970. *Biochem. J.* 117:661–65
399. Srivastava, S. K., Beutler, E. 1968. *Anal. Biochem.* 25:70–76
400. Brehe, J. E., Burch, H. B. 1976. *Anal. Biochem.* 74:189–97
401. Moore, S., Stein, W. H. 1954. *J. Biol. Chem.* 211:893–906
402. Purdie, J. W., Farant, J. P., Gravelle, R. A. 1966. *J. Chromatogr.* 23:242–47
403. Purdie, J. W., Hanafi, D. E. 1971. *J. Chromatogr.* 59:181–84
404. Hsiung, M., Yeo, Y. Y., Itiaba, K., Crawhall, J. C. 1978. *Biochem. Med.* 19:305–17
405. Bowie, L., Crawhall, J. C., Gochman, N., Johnson, K., Schneider, J. A. 1976. *Clin. Chim. Acta* 68:349–53
406. Rabenstein, D. L., Saetre, R. 1977. *Anal. Chem.* 49:1036–39
407. Mefford, I., Adams, R. N. 1978. *Life Sci.* 23:1167–74
408. Takahashi, H., Nara, Y., Meguro, H., Tuzimura, K. 1979. *Agric. Biol. Chem.* 43:1439–45
409. Lankmayr, E. P., Budna, K. W., Muller, K., Nachtmann, F. 1979. *Fresenius Z. Anal. Chem.* 295:371–74
410. Reeve, J., Kuhlenkamp, J., Kaplowitz, N. 1980. *J. Chromatogr.* 194:424–28
411. Reed, D. J., Babson, J. R., Beatty, P. W., Brodie, A. E., Ellis, W. W., Potter, D. W. 1980. *Anal. Biochem.* 106:55–62
412. Murayama, K., Kinoshita, T. 1981. *Anal. Lett.* 14(B15):1221–32
413. Nakamura, H., Tamura, Z. 1981. *Anal. Chem.* 53:2190–93
414. Nakamura, H., Tamura, Z. 1982. *Anal. Chem.* 54:1951–55
415. Newton, G. L., Dorian, R., Fahey, R. C. 1981. *Anal. Biochem.* 114:383–87
416. Kosower, E. M., Pazhenchevsky, B., Hershkowitz, E. 1978. *J. Am. Chem. Soc.* 100:6516–18
417. Modig, H. 1968. *Biochem. Pharmacol.* 17:177–86
418. Harding, J. J. 1970. *Biochem. J.* 117:957–60

419. Modig, H. G., Edgren, M., Revesz, L. 1971. *Int. J. Radiat. Biol.* 22:257–68
420. Harrap, K. R., Jackson, R. C., Riches, P. G., Smith, C. A., Hill, B. T. 1973. *Biochim. Biophys. Acta* 310:104–10
421. Isaacs, J., Binkley, F. 1977. *Biochim. Biophys. Acta* 497:192–204
422. Harisch, G., Eikemeyer, J., Schole, J. 1979. *Experientia* 35:719–20
423. Harisch, G., Mahmoud, M. F. 1980. *Hoppe-Syelers Z. Physiol. Chem.* 361:1859–62
424. Brigelius, R., Lenzen, R., Sies, H. 1982. *Biochem. Pharmacol.* 31:1637–41
425. Krimsky, I., Racker, E. 1952. *J. Biol. Chem.* 198:721–29
425a. Meister, A. 1983. *Science* 220:472–77
426. Loomis, W. F. 1955. *Ann. N Y Acad. Sci.* 62:209–28
427. Tate, S. S., Meister, A. 1976. *Biochem. Biophys. Res. Commun.* 70:500–5
428. Hanai, K. 1981. *J. Comp. Physiol.* 144:503–8
429. Abraham, E. C., Taylor, J. F., Lang, C. A. 1978. *Biochem. J.* 174:819–25
430. Hazelton, G. A., Lang, C. A. 1980. *Biochem. J.* 188:25–30
431. Meister, A. 1980. In *Microorganisms and Nitrogen Sources*, ed. J. W. Payne, pp. 493–509. New York: Wiley
431a. Rennenberg, H. 1982. *Phytochemistry* 21:2771–81
432. Fahey, R. C., Brown, W. C., Adams, W. B., Worsham, M. B. 1978. *J. Bacteriol.* 133:1126–29
433. Murata, K., Tani, K., Kato, J., Chibata, I. 1981. *Eur. J. Appl. Microbiol. Biotechnol.* 11:72–77
434. Rennenberg, H. 1976. *Phytochemistry* 15:1433–34
435. Tabor, H., Tabor, C. W. 1972. *Adv. Enzymol.* 36:203–68
436. Carnegie, P. R. 1963. *Biochem. J.* 89:471–78
437. Buchanan, B. B. 1983. See Ref. 6, pp. 229–40
438. Guddewar, M. B., Dauterman, W. C. 1979. *Phytochemistry* 18:735–40

Ann. Rev. Biochem. 1983. 52:761–99

CELL SURFACE INTERACTIONS WITH EXTRACELLULAR MATERIALS[1]

Kenneth M. Yamada

Membrane Biochemistry Section, Laboratory of Molecular Biology, National Cancer Institute, Bethesda, Maryland 20205

CONTENTS

PERSPECTIVES AND SUMMARY 761
INTRODUCTION 763
MODEL SYSTEMS 764
   General Approaches 764
   Interactions with Plastic and Glass Substrates 765
   Interactions with Collagen Substrates 767
FIBRONECTIN 768
   Properties 768
   Cell Interaction Site 771
   Interactions with Collagen 771
   Interactions with Fibrin 773
   Glycosaminoglycan-Binding Domains 773
   Other Interactions of Fibronectin 776
   Primary Structure of Fibronectin 777
   Structure-Function Relationships 781
LAMININ 782
   Location and General Structure 782
   Function of Laminin 783
   Structural and Functional Domains 783
OTHER CELL ATTACHMENT FACTORS 785
   Chondronectin 785
   Serum Spreading Factors 785
CELLULAR RECEPTORS FOR EXTRACELLULAR MOLECULES 787
SYNTHESIS AND REGULATION 790
FUTURE DIRECTIONS 791

PERSPECTIVES AND SUMMARY

Interactions of cells with extracellular materials are critically important events during embryonic development, growth regulation, and mainten-

[1] The US Government has the right to retain a nonexclusive royalty-free licence in and to any copyright covering this paper.

ance of normal tissue function. For example, the interaction of a cell with either of two purified extracellular molecules, fibronectin or collagen, can alter its adhesion, biosynthetic patterns, and capacity to migrate and proliferate. Considering the biological importance of these interactions, surprisingly little was known until recently about the mechanisms by which these extracellular molecules interact with cells and with other extracellular molecules. Recent progress has resulted from the purification of specific proteins involved in such interactions, and from attempts to explain complex biological events in terms of the combined action of specialized structural and functional domains of these molecules. This review discusses several general mechanisms and systems for studying cell-extracellular material interactions and focuses on several specific molecules involved in these events.

The most intensively studied proteins, fibronectin and laminin, are reviewed here in depth. Fibronectin contains 220,000-dalton subunits linked into dimers and polymers by disulfide bonds. It binds to the cell surface as well as to a series of other extracellular molecules to mediate cell adhesion, binding of native and denatured collagens, and a variety of other interactions with macromolecular ligands. Fibronectin interacts with cells by a specific cell-binding domain to mediate the attachment and spreading of cells. Fibronectin also, however, contains a series of other protease-resistant functional domains arranged in a modular fashion for binding to fibrin, collagen, heparin, hyaluronic acid, and several other macromolecules. Several of these sites have been purified and characterized in terms of specificity and primary structure. By means of these specific domains, fibronectin appears to mediate or modify cell adhesion and several other complex biological events by mechanisms that are under intensive investigation.

Laminin is another adhesive glycoprotein, which differs from fibronectin in structure and in its specificity for type IV collagen and epithelial cells. It exists as a cross-shaped molecule of 1,000,000 daltons consisting of A and B subunits of 200,000 and 400,000 daltons linked by disulfide bonds. Laminin also seems to function by the combined action of independent cell-, collagen-, and heparin-binding domains on each molecule.

Chondronectin is a serum glycoprotein with a high degree of specificity for mediating adhesion of chondrocytes to type II collagen. It is a protein complex of 180,000 daltons stabilized by disulfide bonds, and is structurally distinct from fibronectin and laminin. Chondronectin also binds to chondroitin sulfate; this interaction may be important to its interaction with cartilage extracellular matrix. One or more serum spreading factors of approximately 70,000 daltons can also be critical to the adhesion and growth of a variety of cell types. These spreading factors are monomers, at least one of which differs from the other adhesion factors in its requirement for cellular protein synthesis to mediate cell spreading.

In addition, cells can also interact directly with several types of collagen and with hyaluronic acid by means of plasma membrane binding sites, apparently without the need for an intermediary extracellular molecule. Platelets also bind to native collagen and fibrinogen by receptors, which appear to contain 65,000- and 105,000-dalton glycoproteins, respectively. In general, interactions between cells and extracellular molecules can be mediated by multiple mechanisms, even within the same cell. Each of these systems can be dissected by classical biochemical approaches to aid understanding of complex adhesive and regulatory events in terms of a series of ligand-to-protein domain interactions.

A number of recent reviews are recommended for emphasis on more biological and phenomenological studies (1–9), as well as for coverage of areas not included in this review such as cell-cell adhesion (10), certain interactions with glycosaminoglycans (11), and the physiology and biochemistry of collagen (7, 12).

INTRODUCTION

The composition of extracellular materials interacting with the cell surface has important regulatory and structural consequences for cells, and an extensive literature now documents these biological roles (reviewed in 1–9). Several dramatic examples show the regulation of cell morphology, cytoskeletal organization, and even cell type by purified extracellular macromolecules. Treatment by purified cellular fibronectin of oncogenically transformed cells originally deficient in fibronectin can restore normal cell shape, adhesiveness, cell surface morphology, and actin microfilament organization (13–15); similarly, dissociated epithelial cells cease blebbing and reorganize cortical microfilaments after treatment with purified fibronectin, laminin, or collagen (16). Exposure of cartilage chondrocytes to purified cellular fibronectin alters their morphology and patterns of biosynthesis, causing a reversion from chondrocyte-specific synthesis of type II collagen and proteoglycans to a mesenchymal-cell pattern (17, 18). A particularly striking change was observed when fully differentiated epithelial cells were cultured within a gel of purified collagen. The cells underwent a fundamental alteration in phenotype from epithelial to mesenchymal cell (19). This alteration was accompanied by major changes in cell interactions and morphology, so that cells migrate away from epithelial sheets to form isolated bipolar cells that can invade collagen gels and display altered biosynthetic activity (19). These and other effects of extracellular macromolecules on cell differentiation, motility, and growth underscore the importance of cell surface interactions with the local extracellular microenvironment.

There are two general models for such interactions. Cells can interact either directly with each of these molecules, or secondarily via intermediary linking molecules. For example, cells could interact with collagen directly by means of a specific membrane receptor(s), or indirectly by means of specialized binding molecules that link collagen to the cell surface. In addition, such interactions could be generic or highly specific; i.e. such molecules could bind collagens in general or be specific for each type of collagen. Examples of each type of possible interaction have been reported.

MODEL SYSTEMS

General Approaches

Interactions of extracellular molecules with cells have been analyzed in a variety of cell culture, hemagglutination, phagocytosis, and filter-binding assays. Two particularly fruitful approaches from a biochemical viewpoint, however, have been based on the use of solid supports: cell attachment and spreading on plastic substrates or affinity chromatography of extracellular molecules.

Proteins bind tightly and nonspecifically in low amounts to polystyrene dishes used routinely for tissue culture, and protein adsorption can be confirmed by radioactive and immunological methods. Living cells are plated onto such dishes and examined for attachment and the morphologic effects of the adsorbed molecules. The specific activity and cell-type specificity of adhesion molecules can be examined rapidly in this system; for example, certain cells attach to fibronectin but not to laminin and vice versa (e.g. 20). Proteolytic fragments of proteins and competitive inhibitors can also be examined for their effects on cell attachment and spreading. In addition, collagenous substrates established on these dishes are useful in the analysis of the factors necessary for cell attachment or growth on collagen.

A second approach to studying these interactions is affinity chromatography. A covalently bound extracellular molecule can bind its ligands or receptors; conversely, immobilized ligands can bind and purify specific ligand-binding domains of molecules fragmented by proteases. This approach has permitted the rapid purification of a series of protease-resistant, functional domains for use in primary structure determinations, for mapping these domains and other structural features on the molecules, and for further electron microscopic and spectroscopic characterization.

A complementary approach is to isolate intact extracellular matrices and to selectively ablate specific molecules in order to determine which molecules and complexes of molecules are responsible for complex events such as growth regulation and morphogenesis. Extracellular matrices isolated from cell cultures (21–23) or from tissues (24, 25) are potent stimulators of growth and differentiation. They can act by mechanisms independent of cell attachment activity or adsorbed serum growth factors (26), but the molecules mediating these important activities remain to be identified.

Interactions with Plastic and Glass Substrates

Because of the widespread popularity of cell culture on artificial substrates, the molecules required for cell adhesion to these substrates have been examined in detail. Although cells can attach directly to plastic and glass in protein-free media, such binding is generally considered to be non-physiological; e.g. the adhesive interactions cannot be disrupted by proteases or chelating agents as in normal cell adhesion (2). Cells normally attach to factors provided by the serum present in routine culture media or to factors synthesized by the cells themselves. Successful culture of cells in serum-free medium often requires supplementation of media with such factors as fibronectin (27). Three or more factors have been purified from serum that promote the adhesion and growth of cells on culture substrates: fibronectin, chondronectin, and serum spreading factor(s); these proteins are reviewed separately below.

A general principle of cell-to-substrate adhesion is that attachment and spreading of a cell can be mediated by more than one type of cell surface-ligand interaction. For example, cells attach and spread not only onto substrate-adsorbed attachment factors, but also onto other molecules that interact with cell surface proteins and receptors, such as lectins and anti-plasma membrane antibodies (28–30). Hepatocytes can attach to asialoglycoproteins by means of the asialoglycoprotein receptor, but they do not spread normally (30). Fibroblasts can even attach and spread on substrates composed of glycosidases under conditions restricting enzyme action, presumably by binding carbohydrates on the cell surface (31). Nevertheless, not all such interactions lead to attachment; e.g. binding of the numerous α2 macroglobulin receptors of fibroblasts does not enhance attachment or spreading (32). Many of these interactions are non-physiological models, since cells do not normally encounter sufficient concentrations of ligands such as asialofetuin or, obviously, plant lectins for adhesive interactions to occur. The threshold amounts of ligands required to mediate cell spreading have been determined (29, 33); e.g. a minimum of 45,000 fibronectin molecules per cell are required for fibroblast spreading (29). Defects in adhesion can sometimes be overcome by larger amounts of ligand; e.g. certain cell attachment mutants will respond if provided with unusually large amounts of exogenous fibronectin (34, 35).

IMMUNOLOGICAL APPROACHES   In addition to the analysis of soluble adhesive factors, two other approaches to identifying molecules involved in cell-to-substrate adhesion include: (a) the production and characterization of anti-cell surface antibodies that inhibit adhesion; and (b) the isolation and analysis of substrate-attached adhesive structures. Antibodies that disrupt the adhesion of fibroblasts, myoblasts, and epithelial cells to plastic substrates have been characterized in detail (36–39). Plasma membrane antigens implicated in adhesion by this analysis appear to be glycoproteins with molecular weights of 120,000–160,000. There are an estimated 200,000–800,000 copies per cell (36–39). The mechanisms by which they act are not known; one approach might be to purify and reconstitute these glycoproteins in liposome attachment models of adhesion.

SUBSTRATE-ATTACHED MATERIALS   An alternative approach involves detaching cells from substrates with chelating agents or other methods and examining the adhesive structures remaining attached to the substrate (40, 41). Since electron microscopy reveals that there are at least three classes of adhesive contact with the substrate termed focal, close, and extracellular matrix contacts (42), these substrate-attached adhesive structures are probably complex entities. Cells treated with chelating agents leave behind membrane-enclosed adhesive sites containing intracellular cytoskeletal proteins such as actin and desmin, as well as a variety of extracellular molecules (40, 43). Whether collagen is also present at these sites is debated (44–46). The external components were characterized extensively recently (41, 43, 47–49); they include heparan sulfate proteoglycan, hyaluronic acid, and fibronectin. Pulse-chase isotopic labeling experiments suggest that they exist as two distinct pools of material, a rapidly appearing class of molecules containing heparan sulfate proteoglycan and fibronectin, and a more slowly appearing class containing hyaluronic acid and chondroitin sulfate proteoglycan. The first class may correlate with initial cell attachment and spreading, and the latter with destabilization of adhesive contacts (41, 48). Related approaches include the use of shear forces to obtain "focal contact" sites (50, 51). Besides cytoskeletal proteins such as actin and vinculin, the latter sites contain unidentified components possibly involved in adhesion, such as a major 55,000-dalton protein (51).

Another approach to analyzing cell interactions with substrates is to examine for novel biological activities of molecules released by cells into culture medium (52–54). Such analyses of conditioned media suggest roles for a heparan sulfate proteoglycan in axon extension (52), for a 58,000-

dalton glycoprotein in hepatoma cell attachment to type I collagen (53), and for a 16s complex of extracellular molecules in myoblast attachment (54). These types of approaches should permit the identification of new factors or synergistic combinations of known factors under more physiological conditions than possible using only purified proteins.

Interactions with Collagen Substrates

THREE-DIMENSIONAL GELS   Interactions of cells with collagen can alter cellular differentiation and other important biological processes (reviewed in 1, 3, 7). Cells can interact with collagen either directly or indirectly. Fibroblasts, hepatocytes, and other cells are reported to adhere directly to collagen, especially when it is polymerized into three-dimensional gels to mimic its organization in vivo (55–60). Hepatocytes show little specificity for collagen types, and attach to types I, II, III, IV, and V. Although they bind most rapidly to native collagens, hepatocytes also attach to denatured collagens and even weakly to synthetic collagen-like peptides, which suggests the use of generic cell surface "receptors" for collagenous sequences (58). Hepatocytes also bind directly to fibronectin- and laminin-coated substrates by independent mechanisms, as demonstrated by antibody inhibition studies, somatic cell mutants, and other criteria (58–61). These cells therefore display at least three independent mechanisms for cell-to-collagen interactions. Fibroblast adhesion to native collagen gels can be strengthened by fibronectin (55, 62), but their penetration or locomotion in matrices is inhibited (63, 64). In contrast, neural crest cell and melanoma cell migration is stimulated (64, 64a).

TWO-DIMENSIONAL SUBSTRATES   Under other assay conditions, especially on two-dimensional layers of collagen or on gelatin substrates, there is a strong requirement for an additional cell attachment protein for adhesion of fibroblasts (3, 65), epithelial cells (66), myoblasts (67), and chondrocytes (68) to collagen. Several cell type-specific attachment proteins mediating this type of interaction with collagen have been purified and characterized, including fibronectin, laminin, and chondronectin (see below). The relative biological importance of direct cell-to-collagen compared to cell-linking protein-collagen interactions has not been determined, but both may be of importance depending on the biological event. For example, fibroblasts enmeshed in dense collagenous matrices may not require fibronectin, whereas fibronectin appears important for the organization of collagen in tissue culture models, in which collagen and fibronectin are codistributed in the same fibrils (69, 70), and since disruption of fribronectin with specific antibodies disrupts the deposition of collagen (71).

FIBRONECTIN

Properties

Fibronectin [*fibre* = fiber + *nectere* = to bind, connect] is a cell surface, extracellular matrix, and plasma protein, which has been analyzed in rapidly finer detail (4, 72–81). This glycoprotein is present in large amounts in plasma (0.3 g/liter), where it is termed plasma fibronectin. Fibronectin can be be synthesized by a wide variety of cell types, including fibroblasts and certain epithelial cells; this cell surface form is termed cellular fibronectin (72–81). The plasma form appears to be synthesized by hepatocytes (82, 83; 83a); however, a contribution from endothelial cells is also possible (reviewed in 75, 80).

FUNCTIONS OF FIBRONECTIN   The fibronectins have been implicated in a variety of biological activities, most of which involve adhesive and ligand-binding functions. It is a prototype cell-to-substrate adhesion molecule, and can mediate cell attachment and spreading on collagen, fibrin, and artificial tissue culture substrates (4, 8, 9, 72–81). It has numerous effects on cell morphology and intracellular functions; it also promotes certain phagocytic activities (75) and it binds to a series of ligands including collagen, fibrin, heparin, hyaluronic acid, and actin in events thought to be important to cell interactions with extracellular materials. Fibronectin also stimulates cell movement by mechanisms that are still not understood, but which may be important to cell migratory events in embryonic development and wound healing (reviewed in 80).

TYPES OF FIBRONECTIN   Fibronectins are composed of similar polypeptide subunits of 220–250,000 daltons that are linked by disulfide bonds into dimers and higher polymers (Figure 1). The cellular and plasma forms of fibronectin are very similar in structure and function (84–91), but not identical (88). The two forms have very similar amino acid compositions, carbohydrate structures, and secondary and tertiary structures as evaluated by circular dichroism and tryptophanyl fluorescence measurements and responses to denaturants (84–86, 89–97). In addition, they have the same types and organizations of protease-resistant structural domains (98; reviewed in later sections). Nevertheless, there are several differences in biological specific activity (e.g. in effects on transformed cells) and several minor structural differences (88, 97–100). They differ in solubility (cf 84, 85) and apparent molecular weights (88 and references therein), and plasma fibronectin is a dimer, whereas the cellular form contains dimers and multimers (85, 101, 102). Their N-linked oligosaccharides also differ slightly in the frequent absence of fucose from plasma fibronectin, and in the presence of additional sialic acid in cellular fibronectin, which is also reportedly linked $\alpha 2 \rightarrow 3$ rather than $\alpha 2 \rightarrow 6$ as in the plasma form (93–97).

In addition, recent peptide mapping and monoclonal antibody studies show that at least three regions of difference exist between the two forms, at least one of which is in a region devoid of carbohydrates. All three regions of apparent structural difference can be mapped to the interior of the molecule, which appears to rule out the hypothesis that plasma fibronectin is derived from cellular fibronectin by proteolytic processing (88, 98, 100). Thus, although the two forms of fibronectin are very similar, they are structurally and functionally distinct.

In addition, a third form of fibronectin produced by amniotic cells differs by at least carbohydrate content (103–105). Yet another type of modification is found in the sulfated chains of fibronectin from melanoma cells,

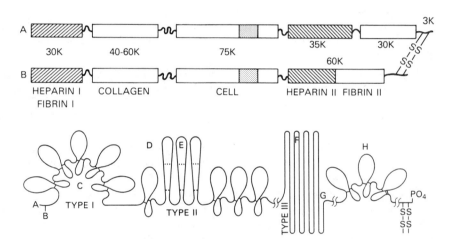

Figure 1 Functional domains and primary structural features of fibronectin. (*Top*) Dimer of plasma fibronectin indicating structural and functional domains. The wavy lines indicate regions of increased susceptibility to proteases between protease-resistant domains; K = kilodaltons. Note that the A and B chains have a site of apparent difference near the carboxyl terminus. (*Bottom*) Schematic representation of homologous repeating structures within the primary sequence of slightly over half of bovine plasma fibronectin, from Petersen et al (125, 201b). A, amino-terminal pyroglutamate; B, site for transglutaminase-mediated cross-linking; C, amino-terminal domain with five "fingers"; D and E, collagen-binding domain with both type I and II homology regions; although the arrangements of disulfide bonds within type II homology regions are not yet known, there are two repeating units of this type in region E; F, type III homology region of four or more repeats containing a cell-binding domain (124, 125); G, heparin-II-binding domain; H, a terminal three-finger repeat, possibly from the fibrin-binding domain; I, carboxyl-terminal 3000-dalton domain with interchain disulfide bonds.

which are 40,000 daltons in apparent size larger than normal fibronectin (106); the nature of this modification is not known. It is important to determine whether the different forms of fibronectin are encoded by separate genes, or are instead produced by different splicing patterns of messenger RNA from the same gene depending on the cell of origin.

GENERAL STRUCTURE   The structure of fibronectin has been evaluated via spectroscopic and sedimentation methods. The circular dichroism (CD) spectra of cellular and plasma fibronectins are indistinguishable and rather unusual, with relatively weak ellipticities in the far ultraviolet (84, 89–91). These CD spectra are not consistent with known elementary structures (89–91), although infrared spectroscopy suggests the possible presence of 35% β structure (107). The existence of domains with tertiary structure is indicated by tryptophanyl fluorescence and circular dichroism studies, which show distinct denaturation curves at different wavelengths and different CD thermal transitions for different proteolytic fragments (89–91, 108). Similarly, distinct endothermal transitions are noted by differential scanning calorimetry (109).

The sedimentation constant of plasma fibronectin is 12–13 at pH 7 with a calculated frictional ratio of 1.7, indicating a nonglobular, relatively asymmetric molecule (91, 110). The sedimentation constant decreases linearly with increasing pH or ionic strength to a value of 8 at pH 11 for both cellular and plasma forms with minimal accompanying spectroscopic changes (89, 91). These results suggest that fibronectin is composed of structured regions of polypeptide linked by flexible regions that permit the molecule to expand or unfold as electrostatic interactions are disrupted.

When visualized directly by electron microscopy, the plasma fibronectin dimer is generally a slender, V-shaped molecule with no large globular regions, but with sites of increased flexibility in the chains that suggest flexible regions of polypeptide between structured domains (108, 111, 112). Proteolytic fragments of fibronectin retain these structural features, and their observed frictional coefficients generally agree with predicted values calculated from models of the shapes observed by electron microscopy (108). The only major discrepancy is for intact fibronectin itself, which should be more compact than observed by microscopy (108); one possibility is that exposure of samples to high ionic strength and glycerol during preparations for electron microscopy causes unfolding of the molecule. In support of this notion, fibronectin lyophilized on grids has a much more globular configuration (107, 113). These microscopy results combined with the spectroscopic and sedimentation findings suggest that fibronectin may normally exist as a folded, but nevertheless asymmetric, molecule that can expand when electrostatic interactions are disrupted. Several suggested configurations are as a molecule in which flexible interdomain regions of random polypeptide can stretch or contract (89), one in which the chains can fold back on themselves and are held in place by complementary electrostic charges on amino and carboxyl domains (79), or one in which the arms of the molecule cross and fold over one another. These considerations of shape are important in light of possible changes in exposure of sites after binding of a ligand or proteolytic cleavage (see section on cellular receptors).

Cell Interaction Site

The binding of fibronectin to the cell surface requires a specific region of the molecule, termed the "cell-binding" region. This region has been purified from among larger proteolytic fragments of fibronectin that do not bind to gelatin affinity columns (114–124). The cell-binding region of fibronectin can mediate the attachment and spreading of cells on plastic substrates, although it lacks collagen- and heparin-binding regions and therefore cannot mediate cell attachment to collagen. A fragment of 11,500 daltons still retaining a significant amount of the cell-binding activity of the intact molecule has been produced through further pepsin treatment, and isolated with a monoclonal antibody that can block cell attachment (123). Analyses of partial digests indicate that this 11,500 fragment and the cell-binding domain are in the central part of the molecule (116–124). The primary sequence of this domain is not unusually hydrophobic, and suggests possible β structure (124, 125). Synthetic peptides spanning this region have been synthesized, and the site of cell-binding activity has been localized to the COOH-terminal 30 amino acids of this region (125a). The existence of other adjacent cell-binding regions, or regions that modulate cell-type specificity, is not yet ruled out (126). Further analyses of this domain in terms of protein conformation, chemical modification, and cell surface protein cross-linking experiments should provide more insight into the nature of its interaction with the cell surface.

Interactions with Collagen

Fibronectin binds to collagen types I, II, III, IV, and V (65, 127–132). The capacity of fibronectin to mediate cell attachment to collagen in vitro has suggested that one important role of fibronectin in vivo is to mediate cell attachment to collagen (3, 65). Cell attachment requires divalent cations and cellular metabolic activity (133). An unusual feature of the interaction of fibronectin with collagen, however, is that it has substantially greater avidity for denatured than for native, triple-helical collagens (128–132). The apparent affinity for nonhelical fragments of collagen is high, with $K_D = 2$–5×10^{-9} M (73). Differences of up to 200-fold in binding between native and denatured collagen type I have been described. There is less difference between binding to native and denatured type III collagen, which is 2- to 6-fold (130–132). These differences are nevertheless significant, and they raise questions about the physiological meaning of the interaction of fibronectin with collagen.

Evidence that fibronectin can actually interact with native collagen rather than at local sites of denaturation includes immunological localization of fibronectin to striated collagen fibrils in tissue culture cells and in vivo (1, 70), as well as a demonstration of binding of purified fibronectin to collagen fibrils and interference with fibrillogenesis (134). These findings suggest that fibronectin can interact with native collagen, especially type III collagen, but further in vivo studies are needed to determine the biological importance of the interaction; even these weaker interactions may be fully sufficient for biological systems. The particularly effective interaction with denatured collagen may be related to the opsonic activity of fibronectin in the clearance of collagenous debris by the reticuloendothelial system (75). Some of the interactions of cells with fibronectin, e.g. during a variety of embryonic cell migratory events, may occur directly with a scaffolding of fibronectin itself containing only limited amounts of striated collagen (e.g. 135, 136).

The collagen-binding domain of fibronectin is approximately 30,000–40,000 daltons and can be isolated after digestion of the intact molecule with a variety of proteases under controlled conditions, even with the broad spectrum protease pronase (98, 114–121, 137–145). These purified fragments can still bind to gelatin or type I collagen affinity columns, although their apparent affinity of binding is decreased substantially from five to 1000-fold depending on the fragment size (117, 137). This domain was mapped to a site 30,000 daltons from the amino terminus (Figure 1; 142–144). There is one such domain per fibronectin monomer (117), but the decrease in affinity of binding to gelatin with increasing proteolysis is not yet ruled out as the result of a loss of adjacent collagen-binding sites. Antibodies to this domain inhibit the binding of fibronectin to gelatin and disrupt the extracellular organization of fibronectin and collagen (71, 146).

The site on collagen to which fibronectin binds is primarily within residues 757–791 in the $\alpha 1(I)$ chain, and similar regions in the $\alpha 2(I)$ chain of type I collagen and in type II collagen (128, 129). This is a less stable region of the collagen triple helix, and it contains the only site on collagen susceptible to vertebrate collagenases (128). Collagen bound to fibronectin by means of its collagen-binding domain can be covalently cross-linked to the adjacent amino-terminal domain by factor XIIIa transglutaminase (140). This cross-linking would stabilize binding, particularly if it involves the more weakly bound native collagens.

Interactions with Fibrin

Fibronectin binds to fibrin and, less avidly, to fibrinogen; the binding appears to be to the COOH-terminal region of the Aα chain of fibrinogen (110, 147–152). Fibronectin can be cross-linked to fibrin or fibrinogen by factor IIIa transglutaminase (149, 150). Such covalent cross-linking increases the stiffness of fibrin clots in vitro (153), although the actual amounts of fibronectin covalently bound to clots in vivo remain to be demonstrated. The fibronectin-fibrin interaction may be important during wound healing to provide a substrate for fibroblasts migrating into wounds to initiate the repair process (81). For example, fibroblasts can adhere to purified fibronectin bound to fibrin and fibrinogen substrates, and they adhere even more readily to fibronectin cross-linked to fibrin by factor XIIIa transglutaminase (154). In addition, macrophage binding of fibrin is stimulated by fibronectin (155).

Fibronectin binds to fibrin or fibrinogen affinity columns, but the binding to fibrin is of significantly higher affinity; binding to fibrinogen, however, occurs at 0°C (148, 151, 152). Complexes of fibronectin and fibrinogen can precipitate in the cold, and plasma fibronectin was originally termed "cold-insoluble globulin" because early preparations contained fibrinogen contaminants and cryoprecipitated (147).

Fibronectin binds to fibrin by at least two domains. The binding to the amino-terminal domain is of highest apparent affinity (119, 156), although binding through a second site close to the carboxyl terminus of one (119) or both subunits also occurs (156a; 156b). An additional weak binding site has also been reported in the gelatin-binding domain (131, 156). None of these fibrin-binding sites is of high affinity at 37°C (148); the covalent cross-linking of fibrin or fibrinogen to fibronectin may therefore be important to stabilize this relatively labile interaction. This transglutaminase-mediated cross-linking of fibrin, as well as of collagen and various amines, occurs primarily in the amino-terminal domain (117, 119, 140, 157–159) to a glutamine residue at position 3 of bovine fibronectin (125). A second, minor cross-linking site near the carboxyl terminus of the molecule can be detected under certain conditions (159).

Glycosaminoglycan-Binding Domains

Fibronectin binds to the sulfated glycosaminoglycans heparin and heparan sulfate (41, 47, 141, 152). This interaction is involved in at least four types of cooperative events influencing the binding of fibronectin to other ligands. The binding of heparin stimulates fibronectin-mediated uptake of gelatinized particles by macrophages (reviewed in 75, 79, 80). In addition, heparin and heparan sulfate enhance the binding of fibronectin to native and denatured collagens types I and III (160–163). Heparin (as macromolecular heparin) disrupts the attachment of fibroblasts to collagen films (164). Heparin also promotes the release of fibronectin from tissues (165). Binding of heparin to fibronectin in chilled plasma also results in a precipitation of fibrinogen (152). Since heparin is normally an intracellular rather than an extracellular molecule, the closely related molecule heparan sulfate may be the molecule actually involved in some of these events in vivo. Heparan sulfate proteoglycans may be capable of establishing transmembrane associations with the plasma membrane (166; A. C. Rapraeger, M. Bernfield, submitted for publication), and may consequently function as binding sites for fibronectin on cells (167). An additional function for the interaction of fibronectin with heparan-sulfate-containing proteoglycans may be to provide structural organization to the extracellular matrix, where fibronectin could function as a cross-linking molecule in stabilizing the matrix around cells.

In binding assays in vitro, the interactions of fibronectin and heparin are of moderately high affinity ($K_D = 10^{-7}$–4×10^{-9} M); the binding appears complex, with at least two components by Scatchard analysis (141). It is not known whether this range of affinities is caused by binding of more than one species of ligand or by the action of more than one binding site on fibronectin. Similarly to heparin-antithrombin binding, the binding of fibronectin to heparin is disrupted by 0.3–0.5 M NaCl, which implies that the interaction is electrostatic (141, 152, 168). However, the binding is specific in that it is saturable and is competitively inhibited by only heparin or heparan sulfate among a series of glycosaminoglycans examined (141).

There are two or three distinct heparin-binding sites on each fibronectin chain, which have been isolated by affinity and ion exchange chromatography. One binding site is located in the amino-terminal domain of the molecule (Figure 1; Table 1; 98, 118, 119, 156, 169, 170). Binding to heparin at this site appears to be of lower affinity than at the carboxyl-terminal domain, and is inhibited by 0.25 M NaCl (159). The interaction of this domain with heparin can be modulated by physiological concentrations of calcium, but not by other divalent cations (169). This modulation by calcium suggests that the binding of heparin-like molecules to this domain are regulated by local concentrations of calcium, e.g. in regions of calcium accumulation near certain glycosaminoglycans.

A second heparin-binding site is located towards the carboxyl terminus of fibronectin (Figure 1; Table 1; 98, 118–120, 122, 141, 159, 169, 171–173). Binding at this site is of higher apparent affinity, and is inhibited by 0.5 M NaCl, but not by divalent cations (159, 169, 171). A third site containing latent heparin-binding activity is normally suppressed in the presence of physiological concentrations of divalent cations (98, 169). The relative biological importance and function of each of these sites in vivo remain to be elucidated. Fibronectin can also bind to DNA by several sites (98, 174), but binding is inhibited by physiological concentrations of divalent cations (169).

Fibronectin also binds to the glycosaminoglycan hyaluronic acid (141, 160, 161, 163), which could influence the interactions of hyaluronic acid with cells or with other constituents of extracellular spaces. In binding studies with purified components, fibronectin binds to hyaluronic acid with moderate affinity and complex kinetics ($K_D = 10^{-7}$ M). The binding sites for hyaluronic acid and heparin are clearly distinct, since the binding of radioactively labeled ligand to one site is not affected by saturating amounts of binding to the other type of site; the binding of hyaluronic acid also differs in its lack of inhibition by even 1–2 M NaCl (141, 175).

Fibronectin interactions with hyaluronic acid have complicated cooperative effects in inhibiting the heparin-stabilized binding of fibronectin to native type III collagen or the heparin stimulation of fibronectin-mediated binding of gelatin by macrophages (160, 163). Although it is specific and saturable (141), the binding of fibronectin to hyaluronic acid is complex and poorly understood, in that hyaluronic acid binding is greater to aggregates of cellular fibronectin than to the same material immobilized on affinity columns (141, 176), and is reportedly minimal to human plasma fibronectin (176). On the other hand, intact porcine plasma fibronectin binds tightly to hyaluronic acid affinity columns. After cleavage by cathepsin D, nearly all porcine fibronectin fragments reportedly bind to hyaluronic acid, although the fragments differ in their sensitivities to elution by increasing ionic strength; these hyaluronate-binding sites are not yet mapped (175). The nature of fibronectin binding sites for hyaluronic acid therefore requires

Table 1 Current summary of properties of fibronectin domains[a]

Domain	Size (kilodaltons)	pI	Disulfides	CHO[b]	Other properties
Amino-terminal	27–31	8.2	10	—	Binds fibrin, heparin, S. aureus, actin, and transglutaminase
Collagen-binding	30–70	4.9–5.3	~13	≥60%	Also binds fibrin weakly
Cell-binding	75–140	?	Few[c]	≤40%	
Heparin-binding II	35–38	8.2–8.9	~1	—	Site of highest affinity
Fibrin-binding II	30–34	5.0–5.8	~11[c]	—	Lower apparent affinity
Heparin II/Fibrin II	58–62	6.4–6.8	?[c]	—	Probable B chain origin
Carboxyl-terminal	3	?	2	—	Interchain disulfides

[a] References concerning each domain are listed by domain in the text; see also Figure 1.
[b] Carbohydrates as percentage of total.
[c] One free sulfhydryl group is also present.

further detailed examination. Hyaluronic acid can also bind to cells with moderately high affinity by means of a receptor that is probably distinct from fibronectin (177).

Chondroitin sulfate proteoglycans also interact with fibronectin (178–181), although free chondroitin sulfate chains do not (141, 180). Chondroitin sulfate proteoglycans inhibit fibronectin-mediated adhesion, whereas free glycosaminoglycan chains do not (178, 179, 181). Proteoglycans can also form complexes that increase the binding of fibronectin to native type I collagen (180). These proteoglycans may therefore modulate fibronectin-cell and fibronectin-extracellular matrix interactions.

Other Interactions of Fibronectin

CYTOSKELETAL PROTEINS   Plasma and cellular fibronectins bind to actin (182–184). The interaction occurs at the amino-terminal domain, but other binding sites are revealed under nonphysiological salt conditions (98, 183). In addition, interactions with myosin, tropomyosin, α-actinin, and vinculin have been reported; the affinities of these interactions are significantly lower than for binding to gelatin (185). The physiological significance of these interactions with many cytoskeletal proteins is not yet clear. The interactions with actin are claimed to require denaturation of the actin, and preliminary studies show that actin can be phagocytosed by macrophages in the presence of fibronectin (184, 186). These results suggest that such interactions are involved in the clearance of cytoskeletal proteins from blood by the reticuloendothelial system, as proposed for collagenous debris (75, 168).

AMINES AND POLYAMINES   Amino sugars and other amines inhibit the hemagglutinating activity of cellular and plasma fibronectins, and fibronectin-mediated cell attachment to collagen (187–189). In addition, amines and especially polyamines such as spermidine inhibit the binding of plasma fibronectin to gelatin affinity columns; this inhibition provides a nondenaturing method for eluting fibronectin from affinity columns during purification (190). Spermidine binds to the fibronectin rather than to the gelatin at physiological salt concentrations. Several domains of fibronectin bind to amines, including the gelatin-binding domain (191). Amines can also promote the polymerization of plasma fibronectin, but the effects are highly dependent on ionic strength, and no effects are observed at physiological ionic strength (192).

SELF-ASSOCIATION   A striking feature of fibronectin on the cell surface is its arrangement into fibrils as a result of self-self associations (72–80). In some cases, fibronectin fibrils can closely parallel intracellular microfilament bundles; the nature of this fibronectin-actin relationship is not known, although the existence of intermediary plasma membrane connections has been postulated (42, 78, 80). The mechanisms of fibronectin polymerization are also obscure. Plasma fibronectin can undergo limited amounts of self-aggregation, particularly if incubated at low ionic strength at 4°C, and will form filaments 2 nm and thicker resembling fibrils observed in vivo (192, 193). Fibronectin may have self-association sites, either at specific points, e.g. in a 60,000-dalton gelatin-binding domain for binding to a second COOH-terminal site (122), or as a result of alternating overall positive and negative charges in adjacent domains of fibronectin for electrostatic linking of chains (79; Table 1). Fibronectin polymerization is enhanced by polyamines as noted above, but also by heparin (160). Moreover, the heparin-induced polymerization occurs at physiological ionic strengths; thus proteoglycans might promote the polymerization of fibronectin in vivo. Fibronectins can also undergo covalent cross-linking at the cell surface to form polymers by disulfide bonding or by factor XIIIa transglutaminase (85, 101, 102, 194). The mechanisms and regulation of fibronectin polymerization remain unsolved problems.

BINDING OF BACTERIA   Fibronectin also interacts with certain bacteria, such as *Staphylococcus* and *Streptococcus*. Both types of bacteria are found at the amino-terminal domain, and the former can be cross-linked covalently to fibronectin by transglutaminase (195–198; P. Speziale, M. Höök, T. Wadström, submitted for publication). These studies also suggest that the bacterial "receptors" for fibronectin could be a protein or lipoteichoic acid (198). These interactions of fibronectin with bacteria might be important for phagocytosis by macrophages (75, 196; but see 199). Alternatively, they might be important in permitting the bacteria to adhere to fibronectin or other extracellular molecules in wounds to initiate infections, as supported by the finding that *Escherichia coli* binds to laminin, another extracellular glycoprotein (199a).

Primary Structure of Fibronectin

Petersen et al (125, 200–201b; Figure 1) recently obtained amino acid sequences for over half of the bovine plasma fibronectin molecule; substantially shorter sequences are also available from other laboratories for human plasma fibronectin (124, 139, 202, 203). The subunits of plasma fibronectin are linked together at the carboxyl terminus to form a dimer (72–80, 84, 193, 202, 204) by two interchain disulfide bonds (125). Plasma fibronectin is composed of A and B chains, which differ slightly in apparent molecular weight on polyacrylamide gels, and are thought to form disulfide-linked heterodimers (84, 119, 122, 159, 172, 173, 205). Differences in primary structure between these chains are not yet found by either peptide mapping or amino acid sequencing (125, 205, 206). Nevertheless, a site of structural difference has been identified by limited proteolytic fragmentation with several proteases (119, 122, 156a, 159, 173); although the differences are not caused by carbohydrates, it is not clear whether they are the result of differences in primary structure or in posttranslational modifications. One speculation is that a difference region exists at this site to ensure complementarity and parallel alignment of fibronectin subunits immediately after synthesis, so that the two interchain disulfide bonds form accurately. Cellular and plasma fibronectins also appear to differ in this region (98, 100); this might account for the greater apparent propensity of cellular fibronectin to form disulfide-linked multimers. These speculations could be tested by further primary structure analyses of these regions.

A striking feature of the primary structure of fibronectin is that it contains small repeating units of 45–90 amino acids (125, 201–201b; Figure 1, bottom). These regions of amino acid sequence homology are of at least three types, termed types I, II, and III (125, 201). A repeating unit of 45 amino acids stabilized by two disulfide bonds is present at least 12 times in the fibronectin molecule near its amino and carboxyl termini. This unit is termed homology type I, or "fingers," because they resemble fingers on a hand in current schematic representations; their secondary and tertiary structure is not yet known (125). The amino-terminal domain contains five such fingers, the collagen-binding domain contains four more, and a carboxyl-terminal domain contains another three; all of these domains have been reported to interact with fibrin, although some interactions may be weak. The extensive intrachain disulfide bonding in type I structures accounts for the marked enrichment of disulfides in these regions of the molecule (Table 1; 125, 138, 139).

Gene cloning experiments indicate that the gene for chicken cellular fibronectin contains at least 48 small exons that are approximately 150 base pairs long; these exons would code for polypeptide units the same size as the type I homology known to exist in bovine plasma fibronectin (206a). It is therefore possible that fibronectin was derived from gene duplications of one to only a few primordial genes during evolution, and that the multidomain structure of the present molecule evolved from a primordial, simple, repeating polypeptide structure. Further amino acid and DNA sequencing studies are obviously required to test these hypotheses.

COLLAGEN-BINDING DOMAIN   The amino acid sequences of collagen-binding domains of bovine and human plasma fibronectin have roughly 80% homology according to the limited comparative sequencing performed to date (125, 203). One distinctive feature of this region is a -Pro-Pro-Pro-sequence near the amino terminus of the collagen-binding domain in both human and bovine fibronectins (125, 203); since a 10,000-dalton poly-peptide region at this site is reportedly crucial for the binding of collagen (170), amino acids at this site could contribute to the binding site for collagen. Both type I and type II homology units are probably present in the smallest biologically active fragments that bind to gelatin (115, 125, 137). Both of these types of unit contain intrachain disulfide bonds, and intact disulfide bonds are found to be required for the binding of fibronectin to gelatin affinity columns (138, 207); these findings suggest that a critical functional site in this region is stabilized by disulfide bonding.

CELL-BINDING DOMAIN   The cell-binding region and at least one free sulfhydryl group are located in regions of type III homology that are deficient in disulfide bonds and that may contain substantial amounts of β structure (124, 125; Table 1). The amino acid sequence of the cell-binding regions of human and bovine plasma fibronectins share 98% homology, i.e. there is a high degree of conservation of primary structure at this site (124, 125). As expected from the sequence data, the biological activity of the cell-binding region does not depend on intact disulfides (208, 209).

FREE SULFHYDRYLS   Human plasma fibronectin appears to contain two free sulfhydryls that are not readily accessible; they are exposed only after treatment with harsh denaturants (84, 117, 210). In contrast, one or two free sulfhydryls are apparently readily accessible in hamster cellular fibro-nectins (207, 211), and their alkylation is reported to interfere with the incorporation of fibronectin into the extracellular matrix (207). One free sulfhydryl is approximately 170,000 from the amino terminus, and the second is in the carboxyl-terminal fibrin-binding domain (117, 142, 201, 207, 210, 156a, 211a). Since cellular fibronectins are often extensively disulfide-bonded into polymers, such free sulfhydryls could be involved in the cross-linking of fibronectin molecules at the cell surface (207). One reservation concerning this hypothesis, however, is that the cleavage patterns of purified, multimeric cellular fibronectin appear more consistent with interchain disulfide bonding at the extreme carboxyl terminus, with no convincing evidence as yet for cross-linking at the more internal sulfhydryl sites, which should result in distinctive disulfide-linked fragments (e.g. 212). These uncertainties about the role and relative accessibilities of free sulf-hydryl groups remain to be resolved, perhaps by a combination of chemical modification, biological, and protease cleavage experiments comparing cellular and plasma fibronectins.

PHOSPHATE AND SULFATE   Fibronectin contains phosphate covalently bound to serine residues, although the molar quantity is uncertain (213–

215). One phosphoserine is located only three amino acids from the carboxyl terminus of bovine plasma fibronectin (125). Proteolytic mapping experiments, however, indicate that the phosphate in hamster and human cellular fibronectins is not located in this extreme COOH-terminal domain, but instead in a protease-labile, highly localized region near, but not at, one end of the molecule (214, 215). The labeling of phosphate on fibronectin has been reported to increase or decrease after transformation of fibroblasts from different species (213, 214). The exact locations, possible mechanisms of modulation, and functions of these phosphoserine residues still remain to be discovered. Fibronectin also contains covalently bound sulfate, but its location and function remain to be determined (106, 216).

CHEMICAL MODIFICATION   Preliminary chemical modification studies suggest that the interaction of fibronectin with cells is not dependent on free sulfhydryl groups, amino groups, or methionine residues (208, 209). Derivatization of fibronectin with carbodiimide, however, markedly inhibits fibronectin-mediated spreading of fibroblasts and hemagglutination, which suggests that carboxyl group(s) are required for fibronectin interactions with the cell surface (208, 217). Intact tyrosine and tryptophan residues may also be necessary for cell spreading, and lysine and arginine residues for binding to gelatin (208, 218, but see 209). In the future, quantitative studies of the effects of modifying individual amino acids within purified domains may provide more specific insights into active sites in fibronectin.

CARBOHYDRATES   Plasma and fibroblast cellular fibronectins contain approximately 5% carbohydrate, which are of the biantennary complex-oligosaccharide type linked to asparagine residues (84–86, 92–97). The role of these oligosaccharides on cellular fibronectin can be examined with the glycosylation inhibitor tunicamycin, which relatively specifically inhibits dolichol pyrophosphate-mediated glycosylation at asparagine residues. Nonglycosylated fibronectin is synthesized and secreted by tunicamycin-treated cells in nearly normal quantities (219), suggesting that the carbohydrate moiety has no major function in the secretion of this glycoprotein, as for many other glycoproteins (220, but see 221). Nonglycosylated fibronectin is two- to threefold more susceptible to proteolysis (222, 223), however, resulting in a decrease in quantities of fibronectin on the cell surface (219, 224). The apparent protective effect of carbohydrates against hydrolysis by proteases is most striking for the normally heavily glycosylated collagen-binding domain (223). In the absence of carbohydrate, this domain is apparently susceptible to cleavage in the middle of a disulfide-linked loop, which is followed by degradation of the domain (225).

The role of carbohydrates in the biological function of fibronectin has been examined with nonglycosylated fibronectin purified from tunicamycin-treated cells. The absence of carbohydrate has no detectable effect on a series of biological activities, including hemagglutination, spreading of cells on plastic substrates, adhesion of cells to collagen, and effects on the morphology of transformed cells (222). In addition, alterations in oligosaccharide branching following oncogenic transformation do not affect the morphological effects of fibronectin (226). It is therefore unlikely that the interactions of this extracellular protein with cells involve its carbohydrates, and they may also not be required for its interactions with ligands (222, 223). Instead, the carbohydrate moiety appears to function to protect specific region(s) of the molecule from proteolytic attack. The mechanisms of this putative protective effect on fibronectin and on other glycoproteins remain to be determined.

Structure-Function Relationships

The existence of specific functional domains arrayed in a modular fashion along the polypeptide subunits of fibronectin begins to explain its many ligand-binding and biological activities. To date, the known interactions of fibronectin with the cell surface can be attributed to binding at the cell-binding domain, although the exact boundaries of this region are not known. Proteolytic fragments containing only this region are sufficient to mediate cell attachment and spreading on plastic substrates, as well as activity promoting the directional migration of fibroblasts and neural crest cells in Boyden chamber assays (123, 227–229). Antibodies against this domain block cell attachment, migration of cells in Boyden chambers, and locomotion of neural crest cells (123, 126, 230, 64a). To obtain cell attachment to a layer of collagen, however, requires the concerted action of both cell-binding and collagen-binding domains linked together in an intact polypeptide fragment (114, 116). If these domains are separated by proteases, they compete instead with intact fibronectin and inhibit fibronectin-mediated attachment of cells to collagen (116). Antibodies against the collagen-binding domain block fibronectin interactions with collagen but not attachment and spreading of cells on plastic substrates (146). To obtain the weak cell-cell adhesive activity of fibronectin requires even larger units; monomeric fibronectin is relatively poor in hemagglutination activity, which requires multimeric fibronectin (116).

Other functional combinations of binding domains are required for the more complex cooperative effects reviewed above, in which glycosaminoglycans such as heparin or hyaluronic acid can modify the binding of fibronectin to fibrinogen and to collagen. The relative importance of these multiple-ligand interactions for interactions with the cell surface or for providing structural organization to the extracellular matrix remains to be determined.

LAMININ

Location and General Structure

Laminin [*lamina* = layer] is a glycoprotein of basement membranes (231–235). It is located in the basement membranes of a variety of tissues, and is absent from other extracellular sites (reviewed in 235). By immunoelectron microscopy, laminin has been localized to the lamina rarae of basement membranes at sites immediately adjacent to cells, which suggests that it might help to mediate interactions between epithelial cells and the basement membrane (reviewed in 235). Since the basement membrane is a very thin structure, it has been difficult to purify proteins directly from isolated basement membranes. However, several tumor cell lines produce large amounts of basement membrane proteins such as laminin and type IV collagen, and tumors such as the EHS sarcoma and parietal endoderm cell lines now provide reliable sources of pure basement membrane proteins (231–234).

Laminin is thought to be composed of three A chains, each 200,000 daltons, and one B chain of 400,000 daltons covalently linked at an end by disulfide bonds to form a molecule of approximately 1,000,000 daltons (Figure 2; 231–239). The A and B chains (also known as α and β chains) are distinct polypeptides, according to immunological, protease sensitivity, and peptide mapping criteria (236, 238). Laminin contains 12–15% carbohydrate, apparently primarily asparagine-linked oligosaccharides (231, 236).

As visualized by rotary shadowing electron microscopy, laminin has a distinctive cruciform shape with at least seven globular domains connected by rod-like elements (111, 237–239; Figure 2). The three A chains comprise the three short arms of the cross, each terminating in two globular domains, and the B chain provides the long arm, terminating in one globular domain (237–239). Since these studies were performed in the presence of 50%–70% glycerol, the native shape of the molecule could be slightly different. In addition, laminin from different cell types may have slightly differing structures. Laminin from teratocarcinoma-derived parietal endoderm cells has a third, nondisulfide-linked polypeptide component of 150,000 daltons, and it may also contain only two B chains (233). Furthermore, these cells produce two types of B chain differing in apparent size by 10,000 daltons (233). Besides soluble laminin, EHS sarcoma cells also produce another class of relatively insoluble basement membrane material that is immunologically related to laminin, but which is probably not identical (240).

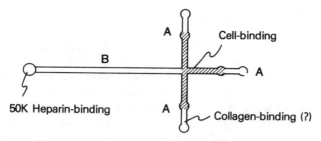

Figure 2 Structural and functional domains of laminin. Laminin consists of A and B chains linked by disulfide bonds and terminating in globular domains. Putative sites of ligand-binding activity are indicated; the central shaded region and some of the globular domains can be obtained as protease-resistant fragments.

Function of Laminin

Laminin functions to mediate the attachment of epithelial cells to type IV collagen of basement membranes (6, 20, 61, 66). Although laminin appears to be specific for binding to type IV collagen in a variety of in vitro assays, and in fact binds preferentially to native rather than denatured type IV collagen (H. K. Kleinman, personal communication), the specificities of laminin are not absolute. For example, laminin can also mediate the attachment of not only epithelial, but also fibroblastic cells to plastic or glass substrates (241), although it may be substantially less effective than fibronectin for fibroblastic cells (241a). It is also important to note that besides laminin, certain epithelial cells can also readily utilize fibronectin for attachment (61, 242). Hepatocytes attach to either protein by function-ally independent receptors (61); on the other hand, epidermal cells appear to have a strong dependence on laminin for attachment to type IV collagen (66, 243). The degree of dependence on laminin for attachment by other epithelial cells requires more extensive investigation; nevertheless, it is likely to be an important adhesive factor for them because of its location in adjacent basement membranes. Similarly, although laminin is a more effective substrate for sensory ganglion neurite outgrowth, fibronectin is also active for such neuronal cell outgrowth (244, 245).

Besides type IV collagen, laminin also binds to heparin and heparan sulfate (246). This interaction may be particularly important in interactions of laminin with heparan sulfate proteoglycans, which are present in sub-stantial amounts in basement membranes (235).

Structural and Functional Domains

Circular dichroism studies suggest that laminin may contain 20%–30% α helix and 15% β structure (237). When laminin is heated, a sharp thermal transition is identified at 58°C, which suggests disruption of a large number of noncovalent interactions at this temperature (237). A similar loss of apparent α helical structure occurs after partial digestion of laminin by elastase, and a thermal transition is no longer demonstrable with the remaining fragments (237). Since a number of domains remain intact after this treatment according to electrophoretic and electron microscopic criteria, laminin appears to be composed of heat- and elastase-resistant domains connected by α-helical regions sensitive to these agents.

Proteases have been used to dissect the laminin molecule into domains that bind to cells or to heparin (232, 237–239; Figure 2). The B subunit comprising the long arm is particularly sensitive to proteases. For example, thrombin and other proteases destroy much of the B subunit, leaving a three-chain complex containing the short arms and the terminal globular domains intact (237–239). This remaining structure can still mediate cell attachment to type IV collagen, and it still binds weakly to heparin (237, 239). This fragment therefore appears to retain the cell-binding, collagen-binding, and possibly weak heparin-binding sites of intact laminin. In contrast, digestion with pepsin degrades about 70% of the laminin molecules into small peptides, cleaving not only the long arm, but also substantial portions of the terminal globular domains, and the capacity for cell attachment to collagen is lost (232, 239, 240). A truncated fragment of about 300,000 daltons is recovered (232, 240) which can nevertheless apparently still bind to cells (239, 239a; Figure 2). This fragment is rich in half-cystine (12% of its residues) and appears homogeneous in the absence of reducing agents, but it dissociates into several cleavage products of 30,000–120,000 after reduction (240). Since this fragment loses the capacity to attach cells to collagen after removal of its terminal globular domains, these domains could be collagen-binding domains, although this hypothesis has not been tested directly (239, 239a).

Another globular domain of 50,000 daltons, which may be derived from the end of the B subunit, can also be recovered from digests and contains a heparin-binding site (237). Although this fragment binds quantitatively to heparin affinity columns, 60%–90% can be eluted at only 0.2 M salt (237); thus binding may be relatively weak at physiological salt concentrations. This heparin-binding domain also contains much of the β structure of laminin as inferred from CD spectra (237).

The functional model for laminin depicted in Figure 2 contains a central cell-binding region and terminal collagen- and heparin-binding domains. The mapping of these sites is still tentative, and it is still not clear whether there are multiple heparin- and collagen-binding sites; for example, additional heparin-binding sites on the short arms are likely from preliminary published data (237). An additional property of laminin is its apparent capability of binding directly to living cells (239) as well as to fixed erythrocytes (241a); if these interactions are of high affinity, they may permit the isolation of a cell-binding domain of laminin and a laminin receptor on cells. Rapid progress is expected in the details of laminin function, as its functional domains and primary structure are elucidated.

OTHER CELL ATTACHMENT FACTORS

Chondronectin

Chondronectin [*chondros* = cartilage + *nectere* = to bind, connect] is a glycoprotein isolated from chicken serum that specifically mediates the attachment of chondrocytes to cartilage type II collagen in vitro (68, 247, 248). It exists as a disulfide-linked multimer of 180,000 daltons composed of subunits estimated as 70,000 by gel electrophoresis (247). Chondronectin acts at relatively low protein concentrations; 50 ng/ml produces 50% attachment of chondrocytes (247), whereas fibronectin and laminin require 100–1000 ng/ml under even ideal assay conditions with other cells (61, 116). Chondronectin is specific for chondrocytes, and displays minimal attachment activity for fibroblasts or epithelial cells. An immunologically and biologically related protein is produced by chondrocytes; by immunofluorescence microscopy, this protein is located at the interface between the cells and the extracellular matrix (247). Chondronectin resembles fibronectin and laminin in binding to glycosaminoglycans; although it has poor avidity for type II collagen alone (68), it binds readily to a complex of cartilage proteoglycan and type II collagen (248). Free chondroitin sulfate chains also interact with chondronectin, whereas several other glycosaminoglycans do not (248). Chondronectin therefore appears to be a specific attachment protein, with a high specificity for mediating chondrocyte-to-cartilage extracellular matrix interactions.

Serum Spreading Factors

Although fibronectin and chondronectin are important cell attachment factors in serum, there is at least one other major glycoprotein attachment factor for adhesion of cells to collagen (249) and plastic substrates (250–253). In fact, during routine cell culturing in 10% serum, fibronectin attaches relatively poorly to plastic substrates as a result of competitive occupation of substrate adsorption sites by other serum proteins (252–254). Other components that attach to substrates more avidly or that interact with cells to stimulate adhesion indirectly therefore appear to be of greater importance than fibronectin for cells under these conditions (252–255); this finding may also explain the effectiveness of gelatin-coated substrates in tissue culture, since they can specifically concentrate plasma fibronectin from serum into the substrate (256). Nevertheless, endogenous cellular fibronectin probably furthers cytoskeletal organization even in 10% serum (35, 42, 63, 80, 167, 255).

A spreading factor for both fibroblastic and epithelial cells of 62,000–70,000 daltons has been characterized recently (252, 257). It exists as a monomer containing 12% carbohydrate, and has an unremarkable amino acid composition (257). Two unusual features of this factor are that it promotes adhesion several-fold more slowly than factors such as fibronectin and laminin, and that it is dependent on protein synthesis by the attaching cells (252). This glycoprotein, therefore, may not function by a simple cell-to-substrate ligand-binding mechanism. Another monomeric spreading factor with a similar size of 70,000 daltons is required for the spreading and growth of a variety of cell types in serum-free medium (258–260). It can be recovered in active form even after disulfide bond reduction and sodium dodecyl sulfate gel electrophoresis (260). Yet a third glycoprotein factor of 65,000 daltons promotes the spreading of epithelial cells, and has been termed epibolin (261). Finally, another attachment and spreading protein of 140,000 daltons from chicken serum that can induce fibroblast close contacts also remains to be characterized (251). The possible overlapping identities, structure, and mechanisms of action of these proteins are still under investigation, and should provide valuable insights into the multiple mechanisms of cell adhesion.

A cellular glycoprotein of 140,000 daltons can be purified from detergent extracts of cultured cells (262–264). This extracellular glycoprotein is extensively disulfide bonded into multimers, and the purified molecule can mediate weak cell attachment to plastic, which is stimulated by fibronectin (263, 264, 264a). Its mechanism of action and relationship to the 140,000-dalton glycoprotein complex implicated in cell-to-substrate adhesion by immunological criteria remain to be determined. Another fibronectin-independent cell-to-substrate attachment mechanism has been identified with somatic cell mutants that attach to substrate-attached cellular matrices but not to fibronectin. A protein of 265,000 daltons may help to mediate this type of adhesion (265).

Even cell-to-substrate interactions in simple model systems on artificial substrates are remarkably complex, and multiple mechanisms that act in concert to mediate normal cell interactions will probably be identified. As increasing numbers of such molecules are purified, the challenge is to determine which adhesive system becomes the most important during each of various cell-to-extracellular material interactions.

CELLULAR RECEPTORS FOR EXTRACELLULAR MOLECULES

Possible plasma membrane receptors for fibronectin, fibrinogen, laminin and collagen have been described. Determining the plasma membrane constituent to which fibronectin binds has been complicated by the inability of plasma fibronectin to bind with measurable affinity to the cell surface unless it has first interacted with a substrate (65, 266); it has therefore been difficult to demonstrate the existence of a discrete "receptor" by classical criteria such as saturability. Possible explanations for this apparent substrate activation include allosteric changes in the cell-binding site after binding to a substrate or a requirement for multivalency by unusually low-affinity receptors. Although cellular fibronectin and its cell-binding fragment appear to bind directly but weakly to cells (116), these molecules may simply be multivalent aggregates that are operationally similar to substrate-bound fibronectin.

Certain interactions of soluble plasma fibronectin with monocytes require fibronectin fragments rather than the intact molecule (267, 268). Monocyte chemotaxis and phagocytosis of particulate activators of the alternative complement pathway are not mediated by fibronectin itself, but instead by certain proteolytic cleavage products of the molecule. Among other possibilities, such cleavages might expose or activate a monocyte cell-binding site; whether binding of such a site would require a unique cellular "receptor," or whether this mechanism provides the equivalent of substrate activation for a soluble molecule is not clear. Another type of receptor for fibronectin appears on platelets activated by thrombin (269, 270); there are 10^5 of these receptors per platelet with a moderate affinity ($K_D = 3 \times 10^{-7}$ M). Whether these receptors are actually platelet-bound fibrin, thrombospondin (271–273), or a unique, fibronectin-specific receptor with an unusually high affinity relative to fibroblast "receptors" remains to be determined.

Several leading possibilities for a "receptor" for fibronectin on fibroblasts are glycoproteins (274–279), glycolipids (280–283), and heparan sulfate proteoglycans (41, 167). Indirect evidence that fibronectin requires a cell surface glycoconjugate to interact with the cell surface derives from studies of ricin-resistant mutants (284) and tunicamycin-treated cells (285, 286), both of which have defects in fibronectin-mediated adhesion. Fibronectin interaction can also be inhibited more directly by protease treatment of cells such as pronase or trypsin in low concentrations of calcium, as assayed by inhibition of cellular binding of fibronectin-coated beads or of cell spreading on fibronectin substrates. In the cases examined, a restoration of cellular capacity to attach to fibronectin requires protein synthesis (274, 275, 278, 279). Treatment of BHK cell surface proteins with trypsin in the presence of calcium does not inhibit cell attachment (279), and the proteins that remain undigested and are immunoprecipitated by an antiserum that inhibits cell attachment include glycoproteins of 80,000 and 120,000 daltons (279). These proteins are therefore candidates for the fibronectin "receptor" or a cell surface component necessary for postreceptor events.

An alternative approach has been to determine which protein becomes cross-linked to fibronectin in nearest-neighbor analyses with chemical cross-linkers. A glycoprotein of 47,000 is specifically cross-linked to fibronectin in cell-spreading assays, although the same protein appears to be cross-linked during cell spreading on lectins (276). Antibodies to partially purified preparations of this glycoprotein block cell spreading (287). This protein could therefore be a receptor for fibronectin that is also involved in cell spreading on lectins. In longer-term cultures of another cell line, however, fibronectin is preferentially cross-linked to sulfated proteoglycans (288).

Gangliosides are also possible receptors for fibronectin. All cell-adhesive interactions of fibronectin tested to date are sensitive to inhibition by gangliosides, although certain phospholipids are also inhibitors (280–283). Laminin-mediated hemagglutination is similarly inhibited by charged lipids (241a). The spreading of fibroblastic cells on lectins and glycosidases. however, is also inhibited by gangliosides (31), although not at high concentrations of lectin (282); high concentrations of fibronectin can also overcome ganglioside inhibition of fibronectin-mediated events (280, 282). These findings suggest that gangliosides competitively inhibit a process common to cell-spreading events in general, rather than only a fibronectin-specific receptor. In addition, ganglioside-mediated spreading of cells on substrate-bound cholera toxin can only partially mimic normal cell spreading (283), which implies that gangliosides alone are not sufficient as a receptor.

Purified gangliosides bind to fibronectin, although the interaction appears weak (281, 283). Nevertheless, fibronectin does not bind effectively to cells deficient in gangliosides according to immunofluorescence microscopy, and the addition of purified gangliosides to these cells restores fibronectin binding in characteristic fibrillar patterns (288a). These experiments suggest that gangliosides or other charged lipids can serve as receptors or binding sites for fibronectin in certain events such as organization of cell surface fibronectin; but they may not be the physiological receptor and may mimic structural features of a protein receptor, such as an unusual configuration of charges.

Another possible candidate is heparan sulfate proteoglycan. Although it is probably not a primary receptor because of the absence of heparin-binding activity in the fibronectin cell-binding site (123) and a lack of cell-spreading activity in a purified heparin-binding domain (171), heparan sulfate may be required for complete spreading of cells on a substrate (167). There may therefore be multiple cell surface components to which fibronectin can bind on the plasma membrane. Unambiguous identification of fibroblast and epithelial cell receptors for fibronectin and laminin is an important goal for the next few years.

A cell surface receptor for laminin has been isolated recently, and it contains a disulfide-complexed protein of 70,000 daltons (239b). This receptor binds to laminin with high affinity ($K_D = 2 \times 10^{-9}$ M), and there are approximately 50,000 binding sites per cell (239a,b). The isolation of this receptor should permit an evaluation of its role in the specificity and regulation of cell attachment to laminin. Saturable receptors for type I collagen have been identified on 3T3 cells; there are 500,000 sites with remarkably high apparent affinity, with $K_D = 10^{-11}$ M (289). The molecule responsible for this receptor activity is not yet identified on fibroblasts. However, platelets provide a particularly favorable system for analyzing cell interactions with collagen. Platelets adhere to native collagen more readily than to denatured collagens (290). A putative plasma membrane receptor for collagen fibrils or the $\alpha 1(I)$ chain of collagen can be purified by chromatography on collagen affinity columns; it is a protein of 65,000 or 75,000 daltons (291, 292). It may exist as a disulfide-linked tetramer, although it remains active after treatment with reducing agent (291, 292). The purified receptor binds to collagen with good affinity ($K_D = 2 \times 10^{-8}$ M), which is similar to affinities estimated with plasma membranes; purified receptor preparations inhibit the binding of platelets to collagen (291, 292). There is considerable uncertainty about the role of fibronectin in platelet adhesion to collagen. Fibronectin can stimulate the spreading of platelets on dried collagen films (reviewed in 80), and a patient with a form of Ehlers-Danlos syndrome has a defect in platelet activation by collagen correctable by the addition of normal fibronectin to assay mixtures (293). Fibronectin can also be cross-linked to the thrombospondin, a platelet lectin implicated in platelet aggregation (273). Whether these observations reflect a role for fibronectin in platelet adhesion is not yet clear. Earlier immunological evidence supporting fibronectin as a collagen receptor on platelets (294) has been shown to be the result of platelet activation by the F_c region of antibody complexes rather than by antibody binding to a fibronectin-containing receptor (295). There is consequently no direct evidence as yet for a role for fibronectin in platelet aggregation, but further studies are clearly needed.

Macrophages can bind fibrin by a nonsaturable mechanism involving fibronectin (155), and by a separate mechanism that binds fibrin monomer at a saturable, high-affinity site reportedly binding to the amino terminus of the fibrin α chain (296). In addition, fibroblasts bind fibrin and fibrinogen in a process probably involving fibronectin (297). The plasma membrane molecules directly or indirectly involved in these fibrin-binding events remain to be identified. Stimulation of platelets by thrombin and other agonists exposes receptors for fibrinogen; there are 4×10^4 receptor sites per platelet with affinities of $K_D = 10^{-7}$ M. According to protein/cross-linking and other studies, this receptor contains the platelet glycoprotein IIIa of M_r 105,000 (298). Platelets also interact with other platelets or substrates by other adhesive glycoproteins including von Willebrand factor (reviewed in 299) and thrombospondin (271–273). The mechanisms of these important platelet interactions still require elucidation.

SYNTHESIS AND REGULATION

The synthesis and processing of collagen have been reviewed previously in detail (12). Fibronectin is synthesized in the rough endoplasmic reticulum (300, 301) as a monomer with high mannose oligosaccharides. It very rapidly becomes a disulfide-linked dimer, and its oligosaccharides are trimmed and processed to the complex form prior to secretion via the Golgi apparatus in 30–60 min (302–305). Immature forms of fibronectin can sometimes exist as chains with two or three slightly different polypeptide sizes (205, 303); whether they represent distinct A and B chains of cellular fibronectin is not known, since the mature molecule migrates as a single band in gel electrophoresis. Fibronectin that is secreted onto the cell surface gradually forms higher disulfide-bonded polymers, whereas the material secreted into solution tends to remain dimeric (303). Limited amounts of plasma fibronectin from culture medium can also become incorporated into a disulfide-bonded cell surface matrix (306, 307). Fibronectin on the cell surface undergoes protein turnover, although varying amounts are lost by sloughing into medium (302).

The metabolism of laminin has not yet been examined extensively, but it also is synthesized as monomers of A (α) and B (β) chains that begin form-ing interchain disulfide bonds in less than 10 minutes; however, sub-stantial pools of unbonded monomers appear to exist in steady state (236). The early biosynthetic products of laminin contain high mannose oligo-saccharides (236), which are presumably processed to form complex oligosaccharides.

An expanding literature too extensive to include comprehensively in this review documents examples of the regulation of quantities of the various extracellular molecules that interact with cells, especially after oncogenic transformation. Such decreases, or occasionally increases or changes in types, of fibronectin, laminin, collagen, and heparan sulfate proteoglycan are complex and can have many sequelae; they may also correlate with the loss of differentiated cell function, and they can vary depending on the cell type (reviewed in 1, 4, 5, 308). Alterations of fibronectin in transformed cells have been studied in detail, and are found to result from multiple causes (4), the most common of which are defects of unknown cause in the capacity of cells to bind fibronectin (e.g. 72, 226, 309) and decreases in biosynthesis resulting from decreased levels of fibronectin messenger RNA (310).

Besides oncogenic transformation, treatment by tumor promoters or even the acquisition of immortality by cell strains often results in substantial decreases in cell surface fibronectin (311–313). Certain hormones and other effectors can increase levels of fibronectin. Glucocorticoids can increase the amounts of fibronectin on hepatocytes and transformed fibroblasts (314, 315); epidermal growth factor has similar effects under certain culture conditions (316); and elevated cyclic AMP (317, 318), vitamin A (319), and butyrate (320, 321) can all stimulate accumulation of cell surface fibronectin in different cell lines, often with accompanying morphological alterations. Therefore, fibronectin levels can be regulated, but the mechanisms of these effects remain to be determined.

As indicated in the introduction, extracellular molecules themselves can regulate the quantities of other extracellular molecules (1, 3, 5, 7). For example, fibronectin can stimulate the synthesis of collagen by hepatocytes (82), and of type I rather than type II collagen by chondrocytes (17, 18). Whether these effects are secondary to changes in cell shape or to direct metabolic effects of fibronectin, e.g. on protein phosphorylation (322, 323), is not yet known. Another series of effects includes reestablishment of a basal lamina (basement membrane) after treatment of tooth epithelium with fibronectin (324), thyroid epithelium by laminin (325), or mammary epithelium by type I collagen (326). The latter effect is attributable to a decreased rate of turnover of heparan sulfate proteoglycan (326), and turnover and remodeling of basal lamina may be regulated by mesenchyme cells (327). Determining the biological roles, specificity, and biochemical mechanisms of these types of interactions should be an area of intense investigation in the next few years.

FUTURE DIRECTIONS

Rapidly increasing numbers of extracellular molecules are being discovered that interact with the cell surface. To evaluate their roles and specificities in modulating cell behavior, the purified molecules need to be compared directly in a series of assays, e.g. for cell attachment and spreading on various substrates, for cell proliferation, and for cell migration and differentiation. The cell-type specificity of these factors, including laminin and the serum spreading factors, also needs further clarification using attachment assays with different types of epithelial cells, fibroblasts, and other cell types. In addition, our current concepts of the functions of these molecules based on model systems must still be verified experimentally in vivo, for example, by antibody inhibition studies and by characterizing cell and animal mutants defective in each specific molecule.

The details of the biochemical mechanisms of interactions of extra-cellular molecules with the cell surface often still remain to be discovered, for example, by purifying and reconstituting each of the pertinent plasma membrane receptors. Characterization of the active sites for binding to these receptors and for binding to many other extracellular macromole-cules should provide rich avenues of investigation. Even when these single-component binding systems are understood, it will still be necessary to determine how cooperative interactions between sites occur.

Finally, another poorly understood area concerns the mechanisms by which extracellular molecules, many of which appear to have adhesive or structural roles, can alter cellular biosynthetic patterns and cytoskeletal organization; step-by-step analysis of the intracellular events following "receptor" binding will be required. The recent purification and initial successes in dissecting apart the functional domains of several new factors with a wide variety of differing specificities suggest that the field will continue its rapid progress in understanding these complex, important processes at the molecular level.

ACKNOWLEDGMENTS

I am grateful to my colleagues in the field for valuable discussions and especially for so generously providing many unpublished manuscripts.

Literature Cited

1. Hay, E. D., ed. 1981. *Cell Biology of Extracellular Matrix.* New York: Plenum. 417 pp.
2. Grinnell, F. 1978. *Int. Rev. Cytol.* 53:65–144
3. Kleinman, H. K., Klebe, R. J., Martin, G. R. 1981. *J. Cell Biol.* 88:473–85
4. Chen, L. B., ed. 1981. *Oncology Overview: Selected Abstr. on Fibronectin and Related Transformation-Sensitive Cell Surface Proteins.* Bethesda: Int. Cancer Res. Data Bank, Nat. Cancer Inst. 95 pp.
5. Yamada, K. M., ed. 1983. *Cell Inter-actions and Development: Molecular Mechanisms.* New York: Wiley. 287 pp.
6. Timpl, R., Martin, G. R. 1982. In *Immunochemistry of the Extracellular Matrix,* ed. H. Furthmayr, 2:119–50. Boca Ratan, FL: CRC.
7. Hay, E. D. 1982. In *Spatial Organization of Eukaryotic Cells,* ed. J. R. McIntosh. New York: Liss. In press
8. Hughes, R. C., Pena, S. D. J., Vischer, P. 1980. In *Cell Adhesion and Motility,* ed. A. S. G. Curtis, J. D. Pitts, pp. 329–56. London: Cambridge Univ. Press
9. Rees, D. A., Badley, R. A., Woods, A. 1980. See Ref. 8, pp. 389–408
10. Frazier, W. A., Glaser, L. 1979. *Ann. Rev. Biochem.* 48:491–523
11. Höök, M., Kjellén, L., Johansson, S.,

Robinson, J. 1984. *Ann. Rev. Biochem.* 53: In press
12. Bornstein, P., Sage, H. 1980. *Ann. Rev. Biochem.* 49:957–1003
13. Yamada, K. M., Yamada, S. S., Pastan, I. 1976. *Proc. Natl. Acad. Sci. USA* 73:1217–21
14. Yamada, K. M., Ohanian, S. H., Pastan, I. 1976. *Cell* 9:241–45
15. Ali, I. U., Mautner, V., Lanza, R., Hynes, R. O. 1977. *Cell* 11:115–26
16. Sugrue, S. P., Hay, E. D. 1981. *J. Cell Biol.* 91:45–54
17. Pennypacker, J. P., Hassell, J. R., Yamada, K. M., Pratt, R. M. 1979. *Exp. Cell Res.* 121:411–15
18. West, C. M., Lanza, R., Rosenbloom, J., Lowe, M., Holtzer, H., Avdalovic, N. 1979. *Cell* 17:491–501
19. Greenburg, G., Hay, E. D. 1982. *J. Cell Biol.* 95:333–39
20. Vlodavsky, I., Gospodarowicz, D. 1981. *Nature* 289:304–6
21. Chen, L. B., Murray, A., Segal, R. A., Bushnell, A., Walsh, M. L. 1978. *Cell* 14:377–91
22. Hedman, K., Kurkinen, M., Alitalo, K., Vaheri, A., Johansson, S., Höök, M. 1979. *J. Cell Biol.* 81:83–91
23. Gospodarowicz, D., Vlodavsky, I., Savion, N. 1980. *J. Supramolec. Struct.* 13:339–72
24. Rojkind, M., Gatmaitan, Z., Mac-Kensen, S., Giambrone, M.-A., Ponce, P., Reid, L. M. 1980. *J. Cell Biol.* 87: 255–63
25. Weiss, R. E., Reddi, A. H. 1980. *Proc. Natl. Acad. Sci. USA* 77:2074–78
26. Gospodarowicz, D., Gonzales, R., Fujii, D. K. 1983. *J. Cell. Physiol.* 114:191–202
27. Barnes, D., Sato, G. 1980. *Cell* 22:649–55
28. Grinnell, F., Hays, D. G. 1978. *Exp. Cell Res.* 116:275–84
29. Hughes, R. C., Pena, S. D. J., Clark, J., Dourmashkin, R. R. 1979. *Exp. Cell Res.* 121:307–14
30. Gjessing, R., Seglen, P. O. 1980. *Exp. Cell Res.* 129:239–49
31. Rauvala, H., Carter, W. G., Hakomori, S. 1981. *J. Cell Biol.* 88:127–37
32. Cassiman, J. J., Marynen, P., Brugmans, M., Van Leuven, F., Van den Berghe, H. 1981. *Cell Biol. Int. Rep.* 5:901–11
33. Aplin, J. D., Hughes, R. C. 1981. *J. Cell Sci.* 50:89–103
34. Klebe, R. J., Rosenberger, P. G., Naylor, S. L., Burns, R. L., Novak, R., Kleinman, H. 1977. *Exp. Cell Res.* 104:119–25
35. Norton, E. K., Izzard, C. S. 1982. *Exp. Cell Res.* 139:463–67

36. Wylie, D. E., Damsky, C. H., Buck, C. A. 1979. *J. Cell Biol.* 80:385–402
37. Damsky, C. H., Knudsen, K. A., Dorio, R. J., Buck, C. A. 1981. *J. Cell Biol.* 89:173–84
38. Damsky, C. H., Knudsen, K. A., Buck, C. A. 1982. *J. Cell Biochem.* 18:1–13
39. Neff, N. T., Lowrey, C., Decker, C., Tovar, A., Damsky, C., et al. 1982. *J. Cell Biol.* 95:654–66
40. Culp, L. A. 1976. *Biochemistry* 15:4094–4104
41. Rollins, B. J., Cathcart, M. K., Culp, L. A. 1982. In *The Glycoconjugates*, ed. M. I. Horowitz, 3:289–329. New York: Academic
42. Chen, W.-T., Singer, S. J. 1982. *J. Cell Biol.* 95:205–22
43. Culp, L. A., Ansbacher, R., Domen, C. 1980. *Biochemistry* 19:5899–907
44. Culp, L. A., Bensusan, H. 1978. *Nature* 273:680–82
45. Bruns, R. R., Gross, J. 1980. *Exp. Cell Res.* 128:1–7
46. Kleinman, H. K., McGarvey, M. L., Martin, G. R. 1982. *Cell Biol. Int. Resp.* 6:591–99
47. Garner, J. A., Culp, L. A. 1981. *Biochemistry* 20:7350–59
48. Lark, M. W., Culp, L. A. 1982. *J. Biol. Chem.* 257:14073–80
49. Rollins, B. J., Culp, L. A. 1979. *Biochemistry* 18:141–48
50. Avnur, Z., Geiger, B. 1981. *J. Mol. Biol.* 153:361–79
51. Bayley, S. A., Rees, D. A. 1982. *Biochim. Biophys. Acta* 689:351–62
52. Lander, A. D., Fujii, D. K., Gospodarowicz, D., Reichardt, L. F. 1982. *J. Cell Biol.* 94:574–85
53. Dickey, W. D., Seals, C. M. 1981. *Cancer Res.* 41:4027–30
54. Schubert, D., LaCorbiere, M. 1982. *J. Cell Biol.* 94:108–14
55. Linsenmayer, T. F., Gibney, E., Toole, B. P., Gross, J. 1978. *Exp. Cell Res.* 116:470–74
56. Grinnell, F., Minter, D. 1978. *Proc. Natl. Acad. Sci. USA* 75:4408–12
57. Schor, S. L., Court, J. 1979. *J. Cell Sci.* 38:267–81
58. Rubin, K., Höök, M., Obrink, B., Timpl, R. 1981. *Cell* 24:463–70
59. Briles, E. B., Haskew, N. B. 1982. *Exp. Cell Res.* 138:436–41
60. Briles, E. I. B. 1982. In *The Extracellular Matrix*, ed. S. P. Hawkes, J. L. Wang, pp. 115–19. New York: Academic
61. Johansson, S., Kjellén, L., Höök, M., Timpl, R. 1981. *J. Cell Biol.* 90:260–64
62. Schor, S. L., Schor, A. M., Bazill, G. W. 1981. *J. Cell Sci.* 49:299–310

63. Couchman, J. R., Rees, D. A., Green, M. R., Smith, C. G. 1982. *J. Cell Biol.* 93:402–10
64. Schor, S. L., Schor, A. M., Bazill, G. W. 1981. *J. Cell Sci.* 48:301–14
64a. Rosavio, R. A., DeLouvee, A., Yamada, K. M., Timpl, R., Thiery, J. P. 1982. *J. Cell Biol.* 96:462–73
65. Klebe, R. J. 1974. *Nature* 250:248–51
66. Terranova, V. P., Rohrbach, D. H., Martin, G. R. 1980. *Cell* 22:719–26
67. Chiquet, M., Puri, E. C., Turner, D. C. 1979. *J. Biol. Chem.* 254:5475–82
68. Hewitt, A. T., Kleinman, H. K., Pennypacker, J. P., Martin, G. R. 1980. *Proc. Natl. Acad. Sci. USA* 77:385–88
69. Vaheri, A., Kurkinen, M., Lehto, V.-P., Linder, E., Timpl, R. 1978. *Proc. Natl. Acad. Sci. USA* 75:4944–48
70. Furcht, L. T., Smith, D., Wendelschafer-Crabb, G., Mosher, D. F., Foidart, J. M. 1980. *J. Histochem. Cytochem.* 28: 1319–33
71. McDonald, J. A., Kelley, D. G., Broekelmann, T. J. 1982. *J. Cell Biol.* 92:485–92
72. Vaheri, A., Mosher, D. F. 1978. *Biochim. Biophys. Acta* 516:1–25
73. Mosher, D. F. 1980. *Prog. Hemostasis. Thromb.* 5:111–51
74. Mosesson, M. W., Amrani, D. L. 1980. *Blood* 56:145–48
75. Saba, T. M., Jaffe, E. 1980. *Am. J. Med.* 68:577–94
76. Pearlstein, E., Gold, L., Garcia-Pardo, A. 1980. *Mol. Cell Biochem.* 29:103–27
77. Ruoslahti, E., Engvall, E., Hayman, E. G. 1981. *Coll. Rel. Res.* 1:95–128
78. Hynes, R. O. 1981. See Ref. 1, pp. 295–334
79. Hörmann, H. 1982. *Klin. Wochenschr* 60:1265–77
80. Hynes, R. O., Yamada, K. M. 1982. *J. Cell. Biol.* 95:369–77
81. Akiyama, S., Yamada, K. M. 1983. Fibronectin and disease. In *Connective Tissue and Diseases of Connective Tissue*, ed. B. Wagner, R. Fleischmajer. Baltimore: Williams & Wilkins. In press
82. Foidart, J.-M., Berman, J. J., Paglia, L., Rennard, S., Abe, S., et al. 1980. *Lab. Invest.* 42:525–32
83. Owens, M. R., Cimino, C. D. 1982. *Blood* 59:1305–9
83a. Tamkun, J. W., Hynes, R. O. 1983. *J. Biol. Chem.* In press
84. Mosesson, M. W., Chen, A. B., Huseby, R. M. 1975. *Biochim. Biophys. Acta* 386:509–24
85. Yamada, K. M., Schlesinger, D. H., Kennedy, D. W., Pastan, I. 1977. *Biochemistry* 16:5552–59
86. Vuento, M., Wrann, M., Ruoslahti, E.

1977. *FEBS Lett.* 82:227–31
87. Pena, S. D. J., Hughes, R. C. 1978. *Cell. Biol. Int. Rep.* 2:339–44
88. Yamada, K. M., Kennedy, D. W. 1979. *J. Cell Biol.* 80:492–98
89. Alexander, S. S., Colonna, G., Yamada, K. M., Pastan, I., Edelhoch, H. 1978. *J. Biol. Chem.* 253:5820–24
90. Colonna, G., Alexander, S. S., Yamada, K. M., Pastan, I., Edelhoch, H. 1978. *J. Biol. Chem.* 253:7787–90
91. Alexander, S. S., Colonna, G., Edelhoch, H. 1979. *J. Biol. Chem.* 254:1501–5
92. Wrann, M. 1978. *Biochem. Biophys. Resp Commun.* 84:269–74
93. Carter, W. G., Hakomori, S. 1979. *Biochemistry* 18:730–38
94. Fukuda, M., Hakomori, S. 1979. *J. Biol. Chem.* 254:5451–57
95. Takasaki, S., Yamashita, K., Suzuki, K., Iwanaga, S., Kobata, A. 1979. *J. Biol. Chem.* 254:8548–53
96. Fisher, S. J., Laine, R. A. 1979. *J. Supramolec. Struct.* 11:391–99
97. Fukuda, M., Levery, S. B., Hakomori, S. 1982. *J. Biol. Chem.* 257:6856–60
98. Hayashi, M., Yamada, K. M. 1981. *J. Biol. Chem.* 256:1292–300
99. Hynes, R. O., Ali, I. U., Destree, A. T., Mautner, V., Perkins, M. E., et al. 1978. *Ann. NY Acad. Sci.* 312:317–42
100. Atherton, B. T., Hynes, R. O. 1981. *Cell* 25:133–41
101. Keski-Oja, J., Mosher, D. F., Vaheri, A. 1977. *Biochem. Biophys. Res. Commun.* 74:699–706
102. Hynes, R. O., Destree, A. 1977. *Proc. Natl. Acad. Sci. USA* 74:2855–59
103. Balian, G., Crouch, E., Click, E. M., Carter, W. G., Bornstein, P. 1979. *J. Supramolec. Struct.* 12:505–16
104. Ruoslahti, E., Engvall, E., Hayman, E. G., Spiro, R. G. 1981. *Biochem. J.* 193:295–99
105. Pande, H., Corkill, J., Sailor, R., Shively, J. E. 1981. *Biochem. Biophys. Res. Commun.* 101:265–72
106. Wilson, B. S., Ruberto, G., Ferrone, S. 1981. *Biochem. Biophys. Res. Commun.* 101:1047–51
107. Koteliansky, V. E., Glukhova, M. A., Bejanian, M. V., Smirnov, V. N., Filimonov, V. V., et al. 1981. *Eur. J. Biochem.* 119:619–24
108. Odermatt, E., Engel, J., Richter, H., Hörmann, H. 1982. *J. Mol. Biol.* 159:109–23
109. Wallace, D. G., Donovan, J. W., Schneider, P. M., Meunier, A. M., Lundblad, J. L. 1981. *Arch. Biochem. Biophys.* 212:515–24
110. Mosesson, M. W., Umfleet, R. A. 1970. *J. Biol. Chem.* 245:5728–36

111. Engel, J., Odermatt, E., Engel, A., Madri, J. A., Furthmayr, H., et al. 1981. *J. Mol. Biol.* 150:97–120
112. Erickson, H. P., Carrell, N., McDonagh, J. 1981. *J. Cell Biol.* 91:673–78
113. Koteliansky, V. E., Bejanian, M. V., Smirnov, V. N. 1980. *FEBS Lett.* 120:283–86
114. Ruoslathi, E., Hayman, E. G. 1979. *FEBS Lett.* 97:221–24
115. Hahn, L.-H. E., Yamada, K. M. 1979. *Proc. Natl. Acad. Sci. USA* 76:1160–63
116. Hahn, L.-H. E., Yamada, K. M. 1979. *Cell* 18:1043–51
117. McDonald, J. A., Kelley, D. G. 1980. *J. Biol. Chem.* 255:8848–58
118. Sekiguchi, K., Hakomori, S. 1980. *Proc. Natl. Acad. Sci. USA* 77:2661–65
119. Sekiguchi, K., Fukuda, M., Hakomori, S. 1981. *J. Biol. Chem.* 256:6452–62
120. Ruoslathi, E., Hayman, E. G., Engvall, E., Cothran, W. C., Butler, W. T. 1981. *J. Biol. Chem.* 256:7277–81
121. Ehrismann, R., Chiquet, M., Turner, D. C. 1981. *J. Biol. Chem.* 256:4056–62
122. Ehrismann, R., Roth, D. E., Eppenberger, H. M., Turner, D. C. 1982. *J. Biol. Chem.* 257:7381–87
123. Pierschbacher, M. D., Hayman, E. G., Ruoslahti, E. 1981. *Cell* 26:259–67
124. Pierschbacher, M. D., Ruoslahti, E., Sundelin, J., Lind, P., Peterson, P. A. 1982. *J. Biol. Chem.* 257:9593–97
125. Petersen, T. E., Thøgersen, H. C., Skorstengaard, K., Vibe-Pedersen, K., Sahl, P., et al. 1983. *Proc. Natl. Acad. Sci. USA* 80:137–41
125a. Pierschbacher, M., Hayman, E. G., Ruoslahti, E. 1983. *Proc. Natl. Acad. Sci. USA.* In press
126. McDonald, J. A., Senior, R. M., Griffin, G. L., Broekelmann, T. J., Prevedel, P. 1982. *J. Cell Biol.* 95:123a (Abstr.)
127. Engvall, E., Ruoslahti, E. 1977. *Int. J. Cancer* 20:1–5
128. Kleinman, H. K., McGoodwin, E. B., Martin, G. R., Klebe, R. J., Fietzek, P. P., Woolley, D. E. 1978. *J. Biol. Chem.* 253:5642–46
129. Dessau, W., Adelmann, B. C., Timpl, R., Martin, G. R. 1978. *Biochem. J.* 169:55–59
130. Jilek, F., Hörmann, H. 1978. *Hoppe-Seyler's Z. Physiol. Chem.* 359:247–50
131. Engvall, E., Ruoslahti, E., Miller, E. J. 1978. *J. Exp. Med.* 147:1584–95
132. Hörmann, H., Jilek, F. 1980. *Artery* 8:482–86
133. Klebe, R. J., Hall, J. R., Rosenberger, P., Dickey, W. D. 1977. *Exp. Cell Res.* 110:419–25
134. Kleinman, H. K., Wilkes, C. M., Martin, G. R. 1981. *Biochemistry* 20:2325–30
135. Newgreen, D., Thiery, J.-P. 1980. *Cell Tissue Res.* 211:269–91
136. Mayer, B. W., Hay, E. D., Hynes, R. O. 1981. *Devel. Biol.* 82:267–86
137. Ruoslahti, E., Hayman, E. G., Kuusela, P., Shively, J. E., Engvall, E. 1979. *J. Biol. Chem.* 254:6054–59
138. Balian, G., Click, E. M., Crouch, E., Davidson, J., Bornstein, P. 1979. *J. Biol. Chem.* 254:1429–32
139. Gold, L. I., Garcia-Pardo, A., Frangione, B., Franklin, E. C., Pearlstein, E. 1979. *Proc. Natl. Acad. Sci. USA* 76:4803–7
140. Mosher, D. F., Schad, P. E., Vann, J. M. 1980. *J. Biol. Chem.* 255:1181–88
141. Yamada, K. M., Kennedy, D. W., Kimata, K., Pratt, R. M. 1980. *J. Biol. Chem.* 255:6055–63
142. Wagner, D. D., Hynes, R. O. 1980. *J. Biol. Chem.* 255:4304–12
143. Balian, G., Click, E. M., Bornstein, P. 1980. *J. Biol. Chem.* 255:3234–36
144. Furie, M. B., Frey, A. B., Rifkin, D. B. 1980. *J. Biol. Chem.* 255:4391–94
145. Vartio, T., Seppä, H., Vaheri, A. 1981. *J. Biol. Chem.* 256:471–77
146. McDonald, J. A., Broekelmann, T. J., Kelley, D. G., Villiger, B. 1981. *J. Biol. Chem.* 256:5583–87
147. Morrison, P. R., Edsall, J. T., Miller, S. G. 1948. *J. Am. Chem. Soc.* 70:3103–8
148. Ruoslahti, E., Vaheri, A. 1975. *J. Exp. Med.* 141:497–501
149. Mosher, D. F. 1975. *J. Biol. Chem.* 250:6614–21
150. Mosher, D. F. 1976. *J. Biol. Chem.* 251:1639–45
151. Stemberger, A., Hörmann, H. 1976. *Hoppe-Seyler's Z. Physiol. Chem.* 357:1003–5
152. Stathakis, N. E., Mosesson, M. W. 1977. *J. Clin. Invest.* 60:855–65
153. Kamykowski, G. W., Mosher, D. F., Lorand, L., Ferry, J. D. 1981. *Biophys. Chem.* 13:25–28
154. Grinnell, F., Feld, M. K., Minter, D. 1980. *Cell* 19:517–25
155. Jilek, F., Hörmann, H. 1978. *Hoppe-Seyler's Z. Physiol. Chem.* 359:1603–5
156. Hörmann, H., Seidl, M. 1980. *Hoppe-Seyler's Z. Physiol. Chem.* 361:1449–52
156a. Hayashi, M., Yamada, K. M. 1983. *J. Biol. Chem.* 258:3332–40
156b. Sekiguchi, K., Hakomori, S. 1983. *J. Biol. Chem.* 258:3967–73
157. Jilek, F., Hörmann, H. 1977. *Hoppe-Seyler's Z. Physiol. Chem.* 358:1165–68
158. Mosher, D. F. 1977. *Biochim. Biophys. Acta* 491:205–10
159. Richter, H., Seidl, M., Hörmann, H.

1981. *Hoppe-Seyler's Z. Physiol. Chem.* 362:399–408
160. Jilek, F., Hörmann, H. 1979. *Hoppe-Seyler's Z. Physiol. Chem.* 360:597–603
161. Ruoslahti, E., Engvall, E. 1980. *Biochim. Biophys. Acta* 631:350–58
162. Johansson, S., Höök, M. 1980. *Biochem. J.* 187:521–24
163. Hörmann, H., Jelinić, V. 1981. *Hoppe-Seyler's Z. Physiol. Chem.* 362:87–94
164. Klebe, R. J., Mock, P. J. 1982. *J. Cell. Physiol.* 112:5–9
165. Bray, B. A., Mandel, I., Turino, G. M. 1981. *Science* 214:793–95
166. Kjellén, L., Pettersson, I., Höök, M. 1981. *Proc. Natl. Acad. Sci. USA* 78:5371–75
167. Laterra, J., Silbert, J. E., Culp, L. A. 1983. *J. Cell Biol.* 96:112–23
168. Molnar, J., Gelder, F. B., Lai, M. Z., Siefring, G. E., Credo, R. B., Lorand, L. 1979. *Biochemistry* 18:3909–16
169. Hayashi, M., Yamada, K. M. 1982. *J. Biol. Chem.* 257:5263–67
170. Smith, D. E., Furcht, L. T. 1982. *J. Biol. Chem.* 257:6518–23
171. Hayashi, M., Schlesinger, D. H., Kennedy, D. W., Yamada, K. M. 1980. *J. Biol. Chem.* 255:10017–20
172. Isemura, M., Yosizawa, Z., Takahashi, K., Kosaka, H., Kojima, N., Ono, T. 1981. *J. Biochem.* 90:1–9
173. Richter, H., Hörmann, H. 1982. *Hoppe-Seyler's Z. Physiol. Chem.* 363:351–64
174. Zardi, L., Siri, A., Carnemolla, B., Santi, L., Gardner, W. D., Hoch, S. O. 1979. *Cell* 18:649–57
175. Isemura, M., Yosizawa, Z., Koide, T., Ono, T. 1982. *J. Biochem.* 91:731–34
176. Laterra, J., Culp, L. A. 1982. *J. Biol. Chem.* 257:719–26
177. Underhill, C. B., Toole, B. P. 1979. *J. Cell. Biol.* 82:475–84
178. Knox, P., Wells, P. 1979. *J. Cell Sci.* 40:77–88
179. Rich, A. M., Pearlstein, E., Weissmann, G., Hoffstein, S. T. 1981. *Nature* 293:224–26
180. Oldberg, Å., Ruoslahti, E. 1982. *J. Biol. Chem.* 257:4859–63
181. Knox, P., Griffiths, S., Wells, P. 1982. See Ref. 60, pp. 103–7
182. Keski-Oja, J., Sen, A., Todaro, G. J. 1980. *J. Cell Biol.* 85:527–33
183. Keski-Oja, J., Yamada, K. M. 1981. *Biochem. J.* 193:615–20
184. Koteliansky, V. E., Glukhova, M. A., Morozkin, A. D., Musatov, A. P., Shirinsky, V. P., et al. 1981. *FEBS Lett.* 133:31–35
185. Koteliansky, V. E., Gneushev, H. N., Glukhova, M. A., Shartava, A. S., Smirnov, V. N. 1982. *FEBS Lett.* 143:168–70
186. Molnar, J., Froehlich, J., Rovin, B. 1981. *J. Supramolec. Struct. Suppl.* 5:310 (Abstr.)
187. Yamada, K. M., Yamada, S. S., Pastan, I. 1975. *Proc. Natl. Acad. Sci. USA* 72:3158–62
188. Pearlstein, E. 1976. *Nature* 262:497–500
189. Vuento, M. 1979. *Hoppe-Seyler's Z. Physiol. Chem.* 360:1327–33
190. Vuento, M., Vaheri, A. 1978. *Biochem. J.* 175:333–36
191. Vartio, T. 1982. *Eur. J. Biochem.* 123:223–33
192. Vuento, M., Vartio, T., Saraste, M., von Bonsdorff, C.-H., Vaheri, A. 1980. *Eur. J. Biochem.* 105:33–42
193. Iwanga, S., Suzuki, K., Hashimoto, S. 1978. *Ann. NY Acad. Sci.* 312:56–73
194. Keski-Oja, J., Mosher, D. F., Vaheri, A. 1976. *Cell* 9:29–35
195. Kuusela, P. 1978. *Nature* 276:718–20
196. Proctor, R. A., Prendergast, E., Mosher, D. F. 1982. *Blood* 59:681–87
197. Mosher, D. F., Proctor, R. A. 1980. *Science* 209:927–29
198. Beachey, E. H., Simpson, W. A. 1982. *Infection* 10:65–69
199. Verbrugh, H. A., Peterson, P. K., Smith, D. E., Nguyen, B.-Y. T., Hoidal, J. R., et al. 1981. *Infect. Immun.* 33:811–19
199a. Speziale, P., Höök, M., Wadström, T., Timpl, R. 1982. *FEBS Lett.* 146:55–8
200. McDonagh, R. P., McDonagh, J., Petersen, T. E., Thøgersen, H. C., Skorstengaard, K., et al. 1981. *FEBS Lett.* 127:174–78
201. Vibe-Pedersen, K., Sahl, P., Skorstengaard, K., Petersen, T. E. 1982. *FEBS Lett.* 142:27–30
201a. Skorstengaard, K., Thøgersen, H. C., Vibe-Pedersen, K., Petersen, T. E., Magnusson, S. 1982. *Eur. J. Biochem.* 128:605–23
201b. Skorstengaard, K., Thøgersen, H. C., Vibe-Pedersen, K., Sahl, P., Petersen, T. E. 1983. In *New Clinical and Biological Aspects on Factor XIII and Fibronectin,* ed. H.-G. Klingemann, R. Egbring. Marburg, W. Germany: Medizinische Verlagsgesellschaft. In press
202. Furie, M. B., Rifkin, D. B. 1980. *J. Biol. Chem.* 255:3134–40
203. Pande, H., Shively, J. E. 1982. *Arch. Biochem. Biophys.* 213:258–65
204. Chen, A. B., Amrani, D. L., Mosesson, M. W. 1977. *Biochim. Biophys. Acta* 493:310–22
205. Kurkinen, M., Vartio, T., Vaheri, A. 1980. *Biochim. Biophys. Acta* 624:490–98
206. Birdwell, C. R., Brasier, A. R., Taylor,

L. A. 1980. *Biochem. Biophys. Res. Commun.* 97:574–81
206a. Hirano, H., Yamada, Y., Sullivan, M., deCrombrugghe, B., Pastan, I., Yamada, K. M. 1982. *Proc. Natl. Acad. Sci. USA* 80:46–50
207. Wagner, D. D., Hynes, R. O. 1979. *J. Biol. Chem.* 254:6746–54
208. Grinnell, F., Minter, D. 1979. *Biochim. Biophys. Acta* 550:92–99
209. Gold, L. I., Pearlstein, E. 1979. *Biochim. Biophys. Acta* 581:237–51
210. Smith, D. E., Mosher, D. F., Johnson, R. B., Furcht, L. T. 1982. *J. Biol. Chem.* 257:5831–38
211. Fukuda, M., Hakomori, S. 1979. *J. Biol. Chem.* 254:5442–50
211a. Sekiguchi, K., Hakomori, S. 1983. *Biochemistry* 22:1415–22
212. Yamada, K. M., Olden, K., Pastan, I. 1978. *Ann. NY Acad. Sci.* 312:256–77
213. Teng, M.-H., Rifkin, D. B. 1979. *J. Cell Biol.* 80:784–91
214. Ali, I. U., Hunter, T. 1981. *J. Biol. Chem.* 256:7671–77
215. Ledger, P. W., Tanzer, M. L. 1982. *J. Biol. Chem.* 257:3890–95
216. Dunham, J. S., Hynes, R. O. 1978. *Biochim. Biophys. Acta* 506:242–55
217. Vuento, M., Salonen, E. 1981. *Z. Naturforsch. Teil C* 36:863–68
218. Vuento, M., Salonen, E., Osterlund, K., Stenman, U. H. 1982. *Biochem. J.* 201:1–8
219. Olden, K., Pratt, R. M., Yamada, K. M. 1978. *Cell* 13:461–73
220. Yamada, K. M., Olden, K. 1982. In *Tunicamycin*, ed. G. Tamura, pp. 119–44. Tokyo: Japan Scientific Societies
221. Housley, T. J., Rowland, F. N., Ledger, P. W., Kaplan, J., Tanzer, M. L. 1980. *J. Biol. Chem.* 255:121–28
222. Olden, K., Pratt, R. M., Yamada, K. M. 1979. *Proc. Natl. Acad. Sci. USA* 76:3343–47
223. Bernard, B. A., Yamada, K. M., Olden, K. 1982. *J. Biol. Chem.* 257:8549–54
224. Duksin, D., Bornstein, P. 1977. *Proc. Natl. Acad. Aci. USA* 74:3433–37
225. Bernard, B. A., Olden, K., Yamada, K. M. 1982. See Ref. 60, pp. 225–29
226. Wagner, D. D., Ivatt, R., Destree, A. T., Haynes, R. O. 1981. *J. Biol. Chem.* 256:1708–15
227. Postlethwaite, A. E., Keski-Oja, J., Balian, G., Kang, A. H. 1981. *J. Exp. Med.* 153:494–99
228. Seppä, H. E. J., Yamada, K. M., Seppä, S. T., Silver, M. H., Kleinman, H. K., Schiffmann, E. 1981. *Cell Biol. Int. Rep.* 5:813–19
229. Greenberg, J. H., Seppä, S., Seppä, H., Hewitt, A. T. 1981. *Devel. Biol.* 87:259–66
230. Schoen, R. C., Bentley, K. L., Klebe, R. J. 1982. *Hybridoma* 1:99–107
231. Chung, A. E., Jaffe, R., Freeman, I. L., Vergnes, J.-P., Braginski, J. E., Carlin, B. 1979. *Cell* 16:277–87
232. Timpl, R., Rhode, H., Robey, P. G., Rennard, S. I., Foidart, J.-M., Martin, G. R. 1979. *J. Biol. Chem.* 254:9933–37
233. Hogan, B. L. M., Cooper, A. R., Kurkinen, M. 1980. *Devel. Biol.* 80:289–300
234. Wewer, U., Albrechtsen, R., Ruoslahti, E. 1981. *Cancer Res.* 41:1518–24
235. Farquhar, M. G. 1981. See Ref. 1, pp. 335–78
236. Cooper, A. R., Kurkinen, M., Taylor, A., Hogan, B. L. M. 1981. *Eur. J. Biochem.* 119:189–97
237. Ott, U., Odermatt, E., Engel, J., Furthmayr, H., Timpl, R. 1982. *Eur. J. Biochem.* 123:63–72
238. Rao, C. N., Margulies, I. M. K., Goldfarb, R. H., Madri, J. A., Woodley, D. T., Liotta, L. A. 1982. *Arch. Biochem. Biophys.* 219:65–70
239. Rao, C. N., Margulies, I. M. K., Tralka, T. S., Terranova, V. P., Madri, J. A., Liotta, L. A. 1982. *J. Biol. Chem.* 257:9740–44
239a. Terranova, V. P., Rao, C. N., Kalebic, T., Margulies, I. M., Liotta, L. A. 1983. *Proc. Natl. Acad. Sci. USA* 80:444–48
239b. Malinoff, H. L., Wicha, M. S. 1983. *J. Cell Biol.* In press
240. Rohde, H., Bächinger, H. P., Timpl, R. 1980. *Hoppe-Seyler's Z. Physiol. Chem.* 361:1651–60
241. Couchman, J. R., Höök, M., Rees, D. A., Timpl, R. 1983. *J. Cell. Biol.* 96:177–83
241a. Kennedy, D. W., Rohrbach, D. H., Martin, G. R., Momoi, T., Yamada, K. M. 1983. *J. Cell. Physiol.* 114:257–62
242. Carlsson, R., Engvall, E., Freeman, A., Ruoslahti, E. 1981. *Proc. Natl. Acad. Sci. USA* 78:2403–6
243. Federgreen, W., Stenn, K. S. 1980. *J. Invest. Dermatol.* 75:261–63
244. Akers, R. M., Mosher, D. F., Lilien, J. E. 1981. *Develop. Biol.* 86:179–88
245. Baron-Van Evercooren, A., Kleinman, H. K., Ohno, S., Marangos, P., Schwartz, J. P., Dubois-Dalcq, M. E. 1983. *J. Neurosci. Res.* 8:179–94
246. Sakashita, S., Engvall, E., Ruoslahti, E. 1980. *FEBS Lett.* 116:243–46
247. Hewitt, A. T., Varner, H. H., Silver, M. H., Dessau, W., Wilkes, C. M., Martin, G. R. 1982. *J. Biol. Chem.* 257:2330–34
248. Hewitt, A. T., Varner, H. H., Silver, M. H., Martin, G. R. 1982. In *Limb Development and Regeneration, Part B,*

ed. R. O. Kelley, P. F. Goetinck, J. B. MacCabe, pp. 25–33. New York: Liss

249. Klebe, R. J., Hall, J. R., Naylor, S. L., Dickey, W. D. 1978. *Exp. Cell Res.* 115:73–78

250. Grinnell, F., Hays, D. G., Minter, D. 1977. *Exp. Cell Res.* 110:175–90

251. Thom, D., Powell, A. J., Rees, D. A. 1979. *J. Cell Sci.* 35:281–305

252. Knox, P., Griffiths, S. 1980. *J. Cell Sci.* 46:97–112

253. Knox, P., Griffiths, S. 1982. *J. Cell Sci.* 55:301–16

254. Grinnell, F., Feld, M. K. 1982. *J. Biol. Chem.* 257:4888–93

255. Virtanen, I., Vartio, T., Badley, R. A., Lehto, V.-P. 1982. *Nature* 298:660–63

256. Klebe, R. J., Bentley, K. L., Sasser, P. J., Schoen, R. C. 1980. *Exp. Cell Res.* 130:111–17

257. Whateley, J. G., Knox, P. 1980. *Biochem. J.* 185:349–54

258. Holmes, R. 1967. *J. Cell Biol.* 32:297–308

259. Barnes, D., Wolfe, R., Serrero, G., McClure, D., Sato, G. 1980. *J. Supramolec. Struct.* 14:47–63

260. Hayman, E. G., Engvall, E., A'Hearn, E., Barnes, D., Pierschbacher, M., Ruoslahti, E. 1982. *J. Cell Biol.* 95:20–23

261. Stenn, K. S. 1981. *Proc. Natl. Acad. Sci. USA* 78:6907–11

262. Lehto, V.-P., Vartio, T., Virtanen, I. 1980. *Biochem. Biophys. Res. Commun.* 95:909–16

263. Carter, W. G., Hakomori, S. 1981. *J. Biol. Chem.* 256:6953–60

264. Carter, W. G. 1982. *J. Biol. Chem.* 257:3249–57

264a. Carter W. G. 1983. *J. Biol. Chem.* 257:13805–15

265. Harper, P. A., Juliano, R. L. 1981. *J. Cell Biol.* 91:647–53

266. Pearlstein, E. 1978. *Int. J. Cancer* 22:32–35

267. Czop, J. K., Kadish, J. L., Austen, K. F. 1981. *Proc. Natl. Acad. Sci. USA* 78:3649–53

268. Norris, D. A., Clark, R. A. F., Swigart, L. M., Huff, J. C., Weston, W. L. 1982. *J. Immunol.* 129:1612–18

269. Ginsberg, M. H., Painter, R. G., Forsyth, J., Birdwell, C., Plow, E. F. 1980. *Proc. Natl. Acad. Sci. USA* 77:1049–53

270. Plow, E. F., Ginsberg, M. H. 1981. *J. Biol. Chem.* 256:9477–82

271. Gartner, T. K., Williams, D. C., Minion, F. C., Phillips, D. R. 1978. *Science* 200:1281–83

272. Jaffe, E. A., Leung, L. L. K., Nachman, R. L., Levin, R. I., Mosher, D. F. 1982.

Nature 295:246–48

273. Lahav, J., Schwartz, M. A., Hynes, R. O. 1982. *Cell* 31:253–62

274. Grinnell, F., Milam, M., Srere, P. A. 1973. *J. Cell Biol.* 56:659–65

275. Grinnell, F. 1980. *J. Cell Biol.* 86:104–12

276. Aplin, J. D., Hughes, R. C., Jaffe, C. L., Sharon, N. 1981. *Exp. Cell Res.* 134:488–94

277. Oppenheimer-Marks, N., Grinnell, F. 1981. *Eur. J. Cell Biol.* 23:286–94

278. Bevilacqua, M. P., Amrani, D., Mosesson, M. W., Bianco, C. 1981. *J. Exp. Med.* 153:42–60

279. Tarone, G., Galetto, G., Prat, M., Comoglio, P. M. 1982. *J. Cell Biol.* 94:179–86

280. Kleinman, H. K., Martin, G. R., Fishman, P. H. 1979. *Proc. Natl. Acad. Sci. USA* 76:3367–71

281. Kleinman, H. K., Hewitt, A. T., Murray, J. C., Liotta, L. A., Rennard, S. I., et al. 1979. *J. Supramolec. Struct.* 11:69–78

282. Yamada, K. M., Kennedy, D. W., Grotendorst, G. R., Momoi, T. 1981. *J. Cell. Physiol.* 109:343–51

283. Perkins, R. M., Kellie, S., Patel, B., Critchley, D. R. 1982. *Exp. Cell Res.* 141:231–43

284. Pena, S. D. J., Hughes, R. C. 1978. *Nature* 276:80–83

285. Pratt, R. M., Yamada, K. M., Olden, K., Ohanian, S. H., Hascall, V. C. 1979. *Exp. Cell Res.* 118:245–52

286. Butters, T. D., Devalia, V., Aplin, J. D., Hughes, R. C. 1980. *J. Cell Sci.* 44:33–58

287. Hughes R. C., Butters, T. D., Aplin, J. D. 1981. *Eur. J. Cell Biol.* 26:198–207

288. Perkins, M. E., Ji, T. H., Hynes, R. O. 1979. *Cell* 16:941–52

288a. Yamada, K. M., Critchley, D. R., Fishman, P. H., Moss, J. 1983. *Exp. Cell Res.* 143:295–302

289. Goldberg, B. 1979. *Cell* 16:265–75

290. Santoro, S. A., Cunningham, L. W. 1979. *Proc. Natl. Acad. Sci. USA* 76:2644–48

291. Saito, Y., Imada, T., Inada, Y. 1982. *Thromb. Res.* 25:143–47

292. Chiang, T. M., Kang, A. H. 1982. *J. Biol. Chem.* 257:7581–86

293. Arneson, M. A., Hammerschmidt, D. E., Furcht, L. T., King, R. A. 1980. *J. Am. Med. Assoc.* 244:144–47

294. Bensusan, H. B., Koh, T. L., Henry, K. G., Murray, B. A., Culp, L. A. 1978. *Proc. Natl. Acad. Sci. USA* 75:5864–68

295. Holderbaum, D., Culp, L. A., Bensusan, H. B., Gershman, H. 1982. *Proc. Natl. Acad. Sci. USA* 79:6537–40

296. Gonda, S. R., Shainoff, J. R. 1982. *Proc. Natl. Acad. Sci. USA* 79:4565–69

297. Colvin, R. B., Gardner, P. I., Roblin, R.

O., Verderber, E. L., et al. 1979. *Lab. Invest.* 41:464–73
298. Bennett, J. S., Vilaire, G., Cines, D. B. 1982. *J. Biol. Chem.* 257:8049–54
299. Hoyer, L. W. 1981. *Blood* 58:1–13
300. Yamada, S. S., Yamada, K. M., Willingham, M. C. 1980. *J. Histochem. Cytochem.* 28:953–60
301. Hedman, K. 1980. *J. Histochem. Cytochem.* 28:1233–41
302. Olden, K., Yamada, K. M. 1977. *Cell* 11:957–69
303. Choi, M. G., Hynes, R. O. 1979. *J. Biol. Chem.* 254:2050–55
304. Olden, K., Hunter, V. A., Yamada, K. M. 1980. *Biochim. Biophys. Acta* 632:408–16
305. Uchida, N., Smilowitz, H., Ledger, P. W., Tanzer, M. L. 1980. *J. Biol. Chem.* 255:8638–44
306. Hayman, E. G., Ruoslahti, E. 1979. *J. Cell Biol.* 83:255–59
307. McKeown-Longo, P. J., Mosher, D. F. 1982. *Fed. Proc.* 41:754 (Abstr.)
308. Bateman, J. F., Peterkofsky, B. 1981. *Proc. Natl. Acad. Sci. USA* 78:6028–32
309. Hayman, E. G., Oldberg, A., Martin, G. R., Ruoslahti, E. 1982. *J. Cell. Biol.* 94:28–35
310. Fagan, J. B., Sobel, M. E., Yamada, K. M., deCrombrugghe, B., Pastan, I. 1981. *J. Biol. Chem.* 256:520–25
311. Blumberg, P. M., Driedger, P. E., Rossow, P. W. 1976. *Nature* 264:446–47
312. Keski-Oja, J., Shoyab, M., De Larco, J. E., Todaro, G. J. 1979. *Int. J. Cancer* 24:218–24
313. Yamada, K. M., Yamada, S. S., Pastan, I. 1977. *J. Cell Biol.* 74:649–54
314. Furcht, L. T., Mosher, D. F., Wendelschafer-Crabb, G., Woodbridge, P. A., Foidart, J. M. 1979. *Nature* 277:393–95
315. Marceau, N., Goyette, R., Valet, J. P., Deschenes, J. 1980. *Exp. Cell Res.* 125:497–502
316. Chen, L. B., Gudor, R. C., Sun, T.-T., Chen, A. B., Mosesson, M. W. 1977. *Science* 197:776–78
317. Nielson, S. E., Puck, T. T. 1980. *Proc. Natl. Acad. Sci. USA* 77:985–89
318. Rajaraman, R., Sunkara, S. P., Rao, P. N. 1980. *Cell Biol. Int. Rep.* 4:897
319. Hassell, J. R., Pennypacker, J. P., Kleinman, H. K., Pratt, R. M., Yamada, K. M. 1979. *Cell* 17:821–26
320. Milhaud, P., Yamada, K. M., Gottesman, M. M. 1980. *J. Cell Physiol.* 104:163–70
321. Hayman, E. G., Engvall, E., Ruoslahti, E. 1980. *Exp. Cell Res.* 127:478–81
322. Rephaeli, A., Spector, M., Racker, E. 1981. *J. Biol. Chem.* 256:6069–74
323. Rieber, M., Rieber, M. S. 1982. *Exp. Cell Res.* 138:441–46
324. Brownell, A. G., Bessem, C. C., Slavkin, H. C. 1981. *Proc. Natl. Acad. Sci. USA* 78:3711–15
325. Garbi, C., Wollman, S. H. 1982. *J. Cell Biol.* 94:489–92
326. David, G., Bernfield, M. 1981. *J. Cell Biol.* 91:281–86
327. Smith, R. L., Bernfield, M. 1982. *Dev. Biol.* 94:378–90

Ann. Rev. Biochem. 1983. 52:801–24

PROTON ATPASES: STRUCTURE AND MECHANISM

L. Mario Amzel and Peter L. Pedersen

Department of Biophysics and Physiological Chemistry, Laboratories for Molecular and Cellular Bioenergetics, Johns Hopkins University School of Medicine, 725 N. Wolfe Street, Baltimore, Md. 21205

CONTENTS

INTRODUCTION 801
   Perspective and Summary 801
STRUCTURE OF THE ATP-SYNTHASE 803
   Subunit Composition 803
   Amino Acid Sequences 804
   Subunit Stoichiometry 808
   Three-Dimensional Structure 809
BINDING AND LABELING STUDIES 813
ASSEMBLY OF THE ATP-SYNTHASE 815
MECHANISM OF THE ATP-SYNTHASE REACTION 816
   Introduction 816
   Proton Translocation 816
   Bond Formation and Hydrolysis 817
   Coupling 820
CONCLUDING REMARKS 821

INTRODUCTION

Perspectives and Summary

The most complex enzyme systems known are those found associated with energy-transducing membranes of mitochondria, chloroplasts, and bacteria. These systems consist of electron transport proteins and a proton ATPase complex (H^+-ATPase). The two systems appear to be coupled by an electrochemical gradient of protons generated by the redox components (1, 2). The role of the H^+-ATPase is to allow protons to cross the lipid bilayer (translocation step), to catalyze the formation of ATP from ADP and P_i (catalytic step) and to direct the protons in such a way as to utilize

801

0066-4154/83/0701-0801$02.00

energy stored in the electrochemical gradient to produce high ATP concentrations (coupling step) (see Figure 1).

As in the case of all enzymatic reactions this "oxidative phosphorylation" process is reversible. In the absence of oxidizable substrates (mitochondria and aerobic bacteria) or in the dark (chloroplasts and photosynthetic bacteria), ATP hydrolysis can induce the formation of an electrochemical gradient of protons. This proton gradient, depending on the energy-transducing membrane in question, can drive reverse electron flow, metabolite transport, pyridine nucleotide reduction, and perhaps in some bacteria chemotaxis. In anaerobic bacteria like *Streptococcus faecalis* the physiological direction of the H^+-ATPase reaction is toward hydrolysis of ATP (3), whereas in mitochondria of animal cells and most aerobic bacteria the physiological direction appears to be primarily toward ATP synthesis.

H^+-ATPases are comprised of three functionally different units called F_0, F_1, and "I." F_0 is a water-insoluble component that spans the energy-transducing membrane and directs protons to or away from F_1. F_1 is a water-soluble catalytic component that participates directly in ATP synthesis or ATP hydrolysis. "I" is a regulator protein whose primary function seems to be to prevent the hydrolysis of newly synthesized ATP. In recent years a number of reviews have been written about proton ATPases from mitochondria (4–8), chloroplasts (9), and bacteria (3, 10–14). Two recent reviews about the regulator protein "I" have appeared (15, 16). For detailed background material the reader is referred to these articles.

It is the purpose of this review to provide an up-to-date appraisal of the

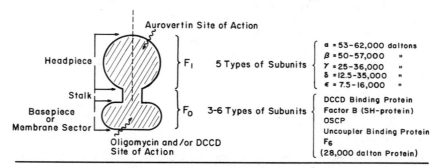

Figure 1 Current view of the structure of H^+-ATPases of bacteria, mitochondria, and chloroplasts considered in this review. In most cases the F_1 moiety consist of the five nonidentical subunits α, β, γ, δ, and ε. In bacteria the F_0 moiety is thought to consist of no more than three to four different types of polypeptides whereas in eukaryotic cells the F_0-moiety appears to be more complex. Common to the F_0-unit of all these H^+-ATPases is the DCCD-binding proteins. Not shown here is the regulator protein "I," which is thought to interact with the β-subunit of the F_1 moiety.

literature on the structure and mechanism of H^+-ATPases. We restrict our coverage to those H^+-ATPases associated with the inner membranes of mitochondria and bacteria and to the analogous enzymes associated with the thylakoid membrane of chloroplasts and photosynthetic bacteria. We do not deal in this review with H^+-ATPases associated with the plasma membranes of *Neurospora*, yeast, and a variety of nonmitochondrial membranes of animal cells. The properties of the plasma membrane type H^+-ATPase of *Neurospora* and yeast is covered in an excellent recent review (17). Suffice it to say that, in contrast to the H^+-ATPases discussed here, the plasma membrane-type enzyme appears to function either directly (18) or indirectly in transport processes (19–22). Also, they are structurally and mechanistically simpler, having one major type of subunit and proceeding through a reaction pathway that involves a covalently bound phosphoenzyme intermediate (17). As yet, sufficient information is not available to discuss critically H^+-ATPases associated with chromaffin granules (19), anterior pituitary granules (22), lysosomes (20), and synaptosomes (22).

Within the past five years over 1000 papers have appeared on the subject of H^+-ATPases. Therefore, we restrict this review to those studies that we believe provide the most informative data about the subject matter at hand. In some cases we have taken the opportunity to speculate where obvious gaps in our knowledge about structure and mechanism of H^+-ATPases exist. It is hoped that such speculations will help stimulate additional thinking about the subject, but most importantly additional experiments to distinguish between two or more apparently conflicting views. For an excellent but different approach to much of the same subject matter the reader is advised to consult the 1981 review by Cross (23) in Volume 50 of this series.

Because a complete understanding of enzyme mechanism relies heavily on structural information, we commence this review by discussing the structure of H^+-ATPases.

STRUCTURE OF THE ATP-SYNTHASE

Subunit Composition

The subunit composition of H^+-translocating ATPases is highly complex (Figure 1). It is best understood in bacterial systems in which the enzyme complex appears to be composed of eight different polypeptide chains: five in the F_1-sector and three in the F_0-sector. In *Escherichia coli* the eight chains are coded by the *atp* or *unc* operon that occurs at the 83-minutes region of the *E. coli* linkage map (see 10, 11, 24, 25). Recently Futai and co-workers (26, 27) found that a transducing phage λasn-5 carried a complete set of structural genes for F_0F_1 (26, 27). In addition they localized them on a 4.5-megadalton segment of $E.\ coli$ DNA carried by the phage. Using this phage Walker and co-workers (28–32) and Futai and co-workers (33–38) sequenced the operon. The unc operon contains nine genes: eight are the structural genes coding for the subunits of the H^+-ATPase while the ninth was postulated to code for a polypeptide that functions, according to Gay & Walker (29), as "a pilot for assembly." Figure 2 shows the organization of the operon (30).

In nonbacterial systems the situation is less clear but all groups agree that the F_1-sector contains five subunits while the number of additional subunits in the F_0F_1-complex varies in the different preparations. The minimum number of subunits necessary to assemble a fully functional ATP-synthase in nonbacterial systems is not known but it probably includes in addition to the F_1-sector five or more subunits (4–9, 39–41). A regulator or inhibitor protein "I" (M_r = 9578 in beef heart) associated with the system is sometimes considered part of the complex.

Amino Acid Sequences

F_1-SECTOR   The sequences of the five subunits of the F_1-sector of the $E.\ coli$ ATPase (EF_1) have been determined by DNA-sequencing techniques. Several of the subunits were partially sequenced in the enzyme of beef heart and that of the thermophilic bacterium PS3. For the $E.\ coli$ enzyme the number of amino acids found in the α-subunit is 513, with 460 in β, 287 in γ, 177 in δ, and 133 in ε. Accurate subunit molecular weights were also obtained from the sequences for the subunits of the EF_1 (30) (α, 55,264; β, 50,157; γ, 31,410; δ, 19,310; ε, 14,194). Whenever available the sequences of the different subunits were compared. The two major subunits from $E.\ coli$ and beef heart were found to be highly homologous (28, 32). However few or no sequence homologies were found between the three small subunits of the same proteins (32). Comparison of the sequences of the α- and β-subunits of $E.\ coli$ showed similarities in sequence that confirmed previous ideas about their relatedness (30, 32, 42). The sequence homologies between the α- and β-subunits are less pronounced than those between α- (or β-) subunits of

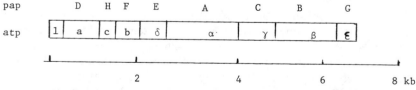

Figure 2   Genetic map of the ATP operon reproduced from (30). The nomenclature proposed by Futai et al (36) is also included.

different sources (32). However, they are consistent and they appear throughout the whole polypeptide chain suggesting that α- and β-subunits are evolutionarily related and that they share the same polypeptide chain folding. In addition, the local sequences in two regions of the α- and the β-subunits have sequence homologies with other ATP-binding proteins, in particular myosin, ATP/ADP translocase, Ca^{2+}-ATPase, adenylate kinase, phosphofructokinase, the *rec* A protein of *E. coli* (32, 37), and asparto-kinase. Table 1 [reproduced from Refs. (32) and (37)] shows some of these similarities. Two amino acids that were shown to be involved in the active sites of ATP-synthases have been identified in the β-subunits. A glutamic acid was found in the β-subunit of the thermophilic bacterium ATPase and shown to be involved in the binding of Mg^{2+}. Sequence comparison of the β-subunit of TF_1 with that of EF_1 showed that the sequence around the glutamic acid residue is highly conserved not only in both bacterial systems but also in the β-subunit of beef heart (30; Table 2).

In beef heart F_1 (BhF_1), inactivation with *p*-fluorosulphonyl-benzoyl-5′-adenosine showed that a tyrosine residue in the β-subunit was the primary site of labeling and binding to this site was responsible for the inactivation of the enzyme (43). The sequence that includes this tyrosine in BhF_1 was found to be highly homologous to that of the *E. coli* F_1 (Table 3) showing that at least part of the active-site residues are also preserved between these two evolutionarily very distant species (30).

Secondary structure predictions were attempted for all the subunits of the EF_1. Although unreliable these predictions produced some interesting patterns: In the β-subunit, residues 240 to 330 are predicted as alternating stretches of α-helices and β-sheets (37) reminiscent of the nucleotide-binding domain (37, 44). The analysis of the δ-subunits show that residues 91 to 140 are predicted as a continuous 50-residue-long helix (34). Interestingly, small-angle X-ray diffraction studies (45) have been inter-preted as suggesting that the δ-subunit of chloroplast F_1 is a 25×90–Å ellipsoid. The high helical content (60% predicted), the possibility of a 50-residue-long helix and the highly elongated shape of the δ-subunit led several authors to suggest an important structural role for this subunit.

INHIBITOR PROTEIN   The complete amino acid sequence of the inhibitor protein "I" was reported for the protein isolated from bovine heart mitochondria (46). The protein consists of 84 amino acids of M_r 9578. It lacks cysteine, proline, methionine, threonine, and tryptophane, while charged amino acids comprise 50% of the sequence. These charged amino acids occur in clusters distributed along the sequence similar to those present in membrane-associated proteins (46). A section of the protein located in the carboxy-terminus contains several duplicated regions.

Table 1 Sequence homologies of selected regions of the α- and β-subunits of F$_1$-ATPases and other ATP-requiring enzymes[a]

Protein	Residue numbers Start	End	Sequence
(a) RecA-protein	265	285	G E G I N F Y G E L V D L G V K E K L I E
E. coli ATPase β	185	205	R E G N D F Y H E M T D S N V I D K U S L
(b) Bovine ATPase β	149	173	K G G K I G L F – G G A G V G K T V F I M E L I N
E. coli ATPase β	142	165	K G G K V G L F – G G A G V G K T V N M M E L I R
E. coli ATPase α	161	184	R G Q R E L I I – G D R Q T G K T A L A I D A I I
Adenylate kinase	6	30	K K S K I I F V V G G P G S G K G T Q C E K I V Q
RecA-protein	57	84	P M G R I V E I Y G P E S S G K T T L T L Q V I A
Myosin, nematode	161	184	E N Q S M L I T – G E S G A G K T E N T K K V I C
Myosin, rabbit	170	193	E N Q S I L I T – G E S G A G K T V N T K R V I Q
ThrA; aspartokinase	671	700	D V A R K L L I L A R E – T G R E L E L A D I E I
(c) Bovine ATPase β	245	264	F R D Q E G Q D V L L F I D N I F R F T
E. coli ATPase β	231	249	F R D – E G R D V L L F V D N I Y R Y T
E. coli ATPase α	269	288	F R D – R G E D A L I I Y D D L S K Q A
ATP/ADP translocase	76	89	– R G M G G A F V L V L Y D E I K K F V
Ca^{2+}-ATPase	(45	64)	H K S – K I V E Y L Z S Y D E I T A M T
Adenylate kinase	106	125	E R K – I G Q P T L L Y – D A G P E T
Phosphofructokinase	86	105	L K K – H G I Q G L V V I G G D G S Y Q

[a] Reproduced with minor modification from Ref. 37 (a) and Ref. 32 (b and c). Single-letter amino acid codes are given as in *Atlas of Protein Sequences and Structure*, vol. 5, 1972, ed. M. O. Dayhoff. Washington DC: Natl. Biomed. Res. Found.

Table 2 Sequences of β-subunits in the vicinity of the glutamic acid residues involved in magnesium binding[a]

Source	Sequence
Beef heart	E L I N N V A K A H G G Y S V F A G V G E R T R E G N D L Y E* H M
E. coli	E L I R N I A I E H S G Y S V F A G V G E R T R E G N D F Y H E M
Bacterium PS3	A G V G E* R

[a] Residues identified in BhF$_1$ and TF$_1$ are indicated. Reproduced from (30).

F$_0$-SECTOR   The three polypeptide chains present in the F$_0$-sector of the *E. coli* ATP-synthase were also sequenced. Some of these proteins had been sequenced previously in other systems but we discuss them in the order they occur in the *E. coli* genome. These three proteins are the second, third, and fourth coding sequences of the *unc* operon [Figure 2; (30)]. Gene 2 in the *E. coli unc* operon codes for protein *a*, a hydrophobic protein [M_r 30,260; (29, 36)] that was shown to be homologous in the carboxy-terminal region to the product of the gene ATP-6 of the human (47), yeast (48), and mouse mitochondrial genome (49). Sequence comparison shows that this protein is fairly well preserved throughout the phylogenetic scale (29).

Gene 3 codes for a highly hydrophobic protein (79 residues) that had been previously identified as the DCCD-binding protein, also called proteolipid or protein *c* of the ATP synthetase (36, 29). This protein has probably been studied in more systems than any other ATPase subunit (see 50). In all systems studied it has 76 to 81 residues and all sequences appear to contain a short central homologous region that contains an aspartyl or glutamyl residue that can be labeled by DCCD (50, 51, 52). All sequences exhibit low polarity; the most striking is that of *E. coli* that contains only 16% hydrophilic residues. The high percentage of small amino acids (glycine plus alanine, 25%) is also striking. The sequences appear to have two clusters of about 25 hydrophobic residues that are thought to form regions embedded in the lipid bilayer. The second of these regions contains a single charged residue (aspartyl or glutamyl) that was shown to be the one to react with DCCD.

Table 3 Sequences of β-subunits of EF$_1$ and BhF$_1$ in the vicinity of the tyrosine residue that was labeled with *p*-fluorosulphonyl-benzoyl-5′-adenosine in BhF$_1$[a]

Source	Sequence
Beef heart	I M D P N I V G S E H Y* D V A R
E. coli	Q L D P L V V G Q E H Y D T A R

[a] Reproduced from (30).

Gene 4 codes for a protein of M_r 17,220 [protein b; (29, 36)]. The sequence of this polypeptide has striking characteristics. The N-terminal portion spanning residues 1 to 32 contains only one charged amino acid (lysine 23) and all other amino acids are hydrophobic. This is in contrast to the rest of the protein that contains a high proportion of charged groups (29). A stretch of nine hydrophobic residues is also found at residues 124 to 132 (36). The first 31 residues probably form a hydrophobic α-helical tail that is buried in the lipid bilayer while the rest of the sequence is proposed to form an α-helix that protrudes from the membrane (31). This charged protrusion is thought to be a central domain around which the F_1-subunits could become membrane bound. Residues 124–132 could interact with the membrane or with a hydrophobic portion of F_1. The type of contact between F_0 and F_1 is not known but helix–helix contacts are postulated to be the main type of interaction probably involving helices in protein b of F_0 and the δ-subunit of F_1 (31).

Most of the sequence data on the H^+-ATPase subunits is very new and has not been fully analyzed. The wealth of information contained in these data is undoubtedly going to provide new insights into the structural and functional characteristics of the system.

Subunit Stoichiometry

It is fair to say that the subunit stoichiometry of the ATP-synthase complex is still unknown (see 53). In the case of F_1, subunit stoichiometries of the type $\alpha_2\beta_2\gamma_2\delta_2\varepsilon_2$ and $\alpha_3\beta_3\gamma\delta\varepsilon$ have been proposed. The sequence data on the E. coli F_1 provided accurate values for the molecular weights of the individual subunits that confirmed the previously reported values for bacterial systems (see above). With these accurate values the first stoichiometry gives M_r 340,000 while the second stoichiometry gives M_r 381,000 for the complete F_1. Both estimates are within the experimental error of the values reported for the molecular weight of F_1. A large number of experiments were performed to determine the subunit stoichiometry of F_1 and both stoichiometries were supported by experimental evidence of one kind or the other. In the last few years, however, strong evidence has accumulated suggesting that the subunit stoichiometry is actually $\alpha_3\beta_3\gamma\delta\varepsilon$ (54–59).

The problem of determining the subunit stoichiometry of the F_0-sector is even more complex. There are really no good values for the number of copies of any of the subunits. It is assumed that some of them occur in multiple copies. In particular, the proteolipid (protein c), is thought to exist in at least six copies per ATPase. Since it appears that the unc operon has a single promoter (29) [other weak promoter sequences could be present (29,

36)] the complete operon is transcribed as a single polycistronic message. The $\alpha_3\beta_3\gamma\delta\varepsilon$ stoichiometry would require a special mechanism to produce different amounts of the different subunits.

It is particularly striking that the order of the genes is such that the β-subunit that has to be produced in three copies per F_1 is at the end of the polycistronic message (Figure 2). It was suggested that, since the different subunits have different codon usage, the utilization of tRNA's of different abundance determines, at the translational level, the number of copies of each of the subunits that is produced. Kanazawa et al (36) attempted a more quantitative analysis. They expressed the frequency of codon usage y as a linear function of x the amount of the corresponding tRNA in *E. coli* reported by Ikemura (60), namely $y = mx + p$. The values of the coefficients m and p for different F_0F_1-chains are given in Table 4. The values are striking in the sense that the coefficient "m" relating tRNA abundance with codon usage appears to correlate well with the expected number of copies of the subunits in the complex.

Three-Dimensional Structure

Figure 3 shows the typical appearance of an ATPase complex. The complex protrudes from the inner mitochondrial membrane into the matrix and in this type of electron micrograph seems to be attached to the membrane by a narrow connecting structure. The large spherical portion is the F_1-sector of the molecule. Structural information relevant to the size and the shape of the ATP-synthase has been obtained in micrographs of preparations of the inner mitochondrial membrane and more recently of vesicular and micellar preparations of F_0F_1-complex. In all these preparations the F_1 portion appears as a protruding sphere 90 Å in diameter connected to the membrane or micelle by a narrow stalk approximately 30 Å in diameter and 40 Å long. Further detail on the structure of the F_1-sector was obtained from electron micrographs of negatively stained purified F_1 particles (61–63). In these micrographs the particles appear as a ring-shaped stain-excluding region surrounding a stained interior. In most fields, the stain-excluding regions appear to be formed by six resolved regions. Additional information was obtained from two-dimensional arrays of TF_1 obtained by ammonium sulphate precipitation (64). These two-dimensional arrays show consistently the central stain region and the surrounding stain-excluding portions forming each particle. Image enhancement techniques were used and the enhanced images showed that the stain-excluding regions appear divided in three parts, each of them probably formed by two smaller regions (64).

The most detailed information so far has been provided by single crystal

Table 4 Coefficients of the equation $y = mx + p$ for some of the *E. coli* proton ATPase subunits[a]

Subunit	m	p	Correlation coefficient (α)
a	5.40	+0.60	0.62
b	6.63	−0.49	0.75
c	7.68	−0.61	—
α	8.07	−1.15	0.93
γ	5.20	+0.50	0.86

[a] In the equation, y is the frequency of codon usage and x is the amount of the corresponding tRNA present in *E. coli*.

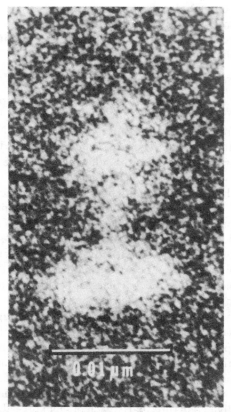

Figure 3   Electron micrograph of a negatively stained F_0F_1-complex. The three regions of the molecule are clearly visible. The F_1-sector is approximately 90 Å in diameter and is connected by a narrow structure to the F_0 portion of the molecule. The size of the F_0 is probably exaggerated by the presence of nonprotein material. Reproduced with permission from (40).

X-ray diffraction studies on the F_1-sector of rat liver mitochondria (rℓ F_1) (65, 66). The structure of crystals of this molecule was determined to 9-Å resolution. Crystallographic data and measurements of the crystal density showed that the asymmetric unit of the crystals contains 180,000 daltons, that is, approximately one-half of an F_1-ATPase (65). Since the crystals have twofold axis of symmetry it was concluded that the molecule was formed by two crystallographically equivalent halves. Each half of the molecule was found to be divided in three regions or masses (A, B, and C, Figure 4) of approximately equal size ($40 \times 50 \times 60$ Å). The complete F_1-ATPase molecule is then a dimer of these three masses $(ABC)_2$ and has dimensions of $120 \times 110 \times 80$ Å. The arrangement of the masses is similar to that observed in the electron micrographs of individual particles since two of the masses lay far from the twofold axis leaving the center of the molecule quite devoid of density. Looking "down" into this central space, one would see the six masses apparently organized in two layers of three masses each. The three masses in each layer are centered at the vertices of approximately equilateral triangles. They make few contacts with each other but interact more strongly with masses in the other layer. Masses A and C interact with their symmetry mates and with the mass B of the other layer while the masses B interact only with the masses A and C of the other layer. The arrangement is such that equivalent masses do not form symmetrical trimers in the complex. The most symmetrical trimeric arrangement in the complex appears to be that formed by an A, B, and C mass in the same layer. However, these masses are in principle not equivalent and have very little contact with each other.

The structure described is directly compatible with a dimeric stoichiometry of the kind $\alpha_2\beta_2\gamma_2\delta_2\varepsilon_2$. Since, as we mentioned, evidence is accumulating that suggests that the subunit stoichiometry is actually $\alpha_3\beta_3\gamma\delta\varepsilon$, the implications of the crystallographic data have to be relaxed in order to accommodate this putative stoichiometry. It was suggested that a structure compatible with both the X-ray model and this stoichiometry should have the general characteristics shown in Figure 5; two pairs of equivalent masses should contain two α- and two β-subunits or combinations of their putative domains. The two masses in the last pair should each contain one of either the α- or the β-subunits combined with some of the minor subunits. Since the α- and β-subunits (32) have very similar polypeptide chain folding, the equivalent masses containing either an α- or a β-subunit could occupy indistinguishably the same crystallographic positions as long as nonidentical residues are not involved in intermolecular contacts. The structure is such that the three α- (and the three β-) subunits are not structurally equivalent in the complex since they seem to occupy positions that cannot be related by threefold symmetry. The smaller

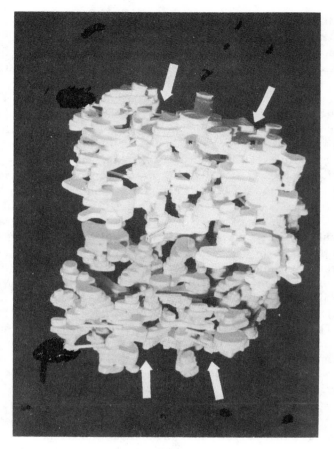

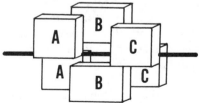

Figure 4   (top) Balsa wood model of F_1-ATPase: The molecule is about $110 \times 120 \times 80$ Å and is composed of two halves (white and grey). Each half of the molecule appears to be formed by three masses of density (the approximate location of the intermass spaces are indicated by the arrows). In the white half the central mass is in the back while in the grey half the masses are in the back.

(bottom) Schematic representation of the model: The six masses in the model are represented as six blocks in the correct three-dimensional arrangement. The twofold axis of symmetry is shown as a thick rod. Reproduced from (66).

a   **b**

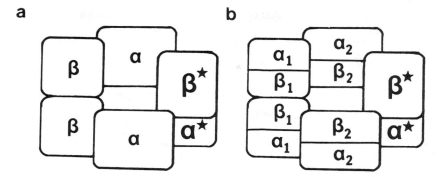

Figure 5 Schematic representation of two of the possible subunit arrangements that are compatible with the $\alpha_3\beta_3\gamma\delta\varepsilon$ stoichiometry. The six blocks corresponding to six regions in the model are drawn to represent approximately the spatial distribution found in the X-ray model. The two figures represent two classes of possible arrangements: (*a*) each of the regions A and B of Figure 4 (*bottom*) contains one subunit; (*b*) each of the regions A and B of Figure 4 (*bottom*) contains domains from both major subunits.

In all cases α^* and β^* represent combinations of subunits including either α or β and some portions of minor subunits. In the case of the $\alpha_2\beta_2\gamma_2\delta_2\varepsilon_2$ stoichiometry (not shown) the regions containing α^* and β^* (region C) would contain each a γ, a δ and an ε subunit. Reproduced from (66).

subunits (present in single copies) probably "tag" the larger subunits for this type of assembly. The γ-subunit appears to be especially important in this respect (see below). In principle two $\alpha\beta$-pairs are almost structurally equivalent in the complex. (The complex would have to be perfectly symmetrical for the pairs to be strictly equivalent) while the third $\alpha\beta$-pair occupies a different environment in the complex. These α- and β-subunits, which are structurally not equivalent, may have different functional properties from the other α- and β-subunits and this arrangement could have special functional significance.

BINDING AND LABELING STUDIES

The F_1-sector of the ATP-synthase complex contains both catalytic and regulatory sites that bind adenine nucleotides. The total number of these sites is not firmly established. However, it appears that up to six adenine nucleotide-binding sites could be present in the F_1-sector. These sites appear to be in the α- and in the β-subunits as shown by binding to isolated subunits (67) and by covalently bound analogs (68–90). The catalytic site(s) is most probably located in the β-subunits. For a structure of this kind ($\alpha_3\beta_3\gamma\delta\varepsilon$) it can be expected that adenine nucleotides would bind with multiplicities of three or six per F_1 depending on whether they bind to one

Table 5 Binding studies on F_1-ATPases substrates, products, and analogs

Compound	Sites/F_1	Source of F_1	Reference
Nucleotides	2 ATP, 1 ADP	Bh (as prepared)	68
Nucleotides	2 ADP, 1 ATP	Bh (as prepared)	69
Nucleotides	3 ATP (non-exch.)	Bh	70
	3 AMP–PNP (exch.)		
Nucleotides	5 AMP–PNP	Bh	69
Nucleotides	3 ATP, 1 AMP–PNP	*M. phlei*	71
Nucleotides	2 ADP, 1 AMP–PNP	*M. phlei*	71
Nucleotides	3 ATP, 2 ADP	Bh	72
Phosphate	1	Bh	109
Mg[a]	1	Bh	73

[a] Dial-ATP(ADP),$2',3'$ dialdehyde derivatives of ATP(ADP).

or to both of the subunits. Most binding studies, however, show that adenine nucleotides (and adenine nucleotide analogs) bind to the enzyme with multiplicities smaller than three sites per F_1. The same is true for inhibitors of the ATPase activity that in most cases were found to inhibit the enzyme completely while binding at one or two moles of inhibitor per mole of F_1. This data is summarized in Tables 5 and 6. The data in both tables can be readily systematized using a model of the F_1-sector described in the previous section. As is clear from Figure 5, such a model could have sites with multiplicities of one, two, or three depending on whether they occur in: (*a*) α^* (or β^*) (multiplicity: one site per F_1); (*b*) interfaces between

Table 6 Binding and labeling studies on F_1-ATPases, inhibitors, and affinity labels

Compound	Sites F_1	Subunit	Sites/subunit	Source	References
Aureovertin	2	β	1	Bh, others	74, 75
Quercetin	2	ND	ND	Chℓ	90
Dial-ATP[a]	1	α	ND	*M. phlei*	71
Dial-ADP	~3	α, β	2:1	*M. phlei*	71
NBD Chloride[b]	2	β	ND	Chℓ	89
NBD Chloride	1	β	ND	Bh	87, 88
8-azido-ATP					
8-azido-ADP	6	α, β	2, 2 (1, 1)	Bh	72, 78
Nucleotides					
8-azido-ATP	1	α	1	*E. coli*	83
ANS[c]	2	ND	ND	Chℓ	90

[a] Dial-ATP(ADP),$2',3'$ dialdehyde derivatives of ATP(ADP).
[b] NBD Chloride, 7-chloro-4-nitrobenzo-2-oxa-1,3-diazole.
[c] ANS, 8-anilino-1-naphtalenesulfonate.

α's (or β's) (multiplicity: one site per F_1); (c) the α-subunits (or the β-subunits) (multiplicity: two sites per F_1); (d) the α-β interfaces (multiplicity: two sites per F_1); or (e) all α's (or all β's) (multiplicity: three sites per F_1).

ASSEMBLY OF THE ATP-SYNTHASE

In bacterial systems it was shown that all eight subunits are needed for assembly of a fully active and coupled ATP synthase (see 11). The three subunits of the F_0-portion are all needed for assembly of the molecule. It was shown that probably the proteolipid can by itself form a proton conducting channel in lipid vesicles (91, 92). Also, the proteolipid is the chain that is labeled when proton conduction is stopped with DCCD. This set of experiments indicates that the proteolipid is the part of the molecule that forms the proton channel and that the other two F_0-polypeptides appear to be involved in the binding of the F_1-sector (see above and the section on Amino Acid Sequences). The interactions between the F_1-sector and the F_0-sector behave in a way which is suggestive of hydrophobic binding. For example, the release of F_1 is favored by lowering the ionic strength of the medium. Also, F_1 can be purified by extraction with chloroform-toluene mixtures. By contrast the protein b to δ-subunit binding discussed by Walker et al (31) appears to be ionic. The binding seems also to be dependent on divalent cations since inclusion of EDTA in the medium strongly enhances the release of the F_1-sector from its membrane-bound form.

The F_1-sector of the molecule of mesophilic organisms is cold labile (11). That is, the ATPase activity is lost when stable solutions containing F_1 are taken to 4°C. It is accepted that inactivation is due to dissociation of the F_1-sector into fragments containing different subunits. The dissociation pattern was found to be quite complex (93) and it appears to be different in different systems and to depend on the composition of the medium. High concentrations of LiCl stabilize the dissociated F_1 and analysis of the different fragments was carried out in some systems (93, 94; see also 24).

Reconstitution experiments using purified F_1-subunits also attempted to determine the requirements for the formation of oligomeric structures and for the reconstitution of ATPase and ATP-synthetic activity (10, 24, 95, 96). The most detailed study was carried out with TF_1 (95, 96). All possible combinations of subunits (31 mixtures): five singles, ten pairs, ten triplets, five quadruplets and one quintuplet were tested. Four functional/structural parameters were monitored as indicative of successful reconstitution: (a) the ATPase activity; (b) the appearance of an oligomeric structure of electrophoretic mobility similar to that of TF_1; (c) the sensitivity of the activity to Na-azide; and (d) the retention of the reconstituted activity at 60°C were measured after attempted reconstitution trials (97). It was shown that the individual subunits do not have ATPase activity and they do not form oligomers of high molecular weight. Combinations of α- and β-subunits also failed to form large oligomers and were devoid of ATPase activity. The minimum combination of subunits that can reconstitute a large oligomeric structure contains α-, β-, and γ-subunits. The reconstituted complex has size and specific activity similar to that of the native enzyme. Experiments in which F_1 or combinations of subunits were rebound to F_1-depleted particles showed that the δ- and ε-subunits bind to the F_0 portion and are essential for restoration of coupled, oligomycin sensitive ATPase activity and of ATP synthetic activity. It is interesting that the isolated α- and β-subunits do not appear to form oligomers in solution. This suggests that neither α- nor β-subunits form closed structures in the complex because these structures would probably also be stable in solution. The combination of α- and β-subunits do not form oligomers either, suggesting that even in combination α- and β-subunits do not form closed structures in the intact complex. All these suggestions from the reconstitution data are in complete agreement with the structural subunit organization described in a previous section.

MECHANISM OF THE ATP-SYNTHASE REACTION

Introduction

The mechanism of the ATP-synthase reaction can be divided conceptually into three processes. The first is the *translocation of protons* across a lipid bilayer; the second corresponds to the *catalysis* of the formation (hydrolysis) of the P–O–P anhydride bond and the third corresponds to the *coupling* of the dissipation of the proton gradient with the formation of P–O–P anhydride bonds at concentrations much higher than those expected for the uncoupled reaction.

The first two processes are carried out by the F_0 and the F_1 portions of the complex respectively. The third process probably requires the correct quaternary arrangement between the F_0- and F_1-sector for tight coupling.

Proton Translocation

Proton conduction by the ATP-synthase was determined to be carried out by an oligomeric form of the proteolipid (12, 91, 92, 95) and it involves side chains of amino acids identified as aspartate or glutamate, arginine, and tyrosine residues (50, 51). Glasser (98) presented a comprehensive description of all known mechanisms of proton conduction.

Two general kinds of mechanisms were proposed for the conduction of protons by a protein embedded in a lipid bilayer. The first mechanism utilizes the ideas of proton conduction in ice. Several authors suggested it as a possible mechanism (98, 99) and lately Nagle, Mille & Morowitz (100, 101) discussed it in more detail, including a proposal of a mechanism for coupling the dissipation of a proton gradient to the production of an otherwise endergonic reaction. The mechanisms involve the movement of protons (or proton defects) through a chain of hydrogen-bonded groups belonging to either the side chains of the polypeptide or water bound to a protein channel.

The other mechanism [proposed by Freund (102)] is drawn by analogy with proton conduction in solid metal hydroxides [i.e. $A\ell(HO)_3$, $Mg(HO)_2$, $Ca(HO)_2$] and it involves the tunneling of protons through a proton conducting band that has an energy much higher than the mean energy of the protons. Not enough experimental information is available to decide which, if any, of these proposals describes the mechanism of proton translocation by F_0. The mechanism of proton conduction by proteins that span biological membranes will probably turn out to be general in nature. In that case it can be expected that information gained in other systems for which there is already available more detailed structural information [i.e. bacteriorodopsin (103–105)] could aid in the elucidation of the mechanism of proton conduction through F_0.

Bond Formation and Hydrolysis

The second process corresponds to the P–O–P anhydride bond formation (or hydrolysis). The mechanism of this process is the best understood of the processes carried out by the ATP-synthase complex (see 23). It has been demonstrated that the reaction occurs by an in-line displacement (106) of a water molecule from the phosphate brought about by a nucleophilic attack by a β-oxygen of ADP (reviewed in 107). Information has accumulated— from isotopic exchange and bound nucleotide-labeling experiments from P. Boyer's laboratory (reviewed in 107)—that is best understood by assuming that the standard free energy of the reaction is approximately zero ($\Delta G^\circ \simeq 0$) when it occurs between enzyme-bound compounds. Recently the laboratory of Penefsky (108) measured the rate and equilibrium constants for many of the steps involved in a minimum mechanism of ATP hydrolysis (or synthesis) by F_1-ATPase. Figure 6 shows the values they obtained. The value for k_{-3} was not reported but using the value of the equilibrium constant for the overall hydrolysis of ATP ($K = e^{-\Delta G^\circ/RT}$, $\Delta G^\circ = 7.5$ Kcal/mol) a value of 0.2–0.3 M is obtained for K_3. This is in apparent agreement with the observation of Penefsky (109) that ADP "inhibits" binding of P_i to F_1. The dissociation constant of ADP from $F_1 \cdot ADP$ (K_4) was found in this and other studies to be around 1 μM [$K_4 = 0.3 \times 10^{-6}$ in Figure 6 from Ref. (108)] while the dissociation constant of P_i from $F_1 \cdot P_i$

(K_4) was found to be of the order of 100 μM [for example 80 μM in Ref. (109)]. Using the data in Figure 6 investigators estimated the K_D for P_i and ADP dissociating from $F_1 \cdot ADP \cdot P_i$ to be approximately 0.7×10^{-6}. This dissociation constant has to be equal to the product of the stepwise dissociation constants for either of the two dissociation pathways. Therefore $K_3 \cdot K_4 = K'_3 \cdot K'_4 = K_D = 0.7 \times 10^{-6}$ and values can be obtained for K_3 and K'_3 of 0.7 and 7×10^{-3} respectively (2.4 and 9×10^{-3} for the values used in Figure 6). These values show that even though P_i and ADP individually bind quite tightly to the enzyme, binding of both P_i and ADP simultaneously to the same molecule is very unfavorable. Boyer et al (110) made an identical statement based on independent information. In other words, if P_i and ADP binding to the enzyme were independent, K_D would be $K_4 \cdot K'_4 \simeq 10^{-10}$. Since K_D is actually 0.7×10^{-6} there is a factor of 1.4×10^{-4} that arises from the unfavorable free energy of interaction between ADP and P_i bound to the same molecule ($\Delta G^\circ = -RT\ell nK = 5.2$ Kcal/mole). Since this unfavorable energy is relieved when enzyme-bound ATP is formed from bound ADP and P_i, it can be considered that this term is partially responsible for the stabilization of the enzyme-bound ATP. If

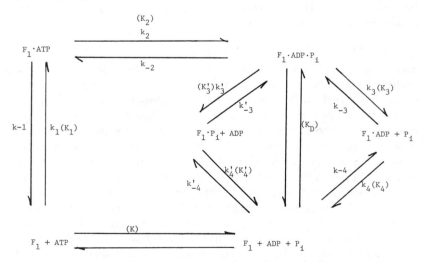

Figure 6   Equilibrium and kinetic constants for the hydrolysis of ATP by F_1. Lower-case indicates kinetic constants and upper case indicates equilibrium constants. The values of some of the constants are given below. Other values are discussed in the text:

$k_1 = 6.4 \times 10^6 \, M^{-1} \, sec^{-1}$	$k_{-1} = 7 \times 10^{-6} \, sec^{-1}$	$K_1 \simeq 10^{12}$
$k_2 = 12 \, sec^{-1}$		$K_2 = 0.5$
$k_3 = 2.7 \times 10^{-3} \, sec^{-1}$		
$k_4 = 3.6 \times 10^{-4} \, sec^{-1}$	$k_{-4} = 1.3 \times 10^3 \, M^{-1} \, sec^{-1}$	$K_4 = 0.3 \times 10^{-6}$
		$K'_4 = 80 \times 10^{-6}$
		$K = 3.6 \times 10^5$

this unfavorable interaction was not present the association constant of binding of ATP to the enzyme, K_1 would have to have been 10^4 times larger ($K_1 \simeq 10^{16}$ instead of 10^{12}). Even for a molecule of the size of ATP a binding constant of that magnitude appears difficult to achieve. So it appears that for the enzyme-bound nucleotides, ATP is the more favorable species ($K_2 = 0.5$) in contrast to the situation in solution where ATP is highly unstable ($K = 3.6 \times 10^5$). This stabilization is accomplished by tightly binding ATP ($K_1 = 10^{12}$) and by destabilizing the simultaneous binding of ADP and P_i ($K_{destabilization} = 1.4 \times 10^4$). Such a thermodynamic behavior should be a direct consequence of the binding characteristic of the catalytic site. The simplest site that would give such a behavior corresponds to one that is highly complementary to all groups in ATP in such a fashion that when ADP and P_i are bound simultaneously to the active site some of the ADP and P_i oxygens are forced to be in too close spatial proximity.

Evidently many other active-site descriptions can be proposed that would be compatible with the thermodynamic data. The major merit of the one just described is that it is probably one of the simplest. The binding data we described correspond to the events occurring at very low substrate concentrations. Under these conditions the enzyme does not behave as a good catalyst for either ATP synthesis or ATP hydrolysis since in both cases it binds too tightly to the products of the reaction. Data from isotope exchange experiments from Boyer's laboratory (reviewed in 23, 107) appear to give the clue as to how this difficulty is solved. Several lines of evidence indicate that ADP binding is necessary for ATP release during synthesis while the release of ADP in hydrolysis is enhanced by ATP binding. The experiments appear to be best understood by assuming that these ATP (ADP) binding sites are not just regulatory sites but that they correspond to the initial binding of the substrates to the active site. That is, for both the forward and back reactions, binding of substrate to an active site enhances product release. This was taken by some researchers as an indication of the existence of a "binding change" mechanism (107). In this mechanism binding of reactants to a weakly binding active site in one β-subunit produces a conformational change that results in the reactant being tightly bound to that site while the product in the active site of another β-subunit becomes loosely bound and is eventually released. This type of mechanism is very attractive and appears to explain a large variety of experimental observations. However, in a general sense other mechanisms were proposed that could explain equally well most of the data. This category includes, for example, the mechanism for F_0F_1 proposed by Mitchell (2) as well as the one proposed by Kozlov & Skulachev (112, 113). Recently, Williams & Coleman (114) suggested a very interesting mechanism based on photo-affinity labeling data. In their scheme, like in that of Kozlov & Skulachev (112, 113), an ATP-binding site in a β-subunit is in close spatial proximity and connected to an ADP-binding site in an α-subunit. Both sites when individually occupied are tight binding sites. When a single nucleotide is bound to such a pair of sites the $ATP + H_2O = ADP + P_i$ equilibrium on the enzyme could have the value suggested by Penefsky et al (108). However in this mechanism hydrolysis/synthesis is accompanied by the translocation of the adenine nucleotide in the direction β to α during hydrolysis or α to β during synthesis. Release of product is enhanced by binding a new substrate molecule to the original catalytic site since the two sites are too close to be comfortably occupied simultaneously. This scheme is in principle very similar to the one proposed by Boyer (107) but in the proposal of Williams & Coleman (114) the site that enhances release of product by binding new substrate is in the same $\alpha\beta$-pair as the original site. The release is effected by destabilization of bound product by the presence of a new substrate bound in close proximity. That is, in the mechanism proposed by Boyer (107), release of substrate is effected allosterically, while in the active-site description of Williams & Coleman (114) the site-to-site interaction is direct. This kind of argument appears difficult to settle in the absence of structural information.

Coupling

The third process catalyzed by the ATP-synthase corresponds to the coupling of P–O–P anhydride bond formation (or hydrolysis) to the translocation of protons across a lipid bilayer. The mechanism of this process is the least understood of those catalyzed by F_0F_1 and it seems to be difficult to address experimentally. Given the thermodynamic description of the bond formation step that is emerging (see above) it appears that the energy of the proton gradient has to be utilized to increase the release of ATP as to produce the high concentrations characteristic of coupled ATP synthesis. Different proposals exist that describe the mechanism of enhanced ATP release brought about by proton translocation. In the mechanism proposed by Boyer and co-workers (107, 115) the protons interact with a conformation of the ATP-synthase that contains tightly bound ATP and loosely bound ADP and P_i. As a result of the proton translocation the conformation is changed to one in which ADP and P_i are tightly bound while the ATP is loosely bound and is eventually released. So in this mechanism the proton gradient affects the steps that were identified as having the most unfavorable standard free energies for ATP formation, namely ADP and P_i binding and ATP release. This is brought about indirectly by a conformational change produced by proton binding to the ATP-synthase complex. In the mechanism proposed by Mitchell (2) the protons are directly involved in the abstraction of an (effective) O^{2-} from bound ADP, P_i and Mg^{2+}. Since in this model (2) formation of ATP in one site is accompanied by ATP release from another site, again in this mechanism protons appear to increment ATP release and ADP and P_i binding. Similar effects are described in the mechanism proposed by Kozlov & Skulachev (112, 113), but in this case, in addition to promoting P–O–P bond formation, the two protons change the charge of the bound Mg^{2+}-nucleotide phosphate complex by $+2$. This change of charge is proposed to promote ATP release. This last proposal is extremely interesting in the sense that independent of the validity of the complete mechanism, the idea that protons effect directly the release of ATP by binding either to ATP or to the active site can be combined with mechanisms like the one proposed by Boyer et al (115) or with mechanisms compatible with the active-site topology presented by Williams & Coleman (114). This last case appears particularly suited for a mechanism of this type since in it the events leading to the release of ATP are localized on a single α-β interface. Protons directed by F_0 to these sites would promote the release of ATP by changing the charge of either ATP (or ADP and P_i) or the side chains of the amino acids involved in nucleotide binding. Titration of groups of this kind is directly contained in the proposal of Boyer and co-workers (107, 115) since the conformational changes they propose could be mainly changes in the ionization of active-site groups. In both cases, the groups titrated should have low enough pK's to be able to take full advantage of the free energy present in the proton gradient (116). In situations where the electrochemical proton gradient ($\Delta\tilde{\mu}_{H^+}$) is mostly given by the membrane potential ($F\Delta\Psi$) the same mechanism would be operational since [as suggested by Mitchell (117; reviewed in 118)] F_0 probably functions as a "proton-specific well" such that the groups in F_1 always "see" the electrochemical proton gradient as a high H^+ concentration. For reasons of brevity we cannot discuss how well all the mechanisms we considered agree with the available experimental data. However, we believe that, in general terms, at least the parts we presented are compatible with most of the published experimental results.

CONCLUDING REMARKS

The ATP-synthase is probably one of the most complex enzyme systems known. The reaction it catalyzes (a coupled vectorial reaction) is mechanistically equally complex. Excellent progress has been achieved in understanding both the structure and the mechanism and regulation of this enzyme. Sequence data is available for both bacterial and eukaryotic systems. A low resolution structure of the F_1-sector has been obtained that provides information about the quaternary arrangement of subunits. Binding and labeling experiments revealed a complex pattern of binding (catalytic and regulatory sites). In conjunction with reconstitution experiments the labeling data permit the assignment of specific functions to each of the individual subunits. Kinetic data and isotope exchange experiments provide invaluable insight into the mechanism of many of the steps involved in ATP synthesis. However, for most steps, several alternative mechanisms still seem to be compatible with most of the available information. It appears then that more experimental data and maybe new experimental approaches will be needed to elucidate fully the mechanism of coupled ATP synthesis by the ATP-synthase complex. In particular it seems clear that many of the problems stated presently as kinetic questions could be most suitably answered by experimental approaches that provide direct structural information. Sequence data and X-ray diffraction will probably make an important contribution, but spectroscopic and resonance techniques, labeling methods, and development of reagents that can map the active site(s) will give many of the clues necessary for the final description of the chemical events.

ACKNOWLEDGMENTS

The work carried out in the author's laboratories was supported by grants GM 25432 and CA 10951 from the National Institutes of Health. We thank Dr. Noreen Williams for critically reviewing the manuscript and Arleen Skaist for skillfully typing and proofreading it.

Literature Cited

1. Mitchell, P. 1966. *Chemiosomotic Coupling in Oxidative and Photosynthetic Phosphorylation*, pp. 1–192. Bodmin Cornwall, England: Glynn Res. Labs.
2. Mitchell, P. 1981. In *Mitochondria and Microsomes in Honor of Lars Ernster*, ed. C. P. Lee, G. Shatz, pp. 427–57. Palo Alto, Calif: Addison-Wesley
3. Abrams, A. 1976. In *The Enzymes of Biological Membranes*, ed. A. Martonosi, pp. 57–73. New York: Plenum
4. Senior, A. E. 1979. In *Membrane Proteins in Energy Transduction*, ed. R. A. Capaldi, M. Dekker, pp. 223–78. New York: Dekker
5. Penefsky, H. S. 1979. *Adv. Enzymol.* 49:223–80
6. Kozlov, I. A., Skulachev, V. P. 1977. *Biochim. Biophys. Acta* 463:29–89
7. Panet, R., Sanadi, R. 1976. *Curr. Top Membr. Transp.* 8:99–150
8. Pedersen, P. L. 1975. *Bioenergetics* 6:243–75
9. Nelson, N. 1981. *Curr. Top. Bioenerg.* 11:1–29
10. Futai, M., Kanazawa, H. 1980. *Curr. Top. Bioenerg.* 10:181–209
11. Fillingame, R. H. 1981. *Curr. Top Bioenerg.* 11:35–100
12. Kagawa, Y. 1978. *Biochim. Biophys. Acta* 505:45–93
13. Wilson, D. B., Smith, J. B. 1978. In *Bacterial Transport*, ed. B. P. Rosen, pp. 495–557. New York: Dekker
14. Harold, F. M. 1977. *Curr. Top. Bioenerg.* 6:83–149
15. Pedersen, P. L., Schwerzmann, K., Cintron, N. 1981. *Curr. Top. Bioenerg.* 11:149–99
16. Ernster, L., Carlsson, C., Hundal, T., Nordenbrand, K. 1979. *Methods Enzymol.* 55:399–407
17. Goffeau, A., Slayman, C. W. 1982. *Biochim. Biophys. Acta* 639:197–223
18. Villalobo, A. 1982. *J. Biol. Chem.* 257:1824–28
19. Apps, D. K., Schatz, G. 1979. *Eur. J. Biochem.* 255:1787–89
20. Schneider, D. L. 1981. *J. Biol. Chem.* 256:3858–64
21. Toll, L., Howard, B. D. 1980. *J. Biol. Chem.* 255:1787–89
22. Carty, S., Johnson, R. G., Scarpa, A. 1982. *J. Biol. Chem.* 257:7269–73
23. Cross, R. L. 1981. *Ann. Rev. Biochem.* 50:681–714
24. Downie, J. A., Gibson, F., Cox, G. B. 1979. *Ann. Rev. Biochem.* 48:103–31
25. Cox, G. B., Downie, J. A. 1978. *Methods Enzymol.* 56:409–15
26. Kanazawa, H., Tamura, F., Mabuchi, K., Miki, T., Futai, M. 1980. *Proc. Natl. Acad. Sci. USA* 77:7005–9

27. Kanazawa, H., Miki, T., Tamura, F., Yura, T., Futai, M. 1979. *Proc. Natl. Acad. Sci. USA* 76:1126–30
28. Gay, N. J., Walker, J. E. 1981. *Nucleic Acids Res.* 9:2187–94
29. Gay, N. J., Walker, J. E. 1981. *Nucleic Acids Res.* 9:3919–26
30. Saraste, M., Gay, N. J., Eberle, A., Runswick, M. J., Walker, J. E. 1981. *Nucleic Acids Res.* 9:5287–96
31. Walker, J. E., Saraste, M., Gay, N. J. 1982. *Nature* 286:867–69
32. Walker, J. E., Eberle, A., Gay, N. J., Runswick, M. J., Saraste, M. 1982. *Biochem. Soc. Trans.* 10:203–6
33. Kanazawa, H., Mabuchi, K., Kayano, T., Tamura, F., Futai, M. 1981. *Biochem. Biophys. Res. Commun.* 100:219–25
34. Mabuchi, K., Kanazawa, H., Kayano, T., Futai, M. 1981. *Biochem. Biophys. Res. Commun.* 102:172–79
35. Kanazawa, H., Kayano, T., Mabuchi, K., Futai, M. 1981. *Biochem. Biophys. Res. Commun.* 103:604–12
36. Kanazawa, H., Mabuchi, K., Kayano, T., Noumi, T., Sekiya, T., Futai, M. 1981. *Biochem. Biophys. Res. Commun.* 103:613–20
37. Kanazawa, H., Kayano, T., Kiyasu, T., Futai, M. 1982. *Biochem. Biophys. Res. Commun.* 105:1257–64
38. Kanazawa, H., Mabuchi, K., Futai, M. 1982. *Biochem. Biophys. Res. Commun.* 107:568–75
39. Galante, Y. M., Wong, S. Y., Hatefi, Y. 1979. *J. Biol. Chem.* 254:12372–78
40. Soper, J. W., Decker, G. L., Pedersen, P. L. 1979. *J. Biol. Chem.* 254:11170–76
41. Glaser, E., Norling, B., Ernster, L. 1981. *Eur. J. Biochem.* 110:225–35
42. Knowles, A. F., Penefsky, H. S. 1972. *J. Biol. Chem.* 247:6624–30
43. Esch, F. S., Allison, W. S. 1978. *J. Biol. Chem.* 253:6100–6
44. Rossmann, M. G., Argos, P. 1981. *Ann. Rev. Biochem.* 50:497–532
45. Schmidt, U. D., Paradies, H. 1977. *Biochem. Biophys. Res. Commun.* 78:1043–52
46. Frangione, B., Rosenwasser, E., Penefsky, H. S., Pullman, M. E. 1981. *Proc. Natl. Acad. Sci. USA* 78:7403–7
47. Anderson, S., Bankier, A. T., Barrell, B. G., deBruijn, M. H. L., Coulson, A. R., et al. 1981. *Nature* 290:457–65
48. Macino, G., Tzagoloff, A. 1980. *Cell* 20:507–17
49. Bibb, M. J., van Etten, R. A., Wright, C. T., Walberg, M. W., Clayton, D. A. 1981. *Cell* 26:167–80
50. Sebald, W., Wachter, E. 1978. In *Energy Conservation in Biological Membranes*, ed. G. Schafer, M. Klingenberg, pp. 228–36. Berlin/Heidelberg/New York: Springer
51. Sebald, W., Machleidt, W., Wachter, E. 1980. *Proc. Natl. Acad. Sci. USA* 77:785–89
52. Hoppe, J., Schairer, H. U., Sebald, W. 1980. *Eur. J. Biochem.* 112:17–24
53. Amzel, L. M. 1981. *J. Bioenerg. Biomembr.* 13:109–21
54. Huberman, M., Salton, M. R. J. 1979. *Biochim. Biophys. Acta* 547:230–40
55. Yoshida, M., Sone, N., Hirata, H., Kagawa, Y. 1978. *Biochem. Biophys. Res. Commun.* 84:117–22
56. Todd, R. D., Griesenbeck, T. A., Douglas, M. G. 1980. *J. Biol. Chem.* 255:5461–67
57. Babakov, A. V., Terekhov, O. P., Yanenko, A. S. 1979. *Biochim. Biophys. Acta* 547:438–46
58. Stutterheim, E., Henneke, M. A. C., Berden, J. A. 1981. *Biochim. Biophys. Acta* 634:271–78
59. Yoshida, M., Sone, N., Hirata, H., Kagawa, Y., Ui, N. 1979. *J. Biol. Chem.* 254:9525–33
60. Ikemura, T. 1981. *J. Mol. Biol.* 146:1–21
61. Catterall, W. A., Pedersen, P. L. 1974. *Biochem. Soc. Spec. Publ.* 4:63–88
62. Kagawa, Y., Racker, H. S. 1966. *J. Biol. Chem.* 241:2475–82
63. Adolfsen, R., Moudrianakis, E. N. 1971. *Biochemistry* 10:2247–53
64. Wakabayashi, T., Kubota, M., Yoshida, M., Kagawa, Y. 1977. *J. Mol. Biol.* 117:515–19
65. Amzel, L. M., Pedersen, P. L. 1978. *J. Biol. Chem.* 253:2067–69
66. Amzel, L. M., McKinney, M., Narayanan, P., Pedersen, P. L. 1982. *Proc. Natl. Acad. Sci. USA* 79:5852–56
67. Ohta, S., Tsuboi, M., Oshima, T., Yoshida, M., Kagawa, Y. 1980. *J. Biochem.* 87:1609–17
68. Harris, D. A., Gomez-Fernandez, J. C., Klungsoyr, L., Radda, G. K. 1978. *Biochim. Biophys. Acta* 504:364–83
69. Garrett, N. E., Penefsky, H. S. 1978. *J. Biol. Chem.* 250:6640–47
70. Cross, R. L., Nalin, C. M. 1982. *J. Biol. Chem.* 257:2874–81
71. Kumar, G., Kalra, V. K., Brodie, A. F. 1979. *J. Biol. Chem.* 254:1964–71
72. Wagenvoord, R. J., van der Kraam, I., Kemp, A. 1977. *Biochim. Biophys. Acta* 460:17–24
73. Senior, A. E. 1979. *J. Biol. Chem.* 254:11319–22
74. Chang, T. M., Penefsky, H. S. 1973. *J. Biol. Chem.* 218:2746–54
75. Verschoor, G. J., van der Sluis, P. R., Slater, E. C. 1977. *Biochim. Biophys. Acta* 462:438–49
76. Wagenvoord, R. J., Verschoor, G. J.,

Kemp, A. 1981. *Biochim. Biophys. Acta* 634:229–36

77. Hulla, F. W., Hockel, M., Rack, M., Risi, S., Dose, K. 1978. *Biochemistry* 17:823–28

78. Wagenvoord, R. J., van der Kraam, I., Kemp, A. 1979. *Biochim. Biophys. Acta* 548:85–95

79. Bragg, P. D., Hou, C. 1980. *Biochem. Biophys. Res. Commun.* 95:952–57

80. Kozlov, I. A., Milgrom, Y. M. 1980. *Eur. J. Biochem.* 106:457–62

81. Drutsa, V. L., Kozlov, I. A., Milgrom, Y. M., Shabarova, Z. A., Sokolova, N. I. 1979. *Biochem. J.* 182:617–19

82. Wagenvoord, R. J., Kemp, A., Slater, E. C. 1980. *Biochim. Biophys. Acta* 593:204–11

83. Verheijen, J. H., Postma, P. W., van Dam, K. 1978. *Biochim. Biophys. Acta* 502:345–53

84. Scheurich, P., Schafer, H. J., Dose, K. 1978. *Eur. J. Biochem.* 88:253–57

85. Schafer, H. J., Scheurich, P., Rathgeber, G., Dose, K. 1980. *Anal. Biochem.* 104:106–11

86. Gregory, R., Recktenwald, D., Hess, B., Schafer, H. J., Scheurich, P., et al. 1980. *FEBS Lett.* 108:253–56

87. Ferguson, S. J., Lloyd, W. J., Lyons, M. H., Radda, G. K. 1975. *Eur. J. Biochem.* 54:117–26

88. Lunardi, J., Vignais, P. V. 1979. *FEBS Lett.* 102:23–28

89. Deters, D. W., Racker, E., Nelson, N., Nelson, H. 1975. *J. Biol. Chem.* 250:1041–47

90. Cantley, L. C., Hammes, G. G. 1976. *Biochemistry* 15:1–15

91. Altendorf, K. 1977. *FEBS Lett.* 73:271–75

92. Criddle, R. S., Packer, L., Shieh, P. 1977. *Proc. Natl. Acad. Sci. USA* 74:4306–10

93. Vogel, G., Steinhart, R. 1976. *Biochemistry* 15:208–16

94. Philosoph, S., Binder, A., Gromet-Elhanan, Z. 1977. *J. Biol. Chem.* 252:8747–52

95. Kagawa, Y., Sone, N., Hirata, H., Okamoto, H. 1979. *J. Bioenerg. Biomembr.* 11:39–78

96. Kagawa, Y. 1978. See Ref. 50, pp. 195–219

97. Yoshida, M., Sone, N., Hirata, H., Kagawa, Y. 1977. *J. Biol. Chem.* 252:3480–85

98. Glasser, L. 1975. *Chem. Rev.* 75:785–89

99. Williams, R. J. P. 1978. *FEBS Lett.* 85:9–19

100. Nagle, J. F., Mille, M., Morowitz, H. J. 1980. *J. Chem. Phys.* 72:3959–71

101. Nagle, J. F., Morowitz, H. J. 1978. *Proc. Natl. Acad. Sci. USA* 75:298–302

102. Freund, F. 1981. *Trends Biochem. Sci.* 6:142–45

103. Stoeckenius, W. 1979. In *Membrane Transduction Mechanisms*, ed. R. A. Cone, J. E. Dowling, pp. 39–47. New York: Raven

104. Ovchinnikov, Y. A., Abdulaev, N. G., Feigina, M. Y., Kiselev, A. V., Lobanov, N. A. 1979. *Biochem. Biophys. Res. Commun.* 82:727–35

105. Unwin, P. N. T., Henderson, R. 1975. *J. Mol. Biol.* 94:425–40

106. Webb, M. R., Grubmeyer, C., Penefsky, H. S., Trentham, D. R. 1980. *J. Biol. Chem.* 255:11,637–39

107. Boyer, P. D. 1980. In *Membrane Bioenergetics*, ed. C. P. Lee, G. Schatz, L. Ernster, pp. 461–79. Palo Alto, Calif: Addison-Wesley

108. Penefsky, H. S., Cross, R. L., Grubmeyer, C. 1982. *Abstr. Second Eur. Bioenerg. Conf.*, pp. 17–18. Lyon, France:LBTM-CNRS

109. Penefsky, H. S. 1977. *J. Biol. Chem.* 252:2891–99

110. Boyer, P. D., Kohlbrenner, W. E., Smith, L. T., Feldman, R. E. 1982. See Ref. 108, pp. 23–24

111. Deleted in proof

112. Kozlov, I. A., Skulachev, V. P. 1977. *Biochim. Biophys. Acta* 463:29–89

113. Kozlov, L. A., Skulachev, V. P. 1982. In *Curr. Top. Membr. Trans.* 16:285–301

114. Williams, N., Coleman, P. S. 1982. *J. Biol. Chem.* 257:2834–41

115. Kayalar, C., Rosing, J., Boyer, P. D. 1977. *J. Biol. Chem.* 252:2486–91

116. Jencks, W. P. 1980. *Adv. Enzym.* 51:75–106

117. Mitchell, P. 1969. *Theor. Exp. Biophys.* 2:159–216

118. Maloney, P. C. 1982. *J. Membr. Biol.* 67:1–12

Ann. Rev. Biochem. 1983. 52:825–69

PENICILLIN-BINDING PROTEINS AND THE MECHANISM OF ACTION OF β-LACTAM ANTIBIOTICS[1]

David J. Waxman[2]

Department of Chemistry, Massachusetts Institute of Technology, Cambridge, Mass. 02139

Jack L. Strominger

Department of Biochemistry and Molecular Biology, Harvard University, Cambridge, Mass. 02138

CONTENTS

PERSPECTIVE 826
PENICILLIN-BINDING PROTEINS (PBPs): GENERAL PROPERTIES 828
PBPs: TARGETS OF β-LACTAM ANTIBIOTICS AND ROLES IN BACTERIAL
    PHYSIOLOGY 829
    High-Molecular-Weight PBPs of E. coli 830
    High-Molecular-Weight PBPs from Other Bacteria 834
    Importance of D-*Alanine Carboxypeptidase (CPase)* 836
BIOCHEMICAL STUDIES OF PURIFIED PBPs 839
    Purification of PBPs 839
    Interactions with Cell Wall–Related Substrates 840
    Interactions with β-Lactam Antibiotics 845
STRUCTURAL STUDIES OF PBPs 851
    Active-Site Peptides and the Mechanism of Action of β-Lactam Antibiotics ... 851
    Homology to β-Lactamases 853
    Other Active-Site Residues 856
    High Resolution Structural Analysis 856
    Relationships Among PBPs 857
PBPs AS MEMBRANE-BOUND ENZYMES 858
β-LACTAMS AS SUBSTRATE ANALOGS 861
CONCLUSION 862

[1] The abbreviations used are: PBPs, penicillin-binding proteins; CPase, D-alanyl-D-alanine carboxypeptidase, referring to low-molecular-weight PBPs that exhibit D,D-alanyl carboxypeptidase and/or transpeptidase activities using simple peptide substrates *in vitro*; and SDS, sodium dodecylsulfate.

[2] Present address: Dana-Farber Cancer Institute, 44 Binney St., Boston, Mass. 02115.

825

0066-4154/83/0701-0825$02.00

PERSPECTIVE

Interest in the mechanism by which penicillin kills bacteria began with its discovery by Fleming in 1929 (68) and intensified with its use during World War II. Three fundamental observations made in this early period were harbingers of the complexities soon to be discovered. Morphological studies by Gardner (83) revealed that low concentrations of penicillin induced filamentation of *Escherichia coli*, with bulge formation and lysis occurring only at higher antibiotic concentrations. Duguid (59) made similar morphological observations and correctly concluded that penicillin was interfering with the synthesis of some cell surface structure (the still undiscovered bacterial cell wall). Several laboratories then recognized that radioactive penicillin G was bound to the cell membrane and that this binding was related to the killing action of penicillin (36). Thus the stage was set for a series of discoveries that led to our present knowledge of the complex interactions of penicillins and other β-lactam antibiotics with the bacterial cell. These included: (*a*) discovery of bacterial cell wall peptidoglycan and of the uridine nucleotide precursor (180, 182; UDP *N*-acetylmuramyl pentapeptide) that accumulates in penicillin-inhibited bacteria; (*b*) elucidation of the detailed chemical structure of peptidoglycan, and the unraveling of the complex pathway that leads to its biosynthesis (reviewed in 91, 92, 151, 197, 250); (*c*) identification of peptidoglycan transpeptidase(s) as the lethal target(s) of β-lactam antibiotics (248, 276); and (*d*) the realization that much of the complexity observed is due to the multiplicity of proteins in the bacterial cell membrane that bind penicillins and related β-lactam antibiotics covalently (the penicillin-binding proteins or PBPs). Thus there are multiple penicillin-sensitive enzymes and also multiple lethal targets (12, 14, 209, 235). In turn, this last discovery has led to a deeper understanding of the intricate mechanisms by which bacterial cell walls are synthesized and the processes involved in the growth, division, and determination of cell shape for bacterial cells.

All *Eubacteria* examined [with the possible exception of mycoplasma, (143) and halobacteria] contain multiple penicillin-binding proteins (PBPs), with three to eight distinct PBPs typically found in a given organism (e.g. 6, 14, 39, 62, 84). This review centers on studies of these PBPs as the targets of β-lactam antibiotics and their roles in bacterial physiology as elucidated by morphological observations, genetic studies, structural analysis, and biochemical investigations of their interactions both with cell wall substrates and with β-lactam antibiotics. Early studies that formed the basis for much of the work in this field during the past ten years were reviewed in 1974 (14) and are not discussed in any detail here. By focusing on the penicillin-binding proteins in our approach to the mechanism of action of β-lactam antibiotics, we hope to address the following questions:

1. Which enzymatic reactions are catalyzed by PBPs and how do these reactions relate to the physiology of microbial growth? Are penicillin-sensitive transpeptidase and D-alanine carboxypeptidase activities (Figure 1) among the reactions catalyzed? Are there penicillin-sensitive enzymes that cannot be detected as PBPs?

2. By which precise mechanisms do PBPs react with penicillins? What accounts for the selectivity of different β-lactam antibiotics for different PBPs? Are β-lactam antibiotics really substrate analogs? Which, if any, of the PBPs are lethal targets of β-lactam antibiotics? Are there multiple killing sites?

3. What are the functions of the multiple PBPs in cell wall biosynthesis? Are all PBPs essential for cell viability?

4. What evolutionary relationships exist among PBPs of a given bacterium and between PBPs of related bacteria? What is their relationship to β-lactamases?

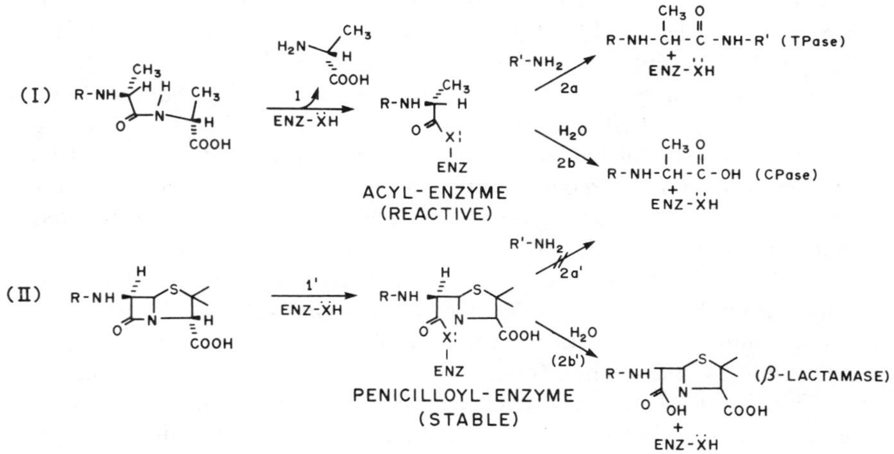

Figure 1 Proposed mechanistic homology between processing of R-D-Ala-D-Ala substrates (I) and β-lactam antibiotics (II) by penicillin-sensitive enzymes. The contrast between the stability of the penicilloyl-enzyme linkage (permitting detection of PBPs) and the high reactivity of the substrate-derived acyl enzyme is an essential feature for the inhibition by β-lactams. Reaction of the acyl enzyme with cell wall amino acceptors (Reaction 2a) results in cross-linking (transpeptidase; TPase) vs reaction with water results in CPase activity (2b). Penicilloyl enzymes are stable to transfer to cell wall nucleophiles (2a'). β-Lactamases could have evolved from penicillin-sensitive enzymes by developing an efficient catalytic mechanism for hydrolysis of the penicilloyl-enzyme linkage (2b').

PENICILLIN-BINDING PROTEINS (PBPS): GENERAL PROPERTIES

PBPs are defined as those bacterial proteins that bind penicillins and other β-lactam antibiotics covalently. PBPs are readily detected and their relative amounts quantitated by incubation of bacterial membranes with [^{14}C]penicillin G, followed by sodium dodecylsulfate (SDS) gel electrophoresis and fluorography (e.g. 12, 210, 261). Use of [^{3}H]penicillin G of high specific activity permits labeling of PBPs in vivo (54, 272). In a given organism, PBPs are numbered in order of decreasing apparent molecular weight, which usually ranges from $M_r \sim 140,000$–$40,000$ (although PBPs as small as $M_r \sim 25,000$ have also been detected; 62). There is no necessary relationship between equivalently numbered PBPs of two unrelated organisms (e.g. PBP-2 of *E. coli* and PBP-2 of *Bacillus subtilis*) although taxonomically related bacteria have similar PBP patterns (39, 84). Although most investigators use [^{14}C]penicillin G (available commercially) to detect PBPs, additional binding proteins are, in general, not detected using other radioactively labeled β-lactams including ^{14}C-labeled mecillinam, cefoxitin, PC-904, furazlocillin, and moxalactam or ^{3}H-labeled propionyl ampicillin (124, 129, 158, 167, 201, 204, 210). Using [^{125}I]-Bolton-Hunter ampicillin, however, an additional PBP (PBP-1C) is detected in *E. coli* membranes (201). Most bacteria contain between 1000 and 10,000 total PBP molecules per cell, with the PBPs comprising approximately 1% of the total membrane protein (14). Particular PBPs vary greatly in their relative abundance: *E. coli* PBP-2 accounts for approximately 1% of the organism's penicillin-binding activity, estimated to correspond to 0.01% of total membrane protein and only 15–20 molecules of PBP-2 per cell (210) as compared to $\sim 90\%$ of the total penicillin binding activity for *Bacillus stearothermophilus* PBP-5 (280).

The affinity of a PBP for a given (nonradiolabeled) β-lactam is usually expressed as the concentration of antibiotic required to reduce [^{14}C]penicillin G binding to the PBP by 50% and is determined after preincubation with the unlabeled β-lactam under given conditions of time, temperature, etc. However, quantitative measurements of a PBP's sensitivity to β-lactams determined in this manner will vary, depending on both the time of incubation with the β-lactam and on the concentration of membrane suspensions used as source of PBP, with lower inhibitory concentrations often obtained using more dilute membrane suspensions. For these reasons, affinities of PBPs for β-lactams are probably more meaningful when determined in vivo (e.g. 54, 189, 272). PBPs of a given organism exhibit widely varying sensitivities to β-lactams, e.g. in *B. stearothermophilus*, where PBPs-1 to -4 are greater than 10,000-fold more sensitive to cephalothin than PBP-5 (280). Examination of the sensitivity of PBPs from several organisms to a wide variety of β-lactam antibiotics suggests two general categories of PBPs, those of low molecular weight (M_r $\sim 40,000$–$50,000$), which are somewhat less sensitive to many penicillins and highly insensitive to most cephalosporins, and those of high molecular weight (M_r $\sim 60,000$–$140,000$), which are generally more sensitive to both penicillins and cephalosporins. This dichotomy, based on sensitivity to β-lactam antibiotics, extends to several other properties of the PBPs, including their relative abundance, in vitro activities, and importance for cell growth and viability. Thus, in general, the low-molecular-weight PBPs are relatively abundant, catalyze efficient penicillin-sensitive D,D-alanine carboxypeptidase (CPase) reactions in vitro and appear to be unnecessary for cell viability. By contrast, high-molecular-weight PBPs are usually minor components, do not catalyze CPase or transpeptidation reactions (Figure 1) using model substrates, and are often essential for cell viability. Most PBPs are integral membrane proteins [found in the inner (cytoplasmic) membrane in gram-negative bacteria] and thus require treatment with nonionic or mild anionic detergents for solubilization in active form. In the case of PBP-5 from $B. subtilis$ (CPase), efficient solubilization is effected by a small number of detergents, all having a hydrophilic-lipophilic balance number (HLB) of 12.5–13.5 (256). A careful study of the optimal HLB required to solubilize high-molecular-weight PBPs has not been undertaken.

PBPs: TARGETS OF β-LACTAM ANTIBIOTICS AND ROLES IN BACTERIAL PHYSIOLOGY

Studies aimed at deducing which PBPs are lethal targets for β-lactam antibiotics have utilized two principal approaches: analysis of mutants with altered PBP patterns and correlations between in vivo and in vitro effects of β-lactams. Thus, the physiological consequences of inactivation of particular PBPs, either by mutation or by treatment with highly selective β-lactams, have been examined and the in vivo functions of the PBPs inferred. Conclusions based on β-lactam binding studies alone must be scrutinized carefully, as there are secondary factors likely to complicate the in vivo situation including permeability problems (24, 53a, 163a, 174), resistance due to β-lactamase production (172a, 189a, 239, 239a) and indirect effects due to treatment with β-lactams, such as activation of the peptidoglycan hydrolase (autolytic) systems (205, 251–253). Questions to be answered in such analyses include the following: 1. What effects does mutation or, better, deletion of the gene encoding a given PBP have on the morphology or β-lactam sensitivity of the mutant? 2. Can mutant strains defective in more than one PBP be constructed? 3. Do certain PBPs assume compensatory functions, acting as "detour" enzymes in the absence of the wild-type complement of PBPs? 4. Does a given β-lactam saturate a particular PBP at subinhibitory growth concentrations? Although such a PBP is probably not essential for cell viability (when all the other PBPs are active), it may nevertheless fill an important role in normal cell growth and division. Conversely, if cell growth or viability is affected by a β-lactam at a concentration that saturates a single PBP in vivo, that PBP might be an important target. 5. Finally, does one expect that induction of lethal effects in vivo will occur at 10% saturation or at 95 + % saturation of an essential PBP? Implicit in this approach is the assumption that formation of the covalent adduct between PBP and β-lactam is directly responsible for loss of enzymatic activity in vivo. However, other possibilities could conceivably include: (a) inhibition due to formation of noncovalent complexes without reaction of the β-lactam bond; (b) indirect effects of β-lactams, e.g. on the biosynthesis of an essential PBP that itself does not exhibit a high sensitivity to the β-lactam in question (104, 271); and (c) unusual modes of action, e.g. the covalent incorporation of the β-lactam MT-141 into E. coli peptidoglycan (149). As most of the detailed studies of the targets of β-lactam antibiotics have focused on E. coli PBPs, these will be described in some detail. PBPs of several gram-negative rods have similar biochemical properties and, most likely, analogous in vivo functions (39, 168, 169, 176, 210, 212, 216), and thus the results obtained from studies of E. coli PBPs [PBPs-1A, -1Bs, -2, -3 (all high-molecular-weight PBPs; M_r ~90,000, 87,000, 66,000, and 60,000, respectively) and PBPs-4, -5, and -6 (low-molecular-weight PBPs; M_r ~49,000, 42,000, and 40,000, respectively)] are likely to be of general importance.

High-Molecular-Weight PBPs of E. coli

PBPS-1A AND -1BS: ENZYMES OF CELL ELONGATION   Many β-lactams induce morphological changes in growing E. coli cells when present at low concentration, with lysis effected only at higher concentrations of antibiotic. Some, e.g. cephaloridine and cefsulodin, cause cell lysis at their minimal inhibitory concentrations, presumably by inhibiting cell elongation. That cephaloridine has the highest affinity for E. coli PBP-1 suggested that this PBP might be involved in cell elongation (209). PBP-1 actually consists of two proteins having distinct affinities for various β-lactam antibiotics (termed PBP-1A and PBP-1Bs; PBP-1Bs itself consists of at least three components distinguishable on SDS-gels; 220, 236, 242). The good correlation between cell lysis and binding of β-lactams to PBP-1Bs suggests that PBP-1Bs is the PBP most important for cell elongation (220).

PBP-1A$^-$ mutants appear to grow normally (220, 236), suggesting that in strains containing normal levels of PBP-1Bs, PBP-1A might be nonessential for normal growth and morphology. The differential effects of PBP-3-selective β-lactams on PBP-1A$^-$ vs PBP-1Bs$^-$ strains (200) as well as the slow rate of β-lactam-induced lysis of a, PBP-1A$^-$ mutant (220) does, however, indicate a role for PBP-1A in peptidoglycan metabolism. As these PBP mutants are identified by the absence of detectable penicillin-binding activity, it is possible that essential PBP-1A activities might be retained in vivo. It is therefore important to establish the absence of the protein moiety of PBP-1A, e.g. using genetic or immunochemical techniques, before firm conclusions can be drawn with regard to the essentiality of PBP-1A. Mutants lacking the penicillin-binding activity of PBP-1Bs have also been isolated (236, 242). Membranes isolated from these mutants have significantly decreased peptidoglycan biosynthetic activity, suggesting that PBP-1Bs catalyzes the majority of this activity in vitro. The loss of penicillin binding by all three components of PBP-1Bs (likely to be related structurally; 161) and the loss of in vitro biosynthetic activity were simultaneously restored in revertants and transductants, suggesting that the changes are the consequence of a single mutation. Although neither the PBP-1A$^-$ nor the PBP-1Bs$^-$ mutation conferred thermosensitive growth, the double mutant PBP-1A^{ts} PBP-1Bs$^-$ is thermosensitive, with lysis occurring at the restrictive temperature (236). The double mutation PBP-1A$^-$ PBP-1Bs$^-$ is apparently lethal. Thus, PBPs-1A and -1Bs are redundant (or at least capable of filling compensatory roles), it being necessary to delete or inhibit both PBPs to prevent cell elongation and thereby induce lysis.

Complete inactivation of PBP-1Bs alone may be sufficient to induce cell lysis, the normal growth of PBP-1A$^+$ PBP-1Bs$^-$ strains (236) possibly reflecting the presence of residual PBP-1Bs activity in vivo (despite the absence of detectable activity in vitro) that is sufficient for normal growth in the presence of a fully active PBP-1A (216). The supersensitivity of PBP-1Bs$^-$ mutants to cephalosporins (236) can be explained by the high sensitivity of PBP-1A to these β-lactams: in the absence of a functional PBP-1Bs, PBP-1A becomes indispensable and thus the mutant organism reflects the hypersensitivity of this PBP to cephalosporins. Although PBP-1A probably catalyzes transpeptidation in vivo, the absence of in vitro peptidoglycan biosynthetic activity using PBP-1A$^+$ PBP-1B$^-$ membranes (236, 242) might reflect inactivation of this PBP [which is significantly more thermolabile than PBP-1Bs (160)] during preparation of the membrane fraction.

PBP-2 AND CELL SHAPE   *E. coli* exhibits an unusual morphological response to mecillinam, a 6β-amidinopenicillanic acid derivative: in its presence, large osmotically stable round cells form and lyse slowly after several hours of active growth (138, 255). Total cell [^{14}C]penicillin G binding activity as well as total CPase, transpeptidase, and endopeptidase activity are unaffected by mecillinam (145, 181), consistent with the highly selective binding of mecillinam to *E. coli* PBP-2, a protein representing only 1% of the total PBPs (209, 219, 218). Similar morphological effects are obtained with thienamycin, clavulanic acid, and various mecillinam analogs, all of which have a high affinity for PBP-2 (211, 212, 219). Several PBP-2 mutants support the importance of this PBP for maintenance of the characteristic rod shape of *E. coli*. These include: (*a*) mecillinam-resistant round cells, in which no PBP-2 can be detected by penicillin binding (236); (*b*) cells temperature sensitive for rod morphology. At the permissive temperature, the cells are mecillinam sensitive and contain a penicillin-binding PBP 2; upon shifting to high temperature, osmotically stable round cells form with loss of PBP-2 activity (116, 212); and (*c*) cells exhibiting partial resistance to mecillinam. The resistance of these cells to a 10- to 15-fold higher concentration of mecillinam [and to other β-lactams that bind exclusively to PBP-2 at their minimal inhibitory concentration (MIC)] is accompanied by a corresponding decrease in the affinity of PBP-2 for all β-lactams tested (214). Acquisition of high levels of resistance is presumably limited by the difficulty of altering the PBP's affinity for mecillinam significantly without impairing its capability for processing its yet-to-be-identified physiologic substrates.

Although PBP-2 might catalyze a novel reaction, it probably mediates a topologically restricted transpeptidase (or carboxypeptidase) reaction, perhaps during a specific portion of the cell cycle. Inactivation of PBP-2 results in a doubling of the anhydromuramic acid content of *E. coli* cell walls (63), suggesting that this PBP might play a role in the regulation of cell wall glycan chain-length determination. That such regulation probably occurs at specific sites along the cell wall is consistent with the observed restricted topographical distribution of PBP-2 in the *E. coli* membrane (25). That partial revertants of a PBP-1Bs^{ts} mutant overproduce both PBPs-1A and -2 (242) suggests that PBP-2 is capable of participating in the incorporation of new glycan strands for cell elongation and is consistent with the suggestion (209) that it may act to insure that new initiation sites for elongation are introduced with the correct rod orientation. In the absence of an active PBP-2, new glycan strands might be utilized preferentially by the septum-forming enzyme system, resulting in growth as round cells. Similarly, the round-cell phenotype of PBP-5-overproducing cells (139) could result from cleavage of the pentapeptide substrates needed for cell elongation, which, at the same time, would increase the pool of tetrapeptides that might be required for septal biosynthesis (17, 139).

Although mecillinam-sensitive transpeptidation has been detected in membranes from PBP-2-overproducing cells (114), the detailed structure of its biosynthetic product(s) is unknown. PBP-2 does not catalyze transpeptidation of the outer membrane lipoprotein to peptidoglycan (21).

PBP-3 AND CELL DIVISION   Each of several β-lactams, including cephalexin, cefuroxime, furazlocillin, piperacillin, and azthreonam, induces filamentation of E. coli by specifically inhibiting cell division without effecting cell lysis under conditions where the antibiotic is bound to only PBP-3 (17, 87, 200, 209, 216). PBP-3^{ts} mutants are temperature sensitive for cell division, and their revertants, growing as normal rods at the high temperature, regain a thermostable PBP-3 (210), which provides strong evidence that inactivation of PBP-3 at the high temperature is the cause of inhibition of cell division. Mutants defective in the proteolytic processing of pre-PBP-3 to the mature PBP-3 grow in chains, probably reflecting an abnormality in septation such that cell separation is delayed (D. Karibian et al, submitted for publication; Y. Hirota, unpublished observations). In addition, recent studies suggest that E. coli PBP 1C might also be necessary for cell division (17).

Cell division can be divided into three stages: initiation of septation, septum formation, and cell separation (207). PBP-3-selective β-lactams appear to uncouple the first stage from the second (178), probably by blocking septal murein synthesis (17, 178). PBP-3 might be activated or synthesized upon completion of DNA replication, or, alternatively, its activity might be regulated by substrate availability at the site of septation. The increase in D-alanine carboxypeptidase activity as well as D-alanine carboxypeptidase II activity (release of D-Ala from a cell wall tetrapeptide yielding a tripeptide) just before cell division (7a, 153; see below) suggests (17, 139) that PBP-3 might utilize tripeptides (or tetrapeptides) as amino acceptors for septal transpeptidation.

Although these studies of the physiological consequences of inactivation of E. coli PBPs-1, -2, or -3 demonstrate that particular β-lactams can kill the bacterium by inactivation of a single PBP (e.g. mecillinam—PBP-2; furazlocillin—PBP-3), many β-lactams have similar affinities for several targets. In such cases the morphological effects produced can be explained by the relative affinity of the β-lactam for its PBP targets (209, 216). The lethal target for a particular β-lactam will, therefore, be the essential PBP with the highest affinity. That some β-lactams kill E. coli by binding to PBP-2 or PBP-3 exclusively demonstrates that inactivation of the major transpeptidase (likely to be PBP-1Bs in E. coli) is not the sole route by which β-lactams kill bacteria.

The structural genes for PBP-1A (designated either pon A or mrc A), PBP

1Bs (*pon B* or *mrc B*), PBP 2 (*pbp A* or *mrd A*) and PBP 3 (*fts I* or *pbpB*) have been mapped at 73.5, 3.3, 14.4, and 1.8 minutes, respectively, on the *E. coli* chromosome (236). The separate location for the genes for PBPs-1A and -1Bs clearly establishes their separate identity. PBP-2 is located in a gene cluster (*lip–dac A–rod A–pbp A–leu S*) that includes a second round morphology gene (*rod A* or *mrd B*) and the gene for PBP-5 (*dac A*) (221, 244). Interestingly, PBP-5 can also be viewed as a cell-shape gene since overproduction of this CPase results in ovoid cells having the same abnormalities in peptidoglycan biosynthesis as do cells with an inactivated PBP-2 (139). Each of the *E. coli* PBPs has been cloned into plasmid and overproduced (>100-fold in some cases) by cloning into "runaway" plasmids or by constructing plasmids that contain a promoter under the control of a thermolabile repressor (133, 164, 201, 221). Under these circumstances the overproduced PBPs appear to be inserted properly into the cell's cytoplasmic membrane (133). Thus, relatively large amounts of *E. coli* PBPs can be obtained for structural and enzymatic studies. Additional genetic studies have revealed further complexities including: (*a*) another gene cluster, near *env B*, which appears to have effects on morphology and may contain genes that regulate the expression of PBPs (149); and (*b*) a mutant described previously (*iap*⁻, isomers of alkaline phosphatase⁻) that has recently been found to have a second phenotypic characteristic; namely it synthesizes only a single species of PBP-1Bs (J. T. Park, unpublished observations).

High-Molecular-Weight PBPs from Other Bacteria

The relative insensitivity of the low-molecular-weight PBPs (CPases) of gram-positive bacteria to many penicillins and most cephalosporins suggests that these proteins are not vital targets for these β-lactam antibiotics (11). In contrast, the cephalosporin-sensitive high-molecular-weight proteins, e.g. PBPs-1, -2, and -4 of *B. subtilis* and *Bacillus megaterium*, are all candidates for killing sites, since their antibiotic sensitivity profiles closely resemble those of the whole organism (12, 14, 32). These high-molecular-weight PBPs probably include multiple killing targets and, as with *E. coli*, the target can vary with the β-lactam antibiotic. Conclusions drawn from the antibiotic sensitivities of PBPs in the absence of supporting genetic evidence should be viewed with caution since, as already noted, several factors can complicate the effects of β-lactams in vivo.

The importance of the high-molecular-weight PBPs of *B. subtilis* was studied by stepwise selection of cloxacillin-resistant mutants (26). PBPs-1ab, -2b, -3, -4, and -5 were each unaltered in their sensitivity to cloxacillin in vitro in each of the stepwise isolates. By contrast, an increase in the cloxacillin resistance of PBP-2a at each step corresponded approximately to the increased cloxacillin resistance observed in vivo ($\sim$180-fold at the fifth step isolate). The mutant PBP-2a did not show an increased resistance to penicillin G, consistent with the unaltered sensitivity of the organism to this β-lactam. Thus, a PBP binding site can selectively mutate with respect to its affinity for particular β-lactams. Although PBP-2a is therefore a likely target for the lethal effects of cloxacillin, its relatively low sensitivity to cephalosporins suggests that another PBP, possibly PBP-2b, is the cephalosporin killing site. Further analysis indicated that the increased resistance of PBP-2a to cloxacillin in the fifth step isolate was also accompanied by: (a) loss of the penicillin-binding activity (or deletion) of PBP-1ab; (b) overproduction of a PBP-2a with an altered electrophoretic mobility; and (c) a small decrease in diameter ($\sim$15%) for the growing cells (26, 125). In addition, transformants obtained from the fifth step mutant and selected for cloxacillin resistance were isolated (at an unexpectedly low frequency), but exhibited all of these phenotypic alterations (C. E. Buchanan, unpublished observations). In these mutants, PBP-2a overproduction might compensate for the deletion of PBP-1ab. Since the ultimate diameter of the organism is probably determined by the length of septum synthesized, PBP-1ab might be involved in septation, as suggested by the reduced diameter of this PBP-1ab$^-$ mutant and by the small diameter of a detergent-resistant B. subtilis mutant (247) having a PBP-1ab of altered mobility (125).

B. megaterium cloxacillin-resistant isolates that filament and have a decreased growth rate have a significantly increased ratio of PBP-3 to PBP-1 (98). In this case, no change in the sensitivity of the PBPs of the mutant was detected. The organism's increased resistance to cloxacillin might result from overproduction of the more cloxacillin-resistant PBP-3, which would partially compensate for the cloxacillin-inhibited PBP-1. That B. megaterium PBP-1 (which is biochemically analogous to B. subtilis PBP-1; 32, 33, 124, 263) probably catalyzes a penicillin-sensitive transpeptidase reaction in vivo is supported by the observation that following penicillin G-inhibition of transpeptidation catalyzed by a membrane-wall fraction from B. megaterium, activity is restored with a half-life of 60 min at 23°C, the same rate at which B. megaterium PBP-1 releases its bound penicilloyl moiety (189).

A variety of additional studies have been carried out with a number of organisms to define the range of PBPs that are present and to use similar techniques to identify those PBPs that are essential for viability and are lethal targets. These include: (a) β-lactam binding studies to identify those PBPs that are saturated below the MIC (and thus are likely not essential for growth) as well as those whose saturation concentrations correspond to the

MIC. The selectivity of different β-lactam antibiotics for different PBPs has also been examined using similar techniques; (*b*) studies of clinical isolates that are resistant to β-lactam antibiotics to identify those PBPs whose decrease in affinity for β-lactams correlates with clinical resistance. In several cases, strains that have increased production of a particular PBP (possibly the basis of the resistance phenotype) have been identified, as have strains with PBP deletions, which evidences that the deleted PBPs cannot be essential for growth under laboratory culture conditions; (*c*) changes in PBP patterns under a variety of growth conditions, e.g. in exponential vs stationary phase cultures (28, 46) as well as in sporulating *B. subtilis* (208). The organisms used in these studies have included, in addition to *E. coli* (8, 9, 87, 111, 123, 129, 144, 167, 170, 171, 175, 177, 200, 213, 215, 219): *B. megaterium* (192), *Staphylococcus aureus* (23, 85, 88, 108, 110, 279), *Streptococcus* sp. (38, 69, 70, 85, 101, 183, 272–274, 284), *Neisseria gonorrhoeae* (5, 54, 55), *Proteus* sp. (176), *Pseudomonas aeruginosa* (190, 191), *Clostridium prefringens* (158, 275), *Klebsiella pneumoniae* (194), *Rhodopseudomonas sphaeroides* (204), *Chlamydia trachomatis* (6), *Caulobacter crescentus* (130), and *Streptomyces* sp. (61, 62, 173). Thus, for example, in *S. aureus* PBPs-1 and -4 are unessential as evidenced by the normal growth of mutants lacking them, PBP-2 probably plays a role in cell wall thickening, and PBP-3 probably catalyzes the major cell wall transpeptidation, although PBP-2 can compensate in cases of PBP-3 inhibition. PBPs-2 and/or -3 were found to be either changed in mobility, increased in relative amount, and/or have decreased affinity for methicillin (in some cases *> 1000-fold*) in methicillin-resistant *Staphylococci* (23, 108, 110, 279). Similarly, in *Streptococcus pneumonia*, PBPs-1a, -1b, and -2a, -2b are altered in parallel with the development of resistance (102, 274, 284).

Importance of D-*Alanine Carboxypeptidase (CPase)*

In certain cocci, such as *S. aureus*, where CPase is a minor PBP (PBP-4), more than 90% of the peptide side chains are involved in cross-links, with uncross-linked peptides present as pentapeptides. By contrast, only 20%–50% of the peptides are cross-linked in gram-positive and gram-negative bacilli, where CPase is often the major PBP, the uncross-linked chains having been converted to tetrapeptides (197). As the tetrapeptides produced by a CPase reaction cannot serve as carboxyl donors for transpeptidation, CPase might regulate the degree of peptidoglycan cross-linking (117). Several observations suggested, however, that CPase might be unnecessary for normal growth in several organisms. Thus, (*a*) an estimated 95% inhibition of *B. subtilis* CPase by penicillin had no apparent effect on cell growth or the extent of cell wall peptide cross-linking in vivo (11, 202); and (*b*) mutants deficient in the penicillin-binding and CPase activity of the three *E. coli* CPases, namely PBP-4 (*dac B* mutants), PBP-5 (*dac A* mutants), PBP-6 (*dac C* mutants), and a double mutant (*dac A dac B*) all appear to grow normally and to synthesize normally cross-linked peptidoglycan in vitro (22, 115, 146, 147, 165, 217). These results are not, however, conclusive, for several reasons: 1. One cannot be certain that CPase was inhibited in vivo to the extent estimated (11). 2. As low as 5% residual CPase activity might be sufficient to impart the normal degree of peptide cross-linking. High-molecular-weight PBPs, likely to function as major cell wall transpeptidases (e.g. *E. coli* PBP-1Bs; see above), are often present in amounts as small as 5% of CPase. In addition, CPase is likely to be distributed throughout the cell surface (100) with the large majority of its molecules functionally inactive, its activity possibly regulated by the availability of free pentapeptides found only at the growing point. 3. Measurements of the extent of cross-linking were rather insensitive in one of the studies ($\pm 7\%$; 202), so that small differences in the overall cross-linking, possibly reflecting important differences in particular cross-links, would not have been detected. 4. *E. coli* PBPs-4, -5, and -6 might be able to fill compensatory functions and thus even the 10% residual CPase activity of the *dac A dac B* double mutant (PBP-4$^-$ PBP-5$^-$) [i.e. that of PBP-6 (2)] might be sufficient for cell viability. The recent isolation of an *E. coli* PBP-6 deletion mutant (*dac C*; 217) should permit an assessment of the morphology and viability of a hypothetical triple mutant, PBP-4$^-$ PBP-5$^-$ PBP-6$^-$.

Recent studies indicate that CPases such as *E. coli* PBP-4 and *S. aureus* PBP-4 might function as secondary transpeptidases of peptidoglycan biosynthesis (41, 49). Whereas a primary transpeptidase effects the insolubilization of nascent peptidoglycan strands by covalently linking them to the peptidoglycan sacculus in an anchorage reaction (151), a secondary transpeptidase catalyzes the further cross-linking of the newly incorporated strands. Studies of the kinetics of cross-linking of newly synthesized *E. coli* cell wall indicate that a slow increase in cross-linking to the levels found in mature peptidoglycan occurs over several generations and is paralleled by a reduction in the number of free pentapeptides in the newly synthesized material (49). Two transpeptidases are apparently involved—a primary transpeptidase engaged in insertion of new subunits into existing wall and a secondary transpeptidase responsible for the subsequent increase in cross-linkage. The two enzymes appear distinct, as the secondary transpeptidase is selectively inhibited by several β-lactams and appears absent in the *E. coli* PBP-4$^-$ (*dac B*) mutant (49). This demonstration of a reduced *rate* of cross-linking constrasts with the absence of a significant effect on *net* cross-linkage in the *dac B* mutant (147). A similar situation occurs in *S. aureus* where mutational loss of PBP-4 (41) (a low-molecular-weight PBP, termed

CPase) or its inhibition by cefoxitin results in a marked reduction in cell wall cross-linking (from 67% oligomers to 26% oligomers in the PBP-4⁻ mutant; 278). The corresponding loss of in vitro transpeptidase activity further evidences that *S. aureus* PBP-4, which exhibits transpeptidase activity in vitro (131) and probably also in a membrane-wall fraction (152), functions as a transpeptidase in vivo. That the PBP-4⁻ strain grows normally (41) indicates that this PBP does not function as a primary transpeptidase, but rather catalyzes a secondary transpeptidation resulting in the high cross-linkage characteristic of this organism. Although mutational loss of either *E. coli* PBP-4 or *S. aureus* PBP-4 is not lethal under laboratory conditions (41, 115, 146), the conservation of these PBPs during evolution suggests the importance of this secondary transpeptidation for cell survival in nature.

In contrast to these CPases that catalyze secondary transpeptidations in vivo, CPases such as *E. coli* PBP-5 appear to function as D-alanine carboxypeptidases in vivo. Thus overproduction of *E. coli* PBP-5 results in a 40% decrease in the amount of free pentapeptide initially present in newly synthesized peptidoglycan (139) while mutational loss of PBP-5 results in a significant increase in the amount of free pentapeptide (48). In this latter case a slow reduction in free pentapeptide level occurs, presumably catalyzed by the less active PBP-6 (2). That *E. coli* PBP-5⁻ mutants exhibit an increased sensitivity to several β-lactams (165, 243) further supports the presence of structural alterations in their cell wall and, in addition, demonstrates that alteration in a "nonessential" PBP can alter an organism's sensitivity to β-lactams.

Studies of peptidoglycan biosynthesis in a thermosensitive division mutant of *E. coli* establish the importance of D-alanine carboxypeptidase activity in the cell cycle of *E. coli* (153, 154). Induction of filamentation either by adding PBP-3-specific β-lactams or by shifting the mutant culture to the nonpermissive temperature resulted in an inhibition of D-alanine carboxypeptidase activity as well as the hypercross-linking of the newly synthesized peptidoglycan. Upon shifting filamentous cultures to the permissive temperature, D-alanine carboxypeptidase activity increased relative to transpeptidase activity, the extent of cross-linking decreased and septum formation resumed. The importance of a proper balance between carboxypeptidase and transpeptidase activity for cell septation was confirmed by the demonstration that these two activities vary significantly during the cell cycle: maximal transpeptidase activity and cross-linkage occurred immediately after cell division and maximal carboxypeptidase activity occurred just before cell division (155), presumably reducing the pool of precursor pentapeptide donors available for continued cell elongation and/or increasing the pool of tetrapeptide (or tripeptide) acceptors that might be required for septum formation (see above).

A unique physiological role for a D-alanine carboxypeptidase activity has been described in *Gaffkya homari*, where a penicillin-*insensitive* transpeptidase incorporates linear, uncross-linked peptidoglycan strands into existing cell wall (107), a role analogous to that of the primary transpeptidase of *E. coli*. This transpeptidase is further distinguished by its high specificity for the tetrapeptide moiety of the nascent peptide unit, which acts as an amino acceptor in the cross-linking reaction. As this tetrapeptide is generated from a corresponding pentapeptide by the action of a penicillin-sensitive CPase, penicillin treatment *indirectly* inhibits the transpeptidase responsible for incorporating newly synthesized glycan strands into existing peptidoglycan. Addition of small amounts of UDP-*N*-acetylmuramyl tetrapeptide reverses the penicillin sensitivity of this incorporation by allowing for formation of nascent strands containing a sufficient number of tetrapeptide units, which are then recognized as substrates for the transpeptidase (107). The PBP corresponding to this D-alanine carboxypeptidase activity, the lethal target of β-lactams in this organism, has not been identified.

BIOCHEMICAL STUDIES OF PURIFIED PBPs

Purification of PBPs

A major advance in the purification of PBPs was made with the introduction of covalent penicillin affinity chromatography (13). As originally described, a penicillin affinity column is constructed by covalently immobilizing a penicillin (e.g. 6-aminopenicillanic acid) to a carboxyl-substituted Sepharose resin in a carbodiimide-catalyzed reaction. Detergent extracts of bacterial membranes are stirred with the penicillin-substituted Sepharose, noncovalently bound proteins removed by extensive washing, and specifically bound proteins (the PBPs) subsequently eluted by treatment with hydroxylamine at neutral pH, which cleaves the covalent bond between penicillin and the PBP. Although one can thus isolate significant amounts of PBPs from many bacterial species. e.g. ~ 20 mg *B. subtilis* PBPs per kg (wet weight) cells, mixtures of PBPs are usually obtained, reflecting the multiplicity of PBPs present in the membrane (13, 31–33, 37, 124, 131, 269).

Several modifications of this technique have been applied to effect purification of individual PBPs. These include: (*a*) variation of the β-lactam ligand to effect isolation of specific PBPs, e.g. the high-molecular-weight, cephalosporin-sensitive PBPs of *Bacillus* sp. on a 7-aminocephalosporanic acid-Sepharose column (124, 269) and *E. coli* PBP 3 on a cephalexin-Sepharose column (246); (*b*) prebinding of a particular β-lactam antibiotic to the membrane extract before incubation with the penicillin affinty column. Only the PBP(s) that are insensitive to the added β-lactam will be free to bind to the affinity column (13). Several low-molecular-weight PBPs, as well as the cefoxitin-insensitive PBP-2 of *E. coli* (43), have been purified by this method; (*c*) saturation of the membranes with penicillin G followed by removal of excess antibiotic and subsequent incubation of a detergent extract with the affinity column. Those PBP(s) that enzymatically release the bound penicillin G are free to bind to the affinity resin; PBP-4 of *S. aureus* (131) and PBP-1 from several species of *Bacillus* (33, 269) have thus been purified; (*d*) utilization of mutants with altered levels of one or more PBPs—deletion mutants have facilitated purification of *E. coli* PBPs-1A, -1Bs, and -3 (246), as have recombinant *E. coli* strains (133, 216, 240) that overproduce particular PBPs; (*e*) variation in affinity column spacer arm, sometimes resulting in an unanticipated selectivity; e.g. *E. coli* PBP-6 binds to the two carbon spacer resin 6-amino penicillanic acid-carboxymethyl Sepharose while PBP-5 does not bind efficiently until a long, hydrophilic spacer arm is employed (2); (*f*) differential extraction of a given PBP(s) from the bacterial membrane, differential hydroxylamine elution from the affinity column, or selective precipitation from a mixture of purified PBPs by dialysis to low ionic strength (2, 31, 113, 246, 269); (*g*) use of conventional biochemical techniques (e.g. ion exchange, hydroxylapatite chromatography, isoelectric focusing) to purify PBPs from a mixture obtained by β-lactam affinity chromatography (31, 199, 269; T. A. O'Brien et al, unpublished results).

Interactions with Cell Wall–Related Substrates

The catalytic activities of purified PBPs have been studied extensively in the hope of discerning their enzymatic functions in vivo. Most studies have utilized either: (*a*) low-molecular-weight PBPs (CPases) purified from membranes of several organisms including *Bacillus sp.* (PBP-5; 257, 280), *E. coli* (PBP-5 and PBP-6; 2, 245), *Salmonella typhimurium* (PBPs-4 and -5; 203), *Proteus mirabilis* (141, 199), *Streptococcus faecalis* (37), and *S. aureus* (PBP 4; 131); (*b*) exocellular (water-soluble) CPases, principally those of *Streptomyces* R61, *Actinomadura* R39, and *Streptomyces albus* G (reviewed by Ghuysen et al. in 90, 95–97); and (*c*) high-molecular-weight PBPs, either in membrane-bound form, in mixtures of PBPs, or as purified proteins.

CATALYTIC PROPERTIES OF CPASES   Low-molecular-weight PBPs, both the membrane-bound and exocellular CPases, catalyze nucleophilic attacks at the carbonyl carbon of the penultimate D-alanine residue of compounds analogous to the cell wall pentapeptide. The nucleophile may be water, yielding the corresponding tetrapeptide in a CPase reaction, or alternatively, an amino compound, R-NH$_2$, with formation of a transpeptidation product (Figure 1), with both activities inhibited by β-lactam antibiotics. Certain low-molecular-weight PBPs also catalyze a penicillin-sensitive endopeptidase reaction, with hydrolysis of the peptide dimers that would be formed by transpeptidation (95, 199, 203, 245). Thus in some bacterial systems, β-lactam treatment might effectively inhibit endopeptidase, a potential autolytic activity. Most CPases are specific for removal of the terminal D-alanine with no detectable cleavage of the pentultimate D-alanine residue (e.g. 245, 257, 280).

The interaction of CPase with acyl-D-alanyl-D-alanine substrates is best described by the three step mechanism shown in Equation 1:

$$E + S \underset{k_{-1}}{\overset{k_1}{\rightleftharpoons}} E \cdot S \overset{k_2}{\underset{\text{DAla}}{\longrightarrow}} E\text{-}P \underset{\substack{\\ RNH_2}}{\overset{H_2O}{<}} \begin{array}{l} E + P\text{—OH} \quad \text{(CPase)} \\ \\ \overset{k_3}{\longrightarrow} E + P\text{—NH—R (TPase)} \end{array} \qquad 1.$$

where enzyme E and substrate S first react to form the noncovalent complex E $\cdot$ S. In the second step, a covalent acyl-enzyme intermediate (see below) E–P forms, with release of the terminal D-alanine. Reaction of E–P with either H$_2$O or RNH$_2$ regenerates free enzyme with formation of either the CPase product (P–OH) or the transpeptidase product (P–NH–R) (also see Figure 1). In some cases binding of RNH$_2$ (amino acceptor) precedes binding of peptide substrate (carbonyl donor; 71, 72) suggesting their binding to distinguishable sites. Alternatively, with other PBPs, the incoming nucleophile and terminal D-alanine might share the same or overlapping sites, in which case one would not expect binding of the amino acceptor to precede release of the terminal D-alanine residue.

Substrates utilized in CPase and transpeptidase assays include the so-called natural substrates (e.g. UDP-N-acetylmuramyl pentapeptide), synthetic analogs [e.g. diacetyl-L-lysyl-D-analyl-D-alanine (183a)] or more complex substrates, such as the linear uncross-linked peptidoglycan polymer (N-acetylglucosamine-β-1,4-N-acetylmuramyl pentapeptide)$_n$ (267). CPase activity can be assayed by D-alanine release, which is quantitated following an electrophoretic or chromatographic separation of the reaction products. With the B. subtilis CPase, no differences in catalytic activity are seen when the linear uncross-linked peptidoglycan polymer is compared as a substrate to the disaccharide pentapeptide monomer derived from it, indicating a lack of specificity for the degree of subunit polymerization (267). The absence of a simple, rapid, and sensitive assay has made detailed kinetic and enzymological studies of purified CPases slow and difficult.

In assays for transpeptidase activity either simple amino acceptors, such as NH_2OH, glycine, or D-alanine, or more complex amino compounds can be used as cosubstrates with the (R)-D-Ala-D-Ala compound. With most CPases, the partitioning of the acyl enzyme intermediate between hydrolysis (CPase reaction) and transpeptidation is influenced by various factors such as pH, amino acceptor concentration, and polarity of the reaction mixture (72, 93). As it is not known which reaction conditions best approximate the situation in vivo, it is impossible to discern the physiologic activities of the low-molecular-weight PBPs (termed CPases in this review) from their observed activities in vitro. For example, purified S. aureus PBP-4 catalyzes CPase (and transpeptidase) activity in vitro even though the presence of pentapeptides and absence of tetrapeptides in the peptidoglycan of this organism gives evidence for the absence of a detectable functional D,D-alanyl carboxypeptidase activity in vivo. Changes in enzyme microenvironment (e.g. saturation with glycine) could provide a mechanism for suppression of this potential carboxypeptidase activity in vivo. CPases from S. typhimurium (PBPs-4 and -5; 203), Streptomyces R61 (77), and Actinomadura R39 (94) catalyze transpeptidase reactions using more complicated amino acceptors with formation of peptide dimers. This suggestion of specific acceptor sites is consistent with the observation that the specificity profiles of the R61 and R39 enzymes for tripeptide donor and for amino acceptor reflect the peptidoglycan structure found in the corresponding bacterial strains (95).

The substrate specificities of several exocellular CPases have been studied in detail using synthetic substrates such as diacetyl-L-Lys-D-Ala-D-Ala and a series of related peptide analogs (95) and can be summarized as follows: 1. These CPases exhibit a high specificity for the penultimate D-alanine residue and substitution by D-Glu, D-Leu, or L-Ala greatly reduces enzyme activity. 2. Specificity for the terminal D-alanine is less rigorous, with significant activity observed using tripeptides terminating in D-Lys, D-Leu, D-Glu, or glycine, but again no activity with L-Ala in this position. 3. Shortening the neutral side chain of the NH_2-terminal L-Lys causes a drastic decrease in activity. The absence of substitution of the L-Lys ε-amino group or substitution with charged groups (e.g. succinylation) may either increase or decrease substrate activity.

REACTIONS CATALYZED BY HIGH-MOLECULAR-WEIGHT PBPS   Attempts to demonstrate either CPase or transpeptidase activity catalyzed by high-molecular-weight PBPs from Bacillus sp. (31, 33, 124, 267), and S. aureus (131; T. A. O'Brien et al, unpublished results) generally have been unsuccessful. Substrates utilized have included a wide variety of synthetic CPase substrates (i.e. R-D-Ala-D-Ala peptides) as well as linear uncross-linked peptidoglycan isolated from penicillin-treated *B. subtilis* (267). The inability to demonstrate activity for the purified PBPs using cell wall–related substances [despite the demonstration of enzyme activity for crude membrane-wall systems using some of the same substrates (e.g. 152, 260)] might be explained in several ways: 1. The appropriate substrates have not been tested. Some of the PBPs might function in vivo to form specific structures which might be found, e.g. at cell poles or septa (64), in which case highly specialized substrates might be required for in vitro activity. 2. The lipid requirements of the PBPs for activity in vitro have not been met. 3. Purification has effected denaturation or, alternatively, removal of an effector or cofactor required for enzymatic activity. 4. Transpeptidation requires the coupled and concerted activity of several membrane enzymes such that in vitro activity cannot be demonstrated using the purified PBPs with synthetic substrates.

In contrast to the inactivity of high-molecular-weight PBPs purified from these gram-positive bacteria, purified *E. coli* PBPs-1A, -1Bs, and -3 have recently been shown to catalyze a (partially coupled) transglycosylase-transpeptidase reaction in vitro using the lipid intermediate, undecaprenol pyrophosphoryl disaccharide pentapeptide, as substrate (112–114, 149, 150, 159, 161, 237, 246, 254); i.e. these PBPs are bifunctional enzymes. The relatively low catalytic activities reported in these in vitro studies (0.7–10 mol product per mol PBP per min) could reflect, at least in part, the low penicillin-binding stoichiometries (0.2 mol per mol) of some of the PBP preparations utilized (161). Confidence that the observed transglycosylase activities are in fact due to the activities of the PBPs themselves, and not to contaminating proteins, was obtained by inhibition of enzymatic activity using PBP-selective β-lactams such as penicillin G (PBP-1A) and apalcillin (PBP-3) (149). Interestingly, purified *E. coli* PBP-1A catalyzed formation of hypercross-linked peptidoglycan (39% cross-linked, as compared to 25% cross-linking for total *E. coli* cell wall or for the peptidoglycan formed by PBP-1Bs in vitro), which contained a significant amount of dissacharide-peptide trimers (254). Such trimers are found in small amounts in *E. coli* cell walls in vivo (99) suggesting that PBP-1A could hypercross-link localized regions of the cell wall for specific, yet to be defined, structural purposes. Using similar assay methods, transglycosylase activities and low level transpeptidase activities have recently been observed with high-molecular-weight PBP mixtures from *Bacillus* sp. (G. Jackson, J. L. Strominger, submitted for publication). In addition, a peptidoglycan transglycosylase (devoid of transpeptidase activity, at least in vitro) purified from *B. megaterium* is apparently identical to PBP-4 of that organism (241). It should be noted that the assays presently performed with the lipid intermediate as substrate must be carried out either by incubation on filter paper (which might serve either as an initial transglycosylase acceptor or to provide a solid support for the reaction) or in reaction mixtures containing organic solvents (e.g. methanol or glycerol, which could also function as transglycosylation acceptors).

ACYL-ENZYME INTERMEDIATES   Acyl enzymes (Figure 1) have been proposed as catalytic intermediates in reactions catalyzed by penicillin-sensitive enzymes (248). Acylation of S. aureus CPase (PBP-4) by diacetyl-L-Lys-D-Ala-D-Ala is more rapid than deacylation ($k_2 > k_3$ in Equation 1) such that the acyl-enzyme E–P accumulates and can be trapped by rapid denaturation (131). Much smaller amounts of acyl-enzyme can be trapped with E. coli PBPs-5 and -6 using the same substrate, suggesting that the rate constants k_2 and k_3 are of the same order of magnitude. Upon treatment of these PBPs with the sulfhydryl inhibitor p-chloromercuribenzoate, deacylation (k_3) is slowed relative to acylation (k_2), favoring accumulation of the acyl-enzyme (42). Alternatively, with an E. coli PBP-5 mutant defective in deacylation (k_3 very small) (148), an acyl-enzyme readily accumulates (H. Amanuma and J. L. Strominger, unpublished observations). Although kinetic studies (166) also indicate acyl-enzymes as catalytic intermediates in reactions catalyzed by CPases from Bacillus sp., attempts to trap acyl-enzymes with these CPases or with the exocellular CPase of Streptomyces R61 using diacetyl-L-Lys-D-Ala-D-Ala have been unsuccessful, suggesting that $k_3 \gg k_2$ and that, consequently, the E–P intermediate exists at low levels in the steady state. However, when the depsipeptide diacetyl-L-Lys-D-Ala-D-lactate (the ester analog of diacetyl-L-Lys-D-Ala-D-Ala and a good substrate for several CPases; 188) is used, k_2 is markedly accelerated relative to k_3 such that acyl-enzymes accumulate and are readily trapped in high yields for several CPases (188, 268, 281, 283). Consistent with the observation that deacylation of the S. aureus CPase is already rate limiting with the amide substrate (131), no rate acceleration is observed with the depsipeptide (188). It is thus well established that penicillin-sensitive CPases form acyl-enzymes as catalytic intermediates.

Attempts to demonstrate acylation of high-molecular-weight PBPs or even to detect their recognition of simple peptide substrates by competition for penicillin binding have been unsuccessful (124, 269). However, a slow acylation of B. subtilis high-molecular-weight PBPs by [^{14}C]diacetyl-L-Lys-D-Ala-D-Lactate has been observed (269). Cephalothin pretreatment or heat or detergent denaturation blocked the acylation, which was also not observed when ovalbumin was used in place of PBPs, suggesting that the acylation reflects specific interactions between PBPs and substrate. Further study is required to clarify the relevance of this exceedingly slow acylation to the processing of physiologically relevant substrates in vivo.

Interactions with β-Lactam Antibiotics

PENICILLIN BINDING: CHEMICAL ASPECTS   The β-lactam-PBP linkage is covalent and sufficiently stable to permit its detection following SDS-gel electrophoresis. This covalent penicillin binding is dependent upon specific structural features of the antibiotic, including the free carboxyl, D-asymmetric center at C-3, and, as an absolute requirement, an intact β-lactam bond. Mild heat or ionic detergent treatment of the PBPs abolishes their binding activity, indicating that binding reflects specific interations between β-lactam and a native enzyme, rather than a facile chemical acylation of a reactive protein nucleophile. Stoichiometric penicillin-PBP complexes have been isolated by gel filtation after saturating CPases (90, 262, 268) or high-molecular-weight PBPs (33, 269) with [^{14}C]penicillin G. With the low-molecular-weight PBPs, loss of CPase activity is directly proportional to the extent of complex formation (15, 135). That a near-stoichiometric complex is recovered after denaturation by any one of several methods (264) indicates that a covalent complex most likely exists in the native enzyme. The β-lactam moiety is most probably bound as a penicilloyl moiety in this native complex as suggested by the PBP-catalyzed enzymatic release of bound antibiotic as a penicilloyl derivative (e.g. penicilloic acid or penicilloyl hydroxamate; see below). Release of penicilloic acid by mild base (281) and nuclear magnetic resonance data (45) evidence a penicilloyl moiety in the denatured complex. The corresponding acyl moiety is probably bound in the case of cephalosporins, but this has not been examined directly. Δ^3-Cephalosporins having a suitable leaving group at C-3 (e.g. an acetoxy side chain) might react to form an exocyclic methylene upon binding to PBPs, as occurs during their reaction with β-lactamases (172). Thus, although not yet rigorously proved, the available data are fully consistent with a covalent penicilloyl-enzyme as the inactive form of CPase.

Early studies suggested that penicillin might be bound as an ester because of its alkali lability (196). Subsequently, neutral hydroxylamine, hydrogen peroxide, or thiols were found to reverse penicillin binding to a crude membrane fraction from B. subtilis [principally PBP-5, and see below (134)]. Because serine ester, threonine ester, and lysine amide bonds would not be susceptible to cleavage under these conditions, these results suggested that penicillin was bound to a cysteine residue via a thioester linkage. The more recent demonstration that hydroxylaminolysis of the bound penicilloyl moiety does not occur after denaturation of the PBP-penicillin complex (15, 132) indicates that the reactivity of the penicilloyl-enzyme linkage only corresponds to that of a thioester—but only in the native complex. Thus denatured [^{14}C]penicilloyl *B. subtilis* PBP-5 is stable to NH_2OH and other nucleophiles at pH 9, and is sensitive to borohydride reduction (pH 9) and high pH (pH 12), consistent with an (oxygen) ester linkage (132). Similar results were obtained with mixtures of high-molecular-weight PBPs purified from *Bacillus* sp. (269). To approach the question of the penicilloyl-enzyme linkage more directly, penicilloyl peptides were isolated from proteolytic digests of [^{14}C]penicillin G-labeled *Streptomyces* R61 CPase and *B. subtilis* PBP-5 (78, 85a). In both cases the penicilloyl moiety was bound to an unidentified serine residue, confirming the originally predicted ester linkage (196). The β-lactam-modified serine has since been identified as residue 36 (numbered from the protein NH_2-terminus) by amino acid sequence analysis of penicilloyl peptides derived from CPases purified from two *Bacillus* sp. (Table 1; see below; 264, 281, 282).

PENICILLIN BINDING: KINETICS   The binding of [^{14}C]penicillin G by PBPs can be assayed: (*a*) by a direct filter binding assay, using purified PBPs (103, 262); (*b*) by densitometry after SDS-gel electrophoresis and fluorography, using purified PBPs or membranes containing a mixture of PBPs (12); (*c*) by measuring loss of CPase activity, in cases where loss of activity parallels the covalent binding of penicillin; or (*d*) by using fluorimetric methods (82, 163). If either of the first two methods is used, binding of nonradioactive β-lactams must be measured indirectly, i.e. by prebinding the unlabeled β-lactam antibiotic followed by rapid saturation of the residual binding sites with [^{14}C]penicillin G.

The kinetics of β-lactam binding and release by CPases suggest a three-step mechanism (11, 74), shown in Equation 2:

$$E + I \underset{k_{-1}}{\overset{k_1}{\rightleftharpoons}} E \cdot I \overset{k_2}{\rightarrow} E-I' \underset{slow}{\overset{k_3}{\rightarrow}} E + \text{degraded } \beta\text{-lactam.} \qquad 2.$$

where $E \cdot I$ is a reversibly formed, noncovalent complex, and $E-I'$ is the covalent penicilloyl-PBP. The overall scheme for processing β-lactams is analogous to the processing of R-D-Ala-D-Ala substrates by CPases (Equation 1, above) excepting that k_3 is large for R-D-Ala-D-Ala substrates, such that E–P corresponds to a reactive acyl-enzyme intermediate, while k_3 is very small for β-lactams such that a relatively stable penicilloyl-enzyme, $E-I'$ accumulates (Figure 1). This proposed kinetic scheme for penicillin binding is the same as for active site–directed enzyme inactivators such as the trypsin inactivator *N*-tosyl lysylchloromethyl ketone (TLCK) (259a). Consistent with this mechanism, β-lactam binding exhibits first-order kinetics (74). The kinetics of CPase inhibition are apparently competitive

Table 1 Active-site sequences of PBPs and β-lactamases

Enzyme	Microorganism	Active-site amino acid sequence[a]	Reference
CPase		31 42 *	
	B. subtilis	Arg-Leu-|Pro|-Ile-|Ala-Ser|-Met-Thr-|Lys|-Met-Met-|Thr|	264
	B. stearothermophilus	-Val-Leu-Gly-Ile-|Ala-Ser|-Met-Thr-|Lys|-Met-	282
	E. coli PBP-5	-Arg-Asp-Pro-|Ala-Ser|-Leu-Thr-|Lys|-Met-Met-|Thr|	J. K. Broome-Smith, B. G. Spratt, unpublished
Class A β-lactamase		34 45 *	
	S. aureus	Arg-Phe-Ala-Tyr-|Ala-Ser|-Thr-Ser-|Lys|-Ala-Ile-Asn-	4
	B. cereus	Arg-Phe-Ala-Phe-|Ala-Ser|-Thr-Tyr-|Lys|-Ala-Leu-Ala-	4
	B. licheniformis	Arg-Phe-Ala-Phe-|Ala-Ser|-Thr-Ile-|Lys|-Ala-Leu-|Thr|	4
	E. coli	Arg-Phe-|Pro|-Met-Met-|Ser|-Thr-Phe-|Lys|-Val-Leu-Leu-	4
Class C β-lactamase[b]		75 86 *	
	E. coli amp C	Leu-Phe-Glu-Leu-Gly-|Ser|-Val-Ser-|Lys|-Thr-Phe-|Thr|	118
	P. aeruginosa	Leu-Phe-Glu-Ile-|Gly|-Ser-Val-Ser-|Lys|	127
Exocellular CPase		*	
	Streptomyces R61	-Val-Gly-|Ser|	78
	Actinomadura R39	-Leu-Pro-Ala-|Ser|-Asn-Gly-Val-	57

[a] Active-site serines (*) are those covalently modified by β-lactams. Residues are numbered from the amino terminus, in cases where known. Boxed residues are homologous to those present in *both* the CPases and class A β-lactamases. Ser 36 has also been implicated as the site of acylation by diacetyl-L-Lys-D-Ala-D-Lactate for the two CPases from *Bacillus* sp.
[b] Possible localized sequence homology is suggested by the 30% identity (9/30 residues) of *amp C* β-lactamase (residues Gly79–Tyr108) with *B. subtilis* CPase and *E. coli* PBP-5 (CPase residues Ala35–Asp64). Seven of the nine identities occur at residues where the two CPases are also homologous.

(75, 258, 280) consistent with the substrate analog hypothesis (248), although at least some of the data are also consistent with noncompetitive inhibition (75). Although detailed kinetic analyses have not been carried out with the high-molecular-weight PBPs, it is likely that these proteins also process β-lactams via the three-step mechanism (Equation 2). The available data (e.g. for *B. subtilis* and *E. coli* PBPs; 12, 210) are, however, also consistent with formation of penicilloyl-PBPs via a simple bimolecular reaction. The biphasic kinetics of penicillin binding originally observed with *B. subtilis* PBP-2 (12) reflect the presence of two PBPs, one of which (PBP-2a) binds penicillin G at ten times the rate of the other (PBP-2b; 124). A similar situation probably accounts for the complex kinetics of binding observed with *E. coli* PBP-4 (210, 221).

A comparison of constants for binding affinity $[K_D = (k_{-1}/k_1)]$ and acylation (k_2) of various CPases by β-lactam antibiotics (268) indicates that in most cases, K_D values range from 0.1 mM to 10 mM (15, 95, 258), with K_D values for good inhibitors (high k_2/K_D) sometimes larger than those of poor inhibitors, e.g. penicillin G versus cephaloglycin with the *Streptomyces* R61 CPase (90, 95). Although recognition of β-lactams by CPase is neither highly efficient nor selective, neither is the analogous process for R-D-Ala-D-Ala substrates, as suggested by the millimolar K_m values generally obtained (e.g. 90, 188). Possibly the effective concentration of physiological substrates is effectively high in vivo as a consequence of the juxtapositioning of a membrane-anchored CPase and nascent cell wall, such that low K_m values (i.e. μM) for substrate, and by extension, low K_D values for β-lactams, did not evolve. Variations in the penicillin side chain can have a significant effect on the acylation rates (k_2) by β-lactams. Thus, ampicillin acylates the *Streptomyces* R61 CPase 200 times slower than does penicillin G, and 6-aminopenicillanic acid, which has no side chain, acylates 10^6-fold more slowly (96). Acylation by R-D-Ala-D-Ala substrates (as reflected by k_{cat} values) is also affected by changes in the R group, although to a lesser extent than with the β-lactams. These results suggest that once bound to the active site, a suitable β-lactam side chain or substrate R group might interact with a specific enzyme grouping so as to correctly position and/or catalytically activate the β-lactam or amide bond for subsequent enzyme acylation (96).

The exocellular CPase of *Streptomyces albus* G is distinguished from other CPases by its uncharacteristically low molecular weight (M_r $\sim$22,000) and by its extremely low sensitivity to inhibition by β-lactams (10^6-fold less reactive than the R61 enzyme) (56, 80). Although noncovalent penicilloyl-CPase complexes form readily, as indicated by K_D values comparable to those of other CPases, acylation by β-lactams is extremely slow, in fact, no more efficient than penicilloylation of bovine serum albumin (96). That this CPase is a zinc enzyme (51) suggests that it probably does not form acyl-enzyme intermediates with peptide substrates, by analogy to other zinc carboxypeptidases (109). Thus, the low levels of penicilloylation observed could be mechanistically irrelevant and might therefore explain the noncompetitive inhibition of CPase activity by penicillin observed with this enzyme (80). Although this CPase is a poor model for a penicillin-sensitive enzyme or PBP, it might be analogous to physiologically important transpeptidases or CPases having very low sensitivities to penicillins and therefore not identified as PBPs.

β-LACTAM RELEASE REACTIONS   Incubation of penicilloyl-PBPs in the absence of excess unbound penicillin can result in release of the bound penicilloyl moiety with concomitant reactivation of the PBP. Reversal of penicillin binding to *B. subtilis* membranes (principally *B. subtilis* CPase) by hydroxylamine yields penicilloyl hydroxamate concomitant with restoration of CPase activity (134, 135). This observation is the basis of the hydroxylamine elution of active CPase from β-lactam affinity columns (13). Denaturation of the penicilloyl-CPase complex prevents the release of penicilloyl-hydroxylamine, indicating that it is an enzyme-catalyzed reaction (15, 132). A native penicilloyl-PBP complex is also required for this reaction in the case of high-molecular-weight PBPs (e.g. 124, 263). In the absence of nucleophiles such as hydroxylamine, degradation of penicilloyl-PBP complexes can occur either by hydrolysis or by fragmentation pathways (Figure 2). With several CPases, including those of *E. coli* (PBPs-5 and -6; 2, 245), *P. mirabilis* (141, 198), and *S. aureus* (PBP-4; 131), the bound penicilloyl moiety is enzymatically hydrolyzed to penicilloic acid. This penicillin release is apparently essential for the continued synthesis of *P. mirabilis* peptidoglycan with near-normal cross-linking during growth in medium containing relatively high concentrations of penicillin G (142). The low turnover number of this penicillinase reaction (~ 0.1 min^{-1} to 10 min^{-1} at 37°C) contrasts with the rapid turnover effected by typical β-lactamases (on the order of 10^5 min^{-1} for good β-lactam substrates; 184). It is therefore essential to establish that the penicilloic acid formed is not due to contamination of the PBPs with traces of β-lactamase, e.g. by confirming that the rate of penicilloic acid formation effected by the isolated penicilloyl-PBP complex is the same as the rate of regain of CPase activity. Possible structural and mechanistic relationships between these PBPs and β-lactamases are discussed below.

Penicillin-inactivated CPases from several species, including *Bacillus*, *Streptomyces*, and *Proteus* sp. regain catalytic activity (15, 73) with fragmentation of the bound penicilloyl moiety to yield phenylacetylglycine (76, 105, 199) and dimethylthiazoline carboxylate or its hydrolysis product, *N*-formyl-D-penicillamine (1, 79, 106), these latter compounds derived from the antibiotic's thiazolidine ring (Figure 2). Fragmentation, also observed with high-molecular-weight PBPs (263), is significantly slower ($k_3 = 0.01$ to $< 0.002 \, min^{-1}$, i.e. $t_{1/2}$ of penicilloyl-PBP complex $= 1 \, hr$ to $> 6 \, hr$ at $37°C$) than release of penicilloic acid by the other CPases ($t_{1/2} \leqq 10 \, min$ at $37°C$) and involves an enzymatic scission of the C-5–C-6 bond of the β-lactam ring as the rate-limiting step. The high stability of the bound penicilloyl moiety to hydrolysis or fragmentation may reflect interactions between the enzyme and the antibiotic's thiazolidine (or dihydrothiazine) ring, as suggested by the rapid hydrolysis of the phenacetylglycyl-enzyme formed upon release of the thiazolidine moiety during the fragmentation reaction [(140); see however (89)]. Stabilization of the penicilloyl-PBP complex by the antibiotic's 6-acyl side chain is also suggested by the more rapid release

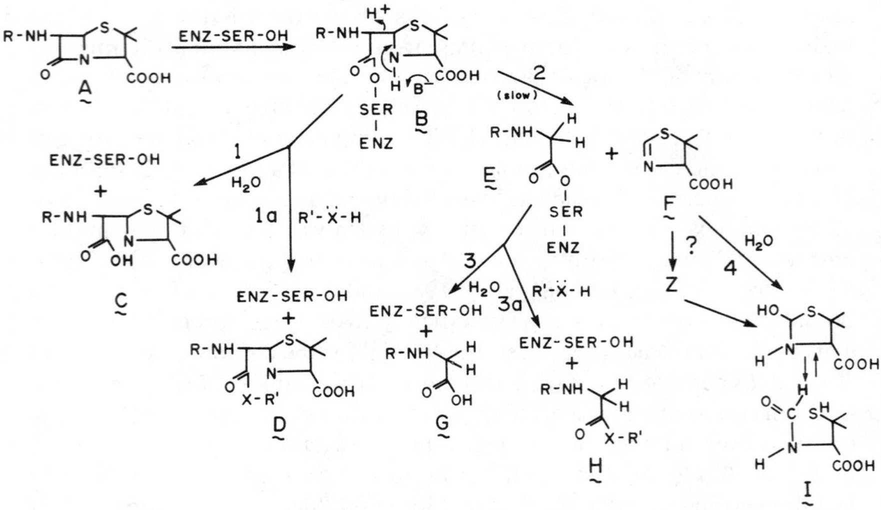

Figure 2 Pathways leading to reactivation of penicillin-inhibited PBPs (β-lactam release reactions). Penicillin (A) binds to and acylates the active-site serine hydroxyl group yielding a penicilloyl enzyme (B). PBPs can regain catalytic activity by direct transfer of the penicilloyl moiety to water (Pathway $1 \rightarrow$ penicilloic acid, C; i.e. β-lactamase activity) or to nucleophiles such as hydroxylamine (Pathway $1a \rightarrow$ penicilloyl hydroxamate, D). Fragmentation via proton abstraction at N-4 and cleavage of C5-C6 (Pathway 2) proceeds more slowly than direct transfer reactions (Pathway 1). The products of this pathway are N-acylglycyl derivatives (G, H) formed via a reactive N-acylglycyl-enzyme (E), in addition to Δ^2-thiazoline carboxylate (F), which is hydrolyzed (possibly enzymatically; possibly via intermediate Z) to yield N-formyl-D-penicillamine (I), shown in equilibrium with its hydroxythiazolide. Some PBPs could conceivably release the bound β-lactam moiety by a combination of fragmentation and hydrolysis pathways (e.g. *E. coli*; PBPs-5 and -6; H. Amanuma, unpublished results).

of 6-aminopenicillanic acid, as compared to penicillin G, by the *B. subtilis* CPase ($t_{1/2}$ of complex = 27 min and 200 min, respectively; 15).

Although the fragmentation pathway probably reflects intrinsic catalytic capabilities of the PBPs, in most cases it proceeds too slowly to be of physiological significance. The mechanism of this novel degradation (Figure 2; 81, 105, 140) may have features in common with release of the substrate-derived acyl group during a CPase or transpeptidase reaction. Partitioning of penicilloyl-PBPs between hydrolysis and fragmentation pathways can be influenced by the nature of the β-lactam (96), reaction conditions (60) and the presence of exogenous nucleophiles. Thus the "reversibility" or "irreversibility" of the inhibition of sensitive enzymes by β-lactams reflects the reactivity of the penicilloyl-PBP complex, as influenced by these factors, rather than fundamental differences in modes of inhibition by β-lactams. Different specificity profiles observed for the transfer of different acyl moieties (phenylacetylglycyl-, penicilloyl-, or substrate-derived R-D-alanyl-) from *Streptomyces* R61 exocellular CPase to a series of exogenous nucleophiles (140) suggest exclusion of certain nucleophiles, possibly by restrictive conformations at an acceptor site. Indeed, if this were not the case, one might expect reactivation of PBPs in vivo by transfer of the penicilloyl moiety to cell wall amino acceptors, and this does not occur to any significant extent (249). That the fragmentation pathway might require catalytic groups not necessary for penicilloyl release is suggested by the contrast between the facile hydroxylaminolysis of the bound penicilloyl moiety of *S. aureus* PBP-3 and its stability to fragmentation (263).

STRUCTURAL STUDIES OF PBPs

Active-Site Peptides and the Mechanism of Action of β-Lactam Antibiotics

It was proposed in the substrate analog hypothesis that penicillin binds to and inactivates peptidoglycan transpeptidase by virtue of its structural similarities to at least some of the possible conformations of the acyl-D-alanyl-D-alanine termini of nascent peptidoglycan strands (248). These structural similarities are most evident when the free carboxyl groups and the terminal asymmetric centers, necessary for activity of both penicillin and peptide substrates, are aligned, resulting in the similar positioning of the highly reactive CO–NH β-lactam bond and the peptide bond cleaved during a transpeptidase (or carboxypeptidase) reaction. The transpeptidase was postulated to react with its peptide substrate to form an acyl-enzyme intermediate, with the elimination of D-alanine (Figure 1). Subsequent reaction with the free amino group of a second cross-bridge would lead to formation of a cross-link and regeneration of the enzyme. Alternatively, hydrolysis of the acyl-enzyme intermediate would effect a D-alanine carboxypeptidase reaction. As a substrate analog, penicillin would bind to the enzyme (transpeptidase or carboxypeptidase) with its β-lactam bond positioned at the active site. A relatively facile acylation of the same active site nucleophile involved in acyl-enzyme formation would then occur with the opening of the β-lactam ring, forming an inactive penicilloyl-enzyme (Figure 1; 248).

This proposal for the mechanism of action of penicillin predicts: (a) the existence of a covalent penicilloyl-enzyme as the inactive form of peptidoglycan transpeptidase; as a corollary, penicillin-sensitive enzymes should be detectable as proteins that bind penicillin covalently; (b) occurrence of acyl-enzymes as catalytic intermediates in the uninhibited reaction; and (c) the substitution of both a penicilloyl moiety and an acyl-D-alanyl moiety derived from substrate on the same amino acid residue after reaction of enzyme with penicillin or with substrate. In addition, it was hypothesized that β-lactamases might have evolved from penicillin-sensitive enzymes by developing an efficient catalytic mechanism for hydrolysis of the penicilloyl enzyme (248; Figure 1). Many of the studies described in the previous sections support the first two predictions of this model. The ability to trap near-stoichiometric acyl-enzyme intermediate using the depsipeptide substrate diacetyl-L-Lys-D-Ala-D-lactate (188, 281) made it feasible to test directly the third prediction that a penicilloyl moiety and an acyl moiety both bind to the same active-site amino acid residue using CPases from *Bacillus* sp. as model penicillin-sensitive enzymes. Radiolabeled peptides were purified from chemical or enzymatic digests of CPase labeled at the antibiotic binding site with [^{14}C]penicillin G or, alternatively, labeled at the catalytic site by rapid denaturation of an acyl enzyme formed using the depsipeptide [^{14}C]-diacetyl-L-lys-D-ala-D-lactate and their amino acid sequences determined (Table 1; 264, 281, 282). Penicillin- and substrate-labeled peptides were shown to have identical primary structures, with acyl moieties derived from antibiotic and from substrate bound in ester linkage to the identical residue, serine 36, in both CPases. Further studies demonstrated that cefoxitin, a 7α-methoxycephalosporin, also binds to serine 36 of the *B. subtilis* CPase (264). These findings provide strong evidence that β-lactam antibiotics are active site–directed acylating agents, as predicted by the substrate analog hypothesis (268). This important conclusion was later extended to include *E. coli* PBP-6 and *Streptomyces* R61 exocellular CPase [for which it had been suggested that penicillin binds at an allosteric site (93, 163)] by comparison of the two sets (i.e. β-lactam- and substrate-labeled) of peptides after partial proteolysis using several cleavage methods (283). The unavailability of near-stoichiometric acyl-enzymes derived from any high-molecular-weight PBPs precludes comparison of substrate and antibiotic binding sites of these PBPs using these methods.

Homology to β-Lactamases

That β-lactamases might have evolved from penicillin-sensitive enzymes of cell wall biosynthesis (185, 248) is supported by the finding that several PBPs catalyze a weak β-lactamase activity (discussed above). Amino acid sequences obtained for the active-site peptides of CPases from *Bacillus* sp. suggested that such an evolutionary relationship might be reflected by sequence homology between the CPases and the class A (4) β-lactamases (264). A computer search for homologous proteins confirmed the existence of significant homology between the active-site region of these CPases and the NH_2-terminal portion of class A β-lactamases, with the amino acid sequences most highly conserved in the vicinity of the active-site residue of CPase, serine 36 (Table 1; 264). In the computer-generated alignments this active-site serine of CPase was aligned correctly with an analogous, catalytically active serine found in the class A β-lactamases (29, 30, 35, 43a, 67, 126, 205a). Insertion of the same gaps necessary for optimal alignment of the various β-lactamase sequences (4) markedly improved the sequence homology between CPase and the β-lactamases (264). This amino acid sequence homology, which is also evident at the level of secondary structure (156) as predicted by the method of Chou & Fasman (34), is unlikely to extend throughout the CPase polypeptide, since CPase ($M_r \simeq 50,000$) is significantly larger than the class A β-lactamases ($M_r \simeq 29,000$).

Sequence analysis of the amino terminal regions of *E. coli* PBPs-5 and -6 suggested homology between these two gram-negative CPases and also evidenced homology (particularly in the case of PBP-5) both to the CPases from *Bacillus* sp. and to the class A β-lactamases (270). The PBP-5 NH_2-terminal sequence has recently been confirmed and extended by DNA sequence analysis of the cloned *E. coli* PBP-5 gene (J. K. Broome-Smith, B. G. Spratt, unpublished observations, Table 1). The sequence determined is 39% identical to the NH_2-terminal 64 residues of *B. subtilis* PBP-5 (264) and $\sim 25\%$ identical to the corresponding region of the class A β-lactamases, with conservation of the active-site serine. Preliminary comparisons of the full-length PBP-5 sequence to the β-lactamase sequences suggests a significant decrease in homology in regions further from the active-site serine.

Although CPase contains an active-site serine and forms an acyl-enzyme, the absence of significant sequence homology to serine proteases (264) suggests that CPase is not a member of this well-studied class of peptidases. In contrast, the formation of acyl-enzymes, alignment of active-site serines, and significant sequence homology suggests that CPases and class A β-lactamases may utilize similar catalytic mechanisms. Relatedness between CPases and β-lactamases can also be seen with regard to their interactions with cephamycins (7α-methoxycephalosporins, e.g. cefoxitin, cefmetazole) and 6α-hydroxyethyl penems. Cefoxitin is a poor substrate for the *E. coli* RTEM β-lactamase and forms a relatively long-lived acyl-enzyme (66) by binding to the serine homologous to the active-site serine 36 of CPase (67), which is also acylated by cefoxitin (264). These 7α-methoxy cephalosporins also bind to several CPases at >100-fold lower concentrations than required for binding the corresponding 7α-H cephalosporins (40, 175). As with the β-lactamases, the 7α-methoxy group is essential for the high stability of the cephamycin-PBP complex, which contrasts to the relative lability of the corresponding penicilloyl-PBP complex. Similarly, β-lactamase-stable penems that have a 6α-hydroxyethyl group bind tightly to all three *E. coli* CPases (PBPs-4, -5, and -6), in contrast to the corresponding 6α-H-analogs (177). The relatively bulky 6α- and 7α-substituents might sterically hinder nucleophilic attack by a water molecule at the antibiotic's α-face in the case of both the CPases and the β-lactamases and thus stabilize the antibiotic-enzyme complexes. Alternatively, the α-substituent might reduce the activity of an enzyme base that facilitates hydrolysis or fragmentation of the bound antibiotic moiety. Abstraction of the C-6 hydrogen by an active-site base has been suggested as an important step in the mechanism of inactivation of β-lactamases by such compounds as β-bromopenicillanic acid, penicillanic acid sulfone, and olivanic acid (65). An active-site base is also indicated by the mechanism proposed for fragmentation of the bound penicilloyl moiety by CPases from *Bacillus* and *Streptomyces* sp. (Figure 2). A homologous active-site base might abstract the C-6 hydrogen leading to inactivation of β-lactamases by β-bromopenicillanic acid, or the N-4 hydrogen of a bound penicilloyl moiety, resulting in its fragmentation with formation of phenylacetylglycine in the case of CPases. Interestingly, mutation of aspartic acid 179 (a conserved residue in Class A β-lactamases) to asparagine in the *S. aureus* β-lactamase results in 99% loss of enzymatic activity (4). Studies of the processing of β-lactamase inactivators by CPases (120) should be useful in further probing the mechanistic homology of these two groups of enzymes.

The apparent homology between CPases and class A β-lactamases raises the question as to whether CPases and β-lactamases each recognize specific substrates of the other enzyme group. That CPases recognize and are acylated by β-lactamase substrates (i.e. β-lactam antibiotics) is not necessarily relevant since the structurally unrelated *Bacillus cereus* β-lactamase II of class B (see below) also processes β-lactams efficiently. Rather, a more meaningful test would involve non-β-lactam-containing compounds, such as phenylpropynal (195), which appear specific for class A β-lactamases. Both *B. subtilis* CPase and *E. coli* PBP-6 are inactivated by phenylpropynal in a pseudo-first-order process ($t_{1/2}$ = 18 min at pH 7, 25°C at 3 mM phenylpropynal); with *E. coli* PBP-5 inactivation is even more rapid ($t_{1/2}$ = 18 min at 0.5 mM phenylpropynal) (D. J. Waxman, H. Amanuma, unpublished observations). These rates are comparable to those seen with several class A β-lactamases (195) and further support the proposed mechanistic homologies. With regard to the hydrolysis of CPase substrates by β-lactamases, simple R-D-Ala-D-Ala-containing peptides are not hydrolyzed by β-lactamases (187). One would, however, anticipate a greater likelihood of hydrolysis of an activated acyclic CPase substrate such as the depsipeptide diacetyl-L-Lys-D-Ala-D-Lactate. Purified preparations of the RTEM β-lactamase from *E. coli* do hydrolyze this CPase substrate, albeit with a turnover number of only $\sim 10^{-3}$ min^{-1}, a rate sufficiently low to suggest that $\sim 0.5\%$ contamination of the β-lactamase preparation by *E. coli* CPases could account for the observed activity (D. J. Waxman, unpublished observations). That the relative order of β-lactamase reactivity (i.e. bicyclic β-lactam > monocyclic β-lactam > constrained peptide) follows the order of susceptibility to nucleophilic cleavage is consistent with the suggestion that β-lactamases lack an effective general acid catalyst found in the homologous CPases (195). One might further speculate that β-lactamase evolution was accompanied by the loss of D-alanyl-D-alanine carboxypeptidase activity to minimize interference by the unregulated periplasmic or exocellular β-lactamases in cell wall biosynthesis.

In contrast to the apparent divergent evolution of CPases and the class A β-lactamases, the *Bacillus cereus* β-lactamase II, a class B β-lactamase distinguished by its distinct primary structure (4) and by its requirement of zinc for catalysis (44), is likely to be related to the class A β-lactamases by convergent evolution. Sequence analysis indicates that this zinc β-lactamase is not homologous to the zinc CPase of low penicillin sensitivity excreted by *Streptomyces albus* G and, in addition, neither enzyme is homologous to the well-studied zinc enzyme carboxypeptidase A (119). Although computer analysis indicates that class C β-lactamases such as the *amp C* enzyme from *E. coli* are not homologous to the class A enzymes (118, 127), it is important to examine by direct comparison the possibility of sequence homology of the penicillin-sensitive CPases from *Bacillus* sp. and *E. coli* to the class C β-lactamases (e.g. see the localized homology suggested in Table 1, note *b*). Penicillin-binding peptides have also been isolated and sequenced for the exocellular CPases from *Streptomyces* R61 and *Actinomadura* R39 (57, 78; Table 1), but the available sequence data is insufficient to determine whether these exoenzymes are homologous either to the class A β-lactamases and CPases from *Bacillus* sp. and *E. coli*, or possibly to the class C β-lactamases. That the β-lactam binding site of the R61 CPase identified by X-ray diffraction analysis might not be in the amino terminal region [see Figure 1b of (121)] makes its homology to the class C β-lactamases a distinct possibility. Although the complete primary structure for *E. coli* PBP-3 has recently been deduced (Y. Hirota, unpublished observations), it is not known whether this high-molecular-weight PBP exhibits sequence homology to any of the above-mentioned enzymes.

Other Active-Site Residues

Nothing is known of the mechanism whereby serine 36 is rendered catalytically active. The demonstration of serine 36 as a catalytic intermediate for CPase does not preclude covalent intermediates involving other residues (e.g. histidine, cysteine). A possible role for cysteines in the catalytic mechanism is suggested by the following experiments. Sulfhydryl agents inhibit release of the bound penicilloyl moiety by *E. coli* PBPs-5 and -6 without significantly affecting their penicillin-binding activity (42, 245). Inhibition of penicillin release is paralleled by inhibition of deacylation of a substrate-derived acyl moiety, such that an acyl-enzyme intermediate accumulates and CPase activity is lost (42). A mutant of *E. coli* PBP-5 (*dac A*) has biochemical properties similar to the sulfhydryl-inhibited wild-type PBP (3, 148; H. Amanuma, J. L. Strominger, submitted for publication), suggesting that the mutation might involve an active-site cysteine. In the case of the CPases from *Bacillus* sp. and from *S. faecalis*, sulfhydryl reagents inhibit penicillin binding and CPase activity (38, 135, 257, 280). Enzymatic release of the bound penicilloyl moiety (either after fragmentation or in the presence of hydroxylamine) is, however, insensitive to sulfhydryl reagents with the CPases from *Bacillus* sp. (D. J. Waxman, unpublished observations). Thus sulfhydryl alkylation interferes with the binding and/or acylation by penicillin and substrate of the CPases from *Bacillus* sp., but not with deacylation either in the presence of hydroxylamine or following fragmentation. A possible role for an active-site arginine in stabilizing the anionic substrates of the R61 CPase has been proposed based on chemical modification studies using α,β-dicarbonyl compounds (86).

High Resolution Structural Analysis

X-ray structural analysis of crystalline PBPs complexed either with β-lactams or with suitable cell wall peptides should help further our understanding of the mechanism of penicillin action. Such studies could potentially provide information relating to: (*a*) the enzymatic mechanisms for processing R-D-Ala-D-Ala substrates; (*b*) the question of whether β-lactams are recognized as substrate analogs; (*c*) the structural relationships among PBPs and between CPases and β-lactamases; and (d) the rational design of new β-lactam antibiotics. As most PBPs are dissolved in detergent micelles and are often insoluble in aqueous media, significant difficulty would be encountered in obtaining these proteins in crystalline form. Studies designed to cleave proteolytically membrane-derived CPases to yield active, water-soluble fragments suitable for crystallization are described below. With the exocellular (water-soluble) CPases from *Streptomyces* sp., such complications are avoided, and crystalline complexes suitable for structural analysis have been obtained (50, 128). The three-dimensional structure of the *Streptomyces* R61 exocellular CPase has recently been determined to 2.8-Å resolution (121). Electron-density difference maps were obtained by Fourier difference synthesis by comparing crystals of the native enzyme with a crystalline cephalosporin C-, o-iodophenylpenicillin- and R-L-Lys-D-Glu-D-Ala-CPase complexes, and suggested the binding of all three ligands in the same region of the molecule (47, 121), consistent with the prediction of the substrate analog hypothesis (248). Studies on the primary structure of this CPase begun recently (58) should be helpful in further defining molecular structure at the active site.

Proton-induced X-ray emission studies of the *Streptomyces albus* G exocellular CPase (51) have led to the discovery of a previously undetected enzyme-bound zinc atom essential for catalysis: activity lost upon its removal by dialysis is restored with its stoichiometric readdition. The three-dimensional structure of this zinc CPase has recently been determined (52, 53) and, together with its primary structure (119), implicate three histidine residues located in a cleft in the carboxyl-terminal domain as ligands for the active-site zinc. Fourier synthesis at 2.8-Å resolution of a zinc CPase-β-iodopenicillanate complex suggested the possibility of acylation of yet another histidine (His 191) as the cause of inhibition by the β-lactam. In addition, roles for either Arg 137 in charge-pairing the substrate's carboxyl group and His 191 as a proton donor were proposed. As noted earlier, this penicillin-resistant zinc enzyme is most likely not related to penicillin-sensitive CPases detected as PBPs.

Relationships Among PBPs

The multiple membrane-bound PBPs of a given organism are independent proteins, not related by, for example, precursor-product relationships. Thus each of the high-molecular-weight PBPs of *E. coli* maps to a different position on the bacterial chromosome (236), and PBPs-5 and -6 have different peptide maps and different NH_2-terminal amino acid sequences (2, 270). Antibody raised to *B. subtilis* PBP-5 (CPase) does not cross-react with any of the high-molecular-weight PBPs of that organism (27), each of which has a distinct [^{14}C]penicilloyl peptide (124). These results do not, however, preclude the existence of important structural similarities or evolutionary relationships among the PBPs of an organism, similarities of which might be reflected by amino acid sequence homology. Proteins exhibiting poor immunological cross-reactivity and having apparently unrelated peptide maps could still have significant (50%) amino acid sequence homology.

Corresponding PBPs in related organisms exhibit structural homologies. Thus, the NH_2-terminal 40 amino acids of the *B. subtilis* and *B. stearothermophilus* CPases are more than 60% identical (264) and the NH_2-terminal 64 residues of *B. subtilis* CPase and *E. coli* PBP-5 39% identical (264; J. K. Broome-Smith, B. G. Spratt unpublished observations). Although structural data are still unavailable for most PBPs, similar biochemical properties of equivalent PBPs in related systems make structural homologies quite likely. For example the rapid release of the bound penicilloyl moiety (as penicilloic acid?) by PBPs-1 from *B. megaterium*, *Bacillus licheniformis* and *B. subtilis* ($t_{1/2} \leq 10$ min for all three enzymes; 33, 263) is suggestive of strong homology. Biochemical similarities between other PBPs in *Bacillus* sp. are also apparent (12, 32, 124) as they are among several gram-negative bacilli, each of which has PBPs of comparable molecular weight which have biochemical properties corresponding to: (*a*) the mecillinam-binding PBP-2 of *E. coli*; (*b*) the *E. coli* PBPs of cell elongation and cell septation; and (*c*) the *E. coli* CPases (PBPs-5 and -6) that catalyze a weak β-lactamase activity (39, 169, 176, 212). These corresponding PBPs are also likely to exhibit distinct structural homologies.

PBPs AS MEMBRANE-BOUND ENZYMES

The structure, biosynthesis, and interactions of PBPs with membrane lipids and with detergents have been studied with the following questions in mind:

1. *Enzymatic activity*—Does the membrane influence the catalytic properties of PBPs or does it simply provide a matrix for anchoring the PBPs in two dimensions? Can one activate detergent-solubilized high-molecular-weight PBPs by readdition of membrane lipids or by reconstitution into lipid vesicles?

2. *PBPs as membrane proteins*—PBPs (other than the exocellular CPases) are intrinsic membrane proteins requiring nonionic detergents or mild ionic detergents for solubilization in active form. How extensive is their membrane association? Is the polypeptide largely membrane embedded [e.g. as is bacteriorhodopsin (122, 259)], or are PBPs essentially water-soluble enzymes anchored to the membrane by a short, hydrophobic segment [e.g. as is microsomal cytochrome b_5 (179, 234)]? Are any PBPs transmembrane proteins? Are their activities modulated by interaction with intracellular or periplasmic effectors?

3. *Water-soluble PBPs*—Several intrinsic membrane proteins can be solubilized by limited proteolysis (e.g. 222, 234). Can PBPs be cleaved to generate active, water-soluble fragments possibly suitable for crystallization and X-ray structural analysis? Can one cleave transglycosylase domains from transpeptidase domains in the case of the bifunctional high-molecular-weight PBPs? How are periplasmic or exocellular CPases related to the corresponding membrane-bound enzymes?

Proteolytic cleavage of the CPases from *B. stearothermophilus* and *B. subtilis* has been used to identify the membrane-binding regions of these PBPs (262, 265). Water-soluble fragments of $M_r \simeq 45,000$ were formed by limited proteolytic cleavage of the *B. stearothermophilus* CPase (M_r 46,500) by loss of a COOH-terminal hydrophobic peptide. Purification of the water-soluble fragments was greatly facilitated by proteolysis of CPase while covalently immobilized on a penicillin affinity resin. The isolated fragments retained full enzymatic activity, became significantly more resistant to thermal inactivation, and no longer bound detergent micelles (262). Active, water-soluble fragments of M_r 47,000 and 35,000 were similarly obtained from the *B. subtilis* CPase (M_r 50,000) (265). In addition, smaller fragments (M_r 15,000–18,000) of both CPases (probably amino-terminal) retained penicillin-binding activity. Native *B. stearothermophilus* CPase, but not its water-soluble fragments could be reconstituted into lipid vesicles in an enzymatically active form with the CPase cleaved (slowly) from the vesicle surface to generate the water-soluble fragments (262). In addition, protease treatment of *B. subtilis* membrane vesicles released the same water-soluble fragments obtained by cleavage of that purified CPase. These studies demonstrate that CPase is composed of a hydrophilic, catalytic domain and a COOH-terminal hydrophobic region that mediates its anchorage to the bacterial membrane.

The generality of these findings is suggested by the following observations. Incubation of *B. megaterium* protoplasts with trypsin releases an active, water-soluble fragment of $M_r \simeq 25,000$ derived from the membrane-bound CPase (PBP-5), M_r 45,000 (P. E. Reynolds, unpublished observations). Similarly, trypsin treatment converts the *S. faecalis* CPase (PBP-6; M_r 43,000) to a 30,000-dalton fragment that retains penicillin binding and CPase activity (38). With regard to high-molecular-weight PBPs, incubation of *S. faecalis* membranes (in the absence of added protease) effects conversion of PBP-4 (M_r 80,000) to a water-soluble, penicillin-binding fragment ($M_r \simeq 73,000$) (38). Trypsin treatment of *S. aureus* PBP-1, -2, and -3ab mixtures ($M_r \simeq 90,000–70,000$) bound covalently to a cephalosporin affinity resin yields a water-soluble, penicillin-binding fragment approximately the same molecular weight as the *S. aureus* CPase (PBP-4), M_r 46,000 (T. A. O'Brien et al, unpublished results). It would be interesting to determine whether these high-molecular-weight PBPs are also membrane-bound via short COOH-terminal segments, i.e. if cleavage of only 1000–3000 daltons is sufficient to effect their conversion to water-soluble form. Exocellular CPases of *Streptomyces* R61 and K15 are likely related to membrane-bound PBPs of similar molecular weight found in the same organisms (137), with immunological cross-reactivity demonstrated for several of these proteins (162). Periplasmic *E. coli* CPase IC (245) is probably related to the membrane-bound PBP-4 (CPase IB) in a similar fashion, with the distinct possibility that these exocellular and periplasmic CPases were formed by cleavage of a COOH-terminal hydrophobic region from their corresponding membrane-bound forms. The exocellular β-lactamase of *B. licheniformis* is derived from a membrane-bound form by cleavage, in this case, of an NH_2-terminal membrane anchor (193, 206, 133a). Periplasmic or exocellular PBPs corresponding to high-molecular-weight PBPs have not been detected.

Amino acid sequence analysis of the COOH-terminal hydrophobic region of the two CPases from *Bacillus* sp. indicated 50% sequence identity in this region of the molecule (266). Identification of the hydrophobic peptides released in the conversion of CPase to its water-soluble fragments established that removal of as few as 24 amino acids is sufficient to cleave CPase from the bacterial surface. The absence of a large grouping of hydrophilic amino acids near the COOH-terminus suggests that, in contrast to several well-studied transmembrane proteins (e.g. glycophorin A), CPase might not have a significant cytoplasmic domain. Important remaining questions relating to the membrane anchoring of CPase include: (*a*) the disposition of particular amino acid residues within the hydrophobic core of the bilayer, e.g. as determined using photoactivatable hydrophobic labeling reagents (7, 10) and whether the enzyme is transmembrane; (*b*) the secondary structure of the membranous segment, proposed to be an α-helix having both a hydrophobic and a hydrophilic face (266); and (*c*) whether CPase interacts with other proteins, including other PBPs, on the membrane surface.

CPase is probably membrane bound with its catalytic domain exposed at the cell surface (or in the periplasm, in gram-negative bacteria). This would permit access to pentapeptide moieties of nascent peptidoglycan, enabling CPase to regulate the extent of cell wall cross-linking via the D-alanine release that continues after attachment of new glycan strands to preexisting peptidoglycan (49). Inhibition of CPase activity by exposure of *B. subtilis* protoplasts to Sepharose-bound penicillin (223) and the trypsin-cleavage of *B. megaterium* CPase from protoplasts cited above are both consistent with this extracytoplasmic orientation of CPase. Presumably, the catalytic sites of the high-molecular-weight PBPs are membrane bound with a similar orientation towards peptidoglycan. The biosynthesis of two *E. coli* CPases (PBPs-5 and -6) (186; B. G. Spratt, unpublished observations) as well as *E. coli* PBP-3 (Y. Hirota, unpublished observations) as precursor forms containing cleavable hydrophobic signal sequences further supports the idea that these membrane proteins require translocation of significant hydrophilic portions of their polypeptide chain to an extracytoplasmic site.

β-LACTAMS AS SUBSTRATE ANALOGS

We have recently attempted to evaluate the structural features that confer on β-lactam antibiotics their potent and specific inhibitory properties (268). It is apparent that β-lactams that are effective active-site inhibitors of cell wall biosynthesis require: (*a*) sufficient structural similarity to physiological substrates to permit recognition by essential cell-wall enzymes; (*b*) a highly reactive β-lactam bond to facilitate rapid acylation of the enzyme's active site; and (*c*) structural features that effectively minimize transfer of the covalently bound penicilloyl moiety to solvent or to cell wall amino acceptors; i.e. the inhibition must be slowly reversible or irreversible. Steric hindrance of incoming nucleophiles by the antibiotic's thiazolidine ring (still bound covalently via the C5–C6 bond after cleavage of the β-lactam) probably is an important factor contributing to stability of the penicilloyl enzyme.

Although important differences in bond angles and bond distances between effective β-lactam antibiotics and peptide substrates exist, some can be understood in terms of the proposed structural analogy between penicillin and a possible transition state structure formed during the enzyme-mediated cleavage of the D-Ala-D-Ala peptide bond (136, 248), a possibility supported by molecular orbital calculations (18, 20). Affinity of the enzyme for such a transition state would facilitate compression of the peptide dihedral angle resulting in the loss of double-bond character and a weakening of the peptide bond. The energy required for this distortion would reduce the net energy of binding substrate to enzyme. In contrast, the nonplanarity of the bicyclic ring system of penicillins (and cephalosporins) already imparts significant single-bond character to the corresponding β-lactam bond, thus favoring binding and acylation by β-lactams.

The question as to whether penicillin-sensitive enzymes bind β-lactams more tightly than normal substrates can be examined by comparing constants for binding ($K_D = k_{-1}/k_1$) and acylation (k_2) by various β-lactam antibiotics to the analogous K_M and k_{cat} values for CPase substrates (268). Such comparisons suggest that CPases do *not* bind β-lactams with significantly higher affinity (lower K_D) than peptide substrates (37, 95, 256, 257). Thus, if certain structural features of β-lactams favor binding by mimicking the tetrahedral transition state of a dipeptide undergoing cleavage, other features, including portions of the thiazolidine (dihydro-thiazine) ring with no analogy in the dipeptide, might decrease enzyme recognition such that, overall, substrate and antibiotic bind with similar affinity. That the side chains of effective β-lactams and active peptide substrates are structurally unrelated points to the complexity of their interactions with penicillin-sensitive enzymes (95). High-resolution X-ray diffraction studies in progress (see above) should help resolve these and related questions.

The highly reactive β-lactam bond essential for the rapid acylation of target PBPs by many penicillins and cephalosporins largely reflects the β-lactam's strained four-membered ring resulting, in part, from the loss of normal amide resonance due to steric constraints that prevent coplanarity of the three substituents of the β-lactam nitrogen. This is seen as a greater $C=O$ stretching frequency, shorter $C=O$ bond length and longer CO–N bond length in active penicillins and cephalosporins when compared to unstrained β-lactams (238, 277). In Δ^3-cephalosporins amide resonance is further decreased by the enamine contribution (157). Substitution of a good leaving group at the C-3 methylene of cephalosporins permits expulsion of the methylene substituent with formation of an exocyclic double bond upon cleavage of the β-lactam thereby increasing both the β-lactam's reactivity and biological activity significantly (19, 157). The low chemical reactivity and reduced biological activity of Δ^2-cephalosporin derivatives [which appear more closely isosteric to penicillin than the active Δ^3-cephalo-sporins (238)] further support the importance of a reactive β-lactam bond for the efficient acylation of essential PBPs. Thus β-lactams such as penicillin G are good inhibitors of CPase as a consequence of their rapid acylation of the enzyme [$k_2 = 20$–6000 times greater than k_2 for the usual CPase substrates, in the case of Streptomyces R61 CPase (95)] and not because of a particularly high affinity for the enzyme (as reflected by a K_D value comparable to other CPase substrates). Thus, the high potency and specific inhibitory properties of β-lactam antibiotics reflect a relatively favorable recognition by the target PBPs coupled with their strong acylating capability. These properties impart a high k_2/K_D ratio which, when coupled with the stability of the covalent β-lactam-enzyme complex (small k_3), enables these antibiotics to inactivate sensitive enzymes of cell wall biosynthesis with high efficiency.

CONCLUSION

A decade has passed since the discovery of multiple penicillin-binding proteins in 1972. During this period much progress has been made, and we now have considerable information about the physiological importance of these PBPs, the nature of their interactions with cell wall substrates and β-lactam antibiotics, and their chemical structures. For the PBPs that have been studied in detail, it has been demonstrated unequivocally that β-lactam antibiotics bind and acylate the catalytic site. The relationship of the PBPs to each other and to another group of proteins that utilize β-lactams as substrates (the β-lactamases) is beginning to emerge. In the next decade rapid development in two important areas can be anticipated. First, the precise enzymatic reactions catalyzed by these PBPs and the relationship of their inhibition to the lethality of β-lactam antibiotics now seem amenable to investigation. Second, protein structural studies, in particular of the crystal structures of the high-molecular-weight PBPs, may lead to an understanding of the detailed molecular interactions between penicillin-sensitive enzymes, cell wall substrates and β-lactam antibiotics. In addition, such studies should help elucidate the basis of the selectivity of different β-lactam antibiotics with different ring structures and/or different side chains for different PBPs. Recent advances in this area made with the low-molecular-weight PBPs are encouraging and suggest that the high-molecular-weight PBPs will be amenable to future study. Genetic approaches that allow for production and manipulation of PBPs to be used in enzymatic, structural, and crystallographic studies will certainly continue to play an extremely important role in the future studies of the mechanisms of interactions of PBPs with β-lactam antibiotics.

ACKNOWLEDGEMENTS

The authors wish to thank their many colleagues who provided unpublished data or manuscripts before publication. Special thanks to Cindy Lehn and Alice Furumoto-Dawson for expert assistance in preparation of this manuscript. Supported in part by research grants from the NIH (AI-09152) and NSF (PCM 78-24129).

Literature Cited

1. Adriaens, P., Meesschaert, B., Frere, J.-M., Vanderhaeghe, H. Degelaen, J., et al. 1978. *J. Biol. Chem.* 253:3660–65
2. Amanuma, H., Strominger, J. L. 1980. *J. Biol. Chem.* 255:1173–80
3. Amanuma, H., Strominger, J. L. 1981. In *Beta-Lactam Antibiotics*, ed. S. Mitsuhashi, pp. 179–84. New York: Springer and Tokyo: Japan Scientific Societies
4. Ambler, R. P. 1980. *Philos. Trans. R. Soc. London Ser. B* 289:321–31
5. Barbour, A. G. 1981. *Antimicrob. Agents Chemother.* 19:316–22
6. Barbour, A. G., Amano, K.-I.,

Hackstadt, T., Perry, L., Harlan, D. C. 1982. *J. Bacteriol.* 151:420–28
7. Bayley, H., Knowles, J. R. 1978. *Biochemistry* 17:2420–23
7a. Beck, B. D., Park, J. T. 1976. *J. Bacteriol.* 126:1250–60
8. Berenguer, J., de Pedro, M. A., Vazquez, D. V. 1982. *Eur. J. Biochem.* 126:155–59
9. Berenguer, J., de Pedro, M. A., Vazquez, D. V. 1982. *Antimicrob. Agents Chemother.* 21:195–200
10. Bercovici, T., Gitler, C. 1978. *Biochem.* 17:1484–89
11. Blumberg, P. M., Strominger, J. L. 1971. *Proc. Natl. Acad. Sci. USA* 68:2814–17

12. Blumberg, P. M., Strominger, J. L. 1972. *J. Biol. Chem.* 247:8107–13
13. Blumberg, P. M., Strominger, J. L. 1972. *Proc. Natl. Acad. Sci. USA* 69:3751–55
14. Blumberg, P. M., Strominger, J. L. 1974. *Bacteriol. Rev.* 38:291–335
15. Blumberg, P. M., Yocum, R. R., Willoughby, E., Strominger, J. L. 1974. *J. Biol. Chem.* 249:6828–35
16. Deleted in proof
17. Botta, G. A., Park, J. T. 1981. *J. Bacteriol.* 145:333–40
18. Boyd, D. B. 1979. *J. Med. Chem.* 22:533–37
19. Boyd, D. B., Lunn, W. H. W. 1979. *J. Med. Chem.* 22:778–84
20. Boyd, D. B. 1982. In *The Chemistry and Biology of β-Lactam Antibiotics*, ed. R. B. Morin, M. Gorman, Vol. 1, Chap. 5. New York: Academic
21. Braun, V., Wolff, H. 1975. *J. Bacteriol.* 123:888–97
22. Broome-Smith, J. K., Spratt, B. G. 1982. *J. Bacteriol.* 152:904–6
23. Brown, D. F. J., Reynolds, P. E. 1980. *FEBS Lett.* 122:275–78
24. Brown, M. R. W. 1975. In *Resistance of P. aeruginosa*, ed. M. R. W. Brown, pp. 71–107. New York: Wiley
25. Buchanan, C. E. 1981. *J. Bacteriol.* 145:1293–98
26. Buchanan, C. E., Strominger, J. L. 1976. *Proc. Natl. Acad. Sci. USA* 73:1816–20
27. Buchanan, C. E., Hsia, J., Strominger, J. L. 1977. *J. Bacteriol.* 131:1008–10
28. Buchanan, C. E., Sowell, M. O. 1977. *J. Bacteriol.* 151:491–94
29. Cartwright, S. J., Coulson, A. F. W. 1980. *Philos. Trans. R. Soc. London. Ser. B* 289:370–72
30. Cartwright, S. J., Fink, A. L. 1982. *FEBS Lett.* 137:186–88
31. Chase, H. A. 1980. *J. Gen. Microbiol.* 117:211–24
32. Chase, H. A., Shepherd, S. T., Reynolds, P. E. 1977. *FEBS Lett.* 76:199–203
33. Chase, H. A., Reynolds, P. E., Ward, J. B. 1978. *Eur. J. Biochem.* 88:275–85
34. Chou, P. Y., Fasman, G. D. 1977. *J. Mol. Biol.* 155:135–75
35. Cohen, S. A., Pratt, R. F. 1980. *Biochemistry* 19:3996–4003
36. Cooper, P. D. 1956. *Bacteriol. Rev.* 20:28–48
37. Coyette, J., Ghuysen, J.-M., Fontana, R. 1978. *Eur. J. Biochem.* 88:297–305
38. Coyette, J., Ghuysen, J.-M., Fontana, R. 1980. *Eur. J. Biochem.* 110:445–56
39. Curtis, N. A. C., Orr, D., Ross, G. W., Boulton, M. G. 1979. *Antimicrob. Agents Chemother.* 16:325–28
40. Curtis, N. A. C., Ross, G. W. 1980. In *Recent Advances in the Chemistry of β-Lactam Antibiotics*, ed. G. I. Gregory. London: Burlington House
41. Curtis, N. A. C., Hayes, M. V., Wyke, A. W., Ward, J. B. 1980. *FEMS Microbiol. Lett.* 9:263–66
42. Curtis, S. J., Strominger, J. L. 1978. *J. Biol. Chem.* 253:2584–88
43. Curtis, S. J., Strominger, J. L. 1981. *J. Bacteriol.* 145:398–403
43a. Dalbadie-McFarland, G., Cohen, L. W., Riggs, A. D., Morin, C., Itakura, K., Richards, J. H. 1982. *Proc. Natl. Acad. Sci. USA* 79:6409–13
44. Davies, R. B., Abraham, E. P. 1974. *Biochem. J.* 143:129–35
45. Degelaen, J., Feeney, J., Roberts, G. C. K., Burgen, A. S. V., Frere, J.-M., Ghuysen, J.-M. 1979. *FEBS Lett.* 98:53–55
46. delaRosa, E. J., dePedro, M. A., Vazquez, D. 1982. *FEMS Microb. Lett.* 14:91–94
47. DeLucia, M. L., Kelly, J. A., Mangion, M. M., Moews, P. C., Knox, J. R. 1980. *Philos. Trans. R. Soc. London Ser. B* 289:374–76
48. dePedro, M. A., Schwartz, U., Nishimura, Y., Hirota, Y. 1980. *FEMS Microbiol. Lett.* 9:219–21
49. dePedro, M. A., Schwarz, U. 1981. *Proc. Natl. Acad. Sci. USA* 78;5856–60
50. Dideberg, O., Frere, J.-M., Ghuysen, J.-M. 1979. *J. Mol. Biol.* 129:677–79
51. Dideberg, O., Joris, B., Frere, J.-M., Ghuysen, J.-M., Weber, G., et al. 1980. *FEBS Lett.* 117:215–18
52. Dideberg, O., Charlier, P., Dupont, L., Vermeirem, M., Frere, J.-M., Ghuysen, J.-M. 1980. *FEBS Lett.* 117:212–14
53. Dideberg, O., Charlier, P., Dive, G., Joris, B., Frere, J.-M., Ghuysen, J.-M. 1982. *Nature* 299:469–70
53a. DiRienzo, J. M., Nakamura, K., Inouye, M. 1978. *Ann. Rev. Biochem.* 47:481–532
54. Dougherty, T. J., Koller, A. E., Tomasz, A. 1980. *Antimicrob. Agents Chemother.* 18:730–37
55. Dougherty, T. J., Koller, A. E., Tomasz, A. 1981. *Antimicrob. Agents Chemother.* 20:109–14
56. Duez, C., Frere, J.-M., Geurts, F., Ghuysen, J.-M., Dierickx, L., Delcambe, L. 1978. *Biochem. J.* 175:793–800
57. Duez, C., Joris, B., Frere, J.-M., Ghuysen, J.-M. 1981. *Biochem. J.* 193:83–86
58. Duez, C., Frere, J.-M., Ghuysen, J.-M., Van Beeumen, J., Vandekerckhove, J. 1981. *Biochim. Biophys. Acta* 671:109–16

59. Duguid, J. P. 1946. *Edinburgh Med. J.* 53:401–12
60. Dusart, J., Leyh-Bouille, M., Ghuysen, J.-M. 1977. *Eur. J. Biochem.* 81:33–44
61. Dusart, J., Leyh-Bouille, M., Nguyen-Disteche, M., Ghuysen, J.-M., Reynolds, P. E. 1980. *Arch. Int. Physiol. Biochim.* 88:327–28
62. Dusart, J., Reynolds, P. E., Ghuysen, J.-M. 1981. *FEMS Microbiol. Lett.* 12:299–303
63. Essig, P., Martin, H. H., Gmeiner, J. 1982. *Arch. Microbiol.* 132:245–50
64. Fan, D. P., Beckman, B. E. 1973. *J. Bacteriol.* 114:790–97
65. Fisher, J., Belasco, J. G., Charnas, R. L., Khosla, S., Knowles, J. R. 1980. *Philos. Trans. R. Soc. London Ser. B* 289:303–19
66. Fisher, J., Belasco, J. G., Khosla, S., Knowles, J. R. 1980. *Biochemistry* 19:2895–901
67. Fisher, J., Charnas, R. L., Bradley, S. M., Knowles, J. R. 1981. *Biochemistry* 20:2726–31
68. Fleming, A. 1929. *Br. J. Exp. Pathol.* 10:226–36
69. Fontana, R., Cerini, R. 1981. In *Current Chemotherapy and Immunology*, ed. P. Penti, G. G. Grass. Florence, Italy: Am. Soc. Microbiol.
70. Fontana, R., Canepari, P., Satta, G., Coyette, J. 1980. *Nature* 287:70–72
71. Frere, J.-M. 1973. *Biochem. J.* 135:469–81
72. Frere, J.-M., Ghuysen, J.-M., Perkins, H. R., Nieto, M. 1973. *Biochem. J.* 135:483–92
73. Frere, J.-M., Leyh-Bouille, M., Ghuysen, J.-M., Perkins, H. R. 1974. *Eur. J. Biochem.* 50:203–14
74. Frere, J.-M., Ghuysen, J.-M., Iwatsubo, M. 1975. *Eur. J. Biochem.* 57:343–51
75. Frere, J.-M., Ghuysen, J.-M., Perkins, H. R. 1975. *Eur. J. Biochem.* 57:353–59
76. Frere, J.-M., Ghuysen, J.-M., Degelaen, J., Loffet, A., Perkins, H. R. 1975. *Nature* 258:168–70
77. Frere, J.-M., Ghuysen, J.-M., Zeiger, A. R., Perkins, H. R. 1976. *FEBS Lett.* 63:112–16
78. Frere, J.-M., Duez, C., Ghuysen, J.-M. 1976. *FEBS Lett.* 70:257–60
79. Frere, J.-M., Ghuysen, J.-M., Vanderhaeghe, H., Adriaens, P., Degelaen, J., DeGraeve, J. 1976. *Nature* 260:451–54
80. Frere, J.-M., Geurts, F., Ghuysen, J.-M. 1978. *Biochem. J.* 175:801–5
81. Frere, J.-M., Ghuysen, J.-M., De-Graeve, J. 1978. *FEBS Lett.* 88:147–50
82. Fuad, N., Frere, J.-M., Ghuysen, J.-M., Duez, C., Iwatsubo, M. 1976. *Biochem. J.* 155:623–29
83. Gardner, A. D. 1940. *Nature* 146:837–38
84. Georgopapadakou, N. H., Liu, F. Y. 1980. *Antimicrob. Agents Chemother.* 18:148–57
85. Georgopapadakou, N. H., Liu, F. Y. 1980. *Antimicrob. Agents Chemother.* 18:834–36
85a. Georgopapadakou, N. H., Hammarström, S., Strominger, J. L. 1977. *Proc. Natl. Acad. Sci. USA* 74:1009–12
86. Georgopapadakou, N. H., Liu, F. Y., Ryono, D. E., Neubeck, R., Ondetti, M. A. 1980. *Eur. J. Biochem.* 115:53–57
87. Georgopapadakou, N. H., Smith, S. A., Sykes, R. B. 1982. *Antimicrob. Agents Chemother.* 21:950–56
88. Georgopapadakou, N. H., Smith, S. A., Bonner, D. P. 1982. *Antimicrob. Agents Chemother.* 22:172–75
89. Georgopapadakou, N. H., Smith, S. A., Cimarusti, C. M. 1982. *Eur. J. Biochem.* 124:507–12
90. Ghuysen, J.-M. 1976. In *The Bacterial DD-Carboxypeptidase-Transpeptidase Enzyme System*, pp. 1–162. Tokyo: Univ. Tokyo Press
91. Ghuysen, J.-M. 1977. *Cell Surf. Rev.* 4:463–596
92. Ghuysen, J.-M., Schockman, G. D. 1973. *Microbiol. Ser.* 1:37–130
93. Ghuysen, J.-M., Leyh-Bouille, M., Frere, J.-M., Dusart, J., Marquet, A., et al. 1974. *Ann. N. Y. Acad. Sci.* 235:236–66
94. Ghuysen, J.-M., Reynolds, P. E., Perkins, H. R., Frere, J.-M., Moreno, R. 1974. *Biochemistry* 13:2539–47
95. Ghuysen, J.-M., Frere, J.-M., Leyh-Bouille, M., Coyette, J., Dusart, J., Nguyen-Disteche, M. 1979. *Ann. Rev. Biochem.* 48:73–101
96. Ghuysen, J.-M., Frere, J.-M., Leyh-Bouille, M., Perkins, H. R., Nieto, M. 1980. *Philos. Trans. R. Soc. London Ser. B* 289:285–301
97. Ghuysen, J.-M., Frere, J.-M., Leyh-Bouille, M., Coyette, J., Duez, C., et al. 1981. See Ref. 3, pp. 185–202
98. Giles, A. F., Reynolds, P. E. 1979. *Nature* 230:167–68
99. Gmeiner, J. 1980. *J. Bacteriol.* 143:510–12
100. Goddell, E. W., Schwarz, U. 1977. *Eur. J. Biochem.* 81:205–10
101. Gutmann, L., Williamson, R., Tomasz, A. 1981. *Antimicrob. Agents Chemother.* 19:872–80
102. Hakenbeck, R., Tarpay, M., Tomasz, A. 1980. *Antimicrob. Agents Chemother.* 17:364–71
103. Hakenbeck, R., Kohiyama, M. 1982. *FEMS Microbiol. Lett.* 14:241–45
104. Hamilton, T. E., Lawrence, P. J. 1975. *J. Biol. Chem.* 250:6578–85

105. Hammarström, S., Strominger, J. L. 1975. *Proc. Natl. Acad. Sci. USA* 72:3463–67
106. Hammarström, S., Strominger, J. L. 1976. *J. Biol. Chem.* 251:7947–49
107. Hammes, W. P. 1976. *Eur. J. Biochem.* 70:107–13
108. Hartman, B., Tomasz, A. 1981. *Antimicrob. Agents Chemother.* 19:726–35
109. Hartsuck, J. A., Lipscomb, W. N. 1971. *Enzymes* 3:1–57
110. Hayes, M. V., Curtis, N. A. C., Wyke, A. W., Ward, J. B. 1981. *FEMS Microbiol. Lett.* 10:119–22
111. Imada, A., Kitano, K., Kintaka, K., Muroi, M., Asai, M. 1981. *Nature* 289:590–91
112. Ishino, F., Matsuhashi, M. 1979. *Agric. Biol. Chem.* 343:2641–42
113. Ishino, F., Matsuhashi, M. 1981. *Biochem. Biophys. Res. Commun.* 101:905–11
113a. Ishino, F., Mitsui, K., Tamaki, S., Matsuhashi, M. 1980. *Biochem. Biophys. Res. Commun.* 97:287–93
114. Ishino, F., Tamaki, S., Spratt, B. G., Matsuhashi, M. 1982. *Biochem. Biophys. Res. Commun.* 109:689–96
115. Iwaya, M., Strominger, J. L. 1977. *Proc. Natl. Acad. Sci. USA* 74:2980–84
116. Iwaya, M., Goldman, R., Tipper, D. J., Feingold, B., Strominger, J. L. 1978. *J. Bacteriol.* 136:1143–58
117. Izaki, K., Matsuhashi, M., Strominger, J. L. 1966. *J. Biol. Chem.* 243:3180–92
118. Jaurin, B., Grundstrom, T. 1981. *Proc. Natl. Acad. Sci. USA* 78:4897–901
119. Joris, B., Casagrande, F., Van Beeumen, J., Gerday, C., Frere, J.-M., Ghuysen, J.-M. 1983. *Eur. J. Biochem.* 130:53–69
120. Kelly, J. A., Frere, J.-M., Klein, D., Ghuysen, J.-M. 1981. *Biochem. J.* 199:129–36
121. Kelly, J. A., Moews, P. C., Knox, J. R., Frere, J.-M., Ghuysen, J.-M. 1982. *Science* 218:479–81
122. Khorana, H. G., Gerber, G. E., Herlihy, W. C., Gray, C. P., Anderegg, R. J., et al. 1979. *Proc. Natl. Acad. Sci. USA* 76:5046–50
123. Kitano, K., Williamson, R., Tomasz, A. 1980. *FEMS Microbiol. Lett.* 7:133–36
124. Kleppe, G., Strominger, J. L. 1979. *J. Biol. Chem.* 254:4856–62
125. Kleppe, G., Yu, W., Strominger, J. L. 1982. *Antimicrob. Agents Chemother.* 21:979–83
126. Knott-Hunziker, V., Waley, S. G., Orlek, B. S., Sammes, P. B. 1979. *FEBS Lett.* 99:59–61
127. Knott-Hunziker, V., Petursson, S., Jayatilake, G. S., Waley, S. G., Jaurin,

B., Grundstrom, T. 1982. *Biochem. J.* 201:621–27
128. Knox, J., DeLucia, M., Murthy, N., Kelly, J., Moews, P., et al. 1979. *J. Mol. Biol.* 127:217–24
129. Komatsu, Y., Nishikawa, T. 1980. *Antimicrob. Agents Chemother.* 17:316–21
130. Koyasu, S., Fukuda, A., Okada, Y. 1980. *J. Biochem.* 87:363–66
131. Kozarich, J. W., Strominger, J. L. 1978. *J. Biol. Chem.* 253:1272–78
132. Kozarich, J. W., Nishino, T., Willoughby, E., Strominger, J. L. 1977. *J. Biol..Chem.* 252:7525–29
133. Kraut, H., Keck, W., Hirota, Y. 1981. *Ann. Rep. Natl. Inst. Genet.* 31:24–29
133a. Lai, J.-S., Sarvas, M., Brammar, W. J., Neugebauer, K., Wu, C. 1981. *Proc. Natl. Acad. Sci. USA* 78:3506–10
134. Lawrence, P. J., Strominger, J. L. 1970. *J. Biol. Chem.* 245:3653–59
135. Lawrence, P. J., Strominger, J. L. 1970. *J. Biol. Chem.* 245:3660–66
136. Lee, B. 1971. *J. Mol. Biol.* 61:463–69
137. Leyh-Bouille, J., Dusart, J., Nguyen-Disteche, M., Ghuysen, J.-M., Reynolds, P. E., Perkins, H. R. 1977. *Eur. J. Biochem.* 81:19–28
138. Lund, F., Tybring, L. 1972. *Nature New Biol.* 236:135–37
139. Markiewicz, Z., Broome-Smith, J. K., Schwarz, U., Spratt, B. G. 1982. *Nature* 297:702–4
140. Marquet, A., Frere, J.-M., Ghuysen, J.-M., Loffet, A. 1979. *Biochem. J.* 177:909–16
141. Martin, H. H., Schilf, W., Maskos, C. 1976. *Eur. J. Biochem.* 71:585–93
142. Martin, H. H., Gmeiner, J. 1979. *Eur. J. Biochem.* 95:487–95
143. Martin, H. H., Schilf, W., Schiefer, H.-G. 1980. *Arch. Microbiol.* 127:297–99
144. Matsubara, N., Minami, S., Matsuhashi, M., Takaoka, M., Mitsuhashi, S. 1980. *Antimicrob. Agents Chemother.* 18:195–99
145. Matsuhashi, S., Kamiyro, T., Blumberg, P. M., Linnett, P., Willoughby, E., Strominger, J. L. 1974. *J. Bacteriol.* 117:578–87
146. Matsuhashi, M., Takagaki, Y., Maruyama, I. N., Tamaki, S., Nishimura, Y., et al. 1977. *Proc. Natl. Acad. Sci. USA* 74:2976–79
147. Matsuhashi, M., Maruyama, I. N., Takagaki, Y., Tamaki, S., Nishimura, Y., Hirota, Y. 1978. *Proc. Natl. Acad. Sci. USA* 75:2631–35
148. Matsuhashi, M., Tamaki, S., Curtis, S. J., Strominger, J. L. 1979. *J. Bacteriol.* 137:644–47
149. Matsuhashi, M., Ishino, F., Tamaki, S.,

Nakajima-Iijima, S., Tomioka, S., et al. 1982. In *Trends in Antibiotics Research—Genetics, Biosyntheses, Actions and New Substances*, ed. H. Umezawa, A. L. Demain, T. Hata, C. R. Hutchinson, pp. 99–114. Tokyo: Japan Antibiotics Res. Assoc.

150. Matsuhashi, M., Nakagawa, J., Tomioka, S., Ishino, F., Tamaki, S. 1982. In *Drug Resistance in Bacteria— Genetics, Biochemistry and Molecular Biology*, ed. S. Mitsuhashi, pp. 297–310. Tokyo: Japan Sci. Soc.

151. Mirelman, D. 1980. In *Bacterial Outer Membranes*, ed. M. Inouye, pp. 115–66. New York: Wiley

152. Mirelman, D., Sharon, N. 1972. *Biochem. Biophys. Res. Commun.* 46:1909–17

153. Mirelman, D., Yashouv-Gan, Y., Schwarz, U. 1976. *Biochemistry* 15:1781–90

154. Mirelman, D., Yashouv-Gan, Y., Schwarz, U. 1977. *J. Bacteriol.* 129:1593–1600

155. Mirelman, D., Yashouv-Gan, Y., Nuchamovitz, Y., Rozenhak, S., Ron, E. Z. 1978. *J. Bacteriol.* 134:458–61

156. Moews, P. C., Knox, J. R., Waxman, D. J., Strominger, J. L. 1981. *Int. J. Pept. Protein Res.* 17:211–18

157. Morin, R. B., Jackson, B. G., Mueller, R. A., Lavagnino, E. R., Scanlon, W. B., Andrews, S. L. 1969. *J. Am. Chem. Soc.* 91:1401–7

158. Murphy, T. F., Barza, M., Park, J. T. 1981. *Antimicrob. Agents Chemother.* 20:809–13

159. Nakagawa, J., Tamaki, S., Matsuhashi, M. 1979. *Agric. Biol. Chem.* 43:1379–80

160. Nakagawa, J., Matsuzawa, H., Matsuhashi, M. 1979. *J. Bacteriol.* 138:1029–32

161. Nakagawa, J., Matsuhashi, M. 1982. *Biochem. Biophys. Res. Commun.* 105:1546–53

162. Nguyen-Disteche, M., Frere, J.-M., Dusart, J., Leyh-Bouille, M., Ghuysen, J.-M., et al. 1977. *Eur. J. Biochem.* 81:29–32

163. Nieto, M., Perkins, H. R., Frere, J.-M., Ghuysen, J.-M. 1973. *Biochem. J.* 135:493–505

163a. Nikaido, H., Nakae, T. 1979. *Adv. Microb. Physiol.* 20:164–250

164. Nishimura, Y., Takeda, Y., Nishimura, A., Suzuki, H., Inouye, M., Hirota, Y. 1977. *Plasmid* 1:67–77

165. Nishimura, Y., Suzuki, H., Hirota, Y., Park, J. T. 1980. *J. Bacteriol.* 143:531–34

166. Nishino, T., Kozarich, J. W., Strominger, J. L. 1977. *J. Biol. Chem.* 252:2934–39

167. Noguchi, H., Matsuhashi, M., Takaoka, M., Mitsuhashi, S. 1978. *Antimicrob. Agents Chemother.* 14:617–24

168. Noguchi, H., Matsuhashi, M., Nikaido, T., Itoh, J., Matsubara, N., et al. 1979. *Microb. Drug. Resis.* 2:361–87

169. Noguchi, H., Matsuhashi, M., Mitsuhashi, S. 1979. *Eur. J. Biochem.* 100:41–49

170. Nozaki, Y., Imada, A., Yoneda, M. 1979. *Antimicrob. Agents Chemother.* 15:20–27

171. Nozaki, Y., Kawashima, F., Imada, A. 1981. *J. Antibiotics* 34:206–11

172. O'Callaghan, C. H., Kirby, S. M., Moris, A., Waller, R. E., Duncombe, R. E. 1972. *J. Bacteriol.* 110:988–91

172a. Ogawara, H. 1981. *Microbiol. Rev.* 45:591–619

173. Ogawara, H., Horikawa, S. 1980. *Antimicrob. Agents Chemother.* 17:1–7

174. Ohkoshi, M. 1970. *Prog. Antimicrob. Anticancer Chemother., Proc., Int. Congr. Chemother., 6th, 1969* 2:717–22

175. Ohya, S., Yamazaki, M., Sugawara, S., Tamaki, S., Matsuhashi, M. 1978. *Antimicrob. Agents Chemother.* 14:780–85

176. Ohya, S., Yamazaki, M., Sugawara, S., Matsuhashi, M. 1979. *J. Bacteriol.* 137:474–79

177. Ohya, S., Utsui, Y., Sugawara, S., Yamazaki, M. 1982. *Antimicrob. Agents Chemother.* 21:492–97

178. Olijhoek, A. J. M., Klencke, S., Pas, E., Nanninga, N., Schwarz, U. 1982. *J. Bacteriol.* 152:1248–54

179. Ozols, J., Gerard, C. 1977. *J. Biol. Chem.* 252:8549–53

180. Park, J. T. 1952. *J. Biol. Chem.* 194:897–904

181. Park, J. T., Burman, L. 1973. *Biochem. Biophys. Res. Commun.* 51:863–68

182. Park, J. T., Strominger, J. L. 1957. *Science* 125:99–101

183. Percheson, P. B., Bryan, L. E. 1980. *Antimicrob. Agents Chemother.* 12:390–96

183a. Perkins, H. R., Nieto, M., Frere, J.-M., Leyh-Bouille, M., Ghuysen, J.-M. 1973. *Biochem. J.* 131:707–18

184. Pollock, M. R. 1965. *Biochem. J.* 94:666–75

185. Pollock, M. R. 1967. *Br. Med. J.* 4:71–77

186. Pratt, J. M., Holland, B., Spratt, B. G. 1981. *Nature* 293:307–9

187. Pratt, R. F., Anderson, E. G., Odeh, I. 1980. *Biochem. Biophys. Res. Commun.* 93:1266–73

188. Rasmussen, J. R., Strominger, J. L. 1978. *Proc. Natl. Acad. Sci. USA* 75:84–88

189. Reynolds, P. E., Shephard, S. T., Chase, H. A. 1978. *Nature* 271:568–70
189a. Richmond, M. H., Curtis, N. A. C. 1974. *Ann. N. Y. Acad. Sci.* 235:533–67
190. Rodriguez-Tebar, A., Rojo, F., Damaso, D., Vazquez, D. 1982. *Antimicrob. Agents Chemother.* 22:255–61
191. Rodriguez-Tebar, A., Rojo, F., Montilla, J. C., Vazquez, D. 1982. *FEMS Microbiol. Lett.* 14:295–98
192. Rodriguez-Tebar, A., Rojo, F., Vazquez, D. 1982. *Eur. J. Biochem.* 126:161–66
193. Sargent, M. G., Lampen, J. O. 1970. *Proc. Natl. Acad. Sci. USA* 65:962–69
194. Satta, G., Canepari, P., Botta, G., Fontana, R. 1980. *J. Bacteriol.* 142:43–51
195. Schenkein, D. P., Pratt, R. F. 1980. *J. Biol. Chem.* 255:45–48
196. Schepartz, S. A., Johnson, M. J. 1956. *J. Bacteriol.* 71:84–90
197. Schleifer, K. H., Kandler, O. 1972. *Bacteriol. Rev.* 36:407–77
198. Schilf, W., Frere, P., Frere, J.-M., Martin, H. H., Ghuysen, J.-M., et al. 1978. *Eur. J. Biochem.* 85:325–30
199. Schilf, W., Martin, H. H. 1980. *Eur. J. Biochem.* 105:361–70
200. Schmidt, L. S., Botta, G., Park, J. T. 1981. *J. Bacteriol.* 145:632–37
201. Schwarz, U., Seeger, K., Wengenmayer, F., Strecker, H. 1981. *FEMS Microbiol. Lett.* 10:107–9
202. Sharpe, A., Blumberg, P. M., Strominger, J. L. 1974. *J. Bacteriol.* 117:926–27
203. Shephard, S. T., Chase, H. A., Reynolds, P. E. 1977. *Eur. J. Biochem.* 78:521–32
204. Shepherd, W. D., Kaplan, S., Park, J. T. 1981. *J. Bacteriol.* 147:354–62
205. Shockman, G. D., Daneo-Moore, L., McDowell, T. D., Wong, W. 1982. In *The Chemistry and Biology of β-Lactam Antibiotics*, ed. R. B. Morin, M. Gorman, 3:304–38. New York: Academic
205a. Sigal, I. S., Harwood, B. G., Arentzen, R. 1982. *Proc. Natl. Acad. Sci. USA* 79:7157–60
206. Simons, K., Sarvas, M., Garoff, H., Helenius, A. 1978. *J. Mol. Biol.* 126:673–90
207. Slater, M., Schaechter, M. 1974. *Bacteriol. Rev.* 38:199–221
208. Sowell, M. O., Buchanan, C. E. 1982. *J. Bacteriol.* 151:491–94
209. Spratt, B. G. 1975. *Proc. Natl. Acad. Sci. USA* 72:2999–3003
210. Spratt, B. G. 1977. *Eur. J. Biochem.* 72:341–52
211. Spratt, B. G. 1977. *Antimicrob. Agents Chemother.* 11:161–66
212. Spratt, B. G. 1977. *J. Antimicrob. Chemother.* 3:(Suppl. B)13–19
213. Spratt, B. G. 1977. *J. Bacteriol.* 131:293–305
214. Spratt, B. G. 1978. *Nature* 274:713–15
215. Spratt, B. G. 1979. *Microb. Drug. Resist.* 2:349–60
216. Spratt, B. G. 1980. *Philos. Trans. R. Soc. London Ser. B* 289:273–83
217. Spratt, B. G. 1980. *J. Bacteriol.* 144:1190–92
218. Spratt, B. G., Pardee, A. B. 1975. *Nature* 254:516–17
219. Spratt, B. G., Jobanputra, V., Zimmermann, W. 1977. *Antimicrob. Agents Chemother.* 12:406–9
220. Spratt, B. G., Jobanputra, V., Schwarz, U. 1977. *FEBS Lett.* 79:374–78
221. Spratt, B. G., Boyd, A., Stoker, N. 1980. *J. Bacteriol.* 143:569–81
222. Springer, T. A., Strominger, J. L. 1976. *Proc. Natl. Acad. Sci. USA* 73:2481–85
223. Storm, D. R., Blumberg, P. M., Strominger, J. L. 1974. *J. Bacteriol.* 1,17:783–85
224–233. Deleted in proof
234. Strittmatter, P., Rogers, M. J., Spatz, L. 1972. *J. Biol. Chem.* 247:7188–94
235. Suginaka, H., Blumberg, P. M., Strominger, J. L. 1972. *J. Biol. Chem.* 247:5279–88
236. Suzuki, H., Nishimura, Y., Hirota, Y. 1978. *Proc. Natl. Acad. Sci. USA* 75:664–68
237. Suzuki, H., van Heijenoort, Y., Tamura, T., Mizoguchi, J., Hirota, Y., van Heijenoort, J. 1980. *FEBS Lett.* 110:245–49
238. Sweet, R. M. 1973. In *Cephalosporins and Penicillins: Chemistry and Biology*, ed. E. H. Flynn, pp. 280–309. New York: Academic
239. Sykes, R. B., Bush, K. 1982. See Ref. 205, Vol. 3, pp. 155–207
239a. Sykes, R. B., Matthew, M. 1976. *J. Antimicrob. Chemother.* 2:115–57
240. Takeda, Y., Nishimura, A., Nishimura, Y., Yamada, M., Yasuda, S. et al. 1981. *Plasmid* 6:86–98
241. Taku, A., Stuckey, M., Fan, D. P. 1982. *J. Biol. Chem.* 257:5018–22
242. Tamaki, S., Nakajima, S., Matsuhashi, M. 1977. *Proc. Natl. Acad. Sci. USA* 74:5472–76
243. Tamaki, S., Nakagawa, J., Maruyama, I. N., Matsuhashi, M. 1978. *Agric. Biol. Chem.* 42:2147–50
244. Tamaki, S., Matsuzawa, H., Matsuhashi, M. 1980. *J. Bacteriol.* 141:52–57
245. Tamura, T., Imae, Y., Strominger, J. L. 1976. *J. Biol. Chem.* 251:414–23
246. Tamura, T., Suzuki, H., Nishimura, Y.,

Mizoguchi, J., Hirota, Y. 1980. *Proc. Natl. Acad. Sci. USA* 77:4499–503

247. Tilby, M. J. 1977. *Nature* 266:450–52

248. Tipper, D. J., Strominger, J. L. 1965. *Proc. Natl. Acad. Sci. USA* 54:1133–41

249. Tipper, D. J., Strominger, J. L. 1968. *J. Biol. Chem.* 243:3169–79

250. Tipper, D. J., Wright, A. 1979. In *The Bacteria*, ed. J. R. Sokatch, L. N. Ornstein, 7:291–426. New York: Academic

251. Tomasz, A. 1979. *Ann. Rev. Microb.* 33:113–37

252. Tomasz, A. 1979. *Rev. Infect. Dis.* 1:434–67

253. Tomasz, A., Kitano, K., Lopez, R., DeFreitas, C. 1979. *Dev. Biochem.* 6:197–222

254. Tomioka, S., Ishino, F., Tamaki, S., Matsuhashi, M. 1982. *Biochem. Biophys. Res. Commun.* 106:1175–82

255. Tybring, L., Melchior, N. H. 1975. *Antimicrob. Agents Chemother.* 8:271–76

256. Umbreit, J. N., Strominger, J. L. 1973. *Proc. Natl. Acad. Sci. USA* 70:2997–3001

257. Umbreit, J. N., Strominger, J. L. 1973. *J. Biol. Chem.* 248:6759–66

258. Umbreit, J. N., Strominger, J. L. 1973. *J. Biol. Chem.* 248:6767–71

259. Unwin, P. N. T., Henderson, R. 1975. *J. Mol. Biol.* 94:425–40

259a. Walsh, C. 1977. *Horizons Biochem. Biophys.* 3:36–81

260. Ward, J. B. 1974. *Biochem. J.* 141:227–41

261. Waxman, D. J. 1982. See Ref. 205, pp. 415–17

262. Waxman, D. J., Strominger, J. L. 1979. *J. Biol. Chem.* 254:4863–75

263. Waxman, D. J., Strominger, J. L. 1979. *J. Biol. Chem.* 254:12056–61

264. Waxman, D. J., Strominger, J. L. 1980. *J. Biol. Chem.* 255:3964–76

265. Waxman, D. J., Strominger, J. L. 1981. *J. Biol. Chem.* 256:2059–66

266. Waxman, D. J., Strominger, J. L. 1981. *J. Biol. Chem.* 256:2067–77

267. Waxman, D. J., Yu, W., Strominger, J. L. 1980. *J. Biol. Chem.* 255:11577–87

268. Waxman, D. J., Yocum, R. R., Strominger, J. L. 1980. *Philos. Trans. R. Soc. London Ser. B* 289:257–71

269. Waxman, D. J., Lindgren, D. M., Strominger, J. L. 1981. *J. Bacteriol.* 148:950–55

270. Waxman, D. J., Amanuma, H., Strominger, J. L. 1982. *FEBS Lett.* 139:159–63

271. White, J. S., Astill, M., Lawrence, P. J. 1979. *Antimicrob. Agents Chemother.* 15:204–8

272. Williamson, R., Hakenbeck, R., Tomasz, A. 1980. *Antimicrob. Agents Chemother.* 18:629–37

273. Williamson, R., Hakenbeck, R., Tomasz, A. 1980. *FEMS Microbiol. Lett.* 7:127–31

274. Williamson, R., Zighelboim, S., Tomasz, A. 1981. In *β-Lactam Antibiotics*, pp. 215–25. New York: Academic

275. Williamson, R., Ward, J. B. 1982. *J. Gen. Microbiol.* 128: In press

276. Wise, E. M., Park, J. T. 1965. *Proc. Natl. Acad. Sci. USA* 54:75–81

277. Woodward, R. B. 1949. In *The Chemistry of Penicillin*, ed. H. T. Clarke, J. R. Johnson, R. Robinson, p. 440. Princeton: Princeton Univ. Press

278. Wyke, A. W., Ward, J. B., Hayes, M. V., Curtis, N. A. C. 1981. *Eur. J. Biochem.* 119:389–93

279. Wyke, A. W., Ward, J. B., Hayes, M. V. 1982. *Eur. J. Biochem.* 127:553–58

280. Yocum, R. R., Blumberg, P. M., Strominger, J. L. 1974. *J. Biol. Chem.* 249:4863–71

281. Yocum, R. R., Waxman, D. J., Rasmussen, J. R., Strominger, J. L. 1979. *Proc. Natl. Acad. Sci. USA* 76:2730–34

282. Yocum, R. R., Rasmussen, J. R., Strominger, J. L. 1980. *J. Biol. Chem.* 255:3977–86

283. Yocum, R. R., Amanuma, H., O'Brien, T. A., Waxman, D. J., Strominger, J. L. 1982. *J. Bacteriol.* 149:1150–53

284. Zighelboim, S., Tomasz, A. 1980. *Antimicrob. Agents Chemother.* 17:434–42

Ann. Rev. Biochem. 1983. 52:871–926

A MOLECULAR DESCRIPTION OF NERVE TERMINAL FUNCTION

Louis F. Reichardt and Regis B. Kelly

Department of Physiology and Department of Biochemistry, School of Medicine, University of California, San Francisco, Calif., 94143

CONTENTS

PERSPECTIVES AND SUMMARY 871
BIOCHEMICAL AND MOLECULAR STUDIES ON ION-SELECTIVE
    CHANNELS AND PUMPS 874
   Sodium Channels 875
   Potassium Channels 878
   Restoration of Na^+ and K^+ Gradients 882
   Voltage-Sensitive Calcium Channels 884
   Ca^{2+}-Removal from the Cytoplasm 889
MOLECULAR BASIS OF EXOCYTOSIS 890
   Description 890
   Kinetics of the Release Process 890
   Concentration-Dependence of Ca^{2+}-Dependent Exocytosis 892
   Possible Ca^{2+} Targets 893
   Mechanism of Vesicle Fusion 895
REGULATION OF NEUROTRANSMITTER LEVELS IN THE NERVE
    TERMINAL 895
   Metabolism of Peptide and Classical Transmitters 895
   Storage of Classical Transmitters 897
MOVEMENTS OF MEMBRANES AND PROTEINS IN THE NEURON 899
MODIFICATION IN NERVE TERMINAL METABOLISM INDUCED BY
    ACTION POTENTIALS AND NEUROTRANSMITTERS 903
   Regulation by cAMP 904
   Regulation by Calcium 907
   Regulation by Protein Kinases 912

PERSPECTIVES AND SUMMARY

The nerve terminal is a specialized region of a neuron, separated from the neuronal soma by an axon that can be exceedingly long, whose function is

0066/4154/81/0701-0871$02.00

to release neurotransmitter when stimulated by an electrical signal carried by the axon. In this review, we describe the enzymes, channels, and other proteins presently thought to be important in nerve terminal function. Recent progress in the area has been so dramatic that it is becoming possible to relate simple changes in behavior to modifications of defined proteins in particular nerve terminals.

Neurotransmitters are stored in synaptic vesicles and are released by fusion of these vesicles to the plasma membrane. Vesicle fusion is triggered by Ca^{2+}-influx through specific Ca^{2+} channels that open in response to depolarization of the plasma membrane and is terminated by the disappearance of Ca^{2+} from the vicinities of the active zones. The voltage signal that opens the Ca^{2+} gates is not constant, but also subject to regulation. The key elements are the Na^+ and K^+ channels in the nerve terminal. These channels are localized to distinct regions of many neurons and different neurons have different quantities of the different channel types. The voltage sensitive Na^+ channel is responsible for depolarizing the membrane. This channel has recently been purified and reconstituted in artificial membranes, so many of its properties are known. K^+ channels are responsible for repolarizing the membrane. Several K^+ channels have been identified: some activated by voltage, some by intracellular Ca^{2+} and others by neurotransmitters. The properties of several of these have recently been shown to be altered by cAMP-dependent protein kinases, resulting in long-term changes in neuronal activity and the efficiency of synaptic transmission. The voltage-dependent Ca^{2+} channels can also be modified by cAMP-dependent protein kinases. Many of the neurotransmitters and hormones that modulate the efficiency of transmitter release apparently do so by modifying the Ca^{2+} channels. Much of the recent progress in molecular studies on the K^+ and Ca^{2+} channels has been made possible by a dramatic new technique called "patch clamping" that permits individual ion channels to be examined on the tip of a microelectrode. In many ways this technology circumvents the need for biochemical isolation and reconstitution.

Maintenance of the Na^+ and K^+ concentrations in neurons requires the classical Na^+,K^+-ATPase that is found in all cell types. Recent experiments suggest that a novel form of this enzyme exists in neural tissues. This pump appears to be localized to anatomically and functionally distinct regions of many neurons. To restore the cytoplasmic Ca^{2+} concentrations to original levels, the nerve terminal has an array of Ca^{2+} removal systems including a $Na^+:Ca^{2+}$ antiporter in the plasma membrane, a Ca^{2+} porter in the mitochondrial inner membrane and several distinct ATP-dependent Ca^{2+} uptake systems in the plasma membrane, smooth endoplasmic reticulum, and synaptic vesicles.

The mechanism of vesicle fusion is poorly understood despite recent progress in isolating highly purified synaptic vesicles and presynaptic plasma membranes. All the elements required for exocytosis—vesicles, Ca^{2+} channels, and components mediating membrane fusion—appear to be localized in small active zones. The Ca^{2+} that enters the nerve terminal must interact with a class of molecules just inside the nerve terminal that regulates vesicle fusion. These Ca^{2+}-binding molecules induce exocytosis within 100–200 μs at Ca^{2+} concentrations of 1–10 μM. Recent research has strongly suggested, but not proven, that calmodulin mediates Ca^{2+} action in exocytosis at the nerve terminal. Identification of the Ca^{2+} target and the components that mediate membrane fusion is a high priority for future research.

Studies on the synthesis and packaging of peptide and classical neurotransmitters have revealed that the two transmitter classes are regulated by fundamentally different mechanisms. Peptide transmitters are synthesized as precursors, packaged in the cell soma, and processed in secretory granules before arrival at neuronal terminals. The components of these granules are brought to the nerve terminal by an ATP-dependent, fast (3–5 μm/sec) transport system that is likely to involve microtubules and actin. Once released by exocytosis, there is no convincing evidence that peptides are retrieved or reused. In contrast, classical neurotransmitters, such as catecholamines, are mostly synthesized in the nerve terminal and can be recycled. These transmitters are retrieved from extracellular space by transmitter-specific Na^+ driven coporters. The uptake of classical transmitters into synaptic vesicles requires an ATP-dependent proton translocase, similar to the mitochondrial F_1 ATPase, that is found in all secretory granules and maintains their interiors at acidic pH. The uptake of catecholamines into chromaffin granules utilizes a proton-catecholamine antiporter. Uptake of other transmitters has not been described with comparable precision. Several different mechanisms couple the rate of synthesis of classical transmitters to their rate of depletion from the nerve terminal.

Nerve terminal functions are regulated by changes in cyclic nucleotide and Ca^{2+} levels in response to membrane depolarization or the binding of transmitters to receptors. Nerve terminals contain high levels of calmodulin and adenylate cyclase. They also contain high levels of cAMP-dependent protein kinase, several Ca^{2+}-calmodulin-dependent protein kinases and a Ca^{2+}-phospholipid activated protein kinase, each of which has a distinct set of protein substrates. Changes in phosphorylation of specific proteins are likely to modulate many of the rapid changes in the rate of energy metabolism, activity of ion pumps and channels, and rates of transmitter uptake and synthesis that are known to occur in individual nerve terminals in response to electrical activity. In several cases there is now direct evidence that regulation of nerve terminal function by second messengers does indeed involve phosphorylation of specific proteins. Most dramatically, recent work with *Aplysia* has shown that changes in the phosphorylation of specific ion channels alter the properties of individual synapses and cells, which in turn cause changes in the animal's behavior.

BIOCHEMICAL AND MOLECULAR STUDIES ON ION-SELECTIVE CHANNELS AND PUMPS

For the most part, transmitter release is modulated by Ca^{2+} entry and removal. Ca^{2+} enters the cytosol through a Ca^{2+}-selective channel that is opened by depolarization of the plasmalemma. Although the conductance of the calcium channel can be controlled directly by covalent modification (1), its important physiological regulator is the membrane potential, which in turn reflects the conductance of channels and the activity of pumps that are selective for other ions, most notably Na^+ and K^+ (e.g. 2). The sensitivity of Na^{2+} and K^+ channels to transmembrane voltage has been known for many years to be responsible for electric signalling in the nerve system. To understand the regulation of transmitter release, therefore, it is first important to examine the regulation of ionic conductances in the nerve terminal.

The electrical properties of excitable cells reflect changes in the selective permeability of the plasmalemma to different ions. Hodgkin & Huxley (3) found that the initial depolarization of the plasmalemma during an action potential in the squid giant axon could be attributed to an initial large increase in Na^+ permeability, depolarizing the cell membrane. The subsequent repolarization to the starting condition required inactivation of the permeability to Na^+ and also an increase in K^+ permeability. The Na^+ and K^+ permeabilities were distinguished experimentally by replacing one of the ions with an equivalent impermeant ion and measuring the permeability of the other. After a step depolarization the Na^+ current rises to a peak quickly but then because of inactivation decays to a low value. The K^+ current differs from the Na^+ in that it rises more slowly, and does not decay. The opening and closing of both these channels is sensitive to membrane potential.

One major motive for pursuing biochemical studies on the channels has been to obtain molecular insights into the aspects of their structure that determine their ion-selectivity properties and voltage sensitivity. A second motive has been to determine the distribution of individual channel types on individual neurons. Na^+, K^+, and Ca^{2+} channels have precise and often distinct distributions on neurons, and these distributions have important functional consequences. Antibodies to purified channel proteins would be powerful reagents in neuroanatomy. Information on channel properties and distribution is crucial to developing models describing properties of neurons or circuits of neurons in the brain.

Sodium Channels

DESCRIPTION Voltage-sensitive Na^+ channels play a key role in conducting the action potential to the nerve terminal and are the one channel type that has been significantly purified and studied with traditional biochemical methods. The channels in different tissues have very similar properties, suggesting they are either a single molecular species or family of closely related molecules (e.g. 4). Recent research on the physiology, biophysics, and toxicology of Na^+ channels have been extensively reviewed (e.g. 4–7), so this review focuses on recent developments. Models describing the Na^+ channel have usually postulated two "gates" or subsections of the channel, both of which must be open to permit ion flow through the channel (e.g. 4, 5). At typical resting potentials, the m gate is closed and h gate open. Depolarization results in opening of the m gate and ion flow. Depolarization also results in closing of the h gate with slower kinetics, stopping ion flow. Channels are not reactivable by depolarization until the h gates have reopened. Both opening and closing of these gates are sensitive to transmembrane potential. Recently, patch clamps have shown that the individual Na^+ channels have a conductance of approximately 5–20 pS (e.g. 8). The channels do not have to open before being inactivated (9), so the m and h gates appear to function independently, confirming the essential feature of the two gate model.

BIOCHEMICAL CHARACTERIZATION   Several toxins have specific high affinity binding sites on the Na^+ channel and have been invaluable reagents for biochemical studies. Two toxins in particular, tetrodotoxin and saxitoxin, inactivate the channel at nanomolar concentrations. These toxins can be radiolabeled and have served as probes to measure channel density (10) or monitor channel purification (e.g. 11). A second set of alkaloid ligands, among which are batrachotoxin and veratridine, bind the channels at a distinct site in its lipophilic domain and induced persistent activation (e.g. 12). A third set of ligands are peptide toxins from certain scorpions and sea anemones (e.g. 13) that show voltage-sensitive binding to the extracellular surface of the Na^+ channel (14). These inhibit channel inactivation and act synergistically with the alkaloids to enhance persistent activation. The peptide toxins can also be radiolabeled, but do not bind the channel in the absence of a substantial transmembrane potential, making them of little use for purification. Beneski & Catterall (15) used arylazido [^{125}I]-scorpion toxin to radiolabel the subunits of the Na^+ channel in intact rat brain synaptosomes. They identified two peptides with apparent molecular weights of 32,000 and 250,000 as putative Na^+-channel subunits. Catterall (16) also used $[^{125}I]$-scorpion toxin to measure the distribution of Na^+-channels on living neuroblastoma cells and spinal cord neurons (see next section, on Localization of Na^+ Channels in Neurons).

Using saxitonin as a probe, investigators estimated the size characteristics of the solubilized Na^+ channel from rat brain by gel filtration and sucrose gradient sedimentation in D_2O and H_2O. Channels solubilized by Triton or cholate give complexes of M_r 600,000 and 1,000,000, respectively (17, 18). Assuming a reasonable protein density, the protein in these complexes would appear to be about 300 kd in size (17), implying that the channel could not contain more than one 250 kd subunit. Similar analysis of the sodium channel in rat muscle suggests that it also has a mass of about 300 kd (18a).

The Na^+ channel has been purified substantially from *Electrophorus* electroplax (11, 19), rat brain (20), and rat muscle (21). The channel from *Electrophorus* appears to contain only one polypeptide of 270 kd that copurified with saxitonin-binding activity (19). Since this channel has recently been shown to contain 30% carbohydrate and to behave anomalously on gels (22), the value of 270 kd is clearly only an approximation of the molecular weight. The Na^+ channel from rat brain, on the other hand, appears to contain three subunits of about 270, 39 and 37 kd (20, 23), two of which are labeled by arylazido $[^{125}I]$-scorpion toxin. The unlabeled M_r 39,000 subunit is linked to the large subunit by disulfide bonds. The Na^+ channel from rat muscle has been purified to approximately the same specific activity and contains a large glycoprotein with anomalous behavior on acrylamide gels and a 38-kd doublet (23a), which are almost certainly closely related to the 270-, 39-, and 37-kd brain channel subunits (23), and an additional protein of 45 kd (23a).

Recently, highly purified preparations of the eel Na^+ channel have been visualized in the electron microscope (24). Negatively stained preparations consist of clusters of 41-Å × 170-Å rods. Assuming the rod represents the protein portion of the glycoprotein, one can calculate the mass of each cylinder to be approximately 180 kd. If the unit contains a single glycoprotein with 30% carbohydrate, the total molecular weight would be about 260,000.

Crude preparations of Na^+ channels from rat brain can be used to reconstitute a voltage-sensitive Na^+ permeability in liposomes. Goldin et al (18) preloaded phospholipid vesicles containing sodium channels with Cs^+. Addition of the channel-specific alkaloid veratridine allowed Cs^+ to escape, reducing the density of the liposomes in a surcrose gradient. These ingenious experiments not only demonstrated functional reconstitution of the channel, but also yielded a 50-fold purification of the vesicles containing Na^+ channels. Tamkun & Catterall (25) reconstituted the Na^+ channels into liposomes that had a Na^+ concentration gradient across their membranes (inside high). Incubation with veratridine generated a trans-membrane diffusion potential that was assayed by the voltage-dependent binding of scorpion toxin. Recently, purified preparations of both the rat brain and the rat muscle sodium channel have been successfully re-constituted in phospholipid vesicles (26, 27). Both preparations exhibit veratridine-induced Na^+ uptake that is sensitive to tetrodotoxin and saxitoxin. The purified eel protein has not yet been successfully re-constituted. Since open channels have very high conductances, the opening of a single Na^+ channel in a phospholipid vesicle will allow that Na^+ to equilibrate within msec. Incorporation into lipid bilayers will be essential for accurate biophysical analysis of Na flux. Recently, Barchi et al (27a) used quenched-flow kinetic techniques to compare the permeability to different alkaline metal cations of batrachotoxin-activated Na^+ channels. The selectivity of the purified, reconstituted channel is very similar to that of the channel in its original membrane, suggesting no subunits have been lost during purification. Normal responsiveness of the reconstituted channel to transmembrane voltage changes has not yet been demonstrated.

The large subunits identified in the electroplax, rat brain, and muscle preparations seem very likely to be homologous. All are similar in apparent size and contain carbohydrate. The presence of small subunits in purified sodium channels from rat brain and muscle, but not electroplax, is not understood.

Modulation of Na^+ channel activity by covalent modification has not been observed in vivo, but the 270-kd subunit is a preferred substrate for the cAMP-kinase in vitro (28), suggesting that such modifications may be demonstrated to be of functional importance in the future.

LOCALIZATION OF Na^+ CHANNELS IN NEURONS   The distribution of Na^+ channels in dendrites, axons, and nerve terminals is functionally important in modulating integration of different synaptic inputs (e.g. 29), initiation and conductance of action potentials (e.g. 30), and neurotransmitter release. In at least some dendrites, Na^+ channels are present, probably in excitable patches (e.g. 29). $[^{125}I]$-scorpion toxin binding has been used to de-monstrate that initial segments of cultured spinal cord neurons, the normal initiation sites for action potentials, have high concentrations of Na^+ channels, suggesting a molecular basis for the lower threshold for action-potential generation often observed at this part of the neuron in vivo (16). In myelinated axons, the density of channels, whether assayed by focal, extracellular recording or density of saxitoxin receptors, is several hundred-fold higher in nodal than internodal segments of axonal membrane (31, 32). In the nodes, the channels appear to be present at very high densities— 3000–5000 channels per μm^2 (4, 31)—and must be a predominant membrane protein. Indeed, the major particles visualized in freeze-fracture replicas of the nodal membrane have dimensions similar to purified channel protomers (24). Channel density is at least 30-fold lower, but more uniformly distributed in unmyelinated axons (30, 31). Recently, monoclonal and polyclonal sera, specific for the 270-kd protein in the *Electrophorus* Na^+ channel, and a new class of Na^+ channel-specific scorpion toxins that do not require a transmembrane potential for binding have been characterized (33, 33a, 34). In myelinated or acutely demyelinated axons, antibody binding is restricted to nodal regions (33a). We anticipate these probes and others specific for the sodium channel polypeptides to be useful for mapping with more precision the position of these channels in neurons, particularly in the nerve terminal.

Potassium Channels

DESCRIPTION   Hodgkin & Huxley (3) showed that a current of K^+ flowing outward from the axon helps restore the membrane potential to its original value after a depolarization induced by Na^+ entry. The K^+ flows through channels that are activated with slower kinetics than the Na^+ channel, do not inactivate, and are resistant to toxins that block Na^+ channels. K^+ channels are found in virtually all excitable cells and probably have even wider distributions than Na^+ channels. While there may be only one type of Na^+ channel, there are clearly many different types of K^+ channels. Typical vertebrate or invertebrate neurons have a mixture of at least three channel types in different ratios: channels that are open at or near the resting membrane potential and are major regulators of the resting potential and excitability (35–37), voltage-sensitive channels, activated by depolarization, that give rise to the fast outward K^+ current seen in the restorative phase of the action potential (3), and Ca^{2+}-dependent K^+ channels, which only open when intracellular Ca^{2+} levels are high (38). Many neurons also have channels whose permeability is regulated by neurotransmitters (e.g. 39–42). Different channels have been shown clearly to regulate different physiological parameters, such as resting potential, shape of the action potential, Ca^{2+}-influx, adaptation to different firing rates and frequency of action potentials (see e.g. 37, 43, 44). Different neurons have different ratios of different types of K^+ channel, so these channels are particularly important generators of diversity between neurons in the nervous system. The different currents in molluscan neurons have been reviewed recently (45).

Even though no representatives of K^+ channels have yet been purified and subjected to traditional methods of biochemical analysis, K^+ channels are important to discuss in this review because they are clearly crucial in regulating exocytosis and changes in synaptic efficiency (e.g. 44). Many different types are clearly regulated by protein kinases and phosphatases (e.g. 44, 46, 47). Finally, biophysical studies using "patch clamp" procedures to isolate single channel types and to modify the environments at the cytosolic and extracellular surfaces have provided a wealth of information that was unimaginable only a few years ago.

ELEMENTARY CHANNEL TYPES

Channels open near rest A variety of K^+ channels active near the resting membrane potential have been described in molluscan neurons (45), muscle (48), and other preparations (49). Single K^+ channel currents open at the resting potential have been studied by patch clamping cultured rat muscle (48). The channels are present at low density, approximately one per μm^2. The conductance (~ 10 pS) and kinetics are not dramatically different from other channel types. In some cells, neurotransmitters, whose activity is mediated through a cAMP-induced kinase, can increase the conductance of this type of channel (46). By increasing K^+ conductance, such neurotransmitters would act to buffer the membrane against changes in transmembrane potential.

Voltage clamp studies on molluscan and vertebrate neurons reveal a distinct class of channels that carries a transient current that is activated at membrane potentials more positive than -60 mV, but is then rapidly inactivated (36, 42). The current that flows through these channels is conventionally called the A-current. This class of channel exerts a major influence on neuronal excitability, slowing the rate of depolarization in response to a stimulus and regulating the rate of action-potential generation in response to sustained stimuli (37). Recently, *Drosophila shaker* mutants have been shown to have altered A-current in muscle and probably nerve (50). Some mutations alter the voltage-dependence of this class of K^+ channel (51). *D. shaker* mutants have abnormally prolonged transmitter release, suggesting this class of channel is important in repolarizing the nerve terminal (52). Genetic and molecular analysis of this locus is likely to contribute significantly to our understanding of K^+ channel function.

Delayed, voltage-sensitive channel The K^+ channel responsible for repolarizing the membrane in the latter phase of the action potential in the squid giant axon (3) is widely distributed in excitable tissues, usually carrying a significant fraction of the K^+ current during later phases of the action potential (e.g. 53). The channel is defined by the speed of its response to a voltage change, its slow inactivation rate and its sensitivity to triethylammonium. The K^+ channel has been analyzed by patch clamping (54). It has a conductance of approximately 10 pS at physiological K^+ concentrations. Its mean open time is 12 msec, although short interruptions of current flow are seen within this open state. Depolarization seems to alter the probability of opening a channel without changing the conductance or mean open time of single channels (54). It is present in muscle at densities about one-twentieth that of the Na^+ channel (55). Recently, a novel scorpion toxin was discovered with submicromolar affinity for this channel in the squid axon (56). One hopes that this presages the isolation of toxins with tighter binding constants or identification of species in which K^+ channels bind this toxin more avidly. At this point, purification attempts have not been made and biochemical studies are a formidable proposition.

Ca^{2+}-activated potassium channel   The Ca^{2+}-activated K^+ channel requires elevated internal Ca^{2+} to open and can remain open for long periods after trains of impulses, reducing neuronal excitability and exocytosis (e.g. 38). The channel is found not only in excitable tissues such as neurons (38, 53), chromaffin cells (57), and muscle (58), but also in hepatocytes and red blood cells (59, 60). Although these channels can carry currents comparable to those of the voltage-sensitive K^+ channels, they have a unit conductance that is tenfold higher, so their density must be much lower, less than one per μm^2. While there are reported exceptions (61), the typical conductance of single channels, determined by patch clamping, is 100–200 pS at physiological K^+ (53, 57, 58). Increased internal Ca^{2+} seems to increase the number of open channels, not the conductance of those channels that are open (57, 58, 61). Although the channel present in red blood cells is reported to be insensitive to calmodulin (59), the Ca^{2+}-induced opening of the channel in dog heart can be blocked by micromolar trifluoperazine, an antagonist of calci-calmodulin action (62). It is thus possible that calmodulin is a tightly bound component of the channel complex. While the channel requires Ca^{2+} for activation, it is sensitive to transmembrane potential at suboptimal Ca^{2+} concentrations (53, 57, 58, 63). Depolarization increases the number of open channels, perhaps by changing the affinity of the channel for Ca^{2+}. Recently, the bee venom toxin apamine, which blocks neuronal Ca^{2+}-dependent K^+ channels (64), has been shown to bind a protease-sensitive receptor in rat brain with a dissociation constant of only 10 pM (65). The density of this receptor is only 1/150 to 1/300 that of Na^+ channels, approximately the density expected for Ca^{2+}-dependent K^+ channels. Apamine also blocks the Ca^{2+}-dependent K^+ channels in cultured rat myoblasts and myotubes, where it binds a receptor with a dissociation constant of 36–60 pM (65a). The density of this receptor is also low, approximately sevenfold lower than the density of Na^+

channels. Although the toxin provides a potential ligand for assay and purification of this channel, the low density of the channel in membranes means that purification will be a formidable task.

In spite of its low density, the Ca^{2+}-dependent K^+ channel has been the subject of several biochemical studies because it is a common target of intracellular kinases, providing a means for second messengers, such as cAMP and Ca^{2+}, to modify ion fluxes in the nerve terminal. In internally perfused *Helix* neurons addition of the catalytic subunit of the cAMP-dependent protein kinase results in an increased Ca^{2+}-dependent K^+ conductance (66). The channel still requires Ca^{2+} to open. One possible consequence of phosphorylation would be to reduce the K_m for Ca^{2+}. Recently, evidence that catecholamines induce the opening of a Ca^{2+}-dependent K^+ channel in mammalian pyramidal cells has been presented (67). Cyclic AMP is also likely to mediate this response (68).

Surprisingly, in other invertebrate neurons, cAMP-dependent phosphorylation seems to reduce, not increase, the conductance of the Ca^{2+}-activated potassium channel. Either bath application of cAMP analogs or microinjection of the catalytic subunit of the cAMP-dependent protein kinase reduces a late K^+ current in *Aplysia* bag cells (47). The channel is suppressed by Co^{2+} or Ni^{2+}, arguing that it is a Ca^{2+}-activated channel (69). The physiological effects of elevated cAMP on the cells are dramatic. A train of impulses in the fibers innervating the bag cells induces a prolonged period of spontaneous activity in these cells. Cyclic AMP analogs have the same effect. If the kinase catalytic subunit can enhance or reduce Ca^{2+}-activated K^+ conductances in different cell types, either the channels or proteins that modify them must differ in some way.

Transmitter-sensitive K^+ channels   Several distinct K^+-selective channels, restricted to particular classes of neurons, have been demonstrated to be opened or closed as a consequence of neurotransmitter action (e.g. 40, 42, 70).

One type of transmitter-sensitive channel, the M channel, has been found in sympathetic neurons (71) and several populations of central neurons (see 42). In frog sympathetic neurons, for example, the M channel is selective for K^+, opens at voltages below the threshold for initiation of action potentials, and does not inactivate with time. Thus, it exerts a strong stabilizing effect on the membrane of cells that are depolarized by other actions. Acetylcholine and muscarinic agonists suppress this current and thereby increase the excitability of sympathetic neurons (42). Closing of M channels does not appear to be mediated by second messengers, but M channels are closed by Ba^{2+} (72).

Another type of channel, the S channel, has been found in *Aplysia* sensory neurons (73, 74). The S channel is open at rest and is not sensitive to voltage or Ca^{2+}. It is closed, however, by serotonin binding to a receptor (40). The action of serotonin is mediated by a cAMP response that activates a cAMP-dependent protein kinase (e.g. 44). Both the M and S channels are distinct from the fast voltage-sensitive (A), delayed voltage-sensitive (K), and Ca^{2+}-activated (C) K^+ channels that coexist in these cells. The M and S channels are only two of a large number of receptor-sensitive K^+ channels. Many cells contain more than one channel of this type (70). K^+ channels are often regulated by peptide transmitters (e.g. 75).

DISTRIBUTION OF K^+ CHANNELS   The existence of so many diverse types of K^+ channels and absence until recently of specific ligands has limited studies on distribution of K^+ channels in neurons. Recent physiological experiments have shown that K^+ channels exist in nodal, paranodal, and internodal regions of the frog myelinated nerve fiber (76). In mammalian myelinated fibers, K^+ channels appear to be excluded from the node (77), but are present in the paranodal axolemma (78). These preliminary results suggest that both the concentration and distribution of different K^+ channel classes will be functionally important in describing the properties of neurons and neuronal circuits. One must hope that specific reagents for K^+ channel polypeptides will be useful for mapping their distribution more precisely.

SUMMARY K^+ channels exist in a tremendous diversity that is only beginning to be appreciated. The different channel types serve as a reservoir for generation of the multitude of different electrophysiological properties seen in the different neurons. These channels seem also to be a frequent substrate for agents that modify synaptic efficiency, animal behavior, and memory (e.g. 44). Diversity in the channels permits very selective responses in different neurons to the same second messenger. More detailed studies on factors regulating the synthesis and distribution of these channels will be very important for future studies on synaptic function. Generating probes specific for these channels, using molecular biology or hybridoma technologies, will be a formidable but rewarding task.

Restoration of Na^+ and K^+ Gradients

DESCRIPTION   To maintain ionic gradients, neurons use a variety of ion pumps and exchange mechanisms, the majority of which are also used by nonneural cells. The major mechanism for removing internal Na^+ and restoring internal K^+ after action potentials is an ATP-driven Na^+-K^+ exchange pump that is detected in biochemical assays as an Na^+,K^+-ATPase (79, 80) and is the same enzyme as is used by other cells in the body. The pump occupies a central role in neuronal function. The Na^+ gradient that it maintains is essential to control intracellular osmolality and cell volume. The Na^+ gradient also drives several exchange reactions. One of these, an Na^+-Ca^{2+} exchange, removes Ca^{2+} from the cytoplasm and contributes to the termination of exocytosis and other consequences of Ca^{2+} action (81, 82). The Ca^{2+}/Na^+ antiporter from bovine heart has been solubilized, partially purified, and reconstituted (83), but the proteins responsible for mediating Na^+/Ca^{2+} exchange have not yet been identified. Several cotransport reactions, using the energy of the Na^+ gradient, are responsible for accumulation of sugars, amino acids, neurotransmitter precursors, and neurotransmitters. The presence of an avid Na^+-dependent uptake mechanism for neurotransmitter precursors, such as choline, or neurotransmitters such as norepinephrine, correlates closely with the transmitter specificity of neurons (84) and in some neurons is specifically localized to the nerve terminal (85).

In neurons, activity of the Na^+ pump is limited by internal Na^+ (86, 87). Increased cytoplasmic Na^+, introduced by action potentials or micro-pipettes, activates this pump, which can actually hyperpolarize the cell, since three internal Na^+ are exchanged for only two K^+ (79, 80). Ouabain-sensitive hyperpolarizations are prominent features in many neurons following trains of action potentials (e.g. 88) and can block action-potential initiation (89).

BIOCHEMICAL CHARACTERIZATION   The structure and function of the Na^+K$^+$-ATPase have been extensively studied (reviewed in 90–92). The enzyme has been purified from several sources including rat brain (93). In each case, two polypeptide chains of M_r 120,000 and 50,000 are found in highly purified preparations. The larger polypeptide has an internal binding site for ATP and external binding site for ouabain (e.g. 94). The smaller subunit is a glycoprotein that cannot be removed without losing enzymatic activity. Definitive evidence that these two subunits are the only proteins required for activity came from reconstitution experiments (e.g. 95), in which it was shown that no other large proteins were present in high enough concentrations to be present in each vesicle with reconstituted active Na^+ and K^+ transport. A small acidic proteolipid has been labeled with a photoaffinity ouabain derivative and hence is a possible additional component of the ATPase (96). The stoichiometry of this proteolipid in purified and reconstituted preparations has not been determined.

There is evidence for multiple forms of Na^+K$^+$-ATPase in neurons, even though the subunit composition, kinetic properties, and sensitivity to cardiac gangliosides of the neuronal enzymes indicate that neurons use the same basic pump as nonneuronal cells. Two forms of the neuronal ATPase have been identified in vertebrates and invertebrates that are closely related, but differ in sensitivity to proteolysis and cardiac ganglioside-mediated inhibition (93). One of these is the standard (α) form of enzyme that was the only form seen in kidney, muscle, adrenal cortex, and astrocytes. The specific neuronal form of the enzyme (α^+) has a lower affinity for strophanthidine and is the only form found in myelinated axons. The α^+ form appears after the standard form in developing brain. It does not appear to be in all neurons, since cultured sympathetic neurons contained only the standard (α) enzyme.

DISTRIBUTION ON NEURONS   Distribution of the Na^+K^+-ATPase has been examined with immunocytochemical methods (97). On neurons, it is restricted to the plasmalemma, where it is distributed over the surface of cell somas and dendrites. It appears to be concentrated in nodes of Ranvier in myelinated axons. Two classes of synaptic terminals were observed, one with and one without high concentrations of the ATPase. Observed differences in concentration between different nerve terminals is likely to be important in regulating the response properties of these terminals to different patterns of stimulation. Concentration in the nodes of Ranvier is also of physiological significance. The antibodies used in these experiments almost certainly recognized both forms of the enzymes.

REGULATION BY HORMONES AND TRANSMITTERS   Hormones and transmitters, such as vasopressin and the catecholamines, regulate activity of the Na^+K^+-ATPase in neuronal and nonneural tissues. Many of these effects seem likely to be mediated by cAMP and cAMP-dependent protein kinase (e.g. 98–102). Direct effects of catecholamines on the ATPase have been seen in brain synaptosomes (99). Thyroid hormone increases the activity of the ATPase by increasing the number of pumps in the plasmalemma (103). These studies suggest that hormones and transmitters will prove to regulate synaptic function by controlling the density and activity of the Na^+K^+-ATPase. Endogenous compounds with ouabain-like binding have also been identified (104, 105). Further studies will be needed to determine whether activity of the Na^+K^+-ATPase is also regulated by such compounds at the ouabain binding site.

Voltage-Sensitive Calcium Channels

DESCRIPTION   Voltage-gated calcium channels have been described in a wide range of neural and nonneural tissues [for recent review see (106)]. Since Ca^{2+} channels in different tissues vary markedly in their kinetic properties and in their sensitivity to inhibitors, it had been assumed that there was not one but many types of calcium channel (106). With the advent of patch clamping this view has been challenged (107) on the grounds that snail, chick and rat channels are remarkably similar to each other and to those found in heart muscle (108) and chromaffin cells (109).

Measurement of single Ca^{2+} channel conductance by patch clamping has not yet been applied to nerve terminal membranes. Voltage-dependent Ca^{2+} channels in heart cells, chromaffin cells, and snail neurons have, however, been recognized by this procedure. These are open for one or two msec with currents of 0.5–2 pA. At high depolarization, channel openings appear in clusters. Reuter et al (108) suggest that the Ca^{2+} channel enters an activated state from which it rapidly flickers open and closed. The heart cell Ca^{2+} channel also inactivates with prolonged depolarization. The Ca^{2+} channel in heart has a conductance of about 25 pS (108). A detailed study of calcium currents in the nerve terminals of the squid stellate ganglion demonstrated some parallels between the Na^+ and Ca^{2+} channels, but also some striking differences (2). The Ca^{2+} channel opens with applied voltage but only after a significant delay. Consequently the Ca^{2+} channel opens during the falling phase of the presynaptic action potential (110). The total Ca^{2+} current is about 1/20 that of the sodium current. When the voltage is removed the channels close with simple exponential kinetics ($\tau = 630$ μsec). Unlike Na^+ channels, however, the Ca^{2+} channels in this synapse stay open during a prolonged depolarization.

During repetitive stimulation, the amount of Ca^{2+} entering the squid nerve terminal per impulse remains constant. This was shown by injection of the Ca^{2+} sensitive dye Arsenazo III into nerve terminals. A single impulse causes detectable increase in light absorption owing to the formation of Ca^{2+}-Arsenazo III complexes. The increase reaches its maximum at the peak of transmitter release (111). During repetitive stimulation the increase in light absorption increased linearly with the number of stimuli (111, 112), showing that the channel is not altered by activity.

The use of Arsenazo III as a calcium-sensitive dye allows the sites of Ca^{2+} entry to be identified. In the squid giant synapse, changes in light absorption were only seen in the nerve terminals and not in preterminal axon regions, suggesting that nerve terminal membranes are enriched in Ca^{2+} channels (111).

Comparison of the change in Arsenazo III absorption induced by injecting a known amount of Ca^{2+} to that induced by a single action potential has made it possible to estimate that approximately 2×10^8 Ca^{2+} ions enter per impulse. The number of ions that flows through a single channel per impulse could be calculated if the density of channels were known. Large intramembranous particles are seen clustered in squid nerve terminals in putative active-zone regions (113). If it is assumed that all large particles in the clusters are Ca^{2+} channels, then each channel passes about 150 Ca^{2+}/impulse and has a conductance of about 0.14–0.21 pS (114). This conductance is in reasonable agreement with that determined from noise measurements on *Helix* neurons (115), but is two orders of magnitude lower than that measured by patch clamping (107). Unless squid nerve terminal Ca^{2+}-selective channels have a smaller conductance, comparatively few of the large intramembranous particles can be functional Ca^{2+} channels.

BIOCHEMICAL NATURE OF THE CA^{2+} CHANNEL   Progress in Ca^{2+} channel purification has been prevented by lack of specific ligands with nanomolar binding constants and the absence of tissues, equivalent to electric organ, that are rich sources of channels. The Ca^{2+} channel activity also decays rapidly in perfused cells (e.g. 116), so purification may require identification of the factors important for maintaining Ca^{2+} channel activity in the intact cell. One potentially useful approach is the separation of reconstituted phospholipid vesicles containing Ca^{2+} transporters on the basis of Ca^{2+} flux-induced density fluxes (117). ATP-driven Ca^{2+} pumps have already been purified by this procedure. Another is the report that batrachotoxin, long known to interact with voltage-sensitive Na^+ channels, can also inhibit Ca^{2+} channels in a neuroblastoma cell line (118). While 40 nM batrachotoxin gave 50% inhibition, the lipophilic nature of this toxin will limit its utility in providing an assay for the Ca^{2+} channel. Another candidate is maitotoxin, which increases Ca^{2+}-influx into pheochromo-cytoma cells (119). Ca^{2+}-influx induced by this toxin is blocked by classical Ca^{2+} channel inhibitors, such as verapamil, Mn^{2+}_2 and tetracaine. Drugs of the dihydropyridine type are thought to block slow-acting calcium channels. ^{3}H-Nitrendipine appears to bind to the target of these drugs. Its binding requires calcium or strontium and is blocked by lanthanum or cobalt, whereas verapamil and D-600, classic Ca^{2+} channel blockers, have no effect (120). Identification of a soluble Ca^{2+} channel-specific ligand, equivalent to tetrodotoxin, would greatly facilitate attempts at purification.

REGULATION OF THE CALCIUM CHANNEL   All types of short-term synaptic plasticity that have been investigated in detail involve modulation of Ca^{2+} levels in the terminals, though the mechanisms differ (121, 122). The amount of Ca^{2+}-influx during an action potential is primarily regulated by the opening and closing of Na^+ and K^+ channels (e.g. 121). In addition internal Ca^{2+} levels depend on steady state Ca^{2+} currents (123), Ca^{2+} accumulation from previous action potentials (122), and direct modification of Ca^{2+} channel activity (124, 125).

Synaptic facilitation is the increase in transmitter release sometimes observed on repetitive stimulation. Facilitation does not appear to be due to changes in the Ca^{2+} channel but to accumulation of Ca^{2+} in the nerve terminal. Recent experiments on the squid stellate ganglion synapse (112) show that an alteration of Ca^{2+} current of sufficient amplitude to cause facilitation would have been readily detected. Experiments on *Aplysia* neuron L10 show a close correspondence between facilitation and internal calcium as measured by the activity of the Ca^{2+}-dependent K^+ channel (122). The kinetics of disappearance of facilitation are thought to be regulated by the time course of Ca^{2+} removal (126).

The steady-state membrane potential also has an important influence on internal Ca^{2+} levels and exocytosis. Studies on *Aplysia* neuron L10 have shown that depolarization of the nerve terminal enhances exocytosis induced by an action potential in two ways: a steady state Ca^{2+} current is increased, resulting in a high concentration of internal Ca^{2+}; and voltage-sensitive K^+ channels, are inactivated, prolonging the duration of the action potential (123). Hormones and transmitters may thus modify synaptic efficiency by inducing small changes in the resting membrane potential near the terminal.

It was noted earlier that the Ca^{2+} channels in the squid stellate ganglion synapse do not inactivate rapidly as a result of prolonged depolarization (2, 127). Under physiological conditions, repetitive stimulation of the synapse gives a linear increase with time in intraterminal Ca^{2+} (111, 112). In many cells, however, a slow inactivation of Ca^{2+} channels is seen (128, 129). Half times of inactivation are typically about 100 msec, but range from 5 to 1000 msec (106). Inactivation has often been associated with the accumulation of high levels of internal Ca^{2+} (reviewed in 106). Indeed, inactivation of Ca^{2+} channels can be induced by injection of Ca^{2+} into the cytoplasm. If internal Ca^{2+} does control inactivation, differences in surface to volume ratio may account for the differences in observed inactivation rates.

In one case, repetitive stimulation has been shown to reduce intra-terminal Ca^{2+} influx and exocytosis. Repetitive stimulation of a sensory neuron in *Aplysia* gives rise to a synaptic potential in a postsynaptic motor neuron that gradually decrements in amplitude. This decrease is thought to explain the habituation observed in the gill-withdrawal reflex in *Aplysia*. By blocking all the known Na^+ and K^+ channels Klein et al (130) could show that the decrease in postsynaptic size correlated with a decrease in Ca^{2+} current in the presynaptic cell. The decrease, however, is not likely to be due to internal Ca^{2+}-induced Ca^{2+} channel inactivation, since it occurred when Ba^{2+} was substituted for Ca^{2+} and the time course of recovery was much slower than that observed after Ca^{2+} injections.

Ca^{2+} channels are also regulated by neurotransmitters and hormones. Ca^{2+}-dependent potentials in cell bodies of the chick dorsal root ganglia (131, 132) and in the rat superior cervical ganglion (133) decrease in duration and amplitude in the presence of noradrenaline, γ-aminobutyric acid, enkephalin, and somatostatin. Voltage clamping of sensory neurons in the absence of Na^+ and K^+ currents has provided evidence that reduced current is carried through Ca^{2+} channels (124). While a second messenger hypothesis is attractive to explain the synergistic actions of so many different transmitters, bath application of dibutyryl cAMP did not reduce Ca^{2+} conductance and no evidence was found for involvement of internal Ca^{2+}. Similar results have been obtained using an *Aplysia* neuron. Stimulation of presynaptic inhibitory neurons reduces both transmitter release and Ca^{2+} currents in voltage-clamped L10 neurons (125). Reduced current through the Ca^{2+} channel appeared to reflect direct transmitter action on the Ca^{2+} channel that was not mediated by changes in membrane potential or internal Ca^{2+}. The mechanism of action of the inhibitory transmitter is not known. Ca^{2+} channels are not regulated exclusively by neurotransmitters. Nanomolar to micromolar concentrations of extracellular ATP are also reported to increase Ca^{2+} currents by 25%–30% (134). The stimulation does not require ATP hydrolysis. Since ATP is released with many neurotransmitters, this effect could be physiologically significant.

The modifications of neuronal Ca^{2+} channels that result in altered conductance are not known. In heart muscle, unlike sensory and sympathetic neurons, epinephrine acts to increase Ca^{2+} current. Patch clamp analysis of single Ca^{2+} channels suggests that epinephrine increases the mean time that channels remain open (108). Injection of the catalytic subunit of protein kinase prolongs and enhances the Ca^{2+} action potential, and epinephrine has no further effect (1). The inward Ca^{2+} current increased about threefold. Injection of the regulatory subunit on the other hand decreased the duration and amplitude of the Ca^{2+} current. It would appear therefore that the positive inotropic effect exerted on heart muscle by epinephrine is due to stimulation of cAMP followed by phosphorylation of a protein that is part of, or affects, the Ca^{2+} channel. Isolated cardiac sarcolemmal membranes retain the capacity to show depolarization-sensitive Ca^{2+} uptake (134a). When such membranes contain both radioactive ATP and the catalytic subunit of protein kinase, only one protein is phosphorylated during Ca^{2+} uptake (M_r 23,000). This membrane protein has been termed calciductin. Proteins of such small size are not likely to be channels but may be regulatory elements. A possible model for these molecules is provided by phospholamban, a small, heart-specific protein that binds the ATP-dependent Ca^{2+} transporter in the sarcoplasmic reticulum. Phosphorylation of phospholamban results in increased ATP-dependent Ca^{2+} sequestration in the sarcoplasmic reticulum (e.g. 135, 135a).

Ca^{2+}-Removal from the Cytoplasm

DESCRIPTION   The Ca^{2+} that enters during an action potential is removed from the cytoplasm by several mechanisms acting in concert. The plasmalemmas of the squid giant axon and synaptosomes contain both a Na^+/Ca^{2+} exchange activity and a Ca^{2+}-dependent ATPase, which extrude Ca^{2+} (81, 82, 136, 137). The proton electrochemical driving force generated by the mitochondrial ATPase provides the energy required to drive a high capacity Ca^{2+}-uptake into mitochondria (reviewed in 138), which has been seen in both squid (139) and synaptosomes (140, 141). In both systems, the cytoplasmic Ca^{2+} level is normally below that which can be effectively removed by mitochondria (e.g. 142), so the mitochondrial system is likely to be only functionally important under conditions of high Ca^{2+} influx (141). Nonmitochondrial sites of internal Ca^{2+} sequestration include smooth endoplasmic reticulum sacs (143) and synaptic vesicles (144).

An ATP-dependent-Ca^{2+} uptake system has also been seen in synaptic vesicles (e.g. 145, 146), but convincing Ca^{2+}-dependent ATP hydrolysis has not been observed (e.g. 147). Vesicles generate a proton electrochemical gradient that is probably responsible for neurotransmitter uptake (148, 148a) and may well supply the energy required for Ca^{2+} uptake. In cholinergic vesicles from Torpedo electroplax nerve terminals, the ATP requirement for ATP-promoted Ca^{2+} uptake has been shown in part to reflect activation of Ca^{2+} transport by a calci-calmodulin-activated endogenous protein kinase (147).

Ca^{2+} must eventually be released by intracellular organelles and transported out of the cell. Ca^{2+}-efflux from brain mitochondria occurs primarily by Na^+/Ca^{2+} exchange (149). The efflux mechanisms of Ca^{2+} from other neuronal organelles, such as synaptic vesicles, have not yet been investigated.

PURIFICATION AND CHARACTERIZATION OF CA^{2+} PUMPS   The major Ca^{2+}-transporting ATPase from brain synaptosomes have been investigated by several groups and shown to be distinct in kinetic properties from the muscle sarcoplasmic reticulum Ca^{2+} transporter (e.g. 150). The major Ca^{2+}-dependent ATPase activity detected in membrane vesicles from synaptosomes is found in the same vesicles as Na^+-Ca^{2+} exchange activity, which has been detected in intact synaptosomes (82), so it is almost certainly in the plasmalemma.

Neuronal ATP-dependent Ca^{2+} translocators have been purified (117, 151). The synaptosomal ATP-dependent uptake system was solubilized in cholate and reconstituted in phospholipid vesicles at high ratios of phospholipid to protein. The vesicles were formed in the presence of oxalate, which forms an insoluble complex with Ca^{2+}. Addition of ATP and Ca^{2+} to the phospholipid vesicles results in Ca^{2+} transport and a shift in density of vesicles incorporating the Ca^{2+}-dependent ATPase caused by the formation of an insoluble Ca^{2+}-oxalate complex. The transporter is purified 100-fold by this procedure. Analysis on gels reveals two bands of M_r 94,000 and 140,000. Both bands are labeled by ATP to form an acyl phosphate (151, 152).

The M_r 140,000 protein from synaptosomes is the same molecular weight as the $(Ca^{2+} + Mg^{2+})$ ATPase in the plasmalemma of red blood cell ghosts and many other tissues (e.g. 153). Recently, the neuronal protein has been purified by calmodulin affinity chromatography (151–152a), the same procedure used to purify the red blood cell enzyme (154). This enzyme is likely to be the same or very similar to the Ca^{2+}-dependent ATPases in the plasmalemma of nonneuronal cells that are strongly stimulated by calmodulin (e.g. 153, 155).

The M_r 94,000 Ca^{2+} transporter is not purified by calmodulin-affinity chromatography (151, 152). Immunological evidence (152) indicates that it may be nervous system specific. Antibodies do not cross react with sarcoplasmic reticulum or red blood cell Ca^{2+} transporters. These antibodies should make it possible to identify the location of this transporter.

In summary, nerve terminals contain at least two Ca^{2+} transporters directly dependent on ATP and others that are dependent on ion gradients. Ca^{2+} that enters the cytoplasm during the action potential is sequestered by virtually every intracellular organelle before eventually being secreted from the cell.

MOLECULAR BASIS OF EXOCYTOSIS

Description

Ca^{2+} entering the nerve terminal by voltage-gated channels triggers fusion of synaptic vesicles to the plasmalemma and release of neurotransmitter. Exocytosis also requires a transport system to carry vesicles to the nerve terminal and a docking apparatus to localize them within a short distance of sites of release. The molecules that mediate exocytosis in the nerve terminal remain unknown. Constraints on the mechanisms, however, come from studies on the morphology and physiology of the synapse.

Kinetics of the Release Process

Electrophysiological recording gives the time course of neurotransmitter release from nerve terminals with a precision unparalleled in any other exocytotic process. Again, the best described synapse is the squid stellate ganglion (156). There is a lag of about one msec between the arrival of the action potential at the nerve terminal and the opening of channels on the postsynaptic membrane. The majority of this time (800 μsec) is due to the lag in opening the Ca^{2+} channels, as described above. Only a few microseconds are required for transmitter to diffuse from the nerve terminals to the very close postsynaptic receptors. This leaves only about 200 μsec for Ca^{2+} to interact with its target and initiate the fusion of synaptic vesicle with presynaptic plasma membrane. Since these experiments were performed at 18°C, it is very likely that in mammalian synapses, the Ca^{2+}-triggered transmitter release takes even less time. The duration of the release process is also very short. The current through the postsynaptic membrane falls with a half-time of approximately one msec. The rapid kinetics of the release process put several constrains on models of neurotransmitter release. First, the brief time involved would allow only those vesicles within a vesicle diameter of the presynaptic membrane to participate in rapid exocytosis. Electron microscopy of the nerve terminal provides evidence for attachment or docking of a small proportion of synaptic vesicles at morphologically specialized attachment sites in the plasmalemma in both central and peripheral synapses (e.g. 157). Even chromaffin granules may make connections with the plasmalemma before exocytosis (158). Affinity chromatography of detergent-solublized chromaffin cell plasma membrane proteins on a column of glutaraldehyde-fixed chromaffin granules results in purification of a 51-kd protein, distinct from actin or tubulin, that is a possible docking protein (159). Proteins of unknown function that are localized specifically to the presynaptic membrane or synaptic vesicle have been identified in neuronal cells (e.g. 160–162) and are thus also candidates for regulation of synaptic vesicle binding. Only those vesicles attached to the presynaptic plasma membranes appear able to participate in normal exocytosis (163, 157), although prolonged exocytosis can result in depletion of almost the entire pool of vesicles (164).

The brief time involved also requires that Ca^{2+} channels be located within 100 nm of the site of exocytosis. The 200 μsec would allow Ca^{2+} to diffuse only 40–100 nm through the cytoplasm (156). In fact, the attachment sites for vesicles are usually associated with arrays of intramembranous particles (165). Vesicle openings ruptured by fast freezing during exocytosis are found almost entirely within 80 nm of this zone (157). Indeed, release takes place within a short distance of these particles even in terminals where the arrays are dispersed by soaking in Ca^{2+}-free solutions (163). Ca^{2+} entry, measured by dye binding, is localized to nerve terminal regions (111). It has been argued that the large particles in the active zones are present in appropriate numbers to be Ca^{2+} channels (114). Unfortunately direct evidence is lacking.

The kinetics of transmitter release seem to follow the kinetics of Ca^{2+} entry quite closely, suggesting that the Ca^{2+} that enters is quickly removed from the exocytotic apparatus (111, 112). Disappearance of intracellular Ca^{2+} however, as monitored by Arsenazo III adsorption, occurs much more slowly (secs) that the cessation of transmitter release (msecs). Also, intracellular Ca^{2+} concentration increases linearly during repeated stimulations, but exocytosis correlates not with the total apparent Ca^{2+} concentration, but with the entry of additional Ca^{2+} during each impulse. To resolve this apparent paradox, it has been proposed that Ca^{2+} entry occurs near release zones, where it promotes exocytosis but then diffuses away from these zones, terminating exocytosis. Binding to Arsenazo III monitors average cytoplasmic Ca^{2+} concentrations. In this model, Ca^{2+} diffusion is primarily responsible for the cessation of exocytosis.

Concentration-Dependence of Ca^{2+}-Dependent Exocytosis

Even though the Ca^{2+} current is linear with external Ca^{2+} below saturation (2), the rate of neurotransmitter release increases with a higher power dependence in the squid giant synapse (110, 112) and frog or crayfish neuromuscular junction (126, 166). The slope of Ca^{2+} vs release varies from between 1 and 2 (squid) to 4 (neuromuscular junction), so activation of the Ca^{2+} target is likely to have a nonlinear dependence on Ca^{2+}.

Estimates of the local Ca^{2+} concentration during release have been made from measurements of Ca^{2+} entry per unit area, based on current or dye binding (112, 156) and suggest a concentration of 10 μM. Alternative calculations have been based on the ratio of spontaneous to induced release. Assuming a fourth power dependence on Ca^{2+} level, internal Ca^{2+} must increase 16-fold from 0.1 to 1.6 μM to account for the 60,000-fold increase in release rate (126).

Direct measurements of the dependence of exocytosis on Ca^{2+} concentration have been made in chromaffin cells and sea urchin eggs that have been permeabilized to molecules of less than 4-nm diameter by brief high voltage electric shocks. The concentration for half-maximal release is about 1 μM in each system and appears to be cooperative with release proportional to the second power of Ca^{2+} concentration (167, 168). The Ca^{2+} requirement can not be replaced by Mg^{2+}. A similar dependence on Ca^{2+} is seen by generating from sea urchin eggs a preparation of cortical granule surfaces attached to polylysine on coverslips (169). The calculations of Ca^{2+} dependence in each system assume that Ca^{2+}-EGTA complexes are biologically inert, an assumption that does not appear to be true for ATP-dependent Ca^{2+} transporters (e.g. 170). The evidence that these quasi–in vitro and in vitro systems measure exocytosis is convincing. In each system, vesicles disappear from the cytoplasm and are fused to the plasma membrane. In permeabilized chromaffin cells, release of catecholamines is accompanied by release of the vesicle protein, dopamine-β-hydroxylase; but not cytoplasmic enzymes.

Possible Ca^{2+} Targets

A molecule of the calmodulin type is an obvious candidate for the Ca^{2+} binding site. Calmodulin is present in all eukaryotic cells, including nervous tissue, at concentrations of approximately 10 μM (171, 172). It has a 0.1–10-μM binding affinity for Ca^{2+} depending on conditions (reviewed in 172), which is in the same range as the intracellular Ca^{2+} concentrations during exocytosis. It has four Ca^{2+} binding sites and the activation of some of its targets such as cyclic nucleotide phosphodiesterase, adenylate cyclase, and ATP-dependent Ca^{2+} transporters, require 3 or 4 Ca^{2+} bound (173–175). Thus, a requirement for more than one bound Ca^{2+} could explain nonlinearities in the dependence of release on Ca^{2+}. Finally, since calmodulin and calmodulin-like proteins bind to a subset of their target enzymes, such as phosphorylase kinase and protein phosphatase 2B, even in the absence of Ca^{2+} (176, 177), calci-calmodulin could initiate vesicle fusion within the 200 μsec required for exocytosis.

Inhibitors of calmodulin action, in particular the phenothiazines, have been shown to inhibit exocytosis in virtually every Ca^{2+}-dependent system examined, including permeabilized adrenal medullary cells (168) and cholinergic synaptosomes (178). Since these drugs shows some interaction with other proteins, for example the phospholipid-sensitive Ca^{2+}-dependent protein kinase (179), and have general membrane perturbant effects, sensitivity to them is not convincing proof of calmodulin involvement. Inhibition of exocytosis in mast cells, though, has been shown to occur at drug concentrations much lower than that required to produce general cell damage (180). Other inhibitory effects, consistent with interference with Ca^{2+} entry, were also observed. In hamster insulinoma cells, concentrations of phenothiazine that block glucose-stimulated, Ca^{2+}-dependent release do not reduce the glucagon-stimulated, cAMP-dependent exocytosis that is independent of extracellular Ca^{2+} (181). In this case, the predominant effect of phenothiazines could be on Ca^{2+} entry rather than on exocytosis, suggesting caution in the interpretation of phenothiazine experiments (182).

Compelling evidence for the involvement of calmodulin in exocytosis has been obtained using sea urchin eggs. Using a preparation of cortical coverslips, Steinhardt & Alderton (183) demonstrated that anticalmodulin antibodies prevented Ca^{2+}-dependent fusion of cortical granules to the plasma membrane. The inhibition could be reversed with excess calmodulin. Extensive endogenous calmodulin was detected on the plasma membrane with immunofluorescence. These experiments strongly suggest that calmodulin will be a general mediator of Ca^{2+}-dependent exocytosis.

Direct evidence, however, for the involvement of calmodulin in exocytosis at the nerve terminal is not nearly as compelling. Calmodulin makes up 0.7% of the protein in isolated nerve terminal preparations (184), certainly a sufficient concentration to play a role in exocytosis. A preparation of brain synaptic vesicles also contains calmodulin that can be removed by washing in chelating agents (184, 185). Since the synaptic vesicle preparation contained coated vesicles (186), which also bind calmodulin (e.g. 187) it will be important to verify that the calmodulin is truly bound to the synaptic vesicles. There are preliminary reports that calmodulin binds to purified electric organ synaptic vesicles (188), chromaffin granules (189, 190) and platelet alpha granules (191). Both Ca^{2+}-independent and Ca^{2+}-dependent binding sites with ~ 30-nM binding constants have been described on the cytosolic surface of chromaffin granule membranes (191a). The former sites appear to represent binding to membrane proteins of M_r 25,000 and 23,000. Binding to additional membrane proteins of M_r 69,000 and 50,000 is seen in the presence of 1 μM Ca^{2+}. Cytosolic proteins of M_r 70,000, 36,000, 34,000, and 32,000 are recruited to the membrane by Ca^{2+}-calmodulin. The latter two proteins appear to be closely related to the clathrin-associated light chains in coated vesicles. Thus, there is no shortage of calmodulin target proteins that could potentially mediate exocytosis. Evidence that the nerve terminal or vesicle-associated calmodulin is involved in exocytosis is scarce, however. While brain synaptic vesicle preparations have been reported to aggregate and lose half their norepinephrine content in the presence of Ca^{2+} and calmodulin (184, 185), this is not compelling evidence that calmodulin is involved in exocytosis at the nerve terminal. Calmodulin clearly regulates additional synaptic vesicle functions not related to exocytosis. In vesicles from electric organ, for example, it regulates activity of the Ca^{2+}-transporter by activating an endogenous protein kinase (147). Thus, it is not clear which, if any, of the calmodulin binding proteins in the chromaffin granule membrane are required for exocytosis.

Another candidate for the calcium-binding site is synexin, a 47-kd soluble protein found in secretory tissues but also liver cells (192, 193). In the presence of Ca^{2+} but not Sr^{2+} or Ba^{2+}, synexin aggregates into rod-like structures (192). Synexin also induces aggregation of chromaffin granules in the presence of Ca^{2+}. This aggregation is blocked by trifluoperazine (193). Synexin may not bind a protein receptor since it enhances the Ca^{2+}-dependent fusion of pure phospholipid vesicles (194). Since synexin is not unique to secretory tissue, works on phospholipid bilayers, causes vesicle-vesicle interaction instead of vesicle-plasma membrane interactions and does not function in the presence of Sr^{2+}, synexin should not be attributed a role in Ca^{2+}-triggered exocytosis until more definitive evidence is available. An alternative hypothesis is that any protein, including calmodulin, that exposes hydrophobic domains on binding Ca^{2+} might either self-polymerize or cross link membranes. These might represent nonspecific analogs of the correct in vivo process.

Mechanism of Vesicle Fusion

In the only case of membrane fusion that is understood, the fusion of a membrane virus attached to the plasma membrane at low pH, one of the viral proteins undergoes a pH-induced change in conformation that exposes a new hydrophobic domain (195), resulting in fusion of the viral and host plasma membranes. A similar Ca^{2+} or calci-calmodulin induced hydrophobic domain in a plasma membrane or synaptic vesicle protein could result in fusion of synaptic vesicles and exocytosis. Different groups have noticed correlations between secretion and phospholipase activation (e.g. 196), phosphorylation (e.g. 197), or methylation (198). On this basis, it has been proposed that one or more enzymatic mechanisms, activated by Ca^{2+}, might be required for exocytosis. The kinetics of release make it impossible for extensive catalysis to be required. For example, phospholipases with conventional turnover numbers could catalyze at most one hydrolytic event per enzyme in the time available (199). Therefore, phosphorylation or phospholipid hydrolysis during transmitter release seem more likely a priori to be involved in mobilization of vesicles to their release sites or in regulatory controls than to be primary mechanisms of exocytosis.

The requirements for exocytosis can be investigated in more detail in preparations of permeabilized cells or cortical granule surfaces that are readily accessible to small molecules. A requirement for Mg^{2+}-ATP has consistently been found (167, 168, 183). The requirement for ATP could not be replaced by S-adenosyl methionine, so there is no evidence for a requirement for methylation (168). The requirement for ATP is consistent with a requirement for phosphorylation, but there are many other explanations including mobilization of secretory granules to the active zone. Detailed analysis of the mechanism of fusion will require identification and characterization of the Ca^{2+}-dependent fusogen.

REGULATION OF NEUROTRANSMITTER LEVELS IN THE NERVE TERMINAL

Metabolism of Peptide and Classical Transmitters

Neurons have evolved several specific mechanisms to maintain neurotransmitter levels in the nerve terminal. The mechanisms employed to regulate the levels of peptide neurotransmitters, which are derived from proteins, are fundamentally different than those for regulating synthesis of classical neurotransmitters, such as catecholamines or acetylcholine, which are derived from metabolites transported into the cell.

To consider the regulation of neuropeptide levels first, the unusual geometry of the neuron restricts protein synthesis to the cell body, so all proteins and peptides have to be transported from there to the nerve terminal (e.g. 200). Recent work has resulted in the cloning of the genes coding for several peptide precursors and identification of the important steps in peptide processing (e.g. 201–206). The enkephalins, corticotropin, and angiotensin are all synthesized as segments of large proteins that appear to be synthesized on membrane-bound ribosomes, glycosylated in the Golgi apparatus, and transported to the nerve terminal in vesicles or other membrane-enclosed compartments. In each case, the precursor protein contains more than one peptide. Neuropeptides are always separated by a pair of dibasic amino acids (lys-lys or lys-arg). Processing of the angiotensin and vasopressin precursors occurs in membrane vesicles during axonal transport (207). Processing occurs by sequential action of a trypsin-like enzyme followed by a carboxypeptidase B-like activity (208). Enzymes in secretory granules that appear to be responsible for each of these steps have been identified and partially purified (209–211). There is evidence suggesting that the enzymes responsible for processing different precursors may differ slightly from each other (reviewed in 200). The corticotropin precursor is processed into different peptides in different cell types (212). This may prove to be true for other precursors also, such as the enkephalin precursor (e.g. 213). Once released by exocytosis, peptides appear to be degraded, not recycled. Characterization of the proteases responsible for peptide degradation is an active, but nascent field (reviewed in 214). Future studies on regulation of peptide stores in response to different rates of exocytosis will almost certainly focus on gene expression and precursor processing.

In contrast, much of the regulation of metabolite-derived neurotransmitter levels occurs in the nerve terminal (e.g. 215). As the transporters and enzymes required for synthesis of norepinephrine and acetylcholine, for example, are found in the nerve terminal, very tight coupling between the rates of transmitter release and replacement are possible. Acetylcholine synthesis is regulated by changes in activity of the choline transporter, which limits precursor availability (215a, 215b). Catecholamine biosynthesis is regulated by changes in tyrosine hydroxylase, the rate-limiting enzyme. The activity of tyrosine hydroxylase is increased within minutes in active nerve terminals by reduced end product catecholamine-mediated inhibition (215) and increased protein kinase activity (e.g. 215c). In contrast to peptides, classical transmitters in granules are in equilibrium with a cytoplasmic pool (e.g. 216). Increased release, therefore, can rapidly affect the cytoplasmic transmitter pool. Ca^{2+} and cAMP increases in response to stimulation also activate tyrosine hydroxylase by protein kinase action (216a, 217, 218). Over hours, high rates of stimulation can result in increased enzyme synthesis in the cell body (219). In terminals containing classic neurotransmitters, vesicle membrane is retrieved after exocytosis and used to form new vesicles without leaving the nerve terminal (220, 221). The terminal plasma membrane contains a Na^+-dependent cotransporter that functions to recover a significant fraction of the released neurotransmitter or its catabolites (222). Proton-driven antiporters are localized in the vesicle membrane where they function to concentrate neurotransmitter stores. Interference with either uptake or new synthesis of transmitter prevents the maintenance of adequate transmitter stores (223, 224).

Storage of Classical Transmitters

The most thoroughly characterized vesicle is the chromaffin granule, which is packed with catecholamines (0.6 M), ATP (0.13 M), protein (0.2 g/ml) and proteoglycans (reviewed in 225). An ideal solution of these components would be hypersomotic and unstable. Direct osmometric measurements, however, have recently shown that a mixture of ATP and catecholamines at these concentrations form a nonideal solution with a reduced osmolality (226). There is no evidence for formation of crystals, however, since NMR measurements have shown that ATP and catecholamines tumble as rapidly in chromaffin granules as in free solution (227).

Catecholamine and ATP uptake into isolated chromaffin granules is driven by pH and membrane potential gradients that are established by an ATP-dependent proton translocase. Isolated chromaffin granules are acidic (pH 5.2), whether measured by methylamine redistribution or protonation of the gamma phosphate of ATP (228, 229). Isolated granules in the absence of ATP have a negative internal charge that reflects a diffusion potential induced by the proton gradient (228). With addition of external ATP, though, an ATP-dependent proton translocase transports positive charge into the granule, which becomes positively charged compared to the outside. The voltage buildup is blocked by proton uncouplers. The translocase can be solubilized by bile salts and reconstituted in phospholipid vesicles (230).

Most experiments suggests that the mitochondrial and granule ATP-dependent proton translocases are closely related, but not identical. Both are sensitive to a similar array of inhibitors (231). Both translocases contain a similar proteolipid, which is the target of DCCD, a proton channel blocker. The proteolipid target of DCCD in chromaffin granules, however, is slightly smaller than in mitochondria (232, 233). Chromaffin granule preparations contain an ATPase that appears to be almost identical to the

F_1-ATPase of mitochondria in size, peptide fragments, and antigenicity of its subunits (234) and an F_1-ATPase-like component with a characteristic lollipop morphology has been visualized in negatively stained chromaffin granules (235). Antibodies to the mitochondrial ATPase block the activity of the granule ATPase and the uptake of monoamines into these granules (234). Although the possibility of contamination by mitochondria or by adsorbed F_1-ATPase has not been completely eliminated, the weight of the evidence favors a close similarity between mitochondrial and secretory granule ATPases.

Chromaffin granule ghosts, depleted of internal triphosphates and proteoglycans, have been prepared and used to characterize the catecholamine and ATP transporters. In the presence of ATP, these ghosts generate transmembrane voltage and pH gradients and accumulate catecholamines by a reaction that is sensitive to reserpine, the well-characterized inhibitor of normal catecholamine accumulation in storage granules (e.g. 148). If a pH gradient is produced artificially by suspending the ghosts in a buffer at a higher pH, catecholamine accumulation is also seen. When a transmembrane, interior positive, potential gradient is generated by suspending ghosts filled with NaCl in an KCl solution and adding an ionophore specific for K^+, catecholamine accumulation also is stimulated. Accumulation of catecholamines in response to either gradient is sensitive to reserpine. These results show that the uptake of catecholamines is promoted by both gradients, but does not depend directly on ATP hydrolysis. Similar conclusions have been reached in studies in which the ATP-induced pH gradient or voltage gradient was selectively discharged with NH_4^+ or thiocyanate, respectively (237–242). The responsiveness of catecholamine transport to membrane potential means that the transporter moves charged, not electroneutral species of catecholamines. Its responsiveness to a pH gradient means that it is a proton-driven antiporter. Measurement of the pH-generated driving force suggests that two protons are expelled for each catecholamine. The antiporter binds reserpine and has been solubilized and reconstituted into liposomes (243). ATP uptake, sensitive to atractylosides, has been seen in chromaffin granules. It has been reported to depend only on the transmembrane potential (244).

Accumulation of hyperosmotic concentrations of catecholamines must require other compounds such as ATP that reduce the osmolarity of the granule interior. At physiological cytoplasmic concentrations (20 μM) of catecholamines, the maximum concentration inside ghosts is only 20 mM (245), less than 10% of the concentration in normal granules. Additional accumulation may well require other vesicle constituents, such as ATP, proteins or proteoglycans (246).

Other vesicles appear to use the proton electrochemical gradient to accumulate small molecules. This has been clearly demonstrated in platelet and mast cell granules. Catecholamine uptake into brain synaptic vesicle fractions is stimulated by ATP and inhibited by most of the same drugs that prevent uptake into chromaffin granules (247). Cholinergic vesicles, purified from electric organ, have an internal pH of ~ 5.5 and contain ~ 0.15 M ATP in free solution (248–251). The purified vesicles contain an ATPase that shows the same drug sensitivity as the ATPase in chromaffin granule ghosts, sensitivity to DCCD and resistance to oligomycin or efrapeptin (252, 253). In the presence of Mg^{2+}-ATP, acetylcholine can be concentrated up to tenfold in these vesicles (254). Bicarbonate converts the ATPase to a high affinity, high velocity form and is required to see significant storage in vitro (253). A similar requirement is seen in cholinergic neurons in vivo (224). Uptake in vitro is blocked by uncouplers of the proton transporter (254), but comparatively little evidence has been obtained for uptake in the absence of ATP (see 255). This may reflect the small size of these vesicles. Alternatively, studies on intact electric organ have suggested that acetylcholine and ATP uptake is restricted to a subpopulation of vesicles with a size and density that differ from those of most vesicles (256–258). If only a small fraction of cholinergic vesicles have a functional uptake system, that could account for many of the difficulties encountered.

Peptide-containing granules can also generate a proton electrochemical gradient. The contents of several granule preparations containing different peptides are also acidic on isolation. Addition of ATP results in generation of an inside-positive membrane potential, which is inhibited by DCCD, but not oligomycin (259–261). An anion-stimulated ATPase with properties similar to the cholinergic ATPase has been identified in these granules (261a). Proton translocases in these granules have several probable functions. An acidic interior may be required for segregation of lysosomal and hormone proteins from other proteins of the Golgi lumen (262, 263) or activity of the proteases that process the peptide precursors (e.g. 210) may require a low pH. A low pH may also promote the formation of peptide aggregates and, of course, permits vesicles with appropriate transporters to accumulate ATP and classical neurotransmitters. A substantial proportion of peptidergic granules contain classic transmitters. The most prominent example is actually the chromaffin granule, which stores both adrenalin and the enkephalins (264).

MOVEMENTS OF MEMBRANES AND PROTEINS IN THE NEURON

Membrane traffic in the nerve terminal has several components. Components of synaptic vesicles are moved to active zones, fused to the plasma membrane by exocytosis, retrieved by endocytosis, and recycled within the nerve terminal. The terminal is the primary site of insertion of plasma membrane components. It is also the site at which receptor-hormone complexes and other material begin transport back to the cell soma.

Mobilization of secretory vesicles can be divided into transport down the axon and further targeting to the active zone. While the two processes are believed to be similar, most experimental evidence derives from studies of movement in the axon. Transport is measured by pulse labeling of newly synthesized proteins and lipids, accumulation of material at ligatures or cold blocks, and direct visualization of moving particles in the light microscope. Membrane components are transported at exceedingly rapid speeds by mechanisms dependent on local energy. The fastest component, traveling at speeds greater than 250 mm/day includes identified plasma membrane components, such as the Na^+K^+-ATPase (265). Mitochondria and at least some components of synaptic vesicles move at somewhat slower rates (265–267). Separate pathways for export of plasma membrane and secretory granule components are seen in nonneuronal peptidergic cells (268), and may be related to some of the different transport components seen in axons. All proteins transported by fast transport appear to be associated with membranous organelles, derived in most cases from the Golgi (268a), and the majority appear to be targeted for the nerve terminal, where many have very fast turnover times (269, 270). A subclass of these proteins does, however, appear to be preferentially deposited along the axon en route (271). Axons also contain membranous elements that are returning to the cell body by retrograde transport, a process typically occurring at 40 mm/day and dependent on local energy sources. About 50% of the membrane protein that reaches the nerve terminal is eventually returned to the cell body (272). Receptors and synaptic vesicle proteins have been identified in this fraction.

Cytosolic proteins not associated with membranous elements are moved to the nerve terminal by a much slower and undirectional process that does not depend on energy in the axon, but does require metabolism in the neuronal cell body. Slow component b moves at 3–6 mm/day and includes actin (273), the spectrin-like protein fodrin (273a), clathrin (274), and several metabolic enzymes (275). It has been pointed out that actin and fodrin move together at a similar speed during lymphocyte capping (275a). An even slower system, slow component a, moves at 0.7–1.1 mm/day. Unlike the other systems, the majority of the proteins in this class have been identified. They are the α-and β-tubulins, the neurofilament proteins (276), and the tau-like microtubule associated proteins (277). The nerve terminal must contain enzymes to degrade proteins transported by slow transport, since movement is unidirectional and the vast majority of these proteins actually reach the terminal region (278).

Spherical vesicles and small membranous tubules appear to contain the material that is moved by fast transport. Neurotransmitters in spherical vesicles and small tubules accumulate at ligatures or cold blocks (279–283). A third membrane system does not participate in fast transport, but is prominent in the axon. An extensively anastomosing network of smooth endoplasmic reticulum is seen in the subaxollemmal regions of neurons, but shows no sign of accumulation on the proximal side or depletion on the distal side of a cold block (283). These experiments make it unlikely that the reticular network participates in fast transport or gives rise to the tubules and vesicles that do accumulate.

The axon contains three major cytoskeletal structures—microtubules, actin filaments and neurofilaments—which could a priori mediate fast and retrograde transport. Morphological evidence favors the involvement of microtubules. Membranous organelles are associated with fascicles of microtubules in axons (284, 285, 285a) and accumulate at ligatures or cold blocks in association with them (281, 282). Movement of membrane particles in vitro, examined by new high resolution microscopy, also follows filaments believed to be microtubule bundles (286). Drugs, such as colchicine, which depolymerize microtubules block fast transport (287), and taxol, which prevents microtubule depolymerization, protect fast transport from colchicine (288).

Some experiments suggest that microtubule-associated dynein may provide the force for fast transport. Low concentrations of erythro-9[3-(2-hydroxy-nonyl)]-adenine, an inhibitor of dynein-dependent motility, block fast transport (288–290). This drug, however, also interferes with other ATPases and thus does not provide unambiguous evidence for the involvement of dynein.

Pharmacological evidence also suggests that actin is involved in fast transport. Injection of DNase I and other actin-depolymerization agents has been observed to block fast transport (290–292). Two agents that should interfere with the function of actin filaments without causing extensive depolymerization, dihydrocytocholasin B and the N-ethylmaleimide-modified S1 fragment of myosin, do not block fast transport (290). These results suggest that actin may play a structural role in the cytoplasm rather than a role in transport force generation (290). To summarize, though, it is not yet clear which proteins are responsible for generating the force used to move organelles by fast transport.

Possible clues to the mechanism of polarized movement might be found in the tail-like appendages on membranous organelles reported to be associated with the filamentous network (285), or the higher concentration of cross-bridges found at the leading edge of membranous organelles (283). There is ample support biochemically for direct interactions between secretory vesicles and actin (293–295) and microtubules (296). The latter interactions may involve microtubules associated proteins (297).

One proposed explanation for the range of transport rates is that the transport vectors might transiently be associated with the linear axonal motor. The rate of the motor could be constant and the transport rate regulated by the fraction of time the transport vector associated with the motor (298, 299). An appealing feature of this model is its ability to explain the saltatory movement of axonal particles seen in the light microscope. Recent improvements in visualizing axonal particles have made it clear that while large membranous organelles may be moving irregularly, small organelles, presumably elements of the tubulo-vesicular system move at fast speeds (3–5 μm/sec) in a continuous fashion along linear elements (286, 300). In addition the slower more intermittent movements of large membranous organelles seem to be not as much saltatory as an "elastic recoil" (286). Thus when particles stop they do not resume at the same speed but at a faster one until they catch up with where they would have been had they not stopped. Such movement suggests stops might come about by transient blocking of movement.

Using organelle movement as an assay of transport, one can remove the permeability barrier of the plasmalemma and examine directly the effects of membrane-impermeable inhibitors on axoplasmic transport. Permeabilization can be brought about by electric shock (289), detergent (288), or extrusion of the axoplasm from a giant squid axon (301). Not surprisingly, ATP is absolutely required for transport and GTP will not substitute (289). Unexpected however is the observation that the movement of large particles, presumably mitochondria, is blocked by 2,4-dinitrophenol even in the presence of ATP (301). Vandate ions at concentrations that inhibit mitosis and ciliary beating also block transport (288, 289). An observation difficult to reconcile with earlier observations (302) is that transport occurs in EGTA buffers in which the free Ca^{2+} ion concentration is varied between 10^{-8} M and 5×10^{-4} (303). The discrepancy might arise because different measures of axonal transport were used, or because Ca^{2+} leaks from intracellular organelles in permeabilized tissues. Further experiments should clarify this point and also indicate which proteins are responsible for generating the force used in fast transport.

We look forward to the extension of these studies to vesicle transport in the nerve terminal. At present, we know that transmitter release is blocked ten min after injection of DNase I into the cell body of *Aplysia* neuron L10 (304). We also know that release in permeabilized adrenal medullary cells is unaffected by vanadate, or by agents that disrupt microtubules or microfilaments (168). Much more detailed information using permeabilized systems, immunoelectron microscopy, and well-characterized inhibitors is clearly needed.

MODIFICATIONS IN NERVE TERMINAL METABOLISM INDUCED BY ACTION POTENTIALS AND NEUROTRANSMITTERS

The activity of the nerve terminal is regulated by two exogenous influences—changes in membrane potential communicated via the axon and direct interactions of neurotransmitters with receptors in the terminal plasma membrane. As discussed earlier, depolarization of the nerve terminal increases Ca^{2+} flow into the terminal. Cytoplasmic Ca^{2+} binds calmodulin and other Ca^{2+}-binding proteins, which directly activate several enzymes and indirectly activate many more through the action of Ca^{2+}-dependent protein kinases.

The binding of agonists to receptors on the nerve terminal membrane can result in changes of membrane potential or activation of second messenger systems (305). Large numbers of receptors, specific for a variety of transmitters, have been discovered in preparations of nerve terminals or neuronal cell lines. Activation of different receptors results in specific changes in cAMP and cGMP levels (306). cAMP and cAMP-dependent protein kinases are found throughout the nervous system (307), and hence are believed to be important modulators of neuronal function. The effects of cAMP appear to be mediated largely and perhaps exclusively through cAMP-dependent protein kinases, which alter the activity of many different enzymes and transporters within the nerve terminal (308).

Current evidence suggests that cGMP is important in regulating the metabolism of a small percentage of the cells in the nervous system. cGMP is more specifically localized than cAMP. cGMP is 10–50-fold more concentrated in the cerebellum than other brain regions and the high levels of cGMP within the cerebellum are found in one cell type, the Purkinje cells (307). cGMP-dependent protein kinase is also found primarily in cerebellar Purkinje cells (309, 310).

Changes in Ca^{2+} or cyclic nucleotide levels have several consequences on nerve terminal metabolism. Cytoplasmic Ca^{2+} stimulates not only exocytosis, but also glycogenolysis (311, 312), mitochondrial respiration (311, 312), endocytosis (313) and neurotransmitter synthesis (314, 315). These changes are basically homeostatic in nature, restoring depleted levels of ATP, neurotransmitters and synaptic vesicles. The actions of elevated cAMP are as pleiotropic as those of Ca^{2+}, but not so simple to summarise. cAMP also regulates energy metabolism (316) and can potentiate neurotransmitter synthesis (315) and release (317). In addition, cAMP exerts actions that lead to long-term changes in synaptic efficiency, the basis of phenomena such as habituation and sensitization (44). Mutations in *Drosophila* that alter the enzymes regulating cAMP synthesis or degradation result in pleiotropic defects in learning, habituation, and sensitization (317a–320).

Regulation by cAMP

BRAIN ADENYLATE CYCLASE   Cellular cAMP levels are determined by the activity of adenylate cyclase, the enzyme responsible for its synthesis, and phosphodiesterase, the enzyme that hydrolyzes it. Adenylate cyclase (321) is associated with the cell membrane in neurons and other cells. Stimulation occurs via receptor-mediated formation of an enzyme complex that requires a GTP/GDP-sensitive coupling protein. Synthesis of cAMP requires GTP binding and stops when GTP is hydrolyzed to GDP at the regulatory site. A plethora of neuronal transmitters, most notably dopamine, activates adenylate cyclase probably by promoting an interaction between their receptors and the GTP/GDP-sensitive regulator protein (322). Other transmitters inhibit adenylate cyclase. Some may act by receptor-mediated binding to a distinct GTP/GDP-sensitive coupling protein that inhibits adenylate cyclase. There is suggestive, but not conclusive, evidence for inhibitory GTP-sensitive regulatory proteins (322a). Opiates act by receptor-mediated stimulation of GTP hydrolysis (322b, 322c). It has been suggested that they reduce cAMP synthesis by reducing GTP occupancy at the regulatory site in the stimulatory coupling protein (322c).

Neural tissues contain an additional form of adenylate cyclase. It is regulated by both the classical GTP/GDP-sensitive regulatory subunit and by Ca^{2+}-calmodulin (323). Recently, Ca^{2+}-calmodulin has been shown to stabilize and activate fourfold the isolated catalytic subunit of this form of the enzyme (324, 325). Addition of the GTP/GDP-sensitive regulatory subunit results in further, additive activation of cAMP synthesis. No evidence for Ca^{2+}-calmodulin binding to this second subunit was found. The cell types that contain this Ca^{2+}-calmodulin form of adenylate cyclase are not known. The results suggest, though, that Ca^{2+} could potentiate the effect of neurotransmitters on cAMP synthesis in those cells that contain this form of the cyclase.

Studies on the time course of sensitization of the gill-withdrawal reflex in *Aplysia*, a process mediated by serotonin-induced cAMP synthesis and cAMP-dependent phosphorylation of a K^+ channel in sensory neurons (44), have shown that the memory for sensitization resides in a persistent elevation of cAMP (326) and not in the slow reversal of a later step in the cAMP-induced sequence of events that results in phosphorylation and closure of the K^+ channels (327). These studies suggest that some neurons may contain forms of adenylate cyclase that are inactivated only slowly after the removal of neurotransmitter. Clearly biochemical studies on the novel forms of adenylate cyclase in nervous tissues will be crucial to understanding the molecular basis of synaptic plasticity.

CAMP-DEPENDENT PROTEIN KINASES   The brain contains the highest levels of cAMP-dependent protein kinases found anywhere in the body, suggesting their importance in neuronal and synaptic function (328). In all systems examined biochemically, the cAMP-dependent kinase consists of a single catalytic (C) subunit that is complexed to one of two classes of regulatory subunit, R_I and R_{II}. The R_I and R_{II} subunits can be distinguished from each other on the basis of size and binding affinities for cAMP and cIMP. In the absence of cAMP, the kinase is a tetramer, consisting of two R and two C subunits. Binding of cAMP to the regulatory unit results in dissociation and activation of the catalytic unit, which phosphorylates protein substrates.

If there is only one protein kinase catalytic unit in brain it is clear that regulation of which protein gets phosphorylated must lie elsewhere. As we shall describe in this section, there are many forms of regulatory subunit, varying in their composition and location in the cell, yet all bind the catalytic subunit. It is therefore a reasonable conjecture that specificity arises because regulatory subunits associate with the correct substrate and thus concentrate the kinase in the vicinity of its target.

Brain cAMP-dependent protein kinases appear to be similar to those in other tissues of the body, but there do appear to be some differences in the regulatory subunits. Antibodies to the R_I and R_{II} subunits, purified from bovine lung and heart, respectively, show weak cross-reactivity with the R_I and R_{II} subunits from bovine brain (329). As assayed by binding to photoactivable $8-N_3$-cAMP, vertebrate brain has significant quantities of both R_I and R_{II} subunits and no significant amounts of other cAMP-binding proteins (330), supporting the proposal that actions of cAMP in brain are mediated exclusively through activation of protein kinases (308). In contrast to other tissues, though, a fraction of the brain R_{II} class cAMP-dependent protein kinase interacts with calmodulin in the presence, but not absence of Ca^{2+} (331, 325). Formation of complexes is promoted by calcineurin (which appears to be protein phosphatase 2B). The Ca^{2+}-calmodulin complex can bind to brain, but not to heart R_{II} subunits. The effect of Ca^{2+}-calmodulin is to inhibit the basal activity of the kinase and lower the enzyme's affinity for cAMP (325). The brain and heart kinases with type II regulatory subunits also have antigenic differences (332). Brain enzyme is poorly bound by antibodies to the heart enzyme.

Labeling cAMP-binding proteins with the photoactivable compound, 8-N_3-cAMP, and subcellular fractionations of these binding proteins also provide evidence for unusual heterogeneity in cAMP-dependent protein kinase regulatory subunits in neuronal tissues. Brain R_{II} consists of a family of molecules with slightly different pI's and, in one case, a slightly different M_r (333). The R_I subunit also is a family with members at more than one pI. Several cAMP-binding proteins, most of which appear to be related to R_I and R_{II}, are also found in *Aplysia* sensory neurons (334). Vertebrate brain and Aplysia sensory neurons also contain both membrane and soluble forms of these subunits (334, 338). Membrane and soluble forms of the R_I and R_{II} subunits have been detected in nonneural tissues, but the brain appears to be the only tissue with both forms of the same subunit. Nonneuronal tissues in vertebrates and *Aplysia* do not have the same heterogeneity in pI or in membrane association. The multiplicity of forms in vertebrate brain could be attributed to the multiplicity of neuronal and supporting cell types in the central nervous system. The results in *Aplysia*, though, demonstrate clearly that single neurons have multiple forms of cAMP-binding proteins distinguished by subcellular distribution and pI.

Recent studies on vertebrate brain tissue have suggested that the cAMP kinase is concentrated in specific positions in the cell by proteins that bind the regulatory subunit. First, analysis of cytosolic and membrane-bound forms of R_{II} subunit indicate that they have the same M_r, pI, and other properties (332). Secondly, experiments in vitro and in vivo have shown that one protein, MAP 2, has a specific binding site for the R_{II} subunit (335, 336). Microtubule-associated protein 2 (MAP 2) consists of two domains, one of which binds the microtubule surface and the other of which appears on a large projection and may mediate interactions with other filaments or oganelles. In neurons, MAP 2 is found on dendritic, but not axonal microtubules (337). The projection portion of MAP 2 binds an R_{II} form of cAMP-dependent protein kinase (335). Exogenous bovine heart R_{II} binds specifically to dendrites in frozen sections of rat brain and the binding is prevented by exogenous MAP 2 (336). It is not clear how many forms of brain R_{II} subunit are bound to MAP 2, but MAP 2 clearly binds the heart form of the enzyme. MAP 2 contains substrates in both of its domains for cAMP-dependent protein kinases. Localized binding may assure preferential phosphorylation in response to cAMP. Phosphorylation would be expected a priori to modify association of MAP 2 with microtubules and other organelles. These experiments suggest that cAMP-dependent protein kinases are localized to specific positions by proteins that bind the regulatory subunit, and raise the possibility that the heterogeneity in regulatory subunits controls the site of kinase binding, thus conferring substrate specificity on a nonspecific catalytic subunit.

Despite the possible heterogeneity in regulatory subunits, suggested by binding to calmodulin and MAP 2, a single catalytic subunit appears to function as the actual kinase in all cell types. No heterogeneity in catalytic subunit has been seen in neural or non-neural tissues (328). Injected vertebrate and endogenous invertebrate catalytic subunits phosphorylate the same spectrum of proteins (338), exert the same influence on synaptic transmission (47, 339), and are sensitive to the same protein inhibitor, the Walsh inhibitor (46, 326, 327).

Regulation by Calcium

SUMMARY OF Ca^{2+} ACTIONS   Changes in cytoplasmic Ca^{2+} have multiple consequences on synaptic function. As discussed earlier, cytoplasmic Ca^{2+} stimulates both exocytosis and endocytosis (156, 164). The sensitivity of each process to calmodulin inhibitors suggests that calmodulin may be the mediator of Ca^{2+} action in each case (168, 340). Ca^{2+}-calmodulin also stimulates activity of several neuronal Ca^{2+} pumps, which would help reduce cytoplasmic Ca^{2+} levels. Ca^{2+}-calmodulin binds directly to the 140-kd ATP-dependent Ca^{2+} transporter (151), which is almost certainly the plasma membrane Ca^{2+}-dependent ATPase (137). Calmodulin also activates a protein kinase that in turn increases activity of a Ca^{2+} transporter in synaptic vesicles (147). Nerve terminal glycogenolysis and respiration are increased in response to activity by a Ca^{2+}-dependent action (311, 312). Increased glycogenolysis is probably mediated in part by Ca^{2+}-calmodulin activation of glycogen phosphorylase b kinase, which activates glycogen phosphorylase (316). The kinase has been detected in brain, but not yet shown to be localized in nerve terminals (341, 342). Cytoplasmic Ca^{2+} also stimulates the synthesis of several neurotransmitters by increasing the activity of rate-limiting transporters or enzymes. Thus, choline transport in cholinergic neurons, tyrosine hydroxylase in catecholaminergic neurons and tryptophan hydroxylase in serotonergic neurons are activated by cytoplasmic Ca^{2+}, the latter two as a result of phosphorylation by Ca^{2+}-calmodulin-sensitive protein kinases. These effects will be discussed in more detail in the next section. These actions are basically homeostatic in nature, restoring cytoplasmic Ca^{2+}, ATP, and neurotransmitters to original levels. Ca^{2+} also has effects on neurons that are less easy to understand. Ca^{2+}-calmodulin interacts with the cyclic AMP pathway at several steps, activating one form of adenylate cyclase, inhibiting the basal activity of a type II kinase, and activating a form of phosphodiesterase with a comparatively high K_m for cyclic nucleotides. These reactions are mediated by Ca^{2+}-dependent binding of calmodulin to the catalytic subunit of cyclase, R_{II} subunit of kinase, and catalytic subunit of phosphodiesterase, respectively (323, 331, 343). Cytoplasmic Ca^{2+} also reduces the activity of certain Ca^{2+} channels and activates Ca^{2+}-dependent K^+ channels, resulting in reductions or increases in synaptic efficiency. Long-term potentiation of synaptic transmission in the hippocampus requires Ca^{2+} and is blocked by trifluoperazine (344), evidence that Ca^{2+} is required for some types of synaptic plastic changes.

Ca^{2+}-calmodulin binds to fodrin, a neuronal cytoskeletal protein that is very similar to spectrin (345), regulates the assembly and disassembly of brain microtubules (346), and controls actin myosin interactions in nonneural cells by activation of a myosin light-chain kinase (347). The role of these activities in short-term synaptic function is not clear. They may be important in generating the changes in synapse number and size that result from long-term facilitation or depression in *Aplysia* and in the mammalian hippocampus (44).

CALMODULIN CALCINEURIN INTERACTIONS   Calmodulin, the virtually ubiquitous Ca^{2+}-binding protein of eukaryotes, is present in high concentrations in nerve terminals (184) and appears to be the major, but not exclusive Ca^{2+}-binding protein and mediator of Ca^{2+} transduction in the synapse. Since it is the subject of frequent and thorough reviews (171, 172, 348, 349), we make no attempt to review its properties here.

The existence in brain of other Ca^{2+}-binding proteins, such as parvalbumin and calcineurin, has suggested the possibility that these proteins may modulate the kinetics of calmodulin activation. Bovine brain contains high levels of calcineurin, originally described as an inhibitor of Ca^{2+}-calmodulin activation of brain cyclic nucleotide phosphodiesterase (350). Calcineurin is predominantly found in the nervous system where it is present at approximately 1 μmole per kg. It contains two subunits of 61 kd and 15 kd, which are tightly linked to each other (351). The small subunit binds 4 moles of Ca^{2+} with submicromolar affinity, i.e. more tightly than calmodulin. The large subunit mediates Ca^{2+}-dependent formation of a calmodulin calcineurin complex. Calcineurin could potentially act as an effective calmodulin buffer—both by serving as a sink for Ca^{2+} and for Ca^{2+}-calmodulin complexes. Computer simulations of Ca^{2+} binding indicate that cytoplasmic Ca^{2+} will be bound to calmodulin for the first few msec after entry into the nerve terminal, but will be largely sequestered by calcineurin over the next 100 msec ($t_{1/2} = 30$ msec) (352). Calcineurin also binds to calmodulin more tightly than to phosphodiesterase, so it would tend to displace calmodulin from other targets. Recent work has shown that the subunit structure and enzymatic properties of calcineurin are identical to those of protein phosphatase 2B, a phosphatase with restricted substrate specificity that dephosphorylates the α-subunit of phosphorylase kinase (353). The enzyme requires Ca^{2+} or Mn^{2+} for activity and is further activated tenfold by Ca^{2+} calmodulin. Immunocytochemical studies in the light and electron microscopes show that it is primarily localized within neurons and is most prominent in dendrites and postsynaptic densities (354). It has been argued that the high concentration of calcineurin in neurons will allow an extremely fast on-off switch for Ca^{2+}-dependent biological processes (352). It now appears that calcineurin can interact with Ca^{2+}-calmodulin at several steps, reducing cytoplasmic Ca^{2+}, sequestering Ca^{2+}-calmodulin, and dephosphorylating proteins activated by Ca^{2+}-sensitive kinases. Further studies will be needed to elucidate the importance of this complex regulatory system.

Ca^{2+}-CALMODULIN-SENSITIVE PROTEIN KINASES   A number of the actions mediated by Ca^{2+} and calmodulin in the nervous system appear to reflect activation of Ca^{2+}-calmodulin-sensitive protein kinases with restricted substrate specificities (355). In contrast to cAMP-dependent protein kinases, which contain only a single catalytic subunit, Ca^{2+}-calmodulin kinases exist as several distinct species with specific targets whose diversity is only now beginning to be explored. These kinases phosphorylate a diverse spectrum of proteins, including some that are also substrates of cAMP-dependent kinases (356, 357).

Separation of a soluble extract of rat brain on Sepharose revealed three distinct peaks of Ca^{2+}-calmodulin-sensitive protein kinase activity with different substrate preferences (358). The kinases that phosphorylated glycogen phosphorylase, tryptophan hydroxylase, and myosin light chain most efficiently were detected at positions corresponding to molecular weights of 1,000,000, 500,000, and 100,000, respectively. Kennedy & Greengard (342) separated four distinct Ca^{2+}-sensitive kinases on DEAE. Two of these appeared to be myosin light-chain kinase and phosphorylase b kinase. The other two kinases were detected by their ability to phosphorylate protein I.

Phosphorylase kinase has been detected in brain, but not actually shown to reside in nerve terminals (341). Since nerve terminal activity increases glycogen breakdown, provided Ca^{2+} is present (311, 312), it seems likely that cytoplasmic Ca^{2+} accelerates glycogenolysis in the same manner in nerve as has been demonstrated in muscle by activating phosphorylase b kinase, which increases the phosphorylation of both glycogen phosphorylase and glycogen synthase (316). As a result, glycogen synthesis is reduced and glycogen breakdown is accelerated. The muscle enzyme contains four subunits of total M_r 1,300,000. One of the subunits is calmodulin, which remains bound without Ca^{2+} (359).

Brain myosin light-chain kinase has been purified and shows Ca^{2+}-dependent binding to calmodulin (331). The purified enzyme is a monomer of 130 kd. A recent study suggests that the brain enzyme may have a broader substrate specificity that myosin light-chain kinases from other sources (359a), but this needs to be confirmed. Myosin light-chain kinase activity has been detected in cultured astrocytes (360), so it is not certain that it is actually present in nerve terminals. It does not appear to phosphorylate efficiently nerve-cytosol proteins in vitro (360a).

One of the two protein I kinases is located in the cytosol and phosphorylates the same region of protein I as cAMP-dependent protein kinase. It is sensitive to low concentrations of trifluoperazine, but not to exogenous calmodulin. To explain these results, Kennedy & Greengard (342) have proposed that it contains calmodulin as a subunit, whether or not Ca^{2+} is present.

The second protein kinase phosphorylates two different sites in protein I and requires exogenous calmodulin to respond to Ca^{2+} (342, 361). This enzyme is found in both membranous and soluble fractions from brain homogenates. The particle enzyme can be solubilized in low ionic strength buffers, though, and once solubilized, does not appear to differ from the soluble enzyme by several criteria (361). Both the soluble and particulate enzymes have been purified about 200-fold using calmodulin affinity chromatography (361). The partially purified preparation is completely dependent on Ca^{2+} and calmodulin. Recently, purification of the enzyme has been completed (M. Kennedy, personal communication). The purified enzyme has a M_r of $\sim 600,000$ and contains both 50-kd and 60-kd subunits in approximately a 3:1 ratio. Both of these subunits are autophosphorylated by the kinase activity. Further studies are needed to determine whether such autophosphorylation is functionally significant. Elevation of cytoplasmic Ca^{2+} results in phosphorylation of protein I at the sites defined by both kinases (362), so there is strong evidence that both protein I kinases are in nerve terminals.

A Ca^{2+}- and calmodulin-dependent protein kinase that seems likely to be related to the calci-calmodulin-dependent protein I kinase has recently been purified 800-fold to homogeneity from rat brain cytosol, using myosin light chain as a substrate (360a). The purified enzyme contains multiple copies of a 49-kd subunit, has a total M_r of 500,000–600,000, and phosphorylates many synaptosomal proteins, including an 80-kd doublet that is probably protein I. This enzyme seems to be slightly smaller than the calci-calmodulin-dependent protein I kinase (M. Kennedy, personal communication) and may well be that enzyme stripped of its 60-kd subunits.

Further research is needed to determine how many additional kinases are required to phosphorylate the diversity of substrates that are detected in brain homogenates (356, 357). At this time, it is not clear which kinases mediate phosphorylation of other known substrates such as tyrosine hydroxylase (358) or the Ca^{2+} transporter in Torpedo synaptic vesicles (147). In the future antibodies to each of the purified kinases should demonstrate the cellular and subcellular localization of these kinases in the nervous system.

Ca^{2+}, PHOSPHOLIPID, AND DIACYLCLYCEROL-ACTIVATED PROTEIN KINASE During the past few years, a new class of kinase has been characterized in many tissues, which has an absolute requirement for Ca^{2+} and phospholipid, but is not activated by calmodulin (363–365). The enzyme is found in higher concentrations in brain than other tissues, but appears to be a virtually ubiquitous enzyme (363, 366). It has been detected in virtually every tissue and species, vertebrate or invertebrate, in which it has been assayed (366). The spectra of proteins phosphorylated by the endogenous kinase in heart, brain, and other tissues are distinct from those phosphorylated by cAMP, cGMP, and Ca^{2+}-calmodulin-dependent kinases (356, 357, 364, 365, 367, 368, 371).

The Ca^{2+}, phospholipid-dependent protein kinase is found mostly in the cytosol (367, 372) and requires unphysiological concentrations of Ca^{2+} for activity. Addition of diacylglycerol, however, reduces the binding constants for both phospholipid and Ca^{2+}, the latter to one μM (373). These experiments suggest that the enzyme is activated by phosphatidylinositol turnover. In the presence of diacylglycerol, the enzyme becomes associated with membranes and phosphorylates endogenous substrates (374). The binding of many different hormones and neurotransmitters to specific classes of receptors results in increased phosphatidylinositol turnover (375), so the enzyme may be an important mediator of hormone and transmitter action.

Studies on platelets have demonstrated that physiological mediators of platelet activation modulate the activity of this kinase in vivo. Thrombin activates both phosphatidylinositol turnover and phosphorylation of a 40-kd protein that is a specific substrate of this kinase in vitro (376, 377). Parallel stimulation of diacylglycerol synthesis and phosphorylation of the 40-kd protein are seen in response to thrombin and phospholipase C. Parallel reductions in each occur in response to PGE_1, and cGMP, respectively. Tumor-promoting phorbol esters replace diacylglycerol in vitro and induce phosphorylation of the 40-kd protein in vivo without increasing diacylglycerol synthesis (378). The kinase is thus likely to mediate many of the effects on cell metabolism of this important class of regulators.

Recently, Ca^{2+}-influx into synaptosomes, induced by depolarization, has been shown to promote phosphorylation of an 87-kd neuronal substrate of this kinase (379). Pursuit of these studies on endogenous substrates should reveal the physiological conditions required for activation of this enzyme in different nerve terminals.

The Ca^{2+}, phospholipid-dependent protein kinase has been purified 15,000 times from bovine heart to near homogeneity and consists of one subunit with an M_r of 83,000–100,000, depending on the procedure used for determination of molecular weight (364). The purified enzyme transfers phosphate from ATP to serine residues and has an absolute requirement for Ca^{2+} and phospholipid. Addition of diacylglycerol reduces the K_m for Ca^{2+} sevenfold and the K_m for phosphatidylserine fourfold. The purified enzyme's spectrum of phosphorylation sites is distinct from those of Ca^{2+}-calmodulin and cAMP-dependent kinases. The size and properties of partially purified enzyme preparations from brain appear to be very similar (366, 380).

The Ca^{2+}, phospholipid-dependent protein kinase contains two distinct functional domains, one with the kinase activity and the second with sites for Ca^{2+}, phospholipid and diacylglycerol binding. Separation of the two domains by proteolysis produces an active kinase (380) that is fully active without Ca^{2+}, phospholipid, or diacylglycerol (381). Calmodulin antagonists, such as trifluoperazine, inactivate the native enzyme but not the proteolytic fragment by competing with phospholipid, but not diacylglycerol (381). Tumor promoting phorbol esters activate the native enzyme by replacing diacylglycerol (378).

The structure of the enzyme has some similarities to enzymes that are regulated by calmodulin. Many of these, including brain phosphodiesterase and the ATP-dependent Ca^{2+} transporter, contain regulatory and catalytic domains that are separable by proteolysis (366, 382). Both classes of enzyme are also directly inhibited by phenothiazines (383, 384). It will be very interesting to sequence the Ca^{2+}-binding site on the Ca^{2+}, phospholipid-dependent kinase and determine whether it represents a domain that could have been derived from a Ca^{2+} binding site on calmodulin.

Regulation by Protein Kinases

The role of protein phosphorylation in the control of cellular metabolism has recently been reviewed with special emphasis on glycogen metabolism (316). In this section, we discuss the regulation by protein phosphorylation of three sets of substrates that are found primarily in nerve terminals: enzymes that regulate transmitter biosynthesis; a protein of unknown function, protein I, that is specifically localized to synaptic vesicles; and ion channels that modify neuronal excitability and synaptic transmission.

REGULATION OF TRANSMITTER SYNTHESIS   Stimulation or depolarization of appropriate nerve terminals results in increased synthesis of acetylcholine, serotonin, and the catecholamines that is not dependent on depletion of release of transmitter stores (215b, 385). Stimulation of transmitter synthesis can require Ca^{2+} entry or cAMP synthesis (215b, 218, 315, 386–388). Depolarization increases the activity of the rate-limiting enzyme in the pathway of biosynthesis for each of these transmitters, either the Na^+-dependent high affinity choline transporter (215b), tyrosine hydroxylase (215, 387), or tryptophan hydroxylase (389). Activation of these enzymes provides one means of maintaining neurotransmitter levels. Since newly synthesized transmitters are in a pool that is preferentially released (390), stimulation of transmitter synthesis may be important for replacing readily releasable transmitter, even in situations where there is not significant depletion of bulk transmitter stores (391). Tyrosine hydroxylase in nerve terminals is activated by stimulation, depolarization, cAMP, and adenosine whether assayed in vivo or in extracts (215, 216a, 315, 385–388, 393, 395).

Tyrosine hydroxylase in brain extracts can be activated by either a cAMP-dependent or calci-calmodulin-dependent protein kinase (218, 315, 358). Tyrosine hydroxylase in adrenal extracts appears to be activated only by the cAMP-dependent kinase (396). While both mechanisms activate the enzyme, the kinetic properties of the two activated enzyme preparations differ, notably in pH optima, suggesting that the different kinases phosphorylate different sites on the enzyme (315).

Purified tyrosine hydroxylase can be phosphorylated by cAMP-dependent kinase in vitro (217, 397–399, 401) and phosphorylation correlates with increased activity (217). There has been some disagreement over the changes in kinetic properties that result in increased activity (217, 297), but more recent studies indicated that differences in assay conditions explain at least some of these observations (400). In appropriate assay conditions, phosphorylation reduces the K_m for pteridine cofactor, increases the V_{max}, and increases the K_i for catecholamines. All these changes are very sensitive to reaction pH (400). In physiological conditions, Lazar & Barchas (391) have argued that these changes will increase tyrosine hydroxylase activity 8–33-fold.

Purified tyrosine hydroxlase can also be activated by a calci-calmodulin-dependent kinase (218, 401). Activation requires a new activator protein in addition to the kinase. This activator protein has been purified and appears to consist of two 33-kd subunits (218). The activator protein has a wide distribution in the nervous system and also is necessary for calci-calmodulin-dependent kinase activation of tryptophan hydroxylase (218). Activation of tyrosine hydroxylase appears to proceed in two steps: phosphorylation of the enzyme precedes activation by the activator protein (401).

While these studies demonstrate that the activity of tyrosine hydroxylase is regulated by protein kinases, the phosphorylation of the enzyme in vivo has not been correlated with enzyme activation. In fact, the only published report of in vivo phosphorylation of tyrosine hydroxylase is in response to treatment of PC12 cells with NGF (402). In view of recent evidence suggesting that insulin and EGF receptors are protein kinases (403, 404), NGF binding may result in phosphorylation and regulation of tyrosine hydroxylase by a third pathway, independent of the cAMP or Ca^{2+}-dependent kinases.

PHOSPHORYLATION OF PROTEIN I   One of the best-characterized synaptic proteins, protein I (synapsin I) has an unknown function but is phosphorylated during nerve terminal stimulation by depolarization or binding of certain neurotransmitters. Protein I consists of a globular head and a proline-rich tail that is sensitive to collagenase (405). Digestion with collagenase releases a protein I fragment from the membrane (406), so the major site of electrostatic interaction with vesicles appears to be in the tail. Protein I is phosphorylated at one serine residue in the collagenase-resistant head by a type II cAMP-dependent kinase (407, 408). A Ca^{2+} dependent kinase in the cytosol also phosphorylates this site (342).

The collagenous tail of protein I is phosphorylated by at least two separate serine residues by a distinct Ca^{2+}-calmodulin-dependent kinase that is found in both particulate and soluble fractions of brain (342, 362). Phosphorylation facilitates release of the membrane-bound protein I by salt extraction (409).

Phosphorylation of protein I responds rapidly to the initiation or cessation of orthograde nerve stimulation in nerve terminals in the posterior pituitary and sympathetic ganglion (410, 411). Increased phosphorylation during stimulation requires external Ca^{2+} and results in phosphorylation of both the head and tail of the protein, implying activation of both Ca^{2+}-dependent kinases. The protein I in nerve terminals is also phosphorylated in response to cAMP or transmitters, such as dopamine and serotonin, that increase cAMP synthesis (410–412). Only the cAMP kinase-sensitive serine in the globular head is phosphorylated in response to these agents.

Protein I appears to be a synaptic vesicle-associated phosphoprotein (413, 414). Much of the protein is clearly bound to vesicles, since it copurifies with them in vitro (409) and is found associated with vesicles in immunocytochemical examination of nerve terminals (414, 415). In vivo transport of protein I occurs more slowly than that of the fastest moving elements (26), which in other systems include synaptic vesicles. Slower transport would be expected if protein I in vivo is in equilibrium between vesicle-bound and soluble pools. Since phosphorylation facilitates release of vesicle-bound protein I in vitro (409), it may also do so in vivo. Alternatively, phosphorylation may change the affinity of protein I for other proteins or organelles without reducing the binding of protein I to vesicles. The affinity of vesicles for other organelles, such as the plasma-lemma and cytoskeleton, might then be altered to change the efficiency of vesicle mobilization or fusion.

REGULATION OF NEURONAL EXCITABILITY OF PHOSPHORYLATION OF POT-ASSIUM CHANNELS   In an earlier section, we discussed the modification of K^+ channels by phosphorylation, and how this phosphorylation affected the probability of channel opening. We end this chapter with a summary of some cases in which such phosphorylation of nerve terminal K^+ channels appears to control behavior. (See also 355.)

Aplysia neuron R_{15} is a large neurosecretory cell, which is believed to be a peptidergic neuron (416, 417). The neuron has a slow oscillation in its transmembrane potential that induces bursts of action potentials during the depolarizing phase of the cycle. The oscillations are clearly due to fluctuations in the comparative activities of Ca^{2+} and K^+ channels. Slow inward Ca^{2+} current depolarizes the cell, raising the internal Ca^{2+} concentrations. There is disagreement over whether this internal Ca^{2+} modifies the current flow by inactivating the Ca^{2+} channel or activating the Ca^{2+}-dependent K^+ channel (W. Adams, personal communication). In either case, K^+ currents become a large fraction of the total current and repolarize the cell. When internal Ca^{2+} is reduced, Ca^{2+} entry again exceeds K^+-efflux and the cycle repeats. Action potentials are controlled by the classical Na^+, Ca^{2+}, and delayed K^+ voltage-sensitive channels. The period between bursts can be shortened or lengthened by excitatory or inhibitory hormones and neurotransmitters. Application of serotonin, an inhibitor, increases the interburst period and, in sufficient quantities, suppresses bursting altogether by increasing the conductance of a K^+ channel (418). R_{15} membranes contain a serotonin receptor that activates an adenylate cyclase (419). The binding of serotonin results in increased levels of intracellular cAMP (420), protein phosphorylation (421), and K^+ conductance. cAMP analogs also increase the K^+ conductance, and inhibitors of cyclic nucleotide phosphodiesterase potentiate the effect of suboptimal concentrations of serotonin (418). So the effect of serotonin on K^+ conductance appears to reflect activation of a serotonin-dependent adenylate cyclase. Indeed, direct injection into R_{15} of cAMP analogs or an activator of adenylate cyclase, guanylphosphoiminodiphosphate, mimics the effect of serotonin (422, 423). Intracellular injection of a specific inhibitor of cAMP-dependent protein kinases, the Walsh inhibitor, pre-vents the response of R_{15} cells to serotonin but not to transmitters acting by other mechanisms (46), providing evidence that phosphorylation by the cAMP-dependent kinase is required to activate the K^+ channel. The affected channel appears to be a channel that opens near rest (46).

Uncertain at this time is the target of kinase action. Lemos et al (421) have injected [^{32}P]ATP into R_{15} and shown that it is retained within the cell. Application of serotonin results in increased phosphorylation of at least four proteins (M_rs of 230,000, 205,000, 135,000, and 26,000) and decreased phosphorylation of at least one (M_r 43,000). The minute amounts of protein available from single cells make it a formidable problem to relate the phosphorylation of individual proteins to particular functional changes.

Modulation of transmitter release by sensory neurons at synapses on motoneurons regulates a monosynaptic gill-withdrawal reflex in *Aplysia* (44). Behavioral changes in gill withdrawal by *Aplysia* have been traced to such modulation. Habituation of the withdrawal reflex to repeated mild sensory stimuli is caused by reduced Ca^{2+}-influx into the sensory neuron terminals (123). The habituated response can be reversed, or "sensitized," by a painful sensory stimulus. The reflex can also be potentiated by a paired stimulus in a classical conditioning training session (424, 425). Sensitization involves an increased Ca^{2+}-influx caused by a reduced K^+ conductance, prolonging the duration of the action potential (130). Sensitization occurs because the painful stimulus causes the release of serotonin by a third neuron onto the sensory neuron terminals (426). Serotonin acts by increasing cAMP synthesis, cAMP-dependent protein kinase activity, and phosphorylation of a protein that results in closing a novel K^+ channel. This K^+ channel is a major channel in *Aplysia* sensory neurons and has been studied by patch clamping (40). These channels are distinguished from the rapid (A), delayed (K), and Ca^{2+}-dependent (C) K^+ channels that are also present in these neurons (427). The serotonin-sensitive channels are not regulated by voltage or internal Ca^{2+}. Both cAMP and serotonin reduce almost to zero the probability of channel opening (40). Intracellular injection of the catalytic subunit of cAMP-dependent protein kinase mimics the effect of serotonin and closes the channels (339). Injection of a specific inhibitor of the cAMP-dependent kinase, the Walsh inhibitor, blocks the effects of cAMP, serotonin, and stimulation of the modulator neurons (428).

Both the strength and duration of the sensitization correlate with increased levels of cAMP in dissected sensory neuron cell somas (429). Brief stimulation of the modulator neurons or application of serotonin result in elevated cAMP levels that persist for 20 min and correlate with closure of the K^+ channel. Injection of the Walsh inhibitor during this interval reverses the effect of elevated cAMP (428), so phosphate is rapidly turned over on the regulatory site of the protein that closes the K^+ channel. The persistance of sensitization thus requires elevated cAMP, which persists long after the stimulus. [Short-term sensitization can be converted to a more permanent increase in efficiency of the same synapse, lasting weeks, by repeated painful sensory stimuli, but the mechanism is not known (44).]

Another dramatic example of changes in neuronal behavior induced by phosphorylation-dependent channel modification is provided by studies on *Aplysia* bag cells (430). Brief stimulation of the innervating nerve generates a burst of action potentials that cause release of an egg-laying hormone. Cyclic AMP plays an important role in generating this response. Cyclic AMP accumulates in the bag cells during the first few minutes of stimulation (431). Addition of cAMP analogs can actually generate bursts of action potentials in the bag cells without electrical stimulation (431). During the burst of action potentials generated by stimulating the nerve, a brief period (< 1 min) of sodium action potentials is followed by 30–40 min of calcium action potentials (432). Cyclic AMP accumulates during the first few minutes of the Ca^{2+} phase of action potentials, but then returns to control levels (431). During the period of elevated cAMP levels, the width of each spike increases (430), because cAMP causes the closure of a K^{+}-channel (41, 69). The K^{+} channel that is closed by cAMP action appears likely to be a Ca^{2+}-dependent K^{+} channel (69). Microinjection of the catalytic subunit of cAMP-dependent protein kinase also decreases the K^{+} conductance (47), suggesting that closure is mediated by phosphorylation.

Activity in bag cells correlates with the level of phosphorylation of specific proteins, measured either by preloading cells with $^{32}P_i$ or labeling extracted proteins with $[^{32}P]ATP$ and the catalytic subunit of the cAMP-dependent protein kinase (338). In the latter case, increased phosphorylation in vivo is revealed as reduced incorporation of ^{32}P in vitro. The phosphorylation of one protein (M_r 33,000) is increased at both early and late times during the discharge. The phosphorylation of a second protein (M_r 21,000) remains low at 2 min, but is dramatically increased after 20 min of discharge. The latter is a bag cell–specific, membrane-associated protein and a prominent substrate of cAMP-dependent phosphorylation in vitro. Since cAMP levels are low at 20 min, phosphorylation in vivo seems more likely to be regulated by a different, perhaps Ca^{2+}-dependent kinase.

The examples of R_{15}, sensory neurons, and bag cells are a few examples of how cells containing different ratios of the basic channel types can have dramatically different resting behavior. These examples also indicate that the cAMP-dependent kinase modifies different channels in different cells. Attention naturally has turned to the possibility that the small phosphorylated proteins, apparently different in each neuron, may act as cell-specific channel modifiers. A possible paradigm is provided by the phosphorylation-dependent activation of Ca^{2+} transport by the heart-specific protein phospholamban (135, 135a). Whatever the target, these examples provide a powerful demonstration of the functional importance of ion channels and protein kinases in controlling the behavior of cells, synapses, and animals.

Literature Cited

1. Osterrieder, W., Brum, G., Hescheler, J., Trautwein, W., Flockerzi, V., Hofmann, F. 1982. *Nature* 298:576–78
2. Llinas, R., Steinberg, I. Z., Walton, K. 1981. *Biophys. J.* 33:289–322
3. Hodgkin, A. L., Huxley, A. F. 1952. *J. Physiol.* 117:500–44
4. Ritchie, J. M. 1979. *Ann. Rev. Neurosci.* 2:341–62
5. Hille, B. 1976. *Ann. Rev. Physiol.* 38:139–52
6. Catterall, W. A. 1980. *Ann. Rev. Pharmacol. Toxicol.* 20:15–43
7. Rogart, R. 1981. *Ann. Rev. Physiol.* 43:711–25
8. Sigworth, F. J., Neher, E. 1980. *Nature* 287:447–49
9. Horn, R., Patlak, J., Stevens, C. R. 1981. *Nature* 291:426–27
10. Colquhoun, D., Henderson, R., Ritchie, J. M. 1972. *J. Physiol.* 227:95–126
11. Agnew, W. S., Levinson, S. R., Brabson, J. S., Raftery, M. A. 1978. *Proc. Natl. Acad. Sci. USA* 75:2606–10
12. Albuquerque, E. X., Daly, J. W. 1976. In *The Specificity of Animal, Bacterial and Plant Toxins*, ed. P. Cuatracasas, pp. 297–338. London: Chapman & Hall
13. Couraud, F. C., Rochat, H., Lissitzky, S. 1978. *Biochem. Biophys. Res. Commun.* 83:1525–30
14. Catterall, W. A. 1977. *J. Biol. Chem.* 252:8669–76
15. Beneski, D., Catterall, W. A. 1980. *Proc. Natl. Acad. Sci. USA* 77:639–43
16. Catterall, W. A. 1981. *J. Neurosci.* 1:777–83
17. Hartshorne, R. P., Coppersmith, J., Catterall, W. A. 1980. *J. Biol. Chem.* 255:10572–75
18. Goldin, S. M., Rhoden, V., Hess, E. J. 1980. *Proc. Natl. Acad. Sci. USA* 77:6884–88
18a. Barchi, R. L., Murphy, L. E. 1981. *J. Neurochem.* 36:2097–2100
19. Agnew, W. S., Moore, A. C., Levinson, S. R., Raftery, M. A. 1980. *Biochem. Biophys. Res. Commun.* 92:860–66
20. Hartshorne, R. P., Catterall, W. A. 1981. *Proc. Natl. Acad. Sci. USA* 78:4620–24
21. Barchi, R. L., Cohen, S. A., Murphy, L. E. 1980. *Proc. Natl. Acad. Sci. USA* 77:1306–10
22. Miller, J. A., Agnew, W. S., Levinson, S. R. 1983. *Biochemistry* 22:462–70
23. Hartshorne, R. P., Messner, D. J., Coppersmith, J. C., Catterall, W. A. 1982. *J. Biol. Chem.* 257:13888–91
23a. Barchi, R. L. 1983. *J. Neurochem.* 40:1377–85
24. Ellisman, M. H., Agnew, W. S., Miller, J. A., Levinson, S. R. 1982. *Proc. Natl. Acad. Sci. USA* 79:4461–65
25. Tamkun, M. M., Catterall, W. A. 1981. *J. Biol. Chem.* 256:11457–63
26. Weigele, J. B., Barchi, R. L. 1982. *Proc. Natl. Acad. Sci. USA* 79:3651–55
27. Talvenheimo, J. A., Tamkun, M. M., Catterall, W. A. 1982. *J. Biol. Chem.* 257:11868–71
27a. Tanaka, J. C., Eccleston, J. F., Barchi, R. L. 1983. *Biophys. J.* 41:50 (abstr.); *J. Biol. Chem.* 258: in press
28. Costa, M. R. C., Casnellie, J. E., Catterall, W. A. 1982. *Soc. Neurosci. Abstr.* 8:727
29. Wong, R. K. S., Prince, D. A., Basbaum, A. I. 1979. *Proc. Natl. Acad. Sci. USA* 76:986–90
30. Huxley, A. F., Stampfli, R. 1949. *J. Physiol.* 108:315–39
31. Ritchie, J. M., Rogart, R. B. 1977. *Proc. Natl. Acad. Sci. USA* 74:211–15
32. Smith, K. J., Hall, S. M. 1980. *J. Neurol. Sci.* 48:201–19
33. Moore, H. P. H., Fritz, L. C., Raftery, M. A., Brockes, J. P. 1982. *Proc. Natl. Acad. Sci. USA* 79:1673–77
33a. Ellisman, M. H., Levinson, S. R. 1982. *Proc. Natl. Acad. Sci. USA* 79:6707–11
34. Couraud, F., Jover, E., Dubois, J. M., Rochat, H. 1982. *Toxicon* 20:9–16
35. Katz, B. 1949. *Arch. Sci. Physiol.* 3:289–300
36. Connor, J. A., Stevens, C. F. 1971. *J. Physiol.* 213:21–30
37. Connor, J. A., Stevens, C. F. 1971. *J. Physiol.* 213:31–53
38. Meech, R. W. 1978. *Ann. Rev. Biophys. Bioeng.* 7:1–18
39. Kehoe, J., Marty, A. 1980. *Ann. Rev. Biophys. Bioeng.* 9:437–65
40. Siegelbaum, S. A., Camardo, J. S., Kandel, E. R. 1982. *Nature* 299:413–16
41. Kaczmarek, L. K., Strumwasser, F. 1981. *J. Neurosci.* 1:626–34
42. Adams, P. R., Brown, D. A., Constanti, A. 1982. *J. Physiol.* 330:537–72
43. Dubois, J. M. 1981. *J. Physiol.* 318:297–316
44. Kandel, E. R., Schwartz, J. H. 1982. *Science* 218:433–42
45. Adams, D. J., Smith, S. J., Thompson, S. H. 1980. *Ann. Rev. Neurosci.* 3:141–67
46. Adams, W. B., Levitan, I. B. 1982. *Proc. Natl. Acad. Sci. USA* 79:3877–80
47. Kaczmarek, L. K., Jennings, K. R., Strumwasser, F., Nairn, A. C., Walter, U., et al. 1980. *Proc. Natl. Acad. Sci. USA* 77:7487–91
48. Ohmori, H., Yoshida, S., Hagiwara, S.

1981. *Proc. Natl. Acad. Sci. USA* 78:4960–64
49. Ohmori, H. 1978. *J. Physiol.* 281:77–99
50. Salkoff, L., Wyman, R. 1981. *Nature* 293:228–30
51. Salkoff, L. 1983. *Nature* 302:249–51
52. Jan, Y. N., Jan, L. Y. 1977. *Proc. R. Soc. London B* 198:87–108
53. Adams, P. R., Constanti, A., Brown, D. A., Clark, R. B. 1982. *Nature* 296:746–49
54. Conti, F., Neher, E. 1980. *Nature* 285:140–43
55. Stefani, E., Chiarandini, D. J. 1982. *Ann. Rev. Physiol.* 44:357–72
56. Carbone, E., Wanke, E., Prestipino, G., Possani, L. D., Maelicke, A. 1982. *Nature* 296:90–91
57. Marty, A. 1981. *Nature* 291:497
58. Pallotta, B. S., Magleby, K. L., Barrett, J. N. 1981. *Nature* 293:471–75
59. Garcia-Sancho, J., Sanchez, A., Herreros, B. 1982. *Nature* 296:744–46
60. Lew, V. L., Muallem, S., Seymour, C. A. 1982. *Nature* 296:742–44
61. Lux, H. D., Neher, E., Marty, A. 1981. *Pfluegers Arch.* 389:293–95
62. Caroni, P., Carafoli, E. 1982. *Proc. Natl. Acad. Aci. USA* 79:5763–67
63. Hermann, A., Hartung, K. 1982. *Pfluegers Arch.* 393:248–53
64. Banks, B. E. C., Brown, C., Burgess, G. M., Burnstock, G., Claret, M., et al. 1979. *Nature* 282:415–17
65. Hugues, M., Duval, D., Kitabgi, P., Lazdunski, M., Vincent, J. P. 1982. *J. Biol. Chem.* 257:2762–69
65a. Hugues, M. Schmid, H., Romey, G., Duval, D., Frelin, C., Lazdunski, M. 1982. *EMBO J.* 1:1039–42
66. De Peyer, J. E., Cachelin, A. B., Levitan, I. B., Reuter, H. 1982. *Proc. Natl. Acad. Sci. USA* 79:4207–11
67. Bernardo, L. S., Prince, D. A. 1982. *Nature* 297:76–79
68. Madison, D. V., Nicoll, R. A. 1982. *Soc. Neurosci. Abstr.* 8:922
69. Kaczmarek, L. K., Strumwasser, F. 1981. *Soc. Neurosci. Abstr.* 7:932
70. Ascher, P., Chesnoy-Marchais, D. 1982. *J. Physiol.* 324:67–92
71. Brown, D. A., Adams, P. R. 1980. *Nature* 283:673–76
72. Constanti, A., Adams, P. R., Brown, D. A. 1981. *Brain Res.* 206:244–50
73. Klein, M., Camardo, J., Kandel, E. R. 1982. *Proc. Natl. Acad. Sci. USA* 79:5713–17
74. Pollack, J. D., Camardo, J. S., Bernier, L., Schwartz, J. H., Kandel, E. R. 1982. *Soc. Neurosci. Abstr.* 8:523
75. Cottrell, G. A. 1982. *Nature* 296:87–89
76. Chiu, S. Y., Ritchie, J. M. 1982. *J.*

Physiol. 322:485–501
77. Chiu, S. Y., Ritchie, J. M., Rogart, R. B., Stagg, D. 1979. *J. Physiol.* 292:149–66
78. Chiu, S. Y., Ritchie, J. M. 1981. *J. Physiol.* 313:415–37
79. Skou, J. C. 1957. *Biochim. Biophys. Acta* 23:294–401
80. Skou, J. C. 1965. *Physiol. Rev.* 45:596–617
81. Baker, P. F., Blaustein, M. P., Hodgkin, A. L., Steinhardt, R. A. 1969. *J. Physiol.* 200:431–58
82. Baker, P. F. 1977. *FEBS Symp.* 42:430–31
83. Miyamoto, H., Racker, E. 1980. *J. Biol. Chem.* 255:2656–58
84. Schon, F., Iverson, L. L. 1974. *Life Sci.* 15:157
85. Suszkiw, J. B., Pilar, G. 1976. *J. Neurochem.* 26:1133–38
86. Thomas, R. C. 1972. *J. Physiol.* 220:55–71
87. Thomas, R. C. 1969. *J. Physiol.* 210:495–514
88. Ritchie, J. M., Straub, R. W. 1957. *J. Physiol.* 136:80–97
89. Van Essen, D. C. 1973. *J. Physiol.* 230:509–34
90. Trachtenberg, M. C., Packey, D. J., Sweeney, T. 1981. *Curr. Top. Cell Regul.* 19:159–217
91. Cantley, L. C. 1981. *Curr. Top. Bioenerg.* 11:201–40
92. Sweadner, K. J., Goldin, S. M. 1980. *N. Engl. J. Med.* 302:777–83
93. Sweadner, K. J. 1979. *J. Biol. Chem.* 254:6060–67
94. Carilli, C. T., Farley, R. A., Perlman, D. M., Cantley, L. C. 1982. *J. Biol. Chem.* 257:5601–6
95. Goldin, S. M. 1977. *J. Biol. Chem.* 252:5630–42
96. Forbush, B., Kaplan, J. H., Hoffman, J. H. 1978. *Biochemistry* 17:3667–76
97. Wood, J. G., Jean, D. H., Whittaker, J. N., McLaughlin, B. M., Albers, R. W. 1977. *J. Neurocytol.* 6:571–81
98. Handler, J. S., Butcher, R. W., Sutherland, R. W., Orloff, J. 1965. *J. Biol. Chem.* 240:4524–26
99. Clausen, T., Flatman, J. A. 1977. *J. Physiol.* 270:455–65
100. Baughler, J. M., Corder, C. N. 1978. *Biochim. Biophys. Acta* 524:455–65
101. Spector, M., O'Neal, S., Racker, E. 1980. *J. Biol. Chem.* 255:5504–7
102. Luly, P., Barnabei, O., Tria, E. 1972. *Biochim. Biophys. Acta* 282:447–52
103. Ismael-Bergi, F., Edelman, I. S. 1971. *J. Gen. Physiol.* 57:710–22
104. Fishman, M. C. 1979. *Proc. Natl. Acad. Sci. USA* 76:4661–63

105. Haupert, G. T. Jr., Sancho, J. M. 1979. *Proc. Natl. Acad. Sci. USA* 76:4658–60
106. Hagiwara, S., Byerly, L. 1981. *Ann. Rev. Neurosci.* 4:69–125
107. Brown, A. M., Camerer, H., Kunze, D. L., Lux, H. D. 1982. *Nature* 299:156–58
108. Reuter, H., Stevens, C. F., Tsien, R. W., Yellen, G. 1982. *Nature* 297:501–4
109. Fenwick, E. M., Marty, A., Neher, E. 1981. *J. Physiol.* 319:100P–1
110. Llinas, R., Sugimori, M., Simon, S. M. 1982. *Proc. Natl. Acad. Sci. USA* 79:2415–19
111. Miledi, R., Parker, I. 1981. *Proc. R. Soc. London Ser. B* 212:197–211
112. Charlton, M. P., Smith, S. J., Zucker, R. S. 1982. *J. Physiol.* 323:173–93
113. Pumplin, D. W., Reese, T. S. 1978. *Neuroscience* 3:685–96
114. Pumplin, D. W., Reese, T. S., Llinas, R. 1981. *Proc. Natl. Acad. Sci. USA* 78:7210–13
115. Krishtal, O. A., Pidoplichko, V. I., Shakhovalov, Y. A. 1981. *J. Physiol.* 310:423–34
116. Byerly, L., Hagiwara, S. 1982. *J. Physiol.* 322:503–28
117. Papazian, D., Rahamimoff, H., Goldin, S. M. 1979. *Proc. Natl. Acad. Sci. USA* 76:3708–12
118. Romey, G., Lazdunski, M. 1982. *Nature* 297:79–80
119. Takahashi, M., Ohizumi, Y., Yasumoto, T. 1982. *J. Biol. Chem.* 257:7287–89
120. Gould, R. J., Murphy, K. M. M., Snyder, S. H. 1982. *Proc. Natl. Acad. Sci. USA* 79:3656–60
121. Kandel, E. R. 1981. *Nature* 293:697–700
122. Kretz, R., Shapiro, E., Kandel, E. R. 1982. *Proc. Natl. Acad. Sci. USA* 79:5430–34
123. Shapiro, E., Castellucci, V. F., Kandel, E. R. 1980. *Proc. Natl. Acad. Sci. USA* 77:1185–89
124. Dunlap, K., Fischbach, G. D. 1981. *J. Physiol.* 317:519–35
125. Shapiro, E., Castellucci, V. F., Kandel, E. R. 1980. *Proc. Natl. Acad. Sci. USA* 77:629–33
126. Parnas, I., Parnas, H., Dudel, J. 1982. *Pfluegers Arch.* 393:232–37
127. Katz, B., Miledi, R. 1971. *J. Physiol.* 216:503–12
128. Tillotson, D. 1979. *Proc. Natl. Acad. Sci. USA* 76:1497–1500
129. Standen, N. B. 1981. *Nature* 293:158–59
130. Klein, M., Shapiro, E., Kandel, E. R. 1980. *J. Exp. Biol.* 89:117–57
131. Dunlap, K., Fischbach, G. D. 1978. *Nature* 276:837–89
132. Mudge, A. W., Leeman, S., Fischbach, G. D. 1979. *Proc. Natl. Acad. Sci. USA* 76:526–30
133. Horn, J. P., McAfee, D. A. 1980. *J. Physiol.* 301:191–204
134. Yatani, A., Tsuda, Y., Akaike, N., Brown, A. M. 1982. *Nature* 296:169–71
134a. Bartschat D. K., Cyr, D. L., Lindenmayer, G. E. 1980. *J. Biol. Chem.* 255:44–47
135. Kranias, E. G., Solaro, R. J. 1982. *Nature* 298:182–84
135a. Louis, C. F., Maffit, M., Jarvis, B. 1982. *J. Biol. Chem.* 257:15182–86
136. Dipolo, R., Beauge, L. 1979. *Nature* 278:271–73
137. Gill, D. L., Grollman, E. F., Kohn, L. D. 1981. *J. Biol. Chem.* 256:184–92
138. Saris, N. E., Akerman, K. E. O. 1980. *Curr. Top. Bioenerg.* 10:103–79
139. Brinley, F. J. Jr., Tiffert, F., Scarpa, A. 1978. *J. Gen. Physiol.* 72:101–27
140. Blaustein, M. P., Ratzlaff, R. W., Kendrick, N. C., Schweitzer, E. S. 1978. *J. Gen. Physiol.* 72:15–41
141. Blaustein, M. P., Ratzlaff, R. W., Schweitzer, E. S. 1978. *J. Gen. Physiol.* 72:43–66
142. Dipolo, R., Requena, J., Brinley, F. J. Jr., Mullins, L. J., Scarpa, A., Tiffert, T. 1976. *J. Gen. Physiol.* 67:433–67
143. McGraw, C. F., Somlyo, A. V., Blaustein, M. P. 1980. *J. Cell Biol.* 85:228–41
144. Schmidt, R., Zimmermann, H., Whittaker, V. P. 1980. *Neuroscience* 5:625–38
145. Michaelson, D. M., Ophir, I., Angel, I. 1980. *J. Neurochem.* 35:117–24
146. Israel, M., Manaranche, R., Marsal, J., Meunier, F. M., Morel, N., et al. 1980. *J. Membr. Biol.* 54:115–26
147. Rephaeli, A., Parsons, S. M. 1982. *Proc. Natl. Acad. Sci. USA* 79:5783–87
148. Kanner, B. I., Sharon, I., Maron, R., Schuldiner, S. 1980. *FEBS Lett.* 111:83–86
148a. Anderson, D. C., King, S. C., Parsons, S. M. 1982. *Biochemistry* 13:3037–43
149. Crompton, M., Moser, R., Ludi, H., Carafoli, E. 1978. *Eur. J. Biochem.* 82:25–31
150. Javors, M. A., Bowden, C. L., Ross, D. H. 1981. *J. Neurochem.* 37:381–87
151. Papazian, D. M., Rahamimoff, H., Goldin, S. M. 1982. *Soc. Neurosci. Abstr.* 8:692
152. Goldin, S. M., Papazian, D., Rahamimoff, H., Chan, S., Hess, E. J. 1983. *Cold Spring Harbor Symp. Quant. Biol.* 38: In press
152a. Goldin, S. M., Moczydlowski, E. G., Papazian, D. M. 1983. *Ann. Rev. Neurosci.* 6:419–46
153. Niggli, V., Adunyah, E. S., Penniston, J.

T., Carafoli, E. 1981. *J. Biol. Chem.* 256:395–401

154. Niggli, V., Penniston, J. T., Carafoli, E. 1979. *J. Biol. Chem.* 254:9955–58
155. Niggli, V., Adunyah, E. S., Carafoli, E. 1981. *J. Biol. Chem.* 256:8588–92
156. Llinas, R., Steinberg, I. Z., Walton, K. 1981. *Biophys. J.* 33:323–51
157. Heuser, J. E., Reese, T. S. 1981. *J. Cell. Biol.* 88:564–80
158. Aunis, D., Hesketh, J. E., Devilliers, G. 1979. *Cell Tissue Res.* 197:433–41
159. Meyer, D. I., Burger, M. M. 1979. *J. Biol. Chem.* 254:9854–59
160. Morell, N., Manaranche, R., Israel, M., Gulik-Krzywicki, T. 1982. *J. Cell Biol.* 93:349–56
161. Matthew, W. D., Tsavaler, L., Reichardt, L. F. 1981. *J. Cell. Biol.* 91:257–69
162. Wagner, J. A., Kelly, R. B. 1979. *Proc. Natl. Acad. Sci. USA* 76:4126–30
163. Ceccarelli, B., Grohovaz, F., Hurlbut, W. P. 1979. *J. Cell Biol.* 81:163–77
164. Ceccarelli, B., Hurlbut, W. P. 1980. *J. Cell Biol.* 87:297–303
165. Dreyer, F., Peper, K., Akert, K., Sandri, C., Moor, H. 1973. *Brain Res.* 62:373–80
166. Dodge, F. A., Rahamimoff, R. 1967. *J. Physiol.* 193:419–32
167. Baker, P. F., Knight, D. E., Whittaker, M. J. 1980. *Proc. R. Soc. London Ser. B* 207:149–61
168. Knight, D. E., Baker, P. F. 1982. *J. Membr. Biol.* 68:107–40
169. Baker, P. F., Whittaker, M. J. 1978. *Nature* 276:513–15
170. Berman, M. C. 1982. *J. Biol. Chem.* 257:1953–57
171. Means, A. R., Dedman, J. R. 1980. *Nature* 285:73–77
172. Klee, C. B., Vanaman, T. C. 1982. *Adv. Protein Chem.* 35:213–321
173. Cox, J. A., Malnoe, A., Stein, E. A. 1981. *J. Biol. Chem.* 256:3218–22
174. Malnoe, A., Cox, J. A., Stein, E. A. 1982. *Biochim. Biophys. Acta* 714:84–92
175. Cox, J. A., Comte, M., Stein, E. A. 1982. *Proc. Natl. Acad. Sci. USA* 79:4265–69
176. Cohen, P. 1980. *Eur. J. Biochem.* 111:563–74
177. Stewart, A. A., Ingrebritsen, T. S., Malnalan, A., Klee, C. B., Cohen, P. 1982. *FEBS Lett.* 137:80–84
178. Schweitzer, E. S., Kelly, R. B. 1982. *Soc. Neurosci. Abstr.* 8:493
179. Schatzman, R. C., Wise, B. C., Kuo, J. F. 1981. *Biochem. Biophys. Res. Commun.* 98:669–76
180. Douglas, W. W., Nemeth, E. F. 1982. *J. Physiol.* 323:229–44
181. Schubart, U. K., Erlichman, J., Fleischer, N. 1980. *J. Biol. Chem.* 255:4120–24
182. Schubart, U. K., Erlichman, J., Fleischer, N. 1982. *Fed. Proc.* 41:2278–82
183. Steinhardt, R. A., Alderton, J. M. 1982. *Nature* 295:154–55
184. DeLorenzo, R. J. 1980. *Ann. NY Acad. Sci.* 356:92–109
185. DeLorenzo, R. J., Freedman, S. D., Yohe, W. B., Mauer, S. C. 1979. *Proc. Natl. Acad. Sci. USA* 76:1838–42
186. DeLorenzo, R. J., Freedman, S. D. 1977. *Biochem. Biophys. Res. Commun.* 77:1036–43
187. Moskowitz, N., Schook, W., Lisanti, M., Hua, E., Puszkin, S. 1982. *J. Neurochem.* 38:1742–47
188. Hooper, J. E., Kelly, R. B. 1982. *Soc. Neurosci. Abstr.* 8:496
189. Burgoyne, R. D., Geisow, M. J. 1981. *FEBS Lett.* 131:127–31
190. Geisow, M. J., Burgoyne, R. D., Harris, A. 1982. *FEBS Lett.* 143:69–72
191a. Geisow, M. J., Burgoyne, R. D. 1983. *Nature* 301:432–35
191. Grinstein, S., Furuya, W. 1982. *FEBS Lett.* 140:49–52
192. Cruetz, C. E., Paxoles, C. J., Pollard, H. B. 1979. *J. Biol. Chem.* 254:553–58
193. Cruetz, C. E., Scott, J. H., Pazoles, C. J., Pollard, H. B. 1982. *J. Cell Biochem.* 18:87–121
194. Hong, K., Duzgunes, N., Papahadjopoulos, D. 1981. *J. Biol. Chem.* 256:3641–44
195. Skehel, J. J., Bayley, P. M., Brown, E. B., Martin, S. R., Waterfield, M. D., et al. 1982. *Proc. Natl. Acad. Sci. USA* 79:968–72
196. Michell, R. H. 1975. *Biochim. Biophys. Acta* 415:81–147
197. DeLorenzo, R. J. 1982. *Fed. Proc.* 41:2265–72
198. Viveros, O. H., Diliberto, E., Axelrod, J. 1977. In *Synapses*, ed. G. A. Cottrell, P. N. R. Usherwood, pp. 368–69. London: Blackie & Son
199. Kelly, R. B., Deutsch, J. W., Carlson, S. S., Wagner, J. A. 1979. *Ann. Rev. Neurosci.* 2:399–446
200. Loh, Y. P., Gainer, H. 1982. In *Brain Peptides*, ed. D. T. Krieger, M. J. Brownstein, J. R. Martin, New York: Academic
201. Nakanishi, S., Inoue, A., Kita, T., Nakamura, M., Chang, A. C. Y., et al. 1979. *Nature* 278:423–27
202. Comb, M., Seeburg, P. H., Adelman, J., Eiden, L., Herbert, E. 1982. *Nature* 295:663–66
203. Land, H., Schutz, G., Schmale, H., Richter, D. 1982. *Nature* 295:299–303
204. Hobart, P., Crawford, R., Shen, L. P.,

Pictet, R., Rutter, W. J. 1980. *Nature* 288:137–41

205. Gubler, U., Seeburg, P., Hoffman, B. J., Gage, L. P., Udenfriend, S. 1982. *Nature* 295:206–8

206. Noda, M., Furutani, Y., Takahashi, H., Toyosato, M., Hirose, T., et al. 1982. *Nature* 295:202–6

207. Brownstein, M. J., Russell, J. T., Gainer, H. 1980. *Science* 207:373–78

208. Tager, H. S., Patzelt, C., Chan, S., Quinn, P. S., Steiner, D. 1979. *Biol. Cell.* 36:127–36

209. Loh, Y. P., Gainer, H. 1982. *Proc. Natl. Acad. Sci. USA* 79:108–12

210. Loh, Y. P., Chang, T.-L. 1982. *FEBS Lett.* 137:57–62

211. Fricker, L. D., Snyder, S. H. 1982. *Proc. Natl. Acad. Sci. USA* 79:3886–90

212. Smyth, D. G., Zaharian, G. 1980. *Nature* 288:613–15

213. Larsson, L.-I., Childers, S., Snyder, S. H. 1979. *Nature* 282:407–10

214. Gorenstein, C., Snyder, S. H. 1980. *Proc. R. Soc. London Ser. B* 210:123–32

215. Weiner, N., Rabadjija, M. 1968. *J. Pharmacol. Exp. Ther.* 160:61–71

215a. Simon, J. R., Kuhar, M. J. 1975. *Nature* 255:162–63

215b. Collier, B., Ilson, D. 1977. *J. Physiol.* 264:489–509

215c. Simon, J. R., Roth, R. H. 1979. *Mol. Pharmacol.* 16:224–33

216. Patterson, P. H., Reichardt, L. F., Chun, L. L. Y. 1976. *Cold Spring Harbor Symp. Quant. Biol.* 40:389–97

216a. Lovenberg, W., Bruckwick, E. A., Hanbauer, I. 1975. *Proc. Natl. Acad. Sci. USA* 72:2955–58

217. Vulliet, P. R., Langan, T. A., Weiner, N. 1980. *Proc. Natl. Acad. Sci. USA* 77:92–96

218. Yamauchi, T., Nakata, H., Fujisawa, H. 1981. *J. Biol. Chem.* 256:5404–9

219. Mueller, R. A., Thoenen, H., Axelrod, J. 1969. *J. Pharmacol. Exp. Ther.* 169:74–79

220. Heuser, J. E., Reese, T. S. 1973. *J. Cell Biol.* 57:315–44

221. Ceccarelli, B., Hurlbut, W. P., Mauro, A. 1973. *J. Cell Biol* 57:499

222. Inverson, L. L. 1971. *Br. J. Pharmacol.* 41:571

223. Birks, R., MacIntosh, F. C. 1961. *Can. J. Biochem. Physiol.* 39:787–827

224. Collier, B., MacIntosh, F. C. 1969. *Can. J. Physiol. Parmacol.* 47:127–35

225. Winkler, H., Carmichael, S. W. 1982. In *The Secretory Granule*, ed. Poisner, Trifaro, pp. 3–79. Amsterdam/New York/Oxford: Elsevier Biomedical

226. Kopell, W. N., Westhead, E. W. 1982. *J. Biol. Chem.* 257:5707–10

227. Sen, R., Sharp, R. R. 1981. *Biochem. J.* 195:329–33

228. Holz, R. W. 1978. *Proc. Natl. Acad. Sci. USA* 75:5190–94

229. Casey, R. P., Njus, D., Radda, G. K., Sehr, P. A. 1977. *Biochemistry* 16:972–77

230. Giraudat, J., Roisin, M. P., Henry, J. P. 1980. *Biochemistry* 19:4499–505

231. Apps, D. K., Pryde, J. G., Phillips, J. H. 1980. *FEBS Lett.* 111:386–90

232. Apps, D. K., Pryde, J. G., Sutton, R., Phillips, J. H. 1980. *Biochem. J.* 190:273–82

233. Sutton, R., Apps, D. K. 1981. *FEBS Lett.* 130:103–6

234. Apps, D. K., Schatz, G. 1979. *Eur. J. Biochem.* 100:411–19

235. Schmidt, W., Winkler, H., Plattner, H. 1982. *Eur. J. Cell Biol.* 27:96–105

236. Kanner, B. I., Sharon, I., Maron, R., Schuldiner, S. 1980. *FEBS Lett.* 111:83–86

237. Knoth, J., Handloser, K., Njus, D. 1980. *Biochemistry* 19:2938–42

238. Knoth, J., Isaacs, J. M., Njus, D. 1981. *J. Biol. Chem.* 256:6541–43

239. Knoth, J., Zallakian, M., Njus, D. 1982. *Fed. Proc.* 41:2742–45

240. Johnson, R. G., Pfister, D., Carty, S. E., Scarpa, A. 1979. *J. Biol. Chem.* 254:10963–72

241. Johnson, R. G., Scarpa, A. 1979. *J. Biol. Chem.* 254:3750–60

242. Scherman, S., Henry, J. P. 1981. *Eur. J. Biochem.* 116:535–39

243. Maron, R., Fishkes, H., Kanner, B. I., Schuldiner, S. 1979. *Biochemistry* 18:4781–85

244. Aberer, W., Kostron, H., Huber, E., Winkler, H. 1978. *Biochem. J.* 172:353–60

245. Phillips, J. H., Apps, D. K. 1980. *Biochem. J.* 192:273–78

246. Kiang, W. L., Krusius, T., Finne, J., Margolis, R. U., Margolis, R. K. 1982. *J. Biol. Chem.* 257:1651–59

247. Toll, L., Howard, B. D. 1980. *J. Biol. Chem.* 255:1787–89

248. Stadler, H., Fauldner, H. H. 1980. *Nature* 286:293–94

249. Fuldner, H. H., Stadler, H. 1982. *Eur. J. Biochem.* 121:519–24

250. Wagner, J. A., Carlson, S. S., Kelly, R. B. 1978. *Biochemistry* 17:1199–1206

251. Breer, H., Morris, S. J., Whittaker, V. P. 1978. *Eur. J. Biochem.* 87:453–58

252. Breer, H., Morris, S. J., Whittaker, V. P. 1977. *Eur. J. Biochem.* 80:313–18

253. Rothlein, J. E., Parsons, S. M. 1982. *J. Neurochem.* 39:1660–68

254. Anderson, D. C., King, S. C., Parsons, S. M. 1982. *Biochemistry* 21:3037–43

255. Giompres, P. E., Luqmani, Y. A. 1980. *Neuroscience* 5:1041–52
256. Zimmermann, H., Denston, C. R. 1977. *Neuroscience* 2:695–714
257. Zimmermann, H., Denston, C. R. 1977. *Neuroscience* 2:715–30
258. Giompres, P. E., Zimmermann, H., Whittaker, V. P. 1981. *Neuroscience* 6:765–75
259. Russell, J. T., Holz, R. W. 1981. *J. Biol. Chem.* 256:5950–53
260. Carty, S. E., Johnson, R. G., Scarpa, A. 1982. *J. Biol. Chem.* 257:7269–77
261. Hutton, J. C. 1982. *Biochem. J.* 204:171–78
261a. Lorenson, M. Y., Lee, Y. C., Jacobs, L. S. 1981. *J. Biol. Chem.* 256:12802–10
262. Gonzales-Noriega, A., Grubb, J. H., Talkad, V., Sly, W. S. 1980. *J. Cell. Biol.* 85:839–52
263. Moore, H.-P., Gumbiner, B., Kelly, R. B. 1983. *Nature* 302:434–36
264. Lewis, R. V., Stern, A. S., Kimura, S., Rossier, J., Stein, S., Udenfriend, S. 1980. *Science* 208:1459–61
265. Baitinger, C., Levine, J., Lorenz, T., Simon, C., Skene, P., Willard, M. 1982. In *Axonal Transport*, ed. D. G. Weiss, pp. 110–20. Berlin/Heidelberg: Springer
266. Dahlstrom, A., Booj, S., Carlson, S. S., Larsson, P. A. 1981. *Acta Physiol. Scand.* 111:217–19
267. Russell, J. T., Brownstein, M. J., Gainer, H. 1981. *Brain Res.* 205:299–311
268. Gumbiner, B., Kelly, R. B. 1982. *Cell* 28:51–59
268a. Hammerschlag, R., Stone, G. C., Bolen, F. A., Lindsey, J. D., Ellisman, M. H. 1982. *J. Cell Biol.* 93:568–75
269. Goodrum, J., Toews, A., Morell, P. 1979. *Brain Res.* 176:255–73
270. Tytell, M., Gulley, R., Wenthold, R., Lasek, R. J. 1980. *Proc. Natl. Acad. Sci. USA* 77:3042–46
271. Levine, J., Willard, M. 1980. *Brain Res.* 194:137–54
272. Bisby, M. A. 1982. In *Axoplasmic Transport*, ed. D. G. Weiss, pp. 193–99. Berlin/Heidelberg: Springer
273. Black, M. N., Lasek, R. J. 1979. *Brain Res.* 171:401–13
273a. Levine, J., Willard, M. 1980. *Brain Res.* 194:137–54
274. Garner, J. A., Lasek, R. J. 1981. *J. Cell Biol.* 88:172–79
275. Brady, S. T., Lasek, R. J. 1981. *Cell* 23:515–23
275a. Levine, J., Willard, M. 1983. *Proc. Natl. Acad. Sci. USA* 80:191–95
276. Hoffman, P. N., Lasek, R. J. 1975. *J. Cell Biol.* 66:351–66
277. Brady, S. T., Lasek, R. J. 1982. See Ref. 272, pp. 206–17
278. Lasek, R. J., Hoffman, P. N. 1976. In *Cell Motility*, ed. R. Goldman, T. Pollard, J. Rosenbaum, pp. 1021–29. Cold Spring Harbor. N. Y.: Cold Spring Harbor Lab.
279. Howes, E. A., McLaughlin, B. J., Heslop, J. P. 1974. *Cell Tissue Res.* 153:545–58
280. Goldman, J. E., Kim, K. S., Schwartz, J. H. 1976. *J. Cell Biol.* 70:304–18
281. Tsukita, S., Ishikawa, H. 1980. *J. Cell Biol.* 84:513–30
282. Smith, R. S. 1980. *J. Neurocytol.* 9:39–65
283. Ellisman, M. H., Lindsey, J. D. 1982. See Ref. 272, pp. 55–63
284. Hirokawa, N. 1982. *J. Cell Biol.* 94:129–42
285. Schnapp, B. J., Reese, T. S. 1982. *J. Cell Biol.* 94:667–79
285a. Papasozomenos, S. C. H., Yoon, M., Crane, R., Autilio-Gambetti, L., Gambetti, P. 1982. *J. Cell Biol.* 95:672–75
286. Allen, R. D., Metuzals, J., Tasaki, I., Brady, S. T., Gilbert, S. P. 1982. *Science* 218:1127–29
287. Brimijoin, S. 1982. *Fed. Proc.* 41:2312–16
288. Forman, D. S. 1982. See Ref. 272, pp. 234–40
289. Adams, R. J. 1982. *Nature* 297:327–29
290. Goldberg, D. J. 1982. *Proc. Natl. Acad. Sci. USA* 79:4818–22
291. Isenberg, G., Schubert, P., Kreutzberg, G. W. 1980. *Brain Res.* 194:588–93
292. Goldberg, D. J., Harris, D. A., Lubit, B. W., Schwartz, J. H. 1980. *Proc. Natl. Acad. Sci. USA* 77:7448–52
293. Burridge, K., Phillips, J. H. 197 *Nature* 254:526–29
294. Wilkins, J. A., Lin, S. 1981. *Biochim. Biophys. Acta* 642:55–66
295. Fowler, V. M., Pollard, H. B. 1982. *Nature* 295:336–39
296. Sherline, P., Lee, Y. C., Jacobs, L. S. 1977. *J. Cell. Biol.* 72:380–89
297. Suprenant, K. A., Dentler, W. L. 1982. *J. Cell Biol.* 93:164–75
298. Ochs, S. 1982. *Fed. Proc.* 41:2301–6
299. Schwartz, J. H. 1979. *Ann. Rev. Neurosci.* 2:467–504
300. Breuer, A. C., Allen, R. D., Lewis, L. J. 1981. *Neurology* 31:118a
301. Brady, S. T., Lasek, R. J., Allen, R. D. 1982. *Science* 218:1129–31
302. Ochs, S., Worth, R. M., Chan, S. Y. 1977. *Nature* 270:748–50
303. Adams, R. J., Baker, P. F., Bray, D. 1982. *J. Physiol.* 326:7–8
304. Shapiro, E., Harris, D. A., Schwartz, J. H. 1981. *Soc. Neurosci. Abstr.* 7:837

305. Miller, R. J., Dawson, G. 1980. *Adv. Biochem. Pharmacol.* 22:11–20
306. Snyder, S. H., Goodman, R. R. 1980. *J. Neurochem.* 35:5–15
307. Greengard, P. 1981. *Harvey Lect.* 75:277–331
308. Greengard, P. 1978. *Science* 199:146–52
309. Walter, U. 1981. *Eur. J. Biochem.* 118:339–46
310. Lohmann, S. M., Walter, U., Miller, P. E., Greengard, P., DeCamilli, P. 1981. *Proc. Natl. Acad. Sci. USA* 78:653–57
311. Landowne, D., Ritchie, J. M. 1971. *J. Physiol.* 212:483–502
312. Landowne, D., Ritchie, J. M. 1971. *J. Physiol.* 212:503–17
313. Ceccarelli, B., Hurlbut, W. P. 1980. *J. Cell. Biol.* 87:297–303
314. Collier, B., Ilson, D. 1977. *J. Physiol.* 264:489–509
315. Simon, J. R., Roth, R. H. 1979. *Mol. Pharmacol.* 16:224–33
316. Cohen, P. 1982. *Nature* 296:613–20
317. Miyamoto, M. D., Breckenridge, B. Mc.L. 1974. *J. Gen. Physiol.* 63:609–24
317a. Livingstone, M. S., Sziber, P. P., Quinn, W. G. 1982. *Soc. Neurosci. Abstr.* 8:384
318. Byers, D., Davis, R. L., Kiger, J. A. 1981. *Nature* 289:79–81
319. Duerr, J. S., Quinn, W. G. 1982. *Proc. Natl. Acad. Sci. USA* 79:3646–50
320. Kauvar, L. M. 1982. *J. Neurosci.* 2:1347–58
321. Ross, E. M., Gilman, A. G. 1980. *Ann. Rev. Biochem.* 49:533–64
322. Limbird, L. E., Gill, D. M., Lefkowitz, R. J. 1980. *Proc. Natl. Acad. Sci. USA* 77:775–79
322a. Hildebrandt, J. D., Hanoune, J., Birnbaumer, L. 1982. *J. Biol. Chem.* 257:14723–25
322b. Koski, G., Streaty, R. A., Klee, W. A. 1982. *J. Biol. Chem.* 257:14035–40
322c. Koski, G., Klee, W. A. 1981. *Proc. Natl. Acad. Sci. USA* 78:4185–89
323. Brostrom, M. A., Brostrom, C. O., Wolff, D. J. 1978. *Arch. Biochem. Biophys.* 191:341–50
324. Salter, R. S., Krinks, M. H., Klee, C. B., Neer, E. J. 1981. *J. Biol. Chem.* 256:9830–33
325. Klee, C. B., Krinks, M. H., Hathaway, D. R., Flockhart, D. A. 1982. In *Calmodulin and Intracellular Ca⁺⁺ Receptors*, ed. S. Kikuichi, H. Hidaka, pp. 303–11. New York: Plenum
326. Bernier, L., Castellucci, V. F., Kandel, E. R., Schwartz, J. H. 1982. *J. Neurosci.* 2:1682–91
327. Castellucci, V. F., Nairn, A., Greengard, P., Schwartz, J. H., Kandel, E. R. 1982. *J. Neurosci.* 2:1673–81
328. Walter, U., Greengard, P. 1981. *Curr. Top. Cell. Regul.* 19:219–55
329. Walter, U., Miller, P., Wilson, F., Menkes, D., Greengard, P. 1980. *J. Biol. Chem.* 255:3757–62
330. Walter, U., Uno, I., Lui, A. Y. C., Greengard, P. 1977. *J. Biol. Chem.* 252:6388–90
331. Hathaway, D. R., Adelstein, R. S., Klee, C. B. 1981. *J. Biol. Chem.* 256:8183–89
332. Rubin, C. S., Rangel-Aldao, R., Sarkar, D., Erlichman, J., Fleischer, N. 1979. *J. Biol. Chem.* 254:3797–805
333. Lohmann, S. M., Walter, U., Greengard, P. 1980. *J. Biol. Chem.* 255:9985–92
334. Eppler, C. M., Palazzolo, M. J., Schwartz, J. H. 1982. *J. Neurosci.* 2:1692–704
335. Vallee, R. B., DiBartolomeis, M. J., Theurkauf, W. E. 1981. *J. Cell Biol.* 90:568–76
336. Miller, P., Walter, U., Theurkauf, W. E., Vallee, R. B., DeCamilli, P. 1982. *Proc. Natl. Acad. Sci. USA* 79:5562–66
337. Ariano, M. A., Matus, A. I. 1981. *J. Cell Biol.* 91:287–92
338. Jennings, K. R., Kaczmarek, L. K., Hewick, R. M., Dreyer, W. J., Strumwasser, F. 1982. *J. Neurosci.* 2:158–68
339. Castellucci, V. F., Kandel, E. R., Schwartz, J. H., Wilson, F. D., Nairn, A. C., Greengard, P. 1980. *Proc. Natl. Acad. Sci. USA* 77:7492–96
340. Salisbury, J. L., Condeelis, J. S., Satir, P., 1980. *J. Cell. Biol.* 87:132–41
341. Osawa, E. 1973. *J. Neurochem.* 20:1487–88
342. Kennedy, M. B., Greengard, P. 1981. *Proc. Natl. Acad. Sci. USA* 78:1293–97
343. Sharma, R. K., Wang, T. H., Wirch, E., Wang, J. H. 1980. *J. Biol. Chem.* 255:5916–23
344. Finn, R. C., Browning, M., Lynch, G. 1980. *Neurosci. Lett.* 19:103–8
345. Glenney, J. R. Jr., Glenney, P., Weber, K. 1982. *Proc. Natl. Acad. Sci. USA* 79:4002–5
346. Marcum, J. M., Dedman, J. R., Brinkley, B. R., Means, A. R. 1978. *Proc. Natl. Acad. Sci. USA* 75:3771–75
347. Walsh, M. P., Vallet, B., Cavadore, J. C., Demaille, J. G. 1980. *J. Biol. Chem.* 255:335–37
348. Cheung, W. Y. 1982. *Fed. Proc.* 41:2253–57
349. Klee, C. B., Crouch, T. H., Richman, P. G. 1980. *Ann. Rev. Biochem.* 49:489–515
350. Wang, J. H., Desai, R. 1977. *J. Biol. Chem.* 252:4175–84
351. Klee, C. B., Crouch, T. H., Krinks, M. H. 1979. *Biochemistry* 18:722–29

352. Klee, C. B., Haiech, J. 1980. *Ann. NY Acad. Sci.* 256:43–54
353. Stewart, A. A., Ingebritsen, T. S., Manalan, A., Klee, C. B., Cohen, P. 1982. *FEBS Lett.* 137:80–83
354. Wood, J. G., Wallace, R. W., Whitaker, J. N., Cheung, W. Y. 1980. *J. Cell Biol.* 84:66–76
355. Kennedy, M. 1983. *Ann. Rev. Neurosci* 6:493–525
356. Walaas, S. I., Nairn, A. C., Greengard, P. 1983. *J. Neurosci.* 3:291–301
357. Walaas, S. I., Nairn, A. C., Greengard, P. 1983. *J. Neurosci.* 3:302–11
358. Yamauchi, T., Fujisawa, H. 1980. *FEBS Lett.* 116:141–44
359. Cohen, P., Burchell, A., Foulkes, J. G., Cohen, P. T. W., Vanaman, T. C., Nairn, A. C. 1978. *FEBS Lett.* 92:287–93
359a. Fukunaga, K., Yamamoto, H., Iwasa, Y., Miyamoto, E. 1982. *Life Sci.* 30:2019–24
360. Scordilis, S. P., Anderson, J. L., Polloch, R., Adelstein, R. S. 1977. *J. Cell Biol.* 74:940–49
360a. Fukunaga, K., Yamamoto, H., Matsui, K., Higashi, K., Miyamoto, E. 1982. *J. Neurochem.* 39:1607–17
361. Kennedy, M., McGuinness, T., Greengard, P. 1983. *J. Neurosci.* 3: 818–31
362. Huttner, W. B., DeGennaro, L. J., Greengard, P. 1981. *J. Biol. Chem.* 256:1482–88
363. Takai, Y., Kishimoto, A., Inoue, M., Nishizuka, Y. 1977. *J. Biol. Chem.* 252:7603–9
364. Wise, B. C., Raynor, R. L., Kuo, J. F. 1982. *J. Biol. Chem.* 257:8481–88
365. Wise, B. C., Glass, D. B., Chou, C. H. J., Raynor, R. L., Katoh, N., et al. 1982. *J. Biol. Chem.* 257:8489–95
366. Kuo, J. F., Anderson, R. G. G., Wise, B. C., Mackerlova, L., Salomonsson, I., et al. 1980. *Proc. Natl. Acad. Sci. USA* 77:7039–43
367. Wrenn, R. W., Katoh, N., Wise, B. C., Kuo, J. F. 1980. *J. Biol. Chem.* 255:12042–46
368. Wrenn, R. W., Katoh, N., Schatzman, R. C., Kuo, J. F. 1981. *Life. Sci.* 29:725–33
369. Deleted in proof
370. Deleted in proof
371. Katoh, N., Wrenn, R. W., Wise, B. C., Shoji, M., Kuo, J. F. 1981. *Proc. Natl. Acad. Sci. USA* 78:4813–17
372. Katoh, N., Kuo, J. F. 1982. *Biochem. Biophys. Res. Commun.* 106:590–95
373. Kishimoto, A., Takai, Y., Mori, T., Kikkawa, U., Nishizuka, Y. 1980. *J. Biol. Chem.* 255:2273–76

374. Takai, Y., Kishimoto, A., Iwasa, Y., Kawahara, Y., Mori, T., Nishizuka, Y. 1979. *J. Biol. Chem.* 254:3692–95
375. Michell, R. H. 1979. *Trends Biochem. Sci.* 4:128–31
376. Kawahara, Y., Takai, Y., Minakuchi, R., Sano, K., Nishizuka, Y. 1980. *Biochem. Biophys. Res. Commun.* 97:309–17
377. Takai, Y., Kaibuchi, K., Matsubara, T., Nishizuka, Y. 1981. *Biochem. Biophys. Res. Commun.* 101:61–67
378. Castagna, M., Takai, Y., Kaibushi, K., Sano, K., Kikkawa, U., Nishizuka, Y. 1982. *J. Biol. Chem.* 257:7847–51
379. Wu, W. C. S., Walaas, S. I., Nairn, A. C., Greengard, P. 1982. *Proc. Natl. Acad. Sci. USA* 79:5249–53
380. Inoue, M., Kishimoto, A., Takai, Y., Nishizuka, Y. 1977. *J. Biol. Chem.* 252:7610–16
381. Mori, T., Takai, Y., Minakuchi, R., Yu, B., Nishizuka, Y. 1980. *J. Biol. Chem.* 255:8378–80
382. Tucker, M. M., Robinson, J. B. Jr., Stellwagon, E. 1981. *J. Biol. Chem.* 256:9051–58
383. Tanaka, T., Hidaka, H. 1980. *J. Biol. Chem.* 255:11078–80
384. Adunyah, E. S., Niggli, V., Carafoli, E. 1982. *FEBS Lett.* 143:65–68
385. Salzman, P. N., Roth, R. H. 1980. *J. Pharmacol. Exp. Ther.* 212:64–73
386. Weiner, N., Lee, F. L., Dreyer, E., Barnes, E. 1978. *Life Sci.* 22:1197–1216
387. Salzman, P. M., Roth, R. H. 1980. *J. Pharmacol. Exp. Ther.* 212:74–83
388. Masserano, J. M., Weiner, N. 1979. *Mol. Pharmacol.* 16:513–28
389. Jequeir, E., Lovenberg, W., Sjoerdsma, A. 1967. *Mol. Pharmacol.* 3:274–78
390. Kopin, I. J., Breese, G. R., Krauss, K. R., Weise, V. K. 1968. *J. Pharmacol. Exp. Ther.* 161:271–78
391. Lazar, M. A., Barchas, J. D. 1982. *Psychopharmacol. Bull.* 18:157–61
392. Morgenroth, V. H., Boadle-Biber, M., Roth, R. H. 1974. *Proc. Natl. Acad. Sci. USA* 71:4283–87
393. Kapatos, G., Zigmond, M. J. 1979. *Brain Res.* 170:299–312
394. Bustos, G., Simon, J., Roth, R. H. 1980. *J. Neurochem.* 35:47–57
395. Erny, R. E., Berezo, M. W., Perlman, R. L. 1981. *J. Biol. Chem.* 256:1335–39
396. Yamauchi, T., Fujisawa, H. 1979. *J. Biol. Chem.* 254:6408–13
397. Joh, T. H., Park, D. H., Reis, D. J. 1978. *Proc. Natl. Acad. Sci. USA* 75:4744–48
398. Yamauchi, T., Fuijisawa, H. 1979. *J. Biol. Chem.* 254:503–7
399. Edelman, A. M., Raese, J. D., Lazar, M.

A., Barchas, J. D. 1981. *J. Pharmacol. Exp. Ther.* 216:647–53
400. Lazar, M. A., Lockfield, A. J., Truscott, R. J. W., Barchas, J. D. 1982. *J. Neurochem.* 39:409–22
401. Yamauchi, T., Fujisawa, H. 1981. *Biochem. Biophys. Res. Commun.* 100:807–13
402. Halegoua, S., Patrick, J. 1980. *Cell* 22:571–81
403. Ushuro, H., Cohen, S. 1980. *J. Biol. Chem.* 255:3863
404. Kasuga, M., Karlsson, F. A., Kahn, C. R. 1982. *Science* 215:185–87
405. Ueda, T., Greengard, P. 1977. *J. Biol. Chem.* 252:5155–63
406. Ueda, T. 1981. *J. Neurochem.* 36:297–300
407. Walter, U., Lohmann, S. M., Sieghardt, W., Greengard, P. 1979. *J. Biol. Chem.* 254:12235–39
408. Huttner, W. B., DeCamilli, P., Schiebler, W., Greengard, P. 1981. *Soc. Neurosci. Abstr.* 7:441
409. Huttner, W. B., Schiebler, W., Greengard, P., DeCamilli, P. 1983. *J. Cell Biol.* 97: In press
410. Tsou, K., Greengard, P. 1982. *Proc. Natl. Acad. Sci. USA* 79:6075–79
411. Nestler, E. J., Greengard, P. 1982. *J. Neurosci.* 2:1011–23
412. Dolphin, A. C., Greengard, P. 1981. *J. Neurosci.* 1:192–203
413. Greengard, P., DeCamilli, P. 1982. In *Disorders of the Motor Unit*, ed. D. L. Schotland, pp. 741–60. New York: Wiley
414. DeCamilli, P., Harris, S. M., Huttner, W. B., Greengard, P. 1983. *J. Cell Biol.* 97: In press
415. Bloom, F. E., Ueda, T., Battenberg, E., Greengard, P. 1979. *Proc. Natl. Acad. Sci. USA* 76:5982–86

416. Loh, Y. P., Gainer, H. 1975. *Brain Res.* 92:193–205
417. Kupfermann, I., Weiss, K. R. 1976. *J. Gen. Physiol.* 67:113–23
418. Drummond, A. H., Benson, J. A., Levitan, I. B. 1980. *Proc. Natl. Acad. Sci. USA* 77:5013–17
419. Levitan, I. B. 1978. *Brain Res.* 154:404–8
420. Cedar, H., Schwartz, J. H. 1972. *J. Gen. Physiol.* 60:570–87
421. Lemos, J. R., Novak-Hofer, I., Levitan, I. B. 1982. *Nature* 298:64–65
422. Treistman, S. N., Levitan, I. B. 1976. *Nature* 261:62–64
423. Treistman, S. N., Levitan, I. B. 1976. *Proc. Natl. Acad. Sci. USA* 73:4689–92
424. Klein, M., Kandel, E. R. 1978. *Proc. Natl. Acad. Sci. USA* 75:3512–16
425. Hawkins, R. D., Abrams, T. W., Carew, T. J., Kandel, E. R. 1983. *Science* 219:400–5
426. Brunelli, M., Castellucci, V., Kandel, E. R. 1976. *Science* 194:1178–81
427. Klein, M., Camardo, J., Kandel, E. R. 1982. *Proc. Natl. Acad. Sci. USA* 79:5713–17
428. Castellucci, V. F., Nairn, A., Greengard, P., Schwartz, J. H., Kandel, E. R. 1982. *J. Neurosci.* 2:1673–81
429. Bernier, L., Castellucci, V. F., Kandel, E. R., Schwartz, J. H. 1982. *J. Neurosci.* 2:1682–91
430. Strumwasser, F., Kaczmarek, L. K., Jennings, K. R., Chiu, A. Y. 1982. In *Neurosecretion*, ed. P. S. Farner, K. Lederis, pp. 249–68. New York: Plenum
431. Kaczmarek, L. K., Jennings, K., Strumwasser, F. 1978. *Proc. Natl. Acad. Sci. USA* 75:5200–4
432. Kaczmarek, L. K., Jennings, K. R., Strumwasser, F. 1982. *Brain Res.* 238:105–115

SUBJECT INDEX

A

Acebutalol-azide
  beta-adrenergic antagonist
    structure of, 164
Acetic anhydride
  plasma proteinase inhibitors
    and, 668
Acetyl coenzyme A
  long-chain fatty acid synthesis
    and, 538
  pyruvate carboxylase activity
    and, 622
Acetyl coenzyme A carboxylase,
    540–43
  animal, 541–43
  fatty acid synthesis regulation
    and, 569–73
  subunits of, 538
  yeast, 543
Acetylglucosamine
  transfer to chitin, 12
Acetyl low-density lipoprotein
    receptor, 227–35
  biochemical properties of,
    227–28
  distribution on cells, 228–29
  function in vivo, 232–35
  ligand specificity of, 229–32
Acetylmuramic acid
  peptidoglycan formation and,
    11
Acetyl transacylase
  fatty acid synthesis and, 556–
    57
Acrosin
  plasma proteinase inhibitors
    and, 667
Actinomycin D
  intestinal calcium transport
    and, 424
Action potentials
  nerve terminal metabolism
    and, 903–17
Active transport
  ATP-linked, 398–402
  free energy coupling and,
    379–405
  ion translocation and, 386–95
Adenosine diphosphate
  see ADP
Adenosine monophosphate
  see AMP
Adenosine triphosphate
  see ATP
Adenoviruses
  DNA methylation and, 98

Adenylate cyclase
  activation mechanisms of,
    174–80
  agonist-specific binding prop-
    erties of, 174–80
  brain, 904–5
Adenylic acid
  macrophage acetyl-LDL up-
    take and, 231
Adenylyl imidodiphosphate
  enzymatic hydrolysis and sol-
    volysis and, 67–68
Adenylyl methylenediphos-
    phonate
  enzymatic hydrolysis and sol-
    volysis and, 67–68
ADP
  fluorosulfonylbenzoyl analog
    of, 76–77
  periodate-oxidized
    isocitrate dehydrogenase
    and, 70
  platelet function and, 86
Adrenal glands
  carbohydrate metabolism and,
    2–3
Adrenergic receptors, 160–62
  properties of, 161
  see also Beta-adrenergic re-
    ceptors
Affinity chromatography
  erythrocyte beta-adrenergic re-
    ceptor purification and,
    172
Affinity labeling
  purine nucleotide, 67–88
Alanine
  gluconeogenesis and, 619
Albumin
  cholesterol excretion by mac-
    rophages and, 251
Alcohol dehydrogenase
  structure and catalysis of, 25–
    26
Alkyl halides
  purine nucleotides and, 72–
    74
Allosteric enzymes
  purine nucleotides and, 67
  structure and catalysis of, 27–31
Alpha-adrenergic agents
  glycogenolysis, glycolysis,
    and gluconeogenesis and,
    641–43
Alpha$_1$-antichymotrypsin, 678–
    80
Alpha$_2$-antiplasmin, 676–78

Alpha-cysteine proteinase inhibi-
    tor, 698–99
Alpha$_2$-macroglobulin, 684–95
α-Amanitin
  influenza virus transcription
    and, 491
Amantidine
  influenza virus and, 484
Amines
  fibronectin and, 776
Amino acids
  nucleophilic side chains of
    alkyl halides and, 72
  sulfur-containing, 187–216
    industrial applications of,
      215–16
  see also Sulfur amino acids
Amino acid sequences
  of proton ATPases, 804–8
  of ribulose bisphosphate car-
    boxylase, 513–17
Amino acid transport
  glutamyl cycle and, 721–25
Aminoaciduria
  glutamylcysteine synthetase
    deficiency and, 747
Aminobutyric acid
  calcium channel regulation
    and, 887–88
Aminooxyacetate
  gluconeogenesis and, 623
3-Aminopicolinate
  phosphoenolpyruvate carboxy-
    kinase and, 624
AMP
  fructose 1,6-bisphosphatase
    and, 627–28
Angiotensin
  glycogenolysis, glycolysis,
    and gluconeogenesis and,
    641–43
  hypertension and, 6–7
Angiotensinogen
  hypertension and, 6–7
Anhydrotrypsin
  aplha$_2$-macroglobulin and, 693
  plasma proteinase inhibitors
    and, 661
Anoxia
  glycolysis regulation and, 619
Anoxygenic photosynthesis,
    127–35
Antibiotics
  see Beta-lactam antibiotics
Antibodies
  E.coli ribosomal protein iden-
    tification and, 51

927

Antichymotrypsin
  alpha$_1$-, 678–80
Anticollagenase
  beta$_1$-, 696–98
Antiplasmin
  alpha$_2$-, 676–78
Anitihrombin III, 674–76
  heparin and, 674–75
  mechanism of, 675–76
  physiological role of, 676
Apamine
  calcium-activated potassium
    channel and, 880
Apoprotein E
  secretion by macrophages,
    249–55
  synthesis and secretion of
    cholesterol loading and,
    252–55
Arachidonic acid
  conjugates of, 356
  dihydroperoxyeicosatetraenoic
    acid formation and, 360
  leukotriene formation and,
    356, 363–64
  malondialdehyde secretion
    and, 233
Arachidonic acid metabolites
  reaction with lysine residues
    of LDL in vivo, 234
Aspartate transcarbamylase
  structure and catalysis of, 27–
    28
Aspartic acid
  nucleophilic side chains of
    alkyl halides and, 72
Aspartic acid amino transferase
  alpha$_2$-macroglobulin and,
    693
Assembly maps
  ribosomal subunits and, 52–55
Asthma
  leukotrienes and, 371–72
  SRS-A and, 357
Atherosclerosis
  foam cell formation in, 255–
    59
Atherosclerotic plaque
  cholesteryl ester/protein com-
    plex receptors from, 241–
    43
  structure of, 224
ATP
  active transport and, 398–402
  2',3'-dealdehyde derivative
    of, 69
  M. phlei ATPase and, 70
  fluorosulfonylbenzoyl analog
    of, 76–77
  periodate-oxidized
    phosphofructokinase in-
      activation and, 71
  structure of, 69

photophosphorylation and,
    126–27
ATP processing
  pathway for, 395–98
ATP-synthase
  assembly of, 815–16
  reaction mechanism of, 816–
    21
ATP-synthetase
  structure of, 803–13
Avian leukosis virus
  B-cell lymphoma induction
    and, 336, 345–46
  erythroblastosis induction and,
    337
Avidin
  acetyl-CoA carboxylase and,
    540
5-Azacytidine
  DNA ethylation and, 104,
    110–12
Azthreonam
  E. coli filamentation and, 833

B

Bacillus stearothermophilus
  phosphofructokinase of
    structure and catalysis of,
      29–30
  three-dimensional ribosomal
    crystals of, 42–43
Bacillus subtilis
  DNA methyltransferases of,
    98
Bacteriophage lambda
  DNA replication of, 606–8
Bacteriophage φ4
  DNA replication of, 608
Bacteriophage T4
  DNA replication of, 588–92
Bacteriophage T7
  DNA replication of, 583–88
Basophils
  leukotrienes in, 365
B-cell lymphomas
  cellular oncogenes and, 336
B-cell tumors
  cellular oncogenes and, 336
Beta-adrenergic affinity ligands
  characteristics of, 166–67
Beta-adrenergic agents
  gluconeogenesis and, 634–39
Beta-adrenergic receptors
  affinity and photoaffinity
    labels for, 164–71
  desensitization and, 181–84
  ligand-binding studies of, 162
  polypeptides of, 170
  purification of, 171–74
  reconstitution studies of, 180–
    81
  structure of, 163–74

Beta$_1$-anticollagenase, 696–98
Beta-lactam antibiotics, 825–63
  mechanism of action of, 851–
    53
  protein-binding proteins and,
    829–39
Bile salts
  vitamin D$_3$ absorption and,
    413
Biosphere
  sulfur flow through, 188
Biotin
  carboxylation-decarboxylation
    of, 541
Bladder carcinoma
  cellular oncogenes and, 332
Blue-green algae
  see Cyanobacteria
Bone
  1,25-dihydroxyvitamin D$_3$
    and, 426–27
Bone calcium mobilization
  1,25-dihydroxyvitamin D$_3$
    analogs and, 430
Bone mineralization
  1,25-dihydroxyvitamin D$_3$
    analogs and, 430
Bromelain
  heagglutinin cleavage and,
    478
Bromoacetate
  pH-independent reaction rate
    of, 74
Bromoethyl-AMP
  pH-independent reaction rate
    of, 74
3-Bromo-2-ketoglutarate
  pH-independent reaction rate
    of, 74
Brown algae
  photosynthetic apparatus of,
    151
Burkitt's lymphoma
  cellular oncogenes and, 332
Buthionine sulfoximine
  glutamylcysteine synthesis
    and, 726

C

Calcineurin
  calmodulin activation and,
    908–9
Calcitroic acid
  1,25-dihydroxyvitamin D$_3$
    and, 416–17
Calcium
  glycogenolysis and, 641–42
  nerve terminal metabolism
    and, 907–12
  potassium channel activation
    and, 880–81
Calcium channels

nerve terminal function and,
884–88
neurotransmitter release and,
872
Calcium metabolism
glutathione and, 744
Calcium transport
intestinal
mechanism of, 424–26
Calmodulin
calcium-activated potassium
channel and, 880
exocytosis and, 893–94
synaptical calcium transduction and, 908–9
Cancer
cellular oncogenes and, 334–
46
glutathione and, 743
Cancer cells
methionine and cyst(e)ine requirements of, 215
Candida lypolytica
acetyl-CoA carboxylase and,
543
Candida utilis
6-phosphogluconate dehyd-
rogenase of
periodate-oxidized NADP
and, 70
Carbamyl phosphate synthetase
fluorosulfonylbenzoyl analogs
and, 77
Carbohydrate
fibronectin and, 780–81
Carbohydrate metabolism
adrenal glands and, 2–3
Carbonic anhydrase
structure and catalysis of, 20–
22
Carboxypeptidase A
alpha$_2$-macroglobulin and,
693
crystalline phase of, 19–20
exterior sidechain and loop
motions in, 287
structure and catalysis of, 22–
24
Carcinogenesis
cellular oncogenes and, 302
Caryophanon latum L
DNA methyltransferases of,
98
Casein
cholesterol excretion by macrophages and, 251
Casein kinase II
fluorosulfonylbenzoyl analogs
and, 77
Catecholamine biosynthesis
regulation of, 896–97
Catecholamines
gluconeogenesis and, 619

physiologic and pharmacologic effects of, 160–61
storage of, 897–99
Cefsulodin
E. coli cell lysis and, 830
Cefuroxime
E. coli filamentation and, 833
Cell surface interactions, 761–
92
collagen substrates and, 767
factors in, 785–86
fibronectin and, 768–82
laminin and, 782–85
plastic and glass substrates
and, 765–67
Cellular oncogenes, 301–48
cancer and, 334–46
DNA-mediated gene transfer
and, 306–8
expression of, 313–14
families of, 317–19
functions of, 314–17
genetic linkages of, 319–20
insertional mutagenesis and,
306
phylogeny of, 309–13
role in normal cells, 331–34
structural characteristics of,
308–9
transduction by retroviruses,
304–6, 320–31
mechanism of, 325–28
Cephalexin
E. coli filamentation and, 833
Cephaloridine
E. coli cell lysis and, 830
Cephalosporins
penicillin-binding proteins
and, 831
Chicken syncytial virus
B-cell lymphoma induction
and, 336
Chitin
acetylglucosamine transfer to,
12
Chloramphenicol
bacteriophage lambda DNA
replication and, 608
DNA synthesis and, 603
E. coli ribosomal binding site
of, 46
Chlorobiaceae
anoxygenic photosynthesis of,
127–32
Chloroflexaceae
anoxygenic photosynthesis of,
127–32
Chlorophyll chromophores, 143–
44
Chlorophyll-protein complexes
light-harvesting, 140–41
in thylakoids, 141–43
Chloroplast ATPase

fluorosulfonylbenzoyl analogs
and, 77
Chlorosomes
of green bacteria, 128–32
Cholecalcioic acid
1,25-dihydroxyvitamin D$_3$
and, 415–16
Cholera toxin
adenylate cyclase activation
and, 177–78
Cholesterol
atherosclerotic plaque and,
224
form in plasma lipoproteins,
225
lipoprotein-bound
macrophage uptake of,
226–43
secretion by macrophages,
249–55
Cholesterol re-esterification reaction
mechanism of, 246
Cholesteryl ester cycle, 247–49
Cholesteryl ester/protein complex receptors
atherosclerotic plaque and,
241–43
Cholesteryl esters
lipoprotein-derived
cytoplasmic re-esterification
and hydrolysis of,
244–47
in macrophages, 225
Chondroitin sulfate
chondronectin and, 785
fibronectin and, 776
Chondronectin
cell surface interactions with,
785
Chromaffin granules
catecholamine storage and,
897–99
Chromatophores
of purple bacteria, 134
Chromophores
chlorophyll, 143–44
Chromosomes
cellular oncogenes and, 319–
20
Chymotrypsin
plasma proteinase inhibitors
and, 667
Citraconic anhydride
plasma proteinase inhibitors
and, 668
Citrate
fatty acid synthesis and, 570
Citrate synthase
domain movement in, 29–30
Cloxacillin
penicillin-binding proteins
and, 834–35

Coagulation
    alpha$_2$-macroglobulin and,
        694
    antithrombin III and, 676
    plasma proteinase inhibitors
        and, 656
Colcemid
    insulin induction of fatty acid
        synthetase and, 574–75
Colchicine
    insulin induction of fatty acid
        synthetase and, 574–75
Collagen
    atherosclerotic plaque and,
        224
    fibronectin and, 771–72
Collagenases
    plasma proteinase inhibitors
        and, 667
Collagen substrates
    cell surface interactions with,
        767
Colon carcinoma
    cellular oncogenes and, 332
Complement activation
    plasma proteinase inhibitors
        and, 656
Condensing enzyme
    fatty acid synthesis and, 558–
        62
Connective tissue turnover
    plasma proteinase inhibitors
        and, 656
Coronary arteries
    leukotrienes and, 373
Cortisol
    pyruvate metabolism and, 621
Coumermycin
    DNA synthesis and, 603
Cryptomonads
    photosynthetic apparatus of,
        150–51
Cyanide
    detoxification of, 204–5
Cyanobacteria
    oxygenic photosynthesis of,
        145–47
Cyanopindolol-azide
    beta-adrenergic antagonist
        structure of, 164
Cyclic AMP
    desensitization mediation and,
        183
    fatty acid synthesis and, 570
    fructose 2,6-bisphosphate and,
        627–28, 641
    gluconeogenesis and, 634–39
    nerve terminal metabolism
        and, 904–7
    phosphoenolpyruvate carboxy-
        kinase and, 625
    phosphofructokinase labeling
        and, 75

pyruvate metabolism and, 621
1,2-Cyclohexanedione
    plasma proteinase inhibitors
        and, 669
Cycloheximide
    ACAT activity in fibroblasts
        and, 246
    intestinal calcium transport
        and, 424
Cystamine
    enzymatic conversions of, 202
    glutamylcysteine synthesis
        and, 726
Cystathionine
    precursor of in microorgan-
        isms, 191
Cystathioninuria, 214
Cysteamine
    pathway to from lanthionine,
        201
Cyst(e)ine
    keto-acid analogs of, 209–11
Cysteine
    nucleophilic side chains of
        alkyl halides and, 72
    pigment formation and, 212–
        13
    transsulfuration pathway to,
        192
Cysteine-carbonyl adducts, 211–
        12
Cysteine metabolism, 198–209
Cysteine sulfur
    precursor of in vertebrates,
        190
Cysteinyldopas
    generation of, 212–13
Cystinosis
    cause of, 214–15
Cytochrome c
    atomic fluctuations in, 275–76
        time dependence of, 278
    electron transfer in, 294
Cytoplasmic factor
    vitamin D metabolism and,
        414
Cytosine methylation
    biochemical consequences of,
        94
Cytoskeletal proteins
    fibronectin and, 776

D

Desulfotomaculum
    reductive dissimilation of sul-
        fate in, 188
Desulfovibrio
    reductive dissimilation of sul-
        fate in, 188
Dextran sulfate
    interferon secretion and, 235
Diabetes

glucokinase and, 634
gluconeogenesis and, 619
phosphoenolpyruvate carboxy-
    kinase and, 625
Diatons
    photosynthetic apparatus of,
        151
1,3-Dibromo-2-propanone
    yeast synthetase inhibition
        and, 562
Diepoxybutane
    RNA-protein cross-linking
        and, 58
Diffraction
    ribosomal particles and, 40–43
Dihydroperoxyeicosatetraenoic
    acid
    formation of, 360
1,25-Dihydroxyvitamin D$_3$
    action in bone, 426–27
    analogs of, 430
    intestinal calcium transport
        and, 424–26
    intestinal phosphate transport
        and, 426
    localization in target tissues,
        420–21
    receptors for, 421–23
Dimethylsulfoxide
    complex with LADH, 26
Dinoflagellates
    light-harvesting pigments of,
        149–50
Dipeptidase
    glutamyl cycle and, 732–33
DNA
    purine nucleotides and, 67
    structural change of
        DNA methylation and, 102
DNA methylation, 93–119
    base modification and, 95–96
    determinations of, 99–100
    biochemistry of, 95–102
    biological function of, 103–17
    DNA structural changes and,
        102
    eukaryotic gene expression
        and, 103–5
    gene inactivation and, 112–
        114
    genetic recombination and,
        117
    Herpes simplex virus latency
        and, 117
    maintenance and de novo, 96–
        99
DNA polymerase
    bacteriophage T7
        DNA strand elongation and,
            584–86
DNA polymerase III holoen-
    zyme, 596–98
DNA replication

bacteriophage lambda, 606–8
bacteriophage φ29, 608
bacteriophage T4, 588–92
bacteriophage T7, 583–88
duplex, 582–83
E. coli, 592–603
plasmid, 603–6
prokaryotic, 581–609
DNase I
DNA sensitivity toward, 101
Dolichol
derivatives of, 13–14
Drosophila DNA
DNA methylation and, 109–10
Dysbetalipoproteinemia
see Familial dysbetalipoproteinemia

E

EDTA
fructose 1,6-biosphosphatase and, 626
Eicosapentaenoic acid
leukotriene formation and, 356
Eicosatrienoic acid
leukotriene formation and, 356
Electron microscopy
ribosomal particles and, 38–40
ribosomal RNA spatial arrangements and, 56
Endocytosis
lysosomal hydrolysis and, 243–44
receptor-mediated
cholesterol uptake in macrophages and, 224
Enkephalin
calcium channel regulation and, 887–88
Enoyl reductase
fatty acid synthesis and, 562–65
Enzymes
allosteric
purine nucleotides and, 67
structure and catalysis of, 27–31
reactivities in solution, 18–20
structure and catalysis of, 17–31
zinc
structure and catalysis of, 20–26
see also specific enzyme
Eosinophils
leukotrienes in, 365
Epinephrine
cyclic AMP concentration in liver and, 635

glycogenolysis and, 12
phosphoenolpyruvate carboxy-kinase and, 625
physiologic and pharmacologic effects of, 160–61
pyruvate metabolism and, 621
Erythroblastosis
cellular oncogenes and, 337
Erythrocytes
cholesterol excretion by macrophages and, 251–52
Erythroleukemia
cellular oncogenes and, 333
Erythroleukemia cells
DNA hypomethylation and, 110
DNA methyltransferases of, 98
Escherichia coli
aspartate transcarbamylase of
structure and catalysis of, 27–28
penicillin-binding proteins of, 830–34
Escherichia coli DNA
polymerase III holoenzyme, 596–98
Escherichia coli helicases, 601
Escherichia coli proteins
DNA replication of, 592–603
Escherichia coli ribosomal proteins
spatial arrangement in situ, 43–55
Escherichia coli ribosomal subunits
assembly maps of, 52–55
fragmentation of, 59
functional domains on, 46
models of, 39
protein binding sites on, 58–59
protein cross-linking in, 49–51
RNA regions in, 57
spatial packing of RNAs in, 59–60
Escerichia coli RNA polymerase
primer synthesis by, 595–96
Escherichia coli SK
DNA methyltransferases of, 98
Escherichia coli 30S ribosomal subunit
protein mapping on, 44
Escherichia coli 50S ribosomal subunit
protein mapping on, 45
Escherichia coli 70S ribosome
crystals of, 42
models of, 41
Estradiol dehydrogenase
fluorosulfonylbenzoyl analogs and, 77

Ethenoadenosine
fluorescence of, 83
Eumalins
tyrosine oxidation and, 213
Exocytosis
molecular basis of, 890–95
Extracellular material
cellular receptors for, 787–90
cell surface interactions with, 761–92

F

Factor XIa
plasma proteinase inhibitors and, 667
Factor Xa
plasma proteinase inhibitors and, 667
Familial dysbetalipoproteinemia
VLDL and, 237
Familial emphysema
proteinase inhibitor deficiency and, 656
Familial hypercholesterolemia
foam cell and, 257–59
low-density lipoprotein receptors and, 226
monocytic production of LDL receptors and, 238
Fatty acids
gluconeogenesis and, 637
leukotriene formation and, 363–64
Fatty acid synthesis, 537–75
animal, 545–51
glycolysis and, 618
regulation of, 569–73
yeast, 552–56
Fatty acid synthetase, 543–45
eukaryotic, 544–45
functional organization of, 556–65
long-chain fatty acid synthesis and, 538
mechanism of action of, 565–69
metabolic control of, 573
structural organization of, 545–56
Fertility
vitamin D and, 431
Fibrin
atherosclerotic plaque and, 224
fibronectin and, 773
Fibrinolysis
alpha$_2$-macroglobulin and, 694
plasma proteinase inhibitors and, 656
Fibroblasts
ACAT activity in, 246
fibronectin and, 787–88

Fibronectin
  cell surface interactions with,
    768–82
  site of, 771
  glycosaminoglycan-binding
    domains of, 773–76
  interactions with collagen,
    771–72
  interactions with fibrin, 773
  primary structure of, 777–81
  properties of, 768–71
  structure-function relationships
    of, 781–82
Fluorescence
  E. coli ribosomal protein
    arrangements and, 48
Fluorosulfonylbenzoyl analogs
  of nucleotides, 76–87
Fluorosulfonylbenzoyl nuc-
    leosides
  reaction of serine in proteins
    and, 79
Foam cells
  atherosclerosis and, 255–59
Fodrin
  calmodulin and, 908
Free energy coupling
  active transport and, 379–405
Fructose 1,6-bisphosphatase
  gluconeogenesis and, 626–29
Fructose 2,6-bisphosphatase
  gluconeogenesis and, 629
Fructose bisphosphate
  light-dependent regulation of,
    525–26
Fructose 2,6-bisphosphate
  biological effects of, 639
  biosynthesis and degradation
    of, 639–40
  cyclic AMP and, 641
  distribution of, 639
  gluconeogenesis and, 640–41
  glycolysis and, 640–41
Fructose 6-phosphate/fructose
    1,6-bisphosphate cycle
  gluconeogenesis and, 646–47
Fucoidin
  acetyl-LDL binding and, 237
Fucoxanthin
  algal light absorption and, 149
Furazlocillin
  E. coli filamentation and, 833

G

Galactose 1-phosphate
  conversion into glucose 1-
    phosphate, 10
Gamma globulin
  cholesterol excretion by mac-
    rophages and, 251

Gangliosides
  fibronectin and, 788
Gene expression
  eukaryotic
    DNA methylation and,
      103–9
Gene transfer
  DNA-mediated
    cellular oncogenes and,
      306–8
Genetic recombination
  DNA methylation and, 117
Globular proteins
  internal motions of, 265–66
Glomerulonephritis
  plasma proteinase inhibitors
    and, 673
Glucagon
  fatty acid synthesis and, 570
  gluconeogenesis and, 619,
    634–39
  phosphoenolpyruvate carboxy-
    kinase and, 625
  pyruvate metabolism and, 621
Glucocorticoids
  1,25-dihydroxyvitamin D_3 re-
    ceptor levels and, 422
  gluconeogenesis and, 619,
    643
  phosphoenolpyruvate carboxy-
    kinase and, 625
Glucokinase
  gluconeogenesis and, 634
Gluconeogenesis
  control of, 619
  liver, 638
  futile cycles in, 644–48
  lactate disposal and, 618
  stimulation of, 642
Glucosamine phosphate
  formation from glutamine and
    hexosephosphate, 12
Glucose
  gluconeogenesis and, 619,
    643–44
  glycolysis regulation and,
    619
  reaction with lysine residues
    of LDL in vivo and, 233
Glucose 1,6-diphosphate
  phospho-glucomutase reaction
    and, 10
Glucose/glucose 6-phosphate
    cycle
  gluconeogenesis and, 644–45
Glucose 1-phosphate
  galactose 1-phosphate conver-
    sion into, 10
Glucose 6-phosphatase
  gluconeogenesis and, 633–34
Glucose 6-phosphate
  formation from glucose 1-
    phosphate, 10
Glutamate

gluconeogenesis and, 635
Glutamate dehydrogenase
  fluorescence of, 83
  fluorosulfonylbenzoyl analogs
    and, 77, 80–84
  GTP binding sites of
    fluorosulfonylbenzoyl ana-
      logs and, 77–78
  purine necleotide sites of, 80–
    81
  reaction with 5'-FSBG, 82
Glutamic acid
  nucleophilic side chains of
    alkyl halides and, 72
Glutamine
  gluconeogenesis and, 635
  glucosamine phosphate forma-
    tion and, 12
Glutamine synthetase
  fluorosulfonylbenzoyl analogs
    and, 77
Glutamyl amino acids
  transport of, 719–21
Glutamyl cycle
  enzymes of, 725–33
  functions of, 721–25
Glutamyl cyclotransferase
  glutamyl cycle and, 729–
    30
Glutamylcysteine synthetase
  glutamyl cycle and, 725–27
Glutamyl transpeptidase
  glutamyl cycle and, 727–29
Glutathione, 711–51
  conjugation of, 739–41
  deficiency and depletion of,
    744–48
  endogenous compounds of,
    740–41
  exogenous compounds of,
    739–40
  function of, 741–44
  metabolism of, 713–15
  oxidation of, 733–35
  pigment synthesis and, 212
  transport of, 715–18
Glutathionedopas
  generation of, 212–13
Glutathionemia
  glutamyl transpeptidase de-
    ficiency and, 747
  transpeptidase deficiency and,
    719
Glutathione peroxidase, 735
  lung disease and, 673
Glutathione reductase, 738–39
  lung disease and, 673
Glutathione synthetase
  glutamyl cycle and, 727
Glutathione S-transferases, 739–
    40
Glutathione transhydrogenases,
    735–38

Glutathionuria
transpeptidase deficiency and,
719
transpeptidase inhibitors and,
715–16
Glycerol
gluconeogenesis and, 619,
623
Glycogenolysis
epinephrine and, 12
stimulation of, 641–42
Glycolysis
control of, 619–20
liver, 638
stimulation of, 642
Glycosaminoglycans
atherosclerotic plaque and,
224
fibronectin and, 773–76
Green algae
chlorophyll chromophores of,
143–44
chlorophyll-protein complexes
in, 141–43
light-harvesting complex of,
140–41
oxygenic photosynthesis of,
136–45
photosystems of, 137–40
Green bacteria
anoxygenic photosynthesis of,
127–32
Guanine nucleotides
acetyl-CoA carboxylase regulation and, 570

H

Haemophilus spp.
DNA methyltransferases of,
98
HDL
see High-density lipoprotein
Hemagglutinin
virus adsorption and, 476–77
Hemagglutinin gene
biosynthesis of, 480
infectivity and, 480–81
structure of, 477–80
Hemoglobin
rigid body motions in, 290–
91, 294
Hemolytic anemia
glutamylcysteine synthetase
deficiency and, 747
Hemolytic disease
glutathione and, 746–47
Heparin
antithrombin III and, 674–
75
fibronectin and, 773–74
Heparin cofactor II, 700
Hepatocellular carcinoma

DNA hypomethylation and,
115
Herpes simplex
virus latency of
DNA methylation and, 117
Herpes simplex virus
thymidine kinase gene of, 445
Hexosephosphate
glucosamine phosphate formation and, 12
High-density lipoprotein
atherosclerosis and, 256
cholesterol excretion by macrophages and, 249–52
cholesterol re-esterification
and, 250–51
High-performance liquid chromatography
leukotrienes and, 369–70
Histidine
fructose 1,6-bisphosphatase
and, 626
nucleophilic side chains of
alkyl halides and, 72
Histone H4
alpha$_2$-macroglobulin and, 693
Histone kinase
affinity labels of, 73
Homocyst(e)ine
keto-acid analogs of, 209–11
Homocysteine
precursor of in vertebrates,
190
Homocysteine-carbonyl adducts,
211–12
Homocystinuria
causes of, 213–14
5-oxoprolinuria and, 747
Homolanthionine
accumulation in E. coli
mutants, 211
Hyaluronic acid
fibronectin and, 775–76
Hydrocortisone
insulin induction of fatty acid
synthetase and, 574
Hydrophobic reactions
alpha$_2$-macroglobulin and, 693
Hydroxacyl dehydratase
fatty acid synthesis and, 565
Hydroxamate
binding to thermolysin, 24
1α-Hydroxylase
vitamin D metabolism and,
427–28
25-Hydroxylation
vitamin D metabolism and,
413–14
Hydroxysteriod dehydrogenase
fluorosulfonylbenzoyl analogs
and, 77
25-Hydroxyvitamin D
metabolism of, 414

Hypercholesterolemia
see also Familial hyper-
cholesterolemia
Hyperlipidemia
abnormal lipoprotein accumulation in, 224
Hypermethionemia, 193
Hypersensitivity reactions
leukotrienes and, 356
Hypertensin, 6
Hypertension
angiotensin and, 6–7
renin and, 5–6
Hypertriglyceridemia
VLDL binding and, 239–40
Hypophosphatemia
vitamin D metabolism and,
427–28

I

Imidazole
fructose 1,6-bisphosphatase
and, 626
2-Iminothiolane
protein-protein cross-linking
and, 49
Immune electron microscopy
E. coli ribosomal proteins
and, 43–46
Immunoaffinity chromatography
erythrocyte beta-adrenergic receptor purification and,
172
Inflammatory reactions
acetyl-LDL receptor and, 235
leukotrienes and, 356
plasma proteinase inhibitors
and, 656
Influenza virus RNA, 467–500
genome structure of, 471–74
morphology of, 469–70
nucleotide sequences of, 474–
90
transcription and replication
of, 490–99
Insulin
acetyl-CoA carboxylase activity and, 572–73
fatty acid synthetase induction
and, 574
gluconeogenesis and, 619
Inter-alpha-trypsin inhibitor,
695–96
Interferon secretion
stimulators of, 235
Intestinal calcium transport
1,25-dihydroxyvitamin D$_3$
analogs and, 430
mechanism of, 424–26
prolactin and, 431
Intestinal phosphate transport
1,25-dihydroxyvitamin D$_3$
and, 426

Iodination
  E. coli ribosomal protein iden-
    tification and, 51–52
Iodoacetamide
  fatty acid synthetase and,
    559–60
  reaction with amino acid side
    chains, 72
Isocitrate dehydrogenase
  ADP activator site of, 73
  periodate-oxidized ADP and,
    70

K

Kethoxal
  E. coli ribosomal protein iden-
    tification and, 51–52
Ketoacyl reductase
  fatty acid synthesis and, 562–
    65
Ketoacyl synthetase
  fatty acid synthesis and, 558–
    62
Kinases
  domain movement in, 29–30
  fluorosulfonylbenzoyl analogs
    and, 77

L

Lactate
  gluconeogenesis and, 619
Lactation
  vitamin D and, 430–31
Lactoperoxidase
  E. coli ribosomal protein iden-
    tification and, 51–52
Laminin
  cell surface interactions with,
    782–85
  function of, 783
  structural and functional do-
    mains of, 783–85
Lanthionine
  pathway to cysteamine from,
    201
LDL
  see Low-density lipoprotein
Leucyl-tRNA synthetase
  affinity labels of, 73
Leukemias
  cellular oncogenes and, 332
  see also specific type
Leukocytes
  leukotrienes and, 373
Leukotriene A_4
  biosynthesis of, 362–63
  stereochemistry of, 361–62
Leukotriene B_4, 359–60
  biosynthesis of, 361
  oxidation of, 367
Leukotriene C_3
  formation of, 364

metabolism of, 366
Leukotriene C_4, 356–59
  biosynthesis of, 361
  structure of, 358–59
Leukotriene C_5
  formation of, 364
Leukotriene D_4
  formation of, 366
Leukotrienes, 355–73
  biological effects of, 371–73
  biosynthesis of, 360–65
  discovery of, 356–60
  distribution and excretion of,
    369
  glutathione and, 740–41
  high-performance liquid chro-
    matography and, 369–70
  mass spectrometry and, 370
  metabolism of, 366
  occurrence of, 365
  radioimmunoassay and, 370–
    71
  structure of, 356–60
Ligand binding
  beta-adrenergic receptors and,
    162
Ligand-protein interactions
  in myoglobin, 283–87
Light-harvesting systems, 125–
    52
Liver alcohol dehydrogenase
  structure and catalysis of, 25–
    26
Liver dysfunction
  methionine toxicity and, 193
Low-density lipoprotein/dextran
    sulfate complexes
  receptor for, 235–37
Low-density lipoprotein recep-
    tors
  familial hypercholesterolemia
    and, 226
Luciferase
  fluorosulfonylbenzoyl analogs
    and, 77
Lumisterol
  vitamin D metabolism and,
    413
Lung carcinoma
  cellular oncogenes and, 332
Lysine
  nucleophilic side chains of
    alkyl halides and, 72
Lysosomal hydrolysis
  endocytosis and, 243–44
Lysozyme
  hinge bending in, 288–90
  molecular conformation in, 19

M

Macroglobulin
  alpha$_2$-, 684–95

Macrophages
  cholesterol and apoprotein E
    secretion by, 249–55
  cholesteryl esters in, 225
  fibronectin and, 789–90
  LDL/dextran sulfate binding
    site on, 236
  lipoprotein-bound cholesterol
    processing and storage
    by, 243–49
  lipoprotein-bound cholesterol
    uptake by, 226–43
  pathogenesis of atherosclerosis
    and, 256–57
Magnesium
  fructose 1,6-bisphosphatase
    and, 626
Malate dehydrogenase
  fluorsulfonylbenzoyl analogs
    and, 77
Maleic anhydride
  plasma proteinase inhibitors
    and, 668
Malonyl coenzyme A
  long-chain fatty acid synthesis
    and, 538
Malonialdihyde
  macrophage uptake of lipopro-
    tein and, 230
  secretion by platelets and
    macrophages, 233
Malonyl transacylase
  fatty acid synthesis and, 556–
    57
Mammary carcinoma
  cellular oncogenes and, 337
Mass spectrometry
  leukotrienes and, 370
Mastocytoma cells
  leukotrienes in, 365
Melanomas
  cellular oncogenes and, 338
3-Mercaptopyruvate-cysteine dis-
    ulfiduria, 213
3-Mercaptopyruvate metabolism,
    207–9
3-Mercaptopyruvate sulfurtrans-
    ferase
  sulfane sulfur transfer and,
    206–7
Methanethiol
  cell division and, 194
Methionine
  keto-acid analogs of, 209–11
  nucleophilic side chains of
    alkyl halides and, 72
Methionine catabolism
  transaminative pathway of,
    193–95
Methionine degradation
  transsulfuration pathway of,
    191–93
Methionine formation

in microorganisms, 190–91
Methionine sulfoxide reductase
plasma proteinase inhibitors
and, 673
Methionine sulfoximine
glutamylcysteine synthesis
and, 726
Methylation
mRNA, 458–59
Methylbuthionine sulfoximine
glutamylcysteine synthesis
and, 726
5-Methylcytosine
crystallization of, 95–96
detection of, 99
distribution of, 100–2
Mitochondrial F₁-ATPase
fluorosulfonylbenzoyl analogs
and, 77
Monocytes
fibronectin and, 787
LDL receptor production and,
238
Monocytic hemopoietic cells
differentiation of
cellular oncogenes and, 332
Mouse mammary tumor virus
mammary carcinoma induction
and, 337
mRNA
cytoplasmic stability of, 460–
62
influenza virus
in vivo transcription of,
494–95
nuclear-cytoplasmic transport
of, 459–60
mRNA formation
eukaryotic
pathway of, 441–62
Mycobacterium phlei
ATPase of
affinity labeling and, 70
Myelin basic protein
alpha₂-macroglobulin and, 693
Myeloblastosis-associated virus
renal carcinoma induction
and, 336–37
Myoglobin
atomic fluctuations in, 289
ligand-protein interaction in,
283–87
Myopathy
glutamylcysteine synthetase
deficiency and, 747
Myosin ATPase
photoaffinity label of, 75

N

NAD
fluorosulfonylbenzoyl analog
of, 76–77

NADH
complex with LADH, 26
fluorosulfonylbenzoyl analog
of, 76–77
NAPD
periodate-oxidized
C. utilis 6-
phosphogluconate de-
hydrogenase and, 70
NADPH dialdehyde deriva-
tives and, 70–71
NADPH
dialdehyde derivatives of
periodate-oxidized NADP
and, 70–71
long-chain fatty acid synthesis
and, 538
vitamin D metabolism and,
414
Nalidixic acid
DNA synthesis and, 603
Nerve terminal function, 871–
917
calcium channels and, 884–88
calcium removal from cytoplasm and, 889–90
exocytosis and, 890–95
membrane and protein movements in neurons and,
899–903
neurotransmitter levels and,
895–99
potassium channels and, 878–
82
sodium channels and, 875–78
sodium-potassium gradient
and, 882–84
Nerve terminal metabolism
induction of modifications in,
903–17
Neuraminidase
biosynthesis of, 483–84
influenza virus, 482–84
Neurons
membrane and protein movements in, 899–903
sodium channels in, 877–78
Neurotransmitters
nerve terminal metabolism
and, 903–17
regulation in nerve terminal,
895–99
release of, 872
Neutron scattering
E. coli ribosomal proteins
and, 46–48
Neutrophil cathepsin G
plasma proteinase inhibitors
and, 667
Neutrophil elastase
plasma proteinase inhibitors
and, 667
Norepinephrine

acetyl-CoA carboxylase inactivation and, 572
calcium channel regulation
and, 887
physiologic and pharmacologic effects of, 160–61
Novobiocin
bacteriophage lambda DNA
replication and, 608
Nucleocapsid protein
influenza virus, 481–82
Nucleoside triphosphate
photoreactive analog of, 76
Nucleotide regulatory protein
adenylate cyclase/beta-
adrenergic receptor in-
teraction and,
174
Nucleotides
fluorosulfonylbenzoyl analogs
of, 76–87
periodate-oxidized, 69–72
purine
affinity labeling and, 67–88
alkyl halide derivatives of,
72–74
Nucleotide sequences
of bacteriophage T7, 587
of influenza virus RNA segments, 474–90

O

Oncogenes
see Cellular oncogenes
Oncogenesis
DNA damage and, 340–43
Oxidants
plasma proteinase inhibitors
and, 670–72
5-Oxoprolinase
fluorosulfonylbenzoyl analogs
and, 77–78
glutamyl cycle and, 730–32
5-Oxoprolinuria
glutathione and, 746–47
Oxygenic photosynthesis, 135–
52
Oxygen toxicity
glutathione and, 742–43

P

Palmitate synthesis
fatty acid synthetase and,
556–65
mechanism of, 567
Palmitoyl thioesterase
fatty acid synthesis and, 565
Pancreatic elastase
plasma proteinase inhibitors
and, 667
Pancreatic trypsin inhibitor

atomic fluctuations in, 273–74
    time dependence of, 278–79
crystal simulation of, 276
tyrosines in, 279–83
Pantetheine
    yeast fatty acid synthetase
        and, 552
Papain
    plasma proteinase inhibitors
        and, 662
Para-azidobenzylcarazolol
    beta-adrenergic antagonist
        structure of, 164
Para-azidobenzylpindolol
    beta-adrenergic antagonist
        structure of, 164
Parvalbumin
    calmodulin activation and,
        908–9
Penicillin-binding proteins, 825–
    63
    beta-lactam antibiotics and,
        829–39
    biochemical studies of, 839–
        51
    properties of, 828–29
    structural studies of, 851–58
Peptidoglycans
    formation of
        intermediates in, 11
Peridinin
    algal light absorption and,
        149–50
Periodate oxidation
    RNA-protein cross-linking
        and, 58
Peripheral neuropathy
    glutamylcysteine synthetase
        deficiency and, 747
Phaeomalins
    tyrosine oxidation and, 213
Phagocytosis
    cholesterol uptake by mac-
        rophages and, 226
Phenothiazines
    exocytosis inhibition and,
        893
Phenoxybenzamine
    acetyl-CoA carboxylase in-
        activation and, 572
Phentolamine
    acetyl-CoA carboxylase in-
        activation and, 572
Phenylephrine
    acetyl-CoA carboxylase in-
        activation and, 572
    cyclic AMP concentration in
        liver and, 635
    pyruvate dehydrogenase activ-
        ity and, 620
Phenylglyoxal hydrate
    plasma proteinase inhibitors
        and, 668–69

Phosphate transport
    intestinal, 426
Phosphoenolpyruvate
    bacterial enzyme catalysis in-
        hibition and, 29
Phosphoenolpyruvate carboxyki-
    nase
    gluconeogenesis and, 623–25
    GTP binding sites of
        fluorosulfonylbenzoyl ana-
            logs and, 77–78
Phosphofructokinase
    allosteric site of
        labeling of, 75
    fluorosulfonylbenzoyl analogs
        and, 77
    gluconeogenesis and, 629–33
    inactivation of
        periodate-oxidized ATP
            and, 71
    structure and catalysis of, 29–
        30
Phospho-glucomutase reaction
    glucose 1,6-diphosphate and,
        10
6-Phosphogluconate dehyd-
    rogenase
    C. utilis
        periodate-oxidized NADP
            and, 70
Phosphoribulokinase
    light-dependent regulation of,
        525–26
Phosphorylase a
    structure and catalysis of, 30–
        31
Phosphorylase b
    affinity labels of, 73
    structure and catalysis of, 30–
        31
Phosphorylase kinase
    detection in brain, 909
Photophosphorylation
    ATP production and, 126–27
Photorespiration
    ribulose bisphosphate carboxy-
        lase and, 509
Photosynthesis
    anoxygenic, 127–35
    fructose 1,6-bisphosphatase
        and, 628
    oxygenic, 135–52
Photosynthetic accessory pig-
    ments
    algal light-harvesting, 147–52
Photosynthetic light-harvesting
    systems, 125–52
Photosystem I
    ATP production and, 126–27
    of green algae, 137–39
    reaction center of, 126
Photosystem II
    of green algae, 139–40

reaction center of, 126
Phycobilisomes
    cyanobacterial, 145–47
Pigment formation
    cysteine and, 212–13
Piperacillin
    E. coli filamentation and, 833
Plasma proteinase inhibitors,
    655–700
    alpha$_1$-, 663–74
        physiological role of, 672–
            74
        polymorphism of, 665–67
    alpha$_1$-antichymotrypsin, 678–
        80
    alpha$_2$-antiplasma, 676–78
    alpha-cysteine, 698–99
    alpha$_2$-macroglobulin, 684–95
    antithrombin III, 674–76
    beta$_1$-anticollagenase, 696–98
    C1-inhibitor, 681–84
    inter-alpha-trypsin inhibitor,
        695–96
    mechanism of action of, 659–
        63
    proteinase associations with,
        658
Plasmid primases
    conjugal transfer and, 606
Plasmid R6K
    DNA replication of, 606
Plasmids
    DNA replication of, 603–6
Plasmin
    plasma proteinase inhibitors
        and, 667
Platelet aggregation
    ADP-induced, 86
    fibronectin and, 789
    purine nucleotides and, 67
Platelets
    ADP receptor protein of, 86–
        87
        fluorosulfonylbenzoyl ana-
            logs and, 77
Polyamines
    fibronectin and, 776
Polycolominic acid
    macrophage acetyl-LDL up-
        take and, 232
Polyglutemic acid
    machrophage acetyl-LDL up-
        take and, 232
Polyguanylic acid
    macrophage acetyl-LDL up-
        take and, 231–32
Polyinosinic acid
    interferon secretion and, 235
Polymorphonuclear leukocytes
    leukotrienes in, 365, 373
Polypurines
    macrophage acetyl-LDL up-
        take and, 231

Polypyrimidines
macrophage acetyl-LDL up-
take and, 232
Polysaccharides
macrophage acetyl-LDL up-
take and, 232
Polyvinyl sulfate
interferon secretion and, 235
Potassium channels
nerve terminal function and,
878–82
neurotransmitter release and,
872
Previtamin D_3
vitamin D production and,
413
Prochlorophyta
oxygenic photosynthesis of,
147
Prolactin
intestinal calcium transport
and, 431
Promyelocytic leukemia
cellular oncogenes and, 332
Propionate
gluconeogenesis and, 635
Prostaglandin synthesis
leukotrienes and, 373
Proteinase inhibitors
see Plasma proteinase inhibitors
Protein complexes
E. coli ribosomal subunits
and, 51
Protein dynamics, 263–97
alpha-helix motion and, 291–
92
atomic fluctuations and, 273–
79
methodology of, 268–73
rigid body motions and, 288–
91
sidechain motions and, 279–
88
Protein kinases
acetyl-CoA carboxylase inactivation and, 572
calcium-calmodulin-sensitive,
909–11
cyclic AMP-dependent, 905–7
affinity labels of, 73
fluorosulfonylbenzoyl analogs
and, 77
nerve terminal metabolism
and, 912–17
Protein-protein cross-linking
E. coli ribosomal subunits
and, 49–51
Proteins
atomic fluctuations in, 273–
79
biological function of, 278–
79

temperature factors and,
273–77
time-dependence of, 277–78
encoded by cellular
oncogenes, 315
rigid body motions in, 288–91
sidechain motions in, 279–88
see also specific type
Prothionine sulfoximine
glutamylcysteine synthesis
and, 726
Proton ATPases, 801–22
assembly of, 815–16
binding and labeling studies
of, 813–15
reaction mechanism of, 816–
21
structure of, 803–13
Proton wire, 390
Proto-oncogene
conversion to oncogene, 338–
40
Pseudomonas aeruginosa elastase
plasma proteinase inhibitors
and, 662
Purine nucleotide affinity labels,
67–88
Purine nucleotide analogs
photoreactive, 74–76
Purine nucleotides
affinity labeling and, 67–88
alkyl halide derivatives of,
72–74
Puromycin
DNA synthesis and, 603
E. Coli ribosomal binding site
of, 46
Purple bacteria
anoxygenic photosynthesis of,
132–35
Putrescine
LDL/dextran sulfate complex
formation and, 237
Pyridoxal phosphate
enoyl reductase activity inhibition and, 563–64
Pyruvate
fatty acid synthesis and, 618
gluconeogenesis and, 619
Pyruvate carboxylase
gluconeogenesis and, 620–23
Pyruvate kinase
fluorosulfonylbenzoyl analogs
and, 84–86
5'-FSBG inactivation of, 85
gluconeogenesis and, 625–26
GTP binding sites of
fluorosulfonylbenzoyl analogs and, 77–78
Pyruvate-phosphoenolpyruvate
cycle
gluconeogenesis and, 647–48

R

Radiation inactivation
beta-adrenergic receptor structure and, 173
Radioimmunoassay
leukotrienes and, 370–71
Receptors
see also specific type
Red algae
photosynthetic apparatus of,
152
Renal carcinoma
cellular oncogenes and, 336–
37
Renin
hypertension and, 5–6
Reproduction
vitamin D and, 430–32
Respiratory function
leukotrienes and, 371–72
Restriction endonucleases
DNA methylation and, 93–94
Retroviruses
cellular oncogene transduction
by, 304–6, 320–31
mechanism of, 325–28
Rheumatoid arthritis
plasma porteinase inhibitors
and
physiological role of, 673
Rhodanese
sulfane sulfur transfer and,
205
Rhodopseudomonas viridis
photosynthetic reaction center
from, 126
Ribosomal particles
diffraction and, 40–43
electron microscopy and, 38–
40
small-angle scattering and,
37–38
Ribosomal RNA
spatial arrangement in situ,
55–60
Ribosomes
architecture of, 35–60
Ribulose bisphosphate carboxylase, 507–32
activation of, 521–26
amino acid sequences of,
513–17
catalysis of, 526–31
genetics of, 510–12
quaternary structure of, 517–
19
subunit composition of, 512–
13
subunit functions of, 519–
20
Ribulose bisphosphate carboxylase-oxygenase

physicochemical properties of, 512–20
Rickets
vitamin D-dependence type II, 423
Rifampicin
bacteriophage lambda DNA replication and, 608
RNA
purine nucleotides and, 67
RNA polymerase
bacteriophage T7
DNA replication and, 583–84
E. coli
primer synthesis by, 595–96
fluorosulfonylbenzoyl analogs and, 77
gene expression and, 106–7
prokaryotic mRNA formation and, 441–42
RNA polymerase II
DNA-dependent
aminitin and, 491
transcription units of, 443–44
RNA-protein cross-links
ribosomal RNA spatial arrangements and, 57–58
RNA-RNA cross-links
ribosomal RNA spatial arrangements and, 56–57
RNA synthesis
influenza virus
control of, 496–97
virion, 496

S

Saccharomyces cerevisiae
acetyl-CoA carboxylase and, 543
fructose 1,6-bisphosphatase and, 628
S-adenosylmethionine metabolism, 195–97
Sarcomas
cellular oncogenes and, 333, 338
Serine proteinases
plasma proteinase inhibitors and, 667
Serratia marcescens metalloproteinase
plasma proteinase inhibitors and, 662
Siphonaxanthin
green algal light absorption and, 149
Slow-reacting substance of anaphylaxis
see Leukotrienes
Small-angle scattering
ribosomal particles and, 37–38

Sodium channels
nerve terminal function and, 875–78
neurotransmitter release and, 872
Sodium-potassium gradient
nerve terminal function and, 882–84
Somatostatin
calcium channel regulation and, 887–88
Sorbitol
gluconeogenesis and, 623
Spermidine
fibronectin and, 776
LDL/dextran sulfate complex formation and, 237
Spermine
LDL/dextran sulfate complex formation and, 237
Spinocerebellar degeneration
glutamylcysteine synthetase deficiency and, 747
SRS-A
see Leukotrienes
Starvation
gluconeogenesis and, 619
phosphoenolpyruvate carboxykinase and, 625
Subtilisin Carlsberg
crystallization and, 19
Sulfane sulfur
enzymes transferring, 204–7
Sulfinoacetaldehyde
stability of, 204
Sulfituria, 213
Sulfur
assimilation into microorganism amino acids, 188–90
Sulfur amino acid metabolism
inborn errors of, 213–15
Sulfur amino acids, 187–216
industrial applications of, 215–16
sulfur cycle and, 188
Superoxide dismutase
lung disease and, 673
Synexin
calcium-triggered exocytosis and, 894–95

T

Tachysterol
vitamin D metabolism and, 413
Taurine
precursors of, 201
Temperature
atomic fluctuations in proteins and, 273–77

Thermolysin
structure and catalysis of, 24–25
Thermus thermophilus
DNA methyltransferases of, 98
Thioredoxin
conversion of PAPS to sulfite in yeast and, 189
fructose 1,6-bisphosphatase and, 628
Thiosulfate reductase
sulfane sulfur transfer and, 205–6
Thiotaurine
precursors of, 201
Thrombin
plasma proteinase inhibitors and, 667
Thrombospondin
platelet interactions and, 790
Thylakoids
chlorophyll-protein complexes in, 141–43
Thyroglobulin
cholesterol excretion by macrophages and, 251
Thyroxine
acetyl-CoA carboxylase regulation and, 573
Tissue kallikrein
plasma proteinase inhibitors and, 667
Transaminative pathway
methionine catabolism and, 193–95
Transpeptidase deficiency
glutathionuria and gluthionemia and, 719
Transsulfuration pathway
methionine degradation and, 191–93
Trehalose-6-phosphate
uridine diphosphate glucose and, 12
Trichochromes
tryosine oxidation and, 213
Triethylammonium
potassium channel and, 879–80
Trifluoperazine
calcium-activated potassium channel and, 880
Triiodothyronine
insulin induction of fatty acid synthetase and, 574
Trypsin
E. coli ribosomal protein identification and, 51
plasma proteinase inhibitors and, 661, 667
Tryptophan
gluconeogenesis and, 623–24

Tryptophanyl-tRNA synthetase
affintiy labels of, 73
Tumorigenesis
cellular oncogenes and, 336–
38
initiation of, 302
Tyrosine hydroxylase
catecholamine biosynthesis
and, 896–97
Tyrosine oxidation
pigment formation and, 213
Tyrosine phosphorylation
mitogenic growth factors and,
315–16
Tyrosines
pancreatic trypsin inhibitor
and, 279–83

U

Ultraviolet irradiation
RNA-protein cross-linking
and, 58
Uridine diphosphate glucose
mechanism of action of, 11–
13
Uridine diphosphate glucose
acetylgalactosamine
isolation from liver, 12

Urinary thiocyanate
origin of, 208–9
Urokinase
plasma proteinase inhibitors
and, 667

V

Vasopressin
glycogenolysis, glycolysis,
and glyuconeogenesis
and, 641–43
pyruvate dehydrogenase activity and, 620
Very low-density lipoprotein receptor, 237–41
biochemical properties of, 238
function in vivo, 240–41
ligand specificity of, 238–40
Vinblastine
insulin induction of fatty acid
synthetase and, 574–75
Viral latency
DNA methylation and, 117
Viral oncogenes
cellular oncogenes and, 334–
35
Virion RNA synthesis, 496
Vitamin D, 411–33

reproduction and, 430–32
Vitamin D metabolism
pathways of, 412–20
regulation of, 427–28
VLDL
see Very low-density lipoprotein
von Willebrand factor
platelet interactions and, 790

Y

Yeast
acetyl-CoA carboxylase and,
543
conversion of PAPS to sulfite
in, 189
Yeast fatty acid synthetase, 552–
56
pantetheine content of, 552
Yeast pyruvate kinase
fluorosulfonylbenzoyl analogs
and, 77

Z

Zinc enzymes
structure and catalysis of, 20–
26

CUMULATIVE INDEXES

CONTRIBUTING AUTHORS, VOLUMES 48–52

A

Abelson, J., 48:1035–69
Adams, E., 49:1005–61
Adelstein, R. S., 49:1005–61
Adhya, P., 49:967–96
Aisen, P., 49:357–93
Amzel, L. M., 48:961–97;
52:801–24
Anderson, M. E., 52:711–60
Argos, P., 50:497–532
Ashwell, G., 51:531–54

B

Badwey, J. A., 49:695–726
Baldwin, R. L., 51:459–89
Barker, H. A., 50:23–40
Barondes, S. H., 50:207–31
Beavo, J. A., 49:923–59
Bell, R. M., 49:459–87
Benkovic, S. J., 49:227–51
Beychok, S., 48:217–50
Bishop, J. M., 52:301–54
Blumenthal, T., 48:525–48
Blundell, T., 51:123–54
Bogomolni, R. A., 51:587–616
Bornstein, P., 49:957–1003
Bowers, B., 51:763–93
Breathnach, R., 50:349–83
Breslow, E., 48:251–74
Bretscher, M. S., 50:85–101
Brown, M. S., 52:223–61
Brownlee, M., 50:385–432
Busch, H., 51:617–54

C

Cabib, E., 51:763–93
Carmichael, G. G., 48:525–48
Caron, M. G., 52:159–86
Carpenter, G., 48:193–216
Cerami, A., 50:384–432
Challberg, M. D., 51:901–34
Chambon, P., 50:349–83
Chaudhuri, A., 51:869–900
Chock, P. B., 49:813–43;
51:935–71
Choi, Y. C., 51:617–54
Choppin, P. W., 52:467–506
Chowdhry, V., 48:293–325
Ciechanover, A., 51:335–64

Cohen, S., 48:193–216
Cohn, M., 51:365–94
Coleman, R. A., 49:459–87
Coligan, J. E., 50:1025–52
Colman, R. F., 52:67–91
Conti-Tronconi, B. M., 51:491–
530
Coon, M. J., 49:315–56
Cooper, A. J. L., 52:187–222
Cooper, P. K., 48:783–836
Cox, G. B., 48:103–31
Coyette, J., 48:73–101
Crawford, I. P., 49:163–95
Cross, R. L., 50:681–714
Crouch, T. H., 49:489–515
Cushman, D. W., 51:283–308

D

Danner, D. B., 50:41–68
Dawid, I. B., 49:727–64
de Haro, C., 48:549–80
Deich, R. A., 50:41–68
DeLuca, H. F., 52:411–39
de Meis, L., 48:275–92
DePamphilis, M. L., 49:481–
532
Doerfler, W., 52:93–124
Downie, J. A., 48:103–31
Dressler, D., 51:727–61
Dusart, J., 48:921–56

E

Engelhardt, W. A., 51:1–19
Englund, P. T., 49:695–726
Ericsson, L. H., 50:261–84
Ewenstein, B. M., 50:1025–52

F

Felsenfeld, G., 49:1115–56
Ferguson, S. J., 51:185–217
Fillingame, R. H., 49:1079–
1113
Fisher, H. W., 48:649–79
Fletterick, R. J., 49:31–61
Folk, J. E., 49:517–31
Foster, D. W., 49:395–420
Frank, L., 49:1005–61
Frazier, W., 48:491–523
Frère, J. M., 48:73–101

Frieden, C., 48:471–89
Friedman, F. I., 48:217–50

G

Gabel, C. A., 50:815–43
Gadian, D. G., 50:69–83
Ganesan, A. K., 48:783–836
Garfinkel, A. S., 49:667–93
Geider, K., 50:233–60
Gelfand, E. W., 50:845–77
Gellert, M., 50:879–910
Ghuysen, J. M., 48:73–101
Gibson, F., 48:103–31
Gilman, A. G., 49:533–64
Glaser, L., 48:491–523
Glazer, A. N., 52:125–57
Goldstein, J. L., 52:223–61

H

Hajduk, S. L., 51:695–726
Hakomori, S., 50:733–64
Hammarström, S., 52:355–77
Hanawalt, P. C., 48:783–836
Harford, J., 51:531–555
Harper, E., 49:1063–78
Heidelberger, M., 48:1–21
Heinrich, P. C., 49:63–91
Hers, H. G., 52:617–53
Hoffmann-Berling, H., 50:233–
60
Hershko, A., 51:335–364
Holcenberg, J. S., 51:795–812
Holzer, H., 49:63–91
Hörz, W., 51:89–121
Huang, C. Y., 51:935–71
Hubbard, S. C., 50:555–83
Hue, L., 52:617–53

I

Igo-Kemenes, T., 51:89–121
Innis, R. B., 48:755–82
Isenberg, I., 48:159–91
Ivatt, R. J., 50:555–83

J

Jackson, C. M., 49:765–811
Jelinek, W. R., 51:813–44
Jones, M. E., 49:253–79

940

Joshi, V. C., 52:537–79

K

Karnovsky, M. L., 49:695–726
Karplus, M., 52:263–300
Kato, I., 49:593–626
Kedes, L. H., 48:837–70
Kelly, R. B., 52:871–926
Kelly, T. J., 51:901–34
Kim, P. S., 51:459–89
Kindt, T. J., 50:1025–52
Klee, C. B., 49:489–515
Klobutcher, L. A., 50:533–54
Knowles, J. R., 49:877–919
Koshland, D. E., Jr., 50:765–82
Krab, K., 50:623–55
Krebs, E. G., 48:923–59
Kreil, G., 50:317–48
Kunkel, T. A., 51:429–57
Kuśmierek, J. T., 51:655–93

L

Lamb, R. A., 52:467–506
LaNoue, K. F., 48:871–922
Laskowski, M., Jr., 49:593–626
Lazarides, E., 51:219–50
Lefkowitz, R. J., 52:159–86
Leloir, L. F., 52:1–15
Lengyel, P., 51:251–283
Leyh-Bouille, M., 48:73–101
Lindahl, T., 51:61–87
Lipscomb, W. N., 52:17–34
Listowsky, I., 49:357–93
Loeb, L. A., 51:429–57
Long, E. O., 49:727–64
Lorimer, G. H., 52:507–35
Luck, J. M., 50:1–22

M

Macino, G., 48:419–41
Madsen, N. B., 49:31–61
Maitra, U., 51:869–900
Malmström, B. G., 51:21–59
Marini, J. C., 51:695–726
Martin, D. W. Jr., 50:845–77
Mazur, S. J., 50:977–1024
McCammon, J. A., 52:263–300
McGarry, J. D., 49:395–420
McGhee, J. D., 49:1115–56
Meister, A., 50:911–68; 52:711–60
Melançon, P., 50:977–1024
Mellman, I. S., 50:1053–86
Miziorko, H. M., 52:507–35
Mortenson, L. E., 48:387–418
Moss, J., 48:581–600

N

Nathenson, S. G., 50:1025–52
Neilands, J. B., 50:715–31

Nemerson, Y., 49:765–811
Nevins, J. R., 52:441–66
Nguyen-Distèche, M., 48:73–101
Nilsson-Ehle, P., 49:667–93
Northrop, D. B., 50:103–31
Nossal, N. G., 52:581–615

O

Ochoa, S., 45:191–216, 48:549–80; 49:1–30
Ogawa, T., 49:421–57
Okazaki, T., 49:241–57
Ondetti, M. A., 51:283–308
Op den Kamp, J. A. F., 48:47–71

P

Parmelee, D. C., 50:261–84
Parsons, T. F., 50:465–95
Pearse, B. M. F., 50:85–101
Pedersen, P. L., 52:801–24
Perutz, M. F., 48:327–86
Petes, T. D., 49:845–76
Pierce, J. G., 50:464–95
Poljak, R. J., 48:961–97
Porter, R. R., 50:433–64
Potter, H., 51:727–61

R

Rabinowitz, J. C., 49:139–61
Radda, G. K., 50:69–83
Raftery, M. A., 51:491–530
Record, M. T. Jr., 50:977–1024
Reddy, R., 51:617–54
Reed, G. H., 51:365–94
Reichard, P., 48:133–58
Reichardt, L. F., 52:871–926
Reid, B. R., 50:969–96
Reid, K. B. M., 50:433–64
Rhee, S. G., 49:765–811; 51:935–71
Richmann, P. G., 49:489–515
Roberts, R., 51:763–93
Roe, J. H., 50:977–1024
Rosenberg, L. E., 50:1053–86
Ross, E. M., 49:533–64
Rossman, M. G., 50:497–532
Rothblum, L., 51:617–54
Ruddle, F. H., 50:533–54

S

Sage, H., 49:957–1003
Salvesen, G. S., 52:655–709
Saraste, M., 50:623–55
Schackmann, R.F W., 50:815–43
Scharff, M. D., 50:657–80
Schimmel, P. R., 48:601–48
Schlesinger, M. J., 50:193–206

Schmid, C. W., 51:813–44
Schnoes, H. K., 52:411–39
Schotz, M. C., 49:667–93
Schoolwerth, A. C., 48:871–922
Schroepfer, G. J. Jr., 50:585–621; 51:555–85
Sebald, W., 48:419–41; 49:281–314
Selzer, G., 48:999–1034
Sennett, C., 50:1053–86
Shanner, S. L., 50:977–1024
Shapiro, B. M., 50:815–43
Shavit, N., 49:111–38
Sherman, M. I., 48:443–70
Shooter, E. M., 51:845–68
Singer, B., 51:655–93
Smith, C. A., 48:783–836
Smith, H. O., 50:41–68
Snyder, S. H., 48:755–82
Söll, D., 48:601–48
Sorgato, M. C., 51:185–217
Stadel, J. M., 52:159–86
Stadtman, E. R., 49:813–43
Stadtman, T. C., 49:93–110
Stauffer, G. V., 49:163–95
Stoeckenius, W., 51:587–616
Stoops, J. K., 52:537–79
Stringer, E. A., 51:869–900
Strobel, G. A., 51:309–33
Strominger, J. L., 52:825–69
Sweeney, W. V., 49:139–61

T

Tanford, C., 52:379–409
Taylor, J. M., 48:681–717
Thelander, L., 48:133–58
Thorneley, R. N. F., 48:387–418
Timasheff, S. N., 49:565–91
Titani, K., 50:261–84
Tolbert, N. E., 50:133–57
Tomizawa, J., 48:999–1034
Travis, J., 52:655–709
Tzagoloff, A., 48:419–41

U

Uehara, H., 50:1025–52
Unger, L., 50:977–1024

V

Vaugham, M., 48:581–600
Vianna, A. L., 48:275–92
von Jagow, G., 49:281–314

W

Wakil, S. J., 52:537–79
Walsh, K. A., 50:261–84
Waring, M. J., 50:159–92
Wassarman, P. M., 49:627–66
Waxman, D. J., 52:825–69

Weinberg, R. A., 49:197–226
Wellner, D., 50:911–68
Westheimer, F. H., 48:293–325
White, R. E., 49:315–56
Wickner, W., 48:23–45
Wikstrom, M., 50:623–55
Williams, R. C., 48:649–79

Wittmann, H. G., 51:155–83; 52:35–65
Wold, F., 50:783–814
Wood, S., 51:123–54
Wool, I. G., 48:719–54

Y

Yamada, K. M., 52:761–99

Yankner, B. A., 51:845–68
Yelton, D. E., 50:657–80
Yuan, R., 50:285–315

Z

Zachau, H. G., 51:89–121
Zimmerman, S. B., 51:395–427

CHAPTER TITLES, VOLUMES 48–52

PREFATORY

A "Pure" Organic Chemist's Downward Path:
Chapter 2—The Years at P. and S.   M. Heidelberger   48:1–21
The Pursuit of a Hobby   S. Ochoa   49:1–30
Confessions of a Biochemist   J. M. Luck   50:1–22
Life and Science   W. A. Engelhardt   51:1–19
Long Ago and Far Away   L. F. Leloir   52:1–15

AMINO ACIDS

Regulation of Tryptophan Biosynthesis   I. P. Crawford, G. V. Stauffer   49:163–95
Metabolism of Proline and the Hydroxyprolines   E. Adams, L. Frank   49:1005–61
Amino Acid Degradation by Anaerobic Bacteria   H. A. Barker   50:23–40
A Survey of Inborn Errors of Amino Acid
Metabolism and Transport in Man   D. Wellner, A. Meister   50:911–68
Biochemistry of Sulfur-Containing Amino Acids   A. J. L. Cooper   52:187–222
Fatty Acid Synthesis and its Regulation   S. J. Wakil, J. K. Stoops, V. C. Joshi   52:537–79

BIONERGETICS (See also Contractile Proteins, Membranes, and Transport)

Energy Transduction in Chloroplasts: Structure and
Function of the ATPase Complex   N. Shavit   49:111–38
b-Type Cytochromes   G. von Jagow, W. Sebald   49:281–314
The Proton-Translocating Pumps of Oxidative
Phosphorylation   R. H. Fillingame   49:1079–1113
Proton-Translocating Cytochrome Complexes   M. Wikström, K. Krab, M. Saraste   50:623–55
The Mechanism and Regulation of ATP Synthesis
by F_1-ATPases   R. L. Cross   50:681–714
Proton Electrochemical Gradients and
Energy-Transduction Processes   S. J. Ferguson, M. C. Sargato   51:185–217
Bacteriorhodopsin and Related Pigments of the
Halobacteria   W. Stoeckenius, R. A. Bogomolni   51:587–616
Comparative Biochemistry of Photo-Synthetic
Light-Harvesting Systems   A. N. Glazer   52:125–57
Mechanism of Free Energy Coupling in Active
Transport   C. Tanford   52:379–409
Proton ATPases: Structure and Mechanism   L. M. Amzel, P. L. Pedersen   52:801–24

CANCER (See Biochemistry of Disease)

CARBOHYDRATES

Lectins: Their Multiple Endogenous Cellular
Functions   S. H. Barondes   50:207–31
Synthesis and Processing of Asparagine-Linked
Oligosaccharides   S. C. Hubbard, R. J. Ivatt   50:555–83
Glycosphingolipids in Cellular Interaction,
Differentiation, and Oncogenesis   S. Hakomori   50:733–64
Carbohydrate-Specific Receptors of the Liver   G. Ashwell, J. Harford   51:531–54
Gluconeogenesis and Related Aspects of
Glycolysis   H. G. Hers, L. Hue   52:617–53
Cell Surface Interactions with Extracellular
Materials   K. M. Yamada   52:761–99

943

CELL ORGANELLES
The Structure and Function of Eukaryotic
  Ribosomes   I. G. Wool   48:719–54
Energy Transduction in Chloroplasts: Structure and
  Function of the ATPase Complex   N. Shavit   49:111–38
Metabolic Pathways in Peroxisomes and
  Glyoxysomes   N. E. Tolbert   50:133–57
Intermediate Filaments: A Chemically
  Heterogeneous, Developmentally Regulated
  Class of Proteins   E. Lazarides   51:219–50
Architecture of Prokaryotic Ribosomes   H. G. Wittman   52:35–65

CELL WALLS
Lectins: Their Multiple Endogenous Cellular
  Functions   S. H. Barondes   50:207–31
Synthesis of the Yeast Cell Wall and its
  Regulation   E. Cabib, R. Roberts, B. Bowers   51:763–93
Penicillin-Binding Proteins and the Mechanism of
  Action of β-Lactam Antibiotics   D. J. Waxman, J. L. Strominger   52:825–69

DEVELOPMENT AND DIFFERENTIATION
Developmental Biochemistry of Preimplantation
  Mammalian Embryos   M. I. Sherman   48:443–70
Surface Components and Cell Recognition   W. Frazier, L. Glaser   48:491–523
Active Oxygen Species and the Functions of
  Phagocytic Leukocytes   J. A. Badwey, M. L. Karnovsky   49:695–726
Molecular Approaches to the Study of
  Fertilization   B. M. Shapiro, R. W. Shackmann,
                                                 C. A. Gabel   50:815–43
The Biology and Mechanism of Action of Nerve
  Growth Factor   B. A. Yankner, E. M. Shooter   51:845–68

DISEASE, BIOCHEMISTRY OF
Integrated Genomes of Animal Viruses   R. A. Weinberg   49:197–226
DNA Modification and Cancer   M. J. Waring   50:149–92
The Biochemistry of the Complications of
  Diabetes Mellitus   M. Brownlee, A. Cerami   50:385–432
Biochemistry of Diseases of Immunodevelopment D. W. Martin, E. W. Gelfand   50:845–77
A Survey of Inborn Errors of Amino Acid
  Metabolism and Transport in Man   D. Wellner, A. Meister   50:911–68
Transmembrane Transport of Cobalamin in
  Prokaryotic and Eukaryotic Cells   C. Sennett, L. E. Rosenberg, I. S.
                                                 Mellman   50:1053–86
Biochemistry of Interferons and Their Actions   P. Lengyel   51:251–82
Enzyme Therapy: Problems and Solutions   J. S. Holcenberg   51:795–812
Lipoprotein Metabolism in the Macrophage:
  Implications for Cholesterol Deposition in
  Atherosclerosis   M. S. Brown, J. L. Goldstein   52:223–61
Cellular Oncogenes and Retroviruses   J. M. Bishop   52:301–54

DNA

General
Mitochondrial Genes and Translation Products   A. Tzagoloff, G. Macino, W. Sebald 48:419–41
Integrated Genomes of Animal Viruses   R. A. Weinberg   49:197–226
Genetic Transformation   H. O. Smith, D. B. Danner, R. A.
                                                 Deich   50:41–68
DNA Modification and Cancer   M. J. Waring   50:159–92
Chemical Mutagenesis   B. Singer, J. T. Kusmierek   51:655–93
The Molecular Biology of Trypanosomes   P. T. Englund, S. L. Hajduk, J. C.
                                                 Marini   51:695–726
DNA Methylation and Gene Activity   W. Doerfler   52:93–124

Recombination
Molecular Mechanisms in Genetic Recombination D. Dressler, H. Potter   51:727–61

Repair
DNA Repair in Bacteria and Mammalian Cells   P. C. Hanawalt, P. K. Cooper, A. K.
                                               Ganesan, C. A. Smith   48:783–836
DNA Repair Enzymes   T. Lindahl   51:61–87

Replication
Initiation of DNA Synthesis in *Escherichia Coli* J. Tomizawa, G. Selzer   48:999–1034
Discontinuous DNA Replication   T. Ogawa, T. Okazaki   49:421–57
Replication of Eukaryotic Chromosomes: A
    Close-up of the Replication Fork   M. L. DePamphilis, P. M.
                                               Wassarman   49:627–66
Fidelity of DNA Synthesis   L. A. Loeb, T. A. Kunkel   51:429–57
Eukaryotic DNA Replication: Viral and Plasmid
    Model Systems   M. D. Challberg, T. J. Kelly   51:901–34
Prokaryotic DNA Replication Systems   N. G. Nossal   52:581–615

Restriction Modification
The Structure and Mechanism of Multifunctional
    Restriction Endonucleases   R. Yuan   50:285–315

Structure
Histone Genes and Histone Messengers   L. H. Kedes   48:837–70
Discontinuous DNA Replication   T. Ogawa, T. Okazaki   49:421–57
Repeated Genes in Eukaryotes   E. O. Long, I. B. Dawid   49:727–64
Nucleosome Structure   J. D. McGhee, G. Felsenfeld   49:1115–56
Proteins Controlling the Helical Structure of
    DNA   K. Geider, H. Hoffmann-Berling   50:233–60
DNA Topoisomerases   M. Gellert   50:879–910
Double Helical DNA: Conformations, Physical
    Properties, and Interactions with Ligands M. T. Record, Jr., S. J. Mazur, P.
                                               Melancon, J. H. Roe, S. L.
                                               Shaner, L. Unger   50:997–1024
Chromatin   T. Igo-Kemenes, W. Hörz, H. G.
                                               Zachau   51:89–121
The Three-Dimensional Structure of DNA   S. B. Zimmerman   51:395–427
Repetitive Sequences in Eukaryotic DNA and
    Their Expression   W. R. Jelinek, C. W. Schmid   51:813–44
DNA Methylation and Gene Actvity   W. Doerfler   52:93–124

DRUGS, ANTIBIOTICS, AND ANTIMETABOLITES
Use of Model Enzymes in the Determination of
    the Mode of Action of Penicillins and
    Δ^3-Cephalosporins   J. M. Ghuysen, J. M. Frère, M.
                                               Leyh-Bouille, J. Coyette, J.
                                               Dusart, M. Nguyen-Disteche   48:73–101
Mechanism of Action of β-Lactam Antibiotics   D. J. Waxman, J. L. Strominger   52:825–69

ENZYMES

Mechanisms and Kinetics
Slow Transitions and Hysteretic Behavior in
    Enzymes   C. Frieden   48:471–89
On the Mechanism of Action of Folate- and
    Biopterin-Requiring Enzymes   S. J. Benkovic   49:227–51
Enzyme-Catalyzed Phosphoryl Transfer Reactions J. R. Knowles   48:877–919
The Expression of Isotope Effects on
    Enzyme-Catalyzed Reactions   D. B. Northrop   50:103–31
The Structure and Mechanism of Multifunctional
    Restriction Endonucleases   R. Yuan   50:285–315
Magnetic Resonance Studies of Active Sites in
    Enzymic Complexes   M. Cohn, G. H. Reed   51:365–94
Enzyme Therapy: Problems and Solutions   J. S. Holcenberg   51:795–812

Subunit Cooperation and Enzymatic Catalysis	C. Y. Huang, S. G. Rhee, P. B. Chock	51:935–71
Structure and Catalysis of Enzymes	W. N. Lipscomb	52:17–34
Mechanism of Action of β-Lactam Antibiotics	D. J. Waxman, J. L. Strominger	352:825–69

Regulation

Activation of Adenylate Cyclase by Choleragen	J. Moss, M. Vaughan	48:581–600
Phosphorylation-Dephosphorylation of Enzymes	E. G. Krebs, J. A. Beavo	48:923–59
Calmodulin	C. B. Klee, T. H. Crouch, P. G. Richman	49:489–515
Interconvertible Enzyme Cascades in Cellular Regulation	P. B. Chock, S. G. Rhee, E. R. Stadtman	49:813–43
Human Plasma Proteinase Inhibitors	J. Travis, G. S. Salvesen	52:655–709

Specific Enzymes and Classes

Membrane Adenosine Triphosphates of Prokaryotic Cells	J. A. Downie, F. Gibson, G. B. Cox	48:103–31
Energy Interconversion by the Ca^{2+}- Dependent ATPase of the Sarcoplasmic Reticulum	L. de Meis, A. L. Vianna	48:275–92
Aminoacyl-tRNA Synthetases: General Features and Recognition of Transfer RNAs	P. R. Schimmel, D. Söll	48:601–48
Selenium-Dependent Enzymes	T. C. Stadtman	49:93–110
Proteins Containing 4Fe-4S Clusters: An Overview	W. V. Sweeney, J. C. Rabinowitz	49:139–61
b-Type Cytochromes	G. von Jagow, W. Sebald	49:281–314
Oxygen Activation by Cytochrome P-450	R. E. White, M. J. Coon	49:315–56
Transglutaminases	J. E. Folk	49:517–31
Biochemical Properties of Hormone-Sensitive Adenylate Cyclase	E. M. Ross, A. G. Gilman	49:533–64
Blood Coagulation	C. M. Jackson, Y. Nemerson	49:765–811
Enzyme-Catalyzed Phosphoryl Transfer Reactions	J. R. Knowles	49:877–919
Regulation and Kinetics of the Actin-Myosin-ATP Interaction	R. S. Adelstein, E. Eisenberg	49:921–56
Collagenases	E. Harper	49:1063–78
DNA Topoisomerases	M. Gellert	50:879–910
Enzymology of Oxygen	B. G. Malmström	51:21–59
DNA Repair Enzymes	T. Lindahl	51:61–87
Enzymes of the Renin-Angiotensin System and Their Inhibitors	M. A. Ondetti, D. W. Cushman	51:283–308
Ribulose-1, 5-Bisphosphate Carboxylase-Oxygenase	H. M. Miziorko, G. H. Lorimer	52:507–35

Structure (Protein)

Structure and Function of Nitrogenase	L. E. Mortenson, R. N. F. Thorneley	48:387–418
The Structures and Related Functions of Phosphorylase α	R. J. Fletterick, N. B. Madsen	49:31–61
Structure and Catalysis of Enzymes	W. N. Lipscomb	52:17–34
DNA Methylation and Gene Activity	W. Doerfler	52:93–124

GENES AND BIOCHEMICAL GENETICS (See also DNA and RNA)

RNA Processing and the Intervening Sequence Problem	J. Abelson	48:1035–69
Repeated Genes in Eukaryotes	E. O. Long, I. B. Dawid	49:727–64
Molecular Genetics of Yeast	T. D. Petes	49:845–76
Genetic Transformation	H. O. Smith, D. B. Danner, R. A. Deich	50:41–68
Organization and Expression of Eucaryotic Split Genes Coding for Proteins	R. Breathnach, P. Chambon	50:349–83
Chromosome Mediated Gene Transfer	L. A. Klobutcher, F. H. Ruddle	50:533–54
A Survey of Inborn Errors of Amino Acid Metabolism and Transport in Man	D. Wellner, A. Meister	50:911–68
Fidelity of DNA Synthesis	L. A. Loeb, T. A. Kunkel	51:429–57

SnRNAs, SnRNPs, and RNA Processing   H. Busch, R. Reddy, L. Rothblum,
                                            Y. C. Choi   51:617–54
Chemical Mutagenesis   B. Singer, J. T. Kuśmierek   51:655–93
The Molecular Biology of Trypanosomes   P. T. Englund, S. L. Hajduk, J. C.
                                            Marini   51:695–726
Repetitive Seqences in Eukaryotic DNA and Their
  Expression   W. R. Jelinek, C. W. Schmid   51:813–44
DNA Methylation and Gene Activity   W. Doerfler   52:93–124
Cellular Oncogenes and Retroviruses   J. M. Bishop   52:301–54
The Gene Structure and Replication of Influenza
  Virus   R. A. Lamb, P. W. Choppin   52:467–506

HORMONES
  Chemistry and Biology of the Neurophysins E. Breslow   48:251–74
  Peptide Neurotransmitters   S. H. Snyder, R. B. Innis   48:755–82
  Biochemical Properties of Hormone-Sensitive
    Adenylate Cyclase   E. M. Ross, A. G. Gilman   49:533–64
  Glycoprotein Hormones: Structure and Function J. G. Pierce, T. F. Parsons 50:465–95
  The Conformation, Flexibility, and Dynamics of
    Polypeptide Hormones   T. Blundelell, S. Wood   51:123–54
  Sterol Biosynthesis   G. J. Schroepfer, Jr.   51:555–85
  Leukotrienes   S. Hammarström   52:355–77

IMMUNOBIOCHEMISTRY
  Three-Dimensional Structure of Immunoglobulins L. M. Amzel, R. J. Poljak 48:961–97
  Active Oxygen Species and the Functions of
    Phagocytic Leukocytes   J. A. Badwey, M. L. Karnovsky   49:695–726
  The Proteolytic Activation Systems of
    Complement   K. B. M. Reid, R. R. Porter   50:433–64
  Monoclonal Antibodies: A Powerful New Tool in
    Biology and Medicine   D. E. Yelton, M. D. Scharff   50:657–80
  Biochemistry of Diseases of Immunodevelopment D. W. Martin Jr., E. W. Gelfand 50:845–77
  Primary Structural Analysis of the Transplantation
    Antigens of the Murine H-2 Major
    Histocompatibility Complex   S. G. Nathenson, H. Uehara, B. M.
                                            Ewenstein, T. J. Kindt, J. E.
                                            Coligan   50:1025–52

LIPIDS
  Lipid Asymmetry in Membranes   J. A. F. Op den Kamp   48:47–71
  Regulation of Hepatic Fatty Acid Oxidation and
    Ketone Body Production   J. D. McGarry, W. Foster   49:395–420
  Enzymes of Glycerolipid Synthesis in Eukaryotes R. M. Bell, R. A. Coleman 49:459–87
  Lipolytic Enzymes in Plasma Lipoprotein
    Metabolism   M. C. Schotz, A. S. Gaarfinkel, P.
                                            Nillson-Ehle   49:667–93
  Proteolipids   M. J. Schlesinger   50:93–206
  Sterol Biosynthesis   G. J. Schroepfer, Jr.   50:585–621
  Glycosphingolipids in Cellular Interaction,
    Differentiation, and Oncogenesis   S. Hakomori   50:733–64
  Sterol Biosynthesis   G. J. Schroepfer   51:555–85
  Lipoprotein Metabolism in the Macrophage:
    Implications for Cholesterol Deposition in
    Atherosclerosis   M. S. Brown, J. L. Goldstein   52:223–61
  Leukotrienes   S. Hammarström   52:355–77
  Fatty Acid Synthesis and its Regulation   S. J. Wakil, J. K. Stoops, V. C.
                                            Joshi   52:537–79

MEMBRANES
  The Assembly of Proteins into Biological
    Membranes: The Membrane Trigger Hypothesis W. Wickner   48:23–45
  Lipid Asymmetry in Membranes   J. A. F. Op den Kamp   48:47–71
  Membrane Recycling by Coated Vesicles   B. M. F. Pearse, M. S. Bretscher 50:85–101

Transfer of Proteins Across Membranes	G. Kreil	50:317–48
Glycosphingolipids in Cellular Interaction, Differentiation, and Oncogenesis	S. Hakomori	50:733–64
Specific Intermediates in the Folding Reactions of Small Proteins and the Mechanism of Protein Folding	P. S. Kim, R. L. Baldwin	51:459–89

METABOLISM

Pyrimidine Nucleotide Biosynthesis in Animals: Genes, Enzymes, and Regulation of UMP Biosynthesis	M. E. Jones	49:253–79
Regulation of Hepatic Fatty Acid Oxidation and Ketone Body Production	J. D. McGarry, D. W. Foster	49:395–420
Enzymes of Glycerolipid Synthesis in Eukaryotes	R. M. Bell, R. A. Coleman	49:459–87
Calmodulin	C. B. Klee, T. H. Crouch, P. G. Richman	49:489–515
Lipolytic Enzymes and Plasma Lipoprotein Metabolism	M. Schotz, A. S. Garfinkel, P. Nilsson-Ehle	49:667–93
Amino Acid Degradation by Anaerobic Bacteria	H. A. Barker	50:23–40
NMR Studies of Tissue Metabolism	D. G. Gadian, G. K. Radda	50:69–83
Transmembrane Transport of Cobalamin in Prokaryotic and Eukaryotic Cells	C. Sennett, L. E. Rosenberg, I. S. Mellman	50:1053–86

METHODOLOGY

Photoaffinity Labeling of Biological Systems	V. Chowdhry, F. H. Westheimer	48:293–325
Electron Microscopic Visualization of Nucleic Acids and of Their Complexes with Proteins	H. W. Fisher, R. C. Williams	48:649–79
NMR Studies of Tissue Metabolism	D. G. Gadian, G. K. Radda	50:69–83
Advances in Protein Sequencing	K. A. Walsh, L. H. Ericsson, D. C. Parmelee, K. Titani	50:261–84
NMR Studies on RNA Structure and Dynamics	B. R. Reid	50:969–96
Magnetic Resonance Studies of Active Sites in Enzymic Complexes	M. Cohn, G. H. Reed	51:365–94

MUSCLE AND CONTRACTILE PROTEINS

Regulation and Kinetics of the Actin-Myosin-ATP Interaction	R. S. Edelstein, E. Eisenberg	49:921–56

NUCLEOTIDES, NUCLEOSIDES, PURINES, AND PYRIMIDINES

Reduction of Ribonucleotides	L. Thelander, P. Reichard	48:133–58
Pyrimidine Nucleotide Biosynthesis in Animals: Genes, Enzymes, and Regulation of UMP Biosynthesis	M. E. Jones	49:253–70
Affinity Labeling of Purine Nucleotide Sites	R. F. Colman	52:67–91

NEUROBIOLOGY AND NEUROCHEMISTRY

Biochemistry of Sensing and Adaptation in a Simple Bacterial System	D. E. Koshland, Jr.	50:765–82
The Biology and Mechanism of Action of Nerve Growth Factor	B. A. Yankner, E. M. Shooter	51:845–68
A Molecular Description of Nerve Terminal Function	L. F. Reichardt, R. B. Kelly	52:871–926

NITROGEN FIXATION

Structure and Function of Nitrogenase	L. E. Mortenson, R. N. F. Thorneley	48:387–418

NUTRITION (See Vitamins, Growth Factors, and Essential Metabolites)

PEPTIDES

Chemistry and Biology of the Neurophysins	E. Breslow	48:251–74
Peptide Neurotransmitters	S. H. Snyder, R. B. Innis	48:755–82
The Conformation, Flexibility, and Dynamics of Polypeptide Hormones	T. Blundell, S. Wood	51:123–54

Enzymes of the Renin-Angiotensin System and
  Their Inhibitors | M. A. Ondetti, D. W. Cushman | 51:283–308
Glutathione | A. Meister, M. E. Anderson | 52:711–60

PHOTOBIOLOGY AND PHOTOSYNTHESIS (See Bioenergetics)
Ribulose-1,5-Bisphosphate Carboxylase-
  Oxygenase | H. M. Miziorko, G. H. Lorimer | 52:507–35

PROTEINS

Binding and Transport Proteins
Iron Transport and Storage Proteins | P. Aisen, I. Listowsky | 49:357–93
Dynamics of Protein: Elements and Function | M. Karplus, J. A. McCammon | 52:263–300

Biosynthesis
Mitochondrial Genes and Translation Products | A. Tzagoloff, G. Macino, W. Sebald | 48:419–41
Regulation of Protein Synthesis in Eukaryotes | S. Ochoa, C. de Haro | 48:549–80
The Structure of Function of Eukaryotic
  Ribosomes | I. G. Wool | 48:719–54
Initiation Factors in Protein Biosynthesis | U. Maitra, E. R. Stringer, A. Chaudhuri | 51:869–900
Architecture of Prokaryotic Ribosomes | H. G. Wittman | 52:35–65

Contractile Proteins
Regulation and Kinetics of the Actin-Myosin-ATP
  Interaction | R. S. Adelstein, E. Eisenberg | 49:921–56

Metabolism
Control of Proteolysis | H. Holzer, P. C. Heinrich | 49:63–91
Mechanisms of Intracellular Protein Breakdown | A. Hershko, A. Ciechanover | 51:335–64

Post-Translational Modification
In Vivo Modification of Proteins | F. Wold | 50:783–814

Special Classes
Histones | I. Isenberg | 48:159–91
Mitochondrial Genes and Translation Products | A. Tzagoloff, G. Macino, W. Sebald | 48:419–41
Histone Genes and Histone Messengers | L. H. Kedes | 48:837–70
Proteins Containing 4Fe-4S Clusters:
  An Overview | W. V. Sweeney, J. C. Rabinowitz | 49:139–61
b-Type Cytochromes | G. von Jagow, W. Sebald | 49:281–314
Calmodulin | C. B. Klee, T. H. Crouch, P. G. Richman | 49:489–515
In Vitro Assembly of Cytoplasmic Microtubules | S. N. Timasheff, L. M. Grisham | 49:565–91
Blood Coagulation | C. M. Jackson, Y. Nemerson | 49:765–811
Structurally Distinct Collagen Types | P. Bornstein, H. Sage | 49:957–1003
Nucleosome Structure | J. D. McGhee, G. Felsenfeld | 49:1115–56
Components of Bacterial Ribosomes | H. G. Wittmann | 51:155–83
Intermediate Filaments: A Chemically
  Heterogeneous, Developmentally Regulated
  Class of Proteins | E. Lazarides | 51:219–50
Biochemistry of Interferons and Their Actions | P. Lengyel | 51:251–82
Human Plasma Proteinase Inhibitors | J. Travis, G. S. Salvesen | 52:655–709

Structure
Probes of Subunit Assembly and Reconstitution
  Pathways in Multisubunit Proteins | F. K. Friedman, S. Beychok | 48:217–50
Regulation of Oxygen Affinity of Hemoglobin:
  Influence of Structure of the Globin on the
  Heme Iron | M. F. Perutz | 48:327–86
Three-Dimensional Structure of Immunoglobulins | L. M. Amzel, R. J. Poljak | 48:961–97
Advances in Protein Sequencing | K. A. Walsh, L. H. Ericsson, D. C. Parmelee, K. Titani | 50:261–84
Protein Folding | M. G. Rossmann, P. Argos | 50:497–532
Specific Intermediates in the Folding Reactions of
  Small Proteins and the Mechanism of Protein
  Folding | P. S. Kim, R. L. Baldwin | 51:459–89

RECEPTORS

Biochemistry of Sensing and Adaptation in a
Simple Bacterial System — D. E. Koshland Jr. — 50:765–82

The Nicotinic Cholinergic Receptor: Correlation of
Molecular Structure with Functional Properties — B. M. Conti-Tronconi, M. A. Raftery — 51:491–530

Carbohydrate-Specific Receptors of the Liver — G. Ashwell, J. Harford — 51:531–54

Adenylate Cyclase–Coupled Beta-Andrenergic
Receptors: Structure and Mechanisms of
Activation and Desensitization — R. J. Lefkowitz, J. M. Stadel, M. G. Caron — 52:159–86

RNA

RNA Replication: Function and Structure of
Qβ-Replicase — T. Blumenthal, G. G. Carmichael — 48:525–48

Aminoacyl-tRNA Synthetases: General Features
and Recognition of Transfer RNAs — P. R. Schimmel, D. Söll — 48:601–48

The Structure and Function of Eukaryotic
Ribosomes — I. G. Wool — 48:719–54

Histone Genes and Histone Messengers — L. H. Kedes — 48:837–70

RNA Processing and the Intervening Sequence
Problem — J. Abelson — 48:1035–69

NMR Studies on RNA Structure and Dynamics — B. R. Reid — 50:969–96

Components of Bacterial Ribosomes — H. G. Wittmann — 51:155–83

Biochemistry of Interferons and Their Actions — P. Lengyel — 51:251–82

SnRNAs, SnRNPs, and RNA Processing — H. Busch, R. Reddy, L. Rothblum, Y. C. Choi — 51:617–54

Architecture of Prokaryotic Ribosomes — H. G. Wittman — 52:35–65

The Pathway of Eukaryotic mRNA Formation — J. R. Nevins — 52:441–66

TOXINS AND TOXIC AGENTS

Activation of Adenylate Cyclase by Choleragen — J. Moss, M. Vaughan — 48:581–600

Phytotoxins — G. A. Strobel — 51:309–33

TRANSPORT

Metabolite Transport in Mitochondria — K. F. LaNoue, A. C. Schoolwerth — 48:871–922

Iron Transport and Storage Proteins — P. Aisen, I. Listowsky — 49:357–93

The Proton-Translocating Pumps of Oxidative
Phosphorylation — R. H. Fillingame — 49:1079–1113

Transfer of Proteins Across Membranes — G. Kreil — 50:317–48

Microbial Iron Compounds — J. B. Neilands — 50:715–31

Transmembrane Transport of Cobalamin in
Prokaryotic and Eukaryotic Cells — C. Sennett, L. E. Rosenberg, and I. S. Mellman — 50:1053–86

The Nicotinic Cholinergic Receptor: Correlation of
Molecular Structure with Functional Properties — B. M. Conti-Tronconi, M. A. Raftery — 51:491–530

Mechanism of Free Energy Coupling in Active
Transport — C. Tanford — 52:379–409

VIRUSES AND BATERIOPHAGES

Integrated Genomes of Animal Viruses — R. A. Weinberg — 49:197–226

Eukaryotic DNA Replication: Viral and Plasmid
Model Systems — M. D. Challberg, T. J. Kelly — 51:901–34

Cellular Oncogenes and Retroviruses — J. M. Bishop — 52:301–54

The Gene Structure and Replication of Influenza
Virus — R. A. Lamb, P. W. Choppin — 52:467–506

VITAMINS, GROWTH FACTORS, ESSENTIAL METABOLITIES

Epidermal Growth Factor — G. Carpenter, S. Cohen — 48:193–216

Selenium-Dependent Enzymes — T. C. Stadtman — 49:93–110

Microbial Iron Compounds — J. B. Neilands — 50:715–31

Transmembrane Transport of Cobalamin in
Prokaryotic and Eukaryotic Cells — C. Sennett, L. E. Rosenberg, I. S. Mellman — 50:1053–86

Vitamin D: Recent Advances — H. F. DeLuca, H. K. Schnoes — 52:411–39

A NONPROFIT SCIENTIFIC PUBLISHER

Annual Reviews Inc.

4139 EL CAMINO WAY • PALO ALTO, CA 94306 USA • (415) 493-4400

Please list the volumes you wish to order by volume number. If you wish a standing order (the latest volume sent to you automatically each year), indicate volume number to begin order. Volumes not yet published will be shipped in month and year indicated. All prices subject to change without notice.

ANNUAL REVIEW SERIES

Annual Review of **ANTHROPOLOGY**

			Prices Postpaid per volume USA/elsewhere	Regular Order Please send: Vol. number	Standing Order Begin with: Vol. number
Vols. 1-10	(1972-1981)	....	$20.00/$21.00		
Vol. 11	(1982)	....	$22.00/$25.00		
Vol. 12	(avail. Oct. 1983)	....	$27.00/$30.00	Vol(s). _____	Vol. _____

Annual Review of **ASTRONOMY AND ASTROPHYSICS**

Vols. 1-19	(1963-1981)	....	$20.00/$21.00		
Vol. 20	(1982)	....	$22.00/$25.00		
Vol. 21	(avail. Sept. 1983)	....	$44.00/$47.00	Vol(s). _____	Vol. _____

Annual Review of **BIOCHEMISTRY**

> Vols. 28-48 $18.00/$18.50
> Price effective through 12/31/82

Vols. 28-50	(1959-1981)	....	$21.00/$22.00		
Vol. 51	(1982)	....	$23.00/$26.00		
Vol. 52	(avail. July 1983)	....	$29.00/$32.00	Vol(s). _____	Vol. _____

Annual Review of **BIOPHYSICS AND BIOENGINEERING**

Vols. 1-10	(1972-1981)	....	$20.00/$21.00		
Vol. 11	(1982)	....	$22.00/$25.00		
Vol. 12	(avail. June 1983)	....	$47.00/$50.00	Vol(s). _____	Vol. _____

Annual Review of **EARTH AND PLANETARY SCIENCES**

Vols. 1-9	(1973-1981)	....	$20.00/$21.00		
Vol. 10	(1982)	....	$22.00/$25.00		
Vol. 11	(avail. May 1983)	....	$44.00/$47.00	Vol(s). _____	Vol. _____

Annual Review of **ECOLOGY AND SYSTEMATICS**

Vols. 1-12	(1970-1981)	....	$20.00/$21.00		
Vol. 13	(1982)	....	$22.00/$25.00		
Vol. 14	(avail. Nov. 1983)	....	$27.00/$30.00	Vol(s). _____	Vol. _____

SEE ORDERING INFORMATION ON PAGE 4.

Annual Review of **ENERGY**		Prices Postpaid per volume USA/elsewhere	Regular Order Please send: Vol. number	Standing Order Begin with: Vol. number
Vols. 1-6	(1976-1981)	$20.00/$21.00		
Vol. 7	(1982)	$22.00/$25.00		
Vol. 8	(avail. Oct. 1983)	$56.00/$59.00	Vol(s). _____	Vol. _____

Annual Review of **ENTOMOLOGY**				
Vols. 7-26	(1962-1981)	$20.00/$21.00		
Vol. 27	(1982)	$22.00/$25.00		
Vol. 28	(avail. Jan. 1983)	$27.00/$30.00	Vol(s). _____	Vol. _____

Annual Review of **FLUID MECHANICS**				
Vols. 1-13	(1969-1981)	$20.00/$21.00		
Vol. 14	(1982)	$22.00/$25.00		
Vol. 15	(avail. Jan. 1983)	$28.00/$31.00	Vol(s). _____	Vol. _____

Annual Review of **GENETICS**				
Vols. 1-15	(1967-1981)	$20.00/$21.00		
Vol. 16	(1982)	$22.00/$25.00		
Vol. 17	(avail. Dec. 1983)	$27.00/$30.00	Vol(s). _____	Vol. _____

Annual Review of **IMMUNOLOGY — New Series 1983**				
Vol. 1	(avail. April 1983)	$27.00/$30.00	Vol(s). _____	Vol. _____

Annual Review of **MATERIALS SCIENCE**				
Vols. 1-11	(1971-1981)	$20.00/$21.00		
Vol. 12	(1982)	$22.00/$25.00		
Vol. 13	(avail. Aug. 1983)	$64.00/$67.00	Vol(s). _____	Vol. _____

Annual Review of **MEDICINE: Selected Topics in the Clinical Sciences**				
Vols. 1-3, 5-15	(1950-1952; 1954-1964)	$20.00/$21.00		
Vols. 17-32	(1966-1981)	$20.00/$21.00		
Vol. 33	(1982)	$22.00/$25.00		
Vol. 34	(avail. April 1983)	$27.00/$30.00	Vol(s). _____	Vol. _____

Annual Review of **MICROBIOLOGY**				
Vols. 15-35	(1961-1981)	$20.00/$21.00		
Vol. 36	(1982)	$22.00/$25.00		
Vol. 37	(avail. Oct. 1983)	$27.00/$30.00	Vol(s). _____	Vol. _____

Annual Review of **NEUROSCIENCE**				
Vols. 1-4	(1978-1981)	$20.00/$21.00		
Vol. 5	(1982)	$22.00/$25.00		
Vol. 6	(avail. March 1983)	$27.00/$30.00	Vol(s). _____	Vol. _____

2

Annual Review of **SOCIOLOGY**		Prices Postpaid per volume USA/elsewhere	Regular Order Please send:	Standing Order Begin with:
			Vol. number	Vol. number
Vols. 1-7	(1975-1981)	$20.00/$21.00		
Vol. 8	(1982)	$22.00/$25.00		
Vol. 9	(avail. Aug. 1983)	$27.00/$30.00	Vol(s). _____ Vol. _____	

SPECIAL PUBLICATIONS

	Prices Postpaid per volume USA/elsewhere	Regular Order Please send:
Annual Reviews Reprints: Cell Membranes, 1975-1977		
(published 1978) Softcover .... $12.00/$12.50		_____ Copy(ies).
Annual Reviews Reprints: Cell Membranes, 1978-1980		
(published 1981) Hardcover .... $28.00/$29.00		_____ Copy(ies).
Annual Reviews Reprints: Immunology, 1977-1979		
(published 1980) Softcover .... $12.00/$12.50		_____ Copy(ies).
History of Entomology		
(published 1973) Clothbound .... $10.00/$10.50		_____ Copy(ies).
Intelligence and Affectivity: Their Relationship During Child Development, by Jean Piaget		
(published 1981) Hardcover .... $8.00/$9.00		_____ Copy(ies).
Telescopes for the 1980s		
(published 1981) Hardcover .... $27.00/$28.00		_____ Copy(ies).
The Excitement and Fascination of Science, Volume 1		
(published 1965) Clothbound .... $6.50/$7.00		_____ Copy(ies).
The Excitement and Fascination of Science, Volume 2		
(published 1978) Hardcover .... $12.00/$12.50 Softcover .... $10.00/$10.50		_____ Copy(ies).

To: ANNUAL REVIEWS INC. 4139 El Camino Way, Palo Alto, CA 94306-9981 USA   Tel. 415-493-4400

Please enter my order for the publications checked above.

Institutional purchase order No. _____

Amount of remittance enclosed $_____

Charge my account ☐ MasterCard ☐ Visa   Acct. No. _____

Exp. Date _____   _____

Individuals: Prepayment required, or charge account below.

California residents, please add applicable sales tax.
Prices subject to change without notice.

Signature

Name _____
       Please print

Address _____
         Please print

_____ Zip Code _____   ☐ Please send free copy of current *Prospectus*